Sixteenth Edition

INQUIRY
INTO LIFE

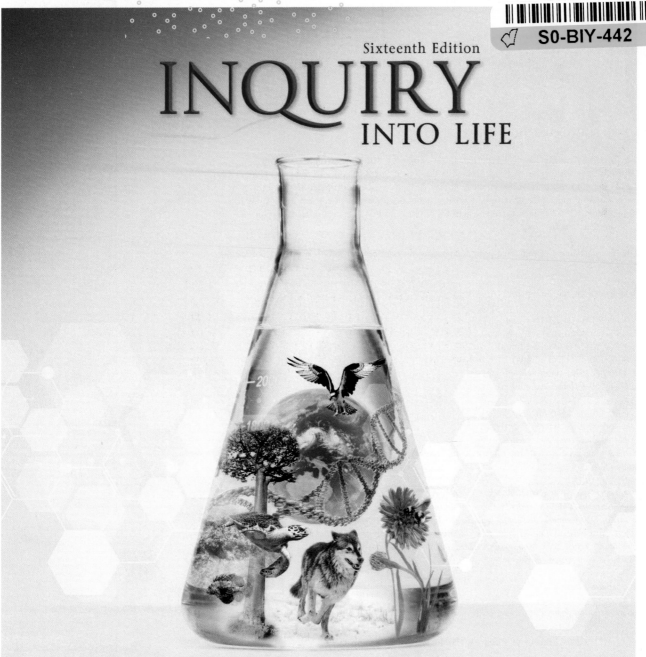

Sylvia S. Mader

Michael Windelspecht

Mc Graw Hill Education

INQUIRY INTO LIFE, SIXTEENTH EDITION

1 2 3 4 5 6 7 8 9 LWI 21 20 19

ISBN 978-1-260-23170-0 (bound edition)
MHID 1-260-23170-4 (bound edition)
ISBN 978-1-260-48259-1 (loose-leaf edition)
MHID 1-260-48259-6 (loose-leaf edition)

Executive Portfolio Manager: *Michelle Vogler*
Product Developers: *Anne Winch, Joan Weber*
Marketing Manager: *Britney Ross*
Content Project Managers: *Kelly Hart, Tammy Juran, Sandy Schnee*
Buyer: *Sandy Ludovissy*
Design: *David W. Hash*
Content Licensing Specialist: *Lori Hancock*
Cover Image: *Scientist adding chemicals: ©Image Source/Getty Images; DNA: ©Comstock/Getty Images; Hawksbill turtle: ©cookelma/123RF.com; Wolves: ©actionsports/123RF.com; Baobab trees: ©Jialiang Gao/ Getty Images; Cornflower: ©Pixtal/SuperStock; Lionfish: ©poweroffforever/Getty Images; Osprey: Source: NASA; Bumblebee: ©Jari Sokka/Shutterstock; Hexagon: ©brkart/Shutterstock*

Compositor: *Aptara® Inc.*

Library of Congress Cataloging-in-Publication Data

Names: Mader, Sylvia S., author. | Windelspecht, Michael, 1963- author.
Title: Inquiry into life / Sylvia S. Mader, Michael Windelspecht.
Description: 16 edition. | New York, NY : McGraw-Hill
  Education, [2020] | Includes bibliographical references and index.
Identifiers: LCCN 2018034426 | ISBN 9781260231700 (alk. paper)
Subjects: LCSH: Biology—Textbooks.
Classification: LCC QH308.2 .M363 2020 | DDC 570—dc23 LC record available at
https://lccn.loc.gov/2018034426

©Jacqueline Baer
Photography

Sylvia S. Mader

Sylvia Mader has authored several nationally recognized biology texts published by McGraw-Hill. Educated at Bryn Mawr College, Harvard University, Tufts University, and Nova Southeastern University, she holds degrees in both biology and education. Over the years she has taught at University of Massachusetts, Lowell; Massachusetts Bay Community College; Suffolk University; and Nathan Mayhew Seminars. Her ability to reach out to science-shy students led to the writing of her first text, *Inquiry into Life*. Highly acclaimed for her crisp and entertaining writing style, her books have become models for others who write in the field of biology.

©Ricochet Creative
Productions LLC

Michael Windelspecht

As an educator, Dr. Windelspecht has taught introductory biology, genetics, and human genetics in the online, traditional, and hybrid environments at community colleges, comprehensive universities, and military institutions. For over a decade he served as the Introductory Biology Coordinator at Appalachian State University, where he directed a program that enrolled over 4,500 students annually.

He received degrees from Michigan State University (BS, zoology-genetics) and the University of South Florida (PhD, evolutionary genetics) and has published papers in areas as diverse as science education, water quality, and the evolution of insecticide resistance. His current interests are in the analysis of data from digital learning platforms for the development of personalized microlearning assets and next-generation publication platforms. He is currently a member of the National Association of Science Writers and several science education associations. He has served as the keynote speaker on the development of multimedia resources for online and hybrid science classrooms. In 2015, he won the DevLearn HyperDrive competition for a strategy to integrate student data into the textbook revision process.

As an author and editor, Dr. Windelspecht has over 20 reference textbooks and multiple print and online lab manuals. He has founded several science communication companies, including Ricochet Creative Productions, which actively develops and assesses new technologies for the science classroom. You can learn more about Dr. Windelspecht by visiting his website at www.michaelwindelspecht.com.

Contributor

Dave Cox serves as Professor of Biology at Lincoln Land Community College, in Springfield, Illinois. He was educated at Illinois College and Western Illinois University. As an educator, Professor Cox teaches introductory biology for nonmajors in the traditional classroom format as well as in a hybrid format. He also teaches biology for majors, and marine biology and biological field studies as study-abroad courses in Belize. He is the co-owner of Howler Publications, a company that specializes in scientific study abroad courses. Professor Cox served as a contributor to both *Inquiry 14e* and *Human Biology 13e*. He also develops educational resources for the ecotourism industry in Belize.

PREFACE

Goals of *Inquiry into Life* 16e

Dr. Sylvia Mader's text, *Inquiry into Life,* was originally developed to reach out to science-shy students. The text now represents one of the cornerstones of introductory biology education. *Inquiry into Life* was founded on the belief that teaching science from a human perspective, coupled with human applications, would make the material more relevant to the student. Interestingly, even though it has been over 40 years since the first edition was published, this style of relevancy-based education remains the focus of the national efforts to increase scientific literacy in the general public.

Our modern society is based largely on advances in science and technology over the past few decades. As we present in this text, there are many challenges facing humans, and an understanding of how science can help analyze, and offer solutions to these problems is critical to our species' health and survival.

The front cover of this text was chosen to indicate not only that humans are the stewards of the planet, but that we also possess the ability to influence, both positively and negatively, all of the inhabitants of our world. Therefore, it is important that we know not only why we are different, but how we are the same as the species we share the planet with. Students in today's world are being exposed, almost on a daily basis, to exciting new discoveries and insights that, in many cases, were beyond our predictions even a few short years ago. It is our task, as instructors, not only to make these findings available to our students, but to enlighten students as to why these discoveries are important to their lives and society. At the same time, we must provide students with a firm foundation in those core principles on which biology is founded, and in doing so, provide them with the background to keep up with the many discoveries still to come.

This text, along with the *Inquiry into Life* 15.1 edition, represents an ongoing project in the development of a continuously updated textbook. As scientists and educators, the authors of this text are well aware that scientific discovery is a dynamic process. Fortunately, the advances in digital publishing are allowing authors to update content on an ongoing basis, which in turn is promoting the ability to update content on a regular basis. This text represents the prototype of those efforts. Specifically, the authors sought to:

1. Update chapter openers and readings within the text to reflect more recent discoveries or topics of interest in the life sciences.
2. Update statistics, maps, and tables to reflect changes in our scientific understanding of the topic.
3. Assess and redesign art to better fit the digital learning environment.
4. Develop a relevancy module to supplement the content of the text. These relevancy modules allow the instructor to engage students with topics that are not normally covered in an introductory textbook.

Relevancy

The use of real-world examples to demonstrate the importance of biology in the lives of students is widely recognized as an effective teaching strategy for the introductory biology classroom. Students want to learn about the topics they are interested in. The development of relevancy-based resources is a major focus for the authors of the Mader series of texts. Some examples of how we have increased the relevancy content of this edition include:

Relevancy Modules

Relevancy modules have been authored to accompany each unit in Mader/Windelspecht *Inquiry into Life*. These modules demonstrate the connections between biological content and topics that are of interest to society as a whole. Each module consists of an introductory video, an overview of basic scientific concepts, and then a closer look at the application of these concepts to the topic. Discussion and assessment questions, specific to the modules, are also available.

These modules are available as a supplementary eBook to the existing text within Connect, and may be assigned by the instructor for use in a variety of ways in the classroom. Examples of topics covered include cancer biology, fermentation science, weed evolution, antibiotic resistance, mega crops, biology of weight gain, and climate change. New topics are planned for launch each year to keep these resources current.

BioNow Videos

The BioNow series of videos, narrated and produced by educator Jason Carlson, provide a relevant, applied approach that allows your students to feel they can actually do and learn biology themselves. While tying directly to the content of your course, the videos help students relate their daily lives to the biology you teach and then connect what they learn back to their lives.

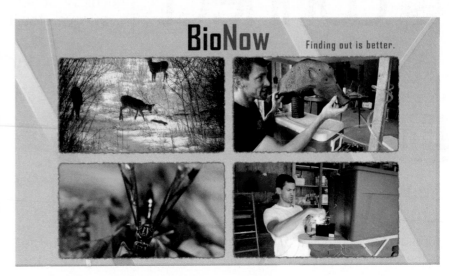

Each video provides an engaging and entertaining story about applying the science of biology to a real situation or problem. Attention is given to using tools and techniques that the average person would have access to, so your students see the science as something they could do and understand.

RicochetScience Website

The RicochetScience.com website, managed by Dr. Michael Windelspecht, provides content that is of interest to students who are not majoring in the sciences. For example, the PopScience articles on this site provide an excellent focus for classroom discussions on topics that are currently being debated in society, such as vaccines. The content is organized by the same topic areas that are the focus of the relevancy modules, making it simple for instructors to find and utilize these resources. The site also features videos and tutorial animations to assist the students in recognizing the relevancy of what they are learning in the classroom.

Students—study more efficiently, retain more and achieve better outcomes. Instructors—focus on what you love—teaching.

SUCCESSFUL SEMESTERS INCLUDE CONNECT

FOR INSTRUCTORS

You're in the driver's seat.

Want to build your own course? No problem. Prefer to use our turnkey, prebuilt course? Easy. Want to make changes throughout the semester? Sure. And you'll save time with Connect's auto-grading too.

65%
Less Time Grading

They'll thank you for it.

Adaptive study resources like SmartBook® help your students be better prepared in less time. You can transform your class time from dull definitions to dynamic debates. Hear from your peers about the benefits of Connect at **www.mheducation.com/highered/connect**

Make it simple, make it affordable.

Connect makes it easy with seamless integration using any of the major Learning Management Systems—Blackboard®, Canvas, and D2L, among others—to let you organize your course in one convenient location. Give your students access to digital materials at a discount with our inclusive access program. Ask your McGraw-Hill representative for more information.

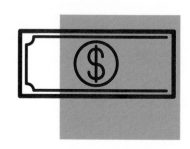

©Hill Street Studios/Tobin Rogers/Blend Images LLC

Solutions for your challenges.

A product isn't a solution. Real solutions are affordable, reliable, and come with training and ongoing support when you need it and how you want it. Our Customer Experience Group can also help you troubleshoot tech problems—although Connect's 99% uptime means you might not need to call them. See for yourself at **status.mheducation.com**

FOR STUDENTS

Effective, efficient studying.

Connect helps you be more productive with your study time and get better grades using tools like SmartBook, which highlights key concepts and creates a personalized study plan. Connect sets you up for success, so you walk into class with confidence and walk out with better grades.

©Shutterstock/wavebreakmedia

"I really liked this app—it made it easy to study when you don't have your textbook in front of you."

—Jordan Cunningham,
Eastern Washington University

Study anytime, anywhere.

Download the free ReadAnywhere app and access your online eBook when it's convenient, even if you're offline. And since the app automatically syncs with your eBook in Connect, all of your notes are available every time you open it. Find out more at **www.mheducation.com/readanywhere.**

No surprises.

The Connect Calendar and Reports tools keep you on track with the work you need to get done and your assignment scores. Life gets busy; Connect tools help you keep learning through it all.

13	14
Chapter 12 Quiz	Chapter 11 Quiz
Chapter 13 Evidence of Evolution	Chapter 11 DNA Technology
	Chapter 7 Quiz
	Chapter 7 DNA Structure and Gene...
	and 7 more...

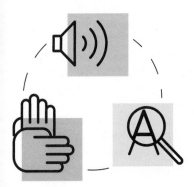

Learning for everyone.

McGraw-Hill works directly with Accessibility Services Departments and faculty to meet the learning needs of all students. Please contact your Accessibility Services office and ask them to email accessibility@mheducation.com, or visit **www.mheducation.com/about/accessibility.html** for more information.

ACKNOWLEDGMENTS

Dr. Sylvia Mader is one of the icons of science education. Her dedication to her students, coupled with her clear, concise writing style, has benefited the education of thousands of students over the past four decades. As an educator, it is an honor to continue her legacy and to bring her message to the next generation of students.

As always, I had the privilege to work with a phenomenal group of people on this edition. I would especially like to thank you, the numerous instructors who have shared e-mails with me or have invited me into your classrooms, both physically and virtually, to discuss your needs as instructors and the needs of your students. You are all dedicated and talented teachers, and your energy and devotion to quality teaching is what drives a textbook revision.

Many dedicated and talented individuals assisted in the development of this edition of *Inquiry into Life*. I am very grateful for the help of so many professionals at McGraw-Hill who were involved in bringing this book to fruition. Therefore, I would like to thank the following:

- The product developer, Anne Winch, for her patience and impeccable ability to keep me focused.
- My brand manager, Michelle Vogler, for her guidance and reminders of why what we do is important.
- My marketing manager, Britney Ross, and market development manager, Beth Bettcher, for placing me in contact with great instructors, on campus and virtually, throughout this process.
- My digital expert, Eric Weber, for helping me envision the possibilities in our new digital world.
- My content project manager, Kelly Hart, and program manager, Angie FitzPatrick, for calmly steering this project throughout the publication process.
- Lori Hancock and Jo Johnson, for the photos within this text. Biology is a visual science, and your contributions are evident on every page.
- Dawnelle Krouse, Michael McGee, and Sharon O'Donnell, who acted as my proofreaders and copyeditors for this edition.
- Jane Peden, for her behind-the-scenes work that keeps us all functioning.

As both an educator and an author, communicating the importance of science represents one of my greatest passions. Our modern society is based largely on advances in science and technology over the past few decades. As I present in this text, there are many challenges facing humans, and an understanding of how science can help analyze, and offer solutions to, these problems is critical to our species' health and survival.

I also want to acknowledge my family and friends for all of their support. My wife, Sandy, who has never wavered in her energy and support of my projects. Over the course of my work with McGraw-Hill, I have watched the natural curiosity of my children, Devin and Kayla, develop. Thank you both for your motivation in making our world a better place.

Michael Windelspecht, PhD
Blowing Rock, NC

Reviewers for This Edition

I would like to thank the individuals below for taking the time to review the content of the previous edition. Your comments and suggestions played an important part of this revision.

Reviewers of the Sixteenth Edition

Sara Albert, *South Louisiana Community College*
Bert Atsma, *Union County College*
Ellen Baker, *Santa Monica College*
Linda Berlin, *Seminole State College*
Seth Bybee, *Brigham Young University*
Erica Cain, *Northeast Lakeview College*
Candice Chatman, *Atlanta Metropolitan State College*
Thomas Chen, *Santa Monica College*
Ryan Combs, *Crowder College*
Sandi Connelly, *Rochester Institute of Technology*
Stella Evans, *Towson University*
Michelle A. Fisher, *Three Rivers College*
Ruchi Gehi, *George Brown College*
Elesha Goodfriend, *Walters State Community College*
John Griffis, *Joliet Junior College*
Daniel B. Hewins, *Rhode Island College*
Laura Houston, *Northeast Lakeview College*
Tonya Huff, *Riverside City College*
Jason Jennings, *Southwest Tennessee Community College*
Michelle L. LaPorte, *St. Louis Community College at Forest Park*
Cheri LaRue, *University of Arkansas*
Julie Littrell, *Kaskaskia College*
Jean Mitchell, *Northwest Florida State College and University of Florida*
Jane Monroe, *Humboldt State University*
Kelly Moore, *Walters State Community College*
Lucas Moyer-Horner, *University of Utah*
Eric Myers, *South Suburban College*
Terina Nusinov, *Seminole State College*
Michael Patterson, *Clackamas Community College*
Michelle Priest, *Santa Ana College*
Darryl Ritter, *Northwest Florida State College*
Peter Rubinelli, *Northwest Arkansas Community College*
Nilesh Sharma, *Western Kentucky University*
Syede Sharmin Maswood, *Millersville University*
Robert Sikes, *University of Delaware*
Heather M. Smith, *Riverside City College*
Marc A. Smith, *Sinclair Community College*
Barbara Stegenga, *The University of North Carolina at Chapel Hill*
Keith M. Vogelsang, *Ivy Tech Community College*
Ryan Wager, *Millersville University of Pennsylvania*
Lynnette Wick, *Walters State Community College*

BRIEF CONTENTS

CONTENTS

Unit 7 Behavior and Ecology

SEM 37,000X ©Thomas Deerinck/Science Source

1

Biology: The Study of Life

Artificial Life

What are the minimal requirements for life? That question has occupied the minds of philosophers and scientists for thousands of years. However, in just the past decade, answers to this question have begun to emerge from a developing field of scientific study called artificial life.

One of the first of these studies occurred in 2010, when a research team led by Craig Venter (a pioneer in genetic research) was successful in removing the genetic information contained within the DNA of a bacterium and replacing it with a synthetic form of DNA.

In 2016, the same group of researchers took their research one step further. This time, they asked what minimal instructions were needed by a cell for it to be considered alive. They constructed a cell that functioned on just 473 genes (humans have around 19,000). In the process, they not only narrowed in on what the minimal requirements for life are, but also created the first example of an artificial species.

The development of artificial life opens up the opportunity for humans to construct cells that perform specific tasks, such as producing insulin, cleaning toxic waste, or producing fuel more efficiently. However, there are concerns about these new endeavors, and some scientists are urging constraint until the risks have been determined.

In this chapter, we are going to explore the concept of life by examining the general characteristics that are shared by all living organisms on our planet.

As you read through this chapter, think about the following questions:
1. What are the basic characteristics that define life?
2. How do the processes of adaptation and evolution shape life over time?
3. How might the use of artificial life help scientists address some of the challenges facing society?

CHAPTER OUTLINE

1.1 The Characteristics of Life
1.2 The Classification of Organisms
1.3 The Process of Science
1.4 Challenges Facing Science

1.1 The Characteristics of Life

Life. Everywhere we look, from the deepest trenches of the oceans to the geysers of Yellowstone, we find that planet Earth is teeming with life. Without life, our planet would be nothing but a barren rock hurtling through space. The variety of life on Earth is staggering. Recent estimates suggest that there are around 8.7 million species on the planet, but this number does not include the bacteria (Fig. 1.1), which historically have been difficult to identify. The variety of living organisms ranges in size from bacteria, much too small to be seen by the naked eye, all the way up to 100-foot-tall giant sequoia trees or 40-ton humpback whales. Humans are just one of those species.

The diversity of life seems overwhelming, and yet all living organisms have certain characteristics in common. Taken together, these characteristics give us insight into the nature of life and help us distinguish living from nonliving organisms. All life generally shares the following characteristics: (1) is organized, (2) requires materials and energy, (3) has the ability to reproduce and develop, (4) responds to its environment, (5) maintains an internal environment, and (6) has the capacity to adapt to its environment. Next, we will explore each of these characteristics in more detail.

Life Is Organized

Life can be organized in a hierarchy of levels (Fig. 1.2). In trees, humans, and all other organisms, **atoms** join together to form

Figure 1.1 Diversity of life on Earth. If aliens ever visit our corner of the universe, they will be amazed at the diversity of life on our planet. Yet despite its diversity, all life shares some common characteristics. (whales): ©iStockphoto/Getty Images; (*E. coli*): Source: USDA-ARS/Eric Erbe & Chris Pooley, photographers; (sequoia): ©Robert Glusic/Getty Images; (Earth): ©Ingram Publishing/Alamy; (mushroom): ©IT Stock/age fotostock; (*Homo sapiens*): ©Blend Images/Ariel Skelley/Getty Images; (*Euglena*): ©blickwinkel/Alamy

Figure 1.2 Levels of biological organization. Life is connected from the atomic level to the biosphere. While the cell is the basic unit of life, it is comprised of molecules and atoms. The sum of all life on the planet is called the biosphere.

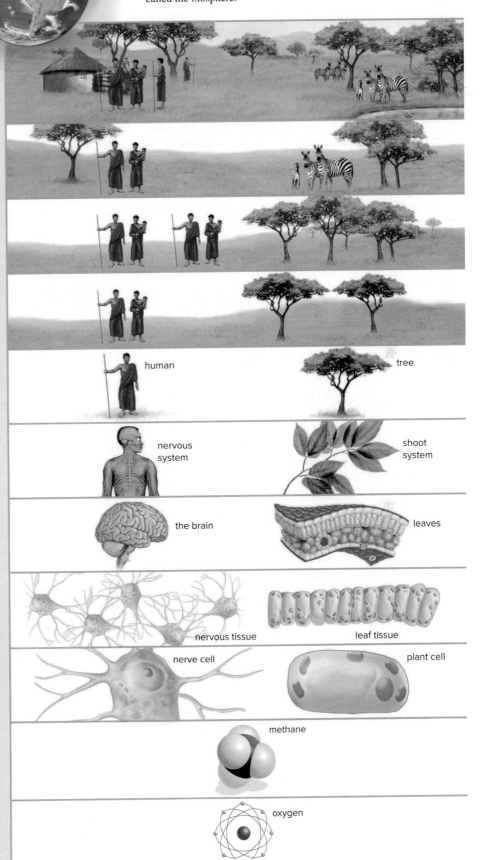

Biosphere
Regions of the Earth's crust, waters, and atmosphere inhabited by living organisms

Ecosystem
A community plus the physical environment

Community
Interacting populations in a particular area

Population
Organisms of the same species in a particular area

Species
A group of similar, interbreeding organisms

Organism
An individual; complex individuals contain organ systems

Organ System
Composed of several organs working together

Organ
Composed of tissues functioning together for a specific task

Tissue
A group of cells with a common structure and function

Cell
The structural and functional unit of all living organisms

Molecule
Union of two or more atoms of the same or different elements

Atom
Smallest unit of an element; composed of electrons, protons, and neutrons

human

tree

nervous system

shoot system

the brain

leaves

nervous tissue

leaf tissue

nerve cell

plant cell

methane

oxygen

molecules, such as DNA molecules that occur within cells. A **cell** is the smallest unit of life, and some organisms are single-celled. In multicellular organisms, a cell is the smallest structural and functional unit. For example, a human nerve cell is responsible for conducting electrical impulses to other nerve cells. A **tissue** is a group of similar cells that perform a particular function. Nervous tissue is composed of millions of nerve cells that transmit signals to all parts of the body. Several tissues then join together to form an **organ.** The main organ that receives signals from nerves is the brain. Organs then work together to form an **organ system.** In the nervous system, the brain sends messages to the spinal cord, which in turn sends them to body parts through spinal nerves. In describing the levels of biological organization, the term **organism** is used to describe an individual that is a collection of multiple organ systems. However, the term is also used generally to indicate an individual member of a species.

The levels of biological organization extend beyond the individual. All the members of one **species** (a group of interbreeding organisms) in a particular area belong to a **population.** A tropical grassland may have a population of zebras, acacia trees, and humans, for example. The interacting populations of the grasslands make up a **community.** The community of populations interacts with the physical environment to form an **ecosystem.** Finally, all the Earth's ecosystems collectively make up the **biosphere.**

Life Requires Materials and Energy

Living organisms need an outside source of materials and energy to maintain their organization and carry on life's other activities. Plants, such as trees, use carbon dioxide, water, and solar energy to make their own food. Humans and other animals acquire materials and energy by eating food.

The food we eat provides nutrients, which cells use as building blocks or for **energy**—the capacity to do work. Cells use energy from nutrients to carry out everyday activities. Some nutrients are broken down completely by chemical reactions to provide the necessary energy to carry out other reactions, such as building proteins. The term **metabolism** is used to describe all of the chemical reactions that occur in a cell. Cells need energy to perform their metabolic functions, and it takes work to maintain the organization of a cell, as well as to maintain an organism.

The ultimate source of energy for nearly all life on Earth is the sun. Plants and certain other organisms are able to capture solar energy and carry on photosynthesis, a process that transforms solar energy into the chemical energy of organic nutrient molecules. All life on Earth acquires energy by metabolizing nutrient molecules made by photosynthesizers. This applies even to plants themselves.

The energy and chemical flow between organisms also defines how an ecosystem functions (Fig. 1.3). Within an ecosystem, chemical cycling and energy flow begin when producers, such as grasses, take in solar energy and inorganic nutrients to produce food (organic nutrients) by photosynthesis. Chemical cycling (aqua arrows in Fig. 1.3) occurs as chemicals move from one population to another in a food chain, until death and decomposition allow inorganic nutrients to be returned to the producers once again. Energy (red arrows in Fig. 1.3), on the other hand, flows from the sun through plants and the other members of the food

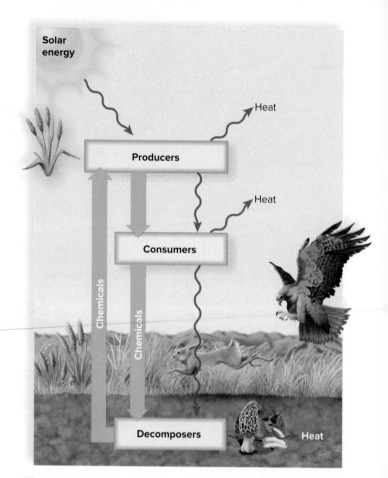

Figure 1.3 Chemical cycling and energy flow in an ecosystem.   In an ecosystem, chemical cycling (aqua arrows) and energy flow (red arrows) begin when plants use solar energy and inorganic nutrients to produce their own food. Chemicals and energy are passed from one population to another in a food chain. Eventually, energy dissipates as heat. With the death and decomposition of organisms, chemicals are returned to living plants once more.

chain as they feed on one another. The energy gradually dissipates and returns to the atmosphere as heat. Because energy does not cycle, ecosystems could not stay in existence without solar energy and the ability of photosynthetic organisms to absorb it.

Energy flow and nutrient cycling in an ecosystem climate largely determine not only where different ecosystems are found in the biosphere, but also what communities are found in the ecosystem. For example, deserts exist in areas of minimal rain, while forests require much rain. The two most biologically diverse ecosystems—tropical rain forests and coral reefs—occur where solar energy is most abundant. One example of an ecosystem in North America is the grasslands, which are inhabited by populations of rabbits, hawks, and various types of grasses, among many others. These populations interact with each other by forming food chains in which one population feeds on another. For example, rabbits feed on grasses, while hawks feed on rabbits and other organisms.

Life Has the Capacity to Reproduce and Develop

Life comes only from life. All forms of life have the capability of **reproduction,** or to make another organism like themselves. Bacteria, protists, and other single-celled organisms simply split in

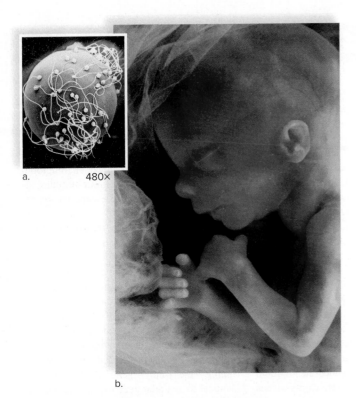

a.　　480×

b.

Figure 1.4 Growth and development define life. Following the (**a**) fertilization of an egg cell by a sperm cell (**b**) humans grow and develop. All life exhibits the characteristics of growth and development.

(a): ©David M. Phillips/Science Source; (b): ©Brand X Pictures/Punchstock

two. In most multicellular organisms, the reproductive process begins with the pairing of a sperm from one partner and an egg from the other partner. The union of sperm and egg (Fig 1.4), followed by many cell divisions, results in an immature stage, which proceeds through stages of **development,** or change, to become an adult.

When living organisms reproduce, their **genes,** or genetic instructions, are passed on to the next generation. Random combinations of sperm and egg, each of which contains a unique collection of genes, ensure that the offspring has new and different characteristics. An embryo develops into a whale, a yellow daffodil, or a human because of the specific set of genes it inherits from its parents. In living organisms, the genes are made of long **DNA (deoxyribonucleic acid)** molecules. DNA provides the blueprint, or instructions, for the organization and metabolism of the particular organism. All cells in a multicellular organism contain the same set of genes, but only certain genes are turned on in each type of specialized cell. You may notice that not all members of a species, including humans, are exactly the same, and that there are obvious differences between species. These differences are the result of **mutations,** or inheritable changes in the genetic information. Mutation provides an important source of variation in the genetic information. However, not all mutations are bad—the observable differences in eye and hair color are examples of mutations.

Life Responds to Its Environment

Right now, your eyes and ears are receiving stimuli from the external environment. If there is a loud noise nearby, your natural tendency is to look in that direction. All organisms respond to information, called stimuli, from their environment. This may be as simple as the movement toward food, or away from a predator. Movement in animals, including humans, is dependent upon their nervous and musculoskeletal systems. Other living organisms use a variety of mechanisms in order to move. The leaves of plants track the passage of the sun during the day, and when a houseplant is placed near a window, hormones help its stem bend to face the sun.

The movement of an organism, whether self-directed or in response to a stimulus, constitutes a large part of its **behavior.** Behavior is largely directed toward minimizing injury, acquiring food, and reproducing.

Life Maintains an Internal Environment

The internal environment of your body, and every cell in your body, is kept relatively stable. For example, human body temperature will show only a slight fluctuation throughout the day. Also, the body's ability to maintain a normal internal temperature is somewhat dependent on the external temperature—we will die if the external temperature becomes too hot or cold. This tendency toward a stable environment is called **homeostasis,** and it is the way that a living organism distinguishes itself from its environment.

Organisms have intricate feedback and control mechanisms that do not require any conscious activity. These mechanisms may be controlled by one or more tissues themselves or by the nervous system. When you are studying and forget to eat lunch, your liver releases stored sugar to keep blood sugar levels within normal limits. Many organisms depend on behavior to regulate their internal environment. In animals, these behaviors are controlled by the nervous system and are usually not consciously controlled. For example, a lizard may raise its internal temperature by basking in the sun, or cool down by moving into the shade.

Life Has the Ability to Adapt

Throughout the nearly 4 billion years that life has been on Earth, the environment has constantly been changing. For example, glaciers that once covered much of the world's surface 10,000–15,000 years ago have since receded, and many areas that were once covered by ice are now habitable. On a smaller scale, a hurricane or fire could drastically change the landscape in an area quite rapidly.

As the environment changes, some individuals of a species (a group of organisms that can successfully interbreed and produce fertile offspring) may possess certain features that make them better suited to the new environment. We call such features **adaptations.** For example, consider a hawk, which can catch and eat a rabbit. A hawk, like other birds, can fly because it has hollow bones, which is an adaptation. Similarly, its strong feet can take the shock of a landing after a hunting dive, and its sharp claws can grab and hold onto prey. As is presented in the Scientific Inquiry feature, "Adapting to Life at High Elevations," humans also exhibit adaptations to their environment.

Individuals of a species that are better adapted to their environment tend to live longer and produce more offspring than other individuals. This differential reproductive success, called **natural selection,** results in changes in the characteristics of a population (all the members of a species within a particular area) through

SCIENCE IN YOUR LIFE ▶ SCIENTIFIC INQUIRY

Adapting to Life at High Elevations

Humans, like all other organisms, have an evolutionary history. This means not only that we share common ancestors with other animals, but also that over time we demonstrate adaptations to changing environmental conditions. One study of populations living in the high-elevation mountains of Tibet (Fig. 1A) demonstrates how the processes of evolution and adaptation influence humans.

Normally, if a person moves to a higher altitude, his or her body responds by making more hemoglobin, the component of blood that carries oxygen, which thickens the blood. For minor elevation changes, this does not present much of a problem. But for people who live at

Figure 1A Humans' adaptations to their environments. Humans have adaptations that allow them to live at high altitudes, such as these individuals in Tibet.
©Michael Freeman/Getty Images

extreme elevations (some people in the Himalayas can live at elevations of over 13,000 ft, or close to 4,000 m), this can present a number of health problems, including chronic mountain sickness, a disease that affects people who live at high altitudes for extended periods of time. The problem is that, as the amount of hemoglobin increases, the blood thickens and becomes more viscous. This can cause elevated blood pressure, or hypertension, and an increase in the formation of blood clots, both of which have negative physiological effects.

Because high hemoglobin levels would be a detriment to people at high elevations, it makes sense that natural selection would favor individuals who produced less hemoglobin at high elevations. Such is the case with the Tibetans in this study. Researchers have identified an allele of a gene that reduces hemoglobin production at high elevations. Comparisons between Tibetans at both high and low elevations strongly suggest that selection has played a role in the prevalence of the high-elevation allele.

The gene is *EPSA1*, located on chromosome 2 of humans. *EPSA1* produces a transcription factor, which basically regulates which genes are turned on and off in the body, a process called gene expression. The transcription factor produced by *EPSA1* has a number of functions in the body. For example, in addition to controlling the amount of hemoglobin in the blood, this transcription factor also regulates other genes that direct how the body uses oxygen.

When the researchers examined the variations in *EPSA1* in the Tibetan population, they discovered that their version greatly reduces the production of hemoglobin. Therefore, the Tibetan population has lower hemoglobin levels than people living at lower altitudes, allowing these individuals to escape the consequences of thick blood.

How long did it take for the original population to adapt to living at higher elevations? Initially, the comparison of variations in these genes between high-elevation and low-elevation Tibetan populations suggested that the event may have occurred over a 3,000-year period. But researchers were skeptical of those data because they represented a relatively rapid rate of evolutionary change. Additional studies of genetic databases yielded an interesting finding—the *EPSA1* gene in Tibetans was identical to a similar gene found in an ancient group of humans called the Denisovans (see Section 32.5). Scientists now believe that the *EPSA1* gene entered the Tibetan population around 40,000 years ago, either through interbreeding between early Tibetans and Denisovans, or from one of the immediate ancestors of this lost group of early humans.

Questions to Consider

1. What other environments do you think could be studied to look for examples of human adaptation?
2. In addition to hemoglobin levels, do you think that people at high elevations may exhibit other adaptations?

---

time. That is, adaptations that result in higher reproductive success tend to increase in frequency in a population from one generation to the next. This change in the frequency of traits in populations and species is called **evolution.**

Evolution explains both the unity and diversity of life. As stated at the beginning of this chapter, all organisms share the same basic characteristics of life because we all share a common ancestor—the first cell or cells—that arose nearly 4 billion years ago. During the past 4 billion years, the Earth's environment has changed drastically, and the diversity of life has been shaped by the evolutionary responses of organisms to these changes.

Check Your Progress 1.1

1. List the common characteristics of all living organisms.
2. Trace the organization of life from the cell to the biosphere.
3. Explain how adaptations relate to evolutionary change.

1.2 The Classification of Organisms

Learning Outcomes

Upon completion of this section, you should be able to
1. Describe how living organisms are classified.
2. Distinguish between the three domains of life.
3. Explain the role of supergroups in the classification of life.

Because life is so diverse, it is helpful to group organisms into categories. **Taxonomy** is the discipline of identifying and grouping organisms according to certain rules. Taxonomy makes sense out of the bewildering variety of life on Earth and is meant to provide valuable insight into evolution. **Systematics** is the study of the evolutionary relationships between species. As systematists learn more about living organisms, the taxonomy often changes. DNA technology is now widely used by systematists to revise current information and to discover previously unknown relationships between organisms.

Several of the basic classification categories, also called *taxa,* are **domain, kingdom, phylum, class, order, family, genus,** and, finally, **species.** These are listed in order from the most inclusive (domains) to the least inclusive (species).

Domains

Domains are the largest classification category. Based upon biochemical and genetic evidence, scientists have identified three domains: **domain Archaea, domain Bacteria,** and **domain Eukarya.** Both domain Archaea and domain Bacteria contain single-celled **prokaryotes,** which lack the membrane-bound nucleus and organelles found in the cells of **eukaryotes** in domain Eukarya.

Prokaryotes are structurally simple (Fig. 1.5) but metabolically complex. *Bacteria* are found almost anywhere—in the water, soil, and atmosphere, as well as on our skin and in our digestive tracts. Although some bacteria cause diseases, others are beneficial, both environmentally and commercially. For example, bacteria can be used to develop new medicines, to clean up oil spills, or to help purify water in sewage treatment plants. *Archaea* live in aquatic environments that lack oxygen or are too salty, too hot, or too acidic for most other organisms. Because these environments are similar to those of the primitive Earth, archaea represent the first cells that evolved on the planet.

Domain Bacteria

- Prokaryotic cells of various shapes
- Adaptations to all environments
- Absorb, photosynthesize, or chemosynthesize food
- Unique chemical characteristics

6,600×

Escherichia coli, a bacterium

a.

Domain Archaea

- Prokaryotic cells of various shapes
- Adaptations to extreme environments
- Absorb or chemosynthesize food
- Unique chemical characteristics

33,200×

Sulfolobus, an archaean

b.

Figure 1.5 Prokaryotic domains.   The two prokaryotic domains of life include (**a**) bacteria and (**b**) archaea. (a): ©A.B. Dowsett/SPL/Science Source; (b): ©Eye of Science/Science Source

Eukaryotic Supergroups and Kingdoms

Historically, the classification of domain Eukarya divided organisms into one of four kingdoms (Fig. 1.6). **Protists** (kingdom Protista), which comprise a very diverse group of organisms, range

Domain Eukarya: Kingdom Protista

- Algae, protozoans, slime molds, and water molds
- Complex single cell (sometimes filaments, colonies, or even multicellular)
- Absorb, photosynthesize, or ingest food

160×

Paramecium, a single-celled protozoan

Domain Eukarya: Kingdom Fungi

- Molds, mushrooms, yeasts, and ringworms
- Mostly multicellular filaments with specialized, complex cells
- Absorb food

Amanita, a mushroom

Domain Eukarya: Kingdom Plantae

- Certain algae, mosses, ferns, conifers, and flowering plants
- Multicellular, usually with specialized tissues, containing complex cells
- Photosynthesize food

Phalaenopsis, orchid, a flowering plant

Domain Eukarya: Kingdom Animalia

- Sponges, worms, insects, fishes, frogs, turtles, birds, and mammals
- Multicellular with specialized tissues containing complex cells
- Ingest food

Vulpes, a red fox

Figure 1.6 Kingdoms of domain Eukarya.   Historically, eukaryotes were divided into four kingdoms. However, recent genetic analyses have added a new level of classification, the supergroup (Table 1.1). (*Paramecium*): ©M. I. Walker/Science Source; (mushroom): ©Tinke Hamming/Ingram Publishing; (orchid): ©Pixtal/age fotostock; (fox): ©Fuse/Getty Images

TABLE 1.1   Eukaryotic Supergroups

Supergroup	Representative Organisms
Archaeplastida	Plants, red and green algae
SAR (stramenopiles, alveolates, and rhizaria)	Brown algae, dinoflagellates, paramecia
Excavata	Euglenoids, trypanosomes
Amoebozoa	Amoeba, slime molds
Opisthokonta	Animals, fungi, choanoflagellates

TABLE 1.2   Classification of Humans

Classification Category	Characteristics
Domain Eukarya	Cells with nuclei
Supergroup Opisthokonta	Possess cells with flagella
Kingdom Animalia	Multicellular, motile, ingestion of food
Phylum Chordata	Dorsal supporting rod and nerve cord
Class Mammalia	Hair, mammary glands
Order Primates	Adapted to climb trees
Family Hominidae	Adapted to walk erect
Genus *Homo*	Large brain, tool use
Species *Homo sapiens**	Body characteristics similar to modern humans

* To specify an organism, you must use the full binomial name, such as *Homo sapiens*.

from single-celled forms to a few multicellular ones. Some use photosynthesis to manufacture food, and some must acquire their own food. Common protists include algae, the protozoans, and the water molds.

The other three kingdoms of eukaryotes in Figure 1.6 (plants, fungi, and animals) all evolved from protists. **Plants** (kingdom Plantae) are multicellular, photosynthetic organisms. Example plants include azaleas, zinnias, and pines. Among the **fungi** (kingdom Fungi) are the familiar molds and mushrooms that, along with bacteria, help decompose dead organisms. **Animals** (kingdom Animalia) are multicellular organisms that must ingest and process their food. Aardvarks, jaguars, and humans are representative animals.

Recently, the development of improved techniques in analyzing the DNA of organisms suggests that not all of the protists share the same evolutionary lineage, meaning that the evolution of the eukaryotes has occurred along several paths. A new taxonomic group, called a **supergroup,** is being developed to explain these evolutionary relationships. There are currently five supergroups for domain Eukarya (Table 1.1). However, as studies continue, the relationship and structure of these groups are being revised. We will explore the structure of the eukaryotic supergroups in more detail in Section 29.1.

Additional Levels of Classification

The other classification categories are phylum, class, order, family, genus, and species. Each classification category is more specific than the one preceding it. For example, the species within one genus share very similar characteristics, while those within the same kingdom share only general characteristics. Modern humans are the only living species in the genus *Homo,* but many different types of animals are in the animal kingdom (Table 1.2). To take another example, all species in the genus *Pisum* (pea plants) look quite similar, while the species in the plant kingdom can be quite different, as is evident when we compare grasses to trees.

Systematics helps biologists make sense out of the bewildering variety of life on Earth because organisms are classified according to their presumed evolutionary relationships. Organisms placed in the same genus are the most closely related, and those in separate domains are the most distantly related. Therefore, all eukaryotes are more closely related to one another than they are to bacteria or archaea. Similarly, all animals are more closely related to one another than they are to plants. As more is learned about evolutionary relationships among species, systematic relationships are changed. Systematists are even now making observations and

performing experiments that will soon result in changes in the classification system adopted by this text.

Scientific Names

Taxonomists assign a binomial, or two-part name, to each species. For example, the scientific name for humans is *Homo sapiens,* and for the garden pea, *Pisum sativum.* The first word is the genus to which the species belongs, and the second word is the specific epithet, or species name. Note that both words are in italics, but only the genus name is capitalized. The genus name can be used alone to refer to a group of related species. Also, a genus can be abbreviated to a single letter if used with the species name (e.g., *P. sativum*).

Scientific names are in a common language—Latin—and biologists use them universally to avoid confusion. Common names, by contrast, tend to overlap across multiple species.

Check Your Progress 1.2

1. Describe the differences among a domain, supergroup, and kingdom.
2. List the levels of taxonomic classification from most inclusive to least inclusive.
3. Explain why scientists assign species to a hierarchical classification system (e.g., kingdom, phylum, class).

1.3 The Process of Science

Learning Outcomes

Upon completion of this section, you should be able to
1. Identify the components of the scientific method.
2. Distinguish between a theory and a hypothesis.
3. Analyze a scientific experiment and identify the hypothesis, experiment, control groups, and conclusions.

The process of science pertains to the study of biology. As you can see from Figure 1.2, the multiple stages of biological organization mean that life can be studied at a variety of levels. Some biological disciplines are cytology, the study of cells; anatomy, the study of structure; physiology, the study of function; botany, the study of plants; zoology, the study of animals; genetics, the study of heredity; and ecology, the study of the interrelationships between organisms and their environments.

Religion, aesthetics, ethics, and science are all ways in which humans seek order in the natural world. The nature of scientific inquiry differs from these other ways of knowing and learning, because the scientific process uses the **scientific method,** a standard series of steps used in gaining new knowledge that is widely accepted among scientists. The scientific method (Fig. 1.7) acts as a guideline for scientific studies. Scientists often modify or adapt the process to suit their particular field of study.

Observation

Scientists believe that nature is orderly and measurable—that natural laws, such as the law of gravity, do not change with time—and that a natural event, or *phenomenon,* can be understood more fully through **observation**—a formal way of watching the natural world.

Scientists use all of their senses in making observations. The behavior of chimpanzees can be observed through visual means, the disposition of a skunk can be observed through olfactory means, and the warning rattles of a rattlesnake provide auditory information of imminent danger. Scientists also extend the ability of their senses by using instruments; for example, the microscope enables us to see objects that could never be seen by the naked eye. Finally, scientists may expand their understanding even further by taking advantage of the knowledge and experiences of other scientists. For instance, they may look up past studies at the library or on the Internet, or they may write or speak to others who are researching similar topics.

Hypothesis

After making observations and gathering knowledge about a phenomenon, a scientist uses inductive reasoning to formulate a possible explanation. **Inductive reasoning** occurs whenever a person uses creative thinking to combine isolated facts into a cohesive whole. In some cases, a chance observation alone may help a scientist arrive at an idea.

One famous case pertains to the antibiotic penicillin, which was discovered in 1928. While examining a petri dish of bacteria that had become contaminated with the mold *Penicillium,* Alexander Flemming observed an area that was free of bacteria. Flemming, an early expert on antibacterial substances, reasoned that the mold might have been producing an antibacterial compound.

We call such a possible explanation for a natural event a **hypothesis.** A hypothesis is not merely a guess; rather, it is an informed statement that can be tested in a manner suited to the processes of science.

All of a scientist's past experiences, no matter what they might be, have the potential to influence the formation of a hypothesis. But a scientist considers only hypotheses that can be tested. Moral and religious beliefs, while very important in the lives of many people, differ between cultures and through time and may not be scientifically testable.

Predictions and Experiments

Scientists often perform an **experiment,** which is a series of procedures, to test a hypothesis. To determine how to test a hypothesis, a

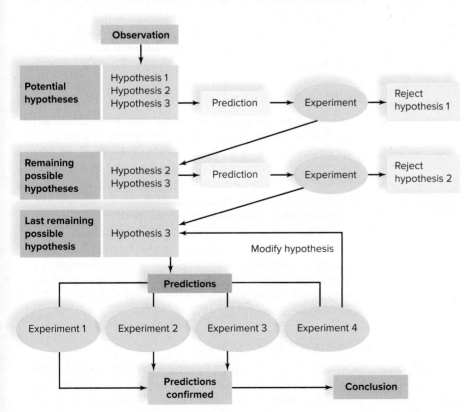

Figure 1.7 Flow diagram for the scientific method.   On the basis of new and/or previous observations, a scientist formulates a hypothesis. The hypothesis is used to develop predictions to be tested by further experiments and/or observations, and new data either support or do not support the hypothesis. Following an experiment, a scientist often chooses to retest the same hypothesis or to test a related hypothesis. Conclusions from many different but related experiments may lead to the development of a scientific theory. For example, studies pertaining to development, anatomy, and fossil remains all support the theory of evolution.

scientist uses deductive reasoning. **Deductive reasoning** involves "if, then" logic. In designing the experiment, the scientist may make a **prediction,** or an expected outcome, based on knowledge of the factors in the experiment.

The manner in which a scientist intends to conduct an experiment is called the **experimental design.** A good experimental design ensures that scientists are examining the contribution of a specific variable, called the **experimental variable,** to the observation. The result is termed the **responding variable,** or dependent variable, because it is due to the experimental variable:

Experimental Variable (Independent Variable)	Response Variable (Dependent Variable)
Factor of the experiment being tested	Result or change that occurs due to the experimental variable

To ensure that the results will be meaningful, an experiment contains both test groups and a **control** group. A test group is exposed to the experimental variable, but the control group is not. If the control group and test groups show the same results, the experimenter knows that the hypothesis predicting a difference between them is not supported.

Scientists often use **model** organisms and model systems to test a hypothesis. Model organisms, such as the fruit fly *Drosophila melanogaster* or the mouse *Mus musculus,* are chosen because they allow the researcher to control aspects of the experiment, such as age and genetic background. Cell biologists may use mice for modeling the effects of a new drug. Like model organisms, model systems allow the scientist to control specific variables and environmental conditions in a way that may not be possible in the natural environment. For example, ecologists may use computer programs to model how human activities will affect the climate of a specific ecosystem. While models provide useful information, they do not always answer the original question completely. For example, medicine that is effective in mice should ideally be tested in humans, and ecological experiments that are conducted using computer simulations need to be verified by field experiments. Biologists, and all other scientists, continuously design and revise their experiments to better understand how different factors may influence their original observation.

Presenting and Analyzing the Data

The **data,** or results, from scientific experiments may be presented in a variety of formats, including tables and graphs. A graph shows the relationship between two quantities. In many graphs, the experimental variable is plotted on the *x*-axis (horizontal), and the result is plotted along the *y*-axis (vertical). Graphs are useful tools to summarize data in a clear and simplified manner. For example, the line graph in Figure 1.8 shows the variation in the concentration of blood cholesterol over a 4-week study. The bars above each data point represent the variation, or standard error, in the results. The title and labels can assist you in reading a graph; therefore, when looking at a graph, first check the two axes to determine what the graph pertains to. By looking at this graph, we know that the blood cholesterol levels were highest during week 2, and we can see to what degree the values varied over the course of the study.

Figure 1.8 **Presentation of scientific data.**   This line graph shows the variation in the concentration of blood cholesterol over a 4-week study. The bars above each data point represent the variation, or standard error, in the results.

Statistical Data

Most scientists who publish research articles use statistics to help them evaluate their experimental data. In statistics, the standard error, or standard deviation, tells us how uncertain a particular value is. Suppose you predict how many hurricanes Florida will have next year by calculating the average number during the past 10 years. If the number of hurricanes per year varies widely, your standard error will be larger than if the number per year is usually about the same. In other words, the standard error tells you how far off the average could be. If the average number of hurricanes is four and the standard error is ± 2, then your prediction of four hurricanes is between two and six hurricanes. In Figure 1.8, the standard error is represented by the bars above and below each data point. This provides a visual indication of the statistical analysis of the data.

Statistical Significance

When scientists conduct an experiment, there is always the possibility that the results are due to chance or to some factor other than the experimental variable. Investigators take into account several factors when they calculate the probability value (p) that their results were due to chance alone. If the probability value is low, researchers describe the results as statistically significant. A probability value of less than 5% (usually written as $p < 0.05$) is acceptable; even so, keep in mind that the lower the p value, the less likely it is that the results are due to chance. Therefore, the lower the p value, the greater the confidence the investigators and you can have in the results. Depending on the type of study, most scientists like to have a p value of < 0.05, but p values of < 0.001 are common in many studies.

Scientific Publications

Scientific studies are customarily published in scientific journals (Fig. 1.9), so that all aspects of a study are available to the scientific community. Before information is published in scientific

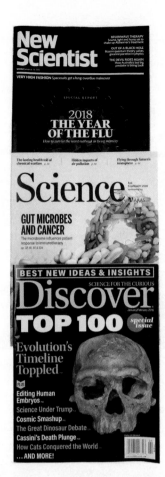

Figure 1.9 Scientific publications. Scientific journals, such as *Science,* are scholarly journals in which researchers share their findings with other scientists. Scientific magazines, such as *Discover* and *New Scientist,* contain articles that are usually written by reporters for a broader audience.
©Ricochet Creative Productions LLC

journals, it is typically reviewed by experts, who ensure that the research is credible, accurate, unbiased, and well-executed. Another scientist should be able to read about an experiment in a scientific journal, repeat the experiment in a different location, and get the same (or very similar) results. Some articles are rejected for publication by reviewers when they believe there is something questionable about the design of an experiment or the manner in which it was conducted. This process of rejection is important in science because it causes researchers to critically review their hypotheses, predictions, and experimental designs, so that their next attempt will more adequately address their hypothesis. Often, it takes several rounds of revision before research is accepted for publication in a scientific journal.

Scientific magazines (Fig. 1.9), such as *Discover,* differ from scientific journals in that they report scientific findings to the general public. The information in these articles is usually obtained from articles first published in scientific journals.

Scientific Theory

The ultimate goal of science is to understand the natural world in terms of scientific **theories,** which are concepts that join together

well-supported and related hypotheses. In ordinary speech, the word *theory* refers to a speculative idea. In contrast, a scientific theory is supported by a broad range of observations, experiments, and data, often from a variety of disciplines. Some of the basic theories of biology are:

Theory	Concept
Cell	All organisms are composed of cells, and new cells only come from preexisting cells.
Homeostasis	The internal environment of an organism stays relatively constant—within a range that is protective of life.
Evolution	A change in the frequency of traits that affect reproductive success in a population or species across generations.

As stated earlier, the theory of evolution is the unifying concept of biology because it pertains to many different aspects of life. For example, the theory of evolution enables scientists to understand the history of life, as well as the anatomy, physiology, and embryological development of organisms, to name a few.

The theory of evolution has been a fruitful scientific theory, meaning it has helped scientists generate new hypotheses. Because this theory has been supported by so many observations and experiments for over 100 years, some biologists refer to the **principle** of evolution, a term sometimes used for theories that are generally accepted by an overwhelming number of scientists. The term **law** instead of principle is preferred by some. For instance, in our discussion of energy in Chapter 6, we will examine the laws of thermodynamics.

An Example of the Scientific Method

We now know that most stomach and intestinal ulcers (open sores) are caused by the bacterium *Helicobacter pylori.* Let's say investigators want to determine which of two antibiotics is best for the treatment of an ulcer. When clinicians do an experiment, they try to vary just the experimental variables—in this case, the medications being tested. A control group is not given the medications, but one or more test groups receive them. If by chance the control group shows the same results as a test group, the investigators immediately know that the results of their study are invalid, because the medications may have had nothing to do with the results. The study depicted in Figure 1.10 shows how investigators may study this hypothesis:

HYPOTHESIS: Newly discovered antibiotic B is a better treatment for ulcers than antibiotic A, which is in current use.

Experimental Design

Next, the investigators might decide to use three experimental groups: one control group and two test groups. It is important to reduce the number of possible variables (differences), such as sex, weight, and other illnesses, among the groups. Therefore, the investigators randomly divide a very large group of volunteers equally into the three groups. The hope is that any differences will be distributed evenly among the three groups. This is possible only if the investigators have a large number of volunteers. The three groups are to be treated like this:

Control group: Subjects with ulcers are not treated with either antibiotic.
Test group 1: Subjects with ulcers are treated with antibiotic A.
Test group 2: Subjects with ulcers are treated with antibiotic B.

After the investigators have determined that all volunteers do have ulcers, they will want the subjects to think they are all receiving the same treatment. This is an additional way to protect the results from any influence other than the medication. To achieve this end, the subjects in the control group can receive a placebo, a treatment that appears to be the same as that administered to the other two groups but actually contains no medication. In this study, the use of a placebo would help ensure the same dedication by all subjects to the study.

Results and Conclusion

After 2 weeks of administering the same amount of medication (or placebo) in the same way, the stomach and intestinal linings of each subject are examined to determine if ulcers are still present. Endoscopy is a procedure that involves inserting an endoscope (a small, flexible tube with a tiny camera on the end) down the throat and into the stomach and the upper part of the small intestine. Then, the doctor can see the lining of these organs and can check for ulcers. Tests performed during an endoscopy can also determine if *Helicobacter pylori* is present.

Because endoscopy is somewhat subjective, it is probably best if the examiner is not aware of which group the subject is in; otherwise, the examiner's prejudice may influence the examination. When neither the patient nor the technician is aware of the specific treatment, it is called a double-blind study.

In this study, the investigators may decide to determine the effectiveness of the medication by the percentage of people who no longer have ulcers. So, if 20 people out of 100 still have ulcers, the medication is 80% effective. The difference in effectiveness is easily read in the graph portion of Figure 1.10.

CONCLUSION: On the basis of their data, the investigators conclude that their hypothesis has been supported.

Check Your Progress 1.3

1. Identify the role of the experimental variable in an experiment.
2. Distinguish between the roles of the test group and the control group in an experiment.
3. Describe the process by which a scientist may test a hypothesis about an observation.

Figure 1.10 Example of a controlled study.   In this study, a large number of people were divided into three groups. The control group received a placebo and no medication. One of the test groups received medication A, and the other test group received medication B. The results are depicted in a graph, and it shows that medication B was a more effective treatment than medication A for the treatment of ulcers.

(students): ©Andrey Popov/Shutterstock

1.4 Challenges Facing Science

As we have learned in this chapter, science is a systematic way of acquiring knowledge about the natural world. Science is a slightly different endeavor than technology. **Technology** is the application of scientific knowledge to the interests of humans. Scientific investigations are the basis for the majority of our technological advances. As is often the case, a new technology, such as your cell phone or a new drug, is based on years of scientific investigations. In this section, we are going to explore some of the challenges facing science, technology, and society.

Climate Change

The term **climate change** refers to changes in the normal cycles of the Earth's climate that may be attributed to human activity. Climate change is primarily due to an imbalance in the chemical cycling of the element carbon. Normally, carbon is cycled within an ecosystem. However, due to human activities, more carbon dioxide is being released into the atmosphere than is being removed. In 1850, atmospheric CO_2 was at about 280 parts per million (ppm); today, it is over 400 ppm (Fig. 1.11). This increase is largely due to the burning of fossil fuels and the destruction of forests to make way for farmland and pasture. Today, the amount of carbon dioxide released into the atmosphere is about twice the amount that remains in the atmosphere. It's believed that most of this dissolves in the oceans, which is increasing their acidity. The increased amount of carbon dioxide (and other gases) in the atmosphere is causing a rise in temperature called **global warming.** These gases allow the sun's rays to pass through, but they absorb and radiate heat back to Earth, a phenomenon called the *greenhouse effect.*

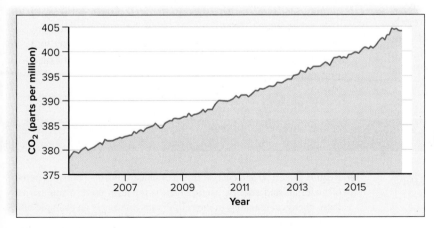

Figure 1.11 **Increases in atmospheric carbon dioxide concentrations.**
The global average carbon dioxide concentration now exceeds 400 ppm and is contributing to climate change and global warming. Source: NOAA, "Global Climate Change: Facts," http://climate.nasa.gov/vital-signs/carbon-dioxide/.

Figure 1.12 **Climate change and extinction.**   The Bramble Cay melomys (*Melomys rubicola*) was native to Australian great barrier reef islands. However, it has not been observed since 2009 and is believed to be the first mammalian species to go extinct as a result of climate change. ©Polaris/Newscom

There is a consensus among scientists from around the globe that climate change and global warming are causing significant changes in many of the Earth's ecosystems and represent one of the greatest challenges of our time. Throughout this text, we will return to how climate change is affecting ecosystems, reducing biodiversity, and contributing to human disease. We will examine climate change in more detail in Section 35.3.

Biodiversity and Habitat Loss

Biodiversity is the total number and relative abundance of species, the variability of their genes, and the different ecosystems in which they live. The biodiversity of our planet has been estimated to be around 8.7 million species (not counting bacteria), and so far, approximately 2.3 million have been identified and named. **Extinction** is the death of a species or larger classification category. It is estimated that presently we are losing hundreds of species every year due to human activities and that as much as 38% of all species, including most primates, birds, and amphibians, may be in danger of extinction before the end of the century. In many cases, these extinctions are associated with climate change (Fig. 1.12). Many biologists are alarmed about the present rate of extinction and hypothesize it may eventually rival the rates of the five mass extinctions that occurred during our planet's history. The last mass extinction, about 65 million years ago, caused many plant and animal species, including the dinosaurs, to become extinct.

The two most biologically diverse ecosystems—tropical rain forests and coral reefs—are home to many organisms. These ecosystems are also threatened by human activities. The canopy of the tropical rain forest alone supports a variety of organisms, including orchids, insects, and monkeys. Coral reefs, which are found just offshore of the continents and islands near the equator, are built up from calcium carbonate skeletons of sea animals called corals. Reefs provide a habitat for many animals, including jellyfish, sponges, snails, crabs, lobsters, sea turtles, moray eels, and some of the world's most colorful fishes. Like tropical rain forests, coral reefs are severely threatened as the human population increases in size. Some reefs are 50 million years old, yet in just a few decades, human activities have destroyed an estimated 25% of all coral reefs and seriously degraded another 30%. At this rate, nearly three-quarters could be destroyed within 40 years. Similar statistics are available for tropical rain forests.

The destruction of healthy ecosystems has many unintended effects. For example, we depend on them for food, medicines, and various raw materials. Draining of the natural wetlands of the Mississippi and Ohio Rivers and the construction of levees have worsened flooding problems, making once fertile farmland undesirable. The destruction of South American rain forests has killed many species that may have yielded the next miracle drug and has decreased the availability of many types of lumber. We are only now beginning to realize that we depend on ecosystems even more for the services they provide. Just as chemical cycling occurs within a single ecosystem, so all ecosystems keep chemicals cycling throughout the biosphere. The workings of ecosystems ensure that the environmental conditions of the biosphere are suitable for the continued existence of humans. In fact, several studies show that ecosystems cannot function properly unless they remain biologically diverse. We will explore the concept of biodiversity in greater detail in later chapters of this text.

Emerging and Reemerging Diseases

Over the past decade, avian influenza (H5N1 and H7N9), swine flu (H1N1), severe acute respiratory syndrome (SARS), and Middle East respiratory syndrome (MERS) have been in the news. These are called **emerging diseases** because they are relatively new to humans. Where do emerging diseases come from? Some may result from new and/or increased exposure to animals or insect populations that act as vectors for disease. Changes in human behavior and use of technology can also result in new diseases. SARS is thought to have arisen in the Guangdong province of China due to the consumption of civets, a type of exotic cat considered a delicacy. The civets were possibly infected by exposure to horseshoe bats sold in open markets. Legionnaires' disease emerged in 1976 due to bacterial contamination of a large air-conditioning system in a hotel. The bacteria thrived in the cooling tower used as the water source for the air-conditioning system.

In addition, globalization results in the transport all over the world of diseases that were previously restricted to isolated communities. The first SARS cases were reported in southern China in November of 2002. By the end of February 2003, SARS had reached nine countries/provinces, mostly through airline travel.

Some pathogens mutate and change hosts, jumping from birds to humans, for example. Before 1997, avian flu was thought to affect only birds. A mutated strain jumped to humans in the 1997 outbreak. To control that epidemic, officials killed 1.5 million chickens to remove the source of the virus. New forms of avian influenza (bird flu) are being discovered every few years.

Reemerging diseases are also a concern. Unlike an emerging disease, a reemerging disease has been known to cause disease in humans for some time, but generally has not been considered a health risk due to a relatively low level of incidence in human populations. Even so, reemerging diseases can cause problems. An excellent example is the Ebola outbreak in West Africa of 2014–2015. Ebola outbreaks have been known since 1976, but generally have affected only small groups of humans. The 2014–2015 outbreak was a much larger event. Although the exact numbers may never be known, it is estimated that over 28,000 people were infected, with over 11,000 fatalities. The outbreak has disrupted the societies of several West African nations.

Both emerging and reemerging diseases have the potential to cause health problems for humans across the globe. Scientists investigate not only the causes of these diseases (for example, the viruses) but also their effects on our bodies and the mechanisms by which they are transmitted. We will take a closer look at viruses in Section 28.5.

Check Your Progress 1.4

1. Explain how a new technology differs from a scientific discovery.
2. Explain why the conservation of biodiversity is important to human society.
3. Summarize how emerging diseases and climate change have the potential for influencing the entire human population.

Conclusion

In the beginning of this chapter, we introduced the idea of artificial life and asked the question of what it meant to be alive. Throughout this chapter, you have explored some of the basic characteristics of life as we currently understand it. By using artificial systems and experimentation, scientists are beginning to determine what it means to be alive, and apply that knowledge to benefit not only human society, but other life that shares our planet with us.

SUMMARIZE

1.1 The Characteristics of Life

Evolution accounts for both the diversity and the unity of life we see about us. All organisms share the following characteristics of life:

- Organization: the levels of biological organization become larger as follows: **atoms** and **molecules** → **cells** → **tissues** → **organs** → **organ systems** → **organisms** → **species** → **populations** → **communities** → **ecosystems** → **biosphere.**

- Require materials and energy: living organisms need an outside source of materials and **energy. Metabolism** is the term used to summarize these chemical reactions in the cell. Photosynthesis is an example of a metabolic process.

- Reproduce and develop: Organisms **reproduce** to pass on their genetic information, included in the **genes** of their **DNA,** to the next generation. **Mutations** introduce variation into the DNA. **Development** is the series of steps that an organism proceeds through to become an adult.

- Respond to stimuli: they detect and react to internal and external events.
- Maintain homeostasis: **homeostasis** is the ability to maintain a stable internal environment.
- Adaptations: **Adaptations** allow an organism to exist in a particular environment. **Evolution** is the accumulation of these changes over multiple generations and the resulting adaptation to the organism's environment.

1.2 The Classification of Organisms

- In **taxonomy,** organisms are assigned an italicized binomial name that consists of the genus and the specific epithet. From the least inclusive to the most inclusive category, each **species** belongs to a **genus, family, order, class, phylum, kingdom,** and, finally, **domain. Systematics** is the study of evolutionary relationships between species.
- The three domains of life are Archaea, Bacteria, and Eukarya. **Domain Archaea** and **domain Bacteria** contain **prokaryotic** organisms that are structurally simple but metabolically complex. **Domain Eukarya** contains the **eukaryotic** protists, fungi, plants, and animals. These are now being organized as **supergroups** to reflect their evolutionary relationships. **Protists** range from single-celled to multicellular organisms and include the protozoans and most algae. Among the **fungi** are the familiar molds and mushrooms. **Plants** are well-known as photosynthetic multicellular organisms, while **animals** are multicellular and ingest their food.
- Each species has a Latin scientific name that consists of the genus and species name. Both names are italicized, but only the genus is capitalized, as in *Homo sapiens*.

1.3 The Process of Science

When studying the natural world, scientists use a process called the **scientific method.**

- **Observations,** along with previous data, are used to formulate a hypothesis. **Inductive reasoning** allows a scientist to combine facts into a **hypothesis.**
- New observations and/or experiments are carried out in order to test the hypothesis. **Deductive reasoning** allows for the development of a **prediction** of what may occur as a result of the experiment. A good **experimental design** includes an **experimental variable** and a **control** group. Scientists may use **models** and model organisms in their experimental design.
- The **data** from the experimental and observational results are analyzed, often using statistical methods. The results are often presented in tables or graphs for ease of interpretation.
- A conclusion is made as to whether the results support the hypothesis or do not support the hypothesis.
- The results may be submitted to a scientific publication for review by the scientific community.
- Over time, multiple conclusions in a particular area may allow scientists to arrive at a **theory** (or **principle** or **law**), such as the cell theory or the theory of evolution. The theory of evolution is a unifying concept of biology.

1.4 Challenges Facing Science

While science investigates the principles of the natural world, **technology** applies this knowledge to the needs of society. Some challenges that scientists are investigating include:

- The loss of **biodiversity** and habitats such as coral reefs and rain forests. This often results in the **extinction** of species.

- **Emerging diseases,** such as avian influenza and SARS, and reemerging diseases, such as Ebola.
- The impact of **climate change** and **global warming.**

ASSESS

Testing Yourself

Choose the best answer for each question.

1.1 The Characteristics of Life

1. Which of these is not a property of all living organisms?
   a. organization
   b. acquisition of materials and energy
   c. care for their offspring
   d. reproduction
   e. responding to the environment

2. Which of the following on this list represents the lowest level of biological organization?
   a. an organ   c. an organ system
   b. a cell   d. an organism

3. The process that involves passing on genetic information between generations is called
   a. natural selection.   c. development.
   b. reproduction.   d. metabolism.

1.2 The Classification of Organisms

4. Which of the following includes prokaryotic organisms?
   a. protists   c. archaea
   b. fungi   d. plants

5. The term *supergroup* is used in the classification of
   a. the archaeans.
   b. the bacteria.
   c. the eukaryotes.
   d. all of the domains of life.

6. The process by which evolution occurs is called
   a. natural selection.   c. reproduction.
   b. development.   d. taxonomy.

1.3 The Process of Science

7. After formulating a hypothesis, a scientist
   a. proves the hypothesis to be true or false.
   b. tests the hypothesis by experimentation.
   c. decides how to best avoid having a control.
   d. makes sure environmental conditions are just right.
   e. formulates a scientific theory.

8. Experiments examine the contribution of the _____ to the observation.
   a. responding variable
   b. control group
   c. standard deviation
   d. experimental variable

9. Which of the following is not correctly linked?
   a. model: a representation of an object used in an experiment
   b. standard deviation: a form of statistical analysis
   c. principle: a theory that is not supported by experimental evidence
   d. data: the results of an experiment or observation

1.4 Challenges Facing Science

10. Which of the following applies scientific knowledge to the needs of society?
    a. evolution
    c. systematics
    b. taxonomy
    d. technology

11. Which of the following represents the permanent loss of a species?
    a. natural selection
    c. extinction
    b. greenhouse effect
    d. climate change

12. H5N1 and SARS are examples of
    a. extinct species.
    c. endangered habitats.
    b. forms of greenhouse gases.
    d. emerging diseases.

ENGAGE

BioNOW

Want to know how this science is relevant to your life? Check out the BioNOW video below:

■ Characteristics of Life

At the end of the day, you head over to the gym for a game of basketball with your friends. Afterward, you go out to eat at a local restaurant. Explain how the characteristics of life apply to you during these activities.

Thinking Critically

1. Explain how model organisms make the study of the dependent variable in an experiment easier.

2. Suppose that we find a form of life on another planet that has a different characteristic than those listed in this chapter. What would this tell us about life on our planet and the process of evolution?

3. You are a scientist working at a pharmaceutical company and have developed a new cancer medication that has the potential for use in humans. Outline a series of experiments, including the use of a model, to test whether the cancer medication works.

©Kim Scott/Ricochet Creative Productions LLC

CHAPTER

2

The Molecules of Cells

Good or Bad Cholesterol?

Without cholesterol, your body would not function well. What is it then about cholesterol that can cause problems in your body? Although blood tests now distinguish between "good" cholesterol and "bad" cholesterol, in reality there is only one molecule called cholesterol. Cholesterol is essential to the normal functioning of cells. It is part of a cell's outer membrane, and it serves as a precursor to hormones, such as the sex hormones testosterone and estrogen. Without cholesterol, your cells would be in major trouble.

So when does cholesterol become "bad"? Your body packages cholesterol in high-density lipoproteins (HDLs) or low-density lipoproteins (LDLs), depending on such factors as genetics, your diet, your activity level, and whether you smoke. Cholesterol packaged as HDL is primarily on its way from the tissues to the liver for recycling. This reduces the likelihood of the formation of deposits called plaques in the arteries; thus, HDL cholesterol is considered "good." LDL is carrying cholesterol to the tissues; high levels of LDL cholesterol contribute to the development of plaques, which can result in heart disease, making LDL the "bad" form. When cholesterol levels are measured, the ratio of LDL to HDL is determined.

Exercising, improving your diet, and not smoking can all lower LDL levels. In some cases, the ability of your body to regulate blood cholesterol levels is genetic; a defect in a gene called *LDLR* lowers the body's ability to move cholesterol into cells. For individuals with high blood cholesterol levels, there are a variety of prescription medications that can help lower LDL levels and raise HDL levels.

These lipoproteins are just one example of the organic molecules that our bodies need to function correctly. In this chapter, you will learn about the structure and function of the major classes of organic molecules, including carbohydrates, lipids, proteins, and nucleic acids.

As you read through this chapter, think about the following questions:

1. What class of biological molecules does cholesterol belong to?
2. Is cholesterol a hydrophobic or hydrophilic molecule?
3. What distinguishes cholesterol from other biological molecules?

CHAPTER OUTLINE

BEFORE YOU BEGIN

Before beginning this chapter, take a few moments to review the following discussions:

Section 1.1 Why must organisms acquire materials and energy?

Section 1.1 How does energy flow through an ecosystem?

Figure 1.2 Where are atoms and molecules located in the levels of biological organization?

2.1 Introduction to Chemistry

Learning Outcomes

Upon completion of this section, you should be able to

1. Describe how protons, neutrons, and electrons relate to atomic structure.
2. Understand how to interpret the periodic table of elements.
3. Describe how variations in an atomic nucleus account for its physical properties.
4. Identify the beneficial and harmful uses of radiation.

Everything—including your computer, the chair you're sitting on, the water you drink, and the air you breathe—is composed of matter. **Matter** refers to anything that takes up space and has mass. Matter has many diverse forms, but it can exist only in four distinct states: solid, liquid, gas, or plasma.

All matter, both nonliving and that found in living organisms, is composed of certain basic substances called **elements.** An element is a substance that cannot be broken down into simpler substances with different properties by ordinary chemical means. Each element has its own unique properties such as density, solubility, melting point, and chemical reactivity. Only 92 naturally occurring elements (see Appendix A) serve as the building blocks of all matter. Other elements have been made in research labs, but they are not biologically important.

The Earth's crust, as well as all organisms, is composed of elements, but they differ as to which elements are predominant (Fig. 2.1). Only six elements—carbon, hydrogen, nitrogen, oxygen, phosphorus, and sulfur—are basic to life and make up about 95%

of the body weight of organisms. The properties of these elements are essential to the uniqueness of cells and organisms. Other elements, including sodium, potassium, calcium, iron, and magnesium, are also important to all organisms.

Atomic Structure

In the early 1800s, the English scientist John Dalton championed the atomic theory, which says that elements consist of tiny particles called **atoms.** An atom is the smallest part of an element that displays the properties of the element. An element and its atoms share the same name. Elements are identified by an atomic symbol that is composed of one or two letters. These letters are usually derived from either the English or Latin names of the element. For example, the symbol H means a hydrogen atom, the symbol Cl stands for chlorine, and the symbol Na (for *natrium* in Latin) is used for a sodium atom.

Physicists have identified a number of subatomic particles that make up atoms. The three best-known subatomic particles are positively charged **protons,** uncharged **neutrons,** and negatively charged **electrons.** Protons and neutrons are located within the nucleus of an atom, and electrons move around the nucleus. Figure 2.2 shows the arrangement of the subatomic particles in a helium atom, which has only two electrons. In Figure 2.2*a*, the shading shows the probable location of electrons, and in Figure 2.2*b*, the circle represents an **electron shell** (or orbital), which indicates the average location of electrons.

The concept of an atom has changed greatly since Dalton's day. If an atom could be drawn the size of a football field, the nucleus would be like a gumball in the center of the field. The electrons would be tiny specks whirling about in the upper stands. Most of an atom is empty space. We should also realize that both of the models in Figure 2.2 indicate only where the electrons are expected to be most of the time. In our analogy, the electrons might very well stray outside the stadium at times.

All atoms of an element have the same number of protons. This is called the **atomic number.** The number of protons in the nucleus makes each atom unique. Generally, atoms are assumed to be electrically neutral, meaning that the number of electrons is the same as the number of protons in the atom. The atomic number tells you not only the number of protons but also the number of electrons.

Each atom also has its own **mass number,** which is the sum of the protons and neutrons in the nucleus. Protons and neutrons are assigned one atomic mass unit (AMU) each. Electrons are so

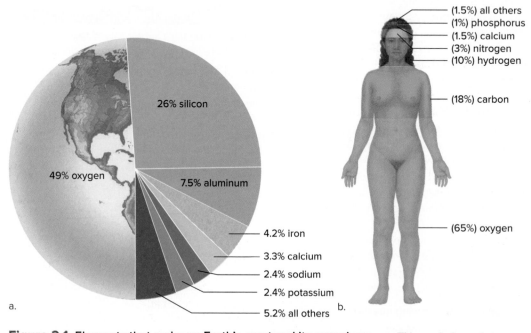

Figure 2.1 Elements that make up Earth's crust and its organisms.   a. This graph shows that Earth's crust primarily contains the elements silicon (Si), aluminum (Al), and oxygen (O). **b.** Organisms primarily contain the elements oxygen (O), nitrogen (N), carbon (C), and hydrogen (H). Biological molecules often also contain the elements sulfur (S) and phosphorus (P).

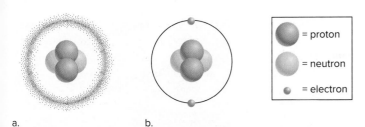

Subatomic Particles			
Particle	Electric Charge	Atomic Mass Unit (AMU)	Location
Proton	+1	1	Nucleus
Neutron	0	1	Nucleus
Electron	−1	0	Electron shell

Figure 2.2 Model of a helium (He) atom.   Atoms contain subatomic particles called protons, neutrons, and electrons. Protons and neutrons are within the nucleus, and electrons are outside the nucleus. **a.** The shading shows the probable location of the electrons in the helium atom. **b.** A circle, called an electron shell or orbital, represents the average location of an electron. **c.** The electric charge and the atomic mass units of the subatomic particles vary as shown.

small that their AMU is considered zero in most calculations (Fig. 2.2*c*). By convention, when we are discussing an atom independently of the periodic table (see below), the atomic number is written as a subscript to the lower left of the atomic symbol. The mass number is written as a superscript to the upper left of the atomic symbol.

$$^{12}_{6}C$$

mass number — 12 — atomic symbol
atomic number — 6 — C

Whereas each atom of an element has the same atomic number (and thus the same number of protons), the number of neutrons may vary slightly. **Isotopes** (Gk. *isos,* "equal") are atoms of the same element that differ in the number of neutrons. For example, the element carbon has three naturally occurring isotopes:

$$^{12}_{6}C \qquad ^{13}_{6}C \qquad ^{14}_{6}C$$

The term **atomic mass** refers to the average mass for all the isotopes of that atom. Because the majority of carbon is carbon 12, the atomic mass of carbon is closer to 12 than to 13 or 14.

6 — atomic number
C — atomic symbol
12.01 — atomic mass

To determine the number of neutrons from the atomic mass, subtract the number of protons from the atomic mass and take the closest whole number.

It is important to note that the term *mass* is used, not *weight,* because mass is constant, while weight changes according to the gravitational force of a body. The gravitational force of the Earth is greater than that of the moon; therefore, substances weigh less on the moon, even though their mass has not changed.

The Periodic Table

Once chemists discovered a number of the elements, they began to realize that even though each element consists of a different atom, certain chemical and physical characteristics are common between groups of elements. The periodic table was constructed as a way to group the elements, and therefore atoms, according to these characteristics. Notice in Figure 2.3 that the periodic table is arranged according to increasing atomic number. The vertical columns in the table are groups. The horizontal rows are periods, which cause each atom to be in a particular group. For example, all the atoms in group VII react with one atom at a time, for reasons we will soon explore. The atoms in group VIII are called the noble gases because they are inert and rarely react with another atom. Helium and krypton are examples of noble gases.

Radioactive Isotopes

Some isotopes of an element are unstable, or radioactive. For example, unlike the other two isotopes of carbon, carbon 14 changes over time into nitrogen 14, which is a stable isotope of the element nitrogen. As carbon 14 decays, it releases various types of energy in the form of rays and subatomic particles. The radiation given off by radioactive isotopes can be detected in various ways. The Geiger counter is an instrument that is commonly used to detect radiation.

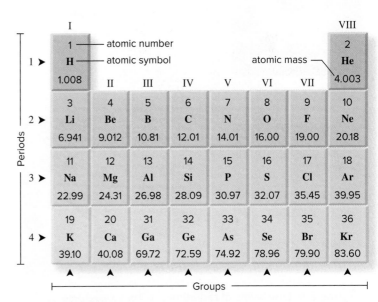

Figure 2.3 A portion of the periodic table.   In the periodic table, the elements are arranged in the order of their atomic numbers and placed in groups (vertical columns) and periods (horizontal rows). All the atoms in a particular group have certain chemical characteristics in common. The first four periods contain the elements most important in biology. The complete periodic table is in Appendix A.

In 1896, the French physicist Antoine-Henri Becquerel discovered that a sample of uranium would produce a bright image on a photographic plate even in the dark, and a similar method of detecting radiation is still in use today. Marie Curie, who worked with Becquerel, coined the term *radioactivity* and contributed much to its study. Today, biologists use radiation to date objects from our distant past, to create images, and to trace the movement of substances in the body.

Low Levels of Radiation

The chemical behavior of a radioactive isotope is essentially the same as that of the stable isotopes of an element. This means you can put a small amount of radioactive isotope in a sample and it becomes a *tracer,* or chemical tag, by which to detect molecular changes.

The importance of chemistry to medicine is nowhere more evident than in the many medical uses of radioactive isotopes. Specific tracers are used in imaging the body's organs and tissues. For example, after a patient drinks a solution containing a minute amount of radioactive iodine 131, it becomes concentrated in the thyroid—the only organ to take up iodine. A subsequent image of the thyroid indicates whether it is healthy in structure and function (Fig. 2.4a). Positron emission tomography (PET) is a way to determine the comparative activity of tissues. Radioactively labeled

a.   b.

Figure 2.5 High levels of radiation.   a. Radiation kills bacteria and fungi. Irradiated peaches (*top*) spoil less quickly (compared to *bottom*) and can be kept for a longer length of time. **b.** Physicians use radiation therapy to kill cancer cells. (a): ©Kim Scott/Ricochet Creative Productions LLC; (b): ©Mark Kostich/Getty Images

glucose, which emits a subatomic particle known as a positron, is injected into the body. The radiation given off is detected by sensors and analyzed by a computer. The result is a color image that shows which tissues took up glucose and are metabolically active (Fig. 2.4b). A PET scan of the brain can help diagnose a brain tumor, Alzheimer disease, epilepsy, or whether a stroke has occurred.

High Levels of Radiation

Radioactive substances in the environment can harm cells, damage DNA, and cause cancer. When researchers, such as Marie Curie, began studying radiation in the nineteenth century, its harmful effects were not known, and many developed cancer. The release of radioactive particles following a nuclear power plant accident, such as occurred in Japan in 2011 following the tsunami (see the Health feature "Japan's Nuclear Crisis") can have far-reaching and long-lasting effects on human health.

However, the effects of radiation can also be put to good use (Fig. 2.5). Radiation from radioactive isotopes has been used for many years to sterilize medical and dental products. Radiation is now used to sterilize the U.S. mail and other packages to free them of possible pathogens, such as anthrax spores. The ability of radiation to kill cells is often applied to cancer cells. Targeted radioisotopes can be introduced into the body so the subatomic particles emitted destroy only cancer cells, with little risk to the rest of the body. X-rays, another form of high-energy radiation, can be used for medical diagnosis.

Electrons

In an electrically neutral atom, the positive charges of the protons in the nucleus are balanced by the negative charges of electrons moving about the nucleus. Various models may be used to illustrate the structure of at atom. One of the more common (Fig. 2.6) is called the Bohr model, and is named after the physicist Niels Bohr. The Bohr model is useful in examining the number of

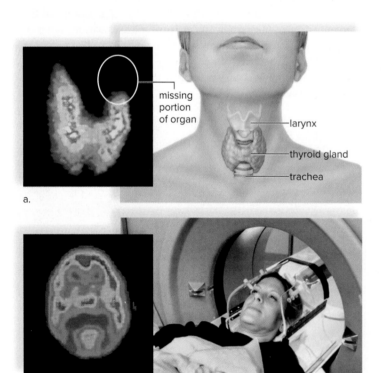

a.

b.

Figure 2.4 Low levels of radiation.   a. The asymmetrical appearance of the thyroid on the PET (positron emission tomography) scan indicates the presence of a tumor that does not take up the radioactive iodine. **b.** A PET scan reveals which portions of the brain are most active (yellow and red colors). (photos) (a): ©Southern Illinois University/Science Source; (b) (PET scan): ©Mazziota et al./Science Source; (woman): Source: National Institutes of Health

missing portion of organ
larynx
thyroid gland
trachea

Japan's Nuclear Crisis

On March 11, 2011, a magnitude 8.9 earthquake struck the coast of Japan, causing the shutdown of 11 of Japan's nuclear power reactors, including reactors 1, 2, and 3 at the Fukushima Daiichi power plant (Fig. 2A). Shortly after the shutdown, the pressure within reactor 1 built up to twice that of normal levels. In an attempt to relieve pressure, steam-containing radiation was vented from the reactors. Over the next several days, the plant continued to have problems and the levels of radiation in the surrounding areas continued to rise. Shortly thereafter, Japanese officials formed a 10-kilometer (km) evacuation zone, which, over the next few days, was increased to 20 km, resulting in the relocation of close to 390,000 people.

The Fukushima disaster released a number of radioactive isotopes into the atmosphere and the surrounding ocean. These included isotopes of iodine, cesium, and xenon. Each of these isotopes is a by-product of the nuclear fission reactions that occur within a nuclear power plant. As the uranium in these plants breaks down, the uranium atoms release energy (such as gamma radiation) and particles (alpha and beta particles). This is called radioactive decay, and the loss of particles effectively transforms the original uranium atom into isotopes of another element. In many cases, the resulting isotopes are also radioactive, and then undergo additional radioactive decay.

Three of the more common isotopes released by the Fukushima disaster are listed next. These isotopes differ considerably in the types of radiation they release, their half-lives, and their potential adverse effects on living organisms, such as humans.

Cesium-134 and Cesium-137

Perhaps the biggest concern following the Fukushima disaster was the release of radioactive cesium. Naturally, cesium (Cs) is a metal with an atomic number of 55 and an atomic mass of approximately 132.9. However, within a nuclear reactor, two isotopes of cesium (^{134}Cs and ^{137}Cs) are formed. Cesium 134 and Cesium 137 have half-lives of 2 years and 30 years, respectively. Both of these isotopes release beta particles and gamma radiation as they decay. Both beta particles and gamma radiation have the ability to penetrate tissues and damage cells and their DNA. Exposure to beta particles and gamma radiation, either internally or externally, can cause burns and greatly increases the chance of cancer.

Following the Fukushima disaster, there were significant concerns of ^{137}Cs entering the atmosphere and the Pacific Ocean. Because of the winds at the time, most of the ^{137}Cs was carried back over Japan, forcing evacuations up to 35 kilometers (22 miles) inland. Large regions of this evacuation zone still remain uninhabitable due to soil contamination by ^{137}Cs.

Small amounts of atmospheric ^{137}Cs were detected on the West Coast of the United States, and some ^{137}Cs has been detected in the waters off the coast of California, but in neither case at levels considered hazardous to humans.

Iodine-131

Radioactive iodine, ^{131}I, was also released by the reactors. Iodine is a water-soluble element that has an atomic number of 53 and an atomic mass of approximately 126.9. There are several forms of iodine isotopes, with ^{131}I being formed mainly by nuclear fission reactions. Like ^{137}Cs, ^{131}I emits beta particles that may damage tissues. It also releases small amounts of gamma radiation.

Radioactive iodine has a half-life of only about 8 days, so its long-term effect on the environment is minimal. In our bodies, iodine is mainly used by the thyroid gland to manufacture the thyroid hormones associated with metabolic functions, including overall metabolic rate. Intense exposure to ^{131}I may cause thyroid problems, including thyroid cancer. Interestingly, isotopes of iodine at low doses are used by the medical profession to diagnose problems with the thyroid gland and treat some forms of thyroid cancer.

Xenon-133

Xenon (Xe) is a noble gas with an atomic number of 54 and an atomic weight of 131.2. One isotope of xenon, ^{133}Xe, has a half-life of only 5 days. Like ^{131}I and ^{137}Cs, ^{133}Xe emits beta particles that may damage tissue, but the short half-life of the isotope limits the chances of severe problems. Further reducing the risk is the fact that, as a noble gas, xenon does not react with other elements, and thus is not easily introduced into the chemical compounds within cells.

Like radioactive iodine, ^{133}Xe is used by the medical profession to diagnose disease. Because it is a gas, it is frequently used in the diagnosis of lung disorders. Studies are now underway to use ^{133}Xe as a form of treatment for certain types of lung cancer, because it easily enters the lungs and the emission of beta particles can help destroy lung cancer cells.

Questions to Consider

1. Describe the differences among ^{137}Cs, ^{131}I, and ^{133}Xe and their nonradioactive isotopes.
2. Explain why ^{137}Cs is more dangerous than ^{131}I or ^{133}Xe.

Figure 2A   The Fukushima Power Plant.   The explosion of the plant following the tsunami released radioactive isotopes into the atmosphere and surrounding water. ©Gamma/Gamma-Rapho/Getty Images

Figure 2.6 Bohr models of atoms.   Electrons orbit the nucleus at particular energy levels (electron shells or orbitals). The first shell contains up to two electrons, and each shell thereafter is most stable when it contains eight electrons. Atoms with an atomic number above 20 may have more electrons in their outer shells. The outermost, or valence, shell helps determine the atom's chemical properties and how many other elements it can interact with.

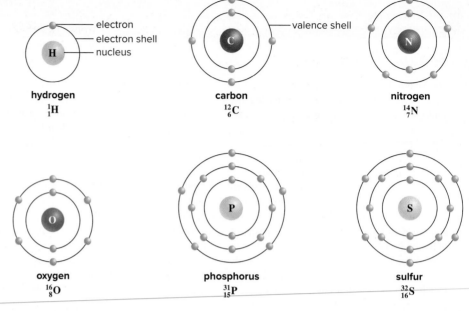

hydrogen
$^{1}_{1}H$

carbon
$^{12}_{6}C$

nitrogen
$^{14}_{7}N$

oxygen
$^{16}_{8}O$

phosphorus
$^{31}_{15}P$

sulfur
$^{32}_{16}S$

electrons in an atom, but it does not accurately depict the exact location of individual electrons at any given moment.

In the Bohr diagrams in Figure 2.6, the energy levels of the electrons, called electron shells or electron orbitals, are drawn as concentric rings about the nucleus. In all atoms, the first shell (closest to the nucleus) can contain two electrons. The next electron shell contains a maximum of eight electrons. In all atoms, the lower shells are filled with electrons before the next higher level contains any electrons.

The sulfur atom, with an atomic number of 16, has two electrons in the first shell, eight electrons in the second shell, and six electrons in the third, or outer, orbital. Revisit the periodic table (see Fig. 2.3) and note that sulfur is in the third period. In other words, the horizontal row tells you how many electron shells an atom has. Also note that sulfur is in group VI. The group tells you how many electrons an atom has in its outer shell.

Regardless of how many shells an atom has, the outermost shell is called the **valence shell.** The valence shell is important, because it determines many of an atom's chemical properties. If an atom has only one shell, the valence shell is complete when it has two electrons. In atoms with more than one shell, the valence shell is most stable when it has eight electrons. This is called the **octet rule.** Each atom in a group within the periodic table has the same number of electrons in its valence shell. As mentioned previously, all the atoms in group VIII of the periodic table have eight electrons in their valence shell. These elements are also called the noble gases, because they do not ordinarily participate in chemical reactions.

The electrons in the valence shells play an important role in determining how an element undergoes chemical reactions. Atoms with fewer than eight electrons in the outer shell react with other atoms in such a way that, after the reaction, each has a stable outer shell. As we will see, the number of electrons in an atom's valence shell determines whether the atom gives up, accepts, or shares electrons to acquire eight electrons in the outer shell.

Check Your Progress 2.1

1. Identify why the key elements in the Earth's crust would differ from those present in living organisms.
2. Predict the number of electrons, protons, and neutrons in calcium (see Fig. 2.3).
3. Explain how radiation can be both beneficial and harmful to humans.
4. Explain the differences between oxygen 16 and oxygen 18.

2.2 Molecules and Compounds

Learning Outcomes

Upon completion of this section, you should be able to

1. Describe how elements are combined into molecules and compounds.
2. List the different types of bonds that occur between elements.
3. Compare the relative strengths of ionic, covalent, and hydrogen bonds.

Atoms, except for noble gases, routinely bond with one another. A **molecule** is formed when two or more atoms bond together. For example, oxygen does not exist in nature as a single atom, O. Instead, two oxygen atoms are joined to form a molecule of oxygen, O_2. When atoms of two or more different elements bond together, the product is called a **compound.** Water (H_2O) is a compound that contains atoms of hydrogen and oxygen. We can also speak of molecules of water because a molecule is the smallest part of a compound that still has the properties of that compound.

Electrons possess energy, and the bonds that exist between atoms also contain energy. Organisms are directly dependent on chemical-bond energy to maintain their organization. When a chemical reaction occurs, electrons shift in their relationship to one another, and energy may be given off or absorbed. This same energy is used to carry on our daily lives.

Ionic Bonding

Ions form when electrons are transferred from one atom to another. For example, sodium (Na), with only one electron in its valence shell, tends to be an electron donor (Fig. 2.7a). Once it gives up this electron, the second shell, with eight electrons, becomes its outer shell. Chlorine (Cl), on the other hand, tends to be an electron acceptor. Its outer shell has seven electrons, so it requires one more electron to have a complete outer shell. When a sodium atom and a chlorine atom come together, an electron is transferred from the sodium atom to the chlorine atom. Now both atoms have eight electrons in their outer shells.

sodium atom (Na)   chlorine atom (Cl)

+   −

sodium ion (Na⁺)   chloride ion (Cl⁻)

a.   sodium chloride (NaCl)

This electron transfer, however, causes a charge imbalance in each atom. The sodium atom has one more proton than it has electrons. Therefore, it has a net charge of +1 (symbolized by Na^+). The chlorine atom has one more electron than it has protons. Therefore, it has a net charge of −1 (symbolized by Cl^-). Such charged particles are called **ions.** Sodium (Na^+) and chloride (Cl^-) are not the only biologically important ions. Some, such as potassium (K^+), are formed by the transfer of a single electron to another atom. Others, such as calcium (Ca^{2+}) and magnesium (Mg^{2+}), are formed by the transfer of two electrons.

Ionic compounds are held together by an attraction between negatively and positively charged ions called an **ionic bond.** When sodium reacts with chlorine, an ionic compound called sodium chloride (NaCl) results. Sodium chloride is a salt, commonly known as table salt because it is used to season our food (Fig. 2.7b). Salts can exist as a dry solid, but when salts are placed in water, they release ions as they dissolve. NaCl separates into Na^+ and Cl^-. In biological systems, because they are 70–90% water, ionic compounds tend to be dissociated (ionized) rather frequently.

Covalent Bonding

A **covalent bond** results when two atoms share electrons in such a way that each atom has an octet of electrons in the outer shell. In a hydrogen atom, the outer shell is complete when it contains two electrons. If hydrogen is in the presence of a strong electron acceptor, it gives up its electron to become a hydrogen ion (H^+). If this is not possible, hydrogen can share with another atom and thereby have a completed outer shell. For example, one hydrogen atom will share with another hydrogen atom. Their two shells overlap, and the electrons are shared between them (Fig. 2.8a). Each atom has a completed outer shell due to the sharing of electrons.

A more common way to symbolize that atoms are sharing electrons is to draw a line between the two atoms, as in the structural formula H—H. In a molecular formula, the line is omitted and the molecule is simply written as H_2.

Sometimes, atoms share more than one pair of electrons to complete their octets. A double covalent bond occurs when two atoms share two pairs of electrons (Fig. 2.8b). To show that oxygen gas (O_2) contains a double bond, the molecule can be written as O=O.

It is also possible for atoms to form triple covalent bonds, as in nitrogen gas (N_2), which can be written as

Na⁺ Cl⁻

b.

Figure 2.7 Formation of sodium chloride (table salt).   a. To form sodium chloride, an electron is transferred from the sodium atom to the chlorine atom. At the completion of the reaction, each atom is an ion containing eight electrons in its outer shell. **b.** In a sodium chloride crystal, ionic bonding between Na⁺ and Cl⁻ causes the atoms to assume a three-dimensional lattice in which each sodium ion is surrounded by six chloride ions, and each chloride ion is surrounded by six sodium ions. This forms crystals, such as in table salt. (b) (crystals): ©Evelyn Jo Johnson; (salt shaker): ©McGraw-Hill Education/Evelyn Jo Johnson, photographer

Electron Model	Structural Formula	Molecular Formula
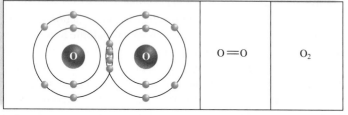	H—H	H_2

a. Hydrogen gas

	O=O	O_2

b. Oxygen gas

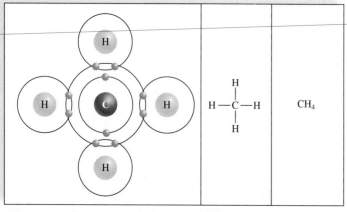

	Structural Formula	Molecular Formula
	$H-\overset{\displaystyle H}{\underset{\displaystyle H}{C}}-H$	CH_4

c. Methane

Ball-and-Stick Model	Space-Filling Model
hydrogen—H carbon—C H covalent bond— H H 109°	

d. Methane—continued

Figure 2.8 Covalently bonded molecules.   In a covalent bond, atoms share electrons, allowing each atom to have a completed valence shell. **a.** A molecule of hydrogen (H_2) contains two hydrogen atoms sharing a pair of electrons with a single covalent bond. **b.** A molecule of oxygen (O_2) contains two oxygen atoms sharing two pairs of electrons. This results in a double covalent bond. **c.** A molecule of methane (CH_4) contains one carbon atom bonded to four hydrogen atoms. **d.** When carbon binds to four other atoms, as in methane, each bond actually points to one corner of a tetrahedron. Ball-and-stick models and space-filling models are three-dimensional representations of a molecule.

$N\equiv N$. Single covalent bonds between atoms are quite strong, but double and triple bonds are even stronger.

The Shape of Molecules

Structural formulas make it seem as if molecules are one-dimensional. Actually, molecules have a three-dimensional shape that often determines their biological function. Molecules consisting of only two atoms are always linear, but a molecule such as methane with five atoms (Fig. 2.8c) has a tetrahedral shape. Why? Because, as shown in the ball-and-stick model, each bond is pointing to the corners of a tetrahedron (Fig. 2.8d, *left*). The space-filling model often comes closest to the actual shape of the molecule (Fig. 2.8d, *right*).

There is a saying in biology that "structure equals function." Frequently, the shape of a molecule is important to the structural and functional roles it plays in an organism. For example, hormones have specific shapes that allow them to be recognized by the cells in the body. Antibodies combine with disease-causing agents, like a key fits a lock, to protect us. Similarly, homeostasis is maintained only when enzymes have the proper shape to carry out their particular reactions in cells.

Nonpolar and Polar Covalent Bonds

When the sharing of electrons between two atoms is fairly equal, the covalent bond is said to be a **nonpolar covalent bond.** All the molecules in Figure 2.8, including methane (CH_4), are nonpolar. In the case of water (H_2O), however, the sharing of electrons between oxygen and each hydrogen is not completely equal (Fig. 2.9a). This is because, in some cases, one atom is able to attract electrons to a greater degree than the other atom. In this case, we say that the atom that has a greater attraction for a shared pair of electrons has a greater **electronegativity.** In the case of water, the oxygen atom is more electronegative than the hydrogen atom and therefore the oxygen atom can attract the electron pair to a greater extent than each hydrogen atom can. It may help to think of electronegativity as where the electron pair chooses to "spend its time." In a water molecule, the shared electron pair spends more time around the nucleus of the oxygen atom than around the nucleus of the hydrogen atom. This causes the oxygen atom to assume a slightly negative charge (δ^-), and it causes the hydrogen atoms to assume slightly positive charges (δ^+). The unequal sharing of electrons in a covalent bond creates a **polar covalent bond.** The bond is called polar because it differs in polarity (the electrical charge) across the molecule.

Hydrogen Bonding

Polarity within a water molecule causes the hydrogen atoms in one molecule to be attracted to the oxygen atoms in other water molecules (Fig. 2.9b). This attraction, although weaker than an ionic or covalent bond, is called a **hydrogen bond.** Because a hydrogen bond is easily broken, it is often represented by a dotted line. Hydrogen bonding is not unique to water. Many biological molecules have polar covalent bonds involving an electropositive hydrogen and usually an electronegative oxygen or nitrogen. In these instances, a hydrogen bond can occur within the same molecule or between different molecules.

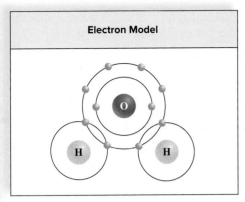

Electron Model

a. Water (H_2O)

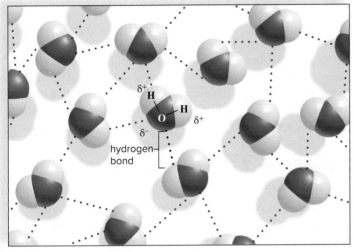

b. Hydrogen bonding between water molecules

Figure 2.9 Water molecule. a. The electron model of water.
b. Hydrogen bonding between water molecules. A hydrogen bond is the
attraction of a slightly positive hydrogen to a slightly negative atom in the
vicinity. Each water molecule can hydrogen-bond to four other molecules in
this manner. When water is in its liquid state, some hydrogen bonds are
forming and others are breaking at all times.

Although a hydrogen bond is more easily broken than a covalent bond, many hydrogen bonds taken together are quite
strong. Hydrogen bonds between cellular molecules help maintain their proper structure and function. For example, hydrogen
bonds hold the two strands of DNA together. When DNA makes
a copy of itself, each hydrogen bond easily breaks, allowing the
DNA to unzip. On the other hand, the hydrogen bonds acting together add stability to the DNA molecule. As we shall see, many
of the important properties of water are the result of hydrogen
bonding.

Check Your Progress 2.2

1. Identify whether carbon dioxide (CO_2) and nitrogen gas (N_2)
   are considered molecules, compounds, or both.
2. Describe why a carbon atom (C) is capable of forming four
   covalent bonds.
3. Explain the difference between a polar and nonpolar covalent bond.

2.3 Chemistry of Water

Learning Outcomes

Upon completion of this section, you should be able to
1. Evaluate which properties of water are important for
   biological life.
2. Identify common acidic and basic substances.
3. Describe how buffers are important to living organisms.

The first cell(s) evolved in water, and organisms are composed of
70–90% water. Water is a polar molecule, and water molecules are
hydrogen-bonded to one another (see Fig. 2.9*b*). Due to hydrogen
bonding, water molecules cling together. Without hydrogen bonding between molecules, water would change from a solid to liquid
state at –100°C and from a liquid to gaseous state at –91°C. This
would make most of the water on Earth steam, and life unlikely.
But because of hydrogen bonding, water is a liquid at temperatures
typically found on Earth's surface. It melts at 0°C and boils at
100°C. These and other unique properties of water make it essential to the existence of life.

Properties of Water

Water Has a High Heat Capacity

A **calorie** is the amount of heat energy needed to raise the temperature of 1 gram (g) of water by 1°C. In comparison, other covalently bonded liquids require input of only about half this amount
of energy to rise 1°C in temperature. The many hydrogen bonds
that link water molecules help water absorb heat without a great
change in temperature.

Converting 1 g of the coldest liquid water to ice requires the
loss of 80 calories of heat energy. Water holds onto its heat, and its
temperature falls more slowly than that of other liquids. This property of water is important not only for aquatic organisms, but for all
organisms. Because the temperature of water rises and falls slowly,
organisms are better able to maintain their normal internal temperatures and are protected from rapid temperature changes.

Water Has a High Heat of Vaporization

Converting 1 g of the hottest water to a gas requires an input of 540
calories of heat energy. Water has a high heat of vaporization because hydrogen bonds must be broken before water boils and
changes to a vaporized state. Water's high heat of vaporization
gives animals in a hot environment an efficient way to release excess body heat (Fig. 2.10). For example, when we sweat, or get
water splashed on us, our body heat is used to vaporize the water,
thus cooling us down.

Temperatures along coasts are moderate due to water's high
heat capacity and high heat of vaporization. During the summer,
the ocean absorbs and stores solar heat, and during the winter, the
ocean slowly releases it. In contrast, the interior regions of continents can experience severe changes in temperature.

Water Is a Solvent

Due to its polarity, water facilitates chemical reactions, inside and
outside living organisms. It dissolves a great number of substances.

Figure 2.10 Water's high heat of vaporization. When body temperature increases, sweat is produced by glands in the dermal layer of the skin. The evaporation of the water off your skin aids in cooling your body.
©PNC/Getty Images

A **solution** contains dissolved substances called **solutes.** When ionic salts—for example, sodium chloride (NaCl)—are put into water, the negative ends of the water molecules are attracted to the sodium ions, and the positive ends of the water molecules are attracted to the chloride ions. This causes the sodium ions and the chloride ions to separate, or dissociate, in water:

An ionic salt dissolves in water.

Water is also a solvent for larger molecules that either contain ionized atoms or are polar molecules.

Molecules that can attract water are said to be **hydrophilic.** When ions and molecules disperse in water, they move about and collide, allowing reactions to occur. Nonionized and nonpolar molecules, such as oil, that cannot attract water are said to be **hydrophobic.**

Water Molecules Are Cohesive and Adhesive

Cohesion refers to the ability of water molecules to cling to each other due to hydrogen bonding. At any moment in time, a water molecule can form hydrogen bonds with at most four other water molecules. Because of cohesion, water exists as a liquid under the conditions of temperature and pressure present at the Earth's surface. The strong cohesion of water molecules is apparent because water flows freely, yet water molecules do not separate from each other.

Adhesion refers to the ability of water molecules to cling to other polar surfaces. This is a result of water's polarity. Multicellular

animals often contain internal vessels in which water assists the transport of nutrients and wastes, because the cohesion and adhesion of water allow blood to fill the tubular vessels of the cardiovascular system. For example, the liquid portion of our blood, which transports dissolved and suspended substances about the body, is 90% water.

Cohesion and adhesion also contribute to the transport of water in plants. The roots of plants absorb water, while the leaves lose water through evaporation. A plant contains a system of vessels that reaches from the roots to the leaves. Water evaporating from the leaves is immediately replaced with water molecules from the vessels. Because water molecules are cohesive, a tension is created that pulls a water column up from the roots. Adhesion of water to the walls of the vessels also helps prevent the water column from breaking apart.

Because water molecules are attracted to each other, they cling together where the liquid surface is exposed to air. The stronger the force between molecules in a liquid, the greater the **surface tension.** This surface tension acts as a barrier between the surface of the water and the atmosphere. Water striders, common insects, use the surface tension of water to walk on the surface without breaking the surface.

Frozen Water (Ice) Is Less Dense than Liquid Water

As liquid water cools, the molecules come closer together. They are densest at 4°C, but they are still moving about, bumping into each other (Fig. 2.11). At temperatures below 4°C, including at 0°C when water is frozen, the water forms a regular crystal lattice that is rigid and has more open space between the water molecules. For this reason, water expands as it freezes, which is why cans of soda burst when placed in a freezer or why frost heaves make northern roads bumpy in the winter. It also means that ice is less dense than liquid water, and floats on liquid water.

If ice did not float on water, it would sink, and ponds, lakes, and perhaps even regions of the ocean would freeze solid. This would make life impossible in the water and also on land. Instead, bodies of water always freeze from the top down. When a body of water freezes on the surface, the ice acts as an insulator to prevent the water below it from freezing. This protects aquatic organisms so they can survive the winter. As ice melts in the spring, it draws heat from the environment, helping to prevent a sudden change in temperature that might be harmful to life.

Acids and Bases

When water ionizes, it releases an equal number of hydrogen ions (H⁺; also called protons[1]) and hydroxide ions (OH⁻) into the solution:

$$H_2O \rightleftharpoons H^+ + OH^-$$
water   hydrogen   hydroxide
           ion   ion

---

[1] A hydrogen atom contains one electron and one proton. A hydrogen ion has only one proton, so it is often simply called a proton.

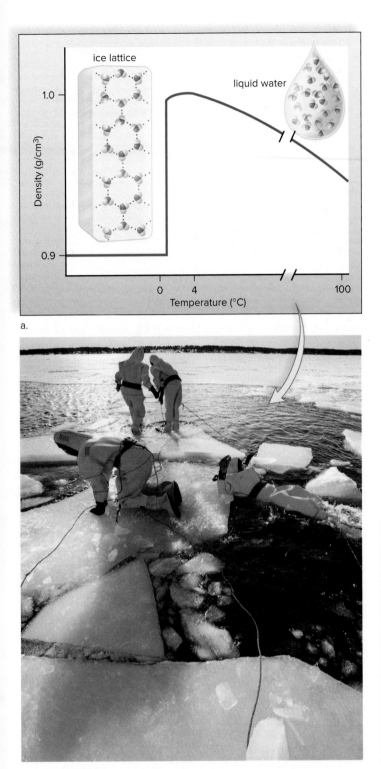

a.

b.

Figure 2.11 Ice is less dense than water.   a. Water is more dense at 4°C than at 0°C. While most substances contract when they solidify, water expands when it freezes because the water molecules in ice form a lattice in which the hydrogen bonds are farther apart than in liquid water. **b.** This allows ice to float on the surface of liquid water. ©Gonzalo Azumendi/Getty Images

Only a few water molecules at a time dissociate, and the actual number of H⁺ and OH⁻ is very small (1×10^{-7} moles/liter).[2]

---
[2] In chemistry, a mole is defined as the amount of matter that contains as many objects (atoms, molecules, ions) as the number of atoms in exactly 12 g of ^{12}C.

Acidic Solutions (High H⁺ Concentrations)

Lemon juice, vinegar, tomatoes, and coffee are all acidic solutions. **Acids** are substances that release hydrogen ions (H⁺) when they dissociate in water. Therefore, they contain a higher concentration of H⁺ than OH⁻. For example, hydrochloric acid (HCl) is an important acid that dissociates in this manner:

$$HCl \longrightarrow H^+ + Cl^-$$

Because dissociation is almost complete, HCl is called a strong acid. If hydrochloric acid is added to a beaker of water, the number of hydrogen ions (H⁺) increases greatly.

Basic Solutions (Low H⁺ Concentrations)

Baking soda and antacids are common basic solutions familiar to most people. **Bases** are substances that either take up hydrogen ions (H⁺) or release hydroxide ions (OH⁻). They contain a higher concentration of OH⁻ than H⁺. For example, sodium hydroxide (NaOH) is an important base that dissociates in this manner:

$$NaOH \longrightarrow Na^+ + OH^-$$

Because dissociation is almost complete, sodium hydroxide is called a strong base. If sodium hydroxide is added to a beaker of water, the number of hydroxide ions increases.

pH Scale

The **pH scale** is used to indicate the acidity or basicity (alkalinity) of solutions.[3] The pH scale (Fig. 2.12) ranges from 0 to 14.

Figure 2.12 The pH scale.   The pH scale ranges from 0 to 14, with 0 being the most acidic and 14 being the most basic. A solution at pH 7 (neutral pH) has equal amounts of hydrogen ions (H⁺) and hydroxide ions (OH⁻). An acidic pH has more H⁺ than OH⁻, while a basic pH has more OH⁻ than H⁺. The pH of familiar solutions can be seen on the scale.

---
[3] pH is defined as the negative log of the hydrogen ion concentration [H⁺]. A log is the power to which ten must be raised to produce a given number.

A pH of 7 represents a neutral state in which the hydrogen ion and hydroxide ion concentrations are equal. A pH below 7 is an acidic solution because the hydrogen ion concentration [H$^+$] is greater than the hydroxide concentration [OH$^-$]. A pH above 7 is basic because [OH$^-$] is greater than [H$^+$]. As we move down the pH scale from pH 14 to pH 0, each unit has ten times the H$^+$ concentration of the previous unit. As we move up the scale from 0 to 14, each unit has ten times the OH$^-$ concentration of the previous unit.

The pH scale was devised to eliminate the use of cumbersome numbers. For example, in Figure 2.12, notice that the exponent in the hydrogen ion concentration (on the left) corresponds to the pH value in the center of the image. As shown next, a number with a smaller negative exponent indicates a greater quantity of hydrogen ions than one with a larger negative exponent. Therefore, a solution with a pH of 6 is more acidic than a solution with a pH of 8.

[H$^+$] (moles per liter)		pH
0.000001	$= 1 \times 10^{-6}$	6
0.0000001	$= 1 \times 10^{-7}$	7
0.00000001	$= 1 \times 10^{-8}$	8

Buffers and pH

A **buffer** is a substance that keeps pH within normal limits. Many commercial products, such as aspirin, shampoos, or deodorants, are buffered as an added incentive for us to buy them. Buffers resist pH changes because they can take up excess hydrogen ions (H$^+$) or hydroxide ions (OH$^-$).

In animals, the pH of body fluids is maintained within a narrow range, or else health suffers. The pH of our blood when we are healthy is always about 7.4—that is, just slightly basic (alkaline). If the blood pH drops to about 7, *acidosis* results. If the blood pH rises to about 7.8, *alkalosis* results. Both conditions can be life threatening because our body needs to operate under a narrow range of internal environmental conditions (see Section 11.5). Normally, pH stability is possible because the body has built-in mechanisms to prevent pH changes. Buffers are the most important of these mechanisms. For example, carbonic acid (H$_2$CO$_3$) is a weak acid that minimally dissociates and then re-forms in the following manner:

$$\underset{\text{carbonic acid}}{H_2CO_3} \underset{\text{re-forms}}{\overset{\text{dissociates}}{\rightleftharpoons}} H^+ + \underset{\text{bicarbonate ion}}{HCO_3^-}$$

Blood always contains a combination of carbonic acid and bicarbonate ions. When hydrogen ions (H$^+$) are added to blood, the following reaction occurs:

$$H^+ + HCO_3^- \longrightarrow H_2CO_3$$

When hydroxide ions (OH$^-$) are added to blood, this reaction occurs:

$$OH^- + H_2CO_3 \longrightarrow HCO_3^- + H_2O$$

These reactions prevent any significant change in blood pH.

Check Your Progress 2.3

1. Compare the difference between water's high heat capacity and high heat of vaporization.
2. Explain the difference between cohesion and adhesion.
3. Explain why a solution with a pH of 3 contains more H$^+$ than a solution with a pH of 6.

2.4 Organic Molecules

Learning Outcomes

Upon completion of this section, you should be able to

1. Compare inorganic molecules to organic molecules.
2. Identify the role of a functional group.
3. Recognize how monomers are joined to form polymers.

Inorganic molecules constitute nonliving matter, but even so, inorganic molecules such as salts (e.g., NaCl) and water play important roles in living organisms. The molecules of life, however, are organic molecules. **Organic molecules** always contain carbon (C) and hydrogen (H). The chemistry of carbon accounts for the formation of the very large variety of organic molecules found in living organisms. Carbon atoms contain four electrons in their outer shell. In order to achieve eight electrons in the outer shell, a carbon atom shares electrons covalently with as many as four other atoms, as in methane (CH$_4$). Carbon atoms often share electrons with other carbon atoms, forming long hydrocarbon chains.

Under certain conditions, a hydrocarbon chain can turn back on itself to form a ring compound. The carbon chain of an organic molecule is often called its skeleton, or backbone. Just as a skeleton accounts for your body's shape, so does the carbon skeleton of an organic molecule account for its shape. It is possible for organic molecules to have the same molecular formula, but different shapes. These molecules are called **isomers.** What gives an organic molecule its unique chemical characteristics are the combinations of atoms, called functional groups, that are attached to the carbon skeleton.

Functional Groups

The chemical reactivity of an organic molecule is determined by the types and locations of functional groups on the organic molecule. A **functional group** is a specific combination of bonded atoms that always has the same chemical properties and therefore always reacts in the same way.

Typically, the carbon skeleton acts as a framework for the positioning of the functional groups. It is important to recognize that how the functional groups are positioned in the organic molecule

TABLE 2.1   Functional Groups

Functional Groups			
Group	*Structure*	*Compound*	*Significance*
Hydroxyl	R—OH	Alcohol as in ethanol	Polar, forms hydrogen bond; present in sugars, some amino acids.
Carbonyl	R—C(=O)H; R—C(=O)—R	(1) Aldehyde as in formaldehyde (2) Ketone as in acetone	(1) Polar; present in sugars. (2) Polar; present in sugars.
Carboxyl (acidic)	R—C(=O)OH	Carboxylic acid as in acetic acid	Polar, acidic; present in fatty acids, amino acids.
Amino	R—N(H)(H)	Amine as in tryptophan	Polar, basic, forms hydrogen bonds; present in amino acids.
Sulfhydryl	R—SH	Thiol as in ethanethiol	Forms disulfide bonds; present in some amino acids.
Phosphate	R—O—P(=O)(OH)—OH	Organic phosphate as in phosphorylated molecules	Polar, acidic; present in nucleotides, phospholipids.

R = remainder of molecule

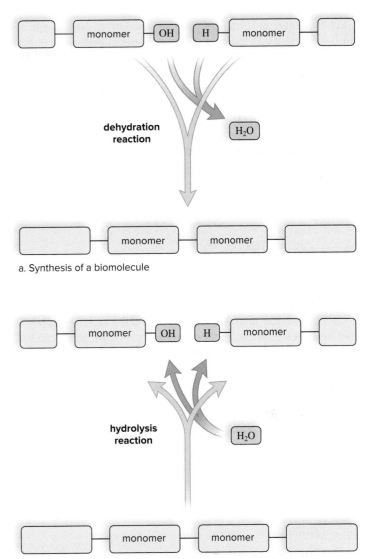

a. Synthesis of a biomolecule

b. Degradation of a biomolecule

Figure 2.13   Synthesis and degradation of polymers.   a. In cells, synthesis often occurs when monomers join (bond) during a dehydration reaction (removal of H_2O). **b.** Degradation occurs when the monomers in a polymer separate during a hydrolysis reaction (addition of H_2O).

determines the chemical properties of the molecule. For example, the addition of an —OH (hydroxyl group) to a carbon skeleton gives that molecule the chemical characteristics of an alcohol.

Table 2.1 lists some of the more common functional groups. The R indicates the "remainder" of the molecule. This is the place on the functional group that attaches to the carbon skeleton.

From Monomers to Polymers

Many of the organic molecules you are familiar with, such as carbohydrates, lipids, proteins, and nucleic acids, are macromolecules (also called biomolecules), meaning they contain smaller subunits joined together. The carbohydrates, proteins, and nucleic acids are referred to as **polymers,** because they are constructed by linking together a large number of the same type of subunit, called a **monomer.** Lipids are not polymers because they contain two different types of subunits (glycerol and fatty acids). Polymers may vary considerably in length. Just as a train increases in length when boxcars are hitched together one by one, so a polymer gets longer as monomers bond to one another.

Cells have a common way of joining monomers to build polymers. During a **dehydration reaction,** an —OH (hydroxyl group) and an —H (hydrogen atom), the equivalent of a water molecule, are removed as the reaction proceeds (Fig. 2.13a). To degrade polymers, the cell uses a **hydrolysis reaction,** in which the components of water are added (Fig. 2.13b).

Check Your Progress 2.4

1. Describe why organic molecules are considered the molecules of life.
2. Compare and contrast dehydration and hydrolysis reactions.

2.5 Carbohydrates

Learning Outcomes

Upon completion of this section, you should be able to

1. Identify the structural components of a carbohydrate.
2. List several examples of important monosaccharides and polysaccharides.

Carbohydrates are almost universally used as an energy source for living organisms, including humans. Carbohydrates also play a structural role in woody plants, bacteria, and animals such as insects. In addition, carbohydrates on cell surfaces are involved in cell-to-cell recognition, as we will learn in Section 4.1.

Carbohydrate molecules are characterized by the presence of the atomic grouping H—C—OH, in which the ratio of hydrogen atoms (H) to oxygen atoms (O) is approximately 2:1. The term *carbohydrate* (literally, carbon-water) includes single sugar molecules and chains of sugars. Chain length varies from a few sugars to hundreds of sugars. The monomer subunits, called monosaccharides, are assembled into long polymer chains called polysaccharides.

Monosaccharides—Simple Sugars

Monosaccharides (*mono*, one; *saccharide*, sugar) consist of only a single sugar molecule and are commonly called simple sugars. A monosaccharide can have a carbon backbone of three to seven carbons. For example, **pentoses** are monosaccharides with five carbons, and **hexoses** are monosaccharides with six carbons. Monosaccharides have a large number of hydroxyl groups (—OH), and the presence of this polar functional group makes them soluble in water.

Glucose is a hexose sugar that plays an important role in living organisms (Fig. 2.14). Like all organisms, our bodies use glucose as an immediate source of energy. Other monosaccharides include fructose, found in fruits, and galactose, the sugar found in milk. All of these are isomers with the molecular formula $C_6H_{12}O_6$. Each also forms a ring structure when placed in a water environment. The exact shape of the ring differs, as does the arrangement of the hydrogen (—H) and hydroxyl (—OH) groups attached to the ring.

Disaccharides

A **disaccharide** (*di*, two) contains two monosaccharides that have joined during a dehydration reaction. Figure 2.15 shows how the disaccharide maltose forms when two glucose molecules bond together. Note the position of this bond. Our hydrolytic digestive juices can break this bond, and the result is two glucose molecules.

When glucose and fructose join, the disaccharide sucrose forms. Sucrose is another disaccharide of special interest because

Figure 2.14 Three ways to represent the structure of glucose. $C_6H_{12}O_6$ is the molecular formula for glucose. **a.** This structure shows all the carbon atoms. **b.** This structure shows only the carbon atoms outside the ring. **c.** This is the simplest way to represent glucose. Note that in **a** and **b,** each carbon has an attached —H and —OH group. Those groups are assumed in **c.**

we use it at the table to sweeten our food. We acquire sucrose from plants such as sugarcane and sugar beets. Lactose is glucose combined with galactose. Individuals who are *lactose intolerant* lack the enzyme (called lactase) that breaks down lactose. To prevent gastrointestinal discomfort, they can either avoid foods that contain lactose (e.g., dairy products) or take nutritional supplements that contain the enzyme.

Polysaccharides—Complex Carbohydrates

Long polymers such as starch, glycogen, and cellulose are **polysaccharides** (*poly*, many) that contain long chains of glucose subunits. Due to their length, they are sometimes referred to as complex carbohydrates.

Energy Storage Polysaccharides

Plants store a large amount of glucose in the form of **starch.** The cells of a potato contain granules where starch resides during winter until energy is needed for growth in the spring. Starch may have either a branched or unbranched structure (Fig. 2.16).

Animals store a large amount of glucose in the form of **glycogen.** In our bodies and those of other vertebrates, liver cells contain granules, where glycogen is stored until needed. The storage and

Figure 2.15 Synthesis and degradation of maltose, a disaccharide. Synthesis of maltose occurs following a dehydration reaction when a bond forms between two glucose molecules and water is removed. Degradation of maltose occurs following a hydrolysis reaction when this bond is broken by the addition of water.

Figure 2.16 Starch structure and function. Starch is composed of chains of glucose molecules. Some chains are branched, as indicated. Starch is the storage form of glucose in plants. The electron micrograph shows starch granules in potato cells. (photo): ©Jeremy Burgess/SPL/Science Source

Figure 2.17 Glycogen structure and function. Glycogen is a highly branched polymer of glucose molecules that serves as the storage form of glucose in animals. The electron micrograph shows glycogen granules in liver cells. (photo): ©Don W. Fawcett/Science Source

release of glucose from liver cells are controlled by hormones. After we eat, the release of the hormone insulin from the pancreas promotes the storage of glucose as glycogen. Notice in Figure 2.17 that glycogen is even more branched than starch.

Polysaccharides serve as storage molecules because they are not as soluble in water and are much larger than a simple sugar. Therefore, as we will learn later (see Section 4.2), polysaccharides cannot easily move in or out of cells.

Structural Polysaccharides

Some types of polysaccharides function as structural components of cells. The polysaccharide **cellulose** is found in plant cell walls, which helps account for the strong nature of these walls.

In cellulose (Fig. 2.18), the glucose units are joined by a slightly different type of linkage than that found in starch or

Figure 2.18 Cellulose structure and function. In cellulose, the linkage between glucose molecules is slightly different from that in starch or glycogen. Plant cell walls contain cellulose, and the rigidity of these cell walls permits nonwoody plants to stand upright as long as they receive an adequate supply of water. (photo): ©Scimat/Science Source

glycogen. Notice the alternating up/down position of the oxygen atoms in the linked glucose units in Figure 2.18. This small difference is significant because it prevents us from digesting foods containing this type of bonding pattern. Therefore, cellulose largely passes through our digestive tract as fiber, or roughage. Fiber in the diet is necessary to the health of our digestive system, and plays a role in preventing some forms of cancer.

Chitin, which is found in the exoskeleton (shell) of crabs and related animals, is another structural polysaccharide. Chitin is used in food processing and in the manufacture of some types of paper.

Check Your Progress 2.5

1. Identify the structural element that all carbohydrates have in common.
2. Explain why starch in plants is a source of glucose for our bodies but cellulose in plants is not.

2.6 Lipids

Learning Outcomes

Upon completion of this section, you should be able to
1. Compare the structures of fats, phospholipids, and steroids.
2. Identify the functions lipids play in our bodies.

A variety of organic compounds are classified as **lipids.** While they differ in structure, all lipids have a common characteristic—they are hydrophobic molecules that are insoluble in water due to the structure of their hydrocarbon chains.

Triglycerides are most often used as long-term energy storage compounds of plants and animals. The phospholipids form a membrane that separates the cell from its environment and forms its

inner compartments as well. The steroids are a large class of lipids that includes, among others, the sex hormones. Waxes are a form of lipid that is used for protection and to prevent water loss.

Triglycerides: Fats and Oils

The most familiar lipids are those found in fats and oils. **Fats** tend to be of animal origin (e.g., lard and butter), and are solid at room temperature. **Oils,** which are usually of plant origin (e.g., corn oil and soybean oil), are liquid at room temperature. Fat has several functions in the body: it is used for long-term energy storage, it insulates against heat loss, and it forms a protective cushion around major organs.

Fats and oils form when one glycerol molecule reacts with three fatty acid molecules (Fig. 2.19). A fat molecule is sometimes called a **triglyceride** because of its three-part structure.

While fats and oils are hydrophobic molecules, the addition of emulsifiers can allow them to mix with water. Emulsifiers contain molecules with a nonpolar end and a polar end. The molecules position themselves about an oil droplet so their nonpolar ends project inward and their polar ends project outward. As a result, the fat or oil disperses in the water. This process is called emulsification.

Emulsification takes place when dirty clothes are washed with soaps or detergents. It explains why some salad dressings are uniform in consistency (emulsified), while others separate into two layers. Also, prior to the digestion of fatty foods, fats are emulsified by bile in the intestines. The liver produces bile, which is then stored in the gallbladder. Individuals who have had their gallbladder (an organ where bile is stored) removed may have trouble digesting fatty foods.

Saturated, Unsaturated, and Trans Fatty Acids

A **fatty acid** is a hydrocarbon chain that ends with the acidic group —COOH (Fig. 2.19). Most of the fatty acids in cells contain 16 or

glycerol   3 fatty acids   3 water molecules   fat molecule

Figure 2.19 Synthesis and degradation of a triglyceride.   When a fat molecule (triglyceride) forms, three fatty acids combine with glycerol, and three water molecules are produced.

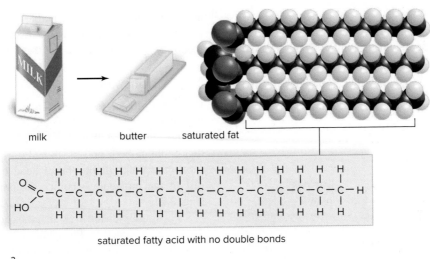

milk   butter   saturated fat

saturated fatty acid with no double bonds

a.

corn   corn oil

unsaturated fatty acid with double bonds (yellow) and a *trans* bond (red)

unsaturated fat

b.

Figure 2.20 Saturated fats, unsaturated fats, and trans fats.   Saturated fats have no double bonds between the carbon atoms, whereas unsaturated fats possess one or more double bonds. In a trans fat, the hydrogen atoms are on opposite sides of the double bond. A fatty acid has a carboxyl group attached to a long hydrocarbon chain. **a.** If there are no double bonds between the carbons in the chain, the fatty acid is saturated. **b.** If there are double bonds between some of the carbons, the fatty acid is unsaturated and a kink occurs in the chain. A trans fat is an unsaturated fat with the type of double bond shown in red.

18 carbon atoms per molecule, although smaller ones with fewer carbons are also known.

Fatty acids are either saturated or unsaturated (Fig. 2.20). **Saturated fatty acids** have no double bonds between carbon atoms. The carbon chain is saturated, so to speak, with all the hydrogens it can hold. Saturated fatty acids account for the solid nature at room temperature of fats such as lard and butter. **Unsaturated fatty acids** have double bonds between carbon atoms wherever the number of hydrogens is less than two per carbon atom. This produces a bend, or kink, in the fatty acid chain. Unsaturated fatty acids account for the liquid nature of vegetable oils at room temperature.

Unsaturated fats are also often referred to as being "cis" or "trans." This terminology refers to the configuration of the hydrogen atoms in the double bond of an unsaturated fat (Fig. 2.20*b*). **Trans fats** are often produced by hydrogenation, or the chemical addition of hydrogen to vegetable oils. The Health feature "Saturated and Trans Fats in Foods" explores why physicians recommend limiting consumption of these types of fats.

Phospholipids

Phospholipids, as their name implies, contain a phosphate group. Essentially, they are constructed like fats, except that in place of the third fatty acid, there is a polar phosphate group or a grouping that contains both phosphate and nitrogen. The phosphate group forms the polar (hydrophilic) head of the molecule, while the rest of the molecule becomes the nonpolar (hydrophobic) tails (Fig. 2.21).

Phospholipids illustrate that the chemistry of a molecule helps determine its function. Phospholipids are the primary components of cellular membranes. They spontaneously form a bilayer in which the hydrophilic heads face outward toward watery solutions and the tails form the hydrophobic interior. Plasma membranes separate extracellular from intracellular environments and are absolutely vital to the form and function of a cell.

Steroids

Steroids have a backbone of four fused carbon rings. Each one differs primarily by the arrangement of the atoms in the rings and the type of functional groups attached to them. Cholesterol is a steroid formed by the body that also enters the body as part of our diet. Cholesterol has several important functions. It is a component of an animal cell's plasma membrane and is the precursor of several other steroids, such as bile salts and the sex hormones testosterone and estrogen (Fig. 2.22).

a. Plasma membrane of a cell

b. Phospholipid structure

Figure 2.21 Phospholipids form membranes. **a.** Phospholipids arrange themselves as a bilayer in the plasma membrane that surrounds cells. **b.** Phospholipids are constructed like fats, except that in place of the third fatty acid, they have a polar phosphate group. The bilayer structure forms because the polar (hydrophilic) head is soluble in water, whereas the two nonpolar (hydrophobic) tails are not. The polar heads interact with the inside and outside of the cell, while the nonpolar tails interact with each other.

Check Your Progress 2.6

1. State the common characteristic of all lipids.
2. Explain how the presence of a double bond in an unsaturated fatty acid affects whether that substance is a solid or liquid.
3. Distinguish between the roles of triglycerides and steroids in humans.

2.7 Proteins

Learning Outcomes

Upon completion of this section, you should be able to
1. Describe the functions of proteins in cells.
2. Explain how a polypeptide is constructed from amino acids.
3. Compare the four levels of protein structure.

Proteins are polymers composed of amino acid monomers. An **amino acid** has a central carbon atom bonded to a hydrogen atom and three functional groups. The name of the molecule is appropriate because one of these groups is an amino group ($-NH_2$) and another is an acidic group ($-COOH$). The third group is called an *R* group and determines the uniqueness of each amino acid.

Saturated and Trans Fats in Foods

You have probably heard that you should limit the amount of saturated fats and trans fats in your diet. But why? We know that saturated fats, which come from animals and are solid at room temperature, have effects in the body that are different from those of unsaturated fats, which come from plants and are liquid at room temperature. Saturated fats are flat molecules that easily stick together in the blood, and too much saturated fat has been shown by scientists to negatively affect heart health, contributing to clogging of arteries and cardiovascular disease (CVD). By comparison, unsaturated fats seem to help prevent CVD, because they don't stick together in the blood and therefore don't clog arteries.

Unsaturated fats might be healthier for you, but plant oils can easily go rancid and aren't solid at room temperature, which makes them more difficult to cook with and to use in solid food products. To get around this problem, food manufacturers hydrogenated unsaturated fatty acids by heating the oil and exposing it to hydrogen gas. This treatment made the otherwise liquid plant oils semisolid at room temperature and gave foods containing partially hydrogenated oils better shelf life.

An unintended consequence of hydrogenation, however, was the formation of trans fats. Many commercially packaged foods contain trans fats, which recently have been shown to increase LDL (sometimes called bad cholesterol) and lower HDL (sometimes called good cholesterol) levels in the blood. Trans fat consumption also appears to increase risk of CVD and heart attack.

At one point, investigators thought that the total amount of lipid in the diet caused coronary and other heart-related diseases. As scientific evidence accumulated showing a distinction between the effects of saturated and unsaturated fats, public perception changed. Until recently, trans fats were of little concern to the general public, and people readily consumed them without much thought. As science has brought the negative effects of trans fats to light, perception has changed once again. Public outcry has prompted changes in the food services industry, with clear labeling of trans fats on all food products and with more restaurants using trans fat–free oils during cooking. This is a good example of how our perceptions change over time based on scientific evidence, and it illustrates the essential role science plays in the common good. Science constantly refines what we know as new evidence provides greater insights into how we function and live.

Questions to Consider

1. What is the chemical structure of a trans fat compared to that of a non-trans fat?
2. How much trans fat do you consume daily?
3. How would you balance the needs of the food industry with the health risks associated with changing the chemical composition of a food?

Figure 2.22 Steroids.   a. Built like cholesterol, (**b**) testosterone and (**c**) estrogen have different effects on the body due to different functional groups attached to the same carbon skeleton. Testosterone is the male sex hormone (*left*), and estrogen is the female sex hormone (*right*). (photo): ©Ingram Publishing

Name	Structural Formula	R Group
alanine (Ala)	$H_3N^+ - \overset{\overset{H}{|}}{\underset{\underset{CH_3}{|}}{C}} - C \overset{O}{\underset{O^-}{\big\langle}}$	R group has a single carbon atom
valine (Val)	$H_3N^+ - \overset{\overset{H}{|}}{\underset{\underset{CH}{|}}{C}} - C \overset{O}{\underset{O^-}{\big\langle}}$ $H_3C \quad CH_3$	R group has a branched carbon chain
cysteine (Cys)	$H_3N^+ - \overset{\overset{H}{|}}{\underset{\underset{CH_2}{|}}{C}} - C \overset{O}{\underset{O^-}{\big\langle}}$ SH	R group contains sulfur
phenylalanine (Phe)	$H_3N^+ - \overset{\overset{H}{|}}{\underset{\underset{CH_2}{|}}{C}} - C \overset{O}{\underset{O^-}{\big\langle}}$ ⬡	R group has a ring structure

Figure 2.23 Representative amino acids.   Amino acids differ from one another by their *R* group. The simplest *R* group is a single hydrogen atom (H). *R* groups (blue) containing carbon vary as shown. These are just 4 of the 20 amino acids found in living organisms.

The *R* group varies from having a single carbon to being a complicated ring structure (Fig. 2.23).

Proteins perform many functions. Some proteins have structural functions. Examples of these are keratin, which makes up hair and nails, and collagen, which lends support to ligaments, tendons, and skin. Other proteins act as enzymes. **Enzymes** act as catalysts, meaning they speed up chemical reactions. They work so quickly that a reaction that normally takes several hours or days without an enzyme takes only a fraction of a second with an enzyme. Therefore, enzymes play an important role in the chemical workings of the cell, and the body.

In addition, some proteins act as hormones, messengers that influence cellular metabolism. Others, such as actin and myosin, account for the movement of cells and the ability of our muscles to contract. Some proteins transport molecules in the blood.

Hemoglobin is a complex protein in our blood that transports oxygen. Antibodies in blood and other body fluids are proteins that combine with foreign substances, preventing them from destroying cells and upsetting homeostasis.

Proteins in the plasma membrane of cells have various functions: Some form channels that allow substances to enter and exit cells; some are carriers that transport molecules into and out of the cell; and some are enzymes.

Peptides

Figure 2.24 shows that a synthesis reaction between two amino acids results in a dipeptide and a molecule of water. A **polypeptide** is a chain of amino acids joined to one another by a **peptide bond.** The atoms associated with a peptide bond unevenly because oxygen is more electronegative than nitrogen. Therefore, the hydrogen attached to the nitrogen has a slightly positive charge, while the oxygen has a slightly negative charge, as shown next.

The polarity of the peptide bond means that hydrogen bonding is possible between the C—O of one amino acid and the N—H of another amino acid in a polypeptide. This hydrogen bonding influences the structure, or shape, of a protein.

Levels of Protein Organization

Proteins can have up to four levels of structural organization (Fig. 2.25). The first level, called the *primary structure,* is the linear sequence of the amino acids joined by peptide bonds. Polypeptides can be quite different from one another. A polypeptide chain may be made from any combination of 20 different amino acids. Each particular polypeptide has its own sequence of amino acids and its own sequence of *R* groups.

The *secondary structure* of a protein comes about when the polypeptide takes on a certain orientation in space. A coiling of the chain results in an α (alpha) helix, or a right-handed spiral, similar to a spiral staircase. Or a folding of the chain results in a β (beta) pleated sheet similar to a handheld fan. Hydrogen bonding between peptide bonds holds the shape in place.

The *tertiary structure* of a protein is its final three-dimensional shape. In muscles, myosin molecules have a rod shape ending in globular (globe-shaped) heads. In enzymes, the polypeptide bends and twists in different ways. Invariably, the hydrophobic portions

Figure 2.24 Synthesis and degradation of a dipeptide.   Following a dehydration reaction, a peptide bond joins two amino acids, and a water molecule is released. Following a hydrolysis reaction, the bond is broken with the addition of water.

Primary Structure

This level of structure is determined by the linear sequence of amino acids, coded for in the genes of the DNA.

amino acid

peptide bond

Secondary Structure

Hydrogen bonding between amino acids causes the polypeptide to form an alpha helix or a pleated sheet.

hydrogen bond

hydrogen bond

α (alpha) helix

β (beta) sheet = pleated sheet

Tertiary Structure

Interactions of amino acid side chains with water, covalent bonding between R groups, and other chemical interactions determine the folded three-dimensional shape of a protein.

disulfide bond

Quaternary Structure

This level of structure occurs when two or more folded polypeptides interact to perform a biological function.

Figure 2.25 Levels of protein organization.   All proteins have a primary structure. Both fibrous and globular proteins have a secondary structure. They are either α (alpha) helices or β (beta) pleated sheets. Globular proteins always have a tertiary structure, and most have a quaternary structure.

are packed mostly on the inside, and the hydrophilic portions are on the outside where they can make contact with water. The tertiary shape of a polypeptide is maintained by various types of bonding between the R groups; covalent, ionic, and hydrogen bonding all occur. One common form of covalent bonding between R groups is a disulfide (S—S) linkage between two cysteine amino acids.

Some proteins have only one polypeptide, and others have more than one polypeptide, each with its own primary, secondary, and tertiary structures. In proteins with multiple polypeptide chains, these separate polypeptides are arranged to give such proteins a fourth level of structure, termed the *quaternary structure.* Hemoglobin is a complex protein having a quaternary structure. Most enzymes also have a quaternary structure. Thus, proteins can differ in many ways, such as in length, sequence, and structure. Each individual protein is chemically unique as well.

The final shape of a protein is very important to its function. As we will discuss in Section 6.3, enzymes cannot function unless they

have their normal shape. When proteins are exposed to extremes in heat and pH, they undergo an irreversible change in shape. This is referred to as being **denatured.** For example, we are all aware that adding acid to milk causes curdling and that heating causes egg white, which contains a protein called albumin, to coagulate. Denaturation occurs because the normal bonding between the R groups has been disturbed. Once a protein loses its normal shape, it can no longer function normally. For example researchers hypothesize that an alteration in protein organization, forming structures called *prions,* is related to the development of Alzheimer disease and Creutzfeldt–Jakob disease (the human form of "mad cow" disease).

Check Your Progress 2.7

1. List some of the functions of proteins.
2. Describe the structure of an amino acid.
3. Compare and contrast the four levels of protein structure.

2.8 Nucleic Acids

The two types of nucleic acids are **DNA (deoxyribonucleic acid)** and **RNA (ribonucleic acid)**. DNA stores genetic information in the cell and in the organism. Further, the cell replicates and transmits this information when the cell copies itself, as well as when the organism reproduces. The discovery of the structure of DNA has had an enormous influence on biology and on society in general. The science of biotechnology (see Section 26.1) is largely devoted to altering the genes in living organisms. New technologies are allowing scientists to more easily edit and modify the genetic material of organisms.

Structure of DNA and RNA

Both DNA and RNA are polymers of nucleotides. Every **nucleotide** is a molecular complex of three subunits (Fig. 2.26)—a phosphate functional group, a pentose sugar, and a nitrogen-containing base.

The nucleotides in DNA contain the sugar deoxyribose, while the nucleotides in RNA contain the sugar ribose. This difference accounts for their respective names (Table 2.2). There are four different types of bases in DNA: **adenine (A), thymine (T), guanine (G),** and **cytosine (C).** The base can have two rings (adenine or guanine) or one ring (thymine or cytosine). In RNA, the base **uracil (U)** replaces the base thymine. These structures are called bases because their presence raises the pH of a solution.

The nucleotides form a linear molecule called a strand, which has a backbone made up of alternating phosphates and sugars, with the bases projecting to one side of the backbone. The nucleotides and their bases occur in a specific order. After many years of work, researchers now know the sequence of the bases in human DNA—

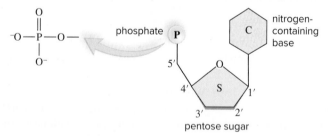

Figure 2.26 Structure of a nucleotide.   Each nucleotide consists of a pentose sugar, a nitrogen-containing base, and a phosphate functional group.

TABLE 2.2	DNA Structure Compared with RNA Structure	
	DNA	**RNA**
Sugar	Deoxyribose	Ribose
Bases	Adenine, guanine, thymine, cytosine	Adenine, guanine, uracil, cytosine
Strands	Double-stranded with base pairing	Single-stranded
Helix	Yes	No

a. Space-filling model

b. Double helix

c. Complementary base pairing

Figure 2.27 Overview of DNA structure.   The structure of DNA is absolutely essential to its ability to replicate and to serve as the genetic material. **a.** Space-filling model of DNA's double helix. **b.** Diagram showing the structure of the double helix **c.** Complementary base pairing between nucleotides. The heredity information stored by DNA is the sequence of its bases, which determines the primary structure of the cell's proteins. (a): ©Molekuul/SPL/age fotostock

the human genome. This breakthrough is expected to lead to improved genetic counseling, gene therapy, and medicines to treat the causes of many human illnesses.

DNA is double-stranded, with the two strands twisted about each other in the form of a **double helix** (Fig. 2.27a, b). In

DNA, the two strands are held together by hydrogen bonds between the bases. When unwound, DNA resembles a ladder. The uprights (sides) of the ladder are made entirely of the alternating phosphate and sugar molecules, and the rungs of the ladder are made only of complementary paired bases. Thymine (T) always pairs with adenine (A), and guanine (G) always pairs with cytosine (C). Complementary bases have shapes that fit together (Fig. 2.27c).

Complementary base pairing allows DNA to replicate in a way that ensures the sequence of bases will remain the same. The base sequence of specific sections of DNA contain a code that specifies the sequence of amino acids in the proteins of the cell.

RNA is single-stranded and is formed by complementary base pairing with one DNA strand. There are several types of RNA. One type of RNA, mRNA, or messenger RNA, carries the information from the DNA strand to the ribosome where it is translated into the sequence of amino acids specified by the DNA.

ATP (Adenosine Triphosphate)

In addition to being the monomers of nucleic acids, nucleotides have other metabolic functions in cells. When adenosine (adenine plus ribose) is modified by the addition of three phosphate groups instead of one, it becomes **ATP (adenosine triphosphate),** an energy carrier in cells. Glucose is broken down in a step-wise fashion so the energy of glucose is converted to that of ATP molecules. ATP molecules serve as small "energy packets" suitable for supplying energy to a wide variety of a cell's chemical reactions. ATP can be said to be the energy "currency" of the cell. Reactions in the cell that need energy require ATP.

ATP is a high-energy molecule because the last two phosphate bonds are unstable and easily broken. In cells, the terminal phosphate bond usually is hydrolyzed, leaving the molecule **ADP (adenosine diphosphate)** and a molecule of inorganic phosphate, ℗ (Fig. 2.28). The cell uses the energy released by ATP breakdown to synthesize macromolecules such as carbohydrates and proteins. Muscle cells use the energy for muscle contraction, while nerve cells use it for the conduction of nerve impulses. After ATP breaks down, it is rebuilt by the addition of ℗ to ADP. Notice in Figure 2.28 that an input of energy is required to reform ATP.

a.   **adenosine**   **triphosphate**

b.   **ATP**

ADP

> **Check Your Progress 2.8**
>
> 1. Explain the difference between DNA and RNA.
> 2. Identify how information is stored in DNA.
> 3. Describe how energy is stored in ATP.

> **Conclusion**
>
> In many cases, our knowledge of scientific concepts, such as nutrition, is based on what we hear in the media. For example, we started this chapter with a discussion of *good* and *bad* *cholesterol.* We learned that what we were actually referring to are lipoproteins, a form of protein that transports lipids and cholesterol in the blood. A high level of LDL—the "bad cholesterol"—means the body has an excess of fat to be transported, and HDL represents empty transport proteins. Ideally, low LDL and high HDL values signify a healthy cardiovascular system and a reduced risk for a number of diet-related diseases.

Figure 2.28 ATP reaction.   ATP, the universal energy "currency" of cells, is composed of adenosine and three phosphate groups (called a triphosphate). When cells require energy, ATP undergoes hydrolysis, producing ADP + ℗, with the release of energy.

SUMMARIZE

2.1 Introduction to Chemistry

■ Both living organisms and nonliving materials are composed of **matter** consisting of **elements.** The most significant elements found in all organisms are carbon, hydrogen, nitrogen, oxygen, phosphorus, and sulfur.

■ Elements contain atoms, and **atoms** are composed of subatomic particles. **Protons** and **neutrons** in the nucleus determine the **atomic mass** of an atom. The **atomic number** indicates the number of protons and the number of **electrons** in electrically neutral atoms. Protons have positive charges, neutrons are uncharged, and electrons have negative charges. **Isotopes** are atoms of a single element that differ in their numbers of neutrons. Radioactive isotopes have many uses, including serving as tracers in biological experiments and medical procedures.

■ Electrons occupy energy levels (**electron shells**) at discrete distances from the nucleus. The number of electrons in the **valence shell** determines the reactivity of an atom. The first shell is complete when it is occupied by two electrons. Shells beyond the first shell are stable when eight electrons are present. The **octet rule** states that atoms react with one another in order to have a stable valence shell. Most atoms do not have filled outer shells, and this causes them to react with one another to satisfy the octet rule.

2.2 Molecules and Compounds

■ **Compounds** and **molecules** are formed when elements associate with each other.

■ **Ions** form when atoms lose or gain one or more electrons to achieve a completed outer orbital. An **ionic bond** is an attraction between oppositely charged ions. A **covalent bond** is one or more shared pairs of electrons.

■ In **nonpolar covalent bonds,** the sharing of electrons between atoms is fairly equal. In **polar covalent bonds,** the sharing of electrons is not equal. One of the atoms exerts greater attraction for the shared electrons than the other, resulting in a difference in **electronegativity.** This creates a slight charge on each atom. A **hydrogen bond** is a weak attraction between a slightly positive hydrogen atom and a slightly negative oxygen or nitrogen atom within the same or a different molecule. Hydrogen bonds help maintain the structure and function of cellular molecules.

2.3 Chemistry of Water

■ Water is a polar molecule. Its polarity allows hydrogen bonding to occur between water molecules. Water's polarity and hydrogen bonding account for its unique properties. These properties include:

• Water has a high heat capacity. Water can absorb heat (measured in **calories**) without a great change in temperature.

• Water has a high heat of evaporation. A large amount of heat is required to cause liquid water to change to a gas.

• Water is a solvent. Water allows the formation of **solutions** with many different **solutes.** Molecules that attract water are **hydrophilic,** whereas molecules that repel water are **hydrophobic.**

• Water is cohesive and adhesive. **Cohesion** allows water molecules to cling together and accounts for the surface tension of water. **Adhesion** allows water to cling to surfaces, such as internal transport vessels.

• Frozen water (ice) is less dense than liquid water. This allows ice to float on liquid water.

■ A small fraction of water molecules dissociate to produce an equal number of hydrogen ions and hydroxide ions. Solutions with equal numbers of H^+ and OH^- are termed neutral. In **acids,** there are more hydrogen ions than hydroxide ions. These solutions have a pH less than 7. In **bases,** there are more hydroxide ions than hydrogen ions. These solutions have a pH greater than 7. Cells are sensitive to pH changes. Biological systems often contain **buffers** that help keep the pH within a normal range.

2.4 Organic Molecules

■ The chemistry of carbon accounts for the chemistry of **organic molecules.** Carbohydrates, lipids, proteins, and nucleic acids are macromolecules with specific functions in cells.

■ **Isomers** are organic molecules with the same molecular formula, but different structures.

■ **Polymers** are formed by the joining together of **monomers.** Long-chain carbohydrates, proteins, and nucleic acids are all polymers. For each bond formed during a **dehydration reaction,** a molecule of water is removed. For each bond broken during a **hydrolysis reaction,** a molecule of water is added.

2.5 Carbohydrates

■ Monosaccharides, disaccharides, and polysaccharides are all **carbohydrates.** Therefore, the term *carbohydrate* includes both the monomers (e.g., glucose) and the polymers (e.g., starch, glycogen, and cellulose).

■ **Monosaccharides** include **hexose** sugars and **pentose** sugars. **Glucose** is a common hexose sugar used for energy.

■ **Disaccharides** consist of two monosaccharides.

■ **Polysaccharides** may serve as energy-storage molecules, such as **starch** (in plants) and **glycogen** (in animals), or as structural molecules, such as **cellulose** and chitin.

2.6 Lipids

■ **Lipids** include a wide variety of compounds that are insoluble in water. The majority of lipids are **triglycerides,** including the **fats** and **oils.** These are involved in long-term energy storage and contain one glycerol and three **fatty acids. Saturated fatty acids** do not have carbon–carbon double bonds, but **unsaturated fatty acids** do have double bonds in their hydrocarbon chain. The double bond causes a kink in the molecule that accounts for the liquid nature of oils at room temperature. **Trans fats** are a form of unsaturated fatty acid.

■ A **phospholipid** replaces one of the fatty acids with a phosphate group. In water, phospholipids form a bilayer, because the head of each molecule is hydrophilic and the tails are hydrophobic. **Steroids,** such as cholesterol, have the same four-ring structure as cholesterol, but each differs by the attached functional groups. Steroids may be used to form hormones, such as testosterone and estrogen.

2.7 Proteins

■ **Proteins** have numerous functions in cells. Some are **enzymes** that speed chemical reactions. The shape of a protein determines its structure.

■ The primary structure of a **polypeptide** is its own particular sequence of the possible 20 types of **amino acids.** The secondary structure is often an alpha (α) helix or a beta (β) pleated sheet. Tertiary structure occurs when a polypeptide bends and twists into a three-dimensional shape. A protein can contain several polypeptides, and this accounts for a possible quaternary structure. When proteins lose their shape, they are said to be **denatured.**

2.8 Nucleic Acids

- Nucleic acids are polymers of **nucleotides.** Each nucleotide has three components: a sugar, a base, and a phosphate functional group. **DNA (deoxyribonucleic acid),** which contains the sugar deoxyribose, is the genetic material that stores information for its own replication and for the sequence of amino acids in proteins. DNA, with the help of **RNA (ribonucleic acid),** specifies protein synthesis.

- There are four different types of bases in DNA: **adenine (A), thymine (T), guanine (G),** and **cytosine (C).** In RNA, the base **uracil (U)** replaces the base thymine.

- DNA molecules consist of two strands arranged as a **double helix.** RNA is usually single-stranded.

- **ATP (adenosine triphosphate),** with its unstable phosphate bonds, is the energy currency of cells. Hydrolysis of ATP to **ADP (adenosine diphosphate) + ℗** releases energy that the cell uses to do metabolic work.

ASSESS

Testing Yourself

Choose the best answer for each question.

2.1 Introduction to Chemistry

1. The atomic number tells you the
   a. number of neutrons in the nucleus.
   b. number of protons in the atom.
   c. atomic mass of the atom.
   d. number of chemical bonds the atom will form.
   e. All of these are correct.

2. Isotopes differ in their
   a. number of protons.
   b. atomic number.
   c. number of neutrons.
   d. number of electrons.

3. The periodic table provides us with what information?
   a. the atomic number, symbol, and mass
   b. how many shells an atom has
   c. how many electrons are in the outer shell
   d. All of these are correct.

2.2 Molecules and Compounds

4. An atom that has two electrons in the valence shell, such as magnesium, would most likely
   a. share to acquire a completed outer shell.
   b. lose these two electrons and become a negatively charged ion.
   c. lose these two electrons and become a positively charged ion.
   d. bind with carbon by way of hydrogen bonds.
   e. bind with another magnesium atom to satisfy its energy needs.

5. When an atom gains electrons, it
   a. forms a negatively charged ion.
   b. forms a positively charged ion.
   c. forms covalent bonds.
   d. forms ionic bonds.
   e. gains atomic mass.

6. An unequal sharing of electrons is a characteristic of a(n)
   a. ionic bond.
   b. polar covalent bond.
   c. nonpolar covalent bond.
   d. All of these are correct.

2.3 Chemistry of Water

7. Hydrogen bonds are formed as a result of which of the following?
   a. ionic bonds
   b. nonpolar covalent bonds
   c. polar covalent bonds
   d. None of these are correct.

8. Which of these properties of water can be attributed to hydrogen bonding between water molecules?
   a. Water stabilizes temperature inside and outside the cell.
   b. Water molecules are cohesive.
   c. Water is a solvent for many molecules.
   d. Ice floats on liquid water.
   e. All of these are correct.

9. The fact that water molecules cling not only to themselves, but also the interior surfaces of vessels, can be attributed to the fact that
   a. water is cohesive and adhesive.
   b. water acts as a solvent.
   c. water has a high heat of vaporization.
   d. All of these are correct.

2.4 Organic Molecules

10. Which of these is not a characteristic of carbon?
    a. forms four covalent bonds
    b. bonds with other carbon atoms
    c. is sometimes ionic
    d. can form long chains
    e. sometimes shares two pairs of electrons with another atom

11. Which of the following reactions combines two monomers to produce a polymer?
    a. dehydration   c. phosphorylation
    b. hydrolysis   d. None of these are correct.

2.5 Carbohydrates

12. The monomers of the carbohydrates are the
    a. polysaccharides.   c. monosaccharides.
    b. disaccharides.   d. waxes.

13. Which of the following polysaccharides is used as an energy-storage molecule in plants?
    a. glycogen   c. starch
    b. chitin   d. cellulose

14. Fructose and galactose are both isomers of
    a. glycogen.   c. starch.
    b. glucose.   d. maltose.

2.6 Lipids

15. A fatty acid is unsaturated if it
    a. contains hydrogen.
    b. contains carbon–carbon double bonds.
    c. contains a carboxyl (acidic) group.
    d. is bound to a glycerol.

16. Which of the following is incorrect regarding phospholipids?
    a. The heads are polar.
    b. The tails are nonpolar.
    c. They contain a phosphate group in place of one fatty acid.
    d. They are energy-storage molecules in the cell.

17. A lipid that contains four fused carbon rings is a
    a. triglyceride.
    c. phospholipid.
    b. wax.
    d. steroid.

2.7 Proteins

18. The chemical differences between one amino acid and another is due to which of the following?
    a. amino group
    d. peptide bond
    b. carboxyl group
    e. carbon atoms
    c. *R* group

19. Which of the following levels of protein structure is determined by interactions of more than one polypeptide chain?
    a. primary
    c. tertiary
    b. secondary
    d. quaternary

20. Which of the following is formed by the linking of two amino acids?
    a. a peptide bond
    c. quaternary structure
    b. a functional group
    d. an ionic bond

2.8 Nucleic Acids

21. Which of the following is incorrect regarding nucleotides?
    a. They contain a sugar, a nitrogen-containing base, and a phosphate functional group.
    b. They are the monomers of fats and polysaccharides.
    c. They join together by alternating covalent bonds between the sugars and phosphate groups.
    d. They are present in both DNA and RNA.

22. Which of the following is correct regarding ATP?
    a. It is an amino acid.
    b. It has a helical structure.
    c. It is a high-energy molecule that can break down to ADP and phosphate.
    d. It is a nucleotide component of DNA and RNA.

ENGAGE

BioNOW

Want to know how this science is relevant to your life? Check out the BioNOW video below:

■ Properties of Water

The elimination of which characteristic of water would have the least impact on a living organism such as yourself?

Thinking Critically

1. On a hot summer day, you decide to dive into a swimming pool. Before you begin your dive, you notice the surface of the water is smooth and continuous. After the dive, you discover that some water droplets are clinging to your skin and that your skin temperature feels cooler. Explain these observations based on the properties of water.

2. In biology, structure equals function. How does the tremendous variability in the structure of proteins relate to their importance (function) in biological systems?

3. Explain why a radioactive isotope of an element reacts chemically the same as a nonradioactive isotope of the same element.

©Steve Gschmeissner/SPL/Getty Images

3

Cell Structure and Function

Tay-Sachs: When Lysosomes Fail to Function

Cells are widely recognized as biological factories capable of performing all of the characteristics of life. While you may think of cells as being simple, they are quite specialized to suit their environment or function in an organism. They are also dynamic and able to change during their life span. One important function of a cell is to break down incoming molecules into their building blocks to provide materials for growth and repair. Breaking down molecules is one of the main activities of the lysosome, due to the digestive enzymes it contains. The lysosome is also involved in the recycling of damaged or unnecessary structures within the cell, and plays an important role in cell differentiation.

Despite their importance in cellular functions, the regulation of lysosomes has not always been well understood. In 2016, the Nobel Prize in Physiology or Medicine was awarded to Yoshinori Ohsumi, who not only determined how lysosomes function, but also identified the genes responsible for regulating their activity. Malfunctioning lysosomes are believed to play a role in a number of human diseases, including Tay-Sachs disease. In Tay-Sachs, a mutation in the *HEXA* gene limits the production of an enzyme called beta-hexosaminidase A. This leads to the accumulation of a fatty acid called ganglioside GM2, which causes the nerve cells in the brain to be malformed, resulting in a deterioration of mental and physical abilities, and death.

In this chapter, you will see that cells are the fundamental building blocks of organisms, organized to carry out basic metabolic functions and adapt to changing environmental conditions.

As you read through this chapter, think about the following questions:

1. What is the role of an organelle in a eukaryotic cell?

2. What other cellular organelles have a similar function to the lysosome?

3.1 The Cellular Level of Organization

3.2 Prokaryotic Cells

3.3 Eukaryotic Cells

3.4 The Cytoskeleton

3.5 Origin and Evolution of the Eukaryotic Cell

Before beginning this chapter, take a few moments to review the following discussions:

Section 1.2 What are the main differences between prokaryotic and eukaryotic cells?

Section 2.3 Why is water so important to cells?

Section 2.6 What role do phospholipids play in forming the cell membrane?

3.1 The Cellular Level of Organization

Learning Outcomes

Upon completion of this section, you should be able to

1. Explain why cells are the basic units of life.
2. List the basic principles of the cell theory.
3. Explain the difference between the surface-area-to-volume ratios for large and small cells.

Cells are the basic units of life. All of the chemistry and biomolecules we have discussed to this point are necessary but insufficient on their own to support life. It is only when these components are brought together and organized into a cell that life is possible. Thus, the answer to what life is must lie within the cell. The smallest living organisms are single-celled, while larger organisms are multicellular—that is, composed of many cells.

There are basically two different types of cells. Prokaryotic cells lack a membrane-bound nucleus. Two of the three domains of life (see Section 1.2) contain prokaryotic cells—bacteria and archaea. While bacteria and archaea were once thought to be closely related because of their similar size and shape, comparisons of DNA and RNA sequences now show these groups to be biochemically distinct from each other. The second type of cell possesses a membrane-bound nucleus, which houses the genetic material. These are called eukaryotic cells, and they belong to the domain Eukarya, which consists of plants, fungi, animals, and protists. In this chapter, we will explore the basic structure of the prokaryotic bacteria and eukaryotic cells. We will visit archaea again in Section 28.3.

The Cell Theory

Today, we are accustomed to thinking of living organisms as being constructed of cells. But the word *cell* didn't enter biology until the seventeenth century. The Dutch scientist Anton van Leeuwenhoek is recognized for making some of the earliest microscopes and observing things no one had seen before. Robert Hooke, an English scientist, confirmed van Leeuwenhoek's observations and was the first to use the term *cell*. The tiny chambers he observed in the honeycomb structure of cork reminded him of the rooms, or cells, in a monastery.

Over 150 years later—in the 1830s—the German microscopist Matthias Schleiden stated that plants are composed of cells. His counterpart, Theodor Schwann, stated that animals are also made up of living units called cells. This was quite a feat, because aside from their own exhausting examination of tissues, both had to take into consideration the studies of many other microscopists. Rudolf Virchow, another German microscopist, later came to the conclusion that cells come from preexisting cells. Sperm fertilizing an egg and a human developing from the zygote is just one example. These observations became the basis of the **cell theory**, which states that all organisms are made up of basic living units called cells, and that all cells come only from previously existing cells. Today, the cell theory is a basic theory of biology.

Microscopes have changed significantly since van Leeuwenhoek's time. Advances in the magnification power of the microscope have allowed scientists to study not only the tremendous variety of cell types, but how these cells function internally. The Scientific Inquiry feature "Modern Microscopy" explores some of the different types of microscopes that scientists use to study cells.

Cell Size

Cells are quite small. A frog's egg, at about 1 millimeter (mm) in diameter, is large enough to be seen by the human eye, but most cells are far smaller than 1 mm. Some are even as small as 1 micrometer (μm)—one thousandth of a millimeter. Cell inclusions and macromolecules are smaller than a micrometer and are measured in terms of nanometers (nm). Figure 3.1 outlines the visual range of the eye, light microscope, and electron microscope. The

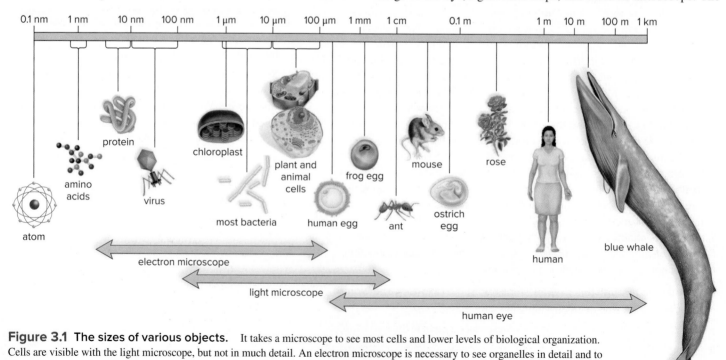

Figure 3.1 The sizes of various objects.   It takes a microscope to see most cells and lower levels of biological organization. Cells are visible with the light microscope, but not in much detail. An electron microscope is necessary to see organelles in detail and to observe viruses and molecules. In the metric system (see Appendix A), each higher unit is ten times greater than the preceding unit. (1 meter = 10^2 cm = 10^3 mm = 10^6 μm = 10^9 /nm)

Modern Microscopy

Microscopes have given scientists a deeper look into how life works than is possible with the naked eye. Today, there are many types of microscopes. A compound light microscope uses a set of glass lenses to focus light rays passing through a specimen to produce an image that is viewed by the human eye. A transmission electron microscope (TEM) uses a set of electromagnetic lenses to focus electrons passing through a specimen to produce an image, which is projected onto a fluorescent screen or photographic film. A scanning electron microscope (SEM) uses a narrow beam of electrons to scan over the surface of a specimen that is coated with a thin metal layer. Secondary electrons given off by the metal are detected and used to produce a three-dimensional image on a television screen. Figure 3A shows these three types of microscopes and their images.

Magnification, Resolution, and Contrast

Magnification is the ratio between the size of an image and an object's actual size. Electron microscopes magnify to a greater extent than do compound light microscopes. A light microscope can magnify objects about a thousand times, but an electron microscope can magnify them hundreds of thousands of times. The difference lies in the means of illumination. The path of light rays and electrons moving through space is wavelike, but the wavelength of electrons is much shorter than the wavelength of light. This difference in wavelength accounts for the electron microscope's greater magnifying capability and its greater ability to distinguish between two points (resolving power).

Resolution is the minimum distance between two objects that allows them to be seen as two separate objects. A microscope with poor resolution might enable a student to see only one cellular granule, while the microscope with the better resolution would show two granules next to each other. The greater the resolving power, the greater the detail seen.

If oil is placed between the sample and the objective lens of the compound light microscope, the resolving power is increased, and if ultraviolet light is used instead of visible light, it is also increased. But typically, a light microscope can resolve down to 0.2 µm, while the transmission electron microscope can resolve down to 0.0002 µm. If the resolving power of the average human eye is set at 1.0, then the typical compound light microscope is about 500, and the transmission electron microscope is 500,000.

The ability to make out, or resolve, a particular object can depend on contrast, a difference in the shading of an object compared to its background. Higher contrast is often achieved by staining cells with colored dyes (light microscopy) or with electron-dense metals (electron microscopy), which make them easier to see. Optical methods such as phase contrast and the use of fluorescently tagged antibodies can also help us visualize subcellular components such as specific proteins.

Questions to Consider

1. What happens to an image if the magnification is increased without increasing resolution?
2. Why is it necessary to artificially color electron micrographs?

Light micrograph   250×

Transmission   10,000×
electron micrograph

Scanning electron micrograph

a. Compound light microscope

b. Transmission electron microscope

c. Scanning electron microscope

Figure 3A Diagram of microscopes with accompanying micrographs. The compound light microscope (**a**) and the transmission electron microscope (**b**) provide an internal view of an organism. The scanning electron microscope (**c**) provides an external view of an organism. (photos)

	One 4-cm cube	Eight 2-cm cubes	Sixty-four 1-cm cubes
Total surface area (height × width × number of sides × number of cubes)	96 cm²	192 cm²	384 cm²
Total volume (height × width × length × number of cubes)	64 cm³	64 cm³	64 cm³
Surface area/Volume **of each cube** (surface area ÷ volume)	1.5:1	3:1	6:1

Figure 3.2 Surface-area-to-volume relationships. As cell size decreases from 4 cm³ to 1 cm³, the surface-area-to-volume ratio increases.

discussion of microscopy in the Scientific Inquiry feature "Modern Microscopy" explains why the electron microscope allows us to see so much more detail than the light microscope.

The fact that cells are so small is a great advantage for multicellular organisms. Nutrients, such as glucose and oxygen, enter a cell, and wastes, such as carbon dioxide, exit a cell at its surface. Therefore, the amount of surface area affects the ability to get material into and out of the cell. A large cell requires more nutrients and produces more wastes than a small cell. However, as cells get larger in volume, the proportionate amount of surface area actually decreases. For example, for a cube-shaped cell, the volume increases by the cube of the sides (height × width × depth), while the surface area increases by the square of the sides and number of sides (height × width × 6). If a cell doubles in size, its surface area increases only fourfold, while its volume increases eightfold. Therefore, small cells, not large cells, are likely to have a more adequate surface area for exchanging nutrients and wastes.

Figure 3.2 illustrates that dividing a large cube into smaller cubes provides a lot more surface area per volume. This relationship is called the **surface-area-to-volume ratio.** Calculations show that a 1-cm cube has a surface-area-to-volume ratio of 6:1, whereas a 4-cm cube has a surface-area-to-volume ratio of 1.5:1. In general, a higher surface-area-to-volume ratio increases the efficiency of transporting materials into and out of the cell.

Most active cells are small. Initially, a frog's egg is not active (and thus fairly large), but once the egg is fertilized and cellular activities begin, the egg divides repeatedly without growth. These cell divisions restore the amount of surface area needed for adequate exchange of materials. Further, cells that specialize in absorption have modifications that greatly increase their surface-area-to-volume ratio. For example, the columnar epithelial cells along the surface of the intestinal wall have surface foldings called microvilli (*sing.,* microvillus) that increase their surface area.

Check Your Progress 3.1

1. Explain why humans are made up of trillions of cells instead of just one.
2. Explain why the cell is the basic unit of life.
3. Describe the metabolic challenges of a large cell compared to a cell smaller in size.

3.2 Prokaryotic Cells

Learning Outcomes

Upon completion of this section, you should be able to

1. Describe the fundamental components of a bacterial cell.
2. Identify the key differences between archaea and bacteria.

As mentioned in Section 3.1, cells can be categorized by the presence or absence of a nucleus. In this section, we will explore the **prokaryotic cells,** or those cells that lack a membrane-bound nucleus. The domains Archaea and Eubacteria consist of prokaryotic cells. Prokaryotes generally exist as single-celled organisms or as simple strings and clusters.

The term *bacteria* is often used to refer to prokaryotic cells that are members of the domain Eubacteria. Many people think of germs when they hear the word *bacteria,* but not all bacteria cause disease. In fact, most bacteria are beneficial and are essential for other living organisms' survival.

Plasma Membrane and Cytoplasm

All cells are surrounded by a **plasma membrane** consisting of a phospholipid bilayer embedded with protein molecules.

The plasma membrane is a boundary that separates the living contents of the cell from the surrounding environment. Inside the cell is a semifluid medium called the **cytoplasm.** The cytoplasm is composed of water, salts, and dissolved organic molecules. The plasma membrane regulates the entrance and exit of molecules into and out of the cytoplasm. We will take a much closer look at the structure and function of the plasma membrane in Section 4.1.

Structure of Bacteria

Bacteria all possess a **cell wall** just outside the plasma membrane. In addition to phospholipids and proteins, the cell wall of bacteria

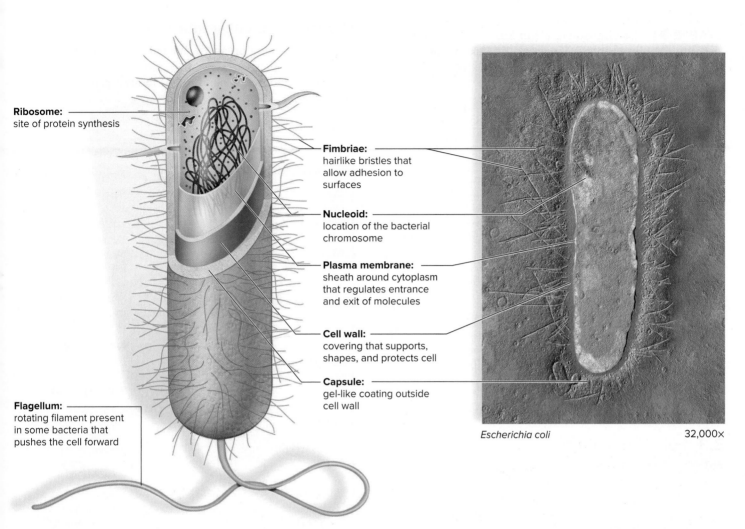

Ribosome:
site of protein synthesis

Fimbriae:
hairlike bristles that
allow adhesion to
surfaces

Nucleoid:
location of the bacterial
chromosome

Plasma membrane:
sheath around cytoplasm
that regulates entrance
and exit of molecules

Cell wall:
covering that supports,
shapes, and protects cell

Capsule:
gel-like coating outside
cell wall

Flagellum:
rotating filament present
in some bacteria that
pushes the cell forward

Escherichia coli   32,000×

Figure 3.3 Prokaryotic cell.   Prokaryotic cells lack a nucleus and other membrane-bound organelles, but they possess a nucleoid region where the DNA is located. (photo): ©Sercomi/Science Source

contains *peptidoglycan,* a complex molecule that is unique to bacteria and composed of chains of disaccharides joined together by peptide chains. The cell wall acts as a form of protection for bacteria. Some classes of antibiotics interfere with the synthesis of the peptidoglycans.

In some bacteria, the cell wall is further surrounded by a gelatinous sheath called a **capsule** (Fig. 3.3). The **flagella** are long, thin protein extensions that rotate like a propeller, allowing the bacterium to move rapidly in a fluid medium. Some bacteria also have **fimbriae,** which are short appendages that help them attach to an appropriate surface. The capsule and fimbriae often increase the ability of pathogenic bacteria to cause disease.

All prokaryotes have a single chromosome (loop of DNA and associated proteins) located within a region of the cytoplasm called the **nucleoid.** In prokaryotes, the nucleoid is not surrounded by a membrane. Many prokaryotes also have small accessory rings of DNA called **plasmids** that are located within the cytoplasm and contain small amounts of genetic information. Plasmids may be moved from one prokaryote to another (see Section 28.4).

Within the cytoplasm are thousands of ribosomes for the synthesis of proteins. The ribosomes of prokaryotic organisms are smaller and structurally different from those of eukaryotic cells, which makes ribosomes a good target for antibacterial drugs. In addition, the photosynthetic cyanobacteria have light-sensitive pigments, usually within a series of internal membranes called *thylakoids.*

Although prokaryotes are structurally simple, they are much more metabolically diverse than eukaryotes. Many of them can synthesize all their structural components from very simple, even inorganic, molecules. Indeed, humans exploit the metabolic capability of bacteria by using them to produce a wide variety of chemicals and products.

Prokaryotes also have adapted to living in almost every environment on Earth. In particular, archaea have been found living under conditions that would not support any other form of life. Archaeal membranes have unique membrane-spanning lipids that help them survive in extremes of heat, pH, and salinity. Table 3.1 compares the major structures of prokaryotes (archaea and bacteria) with those of eukaryotes.

TABLE 3.1 Comparison of Major Structural Features of Archaea, Bacteria, and Eukaryotes

	Archaea	Bacteria	Eukaryotes
Cell wall	Usually present, no peptidoglycan	Usually present, with peptidoglycan	Sometimes present, no peptidoglycan
Plasma membrane	Yes	Yes	Yes
Nucleus	No	No	Yes
Membrane-bound organelles	No	No*	Yes
Ribosomes	Yes	Yes	Yes, larger than prokaryotic

* Some possess internal membranes where chemical reactions may occur.

Check Your Progress 3.2

1. Explain the function of the plasma membrane.
2. Identify the key bacterial structures and their function.
3. Explain the general differences among bacteria, archaea, and eukaryotic cells.

3.3 Eukaryotic Cells

Learning Outcomes

Upon completion of this section, you should be able to

1. Recognize the structure and function of the organelles within eukaryotic cells.
2. Identify the cellular structures unique to both plant and animal cells.
3. Distinguish between the roles of the chloroplast and mitochondria in a cell.

The principal distinguishing feature of **eukaryotic cells** is the presence of a nucleus, which separates the genetic material from the cytoplasm of the cell. Eukaryotes have a higher level of complexity than the prokaryotes, mostly due to the presence of multiple organelles, many of which are surrounded by membranes. The eukaryotes are divided into multiple supergroups, based on their evolutionary relationships (see Section 1.2). Animals, plants, fungi, and protists are all composed of eukaryotic cells.

Cell Walls

The majority of the eukaryotic cells (with the exception of animals and some protists) have a permeable but protective cell wall that surrounds the plasma membrane. The cell walls of plants and algae are composed of the polysaccharide cellulose. Cellulose forms fibrils that lie at right angles to one another for added strength. In some plants, a secondary cell wall, made of a strong compound called lignin, forms inside the primary cell wall. The cell walls of some fungi are composed of cellulose and chitin, the same type of polysaccharide found in the exoskeleton of insects.

Organelles of Eukaryotic Cells

Originally, the term **organelle** referred to only membrane-bound structures within a cell. However, it has come to mean any well-defined subcellular structure that performs a particular function (Table 3.2).

TABLE 3.2 Structures of Eukaryotic Cells

Structure	Composition	Function
Cell wall (plants, fungi, algae, and some protists)	Contains polysaccharides	Support and protection
Plasma membrane	Phospholipid bilayer with embedded proteins	Defines cell boundary; regulates molecule passage into and out of cells
Nucleus	Nuclear envelope, nucleoplasm, chromatin, and nucleoli	Storage of genetic information; synthesis of DNA and RNA
Nucleoli	Concentrated area of chromatin, RNA, and proteins	Ribosomal subunit formation
Ribosomes	Protein and RNA in two subunits	Protein synthesis
Rough endoplasmic reticulum (RER)	Network of folded membranes studded with ribosomes	Folding, modification, and transport of proteins for export or associated with membranes
Smooth endoplasmic reticulum (SER)	Network of folded membranes having no ribosomes	Lipid and carbohydrate synthesis in some cells; detoxification of chemicals
Golgi apparatus	Stack of small membranous sacs	Processing, packaging, and distribution of proteins and lipids
Lysosomes (animal cells only)	Membranous vesicle containing digestive enzymes	Intracellular digestion; recycling of cellular components
Vacuoles and vesicles	Membranous sacs of various sizes	Storage of substances
Peroxisomes	Membranous vesicle containing specific enzymes	Breakdown of fatty acids and other metabolic tasks
Mitochondria	Inner membrane (cristae) bound by an outer membrane	Cellular respiration
Chloroplasts (plant cells and some protists)	Membranous grana bound by two membranes	Photosynthesis
Cytoskeleton	Microtubules, intermediate filaments, actin filaments	Shape of cell and movement of its parts
Cilia and flagella (cilia are rare in plant cells)	Microtubules	Movement of cell
Centriole (animal cells only)	Microtubules	Cell division

We can think of a eukaryotic cell as being a cellular factory. Just as all the assembly lines of a factory operate at the same time, so do all the organelles of a cell. Raw materials enter a factory where different departments turn them into various products. In the same way, the cell takes in chemicals, and then the organelles process them. The factory must also be supplied with energy, and get rid of waste, and the cell performs these functions as well.

In this chapter, we will focus primarily on animal cells (Fig. 3.4) and plant cells (Fig. 3.5). While there are many similarities in these cell types, there are some important differences. For example, both contain mitochondria, but chloroplasts are primarily found in the cells of plants, while centrioles are found in the cells of animals. In the illustrations throughout this text, note that each of the organelles has an assigned color.

nuclear envelope

endoplasmic reticulum

nucleolus

chromatin

a.   10,000×

Plasma membrane: outer surface that regulates entrance and exit of molecules

protein

phospholipid

CYTOSKELETON: maintains cell shape and assists movement of cell parts:

Microtubules: cylinders of protein molecules present in cytoplasm, centrioles, cilia, and flagella

Intermediate filaments: protein fibers that provide support and strength

Actin filaments: protein fibers that play a role in movement of cell and organelles

Centrioles: short, cylinders of microtubules

Centrosome: microtubule organizing center that contains a pair of centrioles

Lysosome: vesicle that digests macromolecules and even cell parts

Vesicle: membrane-bounded sac that stores and transports substances

Cytoplasm: semifluid matrix outside nucleus that contains organelles

NUCLEUS:

Nuclear envelope: double membrane with nuclear pores that encloses nucleus

Chromatin: diffuse threads containing DNA and protein

Nucleolus: region that produces subunits of ribosomes

ENDOPLASMIC RETICULUM:

Rough ER: studded with ribosomes, processes proteins

Smooth ER: lacks ribosomes, synthesizes lipid molecules

Ribosomes: particles that carry out protein synthesis

Mitochondrion: organelle that carries out cellular respiration, producing ATP molecules

Polyribosome: string of ribosomes simultaneously synthesizing same protein

Golgi apparatus: processes, packages, and secretes modified cell products

b.

Figure 3.4 Animal cell anatomy.   **a.** Micrograph of a eukaryotic cell. **b.** Drawing of a generalized animal cell indicating the function of each structure.

(a): ©Alfred Pasieka/Science Source

mitochondrion

nucleus

peroxisome

ribosomes

central vacuole

plasma membrane
cell wall

chloroplast

a.   12,300×

Figure 3.5 Plant cell anatomy.   a. Micrograph of a plant cell. **b.** Drawing of a generalized plant cell indicating the function of each structure. (a): ©Biophoto Associates/Science Source

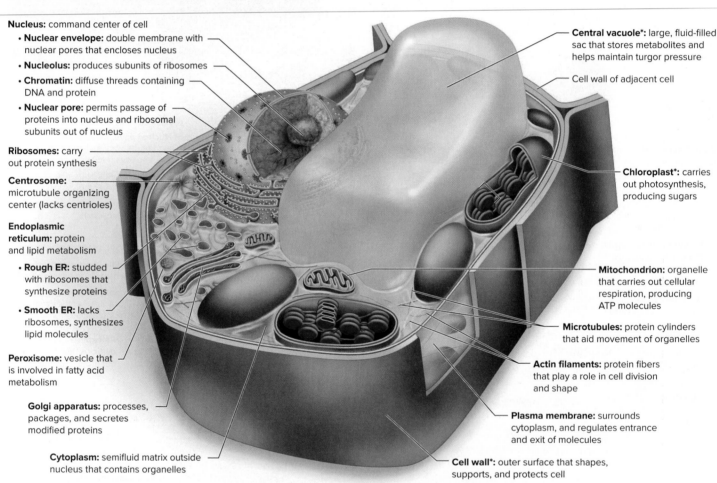

Nucleus: command center of cell
- **Nuclear envelope:** double membrane with nuclear pores that encloses nucleus
- **Nucleolus:** produces subunits of ribosomes
- **Chromatin:** diffuse threads containing DNA and protein
- **Nuclear pore:** permits passage of proteins into nucleus and ribosomal subunits out of nucleus

Ribosomes: carry out protein synthesis

Centrosome: microtubule organizing center (lacks centrioles)

Endoplasmic reticulum: protein and lipid metabolism
- **Rough ER:** studded with ribosomes that synthesize proteins
- **Smooth ER:** lacks ribosomes, synthesizes lipid molecules

Peroxisome: vesicle that is involved in fatty acid metabolism

Golgi apparatus: processes, packages, and secretes modified proteins

Cytoplasm: semifluid matrix outside nucleus that contains organelles

Central vacuole*: large, fluid-filled sac that stores metabolites and helps maintain turgor pressure

Cell wall of adjacent cell

Chloroplast*: carries out photosynthesis, producing sugars

Mitochondrion: organelle that carries out cellular respiration, producing ATP molecules

Microtubules: protein cylinders that aid movement of organelles

Actin filaments: protein fibers that play a role in cell division and shape

Plasma membrane: surrounds cytoplasm, and regulates entrance and exit of molecules

Cell wall*: outer surface that shapes, supports, and protects cell

b.   *Not found in animal cells

The Nucleus

The **nucleus** is one of the defining characteristics of a eukaryotic cell. It is responsible for storing the genetic material, deoxyribonucleic acid (DNA). In the factory analogy, the nucleus is the head office. DNA governs the characteristics of the cell and thus its metabolic functions.

When you look at the nucleus, even in an electron micrograph, you cannot see a DNA molecule (Fig. 3.6). Instead, what you see is **chromatin,** which consists of DNA and associated proteins. Chromatin in most eukaryotic cells is not one continuous strand. During most of the cell's lifetime, chromatin is present, but when the cell is ready to undergo cell division, it will undergo coiling

Figure 3.6 Structure of the nucleus. The nucleus contains chromatin. The nucleolus is where rRNA is produced and ribosomal subunits are assembled. The nuclear envelope contains pores, as shown in the larger micrograph of a freeze-fractured nuclear envelope. Nuclear pores serve as passageways for substances to pass into and out of the nucleus. (photos) (top right): ©E.G. Pollock; (bottom blowup): ©Ron Milligan/Scripps Research Institute

and become highly condensed structures called **chromosomes.** Human cells contain 46 chromosomes, which are immersed in a semifluid medium called the **nucleoplasm.** The nucleoplasm and the cytoplasm of a cell have different compositions.

The **nucleolus** is a region within the nucleus where a type of RNA, called ribosomal RNA (rRNA), is produced. It is here that the rRNA joins with proteins to form the subunits of ribosomes (described below). Under a microscope, the nucleolus often appears as a dark region within the nucleus.

The nucleus is separated from the cytoplasm by a double membrane known as the **nuclear envelope,** which is continuous with the endoplasmic reticulum (discussed later in this section). The nuclear envelope has **nuclear pores** of sufficient size (100 nm) to permit the bidirectional transport of proteins and ribosomal subunits. Regulation of transport through the nuclear pores can control events within the nucleus or the entire cell.

Interestingly, during cell division, the nuclear envelope completely disappears and the contents of the nucleus are mixed with the cytoplasm (see Section 5.3). Following cell division, the nuclear envelope re-forms around the chromosomes.

Ribosomes are responsible for the synthesis of proteins. Ribosomes are composed of two subunits called "large" and "small" because of their relative sizes. Each subunit is a complex of unique ribosomal RNA (rRNA) and protein molecules. Ribosomes can be found individually in the cytoplasm, as well as in groups called **polyribosomes** (several ribosomes associated simultaneously with a single mRNA molecule). Ribosomes can also be found attached to the endoplasmic reticulum, a membranous system of sacs and channels. Proteins synthesized at ribosomes attached to the endoplasmic reticulum have a different destination from that of proteins synthesized at ribosomes free in the cytoplasm.

The Endomembrane System

The endomembrane system consists of the nuclear envelope, the endoplasmic reticulum, the Golgi apparatus, and several small

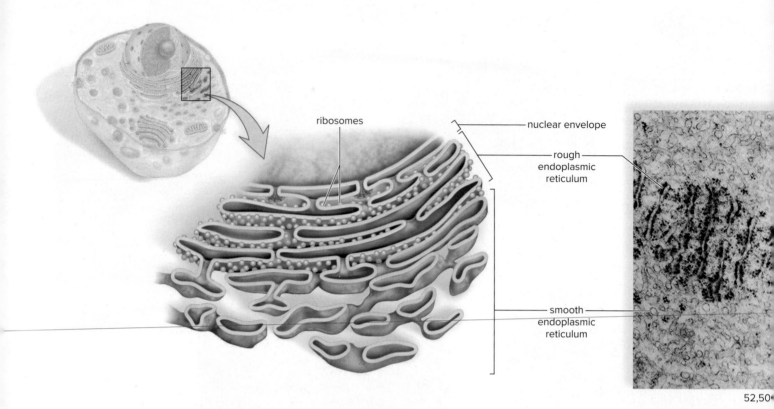

ribosomes

nuclear envelope

rough endoplasmic reticulum

smooth endoplasmic reticulum

52,50•

Figure 3.7 Endoplasmic reticulum (ER).   Ribosomes are present on rough ER, which consists of flattened sacs, but not on smooth ER, which is more tubular. Proteins are synthesized and modified by rough ER, whereas smooth ER is involved in lipid synthesis, detoxification reactions, and several other possible functions. (photo): ©Martin M. Rotker/Science Source

membranous sacs called **vesicles.** Continuing the factory analogy, the endomembrane system is essentially the transportation and product-processing section of the cell. This system compartmentalizes the cell so particular enzymatic reactions are restricted to specific regions. Organelles that make up the endomembrane system are connected either directly or by transport vesicles.

The Endoplasmic Reticulum

The **endoplasmic reticulum (ER)** is an interconnected system of membranous channels and sacs (flattened vesicles) that is physically continuous with the outer membrane of the nuclear envelope.

Rough endoplasmic reticulum (RER) is studded with ribosomes on the side of the membrane that faces the cytoplasm (Fig. 3.7). As proteins are synthesized on these ribosomes, they pass into the interior of the ER, where processing and modification begin. Proteins synthesized here are destined for the membrane of the cell or to be secreted from the cell.

Proper folding, processing, and transport of proteins are critical to the functioning of the cell. For example, in cystic fibrosis, a mutated plasma membrane channel protein is retained in the endoplasmic reticulum because it is folded incorrectly. Without this protein in its correct location, the cell is unable to regulate the transport of the chloride ion, resulting in an inability to regulate water levels in some cells.

Smooth endoplasmic reticulum (SER) is continuous with rough ER but does not have attached ribosomes. Smooth ER synthesizes the phospholipids found in cell membranes, as well as

those that perform various other functions. In the testes, it produces testosterone, and in the liver, it helps detoxify drugs. In muscle cells, smooth ER stores calcium ions. Regardless of any specialized function, smooth ER also forms vesicles in which products are transported to the Golgi apparatus.

The Golgi Apparatus

The **Golgi apparatus** is named for Camillo Golgi, who discovered its presence in cells in 1898. The Golgi apparatus consists of a stack of between 3 and 20 slightly curved sacs whose appearance can be compared to a stack of pancakes (Fig. 3.8). The Golgi apparatus is referred to as the shipping center of the cell because it collects, sorts, packages, and distributes materials such as proteins and lipids. In animal cells, one side of the stack (the inner face) is directed toward the ER, and the other side of the stack (the outer face) is directed toward the plasma membrane. Vesicles can frequently be seen at the edges of the sacs.

The Golgi apparatus receives proteins and also lipid-filled vesicles that bud from the ER. These molecules then move through the Golgi from the inner face to the outer face. During their passage through the Golgi apparatus, proteins and lipids can be modified before they are repackaged in secretory vesicles. Secretory vesicles proceed to the plasma membrane, where they discharge their contents. This action is termed secretion or exocytosis.

The Golgi apparatus is also involved in the formation of lysosomes, vesicles that contain enzymes that remain within the cell. It appears that proteins made at the rough ER have specific

Figure 3.8 Endomembrane system.   The organelles in the endomembrane system work together to carry out the functions noted.

molecular tags that serve as "zip codes" to tell the Golgi apparatus whether they belong inside the cell in some membrane-bound organelle or in a secretory vesicle.

Lysosomes

Lysosomes are membrane-bound vesicles produced by the Golgi apparatus. Lysosomes contain hydrolytic digestive enzymes and act as the garbage disposals of the cellular factory.

Sometimes macromolecules are brought into a cell by vesicle formation at the plasma membrane (Fig. 3.8). When a lysosome fuses with such a vesicle, its contents are digested by lysosomal enzymes into simpler subunits that then enter the cytoplasm. Some white blood cells defend the body by engulfing pathogens via

vesicle formation. When lysosomes fuse with these vesicles, the bacteria are digested. Even parts of a cell are digested by its own lysosomes (called autophagy). For example, the finger webbing found in the human embryo is later dissolved by lysosomes so the fingers are separated.

Lysosomes contain many enzymes for digesting all sorts of molecules. Occasionally, a child inherits the inability to make a lysosomal enzyme, and therefore has a lysosomal storage disease. For example, in Tay-Sachs disease (see chapter opener), the cells that surround nerve cells cannot break down the lipid ganglioside GM2, which then accumulates inside lysosomes and affects the nervous system. At about 6 months, the infant can no longer see and, then, gradually loses hearing and even the ability to move. Death usually occurs by about 3 years of age.

peroxisome

34,000×

Figure 3.9 Peroxisomes. Peroxisomes contain one or more enzymes that can oxidize various organic substances. Peroxisomes also contain the enzyme catalase, which breaks down the hydrogen peroxide (H_2O_2) that builds up after organic substances are oxidized. (photo): ©EM Research Services, Newcastle University

Vacuoles

A **vacuole** is a large membranous sac. A vacuole is larger than a vesicle. Although animal cells have vacuoles, they are much more prominent in plant cells. Typically, plant cells have a large central vacuole filled with a watery fluid that provides added support to the cell (see Fig. 3.5).

Vacuoles store substances. Plant vacuoles contain not only water, sugars, and salts, but also pigments and toxic molecules. The pigments are responsible for many of the red, blue, or purple colors of flowers and some leaves. The toxic substances help protect a plant from herbivorous animals. The vacuoles present in single-celled protozoans are quite specialized. They include contractile vacuoles for ridding the cell of excess water and digestive vacuoles for breaking down nutrients.

Peroxisomes

Peroxisomes, similar to lysosomes, are membrane-bound vesicles that enclose enzymes (Fig. 3.9). These enzymes break down fatty acids, and in the process produce hydrogen peroxide (H_2O_2), a toxic molecule. However, peroxisomes also contain an enzyme called catalase that breaks down the H_2O_2 to water.

The enzymes present in a peroxisome depend on the function of the cell. Peroxisomes are especially prevalent in cells that are synthesizing and breaking down fats. In the liver, some peroxisomes break down fats, while others produce bile salts from

cholesterol. In a disease called adrenoleukodystrophy (ALD), cells lack a carrier protein to transport an enzyme into peroxisomes. As a result, long-chain fatty acids accumulate in the brain, resulting in neurological damage.

Plant cells also have peroxisomes. In germinating seeds, they oxidize fatty acids into molecules that can be converted to sugars needed by the growing plant. In leaves, peroxisomes can carry out a reaction that is opposite to photosynthesis—the reaction uses up oxygen and releases carbon dioxide.

Energy-Related Organelles

Life is possible only because of a constant input of energy. Organisms use this energy for maintenance and growth. Chloroplasts and mitochondria are the two eukaryotic membranous organelles that specialize in converting energy to a form the cell can use. **Chloroplasts** use solar energy to synthesize carbohydrates, which are broken down by the **mitochondria** (*sing.,* mitochondrion) to produce ATP molecules, as shown in Figure 3.10. This diagram shows that chemicals recycle between chloroplasts and mitochondria in the presence of solar energy. When cells use ATP as an energy source, energy dissipates as heat. Most life cannot exist without a constant input of solar energy.

Plants, algae, and cyanobacteria are all capable of carrying on **photosynthesis** in this manner:

solar energy + carbon dioxide + water ⟶ carbohydrate + oxygen

Plants and algae have chloroplasts, while cyanobacteria carry on photosynthesis within independent thylakoids. Solar energy is the ultimate source of energy for most cells because nearly all

solar energy

carbohydrate (high chemical energy)

chloroplast

mitochondrion

$CO_2 + H_2O$ (low chemical energy)

ATP

usable energy for cells

Figure 3.10 Energy-producing organelles in eukaryotic cells. Chloroplasts use sunlight to produce carbohydrates, which in turn are used by mitochondria. The mitochondria then produce carbon dioxide and water, which in turn are used by the chloroplasts.

organisms, either directly or indirectly, use the carbohydrates produced by photosynthesizers as an energy source. We will take a closer look at photosynthesis in Chapter 8.

Many organisms carry on **cellular respiration,** the process by which the chemical energy of carbohydrates is converted to that of ATP (adenosine triphosphate), the common energy carrier in cells. Eukaryotic organisms complete the process of cellular respiration in mitochondria. Cellular respiration can be represented by this equation:

carbohydrate + oxygen $\longrightarrow$ carbon dioxide + water + energy

Here, *energy* is in the form of ATP molecules. When a cell needs energy, ATP supplies it. The energy of ATP is used for all energy-requiring processes in cells. Chapter 7 explores how cellular respiration occurs within cells.

Chloroplasts

Plant and algal cells contain chloroplasts (Fig. 3.11), the organelles that allow them to use solar energy to produce organic molecules. Chloroplasts are about 4–6 μm in diameter and 1–5 μm in length. They belong to a group of organelles known as plastids. Among the plastids are also the *amyloplasts,* common in roots, which store starch, and the *chromoplasts,* common in leaves, which contain red and orange pigments. A chloroplast is green because it contains the green pigment chlorophyll. A typical plant cell in a leaf may contain approximately 50 chloroplasts. Chloroplasts divide by splitting in two in a manner similar to how bacteria divide.

A chloroplast is bounded by two membranes that enclose a fluid-filled space called the **stroma** (Fig. 3.11). The stroma contains a single circular DNA molecule, as well as ribosomes. The chloroplast contains its own genetic material and makes most of its own proteins. The others are encoded by nuclear genes and imported from the cytoplasm. A membrane system within the stroma is organized into interconnected flattened sacs called **thylakoids.** In certain regions, the thylakoids are stacked up in structures called **grana** (*sing.,* granum). There can be hundreds of grana within a single chloroplast (Fig. 3.11).

Chlorophyll, which is located within the thylakoid membranes of grana, captures the solar energy needed to enable chloroplasts to produce carbohydrates. The solar energy excites an electron within the chlorophyll molecule. The energy from that excited electron is used to make high-energy compounds. The stroma contains the enzymes that synthesize carbohydrates from carbon dioxide and water using these high-energy compounds. The chloroplast is a tiny

solar panel for collecting sunlight and a factory for making carbohydrates.

Mitochondria

The majority of eukaryotic cells, with the exception of a few protists, contain mitochondria. Photosynthetic eukaryotes, such as plants and algae, contain *both* chloroplasts and mitochondria. Mitochondria are usually 0.5–1.0 μm in diameter and 2–5 μm in length. Mitochondria are the power plants of the cell. Substrates broken down in the cytoplasm are transported into the mitochondria and converted into ATP to be used by the cell for its various needs. Mitochondria are also involved in cellular differentiation and cell death. It is now recognized that mitochondria play a role in the aging process.

Figure 3.11 Chloroplast structure.
Chloroplasts carry out photosynthesis. **a.** Electron micrograph of a longitudinal section of a chloroplast. **b.** Generalized drawing of a chloroplast in which the outer and inner membranes have been cut away to reveal the grana, each of which is a stack of membranous sacs called thylakoids. In some grana, but not all, it is obvious that thylakoid spaces are interconnected. (photo) (a): ©Science Source

23,000×

a.

double membrane — outer membrane / inner membrane

grana   thylakoid space   stroma   thylakoid

b.

Figure 3.12 Mitochondrion structure. Mitochondria are involved in cellular respiration. **a.** Electron micrograph of a longitudinal section of a mitochondrion. **b.** Generalized drawing in which the outer membrane and portions of the inner membrane have been cut away to reveal the cristae. (photo) (a): ©Keith R. Porter/Science Source

a.

85,000×

double membrane { outer membrane, inner membrane   cristae   matrix

b.

Mitochondria, like chloroplasts, divide by splitting in two. Mitochondria are also bound by a double membrane (Fig. 3.12). In mitochondria, the inner fluid-filled space is called the **matrix.** As with chloroplasts, mitochondria contain their own circular DNA chromosome and encode some, but not all, of their own proteins. The matrix contains ribosomes, and also enzymes that break down carbohydrate products, releasing energy to be used for ATP production.

The inner membrane of a mitochondrion invaginates to form **cristae** (*sing.,* crista). Cristae provide a much greater surface area to accommodate the protein complexes and other participants that produce ATP.

The number of mitochondria per cell can vary considerably. Some cells have only one mitochondrion, while other cells may have thousands. Tissues that need large amounts of energy have more mitochondria per cell than tissues with lower energy demand. Mutations in the mitochondrial DNA usually affect high-energy-demand tissues such as the eye, central nervous system, and muscles.

In humans, the mother's egg supplies the mitochondria. The father's sperm typically does not contribute any mitochondria to the offspring. This has made mitochondrial DNA studies useful for population genetic studies. An interesting example of this is the search for "mitochondrial Eve," which examines different models on how early humans migrated from Africa (see Section 32.5).

Check Your Progress   3.3

1. Explain the function of the cell wall in eukaryotes.
2. Describe how the endomembrane system acts as a transport system.
3. Explain why plant cells need both chloroplasts and mitochondria.

3.4 The Cytoskeleton

Learning Outcomes

Upon completion of this section, you should be able to

1. Compare and contrast the structural differences between actin filaments, intermediate filaments, and microtubules.
2. Identify the cellular structures that are composed of the different cytoskeleton components.
3. Explain the purpose of cilia, flagella, and centrosomes in the cell.

The protein components of the **cytoskeleton** interconnect and extend from the nucleus to the plasma membrane in eukaryotic cells. Advances in microscopy, specifically the invention of the electron microscope, have indicated that the cytoskeleton is a highly organized series of interconnecting proteins. The technique of immunofluorescence microscopy identified the makeup of the protein components within the cytoskeletal network (Fig. 3.13).

The cytoskeleton contains actin filaments, intermediate filaments, and microtubules, which maintain cell shape and allow the cell and its organelles to move. Therefore, the cytoskeleton is often compared with the bones and muscles of an animal. However, the cytoskeleton is dynamic, especially because its protein components can assemble and disassemble as needed.

Actin Filaments

Actin filaments (formerly called microfilaments) are long, extremely thin, flexible fibers (about 7 nm in diameter) that occur in bundles or meshlike networks. Each actin filament contains two chains of globular actin monomers twisted about one another in a helical manner.

Actin filaments play a structural role when they form a dense, complex web just under the plasma membrane, to which they are anchored by special proteins. They are also seen in the microvilli that project from intestinal cells, and their presence accounts for the formation of pseudopods (false feet), extensions that allow certain cells to move in an amoeboid fashion.

a. Actin filaments   850×

b. Intermediate filaments   1,000×

c. Microtubules

Figure 3.13 The cytoskeleton.   The cytoskeleton maintains the shape of the cell and allows its parts to move. Three types of protein components make up the cytoskeleton. **a.** *Left to right:* Fibroblasts in animal tissue contain actin filaments. The drawing shows that actin filaments are composed of a twisted double chain of actin subunits. The giant cells of the green alga *Chara* rely on actin filaments to move organelles from one end of the cell to another. **b.** *Left to right:* Fibroblasts in an animal tissue contain intermediate filaments. The drawing shows that fibrous proteins account for the ropelike structure of intermediate filaments. Human hair is strengthened by the presence of intermediate filaments. **c.** *Left to right:* Fibroblasts in an animal tissue contain microtubules. The drawing shows that microtubules are hollow tubes composed of tubulin subunits. The skin cells of a chameleon rely on microtubules to move pigment granules around so they can take on the color of their environment. (a) (actin filaments): ©Thomas Deerinck/Science Source; (*Chara*): ©Bob Gibbons/Alamy; (b) (intermediate filaments): ©Cultura Science/Alvin Telser, PhD/Getty Images; (girls): ©Amos Morgan/Getty Images; (c) (microtubules): ©Dr. Gopal Murti/ Science Source; (chameleon): ©hlansdown/Getty Images

To produce movement, actin filaments interact with **motor molecules,** which are proteins that can attach, detach, and reattach farther along the actin filament. For example, in muscle cells, the motor molecule myosin pulls actin filaments along in this way using the energy of ATP. Myosin has both a head and a tail. The tails of several muscle myosin molecules are joined to form a thick filament, while the heads interact with ATP and the actin filament.

During animal cell division, the two new cells form when actin, in conjunction with myosin, pinches off the cells from one another.

Intermediate Filaments

Intermediate filaments (8–11 nm in diameter) are intermediate in size between actin filaments and microtubules that will perform a structural role in the cell. They are a ropelike assembly of fibrous polypeptides that vary according to the type of tissue. Some intermediate filaments support the nuclear envelope, whereas others support the plasma membrane and take part in the formation of cell-to-cell junctions. In skin cells, intermediate filaments, made of the protein keratin, provide mechanical strength. They, too, are dynamic structures.

Microtubules

Microtubules are small, hollow cylinders about 25 nm in diameter and 0.2–25 μm in length. Microtubules are made of the globular protein tubulin, which is of two types, called α and β. Microtubules are structured so that the tubulin proteins surround an empty central core.

The regulation of microtubule assembly is controlled by a microtubule organizing center. In most eukaryotic cells, the main microtubule organizing center is in the **centrosome,** which lies near the nucleus. Microtubules radiate from the centrosome, helping to maintain the shape of the cell and acting as tracks along which organelles can move. Whereas the motor molecule myosin is associated with actin filaments, the motor molecules kinesin and dynein are associated with microtubules.

Before a cell divides, microtubules disassemble and then reassemble into a structure called a spindle. The spindle apparatus attaches to the chromosomes and ensures that they are distributed in an orderly manner. It also participates in dividing the cell in half. At the end of cell division, the spindle disassembles, and microtubules reassemble once again into their former array.

Centrioles   Both plant and animal cells contain centrosomes, the major microtubule organizing center for the cell. But in animal cells, a centrosome contains two centrioles lying at right angles to each other. The centrioles may be involved in the process of microtubule assembly and disassembly. **Centrioles** are short cylinders of microtubules surrounding a hollow center (Fig. 3.14).

Before an animal cell divides, the centrioles replicate such that the members of each pair are again at right angles to one another (Fig. 3.14). Then, each pair becomes part of a separate centrosome. During cell division, the centrosomes move apart and may function to organize the mitotic spindle (see Section 5.3).

Cilia and Flagella   Cilia (*sing.,* cilium) and **flagella** (*sing.,* flagellum) are hairlike projections that can move either in an undulating fashion, like a whip, or stiffly, like an oar. Some cells that have these organelles are capable of movement. For example, single-celled organisms called paramecia move by means of cilia, whereas sperm cells move by means of flagella. In the human body, the cells that line our upper respiratory tract have cilia that sweep debris trapped within mucus back up into the throat, where it can be swallowed or ejected. This action helps keep the lungs clean.

one microtubule triplet

Figure 3.14 Centrioles.   In a nondividing animal cell, a single pair of centrioles lies in the centrosome located just outside the nucleus. Before a cell divides, the centrioles replicate, producing two pairs of centrioles. During cell division, centrioles in their respective centrosomes separate so that each new cell has one centrosome containing one pair of centrioles.

In eukaryotic cells, cilia are much shorter than flagella, but they have a similar construction. Both are membrane-bound cylinders. The cylinders are composed of nine microtubule doublets arranged in a circle around two central microtubules. Therefore, they have a 9 + 2 pattern of microtubules. Cilia and flagella move when the microtubule doublets slide past one another (Fig. 3.15).

Check Your Progress 3.4

1. Identify the structural makeup of actin filaments, intermediate filaments, and microtubules.
2. Contrast the function of cilia, flagella, and centrioles.
3. Explain how cilia and flagella are involved in movement.

3.5 Origin and Evolution of the Eukaryotic Cell

Learning Outcomes

Upon completion of this section, you should be able to

1. Define *endosymbiosis*.
2. Describe how the endosymbiotic theory explains eukaryotic cell evolution.

The fossil record, which is based on the remains of ancient life, suggests that the first cells were prokaryotes. Therefore, scientists believe that eukaryotic cells evolved from prokaryotic cells. Biochemical data suggest that eukaryotes are more closely related to archaea than bacteria. The eukaryotic cell probably evolved from a prokaryotic cell in stages. Invagination of the plasma membrane

outer microtubule doublet

central microtubules

dynein side arm

Flagellum cross section   20,000x

The shaft of the flagellum has a ring of nine microtubule doublets anchored to a central pair of microtubules.

The side arms of each doublet are composed of dynein, a motor molecule.

dynein side arms

ATP

In the presence of ATP, the dynein side arms reach out to their neighbors, and bending occurs.

flagellum

Sperm   275x   **Flagellum**

shaft

plasma membrane

Cilia

Figure 3.15 Structure of a flagellum or cilium.   The shaft of a flagellum (or cilium) contains microtubule doublets whose side arms are motor molecules that cause the projection to move. Sperm have flagella. Without the ability of sperm to move to the egg, human reproduction would not be possible. Cilia cover the surface of the cells of the respiratory system where they beat upward to remove foreign matter. (photos) (sperm) ©David M. Phillips/Science Source; (flagellum cross section): ©Steve Gschmeissner/Science Source; (cilia): ©Dr. G. Moscoso/Science Source

might explain the origin of the nuclear envelope and such organelles as the endoplasmic reticulum and the Golgi apparatus. Some believe the other organelles could also have arisen in a similar fashion.

There is evidence that a similar process was involved in the origin of the energy organelles. Observations in the laboratory indicate that an amoeba infected with bacteria can become dependent upon them. Some investigators believe mitochondria and chloroplasts are derived from prokaryotes that were taken up by larger cells. Perhaps mitochondria were originally aerobic heterotrophic bacteria, and chloroplasts were originally cyanobacteria. The eukaryotic host cell would have benefited from an ability to utilize oxygen or synthesize organic food when, by chance, the prokaryote was taken up and not destroyed. After the prokaryote entered the host cell, the two would have begun living together

cooperatively. This is known as the **endosymbiotic theory** (*endo-*, in; *symbiosis*, living together). Some of the evidence supporting this is as follows (Fig. 3.16):

1. Mitochondria and chloroplasts are similar to bacteria in size and structure.
2. Both organelles are bound by a double membrane—the outer membrane may be derived from the engulfing vesicle, and the inner one may be derived from the plasma membrane of the original prokaryote.
3. Mitochondria and chloroplasts contain a limited amount of genetic material and divide by splitting. Their DNA (deoxyribonucleic acid) is a circular loop like that of prokaryotes.
4. Although most of the proteins within mitochondria and chloroplasts are now produced by the eukaryotic host, they do

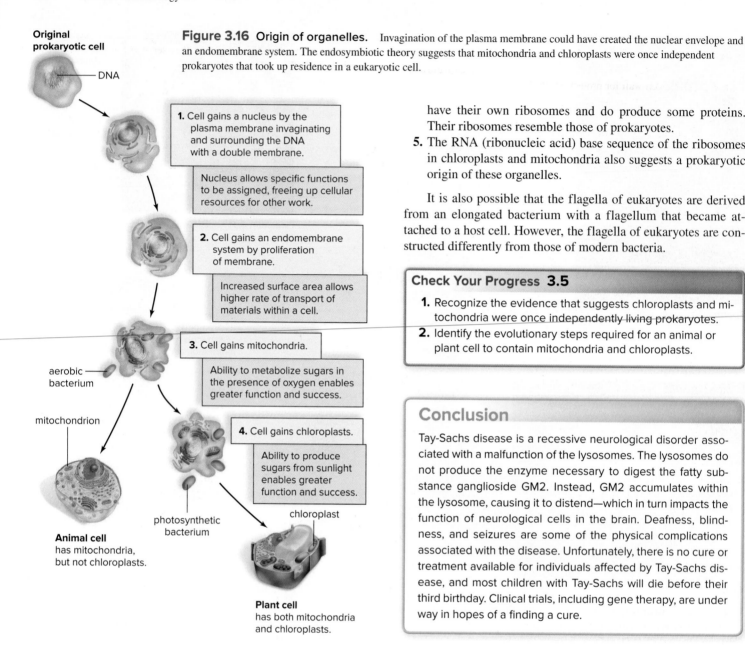

Original prokaryotic cell

DNA

Figure 3.16 Origin of organelles. Invagination of the plasma membrane could have created the nuclear envelope and an endomembrane system. The endosymbiotic theory suggests that mitochondria and chloroplasts were once independent prokaryotes that took up residence in a eukaryotic cell.

1. Cell gains a nucleus by the plasma membrane invaginating and surrounding the DNA with a double membrane.

Nucleus allows specific functions to be assigned, freeing up cellular resources for other work.

2. Cell gains an endomembrane system by proliferation of membrane.

Increased surface area allows higher rate of transport of materials within a cell.

3. Cell gains mitochondria.

aerobic bacterium

Ability to metabolize sugars in the presence of oxygen enables greater function and success.

mitochondrion

4. Cell gains chloroplasts.

Ability to produce sugars from sunlight enables greater function and success.

photosynthetic bacterium

chloroplast

Animal cell
has mitochondria, but not chloroplasts.

Plant cell
has both mitochondria and chloroplasts.

have their own ribosomes and do produce some proteins. Their ribosomes resemble those of prokaryotes.

5. The RNA (ribonucleic acid) base sequence of the ribosomes in chloroplasts and mitochondria also suggests a prokaryotic origin of these organelles.

It is also possible that the flagella of eukaryotes are derived from an elongated bacterium with a flagellum that became attached to a host cell. However, the flagella of eukaryotes are constructed differently from those of modern bacteria.

Check Your Progress 3.5

1. Recognize the evidence that suggests chloroplasts and mitochondria were once independently living prokaryotes.

2. Identify the evolutionary steps required for an animal or plant cell to contain mitochondria and chloroplasts.

Conclusion

Tay-Sachs disease is a recessive neurological disorder associated with a malfunction of the lysosomes. The lysosomes do not produce the enzyme necessary to digest the fatty substance ganglioside GM2. Instead, GM2 accumulates within the lysosome, causing it to distend—which in turn impacts the function of neurological cells in the brain. Deafness, blindness, and seizures are some of the physical complications associated with the disease. Unfortunately, there is no cure or treatment available for individuals affected by Tay-Sachs disease, and most children with Tay-Sachs will die before their third birthday. Clinical trials, including gene therapy, are under way in hopes of a finding a cure.

SUMMARIZE

3.1 The Cellular Level of Organization

■ The **cell theory** states that all organisms are composed of **cells,** the smallest units of living matter, and that new cells come only from pre-existing cells.

■ Cells must remain small in order to maintain an efficient **surface-area-to-volume ratio** for the exchange of molecules with the environment.

3.2 Prokaryotic Cells

■ All cells have a **plasma membrane** consisting of a phospholipid bilayer with embedded proteins. The membrane regulates the movement of molecules into and out of the cell. The inside of the cell is filled with a fluid called **cytoplasm.**

■ **Prokaryotic cells** do not have a nucleus, but they do have a **nucleoid** that is not bound by a nuclear envelope. They also lack most of the other organelles that compartmentalize eukaryotic cells, although some have internal membranes, such as the thylakoids of photosynthetic prokaryotes.

■ Some prokaryotic cells possess a gel-like sheath called a **capsule** and most possess a **cell wall** for protection. Prokaryotic cells may possess **fimbriae** that attach them to surfaces.

■ The DNA of a prokaryote is found primarily in a single, circular chromosome. **Plasmids** in the cytoplasm may contain small amounts of DNA.

■ Some prokaryotic cells move by the use of **flagella.**

3.3 Eukaryotic Cells

■ **Eukaryotic cells** are very complex. Animals, plants, fungi, and protists are all examples of eukaryotes. The common feature is the presence of a **nucleus** that contains the genetic material.

■ Some eukaryotic cells contain cell walls for protection.

■ Eukaryotic cells contain **organelles,** which are subcellular structures that perform a particular function.

■ The nucleus of eukaryotic cells is bound by a **nuclear envelope** containing **nuclear pores.** These pores serve as passageways between the cytoplasm and the **nucleoplasm.** Within the nucleus, the **chromatin** is a complex of DNA and protein. Chromatin is divided into separate structures called **chromosomes.**

■ The **nucleolus** is a special region of the chromatin where rRNA is produced and where proteins from the cytoplasm gather to form ribosomal subunits.

■ **Ribosomes** are organelles that function in protein synthesis. They can be bound to the **endoplasmic reticulum (ER),** or can exist within the cytoplasm singly or in groups called **polyribosomes.**

■ The endomembrane system includes the ER, the **Golgi apparatus,** the **lysosomes,** and other types of **vesicles** and **vacuoles.** The endomembrane system compartmentalizes the cell. The **rough endoplasmic reticulum (RER)** is covered with ribosomes and is involved in the folding, modification, and transport of proteins. The **smooth endoplasmic reticulum (SER)** has various metabolic functions depending on the cell type, but it also forms vesicles that carry products to the Golgi apparatus.

■ The **Golgi apparatus** processes proteins and repackages them into **lysosomes,** which carry out intracellular digestion, or into vesicles for transport to the plasma membrane or other organelles.

■ Vacuoles are large storage sacs, and vesicles are smaller ones. The large, single plant cell vacuole not only stores substances but also lends support to the plant cell.

■ **Peroxisomes** contain enzymes that oxidize molecules by producing hydrogen peroxide, which is subsequently broken down.

■ Cells require a constant input of energy to maintain their structure. **Chloroplasts** capture solar energy to carry on **photosynthesis,** which produces carbohydrates. The **mitochondria** undergo **cellular respiration** to produce ATP. The mitochondria are considered the energy powerhouse of the cell.

■ Chloroplasts contain internal membranes that enclose the **stroma.** These membranes, called **thylakoids,** are organized into stacks called **grana** (*sing.,* granum).

■ Mitochondria contain internal membranes known as **cristae** that enclose a space called the **matrix.**

3.4 The Cytoskeleton

■ The **cytoskeleton** contains **actin filaments, intermediate filaments,** and **microtubules.** These maintain cell shape and allow the cell and its organelles to move. Actin filaments interact with **motor molecules**

such as myosin in muscle cells. Microtubules are present in **centrioles, cilia,** and **flagella.** In the cytoplasm, they serve as tracks along which vesicles and other organelles move due to the action of specific motor molecules.

3.5 Origin and Evolution of the Eukaryotic Cell

■ The first cells were probably prokaryotic cells. Eukaryotic cells most likely arose from prokaryotic cells by a process called the **endosymbiotic theory.**

■ Biochemical data suggest that eukaryotic cells are closer evolutionarily to archaea than to bacteria.

■ The nuclear envelope most likely evolved through invagination of the plasma membrane, but mitochondria and chloroplasts may have arisen through endosymbiotic events.

ASSESS

Testing Yourself

Choose the best answer for each question.

3.1 The Cellular Level of Organization

1. The cell theory states:
   a. Cells form as organelles and molecules become grouped together in an organized manner.
   b. The normal functioning of an organism does not depend on its individual cells.
   c. The cell is the basic unit of life.
   d. Only eukaryotic organisms are made of cells.

2. As the size of a cell decreases, the ratio of its surface area to volume
   a. increases.
   b. decreases.
   c. stays the same.

3.2 Prokaryotic Cells

3. Which of the following would not be found in a prokaryotic cell?
   a. chloroplast
   b. DNA
   c. nucleoid
   d. nucleus

4. The _____ is involved in the movement of a prokaryotic cell.
   a. fimbriae
   b. capsule
   c. nucleoid
   d. flagella

5. Small circular pieces of DNA that are found in the cytoplasm of a prokaryotic cell are called
   a. capsules.
   b. nucleoids.
   c. plasmids.
   d. ribosomes.

3.3 Eukaryotic Cells

6. Eukaryotic cells contain
   a. a nucleus.
   b. DNA.
   c. a plasma membrane.
   d. mitochondria.
   e. All of these are correct.

7. The combination of DNA and protein in the nucleus of a eukaryotic cell is called
   a. nucleoplasm.
   b. plasmids.
   c. nucleoids.
   d. chromatin.

8. This organelle acts as a processing center for vesicles leaving the endoplasmic reticulum.
   a. peroxisome
   b. ribosome
   c. Golgi apparatus
   d. nucleolus

9. This organelle acts as a digestive organelle and may also be used to recycle the internal components of a cell.
   a. ribosome
   b. lysosome
   c. peroxisome
   d. Golgi apparatus

10. The mitochondria of a eukaryotic cell is the site of
    a. photosynthesis.
    b. fatty acid metabolism.
    c. cellular respiration.
    d. protein synthesis.

3.4 The Cytoskeleton

11. This component of the cytoskeleton is located just under the plasma membrane of a eukaryotic cell.
    a. actin filaments
    b. microtubules
    c. centrosomes
    d. intermediate filaments

12. Centrosomes and centrioles act as organizing centers for which component of the cytoskeleton?
    a. actin filaments
    b. microtubules
    c. intermediate filaments
    d. All of these are correct.

3.5 Origin and Evolution of the Eukaryotic Cell

13. The endosymbiotic theory explains the origins of which of the following components of a eukaryotic cell?
    a. mitochondria
    b. chloroplasts
    c. internal membranes
    d. All of these are correct.

14 According to the endosymbiotic theory, _____ were/was originally photosynthetic bacteria.
    a. ribosomes
    b. chloroplasts
    c. mitochondria
    d. the nucleus

ENGAGE

BioNOW

Want to know how this science is relevant to your life? Check out the BioNOW video below:

■ Cell Size

Why would a larger surface-area-to-volume ratio decrease metabolic efficiency?

Thinking Critically

1. One aspect of the new science of synthetic biology involves the laboratory design of cells to perform specific functions. Suppose you wanted to make a protein for use in a drug trial. How would you design a cell that would build the protein for export from the cell?

2. Why does the science of synthetic biology (see the opener for Chapter 1) still need to adhere to the principles of the surface-area-to-volume ratio?

3. How does an understanding of basic cell biology become important as we search for life on other planets, including the moons of our solar system?

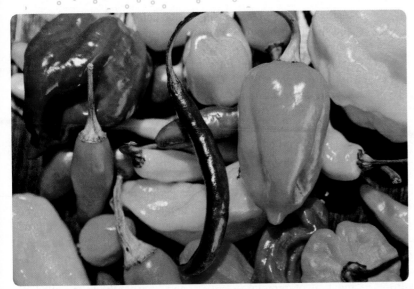
©Tooga Productions, Inc./Getty Images

Membrane Structure and Function

Red Hot Chili Peppers

Have you ever bitten into a hot pepper and had the sensation your mouth is on fire? Your eyes water and you are in real pain! The feelings of heat and pain are due to a membrane protein in your sensory nerves. Capsaicin is the chemical in chili peppers that binds to a channel protein in specialized sensory nerve cell endings called nociceptors (*noci-* means hurtful). One of the important functions of a membrane is to control what molecules move into and out of the cell and when they move. This particular channel protein, when activated, allows calcium ions to flow into the cell. In addition to capsaicin, other factors such as an acidic pH, heat, electrostatic charges, and a variety of chemical agents can activate this channel protein. Once activated by any of these signals, the response is the same. The channel opens, calcium ions flow into the cell, and the nociceptor sends a signal to the brain. The brain then interprets this signal as pain. As long as the capsaicin is present, this pathway will continue to send signals to the brain.

So the quickest way to alleviate the pain is to remove the capsaicin and close the channel protein. While some people drink cold water, this does very little other than cool down their mouth because capsaicin is lipid-soluble and does not dissolve in water. However, drinking milk, or eating rice or bread, usually helps. If you are a true "chili head," you know that if you survive the first bite, the next bite is easier. That is because, within minutes, the pathway becomes desensitized, or fails to respond, to the pain. However, other pathways, such as those in the eyes, may become activated if exposed to the capsaicin!

In this chapter, we will discuss the various functions of proteins embedded in the membranes of your cells and how the membranes control what enters and leaves the cells. We will also describe how cells communicate with each other through signals sent to receptor proteins in the cell membrane.

As you read through this chapter, think about the following questions:

1. What are the roles of the proteins in the plasma membrane of cells?

2. What type of transport is the calcium channel in this story exhibiting?

BEFORE YOU BEGIN

Before beginning this chapter, take a few moments to review the following discussions:

Section 2.6 How does the structure of a phospholipid make it an ideal molecule for the plasma membrane?

Section 2.7 How does a protein's shape relate to its function?

Figures 3.4 and 3.5 What are the key features of animal and plant cells?

4.1 Plasma Membrane Structure and Function

The **plasma membrane** of the cell separates the internal environment of the cell from the external environment. In doing so, it regulates the entrance and exit of molecules to and from the cell. In this way, it helps the cell and the organism maintain a steady internal environment, a process called **homeostasis.** The plasma membrane is made primarily of phospholipids, a type of lipid with both hydrophobic and hydrophilic properties. The phospholipids of the membrane form a bilayer, with the hydrophilic (water-loving) polar heads of the phospholipid molecules facing the outside and inside of the cell (where water is found), and the hydrophobic (water-fearing) nonpolar tails facing each other (Fig. 4.1).

The phospholipid bilayer has a fluid consistency comparable to that of light oil. The fluidity of the membrane is regulated by steroids such as *cholesterol,* which serve to stiffen and strengthen the membrane. Throughout the membrane are numerous proteins, in which protein molecules are either partially or wholly embedded. These proteins are scattered either just outside or within the membrane, and may be either partially or wholly embedded in the phospholipid bilayer. *Peripheral proteins* are associated with only one side of the plasma membrane. Peripheral proteins on the inside of the membrane are often held in place by cytoskeletal filaments. In contrast, *integral proteins* span the membrane, and can protrude from one or both sides. They are embedded in the membrane, but they can move laterally, changing their position in the membrane. The proteins in the membrane form a *mosaic* pattern, and this combination of proteins, steroids, and phospholipids is called the **fluid-mosaic model** of membrane structure (Fig. 4.1).

Figure 4.1 Fluid-mosaic model of plasma membrane structure.   The membrane is composed of a phospholipid bilayer in which proteins are embedded (integral proteins) or associated with the cytoplasmic side (peripheral proteins). Steroids (cholesterol) help regulate the fluidity of the membrane. Cytoskeleton filaments are attached to the inside surface by membrane proteins.

plasma membrane

carbohydrate chain

Outside

extracellular matrix (ECM)

hydrophobic tails

hydrop hea

glycoprotein

phospholipid

glycolipid

phospholipid bilayer

filaments of cytoskeleton

Inside

peripheral protein

integral protein

cholesterol

How Cells Talk to One Another

All organisms are able to sense and respond to specific signals in their environment. A bacterium that has taken up residence in your body is responding to signaling molecules when it finds food and escapes immune cells in order to stay alive. Signaling helps bread mold on stale bread in your refrigerator detect the presence of an opposite mating strain and begin its sexual life cycle. Similarly, the cells of an embryo are responding to signaling molecules when they move to specific locations and assume the shape, and perform the functions, of specific tissues (Fig. 4A). In plants, external signals, such as a change in the amount of light, tell them when it is time to resume growth or flower. Internal signaling molecules enable plants to coordinate the activities of roots, stems, and leaves.

Cell Signaling

The cells of a multicellular organism "talk" to one another by using signaling molecules, called chemical messengers. Some messengers are produced at a distance from a target tissue and, in animals, are carried by the circulatory system to various sites around the body. For example, the pancreas releases a hormone called insulin, which is transported in blood vessels to the liver, and thereafter, the liver stores glucose as glycogen. Failure of the liver to respond appropriately results in a medical condition called diabetes. Growth factors act locally as signaling molecules and cause cells to divide. Overreacting to growth factors can result in a tumor characterized by unlimited cell division. We have learned that cells respond to only certain signaling molecules. Why? Because they must bind to a receptor protein, and cells have receptors for only certain signaling molecules. Each cell has receptors for numerous signaling molecules, and often the final response is due to a summing up of all the various signals received. These molecules tell a cell what it should be doing at the moment, and without any signals, the cell dies.

Signaling not only involves a receptor protein, it also involves a pathway, called a transduction pathway, and a response. To understand the process, consider an analogy. When a TV camera (the receptor) is shooting a scene, the picture is converted to electrical signals (transduction pathway) that are understood by the TV in your house and are converted to a picture on your screen (the response). The process in cells is more complicated because each member of the pathway can turn on the activity of a number of other proteins. As shown in Figure 4A, the cell response to a transduction pathway can be a change in the shape or movement of a cell, the activation of a particular enzyme, or the activation of a specific gene.

Advances in Understanding Cell Communication

The importance of cell signaling causes much research to be directed toward understanding the intricacies of the process, and this research is starting to yield some significant results. For example, researchers have discovered the basis of communication between melanoma cells and neighboring cells of the body. Melanoma is an aggressive form of cancer, in which the cells possess an enhanced ability to move (metastasis) by entering neighboring blood vessels. By identifying the signals generated by the melanoma cells, and their target receptors on the blood vessel cells, researchers have been able to develop chemicals that prevent the movement of the melanoma cells. These chemicals are being used to develop drugs that prevent the spread of melanoma in the body.

Questions to Consider

1. What happens if a cell is missing a receptor for a particular signaling molecule?
2. How does the binding of one type of signaling molecule sometimes result in multiple cellular responses?
3. As a cancer researcher, which segment of the signal transduction pathway would you target to prevent the spread of cancer and why?

Figure 4A Cell signaling and the transduction pathway.   The process of signaling involves three steps: binding of the signaling molecule, transduction of the signal, and response of the cell depending on what type protein is targeted.

Phospholipids may have carbohydrate (sugar) chains attached to them. In this case, they are called **glycolipids.** Proteins with sugar chains attached are called **glycoproteins.** Both of these molecules play an important role in cellular identification. Because the carbohydrate chains occur only on the outside surface, and the peripheral proteins are not distributed equally on the inner and outer faces of the membrane, the two halves of the membrane are not identical.

Allows a particular molecule or ion to cross the plasma membrane freely. Cystic fibrosis, an inherited disorder, is caused by a faulty chloride (Cl⁻) channel; a thick mucus collects in airways and in pancreatic and liver ducts.

a. Channel protein

Selectively interacts with a specific molecule or ion so that it can cross the plasma membrane. The family of GLUT carriers transfers glucose in and out of the various cell types of the body. Different carriers respond differently to blood levels of glucose.

b. Carrier protein

The MHC (major histocompatibility complex) glycoproteins are different for each person, so organ transplants are difficult to achieve. Cells with foreign MHC glycoproteins are attacked by white blood cells responsible for immunity.

c. Cell recognition protein

Shaped in such a way that a specific molecule can bind to it. Some types of dwarfism result not because the body does not produce enough growth hormone, but because the plasma membrane growth hormone receptors are faulty and cannot interact with growth hormone.

d. Receptor protein

Catalyzes a specific reaction. The membrane protein, adenylate cyclase, is involved in ATP metabolism. Cholera bacteria release a toxin that interferes with the proper functioning of adenylate cyclase, which eventually leads to severe diarrhea.

e. Enzymatic protein

Figure 4.2 Examples of membrane protein diversity. These are some of the functions performed by integral proteins found in the plasma membrane.

Functions of the Membrane Proteins

The plasma membranes of various cells and the membranes of various organelles each have their own unique collections of proteins. The peripheral proteins often have a structural role in that they help stabilize and shape the plasma membrane (see Section 3.4). They may also function in signaling pathways. The integral proteins largely determine a membrane's specific functions. Integral proteins can be of the following types:

Channel proteins are involved in the passage of molecules through the membrane. They have a channel that allows a substance to simply move across the membrane (Fig. 4.2a). For example, a channel protein allows hydrogen ions to flow across the inner mitochondrial membrane. Without this movement of hydrogen ions, ATP would never be produced. Channel proteins may contain a gate that must be opened by the binding of a specific molecule to the channel.

Carrier proteins are also involved in the passage of molecules through the membrane. They combine with a substance and help it move across the membrane (Fig. 4.2b). For example, a carrier protein transports sodium and potassium ions across a nerve cell membrane. Without this carrier protein, nerve conduction would not be possible.

Cell recognition proteins are glycoproteins (Fig. 4.2c). Among other functions, these proteins help the body recognize when it is being invaded by pathogens so that an immune reaction can occur.

Receptor proteins have a shape that allows a specific molecule to bind to them (Fig. 4.2d). The binding of this molecule causes the protein to change its shape and thereby bring about a cellular response. The coordination of the body's organs is totally dependent on such signal molecules. For example, the liver stores glucose after it is signaled to do so by insulin. The Scientific Inquiry feature "How Cells Talk to One Another" explores the function of receptor proteins.

Enzymatic proteins carry out metabolic reactions directly (Fig. 4.2e). The integral membrane proteins of the electron transport chain carry out the final steps of aerobic respiration. Without the presence of enzymes, some of which are attached to the various membranes of the cell, a cell would never be able to perform the metabolic reactions necessary for its proper function.

Check Your Progress 4.1

1. Describe the role of proteins, steroids, and phospholipids in the fluid-mosaic model.
2. Distinguish between the roles of the various integral proteins in the plasma membrane.

4.2 The Permeability of the Plasma Membrane

Learning Outcomes

Upon completion of this section, you should be able to

1. Explain why a membrane is selectively permeable.
2. Predict the movement of molecules in diffusion and osmosis.
3. Describe the role of proteins in the movement of molecules across a membrane.

TABLE 4.1   Passage of Molecules into and out of the Cell

	Name	Direction	Requirement	Examples
Energy Not Required	Diffusion	Toward lower concentration	Concentration gradient	Lipid-soluble molecules and gases
	Facilitated transport	Toward lower concentration	Channels or carrier and concentration gradient	Some sugars and some amino acids
Energy Required	Active transport	Toward higher concentration	Carrier plus energy	Sugars, amino acids, and ions
	Exocytosis	Toward outside	Vesicle fuses with plasma membrane	Macromolecules
	Endocytosis	Toward inside	Vesicle formation	Macromolecules

The plasma membrane regulates the passage of molecules into and out of the cell. This function is critical because the life of the cell depends on maintenance of its normal composition. The plasma membrane can carry out this function because it is **selectively permeable,** meaning that certain substances can move across the membrane, while others cannot.

Figure 4.3 illustrates which types of molecules can freely (i.e., passively) cross a membrane and which may require transport by a carrier protein and/or an expenditure of energy. In general, small noncharged molecules, such as carbon dioxide, oxygen, glycerol, and alcohol, can freely cross the membrane. They are able to slip between the hydrophilic heads of the phospholipids and pass through the hydrophobic tails of the membrane. These molecules are said to go "down" their concentration gradient as they move from an area where their concentration is high to an area where their concentration is low. Some molecules are able to go "up" their concentration gradient, or move from an area where their concentration is low to an area where their concentration is high, but this requires energy.

Water, a polar molecule, would not be expected to readily cross the primarily nonpolar membrane. While the small size of the water molecule may allow some water to diffuse across the plasma membrane, the majority of cells have special channel proteins called *aquaporins* that allow water to quickly cross the membrane.

Large molecules and some ions and charged molecules are unable to freely cross the membrane. They can cross the plasma membrane through channel proteins, with the assistance of carrier proteins, or in vesicles. A channel protein forms a pore through the membrane that allows molecules of a certain size and/or charge to pass. Carrier proteins are specific for the substances they transport across the plasma membrane—for example, sodium ions, amino acids, or glucose. Table 4.1 summarizes the movement of particles across the membrane.

Vesicle formation is another way a molecule can exit a cell by exocytosis, or enter a cell by endocytosis. This method of crossing a plasma membrane is reserved for macromolecules or even larger materials, such as a virus.

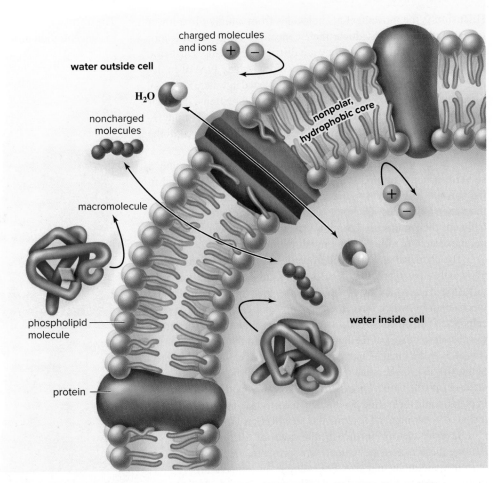

Figure 4.3 How molecules cross the plasma membrane. Molecules that can diffuse across the plasma membrane are shown with long back-and-forth arrows. Substances that cannot diffuse across the membrane are indicated by the curved arrows.

crystal dye

time → time →

a. Crystal of dye is placed in the water   b. Diffusion of water and dye molecules   c. Equal distribution of molecules results

Figure 4.4 Process of diffusion.   Diffusion is spontaneous, and no chemical energy is required to bring it about. **a.** When a dye crystal is placed in water, it is concentrated in one area. **b.** The dye dissolves in the water, and there is a net movement of dye molecules from a higher to a lower concentration. There is also a net movement of water molecules from a higher to a lower concentration. **c.** Eventually, the water and the dye molecules are equally distributed throughout the container.

Diffusion and Osmosis

Diffusion is the movement of molecules from a higher to a lower concentration—that is, down their concentration gradient—until equilibrium is achieved and they are distributed equally. Diffusion is a physical process that can be observed with any type of molecule. For example, when a crystal of dye is placed in water (Fig. 4.4), the dye and water molecules move in various directions, but their net movement, which is the sum of their motion, is toward the region of lower concentration. Eventually, the dye is dissolved in the water, resulting in equilibrium and a colored solution.

A solution contains both a **solute,** usually a solid, and a **solvent,** usually a liquid. The solute is the substance dissolved in the solvent. In Figure 4.4, the solute is the dye, and the solvent is the water molecules. Once the solute and solvent are evenly distributed, their molecules continue to move about, but there is no net movement of either one in any direction.

The chemical and physical properties of the plasma membrane allow only a few types of molecules to enter and exit a cell simply by diffusion. Gases can diffuse through the lipid bilayer. This is the mechanism by which oxygen enters cells and carbon dioxide exits cells when you breathe. Also, consider the movement of oxygen from the alveoli (air sacs) of the lungs to the blood in the lung capillaries (Fig. 4.5). After inhalation (breathing in), the concentration of oxygen in the alveoli is higher than that in the blood. Therefore, oxygen diffuses into the blood. Diffusion also plays an important role in maintaining the resting potential of neurons by using gradients of potassium and sodium ions (see Section 17.1).

Several factors influence the rate of diffusion. Among these factors are temperature, pressure, electrical currents, and molecular size. For example, as temperature increases, the movement of molecules increases, which in turn increases the rate of diffusion.

Osmosis

The diffusion of water across a selectively permeable membrane due to concentration differences is called **osmosis.** To illustrate osmosis,

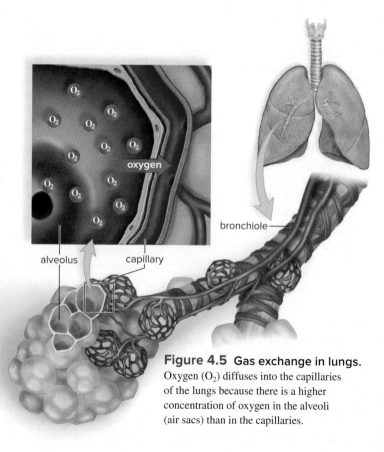

O_2

oxygen

O_2

alveolus   capillary

bronchiole

Figure 4.5 Gas exchange in lungs.
Oxygen (O_2) diffuses into the capillaries of the lungs because there is a higher concentration of oxygen in the alveoli (air sacs) than in the capillaries.

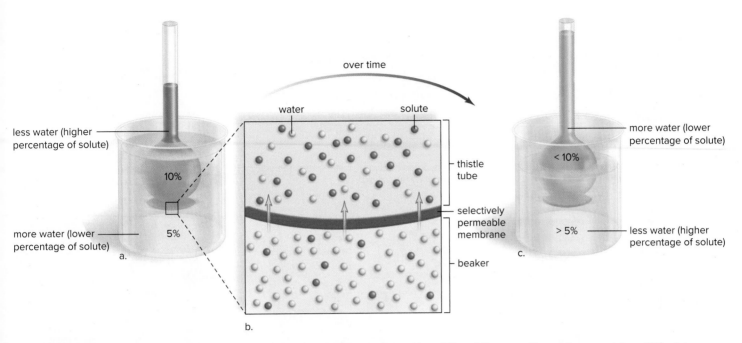

Figure 4.6 Osmosis demonstration.   a. A thistle tube, covered at the broad end by a differentially permeable membrane, contains a 10% solute solution. The beaker contains a 5% solute solution. **b.** The solute (green circles) is unable to pass through the membrane, but the water (blue circles) passes through in both directions. There is a net movement of water toward the inside of the thistle tube, where the percentage of water molecules is lower. **c.** Due to the incoming water molecules, the level of the solution rises in the thistle tube.

a thistle tube containing a 10% solute solution[1] is covered at one end by a selectively permeable membrane and then placed in a beaker containing a 5% solute solution (Fig. 4.6). The beaker has a higher concentration of water molecules (lower percentage of solute), and the thistle tube has a lower concentration of water molecules (higher percentage of solute). Diffusion always occurs from higher to lower concentration. Therefore, a net movement of water takes place across the membrane from the beaker to the inside of the thistle tube.

The solute does not diffuse out of the thistle tube. Why not? Because the membrane is not permeable to the solute. As water enters and the solute does not exit, the level of the solution within the thistle tube rises (Fig. 4.6c). In the end, the concentration of solute in the thistle tube is less than 10%. Why? Because there is now less solute per unit volume of solution. Furthermore, the concentration of solute in the beaker is greater than 5% because there is now more solute per unit volume.

Water enters the thistle tube due to the osmotic pressure of the solution within the thistle tube. **Osmotic pressure**[2] is the pressure that develops in a system due to osmosis. In other words, the greater the possible osmotic pressure, the more likely it is that water will diffuse in that direction. Due to osmotic pressure, water is absorbed by the kidneys and taken up by capillaries in the tissues. Osmosis also occurs across the plasma membrane.

Isotonic Solution   In the laboratory, cells are normally placed in **isotonic solutions.** The prefix *iso-* means "the same as," and the term *tonicity* refers to the osmotic pressure or tension of the solution. In an isotonic solution, the solute concentration and the water concentration both inside and outside the cell are equal, and therefore there is no net gain or loss of water (Fig. 4.7). For example, a 0.9% solution of the salt sodium chloride (NaCl) is known to be isotonic to red blood cells. Therefore, intravenous solutions medically administered usually have this tonicity. Terrestrial animals can usually take in either water or salt as needed to maintain the tonicity of their internal environment. Many animals living in an estuary, such as oysters, blue crabs, and some fishes, are able to cope with changes in the salinity (salt concentrations) of their environment using specialized kidneys, gills, and other structures.

Hypotonic Solution   Solutions that cause cells to swell, or even to burst, due to an intake of water are said to be **hypotonic solutions.** The prefix *hypo-* means "less than" and refers to a solution with a lower concentration of solute (higher concentration of water) than inside the cell. If a cell is placed in a hypotonic solution, water enters the cell. The net movement of water is from the outside to the inside of the cell (Fig. 4.7).

Any concentration of a salt solution lower than 0.9% is hypotonic to red blood cells. Animal cells placed in such a solution expand and sometimes burst or lyse due to the buildup of pressure. The term *cytolysis* is used to refer to disrupted cells. However, if the cell is a red blood cell, the term *hemolysis* is used.

The swelling of a plant cell in a hypotonic solution creates *turgor pressure.* When a plant cell is placed in a hypotonic solution, the cytoplasm expands because the large central vacuole gains water and the plasma membrane pushes against the rigid cell wall. The plant cell does not burst because the cell wall does not give way. Turgor pressure in plant cells is extremely important to

---

[1] Percent solutions are grams of solute per 100 ml of solvent. Therefore, a 10% solution is 10g of sugar with water added to make 100 ml of solution.

[2] Osmotic pressure is measured by placing a solution in an osmometer and then immersing the osmometer in pure water. The pressure that develops is the osmotic pressure of a solution.

Animal cells

15,000×

nucleus — plasma membrane

In an isotonic solution, there is no net movement of water.

15,000×

In a hypotonic solution, water enters the cell, which may burst (lysis).

15,000×

In a hypertonic solution, water leaves the cell, which shrivels (crenation).

Plant cells

nucleus — central vacuole — chloroplast

400×

In an isotonic solution, there is no net movement of water.

cell wall

400×

In a hypotonic solution, the central vacuole fills with water, turgor pressure develops, and chloroplasts are seen next to the cell wall.

plasma membrane

In a hypertonic solution, the central vacuole loses water, the cytoplasm shrinks (plasmolysis), and chloroplasts are seen in the center of the cell.

Figure 4.7 Osmosis in animal and plant cells.   The arrows indicate the movement of water molecules. To determine the net movement of water, compare the number of arrows indicating water molecules being taken into the cell with the number of arrows indicating water being taken out of the cell. In an isotonic solution, a cell neither gains nor loses water; in a hypotonic solution, a cell gains water; and in a hypertonic solution, a cell loses water. (photos): (isotonic animal cell), (hypotonic animal cell), (hypertonic animal cell): ©David M. Phillips/Science Source; (isotonic plant cell): ©Ed Reschke/Photolibrary/Getty Images; (hypotonic plant cell): ©Ray F. Evert, University of Wisconsin; (hypertonic plant cell): ©Ed Reschke/Getty Images

the maintenance of the plant's erect position. If you forget to water your plants, they wilt due to decreased turgor pressure.

Organisms that live in fresh water have to prevent their internal environment from becoming hypotonic. Many protozoans, such as paramecia, have contractile vacuoles that rid the body of excess water. Freshwater fishes have well-developed kidneys that excrete a large volume of dilute urine. Even so, they have to take in salts at their gills. Even though freshwater fishes are good osmoregulators, they would not be able to survive in either distilled water or a marine environment.

Hypertonic Solution   Solutions that cause cells to shrink or shrivel due to loss of water are said to be **hypertonic solutions.** The prefix *hyper-* means "more than" and refers to a solution with a higher percentage of solute (lower concentration of water) than the cell. If a cell is placed in a hypertonic solution, water leaves the cell. The net movement of water is from the inside to the outside of the cell (Fig. 4.7).

Any concentration of a salt solution higher than 0.9% is hypertonic to red blood cells. If animal cells are placed in this solution, they shrink. The term *crenation* refers to the shriveling of a cell in a hypertonic solution. Meats are sometimes preserved by salting them. The salt kills any bacteria present because it makes the meat a hypertonic environment.

When a plant cell is placed in a hypertonic solution, the plasma membrane pulls away from the cell wall as the large central vacuole loses water. This is an example of *plasmolysis,* shrinking of the cytoplasm due to osmosis. The dead plants you may see along a salted roadside died because they were exposed to a hypertonic solution during the winter. Also, when salt water invades coastal marshes due to storms or human activities, coastal plants die. Without roots to hold the soil, it washes into the sea, thereby losing many acres of valuable wetlands.

Marine animals cope with their hypertonic environment in various ways that prevent them from losing water. Sharks increase or decrease the urea in their blood until their blood is isotonic with the environment. Marine fishes and other types of animals excrete salts across their gills. Have you ever seen a marine turtle cry? It is ridding its body of salt by means of glands near the eye.

Transport by Carrier Proteins

The plasma membrane impedes the passage of all but a few substances. Yet, biologically useful molecules are able to enter and exit the cell at a rapid rate because of carrier proteins in the membrane. Carrier proteins are specific. Each can combine with only a certain type of molecule or ion, which is then transported through the membrane. The precise mechanism by which a carrier protein functions varies, but generally after a carrier combines with a molecule, the carrier undergoes a change in shape that moves the molecule across the membrane. Carrier proteins are required for both facilitated transport and active transport (see Table 4.1).

Facilitated Transport

Facilitated transport explains the passage of such molecules as glucose and amino acids across the plasma membrane even though they are not lipid-soluble. The passage of glucose and amino acids is facilitated by their reversible combination with carrier proteins, which in some manner transport them through the plasma membrane. These carrier proteins are specific. For example, various sugar molecules of identical size might be present inside or outside the cell, but glucose can cross the membrane hundreds of times faster than the other sugars. Another example is provided by the aquaporins, which allow water to rapidly move across the plasma membrane of the cell.

A model for facilitated transport (Fig. 4.8) shows that after a carrier has assisted the movement of a molecule to the other side of the membrane, it is free to assist the passage of other similar molecules. Neither diffusion nor facilitated transport requires an expenditure of energy (use of ATP) because the molecules are moving down their concentration gradient.

Active Transport

During **active transport,** molecules or ions move through the plasma membrane, accumulating either inside or outside the cell. For example, iodine collects in the cells of the thyroid gland; glucose is completely absorbed from the gut by the cells lining the digestive tract; and sodium can be almost completely withdrawn from urine by cells lining the kidney tubules. In these instances,

molecules have moved to the region of higher concentration, exactly opposite to the process of diffusion.

Both carrier proteins and an expenditure of energy are needed to transport molecules against their concentration gradient. In this case, chemical energy, usually in the form of ATP, is required for the carrier to combine with the substance to be transported. Therefore, it is not surprising that cells involved primarily in active transport, such as kidney cells, have a large number of mitochondria near membranes where active transport is occurring.

Proteins involved in active transport are often called pumps because, just as a water pump uses energy to move water against the force of gravity, proteins use energy to move a substance against its concentration gradient. One type of pump that is active in all animal cells, but is especially associated with nerve and muscle cells, moves sodium ions (Na^+) to the outside of the cell and potassium ions (K^+) to the inside of the cell. These two events are linked, and the carrier protein is called a **sodium–potassium pump.** A change in carrier shape after the attachment of a phosphate group, and again after its detachment, allows the carrier to combine alternately with sodium ions and potassium ions (Fig. 4.9). The phosphate group is donated by ATP when it is broken down enzymatically by the carrier. The sodium–potassium pump results in both a solute concentration gradient and an electrical gradient for these ions across the plasma membrane.

The passage of salt (NaCl) across a plasma membrane is of primary importance to most cells. The chloride ion (Cl^-) usually crosses the plasma membrane because it is attracted by positively charged sodium ions (Na^+). First, sodium ions are pumped across a membrane, and then chloride ions simply diffuse through channels that allow their passage.

As noted in Figure 4.2a, the genetic disorder cystic fibrosis results from a faulty chloride channel. In cystic fibrosis, Cl^- transport is reduced, and so is the flow of Na^+ and water. Research has shown that the lack of water causes the mucus in the bronchial

Figure 4.8 Facilitated transport.   During facilitated transport, a carrier protein speeds the rate at which the solute crosses the plasma membrane toward a lower concentration. Note that the carrier protein undergoes a change in shape as it moves a solute across the membrane.

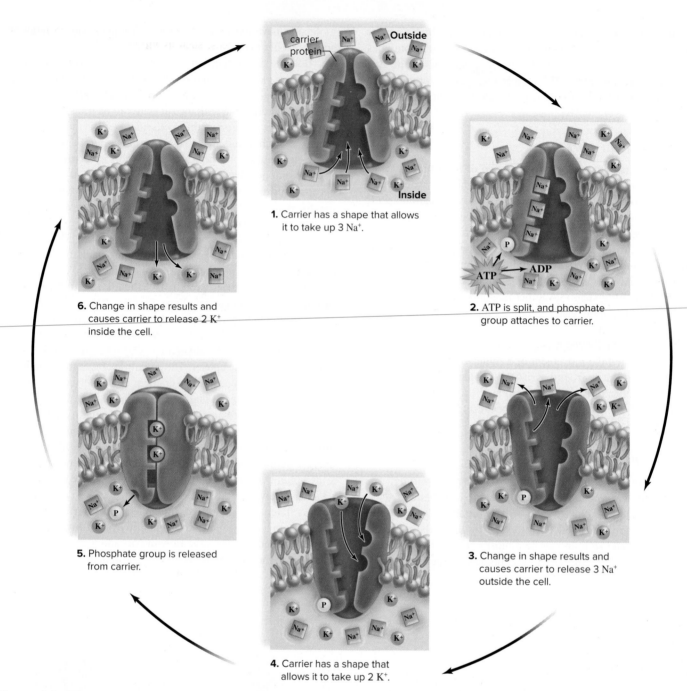

1. Carrier has a shape that allows it to take up 3 Na⁺.

2. ATP is split, and phosphate group attaches to carrier.

3. Change in shape results and causes carrier to release 3 Na⁺ outside the cell.

4. Carrier has a shape that allows it to take up 2 K⁺.

5. Phosphate group is released from carrier.

6. Change in shape results and causes carrier to release 2 K⁺ inside the cell.

Figure 4.9 The sodium–potassium pump.   The same carrier protein transports sodium ions (Na^+) to the outside of the cell and potassium ions (K^+) to the inside of the cell because it undergoes an ATP-dependent change in shape. Three sodium ions are carried outward for every two potassium ions carried inward. Therefore, the inside of the cell is negatively charged compared to the outside.

tubes and pancreatic ducts to be abnormally thick, thus interfering with the function of the lungs and pancreas.

Bulk Transport

How do macromolecules such as polypeptides, polysaccharides, or polynucleotides enter and exit a cell? Because they are too large to be transported by carrier proteins, macromolecules are transported into and out of the cell by vesicle formation. Vesicle formation is called membrane-assisted transport because membrane is needed to form the vesicle. Vesicle formation requires an expenditure of cellular energy, but an added benefit is that the vesicle membrane keeps the contained macromolecules from mixing with molecules within the cytoplasm. Exocytosis is a way substances can exit a cell, and endocytosis is a way substances can enter a cell.

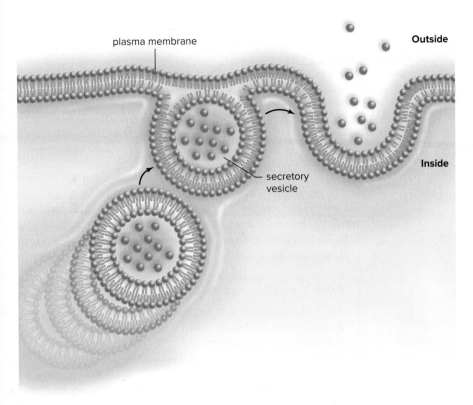

plasma membrane

Outside

secretory
vesicle

Inside

Figure 4.10 Exocytosis.   Exocytosis deposits substances on the outside of the cell and allows secretion to occur.

Exocytosis

During **exocytosis,** a vesicle fuses with the plasma membrane as secretion occurs (Fig. 4.10). Hormones, neurotransmitters, and digestive enzymes are secreted from cells in this manner. The Golgi apparatus often produces the vesicles that carry these cell products to the membrane. Notice that during exocytosis, the membrane of the vesicle becomes a part of the plasma membrane, which is thereby enlarged. For this reason, exocytosis can be a normal part of cell growth. The proteins released from the vesicle adhere to the cell surface or become incorporated in an extracellular matrix.

The process of exocytosis is used by a variety of cell types. For example, pancreatic cells produce digestive enzymes or insulin. In these cells, secretory vesicles accumulate near the plasma membrane, and the vesicles release their contents only when the cell is stimulated by a signal. A rise in blood sugar, for example, signals pancreatic cells to release the hormone insulin. Another example are the sweat glands, which release water and electrolytes when signaled by the nervous system to reduce body temperature.

Endocytosis

During **endocytosis,** cells take in substances by vesicle formation. A portion of the plasma membrane invaginates to envelop the substance, and then the membrane pinches off to form an intracellular vesicle. Endocytosis occurs in one of three ways, as illustrated in

Figure 4.11. Phagocytosis transports large substances, such as viruses, and pinocytosis transports small substances, such as macromolecules, into cells. Receptor-mediated endocytosis is a special form of pinocytosis.

Phagocytosis   When the material taken in by endocytosis is large, such as a food particle or another cell, the process is called **phagocytosis.** Phagocytosis is common in single-celled organisms such as amoebas (Fig. 4.11a). It also occurs in humans. Certain types of human white blood cells are amoeboid—that is, they are mobile like an amoeba, and they are able to engulf debris such as worn-out red blood cells or viruses. When an endocytic vesicle fuses with a lysosome, digestion occurs. We will see that this process is a necessary and preliminary step toward the development of immunity to bacterial diseases.

Pinocytosis   Pinocytosis occurs when vesicles form around a liquid or around very small particles (Fig. 4.11b). White blood cells, cells that line the kidney tubules or the intestinal wall, and plant root cells all use pinocytosis to ingest substances.

Whereas phagocytosis can be seen with the light microscope, an electron microscope is required to observe pinocytic vesicles, which are no larger than 0.1–0.2 μm. Still, pinocytosis involves a significant amount of the plasma membrane because it occurs continuously. However, the loss of plasma membrane due to pinocytosis is balanced by the occurrence of exocytosis.

Receptor-Mediated Endocytosis   Receptor-mediated endocytosis is a form of endocytosis that is quite specific because it uses a receptor protein shaped so that a specific molecule, such as a vitamin, peptide hormone, or lipoprotein, can bind to it (Fig. 4.11c). The receptors for these substances are found at one location in the plasma membrane. This location is called a coated pit because there is a layer of protein on the cytoplasmic side of the pit. Once formed, the vesicle becomes uncoated and may fuse with a lysosome. When empty, used vesicles fuse with the plasma membrane, and the receptors return to their former location.

Receptor-mediated endocytosis is selective and much more efficient than ordinary pinocytosis. It is involved in uptake and also in the transfer and exchange of substances between cells. Such exchanges take place when substances move from maternal blood into fetal blood at the placenta, for example.

The importance of receptor-mediated endocytosis is demonstrated by a genetic disorder called familial hypercholesterolemia. Cholesterol is transported in the blood by a complex of lipids and proteins called low-density lipoprotein (LDL). Ordinarily, body cells take up LDL when LDL receptors gather in a coated pit. But in some individuals, the LDL receptor is unable to properly bind to

a. Phagocytosis

b. Pinocytosis

c. Receptor-mediated endocytosis

d. Micrographs of receptor-mediated endocytosis

Figure 4.11 Three methods of endocytosis. **a.** Phagocytosis occurs when the substance to be transported into the cell is large. Digestion occurs when the resulting vacuole fuses with a lysosome. **b.** Pinocytosis occurs when a macromolecule such as a polypeptide is transported into the cell. **c.** Receptor-mediated endocytosis is a form of bulk transport. Molecules first bind to specific receptor proteins, which migrate to, or are already in, a coated pit. The vesicle that forms contains the molecules and their receptors. **d.** A micrograph of receptor-mediated endocytosis. (d) (both photos): ©Mark Bretscher

the coated pit, and the cells are unable to take up cholesterol. Instead, cholesterol accumulates in the walls of arterial blood vessels, leading to high blood pressure, occluded (blocked) arteries, and heart attacks.

Check Your Progress 4.2

1. Contrast diffusion with facilitated transport.
2. Explain the movement of water between hypotonic and hypertonic environments.
3. Describe the differences between facilitated and active transport.
4. Discuss the potential benefit of receptor-mediated endocytosis.

4.3 Modifications of Cell Surfaces

Learning Outcomes

Upon completion of this section, you should be able to

1. Explain the role of the extracellular matrix in animal cells.
2. Compare the structure and function of adhesion, tight, and gap junctions.

Most cells do not live isolated from other cells. Rather, they live and interact within an external environment that can dramatically affect cell structure and function. This extracellular environment is made of large molecules produced by nearby cells and secreted from their membranes. In plants, prokaryotes, fungi, and most algae, the extracellular environment is a fairly rigid cell wall, which is consistent with a somewhat sedentary lifestyle. Animals, which tend to be more active, have a more varied extracellular environment that can change, depending on the nature of the organism or the tissue type.

Cell Surfaces in Animals

In this section, we will focus on two different types of animal cell surface features: (1) the extracellular matrix (ECM) that is observed outside cells, and (2) junctions that occur between some types of cells. Both of these can connect to the cytoskeleton and contribute to communication between cells, and therefore tissue formation.

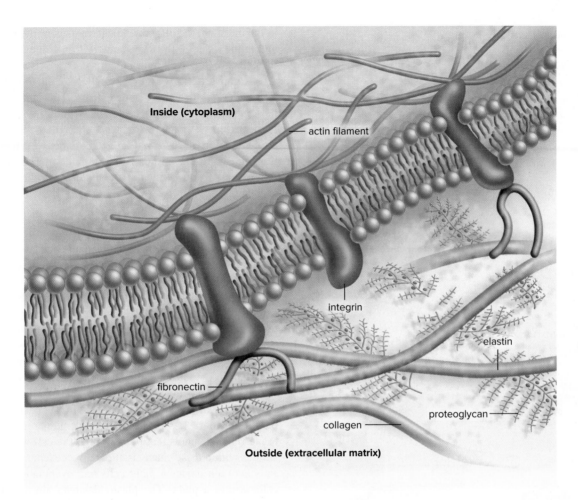

Inside (cytoplasm)

— actin filament

integrin

elastin

fibronectin

collagen —

proteoglycan

Outside (extracellular matrix)

Figure 4.12 Extracellular matrix of an animal cell.   In the extracellular matrix (ECM), collagen and elastin have a support function, while fibronectins bind to integrin, thus assisting communication between the ECM and the cytoskeleton.

Extracellular Matrix

A protective **extracellular matrix (ECM)** is a meshwork of proteins and polysaccharides in close association with the cell that produced them (Fig. 4.12). Collagen and elastin fibers are two well-known structural proteins in the ECM; collagen resists stretching, while elastin gives the ECM resilience.

Fibronectin is an adhesive protein (colored green in Fig. 4.12) that binds to a protein in the plasma membrane called integrin. Integrins are integral membrane proteins that connect to fibronectin externally and to the actin cytoskeleton internally. Through its connections with both the ECM and the cytoskeleton, integrin plays a role in cell signaling, permitting the ECM to influence the activities of the cytoskeleton and, therefore, the shape and activities of the cell.

Amino sugars in the ECM form multiple polysaccharides that attach to a protein and are, therefore, called proteoglycans. Proteoglycans, in turn, attach to a very long, centrally placed polysaccharide. The entire structure, which looks like an enormous bottle brush, resists compression of the extracellular matrix. Proteoglycans assist cell signaling when they regulate the passage of molecules through the ECM to the plasma membrane, where receptors

are located. Thus, the ECM has a dynamic role in all aspects of a cell's behavior.

In Section 11.1, during the discussion of tissues, you'll see that the extracellular matrix varies in quantity and in consistency from being quite flexible, as in loose connective tissue; semiflexible, as in cartilage; and rock solid, as in bone. The extracellular matrix of bone is hard because, in addition to the components mentioned, mineral salts, notably calcium salts, are deposited outside the cell.

Junctions Between Cells

Certain tissues of vertebrate animals are known to have junctions between their cells that allow them to behave in a coordinated manner. Three types of junctions are shown in Figure 4.13. **Adhesion junctions** (Fig. 4.13a) serve to mechanically attach adjacent cells. Desmosomes are one form of adhesion junction. In a desmosome, internal cytoplasmic plaques, firmly attached to the intermediate filament cytoskeleton within each cell, are joined by integral membrane proteins, called cadherins, between cells. The result is a sturdy but flexible sheet of cells. In some organs—such as the heart, stomach, and bladder, where tissues get

a. Adhesion junction

b. Tight junction

c. Gap junction

Figure 4.13 Examples of cell junctions.   **a.** In adhesion junctions, such as the desmosome, adhesive proteins connect two cells. **b.** Tight junctions between cells have joined their adjacent plasma membranes, forming an impermeable layer. **c.** Gap junctions allow communication between two cells by joining plasma membrane channels between the cells.
(photos) (a): ©SPL/Science Source; (b–c): ©David M. Phillips/Science Source

stretched—desmosomes hold the cells together. Adhesion junctions are the most common type of intercellular junction between skin cells.

Another type of junction between adjacent cells are **tight junctions** (Fig. 4.13b), which bring cells even closer than desmosomes. Tight junction proteins actually connect plasma membranes between adjacent cells together, producing a zipperlike fastening. Tissues that serve as barriers are held together by tight junctions;

in the intestine, the digestive juices stay out of the rest of the body, and in the kidneys, the urine stays within kidney tubules, because the cells are joined by tight junctions.

A **gap junction** (Fig. 4.13c) allows cells to communicate. A gap junction is formed when two identical plasma membrane channels join. The channel of each cell is lined by six plasma membrane proteins. A gap junction lends strength to the cells, but it also allows small molecules and ions to pass between them. Gap junctions are important in heart muscle and smooth muscle because they permit a flow of ions that is required for the cells to contract as a unit.

Plant Cell Walls

In addition to a plasma membrane, plant cells are surrounded by a porous cell wall that varies in thickness, depending on the function of the cell.

All plant cells have a cell wall. The primary cell wall contains cellulose fibrils (very fine fibers) in which microfibrils are held together by noncellulose substances. Pectins allow the wall to stretch when the cell is growing, and noncellulose polysaccharides harden the wall when the cell is mature. Pectins are especially abundant in the middle lamella, which is a layer of adhesive substances that holds the cells together.

In a plant, the cytoplasm of living cells is connected by **plasmodesmata** (*sing.,* plasmodesma), numerous narrow, membrane-lined channels that pass through the cell wall. Cytoplasmic strands within these channels allow direct exchange of some materials between adjacent plant cells and eventually connect all the cells within a plant. The plasmodesmata allow only water and small solutes to pass freely from cell to cell.

Check Your Progress 4.3

1. Describe the molecule composition of the extracellular matrix of an animal cell.
2. Explain the differences in functions among an adhesion, gap, and tight junction.
3. Contrast the extracellular matrix of an animal cell with the cell wall of a plant cell.

Conclusion

In the case of the particular channel described in the opening of this chapter, exposure to the capsaicin molecules in a chili pepper opens a channel protein in the membrane of the cells. This allows calcium ions to enter the cell and initiate a cascade of events that results in the sensation of pain. As we have seen, this is a type of facilitated diffusion, because the calcium ions are moving down their concentration gradient with the assistance of a channel protein. It is interesting to note that later in this text (see Section 19.4) you will learn about a different type of voltage-gated ion channel that also allows calcium ions to enter the cell. In cells with this type of channel, muscular contraction or neuron excitation occurs.

SUMMARIZE

4.1 Plasma Membrane Structure and Function

■ The **plasma membrane** plays an important role in isolating the cell from the external environment and in maintaining **homeostasis** within the cell. According to the **fluid-mosaic model** of the plasma membrane, a lipid bilayer is fluid and has the consistency of light oil. The hydrophilic heads of phospholipids form the inner and outer surfaces, and the hydrophobic tails form the interior.

■ Proteins within the membrane are the mosaic portion. The peripheral proteins often have a structural role in that they help stabilize and shape the plasma membrane. They may also function in signaling pathways. The integral proteins have a variety of functions, including acting as **channel proteins, carrier proteins, cell recognition proteins, receptor proteins,** and **enzymatic proteins.**

■ Carbohydrate chains are attached to some of the lipids and proteins in the membrane. These are **glycolipids** and **glycoproteins.**

4.2 The Permeability of the Plasma Membrane

■ The plasma membrane is **selectively permeable,** meaning that some substances, such as gases, freely cross a plasma membrane, while others—particularly ions, charged molecules, and macromolecules—have to be assisted across.

■ Passive ways of crossing a plasma membrane (**diffusion** and **facilitated transport**) do not require an expenditure of chemical energy (ATP). Active ways of crossing a plasma membrane (**active transport** and vesicle formation) do require an expenditure of chemical energy.

■ Lipid-soluble compounds, water, and gases simply diffuse across the plasma membrane by moving down their concentration gradient (high to low concentration).

■ The diffusion of water across a membrane is called **osmosis.** Water (a **solvent**) moves across the membrane into the area of lower water (or higher **solute**) content. When cells are in an **isotonic solution,** they neither gain nor lose water; when they are in a **hypotonic solution,** they gain water; and when they are in a **hypertonic solution,** they lose water. **Osmotic pressure** occurs as a result of differences in tonicity.

■ Some molecules are transported across the membrane by carrier proteins that span the membrane. During **facilitated transport,** a carrier protein assists the movement of a molecule down its concentration gradient. No energy is required.

■ During **active transport,** a carrier protein acts as a pump that causes a substance to move against its concentration gradient. Energy in the form of ATP molecules is required for active transport to occur. The **sodium–potassium pump** is one example of active transport.

■ Larger substances can exit and enter a membrane by **exocytosis** and **endocytosis.** Exocytosis involves secretion. Endocytosis includes **phagocytosis** and **pinocytosis. Receptor-mediated endocytosis,** a type of pinocytosis, makes use of receptor molecules in the plasma membrane and a coated pit, which pinches off to form a vesicle.

4.3 Modifications of Cell Surfaces

■ Animal cells have an **extracellular matrix** (**ECM**) that influences their shape and behavior. The amount and character of the ECM varies by tissue type. Some animal cells have junction proteins that join them to other cells of the same tissue. **Adhesion junctions** and **tight junctions** help hold cells together; **gap junctions** allow passage of small molecules between cells.

■ Plant cells have a freely permeable cell wall, with cellulose as its main component. Also, plant cells are joined by narrow, membrane-lined channels called **plasmodesmata** that span the cell wall and contain strands of cytoplasm that allow materials to pass from one cell to another.

ASSESS

Testing Yourself

Choose the best answer for each question.

4.1 Plasma Membrane Structure and Function

1. In the fluid-mosaic model, the fluid properties are associated with the nature of the _____, and the mosaic pattern is established by the _____.
   a. nucleic acids; phospholipids
   b. phospholipids; embedded proteins
   c. embedded proteins; cholesterol
   d. phospholipids; nucleic acids

2. Which of the following is not a function of proteins present in the plasma membrane?
   a. Proteins assist the passage of materials into the cell.
   b. Proteins interact with and recognize other cells.
   c. Proteins bind with specific hormones.
   d. Proteins carry out specific metabolic reactions.
   e. Proteins produce lipid molecules.

3. The carbohydrate chains projecting from the plasma membrane are involved in
   a. adhesion between cells.
   b. reception of molecules.
   c. cell-to-cell recognition.
   d. All of these are correct.

4.2 The Permeability of the Plasma Membrane

4. When a cell is placed in a hypotonic solution,
   a. solute exits the cell to equalize the concentration on both sides of the membrane.
   b. water exits the cell toward the area of lower solute concentration.
   c. water enters the cell toward the area of higher solute concentration.
   d. there is no net movement of water or solute.

5. When a cell is placed in a hypertonic solution,
   a. solute exits the cell to equalize the concentration on both sides of the membrane.
   b. water exits the cell toward the area of lower solute concentration.
   c. water exits the cell toward the area of higher solute concentration.
   d. there is no net movement of water or solute.

6. Which of the following is incorrect regarding facilitated diffusion?
   a. It is a passive process.
   b. It allows the movement of molecules from areas of low concentration to areas of high concentration.
   c. It may use either channel or carrier proteins.
   d. It allows the rapid transport of glucose across the membrane.

7. The sodium–potassium pump
   a. helps establish an electrochemical gradient across the membrane.
   b. concentrates sodium on the outside of the membrane.
   c. uses a carrier protein and chemical energy.
   d. is present in the plasma membrane.
   e. All of these are correct.

8. Which of the following processes is involved in the bulk transport of molecules out of the cell?
   a. phagocytosis
   b. pinocytosis
   c. receptor-mediated endocytosis
   d. exocytosis
   e. None of these are correct.

9. Which process uses special proteins on the surface of the membrane to identify specific molecules for transport into the cell?
   a. phagocytosis
   b. pinocytosis
   c. receptor-mediated endocytosis
   d. exocytosis

4.3 Modifications of Cell Surfaces

10. The extracellular matrix
    a. assists in the movement of substances across the plasma membrane.
    b. prevents the loss of water when cells are placed in a hypertonic solution.
    c. has numerous functions that affect the shape and activities of the cell that produced it.
    d. contains the junctions that sometimes occur between cells.
    e. All of these are correct.

11. Which of the following junctions allows for cytoplasm-to-cytoplasm communication between cells?
    a. adhesion junctions   c. gap junctions
    b. tight junctions   d. None of these are correct.

ENGAGE

BioNOW

Want to know how this science is relevant to your life? Check out the BioNOW video below:

■ Saltwater Filter

Explain why the branches were able to filter out the bacteria but not the salt from the water sample.

Thinking Critically

1. When a signal molecule such as a growth hormone binds to a receptor protein in the plasma membrane, it stays on the outside of the cell. How might the inside of the cell know that the signal has bound?

2. As mentioned, cystic fibrosis is a genetic disorder caused by a defective membrane transport protein. The defective protein closes chloride channels in membranes, preventing chloride from being exported out of cells. This results in the development of a thick mucus on the outer surfaces of cells. This mucus clogs the ducts that carry digestive enzymes from the pancreas to the small intestine, clogs the airways in the lungs, and promotes lung infections. Why do you think the defective protein results in a thick, sticky mucus outside the cells, instead of a loose, fluid covering?

©Yuri_Arcurs/Getty Images

C H A P T E R

5

Cell Division

CHAPTER OUTLINE

5.1 The Cell Cycle
5.2 Control of the Cell Cycle
5.3 Mitosis: Maintaining the Chromosome Number
5.4 Meiosis: Reducing the Chromosome Number
5.5 Comparison of Meiosis with Mitosis
5.6 The Human Life Cycle

BEFORE YOU BEGIN

Before beginning this chapter, take a few moments to review the following discussions:

Section 2.8 What is the role of DNA in a cell?
Section 3.1 What does the cell theory tell us about the need for cell division?
Section 3.4 What are microtubules and actin filaments?

The Genetics of Cancer

Why don't elephants get cancer? Elephants, like humans, are long-lived organisms. Yet, despite the fact they are long-lived, recent studies have indicated that only about 5% of elephants will develop cancer in their lifetime, compared to up to 25% of humans.

Cancer results from a failure to control the cell cycle, a series of steps that all cells go through prior to initiating cell division. A gene called *p53* is an important component of that control mechanism. This gene belongs to a class of genes called tumor suppressor genes. At specific points in the cell cycle, the proteins encoded by tumor suppressor genes check the DNA for damage. If the cells are damaged, they are not allowed to divide.

Humans have only one pair of *p53* genes (one from each parent). Luckily, you need only a single copy to regulate the cell cycle. But if both genes are not functioning properly, because of inheritance of a defective gene or mutations that inactivate a gene, they can't act as gatekeepers, and instead will allow cells to divide continuously. Unrestricted cell growth is a characteristic of cancer. In humans, mutations in the *p53* gene are associated with increased risk of a number of cancers. Interestingly, elephants have 20 pairs of *p53* genes in their genome, which means it is very unlikely their cells will lose all of this gene's functionality.

In this chapter, we will explore how cell division is regulated, as well as how genes such as *p53* function in the regulation of the cycle.

As you read through this chapter, think about the following questions:

1. What are the roles of the checkpoints in the cell cycle?
2. At which checkpoint do you think *p53* would normally be active?
3. Why would a failure of the checkpoints in the cell cycle result in cancer?

79

5.1 The Cell Cycle

We start life as a single-celled fertilized egg. By the process of cell division, we grow into organisms that contain trillions of cells. The process of cell division serves to increase the number of **somatic cells,** or body cells. However, cell division does not stop when you become an adult. Every day your body produces millions of new red blood cells, skin cells, and cells that line your respiratory and digestive tracts. In addition to growth, cell division is involved in the repair of tissues after an injury.

While cell division increases the number of cells in the body, the process of **apoptosis,** or programmed cell death, acts to decrease the number of cells. Apoptosis occurs during development to remove unwanted tissue—for example, the tail of a tadpole disappears as it matures into a frog. In humans, the fingers and toes of the embryo are initially webbed, but this webbing is removed later in development as a result of apoptosis. Apoptosis also plays an important role in preventing cancer. An abnormal cell that could become cancerous will often die via apoptosis, thus preventing a tumor from developing. Both cell division and apoptosis are normal parts of growth and development.

The Cell Cycle

Cell division is a part of the cell cycle. The **cell cycle** is an orderly set of stages that take place between the time a cell divides and the time the resulting cells also divide.

The Stages of Interphase

As Figure 5.1 shows, most of the cell cycle is spent in **interphase.** This is the time when a cell carries on its usual functions, which are dependent on its location in the body. It also gets ready to divide: It grows larger, the number of organelles doubles, and the amount of DNA doubles. For mammalian cells, interphase lasts for about 20 hours, which is 90% of the cell cycle.

Interphase is divided into three stages: The G_1 stage occurs before DNA synthesis, the S stage includes DNA synthesis, and the G_2 stage occurs after DNA synthesis. Originally, G stood for the "gaps" that occur before and after DNA synthesis during interphase. But now that we know growth occurs during these stages, the G can be thought of as standing for "growth."

During the G_1 stage, a cell doubles its organelles (such as mitochondria and ribosomes), and it accumulates the materials needed for the synthesis of new DNA. Some cells, such as nerve and muscle cells, typically pause during the cell cycle. These cells are said to have entered a G_0 stage. Cells in G_0 phase continue to perform their normal functions, but no longer prepare for cell division. Cells may also enter the G_0 phase if their DNA is damaged. If the repair of the DNA cannot be completed, the cell may undergo apoptosis (see below).

During the S stage, DNA replication occurs. At the beginning of the S stage, each chromosome is composed of one DNA

G₁ checkpoint
Cell cycle main checkpoint. If DNA is damaged, apoptosis may occur. Otherwise, the cell is committed to divide when growth signals are present and nutrients are available.

G₂ checkpoint
Mitosis checkpoint. Mitosis will occur if DNA has replicated properly. Apoptosis may occur if the DNA is damaged and cannot be repaired.

M checkpoint
Spindle assembly checkpoint. Mitosis will not continue if chromosomes are not properly aligned.

Figure 5.1 The cell cycle. Cells normally go through a cycle that consists of four stages: G_1, S, G_2, and M. The G_0 stage represents a holding stage for cells that are not undergoing division. Checkpoints regulate the speed at which a cell moves through the cell cycle.

Figure 5.2 Sister chromatids. The process of DNA replication forms sister chromatids, which are separated by the process of cell division.

molecule, which is called a **chromatid** (Fig. 5.2). At the end of this stage, each chromosome consists of two **sister chromatids** that have identical DNA sequences. Another way of expressing this is to say that DNA replication has resulted in duplicated chromosomes. The sister chromatids remain attached until they are separated during mitosis.

During the G_2 stage, the cell synthesizes the proteins needed for cell division, such as the proteins that make up the microtubules found in the spindle apparatus (see Section 5.3).

Some cells, such as nerve and muscle cells, typically do not complete the cell cycle and are permanently arrested. These cells exit interphase and enter a stage called G_0. While in the G_0 stage, the cells continue to perform normal, everyday processes, but no preparations are being made for cell division. Cells may not leave the G_0 stage without proper signals from other cells and other parts of the body.

The Mitotic Stage

Following interphase, the cell enters the M (for mitotic) stage. This stage includes not only **mitosis,** the division of the nucleus and genetic material, but also **cytokinesis,** the division of the cytoplasm (which may not occur in all cells). During mitosis (covered in Section 5.3), the sister chromatids of each chromosome separate, becoming daughter chromosomes that are distributed to two daughter nuclei. When cytokinesis is complete, two daughter cells are present, each identical to the original mother cell. Mammalian cells usually require only about 4 hours to complete the mitotic stage.

Apoptosis

During apoptosis, the cell progresses through a typical series of events that bring about its destruction. The majority of cells contain enzymes that bring about apoptosis. The enzymes are ordinarily held in check by inhibitors, but they can be unleashed either by internal or external signals. During apoptosis (Fig 5.3), the cell rounds up and loses contact with its neighbors. The nucleus fragments, and the plasma membrane develops blisters. Finally, the cell fragments, and its bits and pieces are engulfed by white blood cells.

Check Your Progress 5.1

1. Describe the cell cycle and explain the major events that occur during each stage.
2. Explain why apoptosis is a necessary process.

5.2 Control of the Cell Cycle

Learning Outcomes

Upon completion of this section, you should be able to

1. Distinguish between internal and external controls of the cell cycle.
2. Describe the purpose of checkpoints in the cell cycle.
3. Differentiate between the role of proto-oncogenes and tumor suppressor genes in regulating the cell cycle.

Eukaryotic cells have evolved a complex system for regulation of the cell cycle. The cell cycle is controlled by both internal and

normal cells				apoptotic cell

blebs

cell fragment

Cell rounds up, and nucleus collapses.

Chromatin condenses, and nucleus fragments.

Plasma membrane blisters, and blebs form.

Cell fragments contain DNA fragments.

2,500×

Figure 5.3 Apoptosis. Apoptosis is a sequence of events that results in a fragmented cell. The fragments are engulfed by white blood cells and neighboring tissue cells. (photo): ©Steve Gschmeissner/Science Source external signals. The events of the cell cycle must occur in the correct order, even if the steps take longer than normal. Internal signals help control the timing of the events. One of these signals is a protein called cyclin, which acts as an internal timekeeper for the cell. The external signals, such as growth factors, tell the cell whether or not to divide.

Checkpoints and Growth Factors

The red stop signs in Figure 5.1 represent three of the checkpoints that control the progress of the cell through the cell cycle. The first checkpoint, during G_1, allows the cell to determine whether conditions are favorable to begin the cell cycle. The cell needs to assess whether there are building blocks available for duplication of the DNA and if the DNA is intact. DNA damage can stop the cell cycle at the G_1 checkpoint. In mammalian cells, the p53 protein stops the cycle at the G_1 checkpoint when DNA is damaged. First, the p53 protein attempts to initiate DNA repair, but if that is not possible, the cell enters G_0 phase and undergoes apoptosis.

The cell cycle stops at the G_2 checkpoint if DNA has not finished replicating. This prevents the initiation of the M stage before completion of the S stage. Also, if DNA is damaged, stopping the cell cycle at this checkpoint allows time for the damage to be repaired. If repair is not possible, apoptosis occurs. Another cell cycle checkpoint occurs during the mitotic (M) stage. The cycle stops if the chromosomes are not going to be distributed accurately to the daughter cells. As we will see in Section 25.5, these checkpoints play an important role in the prevention of cancer.

Mammalian cells tend to enter the cell cycle only when stimulated by an external factor. **Growth factors** are hormones that deliver a signal to the plasma membrane of target cells. These signals set into motion the events that result in the cell entering the cell cycle. For example, when blood platelets release a growth factor, skin fibroblasts in the vicinity are stimulated to finish the cell cycle so an injury can be repaired.

Genes Control the Cell Cycle

Two types of genes control the movement of a cell through the cell cycle: proto-oncogenes and tumor suppressor genes. **Proto-oncogenes** encode proteins that promote the cell cycle and prevent apoptosis. They are often likened to the gas pedal of a car because they

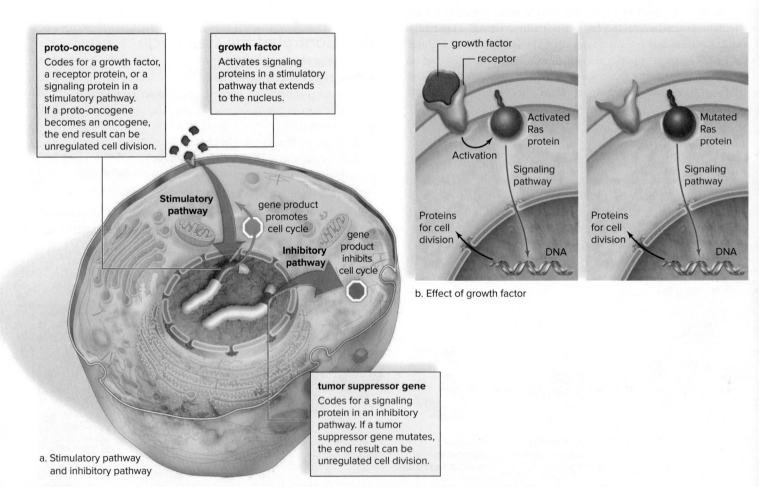

proto-oncogene
Codes for a growth factor, a receptor protein, or a signaling protein in a stimulatory pathway. If a proto-oncogene becomes an oncogene, the end result can be unregulated cell division.

growth factor
Activates signaling proteins in a stimulatory pathway that extends to the nucleus.

Stimulatory pathway

gene product promotes cell cycle

Inhibitory pathway

gene product inhibits cell cycle

tumor suppressor gene
Codes for a signaling protein in an inhibitory pathway. If a tumor suppressor gene mutates, the end result can be unregulated cell division.

a. Stimulatory pathway and inhibitory pathway

growth factor
receptor
Activated Ras protein
Activation
Signaling pathway
Proteins for cell division
DNA

Mutated Ras protein
Signaling pathway
Proteins for cell division
DNA

b. Effect of growth factor

Figure 5.4 Control of the cell cycle.   **a.** Mutated proto-oncogenes and tumor suppressor genes may cause cancer by changing stimulatory and inhibitory pathways. **b.** *RAS* is a proto-oncogene that promotes cell division. When a growth factor binds to the Ras receptor on the cell surface, a signal transduction pathway is activated and relays a signal to the nucleus to express genes whose products promote cell division. Mutations in the *RAS* proto-oncogene cause the Ras protein to be always activated—thus sending a signal for cell division without the presence of a growth factor.

Genetic Testing for Cancer Genes

Genetic tests have become increasingly available for certain cancer genes. If women test positive for defective *BRCA1* and *BRCA2* genes, they have an increased risk for early-onset breast and ovarian cancer. If individuals test positive for the *APC* gene, they are at greater risk for the development of colon cancer. Other genetic tests exist for rare cancers, including retinoblastoma and Wilms tumor.

Advocates of genetic testing say that it can alert those who test positive for these mutated genes to undergo more frequent mammograms or colonoscopies. Early detection of cancer clearly offers the best chance for successful treatment.

Others feel that genetic testing is unnecessary because nothing can presently be done to prevent the disease. Perhaps it is enough for those who have a family history of cancer to schedule more frequent checkups, beginning at a younger age.

Although the U.S. Equal Employment Opportunity Commission (EEOC) prohibits discrimination based on genetic information, recent surveys suggest that the majority of people remain concerned that being predisposed to cancer might threaten one's job or health insurance. Individuals opposed to genetic testing suggest that genetic testing be confined to a research setting, especially because it is not known which particular mutations in the genes predispose a person to cancer. They are afraid, for example, that a woman with a defective *BRCA1* or *BRCA2* gene might make the unnecessary decision to have a bilateral mastectomy. The lack of proper counseling also concerns many. In a study of 177 patients who underwent *APC* gene testing for susceptibility to colon cancer, less than 20% received counseling before the test. Moreover, physicians misinterpreted the test results in nearly one-third of the cases.

Another concern is that testing negative for a particular genetic mutation may give people the false impression they are not at risk for cancer. Such a false sense of security can prevent them from having routine cancer screenings. Regular testing and avoiding known causes of cancer—such as smoking, a high-fat diet, or too much sunlight (see Section 25.5)—are important for everyone.

Questions to Consider

1. Should genetic testing for cancer be available for everyone, or should genetic testing be confined to a research setting?
2. If genetic testing for cancer were offered to you, would you take advantage of it? Why or why not?

---

cause cells to continue through the cell cycle. **Tumor suppressor genes** encode proteins that slow the cell cycle. They are often likened to the brakes of a car because they inhibit cells from progressing through the cell cycle (Fig. 5.4a).

Cancer is unregulated cell growth. Carcinogenesis, or the development of cancer, is a multistage process involving disruption of normal cell division and behavior. This typically occurs due to mutations in either proto-oncogenes or tumor suppressor genes. These genetic changes may be inherited from a parent, the result of an error in replication, or induced by an environmental agent. Later, the altered cell grows and divides to become a population of cells, also called a tumor. The characteristics of cancer cells are discussed more fully in Section 25.5.

Proto-Oncogenes and Tumor Suppressor Genes

When proto-oncogenes mutate (Fig. 5.5), they become cancer-causing genes, called **oncogenes.** For example, the protein product of the *RAS* proto-oncogene is part of a signal transduction pathway, one of a series of relay proteins (see Fig. 5.4b). When a signaling molecule, such as a growth factor, binds to a receptor on the cell surface, the Ras protein is activated. Activated Ras protein conveys a message to the nucleus, resulting in gene expression that brings about cell division. If the *RAS* proto-oncogene is mutated to become an oncogene in such a way that the Ras protein is always activated, no growth factor needs to bind to the cell for the altered Ras protein to send the signal to divide. An altered Ras protein is found in approximately 25% of all tumors.

When tumor suppressor genes mutate, their products no longer inhibit the cell cycle (Fig. 5.6). *p53* and *BRCA1* are both examples of tumor suppressor genes. The p53 protein is such an important protein for regulating the cell cycle that almost half of all human cancers have a mutation in the *p53* gene. Mutations in *BRCA1* are often associated with breast and ovarian cancer. The Bioethical feature "Genetic Testing for Cancer Genes" discusses some of the challenges facing testing for these genes.

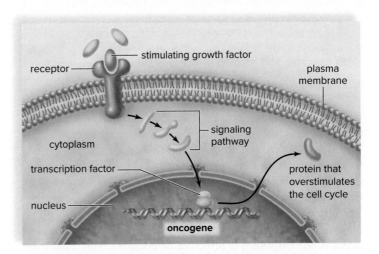

Figure 5.5 Proto-oncogenes mutate to form oncogenes.
An oncogene codes for a protein that either directly or indirectly overstimulates the cell cycle.

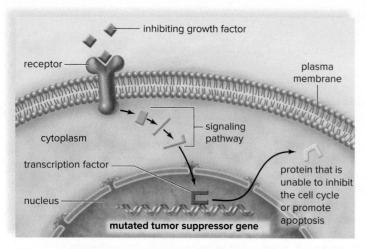

Figure 5.6 Mutations in tumor suppressor genes. A mutated tumor suppressor gene codes for a product that either directly or indirectly fails to inhibit the cell cycle.

Check Your Progress 5.2

1. Explain what factors would cause a cell to stop at each of the three cell cycle checkpoints.
2. Summarize the importance of growth factors in regulating the cell cycle.
3. Distinguish between the action of oncogenes and mutated tumor suppressor genes.

5.3 Mitosis: Maintaining the Chromosome Number

Learning Outcomes

Upon completion of this section, you should be able to

1. Explain the role of mitosis and how it maintains the chromosome number of a cell.
2. Summarize the major events that occur during mitosis and cytokinesis.
3. Compare and contrast mitosis and cytokinesis in plant and animal cells.

Eukaryotic chromosomes are composed of **chromatin,** a combination of both DNA and protein. Some of these proteins are concerned with DNA and RNA synthesis, but a large proportion, termed *histones,* have an important role in chromosome structure. On average, a human cell contains at least 2 m of DNA. Yet all of this DNA is packed into a nucleus that is about 5 μm in diameter. The histones are responsible for packaging the DNA so it can fit into such a small space. When a eukaryotic cell is not undergoing division, the chromatin is dispersed or extended. This makes the DNA available for gene expression (see Section 25.4). At the time of cell division, chromatin coils, loops, and condenses into a highly compacted form.

Each species has a characteristic chromosome number. For instance, human cells contain 46 chromosomes, corn has 20 chromosomes, and the crayfish has 200! This number is called the **diploid (2n)** number because it contains two (a pair) of each type of chromosome. Thus, humans have 23 pairs of chromosomes. One member of each pair originates from the mother, and the other member originates from the father.

Cells with half the diploid number, called the **haploid (n)** number of chromosomes, contain only one of each kind of chromosome. In the life cycle of humans, only sperm and eggs have the haploid number of chromosomes.

Overview of Mitosis

Mitosis is nuclear division in which the chromosome number stays constant. A 2n nucleus divides to produce daughter nuclei that are also 2n. It would be possible to diagram this as 2n → 2n. Figure 5.7 gives an overview of mitosis; for simplicity, only four chromosomes are depicted.

Before nuclear division takes place, DNA replication occurs, duplicating the chromosomes. Each replicated chromosome is composed of two sister chromatids held together in a region called the **centromere.** Sister chromatids are genetically identical, meaning they contain the same DNA sequences. At the completion of mitosis, each of the chromosomes in the daughter cells consists of a single chromatid, sometimes referred to as a daughter chromosome. Notice in Figure 5.7 that each daughter nucleus gets a complete set of chromosomes and has the same number of chromosomes as the parental cell. This makes the daughter cells genetically identical to each other and to the parental cell.

Mitosis is the type of nuclear division that occurs when tissues grow or when repair occurs. Following fertilization, the zygote begins to divide mitotically, and mitosis continues during development and the life span of the individual.

Mitosis in Detail

During mitosis, a structure called a **spindle** assists in distributing the chromosomes to the nucleus of each daughter cell. The spindle contains microtubule fibers. Microtubules are part of the cytoskeleton of the cell, and they have the ability to disassemble and assemble. Prior to mitosis, the centrosome, which is the main microtubule organizing center of the cell, divides. In animal cells, each centrosome contains a pair of barrel-shaped organelles called *centrioles* and an *aster,* which is an array of short microtubules that radiate from the centrosome. The fact that plant cells lack centrioles suggests that centrioles are not required for spindle formation.

Mitosis in Animal Cells

Mitosis is a continuous process that is arbitrarily divided into several phases for convenience of description: prophase, prometaphase, metaphase, anaphase, and telophase (Fig. 5.8).

Prophase **Prophase** starts the process of nuclear division. The centrosomes begin moving away from each other toward opposite

centriole   chromosome

centromere   duplicated chromosome consisting of two sister chromatids

DNA REPLICATION

DURING INTERPHASE

MITOSIS

2n = 4

2n = 4

2n = 4

2n = 4

2n = 4

Figure 5.7 Mitosis overview.   Following DNA replication during interphase, each chromosome in the parental nucleus is duplicated and consists of two sister chromatids. During mitosis, the centromeres divide and the sister chromatids separate, becoming daughter chromosomes that move into the daughter nuclei. Therefore, daughter cells have the same number and kinds of chromosomes as the parental cell. (The blue chromosomes were inherited from one parent, and the red chromosomes were inherited from the other.)

ends of the nucleus. Spindle fibers appear between the separating centrosomes as the nuclear envelope begins to fragment, and the nucleolus begins to disappear. The chromatin condenses and the chromosomes are now visible. Each is duplicated and composed of sister chromatids held together at a centromere. The spindle begins forming during late prophase.

Prometaphase   During **prometaphase,** preparations for sister chromatid separation are evident. Kinetochores appear on each side of the centromere, and these attach sister chromatids to the *kinetochore spindle fibers.* These fibers extend from the poles to the chromosomes, which will soon be located at the center of the spindle.

The kinetochore fibers attach the sister chromatids to opposite poles of the spindle, and the chromosomes are pulled first toward one pole and then toward the other before the chromosomes come into alignment. Notice that even though the chromosomes are attached to the spindle fibers in prometaphase, they are still not in alignment.

Metaphase   By the time of **metaphase,** the fully formed spindle consists of poles, asters, and fibers. The *metaphase plate* is a plane perpendicular to the axis of the spindle and equidistant from the poles. The chromosomes attached to centromeric spindle fibers line up at the metaphase plate during metaphase. Polar spindle fibers reach beyond the metaphase plate and overlap.

Anaphase   At the beginning of **anaphase,** the centromeres uniting the sister chromatids divide. The sister chromatids then separate, becoming daughter chromosomes that move toward the opposite poles of the spindle. Daughter chromosomes have a centromere and a single chromatid.

What accounts for the movement of the daughter chromosomes? First, the kinetochore spindle fibers shorten, pulling the daughter chromosomes toward the poles. Second, the polar spindle

fibers push the poles apart as they lengthen and slide past one another.

Telophase   During **telophase,** the spindle disappears, and the nuclear envelope components reassemble around the daughter chromosomes. Each daughter nucleus contains the same number and kinds of chromosomes as the original parental cell. Remnants of the polar spindle fibers are still visible between the two nuclei.

Following telophase, the chromosomes once again decondense, and a nucleolus appears in each daughter nucleus. Cytokinesis, or division of the cytoplasm, often begins before telophase is complete.

Mitosis in Plant Cells

As with animal cells, mitosis in plant cells permits growth and repair. A particular plant tissue, called meristematic tissue, retains the ability to divide throughout the life of a plant. Meristematic tissue is found at the root tip and also at the shoot tip of stems. One form of meristematic tissue accounts for the ability of trees to increase their girth each growing season.

Figure 5.8 also illustrates mitosis in plant cells. Exactly the same phases occur in plant cells as in animal cells. Although plant cells have a centrosome and spindle, there are no centrioles or asters during cell division. However, this does not have an influence on the ability of the spindle to distribute the chromosomes to each daughter cell.

Cytokinesis in Animal and Plant Cells

Cytokinesis, or cytoplasmic cleavage, usually accompanies mitosis, but they are separate processes. Division of the cytoplasm begins in anaphase and continues in telophase but does not reach completion until just before the next interphase. By that time, the

MITOSIS

Animal Cell at Interphase

centrosome has centrioles

aster 900×

duplicated chromosome 900×

spindle pole 900×

nuclear envelope fragments

chromatin condenses

nucleolus disappears

spindle fibers forming

centromere

kinetoc

kinetochore spindle fiber

polar spindle fiber

Early Prophase
Centrosomes have duplicated. Chromatin is condensing into chromosomes, and the nuclear envelope is fragmenting.

Prophase
Nucleolus has disappeared, and duplicated chromosomes are visible. Centrosomes begin moving apart, and spindle is in process of forming.

Prometaphase
The kinetochore of each chromatid is attached to a kinetochore spindle fiber. Polar spindle fibers stretch from each spindle pole and overlap.

centrosome lacks centrioles

Plant Cell at Interphase

900×

cell wall   chromosomes   900×

spindle pole lacks centrioles and aster   900×

Figure 5.8 Phases of mitosis in animal and plant cells. Following interphase, the steps of mitosis include prophase, prometaphase, metaphase, anaphase, and telophase. The blue chromosomes were inherited from one parent, and the red from the other parent. (photos) (animal early prophase, prophase, metaphase, anaphase, telophase): ©Ed Reschke; (animal prometaphase): ©Michael Abbey/Science Source; (plant early prophase, plant prometaphase): ©Ed Reschke; (plant prophase, metaphase, anaphase, telophase): ©Kent Wood/Science Source

newly forming cells have received a share of the cytoplasmic organelles that duplicated during the previous interphase.

Cytokinesis in Animal Cells

By the start of telophase in animal cells, a **cleavage furrow,** which is an indentation of the membrane between the two daughter nuclei, begins to form. The cleavage furrow deepens when a band of actin filaments, called the contractile ring, slowly forms a constriction between the two daughter cells. The action of the contractile

ring can be likened to pulling a drawstring ever tighter about the middle of a balloon, causing the balloon to constrict in the middle.

A narrow bridge between the two cells can be seen during telophase, and then the contractile ring continues to separate the cytoplasm until there are two daughter cells (Fig. 5.9).

Cytokinesis in Plant Cells

Cytokinesis in plant cells occurs by a process different from that seen in animal cells (Fig. 5.10). The rigid cell wall that surrounds

chromosomes at metaphase plate   900×

daughter chromosome   900×

cleavage furrow   900×

nucleolus

kinetochore spindle fiber

Metaphase
tromeres of duplicated chromosomes
aligned at the metaphase plate (center
ly formed spindle). Kinetochore spindle
ers attached to the sister chromatids
come from opposite spindle poles.

Anaphase
Sister chromatids part and become daughter
chromosomes that move toward the spindle
poles. In this way, each pole receives the same
number and kinds of chromosomes as the parent cell.

Telophase
Daughter cells are forming
as nuclear envelopes and
nucleoli reappear. Chromosomes will
become indistinct chromatin.

spindle fibers   900×

900×

cell plate   900×

plant cells does not permit cytokinesis by furrowing. Instead, cytokinesis in plant cells involves building new cell walls between the daughter cells.

Cytokinesis is apparent when a small, flattened disk appears between the two daughter plant cells. Electron micrographs reveal that the disk is at right angles to a set of microtubules. The Golgi apparatus produces vesicles, which move along the microtubules to the region of the disk. As more vesicles arrive and fuse, a **cell plate** can be seen. The membrane of the vesicles completes the plasma membrane for both cells, and they release

molecules that form the new plant cell walls. The cell walls are later strengthened by the addition of cellulose fibrils.

Check Your Progress 5.3

1. Distinguish between the number of chromatids each chromosome contains before and after replication.
2. Summarize the events that are occurring in each stage of mitosis.
3. Compare and contrast cytokinesis in plant and animal cells.

4,000×

cleavage furrow

contractile ring

4,000×

Figure 5.9 Cytokinesis in animal cells. A single cell becomes two cells by a furrowing process. A contractile ring composed of actin filaments gradually gets smaller, and the cleavage furrow pinches the cell into two cells. (photos) (top): Source: National Institutes of Health (NIH)/USHHS; (bottom): ©Steve Gschmeissner/SPL/Getty Images

5.4 Meiosis: Reducing the Chromosome Number

Learning Outcomes

Upon completion of this section, you should be able to

1. Summarize the purpose of meiosis.
2. Explain what is meant by the term *homologous chromosomes.*
3. Describe the events of meiosis.

Meiosis occurs in any life cycle that involves sexual reproduction. **Meiosis** reduces the chromosome number in such a way that the

24,000×

Figure 5.10 Cytokinesis in plant cells. During cytokinesis in a plant cell, a cell plate forms midway between two daughter nuclei and extends to the plasma membrane. ©Biophoto Associates/Science Source

daughter nuclei receive only one of each kind of chromosome. In other words, meiosis reduces the chromosome number of a cell from diploid (2n) to haploid (n). The process of meiosis ensures that, as a result of sexual reproduction, the next generation of individuals will have the diploid number of chromosomes and a combination of traits different from that of either parent.

Overview of Meiosis

At the start of meiosis, the parental cell has the diploid number of chromosomes. In Figure 5.11, the diploid number is 4, which you can verify by counting the number of centromeres. Notice that meiosis requires two cell divisions, meiosis I and meiosis II, and that the four daughter cells have the haploid number of chromosomes. DNA replication occurs prior to meiosis I.

Recall that when a cell is 2n, or diploid, the chromosomes occur in pairs. For example, the 46 chromosomes of humans occur in 23 pairs. In Figure 5.11, there are two pairs of chromosomes. The members of a pair are called **homologous chromosomes** (or homologues). When we are discussing cell division, homologous chromosomes will be the same size (long or short) but are indicated by different colors—the blue chromosomes are inherited from one parent, and the red are inherited from the other.

Meiosis I

During meiosis I, the homologous chromosomes come together and line up side by side. This process is called **synapsis** and it

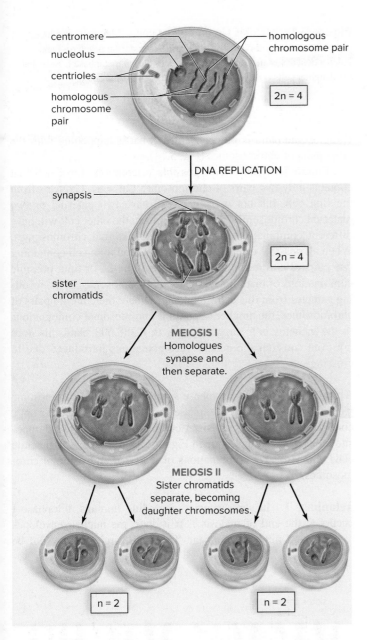

Figure 5.11 Overview of meiosis. Following DNA replication, each chromosome is duplicated. During meiosis I, the homologous chromosomes pair during synapsis and then separate. During meiosis II, the centromeres divide and the sister chromatids separate, becoming daughter chromosomes that move into the daughter nuclei. This diagram shows the process in an animal cell.

results in an association of four chromatids that stay in close proximity during the first two phases of meiosis I.

Because of synapsis, there are pairs of homologous chromosomes at the metaphase plate during meiosis I. Notice that only during meiosis I is it possible to observe paired chromosomes at the metaphase plate. When the members of these pairs separate, each daughter nucleus receives one member of each pair. Therefore, each daughter cell now has the haploid number of chromosomes, as you can verify by counting its centromeres.

Meiosis II and Fertilization

Replication of DNA does not occur between meiosis I and meiosis II because the chromosomes are still duplicated (composed of two sister chromatids). During meiosis II, the centromeres divide and the sister chromatids separate, becoming daughter chromosomes that are distributed to daughter nuclei. In the end, each of four daughter cells has the haploid number of chromosomes, and each chromosome consists of one chromatid.

In some life cycles, such as that of humans (see Fig. 5.17), the daughter cells mature into **gametes** (sex cells, also called the sperm and egg) that fuse during fertilization. **Fertilization** restores the diploid number of chromosomes in a cell that will develop into a new individual. If the gametes carried the diploid instead of the haploid number of chromosomes, the chromosome number would double with each fertilization.

Meiosis in Detail

Meiosis, which requires two nuclear divisions, results in four daughter nuclei, each having one of each kind of chromosome and therefore half the number of chromosomes as the parental cell.

Meiosis I

The phases of meiosis I for an animal cell are diagrammed in Figure 5.12. Meiosis helps ensure that genetic variation of the parental genes occurs through two key events: crossing-over and independent assortment of chromosome pairs. Because the members of a homologous pair can carry slightly different instructions for the same genetic trait (see Section 23.1), these events are significant. For example, one homologue may carry instructions for brown eyes, while the corresponding homologue may carry instructions for blue eyes. Fertilization ensures the offspring will have different combinations of genes from either parent.

Prophase I In prophase I, synapsis occurs, and then the spindle appears while the nuclear envelope fragments and the nucleolus disappear. Figure 5.13 shows that during synapsis, the homologous chromosomes come together and line up side by side. Now an exchange of genetic material may occur between the *nonsister* chromatids of the homologues. This exchange is called **crossing-over.** Crossing-over means that the chromatids held together by a centromere are no longer identical. As a result of crossing-over, the daughter cells receive chromosomes with recombined genetic material.

Notice in Figure 5.13 that crossing-over is represented by an exchange of color between the chromosomes. Because of crossing-over, the chromosomes in the daughter cells have a different combination of genetic material than the parental cell.

Metaphase I During metaphase I, the homologous chromosome pairs align at the metaphase plate. Depending on how they align, the maternal or paternal member of each pair may be oriented toward either pole (see Fig. 5.12). This is called **independent assortment** because the orientation of each of the homologous

2n = 4

↓ DNA REPLICATION

MEIOSIS I

Prophase I
Homologous chromosomes pair during synapsis allowing for crossing-over to occur.

Metaphase I
Independent assortment occurs as homologous chromosomes pair along the metaphase plate.

Anaphase I
Homologous chromosomes separate, pulled to opposite poles by centromeric spindle fibers.

Telophase I
Daughter cells have one chromosome from each homologous pair.

n = 2

Interkinesis
Chromosomes still consist of two chromatids.

Figure 5.12 Meiosis I in an animal cell. Meiosis I introduces genetic variation by the process of crossing-over during prophase I and independent assortment of the chromosomes during metaphase I. Note that the daughter cells possess only one copy of each chromosome.

chromosome pairs is independent of what is happening with the other pairs of chromosomes (Fig. 5.14).

Figure 5.14 shows two possible orientations for a cell that contains only two pairs of chromosomes (2n = 4). Gametes from the first cell, but not the other, will have two paternal or two maternal chromosomes. The gametes from the other cell will have different combinations of maternal and paternal chromosomes. When all possible orientations are considered for a cell containing two pairs of chromosomes, the result will be 2^2, or four, possible combinations of maternal and paternal chromosomes in the resulting gametes from this cell. In humans, where there are 23 pairs of chromosomes, the number of possible chromosomal combinations in the gametes is a staggering 2^{23}, or 8,388,608. And this does not even consider the genetic recombinations introduced due to crossing-over.

Anaphase I During anaphase I, the homologous chromosome pair that aligned during metaphase I separates. Each chromosome, consisting of two identical sister chromatids, moves toward opposite poles of the cell. Notice in Figure 5.12 that this will generate cells with different combinations of maternal and paternal chromosomes.

Telophase I In some species, including humans, telophase I occurs at the end of meiosis I. If it does, the nuclear envelopes re-form, and nucleoli reappear. This phase may or may not be

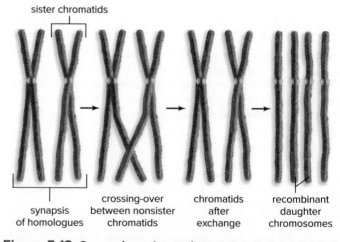

sister chromatids

synapsis of homologues | crossing-over between nonsister chromatids | chromatids after exchange | recombinant daughter chromosomes

Figure 5.13 Synapsis and crossing-over. During prophase I, *from left to right,* duplicated homologous chromosomes undergo synapsis when they line up with each other. During crossing-over, nonsister chromatids break and then rejoin. The two resulting daughter chromosomes will have a different combination of genetic material than they had before.

Figure 5.14 Independent assortment. Two possible orientations of homologous chromosome pairs at the metaphase plate are shown for metaphase I. Each of these will result in daughter nuclei with a different combination of parental chromosomes (independent assortment). In a cell with two pairs of homologous chromosomes, there are 2^2 possible combinations of parental chromosomes in the daughter nuclei.

accompanied by cytokinesis, which is separation of the cytoplasm (see Section 5.3).

Interkinesis The period of time between meiosis I and meiosis II is called **interkinesis.** Because the chromosomes are already duplicated, there is no replication of DNA during interkinesis. The daughter cells of meiosis I then proceed directly into meiosis II.

Meiosis II

Phases of meiosis II for an animal cell are diagrammed in Figure 5.15. At the beginning of prophase II, a spindle appears while the nuclear envelope disassembles and the nucleolus disappears. Each duplicated chromosome attaches to the spindle, and then they align at the metaphase plate during metaphase II. During anaphase II, sister chromatids separate, becoming daughter chromosomes that move into the daughter nuclei. In telophase II, the spindle disappears as the nuclear envelope re-forms.

During the cytokinesis that can follow meiosis II, the plasma membrane furrows to produce two complete cells, each of which has the haploid number of chromosomes. Because each cell from meiosis I undergoes meiosis II, there are four daughter cells altogether.

The Importance of Meiosis

Meiosis produces haploid cells that are genetically different than the original diploid parent cell. The introduction of this variation occurs in two ways. First, during prophase I, crossing-over between nonsister chromatids of the paired homologous chromosomes rearranges genes, with the result that the sister chromatids of each homologue may no longer be identical. Second, the independent assortment of chromosomes during anaphase I means that the gametes produced by meiosis II may have different combinations of chromosomes than the parental cell.

Upon fertilization, the combining of chromosomes from genetically different gametes, even if the gametes are from a single individual (as in many plants), helps ensure that offspring are not identical to their parents. This genetic variability is the main advantage of sexual reproduction.

The staggering amount of genetic variation achieved through meiosis is particularly important to the long-term survival of a species because it increases genetic variation within a population. If the environment changes, genetic variability among offspring introduced by sexual reproduction may be advantageous. Under the new conditions, some offspring may have a better chance of survival and reproductive success than others in a population.

Check Your Progress 5.4

1. Explain why there are two divisions in meiosis and why the DNA is not replicated between meiosis I and meiosis II.
2. Outline the major events that occur during meiosis I and II, focusing on the activities of the chromosomes.
3. Explain how crossing-over and independent assortment introduce genetic variation.

MEIOSIS II

Prophase II
Cells have one chromosome
from each homologous pair.

Metaphase II
Chromosomes align
at the metaphase plate.

Anaphase II
Daughter chromosomes
move toward the poles.

Telophase II
Spindle disappears,
nuclei form, and
cytokinesis takes place.

Daughter Cells
Meiosis results
in four haploid
daughter cells.

n = 2

n = 2

n = 2

n = 2

Figure 5.15 Meiosis II in an animal cell.   During meiosis II, sister chromatids separate, becoming daughter chromosomes that are distributed to the daughter nuclei. Following meiosis II, there are four haploid daughter cells. Comparing the number of chromosomes in each daughter cell with the number in the parental cell at the start of meiosis I (see Fig. 5.13) verifies that each daughter cell is haploid.

5.5 Comparison of Meiosis with Mitosis

Learning Outcomes

Upon completion of this section, you should be able to
1. Compare and contrast the processes of meiosis and mitosis.
2. Identify the differences in the behavior of homologous chromosomes in meiosis and mitosis.

You should have already noted some similarities and differences between the processes of mitosis and meiosis. The differences include:

■ DNA replication takes place only once prior to either meiosis or mitosis. However, meiosis requires two nuclear divisions, whereas mitosis requires only one nuclear division.

■ Meiosis followed by cytokinesis produces four daughter cells. Mitosis followed by cytokinesis results in two daughter cells.

Figure 5.16 Meiosis compared to mitosis. The processes of meiosis and mitosis differ in the processes that introduce genetic variation in meiosis, and in the chromosome number and number of daughter cells produced.

- The four daughter cells following meiosis are haploid and have half the chromosome number as the parental cell. The daughter cells following mitosis have the same chromosome number as the parental cell.
- The daughter cells resulting from meiosis are not genetically identical to each other or to the parental cell. The daughter cells resulting from mitosis are genetically identical to each other and to the parental cell.

The differences in the individual stages of meiosis and mitosis are shown in Figure 5.16.

In addition, meiosis and mitosis differ in the times that each occurs during the life cycle of the organism, and in the internal processes used to generate variation during meiosis.

Occurrence

Meiosis occurs only at certain times in the life cycle of sexually reproducing organisms. In humans, meiosis occurs only in the reproductive organs and produces the gametes. Mitosis occurs almost continuously in all the tissues of an organism for the purpose of growth and repair.

TABLE 5.1   Comparison of Meiosis I with Mitosis

Meiosis I	Mitosis
Prophase I Pairing of homologous chromosomes	*Prophase* No pairing of chromosomes
Metaphase I Homologous duplicated chromosomes at metaphase plate	*Metaphase* Duplicated chromosomes at metaphase plate
Anaphase I Homologous chromosomes separate.	*Anaphase* Sister chromatids separate, becoming daughter chromosomes that move to the poles.
Telophase I Two haploid daughter cells	*Telophase/Cytokinesis* Two diploid daughter cells, identical to the parental cell

TABLE 5.2   Comparison of Meiosis II with Mitosis

Meiosis II	Mitosis
Prophase II No pairing of chromosomes	*Prophase* No pairing of chromosomes
Metaphase II Haploid number of duplicated chromosomes at metaphase plate	*Metaphase* Duplicated chromosomes at metaphase plate
Anaphase II Sister chromatids separate, becoming daughter chromosomes that move to the poles.	*Anaphase* Sister chromatids separate, becoming daughter chromosomes that move to the poles.
Telophase II Four haploid daughter cells, not identical to parental cell	*Telophase/Cytokinesis* Two diploid daughter cells, identical to the parental cell

Processes

The stages of meiosis also differ from those of mitosis in several important aspects.

Comparison of Meiosis I to Mitosis

Notice that these events distinguish meiosis I from mitosis (Table 5.1):

- Homologous chromosomes pair and undergo crossing-over during prophase I of meiosis, but not during mitosis.
- Paired homologous chromosomes align at the metaphase plate during metaphase I in meiosis. These paired chromosomes have four chromatids altogether. Individual chromosomes align at the metaphase plate during metaphase in mitosis. They each have two chromatids.
- Homologous chromosomes (with centromeres intact) separate and move to opposite poles during anaphase I in meiosis. Centromeres split, and sister chromatids, now called daughter chromosomes, move to opposite poles during anaphase in mitosis.

Comparison of Meiosis II to Mitosis

The events of meiosis II are just like those of mitosis, except that in meiosis II the nuclei contain the haploid number of chromosomes (Table 5.2).

Check Your Progress 5.5

1. Describe the differences in the activity of the homologous chromosomes in prophase and metaphase of meiosis I and mitosis.
2. Compare and contrast the outputs of meiosis and mitosis with regard to the number of cells and the chromosome complement of each cell.

5.6 The Human Life Cycle

Learning Outcomes

Upon completion of this section, you should be able to

1. Describe the human life cycle in terms of haploid and diploid cells.
2. Explain the process of gamete production in both males and females.

The human life cycle requires both meiosis and mitosis (Fig. 5.17). A haploid sperm (n) and a haploid egg (n) join at fertilization, and the resulting **zygote** has the full, or diploid (2n), number of chromosomes. During development of the fetus before birth, mitosis keeps the chromosome number constant in all the cells of the body. After birth, mitosis is involved in the continued growth of the child and repair of tissues at any time. As a result of mitosis, each somatic cell in the body has the same number of chromosomes.

Spermatogenesis and Oogenesis in Humans

In human males, meiosis is a part of **spermatogenesis,** which occurs in the testes and produces sperm. In human females, meiosis is a part of **oogenesis,** which occurs in the ovaries and produces eggs.

Spermatogenesis

Following puberty, the time of life when the sex organs mature, spermatogenesis is continual in the testes of human males. As many as 300,000 sperm are produced per minute, or 400 million per day.

During spermatogenesis in mammals (Fig. 5.18), primary spermatocytes, which are diploid, divide during the first meiotic division to form two *secondary spermatocytes,* which are haploid. Secondary spermatocytes divide during the second meiotic division to produce four *spermatids,* which are also haploid. What is the difference between the chromosomes in haploid secondary spermatocytes and those in haploid spermatids? The chromosomes

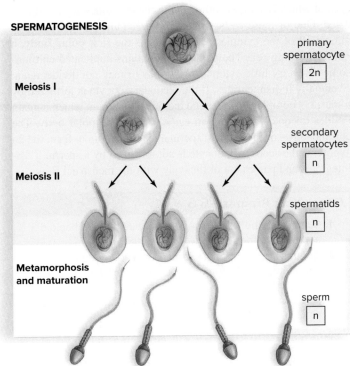

Figure 5.18 Spermatogenesis in mammals.   Spermatogenesis produces four viable haploid sperm.

Figure 5.17 Life cycle of humans.   Meiosis in males is a part of sperm production, and meiosis in females is a part of egg production. When a haploid sperm fertilizes a haploid egg, the zygote is diploid. The zygote undergoes mitosis as it develops into a newborn child. Mitosis continues throughout life during growth and repair.

in secondary spermatocytes are duplicated and consist of two chromatids, whereas those in spermatids consist of only one chromatid. Spermatids then mature into **sperm** (spermatozoa).

Oogenesis

Meiosis in human females begins in the fetus. In the ovaries of the fetus, all of the primary oocytes (diploid cells) begin meiosis but become arrested in prophase I (Fig. 5.19). Following puberty and the initiation of the menstrual cycle, one primary oocyte begins to complete meiosis. It finishes the first meiotic division as two cells,

Figure 5.19 Oogenesis in mammals.   Oogenesis produces one egg and at least two polar bodies, whereas spermatogenesis produces four viable sperm. In humans, both sperm and egg have 23 chromosomes; therefore, following fertilization, the zygote has 46 chromosomes.

each of which is haploid, although the chromosomes are still duplicated. One of these cells, termed the *secondary oocyte,* receives almost all of the cytoplasm. The other is the first **polar body,** a nonfunctioning cell. The polar body contains duplicated chromosomes but very little cytoplasm and may or may not divide again. Eventually it disintegrates. If the secondary oocyte is fertilized by a sperm, it completes the second meiotic division, in which it again divides unequally, forming an **egg** and a second polar body. The chromosomes of the egg and sperm nuclei then join to form the 2n zygote. If the secondary oocyte is not fertilized by a sperm, it disintegrates and passes out of the body with the menstrual flow.

Check Your Progress 5.6

1. Summarize the cells in the human body that are haploid and the role these cells play in the life cycle.
2. Contrast the number of sperm produced from one primary spermatocyte with the number of eggs produced from one oocyte.

Conclusion

In the chapter opener, we asked why the rate of cancer in elephants is significantly lower than in humans. The answer is in the number of copies of a gene, called *p53,* in the elephant genome. *p53* belongs to a class of genes called tumor suppressor genes, which act as gatekeepers for the cell cycle and thus control the rate at which cells divide.

Also called the "Guardian of the Genome," *p53* produces a protein that is responsible for not only activating DNA repair enzymes, but also stopping the cell cycle and placing the cell in the G_0 phase. In doing so, it helps regulate the rate at which cells divide, and limit the probability that cells will become cancerous. While having extra copies of genes is not always advantageous, in this case, elephants have evolved a mechanism of decreasing their rates of cancer.

SUMMARIZE

5.1 The Cell Cycle

- Cell division increases the number of **somatic cells** in the body, and **apoptosis** reduces this number when appropriate. Cells go through a **cell cycle** that includes (1) **interphase** and (2) cell division, consisting of **mitosis** and **cytokinesis.** Interphase, in turn, includes G_1 (growth as certain organelles double), S (DNA synthesis), and G_2 (growth as the cell prepares to divide). Cell division occurs during the mitotic stage (M) when daughter cells receive a full complement of chromosomes.

- During S phase, DNA replication occurs. Each **chromatid** is copied, forming **sister chromatids.**

5.2 Control of the Cell Cycle

- Internal and external signals control the cell cycle, which can stop at any of three checkpoints: in G_1 prior to the S stage, in G_2 prior to the M stage, and near the end of mitosis. DNA damage is one reason the cell cycle stops. The p53 protein is active at the G_1 checkpoint, and if DNA is damaged and can't be repaired, this protein initiates apoptosis.

- **Growth factors** are examples of external signals that control the progress of the cell through the cell cycle.

- Both **proto-oncogenes** and **tumor suppressor genes** produce proteins that control the cell cycle. Proto-oncogenes encode proteins that promote the cell cycle and prevent apoptosis. *RAS* is an example of a proto-oncogene. Mutations in proto-oncogenes create **oncogenes,** which may cause cancer. When tumor suppressor genes mutate, their products no longer inhibit the cell cycle. *p53* and *BRCA1* are tumor suppressor genes.

5.3 Mitosis: Maintaining the Chromosome Number

- Each species has a characteristic number of chromosomes. The total number is the **diploid (2n)** number, and half this number is the **haploid (n)** number. Among eukaryotes, cell division involves nuclear division and division of the cytoplasm (cytokinesis).

- Replication of the chromatids precedes cell division. The duplicated chromosome is composed of two sister chromatids held together at a **centromere.** During mitosis, the centromeres divide, and the daughter chromosomes go into each new nucleus.

- Mitosis has the following phases: **prophase,** when chromosomes condense and the nuclear envelope dissolves; **prometaphase,** when the chromosomes are attached to **spindle** fibers; **metaphase,** when the chromosomes are aligned at the metaphase plate; **anaphase,** when the chromatids separate, becoming daughter chromosomes that move toward the poles; and **telophase,** when new nuclear envelopes form around the daughter chromosomes and cytokinesis is well underway.

- Cytokinesis differs between plant and animal cells. Plant cells form a **cell plate** between the daughter cells, while animal cells develop a **cleavage furrow** that separates the daughter cells.

5.4 Meiosis: Reducing the Chromosome Number

- **Meiosis** occurs in any life cycle that involves sexual reproduction. The end result of meiosis is daughter cells with the haploid number of **homologous chromosomes** (homologues). In some life cycles, the daughter cells become **gametes,** and upon **fertilization,** the offspring have the diploid number of chromosomes, the same as their parents. **Crossing-over** and **independent assortment** of chromosomes during meiosis I ensure genetic variation in daughter cells.

■ Meiosis utilizes two nuclear divisions. During meiosis I, homologous chromosomes undergo **synapsis,** and crossing-over between nonsister chromatids occurs. When the homologous chromosomes separate during meiosis I, each daughter nucleus receives one member from each pair of chromosomes. Therefore, the daughter cells are haploid. Interkinesis during meiosis I and meiosis II does not include DNA replication. Distribution of daughter chromosomes derived from sister chromatids during meiosis II then leads to a total of four new cells, each with the haploid number of chromosomes.

5.5 Comparison of Meiosis with Mitosis

■ While mitosis and meiosis differ in the number of cells produced, and the chromosome complement of each cell, there are many similarities in the processes. The primary difference is the behavior of the homologous chromosomes.

5.6 The Human Life Cycle

■ The human life cycle involves both mitosis and meiosis. Mitosis ensures that each somatic cell has the diploid number of chromosomes. In humans, meiosis is a part of **spermatogenesis** and **oogenesis.** Spermatogenesis in males produces four viable **sperm,** whereas oogenesis in females produces one **egg** and **polar bodies.** Oogenesis does not go on to completion unless a sperm fertilizes the secondary oocyte. Fertilization produces a **zygote,** which grows by mitosis.

ASSESS

Testing Yourself

Choose the best answer for each question.

5.1 The Cell Cycle

1. Label this drawing of the cell cycle and briefly summarize what is happening at each stage.

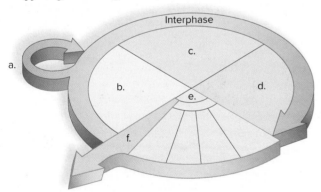

2. Apoptosis
   a. decreases the number of cells in the body.
   b. is programmed cell death.
   c. is controlled by enzymes in the cell.
   d. All of these are correct.

5.2 Control of the Cell Cycle

3. At which of the following checkpoints is the DNA checked for damage and, if damage is present, the cell is placed in G_0 phase?
   a. M
   b. G_1
   c. G_2
   d. None of these are correct.

4. This checkpoint determines if the DNA has been replicated properly prior to cell division.
   a. G_1
   b. S
   c. G_2
   d. M

5. Which of the following act as the brakes of the cell cycle and prevent cells from dividing too quickly?
   a. oncogenes
   b. proto-oncogenes
   c. cyclins
   d. tumor suppressor genes

5.3 Mitosis: Maintaining the Chromosome Number

6. Which of the following terms is described incorrectly?
   a. prometaphase—the kinetochores become attached to spindle fibers
   b. anaphase—daughter chromosomes migrate toward spindle poles
   c. prophase—the chromosomes condense and the nuclear envelope disintegrates
   d. metaphase—the chromosomes are aligned at the metaphase plate
   e. telophase—a resting phase between cell division cycles

7. If a parent cell has a diploid number of 24 chromosomes before mitosis, how many chromosomes will the daughter cells have?
   a. 12
   b. 24
   c. 30
   d. 36
   e. 64

8. Cytokinesis in animal cells involves the formation of a/an
   a. oncogene.
   b. cell plate.
   c. cleavage furrow.
   d. sister chromatid.

5.4 Meiosis: Reducing the Chromosome Number

9. If a parent cell has 22 chromosomes, the daughter cells following meiosis II will have
   a. 22 chromosomes.
   b. 44 chromosomes.
   c. 11 chromosomes.
   d. All of these are correct.

10. Crossing-over occurs between
    a. sister chromatids of the same chromosome.
    b. chromatids of nonhomologous chromosomes.
    c. nonsister chromatids of a homologous pair.
    d. None of these are correct.

11. Which of these helps provide genetic diversity?
    a. independent alignment during metaphase I
    b. crossing-over during prophase I
    c. random fusion of sperm and egg nuclei during fertilization
    d. All of these are correct.

5.5 Comparison of Meiosis with Mitosis

12. The pairing of homologous chromosomes occurs during which of the following?
    a. mitosis
    b. meiosis I
    c. meiosis II
    d. All of these are correct.

13. Sister chromatids separate during anaphase of which of the following?
   a. mitosis
   b. meiosis I
   c. meiosis II
   d. Both a and c are correct.

5.6 The Human Life Cycle

14. Polar bodies are produced during
   a. DNA replication.
   b. mitosis.
   c. spermatogenesis.
   d. oogenesis.
   e. None of these are correct.

15. Following fertilization, the somatic cells of humans divide by
   a. apoptosis.
   b. meiosis.
   c. mitosis.
   d. DNA replication.

ENGAGE

BioNOW

Want to know how this science is relevant to your life? Check out the BioNOW video below:

■ Cell Division

What was the purpose of the rooting hormone in this experiment, and what part of the cell cycle do you think it was targeting?

Thinking Critically

1. BPA is a chemical compound that has historically been used in the manufacture of plastic products. However, cells often mistake BPA compounds for hormones that accelerate the cell cycle. Because of this, BPA is associated with an increased risk of certain cancers.
   a. How might BPA interact with the cell cycle and its checkpoints?
   b. Why do you think very small concentrations of BPA might have a large effect on the cell?

2. In a type of skin cancer known as melanoma, skin cells divide uncontrollably. This out-of-control division can be deadly, but a lack of cell division could be harmful as well. Describe why and when skin cells do need to replicate.

3. From an evolutionary standpoint, why is sexual reproduction advantageous for the continuation of a species?

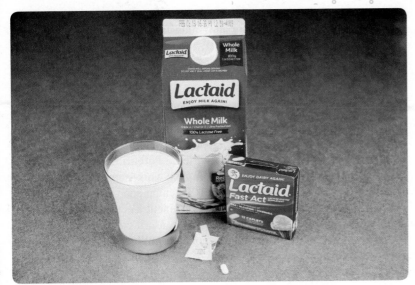

©Ricochet Creative Productions LLC

6

Metabolism: Energy and Enzymes

Lactose Intolerance

For between 65% and 75% of the world's population, a meal containing milk or milk products can result in painful cramping of the intestines, gas, and diarrhea. This condition is called lactose intolerance and it is due to the lack of an enzyme called lactase. Normally, in the small intestine lactase breaks down lactose, the sugar in milk, to glucose and galactose, both of which are easily absorbed by the small intestine. However, if the lactase enzyme is not present, then the lactose sugar enters the large intestine. There, the bacteria of the large intestine digest the sugar, producing the symptoms of lactose intolerance.

Lactose intolerance may be caused either by genetic factors that turn off the lactase gene early in life, or by diseases that affect the small intestine and limit its ability to produce lactase. Fortunately, food manufacturers have developed ways to produce products that do not contain lactose sugar (lactose-free). Supplements of lactase enzyme are also available to provide temporary relief of the condition.

This chapter describes the general characteristics and functions of enzymes, and how enzymes function in the flow of energy and metabolism. Each enzyme in our body is responsible for a unique metabolic reaction, as illustrated by the lactase enzyme, which is responsible for allowing the lactose sugar to be used as an energy source.

As you read through this chapter, think about the following questions:

1. What is the role of an enzyme in a metabolic pathway?
2. Why are enzymes specific in the types of substrates they will interact with?
3. What environmental factors influence the rate of enzyme activity?

BEFORE YOU BEGIN

Before beginning this chapter, take a few moments to review the following discussions:

Section 1.1 What is the relationship between energy and the basic characteristics of life?

Section 2.7 How does the shape of a protein relate to its function?

Section 4.2 In what ways does energy play a role in the movement of molecules across a plasma membrane?

6.1 Life and the Flow of Energy

Energy is the ability to do work or bring about a change. In order to maintain their organization and carry out metabolic activities, cells, as well as organisms, need a constant supply of energy. This energy allows living organisms to carry on the processes of life, including growth, development, response to stimuli, metabolism, and reproduction.

With the exception of some organisms living in the deepest parts of the ocean, the majority of organisms get their energy from organic nutrients produced by photosynthetic organisms, such as algae, plants, and some forms of bacteria. Therefore, life on Earth is ultimately dependent on an input of solar energy.

Forms of Energy

Energy occurs in two forms: kinetic and potential. **Kinetic energy** is the energy of motion, as when a ball rolls down a hill or as you walk through a field. **Potential energy** is stored energy—its capacity to accomplish work is not being used at the moment. The food we eat has potential energy because it can be converted into various types of kinetic energy. Food is specifically called **chemical energy** because it contains energy in the chemical bonds of organic molecules. When we walk, we convert chemical energy into a type of kinetic energy called **mechanical energy** (Fig. 6.1).

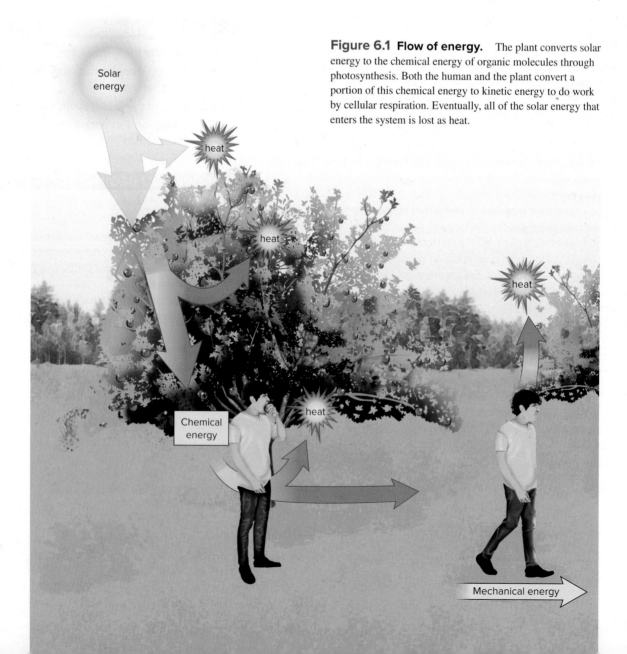

Figure 6.1 Flow of energy.   The plant converts solar energy to the chemical energy of organic molecules through photosynthesis. Both the human and the plant convert a portion of this chemical energy to kinetic energy to do work by cellular respiration. Eventually, all of the solar energy that enters the system is lost as heat.

Two Laws of Thermodynamics

Figure 6.1 illustrates the flow of energy in a terrestrial ecosystem. Through photosynthesis, plants capture a small portion of the incoming solar energy. This is stored in the chemical bonds of organic molecules, such as carbohydrates. When plants use their carbohydrates (through cellular respiration), some of the energy dissipates as heat. In an ecosystem, plants serve as the food source for other organisms, including humans. The cellular respiration pathways in our bodies use these carbohydrates as energy, once again releasing some of the energy as heat. All living organisms metabolize organic nutrient molecules, and eventually, all the captured solar energy dissipates as heat. Therefore, energy flows through an ecosystem. It does not recycle.

Two **laws of thermodynamics** explain why energy flows in ecosystems and in cells. The first law of thermodynamics explains the ability of organisms to convert chemical energy to mechanical energy:

> The first law of thermodynamics, also called the law of conservation of energy, states that energy cannot be created or destroyed, but it can be changed from one form to another.

During photosynthesis, plant cells use solar energy to convert energy-poor molecules, such as carbon dioxide and water, to energy-rich molecules, such as carbohydrates. As we have seen, not all of the captured solar energy is used to form carbohydrates; some becomes heat:

solar energy   carbohydrate synthesis

Obviously, plant cells do not create the energy they use to produce carbohydrate molecules. That energy comes from the sun. The energy from the sun is not destroyed in the process because heat is also a form of energy. Similarly, a moose uses the energy derived from carbohydrates to power its muscles. And as its cells use this energy, none is destroyed, but some becomes heat, which dissipates into the environment:

carbohydrate   muscle contraction

The second law of thermodynamics, therefore, applies to living systems.

> The second law of thermodynamics states that energy cannot be changed from one form to another without a loss of usable energy.

The heat released during photosynthesis (from the plant) and cellular respiration (from both the plant and moose) dissipates into the environment. This energy is no longer usable—that is, it is not available to do work. With transformation upon transformation, eventually all usable forms of energy become heat that is lost to the environment. Heat that dissipates into the environment cannot be captured and converted to one of the other forms of energy.

As a result of the second law of thermodynamics, no process requiring a conversion of energy is ever 100% efficient. Much of the energy is lost in the form of heat. In automobiles, the gasoline engine is between 20% and 30% efficient in converting chemical energy into mechanical energy. The majority of energy is obviously lost as heat. Cells are capable of about 40% efficiency, with the remaining energy given off to their surroundings as heat.

Cells and Entropy

The second law of thermodynamics can be stated another way: Every energy transformation makes the universe less organized and more disordered. The term **entropy** is used to indicate the relative amount of disorganization. Because the processes that occur in cells are energy transformations, the second law means that every process that occurs in cells always does so in a way that increases the total entropy of the universe. Then, too, any one of these processes makes less energy available to do useful work in the future.

Figure 6.2 illustrates the second law of thermodynamics. The second law of thermodynamics tells us that glucose tends to break apart into carbon dioxide and water (Fig. 6.2a). Why? Because glucose is more organized, and therefore less stable, than its breakdown products. Also, hydrogen ions on one side of a membrane tend to move to the other side unless they are prevented from doing so (Fig. 6.2b). Why? Because when they are distributed randomly, entropy has increased. As an analogy, you know from experience that a neat room (Fig. 6.2c) is more organized but less stable than a messy room, which is disorganized but more stable. How do you know a neat room is less stable than a messy room? Consider that a neat room always tends to become more messy.

Cellular processes, such as the synthesis of glucose and ion transport across a membrane, are possible because cells have the ability to obtain an input of energy from an outside source. This energy ultimately comes from the sun. Life depends on a constant supply of energy from the sun because the ultimate fate of all solar energy in the biosphere is to become randomized in the universe as heat. A living cell remains organized because it functions to maintain a constant flow of energy.

> ### Check Your Progress 6.1
> **1.** Explain why energy is not recycled in an ecosystem.
> **2.** Summarize how the two laws of thermodynamics are involved as you consume your daily meals.

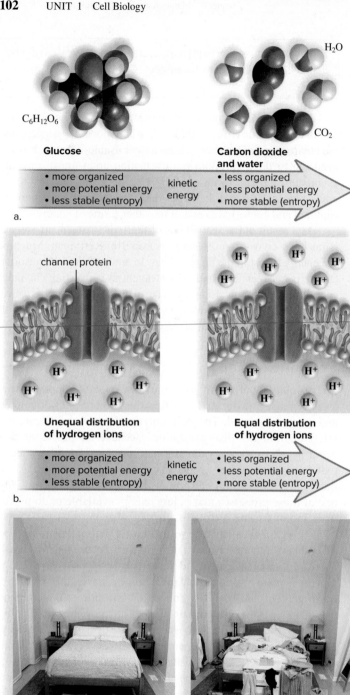

a.

b.

c.

Figure 6.2   Cells and entropy.   The second law of thermodynamics tells us that (**a**) glucose, which is more organized, tends to break down to carbon dioxide and water, which are less organized. **b.** Similarly, hydrogen ions (H^+) on one side of a membrane tend to move to the other side so that the ions are randomly distributed. **c.** Entropy also affects nonliving structures, such as this room. All of these are examples of a loss of potential energy and an increase in entropy. (c, both photos): ©Keith Eng, 2008

6.2 Energy Transformations and Metabolism

Learning Outcomes

Upon completion of this section, you should be able to

1. Identify how the terms *anabolic, catabolic, endergonic,* and *exergonic* relate to metabolic reactions.
2. Summarize the ATP cycle and the role of ATP in the cell.

One of the defining characteristics of life is its ability to use materials and energy to sustain their structure and function. In living organisms, this is accomplished through chemical reactions. The term **metabolism** is used to describe the sum of all the chemical reactions that occur in a cell.

Overview of Metabolic Reactions

A significant part of cellular metabolism involves the breaking down and the building up of molecules. In a chemical reaction, **reactants** are substances that participate in a reaction, while **products** are substances that form as a result of a reaction. For example, in the reaction $A + B \rightarrow C + D$, A and B are the reactants while C and D are the products. It is important to note that most chemical reactions will occur in both directions unless regulated. The term **catabolism** is used to refer to the breaking down of molecules, and the term **anabolism** is used to refer to the building up (synthesis) of molecules.

In cell biology, which occurs on a small scale, we are less concerned about the entire universe, which is vast. In such specific instances, cell biologists use the concept of free energy instead of entropy. **Free energy** (also called ΔG) is the amount of energy available—that is, energy that is still "free" to do work— after a chemical reaction has occurred. The change in free energy after a reaction occurs is calculated by subtracting the free energy content of the reactants from that of the products. A negative result (negative ΔG) means that the products have less free energy than the reactants, and the reaction will go forward. In our reaction, if C and D have less free energy than A and B, the reaction will occur.

Chemical reactions are also often described as whether they require an input of energy, or release energy, during the course of the reaction. **Exergonic reactions** are spontaneous and release energy, while **endergonic reactions** require an input of energy to occur. In the body, many reactions, such as protein synthesis, nerve impulse conduction, or muscle contraction, are endergonic, and they are driven by the energy released by exergonic reactions. ATP is a carrier of energy between exergonic and endergonic reactions.

ATP: Energy for Cells

Adenosine triphosphate (ATP) is the common energy currency of cells. When cells require energy, they "spend" ATP. The more active the organism, the greater the demand for ATP. However, cells do not keep an unlimited supply of ATP on hand. Instead,

Figure 6.3 The ATP cycle.   In cells, ATP carries energy between exergonic reactions and endergonic reactions. When a phosphate group is removed by hydrolysis, ATP releases the appropriate amount of energy for most metabolic reactions.

ATP is constantly being generated from **adenosine diphosphate (ADP)** and a molecule of inorganic phosphate, ℗ (Fig. 6.3).

This cycle is powered by the breakdown of glucose and other biomolecules during cellular respiration (see Section 7.1). However, according to the second law of thermodynamics, this process is not very efficient. Only 39% of the free energy stored in the chemical bonds of a glucose molecule is transformed to ATP; the rest is lost as heat.

There are many biological advantages in using ATP as an energy carrier in living organisms. ATP is a common and universal energy currency because it can be used in many different types of reactions. Also, when ATP is converted to energy, ADP and ℗, the amount of energy released is sufficient for a particular biological function, and little energy is wasted. In addition, ATP breakdown can be coupled to endergonic reactions in such a way that it minimizes energy loss.

Structure of ATP

ATP is a nucleotide composed of the nitrogen-containing base adenine and the 5-carbon sugar ribose (together called adenosine) and three phosphate groups (see Fig. 6.3). ATP is called a "high-energy" compound because of the energy stored in the chemical bonds of the phosphates. This energy is sufficient to power many of the functions of the cell.

Function of ATP

In living systems, ATP can be used for the following:

■ Chemical work. ATP supplies the energy needed to synthesize macromolecules (anabolism) that make up the cell, and therefore the organism.

■ Transport work. ATP supplies the energy needed to pump substances across the plasma membrane.

■ Mechanical work. ATP supplies the energy needed to permit muscles to contract, cilia and flagella to beat, chromosomes to move, and so forth.

Coupled Reactions

In **coupled reactions,** the energy released by an exergonic reaction is used to drive an endergonic reaction. ATP breakdown is often coupled to cellular reactions that require an input of energy. Coupling, which requires that the exergonic reaction and the endergonic reaction be closely tied, can be symbolized like this:

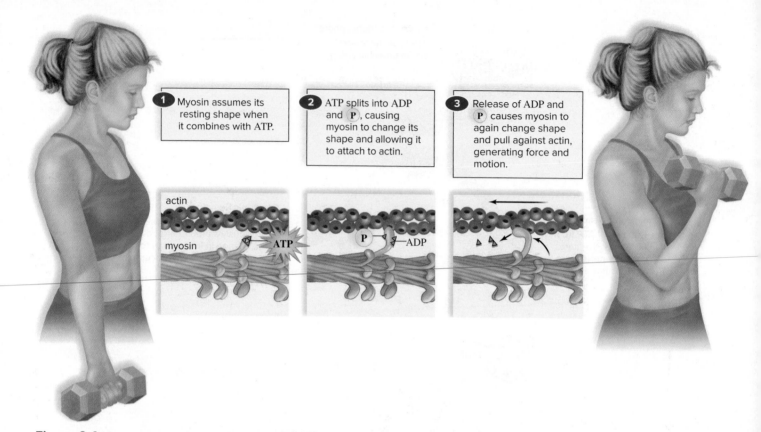

1 Myosin assumes its resting shape when it combines with ATP.

2 ATP splits into ADP and Ⓟ, causing myosin to change its shape and allowing it to attach to actin.

3 Release of ADP and Ⓟ causes myosin to again change shape and pull against actin, generating force and motion.

actin

myosin

ATP

P ADP

Figure 6.4 Coupled reactions. Muscle contraction occurs only when it is coupled to ATP breakdown.

This reaction tells you that coupling occurs, but it does not show how coupling is achieved. A cell has two main ways to couple ATP hydrolysis to an energy-requiring reaction: ATP is used to energize a reactant, or ATP is used to change the shape of a reactant. Both can be achieved by transferring a phosphate group to the reactant, so that the product is phosphorylated.

An example of a coupled reaction is when a polypeptide is synthesized at a ribosome. There, an enzyme transfers a phosphate group from ATP to each amino acid in turn, and this transfer supplies the energy needed to overcome the energy cost associated with bonding one amino acid to another.

Through coupled reactions, ATP allows chemical reactions that may not be favorable from an energy perspective for the cell. However, even though they are unfavorable, these processes must occur to create the high degree of order and structure essential for life. Macromolecules must be made and organized to form cells and tissues; the internal composition of the cell and the organism must be maintained; and movement of cellular organelles and the organism must occur if life is to continue.

Figure 6.4 shows that ATP breakdown provides the energy necessary for muscular contraction to occur. During muscle contraction, myosin filaments pull actin filaments to the center of the cell, and the muscle shortens. First, the myosin head combines with ATP (three connected green triangles) and takes on its resting shape. Next, ATP breaks down to ADP (two connected green triangles) plus Ⓟ (one green triangle). A resulting change in shape

allows myosin to attach to actin. Finally, the release of ADP and Ⓟ from the myosin head causes it to change its shape again and pull on the actin filament. The cycle then repeats. During this cycle, chemical energy has been transformed to mechanical energy, heat has been released, and entropy has increased.

Check Your Progress 6.2

1. Determine whether an anabolic reaction is more likely to be exergonic or endergonic.
2. Explain how the ATP pathway is similar to a rechargeable battery.

6.3 Enzymes and Metabolic Pathways

Learning Outcomes

Upon completion of this section, you should be able to

1. Explain the purpose of a metabolic pathway and how enzymes help regulate it.
2. Recognize how enzymes influence the activation energy rates of a chemical reaction.
3. Identify how environmental conditions influence the activity of an enzyme.

Chemical reactions do not occur haphazardly in healthy cells; they are usually part of a series of linked reactions called a **metabolic pathway.** Metabolic pathways begin with a particular reactant and end with a final product. Many specific steps can be involved in a metabolic pathway, and each step is a chemical reaction catalyzed by an enzyme. An **enzyme** is a molecule that speeds a chemical reaction without itself being affected by the reaction. While most enzymes are proteins, some are made of RNA. These are called **ribozymes,** and they are involved in the synthesis of RNA and the synthesis of proteins at ribosomes.

The diagram below shows an example of a simple metabolic pathway:

Metabolic pathways begin with a particular reactant (such as A in the diagram above) and end with a final product (D in the diagram above). Many specific steps can be involved in a metabolic pathway, and each step is a chemical reaction catalyzed by an enzyme (E_1, E_2, E_3). The substrates for the first reaction are converted into products, and those products then serve as the substrates for the next enzyme-catalyzed reaction. One reaction leads to the next reaction in an organized, highly regulated manner.

It is important to note that in biological systems, any one of the molecules (A–D) in this linear pathway could also be a substrate for an enzyme in another pathway. The presence or absence of an active enzyme determines which reaction takes place. A diagram of the metabolic pathways of a cell would look much like a road map, with many intersecting points.

Energy of Activation

Molecules frequently do not react with one another unless they are activated in some way. In the lab, for example, in the absence of an enzyme, activation is very often achieved by heating a reaction flask to increase the number of effective collisions between molecules. The energy that must be added to cause molecules to react with one another is called the **energy of activation (E_a)** (Fig. 6.5). Even though the reaction will proceed (ΔG is negative), the energy of activation must be overcome. The burning of firewood is a very exergonic reaction, but firewood in a pile does not spontaneously combust. The input of some energy, perhaps a lit match, is required to overcome the energy of activation.

Figure 6.5 shows E_a when an enzyme is not present, compared to when an enzyme is present, illustrating that enzymes lower the amount of energy required for activation to occur. Nevertheless, the addition of the enzyme does not change the end result of the reaction. Notice that the energy of the products is less than the energy of the reactants. This results in a negative ΔG, so the reaction will proceed. But the reaction will not go at all unless the energy of activation is overcome. Without the enzyme, the reaction rate will be very slow. By lowering the energy of activation, the enzyme increases the rate of the reaction.

Sugar in your kitchen cupboard will break down to carbon dioxide and water because the energy of the products (carbon

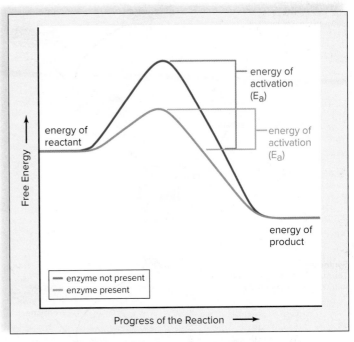

Figure 6.5 Energy of activation (E_a). Enzymes speed the rate of reactions because they lower the amount of energy required for the reactants to react.

dioxide and water) is much less than the free energy of the reactant (sugar). However, the rate of this reaction is so slow that you never see it. If you eat the sugar, the enzymes in your digestive system greatly increase the speed at which the sugar is broken down. However, the end result (carbon dioxide and water) is still the same.

How Enzymes Function

The following equation is often used to indicate that an enzyme forms a complex with its substrate:

S	+	E	$\longrightarrow$	ES	$\longrightarrow$	E	+	P
substrate		enzyme		enzyme-substrate complex				product

As shown in Figure 6.6, in most instances, only one small part of the enzyme, called the **active site,** complexes with the substrate(s). It is here that the enzyme and substrate fit together, seemingly like a key fits a lock.

However, cell biologists now know that the active site undergoes a slight change in shape in order to accommodate the substrate(s). This is called the **induced fit model** because the enzyme is induced to undergo a slight alteration to achieve optimum fit (Fig. 6.7).

The change in shape of the active site facilitates the reaction that now occurs. After the reaction has been completed, the product(s) is released, and the active site returns to its original state, ready to bind to another substrate molecule. Only a small amount of enzyme is actually needed in a cell because enzymes are not used up by the reaction.

Degradation
The substrate is broken down to smaller products.

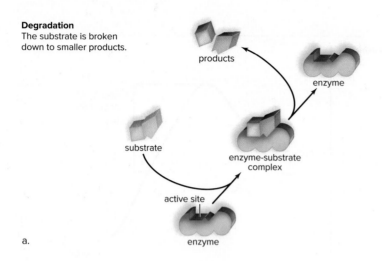

a.

Synthesis
The substrates are combined to produce a larger product.

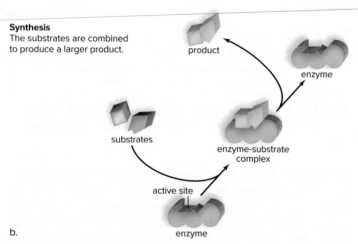

b.

Figure 6.6 Enzymatic action. An enzyme has an active site where the substrates and enzyme fit together in such a way that the substrates react. Following the reaction, the products are released, and the enzyme is free to act again. **a.** The enzymatic reaction can result in the degradation of a substrate into multiple products (catabolism) or, (**b**) the synthesis of a product from multiple substrates (anabolism).

Some enzymes do more than simply complex with their substrate(s); they participate in the reaction. For example, trypsin digests protein by breaking peptide bonds. The active site of trypsin contains three amino acids with *R* groups that actually interact with members of the peptide bond—first to break the bond and then to introduce the components of water. This illustrates that the formation of the enzyme-substrate complex is very important in speeding up the reaction.

Every reaction in a cell requires that its specific enzyme be present. Because enzymes complex only with their substrates, they are often named for their substrates with the suffix *-ase*, as in the following examples:

Substrate	Enzyme
Lipid	Lipase
Urea	Urease
Maltose	Maltase
Ribonucleic acid	Ribonuclease
Lactose	Lactase

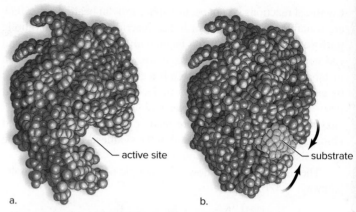

a. b.

Figure 6.7 Induced fit model. These computer-generated images show an enzyme called lysozyme that hydrolyzes its substrate, a polysaccharide that makes up bacterial cell walls. **a.** Shape of enzyme when no substrate is bound to it. **b.** After the substrate binds, the shape of the enzyme changes so that hydrolysis can better proceed.

Factors Affecting Enzymatic Speed

Enzymatic reactions proceed quite rapidly. Consider, for example, the breakdown of hydrogen peroxide (H_2O_2) as catalyzed by the enzyme catalase: $2\ H_2O_2 \rightarrow 2\ H_2O + O_2$. The breakdown of hydrogen peroxide can occur 600,000 times a second when catalase is present! In order to achieve maximum product per unit time, there should be enough substrate to fill the enzyme's active sites most of the time. In addition to substrate concentration, the rate of an enzymatic reaction can also be affected by environmental factors (temperature and pH) or by cellular mechanisms, such as enzyme activation, enzyme inhibition, and cofactors.

Substrate Concentration

Generally, enzyme activity increases as substrate concentration increases because there are more collisions between substrate molecules and the enzyme. As more substrate molecules fill active sites, more product results per unit time. But when the enzyme's active sites are filled almost continuously with substrate, the enzyme's rate of activity cannot increase any more. The maximum rate has been reached.

Temperature and pH

As the temperature rises, enzyme activity increases (Fig. 6.8). This occurs because higher temperatures cause more effective collisions between enzyme and substrate. However, if the temperature rises beyond a certain point, enzyme activity eventually levels out and then declines rapidly because the enzyme is **denatured.** Most enzymes are proteins, and the shape of the protein responds to temperature. Temperature often influences the protein's secondary and tertiary structure (see Fig. 2.25), preventing it from binding its substrate(s) efficiently. Therefore, as the structure of the enzyme changes, its activity decreases.

a. Rate of reaction as a function of temperature.

b. Body temperature of ectothermic animals often limits rates of reactions.

c. Body temperature of endothermic animals promotes rates of reactions.

Figure 6.8 The effect of temperature on rate of reaction. a. Usually, the rate of an enzymatic reaction doubles with every 10°C rise in temperature. This enzymatic reaction is maximum at about 40°C. Then, it decreases until the reaction stops altogether, because the enzyme has become denatured. **b.** The body temperature of ectothermic animals, which require an environmental source of heat, often limits rates of reactions. **c.** The body temperature of endothermic animals, which generate heat through their own metabolism, promotes rates of reaction. (b): ©Brand X Pictures/PunchStock; (c): ©Digital Vision/Getty Images

Each enzyme also has a preferred pH at which the rate of the reaction is highest. Figure 6.9 shows the preferred pH for the enzymes pepsin and trypsin. At this pH value, these enzymes have their normal configurations. The globular structure of an enzyme is dependent on interactions, such as hydrogen bonding, between *R* groups (see Fig. 2.25). A change in pH can alter the ionization of *R* groups and disrupt normal interactions, and under extreme pH conditions, denaturation eventually occurs. If the enzyme's shape is altered, it is then unable to combine efficiently with its substrate.

Enzyme Activation

Not all enzymes are needed by the cell all the time. Genes can be turned on to increase the concentration of an enzyme in a cell or turned off to decrease the concentration. But enzymes can also be present in the cell in an inactive form. Activation of enzymes occurs in many different ways. Some enzymes are covalently modified by the addition or removal of phosphate groups. An enzyme called a kinase adds phosphates to proteins, as shown below, and an enzyme called a phosphatase removes them. In some proteins, adding phosphates activates them; in others, removing phosphates activates them. Enzymes can also be activated by cleaving or removing part of the protein, or by associating with another protein or cofactor.

inactive protein   kinase   active protein

Enzyme Inhibition

Enzyme inhibition occurs when the substrate is unable to bind to the active site of an enzyme. The activity of almost every enzyme in a cell is regulated by feedback inhibition. In the simplest case, when there is plenty of product, it binds to the enzyme's active site, and then the substrate is unable to bind. As the product is used up, inhibition is reduced, and more product can be produced. In this way, the concentration of the product is always kept within a certain range.

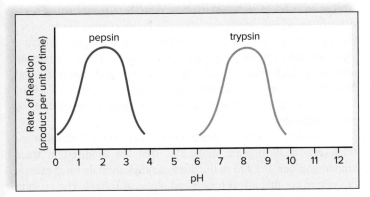

Figure 6.9 The effect of pH on rate of reaction. The preferred pH for pepsin, an enzyme that acts in the stomach, is about 2, while the preferred pH for trypsin, an enzyme that acts in the small intestine, is about 8. At the preferred pH, an enzyme maintains its shape so it can bind with its substrates.

SCIENCE IN YOUR LIFE ▶ **HEALTH**

Enzyme Inhibitors Can Spell Death

Cyanide gas was formerly used to execute people. How did it work? Cyanide can be fatal because it binds to a mitochondrial enzyme necessary for the production of ATP. MPTP (1-methyl-4-phenyl-1,2,3,6-tetrahydropyridine) is another enzyme inhibitor that stops mitochondria from producing ATP. The toxic nature of MPTP was discovered in the early 1980s, when a group of intravenous drug users in California suddenly developed symptoms of Parkinson disease, including uncontrollable tremors and rigidity. All of the drug users had injected a synthetic form of heroin that was contaminated with MPTP. Parkinson disease is characterized by the death of brain cells, the very ones that are also destroyed by MPTP.

Sarin is a chemical that inhibits an enzyme at neuromuscular junctions, where nerves stimulate muscles. When the enzyme is inhibited, the signal for muscle contraction cannot be turned off, so the muscles are unable to relax and become paralyzed. Sarin can be fatal if the muscles needed for breathing become paralyzed. In 2017, the Syrian government used sarin gas against the inhabitants of Khan Sheikhoun, a town in northern Syria (Fig. 6A). It is estimated that the attack killed over 80 civilians.

A fungus that contaminates and causes spoilage of sweet clover produces a chemical called warfarin. Cattle that eat the spoiled feed die from internal bleeding because warfarin inhibits a crucial enzyme for blood clotting. Today, warfarin is widely used as a rat poison. Unfortunately, it is not uncommon for warfarin to be mistakenly eaten by pets and even very small children, with tragic results.

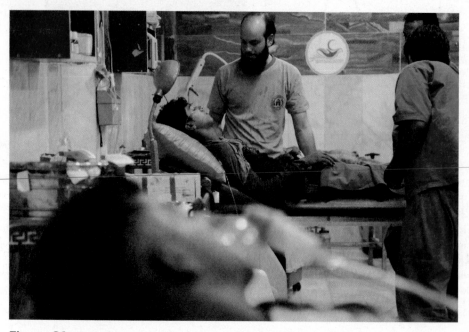

Figure 6A Sarin gas. The aftermath when sarin, a nerve gas that results in the inability to breathe, was used against a Syrian village in 2017. ©Anadolu Agency/Contributor/Getty Images

Many people are prescribed a medicine called warfarin (Coumadin), which acts as an anticoagulant to prevent inappropriate blood clotting. It is often prescribed for people who have received an artificial heart valve or have irregular heartbeats.

These examples all show how our understanding of science can have positive or negative consequences, and emphasize the role of ethics in scientific investigation.

Questions to Consider

1. Do you feel it is ethical to use dangerous chemicals to treat diseases?
2. Many drug companies are actively looking for species that might contain chemical compounds that are new to humans. How might this have both beneficial and harmful side effects?

Most metabolic pathways in cells are regulated by a more complicated type of feedback inhibition (Fig. 6.10). In these instances, the end product of an active pathway binds to a site other than the active site of the first enzyme. The binding changes the shape of the active site so the substrate is unable to bind to the enzyme, and the pathway shuts down (inactive). Therefore, no more product is produced.

As discussed in the Health feature "Enzyme Inhibitors Can Spell Death" many poisons, such as cyanide and sarin, are enzyme inhibitors.

Enzyme Cofactors

Many enzymes require the presence of an inorganic ion or a nonprotein organic molecule at the active site in order to work properly. These necessary ions or molecules are called **cofactors.** The inorganic ions include metals such as copper, zinc, or iron. The nonprotein organic molecules are called **coenzymes.** These cofactors participate in the reaction and may even accept or contribute atoms to the reactions. Examples of these are NAD$^+$ (nicotinamide adenine dinucleotide), FAD (flavin adenine dinucleotide), and NADP$^+$ (nicotinamide adenine dinucleotide phosphate), each of which plays a significant role in either cellular respiration or photosynthesis.

Vitamins are often components of coenzymes. **Vitamins** are relatively small organic molecules that are required in trace amounts in our diet and in the diets of other animals for synthesis of coenzymes that affect health and physical fitness. The vitamin becomes a part of the coenzyme's molecular structure. For example, the vitamin niacin is part of the coenzyme NAD, and riboflavin (B$_2$) is part of the coenzyme FAD.

A deficiency of any one of these vitamins results in a lack of the coenzyme and therefore a lack of certain enzymatic actions. In

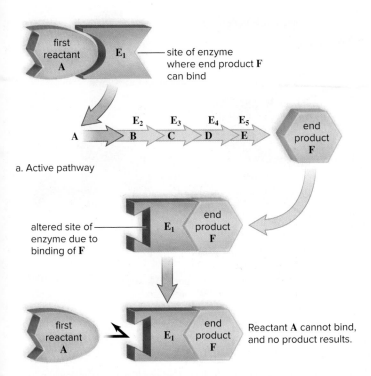

a. Active pathway

b. Inactive pathway

Figure 6.10 Feedback inhibition. **a.** In an active pathway, the first reactant (A) is able to bind to the active site of enzyme E_1. **b.** Feedback inhibition occurs when the end product (F) of the metabolic pathway binds to the first enzyme of the pathway—at a site other than the active site. This binding causes the active site to change its shape. Now, reactant A is unable to bind to the enzyme's active site, and the whole pathway shuts down.

humans, this eventually results in vitamin-deficiency symptoms. For example, niacin deficiency results in a skin disease called pellagra, and riboflavin deficiency results in cracks at the corners of the mouth.

Check Your Progress 6.3

1. Summarize why enzymes are needed in biochemical pathways and how cells may regulate their activity.
2. Describe how enzymes accelerate chemical reactions.
3. Discuss why the three-dimensional shape of an enzyme is important to its function.

6.4 Oxidation-Reduction Reactions and Metabolism

Learning Outcomes

Upon completion of this section, you should be able to

1. Explain how equations for photosynthesis and cellular respiration represent oxidation-reduction reactions.
2. Summarize the relationship between the metabolic reactions of photosynthesis and cellular respiration.

In the next two chapters, you will explore two important metabolic pathways: cellular respiration (see Section 7.1) and photosynthesis (see Section 8.1). Both of these pathways are based on the use of special enzymes to facilitate the movement of electrons. The movement of these electrons plays a major role in the energy-related reactions associated with these pathways.

Oxidation-Reduction Reactions

When oxygen (O) combines with a metal such as iron or magnesium (Mg), oxygen receives electrons and becomes an ion that is negatively charged. The metal loses electrons and becomes an ion that is positively charged. When magnesium oxide (MgO) forms, it is appropriate to say that magnesium has been oxidized. On the other hand, oxygen has been reduced because it has gained negative charges (i.e., electrons). Reactions that involve the gain and loss of electrons are called **oxidation-reduction reactions.** Sometimes, the terms *oxidation* and *reduction* are applied to other reactions, whether or not oxygen is involved. When discussing metabolic reactions, oxidation represents the loss of electrons, and reduction is the gain of electrons. Because oxidation and reduction go hand-in-hand, the entire reaction is often called a **redox reaction.** One easy way to remember what is happening in redox reactions is to remember the term *OIL RIG*:

OIL RIG

Oxidation Is Loss   Reduction Is Gain

An example is the chemical reaction that produces sodium chloride.

$$Na + Cl \rightarrow NaCl$$

In this reaction, $Na + Cl \rightarrow NaCl$, sodium has been oxidized (loss of electron), and chlorine has been reduced (gain of electrons).

The terms *oxidation* and *reduction* also apply to covalent reactions in cells. In this case, however, oxidation is the loss of hydrogen atoms ($e^- + H^+$), and reduction is the gain of hydrogen atoms. Notice that when a molecule loses a hydrogen atom, it has lost an electron, and when a molecule gains a hydrogen atom, it has gained an electron. As we will see, this form of oxidation-reduction is best shown using the overall equations for photosynthesis and cellular respiration.

Chloroplasts and Photosynthesis

The chloroplasts in plants capture solar energy and use it to convert water and carbon dioxide into a carbohydrate. Oxygen is a by-product that is released. The overall equation for photosynthesis can be written like this:

$$\text{energy} + 6\,CO_2 + 6\,H_2O \longrightarrow C_6H_{12}O_6 + 6\,O_2$$

carbon   water   glucose   oxygen
dioxide

Figure 6.11 Relationship of chloroplasts to mitochondria. Chloroplasts produce energy-rich carbohydrate. Carbohydrate is broken down in mitochondria, and the energy released is used for the buildup of ATP. Mitochondria can also respire molecules derived from fats and amino acids for the buildup of ATP. Usable energy is lost as heat due to the energy conversions of photosynthesis, cellular respiration, and the use of ATP in the body. (photos) (leaves): ©Ollirg/Shutterstock; (runner): ©Photodisc/Getty Images

This equation shows that during photosynthesis, hydrogen atoms are transferred from water to carbon dioxide as glucose forms. In this reaction, therefore, carbon dioxide has been reduced and water has been oxidized. It takes energy to reduce carbon dioxide to glucose, and this energy is supplied by solar energy. Chloroplasts are able to capture solar energy and convert it to the chemical energy of ATP, which is used along with hydrogen atoms to reduce carbon dioxide.

The reduction of carbon dioxide to form a mole of glucose stores 686 kcal in the chemical bonds of glucose. This is the energy that living organisms utilize to support themselves only because carbohydrates (and other nutrients) can be oxidized in mitochondria.

Mitochondria and Cellular Respiration

Mitochondria, present in both plants and animals, oxidize carbohydrates and use the released energy to build ATP molecules. Cellular respiration, therefore, consumes oxygen and produces carbon dioxide and water, the very molecules taken up by chloroplasts. The overall equation for cellular respiration is the opposite of the one we used to represent photosynthesis:

$$\underset{\text{glucose}}{C_6H_{12}O_6} + \underset{\text{oxygen}}{6\,O_2} \longrightarrow \underset{\substack{\text{carbon} \\ \text{dioxide}}}{6\,CO_2} + \underset{\text{water}}{6\,H_2O} + \text{energy}$$

In this reaction, glucose has lost hydrogen atoms (been oxidized), and oxygen has gained hydrogen atoms (been reduced). When oxygen gains electrons, it becomes water. The complete oxidation of a mole of glucose releases 686 kcal of energy, and some of this energy is used to synthesize ATP molecules. If the energy within glucose were released all at once, most of it would dissipate as heat instead of some of it being used to produce ATP. Instead, cells oxidize glucose step by step. The energy is gradually stored and then converted to that of ATP molecules, which is used by organisms in a variety of ways.

Figure 6.11 shows us very well that chloroplasts and mitochondria are involved in a cycle. Carbohydrate produced within chloroplasts becomes a fuel for cellular respiration in mitochondria, while carbon dioxide released by mitochondria becomes a substrate during photosynthesis in chloroplasts. These organelles are involved in a redox cycle because carbon dioxide is reduced

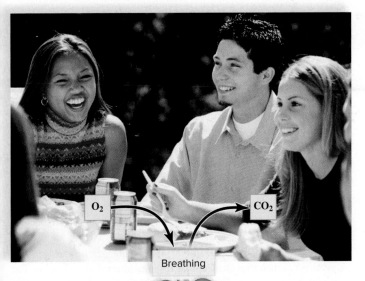

Figure 6.12 Relationship between breathing, eating, and cellular respiration.   The O_2 we inhale and the nutrients resulting from the digestion of our food are carried by the bloodstream to our cells, where they enter mitochondria and undergo cellular respiration. Following cellular respiration, the ATP stays in the cell, but the CO_2 is carried to the lungs for exhalation. (photo): ©Brand X Pictures/Getty Images

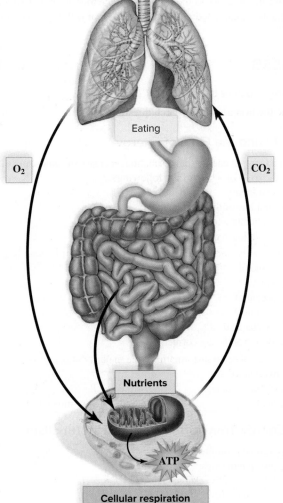

during photosynthesis and carbohydrate is oxidized during cellular respiration. Note that energy does not cycle between the two organelles; instead, it flows from the sun through each step of photosynthesis and cellular respiration until it eventually becomes unusable heat as ATP is used by the cell.

Cellular Respiration and Humans

Humans, like all eukaryotic organisms, are involved in the cycling of molecules between chloroplasts and mitochondria. Our food is derived from plants or we eat other animals that have eaten plants. Also, we take in oxygen released by plants. Nutrients from our food and oxygen enter our mitochondria, which produce ATP (Fig. 6.12).

Without a supply of energy-rich molecules, ultimately derived from plants, we could not produce the ATP molecules needed to maintain our bodies. On the other hand, our mitochondria release carbon dioxide and water. The carbon dioxide is exhaled and it enters the atmosphere where it is accessible to plants once again. In Section 7.4, we will explore how our bodies use energy sources other than carbohydrates for fuel.

Check Your Progress 6.4

1. Explain why overall equations for photosynthesis and cellular respiration represent redox reactions.
2. Summarize how the inputs and outputs of cellular respiration and photosynthesis are related.

Conclusion

Normally, the lactase gene in humans is turned off after the first few years of life. However, a mutation, believed to have originated in what is now Turkey around 8,000 years ago and subsequently spread to European populations, allows some people to consume milk products into their adult years. For this reason, lactose intolerance is more prevalent in Asian and Hispanic populations where the incidence of the mutation is lower.

SUMMARIZE

6.1 Life and the Flow of Energy

■ **Energy** is the ability to do work. It may exist in several forms, including **potential energy** (**chemical energy**) and **kinetic energy** (**mechanical energy**).

■ Two **laws of thermodynamics** are basic to understanding how living organisms relate to energy. The first law of thermodynamics states that energy cannot be created or destroyed but can only be changed from one form to another. The second law of thermodynamics states that one usable form of energy cannot be converted into another form without loss of usable energy. Therefore, every energy transformation makes the universe less organized. As a result of these laws, we know that the **entropy** of the universe is increasing and that only a constant input of energy maintains life's organization.

6.2 Energy Transformations and Metabolism

■ The term **metabolism** encompasses all the chemical reactions occurring in a cell. **Catabolism** involves breaking down **reactants,** whereas **anabolism** involves building up **products. Free energy** is the amount of energy that is actually available to do work.

■ Considering individual reactions, only those that result in products that have less usable energy than the reactants go forward. Such reactions, called **exergonic reactions,** release energy. **Endergonic reactions,** which require an input of energy, occur in cells as **coupled reactions** with an exergonic process. For example, glucose breakdown is an exergonic metabolic pathway that drives the buildup of many **adenosine triphosphate** (**ATP**) molecules. These ATP molecules then supply energy for cellular work. ATP goes through a cycle of constantly being built up from, and then broken down to, **adenosine diphosphate** (**ADP**) + ℗.

6.3 Enzymes and Metabolic Pathways

■ A **metabolic pathway** is a series of reactions that proceed in an orderly, step-by-step manner. Each reaction requires a specific **enzyme.** Enzymes function by reducing the **energy of activation** (E_a) needed for the reaction to proceed. Most enzymes are proteins, but some, called **ribozymes,** are made of RNA.

■ Reaction rates increase when enzymes form a complex with their substrates, called an **induced fit model.** Generally, enzyme activity increases as substrate concentration increases. Once all **active sites** are filled, the maximum rate has been achieved. Any environmental factor, such as temperature or pH, affects the shape of a protein and its ability to function.

■ Cellular mechanisms regulate enzyme quantity and activity. The activity of most metabolic pathways is regulated by **enzyme inhibition,** usually associated with feedback. Many enzymes have **cofactors** or **coenzymes,** such as **vitamins,** that help them carry out a reaction.

6.4 Oxidation-Reduction Reactions and Metabolism

■ The overall equation for photosynthesis is the opposite of that for cellular respiration. Both processes involve **oxidation-reduction reactions. Redox reactions** are a major way in which energy is transformed in cells.

■ During photosynthesis, carbon dioxide is reduced to glucose, and water is oxidized. Glucose formation requires energy, and this energy comes from the sun. Chloroplasts capture solar energy and convert it to the chemical energy of ATP molecules, which are used along with hydrogen atoms to reduce carbon dioxide to glucose. During cellular respiration, glucose is oxidized to carbon dioxide, and oxygen is reduced to water. This reaction releases energy, which is used to synthesize ATP molecules in all types of cells.

■ Energy flows through all living organisms. Photosynthesis is a metabolic pathway in chloroplasts that transforms solar energy to the chemical energy within carbohydrates, and cellular respiration is a metabolic pathway completed in mitochondria that transforms this energy into that of ATP molecules. Eventually, the energy within ATP molecules becomes heat.

■ Cellular respiration is an aerobic process that requires oxygen and gives off carbon dioxide. Cellular respiration involves oxidation. The air we inhale contains the oxygen, and the food we digest after eating contains the carbohydrate glucose needed for cellular respiration.

ASSESS

Testing Yourself

Choose the best answer for each question.

6.1 Life and the Flow of Energy

1. The fact that energy transformations increase the amount of entropy is the basis of which of the following?
   a. cell theory
   b. first law of thermodynamics
   c. second law of thermodynamics
   d. oxidation-reduction reactions

2. The energy stored in the carbon–carbon bonds of glucose is an example of _____ energy.
   a. kinetic
   b. potential
   c. chemical
   d. mechanical
   e. Both b and c are correct.

3. During energy transformations, the majority of energy is converted to
   a. chemical bonds.   c. ATP.
   b. heat.   d. glucose molecules.

6.2 Energy Transformations and Metabolism

4. Exergonic reactions
   a. are spontaneous.
   b. have a negative ΔG value.
   c. release energy.
   d. All of these are correct.

5. Which of the following is incorrect regarding ATP?
   a. It is the energy currency of the cell.
   b. It is stable.
   c. It is recycled using ADP and inorganic phosphate.
   d. Cells keep only small amounts of ATP on hand.

6. The sum of all the chemical reactions in a cell is called
   a. free energy.   c. metabolism.
   b. entropy.   d. oxidation-reduction reactions.

6.3 Enzymes and Metabolic Pathways

7. Which of the following is incorrect regarding the active site of an enzyme?
   a. is unique to that enzyme
   b. is the part of the enzyme where its substrate can fit
   c. can be used over and over again
   d. is not affected by environmental factors, such as pH and temperature

8. Which of the following environmental conditions may have an influence on enzyme activity?
   a. substrate concentration
   b. temperature
   c. pH
   d. All of these are correct.

9. In which of the following does an inhibitor bind to a site other than the active site of the enzyme?
   a. competitive inhibition
   b. noncompetitive inhibition
   c. redox reactions
   d. None of these are correct.

10. Enzymes catalyze chemical reactions by which of the following?
    a. lowering the energy of activation in the reaction
    b. raising the energy of activation in the reaction
    c. increasing entropy
    d. increasing the free energy of the products

6.4 Oxidation-Reduction Reactions and Metabolism

11. The gain of electrons by a molecule is called
    a. inhibition.
    b. entropy.
    c. oxidation.
    d. reduction.

12. In which of the following processes is carbon dioxide reduced to form carbohydrate?
    a. cellular respiration
    b. noncompetitive inhibition
    c. photosynthesis
    d. induced fit model

ENGAGE

BioNOW

Want to know how this science is relevant to your life? Check out the BioNOW video below:

■ Energy Part I : Energy Transfers

Explain how both laws of thermodynamics apply to the experiments in this video.

Thinking Critically

1. Entropy is increased when nutrients break down, so why are enzymatic metabolic pathways required for cellular respiration? (*Hint:* See Fig. 6.5.)

2. Plasminogen is an enzyme that needs to be converted into plasmin so it can dissolve a fibrin blood clot. Vampire bats produce an enzyme that is 150 times more potent at converting plasminogen to plasmin. Explain what information researchers would need to know about the vampire bat enzyme in order to synthesize it in the laboratory.

3. How might a knowledge of the structure of an active site of an enzyme allow you to build a drug to regulate a metabolic pathway?

CHAPTER

7

Cellular Respiration

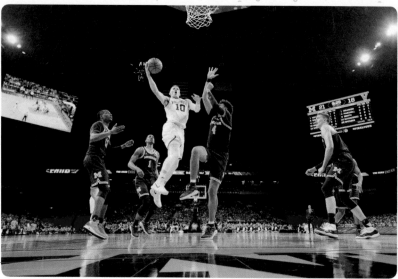

©Brett Wilhelm /NCCA Photos/Getty Images

BEFORE YOU BEGIN

Before beginning this chapter, take a few moments to review the following discussions:

Section 2.5 What is the role of carbohydrates in the body?

Section 3.3 What is the structure and function of the mitochondria in a cell?

Figure 6.3 How does the ATP cycle resemble a rechargeable battery?

Metabolic Demands on Athletes

During a typical basketball game, such as the 2018 NCAA championship game between Michigan and Villanova, the starting players run an average of 4–5 kilometers (around 3 miles) during a 40-minute game. However, unlike the endurance running experienced by marathoners, basketball players experience periods of intense activity (sprinting), followed by brief periods of rest. This start-and-stop nature of the game means that the muscles of the athlete are constantly switching between aerobic and anaerobic metabolism.

During aerobic metabolism, the muscle cells use oxygen in order to completely break down glucose, producing more ATP, a high-energy molecule used for muscle contraction, than otherwise. The breakdown of glucose with oxygen to produce carbon dioxide and water in the cytoplasm and mitochondria of the cell is called cellular respiration. However, running short, fast sprints quickly depletes oxygen levels and drives the muscles into anaerobic metabolism. Without oxygen, glucose cannot be broken down completely. It is changed into lactate, which is responsible for that muscle burn we sometimes feel after strenuous exercise. Once oxygen is restored to the muscles, the body is able to return to aerobic metabolism and dispose of the lactate.

In this chapter, we will discuss the metabolic pathways of cellular respiration that allow the energy within a glucose molecule, and other organic nutrients, to be converted into ATP.

As you read through this chapter, think about the following questions:

1. What are the differences between the aerobic and anaerobic pathways?

2. How is the energy of a glucose molecule harvested by a cell?

3. How are other organic nutrients, such as proteins and fats, used as energy?

7.1 Overview of Cellular Respiration

Learning Outcomes

Upon completion of this section, you should be able to

1. Describe the overall equation for cellular respiration.
2. Explain the role of electron carriers in respiration.
3. Summarize the phases of cellular respiration and indicate where they occur in a cell.

Cellular respiration is the process by which cells release the energy stored within organic molecules, such as glucose, and use this energy to synthesize ATP molecules. The reactions of cellular respiration can be summarized in the following formula:

$$C_6H_{12}O_6 + 6O_2 \longrightarrow 6CO_2 + 6H_2O$$

glucose oxygen   **carbon water
                                 dioxide**

Overall, cellular respiration is an aerobic process, meaning that it requires oxygen (O_2), and in the process, releases carbon dioxide (CO_2). Glucose is a high-energy molecule, but its breakdown products, CO_2 and H_2O, are low-energy molecules. Therefore, as the equation shows, energy is released. This is the energy that will be used to produce ATP molecules. The cell carries out cellular respiration in order to build up ATP molecules.

The pathways of cellular respiration allow the energy within a glucose molecule to be released slowly so that ATP can be produced gradually. Cells would lose a tremendous amount of energy if glucose breakdown occurred all at once—most of the energy would become nonusable heat. The step-by-step breakdown of glucose to CO_2 and H_2O usually produces a maximum yield of 36 to 38 ATP molecules, although the yield is usually less. The energy in these ATP molecules is equivalent to about 39% of the energy that was available in glucose. Even though it might seem less efficient, this conversion is more efficient than many others; for example, only between 20% and 30% of the energy within gasoline is converted to the motion of a car.

The cellular respiration pathways allow the energy within a glucose molecule to be released slowly so the ATP can be produced gradually. The energy in the ATP can be used for most cellular reactions, such as joining one amino acid to another during protein synthesis or joining actin to myosin during muscle contraction. This energy is released by a relatively simple procedure: the removal of a single phosphate group. The cell uses the reverse reaction for building up ATP again.

The NAD$^+$ and FAD Coenzymes

Cellular respiration involves many individual metabolic reactions, each one catalyzed by its own enzyme. Enzymes of particular significance are those that use **NAD$^+$ (nicotinamide adenine dinucleotide),** a coenzyme used in oxidation-reduction reactions). When a metabolite is oxidized, NAD$^+$ accepts two electrons

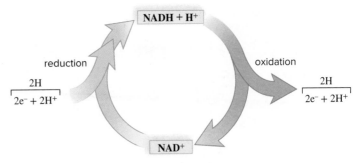

Figure 7.1 The NAD$^+$ cycle. The coenzyme NAD$^+$ accepts two hydrogen atoms (H$^+$ + e$^-$), and NADH + H$^+$ results. When NADH passes on electrons, NAD$^+$ results. Only a small amount of NAD$^+$ need be present in a cell, because each NAD$^+$ molecule is recycled.

plus a hydrogen ion (H$^+$), and NADH results (Fig 7.1). The electrons received by NAD$^+$ are high-energy electrons that are usually carried to the electron transport chain:

Only a small amount of NAD$^+$ needs to be present in a cell, because each NAD$^+$ molecule is used over and over again. **FAD (flavin adenine dinucleotide),** another coenzyme of oxidation-reduction, is sometimes used instead of NAD$^+$. FAD accepts two electrons and two hydrogen ions (H$^+$) to become FADH$_2$.

NAD$^+$ and FAD pick up electrons at specific points in the cellular respiration pathways and then carry these high-energy electrons to an electron transport chain in mitochondria, where they drop them off. The empty NAD$^+$ or FAD is then free to go back and pick up more electrons.

Phases of Cellular Respiration

The metabolic pathways of cellular respiration couple the release of energy within a glucose molecule to the production of ATP. The coupling of these reactions reduces the amount of energy lost as heat, which would be significant if glucose breakdown occurred all at once. Cellular respiration involves four phases (Fig. 7.2).

The first phase, glycolysis, takes place outside the mitochondria and does not utilize oxygen. Therefore, glycolysis is **anaerobic.** The other phases take place inside the mitochondria, where oxygen acts as the final acceptor of electrons. These phases are **aerobic.** The four phases include:

- Glycolysis is the breakdown of glucose ($C_6H_{12}O_6$) to two molecules of pyruvate, a C_3 molecule. Oxidation by removal of electrons (e^-) and hydrogen ions (H$^+$) provides enough energy for the immediate buildup of two ATP.
- During the preparatory (prep) reaction, pyruvate is oxidized to a C_2 acetyl group carried by CoA (coenzyme A), and CO_2 is removed. Because glycolysis ends with two molecules of pyruvate, the prep reaction occurs twice per glucose molecule.
- The citric acid cycle is a cyclical series of oxidation reactions that give off CO_2 and produce one ATP. The citric acid cycle used to be called the Krebs cycle in honor of the researcher who worked out most of the steps. The cycle is now named for citric acid (or citrate), the first molecule in the cycle. The citric acid cycle turns twice because two acetyl CoA molecules enter the cycle per glucose molecule. Altogether, the citric

Figure 7.2 Overview of Cellular Respiration. The complete breakdown of glucose during cellular respiration consists of four phases. Glycolysis in the cytoplasm produces pyruvate, which enters mitochondria if oxygen is available. The preparatory reaction and the citric acid cycle that follow occur inside the mitochondria. Also inside mitochondria, the electron transport chain receives the electrons removed from glucose breakdown products. The result of glucose breakdown is a maximum of 36 to 38 ATP, depending on the particular cell.

acid cycle accounts for two immediate ATP molecules per glucose molecule.

■ The electron transport chain is a series of membrane-bound carriers that pass electrons from one carrier to another. High-energy electrons are delivered to the chain, and low-energy electrons leave it (Fig. 7.3). The electron transport chain is like a flight of stairs. As something bounces down the stairs it loses potential energy. Similarly, as electrons pass through the carriers from a higher-energy to a lower-energy state, energy is released and used for ATP synthesis. The electrons from one glucose molecule passing down the electron transport chain contribute to the production of the majority of the ATP (between 32 and 34, depending on the type of cell).

In cellular respiration, the low-energy electrons are finally received by O_2, which then combines with H^+ and becomes water. Pyruvate is a pivotal metabolite in cellular respiration. If oxygen is not available to the cell, fermentation occurs in the cytoplasm instead of continued aerobic (with oxygen) cellular respiration. During fermentation, pyruvate is reduced to lactate or to carbon dioxide and alcohol, depending on the organism.

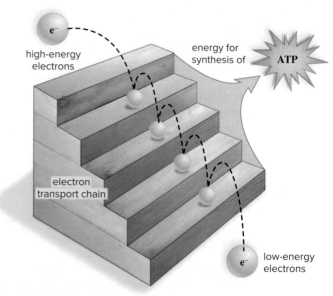

Figure 7.3 Electron transport chain. High-energy electrons are delivered to the chain. As they pass from carrier to carrier, energy is released and used for ATP production.

7.2 Outside the Mitochondria: Glycolysis

Learning Outcomes

Upon completion of this section, you should be able to
1. Describe the location and inputs and outputs of glycolysis.
2. Explain why ATP is both an input and output of glycolysis.

Glycolysis is the initial stage of cellular respiration (Fig. 7.4). It takes place within the cytoplasm, and results in the breakdown of glucose into two pyruvate molecules and a small amount of energy. Because glycolysis occurs universally in all organisms, it most likely evolved before the citric acid cycle and the electron transport chain. Glycolysis most likely evolved when environmental conditions were anaerobic and before cells had mitochondria. This may be why glycolysis does not require oxygen and occurs in the cytoplasm.

Energy-Investment Steps

As glycolysis begins, two ATP are used to activate glucose, and the molecule that results splits into two 3-carbon (C₃) molecules (*G3P*), each of which has an attached phosphate group. From

this point on, each C_3 molecule undergoes the same series of reactions (Fig. 7.5).

Energy-Harvesting Steps

Oxidation of G3P occurs by the removal of hydrogen atoms (H⁺ + e⁻). The hydrogen atoms are picked up by NAD⁺, and NADH + H⁺ results. Later, NADH will pass electrons on to the electron transport chain. Oxidation of G3P and subsequent substrates results in four high-energy phosphate groups, which are used to synthesize four ATP. This is **substrate-level ATP synthesis,** in which an enzyme passes a high-energy phosphate to ADP, and ATP results:

Subtracting the two ATP that were used to get started, glycolysis yields a net gain of two ATP (Fig. 7.5).

Inputs and Outputs of Glycolysis

All together, the inputs and outputs of glycolysis are as follows:

Notice that, so far, we have accounted for only 2 of the 36 to 38 ATP molecules theoretically possible when glucose is completely broken down. When oxygen is available, the end product, pyruvate, enters the mitochondria, where it undergoes further breakdown. If oxygen is not available, pyruvate can be used for fermentation, as described in Section 7.3. For each glucose that enters glycolysis, two ATP, two NADH + H⁺, and two pyruvate are formed.

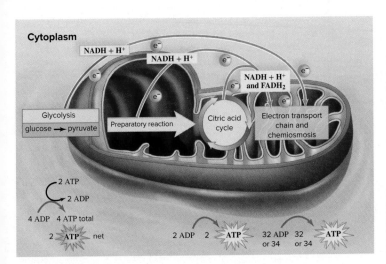

Figure 7.4 Overview of glycolysis. Glycolysis is the first stage of cellular respiration and occurs in the cytoplasm of the cell.

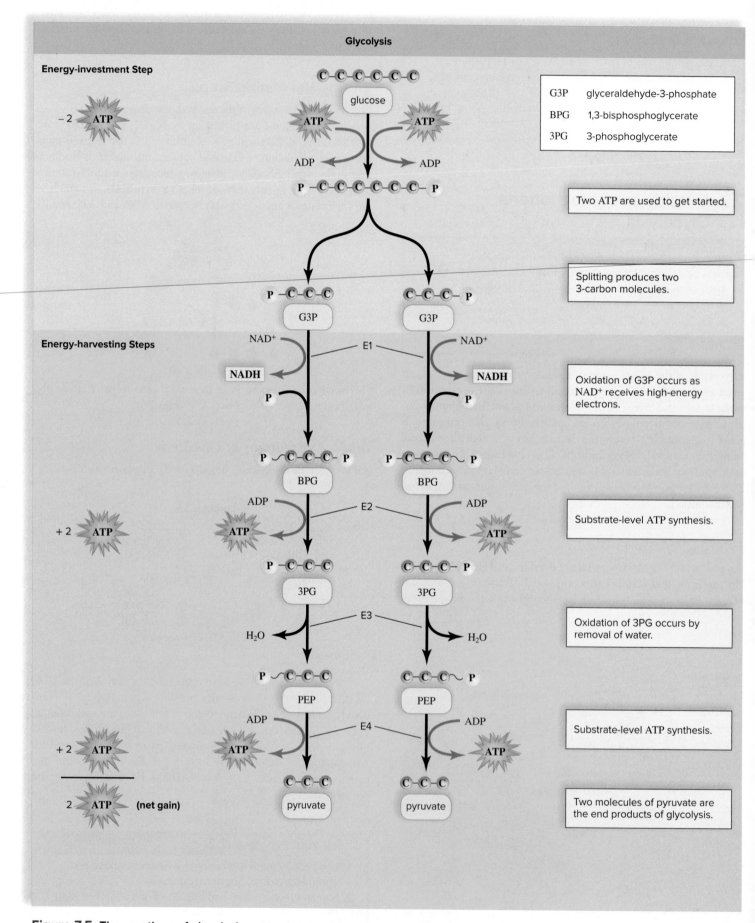

Figure 7.5 The reactions of glycolysis.   Glycolysis begins with glucose and ends with two pyruvate molecules. There is a gain of two NADH + H⁺, and a net gain of two ATP from glycolysis.

7.3 Outside the Mitochondria: Fermentation

Learning Outcomes

Upon completion of this section, you should be able to
1. Explain how ATP can continue to be produced in the absence of oxygen.
2. Describe the advantages and disadvantages of fermentation.

The complete breakdown of glucose requires an input of oxygen to keep the electron transport chain working. If oxygen is limited, pyruvate molecules accumulate in the cell, and intermediates, such as NAD⁺ and FAD, cannot be recycled. To correct for this, cells may utilize anaerobic pathways, such as **fermentation.** There are two basic forms of fermentation—lactic acid fermentation and alcohol fermentation.

Lactic Acid Fermentation

If human cells, or animal cells in general, experience a low-oxygen environment, the pyruvate formed by glycolysis is reduced to form lactate, which is converted to lactic acid in the water environment of the cell.

The electrons needed to reduce pyruvate to lactic acid are supplied by the NADH molecules from glycolysis (Fig. 7.6). Normally, NADH contributes its electrons to the electron transport chain and is then recycled as NAD⁺ for use in glycolysis (see Fig. 7.2). But in the absence of oxygen, the NAD⁺ is regenerated when pyruvate is reduced to lactic acid. Therefore, the lactic acid pathway helps to recycle NAD⁺ molecules and allow glycolysis to proceed with ATP production.

Notice that the fermentation pathway yields only 2 ATP per glucose molecule. Despite the low amounts of ATP produced, fermentation is essential to humans. It is commonly used as a pathway in muscle cells. When our muscles are working vigorously over a short period of time, as when we run, fermentation is a way to produce ATP even though oxygen is temporarily in limited supply.

Lactate, however, is toxic to cells. At first, blood carries away all the lactate formed in muscles. But eventually lactate begins to build up, lowering the pH and causing the muscles to "burn." When we stop running, our bodies are in *oxygen debt,* as signified by the fact that we continue to breathe very heavily for a time. Recovery is complete when the lactate is transported to the liver, where it is reconverted to pyruvate. Some of the pyruvate reenters the cellular respiration pathways, and the rest is converted back to glucose.

Alcohol Fermentation

In other organisms, the pyruvate is reduced to produce alcohol. As was the case with lactic acid fermentation, the electrons needed to reduce the pyruvate are supplied by NADH molecules. In the process, NAD⁺ molecules are regenerated for use in glycolysis (Fig. 7.6). However, unlike lactic acid fermentation, alcohol fermentation releases small amounts of CO₂.

Yeasts are good examples of organisms that use alcohol fermentation. Yeasts are eukaryotic organisms, and thus can carry out

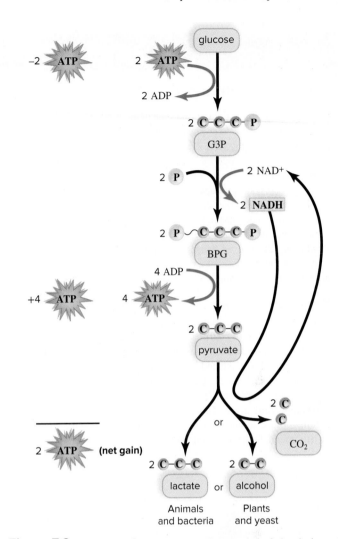

Figure 7.6 Fermentation. Fermentation consists of glycolysis followed by a reduction of pyruvate by NADH + H⁺. The resulting NAD⁺ returns to the glycolytic pathway to pick up more hydrogen atoms.

cellular respiration. But when they are in an anaerobic environment, they can use the fermentation process to produce small amounts of ATP. In the process, they generate ethyl alcohol and CO₂ as a result of fermentation. When yeast is used to leaven bread, the CO₂ produced makes bread rise. Fermentation produces the ethanol found in many alcoholic beverages, from beer to wine. However, it is important to note that the alcohol produced is a waste product for the yeast; it is not used as an energy source. Eventually, yeasts are killed by the very alcohol they produce. Other organisms, such as the bacteria, vary as to whether they produce an organic acid, such as lactate, or an alcohol and CO₂. The Scientific Inquiry feature "Fermentation and Food" gives more information on how fermentation produces many of the food items you consume on a daily basis.

Energy Yield of Fermentation

The ATP produced during fermentation are actually the products of glycolysis. This is because fermentation follows glycolysis (Fig. 7.5). Therefore, the anaerobic pathways produce only two ATP by substrate-level ATP synthesis. These two ATP represent

Fermentation and Food

At the grocery store, you will find such items as bread, yogurt, soy sauce, pickles, and maybe even beer or wine (Fig. 7A). These are just a few of the many foods produced when microorganisms ferment (break down sugar in the absence of oxygen). Foods produced by fermentation last longer, because the fermenting organisms have removed many of the nutrients that would attract other organisms. The products of fermentation can even be dangerous to the very organisms that produced them, as when yeast are killed by the alcohol they produce.

Yeast Fermentation

Baker's yeast, *Saccharomyces cerevisiae,* is added to bread for the purpose of leavening—the dough rises when the yeast give off CO_2. The ethyl alcohol produced by the fermenting yeast evaporates during baking. The many varieties of sourdough breads obtain their leavening from a starter composed of fermenting yeast, along with bacteria from the environment. Depending on the community of microorganisms in the starter, the flavor of the bread may range from sour and tangy, as in San Francisco–style sourdough, to a milder taste, such as that produced by most Amish friendship bread recipes.

Ethyl alcohol in beer and wine is produced when yeast ferment carbohydrates. When yeast ferment fruit carbohydrates, the end result is wine. If they ferment grain, beer results. A few specialized varieties of beer, such as traditional wheat beers, have a distinctive, sour taste because they are produced with the assistance of lactic acid–producing bacteria, such as those of the genus *Lactobacillus.* Stronger alcoholic drinks (e.g., whiskey and vodka) require distillation to concentrate the alcohol content.

Bacteria that produce acetic acid, including *Acetobacter aceti,* spoil wine. These bacteria convert the alcohol in wine or cider to acetic acid (vinegar). Until the renowned nineteenth-century scientist Louis Pasteur invented the process of pasteurization, acetic acid bacteria commonly caused wine to spoil. Although today we generally associate the process of pasteurization with making milk safe to drink, it was originally developed to reduce bacterial contamination in wine, so that limited acetic acid would be produced. The discovery of pasteurization is another example of how the pursuit of scientific knowledge can positively affect our lives.

Bacterial Fermentation

Yogurt, sour cream, and cheese are produced through the action of various lactic acid bacteria that cause milk to sour. Milk contains lactose, which these bacteria use as a carbohydrate source for fermentation. Yogurt, for example, is made by adding lactic acid bacteria, such as *Streptococcus thermophilus* and *Lactobacillus bulgaricus,* to milk and then incubating it to encourage the bacteria to convert the lactose. During the production of cheese, an enzyme called rennin must also be added to the milk to cause it to coagulate and become solid.

Old-fashioned brine cucumber pickles, sauerkraut, and kimchi are pickled vegetables produced by the action of acid-producing, fermenting bacteria that can survive in high-salt environments. Salt is used to draw liquid out of the vegetables and to aid in their preservation. The bacteria need not be added to the vegetables, because they are already present on the surfaces of the plants.

Soy Sauce Production

Soy sauce is traditionally made by adding a mold, *Aspergillus,* and a combination of yeast and fermenting bacteria to soybeans and wheat. The mold breaks down starch, supplying the fermenting microorganisms with sugar they can use to produce alcohol and organic acids.

As you can see from each of these examples, fermentation is a biologically and economically important process that scientists use for the betterment of our lives.

Questions to Consider

1. How would the production of fermentation products differ from that of other food products?
2. What products of fermentation do you use on a daily basis?

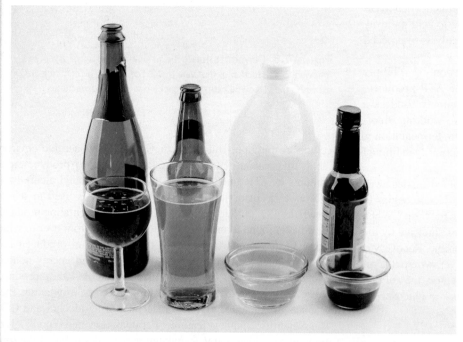

Figure 7A Products from fermentation.   Fermentation of different carbohydrates by microorganisms like bacteria and yeast helps produce the products shown. ©Ricochet Creative Productions LLC only a small fraction of the potential energy stored in a glucose molecule. As noted earlier, complete glucose breakdown during cellular respiration results in a maximum of 36 to 38 ATP. Therefore, following fermentation, most of the potential energy a cell can capture from the respiration of a glucose molecule is still waiting to be released.

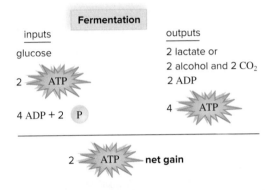

Fermentation

inputs	outputs
glucose	2 lactate or
	2 alcohol and 2 CO_2
2 ATP	2 ADP
4 ADP + 2 P	4 ATP

2 ATP **net gain**

Check Your Progress 7.3

1. Describe the environmental conditions that would cause a muscle cell to undergo fermentation.

2. Explain how fermentation acts as a NAD^+ recycling system.

7.4 Inside the Mitochondria

Learning Outcomes

Upon completion of this section, you should be able to

1. Recognize the role of the mitochondria in cellular respiration.

2. Summarize the inputs and outputs of the preparatory reaction, the citric acid cycle, and the electron transport chain.

3. Identify how each stage of the aerobic pathway contributes to the generation of ATP in a cell.

The remaining reactions of cellular respiration—the preparatory reaction, the citric acid cycle, and the electron transport chain—all occur within the mitochondria (Fig. 7.7). The structure of a mitochondrion includes folds of an inner membrane, called the *cristae,* that jut out into the *matrix,* an innermost compartment filled with a gel-like fluid. The preparatory reaction and the citric acid cycle are located in the matrix, while the electron transport chain occurs along the cristae.

Preparatory Reaction

The **preparatory (prep) reaction** is named because it produces the molecule that can enter the citric acid cycle. In this reaction, pyruvate is converted to a 2-carbon (C_2) *acetyl group* attached to *coenzyme A (CoA)*, and CO_2 is given off. This is an oxidation reaction in which hydrogen atoms ($H^+ + e^-$) are removed from pyruvate by NAD^+, and $NADH + H^+$ results. This reaction occurs twice per glucose molecule:

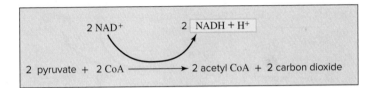

2 NAD^+ → 2 $NADH + H^+$

2 pyruvate + 2 CoA ⟶ 2 acetyl CoA + 2 carbon dioxide

Citric Acid Cycle

The **citric acid cycle** (also called the Krebs cycle) is a cyclical metabolic pathway located in the matrix of mitochondria (Fig. 7.8). At the start of the citric acid cycle, the C_2 acetyl group carried by CoA joins with a C_4 molecule, and a C_6 citrate molecule results. Note that because there are two acetyl CoA molecules entering the cycle for each glucose, the cycle will turn twice.

The CoA returns to the preparatory reaction to pick up another acetyl group. During the citric acid cycle, each acetyl group received from the preparatory reaction is oxidized to two CO_2 molecules. As the reactions of the cycle occur, oxidation is carried out by the removal of hydrogen atoms ($H^+ + e^-$). In three

outer membrane

inner membrane

intermembrane space

matrix

cristae

Cristae: location of the electron transport chain (ETC)

Matrix: location of the prep reaction and the citric acid cycle

Figure 7.7 Structure of a mitochondrion.   Mitochondria are double-membrane organelles. The location of the reactions of cellular respiration are shown here.

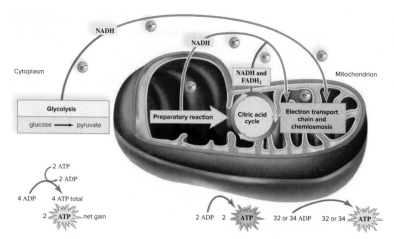

Figure 7.8 Citric acid cycle. The net result of this cycle of reactions is the oxidation of an acetyl group to two molecules of CO_2, along with a transfer of electrons to NAD^+ and FAD and a gain of one ATP. The citric acid cycle turns twice per glucose molecule.

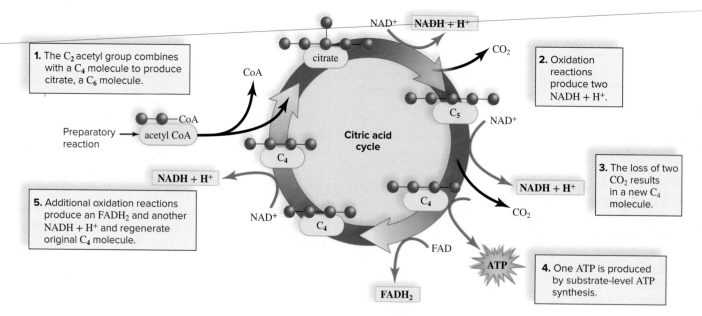

1. The C_2 acetyl group combines with a C_4 molecule to produce citrate, a C_6 molecule.

2. Oxidation reactions produce two $NADH + H^+$.

3. The loss of two CO_2 results in a new C_4 molecule.

4. One ATP is produced by substrate-level ATP synthesis.

5. Additional oxidation reactions produce an $FADH_2$ and another $NADH + H^+$ and regenerate original C_4 molecule.

instances, $NADH + H^+$ results, and in one instance, $FADH_2$ is formed.

Substrate-level ATP synthesis is also an important event of the citric acid cycle. In substrate-level ATP synthesis, an enzyme passes a high-energy phosphate to ADP, and ATP results.

Because the citric acid cycle turns twice for each original glucose molecule, the inputs and outputs of the citric acid cycle per glucose molecule are as follows:

Citric acid cycle

inputs	outputs
2 acetyl groups	4 CO_2
6 NAD^+	6 $NADH + H^+$
2 FAD	2 $FADH_2$
2 ADP + 2 ℗	2 ATP

The six carbon atoms originally located in the glucose molecule have now become CO_2. The preparatory reaction produces two CO_2, and the citric acid cycle produces four CO_2 per glucose molecule.

The Electron Transport Chain

The **electron transport chain (ETC),** located in the cristae of the mitochondria, is a series of protein carriers that pass electrons from one to the other. Some of the electron carriers of the system are called *cytochrome* molecules, which as iron-containing molecules play an important role in redox reactions.

The electron transport chain receives NADH and $FADH_2$ molecules from glycolysis, the preparatory reactions, and the citric acid cycle (Fig. 7.9a). Its role in the process of cellular respiration is to remove the high-energy electrons from these molecules (thus oxidizing them) and then harvest the energy in these electrons to form ATP. Oxygen is used in the electron transport chain as a final electron acceptor.

The electron transport chain is arranged so that high-energy electrons enter the system and low-energy electrons leave the system (Fig. 7.9b). Notice that when NADH gives up its electrons to the chain, it releases H^+ ions and becomes NAD^+. Similarly, when $FADH_2$ gives up its electrons to the chain, it becomes FAD and releases H^+ ions. The next carrier gains the electrons and is reduced. Then, each of the carriers, in turn, becomes reduced and then oxidized as the electrons move down the system.

Figure 7.9 Cellular organization of the electron transport chain. **a.** The electron transport chain is the final stage of cellular respiration. **b.** The pathway is located in the cristae of the mitochondria. As electrons move from one protein complex to the other, hydrogen ions (H⁺) are pumped from the mitochondrial matrix into the intermembrane space. As hydrogen ions flow down a concentration gradient from the intermembrane space into the matrix, ATP is synthesized by the enzyme ATP synthase. ATP leaves the matrix by way of a channel protein.

As the electrons pass from one carrier to the next, energy that will be used to produce ATP molecules is captured and stored as a hydrogen ion gradient (Fig. 7.9*b*). Oxygen receives the energy-spent electrons from the last of the carriers. After receiving electrons, oxygen combines with hydrogen ions and forms water.

Once NADH has delivered electrons to the electron transport chain, the NAD⁺ that results can pick up more hydrogen atoms. In like manner, once FADH₂ gives up electrons to the chain, the FAD that results can pick up more hydrogen atoms. The recycling of coenzymes and ADP increases cellular efficiency because it does away with the necessity to synthesize NAD⁺ and FAD every time.

Figure 7.10 The energy yield of cellular respiration.   This diagram illustrates the energy yield per glucose molecule. Substrate-level ATP synthesis during glycolysis and the citric acid cycle accounts for four ATP. Chemiosmosis accounts for a maximum of 32 or 34 ATP, depending on the shuttle mechanism involved in transporting cytoplasmic $NADH + H^+$ into the mitochondrion. The theoretical maximum total of ATP is therefore 36 to 38 ATP, although most cells produce fewer ATP.

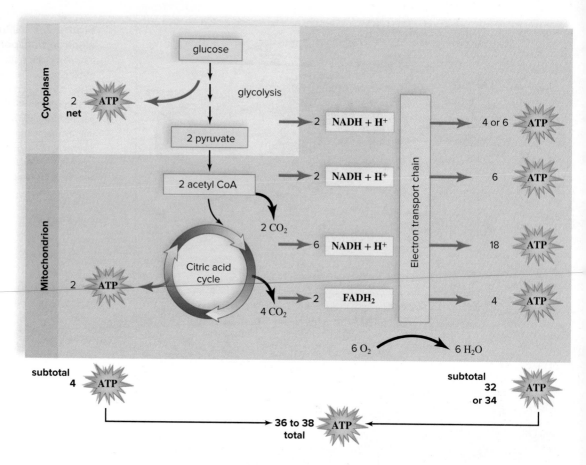

Generating ATP

We have been stressing that the carriers of the electron transport chain accept electrons, which they pass from one to the other. What happens to the hydrogen (H^+) ions from $NADH + H^+$? The complexes in the cristae use the energy released by electrons as they move down the electron transport chain to pump H^+ from the mitochondrial matrix into the space between the outer and inner membrane of a mitochondrion. This space is called the *intermembrane space*. The pumping of H^+ into the intermembrane space establishes an unequal distribution of H^+ ions; there are many H^+ in the intermembrane space and few in the matrix of a mitochondrion. This means the intermembrane space is positively charged in relation to the matrix, as well as more acidic.

The cristae also contain an *ATP synthase complex*. The H^+ ions flow through an ATP synthase complex from the intermembrane space into the matrix. The flow of H^+ through an ATP synthase complex brings about a change in shape, which causes the enzyme ATP synthase to synthesize ATP from ADP + ℗. Mitochondria produce ATP by **chemiosmosis,** which indicates that ATP production is tied to an electrochemical gradient, namely the unequal distribution of H^+ across the cristae. Once formed, ATP molecules pass into the cytoplasm.

Most ATP is produced by the electron transport chain and chemiosmosis. Per glucose molecule, ten NADH and two $FADH_2$ take electrons to the electron transport chain. For each NADH formed *inside* the mitochondria by the citric acid cycle, three ATP result, but for each $FADH_2$, only two ATP are produced. Figure 7.9*b* explains the reason for this difference: $FADH_2$ delivers its

electrons to the transport chain after NADH, and therefore these electrons do not participate in as many redox reactions and don't pump as many H^+ as NADH. Therefore, $FADH_2$ cannot account for as much ATP production.

Chemiosmosis is similar to using water behind a dam to generate electricity. The pumping of H^+ out of the matrix into the intermembrane space is like pumping water behind the dam. The floodgates of the dam are like the ATP synthase complex. When the floodgates are open, the water rushes through, generating electricity. In the same way, H^+ rushing through the ATP synthase complex is used to produce ATP.

Energy Yield from Cellular Respiration

Figure 7.10 provides the theoretical ATP for each stage of cellular respiration. However, we know now that cells rarely ever achieve these theoretical values. Several factors can lower the ATP yield for each molecule of glucose entering the pathway:

- In some cells, NADH cannot cross mitochondrial membranes, but a "shuttle" mechanism allows its electrons to be delivered to the electron transport chain inside the mitochondria. The cost to the cell is one ATP for each NADH that is shuttled to the ETC. This reduces the overall count of ATP produced as a result of glycolysis, in some cells, to four instead of six ATP.

- At times, cells need to expend energy to move ADP molecules and pyruvate into the cell and to establish protein gradients in the mitochondria.

Metabolic Fate of Pizza

Obviously, our diets do not solely consist of carbohydrates. Because fats and proteins are also organic nutrients, it makes sense that our bodies can utilize the energy found in the bonds of these molecules. In fact, the metabolic pathways we have discussed in this chapter are more than capable of accessing the energy of fats and proteins. For example, let's trace the fate of a pepperoni pizza, which contains carbohydrates (crust), fats (cheese), and protein (pepperoni).

We already know that the glucose in the carbohydrate crust is broken down during cellular respiration. When the cheese in the pizza (a fat) is used as an energy source, it breaks down to glycerol and three fatty acids. As Figure 7B indicates, glycerol can be converted to pyruvate and enter glycolysis. The fatty acids are converted to 2-carbon acetyl CoA that enters the citric acid cycle. An 18-carbon fatty acid results in nine acetyl CoA molecules. Calculation shows that respiration of these can produce a total of 108 ATP molecules. This is why fats are an efficient form of stored energy—the three long fatty acid chains per fat molecule can produce considerable ATP when needed.

Proteins are less frequently used as an energy source, but are available as necessary. The carbon skeleton of amino acids can enter glycolysis, be converted to acetyl groups, or enter the citric acid cycle at some other juncture. The carbon skeleton is produced in the liver when an amino acid undergoes deamination, or the removal of the amino group. The amino group becomes ammonia (NH_3), which enters the urea cycle and becomes part of urea, the primary excretory product of humans. Just where the carbon skeleton begins degradation depends on the length of the R group, because this determines the number of carbons left after deamination.

In Section 14.4, we will take a more detailed look at the nutritional needs of humans, including discussions on how vitamins and minerals interact with metabolic pathways, and the dietary guidelines for proteins, fats, and carbohydrates.

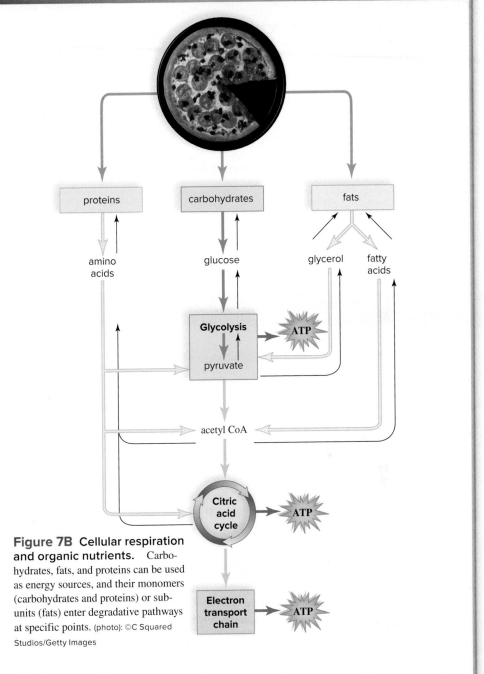

Figure 7B Cellular respiration and organic nutrients.   Carbohydrates, fats, and proteins can be used as energy sources, and their monomers (carbohydrates and proteins) or subunits (fats) enter degradative pathways at specific points. (photo): ©C Squared Studios/Getty Images

Questions to Consider

1. How might a meal of a cheeseburger and fries be processed by the cellular respiration pathways?

2. While Figure 7B does not indicate the need for water, it is an important component of our diet. Where would water interact with these pathways?

There is still considerable research into the precise ATP yield per glucose molecule. However, most estimates place the actual yield at around 30 ATP per glucose. Using this number, we can calculate that only between 32% and 39% of the available energy is usually transferred from glucose to ATP. The rest of the energy is lost in the form of heat.

It is also important to note that while glucose is the preferred energy source for the cellular respiration pathways, our diets do not consist entirely of carbohydrates. The Health feature "Metabolic Fate of Pizza" explores how proteins and lipids (fats) may be used to generate energy by cellular respiration.

Check Your Progress 7.4

1. Explain the relationship between the metabolic pathways within the mitochondria and glycolysis.
2. Calculate the number of NADH, $FADH_2$, and ATP molecules produced by each stage of cellular respiration per glucose molecule.
3. Discuss why there is variation in the number of ATP molecules produced per glucose.

Conclusion

Whether you are playing basketball, studying for an exam, or even sleeping, your metabolic pathways have the ability to break down energy nutrients, primarily carbohydrates, and produce ATP. The specific pathway followed is dependent on the availability of oxygen. During aerobic exercise, oxygen is available for the muscles to completely break down glucose to carbon dioxide and water. If oxygen supplies are low, then muscle cells may switch to fermentation pathways to produce small amounts of ATP to sustain muscle function.

SUMMARIZE

7.1 Overview of Cellular Respiration

■ The metabolic pathways that produce ATP may be either **aerobic** or **anaerobic,** depending on the availability of oxygen.

■ During aerobic **cellular respiration,** glucose is oxidized to CO_2 and H_2O. Four phases are required for glucose to be metabolized to carbon dioxide and water. These include glycolysis, the preparatory reactions, the citric acid cycle, and the electron transport chain. During oxidation of substrates, electrons are removed along with hydrogen ions (H^+). **NAD$^+$ (nicotinamide adenine dinucleotide)** and **FAD (flavin adenine dinucleotide)** act as electron carriers. In the absence of oxygen, fermentation reactions occur.

7.2 Outside the Mitochondria: Glycolysis

■ **Glycolysis,** the breakdown of glucose to two pyruvates, is a series of anaerobic enzymatic reactions that occur in the cytoplasm.

■ Oxidation by NAD$^+$ releases enough energy immediately to give a net gain of two ATP by **substrate-level ATP synthesis.** Two NADH + H^+ are formed.

7.3 Outside the Mitochondria: Fermentation

■ **Fermentation** involves glycolysis, followed by the reduction of pyruvate by NADH + H^+ to either lactate or alcohol and CO_2. The reduction process regenerates NAD$^+$ so it can accept more electrons during glycolysis.

■ Although fermentation results in only two ATP, it still serves a purpose: In humans, it provides ATP energy for short-term, strenuous muscular activity. The accumulation of lactate puts the individual in oxygen debt because oxygen is needed to completely metabolize lactate to CO_2 and H_2O.

7.4 Inside the Mitochondria

■ Pyruvate from glycolysis enters a mitochondrion, where the **preparatory (prep) reaction** takes place. During this reaction, oxidation

occurs as CO_2 is removed. NAD$^+$ and FAD are reduced, and CoA receives the C_2 acetyl group that remains. Because the reaction must take place twice per glucose, two NADH + H^+ result.

■ The acetyl group enters the **citric acid cycle,** a series of reactions located in the mitochondrial matrix. Complete oxidation follows, as two CO_2, three NADH + H^+, and one $FADH_2$ are formed. The cycle also produces one ATP. The entire cycle must turn twice per glucose molecule.

■ The final stage of glucose breakdown involves the **electron transport chain** located in the cristae of the mitochondria. The electrons received from NADH and $FADH_2$ are passed down a chain of electron carriers until they are finally received by O_2, which combines with H^+ to produce H_2O. As the electrons pass down the chain, ATP is produced.

■ The carriers of the electron transport chain are located in protein complexes on the cristae of the mitochondria. Each protein complex receives electrons and pumps H^+ into the intermembrane space, setting up an electrochemical gradient. When H^+ ions flow down this gradient through the ATP synthase complex, energy is released and used to form ATP molecules from ADP and ℗. This is ATP synthesis by **chemiosmosis.**

■ To calculate the total number of ATP produced per glucose molecule, consider that for each NADH + H^+ formed inside the mitochondrion, three ATP are produced. Each molecule of $FADH_2$ results in the formation of only two ATP because the electrons enter the electron transport chain at a lower energy level than the electrons delivered by NADH. In cells that cannot transfer NADH across the mitochondrial membrane, the ATP counts are reduced by two because the electrons from NADH generated in the cytoplasm during glycolysis must be transferred from outside to inside the mitochondria, using two ATP.

■ Of the maximum number of 36 or 38 ATP formed by cellular respiration, 4 are produced outside the electron transport chain: 2 by glycolysis and 2 by the citric acid cycle. The rest are produced by the electron transport chain. The actual yield is closer to 30–32 ATP.

ASSESS

Testing Yourself

Choose the best answer for each question.

7.1 Overview of Cellular Respiration

For questions 1–5, match the letters in the diagram to the appropriate statement.

Key:

1. Preparatory reaction
2. Electron transport chain
3. Glycolysis
4. Citric acid cycle
5. NADH + H$^+$ and FADH$_2$

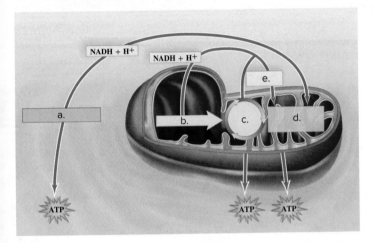

7.2 Outside the Mitochondria: Glycolysis

6. During glycolysis, what is the net production of ATP per glucose molecule?
   a. 0
   b. 1
   c. 2
   d. 8
   e. 32–34
7. Which of the following is not a product or reactant of glycolysis?
   a. NADH
   b. ATP
   c. pyruvate
   d. oxygen

7.3 Outside the Mitochondria: Fermentation

8. Which of these is incorrect regarding fermentation?
   a. There is a net gain of only two ATP per glucose.
   b. It occurs in the cytoplasm.
   c. The process starts with glucose entering glycolysis.
   d. NADH donates electrons to the electron transport chain.
   e. It is anaerobic.

9. Fermentation is primarily involved in the recycling of
   a. ADP.
   b. oxygen.
   c. pyruvate.
   d. NAD$^+$.

7.4 Inside the Mitochondria

10. The greatest contributor of electrons to the electron transport chain is
    a. oxygen.
    b. the prep reaction.
    c. glycolysis.
    d. fermentation.
    e. the citric acid cycle.
11. Which of these is not true of the citric acid cycle?
    a. The citric acid cycle includes the prep reaction.
    b. The citric acid cycle produces ATP by substrate-level ATP synthesis.
    c. The citric acid cycle occurs in the mitochondria.
    d. The citric acid cycle produces two ATP per glucose molecule.
12. Which of these is not true of the electron transport chain?
    a. The electron transport chain is located on the cristae of the mitochindria.
    b. The electron transport chain produces more NADH than any metabolic pathway.
    c. The electron transport chain contains cytochrome molecules.
    d. The electron transport chain ends when oxygen accepts electrons.
13. The oxygen required by cellular respiration is reduced and becomes part of which molecule?
    a. ATP
    b. pyruvate
    c. H$_2$O
    d. CO$_2$

ENGAGE

BioNOW

Want to know how this science is relevant to your life? Check out the BioNOW video below:

- Energy Part III: Cellular Respiration

How does this experiment relate to the relationship between autotrophs and heterotrophs on a global scale?

Thinking Critically

1. Different varieties of beer and wine contain different amounts of alcohol produced through fermentation by yeast. Among the reactants required for fermentation, which will determine the amount of alcohol produced?
2. Rotenone is a broad-spectrum insecticide that inhibits the electron transport chain. Why might it be toxic to humans?
3. Some fat-burning compounds accelerate the movement of fatty acids into the cellular respiration pathways. Explain how these compounds may work.

CHAPTER

8

Photosynthesis

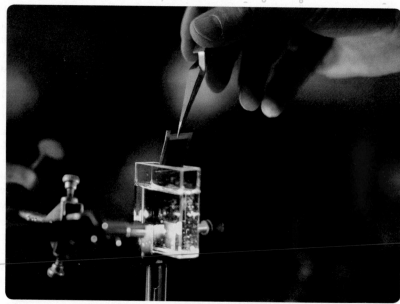

©Deanne Fitzmaurice/National Geographic Creative

BEFORE YOU BEGIN

Before beginning this chapter, take a few moments to review the following discussions:

Section 3.3 Which cellular structures are necessary for photosynthesis?

Figure 6.1 How does energy flow in biological systems?

Section 6.3 What role do enzymes play in regulating metabolic processes?

Artificial Leaves May Power the Future

In nature, the leaves of plants act as metabolic factories that use the energy from the sun, and carbon dioxide from the atmosphere, to manufacture carbohydrates—the basis of all life on Earth.

Scientists have been interested for some time in the development of an artificial leaf that would mimic the natural processes. Various models have been built that model a natural leaf, but usually at a very low efficiency. Normally, a leaf is capable of processing about 1% of the solar energy that strikes its surface. While this may not seem efficient, the sheer number of leaves on a plant, and the ample supply of sunlight, provide enough energy for a plant to manufacture the carbohydrates it needs.

Recently, a Harvard research team developed an artificial leaf that uses solar energy and carbon dioxide to produce hydrocarbons that can be used to produce biofuels. The technological advances in this new artificial leaf are believed to be able to boost efficiency to around 10%, making it an attractive option for commercial and industrial use.

With carbon dioxide levels in the atmosphere increasing annually, scientists have been looking for ways to not only reduce the use of fossil fuels, which add to these levels, but also capture the carbon dioxide that is already present in the atmosphere. The artificial leaf would actually perform these functions well and, in the process, produce a sustainable fuel source.

In this chapter, we will explore how plants use the photosynthetic pathways to capture the energy of the sun and store it within the chemical bonds of carbohydrates.

As you read through this chapter, think about the following questions:

1. Which pigments provide the maximum efficiency for a plant as it conducts photosynthesis?

2. Why do leaves appear green in the spring and summer and then turn to red or yellow in the fall?

8.1 Overview of Photosynthesis

Learning Outcomes

Upon completion of this section, you should be able to

1. Compare and contrast autotrophs and heterotrophs.
2. Explain the role of photosynthesis for all organisms on Earth.
3. Recognize the overall chemical equation for photosynthesis.
4. Describe the process of photosynthesis in terms of two sets of reactions that take place in a chloroplast.

Photosynthesis converts solar energy into the chemical energy of a carbohydrate. Photosynthetic organisms, including plants, algae, and cyanobacteria, are called **autotrophs** because they produce their own food (Fig. 8.1). Each year, photosynthesizing organisms produce over 150 billion metric tons of carbohydrates. This metabolic output allows not only photosynthetic organisms to sustain themselves but also the majority of all the other organisms on Earth.

With some exceptions, it is possible to trace the majority of food chains back to the photosynthetic organisms. In other words, producers, which have the ability to synthesize carbohydrates, feed not only themselves but also consumers, which must take in preformed organic molecules. Collectively, consumers are called **heterotrophs.** Both autotrophs and heterotrophs use organic molecules produced by photosynthesis as a source of building blocks for growth and repair and as a source of chemical energy for cellular work.

Photosynthesis in the Flowering Plants

Although the majority of photosynthesis occurs in bacteria and algae, we are going to focus our exploration of photosynthesis on the flowering plants, which exist mostly on land.

Photosynthesis takes place in the green portions of plants. The leaves of a flowering plant contain mesophyll tissue, in which cells are specialized for photosynthesis (Fig. 8.2a). The raw materials for photosynthesis are water and carbon dioxide. The roots of a plant absorb water, which then moves in vascular tissue up the

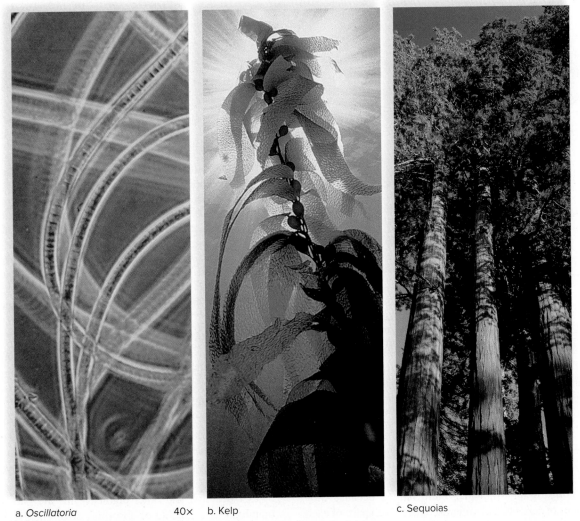

a. *Oscillatoria*   40×   b. Kelp   c. Sequoias

Figure 8.1 Photosynthetic organisms.   Photosynthetic organisms include (**a**) cyanobacteria such as *Oscillatoria,* which are a type of bacterium; (**b**) algae such as kelp, which typically live in water and can range in size from microscopic to macroscopic; and (**c**) plants such as the sequoia, which typically live on land. (a): ©Sinclair Stammers/Getty Images; (b): ©Chuck Davis/Stone/Getty Images; (c): ©Dynamic Graphics Group/Creatas/Alamy

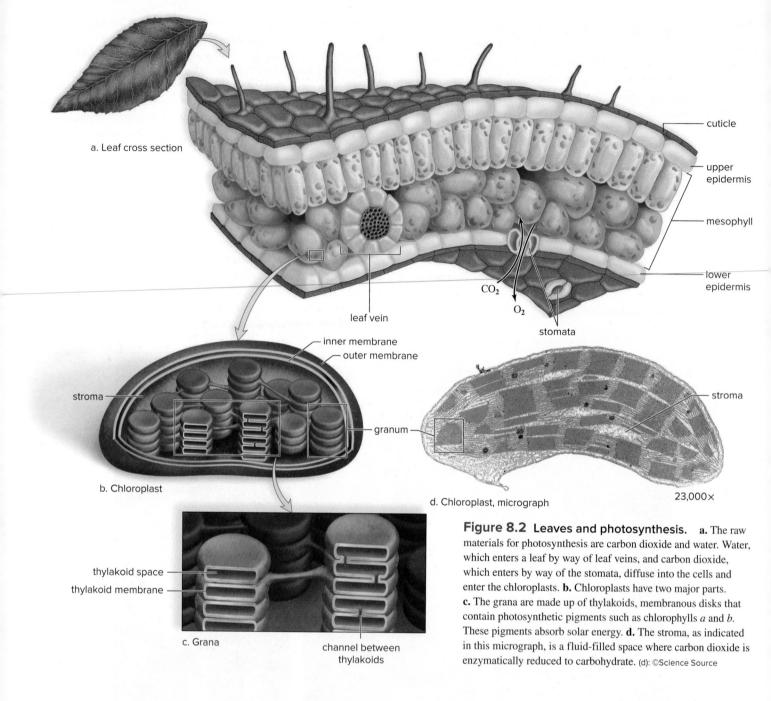

a. Leaf cross section

cuticle

upper epidermis

mesophyll

lower epidermis

CO_2

O_2

stomata

leaf vein

inner membrane
outer membrane

stroma

granum

stroma

b. Chloroplast

d. Chloroplast, micrograph

23,000×

thylakoid space

thylakoid membrane

c. Grana

channel between thylakoids

Figure 8.2 Leaves and photosynthesis. **a.** The raw materials for photosynthesis are carbon dioxide and water. Water, which enters a leaf by way of leaf veins, and carbon dioxide, which enters by way of the stomata, diffuse into the cells and enter the chloroplasts. **b.** Chloroplasts have two major parts. **c.** The grana are made up of thylakoids, membranous disks that contain photosynthetic pigments such as chlorophylls *a* and *b*. These pigments absorb solar energy. **d.** The stroma, as indicated in this micrograph, is a fluid-filled space where carbon dioxide is enzymatically reduced to carbohydrate. (d): ©Science Source

stem to a leaf by way of the leaf veins. Carbon dioxide in the air enters a leaf through small openings called **stomata** (*sing.,* stoma). After entering a leaf, carbon dioxide and water diffuse into **chloroplasts** (Gk. *chloros,* "green"; *plastos,* "formed, molded"), the organelles that carry out photosynthesis.

A double membrane surrounds a chloroplast and its fluid-filled interior, which is called the *stroma* (Fig. 8.2*b*). A different membrane system within the stroma forms flattened sacs called *thylakoids* (Fig. 8.2*c*). In some places, thylakoids are stacked to form *grana* (*sing.,* granum), which were named because they looked like piles of seeds to early microscopists. The space of each thylakoid is connected to the space of every other thylakoid within a chloroplast, thereby forming a continuous inner compartment within the chloroplast, called the thylakoid space.

The ability of organisms to carry out photosynthesis is due to the presence of special photosynthetic pigments, which allow them to capture solar energy. Most photosynthetic organisms contain **chlorophyll,** the pigment that gives them a green color (Fig. 8.2*d*). However, the green of chlorophyll can be masked by other pigments. For example, the **carotenoids** give photosynthesizing cells a yellow to red color. In addition, the *phycobilins* give red algae their red color and cyanobacteria a bluish color.

Chlorophyll and other pigments that are part of a thylakoid membrane are capable of absorbing solar energy. This is the energy that drives photosynthesis. The stroma, filled with enzymes, is where carbon dioxide is first attached to an organic compound and is then reduced to a carbohydrate. Therefore, it is proper to associate the absorption of solar energy with the thylakoid membranes making up the grana, and to associate the reduction of carbon dioxide to a carbohydrate with the stroma of a chloroplast.

Humans, and indeed nearly all organisms, release carbon dioxide into the air as a waste product from metabolic processes such as cellular respiration. This is some of the same carbon dioxide that enters a leaf through the stomata and is converted to carbohydrate. Carbohydrate, in the form of glucose, is the chief energy source for most organisms.

Photosynthetic Reaction

For convenience, the overall equation for photosynthesis is sometimes simplified in this manner:

$$6\,CO_2 + 6\,H_2O \xrightarrow[\text{pigments}]{\text{solar energy}} C_6H_{12}O_6 + 6\,O_2$$

The oxygen given off by photosynthesis comes from water. This was proven experimentally by exposing plants first to CO_2 and then to H_2O that contained an isotope of oxygen called heavy oxygen (^{18}O). Only when heavy oxygen was a part of water did this isotope appear in O_2 given off by the plant. Therefore, O_2 released by chloroplasts comes from H_2O, not from CO_2.

The overall equation tells us that during photosynthesis, H_2O is oxidized and CO_2 is reduced to form a carbohydrate. Reduction of any molecule requires energy. During photosynthesis, this energy is ultimately provided by the sun.

The overall chemical equation for photosynthesis tells us why plants are so important to our lives. Plants provide us with food in the form of carbohydrates, and much of the oxygen we breathe and use for cellular respiration.

Two Sets of Reactions

The word *photosynthesis* suggests that the process requires two sets of reactions: *photo,* which means light, refers to the reactions that capture solar energy, and *synthesis* refers to the reactions that produce carbohydrate.

The two sets of reactions are called the **light reactions** (light-dependent reactions) and the **Calvin cycle reactions** (light-independent reactions) (Fig. 8.3). We will see that the light reactions release oxygen and provide the molecules that allow the Calvin cycle reactions to reduce carbon dioxide to a carbohydrate. The coenzyme **NADP⁺ (nicotinamide adenine dinucleotide phosphate)** carries hydrogen atoms from the light reactions to the Calvin cycle reactions. When NADP⁺ accepts hydrogen atoms, it becomes NADPH. Also, ATP carries energy from the light reactions to the Calvin cycle reactions.

The light reactions are sometimes called the light-dependent reactions because they cannot occur unless light is present. The Calvin cycle reactions are sometimes called the light-independent reactions because they will occur whether light is present or not.

Figure 8.3 Overview of photosynthesis.   The process of photosynthesis consists of the light reactions and the Calvin cycle reactions. The light reactions, which produce ATP and NADPH, occur in the thylakoid membrane. These molecules are used in the Calvin cycle reactions in the stroma to reduce carbon dioxide to a carbohydrate.

Check Your Progress 8.1

1. Name the primary source of energy for carbohydrate production in plants.
2. Identify the major components in the structure of a chloroplast.
3. State the overall equation for photosynthesis.
4. Describe the two sets of reactions involved in photosynthesis.

8.2 Plants as Solar Energy Converters

Learning Outcomes

Upon completion of this section, you should be able to

1. Identify the photosynthetic pigments required to absorb the various wavelengths of light necessary for photosynthesis.
2. Explain the role of the noncyclic electron pathway and the cyclic electron pathway.
3. Describe the organization of the thylakoid and how this organization is critical to the production of ATP during photosynthesis.

During the light reactions, the photosynthetic pigments within the thylakoid membranes absorb solar energy. Solar energy (radiant energy from the sun) can be described in terms of its wavelength and its energy content. The electromagnetic spectrum (Fig. 8.4) describes all of the forms of energy emitted from the sun. The wavelengths characterized as visible light, those we can see and

that are used for photosynthesis, are only a small portion of this spectrum.

Visible Light

Visible light itself contains various wavelengths of light. When it is passed through a prism, we see all the different colors that make up visible light. The sensory areas of our brains identify these wavelengths as distinct colors. The colors in visible light (Fig. 8.4) range from violet (the shortest wavelength) to indigo, blue, green, yellow, orange, and red (the longest wavelength). The energy content is highest for violet light and lowest for red light.

Only about 45% of the solar radiation that hits Earth's atmosphere ever reaches the surface of Earth, and most of this radiation is within the visible-light range. Higher-energy wavelengths are screened out by the ozone layer in the atmosphere, and lower-energy wavelengths are screened out by water vapor and carbon dioxide (CO_2) before they reach the Earth's surface. Our eyes are adapted to detecting the wavelengths of visible light that reach the surface, as are the photosynthetic pigments of plants, algae, and bacteria.

The pigments found within most types of photosynthesizing cells, the chlorophylls a and b and the carotenoids, are capable of absorbing various portions of visible light. The absorption spectrum for these pigments is shown in Figure 8.5. Both chlorophyll a and chlorophyll b absorb violet, indigo, blue, and red light better than the light of other colors. Because green light is reflected and only minimally absorbed, leaves appear green to us. The yellow or orange carotenoids are able to absorb light in the violet-blue-green range. These pigments and others become noticeable in the fall when chlorophyll breaks down and the other pigments are uncovered. Photosynthesis begins when the pigments within thylakoid membranes absorb solar energy.

Figure 8.4 The electromagnetic spectrum.   The electromagnetic spectrum extends from the very short wavelengths of gamma rays through the very long wavelengths of radio waves. Visible light, which drives photosynthesis, is expanded to show its component colors. Each color varies in its wavelength and energy content.

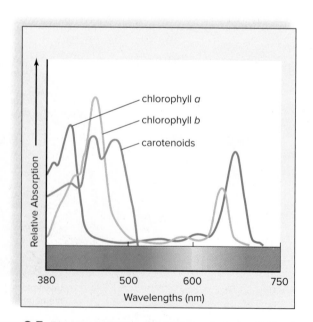

Figure 8.5 Photosynthetic pigments and photosynthesis.
The photosynthetic pigments in chlorophylls a and b and the carotenoids absorb certain wavelengths within visible light. This is their absorption spectrum. Notice that they do not absorb green light. That is why the leaves of plants appear green to us. You observe the color not absorbed by the leaf.

Figure 8.6 Noncyclic electron pathway: Electrons move from water to NADP⁺.   Energized electrons (taken from water, which splits, releasing oxygen) leave photosystem II and pass down an electron transport chain, leading to the formation of ATP. Energized electrons (replaced by photosystem II) leave photosystem I and pass to $NADP^+$, which then combines with H^+, becoming NADPH.

Light Reactions

The light reactions that occur in the thylakoid membrane consist of two electron pathways called the noncyclic electron pathway and the cyclic electron pathway. Both electron pathways produce ATP, but only the noncyclic pathway also produces NADPH.

The Noncyclic Electron Pathway

In the **noncyclic electron pathway,** the electron flow can be traced from water to a molecule of NADP⁺ (Fig. 8.6). This pathway uses two photosystems, called photosystem I (PS I) and photosystem II (PS II). The photosystems are named for the order in which they were discovered, not for the order in which they participate in the photosynthetic process. A **photosystem** consists of a pigment complex (molecules of chlorophyll *a*, chlorophyll *b*, and the carotenoids) and an electron acceptor within the thylakoid membrane. The pigment complex serves as an "antenna" for gathering solar energy.

The noncyclic pathway begins with photosystem II. The pigment complex absorbs solar energy, which is then passed from one pigment molecule to another until it is concentrated in a particular pair of chlorophyll *a* molecules called the *reaction center.* Electrons (e⁻) in the reaction center chlorophyll become so energized that they escape from the reaction center and move to a nearby electron acceptor.

Without replacement electrons, photosystem II would quickly cease to function. The replacement electrons are provided by splitting water (Fig. 8.6), which also releases hydrogen ions (H⁺) and oxygen. This is the same oxygen used by plants and animals in their cellular respiration pathways (see Section 7.4). The H⁺ ions remain within the thylakoid space and contribute to the formation of a hydrogen ion gradient.

The electron acceptor that received the energized electrons from the reaction center will send the electrons down a series of carriers that pass electrons from one to the other, known as the electron transport chain. As the electrons pass from one carrier to

Figure 8.7 Organization of a thylakoid.   Each thylakoid membrane within a granum produces NADPH and ATP. Electrons move through photosystem II, photosystem I, and electron transport chains within the thylakoid membrane. Electrons pass to NADP⁺, after which it becomes NADPH. A carrier at the start of the electron transport chain pumps hydrogen ions from the stroma into the thylakoid space. When hydrogen ions flow back out of the space into the stroma through the ATP synthase complex, ATP is produced from ADP + ℗.

the next, the energy released is used to move hydrogen ions (H⁺) from the stroma into the thylakoid space, forming a hydrogen ion gradient. When these hydrogen ions flow down their electrochemical gradient through **ATP synthase** complexes, ATP production occurs, as will be described shortly. Notice that this ATP will be used in the Calvin cycle reactions in the stroma to reduce carbon dioxide to a carbohydrate.

Similarly, when the photosystem I pigment complex absorbs solar energy, energized electrons leave its reaction center and are captured by a different electron acceptor. After going through the electron transport chain, the electrons from photosystem II are now low-energy electrons. These electrons are used to replace those lost by photosystem I. The electron acceptor in photosystem I passes its electrons to NADP⁺ molecules. Each NADP⁺ accepts two electrons and an H⁺ to become a reduced form of the

molecule—that is, NADPH. This NADPH will also be used by the Calvin cycle reactions in the stroma to reduce carbon dioxide to a carbohydrate.

Photosynthesis and the Thylakoid Membrane

The following molecular complexes are present in the thylakoid membrane and are involved in the noncyclic electron pathway (Fig. 8.7):

- Photosystem II, which consists of a pigment complex and an electron acceptor molecule, receives electrons from water, which splits, releasing oxygen.
- The electron transport chain carries electrons from photosystem II to photosystem I, and pumps H⁺ from the stroma into the thylakoid space.

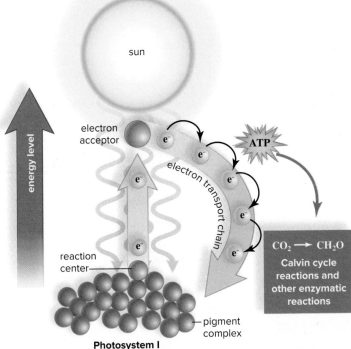

Figure 8.8 Cyclic electron pathway: Electrons leave and return to photosystem I.   Energized electrons leave the photosystem I reaction center and are taken up by an electron acceptor, which passes them down an electron transport chain before they return to photosystem I. Only ATP production occurs as a result of this pathway.

- Photosystem I, which also consists of a pigment complex and an electron acceptor molecule, is adjacent to NADP reductase, which reduces NADP$^+$ to NADPH.
- The ATP synthase complex crosses the thylakoid membrane and contains an interior channel and a protruding ATP synthase, an enzyme that joins ADP + ℗.

Cyclic Electron Pathway

Under some environmental conditions, such as high oxygen levels, NADPH levels may accumulate in the cell, and therefore photosynthetic cells may enter the **cyclic electron pathway** (Fig. 8.8). This pathway is also found in many prokaryotic cells. It differs from the noncyclic reactions in that the electrons are recycled back to photosystem I. The pathway begins when the photosystem I pigment complex absorbs solar energy that is passed from one

pigment to the next until it becomes concentrated in a reaction center. As with photosystem II, electrons (e$^-$) become so energized that they escape from the reaction center and move to nearby electron acceptor molecules.

This time, instead of the electrons moving on to NADP$^+$, energized electrons (e$^-$) taken up by an electron acceptor are sent down an electron transport chain. As the electrons pass from one carrier to the next, their energy becomes stored as a hydrogen (H$^+$) gradient. Again, the flow of hydrogen ions down their electrochemical gradient through ATP synthase complexes produces ATP.

The spent electrons return to photosystem I after the electron transport chain. This is how photosystem I receives replacement electrons and why this electron pathway is called cyclic. It is also why the cyclic pathway produces ATP but does not produce NADPH.

ATP Production

The thylakoid space acts as a reservoir for hydrogen ions (H$^+$). Each time oxygen is removed from water, two H$^+$ remain in the thylakoid space. As the electrons move from carrier to carrier along the electron transport chain, the electrons give up energy, which is used to pump H$^+$ from the stroma into the thylakoid space. This is like pumping water into a reservoir. The end result is that there are more H$^+$ in the thylakoid space than in the stroma. Because H$^+$ is charged, this is an electrochemical gradient. Not only are there more hydrogen ions, but there are more positive charges within the thylakoid space than the stroma. When a channel is opened in the ATP synthase complex, H$^+$ flows from the thylakoid space into the stroma, just like water flowing out of the reservoir. The flow of H$^+$ from high to low concentration across the thylakoid membrane provides the energy that allows the ATP synthase enzyme to enzymatically produce ATP from ADP + ℗. This method of producing ATP is called chemiosmosis because ATP production is tied to the establishment of an H$^+$ gradient.

Check Your Progress 8.2

1. Explain why leaves appear green.
2. Compare the production of NADPH to ATP in noncyclic photosynthesis.
3. Identify which part of a thylakoid will contain the photosystems, electron transport chain, and the ATP synthase complex.
4. Describe why the H$^+$ gradient across a thylakoid membrane is referred to as a storage of energy.

8.3 Plants as Carbon Dioxide Fixers

Learning Outcomes

Upon completion of this section, you should be able to

1. Describe the three phases of the Calvin cycle.
2. Explain how the products of the Calvin cycle are used to form the other molecules found in plants.

The Calvin cycle reactions take place after the light reactions. This is a series of reactions that produce a carbohydrate before returning to the starting point. The cycle is named for Melvin Calvin, who, with colleagues, used the radioactive isotope ^{14}C as a tracer to discover the reactions making up the cycle.

The Calvin cycle is a series of reactions that uses atmospheric carbon dioxide and the solar energy captured during the light-dependent reactions to produce carbohydrate molecules (Fig. 8.9). There are several stages to the Calvin cycle. These include (1) carbon dioxide fixation; (2) carbon dioxide

reduction; and (3) regeneration of RuBP, the starting material of the cycle.

Fixation of Carbon Dioxide

Carbon dioxide fixation is the first step of the Calvin cycle. During this reaction, three molecules of carbon dioxide from the atmosphere are attached to three molecules of RuBP, a 5-carbon molecule. The result is three 6-carbon molecules. Each 6-carbon

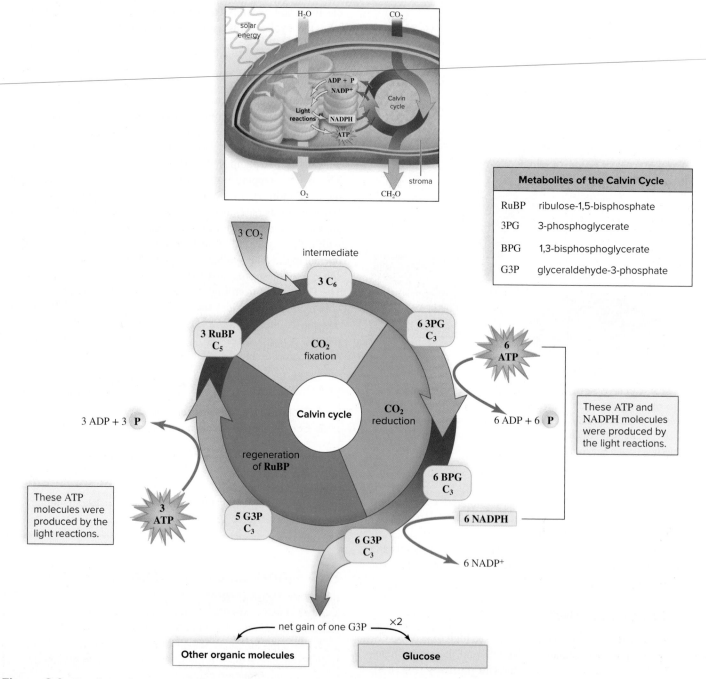

Figure 8.9 The Calvin cycle reactions.   The Calvin cycle is divided into three portions: CO_2 fixation, CO_2 reduction, and regeneration of RuBP. Because five G3P are needed to re-form three RuBP, it takes three turns of the cycle to have a net gain of one G3P. Two G3P molecules are needed to form glucose.

molecule then splits in half, forming a total of six 3-carbon molecules. This first 3-carbon molecule is called 3-phosphoglycerate or 3PG.

The enzyme that speeds this reaction, called **RuBP carboxylase,** is a protein that makes up about 20% to 50% of the protein content in chloroplasts. The reason for its abundance may be that it is unusually slow (it processes only a few molecules of substrate per second compared to thousands per second for a typical enzyme), and so there has to be a lot of it to keep the Calvin cycle going.

Reduction of Carbon Dioxide

Each of two 3PG molecules undergoes reduction to G3P in two steps. The first step converts 3PG into 1,3-bisphosphoglycerate (BPG) using ATP, and the second reaction converts BPG into glyceraldehyde-3-phosphate (G3P) using NADPH.

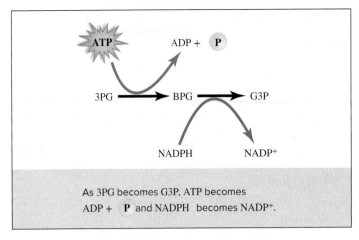

As 3PG becomes G3P, ATP becomes ADP + P and NADPH becomes NADP⁺.

This sequence of reactions uses ATP and NADPH from the light reactions. These reactions also result in the reduction of

---

Diesel Power from Algae

Algae are at the base of many food chains. But researchers are now using algae to produce a variety of different fuel types. Solazyme, a San Francisco–based company, has discovered the methods necessary to produce crude oil from various plant-based sugars with the help of microalgae (Fig. 8A). The process it has developed does not depend on the photosynthetic ability of the algae. It uses genetically modified algae grown in large vats. Instead of being exposed to sunlight, the algae are fed sugar that they convert into various types of oil. Different types of algae produce different types of oil. Because the algae are grown in the dark, the genes for photosynthesis have been switched off. With these genes turned off, the algae actually make more oil.

Algae-based biofuels are considered "green" fuels. These fuels can be used for a variety of purposes, but one of the more common is the production of biodiesel that can be used in cars and mass transportation systems. Some of the benefits of Solazyme-based fuels over traditional petroleum-based fuels are that they meet industry standards and can be used with factory-standard diesel engines without any type of modifications. Not just automobiles may be powered by these biofuels. In 2010, Solazyme delivered 80,000 liters of marine diesel and jet fuel to the U.S. Navy that was derived from algae-based production. In 2011, a United Airways jet conducted the first commercial flight powered by biofuels.

The oils produced by the algae are not only used for the production of fuel, but may

Figure 8A Production of biodiesel from algae.   The algae in these containers have the ability to produce a form of oil that can be used as a fuel. ©Agencja Fotograficzna Caro/Alamy

also be used to manufacture cosmetics and food additives.

With the increasing concern over the effects of global climate change on the world's populations and ecosystems, many governments (both local and national) and private industries are looking to use biofuels as an alternative "green-energy" source. Currently, demand still outpaces the production abilities of the companies, but advances in technology

and techniques will increasingly make biofuels an attractive alternate energy source.

Questions to Consider

1. Why would growing algae help reduce the carbon footprint of a city or town?
2. Fossil fuels are still relatively cheap to extract and produce. What technology advances do you think are needed to make biodiesel cheaper to produce?

carbon dioxide to a carbohydrate because R—CO_2 has become R—CH_2O. Energy and electrons are needed for this reduction reaction, and they are supplied by ATP and NADPH.

Regeneration of RuBP

Notice that the Calvin cycle reactions in Figure 8.9 are multiplied by three because it takes three turns of the Calvin cycle to allow one G3P to exit. For every three turns of the Calvin cycle, five molecules of G3P are used to re-form three molecules of RuBP and the cycle continues. Notice that 5×3 (carbons in G3P) $= 3 \times 5$ (carbons in RuBP):

5 G3P   3 RuBP

3 ATP   3 ADP + **P**

As five molecules of G3P become three molecules of RuBP, three molecules of ATP become three molecules of ADP + **P**.

This reaction also uses some of the ATP produced by the light reactions.

The Importance of the Calvin Cycle

G3P (glyceraldehyde-3-phosphate) is the product of the Calvin cycle that can be converted to other molecules a plant needs. Compared to animal cells, algae and plants have enormous biochemical capabilities. They use G3P for glucose, sucrose, starch, cellulose, fatty acid, and amino acid synthesis.

Glucose phosphate can be combined with fructose (and the phosphate removed) to form sucrose, the molecule that plants use to transport carbohydrates from one part of the plant to the other. Glucose phosphate is also the starting point for the synthesis of starch and cellulose. Starch is the storage form of glucose. Some starch is stored in chloroplasts, but most starch is stored in amyloplasts in roots. Cellulose is a structural component of plant cell walls and becomes fiber in our diet, because we are unable to digest it.

A plant can use the hydrocarbon skeleton of G3P to form fatty acids and glycerol, which are combined in plant oils. We are all familiar with corn oil, sunflower oil, and olive oil used in cooking. When nitrogen is added to the hydrocarbon skeleton derived from G3P, amino acids are formed, allowing the plant to produce protein. As described in the Ecology feature "Diesel Power from Algae," these oils are being used as a new source of fuel called biodiesel.

Check Your Progress 8.3

1. Describe how carbon dioxide is fixed and then reduced to a carbohydrate.
2. Illustrate why it takes three turns of the Calvin cycle to produce one glucose molecule.
3. Explain why G3P is an important molecule in plant metabolism.

8.4 Alternative Pathways for Photosynthesis

Learning Outcomes

Upon completion of this section, you should be able to

1. Contrast the processes of C_3, C_4, and CAM photosynthesis.
2. Explain how different photosynthetic modes allow plants to adapt to a particular environment.

Plants are able to live under all sorts of environmental conditions. One reason for this is because various modes of photosynthesis have evolved.

C_3 Photosynthesis

The majority of plants, such as azaleas, maples, and tulips, carry on photosynthesis as described in Section 8.2 and are called C_3 plants (Fig. 8.10). This is because in C_3 plants the CO_2 coming into the Calvin cycle is fixed to form a 3-carbon molecule (3PG; see Fig 8.9).

As shown in Figure 8.2, leaves have small openings called stomata, through which water can leave and carbon dioxide (CO_2) can enter. If the weather is hot and dry, the stomata close, conserving water. If not, then the water loss might cause the plant to wilt and die. Now the concentration of CO_2 decreases in leaves, while O_2, a by-product of photosynthesis, increases. When O_2 rises in C_3 plants, RuBP carboxylase combines it with RuBP instead of CO_2. The result is one molecule of 3PG and the eventual release of CO_2. This is called *photorespiration,* because in the presence of light (photo-), oxygen is taken up and CO_2 is released (-respiration).

An adaptation called C_4 photosynthesis enables some plants to avoid photorespiration.

C_4 Photosynthesis

In a C_3 plant, the mesophyll cells contain well-formed chloroplasts and are arranged in parallel layers. In a C_4 leaf, the bundle sheath

Figure 8.10 Carbon dioxide fixation in C_3 plants. In C_3 plants, like these wildflowers, CO_2 is taken up by the Calvin cycle directly in mesophyll cells. (photo): ©Don Paulson Photography/Purestock/Superstock RF

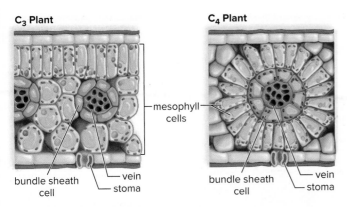

Figure 8.11 Leaf structure of C₃ and C₄ plants.   C₃ and C₄ plants differ in their arrangement of photosynthetic mesophyll cells.

CO₂ fixation in a CAM plant, pineapple, *Ananas comosus*

Figure 8.13 Carbon dioxide fixation in a CAM plant.   CAM plants, such as pineapple, fix CO₂ at night, forming a C₄ molecule that is released to the Calvin cycle during the day. (photo): ©Pixtal/age fotostock

cells, as well as the mesophyll cells, contain chloroplasts (Fig. 8.11). Further, the mesophyll cells are arranged concentrically around the bundle sheath cells.

In the mesophyll of **C₄ plants,** such as sugarcane and corn, the CO₂ is initially fixed by forming a 4-carbon molecule. The formation of this molecule accounts for the terms C₄ plant and C₄ photosynthesis. The 4-carbon molecule releases CO₂ to the Calvin cycle and carbohydrate production follows (Fig. 8.12).

Because of the need to transport molecules from where CO₂ is fixed to where the Calvin cycle is located, it would appear that the C₄ pathway would be little utilized among plants. However, when the weather is hot and dry, C₄ plants have an advantage. This is because when the stomata close in order to conserve water, and oxygen increases in leaf air space, RuBP carboxylase is not exposed to this oxygen in C₄ plants and photorespiration does not occur. Instead, in C₄ plants, the carbon dioxide is delivered to the Calvin cycle, which is located in bundle sheath cells that are sheltered from the leaf air spaces.

When the weather is moderate, C₃ plants ordinarily have the advantage, but when the weather becomes hot and dry, C₄ plants have the advantage, and we can expect them to predominate. In the early summer, C₃ plants such as Kentucky bluegrass and creeping bent grass predominate in lawns in the cooler parts of the United States, but by mid-summer, crabgrass, a C₄ plant, begins to take over.

CAM Photosynthesis

Another alternative pathway, called the **CAM** pathway, has also evolved because of environmental pressures. CAM stands for crassulacean-acid metabolism. Crassulaceae is a family of flowering succulent plants that live in warm, arid regions of the world. CAM photosynthesis is prevalent among most succulent plants that grow in desert environments, including the cacti, but it is now known to occur in a variety of other plants.

While the C₄ pathway separates components of photosynthesis by location, the CAM pathway separates them in time (Fig. 8.13). CAM plants fix CO₂ into a 4-carbon molecule at night, when they can keep their stomata open without losing much water. The 4-carbon molecule is stored in large vacuoles in their mesophyll cells until the next day. During the day, the 4-carbon molecule releases the CO₂ to the Calvin cycle within the same cell.

Plants are capable of fixing carbon by more than one pathway. These pathways appear to be the result of adaptation to different climates.

Figure 8.12 Carbon dioxide fixation in C₄ plants.   C₄ plants like corn form a C₄ molecule in mesophyll cells prior to releasing CO₂ to the Calvin cycle in bundle sheath cells. (photo): Source: USDA/Doug Wilson, photographer

Check Your Progress 8.4

1. Identify various plants that use a method of photosynthesis other than C₃ photosynthesis.
2. Explain why C₄ photosynthesis is advantageous in hot, dry conditions.

SCIENCE IN YOUR LIFE ▶ ECOLOGY

The New Rice

According to international government leaders who met in Japan in July of 2008, a global food crisis is developing. The world's population continues to grow, and grain is being diverted for biofuels and to feed livestock. The world's farmers are not keeping up with the demand for wheat, sorghum, maize, and rice. But according to the International Rice Research Institute (IRRI) in Los Baños, Philippines, two-thirds of the world's poorest people subsist primarily on rice (Fig. 8B). So a shortage of rice affects world hunger more than the other grains. More rice is being consumed than grown, and world rice stockpiles have been declining since 2001. In order to keep up with demand, the world must grow 50 million tons more rice per year than it did in 2005. This requires an increase globally of 1.2% per year.

Several environmental factors are causing these declines, including global climate change and regional flooding. For example, most varieties of rice will die if they are submerged for 3 or more days. However, one variety, called flood resistant (FR), could survive even if submerged for 3 weeks. A single gene, called the *Sub1A* gene, appeared to confer a significant degree of submergence tolerance to the FR variety. IRRI scientists were able to introduce the

Figure 8B Rice to feed the world.
Rice provides food for two-thirds of the world's poorest people. New technologies may increase the production of rice. ©Doable/A.collection/Getty Images

gene into commercial rice strains via hybridization, and four varieties are currently in field trials.

China plants approximately 57% of its rice crop in hybrid varieties. Hybrid rice produces approximately 20% more per acre than the traditional inbred varieties. The hybrids

developed by the Chinese are best suited for growth in the tropics and are not effective in temperate zones. However, the rice lacks in flavor compared to the inbred varieties.

One of the more long-term research programs with rice is trying to convert rice from a C_3 into a C_4 plant! C_4 plants are 50% more efficient at turning sunlight into food than C_3 plants. There are two aspects of a C_4 plant that would need to be introduced into rice. The first would be the actual enzymes involved in fixing carbon dioxide. The second would be the leaf anatomy of the mesophyll and bundle sheath cells. Researchers have succeeded in introducing the appropriate enzymes from maize, a C_4 plant, into rice plants. Without the proper leaf anatomy, however, there is no guarantee the cloned enzymes will increase the efficiency of photosynthesis.

Questions to Consider

1. If the poorest people in the world go hungry due to the lack of grain, how much difference would it make to solving world hunger if you reduced your meat and fuel consumption?
2. Are genetically modified organisms the only way to feed the world's growing population?

8.5 Photosynthesis Versus Cellular Respiration

Learning Outcomes

Upon completion of this section, you should be able to

1. Compare the overall chemical equation for photosynthesis and cellular respiration.
2. Describe the similarities and differences between cellular respiration and photosynthesis.

Both plant and animal cells carry on cellular respiration, but animal cells cannot photosynthesize. Plants, algae, and cyanobacteria are capable of photosynthesis. The organelle for cellular respiration is the mitochondrion, whereas the organelle for photosynthesis is the chloroplast. Note that in some bacteria, such as the cyanobacteria, photosynthesis occurs without chloroplasts—this is because these bacteria do have thylakoid-like structures in the cell. Photosynthesis is the formation of glucose, whereas cellular respiration is the breaking down of glucose. Figure 8.14 compares the two processes.

The following overall chemical equation for photosynthesis is the opposite of that for cellular respiration. The reaction in the forward direction represents photosynthesis, and the word *energy* stands for solar energy. The reaction in the opposite direction represents cellular respiration, and the word *energy* then stands for ATP:

$$\text{energy} + 6\,CO_2 + 6\,H_2O \underset{\text{cellular respiration}}{\overset{\text{photosynthesis}}{\rightleftharpoons}} C_6H_{12}O_6 + 6\,O_2$$

Both photosynthesis and cellular respiration are metabolic pathways within cells, and therefore consist of a series of reactions that the overall equation does not indicate. Both pathways, which use an electron transport chain located in membranes, produce ATP by chemiosmosis. Both also use an electron carrier; photosynthesis uses $NADP^+$, and cellular respiration uses NAD^+.

Both pathways utilize the following reaction, but in opposite directions. For photosynthesis, read the reaction from left

Figure 8.14 Photosynthesis versus cellular respiration.   Both photosynthesis and cellular respiration have an electron transport chain located within membranes, where ATP is produced. Both processes have enzyme-catalyzed reactions located within the fluid interior of respective organelles. In photosynthesis, hydrogen atoms are donated by NADPH + H⁺ when CO_2 is reduced in the stroma of a chloroplast. During cellular respiration, NADH forms when glucose is oxidized in the cytoplasm and glucose breakdown products are oxidized in the matrix of a mitochondrion.

to right; for cellular respiration, read the reaction from right to left:

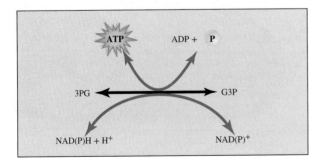

Both photosynthesis and cellular respiration occur in plant cells. In plants, both processes occur during the daylight hours, whereas only cellular respiration occurs at night. During daylight hours, the rate of photosynthesis exceeds the rate of cellular respiration, resulting in a net increase and storage of glucose. The stored glucose is used for cellular metabolism, which continues during the night.

Check Your Progress 8.5

1. Explain the similarities, and differences, between cellular respiration and photosynthesis. Use the chemical equation for these reactions in your answer.
2. Explain why a plant cell must contain both chloroplasts and mitochondria.

Conclusion

The natural process of photosynthesis has been powering life on Earth for billions of years. By understanding these pathways, scientists have been able to develop the technology behind an artificial leaf that is potentially more efficient than the natural system.

These advances are not just limited to the production of biofuels. Scientists are now exploring ways to use artificial leaves to produce other compounds such as plastics. Modifications to these artificial pathways may also one day enable the production of fertilizers, as well as chemicals that can be used to produce medication.

SUMMARIZE

8.1 Overview of Photosynthesis

■ **Photosynthesis** is absolutely essential for the continuance of life, because via this process, **autotrophs,** such as plants, manufacture the food needed for all life, including the **heterotrophs,** to survive.

■ Chloroplasts carry on photosynthesis. During photosynthesis, solar energy is converted to chemical energy within carbohydrates. A chloroplast contains two main portions: the stroma and membranous grana made up of thylakoid sacs. **Chlorophyll** and other pigments like the **carotenoids** that are located within the thylakoid membrane absorb solar energy, and enzymes in the fluid-filled stroma reduce CO_2. CO_2 enters the leaf through small openings called **stomata.**

■ The overall equation for photosynthesis is $6\,CO_2 + 6\,H_2O \rightarrow C_6H_{12}O_6 + 6\,O_2$. Photosynthesis consists of two reactions: the **light reactions** and the **Calvin cycle reactions.** The Calvin cycle reactions, located in the stroma, use **NADP⁺ (nicotinamide adenine dinucleotide phosphate)** and ATP to reduce carbon dioxide. These molecules are produced by the light reactions located in the thylakoid membranes of the grana, after chlorophyll captures the energy of sunlight.

8.2 Plants as Solar Energy Converters

■ Photosynthesis uses solar energy in the visible-light range. Chlorophylls *a* and *b* and the carotenoids largely absorb violet, indigo, blue, and red wavelengths and reflect green wavelengths. This causes leaves to appear green to us.

■ Photosynthesis begins when pigment complexes within **photosystem** I and photosystem II absorb radiant energy. In the **noncyclic electron pathway,** electrons are energized in photosystem II before they enter an electron transport chain. Electrons from H_2O replace those lost in photosystem II. As the electrons pass through the electron transport chain, they help establish a hydrogen ion gradient. The electrons energized in photosystem I pass to NADP⁺, which becomes NADPH. Electrons from the electron transport chain replace those lost by photosystem I.

■ In the **cyclic electron pathway** of the light reactions, electrons energized by the sun leave photosystem I and enter an electron transport chain that produces a hydrogen ion gradient. Then, the energy-spent electrons return to photosystem I.

■ The hydrogen ion gradient across the thylakoid membrane is used to synthesize ATP using an **ATP synthase** enzyme complex.

8.3 Plants as Carbon Dioxide Fixers

■ The ATP and NADPH made in thylakoid membranes pass into the stroma, where carbon dioxide is reduced during the Calvin cycle reactions.

■ **Carbon dioxide fixation** attaches CO_2 to a 5-carbon molecule named RuBP by the enzyme **RuBP carboxylase.** The resulting 6-carbon molecule splits into two molecules of 3-carbons each, called 3PG. ATP and NADPH from the light reactions are then used to reduce 3PG to G3P.

■ G3P is used to synthesize various molecules, including carbohydrates such as glucose.

8.4 Alternative Pathways for Photosynthesis

■ In C_3 plants, the first molecule observed following carbon dioxide fixation is a 3-carbon molecule. The structure of these plants allows RuBP carboxylase to bind with oxygen during photorespiration.

■ Plants utilizing C_4 photosynthesis have a different construction than plants using C_3 photosynthesis. **C_4 plants** fix CO_2 in mesophyll cells and then deliver the CO_2 to the Calvin cycle in bundle sheath cells. Now, O_2 cannot compete for the active site of RuBP carboxylase when stomata are closed due to hot and dry weather. This represents a partitioning of pathways in space: Carbon dioxide fixation occurs in mesophyll cells, and the Calvin cycle occurs in bundle sheath cells.

■ **CAM** plants fix carbon dioxide at night when their stomata remain open. This represents a partitioning of pathways in time: Carbon dioxide fixation occurs at night, and the Calvin cycle occurs during the day.

8.5 Photosynthesis Versus Cellular Respiration

■ Both photosynthesis and cellular respiration utilize an electron transport chain and ATP synthesis. However, photosynthesis reduces CO_2 to a carbohydrate. The oxidation of H_2O releases O_2. Cellular respiration oxidizes carbohydrate, and CO_2 is given off. Oxygen is reduced to water.

ASSESS

Testing Yourself

Choose the best answer for each question.

8.1 Overview of Photosynthesis

1. Label each of the items in the diagram below:

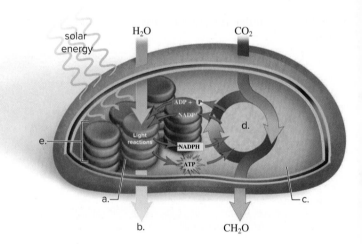

2. The function of light reactions is to
 a. obtain CO_2.
 b. make carbohydrate.
 c. convert light energy into a usable form of chemical energy.
 d. regenerate RuBP.

3. The Calvin cycle reactions
   a. produce carbohydrate.
   b. convert one form of chemical energy into a different form of chemical energy.
   c. regenerate more RuBP.
   d. use the products of the light reactions.
   e. All of these are correct.

8.2 Plants as Solar Energy Converters

4. The final acceptor of electrons during the light reactions of the non-cyclic electron pathway is
   a. PS I.
   b. PS II.
   c. ATP.
   d. $NADP^+$.
   e. water.

5. The oxygen given off by photosynthesis comes from
   a. H_2O.
   b. CO_2.
   c. glucose.
   d. RuBP.

6. The noncyclic electron pathway, but not the cyclic pathway, generates
   a. 3PG.
   b. chlorophyll.
   c. ATP.
   d. NADPH.

8.3 Plants as Carbon Dioxide Fixers

7. The ATP and NADPH from the light reactions are used to
   a. split water.
   b. cause RuBP carboxylase to fix CO_2.
   c. re-form the photosystems.
   d. cause electrons to move along their pathways.
   e. convert 3PG to G3P.

8. In the absence of sunlight, plants are not able to engage in the Calvin cycle due to a lack of
   a. ATP.
   b. oxygen.
   c. NADPH.
   d. Both a and c are correct.

8.4 Alternative Pathways for Photosynthesis

9. CAM photosynthesis
   a. is the same as C_4 photosynthesis.
   b. is an adaptation to cold environments in the Southern Hemisphere.
   c. is prevalent in desert plants that close their stomata during the day.
   d. occurs in plants that live in marshy areas.
   e. stands for "chloroplasts and mitochondria."

10. C_4 photosynthesis
    a. is the same as C_3 photosynthesis, because it takes place in chloroplasts.
    b. occurs in plants whose bundle sheath cells contain chloroplasts.
    c. takes place in plants such as wheat, rice, and oats.
    d. is an advantage when the weather is hot and dry.
    e. Both b and d are correct.

8.5 Photosynthesis Versus Cellular Respiration

11. Cellular respiration and photosynthesis both
    a. use oxygen.
    b. produce carbon dioxide.
    c. contain an electron transport chain.
    d. occur in the chloroplast.

12. Which of the following does not occur in photosynthesis?
    a. reduction of CO_2
    b. oxidation of water
    c. reduction of oxygen
    d. All of these are correct.

ENGAGE

BioNOW

Want to know how this science is relevant to your life? Check out the BioNOW video below:

■ Energy Part II: Photosynthesis

Why does this experiment need to produce light in the wavelengths only at the two ends of the visible light spectrum?

Thinking Critically

1. In biofuel production, plants are grown in order to create fuels that are eventually burned in vehicles like automobiles. Overall, does the process of biofuel production add carbon dioxide to the atmosphere, remove it, or both?

2. In the image below, the alga is the autotroph and undergoes photosynthesis, whereas the bacteria (small, black dots) are the heterotrophs. Explain why the bacteria are clustered at the far ends after the alga was exposed to different color wavelengths of light.

©Stephen Durr

3. In the 17th century, a Belgian doctor, Jan Baptista van Helmont, planted a small willow tree in a pot of soil. He weighed the tree and the soil. The tree was watered for 5 years and weighed 74.4 kg more than when he began, but the soil had lost only 57 g of mass. Explain what accounts for the plant's increase in biomass.

CHAPTER

9

Plant Organization and Function

(tree): ©Dinodia Photos/Alamy; (products): ©bdspn/iStock/360/Getty Images

BEFORE YOU BEGIN

Before beginning this chapter, take a few moments to review the following discussions:

Section 2.3 How do the properties of water make it possible for the movement of nutrients in plants?

Section 3.5 Which cell structures are unique to plants?

Section 8.1 Which parts of the plant are responsible for conducting photosynthesis?

The Amazing Neem Tree

If you were to walk into a health-food store looking for neem products, you would find oils, toothpaste, soap, and facial washes. At a plant nursery, you would find neem insecticide and fungicide. But what exactly is neem? The neem tree (*Azadirachta indica*) is found in India and Pakistan and is one of the oldest and most widely used plants in the world. It is locally known as the "village pharmacy," and the United Nations declared neem the "tree of the 21st century" due to its many uses in health and agriculture.

Neem's nontraditional medical uses are many. Bark and roots act as an analgesic and diuretic, and provide flea and tick protection to dogs. The sap effectively treats skin diseases, such as psoriasis. Gum, a sticky substance exuded by stems, is used to combat scabies (an itch mite) and treat surface wounds. And in some countries, the tree's woody twigs aid dental hygiene, in the form of homemade toothbrushes.

The neem tree also plays an important role in agriculture and pest control. In India and Pakistan, for example, neem leaves have traditionally been mixed with stored grains or placed in drawers with clothing, keeping insects at bay. Not surprisingly, the U.S. Department of Agriculture funds a significant amount of research on neem, due to its potential as an all-natural insecticide and fungicide.

In this chapter, we consider the structure of roots, stems, and leaves, and discuss the specialized cells and tissues that make up these organs.

As you read through this chapter, think about the following questions:

1. How do each of the vegetative plant organs contribute to the success of flowering plants?

2. How do modifications of vegetative organs increase fitness?

9.1 Cells and Tissues of Plants

Learning Outcomes

Upon completion of this section, you should be able to
1. Explain the function of meristematic tissue.
2. Describe the types of tissues found in plants.

Plants are an exceptionally diverse classification of eukaryotes (see Fig. 1.6). Scientists have currently identified approximately 391,000 species of plants, the majority of which are the flowering plants, or **angiosperms.** For this reason, we will use the angiosperms as our example as we examine the structure and function of a plant in this chapter. We will take a deeper look at the diversity of plants again in Chapter 30.

Meristematic Tissue

Plants have levels of biological organization similar to animals (see Fig. 1.2). As in animals, a tissue is composed of specialized cells that perform a particular function, and an organ is a structure made up of multiple tissues. In plants, simple tissues are made up of a single cell type, whereas complex tissues contain a variety of cell types. All the tissue types in a plant arise from **meristematic tissue** (Fig. 9.1). Meristematic tissue allows a plant to grow its

entire life because it retains cells that forever have the ability to divide and produce more tissues. Plants can grow their entire lives. Even a 5,000-year-old tree is still growing!

Types of Tissue in Plants

Apical meristem is the type of meristematic tissue that is present in the tips of all growing stems and roots of a plant. Apical meristem continually produces three types of primary meristem that develop into the three types of specialized primary tissues in the body of a plant: protoderm gives rise to epidermal tissue; ground meristem produces ground tissue; and procambium produces vascular tissue. The functions of these three specialized tissues include the following:

1. **Epidermal tissue** forms the outer protective covering of a plant.
2. **Ground tissue** fills the interior of a plant.
3. **Vascular tissue** transports water and nutrients in a plant and provides support.

A summary of each of these tissue types is presented in Table 9.1.

Epidermal Tissue

The entire body of both nonwoody (herbaceous) and young woody plants is covered by a layer of a complex tissue called the **epidermis.**

Figure 9.1 Role of meristematic tissue.   Meristematic tissue gives rise to a variety of different tissue types, including epidermal, ground, and vascular tissue.

TABLE 9.1 Specialized Tissues in Plants

Tissue	Cell Types	Description	Function
Epidermal tissue	Epidermis	Tissue that covers roots, leaves, and stems	Protection, prevention of water loss
	Periderm (older woody plants)	Composed of waterproof cork cells	Protection against attacks by fungi, bacteria, and animals
Ground tissue	Parenchyma	Most abundant, corresponds to the typical plant cell	Gives rise to specialized, photosynthesis, and storage cells
	Collenchyma	Contains thick primary walls	Gives flexible support to immature regions of plant body
	Sclerenchyma	Contains thick secondary walls, contains lignin making the wall tough and hard	Supports the mature regions of the plant
Vascular: xylem	Tracheids	Elongated with tapered ends	Transport water and minerals from roots to leaves
	Vessel elements	Short and wide, contain perforated plates	Transport water and minerals from roots to leaves
Vascular: phloem	Sieve-tube members	Lack nuclei, associated with a companion cell, sieve plate	Transport sugar and organic compounds throughout the plant

Figure 9.2 Modifications of epidermal tissue. **a.** Root epidermis has root hairs to absorb water. **b.** Leaf epidermis contains stomata (*sing.,* stoma) for gas exchange. **c.** Periderm includes cork and collenchyma cells. Lenticels in cork are important in gas exchange. (a): ©Nigel Cattlin/Alamy; (b): ©Biophoto Associates/Science Source; (c): ©Kingsley Stern

a. Root hairs

cabbage seedling

root hairs

elongating tip of root

b. Stoma of leaf

chloroplasts

stoma

guard cell

epidermal cells

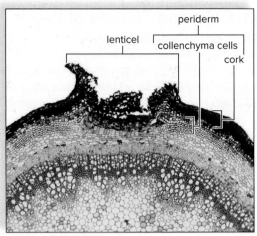

periderm

lenticel

collenchyma cells

cork

c. Cork of older stem

The walls of epidermal cells exposed to air are covered with a waxy **cuticle** to minimize water loss. The cuticle also protects against bacteria and other organisms that might cause disease.

In roots, certain epidermal cells have long, slender projections called **root hairs** (Fig. 9.2*a*). The hairs increase the surface area of the root for absorption of water and minerals.

On stems, leaves, and reproductive organs, epidermal cells produce small hairs that have two important functions: protecting the plant from too much sun and conserving moisture. Sometimes these hairs also help protect a plant from herbivores by producing toxic substances. Under the slightest pressure, the stiff hairs of the stinging nettle lose their tips, forming "hypodermic needles" that inject an intruder with a stinging secretion.

A waxy cuticle reduces the gas exchange in leaves. Leaves will often contain specialized cells called *guard cells* and microscopic pores called *stomata* (*sing.,* stoma) that assist in gas exchange. Guard cells, which are epidermal cells with chloroplasts, surround stomata (Fig. 9.2*b*). When the stomata are open, gas exchange is possible but water loss also occurs.

In older woody plants, the epidermis of the stem is replaced by boxlike **cork** cells. At maturity, cork cells can be sloughed off, and new cork cells are made by a meristem called cork cambium

(Fig. 9.2*c*). As the new cork cells mature, they increase slightly in volume, and their walls become encrusted with a lipid material that makes them waterproof and chemically inert. These nonliving cells protect the plant and make it resistant to attack by fungi, bacteria, and animals. Some cork tissues are commercially used for bottle corks and other products.

The cork cambium overproduces cork in certain areas of the stem surface, causing ridges and cracks to appear. The areas where the cork layer is thin and the cells are loosely packed are called *lenticels*. Lenticels are important in gas exchange between the interior of a stem and the air.

Ground Tissue

Ground tissues form the bulk of a plant and fill the space between the epidermal and the vascular tissue. Most of the photosynthesis and carbohydrate storage takes place in ground tissue. Ground tissue is also responsible for producing hormones, toxins, pigments, and other specialized chemicals. The cell types present in ground tissue are parenchyma, collenchyma, and sclerenchyma cells.

Parenchyma cells are the most abundant and correspond best to the typical plant cell (Fig. 9.3*a*). These are the least specialized

a. Parenchyma cells   100×

b. Collenchyma cells   255×

c. Sclerenchyma cells   340×

Figure 9.3 Ground tissue cells. **a.** Parenchyma cells are the least specialized of the plant cells. **b.** The walls of collenchyma cells are much thicker than those of parenchyma cells. **c.** Sclerenchyma cells have very thick walls and are nonliving at maturity—their primary function is to give strong support.
(a–c): ©Biophoto Associates/Science Source

of the cell types and are found in all the organs of a plant. Parenchyma cells can divide and give rise to more specialized cells such as roots. They have relatively thin walls, are alive at maturity, and may contain chloroplasts and carry out photosynthesis, or they may contain colorless plastids that store the products of photosynthesis. The fleshy part of an apple is mostly storage parenchyma cells.

Collenchyma cells are like parenchyma cells except they have thicker primary walls (Fig. 9.3*b*). The thickness is irregular, with the corners of the cell being thicker than other areas. Collenchyma cells often form bundles just beneath the epidermis and give flexible support to immature regions of a plant body. The familiar strands in celery stalks (leaf petioles) are composed mostly of collenchyma cells.

Sclerenchyma cells have a thick secondary cell wall that forms between the primary cell wall and the plasma membrane. The secondary cell wall is impregnated with **lignin,** a highly resistant organic substance that makes the walls tough and hard (Fig. 9.3*c*). If we compare a cell wall to reinforced concrete, cellulose fibrils would play the role of steel rods, and lignin would be analogous to the cement. Most sclerenchyma cells are dead at maturity. Their primary function is to support the mature regions of a plant. Tracheids and vessel elements are types of sclerenchyma cells that not only function in support but also transport water. Two other types of sclerenchyma cells are fibers and sclereids. Fibers are long and slender, and may be grouped in bundles that can be used commercially. Hemp fibers can be used to make rope, and flax fibers can be woven into linen. Flax fibers, however, are not lignified, which is why linen is soft. Sclereids, which are shorter than fibers and have variable shapes, are found in seed coats and nutshells. Sclereids, or "stone cells," are responsible for the gritty texture of pears, as well as the hardness of nuts and peach pits.

Vascular Tissue

There are two types of vascular (transport) tissue. **Xylem** transports water and minerals from the roots to the leaves, and **phloem** transports sugar and other organic compounds, including hormones, throughout the plant. Both xylem and phloem are considered complex tissues because they are composed of two or more kinds of cells. Xylem contains two types of conducting cells: tracheids and vessel elements (Fig. 9.4*a*). Both types of conducting cells are hollow and nonliving, but the **vessel elements** are shorter and wider. Vessel elements have plates with perforations in their end walls and are arranged to form a continuous vessel for water and mineral transport. The elongated **tracheids,** with tapered ends, form a less efficient means of transport, but water can move across the end walls and side walls because there are pits, or depressions, where the secondary wall does not form. In addition to vessel elements and tracheids, xylem contains parenchyma cells that store various substances and fibers that lend support.

The conducting cells of phloem are **sieve-tube members** arranged to form a continuous sieve tube (Fig. 9.4*b*). Sieve-tube members contain cytoplasm but no nuclei. The term *sieve* refers to a cluster of pores in the end walls, collectively called a sieve plate. Each sieve-tube member has a **companion cell,** which contains a nucleus. The two are connected by numerous **plasmodesmata,**

a. Xylem

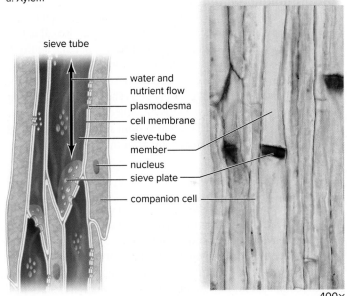

400×

b. Phloem

Figure 9.4 Structure of xylem and phloem.   a. Photomicrograph of xylem vascular tissue and drawing showing tracheids and vessel elements. **b.** Photomicrograph of phloem vascular tissue and drawing showing sieve tubes and companion cells. (photos) (a): ©SUNY, N.C. Brown Center for Ultrastructure Studies, College of Environmental Science & Forestry, Syracuse, NY; (b): ©Dr. Michael Clayton, University of Wisconsin, Madison, Dept. of Botany

cytoplasmic connections between neighboring cells. In the phloem, the plasmodesmata extend from one sieve-tube member to another, through the sieve plate. The nucleus of the companion cell controls and maintains the life of both cells. The companion cells are also believed to be involved in phloem's transport function.

It is important to realize that vascular tissue (xylem and phloem) extends from the root through stems to the leaves and vice versa (see Fig. 9.5). In the roots, the vascular tissue is located in the **vascular cylinder.** In the stem, it forms **vascular bundles,** and in the leaves, it is found in **leaf veins.**

9.2 Plant Organs and Systems

The cells and tissues of plants are organized into organs and organ systems in the same manner as animals (see Fig. 1.2). Recall that an organ is a structure that contains different tissues and the ability to perform a specific function. In plants, the roots, stems, and leaves are called the *vegetative organs*. Vegetative organs are involved in growth and nutrition, but not reproduction. The organs of flowering plants are organized into organ systems called the root system and the shoot system (Fig. 9.5). The **root system** simply consists of the roots, whereas the **shoot system** consists of the stem and leaves.

Root System

Roots have various adaptations and associations to perform the functions of anchorage, absorption of water and minerals, and storage of carbohydrates. The root system in the majority of plants is located underground. The depth and distribution of plant roots depends on the type of plant, the timing and amount of rainfall, and the soil composition. For example, plants growing in deserts tend to have deeper roots than those growing in temperate grasslands. A common misconception is that the size and distribution of tree roots reflect the aboveground trunk and branches of the tree. In reality, 90% of a tree's roots are located within 1 meter (m) of the surface. However, a tree's roots extend out much farther than the crown of the tree. On average, a tree's roots will extend two to four times the diameter of the aboveground portion of the tree.

Extensive root systems of plants help anchor them in the soil and provide support (Fig. 9.6a). The root system absorbs water and minerals from the soil for the entire plant. The cylindrical shape of a root allows it to penetrate the soil as it grows and to absorb water from all sides. The absorptive capacity of a root is also increased by its many branch (lateral) roots and root hairs located in a special zone near a root tip. Root hairs, which are projections from epidermal root-hair cells, are so numerous that they increase the absorptive surface of a root tremendously. In a classic study completed in 1937, H. J. Dittmer counted 13,800,000 root hairs in a single, 20-inch-high rye plant. Because root-hair cells are constantly being replaced, a typical rye plant may form about 100 million new

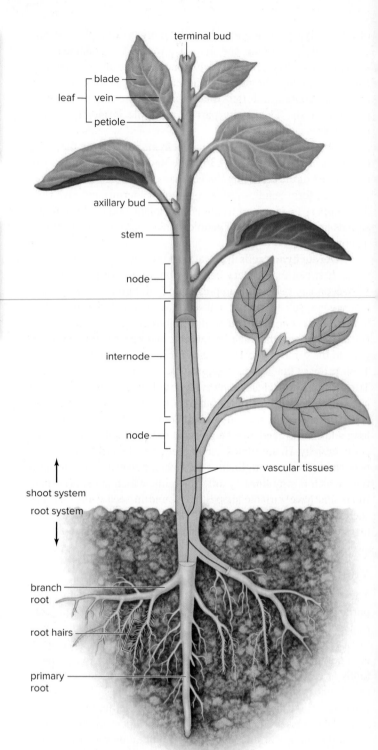

Figure 9.5 Organization of a plant body. The body of a plant consists of a root system and a shoot system. The shoot system contains the vegetative organs of the stem and leaves. Axillary buds can develop into branches or flowers, the reproductive organs of a plant. The root system and shoot system are connected by vascular tissue that extends from the roots to the leaves.

root-hair cells every day. Roughly pulling a plant out of the soil damages the small lateral roots and root hairs, and these plants do not fare well when transplanted. Transplantation is usually more successful if part of the surrounding soil is moved along with the

a. Root system, dandelion

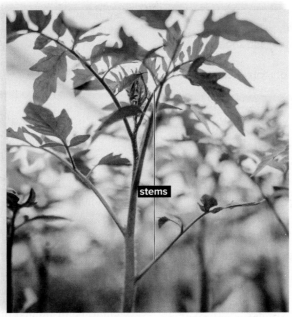

b. Shoot system, tomato plant

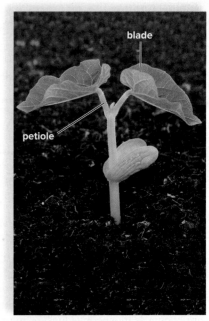

c. Leaves, bean seedling

Figure 9.6 **Vegetative organs of a eudicot.** **a.** The root system anchors the plant and absorbs water and minerals. **b.** The shoot system, the stem and branches, support the leaves and transport water and organic nutrients. **c.** The leaves carry out photosynthesis. (a): ©Dorling Kindersley/Getty Images; (b): ©Snowboy/Shutterstock; (c): ©Emilio100/Shutterstock

plant, because this leaves many of the branch roots and the root hairs intact.

Roots also have other functions. In certain plants, the roots are modified for food storage. For example, branch roots form storage organs for carbohydrates in sweet potatoes. Roots can also store water. Some plants in the pumpkin family store large amounts of water in their roots. Roots also produce hormones that stimulate the growth of stems and coordinate their size with the size of the root.

Shoot System: Stems

The shoot system of a plant is composed of the stem, the branches, and the leaves. A **stem,** the main axis of a plant, terminates in tissue that allows the stem to elongate and produce leaves (Fig. 9.6*b*). A stem can sometimes expand in girth as well as length. As trees grow taller each year, they accumulate woody tissue that adds to their girth. The increase in girth strengthens the stem of the plant. Most stems are upright and support leaves in such a way that each leaf is exposed to as much sunlight as possible. A **node** occurs where leaves are attached to the stem, and an **internode** is the region between the nodes (see Fig. 9.5). The presence of nodes and internodes is the characteristic used to identify a stem, even if it happens to be an underground stem. Some stems are horizontal, and in these cases the nodes may give rise to new plants.

In addition to supporting the leaves, a stem has vascular tissue that transports water and minerals from the roots through the stem to the leaves. Vascular tissue also transports the organic products of photosynthesis, usually in the opposite direction. Xylem cells consist of nonliving cells that form a continuous pipeline for water and mineral transport, whereas phloem cells consist of living cells that join end to end for organic nutrient transport. Stems may have functions other than transport. In some plants (e.g., cactus), the stem is the primary photosynthetic organ. In succulent plants, the stem is a water reservoir. Tubers are enlarged underground horizontal stems that store nutrients.

Shoot System: Leaves

Leaves are the major component of the plant that require H_2O, CO_2, and sunlight to carry on photosynthesis. With few exceptions, their cells are living, and the bulk of a leaf contains tissue specialized to carry on photosynthesis. Leaves receive water from the root system by way of the stem. The position of the leaves on a plant maximizes the plant's photosynthetic efforts. The shape of the leaf also influences its function. For example, leaves that are broad and flat have the maximum surface area for the absorption of carbon dioxide and the collection of solar energy needed for photosynthesis. Also, unlike stems, leaves are almost never woody. When describing a leaf, the wide portion of a leaf is called the **blade,** and the stalk that attaches the blade to the stem is the **petiole** (Fig. 9.6*c*).

The size, shape, color, and texture of leaves can be highly variable. These characteristics are fundamental in plant identification. The leaves of some aquatic duckweeds may be less than 1 millimeter (mm) in diameter, while some palms may have leaves that exceed 6 m in length. The shapes of leaves can vary from cactus spines to deeply lobed oak leaves. Leaves can exhibit a variety of colors, from many shades of green to deep purple. The texture of leaves varies from smooth and waxy like a magnolia to coarse like a sycamore. Not all leaves are foliage leaves. Some are specialized to protect buds, attach to objects (tendrils), store food (bulbs), or even capture insects.

The upper acute angle between the petiole and stem is the leaf axil, where an **axillary bud** (or lateral bud) originates. This bud may become a branch or a flower. Plants that lose their leaves every year are called **deciduous.** This is in contrast to most of the gymnosperm plants (conifers), which usually retain their leaves for 2 to 7 years. These plants are called **evergreens.**

Check Your Progress 9.2

1. List the organs found in the root system and shoot system of a plant.
2. Compare the structure and function of roots, stems, and leaves.

9.3 Monocot Versus Eudicot Plants

Learning Outcome

Upon completion of this section, you should be able to

1. List and describe the key features of monocots and eudicots.

Flowering plants are divided into two groups, depending on the number of **cotyledons,** or seed leaves, in the embryonic plant (Fig. 9.7). Some cotyledons emerge, grow larger, and become green when the seed germinates. Some plants have one cotyledon, and are known as monocotyledons, or **monocots.** Other embryos have two cotyledons, and are known as eudicotyledons, or **eudicots.**

The vascular (transport) tissue is organized differently in monocots and eudicots. In the monocot root, vascular tissue occurs in a ring. In the eudicot root, phloem, which transports organic nutrients, is located between the arms of xylem, which transports water and minerals, and has a star shape. In the monocot stem, the vascular bundles, which contain vascular tissue surrounded by a bundle sheath, are scattered. In a eudicot stem, the vascular bundles occur in a ring.

Leaf veins are vascular bundles within a leaf. Monocots exhibit parallel venation, and eudicots exhibit netted venation, which may be either pinnate or palmate. Pinnate venation means that major veins originate from points along the centrally placed main vein, and palmate venation means that the major veins all originate at the point of attachment of the blade to the petiole:

Netted venation: pinnately veined   palmately veined

Adult monocots and eudicots have other structural differences, such as the number of flower parts and the number of apertures (pores or slits) of pollen grains. The flower parts of monocots are arranged in multiples of three, and the flower parts of eudicots are arranged in multiples of four or five. Eudicot pollen grains usually have three apertures, and monocot pollen grains usually have one aperture.

Although the division between monocots and eudicots may seem to be of limited importance, it does in fact affect many aspects of their structure. The eudicots are the larger group and include some of our most familiar flowering plants—from dandelions to oak trees. The monocots include grasses, lilies, orchids, and palm trees, among others. Some of our most significant food sources are monocots, such as rice, wheat, and corn.

Check Your Progress 9.3

1. Compare the following in monocots and eudicots: number of cotyledons, leaf venation, and flower parts.
2. List examples of plants that are monocots and eudicots.

	Seed	Root	Stem	Leaf	Flower	Pollen
Monocots	One cotyledon in seed	Root xylem and phloem in a ring	Vascular bundles scattered in stem	Leaf veins form a parallel pattern	Flower parts in threes and multiples of three	One pore or slit
Eudicots	Two cotyledons in seed	Root phloem between arms of xylem	Vascular bundles in a distinct ring	Leaf veins form a net pattern	Flower parts in fours or fives and their multiples	Three pores or slits

Figure 9.7 Flowering plants are either monocots or eudicots. The features used to distinguish monocots from eudicots: the number of cotyledons in the seed; the arrangement of vascular tissue in roots, stems, and leaves; the number of flower parts; and the number of pores in the pollen grain.

9.4 Organization of Roots

Learning Outcomes

Upon completion of this section, you should be able to

1. List the zones in the root involved in primary growth.
2. Describe the anatomy of eudicot and monocot roots.
3. Identify different types of roots and root specializations.

Primary growth, which causes a plant to grow lengthwise, is centered in the apex (tip) of the shoot and the root systems. The growth of many roots is continuous, pausing only when temperatures become too cold or when water becomes too scarce. Figure 9.8a, a longitudinal section of a eudicot root, reveals zones where cells are in various stages of differentiation as primary growth occurs. These zones are called the *zone of cell division,* the *zone of elongation,* and the *zone of maturation.*

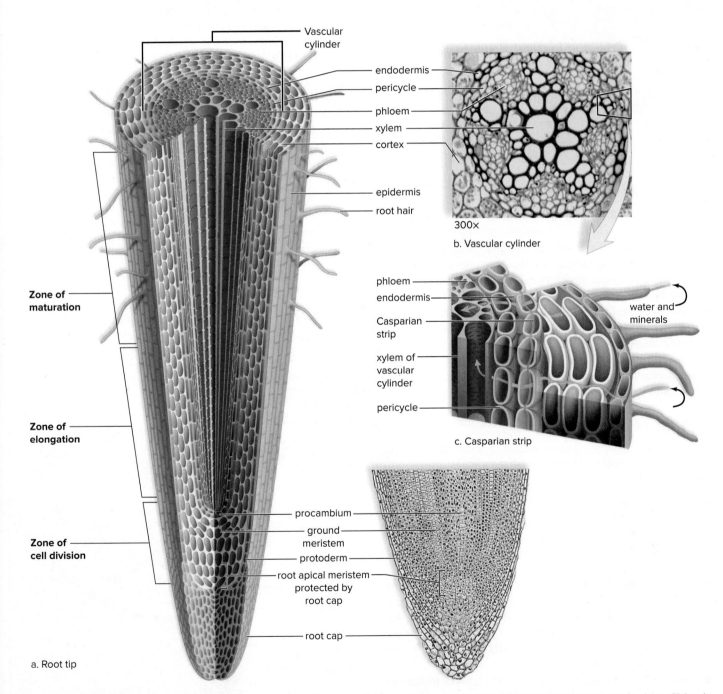

Figure 9.8 Eudicot root tip.   **a.** The root tip is divided into three zones. **b.** The vascular cylinder of a eudicot root contains the vascular tissue. Xylem is typically star-shaped, and phloem lies between the points of the star. **c.** Water and minerals either pass between cells, or they enter root cells directly. In any case, because of the Casparian strip, water and minerals must pass through the cytoplasm of endodermal cells in order to enter the xylem. In this way, endodermal cells regulate the passage of minerals into the vascular cylinder. (photos) (a): ©Ray F. Evert; (b): ©Carolina Biological Supply Company/Medical Images

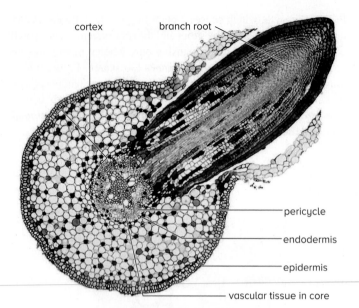

Figure 9.9 Branching of eudicot root. This cross-section shows the origination and growth of a branch root from the pericycle. ©Lee Wilcox

The zone of cell division is protected by the **root cap,** which is composed of parenchyma cells and protected by a slimy sheath. As the root grows, root cap cells are constantly removed by rough soil particles and replaced by new cells. The zone of cell division contains the root apical meristem, where mitosis produces relatively small, many-sided cells having dense cytoplasm and large nuclei. These meristematic cells give rise to the primary meristems, called protoderm, ground meristem, and procambium. Eventually, the primary meristems develop, respectively, into the three mature tissue types discussed previously: epidermis, ground tissue (i.e., cortex), and vascular tissue.

Anatomy of a Eudicot Root

Figure 9.8*a, b* also shows a cross-section of a root at the zone of maturation. These specialized tissues are identifiable:

- *Epidermis.* The epidermis, which forms the outer layer of the root, consists of only a single layer of cells. The majority of epidermal cells are thin-walled and rectangular. The epidermal cells in the zone of maturation have root hairs that project as far as 5–8 mm into the soil.
- *Cortex.* Underneath the epidermis is a layer of large, thin-walled parenchyma cells that make up the **cortex,** a type of ground tissue. These irregularly shaped cells are loosely packed, so that water and minerals can move through the cortex without entering the cells. The cells contain starch granules, which function in food storage.
- *Endodermis.* The **endodermis** is a single layer of rectangular cells that forms a boundary between the cortex and the inner vascular cylinder. The endodermal cells fit snugly together and are bordered on four sides by a layer of impermeable lignin and suberin known as the **Casparian strip** (Fig. 9.8*c*).

This strip prevents the passage of water and mineral ions between adjacent cell walls. The two sides of each cell that contact the cortex and the vascular cylinder, respectively, remain permeable. Therefore, the only access to the vascular cylinder is through the endodermal cells, as shown by the arrows in Figure 9.8*c*.

- *Vascular tissue.* The first layer of cells within the vascular cylinder is the **pericycle.** This tissue can become meristematic and start the development of branch roots (Fig. 9.9). The main portion of the vascular cylinder contains xylem and phloem. The xylem appears star-shaped in eudicots because several arms of tissue radiate from a common center (see Fig. 9.8). The phloem is found in separate regions between the arms of the xylem.

Anatomy of Monocot Roots

Monocot roots (Fig. 9.10) have the same growth zones as eudicot roots, but they do not undergo secondary growth as many eudicot roots do. A monocot root contains **pith,** a type of ground tissue, which is centrally located. The pith is surrounded by a vascular ring composed of alternating xylem and phloem bundles.

Root Diversity

Roots tend to be of two varieties. In eudicots, the first (primary) root grows straight down and is called a **taproot.** Monocots have a

Figure 9.10 Monocot root. a. This overall cross-section of a monocot root shows that a vascular ring surrounds a central pith. **b.** The enlargement shows the placement of various tissues.

(a): ©McGraw-Hill Education/Al Telser, photographer; (b): ©George Ellmore/Tufts University

fibrous root system, which may have large numbers of fine roots of similar diameter. Many mature plants have a combination of taproot and fibrous root systems.

Taproot   Fibrous root

Some plants have roots that possess a variety of adaptations to better perform their functions. Root adaptations can improve anchorage to the ground or the storage of carbohydrates. Other root systems need to increase absorption of water, minerals, oxygen, or sunlight.

Most roots store some food, but in certain plants, such as the sweet potato or carrot, the roots are enlarged and store large quantities of starch. Perennials rely on this stored carbohydrate to grow year after year (Fig. 9.11*a*). When roots develop from organs of the shoot system instead of the root system, they are known as *adventitious roots.* One style of adventitious root is typically found in corn plants. These are called *prop roots* because they emerge above the soil and act as anchors for the plant. Specialized corn prop roots, produced toward the base of stems, support these plants in high wind (Fig. 9.11*b*).

Roots, like all organs, need oxygen to perform cellular respiration. Roots normally get their oxygen from the air pockets in the soil, but oxygen becomes scarce when the plant grows exclusively in water. Large trees growing in water, such as mangrove and bald cypress, have evolved spongy roots called *pneumatophores,* which extend above the water's surface and enhance gas exchange (Fig. 9.11*c*). Many *epiphytes* (plants that live in or on trees) have aerial roots for a variety of reasons. English ivy uses aerial roots to climb the bark of trees. Orchids use their roots to capture moisture in the air and support their weight, and in some instances, green aerial roots perform photosynthesis (Fig. 9.11*d*).

Plants such as dodders and broomrapes are parasitic on other plants. Their stems have rootlike projections called haustoria (*sing.,* haustorium) that grow into the host plant and make contact with vascular tissue from which they extract water and nutrients.

Symbiotic Relationships in Roots

Two symbiotic relationships assist roots in taking up mineral nutrients. In the first type, legumes (soybeans and alfalfa) have roots infected by nitrogen-fixing *Rhizobium* bacteria. These bacteria can fix atmospheric nitrogen (N_2) by breaking the $N \equiv N$ bond and

a.   b.
c.   d.

Figure 9.11 Root diversity and specialization.   a. Sweet potato plants have food storage roots. **b.** Prop roots are specialized for support. **c.** The pneumatophores of this tree allow it to acquire oxygen even though it lives in water. **d.** The aerial roots of orchids offer physical support, water and nutrient uptake, and in some cases, even photosynthesis. (a): ©Brett Stevens/ Getty Images; (b): ©NokHoOkNoi/iStock/360/Getty Images; (c): ©FLPA/Mark Newman/ age fotostock; (d): ©DEA/S Montanari/age fotostock

reducing nitrogen to NH_4^+ for incorporation into organic compounds. The bacteria live in **root nodules** and are supplied with carbohydrates by the host plant (Fig. 9.12*a*). The bacteria, in turn, furnish their host with nitrogen compounds.

The second type of symbiotic relationship, called a mycorrhizal association, involves fungi and almost all plant roots (Fig. 9.12*b*). Only a small minority of plants do not have mycorrhizae (*sing.,* mycorrhiza). Ectomycorrhizae form a mantle that is exterior to the root that grows between the cell walls. Endomycorrhizae can penetrate cell walls. In any case, the fungus increases the surface area available for mineral and water uptake and breaks down organic matter, releasing nutrients for the plant to use. In return, the root furnishes the fungus with sugars and amino acids. Plants are extremely dependent on mycorrhizae for maximum growth (Fig. 9.12*b*). Orchid seeds, which are quite small and contain

a. Root nodules

plant without mycorrhizae

plant with mycorrhizae

mycorrhizae in root cells

100×

b. Mycorrhizae

Figure 9.12 Root nodules and mycorrhizae.
a. Nitrogen-fixing bacteria live in nodules on the roots of plants, particularly legumes. Plants grown with mycorrhizae (*plant on right*) grow much larger than a plant (*left*) grown without mycorrhizae. **b.** Ectomycorrhizae on red pine roots. (a): ©DS7O/Getty Images; (photo) (b): ©Dr. Keith Wheeler/Science Source

limited nutrients, do not germinate until a mycorrhizal fungus has invaded their cells. Nonphotosynthetic plants, such as Indian pipe, use their mycorrhizae to extract nutrients from nearby trees. Plants without mycorrhizae are usually limited as to the environment in which they can grow.

Check Your Progress 9.4

1. Explain how a plant controls what enters the vascular bundle.
2. Contrast eudicot and monocot roots.
3. List several specializations of roots and identify which of the three main functions of a root is involved in each specialization.

9.5 Organization of Stems

Learning Outcomes

Upon completion of this section, you should be able to

1. Identify the structures of a woody twig.
2. Describe the anatomy of eudicot and monocot stems.
3. Explain the secondary growth of stems.
4. Identify the adaptations that lead to stem diversity.

The anatomy of a woody twig helps us review the organization of a stem (Fig. 9.13). The **terminal bud** contains the shoot tip protected by bud scales, which are modified leaves. Leaf scars and bundle scars mark the location of leaves that have dropped. Dormant axillary buds that can give rise to branches or flowers are also found here. Each spring when growth resumes, bud scales fall off and leave a scar. You can tell the age of a stem by counting these groups of bud scale scars because there is one for each year's growth.

The terminal bud includes the shoot apical meristem and also leaf primordia (young leaves) that differentiate from cells produced by the shoot apical meristem (Fig. 9.14*a*). The activity of the terminal bud would be hindered by a protective covering

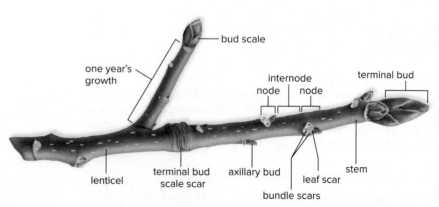

bud scale

one year's growth

internode
node   node

terminal bud

lenticel

terminal bud scale scar

axillary bud

leaf scar

stem

bundle scars

Figure 9.13 Woody twig. The major parts of a stem are illustrated by a woody twig collected in winter.

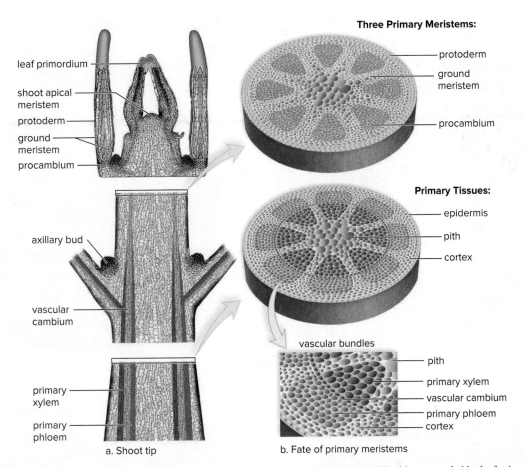

Three Primary Meristems:

- protoderm
- ground meristem
- procambium

Primary Tissues:

- epidermis
- pith
- cortex

vascular bundles

- pith
- primary xylem
- vascular cambium
- primary phloem
- cortex

leaf primordium

shoot apical meristem

protoderm

ground meristem

procambium

axillary bud

vascular cambium

primary xylem

primary phloem

a. Shoot tip

b. Fate of primary meristems

Figure 9.14 Shoot tip and primary meristems.   **a.** The shoot apical meristem within a terminal bud is surrounded by leaf primordia. **b.** The shoot apical meristem produces the primary meristems. Protoderm gives rise to epidermis; ground meristem gives rise to pith and cortex; and procambium gives rise to vascular tissue, including primary xylem, primary phloem, and vascular cambium. (a): ©Steven P. Lynch

comparable to the root cap. Instead, the leaf primordia fold over the apical meristem, providing protection. At the start of the season, the leaf primordia are, in turn, covered by terminal bud scales (scalelike leaves), but these drop off, or abscise, as growth continues.

The shoot apical meristem produces everything in a shoot: leaves, axillary buds, additional stem, and sometimes flowers. In the process, it gives rise to the same primary meristems as in the root. These primary meristems, in turn, develop into the differentiated tissues of a shoot system. The protoderm becomes the epidermis of the stem and leaves. Ground meristem produces parenchyma cells that become the cortex and pith in the stem and mesophyll in the leaves.

Procambium differentiates into the xylem and phloem of a vascular bundle. Certain cells become tracheids and others become vessel elements. The first sieve-tube members of a vascular bundle do not have companion cells and are short-lived (some live only a day before being replaced). Mature vascular bundles contain fully differentiated xylem and phloem, and a lateral meristem called **vascular cambium,** which is responsible for secondary growth. Vascular cambium is discussed more fully later in this section.

Herbaceous Stems

Mature nonwoody stems, called herbaceous stems, exhibit only primary growth. The outermost tissue of herbaceous stems is the epidermis, which is covered by a waxy cuticle to prevent water loss. In the vascular bundle, xylem is typically found toward the inside of the stem, and phloem is found toward the outside.

In a herbaceous eudicot stem, such as a sunflower, the vascular bundles are arranged in a distinct ring that separates the cortex from the central pith, where water and the products of photosynthesis are stored (Fig. 9.15). The cortex is sometimes green and carries on photosynthesis. In a monocot stem such as corn, the vascular bundles are scattered throughout the stem, and often there is no well-defined cortex or well-defined pith (Fig. 9.16).

Woody Stems

A woody plant, such as an oak tree, has both primary and secondary tissues. Primary tissues are those new tissues formed each year from primary meristems right behind the shoot apical meristem. Secondary tissues develop during the first and subsequent years of growth from lateral meristems: vascular cambium and cork

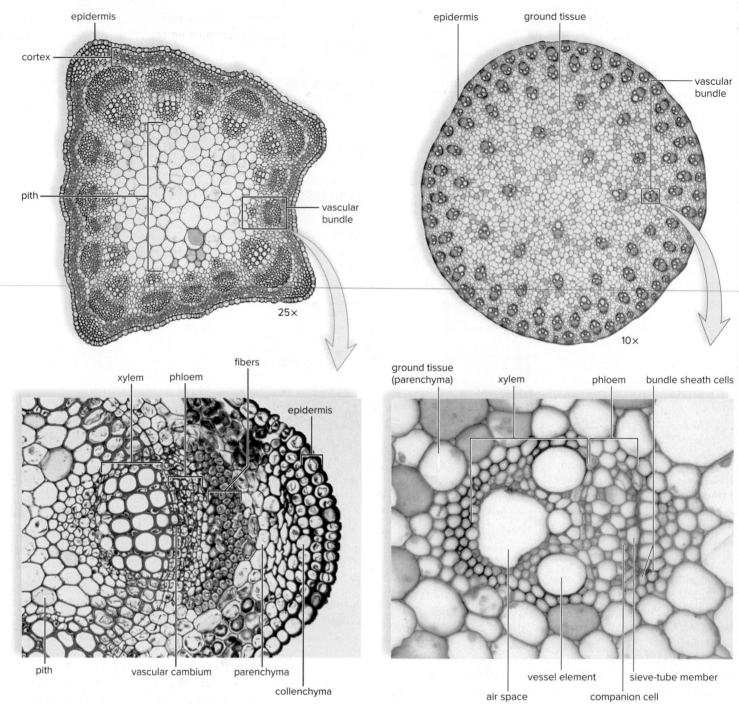

Figure 9.15 Herbaceous eudicot stem.   In eudicot stems, the vascular bundles are arranged in a ring around the pith. (top): ©Ed Reschke; (bottom): ©Ray F. Evert/University of Wisconsin Madison

Figure 9.16 Monocot stem.   In the monocot stem, the vascular bundles are scattered throughout the stem. (top): ©Carolina Biological Supply Company/Medical Images; (bottom): ©Kingsley Stern

cambium. Primary growth occurs in all plants. *Secondary growth,* which occurs only in conifers and woody eudicots, increases the girth of trunks, stems, branches, and roots.

Trees and shrubs undergo secondary growth due to activities of the vascular cambium (Fig. 9.17). In herbaceous plants, vascular cambium is present between the xylem and phloem of each vascular bundle. In woody plants, the vascular cambium develops to form a ring of meristem that divides parallel to the

surface of the plant, and produces new xylem and phloem each year. Eventually, a woody eudicot stem has an entirely different organization from that of a herbaceous eudicot stem. A woody stem has no distinct vascular bundles and instead has three distinct areas: the bark, the wood, and the pith. The pith is a collection of parenchyma cells located at the center of the stem. Vascular cambium lies between the bark and the wood, which are discussed next.

pith
primary xylem
primary phloem
cortex
epidermis

a.

Vascular cambium: Lateral meristem that will produce secondary xylem and secondary phloem in each succeeding year.

Periderm: As a stem becomes woody, epidermis is replaced by the periderm.

pith
primary xylem
secondary xylem
vascular cambium
secondary phloem
primary phloem
cortex
cork cambium
cork

b.

lenticel

Bark: Includes periderm and also living secondary phloem.
Wood: Increases each year; includes annual rings of xylem.

xylem ray
phloem ray

secondary xylem
vascular cambium
secondary phloem
primary phloem

cork cambium
cork

c.

Figure 9.17 Secondary growth of stems. **a.** Diagram showing a eudicot herbaceous stem just before secondary growth begins. **b.** Secondary growth has begun, and periderm has replaced the epidermis. Vascular cambium produces secondary xylem and secondary phloem each year. **c.** In a two-year-old stem, the primary phloem and cortex have disappeared, and only the secondary phloem (within the bark) produced by vascular cambium will be active that year. Secondary xylem builds up to become the annual rings of a woody stem.

Notice in Figure 9.17*c* the xylem rays and phloem rays visible in the cross-section of a woody stem. Rays consist of parenchyma cells that permit lateral conduction of nutrients from the pith to the cortex, as well as some storage of food. A phloem ray is actually a continuation of a xylem ray. Some phloem rays are much broader than other phloem rays.

Bark

The **bark** of a tree contains both periderm (cork, cork cambium, and a single layer of cork cells filled with suberin) and phloem.

Although secondary phloem is produced each year by vascular cambium, phloem does not build up from season to season. The bark of a tree can be removed. However, doing so is very harmful because, without phloem, organic nutrients cannot be transported.

Cork cambium develops beneath the epidermis. When cork cambium first begins to divide, it produces tissue that disrupts the epidermis and replaces it with cork cells. Cork cells are impregnated with suberin, a waxy layer that makes them waterproof but also causes them to die. This is protective, because now the stem is less edible. But an impervious barrier means that gas exchange is impeded except at lenticels, which are pockets of loosely arranged cork cells not impregnated with suberin.

Wood

Wood is secondary xylem that builds up year after year, thereby increasing the girth of trees. In trees that have a growing season, vascular cambium is dormant during the winter. In the spring, when moisture is plentiful and leaves require much water for growth, the secondary xylem contains wide vessel elements with thin walls. In this so-called *spring wood,* wide vessels transport sufficient water to the growing leaves. Later in the season, moisture is scarce, and the wood at this time, called *summer wood,* has a lower proportion of vessels (Fig. 9.18). Strength is required

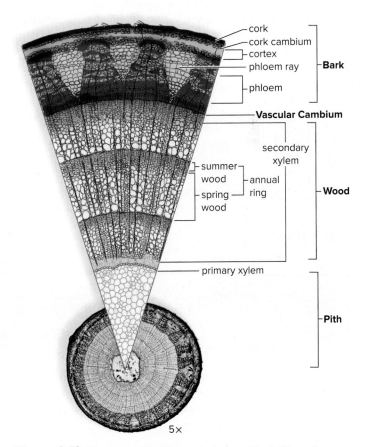

cork
cork cambium
cortex
phloem ray
phloem

Bark

Vascular Cambium

secondary xylem

summer wood
spring wood
annual ring

Wood

primary xylem

Pith

5×

Figure 9.18 Three-year-old woody twig. The buildup of secondary xylem in a woody stem results in annual rings, which tell the age of the stem. The rings can be distinguished because each one begins with spring wood (large vessel elements) and ends with summer wood (smaller and fewer vessel elements). (photo) (cross-section): ©Ed Reschke/Getty Images

The Many Uses of Bamboo

Due to its resilience, amazing rate of growth, and ability to grow in a variety of climates, bamboo is quickly becoming a valuable crop. Certain varieties of bamboo are capable of growing up to a foot per day and can reach their full height within 1 year in the right conditions.

Bamboo is classified as a grass that contains varieties ranging in height from 1 foot to over 100 feet. Globally, there are over 1,400 species of bamboo. Approximately 900 can be found in tropical climates, whereas the remaining 500 are found in temperate environments. Several varieties of bamboo are even found in the eastern and southeastern United States.

Bamboo is recognized for its versatility as a building material, commercial food product, and clothing material. It can be processed into roofing material, flooring, and support beams, as well as a variety of construction materials. The mature stalks can be used as support columns in "Green" construction. It can also be used to replace steel reinforcing rods typically used in concrete-style construction. The bamboo goods industry started increasing in popularity in the United States during the mid-1990s and is expected to exceed $25 billion in the next several years. Bamboo products are three times harder than oak.

While not used extensively in the commercial food market, bamboo does have a variety of uses (Fig. 9A). The shoots are often used in various Asian dishes as a vegetable. They can be boiled and added to many dishes or eaten raw. In China, bamboo is made into a variety of alcoholic drinks; other Asian countries make soups, pancakes, and broths. The hollow bamboo stalk can be used to cook rice and soups within it. This will give the foods a subtle but distinct taste. Additional uses include modifying bamboo into cooking utensils, most notably chopsticks.

Clothing products are now being made out of bamboo. When turned into clothing, bamboo is reported as being very light and extremely soft. It also has the ability to wick moisture away from the skin, making it ideal to wear during exercise. Several lines of baby clothing are being created from bamboo.

Bamboo is also being recognized as one of the most eco-friendly crops. It requires lesser amounts of chemicals or pesticides to grow, and removes nearly five times more greenhouse gases and produces nearly 35% more oxygen than an equivalent stand of trees. Harvesting can be done starting at the second to third year of growth, all the way through to the fifth to seventh year of growth. Because bamboo is a perennial, it allows for a yearly regrowth of the stand after harvesting instead of requiring yearly replanting.

Versatility, hardiness, ease of growing, and greater awareness of environmentally friendly products are quickly making bamboo an ideal product that may someday replace metal, wood, and plastics.

Questions to Consider

1. Would you purchase products made from bamboo, even if they cost more than products made from nonrenewable materials?
2. Should the government provide financial incentives for farmers to switch their crops to bamboo?
3. What are the negative consequences of growing and using bamboo products?

Figure 9A The many uses of bamboo.   Bamboo is quickly becoming a multifunctional product in today's society. Uses range from building materials to food and clothing. (left): ©Roman Borodaev/Alamy; (center): ©Koki Iino/Getty Images; (right): ©Michele Westmorland/Getty Images because the tree is growing larger and summer wood contains numerous thick-walled tracheids. At the end of the growing season, just before the cambium becomes dormant again, only heavy fibers with especially thick secondary walls may develop. When the trunk of a tree has spring wood followed by summer wood, the two together make up one year's growth, or an **annual ring.** You can tell the age of a tree by counting the annual rings (Fig. 9.19*a*). The outer annual rings, where transport occurs, are called sapwood.

In older trees, the inner annual rings, called the heartwood, no longer function in water transport. The cells become plugged with deposits such as resins, gums, and other substances that inhibit the growth of bacteria and fungi. Heartwood may help support a tree, although some trees stand erect and live for many years after the heartwood has rotted away. Figure 9.19*b* shows the layers of a woody stem in relation to one another.

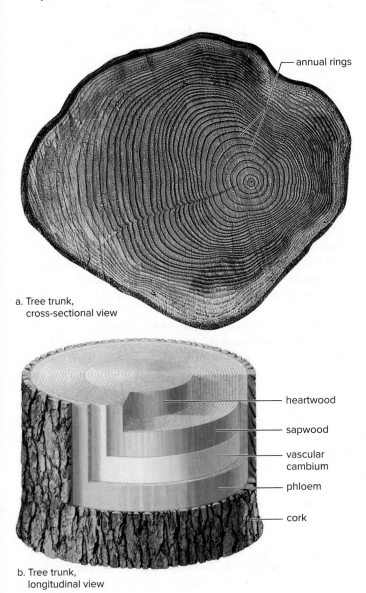

a. Tree trunk, cross-sectional view

— annual rings

— heartwood

— sapwood

— vascular cambium

— phloem

— cork

b. Tree trunk, longitudinal view

Figure 9.19 Tree trunk.   a. A cross-section of a 39-year-old tree trunk. The xylem within the darker heartwood is inactive. The xylem within the lighter sapwood is active. **b.** The relationship of bark, vascular cambium, and wood is retained in a mature stem. The pith has been buried by the growth of layer after layer of new secondary xylem. (a): ©Ardea London/Ardea.com

The annual rings are important not only in telling the age of a tree, but also in serving as a historical record of tree growth. For example, if rainfall and other conditions were extremely favorable during a season, the annual ring may be wider than usual.

Woody Plants

Is it advantageous to be woody? With adequate rainfall, woody plants can grow taller than herbaceous plants and increase in girth because they have adequate vascular tissue to support and service their leaves. However, it takes energy to produce secondary growth and prepare the body for winter if the plant lives in the temperate zone. Also, woody plants need more defense mechanisms because a long-lasting plant that stays in one spot is likely to be attacked by herbivores and parasites. Then, too, trees don't usually reproduce until they have grown for several seasons, by which time they may have succumbed to an accident or disease. In certain habitats, it is more advantageous for a plant to put most of its energy into simply reproducing rather than being woody.

Stem Diversity

Stems exist in diverse forms, or modifications (Fig. 9.20). Aboveground horizontal stems, called **stolons** or runners, produce new plants where the nodes touch the ground. The strawberry plant contains this type of stem, which functions in vegetative reproduction.

Aboveground vertical stems can also be modified. For example, cacti have succulent stems specialized for water storage, and the tendrils of grape plants (which are stem branches) allow them to climb. Morning glory and its relatives have stems that twine around support structures. Such tendrils and twining shoots help plants expose their leaves to the sun.

Underground horizontal stems, called **rhizomes,** may be long and thin, as in sod-forming grasses, or thick and fleshy, as in irises. Rhizomes survive the winter and contribute to asexual reproduction because each node bears a bud. Some rhizomes have enlarged portions called tubers, which function in food storage. Potatoes are an example of tubers.

Corms are bulbous underground stems that lie dormant during the winter, just as rhizomes do. They have thin, papery leaves and a thick stem. They also produce new plants the next growing season. Gladiolus corms are referred to as bulbs by laypersons, but the botanist reserves the term *bulb* for a structure composed of thick modified leaves attached to a short vertical stem. An onion is a bulb.

Humans make use of stems in many ways. The stem of the sugarcane plant is a primary source of table sugar. The spice cinnamon and the drug quinine are derived from the bark of *Cinnamomum verum* and various *Cinchona* species, respectively.

Check Your Progress 9.5

1. Compare and contrast eudicot and monocot stems.
2. Explain why there are two portions to every ring in a tree's annual rings.
3. Describe various stem modifications.

a. Stolon

b. Rhizome

c. Tuber

d. Corm

Figure 9.20 Stem diversity. **a.** A strawberry plant has aboveground horizontal stems called stolons. Every other node produces a new shoot system. **b.** The underground horizontal stem of an iris is a fleshy rhizome. **c.** The underground stem of a potato plant has enlargements called tubers. We call the tubers potatoes. **d.** The corm of a gladiolus is a thick stem covered by papery leaves.
(photos) (a): ©McGraw-Hill Education/Evelyn Jo Johnson, photographer; (b): ©Science Pictures Limited/Science Source; (c–d): McGraw-Hill Education/Carlyn Iverson, photographer

9.6 Organization of Leaves

Learning Outcomes

Upon completion of this section, you should be able to
1. Identify the structures of a leaf.
2. Describe some types of leaf modifications.

Figure 9.21 shows a cross-section of a generalized foliage leaf of a temperate-zone eudicot plant. The functions of a foliage leaf are to carry on photosynthesis, regulate water loss, and be protective against parasites and predators.

Epidermal tissue is located on the leaf's upper and lower surfaces, where it is well situated to be protective. Epidermis contains trichomes, little hairs that are highly protective, especially when they secrete irritating substances against predators. Epidermal cells play a significant role in regulating water loss. Closely packed epidermal cells secrete an outer, waxy cuticle that helps prevent water loss but does not allow the uptake of CO_2 needed for photosynthesis. However, the stomata, located particularly in the lower epidermis, do allow the uptake of CO_2 and can also be closed by guard cells to prevent water loss. Trichomes can also help prevent parasites from entering stomata, and their presence cuts down on water loss when stomata are open.

Leaves are the chief photosynthesizing organs of a plant. Therefore, leaves must absorb solar energy, take up CO_2, and receive water by way of leaf veins. For the reception of solar energy, foliage leaves are generally flat and thin—this shape allows solar energy to penetrate the entire width of the leaf. The body of a leaf is composed of **mesophyll.** The elongated cells of the palisade mesophyll carry on most of the photosynthesis and are situated to allow its chloroplasts to efficiently absorb solar energy. The irregular cells of spongy mesophyll are bounded by air spaces that receive CO_2 when stomata are open. The loosely packed arrangement of the cells in spongy mesophyll increases the amount of surface area for absorption of CO_2 and for water loss. Evaporation of water is actually the means by which water rises in leaf veins, as discussed in Section 9.7.

Leaf veins branch, and water needed for photosynthesis diffuses from narrow terminations that also take up the products of photosynthesis for distribution to other parts of the plant. **Bundle sheaths** are layers of cells surrounding vascular tissue. Most bundle sheaths are parenchyma cells, sclerenchyma cells, or a combination of the two. The parenchyma cells help regulate the entrance and exit of materials into and out of the leaf vein.

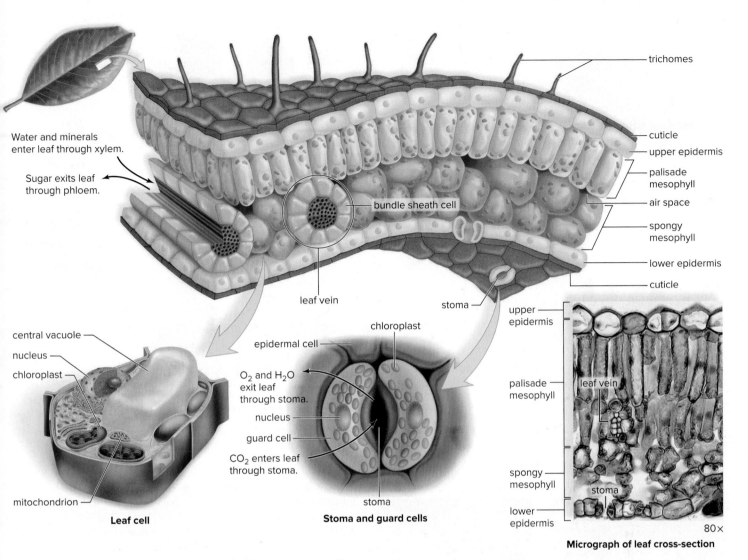

Water and minerals enter leaf through xylem.

Sugar exits leaf through phloem.

trichomes

cuticle

upper epidermis

palisade mesophyll

air space

spongy mesophyll

lower epidermis

cuticle

bundle sheath cell

leaf vein

stoma

upper epidermis

palisade mesophyll

leaf vein

spongy mesophyll

stoma

lower epidermis

80×

Micrograph of leaf cross-section

central vacuole

nucleus

chloroplast

mitochondrion

Leaf cell

epidermal cell

chloroplast

O_2 and H_2O exit leaf through stoma.

nucleus

guard cell

CO_2 enters leaf through stoma.

stoma

Stoma and guard cells

Figure 9.21 Leaf structure.   Photosynthesis takes place in the mesophyll tissue of leaves. The leaf is enclosed by epidermal cells covered with a waxy layer, the cuticle. The veins contain xylem and phloem for the transport of water and solutes. Stoma in the epidermis permit the exchange of gases. (photo): ©Ray F. Evert, University of Wisconsin, Madison

Leaf Diversity

The blade of a leaf can be simple or compound (Fig. 9.22a). A simple leaf has a single blade in contrast to a compound leaf, which is divided into leaflets. For example, a magnolia tree has simple leaves, and a buckeye tree has compound leaves. In pinnately compound leaves, the leaflets occur in pairs, as in a black walnut tree, while in palmately compound leaves, all of the leaflets are attached to a single point, as in a buckeye tree. Plants such as the mimosa have bipinnately compound leaves, with leaflets subdivided into even smaller leaflets.

Leaves can be arranged on a stem in three ways: alternate, opposite, or whorled (Fig. 9.22b). The American beech has alternate leaves; the maple has opposite leaves, being attached to the same node; and bedstraw has a whorled leaf arrangement, with several leaves originating from the same node.

Leaves are adapted to environmental conditions. Shade plants tend to have broad, wide leaves, and desert plants tend to have reduced leaves with sunken stomata. The leaves of a cactus are the spines attached to the succulent (fleshy) stem (Fig. 9.23a). Other succulents have leaves adapted to hold moisture.

An onion bulb is composed of leaves surrounding a short stem. In a head of cabbage, large leaves overlap one another. The petiole of a leaf can be thick and fleshy, as in celery and rhubarb. Leaves of climbing plants, such as those of peas and cucumbers, are modified into tendrils that can attach to nearby objects (Fig. 9.23b).

Some plants have leaves that are specialized for catching insects. A sundew has sticky trichomes that trap insects and others that secrete digestive enzymes. The Venus flytrap has hinged leaves that snap shut and interlock when an insect triggers sensitive trichomes that project from inside the leaves (Fig. 9.23c). Insectivorous plants commonly grow in marshy regions, where the supply of soil nitrogen is severely limited. The digested insects provide the plants with a source of organic nitrogen.

Check Your Progress 9.6

1. Describe the basic structure of a leaf, including the tissues where photosynthesis occurs and those that regulate the movement of water and gases.

2. Identify how leaves are adapted to their environment.

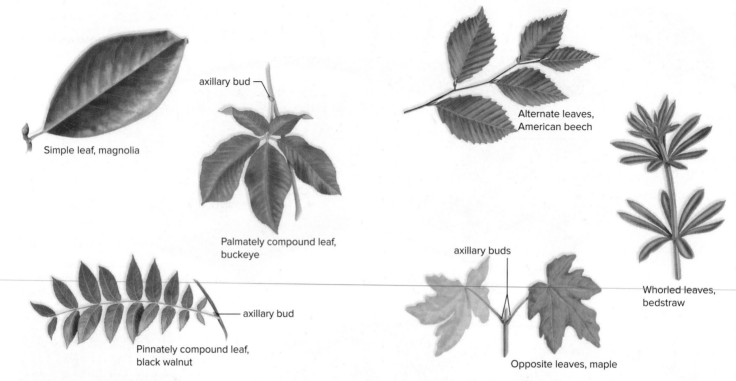

a. Simple versus compound leaves

b. Arrangement of leaves on stem

Figure 9.22 Classification of leaves. **a.** Leaves are simple or compound, being either pinnately compound or palmately compound. Note the one axillary bud per compound leaf. **b.** The leaf arrangement on a stem can be alternate, opposite, or whorled.

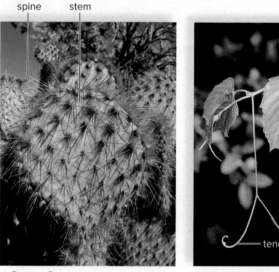

a. Cactus, *Opuntia*

b. Cucumber, *Cucumis*

c. Venus flytrap, *Dionaea*

Figure 9.23 Leaf diversity. **a.** The spines of a cactus are modified leaves that protect the fleshy stem from animal predation. **b.** The tendrils of a cucumber are modified leaves that attach the plant to a physical support. **c.** The modified leaves of the Venus flytrap serve as a trap for insect prey. When triggered by an insect, the leaf snaps shut. Once shut, the leaf secretes digestive juices that break down the soft parts of the insect's body. (a): Source: National Park Service/Robb Hannawacker, photographer; (b–c): ©Steven P. Lynch

9.7 Uptake and Transport of Nutrients

Learning Outcomes

Upon completion of this section, you should be able to

1. Explain the movement of water in a plant according to the cohesion-tension model of xylem transport.
2. Explain the movement of organic nutrients in a plant according to the pressure-flow model of phloem transport.

In order to produce a carbohydrate, a plant requires carbon dioxide from the air and water from the soil. The transport system of the plant consists of the vascular tissue: the xylem and phloem. Water and minerals are transported through a plant in xylem; the products of photosynthesis are transported in phloem.

Water Uptake and Transport

Water and minerals enter the plant through the root, primarily through the root hairs. Osmosis of water and diffusion of minerals

Figure 9.24 **Cohesion-tension model of xylem transport.**   Tension created by evaporation (transpiration) at the leaves pulls water along the length of the xylem from the root hairs to the leaves.

xylem

mesophyll cells

stoma

intercellular space

H_2O

Leaves
• Transpiration creates tension.
• Tension pulls the water column upward from the roots to the leaves.

cohesion due to hydrogen bonding between water molecules

adhesion due to polarity of water molecules

cell wall

water molecule

Stem
• Cohesion makes water column continuous.
• Adhesion keeps water column in place.

xylem

root hair

Roots
• Water enters xylem at root.
• Water column extends from leaves to root.

xylem

cells creates a positive pressure called root pressure that tends to push xylem sap upward. Atmospheric pressure can support a column of water to a maximum height of approximately 10.3 m. However, some trees can exceed 90 m in height, so other factors must be involved in causing water to move from the roots to the leaves. Figure 9.24 illustrates the accepted model for the transport of water and, therefore, minerals in a plant. It is called the **cohesion-tension model.**

The Cohesion-Tension Model of Xylem Transport

The vascular tissue of xylem contains the hollow conducting cells called tracheids and vessel elements (see Fig. 9.6). The vessel elements are larger than the tracheids, and they are stacked one on top of the other to form a pipeline that stretches from the roots to the leaves. It is an open pipeline because the vessel elements have no end walls, just perforation plates, separating one from the other. The tracheids, which are elongated with tapered ends, form a less obvious means of transport. Water can move across the end and side walls of tracheids because of pits, or depressions, where the secondary wall does not form.

Explanation of the Model   Unlike animals, which rely on a pumping heart to move blood through their vessels, plants utilize a passive, not active, means of transport to move water in xylem. The cohesion-tension model of xylem transport relies on the properties of water (see Section 2.3). The term **cohesion** refers to the tendency of water molecules to cling together. Because of hydrogen bonding, water molecules interact with one another, forming a continuous water column in xylem from the roots to the leaves that is not easily broken. **Adhesion** refers to the ability of water, a polar molecule, to interact with the molecules making up the walls of the vessels in xylem. Adhesion gives the water column extra strength and prevents it from slipping downward.

What causes the continuous water column to move passively upward? Consider the structure of a leaf. When the sun rises, stomata open, and carbon dioxide enters a leaf. Within the leaf, the mesophyll cells—particularly the spongy layer—are exposed to the air, which can be quite dry. Water now evaporates from mesophyll cells. Evaporation of water from leaf cells is called **transpiration.** At least 90% of the water taken up by the roots is eventually lost by transpiration. This means the total amount of water lost by a plant over a long period of time is surprisingly large. A single *Zea mays* (corn) plant loses somewhere between 135 and 200 liters (l) of water through transpiration during a growing season.

aid the entrance of water and minerals from the soil into the plant. Eventually, however, a plant uses active transport to bring minerals into root cells. From there, water, along with minerals, moves across the tissues of a root until it enters xylem.

Once water enters xylem, it must be transported upward to the leaves of the plant. This can be a daunting task. Water entering root

Using Plants to Clean Up Toxic Messes

Phytoremediation uses plants like poplar, mustard, and mulberry to clean up lead, uranium, and other environmental pollutants. The genetic makeup of these plants allows them to absorb, store, degrade, or transform substances that normally kill or harm other plants and animals. "It's an elegantly simple solution to pollution problems," says Louis Licht, who runs Ecolotree, an Iowa City phytoremediation company. The idea behind phytoremediation is not new, but the idea of using these plants on contaminated sites has just gained support in the last decade. Different plants work on different contaminants. The mulberry bush, for instance, is effective on industrial sludge; some grasses attack petroleum wastes; and sunflowers (together with soil additives) remove lead. The plants clean up sites in different ways, depending on the substance involved. The plants or microbes around their roots break down organic substances, allowing the remains to be absorbed by the plant or left in the soil or water. Inorganic contaminants are trapped within the body of the plant, which must then be harvested and disposed of, or processed to reclaim the trapped contaminant.

Figure 9B Phytoremediation of nitrates. Poplars and various other plants can be used to absorb excess nitrates that run off from farm fields. ©Danita Delimont/ Gallo Images/Getty Images

Poplars Take Up Excess Nitrates

The poplars act like vacuum cleaners, sucking up nitrate-laden runoff from a fertilized cornfield before this runoff reaches nearby waterways. Nitrate runoff into the Mississippi River from Midwest farms, after all, is a major cause of the large "dead zone" of oxygen-depleted water that develops each summer in the Gulf of Mexico. Poplars planted along the edge of farm fields have been shown to reduce the nitrate levels of streams by more than 90% (Fig. 9B).

Mustard Plants Take Up Uranium

Phytoremediation has also helped clean up badly polluted sites, in some cases at a fraction of the usual cost. Uranium was removed from a Superfund site on an Army firing range in Aberdeen, Maryland. Mustard plants removed the uranium at as little as 10% of the cost of traditional cleanup methods.

Limitations of Phytoremediation

One of the main limitations to phytoremediation is the pace. Depending on the contaminant, it can take several growing seasons to clean a site—much longer than conventional methods. Phytoremediation is also effective only at depths that plant roots can reach, making it useless against deep-lying contamination unless the soil is excavated. Phytoremediation will not work on lead and other metals unless chemicals are added to the soil. Plants will also vary in the amount of pollutant they can absorb.

Despite its shortcomings, experts see a bright future for this technology because, for one reason, the costs are relatively small compared to those of traditional remediation technologies. Traditional methods of cleanup require much energy input and therefore have higher cost. In general, phytoremediation is a low-cost alternative to traditional methods because less energy is required for operation and maintenance.

Questions to Consider

1. What happens to the pollutants when the plant dies?
2. Why would one plant be more adapted to absorbing a particular nutrient than another?
3. How does this process relate to the cohesion-tension and pressure-flow models?

The water molecules that evaporate are replaced by other water molecules from the leaf veins. In this way, transpiration exerts a driving force—that is, a *tension*—which draws the water column up in vessels from the roots to the leaves. As transpiration occurs, the water column is pulled upward, first within the leaf, then from the stem, and finally from the roots.

Tension can reach from the leaves to the root only if the water column is continuous. What happens if the water column within xylem is broken, as by cutting a stem? The water column "snaps back" in the xylem vessel, away from the site of breakage, making it more difficult for conduction to occur. This is why it is best to maximize water conduction by cutting flower stems under water. This effect has also allowed investigators to measure the tension in stems. A device called the pressure bomb measures how much pressure it takes to push the xylem sap back to the cut surface of the stem.

There is an important consequence to the way water is transported in plants. When a plant is under water stress, the stomata close. Now the plant loses little water because the leaves are protected against water loss by the waxy cuticle of the upper and lower epidermis. When stomata are closed, however, carbon dioxide cannot enter the leaves, and plants are unable to photosynthesize. (CAM plants are a notable exception. See Section 8.4.) Photosynthesis, therefore, requires an abundant supply of water so that the stomata remain open and allow carbon dioxide to enter.

Opening and Closing of Stomata

Each stoma, a small pore in the leaf epidermis, is bordered by two **guard cells** (Fig. 9.25). When water enters the guard cells and turgor pressure increases, the stoma opens. When water exits the

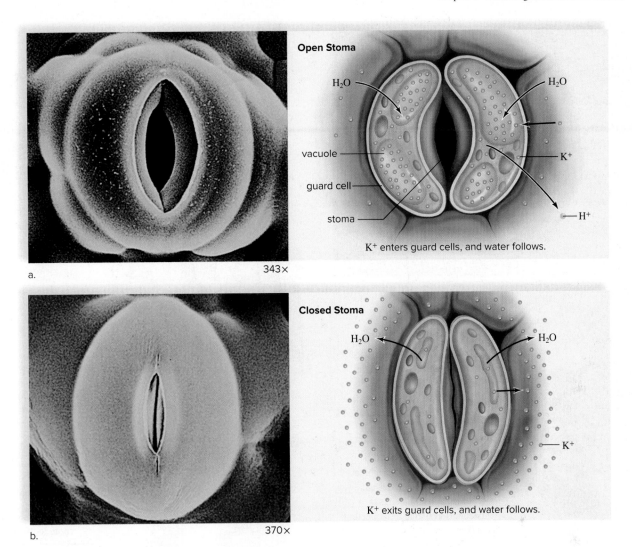

Figure 9.25 Opening and closing of stomata. **a.** A stoma opens when turgor pressure increases in guard cells due to the entrance of K^+ followed by the entrance of water. **b.** A stoma closes when turgor pressure decreases due to the exit of K^+ followed by the exit of water. (photos, both): ©Jeremy Burgess/SPL/ Science Source

guard cells and turgor pressure decreases, the stoma closes. Notice in Figure 9.25 that the guard cells are attached to each other at their ends and that the inner walls are thicker than the outer walls. When water enters, a guard cell's radial expansion is restricted because of cellulose microfibrils in the walls, but lengthwise expansion of the outer walls is possible. When the outer walls expand lengthwise, they buckle out from the region of their attachment, and the stoma opens.

Since about 1968, it has been clear that potassium ions (K^+) accumulate within guard cells when stomata open. In other words, active transport of K^+ into guard cells causes water to follow by osmosis and stomata to open.

If plants are kept in the dark, stomata open and close about every 24 hours, just as if they were responding to the presence of sunlight in the daytime and the absence of sunlight at night. This means some sort of internal biological clock must be keeping time. Circadian rhythms (behaviors that occur nearly every 24 hours) and biological clocks are areas of intense investigation at this time. Other factors that influence the opening and closing of stomata include temperature, humidity, and stress.

Organic Nutrient Transport

Plants transport water and minerals from the roots to the leaves. They also transport organic nutrients from the leaves to the parts of plants that need them. This includes young leaves that have not yet reached their full photosynthetic potential, flowers that are in the process of making seeds and fruits, and the roots, whose location in the soil prohibits them from carrying on photosynthesis.

The Role of Phloem

In the 1600s, Marcello Malpighi suggested that bark is involved in moving sugars from leaves to roots. He observed that if a strip of bark is removed below the level of the majority of the tree's leaves, the bark swells just above the cut, and sugar accumulates in the swollen tissue. We know today this is because the phloem is being removed, but the xylem is left intact. Therefore, the results suggest that phloem is the tissue that transports sugars.

Radioactive tracer studies with carbon 14 (^{14}C) have confirmed that phloem transports organic nutrients. When ^{14}C-labeled carbon dioxide (CO_2) is supplied to mature leaves,

a. An aphid feeding on a plant stem

b. Aphid stylet in place

Figure 9.26 Acquiring phloem sap. Aphids are small insects that remove nutrients from phloem by means of a needlelike mouthpart called a stylet. **a.** Excess phloem sap appears as a droplet after passing through the aphid's body. **b.** Micrograph of a stylet in plant tissue. When an aphid is cut away from its stylet, phloem sap becomes available for collection and analysis. (a): ©Dr. M.H. Zimmerman/ Harvard University; (b): ©Steven P. Lynch

radioactively labeled sugar is soon found moving down the stem into the roots. It's difficult to get samples of sap from phloem without injuring the phloem, but this problem is solved by using aphids, small insects that are phloem feeders. The aphid drives its stylet, a sharp mouthpart that functions like a hypodermic needle, between the epidermal cells, and sap enters its body from a sieve-tube member (Fig. 9.26). If the aphid is anesthetized using ether, its body can be carefully cut away, leaving the stylet. Phloem can then be collected and analyzed. The use of radioactive tracers and aphids has revealed that the movement of nutrients through phloem can be as fast as 60–100 cm per hour and possibly up to 300 cm per hour.

The Pressure-Flow Model of Phloem Transport

The **pressure-flow model** is the current explanation for the movement of organic materials in phloem. Consider the following experiment in which two bulbs are connected by a glass tube. The left-hand bulb contains solute at a higher concentration than the right-hand bulb. Each bulb is bounded by a differentially

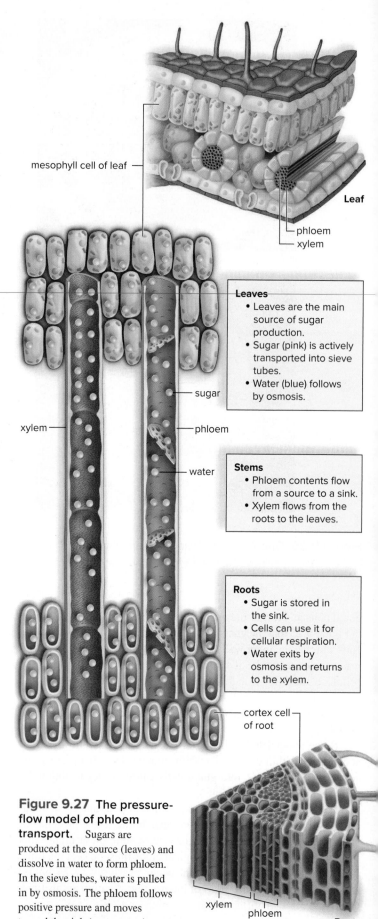

mesophyll cell of leaf

Leaf

phloem
xylem

Leaves
• Leaves are the main source of sugar production.
• Sugar (pink) is actively transported into sieve tubes.
• Water (blue) follows by osmosis.

sugar

xylem

phloem

water

Stems
• Phloem contents flow from a source to a sink.
• Xylem flows from the roots to the leaves.

Roots
• Sugar is stored in the sink.
• Cells can use it for cellular respiration.
• Water exits by osmosis and returns to the xylem.

cortex cell of root

Figure 9.27 The pressure-flow model of phloem transport. Sugars are produced at the source (leaves) and dissolve in water to form phloem. In the sieve tubes, water is pulled in by osmosis. The phloem follows positive pressure and moves toward the sink (root system).

xylem
phloem

Root

permeable membrane, and the entire apparatus is submerged in distilled water:

Distilled water flows into the left-hand bulb to a greater extent because it has the higher solute concentration. The entrance of water creates a positive *pressure,* and water *flows* toward the second bulb. This flow not only drives water toward the second bulb, but it also provides enough force for water to move out through the membrane of the second bulb—even though the second bulb contains a higher concentration of solute than the distilled water.

In plants, sieve tubes are analogous to the glass tube that connects the two bulbs. Sieve tubes are composed of sieve-tube members, each of which has a companion cell. It is believed that the companion cells assist the sieve-tube members in some way. The sieve-tube members align end to end, and strands of plasmodesmata (cytoplasm) extend through sieve plates from one sieve-tube member to the other. Sieve tubes, therefore, form a continuous pathway for organic nutrient transport throughout a plant.

During the growing season, photosynthesizing leaves are producing sugar (Fig. 9.27). Therefore, they are a **source** of sugar. This sugar is actively transported into phloem. Again, transport is

dependent on an electrochemical gradient established by a proton pump, a form of active transport. Sugar is carried across the membrane in conjunction with hydrogen ions (H^+), which are moving down their concentration gradient. After sugar enters sieve tubes, water follows passively by osmosis. The buildup of water within sieve tubes creates the positive pressure that starts a flow of phloem contents. The roots (and other growth areas) are a **sink** for sugar, meaning they are removing sugar, storing it, or using it for cellular respiration. After sugar is actively transported out of sieve tubes, water exits phloem passively by osmosis (which reduces pressure at the sink) and is taken up by xylem, which transports water to leaves, where it is used for photosynthesis. Now, phloem contents continue to flow from the leaves (source) to the roots (sink).

The pressure-flow model of phloem transport can account for any direction of flow in sieve tubes if we consider that the direction of flow is always from source to sink.

Check Your Progress 9.7

1. Describe cohesion and adhesion and how they are involved in moving water up a plant.
2. Explain how transpiration is involved in moving water throughout a plant.
3. Explain how osmosis is involved in moving organic nutrients in the plant.

Conclusion

The preservation of biodiversity is important not only to preserve the health of ecosystems, such as rain forests, but the health of humans as well. Plants, such as bamboo and the neem tree, play an important role in the health and welfare of humans. In addition to providing wood and building materials, plants can be a source of chemicals that may be used to fight diseases, from hypertension and pain to heart disease.

SUMMARIZE

9.1 Cells and Tissues of Plants

■ There are over 391,000 known species of plants. Of these, the majority are the flowering plants, or **angiosperms.**

■ A plant has the ability to grow its entire life because it possesses **meristematic tissue. Apical meristems** are located at or near the tips of stems and roots, where they increase the length of these structures. This increase in length is called *primary growth.* The apical meristems continually produce three types of meristem: protoderm, ground meristem, and procambium.

■ The entire body of both nonwoody (herbaceous) and young woody plants is covered by a layer of **epidermis,** which in most plants contains a single layer of closely packed epidermal cells. The walls of

epidermal cells exposed to air are covered with a waxy **cuticle** to minimize water loss.

■ **Ground tissue** forms the bulk of a plant and contains **parenchyma, collenchyma,** and **sclerenchyma** cells. Parenchyma cells are thin-walled and capable of photosynthesis when they contain chloroplasts. Collenchyma cells have thicker walls for flexible support. Sclerenchyma cells are hollow, nonliving support cells with secondary walls fortified by **lignin.**

■ **Vascular tissue** consists of **xylem** and **phloem.** Xylem contains two types of conducting cells: **vessel elements** and **tracheids.** Xylem transports water and minerals. In phloem, sieve tubes are composed of **sieve-tube members,** each of which has a **companion cell.** Phloem transports sugar and other organic compounds, including hormones.

9.2 Plant Organs and Systems

■ A flowering plant has two main organ systems, the **shoot system** and the **root system.**

■ Vegetative organs are those involved in growth and nutrition. **Roots** are the organ that anchor a plant, absorb water and minerals, and store the products of photosynthesis. **Stems** are organs that support leaves, conduct materials to and from roots and leaves, and help store plant products. **Leaves** are organs that are specialized for gas exchange, and they carry on most of the photosynthesis in the plant.

9.3 Monocot Versus Eudicot Plants

■ Flowering plants are divided into **monocots** and **eudicots** according to the number of **cotyledons** in the seed; the arrangement of vascular tissue in roots, stems, and leaves; number of flower parts; and number of pores or slits in the pollen grains.

9.4 Organization of Roots

■ Primary growth occurs in both the shoot and root systems. In the root system, a root tip has a zone of cell division (protected by a **root cap**), a zone of elongation, and a zone of maturation.

■ A cross-section of a herbaceous eudicot root reveals the epidermis, which protects; the **cortex,** which stores food; the **endodermis,** which regulates the movement of minerals; and the vascular cylinder, which is composed of vascular tissue. The **Casparian strip** regulates the movement of water and minerals.

■ The first layer of cells in the vascular cylinder is called the **pericycle.** In the vascular cylinder of a eudicot, the xylem appears star-shaped, and the phloem is found in separate regions, between the arms of xylem. In contrast, a monocot root has a ring of vascular tissue with alternating bundles of xylem and phloem surrounding the **pith.**

■ Roots are diversified. **Taproots** are specialized to store the products of photosynthesis. A **fibrous root system** covers a wider area. Prop roots are adventitious roots specialized to provide increased anchorage.

■ Roots may enter into symbiotic relationships with bacteria (**root nodules**) and fungi.

9.5 Organization of Stems

■ The activity of the shoot apical meristem within a **terminal bud** accounts for the primary growth of a stem. A terminal bud contains internodes and leaf primordia at the nodes. When stems grow, the internodes lengthen.

■ In a cross-section of a nonwoody eudicot stem, epidermis is the outermost layer of cells, followed by cortex tissue, vascular bundles in a ring, and an inner pith. Monocot stems have scattered vascular bundles, and the cortex and pith are not well defined.

■ Secondary growth of a woody stem is due to activities of the **vascular cambium,** which produces new xylem and phloem every year, and **cork cambium,** which produces new cork cells when needed. Cork, a part of the **bark,** replaces epidermis in woody plants.

■ In a cross-section of a woody stem, the bark is composed of all the tissues outside the vascular cambium: secondary phloem, cork cambium, and cork. **Wood** consists of secondary xylem, which builds up year after year and forms **annual rings.**

■ Stems are diverse. Modified stems include some tendrils, horizontal aboveground **stolons** and underground stems, **rhizomes,** and corms.

9.6 Organization of Leaves

■ The bulk of a leaf is **mesophyll** tissue bordered by an upper and lower layer of epidermis. The epidermis is covered by a cuticle and may bear trichomes. Stomata tend to be in the lower layer. Vascular tissue is present within **leaf veins.** The leaf veins conduct water and the products of photosynthesis to various parts of the plant body. **Bundle sheaths** are the cells surrounding the vascular tissue.

■ Leaves are diverse. The spines of a cactus are leaves. Other succulents have fleshy leaves. An onion is a bulb with fleshy leaves, and the tendrils of cucumbers are leaves. The Venus flytrap has leaves that trap and digest insects.

9.7 Uptake and Transport of Nutrients

■ Water transport in plants occurs within xylem. The **cohesion-tension model** of xylem transport states that transpiration (evaporation of water at stomata) creates tension, which pulls water upward in xylem. This method works only because water molecules are **cohesive** and **adhesive. Transpiration** and carbon dioxide uptake occur when stomata are open.

■ Stomata open when **guard cells** take up potassium (K^+) ions and water follows by osmosis. Stomata open because the entrance of water causes the guard cells to buckle out.

■ Transport of organic nutrients in plants occurs within phloem. The **pressure-flow model** of phloem transport states that sugar is actively transported into phloem at a **source,** and water follows by osmosis. The resulting increase in pressure creates a flow, which moves water and sugar to a **sink.**

ASSESS

Testing Yourself

Choose the best answer for each question.

9.1 Cells and Tissues of Plants

1. All of the tissues of a plant originate from
   a. epidermal tissue.
   b. meristem tissue.
   c. ground tissue.
   d. vascular tissue.

2. _____ forms the outer protective covering of a plant.
   a. Meristem tissue
   b. Ground tissue
   c. Vascular tissue
   d. Epidermal tissue

3. Tracheids and vessel elements would be found in
   a. vascular tissue.
   b. ground tissue.
   c. epidermal tissue.
   d. meristem tissue.

9.2 Plant Organs and Systems

4. Which of the following would not be found in the shoot system of a plant?
   a. stem
   b. leaves
   c. roots
   d. All of these are correct.

5. Nodes and internodes are found on the _____ of a plant.
   a. roots
   b. leaves
   c. stems
   d. All of these are correct.

6. All of the following are vegetative organs except
   a. leaves.
   b. stems.
   c. flowers.
   d. roots.
   e. All of these are correct.

9.3 Monocot Versus Eudicot Plants

7. Which of these is an incorrect contrast between monocots (stated first) and eudicots (stated second)?
   a. one cotyledon—two cotyledons
   b. leaf veins parallel—net veined
   c. pollen with three pores—pollen with one pore
   d. flower parts in threes—flower parts in fours or fives
   e. All of these are correct.

9.4 Organization of Roots

8. Root hairs are found in the zone of
   a. cell division.
   b. elongation.
   c. maturation.
   d. apical meristem.
   e. All of these are correct.

9. The Casparian strip is found
   a. between all epidermal cells.
   b. between xylem and phloem cells.
   c. surrounding endodermal cells.
   d. within the secondary wall of parenchyma cells.
   e. in both endodermis and pericycle.

9.5 Organization of Stems

10. Between the bark and the wood in a woody stem, there is a layer of meristem called
    a. cork cambium.
    b. vascular cambium.
    c. apical meristem.
    d. the zone of cell division.
    e. procambium preceding bark.

11. Which of these is a stem?
    a. carrot
    b. stolon of strawberry plants
    c. spine of cacti
    d. sweet potato
    e. pneumatophore

9.6 Organization of Leaves

12. Which part of a leaf carries out most of the photosynthesis of a plant?
    a. vascular bundle
    b. mesophyll
    c. epidermal layer
    d. guard cells
    e. trichomes

13. Gases, such as CO_2, enter the leaf through
    a. stomata.
    b. bundle sheath cells.
    c. the mesophyll.
    d. the phloem.

9.7 Uptake and Transport of Nutrients

14. What role do cohesion and adhesion play in xylem transport?
    a. Like transpiration, they create a tension.
    b. Like root pressure, they create a positive pressure.
    c. Like sugars, they cause water to enter xylem.
    d. They create a continuous water column in xylem.

15. The sugar produced by mature leaves moves into sieve tubes by way of _____, while water follows by _____.
    a. osmosis, osmosis
    b. active transport, active transport
    c. osmosis, active transport
    d. active transport, osmosis

ENGAGE

BioNOW

Want to know how this science is relevant to your life? Check out the BioNOW video below:

■ Saltwater Filter

Discuss how the vascular tissues xylem and phloem are involved in the experimental process in this video.

Thinking Critically

1. If you were given an unfamiliar vegetable, how could you tell if it was a root or a stem, based on its external features and microscopic examination of its cross-section?

2. What are some of the potential evolutionary advantages of being woody?

3. Compare and contrast the transport of water and organic nutrients in plants with the transport of blood in humans.

4. *Welwitschia* is a genus of plant that lives in the Namib and Mossamedes Deserts in Africa. Annual rainfall averages only 2.5 cm (1 inch) per year. *Welwitschia* plants contain a large number of stomata (22,000 per cm^2), which remain closed most of the time. Can you suggest how a large number of stomata would be beneficial to these desert plants?

CHAPTER

10

Plant Reproduction, Growth, and Response

Source: USDA-ARS/Bruce Fritz, photographer

BEFORE YOU BEGIN

Before beginning this chapter, take a few moments to review the following discussions:

Section 5.4 What role does meiosis play in sexual reproduction?

Section 5.5 What are the differences between mitosis and meiosis?

Section 9.3 What is the difference between a monocot and eudicot plant?

Seedless Plants

The observation that sunflowers track the sun as it moves through the sky is a striking example of a flowering plant's ability to respond to environmental stimuli. Other responses to light can take longer than sun-tracking because they involve hormones and an alteration in growth. For example, flowering plants will exhibit a bend toward the light within a few hours, because a hormone produced by the growing tip has moved from the sunny side to the shady side of the stem. Hormones also help flowering plants respond to stimuli in a coordinated manner. In the spring, seeds germinate and growth begins if the soil is warm enough to contain liquid water. In the fall, when temperatures drop, shoot and root apical growth ceases. Some plants also flower according to the season. The pigment phytochrome is instrumental in detecting the photoperiod and bringing about changes in gene expression, which determine whether a plant flowers or does not flower.

The fundamentals of agriculture and supplying plant-based products to the human population include a better understanding of how flowering plant reproduction occurs, seed production and development, and how plants operate under natural conditions. This chapter discusses the variety of ways flowering plants can respond to their environment, including other organisms.

As you read through this chapter, think about the following questions:

1. What hormones are associated with the normal growth and development of flowering plants?
2. What are some examples of how humans have manipulated plants to better serve our needs?
3. What environmental stimuli trigger the various plant responses?

10.1 Sexual Reproduction in Flowering Plants

Like in animals, sexual reproduction in plants is advantageous because it generates variation among the offspring through the process of meiosis and random fertilization. When plants reproduce sexually, they undergo an **alternation of generations,** in which they alternate between two multicellular stages, one diploid (2n) and one haploid (n). In this chapter, we will use the life cycle of a flowering plant, or angiosperm, as our example (Fig. 10.1).

In the plant life cycle there are two stages. The diploid stage of the life cycle is called the **sporophyte.** The sporophyte produces haploid spores by the process of meiosis. These spores then divide by mitosis to become the **gametophyte.** The gametophyte is haploid and it divides by mitosis to produce haploid gametes (egg and sperm). Fertilization of an egg by a sperm produces a new diploid sporophyte.

Alternation of Generations in Flowering Plants

In flowering plants, the sporophyte is dominant. Dominant simply means that this is the stage where the plant spends the majority of its life cycle and conducts the majority of photosynthesis. In the case of the angiosperms, this is the stage of the life cycle that bears flowers (Fig. 10.1). A **flower** is a reproductive structure and it produces two types of spores: microspores and megaspores. A **microspore** undergoes mitosis and becomes a pollen grain, which is the male gametophyte. Meanwhile, the **megaspore** has undergone mitosis to become a microscopic embryo sac, which is the female gametophyte. The female gametophyte is retained within the flower.

A pollen grain is either windblown or carried by an animal to the vicinity of the embryo sac. At maturity, a pollen grain contains two nonflagellated sperm that travel down a pollen tube to the embryo sac. Upon fertilization, the structure surrounding the embryo sac develops into a **seed.** The seeds are enclosed by a **fruit,** which aids in dispersing the seeds. When a seed germinates, a new sporophyte emerges and develops into a mature organism.

Adaptation to a Land Environment

The life cycle of flowering plants is adapted to a land existence because all stages of the life cycle are protected from drying out (see Section 30.1). For example, the microscopic gametophytes develop within the sporophyte. Pollen grains are not released

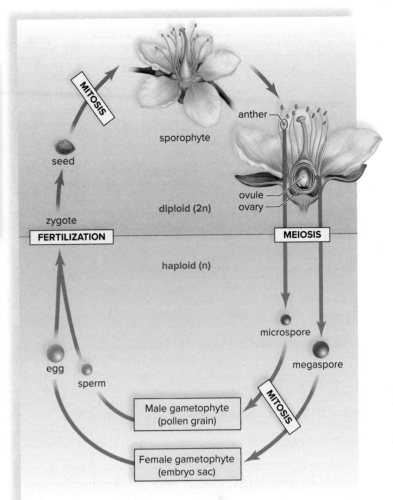

Figure 10.1 Alternation of generations in flowering plants. The sporophyte bears flowers. The flower produces microspores within anthers and megaspores within ovules by meiosis. A megaspore becomes a female gametophyte, which produces an egg within an embryo sac, and a microspore becomes a male gametophyte (pollen grain), which produces sperm. Fertilization results in a seed-enclosed zygote and stored food.

until they have a thick wall, and they are carried by wind or an animal (usually an insect) to another flower, where they develop a pollen tube that carries the sperm to the egg. Following fertilization, the seed coat protects the embryo until conditions favor regrowth.

Flowers

The flower is unique to the group of plants known as the angiosperms. The evolution of the flower was a major factor leading to the success of angiosperms, which now number over 369,000 described species. Aside from producing the spores and protecting the gametophytes, flowers often attract pollinators. Flowers also produce the fruits that enclose the seeds.

A flower develops in response to various environmental signals (see Section 10.4). In many plants, a shoot apical meristem stops producing leaves and starts producing a bud that contains

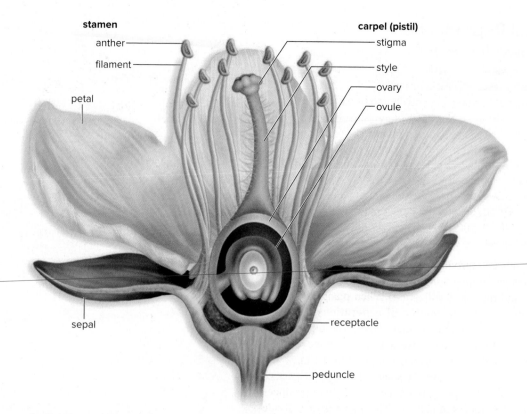

Figure 10.2 Anatomy of a flower.   A complete flower has all flower parts: sepals, petals, stamens, and at least one carpel.

a flower. In other plants, axillary buds develop directly into flowers.

A typical flower has four whorls of modified leaves attached to a receptacle at the end of a flower stalk. The receptacle bearing a single flower is attached to a structure called a peduncle, while the structure that bears one of several flowers is called a pedicel. Figure 10.2 shows the following basic floral structures:

1. **Sepals** are the most leaflike of all the flower parts, are usually green, and protect the bud as the flower develops within. Sepals can also be the same color as the flower petals.
2. **Petals** are the structures used to attract a variety of pollinators. Wind-pollinated flowers may have no petals at all.
3. **Stamens** are the "male" portion of the flower. Each stamen has two parts: the **anther,** a saclike container, and the **filament,** a slender stalk. Pollen grains develop from the microspores produced in the anther.
4. At the very center of a flower is the **carpel** (also called the pistil), a vaselike structure that represents the "female" portion of the flower. A carpel usually has three parts: the **stigma,** an enlarged sticky knob; the **style,** a slender stalk; and the **ovary,** an enlarged base that encloses one or more ovules.

In monocots, the petals and sepals occur in threes and multiples of three. In eudicots, the petals and sepals are in fours or fives and multiples of four or five (Fig. 10.3).

A flower can have a single carpel or multiple carpels. Sometimes several carpels are fused into a single structure, in which case this compound ovary has several chambers containing **ovules,** each of which will house a megaspore. For example, an orange develops from a compound ovary, and every section of the orange is a chamber.

Not all flowers have sepals, petals, stamens, and carpels. Those that do, such as the flower in Figure 10.2, are said to be *complete,* and those that do not are said to be *incomplete.* Flowers that have both stamens and carpels are called *perfect* (bisexual) flowers. Those with only stamens or only carpels are *imperfect* (single-sex) flowers. If staminate and carpellate (pistillate) flowers are on one plant, the plant is *monoecious* (one house) (Fig. 10.4). If staminate and carpellate flowers are on separate plants, the plant is *dioecious* (two houses).

Life Cycle of Flowering Plants

In plants, the sporophyte produces haploid spores by meiosis. The haploid spores grow and develop into haploid gametophytes, which produce gametes by mitotic division.

Flowering plants are heterosporous, meaning they produce microspores and megaspores. Microspores become mature male gametophytes (sperm-bearing pollen grains), and megaspores become mature female gametophytes (egg-bearing embryo sacs).

stamen

p2

s1

s2

carpel

petal

p1

p3

sepal

s3

a. Daylily

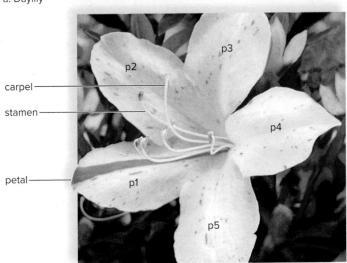

p3

p2

carpel

stamen

p4

petal

p1

p5

b. Festive azalea

Figure 10.3 Monocot versus eudicot flowers.   a. Monocots, such as daylilies, have flower parts usually in threes. In particular, note the three petals and three sepals. **b.** Azaleas are eudicots. They have flower parts in fours or fives; note the five petals of this flower. (p = petal; s = sepal)

(a): ©Gilles Delacroix/Garden World Images/age fotostock; (b): ©Pat Pendarvis

Development of the Male Gametophyte

Microspores are produced in the anthers of flowers (Fig. 10.5). An anther has four pollen sacs, each containing many microspore mother cells. A microspore mother cell undergoes meiosis to produce four haploid microspores. In each, the haploid nucleus divides mitotically, followed by unequal cytokinesis, and the result is two cells enclosed by a finely sculptured wall. This structure, called the **pollen grain,** is at first an immature male gametophyte that consists of a tube cell and a generative cell. The larger tube cell will eventually produce a *pollen tube.* The smaller generative cell divides mitotically either now or later to produce two sperm. Once these events take place, the pollen grain has become the mature male gametophyte.

a. Staminate flowers

b. Carpellate flowers

Figure 10.4 Corn plants are monoecious.   A corn plant has **(a)** clusters of staminate flowers and **(b)** clusters of carpellate flowers. Staminate flowers produce the pollen that is carried by wind to the carpellate flowers, where an ear of corn develops. (a): ©Adam Hart-Davis/SPL/Science Source; (b): ©Design Pics Inc./Alamy

Development of the Female Gametophyte

The ovary contains one or more ovules. An ovule has a central mass of parenchyma cells almost completely covered by layers of tissue called integuments, except where there is an opening called a micropyle. One parenchyma cell enlarges to become a megaspore mother cell, which undergoes meiosis, producing four haploid megaspores (Fig. 10.5). Three of these megaspores are nonfunctional; one is functional. Typically, the nucleus of the functional megaspore divides mitotically until there are eight nuclei in the *female gametophyte.* When cell walls form later, seven cells are formed, one of which contains two nuclei (binucleate). This collection of seven cells is collectively called the female gametophyte, or **embryo sac.**

Pollination and Fertilization

The walls separating the pollen sacs in the anther break down when the pollen grains are ready to be released (Fig. 10.6). **Pollination** is simply the transfer of pollen from an anther to the stigma of a carpel. Self-pollination occurs if the pollen is from the same plant, and cross-pollination occurs if the pollen is from a different plant of the same species.

Angiosperms often have adaptations to foster cross-pollination. For example, the carpels may mature only after the anthers have released their pollen. Cross-pollination may also be brought about with the assistance of a particular pollinator. The Ecology feature "The Coevolution of Plants and Their Pollinators" explores the importance of insects and other pollinators to the life cycle of a flowering plant. Plants have evolved mechanisms, such as color, odors, or production of nectar, to attract specific pollinators to their flowers. If

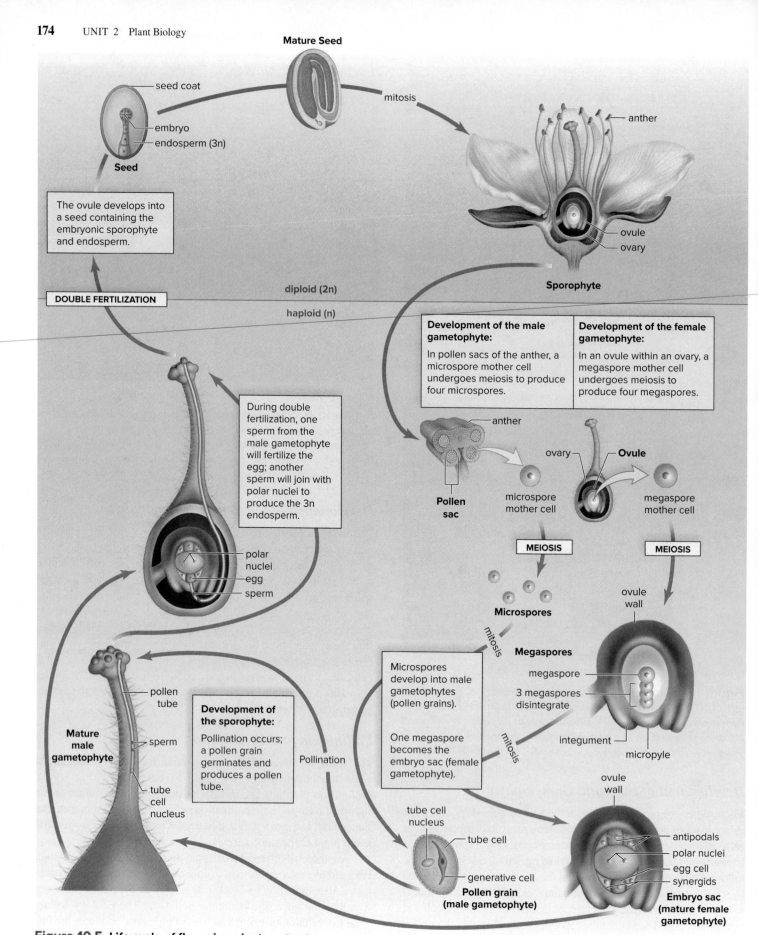

Figure 10.5 Life cycle of flowering plants.   *Development of gametophyte (bottom):* A pollen sac in the anther contains microspore mother cells, which produce microspores by meiosis. A microspore develops into a pollen grain, which germinates and has two sperm. An ovule in an ovary contains a megaspore mother cell, which produces a megaspore by meiosis. A megaspore develops into an embryo sac containing seven cells, one of which is an egg. *Development of sporophyte (top):* A pollen grain contains two sperm by the time it germinates and forms a pollen tube. During double fertilization, one sperm fertilizes the egg to form a diploid zygote, and the other fuses with the polar nuclei to form a triploid (3n) endosperm cell. A seed contains the developing sporophyte embryo plus stored food.

The Coevolution of Plants and Their Pollinators

Plants and their pollinators have adapted to one another. They have a mutualistic relationship in which each benefits—the plant uses its pollinator to ensure that cross-pollination takes place, and the pollinator uses the plant as a source of food. This mutualistic relationship came about through the process of coevolution—that is, the interdependency of the plant and the pollinator is the result of suitable changes in the structure and function of each.

Bee- and Wasp-Pollinated Flowers

There are 20,000 known species of bees that pollinate flowers. The best-known pollinators are the honeybees (Fig. 10A*a*). As noted in the text, bee eyes see ultraviolet (UV) wavelengths. Therefore, bee-pollinated flowers are usually brightly colored and are predominantly blue or yellow; they are not entirely red. They may also have ultraviolet shadings called nectar guides, which highlight the portion of the flower that contains the reproductive structures. The mouthparts of bees are fused into a long tube that contains a tongue. This tube is an adaptation for sucking up nectar provided by the plant, usually at the base of the flower. Bees also collect pollen as a food.

Bee-pollinated flowers are delicately sweet and fragrant, advertising that nectar is present. The nectar guides often point to a narrow floral tube large enough for the bee's feeding apparatus but too small for other insects to reach the nectar. Bee-pollinated flowers may be irregular in shape, and are sturdy because they often have a landing platform where the bee can alight. The flower structure requires the bee to brush up against the anther and stigma as it moves toward the floral tube to feed.

One type of orchid, *Ophrys,* has evolved a unique adaptation. The flower resembles a female wasp, and when the male of that species attempts to copulate with the flower, the flower spring-loads pollen on the wasp's head. When the frustrated wasp attempts to "copulate" with another flower, the pollen is perfectly positioned to come in contact with the stigma of the second flower.

Moth- and Butterfly-Pollinated Flowers

Both moths and butterflies have a long, thin, hollow proboscis, but they differ in other characteristics. Moths usually feed at night and have a well-developed sense of smell. The flowers they visit are visible at night, because they are lightly shaded (white, pale yellow, or pink), and they have strong, sweet perfume, which helps

Figure 10A **Types of pollinators.** **a.** A bee-pollinated flower is a color other than red. **b.** A butterfly-pollinated flower is often a composite, containing many individual flowers. **c.** Hummingbird-pollinated flowers are curved back, allowing the bird to insert its beak to reach the rich supply of nectar. **d.** Bat-pollinated flowers are large, sturdy flowers that can take rough treatment. (a): ©Steven P. Lynch; (b): ©Purestock/Getty Images; (c): ©Daniel Dempster Photography/Alamy; (d): ©Dr. Merlin D. Tuttle/Bat Conservation International/Science Source

attract moths. Moths hover when they feed, and their flowers have deep tubes with open margins that allow the hovering moths to reach the nectar with their long proboscis.

Butterflies, in comparison, are active in the daytime and have good vision but a weak sense of smell. Their flowers have bright colors—even red, because butterflies can see the color red—but the flowers tend to be odorless. Unable to hover, butterflies need a place to land. Flowers that are visited by butterflies often have flat landing platforms (Fig. 10A*b*). Composite flowers (composed of a compact head of numerous individual flowers) are especially favored by butterflies. Each flower has a long, slender floral tube, accessible to the long, thin butterfly proboscis.

Bird- and Bat-Pollinated Flowers

In North America, the most well-known bird pollinators are the hummingbirds. These small animals have good eyesight but do not have a well-developed sense of smell. Like moths, they hover when they feed. Typical flowers pollinated by hummingbirds are red, with a slender floral tube and margins that are curved back

and out of the way. And although they produce copious amounts of nectar, the flowers have little odor. As a hummingbird feeds on nectar with its long, thin beak, its head comes in contact with the stamens and pistil (Fig. 10A*c*).

Bats are adapted to gathering food in various ways, including feeding on the nectar and pollen of plants. Bats are nocturnal and have an acute sense of smell. Those that are pollinators also have keen vision and a long, extensible, bristly tongue. Typically, bat-pollinated flowers open only at night and are light-colored or white. They have a strong, musky smell similar to the odor that bats produce to attract one another. The flowers are generally large and sturdy and are able to hold up when a bat inserts part of its head to reach the nectar. While the bat is at the flower, its head becomes dusted with pollen (Fig. 10A*d*).

Questions to Consider

1. Why do you think pollinators are often very specific to the plants they pollinate?
2. What are the potential consequences if honeybees were to go extinct?
3. What are some other examples of coevolution you can think of?

Figure 10.6 Pollen grains. **a.** Grass releasing pollen. **b.** Pollen grains of Canadian goldenrod, *Solidago canadensis*. **c.** Pollen grains of asparagus, *Asparagus officinalis*. The shape and pattern of pollen grain walls are quite distinctive, and experts can use them to identify the genus, and sometimes even the species, that produced a particular pollen grain. (a): ©Tim Gainey/Alamy; (b): ©Medical-on-line/Alamy; (c): ©Steve Gschmeissner/Science Photo Library/Getty Images

a pollinator goes from flower to flower of only one type of plant, cross-pollination is more likely to occur in an efficient manner. Over time, certain pollinators have become adapted to reach the nectar of only one type of flower. The process by which the evolution of two species is connected is called **coevolution,** and it is considered to be a major factor in the success of the angiosperms.

When a pollen grain lands on the stigma of the same species, it germinates, forming a pollen tube (see Fig. 10.5). The germinated pollen grain, containing a tube cell and two sperm, is the mature male gametophyte. As it grows, the pollen tube passes between the cells of the stigma and the style to reach the micropyle, a pore of the ovule. Now **double fertilization** occurs. One sperm nucleus unites with the egg nucleus, forming a 2n zygote, and the other sperm nucleus migrates and unites with the polar nuclei of the central cell, forming a 3n endosperm cell. The zygote divides mitotically to become the **embryo,** a young sporophyte, and the endosperm cell divides mitotically to become the endosperm. **Endosperm** is the tissue that will nourish the embryo and seedling as they develop.

Check Your Progress 10.1

1. Identify which generation produces two types of spores and what will happen to the spores.
2. List the reproductive parts of a flower and give a function for each part.
3. Describe what is meant by the term *heterosporous*.
4. Describe the process of double fertilization.

10.2 Growth and Development

Learning Outcomes

Upon completion of this section, you should be able to

1. Recognize the developmental steps of a eudicot embryo and compare the function of its cotyledons to that of a cotyledon in monocots.
2. Identify different types of fruits.
3. Label seed structure and describe germination and dispersal.

Development of the Eudicot Embryo

The endosperm cell, shown in Figure 10.7a, divides to produce the endosperm tissue, shown in Figure 10.7b. The zygote also divides, but asymmetrically. One of the resulting cells is small, with dense cytoplasm. This cell is destined to become the embryo, and it divides repeatedly in different planes, forming a ball of cells. The other, larger cell also divides repeatedly, but it forms an elongated structure called a suspensor, which has a basal cell. The suspensor pushes the embryo deep into the endosperm tissue. The suspensor then disintegrates as the seed matures.

During the globular stage, the embryo is a ball of cells (Fig. 10.7c). The root-shoot axis of the embryo is already established at this stage because the embryonic cells near the suspensor will become a root, whereas those at the other end will ultimately become a shoot.

The embryo has a heart shape when the **cotyledons,** or seed leaves, appear (Fig. 10.7d). As the embryo continues to enlarge and elongate, it takes on a torpedo shape (Fig. 10.7e). Now the root tip and shoot tip are distinguishable. The shoot apical meristem in the shoot tip is responsible for aboveground growth, and the root apical meristem in the root tip is responsible for belowground growth. In a mature embryo, the epicotyl is the portion between the cotyledon(s) that contribute(s) to shoot development (Fig. 10.7f). The hypocotyl is that portion below the cotyledon(s) that contributes to stem development. The radicle contributes to root development. Now, the embryo is ready to develop the two main parts of a plant: the shoot system and the root system.

The cotyledons are quite noticeable in a eudicot embryo and may fold over. As the embryo develops, the wall of the ovule becomes the seed coat.

Monocot Versus Eudicot Cotyledons

Eudicots have two cotyledons; monocots have only one. In monocots, the cotyledon rarely stores food. Instead, it absorbs food molecules from the endosperm and passes them to the embryo. In eudicots, the cotyledons usually store the nutrient molecules the embryo uses. Therefore, the endosperm disappears because it has been taken up by the two cotyledons.

Fruit Types and Seed Dispersal

A fruit is derived from an ovary and sometimes other flower parts. It serves to protect and help disperse the offspring. As a fruit develops, the ovary wall thickens to become the pericarp

Figure 10.7 Development of a eudicot embryo.   a. The singled-celled zygote lies beneath the endosperm nucleus. **b, c.** The endosperm is a mass of tissue surrounding the embryo. The embryo is located above the suspensor. **d.** The embryo becomes heart-shaped as the cotyledons begin to appear. **e.** There is progressively less endosperm as the embryo differentiates and enlarges. As the cotyledons bend, the embryo takes on a torpedo shape. **f.** The embryo consists of the epicotyl, the hypocotyl, and the radicle.

(Fig. 10.8*a*). The pericarp can contain up to three layers: exocarp, mesocarp, and endocarp.

Simple fruits are derived from a simple ovary of a single carpel or from a compound ovary of several fused carpels. A pea pod is an example of a simple fruit. At maturity, the pea pod breaks open on both sides of the pod to release the seeds. Peas and beans are legumes. A *legume* is a fruit that splits along two sides when mature.

Like legumes, cereal grains of wheat, rice, and corn are *dry fruits*. Sometimes, these grainlike fruits are mistaken for seeds because a dry pericarp adheres to the seed within. These dry fruits are indehiscent, meaning they don't split open. Humans gather grains before they are released from the plant and then process them to acquire their nutrients.

In some simple fruits, the mesocarp becomes fleshy. When ripe, *fleshy fruits* often attract animals and provide them with food (Fig. 10.8*b*). Peaches and cherries are examples of fleshy fruits that have a hard endocarp. This type of endocarp protects the seed so it can pass through the digestive system of an animal and remain unharmed. In a tomato, the entire pericarp is fleshy. If you cut open a tomato, you see several chambers because the flower's carpel is composed of several fused carpels.

Among fruits, apples are an example of an *accessory fruit* because the bulk of the fruit is not from the ovary, but from the receptacle. Only the core of an apple is derived from the ovary. If you cut an apple crosswise, it is obvious that an apple, like a tomato, came from a compound ovary with several chambers.

Compound fruits develop from several individual ovaries. For example, each little part of a raspberry or blackberry is derived from a separate ovary. Because the flower had many separate carpels, the resulting fruit is called an *aggregate fruit* (Fig. 10.8*c*). The strawberry is also an aggregate fruit, but each ovary becomes a one-seeded fruit called an achene. The flesh of a strawberry is from the receptacle. In contrast, a pineapple comes from many different carpels that belong to separate flowers. As the ovaries mature, they fuse to form a large, *multiple fruit* (Fig. 10.8*d*).

Dispersal of Seeds

To increase a plant's chance of success, it needs to be widely distributed. In order to achieve success, plant seeds need to be dispersed away from the parent plant.

Figure 10.8 Structure and function of fruits.   **a.** Maple trees produce a winged fruit. The wings rotate in the wind and keep the fruit aloft. **b.** Strawberry plants produce an accessory fruit. Each "seed" is actually a fruit on a fleshy, expanded receptacle. **c.** Like strawberries, blackberries are aggregate fruits. Each "berry" is derived from an ovary within the same flower. **d.** A pineapple is a multiple fruit derived from the ovaries of many flowers. (a): ©Tonarinokeroro/123RF; (b): ©MIXA/Getty Images; (c): ©David Marsden/Getty Images; (d): ©Ingram Publishing/Alamy

seed covered by pericarp   wing

a.

b.

one fruit

fruits from ovaries of one flower

c.

one fruit

fruits from ovaries of many flowers

d. A multiple fruit contains many fused produced from simple ovaries of inc flowers.

Plants have various means of ensuring that dispersal takes place. The hooks and spines of clover, bur, and cocklebur attach to the fur of animals and the clothing of humans. Birds and mammals sometimes eat fruits, including the seeds, which are then defecated (passed through the digestive tract with the feces) some distance from the parent plant. Squirrels and other animals gather seeds and fruits, which they bury some distance away.

Some plants have fruits with trapped air or seeds with inflated sacs that help them float in water. The fruit of the coconut palm, for example, which can be dispersed by ocean currents, may land hundreds of kilometers away from the parent plant. Many plants have seeds with structures such as woolly hairs, plumes, and wings that adapt them for dispersal by wind. The seeds of an orchid, however, are so small and light they need no special adaptation to carry them far away. The somewhat heavier dandelion fruit uses a tiny "parachute" for dispersal. The winged fruit of a maple tree, which contains two seeds, has been known to travel up to 10 kilometers (km) from its parent (Fig. 10.8a). A touch-me-not plant has seed pods that swell as they mature. When the pods finally burst, the ripe seeds are hurled out.

Germination of Seeds

Provided the environmental conditions are right, seeds **germinate**—that is, they begin to grow so that a seedling appears (Fig. 10.9).

Some seeds do not germinate until they have been dormant for a specific period of time. For seeds, **dormancy** is the time during which no growth occurs, even though conditions may be favorable for growth. In the temperate zone, seeds often have to be exposed to a period of cold weather before dormancy is broken. In deserts, germination does not occur until there is adequate moisture. This requirement helps ensure that seeds do not germinate until the most favorable growing conditions are present. Germination takes place if there is sufficient water, warmth, and oxygen to sustain growth.

Germination requires regulation, and both inhibitors and stimulators are known to exist. Fleshy fruits (e.g., apples and tomatoes) contain inhibitors so that germination does not occur until the seeds are removed and washed. In contrast, stimulators are present in the seeds of some temperate-zone woody plants. Mechanical action may also be required. Water, bacterial action, and even fire can act on the seed coat, allowing it to become permeable to water. The uptake of water causes the seed coat to burst.

Eudicot Versus Monocot Seed Germination

Eudicot embryos have two cotyledons that have absorbed the endosperm (Fig. 10.9). The cotyledons supply nutrients to the embryo and seedling and eventually shrivel and disappear. If the two cotyledons of a bean seed are parted, you can see a rudimentary plant. The epicotyl bears young leaves and is called a *plumule.* As the eudicot seedling emerges from the soil, the shoot is hook-shaped to protect the delicate plumule. The hypocotyl becomes part of the stem, and the radicle develops into the roots.

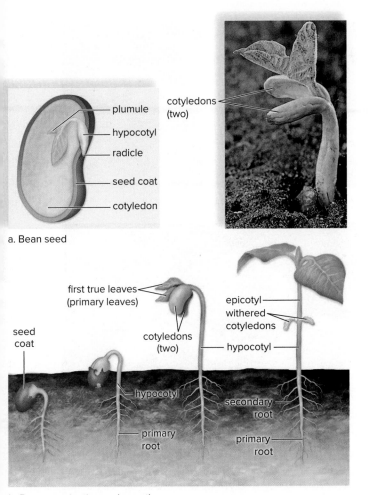

Figure 10.9 Eudicot seed structure and germination.   The common garden bean is a eudicot. **a.** Eudicot seed structure. **b.** Germination and development of the seedling. (photo) (a): ©Ed Reschke

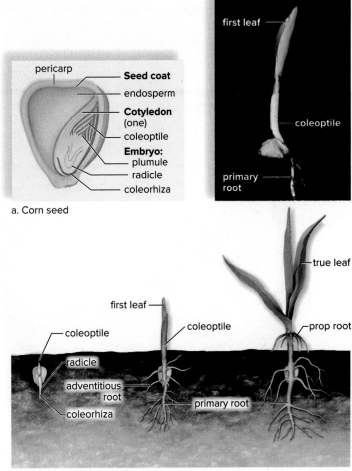

Figure 10.10 Monocot seed structure and germination.   A corn plant is an example of a monocot. **a.** Seed structure. **b.** Germination and development of the seedling. (photo) (a): ©Ed Reschke/Getty Images

In monocots, the endosperm is the food-storage tissue, and the single cotyledon does not have a storage role (Fig. 10.10). Corn is a monocot, and its kernels are actually fruits; therefore, the outer covering is the pericarp. The plumule and radicle are enclosed in protective sheaths called the coleoptile and the coleorhiza, respectively. The plumule and the radicle burst through these coverings during germination.

Check Your Progress 10.2

1. Compare the function of endosperm in monocots versus eudicots.
2. Identify examples of plant adaptations for seed dispersal by wind.
3. Explain why seeds don't germinate immediately after dispersal.
4. Describe the differences between eudicot and monocot seed germination.

10.3 Asexual Reproduction in Plants

Learning Outcomes

Upon completion of this section, you should be able to
1. Identify the asexual methods of reproduction in plants.
2. Describe how tissue culture can be used to clone plants with desirable traits.

Asexual reproduction is the production of an offspring from a single parent. Because the offspring is genetically identical to the parent, it is called a clone. Asexual reproduction is less complicated in plants, because pollination and seed production are not required. Therefore, it can be advantageous when the parent is already well adapted to a particular environment and the production of genetic variations is not an apparent necessity. In addition, clones are desirable for plant sellers, because they know exactly how the plant will grow and look.

Figure 10.11 Asexual reproduction through tissue culture.   a. Meristem tissue is placed on sterile media, and an undifferentiated mass, called a callus, develops. **b.** From the callus, organogenesis takes place and leaves or roots develop. **c.** The callus develops into a plantlet. **d.** Plantlets can be stored, and then shipped in sterile containers and transferred to soil for growth into adult plants. (a): ©Biophoto Associates/Science Source; (b–c): ©Prof. Dr. Hans-Ulrich Koop, from Plant Cell Reports, 17:601-604; (d): ©Edwin Remsberg/Alamy

Asexual Reproduction from Stems, Roots, and Cuttings

Because plants contain nondifferentiated meristem tissue, they routinely reproduce asexually by vegetative propagation. For example, complete strawberry plants will grow from the nodes of stolons (see Fig. 9.20*a*) and irises will grow from the nodes of rhizomes (see Fig. 9.20*b*). White potatoes are actually portions of belowground stems, and each eye is a bud that will produce a new potato plant if it is planted with a portion of the swollen tuber. Sweet potatoes are modified roots. They can be propagated by planting sections of the root. You may have noticed that the roots of some fruit trees, such as cherry and apple trees, produce "suckers," small plants that can grow into new trees.

Many types of plants are now propagated from stem cuttings. The discovery that the plant hormone auxin can cause roots to develop from a stem, as discussed in Section 10.4, has expanded the list of plants that can be propagated from stem cuttings.

Propagation of Plants in Tissue Culture

One of the major disadvantages of most asexual propagation techniques is that they also propagate pathogens. Plant pathogens can be viruses, bacteria, or fungi; clones created from an infected parent will also be infected. However, it is possible to maintain plants in a disease-free status if clones from an uninfected parent are made in sterile conditions through tissue culture. Hence, **tissue culture** is simply plant propagation done in a laboratory under sterile conditions.

One of the key reasons for success in plant tissue culture, as opposed to animal cell culture, is the totipotency of plant cells. **Totipotency** is the ability of individual plant cells to develop into an entire plant. The only exceptions are the plant cells that lose their nuclei (sieve-tube members) or that are dead at maturity (xylem, sclerenchyma, and cork cells).

Techniques for tissue culture vary, but most begin with cells from the *meristem* of the parent plant. Meristematic cells, which are already actively dividing, are grown on an agar medium in flasks, tubes, or petri dishes. The medium contains auxin, cytokinins, various nutrients, and water. Initially, a mass of undifferentiated (unspecialized) cells, called a callus, forms (Fig. 10.11*a*).

The addition of different combinations of plant hormones (see Section 10.4) will signal *organogenesis* (organ formation) and initiate the formation of roots and shoots, until a fully developed plantlet is formed (Fig. 10.11*b, c*). Plantlets can be shipped in their sterile containers to growers for transplantation into a greenhouse or field (Fig. 10.11*d*).

Tissue culture is an important technique for propagating many fruits and vegetables. A notable example is the banana, which is a sterile hybrid that cannot produce seeds. The only way to produce this commercially important fruit is through tissue culture. As another example, asparagus is a dioecious plant, and all commercial stalks are male. The female stalks favor the production of flowers and are undesirable for eating. Tissue culture is a more efficient means of producing disease-free male asparagus for growers.

Many botanical gardens and universities use tissue culture for plant conservation. The Atlanta Botanical Garden has a tissue culture lab that propagates native species of orchids and lilies. The propagated plants are shared with local nurseries, resulting in fewer plants being removed from the wild by collectors. Tissue culture of rare species is also used to help increase the population in the wild, because they are replanted in native habitats

A technique called cell suspension culture allows scientists to extract chemicals (secondary metabolites) from plant cells, which may have been genetically modified. This technique allows scientists to avoid overcollection of wild plants from their natural environments. These cells produce the same chemicals the plant produces. For example, cell suspension cultures of *Cinchona ledgeriana* produce quinine, which is used to treat leg cramping, a major symptom of malaria. Several *Digitalis* species produce digitalis, digitoxin, and digoxin, which are useful in the treatment of heart disease.

Check Your Progress 10.3

1. List two ways in which plants can reproduce asexually.
2. Identify how plants are grown in tissue culture.
3. Identify the possible benefits of asexual reproduction.

10.4 Control of Growth and Responses

Learning Outcomes

Upon completion of this section, you should be able to

1. Explain the importance of plant hormones.
2. Identify the types of plant hormones and describe their function.
3. Recognize how plants respond to stimuli.

Plants respond to environmental stimuli such as light, gravity, and seasonal changes, usually by a change in their pattern of growth. Hormones are involved in these responses.

Plant Hormones

Plant **hormones** are small organic molecules produced by the plant that serve as chemical signals between cells and tissues. The commonly recognized groups of plant hormones include the auxins, gibberellins, cytokinins, abscisic acid, ethylenes, and brassinosteroids. Other chemicals, some of which differ only slightly from the natural hormones, also affect the growth of plants. These and the naturally occurring hormones are sometimes grouped together and called plant growth regulators.

Plant hormones bring about a physiological response in target cells after binding to a specific receptor protein in the plasma membrane. As an example, Figure 10.12 shows how the hormone

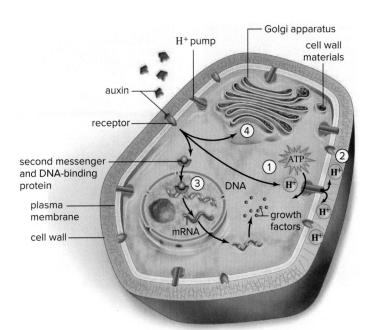

Figure 10.12 Auxin mode of action. (**1**) The binding of auxin to a receptor stimulates an H⁺ pump, moving hydrogen ions into the cell wall. (**2**) The resulting acidity causes the cell wall to weaken, and water enters the cell. (**3**) At the same time, a second messenger stimulates production of growth factors, and (**4**) new cell wall materials are synthesized. The result of all these actions is that the cell elongates, causing the stem to elongate.

auxin brings about elongation of a plant cell, a necessary step toward differentiation and maturation.

Auxins

The most common naturally occurring **auxin** is indoleacetic acid (IAA). It is produced in shoot apical meristem and is found in young leaves, flowers, and fruits. Therefore, you would expect auxin to affect many aspects of plant growth and development.

Effects of Auxin   Auxin is present in the apical meristem of a plant shoot, where it causes the shoot to grow from the top, a phenomenon called **apical dominance.** Only when the terminal bud is removed, deliberately or accidentally, are the axillary buds able to grow, allowing the plant to branch (Fig. 10.13). Pruning the top (apical meristem) of a plant generally achieves a fuller look because of increased branching of the main body of the plant.

The application of a weak solution of auxin to a woody cutting causes adventitious roots to develop more quickly than they would otherwise. Auxin production by seeds also promotes the growth of fruit. As long as auxin is concentrated in leaves or fruits rather than in the stem, leaves and fruits do not fall off. Therefore, trees can be sprayed with auxin to keep mature fruit from falling to the ground.

How Auxins Work   When a plant is exposed to unidirectional light, auxin moves to the shady side, where it binds to receptors and activates an ATP-driven pump that transports hydrogen ions (H⁺) out of the cell (see Fig. 10.12). The acid environment weakens cellulose fibrils, and activated enzymes further degrade the cell wall. Water now enters the cell, and the resulting increase in turgor pressure causes the cell to elongate and the stem to bend

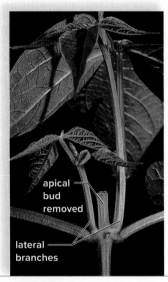

a. Plant with apical bud intact   b. Plant with apical bud removed

Figure 10.13 Apical dominance.   a. Auxin in the apical bud inhibits axillary bud development, and the plant exhibits apical dominance. **b.** When the apical bud is removed, lateral branches develop. (both photos): ©Prof. Malcolm B. Wilkins

a.   b.

Figure 10.14 Effect of gibberellins.   a. The plant on the right was treated with gibberellins; the plant on the left was not treated. **b.** The grapes are larger on the right, because gibberellins caused an increase in the space between the grapes, allowing them to grow larger. (a): ©Science Source; (b): ©Amnon Lichter, The Volcani Center

toward the light. This growth response to unidirectional light is called **phototropism.** Besides phototropism, auxins are also involved in **gravitropism,** in which roots curve downward and stems curve upward in response to gravity, as discussed later in this section.

Synthetic auxins are used today in a number of applications. These auxins are sprayed on plants, such as tomatoes, to induce the development of fruit without pollination, creating seedless varieties. Synthetic auxins have been used as herbicides to control broadleaf weeds, such as dandelions and other plants. These substances have little effect on grasses. Agent Orange is a powerful synthetic auxin that was used in extremely high concentrations to defoliate the forests of Vietnam during the Vietnam War. This powerful auxin proved to be carcinogenic and harmed not only many of the local people, but also any soldiers who came in contact with the compound.

Gibberellins

Gibberellins were discovered in 1926 while a Japanese scientist was investigating a fungal disease of rice plants called "foolish seedling disease." The plants elongated too quickly, causing the stem to weaken and the plant to collapse. The fungus infecting the plants produced an excess of a chemical called gibberellin, named after the fungus *Gibberella fujikuroi.* It was not until 1956 that gibberellic acid was able to be isolated from a flowering plant. Sources of gibberellin in flowering plant parts are young leaves, roots, embryos, seeds, and fruits.

Gibberellins are growth-promoting hormones that bring about elongation of the resulting cells. We know of about 136 gibberellins that differ slightly in their chemical makeup. The most common of these is gibberellic acid, GA_3 (the subscript designation distinguishes it from other gibberellins).

Effects of Gibberellins   When gibberellins are applied externally to plants, the most obvious effect is stem elongation (Fig. 10.14). Gibberellins can cause dwarf plants to grow, cabbage plants to become 2 m tall, and bush beans to become pole beans.

During dormancy a plant does not grow, even though conditions may be favorable for growth. The dormancy of seeds and buds can be broken by applying gibberellins, and research with barley seeds has shown how GA_3 influences the germination of seeds. Barley seeds have a large, starchy endosperm that must be broken down into sugars to provide energy for the embryo to grow. It is hypothesized that after GA_3 attaches to a receptor in the plasma membrane, calcium ions (Ca^{2+}) combine with a protein. This complex is believed to activate the gene that codes for amylase. Amylase then acts on starch to release the sugars needed for the seed to germinate. Gibberellins have been used to promote barley seed germination for the beer brewing industry.

Cytokinins

Cytokinins were discovered as a result of attempts to grow plant tissue and organs in culture vessels in the 1940s. It was found that cell division occurred when coconut milk (a liquid endosperm) and yeast extract were added to the culture medium. Although the effective agent or agents could not be isolated, they were collectively called cytokinins because cytokinesis refers to division of the cytoplasm. A naturally occurring cytokinin was not isolated until 1967. Because it came from the kernels of maize (*Zea*), it was called zeatin.

Effects of Cytokinins   The cytokinins promote cell division. These substances, which are derivatives of adenine, one of the purine bases in DNA and RNA, have been isolated from actively dividing tissues of roots and also from seeds and fruits. A synthetic cytokinin, called kinetin, also promotes cell division.

Researchers have found that **senescence** (aging) of leaves can be prevented by applying cytokinins. When a plant organ, such as a leaf, loses its natural color, it is most likely undergoing senescence. During senescence, large molecules within the leaf are broken down and transported to other parts of the plant. Senescence does not always affect the entire plant at once. For example, as some plants grow taller, they naturally lose their lower leaves. Not only can cytokinins prevent the death of leaves, but they can also initiate leaf growth. Axillary buds begin to grow despite apical dominance when cytokinin is applied to them.

Researchers are well aware that the ratio of auxin to cytokinin and the acidity of the culture medium determine whether a plant tissue forms an undifferentiated mass, called a callus, or differentiates to form roots, vegetative shoots, leaves, or floral shoots. Researchers have reported that chemicals called oligosaccharins (chemical fragments released from the cell wall) are effective in directing differentiation. They hypothesize that reception of auxin and cytokinins, which leads to the activation of enzymes, releases these fragments from the cell wall.

Abscisic Acid

Abscisic acid (ABA) is produced by any "green tissue" with chloroplasts, but also by the monocot endosperm, and in roots. Abscisic acid was once thought to function in **abscission,** the dropping of leaves, fruits, and flowers from a plant. But even though the external application of abscisic acid promotes abscission, researchers no longer believe this hormone functions naturally in this process. Instead, they think the hormone ethylene, discussed next, brings about abscission.

Effects of Abscisic Acid   Abscisic acid is sometimes called the stress hormone because it initiates and maintains seed and bud dormancy and brings about the closure of stomata when a plant is under water stress (Fig. 10.15). Dormancy has begun when a plant stops growing and prepares for adverse conditions (even though conditions at the time are favorable for growth). For example, researchers believe that abscisic acid moves from leaves to vegetative buds in the fall, and thereafter, these buds are converted to winter buds. A winter bud is covered by thick, hardened scales. A reduction in the level of abscisic acid and an increase in the level of gibberellins are believed to break seed and bud dormancy. Then seeds germinate, and buds send forth leaves.

Ultimately, abscisic acid causes potassium ions (K^+) to leave guard cells, causing the guard cells to lose water, and the stomata to close.

Ethylene

Ethylene is a gas that works with other hormones to bring about certain effects.

Effects of Ethylene   Ethylene is involved in abscission. Low levels of auxin and perhaps gibberellin in the leaf, compared to the stem, probably initiate abscission. But once the process of abscission has begun, ethylene stimulates certain enzymes, such as cellulase, which cause leaf, fruit, or flower drop (Fig. 10.16a, b). Cellulase hydrolyzes cellulose in plant cell walls.

In the early 1900s, it was common practice to prepare citrus fruits for market by placing them in a room with a kerosene stove. Only later did researchers realize that ethylene, an incomplete

a. No abscission

b. Abscission

c. Ripening

Figure 10.16 Functions of ethylene.   **a.** Normally, there is no abscission when a holly twig is placed under a glass jar for a week. **b.** When an ethylene-producing ripe apple is also under the jar, abscission of the holly leaves occurs. **c.** Similarly, ethylene given off by this one ripe tomato will cause the others to ripen. (a–b): ©Kingsley Stern; (c): ©Westend61 Premium/Shutterstock

Figure 10.15 Abscisic acid control of stoma closing. K^+ is concentrated inside the guard cells, and the stoma is open (*left*). When ABA binds to its receptor in the guard cell plasma membrane, calcium ions (Ca^{2+}) enter (*middle*). Then, potassium (K^+) channels open and K^+ exits the cell. Water follows and the stoma closes (*right*).

combustion product of kerosene, was ripening the fruit (Fig. 10.16*c*). Because it is a gas, ethylene can act from a distance. A barrel of ripening apples can induce the ripening of a bunch of bananas, even if they are in different containers. Ethylene is released at the site of a plant wound due to physical damage or infection (which is why one rotten apple spoils the whole bushel).

Brassinosteroids

Although **brassinosteroids** were first discovered almost four decades ago, they have only recently been recognized as a separate class of plant hormones. These steroid compounds function in the growth and development of the plant, specifically in the specialization of the vascular tissue. They are also recognized in helping with the response of the plant to stress, such as cold or drought conditions.

Plant Responses to Environmental Stimuli

Environmental signals determine the seasonality of growth, reproduction, and dormancy in plants. Plant responses are strongly influenced by such environmental stimuli as light, day length, gravity, and touch. The ability of a plant to respond to environmental signals fosters the survival of the plant and the species in a particular environment.

Plant responses to environmental signals can be rapid, as when stomata open in the presence of light, or they can take some time, as when a plant flowers in season. Despite their variety, most plant responses to environmental signals are due to growth and sometimes differentiation, brought about at least in part by particular hormones.

Plant Tropisms

Plant growth toward or away from a directional stimulus is called a **tropism.** Tropisms are due to differential growth—one side of an organ elongates faster than the other, and the result is a curving toward or away from the stimulus. The following three well-known tropisms were each named for the stimulus that causes the response:

phototropism	growth in response to a light stimulus
gravitropism	growth in response to gravity
thigmotropism	growth in response to touch

Growth toward a stimulus is called a positive tropism, and growth away from a stimulus is called a negative tropism. For example, in positive phototropism, stems curve toward the light, and in negative gravitropism, stems curve away from the direction of gravity (Fig. 10.17). Roots, of course, exhibit positive gravitropism.

The role of auxin in the positive phototropism of stems has been studied for quite some time. Because blue light, in particular, causes phototropism to occur, researchers believe a yellow pigment related to the vitamin riboflavin acts as a photoreceptor for light. Following reception, auxin migrates from the bright side to the shady side of a stem. The cells on that side elongate faster than those on the bright side, causing the stem to curve toward the light. Negative gravitropism of stems occurs when auxin moves to the lower part of a stem when a plant is placed on its side. Figure 10.12 explains how auxin brings about elongation.

Flowering

The flowering of angiosperms is a striking response to environmental seasonal changes. In some plants, flowering occurs according to the **photoperiod,** which is the ratio of the length of day to the length of night over a 24-hour period. Plants can be divided into the following three groups:

1. **Short-day plants** flower when the day length is shorter than a critical length. (Examples are cocklebur, poinsettia, and chrysanthemum.)
2. **Long-day plants** flower when the day length is longer than a critical length. (Examples are wheat, barley, clover, and spinach.)
3. **Day-neutral plants** do not depend on day length for flowering. (Examples are tomato and cucumber.)

a.

b.

c.

Figure 10.17 Tropisms.   a. In positive phototropism, the stem of a plant curves toward the light. This response is due to the accumulation of auxin on the shady side of the stem. **b.** In negative gravitropism, the stem of a plant curves away from the direction of gravity 24 hours after the plant was placed on its side. This response is due to the accumulation of auxin on the lower side of the stem. **c.** In thigmotropism, contact directs the growth of the plant, as is shown in this morning glory plant. (a): ©Dorling Kindersley/Getty Images; (b): ©Kingsley Stern; (c): ©Alison Thompson/Alamy

Figure 10.18 Photoperiodism and flowering.   a. Short-day plant. When the day is shorter than a critical length, this type of plant flowers. The plant does not flower when the day is longer than the critical length. It also does not flower if the longer-than-critical-length night is interrupted by a flash of light. **b.** Long-day plant. The plant flowers when the day is longer than a critical length. When the day is shorter than a critical length, this type of plant does not flower. However, it does flower if the slightly longer-than-critical-length night is interrupted by a flash of light.

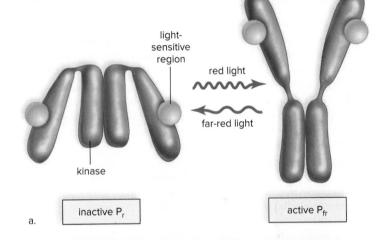

Experiments have shown that the length of continuous darkness, not light, controls flowering. For example, the cocklebur does not flower if a suitable length of darkness is interrupted by a flash of light. In contrast, clover does flower when an unsuitable length of darkness is interrupted by a flash of light (Fig. 10.18). Interrupting the light period with darkness has no effect on flowering.

Phytochrome and Plant Flowering

If flowering is dependent on day and night length, plants must have some way to detect these periods. This appears to be the role of **phytochrome,** a blue-green leaf pigment that alternately exists in two forms—P_r and P_{fr} (Fig. 10.19a).

Direct sunlight contains more red light than far-red light; therefore, P_{fr} (far red) is apt to be present in plant leaves during the day. In the shade and at sunset, there is more far-red light than red light. Therefore, P_{fr} is converted to P_r (red) as night approaches. There is also a slow metabolic replacement of P_{fr} by P_r during the night.

It is possible that phytochrome conversion is the first step in a signaling pathway that results in flowering. A flowering hormone has never been discovered.

Other Functions of Phytochrome

The $P_r \rightarrow P_{fr}$ conversion cycle has other functions. The presence of P_{fr} indicates to seeds of some plants that sunlight is present and conditions are favorable for germination. Such seeds must be only partly covered with soil when planted. Following germination, the presence of P_r indicates that stem elongation may be needed to reach sunlight. Seedlings grown in the dark etiolate—that is, the stem increases in length, and the leaves remain small (Fig. 10.19b). Once the seedling is exposed to sunlight and P_r is converted to P_{fr}, the seedling begins to grow normally—the leaves expand, and the stem branches.

Figure 10.19 Phytochrome control of growth pattern.   a. The inactive form of phytochrome (P_r) is converted to the active form (P_{fr}) in the presence of red light. **b.** In sunlight, phytochrome is active and normal growth occurs. In the shade, etiolation occurs. (b): ©Nigel Cattlin/Alamy

Check Your Progress 10.4

1. Identify the functions of auxins, gibberellins, cytokinins, abscisic acid, ethylene, and brassinosteroids.
2. Recognize the different types of plant growth tropism.
3. Compare the differences between short-day and long-day plants.
4. Identify the potential role of phytochrome in plant flowering.

Conclusion

Like other organisms, plants have the ability to respond to their environment. By studying these responses, scientists are developing a better understanding of how plants grow, seeds develop, and plants respond to stimuli that has benefits for human society.

SUMMARIZE

10.1 Sexual Reproduction in Flowering Plants

- Flowering plants have an **alternation-of-generations** life cycle.
- **Flowers** borne by the **sporophyte** produce **microspores** and **megaspores** by meiosis. Microspores develop into a male gametophyte, and megaspores develop into the female gametophyte. **Gametophytes** produce gametes by mitosis.
- Flowers contain sepals that form an outer whorl; **petals,** which are the next whorl; and **stamens** that form a whorl around the base of at least one carpel. The carpel, in the flower's center, consists of a **stigma, style,** and **ovary,** which contains **ovules.**
- **Anthers** contain microspore mother cells, each of which divides meiotically to produce four haploid microspores. Each microspore divides mitotically to produce a two-celled **pollen grain**—a tube cell and a generative cell. The generative cell later divides mitotically to produce two sperm cells.
- The pollen grain is the male gametophyte. After **pollination,** the pollen grain germinates, and as the pollen tube grows, sperm cells travel to the **embryo sac.** Plants have undergone coevolution with many of their pollinators to ensure fertilization occurs.
- Each ovule contains a megaspore mother cell, which divides meiotically to produce four haploid megaspores, only one of which survives. This megaspore divides mitotically to produce the female gametophyte (embryo sac), which usually has seven cells.
- Flowering plants undergo **double fertilization.** One sperm nucleus unites with the egg nucleus, forming a 2n zygote, and the other unites with the polar nuclei of the central cell, forming a 3n **endosperm** cell. The zygote becomes the sporophyte **embryo,** and the endosperm cell divides to become endosperm tissue. Collectively, these structures are known as the **seed,** which is often found within a fruit.

10.2 Growth and Development

- Prior to seed formation, the zygote undergoes growth and development to become an embryo.
- The seeds enclosed by a fruit contain the embryo (hypocotyl, epicotyl, plumule, radicle) and stored food (endosperm and/or **cotyledons**).
- Following dispersal, a seed **germinates. Dormancy** is the period in which no growth occurs.

10.3 Asexual Reproduction in Plants

- Many flowering plants reproduce asexually (e.g., buds give rise to entire plants, or roots produce new shoots). Traditionally, hybridization produced new varieties of plants.
- Asexual reproduction allows for clonal propagation in tissue culture. Plant **tissue culture** allows production of new plants because most plant cells are **totipotent.**

10.4 Control of Growth and Responses

- Plant **hormones** are chemical signals involved in plant growth and responses to environmental stimuli.
- There are five commonly recognized groups of plant hormones:
  - **Auxins** cause **apical dominance,** the growth of adventitious roots, and two types of **tropisms—phototropism** and **gravitropism.**
  - **Gibberellins** promote stem elongation and break seed dormancy so that germination occurs.
  - **Cytokinins** promote cell division, prevent **senescence** of leaves, and along with auxin, influence differentiation of plant tissues.
  - **Abscisic acid (ABA)** initiates and maintains seed and bud dormancy and closes stomata under water stress.
  - **Ethylene** causes **abscission** of leaves, fruits, and flowers. It also causes some fruits to ripen.
  - **Brassinosteroids** are involved in the growth of the plant, especially in the formation of vascular tissue
- Tropisms are growth responses toward or away from unidirectional stimuli. When a plant is exposed to light, auxin moves laterally from the bright to the shady side of a stem. Then, cells on the shady side elongate, and the stem bends toward the light. Similarly, auxin is responsible for negative gravitropism and growth upward, opposite gravity.
- Flowering is a striking response to seasonal changes. **Phytochrome,** a plant pigment that responds to daylight, is believed to be part of a biological clock system that in some unknown way brings about flowering. Phytochrome has various other functions, such as seed germination, leaf expansion, and stem branching.
- The **photoperiods** of plants can be divided into three groups:
  - **Short-day plants** flower when days are shorter than a critical length.
  - **Long-day plants** flower when days are longer than a critical length.
  - **Day-neutral plants** do not depend on day length for flowering.

ASSESS

Testing Yourself

Choose the best answer for each question.

10.1 Sexual Reproduction in Flowering Plants

1. Which of the following contains the diploid cells that will produce the female gametophyte?
   a. sepal
   b. stamen
   c. ovary
   d. style

2. Label the diagram below of alternation of generations in flowering plants.

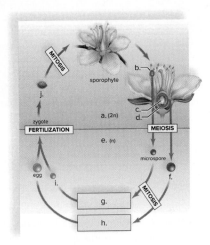

3. Label the following diagram of a flower.

10.2 Growth and Development

4. Fruits
   a. nourish embryo development.
   b. help with seed dispersal.
   c. signal gametophyte maturity.
   d. attract pollinators.
   e. stay where they are produced.

5. The seed leaves of a plant are called the
   a. meristem.
   b. endosperm.
   c. cotyledons.
   d. zygote.

6. A seed contains
   a. a seed coat.
   b. cotyledon(s).
   c. an embryo.
   d. stored food.
   e. All of these are correct.

10.3 Asexual Reproduction in Plants

7. Which term describes a plant cell that has the ability to develop into an entire plant?
   a. totipotent
   b. tissue
   c. stolon
   d. transgenic

8. The growing of plants in a laboratory environment under specific conditions is called
   a. organogenesis.
   b. hybridization.
   c. tissue culture.
   d. None of these are correct.

10.4 Control of Growth and Responses

For questions 9–14, match each statement with a hormone in the key. Answers can be used more than once.

KEY:
   a. auxin
   b. gibberellin
   c. cytokinin
   d. ethylene
   e. abscisic acid
   f. brassinosteroids

9. It is present in a gaseous form.
10. Grapes can grow larger and exhibit stem elongation.
11. Stomata close when a plant is water-stressed.
12. Stems bend toward the sun.
13. It inhibits plant senescence.
14. It causes leaf abscission.
15. It participates in the growth and development of the plant.

ENGAGE

Thinking Critically

1. Explain the similarities and differences in how mitosis and meiosis are used in the plant and human (animal) life cycles.

2. How might a plant benefit from having just one type of pollinator?

3. You hypothesize that abscisic acid (ABA) is responsible for the turgor pressure changes that permit a plant to track the sun (such as a patch of sunflowers on a sunny day). What observations could you make to support your hypothesis?

4. In most parts of the world, commercial potato crops are produced asexually by planting tubers. However, in some regions, such as Southeast Asia and the Andes, some potatoes are grown from true seeds. Discuss the advantages and disadvantages of growing potatoes from true seeds.

C H A P T E R

11

Human Organization

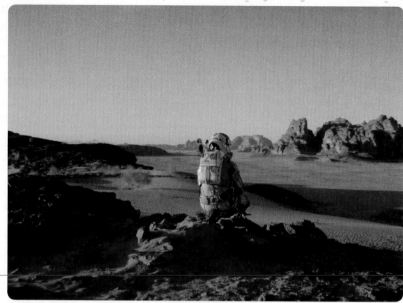

©Moviestore Collection Ltd/Alamy

CHAPTER OUTLINE

BEFORE YOU BEGIN

Before beginning this chapter, take a few moments to review the following discussions:

Section 1.1 Why is homeostasis a characteristic of life?

Figure 1.2 What levels of biological organization are found in humans?

Section 3.3 What major organelles make up animal cells?

Living on Mars

In the 2015 movie *The Martian*, botanist Mark Whatley (Matt Damon) almost dies when an explosion damages his space suit while on the surface of Mars. The fact that an astronaut needs to wear a special suit to survive in a hostile environment is a reminder that the internal environment of our body must stay within normal limits. For example, our enzymes function best at around 37°C; a moderate blood pressure helps blood perfuse the tissues; and a sufficient oxygen concentration facilitates ATP production. This concept is known as homeostasis, a dynamic equilibrium of the internal environment. Climb to the top of Everest, cross the Sahara Desert by camel, visit the South Pole, or visit space—a variety of internal conditions in your body will stay within fairly narrow ranges, as long as you take proper precautions. An astronaut depends on artificial systems in addition to natural systems to maintain homeostasis.

This chapter discusses homeostasis after a look at the body's organization. Just as in other complex animals, each organ system of the human body contains a particular set of organs. For example, the circulatory system contains the heart and blood vessels, and the nervous system contains the brain and nerves. Organs are composed of tissues, and each type of tissue has cells that perform specific functions. We begin the chapter by examining several of the major types of tissues.

As you read through this chapter, think about the following questions:

1. Why is the maintenance of a constant temperature and oxygen concentration important?
2. What types of tissues make up each of the human body's major organ systems?
3. How do the different body systems interact with, and depend on, each other?

11.1 Types of Tissues

Recall from the biological levels of organization (see Section 1.1) that cells are composed of molecules; a tissue has similar types of cells; an organ contains several types of tissues; and several organs make up an organ system. In this chapter, we will take a closer look at the tissue, organ, and organ system levels of organization.

Tissues are composed of similarly specialized cells that perform a common function in the body. The tissues of the human body can be categorized into the following four major types:

- *Epithelial tissue* covers body surfaces and lines body cavities.
- *Connective tissue* binds and supports body parts.
- *Muscular tissue* moves the body and its parts.
- *Nervous tissue* receives stimuli, processes information, and conducts nerve impulses.

We will now examine the structure and function of each of these four types of tissues.

Epithelial Tissue

Epithelial tissue, also called *epithelium,* consists of tightly packed cells that form a continuous layer. Epithelial tissue covers surfaces and lines body cavities. Epithelial tissue has numerous functions in the body. Usually, it has a protective function, but it can also be modified to carry out secretion, absorption, excretion, and filtration.

On the external surface, epithelial tissue protects the body from injury, drying out, and possible invasion by microbes such as bacteria and viruses. On internal surfaces, modifications help epithelial tissue carry out both its protective and specific functions. Epithelial tissue secretes mucus along the digestive tract and sweeps up impurities from the lungs by means of *cilia (sing.,* cilium). It efficiently absorbs molecules from kidney tubules and from the intestine because of minute cellular extensions called *microvilli (sing.,* microvillus).

A *basement membrane* usually joins an epithelium to underlying connective tissue. We now know that the basement membrane consists of glycoprotein secreted by epithelial cells and collagen fibers that belong to the connective tissue.

Epithelial tissue is classified according to the shape of its cells (squamous, cuboidal, or columnar) and the number of layers in the tissue (Fig. 11.1). **Squamous epithelium** is characterized by flattened cells and is found lining the lungs and blood vessels. **Cuboidal epithelium** contains cube-shaped cells and lines the kidney tubules. **Columnar epithelium** has cells resembling rectangular pillars or columns, with nuclei usually located near the bottom of each cell. This epithelium lines the digestive tract. Ciliated columnar epithelium lines the uterine tubes, where it propels the egg toward the uterus, or womb.

Classification is also based on the number of layers in the tissue. *Simple epithelium* has a single layer of cells, whereas *stratified epithelium* has layers of cells piled one on top of another. The walls of the smallest blood vessels, called *capillaries,* are composed of simple squamous epithelium. The permeability of this single layer of cells allows exchange of substances between the blood and tissue cells. Simple cuboidal epithelium lines kidney tubules and the cavities of many internal organs. Stratified squamous epithelium lines the nose, mouth, esophagus, anal canal, and vagina. The outer layer of skin is also stratified squamous epithelium, but the cells have been reinforced by keratin, a protein that provides strength.

Pseudostratified epithelium appears to be layered, but true layers do not exist because each cell touches the basement membrane. The lining of the windpipe, or trachea, is pseudostratified ciliated columnar epithelium. A secreted covering of mucus traps foreign particles, and the upward motion of the cilia carries the mucus to the back of the throat, where it either may be swallowed or expelled. Smoking can cause a change in mucous secretion and inhibit ciliary action, and the result is a chronic inflammatory condition called bronchitis.

When an epithelium secretes a product, it is said to be glandular. A **gland** can be a single epithelial cell, as are the mucus-secreting goblet cells within the columnar epithelium lining the digestive tract, or a gland can contain many cells. Glands that secrete their product into ducts are called exocrine glands, and those that secrete their product into the bloodstream are called endocrine glands. The pancreas is both an exocrine gland, because it secretes digestive juices into the small intestine via ducts, and an endocrine gland, because it secretes insulin into the bloodstream.

Junctions Between Epithelial Cells

The cells of a tissue can function in a coordinated manner when the plasma membranes of adjoining cells interact. Three common types of junctions link epithelial cells (see Section 4.3). These are the *tight junctions,* which form an impermeable barrier between the cells; *gap junctions,* which serve to strengthen connections while allowing small molecules to pass; and *adhesion junctions,* which act like rivets or "spot welds" to anchor tissues in place, increasing their overall strength.

Connective Tissue

Connective tissue binds organs together, provides support and protection, fills spaces, produces blood cells, and stores fat. As a rule, connective tissue cells are widely separated by a *matrix,* consisting of a noncellular material that varies in consistency from solid to jellylike to fluid. A nonfluid matrix may have fibers of three possible types. White **collagen fibers** contain collagen, a protein that gives them flexibility and strength. **Reticular fibers** are very thin collagen fibers that are highly branched and form delicate supporting networks. Yellow **elastic fibers** contain elastin, a protein that is not as strong as collagen but more elastic.

Figure 11.1 Epithelial tissue. Certain types of epithelial tissue—squamous, cuboidal, and columnar—are named for the shapes of their cells. They all have a protective function in addition to other specific functions. (all photos): ©Ed Reschke

Loose Fibrous and Dense Fibrous Tissues

Loose fibrous connective tissue supports epithelium and also many internal organs (Fig. 11.2*a*). Its presence in lungs, arteries, and the urinary bladder allows these organs to expand. It forms a protective covering enclosing many internal organs, such as muscles, blood vessels, and nerves.

Dense fibrous connective tissue contains many collagen fibers that are packed together (Fig. 11.2*b*). This type of tissue has more specific functions than does loose connective tissue. For example, dense fibrous connective tissue is found in **tendons,** which connect muscles to bones, and in **ligaments,** which connect bones to other bones at joints.

Both loose fibrous and dense fibrous connective tissues have cells called **fibroblasts** located some distance from one another and separated by a jellylike matrix containing white collagen fibers and yellow elastic fibers.

Loose fibrous connective tissue
- has space between components.
- occurs beneath skin and most epithelial layers.
- functions in support and binds organs.

fibroblast
elastic fiber   collagen fiber   250×

a.

Dense fibrous connective tissue
- has collagenous fibers closely packed.
- in dermis of skin, tendons, ligaments.
- functions in support.

collagen fibers   nuclei of fibroblasts   400×

b.

Adipose tissue
- cells are filled with fat.
- occurs beneath skin, around heart and other organs.
- functions in insulation, stores fat.

nucleus   250×

c.

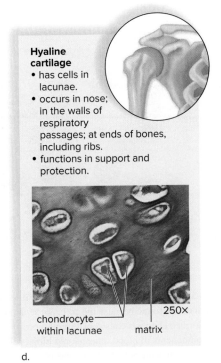

Hyaline cartilage
- has cells in lacunae.
- occurs in nose; in the walls of respiratory passages; at ends of bones, including ribs.
- functions in support and protection.

chondrocyte within lacunae   matrix   250×

d.

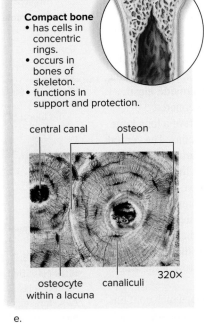

Compact bone
- has cells in concentric rings.
- occurs in bones of skeleton.
- functions in support and protection.

central canal   osteon
osteocyte within a lacuna   canaliculi   320×

e.

Figure 11.2 Connective tissue examples. a. In loose fibrous connective tissue, cells called fibroblasts are separated by a jellylike matrix, which contains both collagen and elastic fibers. **b.** Dense fibrous connective tissue contains tightly packed collagen fibers for added strength. **c.** Adipose tissue cells (adipocytes) have nuclei pushed to one side because the cells are filled with fat. **d.** In hyaline cartilage, the flexible matrix has a white, translucent appearance. **e.** In compact bone, the hard matrix contains calcium salts. Concentric rings of cells in lacunae form an elongated cylinder called an osteon. An osteon has a central canal that contains blood vessels and nerve fibers.

(photos) (a, c–e): ©Ed Reschke; (b): ©McGraw-Hill Education/Dennis Strete, photographer

Adipose Tissue and Reticular Connective Tissue

In **adipose tissue** (Fig. 11.2c), the fibroblasts enlarge and store fat. These cells are commonly called adipocytes. The body uses this stored fat for energy, insulation, and organ protection. Adipose tissue is found beneath the skin, around the kidneys, and on the surface of the heart. **Reticular connective tissue** forms the supporting meshwork of lymphoid tissue in lymph nodes, the spleen, the thymus, and the bone marrow. All types of blood cells are produced in red bone marrow, but a certain type of lymphocyte (the T lymphocyte) completes its development in the thymus. The lymph nodes are sites of lymphocyte responses (see Section 13.1).

Cartilage

Cartilage is a specialized form of dense fibrous connective tissue, which most commonly forms the smooth surfaces that allow bones to slide against each other in joints. Cartilage is found in many other locations, however. The cells in cartilage lie in small chambers called **lacunae** (*sing.,* lacuna), separated by a matrix that is solid yet flexible. Unfortunately, because this tissue lacks a direct blood supply, it heals very slowly. There are three types of cartilage, distinguished by the main type of fiber in the matrix.

Hyaline cartilage (Fig. 11.2d), the most common type of cartilage, contains only very fine collagen fibers. The matrix has a white, translucent appearance. Hyaline cartilage is found in the nose and at the ends of the long bones and the ribs, and it forms rings in the walls of respiratory passages. The fetal skeleton is also made of this type of cartilage. Later, the cartilaginous fetal skeleton is replaced by bone.

Elastic cartilage has more elastic fibers than hyaline cartilage. For this reason, it is more flexible and is found, for example, in the framework of the outer ear.

Fibrocartilage has a matrix containing strong collagen fibers. Fibrocartilage is found in structures that withstand tension and pressure, such as the pads between the vertebrae in the backbone and the wedges in the knee joint.

Bone

Bone is the most rigid connective tissue. It consists of an extremely hard matrix of inorganic salts, notably calcium salts, deposited around protein fibers, especially collagen fibers. The inorganic salts give bone rigidity, and the protein fibers provide elasticity and strength, much as steel rods do in reinforced concrete.

Compact bone (Fig. 11.2e) makes up the shaft of a long bone. It consists of cylindrical structural units called osteons. The central canal of each osteon is surrounded by rings of hard matrix. Bone cells are located in spaces called lacunae between the rings of matrix. In the central canal, nerve fibers carry nerve impulses and blood vessels carry nutrients that allow bone to renew itself. Nutrients can reach all of the bone cells because they are connected by thin processes within canaliculi (minute canals) that also reach to the central canal. The shaft of long bones such as the femur (thigh bone) have a hollow chamber filled with bone marrow, a site where blood cells develop.

The ends of a long bone contain **spongy bone,** which contains numerous bony bars and plates, separated by irregular spaces. Al-though lighter than compact bone, spongy bone is still designed for strength. Just as braces are used for support in buildings, the solid portions of spongy bone follow lines of stress. Section 19.1 provides more information about bone and cartilage.

Blood

Blood is unlike other types of connective tissue in that the matrix (i.e., plasma) is not made by the cells. The internal environment of the body consists of blood and the fluid between the body's cells. The systems of the body help keep the composition and chemistry of blood within normal limits, and blood, in turn, creates **interstitial fluid,** which is the fluid between the cells. Blood transports nutrients and oxygen to the interstitial fluid and removes carbon dioxide and other wastes. It helps distribute heat and also plays a role in fluid, ion, and pH balance. Various components of blood help protect us from disease, and blood's ability to clot prevents fluid loss.

If blood is transferred from a person's vein to a test tube and prevented from clotting, it separates into two layers (Fig. 11.3a). The upper, liquid layer, called **plasma,** represents about 55% of the volume of whole blood and contains a variety of inorganic and organic substances dissolved or suspended in water (Table 11.1). The lower layer consists of red blood cells (erythrocytes), white

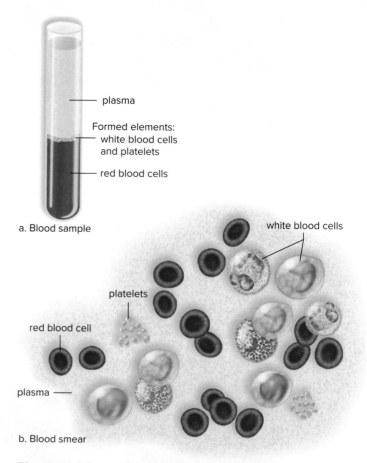

plasma

Formed elements:
white blood cells
and platelets

red blood cells

a. Blood sample

white blood cells

platelets

red blood cell

plasma

b. Blood smear

Figure 11.3 Blood, a liquid connective tissue. a. Blood is often classified as connective tissue because the cells are separated by a matrix—plasma. The formed elements consist of several types of cells. **b.** A drawing of the components of blood: plasma, red blood cells, white blood cells, and platelets (which are actually fragments of a larger cell).

TABLE 11.1 Components of Blood Plasma	
Water (90–92% of total) **Solutes (8–10% of total)**	
Inorganic ions (electrolytes)	Na^+, Ca^{2+}, K^+, Mg^{2+}, Cl^-, HCO_3^-, HPO_4^{2-}, SO_4^{2-}
Gases	O_2, CO_2
Plasma proteins	Albumin, globulins, fibrinogen, transport proteins
Organic nutrients	Glucose, lipids, phospholipids, amino acids, etc.
Nitrogen-containing waste products	Urea, ammonia, uric acid
Regulatory substances	Hormones, enzymes

blood cells (leukocytes), and blood platelets (thrombocytes) (Fig. 11.3*b*). Collectively, these are called the formed elements, and they represent about 45% of the volume of whole blood. Formed elements are manufactured in the red bone marrow of the skull, ribs, vertebrae, and ends of the long bones.

The **red blood cells** are small, biconcave, disk-shaped cells without nuclei. The presence of the red pigment hemoglobin makes the cells red and, in turn, makes the blood red. Hemoglobin is composed of four units. Each contains the protein globin and a complex, iron-containing structure called heme. The iron forms a loose association with oxygen, and in this way, red blood cells transport oxygen.

White blood cells differ from red blood cells in that they are usually larger, have a nucleus, and without staining would appear translucent. When smeared onto microscope slides and stained, the nuclei of white blood cells typically look bluish (Fig. 11.3*b*). White blood cells fight infection in a number of different ways. Some white blood cells are phagocytic and engulf infectious pathogens, while others are responsible for the adaptive immunity that develops after an individual is exposed to various pathogens or toxins, either through natural infection or by vaccination. One important feature of acquired immunity is the production of antibodies, molecules that combine with foreign substances to inactivate them.

Platelets are not complete cells. Rather, they are fragments of large cells present only in bone marrow. When a blood vessel is damaged, platelets help to form a plug that seals the vessel, and injured tissues release molecules that stimulate the clotting process.

We will explore the role of blood cells in human physiology in more detail in Section 12.2 and Chapter 13.

Muscular Tissue

Muscular tissue is composed of cells called muscle fibers. Muscle fibers contain actin filaments and myosin filaments, whose interaction accounts for movement. The three types of muscle tissue are skeletal, smooth, and cardiac.

Skeletal muscle (Fig. 11.4*a*) is usually attached by tendons to the bones of the skeleton, and when it contracts, body parts move. Contraction of skeletal muscle is under voluntary control. Skeletal muscle fibers are cylindrical and quite long—sometimes they run the length of the muscle. They arise from fusion of several cells,

Skeletal muscle
- has striated cells with multiple nuclei.
- occurs in muscles attached to skeleton.
- functions in voluntary movement of body.

striation   nucleus   250×

Smooth muscle
- has spindle-shaped cells, each with a single nucleus.
- cells have no striations.
- occurs in blood vessel walls and walls of the digestive tract.
- functions in movement of substances in lumens of body.
- is involuntary.

smooth muscle cell   nucleus   400×

Cardiac muscle
- has branching, striated cells, each with a single nucleus.
- occurs in the wall of the heart.
- functions in the pumping of blood.
- is involuntary.

intercalated disc   nucleus   250×

a.   b.   c.

Figure 11.4 Muscular tissue.   a. Skeletal muscle is voluntary and striated. **b.** Smooth muscle is involuntary and nonstriated. **c.** Cardiac muscle is involuntary and striated. Cardiac muscle cells branch and fit together at intercalated discs. (photos) (a, c): ©Ed Reschke; (b): ©McGraw-Hill Education/Dennis Strete, photographer resulting in one fiber with multiple nuclei. The fibers have alternating light and dark bands that give them a *striated* appearance.

Smooth muscle is so named because the cells lack striations. The spindle-shaped fibers, each with a single nucleus, form layers in which the thick middle portion of one cell is opposite the thin ends of adjacent cells. Consequently, the nuclei form an irregular pattern in the tissue (Fig. 11.4b). Smooth muscle is not under voluntary control, and therefore is said to be involuntary. Smooth muscle is found in the walls of viscera (intestine, stomach, and other internal organs) and blood vessels.

Cardiac muscle (Fig. 11.4c) is found only in the walls of the heart. Cardiac muscle combines features of both smooth muscle and skeletal muscle. Like skeletal muscle, it has striations, but the contraction of the heart is involuntary. Cardiac muscle cells also differ from skeletal muscle cells in that they usually have a single, centrally placed nucleus. The cells are branched and seemingly fused one with the other, and the heart appears to be composed of one large interconnecting mass of muscle cells. Actually, cardiac muscle cells are separate and individual, but they are bound end to end at *intercalated discs,* areas where folded plasma membranes between two cells contain adhesion junctions and gap junctions. These areas promote the flow of electrical current when the heart muscle contracts by allowing ions to flow freely between cells. See Section 19.4 to learn more about the mechanism of muscle contraction.

Nervous Tissue

Nervous tissue, which contains nerve cells called neurons (Fig. 11.5a), is present in the brain and spinal cord. A **neuron** is a specialized cell that has three parts: a cell body, dendrites, and an axon (Fig. 11.5a, b). The cell body contains the major concentration of the cytoplasm and the nucleus of the neuron. Dendrites are the receptive part of the neuron, meaning they generally conduct signals toward the cell body. Axons are conductive, meaning they transmit nerve impulses away from the cell body. Long axons are covered by myelin, a white fatty substance, which increases the speed of nerve impulses. The term *fiber*[1] is used here to refer to an axon along with its myelin sheath, if it has one. Outside the brain and spinal cord, fibers bound by connective tissue form **nerves.**

The nervous system has just three functions: sensory input, integration of data, and motor output. Nerves conduct impulses from sensory receptors to the spinal cord and the brain where integration occurs. The phenomenon called sensation occurs only in the brain, however. Nerves also conduct nerve impulses away from the spinal cord and brain to the muscles and glands, causing them to contract and secrete, respectively. In this way, a coordinated response to the stimulus is achieved.

Neuroglia

In addition to neurons, nervous tissue contains neuroglia. **Neuroglia** are cells that outnumber neurons nine to one and take up more than half the volume of the brain (Fig. 11.5a). Although the primary function of neuroglia is to support and nourish neurons, research is currently being conducted to determine how much they directly

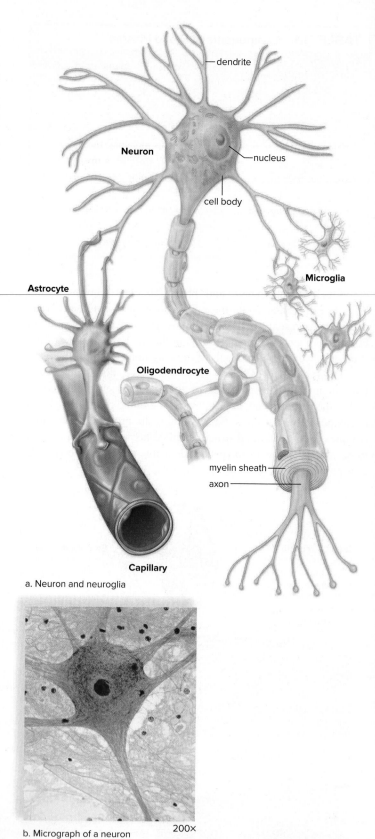

a. Neuron and neuroglia

b. Micrograph of a neuron   200×

Figure 11.5 Neurons and neuroglia. **a.** Neurons conduct nerve impulses. Neuroglia consist of cells that support and nourish neurons and have various functions. Astrocytes lie between neurons and a capillary. Therefore, nutrients entering neurons from the blood must first pass through astrocytes. Oligodendrocytes form the myelin sheaths around fibers in the brain and spinal cord. **b.** A neuron cell body as seen with a light microscope. (b): ©Ed Reschke

---

[1]In connective tissue, a fiber is a component of the matrix; in muscle tissue, a fiber is a muscle cell; in nervous tissue, a fiber is an axon and its myelin sheath.

contribute to brain function. The four types of neuroglia in the brain are microglia, astrocytes, oligodendrocytes, and ependymal cells. *Microglia,* in addition to supporting neurons, engulf bacterial and cellular debris. *Astrocytes* provide nutrients to neurons and produce a hormone known as glia-derived growth factor, which has potential as a treatment for Parkinson disease and other diseases caused by neuron degeneration. *Oligodendrocytes* form myelin sheaths. Outside the brain, *Schwann cells* are the type of neuroglia that encircle long nerve fibers and form a myelin sheath. *Ependymal cells* line the fluid-filled spaces of the brain and spinal cord. Neuroglia don't have long processes, but even so, researchers are now beginning to gather evidence that they do communicate among themselves and with neurons. To learn more about the nervous system, see Section 17.1.

Check Your Progress 11.1

1. List the five types of epithelium, and identify a location where each could be found in the human body.
2. Compare and contrast the three types of connective tissue.
3. Describe the structure and function of skeletal, smooth, and cardiac muscle.
4. Distinguish between a neuron and the neuroglia.

11.2 Body Cavities and Body Membranes

Learning Outcomes

Upon completion of this section, you should be able to

1. List the two main cavities of the human body, and the major organs found in each.
2. Compare and contrast the location(s) and function(s) of the different types of body membranes.

When referring to anatomical parts of humans, certain standard terms are used. Many of the same terms apply to other organisms, although there is some variation because the human terms always refer to a body that is in the upright, standing position, as in Figure 11.6. The term *ventral,* which means the same thing as *anterior* in humans, refers to the front, and *dorsal* or *posterior* both mean toward the back. *Superior* means toward the head, and *inferior* means toward the feet. Other anatomical terms are relative to other body parts; something that is *medial* is closer to the midline of the body, whereas *lateral* means away from the midline. Similarly, when referring to an appendage like an arm or a leg, *proximal* means closer to the trunk of the body, while *distal* means away from the trunk.

Keeping these definitions in mind, the human body can be divided into two main **body cavities:** the ventral cavity and the dorsal cavity (Fig. 11.6a). The **ventral cavity,** which is called a *coelom* during development, becomes divided into the thoracic, abdominal, and pelvic cavities (the latter two are sometimes grouped together as the abdominopelvic cavity). The *thoracic cavity* contains the right and left lungs and the heart. It is separated from the *abdominal cavity* by a horizontal muscle called the diaphragm. The stomach, liver, spleen, gallbladder, and most of the

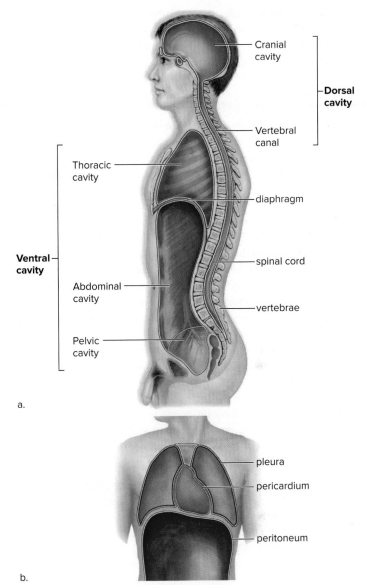

Figure 11.6 Mammalian body cavities. **a.** Side view. The dorsal (toward the back) cavity contains the cranial cavity and the vertebral canal. The brain is in the cranial cavity, and the spinal cord is in the vertebral canal. In the ventral (toward the front) cavity, the diaphragm separates the thoracic cavity and the abdominal cavity. The heart and lungs are in the thoracic cavity, and most other internal organs are in the abdominal cavity. **b.** Types of serous membranes.

small and large intestines are in the upper portion of the abdominal cavity. The *pelvic cavity* contains the rectum, the urinary bladder, the internal reproductive organs, and the rest of the large intestine. Males have an external extension of the abdominal wall, called the scrotum, containing the testes.

The **dorsal cavity** also has two parts: the *cranial cavity* within the skull contains the brain; the *vertebral canal,* formed by the vertebrae, contains the spinal cord.

Body Membranes

Body membranes line cavities and the internal spaces of organs and tubes that open to the outside.

Mucous membranes line the tubes of the digestive, respiratory, urinary, and reproductive systems. They are composed of an epithelium overlying a loose fibrous connective tissue layer. The epithelium contains goblet cells that secrete mucus. This mucus ordinarily protects the body from invasion by bacteria and viruses. More mucus is secreted and expelled when a person has a cold and has to blow her or his nose.

Serous membranes, which line the thoracic and abdominal cavities and cover the organs they contain, are also composed of epithelium and loose fibrous connective tissue. They secrete a watery fluid that keeps the membranes lubricated. Serous membranes support the internal organs and compartmentalize the large thoracic and abdominal cavities. This helps hinder the spread of any infection.

Serous membranes have specific names according to their location (Fig. 11.6b). The *pleurae* (*sing.,* pleura) line the thoracic cavity and cover the lungs; the *pericardium* covers the heart; the *peritoneum* lines the abdominal cavity and covers its organs. A double layer of peritoneum, called mesentery, supports the abdominal organs and attaches them to the abdominal wall. Peritonitis is a potentially life-threatening infection of the peritoneum that may occur if an inflamed appendix bursts before it is removed, or if the digestive tract is perforated for any other reason.

Synovial membranes, composed only of loose connective tissue, line freely movable joint cavities. They secrete synovial fluid that lubricates the cartilage at the ends of the bones so they can move smoothly in the joint cavity.

The **meninges** are membranes within the dorsal cavity. They are composed only of connective tissue and serve as a protective covering for the brain and spinal cord.

Check Your Progress 11.2

1. Identify the two cavities separated by the diaphragm.
2. Describe the function of the fluids produced by various body membranes.

11.3 Organ Systems

Learning Outcomes

Upon completion of this section, you should be able to

1. List the major organs that make up each organ system.
2. Describe the general function(s) of each organ system.

Figure 11.7 illustrates the organ systems of the human body. Just as organs work together in an organ system, organ systems work together in the body. Therefore, while a particular organ may be assigned to one system, it may assist in the function of other organ systems.

Integumentary System

The **integumentary system** (see Section 11.4) contains skin, which is made up of two main types of tissue: the epidermis is composed of stratified squamous epithelium, and the dermis is composed of fibrous connective tissue. The integumentary system also includes nails, located at the ends of the fingers and toes (collectively called the digits); hairs and muscles that move hairs; the oil and sweat glands; blood vessels; and nerves leading to sensory receptors. Besides having a protective function, skin also synthesizes vitamin D, collects sensory data, and helps regulate body temperature.

Cardiovascular System

In the **cardiovascular system** (see Chapter 12), the heart pumps blood and sends it under pressure into the blood vessels. While blood is moving throughout the body, it distributes heat produced by the muscles. Blood transports nutrients and oxygen to the cells and removes their waste molecules, including carbon dioxide. Despite the movement of molecules into and out of the blood, it has a fairly constant volume and pH. The blood is also a route by which cells of the immune system can be distributed throughout the body.

Lymphatic and Immune Systems

The **lymphatic system** (see Chapter 13) consists of lymphatic vessels (which transport lymph), lymph nodes, and other lymphatic (lymphoid) organs. This system protects the body from disease by purifying lymph and storing lymphocytes, the white blood cells responsible for adaptive immunity. Lymphatic vessels absorb fat from the digestive system and collect excess interstitial fluid, which is returned to the cardiovascular system.

The **immune system** (see Chapter 13) consists of all the cells in the body that protect us from disease, especially those caused by infectious agents. The lymphocytes, in particular, belong to this system.

Digestive System

The **digestive system** (see Chapter 14) includes the mouth, esophagus, stomach, small intestine, and large intestine (colon), along with associated organs such as the teeth, tongue, salivary glands, liver, gallbladder, and pancreas. This system receives food and digests it into nutrient molecules, which can enter the body's cells. The nondigested remains are eventually eliminated.

Respiratory System

The **respiratory system** (see Chapter 15) consists of the lungs and the tubes that take air to and from them. The respiratory system brings oxygen into the body for cellular respiration and removes carbon dioxide from the body at the lungs, restoring pH.

Urinary System

The **urinary system** (see Chapter 16) contains the kidneys, the urinary bladder, and the tubes that carry urine. This system rids the body of metabolic wastes, and helps regulate the fluid balance and pH of the blood.

Nervous System

The **nervous system** (see Chapters 17 and 18) consists of the brain, spinal cord, and associated nerves. The nerves conduct nerve impulses from sensory receptors to the brain and spinal cord, where integration occurs. Nerves also conduct nerve impulses from the brain and spinal cord to the muscles and glands, allowing us to respond to both external and internal stimuli.

Integumentary system

- protects body
- provides temperature homeostasis
- synthesizes vitamin D
- receives sensory input
Organ: Skin

Cardiovascular system

- transport system for nutrients, waste
- provides temperature, pH, and fluid homeostasis
Organ: Heart

Lymphatic and immune systems

- defends against infectious diseases
- provides fluid homeostasis
- assists in absorption and transport of fats
Organs: Lymphatic vessels, lymph nodes, spleen

Digestive system

- ingests, digests, and processes food
- absorbs nutrients and eliminates waste
- involved in fluid homeostasis
Organs: Oral cavity, esophagus, stomach, small intestine, large intestine, salivary glands, liver, gallbladder, pancreas

Respiratory system

- exchanges gases at both lungs and tissues
- assists in pH homeostasis
Organs: Lungs

Urinary system

- excretes metabolic wastes
- provides pH and fluids homeostasis
Organs: Kidneys, urinary bladder

Skeletal system

- provides support and protection
- assists in movement
- stores minerals
- produces blood cells
Organ: Bones

Muscular system

- assists in movement and posture
- produces heat
Organ: Muscles

Nervous system

- receives, processes, and stores sensory input
- provides motor output
- coordinates organ systems
Organs: Brain, spinal cord

Endocrine system

- produces hormones
- coordinates organ systems
- regulates metabolism and stress responses
- involved in fluid and pH homeostasis
Organs: Testes, ovaries, adrenal glands, pancreas, thymus, thyroid, pineal gland

Reproductive system

- produces and transports gametes
- nurtures and gives birth to offspring in females
Organs: Testes, penis, ovaries, uterus, vagina

Figure 11.7 Organ systems of the body.   Many of the organs of the body may have functions that are associated with more than one organ system.

Musculoskeletal System

Within the **musculoskeletal system** (see Chapter 19), the bones provide a scaffolding that helps hold and protect body parts. For example, the skull forms a protective encasement for the brain. The skeleton also helps move the body because it serves as a place of attachment for the skeletal muscles. It stores minerals, and it produces blood cells within the bone marrow. Skeletal muscle contraction maintains posture and accounts for the movement of the body and its parts. Cardiac muscle contraction results in the heartbeat. The walls of internal organs contract due to the presence of smooth muscle.

Endocrine System

The **endocrine system** (see Chapter 20) consists of the hormonal glands, which secrete chemical messengers called hormones. Hormones have a wide range of effects, including regulating cellular metabolism, regulating fluid and pH balance, and helping us respond to stress. Both the nervous and endocrine systems coordinate and regulate the functioning of the body's other systems. The endocrine system also helps maintain the functioning of the male and female reproductive organs.

Reproductive System

The **reproductive system** (see Chapter 21) has different organs in the male and female. The male reproductive system consists of the testes, other glands, and various ducts that conduct semen to and through the penis. The testes produce sex cells called sperm. The female reproductive system consists of the ovaries, uterine tubes, uterus, vagina, and external genitals. The ovaries produce sex cells called eggs, or oocytes. When a sperm fertilizes an oocyte, an offspring begins to develop.

Check Your Progress 11.3

1. List a major organ that is found in each organ system.
2. Identify two organ systems that protect the body from disease.

11.4 Integumentary System

Learning Outcomes

Upon completion of this section, you should be able to

1. Identify the two main regions of skin, and how these are distinguished from the subcutaneous layer.
2. Describe the makeup and function of the accessory structures of human skin.
3. List some common disorders of human skin, and how these may be treated.

The skin and its accessory organs (nails, hair, oil glands, and sweat glands) are collectively called the **integumentary system. Skin** protects underlying tissues from physical trauma, pathogen invasion, and water loss. It also participates in homeostasis by helping to regulate body temperature. The skin even synthesizes vitamin D with the aid of ultraviolet radiation. Skin also contains sensory receptors, which help us to be aware of our surroundings and to communicate through touch.

Regions of the Skin

The skin has two regions: the epidermis and the dermis (Fig. 11.8). A subcutaneous layer is present between the skin and any underlying

Figure 11.8 Human skin anatomy.   Skin consists of two regions, the epidermis and the dermis. A subcutaneous layer lies below the dermis.

SCIENCE IN YOUR LIFE ▶ HEALTH

UV Rays: Too Much Exposure or Too Little?

The sun is the major source of energy for life on Earth. Without the sun, most organisms would quickly die out. But the sun's energy can also be damaging. In addition to visible light, the sun emits ultraviolet (UV) radiation, which has a shorter wavelength (and thus a higher energy) than visible light. Based on wavelength, UV radiation can be grouped into three types: UVA, UVB, and UVC, but only UVA and UVB reach the Earth's surface. Scientists have developed a UV index to determine how powerful the solar rays are in different U.S. cities. In general, the more southern the city, the higher the UV index and the greater the risk of skin cancer. Maps indicating the daily risk levels for various U.S. regions can be viewed at http://www.nws.noaa.gov/om/heat/uv.shtml.

Both UVA and UVB rays can damage skin cells. Tanning occurs when melanin granules increase in keratinized cells at the surface of the skin as a way to prevent further damage by UV rays. Too much exposure to UVB rays can directly damage skin cell DNA, leading to mutations that can cause skin cancer. UVB is also responsible for the pain and redness characteristic of a sunburn. In contrast, UVA rays do not cause sunburn, but penetrate more deeply into the skin, damaging collagen fibers and inducing premature aging of the skin, as well as certain types of skin cancer. UVA rays can also induce skin cancer by causing the production of highly reactive chemicals that can indirectly damage DNA.

Skin cancer is the most commonly diagnosed type of cancer in the United States, outnumbering cancers of the lung, breast, prostate, and colon combined. The most common type of skin cancer is basal cell carcinoma (Fig. 11A*a*), which is rarely fatal, but can be disfiguring if allowed to grow. Squamous cell carcinoma (Fig. 11A*b*), the second most common type, is fatal in about 1% of cases. The most deadly form is melanoma (Fig. 11A*c*), which occurs in adolescents and young adults, as well as in older people. If detected early, over 95% of patients survive at least five years, but if the cancer cells have already spread throughout the body, only 10% to 20% can expect to live this long.

Melanoma affects pigmented cells and often has the appearance of an unusual mole (Fig. 11A*c*). Any moles that become malignant are removed surgically. If the cancer has

a. Basal cell carcinoma

b. Squamous cell carcinoma

c. Melanoma

Figure 11A Skin cancer.   a. Basal cell carcinomas are the most common type of skin cancer. b. Squamous cell carcinomas arise from the epidermis, and usually occur in areas exposed to the sun. c. Malignant melanomas result from a proliferation of pigmented cells. Warning signs include a change in the shape, size, or color of a normal mole, as well as itching, tenderness, or pain.
(a–b): ©Dr. P. Marazzi/Science Source; (c): ©James Stevenson/SPL/Science Source

spread, chemotherapy and a number of other treatments are also available. In March 2007, the USDA approved a vaccine to treat melanoma in dogs, which was the first time a vaccine was approved as a treatment for any cancer in animals or humans. Clinical trials are underway to test a similar vaccine for use in humans. Advances have also been made in the field of immunotherapy, where the cells of a person's immune system (see Chapter 13) are directed to attack melanoma cells.

According to the Skin Cancer Foundation, about 90% of nonmelanoma skin cancers, and 65% of melanomas, are associated with exposure to UV radiation from the sun. So how can we protect ourselves? First, try to minimize sun exposure between 10 A.M. and 4 P.M. (when the UV rays are most intense) by wearing protective clothing, hats, and sunglasses. Second, use sunscreen. Sunscreens generally do a better job of blocking UVB than UVA rays. In fact, the SPF, or sun protection factor printed on sunscreen labels, refers only to the degree of protection against UVB. Many sunscreens don't provide as much protection against UVA, and because UVA doesn't cause sunburn, people may have a false sense of security. Some sunscreens do a better job of blocking UVA—look for those that contain zinc oxide, titanium dioxide, avobenzone, or Mexoryl SX. Unfortunately, tanning salons use lamps that emit UVA rays that are two to three times more powerful than the UVA rays emitted by the sun. Because of the potential damage to deeper layers of skin, most medical experts recommend avoiding indoor tanning salons altogether.

Because UV light is potentially damaging, why haven't all humans developed the more

protective dark skin that is common to humans living in tropical regions? It turns out that vitamin D is produced in the body only when UVB rays interact with a form of cholesterol found mainly in the skin. This "sunshine vitamin" serves several important functions in the body, including keeping bones strong, boosting the immune system, and reducing blood pressure. Certain foods also contain vitamin D, but it can be difficult to obtain sufficient amounts through diet alone. Therefore, in more temperate areas of the planet, lighter-skinned individuals have the advantage of being able to synthesize sufficient vitamin D. Interestingly, dark-skinned people living in such regions may be at increased risk for vitamin D deficiency.

So how much sun exposure is enough in temperate regions of the world? During the summer months, an average fair-skinned person will synthesize plenty of vitamin D after exposure to 10–15 minutes of midday sun. During winter months, however, those living north of Atlanta probably receive too few UV rays to stimulate vitamin D synthesis, and therefore they must fulfill their requirement through their diet.

Questions to Consider

1. How would skin cancer negatively affect homeostasis?
2. If caught early, melanoma is rarely fatal. What are some factors that could delay detection of a melanoma?
3. There is considerable controversy about the level of dietary vitamin D intake that should be recommended. Why is this the case?

structures, such as muscle or bone. The subcutaneous layer is not a part of the skin.

The **epidermis** is made up of stratified squamous epithelium. New cells derived from *basal cells* become flattened and hardened as they are pushed to the surface by cells forming underneath them. Hardening takes place because the cells produce keratin, a water-proof protein. A thick layer of dead keratinized cells, arranged in spiral and concentric patterns, forms fingerprints and footprints. Specialized cells in the epidermis called **melanocytes** produce melanin, the main pigment responsible for skin color. Different amounts of melanin in the skin provide protection from damage by excessive ultraviolet (UV) radiation, while allowing sufficient UV activation of vitamin D. As the Health feature "UV Rays: Too Much Exposure or Too Little?" explains, too much exposure to UV light may be dangerous.

The **dermis** is a region of fibrous connective tissue beneath the epidermis. The dermis contains collagen and elastic fibers. The collagen fibers are flexible but offer great resistance to over-stretching. They prevent the skin from being torn. The elastic fibers maintain normal skin tension but also stretch to allow move-ment of underlying muscles and joints. The number of collagen and elastic fibers decreases with exposure to the sun, causing the skin to become less supple and more prone to wrinkling. The der-mis also contains blood vessels that nourish the skin. When blood rushes into these vessels, a person blushes, and when blood flow is reduced, or if there is not adequate oxygen in the blood, the skin turns "blue."

Sensory receptors are specialized free nerve endings in the dermis that respond to external stimuli. There are sensory recep-tors for touch, pressure, pain, and temperature. The fingertips con-tain the most touch receptors, and these add to our ability to use our fingers for delicate tasks.

The *subcutaneous layer* is composed of loose connective tis-sue and adipose tissue, which stores fat. Fat is a stored source of energy in the body. Adipose tissue also helps thermally insulate the body from either gaining heat from the outside or losing heat from the inside.

Accessory Organs of the Skin

Nails, hair, and glands are structures of epidermal origin, even though some parts of hair and glands are largely found in the dermis.

Nails are a protective covering of the distal part of the digits. Nails can help pry things open or pick up small objects. Nails grow from special epithelial cells at the base of the nail in the portion called the nail root. These cells become keratinized as they grow out over the nail bed. The visible portion of the nail is called the nail body. The cuticle is a fold of skin that hides the nail root. The whitish color of the half-moon-shaped base, or lunula, results from the thick layer of cells in this area (Fig. 11.9).

Hair follicles are in the dermis and continue through the epi-dermis where the hair shaft extends beyond the skin. Epidermal cells form the root of hair, and their division causes a hair to grow. As the cells become keratinized and die, they are pushed farther from the root. Interestingly, chemical substances in the body, such as illicit drugs and by-products of alcohol metabolism, are

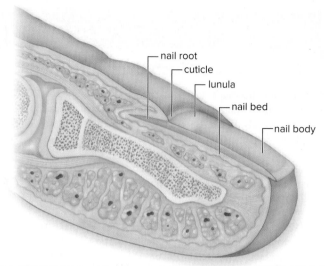

Figure 11.9 Nail anatomy.   Cells produced by the nail root become keratinized, forming the nail body.

incorporated into growing hair shafts, where they can be detected by laboratory tests. The Bioethical feature "Just a Snip, Please: Testing Hair for Drugs" explores how hair can be used to diagnose for drug use.

When we are scared or cold, contraction of the arrector pili muscles attached to hair follicles may cause the hairs to "stand on end" and goose bumps to develop. This trait has no real function in humans, although from an evolutionary perspective it may have helped fur-covered mammals to look larger, as well as to keep warm by trapping air within the fur. The color of hair is due to the variable presence of different forms of melanin. In general, the more melanin, the darker the hair. A decreasing melanin content with age results in the various shades of gray hair. White or gray hair contains little or no melanin. Hair loss is most often age-related, but also may occur following many types of illnesses.

Each hair follicle has one or more **oil glands** (also called seba-ceous glands), which secrete *sebum,* an oily substance that lubri-cates the hair within the follicle and the skin itself. With the exception of the palms of the hands and soles of the feet, all areas of human skin have oil glands.

The skin of an average adult also has about 3,000 **sweat glands** (also called sudoriferous glands) per square inch. A sweat gland is a coiled tubule within the dermis that straightens out near its opening, or pore. Some sweat glands open into hair follicles, but most open onto the surface of the skin. Sweat glands play a role in modifying body temperature. When body temperature starts to rise, sweat glands become active. Sweat absorbs body heat as it evaporates. Once body temperature lowers, sweat glands are no longer active.

Disorders of the Skin

The integumentary system is susceptible to a number of diseases. Perhaps more than animals with skin covered by tough scales or a thick fur, human skin can be easily traumatized. It is also prone to certain infections, although its dryness, mildly acidic pH, and the presence of dead cells on its outermost layers renders skin resistant

SCIENCE IN YOUR LIFE ▶ **BIOETHICAL**

Just a Snip, Please: Testing Hair for Drugs

Imagine you have just started your first real full-time job, with health insurance, a pension plan, and other benefits. The first day, after you settle into your cubicle, a manager comes by and asks for a sample of your hair. "We normally just snip a bit off in the back, where it will hardly show," he says, smiling. When you ask why, he replies that it is company policy to screen employees for drug use every 6 months.

Tests that detect illicit drugs or their metabolic products in urine or saliva are of limited value, because most of these chemicals are secreted by the body for only a few days. In contrast, many illicit drugs and/or their metabolites are incorporated into growing hair shafts throughout the body, where they remain indefinitely. Especially in the first 1.5 inches of hair growth from the scalp, each 0.5 inch is considered to represent 30 days' worth of growth (and thus, potential drug use). Hair from anywhere on the body can be used, although the growth of body hair is usually slower, so the time of any drug use cannot be determined. It generally takes 4 to 5 days from the time a drug is taken into the body until it begins to appear in hair.

Many commercial laboratories now offer hair testing (Fig. 11B). In general, these labs contend that they can distinguish between environmental exposure to a particular drug—from being in the vicinity of someone smoking marijuana, for example—and actual drug use by an individual. In recent years, several court decisions have supported the idea that hair testing can accurately distinguish actual drug use from such "passive" exposure. This has led to the marketing of several shampoos for cleaning

or "detoxifying" the hair shafts, but these may not be effective. And even if they were, a lab could conceivably test hair for common contaminants expected to be found in everyone's hair. If these contaminants were not found, this could be used as evidence that a person has attempted to hide prior drug usage.

Even if one accepts that hair testing is an accurate way to prove that a person has used drugs, is it ethical for businesses to require their employees to undergo such tests? Does it matter what type of job a person has? For example, would it be more difficult to argue against mandatory drug testing for school bus drivers than for stockbrokers? And for those of you still living with your parents, how upset would you be to find out that your mom or dad had slipped into your bedroom at night, snipped off a small bit of hair from the back of your head, and mailed it to one of several companies that now offer hair testing to the general public?

Questions to Consider

1. The Fourth Amendment to the U.S. Constitution guards against "unreasonable searches and seizures" by the government. Do you believe that the types of drug testing described here are unconstitutional?
2. In which of the following additional situations would you support mandatory drug

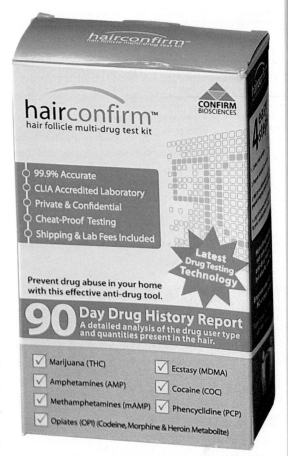

Figure 11B Hair testing kit. Kits for testing hair for illicit or prescription drugs can be purchased at retail stores or online. ©MCT via Getty Images

testing: to test airline pilots for hallucinogens; to test high school athletes for steroids; to test NBA players for marijuana?

---

to many pathogens. Several types of cancer can arise from the skin, often secondary to damage from UV rays (see the Health feature "UV Rays: Too Much Exposure or Too Little?"). Allergic reactions and various irritating chemicals can cause *dermatitis,* or inflammation of the skin.

Acne is a common skin disorder that usually first appears during adolescence. It occurs mostly on the face, shoulders, chest, and back—areas where the skin has the highest density of sebaceous glands. Around the time of puberty, increasing levels of certain hormones cause an increased production of sebum. This, in turn, can lead to a blockage of the exit of the sebum from the gland, and an increase in the growth of *Propionibacterium acnes,* a bacterium that is almost universally found on human skin. This

causes an inflammatory response, commonly known as a pimple. In many cases, white blood cells called neutrophils will migrate into the area, forming pus (see Section 13.1), and sometimes deeper tissues become involved, which increases the possibility of permanent scarring.

Most medications for acne work by drying out and unclogging blocked pores, killing the bacteria, and/or reducing inflammation. Over-the-counter medications can be useful for mild acne, but stronger, prescription medications may be needed for more severe cases. In 2005, the FDA approved Zeno, a handheld medical device that kills *P. acnes* by heating the skin to 120 degrees for 2.5 minutes. Other available treatments rely on lasers or light to kill the bacteria and/or reduce the production of sebum.

Warts are small areas of skin proliferation caused by the human papillomavirus. Though they can occur at any age, non-genital warts most commonly occur between the ages of 12 and 16. Though sometimes embarrassing, these are generally harmless and disappear without treatment—more than half do so within 2 years. Warts that cause cosmetic disfigurement or are painful (such as plantar warts on the foot) can be surgically removed, frozen with liquid nitrogen, or treated with various pharmaceutical compounds. One example is cantharidin, an extract from the blister beetle. Genital warts are discussed in Section 21.5.

> **Check Your Progress 11.4**
> **1.** Compare the structure of the epidermal and dermal layers of the skin.
> **2.** Discuss why a dark-skinned individual living in northern Canada might develop bone problems.
> **3.** List three accessory organs of the skin, and describe the major function(s) of each.

11.5 Homeostasis

> **Learning Outcomes**
> Upon completion of this section, you should be able to
> **1.** Define homeostasis, and describe why it is essential to living organisms.
> **2.** Differentiate between positive and negative feedback mechanisms, and list specific examples of each.
> **3.** Discuss the roles of specific organ systems in homeostasis, and the health consequences if homeostasis is disrupted.

Homeostasis is the maintenance of a relatively constant internal environment by an organism, or even by a single cell. Even though external conditions may change dramatically, internal conditions stay within a narrow range. For example, regardless of how cold or hot it gets, the temperature of the body stays around 98.6°F (37°C). Even if you consume an acidic food such as yogurt (with a pH around 4.5), the pH of your blood is usually about 7.4, and even if you eat a candy bar, the amount of sugar in your blood stays between 0.05% and 0.08%.

It is important to realize that internal conditions are not absolutely constant. They tend to fluctuate above and below a particular value. Therefore, the internal state of the body is often described as one of *dynamic* equilibrium. If internal conditions change to any great degree, illness results. This makes the study of homeostatic mechanisms medically important.

Negative Feedback

Negative feedback is the primary homeostatic mechanism that keeps a variable close to a particular value, or set point. A homeostatic mechanism has at least two components: a sensor and a control center (Fig. 11.10). The sensor detects a change in internal conditions. The control center then directs a response that

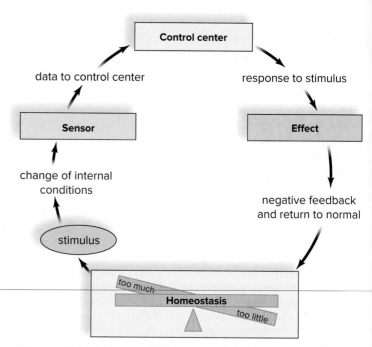

Figure 11.10 Negative feedback mechanism. The sensor and control center of a feedback mechanism work together to keep a variable close to a particular value.

brings conditions back to normal again. Now, the sensor is no longer activated. In other words, a negative feedback mechanism is present when the output of the system dampens the original stimulus.

Let's take a simple example. When the pancreas detects that the blood glucose level is too high, it secretes insulin, a hormone that causes cells to take up glucose. Now the blood sugar level returns to normal, and the pancreas is no longer stimulated to secrete insulin.

Mechanical Example

A home heating system is often used to illustrate how a more complicated negative feedback mechanism works (Fig. 11.11).

You set the thermostat at, say, 68°F. This is the *set point*. The thermostat contains a thermometer, a sensor that detects when the room temperature is above or below the set point. The thermostat also contains a control center. It turns the furnace off when the room is too hot and turns it on when the room is too cold. When the furnace is off, the room cools, and when the furnace is on, the room warms. In other words, typical of negative feedback mechanisms, there is a fluctuation above and below normal.

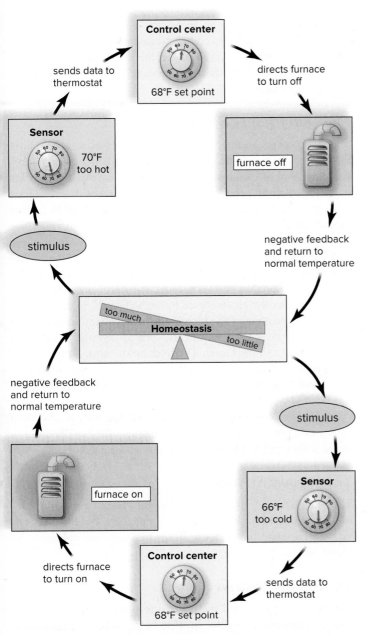

Figure 11.11 Complex negative feedback mechanism.   When a room becomes too warm, negative feedback allows the temperature to return to normal. A contrary cycle, in which the furnace turns on and gives off heat, returns the room temperature to normal when the room becomes too cool.

Human Example: Regulation of Body Temperature

The sensor and control center for body temperature are located in a part of the brain called the hypothalamus. When the body temperature is above normal, the control center directs the blood vessels of the skin to dilate (Fig. 11.12). This allows more blood to flow near the surface of the body, where heat can be lost to the environment. In addition, the nervous system activates the sweat glands, and the evaporation of sweat helps lower body temperature. Gradually, body temperature decreases to 98.6°F (37°C). When the body temperature falls below normal, the control center (via nerve impulses) directs the blood vessels of the skin to constrict. This conserves heat. If body temperature falls even lower,

the control center sends nerve impulses to the skeletal muscles, and shivering occurs. Shivering generates heat, and gradually body temperature rises to 98.6°F. When the temperature rises to normal, the control center is inactivated.

Notice that a negative feedback mechanism prevents change in the same direction; that is, body temperature does not get warmer and warmer because warmth brings about a change toward a lower body temperature. Also, body temperature does not get colder and colder because a body temperature below normal brings about a change toward a warmer body temperature.

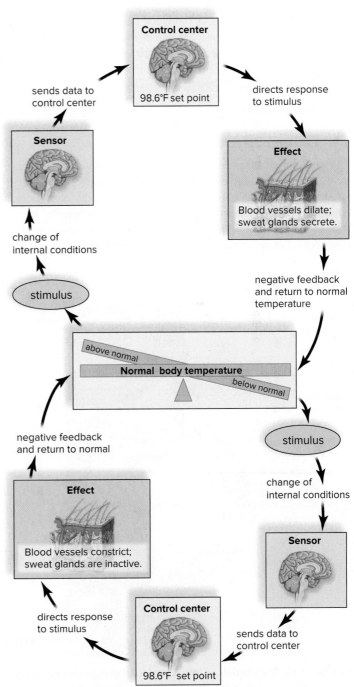

Figure 11.12 Regulation of body temperature.   Normal body temperature is maintained by a negative feedback system.

Positive Feedback

Positive feedback is a mechanism that brings about an ever greater change in the same direction. One example is the process of blood clotting, during which injured tissues release chemical factors that activate platelets. These activated platelets initiate the clotting process, and also release factors that stimulate further clotting (see Section 12.2).

Positive feedback loops tend to be involved in processes that have a definite cutoff point. Consider that when a woman is giving birth, the head of the baby begins to press against the cervix, stimulating sensory receptors there. When nerve impulses reach the brain, the brain causes the pituitary gland to secrete the hormone oxytocin. Oxytocin travels in the blood and causes the uterus to contract. As labor continues, the cervix is ever more stimulated, and uterine contractions become ever stronger until birth occurs.

Homeostasis and Body Systems

All systems of the body contribute toward maintaining homeostasis.

The Transport Systems

The cardiovascular system conducts blood to and away from capillaries, where exchange of gases, nutrients, and wastes occurs. Interstitial fluid, which bathes all the cells of the body, is refreshed when molecules such as oxygen and nutrients move into the interstitial fluid from the blood and when carbon dioxide and wastes move from the interstitial fluid into the blood (Fig. 11.13).

Figure 11.13 Regulation of interstitial fluid composition.   Cells are surrounded by interstitial fluid, which is continually refreshed because oxygen and nutrient molecules constantly exit, and carbon dioxide and waste molecules continually enter the bloodstream.

The lymphatic system is an accessory to the cardiovascular system. Lymphatic capillaries collect excess interstitial fluid, which is returned via lymphatic vessels to the cardiovascular system. Lymph nodes are sites where the immune system responds to invading microorganisms.

The Maintenance Systems

The respiratory system adds oxygen to and removes carbon dioxide from the blood. It also plays a role in regulating blood pH because removal of CO_2 causes the pH to rise, just as CO_2 retention helps to lower the pH. The digestive system takes in and digests food, providing nutrient molecules that enter the blood to replace the nutrients that are constantly being used by the body cells. The liver, an organ that assists the digestive process by producing bile, also plays a significant role in regulating blood composition. Immediately after glucose enters the blood, any excess is removed by the liver and stored as glycogen. Later, the glycogen can be broken down to replace the glucose used by the body cells. In this way, the glucose composition of the blood remains relatively constant. The liver also removes toxic chemicals, such as ingested alcohol and other drugs, from the blood. The liver makes urea, a nitrogenous end product of protein metabolism. Urea and other metabolic waste molecules are excreted by the kidneys, which are a part of the urinary system. Urine formation by the kidneys is extremely critical to the body, not only because it rids the body of unwanted substances, but also because urine formation offers an opportunity to carefully regulate blood volume, salt balance, and pH.

The Support Systems

The integumentary and musculoskeletal systems protect the internal organs. In addition, the integumentary system produces vitamin D, while the skeleton stores minerals and produces the blood cells.

The Control Systems

The nervous system and the endocrine system work together to control other body systems so that homeostasis is maintained. We have already seen that, in negative feedback mechanisms, sensory receptors send nerve impulses to control centers in the brain, which then rapidly direct effectors to become active. Effectors can be muscles or glands. Muscles bring about an immediate change. Endocrine glands secrete hormones that bring about a slower, more lasting change that keeps the internal environment relatively stable.

Disease

A **disease** is an abnormality in the body's normal processes that significantly impairs homeostasis. As will be seen in the physiology chapters of this text, diseases (disorders) can affect virtually every part of the human body. The major causes of human diseases include blood vessel problems, cancers, infections, and inflammatory conditions.

A particular disease may be described as *systemic,* meaning it affects the entire body or at least several organ systems. Other diseases, including many infections or inflammatory diseases such as dermatitis or arthritis, are more *localized* to a specific part of the body. Diseases may also be classified on the basis of their severity and duration. *Acute* diseases, such as poison ivy dermatitis or influenza, tend to occur suddenly and generally last a short time, although some can be life-threatening. *Chronic* diseases, such as multiple sclerosis, AIDS, and most cancers, tend to develop slowly and last a long time, even for the rest of a person's life, unless an effective cure is available.

The term *cancer* refers to a group of disorders in which the usual controls on cell division fail, resulting in the production of abnormal cells that invade and destroy healthy body tissue. Cancers are classified according to the type of tissue from which they arise. *Carcinomas,* the most common type, are cancers of epithelial tissue. The Health feature "UV Rays: Too Much Exposure or Too Little?" in Section 11.4 describes one type of carcinoma, the melanoma. *Sarcomas* are cancers arising in muscle or connective tissue, especially bone or cartilage. *Leukemias* are cancers of the blood cells, and *lymphomas* are cancers that originate in lymph nodes. The chance of developing cancer in a particular tissue shows a positive correlation to the rate of cell division. Epithelial cells reproduce at a high rate, and 2.5 million new blood cells appear each second. Therefore, carcinomas and leukemias are common types of cancers.

Check Your Progress 11.5

1. Explain how positive feedback differs from negative feedback.
2. Describe how several body systems can interact to maintain homeostasis.
3. List several specific diseases that result when a particular body system fails to perform its function.

Conclusion

The human body consists of an incredibly complex system of molecules, cells, tissues, organs, and organ systems functioning together. Ultimately, the interaction of these components provides humans with the ability to achieve the highest levels of physical and mental functions, as well as maintaining general homeostasis. We are aware of only a few of these processes, such as the activity of our muscles when we are trying to learn a new skill or perform a challenging physical task. When these homeostatic processes fail, various types of diseases can result.

SUMMARIZE

11.1 Types of Tissues

In the human body, most cells are organized into **tissues.** Human tissues are categorized into four groups:

■ **Epithelial tissue** covers the body and lines its cavities. **Squamous epithelium, cuboidal epithelium,** and **columnar epithelium** can be simple, stratified, or pseudostratified and may also have cilia or microvilli. Epithelial cells sometimes form **glands** that secrete either into ducts or into the blood.

■ **Connective tissue,** in which cells are separated by a matrix, often binds body parts together. In most cases, the matrix contains **collagen fibers, reticular fibers,** and **elastic fibers.**

- **Loose fibrous connective tissue** supports epithelium and encloses organs. **Dense fibrous connective tissue,** such as that of tendons and ligaments, contains a closely packed matrix. **Cartilage,** a type of dense fibrous connective tissue in which cells are found in **lacunae,** can be of three types: **hyaline cartilage, elastic cartilage,** or **fibrocartilage.**

- **Bone** has a hard matrix due to the presence of calcium salts. **Compact bone** is found in the shafts of long bones; **spongy bone** is found on the ends.

- **Adipose tissue** stores fat, while **reticular connective tissue** forms a supportive mesh in some organs.

- **Blood** is a connective tissue in which the matrix is a liquid. Blood consists of the liquid part, called **plasma,** and the formed elements, which include the **red blood cells, white blood cells,** and **platelets. Interstitial fluid** is created when the liquid portions of the blood surround the individual cells of the body.

■ **Muscular tissue** is of three types. **Skeletal muscle** is found in muscles attached to bones, and **smooth muscle** is found in internal organs. Both skeletal muscle and **cardiac muscle** are striated. Both cardiac and smooth muscle are involuntary.

■ **Nervous tissue** has one main type of conducting cell, the **neuron,** and several types of supporting cells, the **neuroglia.** Each neuron has a cell body, dendrites, and an axon. Axons are specialized to conduct nerve impulses. Bundles of axons bound by connective tissue are called **nerves.**

11.2 Body Cavities and Body Membranes

The internal organs occur within two main **body cavities.**

■ The **dorsal cavity** includes the cranial cavity and vertebral canal.

■ The **ventral cavity** contains the thoracic, abdominal, and pelvic cavities, which house organs of the respiratory, digestive, urinary, and reproductive systems, among others.

■ Four types of membranes line body cavities and the internal spaces of organs. **Mucous membranes** line the tubes of the digestive system; **serous membranes** line the thoracic and abdominal cavities and cover the organs they contain; **synovial membranes** line movable joint cavities; and **meninges** cover the brain and spinal cord.

11.3 Organ Systems

The organ systems of the human body can be grouped according to their major function(s):

■ The **integumentary system** protects deeper tissues and also makes vitamin D, collects sensory data, and helps regulate body temperature.

■ The **cardiovascular, lymphatic, digestive, respiratory,** and **urinary systems** perform processing and transport functions that maintain the normal conditions of the body.

■ The **immune system** defends against infectious disease.

■ The **musculoskeletal system** supports and protects the body and permits movement.

■ The **nervous system** receives sensory input and directs muscles and glands to respond.

■ The **endocrine system** produces hormones, which influence the functions of the other systems.

■ The **reproductive system** produces sex cells (gametes), and ultimately, offspring.

11.4 Integumentary System

■ The **integumentary system** is composed of **skin** and the accessory organs.

■ The main functions of the integumentary system are to protect underlying tissues while providing a barrier to pathogen invasion and water loss. Skin also functions in the sense of touch.

■ Human skin has two regions: (1) the **epidermis** contains basal cells that produce new epithelial cells, which become keratinized as they move toward the surface; and (2) the **dermis** contains glands, hair follicles, blood vessels, and sensory receptors.

■ The accessory organs of skin include the **nails, hair follicles, oil glands** (sebaceous glands), and **sweat glands.**

■ The skin is susceptible to several types of disorders, including infections, cancers, and various inflammatory conditions.

11.5 Homeostasis

■ The body's internal environment consists of blood and interstitial fluid.

■ **Homeostasis** is the maintenance of a relatively constant internal environment, mainly by two mechanisms: (1) **negative feedback** mechanisms keep the environment relatively stable (when a sensor detects a change above or below a set point, a control center brings about an effect that reverses the change and brings conditions back to normal again); and (2) **positive feedback** mechanisms bring about rapid change in the same direction as the stimulus. **Disease** results when one or more body systems fail.

ASSESS

Testing Yourself
Choose the best answer for each question.

11.1 Types of Tissues

1. Tissues are formed from _____ and are arranged together to form _____.
   a. organs; organ systems
   b. cells; molecules
   c. cells; organs
   d. molecules; cells

2. Which choice is true of both cardiac and skeletal muscle?
   a. striated
   b. single nucleus per cell
   c. multinucleated cells
   d. involuntary control

3. Which of the following is not a connective tissue?
   a. blood
   b. bone
   c. cartilage
   d. adipose
   e. All of these are connective tissues.

11.2 Body Cavities and Body Membranes

4. Label the diagram of the body cavities below.

11.3 Organ Systems

5. Which system(s) help(s) control pH balance?
   a. digestive
   b. respiratory
   c. urinary
   d. Both b and c are correct.

6. Which system plays the biggest role in fluid balance?
   a. cardiovascular
   b. urinary
   c. digestive
   d. integumentary

11.4 Integumentary System

7. Label the indicated structures in the diagram of human skin below:

8. Keratinization of epithelial cells occurs in which layer of the skin?
   a. subcutaneous layer
   b. dermis
   c. epidermis
   d. All of these are correct.

11.5 Homeostasis

9. Which of the following allows rapid change in one direction but does not achieve stability?
   a. homeostasis
   b. positive feedback
   c. negative feedback
   d. All of these are correct.

10. Which of the following is an example of negative feedback?
    a. Uterine contractions increase as labor progresses.
    b. Insulin decreases blood sugar levels after a meal is eaten.
    c. Sweating increases as body temperature drops.
    d. Platelets continue to plug an opening in a blood vessel until blood flow stops.

ENGAGE

Thinking Critically

1. In what way(s) is blood like a tissue?

2. Which of these homeostatic mechanisms in the body are examples of positive feedback, and which are examples of negative feedback? Why?
   a. The adrenal glands produce epinephrine in response to a hormone produced by the pituitary gland in times of stress; the pituitary gland senses the epinephrine in the blood and stops producing the hormone.
   b. As the bladder fills with urine, pressure sensors send messages to the brain with increasing frequency, signaling that the bladder must be emptied. The more the bladder fills, the more messages are sent.
   c. When you drink an excess of water, specialized cells in your brain, as well as stretch receptors in your heart, detect the increase in blood volume. Both signals are transmitted to the kidneys, which increase the production of urine.

3. Explain how a failure of homeostasis leads to death.

C H A P T E R
12
Cardiovascular System

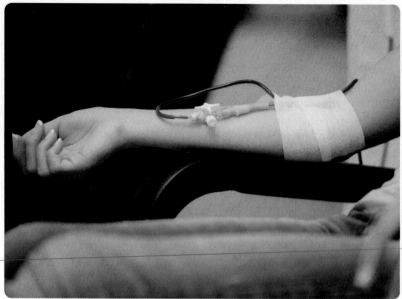

©Wavebreak Media Ltd/123RF

BEFORE YOU BEGIN

Before beginning this chapter, take a few moments to review the following discussions:

Section 4.2 What molecules must diffuse across the plasma membrane of cells lining capillaries?

Section 11.1 How does the structure of cardiac muscle differ from skeletal and smooth muscle?

Section 11.5 The cardiovascular system interacts with which other body systems?

Synthetic Blood

Every 2 seconds, someone in the United States needs a transfusion of blood. The need may have been caused by an injury, or a surgical procedure, or even a disease that requires regular transfusions. To answer these needs, blood donors provide over 36,000 units of blood every day, or 15 million units of blood every year.

Historically, the need for blood transfusions has been met by blood donors. Unfortunately, supply does not always meet demand. Natural disasters may place a strain on supplies, especially considering that blood can be stored for only about 42 days, which makes stockpiling difficult. Furthermore, some people are not able to receive blood from donors due to religious beliefs or medical conditions.

To meet the demand for transfusions, scientists have been developing synthetic, or artificial, blood. The use of an oxygen-carrying blood substitute, called oxygen therapeutics, has the goal of providing a patient's tissues with enough oxygen. In some cases, the blood substitute is completely synthetic, and contains chemicals that mimic the oxygen-carrying hemoglobin found in red blood cells. In other cases, scientists are applying biotechnology to manufacture replacement red blood cells. While several biotech companies are currently conducting clinical trials using synthetic blood, most medical professionals believe we are still several years from being able to produce a synthetic blood supply to meet the needs of society.

In this chapter, we will explore the structure and function of the cardiovascular system in the human body and explore why this organ system is so essential to life.

As you read through this chapter, think about the following questions:

1. What is the role of blood in our body?
2. Besides red blood cells, what other elements need to be present in synthetic blood?

12.1 The Blood Vessels

Learning Outcomes

Upon completion of this section, you should be able to

1. Identify the structural features and function of each of the three main types of blood vessels.
2. Explain the process of capillary exchange.

The **cardiovascular system** has three types of blood vessels: the **arteries** (and arterioles), which carry blood away from the heart to the capillaries; the **capillaries,** which permit exchange of material with the tissues; and the **veins** (and venules), which return blood from the capillaries to the heart. The structure of each of these blood vessels is well adapted to its function in the body.

The Arteries

An arterial wall has three layers (Fig. 12.1*a*). As is the case with all of the blood vessels, arteries have an inner *endothelium,* a simple squamous epithelium attached to a connective tissue basement membrane that contains elastic fibers (see Section 11.1). The middle layer is the thickest layer and consists of smooth muscle that can contract to regulate blood flow and blood pressure. The outer layer is fibrous connective tissue near the middle layer, but it becomes loose connective tissue at its periphery. The largest artery in the human body is the **aorta.** It is approximately 25 mm wide and carries O_2-rich blood from the heart to other parts of the body.

Smaller arteries branch off from the aorta, eventually forming a large number of arterioles. **Arterioles** are small arteries just visible to the naked eye, averaging about 0.5 mm in diameter. The inner layer of arterioles is endothelium. The middle layer is composed of some elastic tissue, but mostly of smooth muscle with fibers that encircle the arteriole. When these muscle fibers are contracted, the vessel has a smaller diameter (is constricted). When these muscle fibers are relaxed, the vessel has a larger diameter (is dilated). Whether arterioles are constricted or dilated affects blood pressure. The greater the number of vessels dilated, the lower the blood pressure.

The Capillaries

Capillaries join arterioles to venules (Fig. 12.1*b*). Capillaries are extremely narrow—only 8–10 μm wide—and have thin walls composed only of a single layer of endothelium with a basement membrane. Although each capillary is small, they form vast networks. Their total surface area in a human body is about 6,300 square meters (m²). Capillary beds (networks of many capillaries) are present in nearly all regions of the body. Consequently, a cut to almost any body tissue draws blood. One region of the body that is nearly capillary-free is the cornea of the eye, so that light can pass through. Therefore, the cells of the cornea must obtain nutrients by diffusion from the tears on the outside surface, and from the aqueous humor on the inside surface.

Capillaries play a very important role in homeostasis because an exchange of substances takes place across their thin walls. Oxygen and nutrients, such as glucose, diffuse out of a capillary into

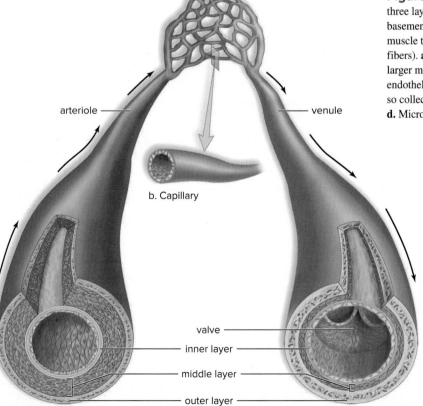

a. Artery

b. Capillary

c. Vein

arteriole

venule

valve

inner layer

middle layer

outer layer

Figure 12.1 Blood vessels.   The walls of arteries and veins have three layers. The inner layer is composed largely of endothelium, with a basement membrane that has elastic fibers; the middle layer is smooth muscle tissue; the outer layer is connective tissue (largely collagen fibers). **a.** Arteries have a thicker wall than veins because they have a larger middle layer than veins. **b.** Capillary walls are one-cell-thick endothelium. **c.** Veins are generally larger in diameter than arteries, so collectively, veins have a larger holding capacity than arteries. **d.** Micrograph of an artery and a vein. (d): ©Ed Reschke

outer layer

middle layer

inner layer

outer layer

middle layer

inner layer

artery

vein

d. Micrograph of artery and vein

100×

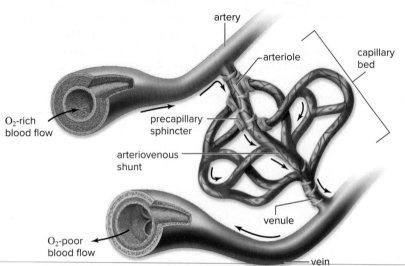

Figure 12.2 Anatomy of a capillary bed. A capillary bed forms a maze of capillary vessels that lies between an arteriole and a venule. When precapillary sphincter muscles are relaxed, the capillary bed is open, and blood flows through the capillaries. When sphincter muscles are contracted, blood flows through the arteriovenous shunt that carries blood directly from an arteriole to a venule. As blood passes through a capillary in the tissues, it gives up its oxygen (O_2). Therefore, blood goes from being O_2-rich in the arteriole (red color) to being O_2-poor in the vein (blue color).

the interstitial fluid that surrounds cells. Wastes, such as carbon dioxide, diffuse into the capillary. Some water also leaves a capillary. Any excess is picked up by the vessels of the lymphatic system (see Figure 12.9). The relative constancy of interstitial fluid is absolutely dependent upon capillary exchange.

Because capillaries serve the cells, the heart and the other vessels of the cardiovascular system can be thought of as the means by which blood is conducted to and from the capillaries. Only certain capillary beds are completely open at any given time. Rings of muscles, called *precapillary sphincters,* control the flow of blood into a capillary bed. When the precapillary sphincters are closed, blood flow into that capillary bed is reduced. For example, after eating, the capillary beds that serve the digestive system are mostly open, and those that serve the muscles are mostly closed. Each capillary bed has *arteriovenous shunts* that allow blood to go directly from arterioles to venules, bypassing the bed (Fig. 12.2).

Contrary to a popular misconception, it is not a decrease in blood flow to the brain after meals that causes "postprandial somnolence," or the sleepiness that many people feel after eating. The blood supply to the brain is maintained under most physiological conditions, including ingestion of a heavy meal. Instead, it is recognized that hormones released by the digestive tract may instead be the culprit.

The Veins

Veins and venules take blood from the capillary beds to the heart. First, the **venules** (small veins) drain blood from the capillaries and then join to form a vein. The walls of veins (and venules) have the same three layers as arteries, but there is less smooth muscle and connective tissue (see Fig. 12.1c). Therefore, the wall of a vein is thinner than that of an artery.

Veins often have *valves,* which allow blood to flow only toward the heart when open and prevent blood from flowing backward when closed. Valves are found in the veins that carry blood against the force of gravity, especially the veins of the lower limbs. Unlike blood flow in the arteries and arterioles, which is kept moving by the pumping of the heart, blood flow in veins is primarily due to skeletal muscle contraction. If these valves become damaged by disease or through the normal wear-and-tear of aging, blood may begin pooling in the veins, causing them to enlarge and be visible as *varicose veins.* These most commonly occur in the lower legs of older individuals. Varicose veins of the anal canal are known as hemorrhoids.

Because the walls of veins are thinner (see Fig. 12.1d), they can expand to a greater extent. At any one time, about 70% of the blood is in the veins. In this way, the veins act as a blood reservoir. If blood is lost due to hemorrhaging, nervous stimulation causes the veins to constrict, providing more blood to the rest of the body. The largest veins are the **venae cavae,** which include the *superior vena cava* (20 mm wide) and the *inferior vena cava* (35 mm wide). Both veins deliver O_2-poor blood into the heart.

Check Your Progress 12.1

1. Describe how blood flow is controlled in each of the three major types of blood vessels.
2. List several specific substances that diffuse across capillary walls.

12.2 Blood

Learning Outcomes

Upon completion of this section, you should be able to

1. List the major types of blood cells, and summarize their functions.
2. Identify the major molecular and cellular events that result in a blood clot.
3. Define capillary exchange, and describe the two major forces involved.

The body contains a number of different types of fluids, some of which are noted in Table 12.1. In this section, we will explore the function of **blood,** a liquid connective tissue.

TABLE 12.1 Body Fluids Related to Blood

Name	Composition
Blood	Formed elements and plasma
Plasma	Liquid portion of blood
Serum	Plasma minus fibrinogen
Interstitial fluid	Plasma minus most proteins
Lymph	Interstitial fluid within lymphatic vessels

Not only does blood have transport functions, it also acts in regulating homeostasis and as an important component of the immune system. Blood also acts as a transport medium for nutrients, wastes, and hormones. Blood helps regulate body temperature by dispersing body heat and helps regulate blood pressure because the plasma proteins contribute to the osmotic pressure of blood. Buffers in blood help maintain blood pH at about 7.4. Blood helps protect the body against invasion by disease-causing pathogens. Clotting mechanisms protect the body against potentially life-threatening loss of blood.

Components of Blood

If blood is collected from a person's vein into a test tube and prevented from clotting, it separates into three layers (Fig. 12.3a). The upper layer is **plasma,** the liquid portion of blood. The two lower layers together are the **formed elements,** which include the white blood cells and platelets (middle layer), and the red blood cells (bottom layer).

Plasma

Plasma contains a variety of inorganic and organic substances dissolved or suspended in water (Fig. 12.3b). Plasma proteins, which make up 7–8% of plasma, assist in transporting large organic molecules in blood. For example, *albumin* transports bilirubin, a breakdown product of hemoglobin. Lipoproteins transport cholesterol. Certain plasma proteins have specific functions. As discussed later in this section, fibrinogen is necessary to blood clotting, and immunoglobulins are antibodies that help fight infection. Plasma proteins also maintain blood volume because they are too large to leave capillaries. Therefore, blood in capillaries normally has a higher solute concentration than does interstitial fluid, and water automatically diffuses into them.

The Red Blood Cells

Red blood cells, also called RBCs or **erythrocytes,** are continuously manufactured in the red bone marrow of the skull, the ribs, the vertebrae, and the ends of the long bones. Normally, there are 4 to 6 million red blood cells per microliter (μl) of whole blood (Fig. 12.3d).

Mature red blood cells don't have a nucleus and are biconcave disks (Fig. 12.4a, b). Their shape increases their flexibility for moving through capillary beds and their surface area for diffusion of gases. Red blood cells carry oxygen because they contain **hemoglobin,** the respiratory pigment. Because hemoglobin is a red pigment, the cells are red. A hemoglobin molecule contains four polypeptide chains (Fig. 12.4c). Each chain is associated with heme, a complex iron-containing group. The iron portion of hemoglobin acquires oxygen in the lungs and gives it up in the tissues.

Red blood cells lack nuclei and most other internal organelles. They produce ATP by the process of anaerobic fermentation (see Section 7.3). Due to their lack of internal structure, red blood cells live only about 120 days. They are destroyed chiefly in the liver and the spleen, where they are engulfed by large phagocytic cells. The iron is mostly salvaged and reused, while the heme portion undergoes chemical degradation, and the liver excretes it into the bile as bile pigments.

When the body has an insufficient number of red blood cells or the red blood cells do not contain enough hemoglobin, an individual suffers from **anemia** and has a tired, rundown feeling. There are three basic causes of anemia: (1) decreased production of red blood cells, (2) loss of red blood cells from the body, and (3) destruction of red blood cells within the body. In the most common type of anemia, iron-deficiency anemia, red blood cell production is decreased, most often due to a diet that does not contain enough iron.

Whenever arterial blood carries a reduced amount of oxygen, as in chronic anemia or when an individual first takes up residence at a high altitude, the kidneys increase their production of a hormone called *erythropoietin,* which speeds the maturation of red blood cells in the bone marrow.

The White Blood Cells

White blood cells, also called WBCs or **leukocytes,** differ from red blood cells in that they are usually larger, have a nucleus, lack hemoglobin, and without staining appear translucent (see Fig. 12.3c, d). White blood cells are not as numerous as red blood cells, with only 5,000–11,000 cells per microliter. White blood cells fight infection and play a role in the development of immunity, the ability to resist disease.

On the basis of structure, it is possible to divide white blood cells into *granular leukocytes* and *agranular leukocytes,* the latter also being known as *mononuclear cells.* Granular leukocytes (neutrophils, basophils, and eosinophils) are filled with spheres that contain enzymes and proteins, which help white blood cells defend the body against microbes. **Neutrophils** are granular leukocytes with a multilobed nucleus joined by nuclear threads. They are the most abundant of the white blood cells and are able to phagocytize and digest bacteria. **Basophils** stain a deep blue and release histamine, which can cause inflammation. **Eosinophils** stain a deep red and are involved in protecting against parasitic worms, although they are also involved in some allergies.

Agranular leukocytes (monocytes and lymphocytes) typically have a kidney-shaped or spherical nucleus. **Monocytes** are the largest of the white blood cells, and they differentiate into phagocytic dendritic cells and macrophages. **Dendritic cells** are present in tissues that are in contact with the environment: skin, nose, lungs, and intestines. Once they have captured a microbe with their long, spiky arms, called dendrites, they stimulate other white blood cells to defend the body. **Macrophages,** well known as ferocious phagocytes (Fig. 12.5), play a similar role in other organs, such as the liver, kidney, and spleen. The **lymphocytes** are of two major types: B lymphocytes (B cells) and T lymphocytes (T cells). B cells produce antibodies, whereas T cells occur as two main types: helper T cells that regulate the responses of other cells, and cytotoxic T cells that are able to kill other cells.

If the number of any type of white blood cell increases or decreases beyond normal, disease may be present. For example, neutrophil numbers tend to increase in response to many different bacterial infections. A person with *infectious mononucleosis,* caused by the Epstein–Barr virus, often has an increased number of B cells. In an HIV-infected person, AIDS is defined, in part, by having an abnormally low number of T cells. *Leukemia* is a form of cancer characterized by uncontrolled production of abnormal white blood cells.

White blood cells live different lengths of time. Many live only a few days or may die combating invading pathogens. Others live for months or even years. How white blood cells interact with the immune system is discussed more extensively in Chapter 13.

Figure 12.3 Composition of blood. a. When blood is collected into a test tube containing an anticoagulant to prevent clotting and then centrifuged, it consists of three layers. The transparent, straw-colored or yellow top layer is the plasma, the liquid portion of the blood. The thin, middle layer consists of leukocytes and platelets. The bottom layer contains the erythrocytes. **b.** Breakdown of the components of plasma. **c.** Micrograph of the formed elements, which are also listed in (**d**). (c): ©Ed Reschke

Plasma (about 55% of whole blood)

Formed elements
{ Leukocytes and platelets (<1% of whole blood)
Erythrocytes (about 45% of whole blood)

a.

Formed Elements		
Type	**Function and Description**	**Source**
Red blood cells (erythrocytes)	Transport O_2 and help transport CO_2	Red bone marrow
4 million–6 million per mm³ blood	7–8 µm in diameter Bright-red to dark-purple biconcave disks without nuclei	
White blood cells (leukocytes) 5,000–11,000 per mm³ blood	Fight infection	Red bone marrow
Granular leukocytes — Neutrophils* 40–70%	10–14 µm in diameter Spherical cells with multilobed nuclei; fine, pink granules in cytoplasm; phagocytize pathogens	
Eosinophils* 1–4%	10–14 µm in diameter Spherical cells with bilobed nuclei; coarse, deep-red, uniformly sized granules in cytoplasm; phagocytize antigen-antibody complexes and allergens	
Basophils* 0–1%	10–12 µm in diameter Spherical cells with lobed nuclei; large, irregularly shaped, deep-blue granules in cytoplasm; release histamine, which promotes blood flow to injured tissues	
Agranular leukocytes — Lymphocytes* 20–45%	5–17 µm in diameter (average 9–10 µm) Spherical cells with large, round nuclei; responsible for specific immunity	
Monocytes* 4–8%	10–24 µm in diameter Large spherical cells with kidney-shaped, round, or lobed nuclei; become macrophages that phagocytize pathogens and cellular debris	
Platelets (thrombocytes)	Aid clotting	Red bone marrow
150,000–300,000 per mm³ blood	2–4 µm in diameter Disk-shaped cell fragments with no nuclei; purple granules in cytoplasm	

d.   *Appearance with Wright's stain.

Plasma		
Type	**Function**	**Source**
Water (90–92% of plasma)	Maintains blood volume; transports molecules	Absorbed from intestine
Plasma proteins (7–8% of plasma)	Maintain blood osmotic pressure and pH	
Albumin	Maintains blood volume and pressure, transport	Liver
Antibodies	Fight infection	B lymphocytes
Fibrinogen	Clotting	Liver
Salts (less than 1% of plasma)	Maintain blood osmotic pressure and pH; aid metabolism	Absorbed from intestine
Gases Oxygen Carbon dioxide	Cellular respiration End product of metabolism	Lungs Tissues
Nutrients Lipids Glucose Amino acids	Food for cells	Absorbed from intestine
Nitrogenous wastes Urea Uric acid	Excretion by kidneys	Liver
Other Hormones, vitamins, etc.	Aid metabolism	Varied

b.

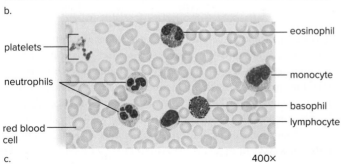

platelets

neutrophils

red blood cell

eosinophil

monocyte

basophil

lymphocyte

c.   400×

a. Blood capillary   400×   b. Red blood cells   SEM 4,175×   c. Hemoglobin molecule

Figure 12.4 Red blood cells and the structure of hemoglobin.   **a.** Red blood cells move in single file through the capillaries. **b.** Each red blood cell is a biconcave disk containing many molecules of hemoglobin, the respiratory pigment. **c.** Hemoglobin contains two types of polypeptide chains (green, purple). There is an iron-containing heme group in the center of each chain. Oxygen combines loosely with iron when hemoglobin is oxygenated. Oxyhemoglobin is bright red, and deoxyhemoglobin is a dark maroon color. (a): ©Ed Reschke/Getty Images (b): ©Andrew Syred/Science Source

Figure 12.5 Macrophage engulfing bacteria.   Monocyte-derived macrophages (red in this image) are the body's scavengers. They engulf microbes (yellow rods above) and debris in the body's fluids and tissues, as illustrated in this colorized scanning electron micrograph. ©Dennis Kunkel/Medical Images

The Platelets and Blood Clotting

Platelets, also called **thrombocytes,** result from fragmentation of certain large cells, called *megakaryocytes,* in the red bone marrow. Platelets are produced at a rate of 200 billion a day, and the blood contains 150,000–300,000 per microliter. These formed elements are involved in the process of blood **clotting,** or coagulation.

There are at least 12 clotting factors in the blood that participate with platelets in the formation of a blood clot. We will discuss the roles played by *fibrinogen* and *prothrombin,* which are proteins manufactured by the liver. Vitamin K, found in green vegetables and also formed by intestinal bacteria, is needed for prothrombin production. Vitamin K deficiency can result in clotting disorders.

Blood Clotting

When a blood vessel in the body is damaged, the process of clotting begins (Fig. 12.6a). First, platelets clump at the site of the puncture and partially seal the leak. Platelets and damaged tissue release *prothrombin activator,* which converts the plasma protein prothrombin to thrombin. This reaction requires calcium ions (Ca^{2+}). **Thrombin,** in turn, acts as an enzyme that severs two short amino acid chains from each fibrinogen molecule, activating it. The activated fragments then join end to end, forming long threads of **fibrin.** Fibrin threads wind around the platelet plug in the damaged area of the blood vessel and provide the framework for the clot. Red blood cells are also trapped within the fibrin threads. These cells make a clot appear red (Fig. 12.6b).

A fibrin clot is only temporary. As soon as blood vessel repair is initiated, an enzyme called plasmin destroys the fibrin network and restores the fluidity of the plasma. Sometimes it is desirable to inhibit the clotting process. This can be done with medications such as heparin, or even with medicinal leeches (see the Health feature "The Use of Leeches in Medicine"). If blood is allowed to clot in a test tube, a yellowish fluid develops above the clotted material. This fluid is called *serum,* and it contains all the components of plasma except fibrinogen.

1. Blood vessel is punctured.

2. Platelets congregate and form a plug.

3. Platelets and damaged tissue cells release prothrombin activator, which initiates a cascade of enzymatic reactions.

Prothrombin activator

Prothrombin $\longrightarrow$ Thrombin
Ca^{2+}

Fibrinogen $\longrightarrow$ Fibrin threads
Ca^{2+}

4. Fibrin threads form and trap red blood cells.

a. Blood-clotting process

fibrin threads

red blood cell

b. Blood clot   4,400×

Figure 12.6 Blood clotting.   a. Platelets and damaged tissue cells release prothrombin activator, which acts on prothrombin in the presence of Ca^{2+} (calcium ions) to produce thrombin. Thrombin acts on fibrinogen in the presence of Ca^{2+} to form fibrin threads. **b.** A scanning electron micrograph of a blood clot shows red blood cells caught in the fibrin threads. (b): ©Eye of Science/Science Source

---

SCIENCE IN YOUR LIFE ▶ HEALTH

The Use of Leeches in Medicine

Although it may seem more like an episode of a popular TV show than a real-life medical treatment, the U.S. Food and Drug Administration has approved the use of leeches as medical "devices" for treating conditions involving poor blood supply to various tissues.

Leeches are bloodsucking, aquatic creatures whose closest living relatives are earthworms (Fig. 12A). Prior to modern times, medical practitioners frequently applied leeches to patients, mainly in an attempt to remove the bad "humors" that they thought were responsible for many diseases. This practice was abandoned, thankfully, in the nineteenth century when it was realized that the "treatment" often harmed the patient.

True to their tenacious nature, however, leeches are making a comeback in twenty-first-century medicine. By applying leeches to tissues that have been injured by trauma or disease, blood supply can be improved. When reattaching a finger, for example, it is easier to suture together the thicker-walled arteries than the veins. Poorly draining blood from the veins

Figure 12A Leeches. Leeches can attach to human skin and suck out blood. Their use in medicine seems to be making a comeback. ©Astrid & Hanns-Frieder Michler/ Science Source

can pool in the appendage and threaten its survival. It turns out that leech saliva contains chemicals that dilate blood vessels and prevent blood from clotting by blocking the activity of thrombin. These effects can improve the circulation to the body part. Another substance in leech saliva actually anesthetizes the bite wound. In a natural setting, this allows the leech to feast on the blood supply of its victim undetected, but in a medical setting, it makes

the whole experience more tolerable, at least physically. Mentally, however, the application of leeches can still be a rather unsettling experience, and patient acceptance is a major factor limiting their more widespread use.

Questions to Consider

1. If you had an injury and your doctor said it might help, would you be willing to let leeches feast for a few minutes, say, on your hand? On your face?

2. Leeches were historically used for bloodletting, one of the oldest medical practices. Can you think of several reasons why, prior to the mid-nineteenth century, early physicians might have gotten the idea that removing blood could help cure various diseases?

3. The use of actual living organisms in medical treatments is sometimes called "biotherapy." Can you think of other situations in which living creatures might be used for human health benefits, either directly or indirectly?

Hemophilia

Hemophilia refers to a group of inherited clotting disorders caused by a deficiency in a clotting factor. The most common type, hemophilia A, accounts for about 90% of all cases, and almost always occurs in males because the faulty gene is found on the X chromosome. (Females have two Xs; therefore, they have a backup copy of the gene.) The slightest bump to an affected person can cause bleeding into the joints. Cartilage degeneration in the joints and resorption of underlying bone can follow. Bleeding into muscles can lead to nerve damage and muscular atrophy. Death can result from bleeding into the brain, with accompanying neurological damage. Hemophiliacs usually require frequent blood transfusions, although they may also be treated with injections of the specific clotting factor they are lacking.

Bone Marrow Stem Cells

A **stem cell** is a cell that is capable of continuously dividing and producing new cells that go on to differentiate into particular types of cells. It's been known for some time that the bone marrow has stem cells, which have the potential to give rise to other stem cells for the various formed elements (Fig. 12.7). Recent research has shown that bone marrow stem cells are also able to differentiate into other types of cells, including liver, bone, fat, cartilage, heart, and even neurons. A person's bone marrow stem cells may be used

for curing certain conditions that might develop, such as diabetes, heart disease, liver disease, or even brain disorders (e.g., Alzheimer disease or Parkinson disease).

The use of a person's own stem cells does away with the problem of possible rejection. Among all the various types of *adult stem cells* in the body, bone marrow stem cells are particularly attractive because they are the most accessible. Adult stem cells are usually multipotent, meaning they can develop into a limited number of cell types. Some researchers prefer to work with *embryonic stem cells,* because these cells are *totipotent,* meaning they are more likely to develop into any type of cell. Embryonic stem cells are available because many early-stage embryos remain unused in fertility clinics, although this issue has become controversial. Scientists have developed a form of stem cell called *induced pluripotent stem cell* that converts adult stem cells to embryonic stem cells. These stem cells give scientists the

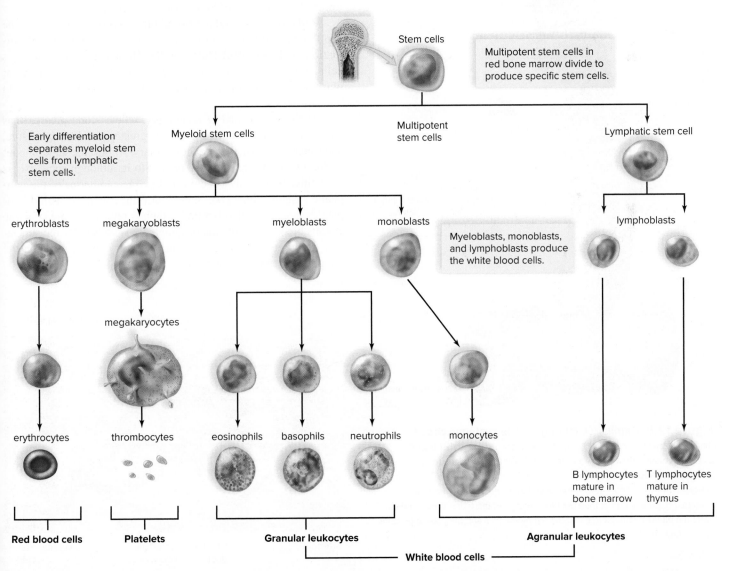

Figure 12.7 Blood cell formation in red bone marrow.   Multipotent stem cells give rise to two specialized types of stem cells. The myeloid stem cells give rise to still other cells, which become red blood cells, platelets, and all the white blood cells except lymphocytes. The lymphoid stem cells give rise to lymphoblasts, which become lymphocytes.

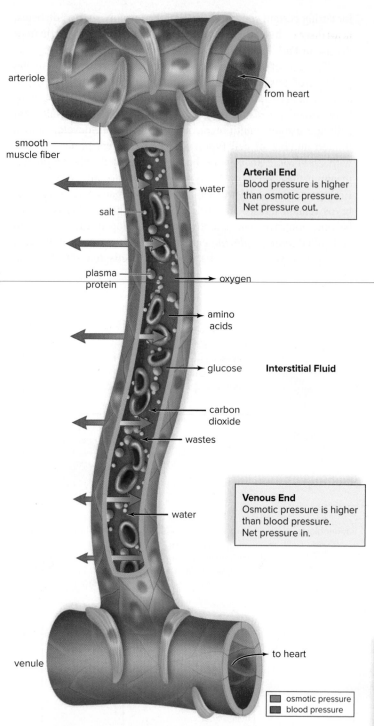

arteriole

from heart

smooth
muscle fiber

water

Arterial End
Blood pressure is higher
than osmotic pressure.
Net pressure out.

salt

plasma
protein

oxygen

amino
acids

glucose

Interstitial Fluid

carbon
dioxide

wastes

Venous End
Osmotic pressure is higher
than blood pressure.
Net pressure in.

water

venule

to heart

☐ osmotic pressure
☐ blood pressure

Figure 12.8 Capillary exchange in the systemic circuit. At
the arterial end of a capillary (*top*), the blood pressure is higher than the
osmotic pressure; therefore, water tends to leave the bloodstream. In the
midsection, molecules, including oxygen and carbon dioxide, follow their
concentration gradients. At the venous end of a capillary (*bottom*), the
osmotic pressure is higher than the blood pressure; therefore, water tends to
enter the bloodstream. Notice that the red blood cells and the plasma proteins
are too large to exit a capillary.

ability to generate new cells outside the body, and are quickly
changing the practice of medicine.

Capillary Exchange

Two forces primarily control movement of fluid through the capillary wall: (1) osmotic pressure, created by salts and plasma proteins, which tends to cause water to move from the interstitial fluid
to the blood; and (2) blood pressure, which tends to cause water to
move in the opposite direction. At the arterial end of a capillary,
blood pressure is higher than the osmotic pressure of blood
(Fig. 12.8), so water exits a capillary at this end.

Midway along the capillary, where blood pressure is lower,
the two forces essentially cancel each other, and there is no net
movement of water. Solutes now diffuse according to their concentration gradient—nutrients (glucose and amino acids) and oxygen
diffuse out of the capillary, and wastes (carbon dioxide) diffuse
into the capillary. In the pulmonary circuit, where the O_2 concentration is higher in the lung tissues, and the CO_2 concentration is
lower, the movement of these gases is reversed.

Red blood cells and almost all plasma proteins remain in the
capillaries, but small substances leave, contributing to the formation of **interstitial fluid,** or the fluid between the body's
cells. Interstitial fluid tends to contain all the components of
plasma but lesser amounts of protein, which generally stays in
the capillaries.

At the venous end of a capillary, where blood pressure has
fallen even more, osmotic pressure is greater than blood pressure,
and water tends to move into the capillary. Almost the same amount
of fluid that left the capillary returns to it, although some excess
interstitial fluid is always collected by the capillaries of the lymphatic system (Fig. 12.9). Interstitial fluid contained within lymphatic vessels is called **lymph.** Lymph is returned to the systemic
venous blood when the major lymphatic vessels enter the subclavian veins in the shoulder region. We will explore more on the
lymphatic system in Chapter 13.

blood capillary   excess interstitial fluid

arteriole   lymphatic lymphatic   venule tissue
             duct   capillaries   cells

Figure 12.9 Return of excess interstitial fluid. A lymphatic
capillary bed (shown here in green) lies near a blood capillary bed. When
lymphatic capillaries take up excess interstitial fluid, it becomes lymph.

12.3 The Human Heart

Learning Outcomes

Upon completion of this section, you should be able to
1. Identify the structural components of the human heart.
2. Trace the path of blood through the heart and lungs.
3. Describe how the heartbeat is controlled.

The **heart** is a cone-shaped, muscular organ about the size of a fist (Fig. 12.10). It is located between the lungs, directly behind the sternum (breastbone), and is tilted so that the apex (the pointed end) is oriented to the body's left. The major portion of the heart, called the *myocardium,* consists largely of cardiac muscle tissue. The muscle fibers of the myocardium are branched and tightly joined to one another. The heart lies within the *pericardium,* a thick, serous membrane that secretes a small quantity of lubricating liquid. The inner surface of the heart is lined with *endocardium,* a membrane composed of connective tissue and endothelial tissue.

Internally, a wall called the *septum* separates the heart into a right side and a left side (Fig. 12.11). The heart has four chambers. The two upper, thin-walled **atria** (*sing.,* **atrium**) are located above the two lower, thick-walled **ventricles.** The ventricles pump the blood to the lungs and the body.

The heart also has four valves that direct the flow of blood and prevent its backward movement. The two valves that lie between the atria and the ventricles are called the **atrioventricular valves.** These valves are supported by strong fibrous strings called *chordae tendineae.* The chordae, which are attached to muscular projections of the ventricular walls, support the valves and prevent them from inverting when the heart contracts. The atrioventricular valve on the body's right side is called the *tricuspid valve* because it has three flaps, or cusps. The valve on the left side is called the *bicuspid (mitral) valve* because it has two flaps. The remaining two valves, between the

Figure 12.10 External heart anatomy. a. The venae cavae and the pulmonary trunk are attached to the right side of the heart. The aorta and the pulmonary veins are attached to the left side of the heart. **b.** The coronary arteries and cardiac veins service the cardiac muscle. The coronary arteries bring oxygen and nutrients to cardiac cells, which derive no benefit from blood coursing through the heart. The cardiac veins drain blood into the right atrium.

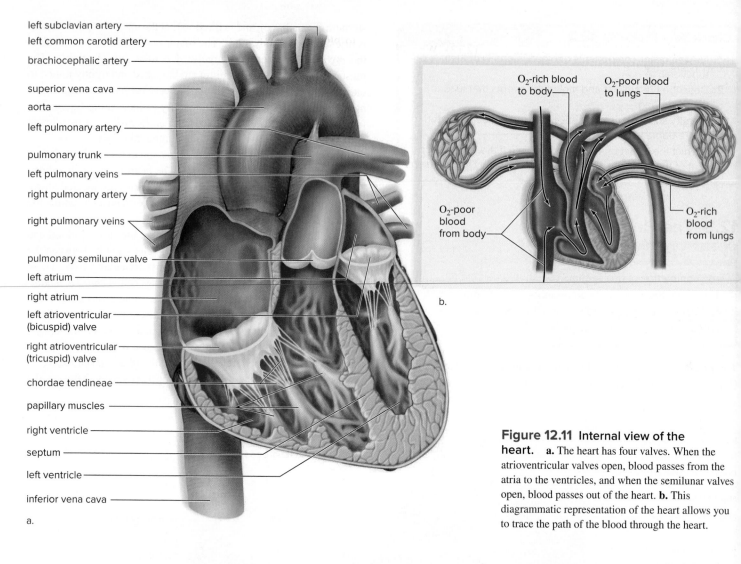

left subclavian artery
left common carotid artery
brachiocephalic artery
superior vena cava
aorta
left pulmonary artery
pulmonary trunk
left pulmonary veins
right pulmonary artery
right pulmonary veins
pulmonary semilunar valve
left atrium
right atrium
left atrioventricular (bicuspid) valve
right atrioventricular (tricuspid) valve
chordae tendineae
papillary muscles
right ventricle
septum
left ventricle
inferior vena cava

a.

O_2-rich blood to body
O_2-poor blood to lungs
O_2-poor blood from body
O_2-rich blood from lungs

b.

Figure 12.11 Internal view of the heart.   a. The heart has four valves. When the atrioventricular valves open, blood passes from the atria to the ventricles, and when the semilunar valves open, blood passes out of the heart. **b.** This diagrammatic representation of the heart allows you to trace the path of the blood through the heart.

ventricles and their attached vessels, are the **semilunar valves,** whose flaps resemble half-moons. The *pulmonary semilunar valve* lies between the right ventricle and the pulmonary trunk. The *aortic semilunar valve* lies between the left ventricle and the aorta.

Path of Blood Through the Heart

By referring to Figure 12.11, we can trace the path of blood through the heart in the following manner:

- The superior vena cava and the inferior vena cava, which carry O_2-poor blood that is relatively high in carbon dioxide, enter the right atrium.
- The right atrium sends blood through the tricuspid valve to the right ventricle.
- The right ventricle sends blood through the pulmonary semilunar valve into the pulmonary trunk and through the two **pulmonary arteries** to the lungs.
- Four **pulmonary veins,** which carry O_2-rich blood, enter the left atrium.
- The left atrium sends blood through the bicuspid (mitral) valve to the left ventricle.
- The left ventricle sends blood through the aortic semilunar valve into the aorta to the rest of the body.

From this description, you should see that O_2-poor blood never mixes with O_2-rich blood, and that blood must go through the lungs in order to pass from the right side to the left side of the heart. In fact, the heart is a double pump because the right ventricle sends blood into the lungs and the left ventricle sends blood into the rest of the body. Because the left ventricle has the harder job of pumping blood to the entire body, its walls are thicker than those of the right ventricle, which pumps blood a relatively short distance to the lungs. The volume of blood that the left ventricle pumps per minute is called the **cardiac output.** In a person with an average heart rate of 70 beats per minute, the cardiac output is about 5.25 liters (l) of blood per minute. This is about equal to the total amount of blood in the body, and adds up to about 750 liters (2,000 gallons) of blood per day. During heavy exercise, the cardiac output can increase as much as fivefold, or perhaps more in trained athletes.

The pumping of the heart sends blood out under pressure into the arteries. Because the left side of the heart is the stronger pump, blood pressure is greatest in the aorta. Blood pressure then decreases as the cross-sectional area of arteries and then arterioles increases. The **pulse** is a wave effect that passes down the walls of arteries when the aorta expands and then recoils with each ventricular contraction. The arterial pulse can be used to determine the heart rate, and a weak or "thready" pulse may indicate a weak heart or low blood pressure.

The Heartbeat

Each heartbeat is called a **cardiac cycle** (Fig. 12.12). When the heart beats, first the two atria contract at the same time, then the two ventricles contract at the same time. Then, all of the chambers relax. The word **systole** refers to contraction of heart muscle, and the word **diastole** refers to relaxation of heart muscle.

The heart contracts, or beats, about 70 times a minute, and each heartbeat lasts about 0.85 second, apportioned as follows:

Cardiac Cycle		
Time	**Atria**	**Ventricles**
0.15 sec	Systole	Diastole
0.30 sec	Diastole	Systole
0.40 sec	Diastole	Diastole

A normal adult heart rate at rest can vary from 60 to 80 beats per minute, which adds up to about 2.5 million beats in a lifetime!

When the heart beats, the familiar "lub-dub" sound occurs. This is best heard using a *stethoscope,* an instrument used to isolate and amplify the body's internal sounds. The longer and lower-pitched "lub" is caused by vibrations occurring when the atrioventricular valves close due to ventricular contraction. The shorter and sharper "dub" results when the semilunar valves close due to back pressure of blood in the arteries. A heart murmur, or a slight whooshing sound after the "lub," is most commonly due to blood flowing back through an ineffective mitral valve (see Section 12.5).

Intrinsic Control of Heartbeat

The rhythmic contraction of the atria and ventricles is due to the intrinsic (or internal) conduction system of the heart that is made possible by the presence of nodal tissue, a unique type of cardiac muscle. Nodal tissue, which has both muscular and nervous characteristics, is located in two regions of the heart. The *SA (sinoatrial) node* is located in the upper dorsal wall of the right atrium. The *AV (atrioventricular) node* is located in the base of the right atrium very near the septum (Fig. 12.13a). The SA node initiates the heartbeat and sends out an excitation impulse every 0.85 seconds. This causes the atria to contract. When impulses reach the AV node, a slight delay allows the atria to finish contraction before the ventricles begin to contract. The signal for the ventricles to contract travels from the AV node through specialized cardiac muscle fibers called the *atrioventricular bundle* (*AV bundle*) before reaching the numerous and smaller *Purkinje fibers.*

The SA node is also called the **pacemaker** because it usually keeps the heartbeat regular. If the SA node fails to work properly, the heart still beats due to impulses generated by the AV node. But the beat is slower (40 to 60 beats per minute). To correct this condition, an artificial pacemaker may be implanted that gives an electrical stimulus to the heart every 0.85 seconds.

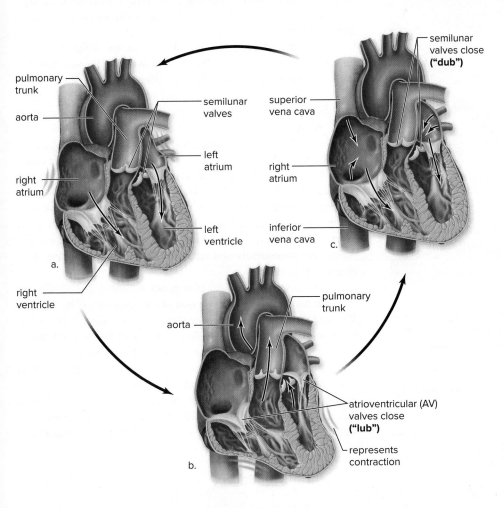

Figure 12.12 The cardiac cycle.
a. When the atria contract, the ventricles are relaxed and filling with blood. **b.** When the ventricles contract, the atrioventricular valves close, preventing blood from flowing back into the atria and producing the "lub" sound of the heartbeat. **c.** After the ventricles contract, the "dub" sound of the heartbeat results from the closing of the semilunar valves to prevent arterial blood from flowing back into the ventricles.

SA node

AV node

branches of
atrioventricular
bundle

Purkinje fibers

a.

R

P

T

Q

S

b.

c.

Figure 12.13 Conduction system of the heart. **a.** The SA node
sends out a stimulus (black arrows), which causes the atria to contract. When
this stimulus reaches the AV node, it signals the ventricles to contract.
Impulses pass down the two branches of the atrioventricular bundle to the
Purkinje fibers, and thereafter the ventricles contract. **b.** A normal ECG
usually indicates that the heart is functioning properly. The P wave occurs just
prior to atrial contraction; the QRS complex occurs just prior to ventricular
contraction; and the T wave occurs when the ventricles are recovering from
contraction. **c.** Ventricular fibrillation produces an irregular electrocardiogram
due to irregular stimulation of the ventricles. (b, c): ©Ed Reschke

Extrinsic Control of Heartbeat

The body also has extrinsic (or external) ways to regulate the
heartbeat. A cardiac control center in the medulla oblongata, a portion of the brain that controls internal organs, can alter the beat of
the heart by way of the autonomic system, a portion of the nervous
system. The autonomic system has two subdivisions: the parasympathetic division, which promotes functions of a resting state, and
the sympathetic division, which brings about responses to increased activity or stress. The parasympathetic division decreases
SA and AV nodal activity when we are inactive, and the sympathetic system increases these nodes' activity when we are active or
excited.

The hormones epinephrine and norepinephrine, which are released by the adrenal medulla (inner portion of the adrenal gland),
also stimulate the heart. When we are frightened, for example, the
heart pumps faster and stronger due to sympathetic stimulation and
because of the release of epinephrine and norepinephrine.

The Electrocardiogram

An **electrocardiogram** (**ECG**) is a recording of the electrical
changes that occur in the myocardium during a cardiac cycle. Body
fluids contain ions that conduct electrical currents, and therefore
the electrical changes in the myocardium can be detected on the
skin's surface. Electrodes placed on or near the chest are connected
by wires to an instrument that detects and records the myocardium's
electrical changes. Figure 12.13*b* depicts the electrical changes
during a normal cardiac cycle.

When the SA node triggers an impulse, the atrial fibers produce an electrical change called the P wave. The P wave indicates
that the atria are about to contract. After that, the QRS complex
signals that the ventricles are about to contract. The electrical
changes that occur as the ventricular muscle fibers recover produce
the T wave.

Various types of abnormalities, known as *arrhythmias,* can be
detected by an electrocardiogram. The most common type is called
atrial fibrillation (AF). Instead of the heart generating single, regular impulses from the AV node, multiple, chaotic impulses are
generated, resulting in an irregular, fast heartbeat. Symptoms of
AF may include *palpitations,* meaning a fluttering sensation in the
heart, along with dizziness, weakness, or chest pain. Sometimes,
AF occurs infrequently and resolves on its own; other cases may
require medication.

A more serious medical emergency is *ventricular fibrillation
(VF)*, an uncoordinated contraction of the ventricles (Fig. 12.13*c*).
VF most commonly follows a heart attack, but it also can be caused
by an injury or drug overdose. Because a heart in VF is not pumping blood, it must be defibrillated by applying a strong electrical
current for a short period of time. Then, the SA node may be able
to reestablish a coordinated beat. Many public places now have
automatic external defibrillators (AEDs), which are small devices
that can be used to determine whether a person is suffering from
ventricular fibrillation—and if so, administer an appropriate electrical shock to the chest.

12.4 Two Cardiovascular Pathways

The cardiovascular system includes two circuits (Fig. 12.14). The **pulmonary circuit** circulates blood through the lungs, and the **systemic circuit** serves the needs of body tissues.

The Pulmonary Circuit

Blood from all regions of the body first collects in the right atrium and then passes into the right ventricle, which pumps it into the pulmonary trunk. The pulmonary trunk divides into the right and left pulmonary arteries, which branch as they approach the lungs. The arterioles take blood to the pulmonary capillaries, where gas exchange occurs. Blood then passes through the pulmonary venules, which lead to the four pulmonary veins (two from each lung) that enter the left atrium. Because blood in the pulmonary arteries is O_2-poor but blood in the pulmonary veins is O_2-rich, it is not correct to say that all arteries carry blood high in oxygen and all veins carry blood low in oxygen. It is just the reverse in the pulmonary circuit, as well as in the umbilical arteries and vein of the developing fetus (see Section 22.3).

The Systemic Circuit

The systemic circuit includes the major arteries and veins shown in Figure 12.15. As mentioned, the largest artery in the systemic circuit is the *aorta;* the largest veins are collectively called the *venae cavae.* The superior vena cava collects blood from the head, the chest, and the arms, while the inferior vena cava collects blood from the lower body regions. Both enter the right atrium. The aorta and the venae cavae serve as the major pathways for blood in the systemic circuit.

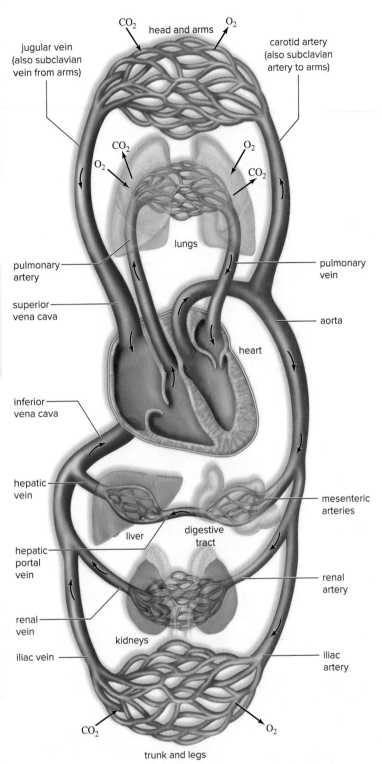

Figure 12.14 Path of blood.   This symbolic and not-to-scale drawing shows the path of blood in the pulmonary and systemic circuits. The pulmonary arteries and veins take blood from the right (blue) to the left (red) side of the heart. Tracing blood from the digestive tract to the right atrium in the systemic circuit involves the hepatic portal vein, the hepatic vein, and the inferior vena cava. The blue-colored vessels carry O_2-poor blood, and the red-colored vessels carry O_2-rich blood; the arrows indicate the direction of blood flow.

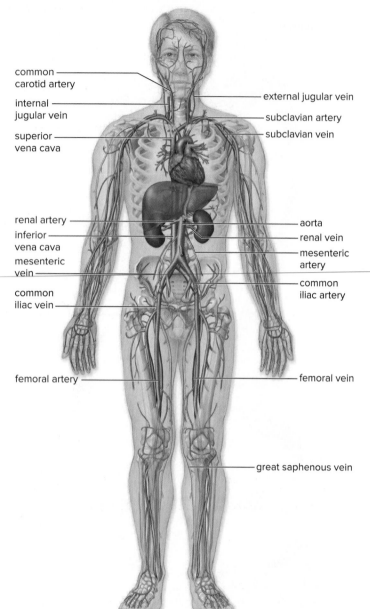

Figure 12.15 Major arteries (red) and veins (blue) of the systemic circuit. This representation of the major blood vessels of the systemic circuit shows how the systemic arteries and veins are arranged in the body. The superior and inferior venae cavae take their names from their relationship to the heart.

The path of systemic blood to any organ in the body begins in the left ventricle. For example, trace the path of blood to and from the legs in Figure 12.15:

left ventricle → aorta → common iliac artery → femoral artery → leg capillaries →

femoral vein → common iliac vein → inferior vena cava → right atrium

Notice that, when tracing blood, you need only mention the aorta, the proper branch of the aorta, the region, and the vein returning blood to the vena cava. In most instances, the artery and the vein that serve the same region are given the same name (Fig. 12.15).

Figure 12.16 Blood velocity and blood pressure. In capillaries, blood is under minimal pressure and has the least velocity. Blood pressure and velocity drop off because capillaries have a greater total cross-sectional area than arterioles.

The *coronary arteries* (see Fig. 12.10) serve the muscles of the heart because these muscles are not nourished by the blood within the chambers. The coronary arteries are the first branches off the aorta. They originate just above the aortic semilunar valve, and they lie on the exterior surface of the heart, where they divide into diverse arterioles. Because they have a very small diameter, the coronary arteries may become clogged, as discussed in Section 12.5. The coronary capillary beds join to form venules. The venules converge to form the cardiac veins, which empty into the right atrium.

A **portal system** in blood circulation begins and ends in capillaries. One such system, the *hepatic portal system,* is associated with the liver. Capillaries that occur in the villi of the small intestine pass into venules that join to form the *hepatic portal vein.* This vein carries the blood to a set of capillaries in the liver, an organ that monitors the makeup of the blood (see Fig. 12.14). The *hepatic vein* leaves the liver and enters the inferior vena cava.

Blood Pressure

When the left ventricle contracts, blood is forced into the aorta and then into other systemic arteries under pressure. *Systolic pressure* results from blood being forced into the arteries during ventricular systole, and *diastolic pressure* is the pressure in the arteries during ventricular diastole.

As blood flows from the aorta into the arteries and arterioles, blood pressure falls. Also, the difference between systolic and diastolic pressure gradually diminishes. In the capillaries, blood flow is slow and fairly even. This may be related to the very high total cross-sectional area of the capillaries (Fig. 12.16). It has been calculated that if all the blood vessels in a single human were connected end to end, the total distance would reach around Earth at the equator two times! Most of this distance would be due to the large number of capillaries.

Blood pressure can be measured with a *sphygmomanometer,* which has a pressure cuff that determines the amount of pressure required to stop the flow of blood through an artery. Blood pressure is normally measured on the *brachial artery,* an artery in the upper arm. Today, automated manometers are often used to take one's blood pressure instead. Blood pressure is expressed in millimeters of mercury (mm Hg). A blood pressure reading consists of two numbers that represent systolic and diastolic pressures, respectively—for example, 120/80 mm Hg. Blood pressure that is too low or—more commonly—too high can be a significant health problem (see Section 12.5).

Blood pressure in the veins is low and by itself is an inefficient means of moving blood back to the heart, especially from the limbs. When skeletal muscles near veins contract, they put pressure on the veins and the blood they contain. Valves prevent the backward flow of blood in veins, and therefore muscle contraction is sufficient to move blood toward the heart (Fig. 12.17). During

long periods of sitting or other inactivity, clots may form in the deep veins of the legs. Because these can cause serious problems, especially if they break free and become lodged in the lungs, take frequent stretch breaks on long car or plane trips.

Check Your Progress 12.4

1. Identify which arteries carry O$_2$-poor blood.
2. Name the major blood vessels through which blood flows to the lungs and other parts of the body.
3. Describe why blood flows in one direction in an artery versus a vein.

12.5 Cardiovascular Disorders

Learning Outcomes

Upon completion of this section, you should be able to

1. Describe the major categories of cardiovascular disease that occur in the United States.
2. Define *hypertension* and explain its most common causes.
3. List three possible treatments for a blocked coronary artery.

Cardiovascular disease (CVD) is the leading cause of untimely death in most Western countries. Modern research efforts have resulted in improved diagnosis, treatment, and prevention. This section discusses some of the major advances that have been made in these areas. The Health feature "Prevention of Cardiovascular Disease" describes ways to maintain cardiovascular health.

Atherosclerosis

Atherosclerosis is an accumulation of soft masses of fatty materials, particularly cholesterol, beneath the inner linings of arteries. Such deposits are called *plaque.* Plaque tends to protrude into the lumen of the vessel and interfere with the flow of blood (see Fig. 12B). In most instances, atherosclerosis begins in early adulthood and develops progressively through middle age, but symptoms may not appear until an individual is 50 or older.

Plaque can cause platelets to adhere to the irregular arterial wall, forming a clot. As long as the clot remains stationary, it is called a *thrombus,* but if it dislodges and moves along with the blood, it is called an *embolus. Thromboembolism,* a clot that has been carried in the bloodstream but is now lodged in a blood vessel, must be treated, or serious complications can arise.

Hypertension

Normal blood pressure values vary somewhat among different age groups, body sizes, and levels of athletic conditioning. However, it is estimated that about 33% of American adults have **hypertension,** which is high blood pressure. Another 30% are thought to have a condition called *prehypertension,* often a precursor to hypertension

a. Contracted skeletal muscle pushes blood past open valve.

b. Closed valve prevents backward flow of blood.

Figure 12.17 Cross-section of a valve in a vein. a. Pressure on the walls of a vein, exerted by skeletal muscles, increases blood pressure within the vein and forces the valve open. **b.** When external pressure is no longer applied to the vein, blood pressure decreases, and back pressure forces the valve closed. Closure of the valve prevents the blood from flowing in the opposite direction.

Prevention of Cardiovascular Disease

Many of us are predisposed to cardiovascular disease (CVD) due to factors beyond our control. Having a family history of heart attacks under age 55, or being male or African American increases the risk. Other risk factors for CVD are related to our behavior.

The Don'ts

Smoking

When a person smokes, the drug nicotine, present in cigarette smoke, enters the bloodstream. Nicotine causes arterioles to constrict, including those that supply the heart itself. It also increases blood pressure, and the tendency of blood to clot. These factors may explain why about 20% of deaths from CVD are directly related to cigarette smoking.

Drug Abuse

Stimulants, such as cocaine and amphetamines, can cause an irregular heartbeat and lead to heart attacks even when using the drugs for the first time.

Obesity

People who are obese have more tissues that need to be supplied with blood. To meet this demand, the heart pumps blood out under greater pressure. Being overweight also increases the risk of type 2 diabetes, in which glucose damages blood vessels and makes them more prone to the development of plaque.

The Do's

Healthy Diet

Diet influences the amount of cholesterol in the blood. Cholesterol is ferried by two types of plasma proteins, called LDL (low-density lipoprotein) and HDL (high-density lipoprotein). LDL (the "bad" lipoprotein) takes cholesterol from the liver to the tissues, and HDL (the "good" lipoprotein) transports cholesterol out of the tissues to the liver. When the LDL level in blood is high or the HDL level is too low, plaque accumulates on arterial walls (Fig. 12B).

Eating foods high in saturated fat (red meat, cream, and butter) or trans fats (most margarines, commercially baked goods, and deep-fried foods) increases LDL cholesterol. Replacement of these harmful fats with monounsaturated fats (olive and canola oil) and polyunsaturated fats (corn, safflower, and soybean oil) is recommended. Most nutritionists also suggest eating at least five servings of antioxidant-rich fruits and vegetables a day to protect against cardiovascular disease. Antioxidants protect the body from free radicals that can damage blood vessels.

The American Heart Association (AHA) recommends eating at least two servings of cold-water fish (e.g., halibut, sardines, tuna, and salmon) each week. These fish contain omega-3 polyunsaturated fatty acids that can reduce plaque. However, some studies suggest that farm-raised fish (such as tilapia) contain low levels of these beneficial fatty acids and may, in fact, be detrimental to heart health.

Resveratrol

The "French paradox" refers to the observation that levels of CVD are relatively low in France, despite the common consumption of a high-fat diet. One possible explanation is that wine is frequently consumed with meals. In addition to its alcohol content, red wine contains an antioxidant called resveratrol. This chemical is mainly produced in the skin of grapes, so it is also found in grape juice, and supplements are available at health food stores.

Cholesterol Profile

Starting at age 20, all adults are advised to have their cholesterol levels tested at least every 5 years. Even in healthy individuals, an LDL level above 160 mg/100 ml and an HDL level below 40 mg/100 ml are matters of concern. Cholesterol-lowering medications are available for those who do not meet these minimum guidelines.

Exercise

People who exercise are less apt to have cardiovascular disease. Exercise not only helps keep weight under control, but may also help minimize stress and reduce hypertension. In fact, short bursts of exercise may be superior to longer sessions. In one study, as few as three 10-minute workout sessions a day reduced triglyceride levels in blood better than one 30-minute session.

Anxiety and Stress

Mental stress can increase the odds of a heart attack. Within an hour of a strong earthquake that struck near Los Angeles in 1994, 16 people died of sudden heart failure (compared to the average of about four per day). Over the next several days, the number of heart-related deaths declined, suggesting that emotional stress had triggered fatal complications in those who were already predisposed to them. Obviously, it is difficult to avoid earthquakes, but we can learn healthy ways to avoid and manage stress.

Can Alcohol Benefit the Heart?

Alcohol abuse can destroy just about every organ in the body, the heart included. Based on several large research studies, however, the AHA notes that people who consume one or two drinks per day have a 30% to 50% reduced risk of CVD compared to nondrinkers. Importantly, because of the potential downsides of alcohol consumption, the AHA does not recommend that nondrinkers start using alcohol.

Questions to Consider

1. Review the risk factors discussed in this reading. Which, if any, could be affecting you right now?
2. Some scientists suspect that infections with certain types of bacteria and viruses may contribute to cardiovascular disease. What are some specific ways infections might do this?
3. What are other possible explanations for the "French paradox"?

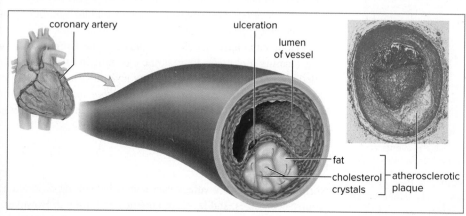

Figure 12B Coronary arteries and plaque. Atherosclerotic plaque is an irregular accumulation of cholesterol and fat. When plaque is present in a coronary artery, a heart attack is more likely to occur because of restricted blood flow. (photo): ©Biophoto Associates/Science Source

TABLE 12.2	Normal Values for Adult Blood Pressure	
	Top Number (Systolic)	**Bottom Number (Diastolic)**
Normal	Below 120	Below 80
Elevated	120–129	Below 80
Stage 1 hypertension	130–139	80–89
Stage 2 hypertension	140 or more	90 or more

Source: Blood pressure values established by the American Heart Association.

(Table 12.2). Under age 45, a blood pressure reading above 130/90 mm Hg is considered to be abnormally high, as is a reading above 140/95 mm Hg in older people.

Hypertension is sometimes called "the silent killer" because it may not be detected until a stroke or heart attack occurs. It most often occurs secondary to a narrowing of a person's arteries from atherosclerosis. Consider trying to force a gallon of water through a straw, versus a garden hose, in the same amount of time. You would need to apply much more force to get the water through the straw. In the same way, forcing blood through narrowed arteries, over time, creates additional pressure on the cardiovascular system that can damage the blood vessels, heart, and other organs.

Many medications are used to treat hypertension, including diuretics (which reduce the blood volume by increasing urine output), vasodilators (which dilate the blood vessels), and various drugs that improve heart function. Only about two-thirds of people with hypertension seek medical help for their condition, and it is likely that many people with high blood pressure are unaware of it.

Heart Valve Disease

Each year in the United States, around 100,000 people undergo surgery to have faulty heart valves repaired or replaced. In some patients, the heart valves are malformed at birth, but more commonly they degenerate due to age or infections, to the point where they no longer prevent the backflow of blood. A narrowing (stenosis) of the aortic valve opening is the most common, followed by mitral valve prolapse, in which abnormally thickened "leaflets" of the mitral valve protrude back into the left ventricle. In some cases, the faulty valves can be repaired during open-heart surgery, but more commonly they are replaced, using either artificial valves or valves removed from an animal (usually a pig) or a deceased human. Surgeons can also replace a defective heart valve by threading a compact artificial valve through an artery in the leg, thus avoiding open-heart surgery.

Stroke, Heart Attack, and Aneurysm

A cerebrovascular accident, also called a **stroke,** often results when an arteriole in the brain bursts or is blocked by an embolus. Lack of oxygen causes a portion of the brain to die, and paralysis or death can result. A person is sometimes forewarned of a stroke by a feeling of numbness in the hands or the face, difficulty in speaking, or temporary blindness in one eye. Cerebrovascular accidents become more common with age, but are certainly not

limited to older people: In 2006, former Minnesota Twins baseball star Kirby Puckett died of a stroke at age 45.

If a coronary artery becomes partially blocked, the individual may suffer from **angina pectoris,** characterized by a squeezing or burning sensation in the chest. Nitroglycerin or related drugs dilate blood vessels and help relieve the pain. When a coronary artery is completely blocked, a portion of the heart muscle dies due to a lack of oxygen, and a myocardial infarction, or **heart attack,** occurs.

An **aneurysm** is the ballooning of a blood vessel, most often the abdominal aorta or the arteries leading to the brain. Atherosclerosis and high blood pressure can weaken the wall of an artery to the point that an aneurysm develops. If a major vessel such as the aorta should burst, about half of victims die before reaching a hospital. It is often possible to replace a damaged or diseased portion of a vessel, such as an artery, with a synthetic graft. Several companies are developing wireless sensors that can be implanted along with these grafts, to help doctors monitor changes in pressure that may signal an aneurysm that is about to burst.

Coronary Bypass Operations

Since the late 1960s, coronary bypass surgery has been a common way to treat an obstructed coronary artery. During this operation, a surgeon usually takes a segment from another blood vessel and stitches one end to the aorta and the other end to a coronary artery past the point of obstruction. Figure 12.18 shows a heart in which three blocked coronary arteries have been bypassed using grafted arteries. Note that if veins are used as replacements, they must be placed in reverse orientation, due to the presence of valves. Even after successful bypass surgery, it is common for blockages to recur in the transplanted veins within 5 to 7 years.

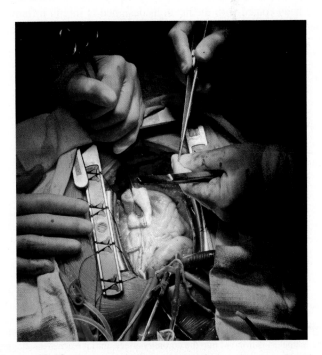

Figure 12.18 Coronary bypass operation. Close-up of hands of a surgeon performing coronary artery bypass surgery. The surgeon has bypassed two blocked arteries using vessels removed from another part of the body. ©aaM Photography, Ltd./Getty Images

Some recent research suggests that heart muscle that has been damaged by a heart attack can be regenerated using stem cells. Studies using models of heart attacks in mice and rats have shown that stem cells injected directly into the damaged heart muscle will differentiate into new heart muscle cells, as well as new blood vessels. A few clinical trials are in progress, in which human patients have been injected with stem cells isolated from their own bodies, but the long-term results are still unknown. Scientists at the University of Minnesota went a step further in April 2011 when they announced they had grown a human heart by starting with a cadaver heart, removing the cardiac muscle cells, and then exposing this "scaffold" of blood vessels and connective tissues to human stem cells. Although these researchers are working to show that the resulting "bioartificial" heart can function normally, the ability to grow such organs would be a breakthrough for transplant medicine.

Clearing Clogged Arteries

Over the past decade, the proportion of heart attack patients receiving bypass surgery has decreased, mainly due to the increased usage of angioplasty and stent placement. In *angioplasty,* a cardiologist threads a catheter into an artery in the groin or upper part of the arm and guides it through a major blood vessel toward the heart. When the tube arrives at the region of plaque in an artery, a balloon attached to the end of the tube is inflated, forcing the vessel open, followed by removal of the tube. However, the artery may not always remain open because the trauma can cause smooth muscle cells in the wall of the artery to proliferate and close it.

More often, the same operation utilizes a cylinder of expandable metal mesh called a *stent.* Once the stent is in place, a balloon inside the stent is inflated, expanding it, and locking it in place (Fig. 12.19). In recent years, the most common types of stents implanted in patients' arteries are drug-eluting stents, because they have been coated with medications designed to inhibit inflammation and scar formation. Since 2003, these drug-coated stents have

been implanted in over 6 million people worldwide. Compared to coronary bypass surgery, patients receiving stents generally have an easier recovery and shorter hospitalization, although there is some controversy about whether stents are now being overused.

Dissolving Blood Clots

Medical treatment for thromboembolism often includes the use of *tissue plasminogen activator (tPA).* This drug converts plasminogen into plasmin, an enzyme that dissolves blood clots, as mentioned in Section 12.2. In fact, tPA is the body's own way of converting plasminogen to plasmin. tPA is also being used for thrombolytic stroke patients but with limited success because some patients experience life-threatening bleeding in the brain.

If a person has symptoms of angina or a stroke, aspirin may be prescribed. Aspirin reduces the stickiness of platelets, and thereby lowers the probability that a clot will form. Evidence indicates that aspirin protects against first heart attacks, but there is no clear support for taking aspirin every day to prevent strokes in symptom-free people. Even so, some physicians recommend taking a low dosage of aspirin every day.

Heart Transplants and Artificial Hearts

Heart transplants are usually successful today, but unfortunately, the need for hearts to transplant is greater than the supply because of a shortage of human donors. Therefore, a left ventricular assist device (LVAD), implanted in the abdomen, may help patients waiting for a transplant. A tube passes blood from the left ventricle to the device, which pumps it to the aorta. A cable passes from the device through the skin to an external battery the patient totes around.

Even fewer patients have received a device called a total artificial heart (TAH), such as that shown in Figure 12.20. An internal battery and a controller regulate the pumping speed, and an external battery powers the device by passing electricity through the

catheter

arterial wall

stent

a. Artery is closed.   b. Stent is placed.   c. Balloon is inflated.

Figure 12.19 Angioplasty with stent placement. **a.** A plastic tube (catheter) is inserted into the coronary artery until it reaches the clogged area. **b.** A metal stent with a balloon inside it is pushed out the end of the plastic tube into the clogged area. **c.** When the balloon is inflated, the vessel opens, and the stent is left in place to keep the vessel open.

replacement heart

wireless energy-transfer system | external wireless driver | internal controller | external battery pack | rechargeable internal battery | photograph of artificial heart

Figure 12.20 A total artificial heart. The SynCardia temporary Total Artificial Heart moves blood in the same manner as a natural heart. The controller is implanted into the patient's abdomen. The heart is powered by an internal rechargeable battery, with an external battery as a backup system.
(photo: ©SynCardia Systems, Inc.

skin via external and internal coils. A rotating centrifugal pump moves silicon hydraulic fluid between left and right sacs to force blood out of the heart into the pulmonary trunk and the aorta. Initially, recipients were near death, and most survived for only a few days. However, medical advances have increased the survival time to months for many recipients, making the TAH a temporary treatment for people who are experiencing heart failure. Over 1,300 TAHs have been implanted into patients.

Check Your Progress 12.5

1. Summarize the major forms of cardiovascular disease.
2. Explain the thresholds for determining if a person has hypertension.
3. Identify some treatment options, as well as potential limitations, for patients who need their heart replaced.

Conclusion

Blood plays an important role in our bodies. During an injury or operation, the loss of blood can cause a life-threatening condition and therefore must be resupplied by a transfusion. While blood donation programs are usually capable of supplying blood during normal circumstances, events such as natural disasters, mass transit accidents, and instances of terrorism can place a high demand on the supply of blood. The development of artificial blood, with its ability to be easily stored, may eventually provide an alternative source of blood for hospitals and first responders.

SUMMARIZE

12.1 The Blood Vessels

There are three types of blood vessels in the **cardiovascular system:**

- **Arteries** (and **arterioles**) take blood away from the heart. The largest artery is the **aorta.**
- **Capillaries** are the smallest vessels, where exchange of substances with the tissues occurs.
- **Veins** (and **venules**) take blood to the heart. Of these three types of vessels, only veins have valves. The largest veins are the **venae cavae.**

12.2 Blood

Blood is a connective tissue that functions in transport, homeostasis, protection, and temperature regulation. It is composed of the plasma and the formed elements:

- **Plasma** contains mostly water and proteins, as well as nutrients and wastes. The plasma proteins prothrombin and fibrinogen are well known for their contribution to the **clotting** process.
- The **formed elements** are red blood cells, white blood cells, and platelets. **Red blood cells (erythrocytes)** contain **hemoglobin,** which functions in oxygen transport. A decrease in red blood cell numbers is known as **anemia. White blood cells (leukocytes)** help defend against infections. **Neutrophils, monocytes,** and **dendritic cells** are phagocytic. **Basophils** release histamine, and **eosinophils** fight parasitic infestations. **Lymphocytes,** which include B and T cells, are involved in the development of adaptive immunity to infections. **Platelets (thrombocytes)** are involved in blood clotting, along with proteins like **thrombin** and **fibrin. Hemophilia** is a clotting deficiency caused by a genetic defect.
- **Stem cells** can give rise to many types of cells. The bone marrow contains multipotent cells that not only produce all types of blood cells, but may also be useful in treating many types of diseases.
- When blood reaches a capillary, water moves out at the arterial end due to blood pressure, forming **interstitial fluid.** Some of this fluid moves back into capillaries due to osmotic pressure. Excess fluid is collected by capillaries of the lymphatic system, forming **lymph.**

12.3 The Human Heart

The **heart** is made up largely of myocardium, and it is surrounded by a membrane called the pericardium. It has a right and left side and four chambers:

- On the right side, the venae cavae deliver O_2-poor blood from the body into the right atrium, and the right ventricle pumps it via the **pulmonary arteries** into the pulmonary circuit.
- On the left side, the **pulmonary veins** bring O_2-rich blood from the lungs into the left atrium, and the left ventricle pumps it through the aorta into the systemic circuit.
- Blood from each **atrium** (*pl.,* **atria**) passes through **atrioventricular valves** into each **ventricle.** Blood leaving the ventricles passes through **semilunar valves.** The heart sounds, "lub-dub," are due to the closing of the atrioventricular valves, followed by the closing of the semilunar valves. **Systole** refers to heart contraction; **diastole** refers to relaxation.
- The volume of blood leaving the heart each minute is the **cardiac output,** and the pumping action of the heart results in a **pulse** that can be felt in the arteries.
- During the **cardiac cycle,** the SA node (**pacemaker**) initiates the heartbeat by causing the atria to contract. The AV node conveys the stimulus to the ventricles, causing them to contract. These electrical impulses can be monitored using an **electrocardiogram (ECG).**

12.4 Two Cardiovascular Pathways

The cardiovascular system is divided into the pulmonary circuit and the systemic circuit:

- In the **pulmonary circuit,** the pulmonary trunk from the right ventricle of the heart and the two pulmonary arteries take O_2-poor blood to the lungs, and four pulmonary veins return O_2-rich blood to the left atrium of the heart.
- In the **systemic circuit,** O_2-rich blood is pumped from the left ventricle into the aorta, which branches off to form arteries going to specific organs.
- **Portal systems** begin and end in capillaries. For example, the hepatic portal vein originates from intestinal capillaries and finishes in hepatic capillaries.

■ The left ventricle forces blood into the aorta, creating **blood pressure.** Blood pressure is higher in arteries than in veins, where muscular movement and the presence of one-way valves help return venous blood to the heart.

12.5 Cardiovascular Disorders

Although disorders of the cardiovascular system are the leading cause of death in many countries, medical science has developed many effective ways to treat, and preferably to prevent them:

■ **Atherosclerosis** is an accumulation of cholesterol and other substances in arteries. This condition can lead to **hypertension,** or high blood pressure.

■ Diseases of the heart valves are also common, and usually treatable by surgery.

■ Strokes, heart attacks, and aneurysms are very common disorders of the blood vessels. A **stroke** results when the blood supply to the brain is disrupted. An **aneurysm** is a ballooning of any blood vessel, but aneurysms of the aorta or blood vessels supplying the brain are especially significant. A **heart attack** usually occurs when one or more of the coronary arteries supplying the heart becomes obstructed.

ASSESS

Testing Yourself

Choose the best answer for each question.

12.1 The Blood Vessels

For questions 1–5, match the descriptions to the blood vessels in the key. Answers may be used more than once.

Key:

    a. venules   **d.** arterioles
    b. arteries   **e.** capillaries
    c. veins

1. drain blood from capillaries
2. empty into capillaries
3. may contain valves
4. muscular vessels that move blood away from the heart
5. sites for the exchange of substances between blood and interstitial fluid

12.2 Blood

6. Which of the following is not a function of blood?
    a. regulation of body temperature
    b. defense of the body
    c. transport of oxygen and carbon dioxide
    d. manufacture of hormones

7. Which of the following is/are formed from megakaryocytes?
    a. basophils
    b. platelets
    c. erythrocytes
    d. fibrinogen

8. Which of the following is not a white blood cell?
    a. lymphocyte
    b. neutrophil
    c. eosinophil
    d. basophil
    e. All of these are white blood cells.

12.3 The Human Heart

9. Which of the following lists the events of the cardiac cycle in the correct order?
    a. both atria contract, then both ventricles
    b. both ventricles contract, then both atria
    c. left atria and left ventricle contract, then right atria and right ventricle
    d. right atria and right ventricle contract, then left atria and left ventricle

10. An electrocardiogram measures
    a. chemical signals in the brain and heart.
    b. electrical activity in the brain and heart.
    c. chemical signals in the heart.
    d. electrical changes in the wall of the heart.

11. Blood moving from the left atrium to the left ventricle passes through which of the following valves?
    a. bicuspid valve
    b. tricuspid valve
    c. aortic semilunar valve
    d. pulmonary semilunar valve

12.4 Two Cardiovascular Pathways

12. Label the blood vessels in the following diagram:

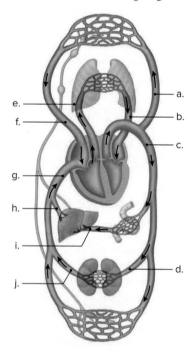

12.5 Cardiovascular Disorders

13. In which of the following does a blood vessel burst, sending blood into the surrounding tissues, such as the brain.
    a. angina pectoris
    b. a heart attack
    c. atherosclerosis
    d. a stroke

14. Which of the following conditions is characterized by the accumulation of plaque in a blood vessel?
    a. hypertension
    b. stroke
    c. atherosclerosis
    d. heart attack

ENGAGE

Thinking Critically

1. A few specialized tissues do not contain any blood vessels, including capillaries. Can you think of two or three? How might these tissues survive without a direct blood supply?

2. Assume your heart rate is 70 beats per minute (bpm), and each minute your heart pumps 5.25 l of blood to your body. Based on your age to the nearest day, about how many times has your heart beat so far, and what volume of blood has it pumped?

3. Examine the following abnormal ECG patterns. Compared with the normal pattern shown in Figure 12.13*b*, what is your best guess as to the type of problem each patient may have? What type of device can be implanted in a patient's heart to correct some abnormalities of the heart, and how does it generally work?

a.

b.

4. There are not enough living hearts available to meet the need for heart transplants. What are some of the major problems a manufacturer would have to overcome when attempting to design an artificial heart that will last for years inside a patient's body?

13

Lymphatic and Immune Systems

©AuntSpray/Shutterstock

CHAPTER OUTLINE

BEFORE YOU BEGIN

Before beginning this chapter, take a few moments to review the following discussions:

Section 2.7 What are the four levels of protein organization?

Section 4.1 What are the roles of proteins in the plasma membrane?

Section 11.5 What is the role of the immune system in maintaining homeostasis?

The Search for a Zika Vaccine

Although Zika virus was first reported in Africa in 1952, the virus did not make an appearance in the Western Hemisphere until 2015, when cases occurred in Brazil. The most common way the virus is transmitted between people is via an infected *Aedes* mosquito. However, it can be sexually transmitted from infected males to females. In a relatively short period of time, the virus has spread throughout South and Central America, and there have already been cases of infected travelers returning to the United States.

For most people, infection with the Zika virus produces mild symptoms, such as fevers, rashes, or joint pain. Some individuals do not experience any symptoms at all, and thus may not know they have been infected. However, in a small number of cases, pregnant females who have been infected with the Zika virus have given birth to children with microcephaly. Microcephaly is a birth defect that causes the head and brain of an infant to be much smaller than normal. This can cause a number of developmental problems, including seizures, intellectual disabilities, and vision problems. Because there is no cure for microcephaly, researchers have been actively looking at ways to develop a vaccine against the virus.

In order to create a vaccine, researchers must identify the parts of a virus that will cause our immune system to react as if it has been infected and build up antibodies to the actual virus. Later, if an individual is exposed to the virus, these antibodies are used by the body to provide immunity. In this chapter, we will explore how our immune system protects us, not only from viruses such as Zika, but from a wide variety of pathogens.

As you read through this chapter, think about the following questions:

1. How does your body respond to its first exposure to a new pathogen?
2. How do vaccines provide immunity?
3. What is the difference between an antibody and an antibiotic?

13.1 The Lymphatic System

Learning Outcomes

Upon completion of this section, you should be able to

1. Describe the major functions of the lymphatic system.
2. Distinguish between the roles of primary and secondary lymphoid organs, and list examples of each.

The **lymphatic system** consists of lymphatic vessels and the lymphoid organs (Fig. 13.1). This system, which is closely associated with the cardiovascular system, has three main functions that contribute to homeostasis:

- Lymphatic capillaries absorb excess interstitial fluid and return it to the bloodstream.
- Lymphatic capillaries absorb fats from the digestive tract and transport them to the bloodstream.
- Lymphoid organs help to defend the body against disease. This function is mainly carried out by the white blood cells present in lymphatic vessels and lymphoid organs, as well as in the blood.

Lymphatic Vessels

The **lymphatic vessels** form a one-way system that begins with **lymphatic capillaries**—tiny, closed-ended vessels whose walls consist of simple squamous epithelium (Fig. 13.1). Most regions of the body are richly supplied with lymphatic capillaries, whose primary purpose is to absorb excess interstitial fluid. Interstitial fluid is mostly water, but it also contains solutes (i.e., nutrients, electrolytes, and oxygen) derived from plasma and cellular products (i.e., hormones, enzymes, and wastes) secreted by cells. The fluid inside lymphatic vessels is called **lymph.**

The lymphatic capillaries join to form lymphatic vessels that merge before entering

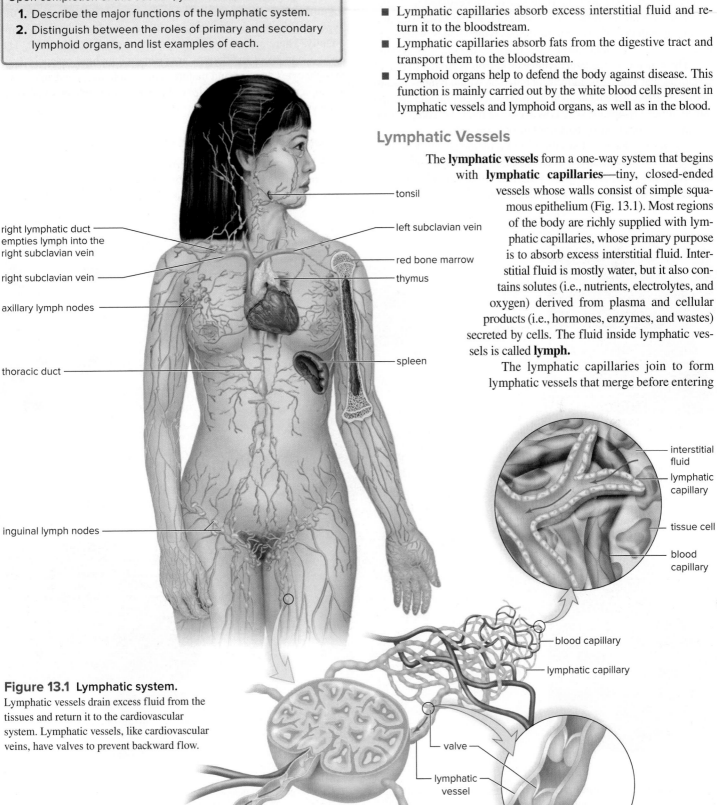

Figure 13.1 Lymphatic system.
Lymphatic vessels drain excess fluid from the tissues and return it to the cardiovascular system. Lymphatic vessels, like cardiovascular veins, have valves to prevent backward flow.

tonsil

left subclavian vein

red bone marrow

thymus

right lymphatic duct empties lymph into the right subclavian vein

right subclavian vein

axillary lymph nodes

spleen

thoracic duct

inguinal lymph nodes

interstitial fluid

lymphatic capillary

tissue cell

blood capillary

blood capillary

lymphatic capillary

valve

lymphatic vessel

lymph node

one of two ducts: the thoracic duct or the right lymphatic duct. The larger, thoracic duct returns lymph collected from the body below the thorax and the left arm and the left side of the head and neck into the left subclavian vein. The right lymphatic duct returns lymph from the right arm and right side of the head and neck into the right subclavian vein. On its way back to the cardiovascular system, lymph percolates through various lymph nodes, where any foreign material present can be recognized by the immune system.

The structure of the larger lymphatic vessels is similar to that of cardiovascular veins. The movement of lymph within lymphatic capillaries is largely dependent upon skeletal muscle contraction. Lymph forced through lymphatic vessels as a result of muscular compression is prevented from flowing backward by one-way valves.

Edema is a localized swelling caused by the accumulation of interstitial fluid that has not been collected by the lymphatic system. It occurs if either too much interstitial fluid is made and/or if not enough of it is drained away. Edema can lead to tissue damage

and even death, illustrating the importance of the fluid-absorbing function of the lymphatic system.

Lymphoid Organs

Lymphoid organs contain large numbers of **lymphocytes,** the type of white blood cell that is mainly responsible for adaptive immunity. There are two types of lymphoid organs (Fig. 13.2). The red bone marrow and thymus are primary lymphoid organs, where lymphocytes develop and mature. The lymph nodes and spleen are examples of secondary lymphoid organs—sites where some lymphocytes become activated.

Primary Lymphoid Organs

Red bone marrow (Fig. 13.2*b*) contains a network of connective tissue fibers, along with stem cells that are ever capable of di-

a. Thymus

lobule
cortex
medulla

tonsil

lymphatic vessel

lymphocyte

monocyte

b. Red bone marrow   250 ×

medulla
cortex
capsule

c. Lymph node   20 ×

capsule
red pulp
white pulp

40 ×

d. Spleen

Figure 13.2 The lymphoid organs.   The thymus (**a**) and red bone marrow (**b**) are the primary lymphoid organs. Blood cells, including lymphocytes, are produced in red bone marrow. B cells mature in the bone marrow. T cells mature in the thymus. The lymph nodes (**c**) and the spleen (**d**) are secondary lymphoid organs. Lymph is cleansed in the nodes, and blood is cleansed in the spleen. (photos) (a): ©McGraw-Hill Education/Dennis Strete, photographer; (b): ©Ed Reschke/Photolibrary/Getty Images; (c–d): ©McGraw-Hill Education/Al Telser, photographer viding and producing blood cells. Some of these cells become the various types of white blood cells: neutrophils, eosinophils, basophils, lymphocytes, and monocytes (see Fig. 12.3). The cells are packed around thin-walled sinuses filled with venous blood. Differentiated blood cells enter the bloodstream at these sinuses.

In children, most bones contain red bone marrow, but in adults it is present mainly in the bones of the skull, sternum, ribs, clavicle, pelvis, and vertebral column. Although all lymphocytes begin their development in the bone marrow, the two major types of lymphocytes finish maturing in different locations. B lymphocytes, also called B cells, remain in the bone marrow until mature. Immature T lymphocytes (T cells) migrate from the bone marrow through the bloodstream to the thymus, where they mature.

The soft, bilobed **thymus** (Fig. 13.2*a*) is located in the thoracic cavity between the trachea and the sternum, above the heart. The thymus grows largest around the time of puberty, and shrinks as we get older. Connective tissue divides the thymus into lobules, which are filled with T cells and supporting cells. Only about 5% of T cells ever exit the thymus. T cells that are capable of reacting to the body's own cells undergo *apoptosis,* which involves a cascade of specific cellular events leading to the death and destruction of the cell (see Section 5.1). Those that leave the thymus are capable of reacting to foreign molecules (see Section 13.4).

Secondary Lymphoid Organs

Lymphocytes frequently migrate from the bloodstream into the secondary lymphoid organs. Here, they may encounter foreign molecules or cells, after which they proliferate and become activated. These activated cells usually reenter the bloodstream, where they search for sites of infection or inflammation, like a team of highly trained military personnel seeking to destroy a specific enemy.

The **spleen** (Fig. 13.2*d*) is located in the upper left side of the abdominal cavity, behind the stomach. Most of the spleen is *red pulp,* which consists of blood vessels and sinuses, where macrophages remove old and defective blood cells. The spleen also has *white pulp* that consists of small areas of lymphoid tissue, where lymphocytes can react to foreign invaders that have gained access to the blood.

Occasionally, the spleen may need to be surgically removed due to trauma or disease. Although the spleen's functions can be mostly replaced by other organs, a person with no spleen is more susceptible to certain types of infections. Because the spleen also removes older red blood cells and platelets, the numbers of both types of cells may increase after spleen removal.

Lymph nodes (Fig. 13.2*c*) are small (from 1 to 25 mm in diameter), ovoid structures occurring along lymphatic vessels, through which the lymph must pass. Connective tissue divides the organ into nodules. Each of these is packed with B and T cells and contains a sinus. As lymph courses through the many sinuses, resident macrophages engulf **pathogens,** which are disease-causing agents such as viruses and bacteria, as well as any debris present in the lymph.

Sometimes incorrectly called "glands," lymph nodes are named for their location. For example, *inguinal* nodes are in the groin, and *axillary* nodes are in the armpits. Physicians often feel for the presence of swollen or tender lymph nodes as evidence that the body is fighting an infection.

Unfortunately, cancer cells sometimes enter lymphatic vessels and move undetected to other regions of the body, where they may produce secondary tumors. In this way, the lymphatic system sometimes inadvertently assists *metastasis,* the spreading of cancer far from its place of origin (see Section 25.5).

Check Your Progress 13.1

1. Summarize the three functions of the lymphatic system.
2. Explain how lymph is different than blood.
3. Distinguish between the function of a primary and secondary lymphoid organ.

13.2 Innate Immunity

Learning Outcomes

Upon completion of this section, you should be able to

1. List examples of physical and chemical barriers to infection.
2. Describe the purpose of the inflammatory response.
3. Explain the function of phagocytes and natural killer cells.
4. Describe the three main functions of the complement system.

The lymphatic system works with the **immune system** to protect the body from pathogens, toxins, and other invaders. The term **immunity** refers to a condition where the body is protected from various threats, like pathogens, toxins, and cancer cells. There are two main types of immunity. Mechanisms of **innate immunity** are fully functional without previous exposure to these substances, whereas **adaptive immunity** is initiated and amplified after specific recognition of these substances (see Section 13.3). In this section, we will examine the mechanisms of innate immunity. These can be divided into at least four types: physical and chemical barriers; inflammation, phagocytes, and natural killer cells; and protective proteins.

Physical and Chemical Barriers

Skin and the mucous membranes lining the respiratory, digestive, and urinary tracts serve as mechanical barriers to entry of pathogens. The upper respiratory tract is lined with ciliated cells that sweep mucus and trapped particles up into the throat, where they can be swallowed or expelled.

The secretions of oil glands in the skin contain chemicals that weaken or kill certain bacteria on the skin. The stomach has an acid pH, which kills many types of bacteria or inhibits their growth. The various bacteria that normally reside in the intestine and other areas, such as the vagina, remove nutrients and block binding sites that potentially could be used by pathogens.

Inflammatory Response

Damage to tissues—whether by physical trauma, chemical agents, or pathogens—initiates a series of events known as the **inflammatory response.** Inflammation tends to wall off infections and increase the exposure and access of the immune system to the inciting agent. Figure 13.3 illustrates the main participants in an inflammatory response.

An inflamed area has several typical characteristics, including redness, heat, swelling, and pain. Most of these signs are due to capillary changes in the damaged area. At least three types of cells that reside in the skin and connective tissues promote inflammatory responses to various types of damage. **Mast cells,** a type of immune cell found especially in the skin, lungs, and intestinal tract, respond to damage by releasing **histamine,** a chemical that binds to receptors present on endothelial cells lining the blood vessels. This causes capillaries in the area to dilate and become more permeable, allowing fluids to escape into the tissues, resulting in swelling. The swollen area may stimulate free nerve endings, causing the sensation of pain. The increased blood flow also causes the skin to redden and feel warm.

Two other cell types that play an important role in inflammation are **macrophages** and **dendritic cells.** Both are phagocytic cells that reside in almost all types of tissues. Especially when they are exposed to invading microbes, both cell types can release **cytokines,** chemical messengers that influence the activities of other immune cells. For example, after ingesting a bacterial pathogen, macrophages release a cytokine called *tumor necrosis factor alpha,* which acts on endothelial cells, and another known as *interleukin-8,* which attracts other types of immune cells to the scene. Some cytokines secreted by macrophages and dendritic cells act on the brain to induce a fever response; others, called *colony-stimulating factors,* cause the bone marrow to produce more white cells.

If the cause of inflammation cannot be eliminated, as occurs in tuberculosis and some types of arthritis, the inflammatory reaction may persist and become harmful rather than helpful. In these cases, anti-inflammatory medications such as aspirin, ibuprofen, or cortisone may be used to minimize the detrimental effects of chronic inflammation.

Phagocytes and Natural Killer Cells

At sites of inflammation, various types of white blood cells migrate through the walls of dilated capillaries, into the inflamed area. Among these are two types of **phagocytes,** which literally means "eating cells." **Neutrophils** are usually the first white blood cells to enter an inflamed area from the blood, and they may accumulate to form *pus,* which has a whitish appearance mainly due to their presence. If the inflammatory reaction continues, **monocytes** will migrate from the blood to the tissues, where they are then called **macrophages** (meaning, literally, "large eaters"). When a neutrophil or a macrophage encounters a pathogen, especially a bacterial cell, it will engulf the pathogen by phagocytosis into a vesicle (see Fig. 4.11) within the cell. This then fuses with a lysosome. Within the lysosome is an acid pH that activates hydrolytic enzymes and various reactive oxygen compounds, which can usually destroy the pathogen.

Natural killer (NK) cells are large, granular, lymphocyte-like cells that kill some virus-infected and cancer cells by cell-to-cell contact. Interestingly, NK cells and cytotoxic T cells (see Section 13.3) use very similar mechanisms to kill their target cells, by inducing them to undergo apoptosis. NK cells, however, seek out and kill cells that lack a particular type of "self" molecule, called MHC-I

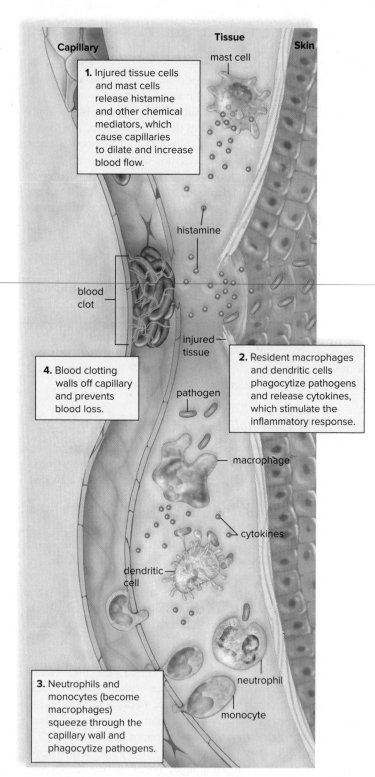

Figure 13.3 Inflammatory response.   Due to capillary changes in a damaged area and the release of chemical mediators, such as histamine by mast cells, an inflamed area exhibits redness, heat, swelling, and pain. The inflammatory reaction can be accompanied by other reactions to the injury. Macrophages and dendritic cells, present in the tissues, phagocytize pathogens, as do neutrophils, which squeeze through capillary walls from the blood. Macrophages and dendritic cells release cytokines, which stimulate the inflammatory and other immune reactions. A blood clot can form to seal a break in a blood vessel.

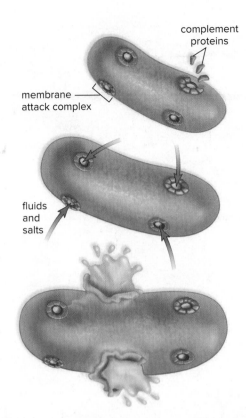

Figure 13.4 Action of the complement system against a bacterium. When complement proteins in the blood plasma are activated in an immune response, they form a membrane attack complex. This complex makes holes in bacterial cell walls and plasma membranes, allowing fluids and salts to enter until the cell eventually bursts.

(major histocompatibility class I), on their surface. Because some virus-infected and cancer cells may lack these MHC-I molecules, they often become susceptible to being killed by NK cells. However, because NK cells do not recognize specific viral or tumor antigens, and do not proliferate when exposed to a particular antigen, they are considered a part of the innate immune system.

Protective Proteins

The **complement system** (often simply called complement) is composed of a number of blood plasma proteins designated by the letter *C* and a number (e.g., C3). The complement proteins "complement" certain immune responses, which accounts for their name. For example, they are involved in and amplify the inflammatory reaction because certain complement proteins can bind to mast cells and trigger histamine release, and others can attract phagocytes to the scene. Some complement proteins bind to the surface of pathogens, which ensures that the pathogens will be phagocytized by a neutrophil, dendritic cell, or macrophage.

Certain other complement proteins join to form a *membrane attack complex* that produces holes in the surface of bacteria and some viruses. Fluids and salts then enter the bacterial cell or virus to the point that they burst (Fig. 13.4).

Interferons are proteins produced by virus-infected cells as a warning to noninfected cells in the area. Interferons can bind to receptors of noninfected cells, causing them to prepare for possible attack by producing substances that interfere with viral replication.

Interferons may be administered to treat certain viral infections, such as hepatitis C.

13.3 Adaptive Immunity

When innate defenses have failed to prevent an infection, adaptive immunity comes into play. The adaptive immune system recognizes, responds to, and usually eliminates antigens from the body. An **antigen** is any molecule, usually a protein or carbohydrate, that stimulates an adaptive immune response. If mechanisms of innate immunity resemble a general police force, with members who are skilled at dealing with many different types of criminals, then the players in adaptive immunity are more like bounty hunters trained to seek out a very specific perpetrator. Adaptive defenses usually take 5 to 7 days to become fully activated, and they may last for many years.

Antigen Receptor Diversity and Clonal Selection

The adaptive immune system depends primarily on the activity of B cells and T cells. Both cell types are capable of recognizing antigens because they have specific *antigen receptors*—plasma membrane proteins whose shape allows them to bind to particular antigens. Each lymphocyte has only one type of receptor. Because we could potentially encounter millions of different antigens during our lifetime, the population of B and T cells in our body requires a diversity of antigen receptors to protect us against them. Remarkably, diversification occurs to such an extent during the maturation process that there are specific B cells and/or T cells for almost any possible antigen.

Although the mechanisms accounting for the generation of antigen receptor diversity are complex, four major mechanisms account for this phenomenon, as follows:

1. The genes that code for antigen receptors of B and T cells contain a variety of gene segments: small DNA sequences (see Section 25.1) that each code for the portion of the antigen receptor that will eventually bind to the antigen.

2. In developing B and T cells, an enzyme cuts out some of these gene segments and links them back together in different combinations.

3. As the gene segments are being linked, another enzyme introduces mutations, thereby greatly increasing the variety in these genes.

4. Each antigen receptor is made up of at least two protein chains, each of which undergoes steps 1–3 independently, resulting in various combinations of the chains.

The **clonal selection** model states that in a large population of B and T cells, only a few cells will have antigen receptors that can react with any particular antigen in the body (be it an infection, toxin, cancer cell, etc.). However, when an antigen binds to receptors on a particular lymphocyte, that cell is "selected" (i.e., it divides, forming "clones" of itself).

Because our various immune defenses do not ordinarily react to our own normal cells, it is said that the immune system is able to distinguish "self" from "nonself." However, because the generation of antigen receptors is a random genetic process, some of these receptors will be capable of reacting with self-antigens (i.e., molecules normally found in the body). Such cells could attack our own tissues, a situation that would be incompatible with life. Fortunately, this disaster is avoided, because any lymphocytes that react with self-antigens early in their development (in the primary lymphoid tissues) undergo apoptosis.

B Cells and Antibody-Mediated Immunity

B cells are usually activated in a lymph node or the spleen when their BCRs bind to specific antigens. This can be seen in Figure 13.5, where each B cell has a specific antigen receptor, or **B-cell receptor (BCR),** represented by shape. Only the B cell with a BCR that fits the antigen (green circles) undergoes clonal expansion. During clonal expansion, cytokines secreted by helper T cells stimulate B cells to divide. Most of the resulting clones become **plasma cells,** which are specialized for the secretion of antibodies. **Antibodies** are the secreted form of the BCR of the B cell that was activated. Some cloned B cells become **memory B cells,** which are the means by which long-term immunity is possible. If the same antigen enters the system again, memory B cells quickly divide and give rise to more plasma cells capable of rapidly producing the correct type of antibody. Once an infection has been cleared from the body, new plasma cells cease to develop, and most of those present undergo apoptosis.

Defense by B cells is called **antibody-mediated immunity.** It is also called *humoral immunity* because these antibodies are present in blood and lymph. (Historically, a humor is any fluid normally occurring in the body.)

Antibody Structure   Antibodies are also called **immunoglobulins (Igs).** They are typically Y-shaped molecules with two arms. Each arm has a "heavy" (long) polypeptide chain and a "light" (short) polypeptide chain. These chains have constant (C) regions, where the sequence of amino acids is set, and variable (V) regions, where the sequence of amino acids varies between antibodies (Fig. 13.6). The constant regions are the same within a particular type or class of antibodies (see below). The variable regions form the antigen-binding sites. The antigen combines with the antibody at the antigen-binding site in a lock-and-key manner.

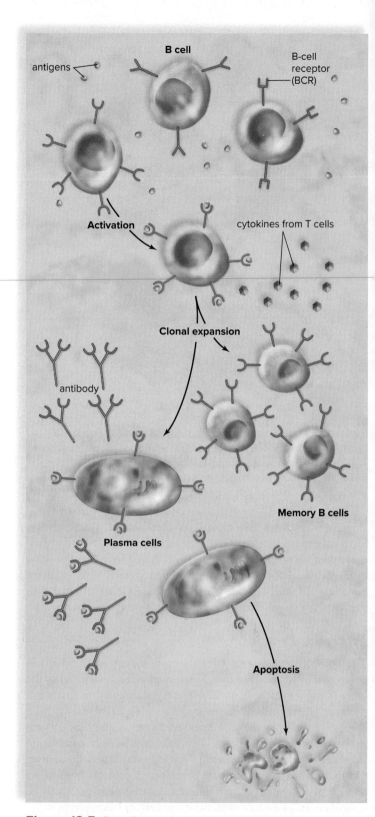

Figure 13.5 B-cell clonal selection.   Each B cell has a B-cell receptor (BCR) designated by shape that will combine with a specific antigen. Activation of a B cell occurs when its BCR can combine with an antigen (colored green). In the presence of cytokines, the B cell undergoes clonal expansion, producing many plasma cells and memory B cells. These plasma cells secrete antibodies specific to the antigen, and memory B cells immediately recognize the antigen in the future. After the infection passes, plasma cells undergo apoptosis, also called programmed cell death.

Figure 13.6 Structure of an antibody. **a.** An antibody contains two heavy (long) polypeptide chains and two light (short) chains arranged so there are two variable regions, where a particular antigen is capable of binding with an antibody (*V* = variable region, *C* = constant region). The shape of the antigen fits the shape of the binding site. **b.** Computer model of an antibody molecule. The antigen combines with the two side branches.

(b): ©Dr. Arthur J. Olson, Scripps Institute

TABLE 13.1 Classes of Human Antibodies

Class	Distribution	Function
IgG	Main antibody type in circulation	Binds to pathogens, activates complement, and enhances phagocytosis
IgM	Antibody type found in circulation; largest antibody	Activates complement; clumps cells
IgA	Main antibody type in secretions such as saliva and milk	Prevents pathogens from attaching to epithelial cells in digestive and respiratory tract
IgD	Antibody type found on surface of immature B cells	Presence signifies readiness of B cell to respond to antigens
IgE	Antibody type found as antigen receptors on eosinophils in blood and on mast cells in tissues	Responsible for immediate allergic response and protection against certain parasitic worms

The antigen–antibody reaction can have several outcomes, but often the reaction produces complexes of antigens combined with antibodies. These complexes, also called *immune complexes,* may mark the antigens for destruction. For example, an immune complex may be engulfed by a neutrophil or macrophage, or it may activate complement. Antibodies may also "neutralize" viruses or toxins by preventing their binding to cells.

Types of Antibodies There are five major classes of antibodies in humans (Table 13.1). Immunoglobulin G (IgG) antibodies are the major type in blood, lymph, and interstitial fluid. IgG antibodies bind to pathogens and their toxins and can activate complement. IgG is the only antibody class that can cross the placenta. Along with IgA, IgG is found in breast milk. IgM antibodies are pentamers, meaning they contain five of the Y-shaped structures shown in Figure 13.6*a*. These antibodies appear in blood soon after a vaccination or infection and disappear relatively quickly. They are good complement activators. IgA antibodies are monomers in blood and lymph or dimers in tears, saliva, gastric juice, and mucous secretions. They are the main type of antibody found in body secretions. IgD molecules appear on the surface of B cells when they are ready to be activated. IgE antibodies are important in the

immune system's response to parasites (e.g., parasitic worms). They are best known, however, for immediate allergic responses, including anaphylactic shock (discussed in Section 13.5). IgD antibodies are found mainly on the surface of immature B cells.

T Cells and Cell-Mediated Immunity

As a T cell leaves the thymus, it has a unique **T-cell receptor (TCR)** similar to the BCR on B cells. Unlike B cells, however, there is no secreted form of the TCR, and T cells are unable to recognize an antigen without help. The antigen must be displayed, or "presented," to the TCR by an **MHC (major histocompatibility complex) protein** on the surface of another cell.

There are two major types of T cells: **helper T cells** (T_H cells) and **cytotoxic T cells** (T_C cells, or CTLs). Each type has a TCR that can recognize an antigen fragment in combination with an MHC molecule. However, T_H cells only recognize antigen presented by specialized **antigen-presenting cells (APCs)** with MHC class II proteins on their surface, whereas the T_C cells only recognize antigen presented by various cell types with MHC class I proteins on their surface (Fig. 13.7).

APCs such as dendritic cells or macrophages play an important role in initiating most adaptive immune responses. After ingesting and destroying a pathogen, APCs travel to a lymph node or the spleen, where T cells also congregate. After breaking the pathogen apart with lysosomal enzymes, the APC displays a piece of the pathogen on its surface, in the binding groove of an MHC class II protein. After they are stimulated by APCs, T_H cells proliferate and secrete various cytokines. Cytokines produced by T_H cells help activate T_C cells, as well as B cells. Other cytokines, produced by types of T_H cells called suppressor T cells, inhibit certain responses, thereby helping to control or regulate adaptive immunity.

In contrast to T_H cells, T_C cells often kill the cells they recognize. One method of killing used by T_C cells depends on storage vacuoles that contain many molecules of *perforin* and enzymes called *granzymes*. After an activated T_C cell binds to a virus-infected or cancer cell that is presenting foreign antigen on its MHC class I proteins, the T_C cell releases perforin, which forms pores in the plasma membrane of the abnormal cell. This allows the granzymes to enter the target cell, inducing it to undergo apoptosis (Fig. 13.8*a*).

Figure 13.7 T-cell clonal selection. Each T cell has a T-cell receptor (TCR) designated by a shape that will combine only with a specific antigen. Activation of a T_C cell occurs when its TCR can combine with an antigen such as a virus (colored green). A dendritic cell presents fragments of the antigen in the groove of its MHC class I (MHC-I) molecules. Thereafter, the T_C cell undergoes clonal expansion, producing many T_C cells with the same antigen specificity. After the antigen has been removed, the majority of T_C cells undergo apoptosis, but a small number of memory T_C cells remain. Memory T_C cells provide protection should the same antigen enter the body again at a future time.

Figure 13.8 Cell-mediated immunity. a. How a T_C cell destroys a virus-infected cell or cancer cell. **b.** The scanning electron micrograph shows cytotoxic T cells attacking and destroying a cancer cell (target cell). (b): ©Steve Gschmeissner/Science Source

SEM 1,250×

The T$_C$ cell is then capable of moving on to kill another target cell (Fig. 13.8*b*).

Notice also in Figure 13.7 that, similar to B cells, some of the clonally expanded T cells become **memory T cells.** These may live for many years and can quickly jump-start an immune response to an antigen previously present in the body. Immunity mediated by T$_H$ cells and T$_C$ cells is sometimes referred to as **cell-mediated immunity.**

Because HIV, the virus that causes AIDS, infects helper T cells and other cells of the immune system, it suppresses many components of acquired immune responses and makes HIV-infected individuals susceptible to opportunistic infections (see the Health feature "Opportunistic Infections and HIV" in Section 13.6). Infected macrophages and dendritic cells also serve as reservoirs for HIV.

Check Your Progress 13.3

1. Describe four mechanisms that result in the generation of antigen receptor diversity.
2. Discuss the major tenets (principles) of the clonal selection theory.
3. List the five classes of human antibodies and their main functions.
4. Contrast the specific functions of helper T cells, suppressor T cells, and cytotoxic T cells.

13.4 Active Versus Passive Immunity

Learning Outcomes

Upon completion of this section, you should be able to

1. Distinguish between active and passive immunity.
2. Describe some specific clinical applications of cytokine therapies.
3. Explain the major steps in the production of monoclonal antibodies, and some of their practical uses.

Immunity is the condition where the body is protected from various threats—usually by the adaptive immune system. Adaptive immunity can be induced actively or passively. In **active immunity,** the individual alone produces an immune response against an antigen. In **passive immunity,** the individual receives antibodies or cells from another individual.

Active Immunity

Active immunity usually develops naturally after a person is infected with a pathogen. However, active immunity can also be induced artificially when a person is well, so that future infection will not take place. To prevent infections, people can be immunized (vaccinated) against them. The United States is committed to immunizing all children against the common types of childhood disease (Fig. 13.9*a*). The Bioethical feature "Should Parents Be Required to Have Their Kids Vaccinated?" examines the issue of parents who choose not to vaccinate their children.

Immunization involves the use of **vaccines,** preparations that contain an antigen to which the immune system responds.

Traditionally, vaccines are the pathogens themselves, which have either been killed, or treated so they are no longer *virulent* (able to cause disease). Today, it is possible to genetically engineer bacteria or other microbes to mass-produce individual antigens, which can be used as *subunit* vaccines. This method has now produced a vaccine against hepatitis B, a viral disease, and is being used to develop a vaccine against malaria, a protozoal disease.

Most recently, animals have been successfully immunized by injecting them with DNA coding for a foreign protein. The DNA is taken up by the animal's cells, which produce the protein, to which the immune system responds. Several of these *DNA vaccines* are already approved for use in veterinary medicine, and DNA vaccines against human diseases are likely to follow.

After a vaccine is given, it is possible to monitor the amount of antibody (also called the *antibody titer*) present in samples of plasma over time. After the first vaccination, a *primary response* occurs. Typically, the antibody titer rises slowly, levels off, and gradually declines as the antibodies bind to the antigen or simply break down (Fig. 13.9*b*). Following a second vaccination, a *secondary response* occurs. The titer rises more rapidly, to a level much greater than before, then slowly declines. The second exposure is called a "booster" vaccination because it boosts the immune response to a high level. The high levels of antigen-specific antibodies and T cells usually prevent disease symptoms even if the individual is exposed to the disease-causing agent.

After exposure to a particular antigen, whether naturally or by immunization, active immunity is dependent upon the generation of a population of memory B cells and memory T cells that are capable of responding quickly to the same antigen, if encountered later. As to the question of how long this immunological memory can last, a 2008 study revealed that most survivors of the 1918 influenza epidemic had memory B cells that could produce antibodies that reacted with the 90-year-old virus strain, but not against more recent strains. The researchers also showed that these antibodies could protect mice from a lethal dose of that same, nearly century-old virus. While the mechanisms of immunological memory are not completely understood, it is possible that certain memory cells can last a lifetime. Still, even though you are immunized

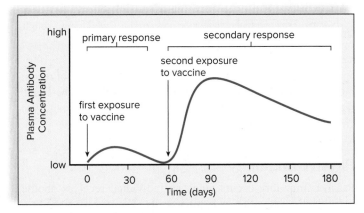

Figure 13.9 Active immunity due to immunizations. During immunization, the primary response, after the first exposure to a vaccine, is minimal. The secondary response, which occurs after the second exposure, shows a dramatic rise in the amount of antibody present in plasma.

SCIENCE IN YOUR LIFE ▶ BIOETHICAL

Should Parents Be Required to Have Their Kids Vaccinated?

The development of vaccines has been one of the greatest medical achievements in the last 200 years. The World Health Organization estimates that vaccination saves about 3 million lives each year. Vaccination against smallpox eradicated that dreaded disease from the planet, and other feared diseases like measles, polio, diphtheria, whooping cough, and tetanus are now rare in countries where vaccination is common. Paradoxically, because vaccination is so effective, most parents of young children today have never seen a child suffering from these diseases, leading some of them to question whether their children need to be vaccinated at all.

Most often, these parents are worried about the possible side effects of vaccination. Indeed, a Google search on "vaccine risks" will turn up some reasonable medical advice, mixed in with a good dose of overhyped hysteria. As with any other medical treatment, vaccines can produce unintended side effects. In one example, a rotavirus vaccine called Rota Shield was withdrawn from the U.S. market in 1999, less than a year after it had been approved by the FDA for use in protecting children from a serious intestinal infection. Out of about 1.5 million doses administered, 15 cases of life-threatening bowel obstructions were reported within 1–2 weeks of vaccination. This resulted in the manufacturer removing the product from the market.

A famous incidence of alleged vaccine side effects occurred in 1976, after an outbreak of "swine flu" at Fort Dix military base. Concerned about a nationwide influenza outbreak, President Gerald Ford recommended that every U.S. citizen should be vaccinated for this new, and possibly fatal, strain of virus. However, about 2 months after the mass vaccination was initiated, the government halted the program in

response to hundreds of lawsuits alleging the vaccine caused side effects ranging from a rare neurological syndrome to sudden death. In all, after about 40 million people received the swine flu vaccine, 25 deaths and hundreds of illnesses were blamed on the vaccine. The feared influenza outbreak never occurred, and only one person died from the virus (a soldier at Fort Dix). Experts still debate whether the vaccine was to blame for all the medical issues. Regardless, some antivaccine activists point to the "swine flu fiasco" as a general argument against vaccination.

In February 1998, a paper was published in the British medical journal *The Lancet,* claiming that an autism-type condition developed in several children who received the measles-mumps-rubella (MMR) vaccine. The article was immediately controversial, and dozens of subsequent studies failed to turn up any vaccine–autism link. In February 2010, *The Lancet* officially retracted the article, accusing the paper's lead author, Andrew Wakefield, of several types of scientific fraud. However, by then, considerable damage had been done to the reputation of vaccines in general, and the MMR in particular. In the years following the Wakefield paper, MMR vaccination rates in Britain dropped from 92% to 73%. In the United States, more cases of measles were reported in 2014 than any year since 1997, and most of these cases were in unvaccinated children.

Ironically, even parents who don't vaccinate their children benefit from the fact that a large percentage of children are vaccinated, and thus are less likely to transmit infectious diseases to their unvaccinated child. This concept of "herd immunity" has been shown to be relevant in recent outbreaks of measles and whooping cough in communities where

increasing numbers of parents are not vaccinating their children. On the front lines of the vaccine issue are pediatricians, many of whom feel so strongly about the need for vaccination that, in a 2011 survey, 25% of them stated they had turned away kids whose parents refused vaccination (compared to 18% in 2005).

The concept that vaccinating a high percentage of children can protect entire communities has led to immunization requirements. Most states require that a parent provide written proof of a child's immunizations prior to registering for school. If the child is not immunized, he or she will not be allowed to attend public school, unless the parent can demonstrate a medical reason, or a religious objection, that explains why the child is not vaccinated. As mentioned, there is no shortage of Internet sites voicing objections to any vaccine requirements, with some going so far as to question the benefit of all vaccinations. While vaccines are not risk-free, these risks seem to be greatly outweighed by the benefits, especially if one considers how infectious diseases were a major cause of death and suffering in our very recent past.

Questions to Consider

1. Do you think children should be prevented from attending school if they are not vaccinated against diseases that may harm their classmates?

2. Do you think it is appropriate to allow parents to forgo the required vaccinations for their children because of religious objections?

3. What is your opinion of pediatricians who refuse to see children if they are not vaccinated? Can you clearly articulate both sides of this issue?

---

against many diseases at a young age, there are a number of recommended vaccinations you should receive as an adult. The Health feature "Adult Vaccinations" explores the schedule for these vaccinations.

Passive Immunity

Passive immunity occurs when an individual receives another individual's antibodies or immune cells. The passive transfer of antibodies is a common natural process. For example, newborn infants are passively immune to some diseases because IgG antibodies have crossed the placenta from the mother's blood (Fig. 13.10a). These antibodies soon disappear, however, so within a few months, infants become more susceptible to infections.

Breast-feeding may prolong the natural passive immunity an infant receives from the mother because IgG and IgA antibodies are present in the mother's milk (Fig. 13.10b). The "first milk," or *colostrum,* produced within the first few days after giving birth, is particularly rich in antibodies.

Even though passive immunity does not last, it is sometimes used to prevent illness in a patient who has been exposed to an infectious disease. Usually, the patient receives a gamma globulin injection (serum that contains antibodies), perhaps taken from individuals who have recovered from the illness (Fig. 13.10c). In the past, horses were immunized, and serum was taken from them to provide the needed antibodies against such diseases as diphtheria, botulism, and tetanus. Unfortunately, about 50% of patients who received these antibodies became ill because their immune system

Adult Vaccinations

Many people mistakenly believe that you have received your full complement of vaccinations by the time you leave high school. In reality, being vaccinated is a lifelong activity. The Centers for Disease Control and Prevention (CDC) has identified a series of vaccinations that are recommended after the age of 18 (Table 13A). In many cases, vaccinations are recommended when an individual is determined to be at risk for a specific disease or condition. For example, while the vaccination for hepatitis B (HepB) may not be required, it is recommended for individuals who have had more than one sexual partner during a 6-month period, have been diagnosed with a sexually transmitted disease (STD; see Section 21.5), use injection drugs, or may have been exposed to blood or infected body fluid. In most cases, vaccinations are recommended as a protective measure even if you are not in an at-risk category.

As always, if you have questions regarding any of these diseases, or about your personal need for vaccinations, consult with your health-care provider. For more information on vaccination schedules from birth through adulthood, visit the CDC's website at www.cdc.gov/vaccines or the Immunization Action Coalition (www.immunize.org).

TABLE 13A   Recommended Immunization Schedule for Adults

Vaccine	19–21 years	22–26 years	27–49 years	50–59 years	60–64 years	≥ 65 years
Influenza	← 1 dose annually →					
Tetanus, diphtheria, pertussis (Td/Tdap)	Substitute 1-time dose of Tdap for Td booster; then boost with Td every 10 years					
Varicella	← 2 doses →					
Human papillomavirus (HPV) female	← 3 doses →					
Human papillomavirus (HPV) male	← 3 doses →					
Shingles					← 1 dose →	
Measles, mumps, rubella (MMR)	← 1 or 2 doses →					
Pneumococcal 13-valent conjugate (PCV13)			← 1 dose →			← 1 dose →
Pneumococcal polysaccharide (PPSV23)			← 1 or 2 doses →			← 1 dose →
Meningococcal	← 1 or more doses →					
Hepatitis A	← 2 doses →					
Hepatitis B	← 3 doses →					
Haemophilus influenzae type b (Hib)	← 1 or 3 doses →					

■ For all persons in this category who meet the age requirements and who lack documentation of vaccination or have no evidence of previous infection; shingles vaccine recommended regardless of prior episode of shingles

■ Recommended if some other risk factor is present (e.g., on the basis of medical, occupational, lifestyle, or other indication)

□ No recommendation

a. Antibodies (IgG) cross the placenta.

b. Antibodies (IgG, IgA) are secreted into breast milk.

c. Antibodies can be injected by a physician.

Figure 13.10 Passive immunity. During passive immunity, antibodies are received (**a**) by crossing the placenta, (**b**) in breast milk, or (**c**) by injection. Because the body is not producing the antibodies, passive immunity is short-lived. (b): JGI/Blend Images LLC; (c): ©Aaron Haupt/Science Source recognized horse protein as foreign. This is called *serum sickness*. Passive immunity using horse antibodies against poisonous snake and spider venom is still used today, but human gamma globulin is used to treat certain infections and immunodeficiencies.

Instead of antibodies, cells of the immune system may be passively transferred into a patient, although this practice is less common. The best example is a bone marrow transplant, in which stem cells that produce blood cells are replenished after a cancer patient's own bone marrow has been intentionally destroyed by radiation or chemotherapy.

Immune Therapies

Cytokines and Immunity

As mentioned, *cytokines* are chemical messengers produced by T cells, macrophages, and other cells. Because cytokines regulate white blood cell formation and/or function, they have multiple uses in medicine. Two of the most common applications of cytokine therapy are in the stimulation of blood cell production, and the treatment of cancer.

Several cytokines stimulate the bone marrow to produce more blood cells. One example is granulocyte-macrophage colony-stimulating factor, a cytokine produced by lymphocytes, macrophages, and other cells. Under the generic name sargramostim, it is mainly used to stimulate the production of white blood cells in patients receiving chemotherapy for cancer, as well as following bone marrow transplantation. It can also be useful in treating certain genetic deficiencies in white blood cell production.

Certain cytokines are used in treating cancer. Although surgery is the main treatment for melanoma, interferon is used to decrease the risk of the cancer recurring after surgery. Interferon may work by directly inhibiting the growth of cancer cells, and/or by stimulating the immune response to the tumor. Due to its antiviral activity, interferon may inhibit viruses that are involved in causing some cancers. The FDA has also approved another cytokine, called interleukin-2 (IL-2) or aldesleukin, for the treatment of kidney cancer, melanoma, and certain immunodeficiencies. IL-2 works mainly by stimulating activities of T cells and NK cells.

One serious drawback with cytokine therapy is the risk of side effects, which can be life-threatening. One approach to this problem is to block the activity of the cytokine until it is needed. In a study published in March 2011, researchers at the University of Rochester (New York) described how they attached another protein to IL-2, effectively blocking it from attaching to its receptor. They designed the protein, however, so it would be cleaved by enzymes commonly present in a particular type of cancer. That way, the cytokine would be active only in the immediate vicinity of the cancer cells, not throughout the body, resulting in fewer side effects. Preliminary studies indicated that the approach works well in mice, but more work must be done before human trials can begin.

Monoclonal Antibodies

Every plasma cell derived from the same B cell secretes antibodies against a specific antigen. These are **monoclonal antibodies** because all of them are the same type and because they are produced by plasma cells derived from the same B cell. One method of

producing monoclonal antibodies in vitro (outside the body in glassware) is depicted in Figure 13.11. Following immunization of an animal (usually lab mice are used) with an antigen of interest, spleen cells are removed and fused with myeloma cells (malignant plasma cells that live and divide indefinitely). The fused cells are called hybridomas—*hybrid-* because they result from the fusion of two different cells,

Figure 13.11 Production of monoclonal antibodies.   a. Spleen cells derived from mice immunized with a specific antigen are fused with myeloma (cancerous) cells. **b.** This produces hybridoma cells that are "immortal." **c.** Hybridoma cells are cultured and clones producing the desired antibody are selected and expanded (**d**), producing monoclonal antibodies (**e**).

and *-oma* because one of the cells is a cancer cell. The cells are cultured in a medium that kills any unfused myeloma cells, while unfused spleen cells die naturally after a few days. The cell cultures are then screened to see if any contain hybridomas that are producing the desired antibody. If so, these specific hybridomas are clonally expanded to produce a large number of cells that secrete the desired antibody.

Because they react very specifically with only one particular antigen, monoclonal antibodies can be used for quick and accurate diagnosis of various conditions. For example, a particular hormone called human chorionic gonadotropin (HCG) is present in the urine of a pregnant woman. A monoclonal antibody can be used to detect this hormone. If it is present, the woman is pregnant. Monoclonal antibodies are also used to identify infections. And because they can distinguish between cancerous and normal tissue cells, they are used to carry radioactive isotopes or toxic drugs to tumors, which can then be selectively destroyed. Trastuzumab (Herceptin) is a monoclonal antibody used to treat breast cancer. Given intravenously, it binds to a protein receptor found on some breast cancer cells and prevents the cancer cells from dividing so fast. Antibodies that bind to cancer cells also can activate complement, increase phagocytosis by macrophages and neutrophils, and attract NK cells.

Check Your Progress 13.4

1. Describe three different types of vaccines.
2. Explain why the passive transfer of antibodies is of great importance for the newborn.
3. Explain how the interaction of cytokines with receptors on target cells may be blocked.
4. Summarize the advantage of using a monoclonal antibody for diagnosis or treatment.

13.5 Adverse Effects of Immune Responses

Learning Outcomes

Upon completion of this section, you should be able to

1. Discuss the most common immunological mechanisms responsible for allergies.
2. Compare the adverse reactions involving the ABO blood system with those involving the Rh system.
3. Explain the types of precautions that must be taken when transplanting organs.

Sometimes the immune system responds to harmless antigens in a manner that damages the body, as when individuals develop allergies, receive an incompatible blood type, or suffer tissue rejection.

Allergies

Allergies are hypersensitivities to substances, such as pollen, food, or animal hair, that ordinarily would do no harm to the body. The response to these antigens, called *allergens,* usually includes some degree of tissue damage.

An **immediate allergic response** can occur within seconds of contact with the antigen. The response is caused by antibodies known as IgE (see Table 13.1). IgE antibodies are attached to receptors on the plasma membrane of mast cells in the tissues, and also to eosinophils and basophils in the blood. When an allergen attaches to the IgE antibodies, these cells release histamine and other substances that bring about the allergic symptoms. When an allergen such as pollen is inhaled, histamine stimulates the mucous membranes of the nose and eyes, causing the runny nose and watery eyes typical of *hay fever.* In **asthma,** the airways leading to the lungs constrict, resulting in difficult breathing characterized by wheezing. When food contains an allergen, nausea, vomiting, and diarrhea often result.

Anaphylactic shock is an immediate allergic response that occurs because the allergen has entered the bloodstream. Bee stings and penicillin shots are known to cause this reaction. Anaphylactic shock is characterized by a sudden and life-threatening drop in blood pressure due to increased permeability of the capillaries caused by histamine. Injecting epinephrine can counteract this reaction until medical help is available.

Treating mild to moderate allergies usually involves antihistamines to relieve the symptoms. In more serious cases, injections of the allergen can be given so that the body will build up high quantities of IgG antibodies (see the Health feature "Immediate Allergic Responses"). The hope is that IgG antibodies will then combine with allergens before they have a chance to reach the IgE antibodies. A monoclonal antibody called Xolair is also available, which binds to and blocks the binding of IgE to the receptor found on inflammatory cells.

A **delayed allergic response** is initiated by memory T cells at the site of allergen contact in the body. The allergic response is regulated by the cytokines secreted by "sensitized" T cells at the site. A classic example of a delayed allergic response is the skin test for tuberculosis (TB). When the test result is positive, the tissue where the antigen was injected becomes red and hardened (Fig. 13.12), indicating prior exposure to tubercle bacilli, the cause of TB.

Figure 13.12 Skin testing for tuberculosis. The formation of a swollen, red area two to three days after injection of *Mycobacterium tuberculosis* antigens into the skin indicates previous exposure. ©Bart's Medical Library/Medical Images

Immediate Allergic Responses

The runny nose and watery eyes of hay fever are often caused by an allergic reaction to the windblown pollen of trees, grasses, and ragweed, particularly in the spring and fall (Fig. 13A). Worse, if a person has asthma, the airways leading to the lungs constrict, resulting in difficult breathing characterized by wheezing. Most people can inhale pollen with no ill effects. But others have developed a hypersensitivity, meaning their immune system responds in a deleterious manner. The problem stems from a type of antibody called immunoglobulin E (IgE), which causes the release of histamine from mast cells and also basophils whenever they are exposed to an allergen. Histamine causes the mucous membranes of the nose and eyes to release fluid as a defense against pathogen invasion. But in the case of allergies, copious fluid is released even though no real danger is present.

Many food allergies are also due to the presence of IgE antibodies, which usually bind to a protein in the food. The symptoms, such as nausea, vomiting, and diarrhea, are due to the mode of entry of the allergen. Skin symptoms may also occur, however. Adults are often allergic to shellfish, nuts, eggs, cows' milk, fish, and soybeans. Peanut allergy is a common food allergy in the United States, possibly because peanut butter is a staple in the diet. People seem to outgrow allergies to cow's milk and eggs more often than allergies to peanuts and soybeans.

Celiac disease occurs in people who are allergic to wheat, rye, barley, and sometimes oats—in short, any grain that contains gluten proteins. It is thought that the gluten proteins elicit a delayed cell-mediated immune response by T cells, with the resultant production of cytokines. The symptoms of celiac disease include diarrhea, bloating, weight loss, anemia, bone pain, chronic fatigue, and weakness.

People can reduce their reactions to airborne and food allergens by avoiding the offending substances. Airlines are now required to provide a peanut-free zone in their planes for those who are allergic. Taking antihistamines can also be helpful.

If these precautions are inadequate, patients can be tested to measure their susceptibility to any number of possible allergens. A small quantity of a suspected allergen is injected just beneath the skin, and the strength of the subsequent reaction is noted. The appearance of a raised, red, so-called wheal-and-flare response at the skin prick site within a few minutes demonstrates that IgE antibodies attached to mast cells have reacted to an allergen. In an immunotherapy called hyposensitization, ever-increasing doses of the allergen are periodically injected subcutaneously with the hope that the body will produce an IgG response against the allergen. IgG, in contrast to IgE, does not cause the release of histamine after it combines with the allergen. If IgG combines first, the allergic response does not occur. Patients know they are cured when the allergic symptoms go away. Therapy may have to continue for as long as 2 to 3 years.

Allergic-type reactions can occur without involving the immune system. Wasp and bee stings contain substances that cause swelling, even in those whose immune system is not sensitized to substances in the sting. Also, jellyfish tentacles and certain foods (e.g., fish that is not fresh and strawberries) contain histamine or closely related substances that can cause a reaction. Furthermore, immunotherapy is not possible in people who are allergic to penicillin and bee stings. Upon the first exposure, high sensitivity has built up so that when reexposed, anaphylactic shock can occur. Among its many effects, histamine causes increased permeability of the capillaries, the smallest blood vessels. When this reaction occurs throughout the body, these individuals experience a drastic decrease in blood pressure that can be fatal within a few minutes. People who know they are allergic to bee stings can obtain a syringe of epinephrine to carry with them. This medication can counteract the symptoms of anaphylactic shock until medical help is reached.

Questions to Consider

1. Based on what you have read in this feature and the chapter, what are some factors that might determine a person's tendency to develop an immediate allergic response to an allergen?

2. An increase in allergic diseases has been reported in developed countries, compared to lower rates in less-developed countries. What are some possible reasons for this difference?

3. What are the specific effects of epinephrine that can save the life of a person who is in anaphylactic shock?

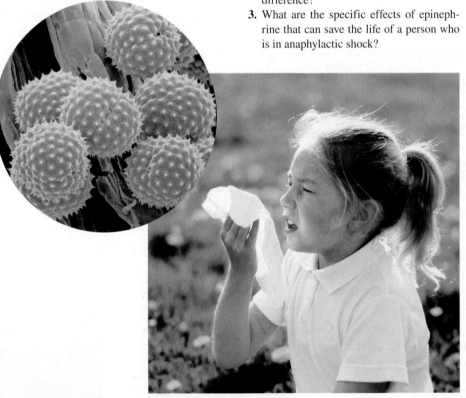

Figure 13A Pollen and allergies. When people are allergic to pollen, they develop symptoms that include watery eyes, sinus headaches, increased production of mucus, labored breathing, and sneezing. (girl): ©Damien Lovegrove/SPL/Science Source; (pollen grains): ©David Scharf/SPL/Science Source

Blood-Type Reactions

Several blood typing systems are currently in use. The most important is the ABO system. People often carry information about their ABO type in their wallets in case an accident requires them to need blood.

ABO System

In the ABO system, the presence or absence of type A and type B antigens on red blood cells determines a person's blood type. For example, a person with type A blood has the A antigen on his or her red blood cells. Because it is considered "self," this molecule is not recognized as an antigen by this individual's body, although it can be an antigen to a recipient who does not have type A blood.

The ABO system identifies four types of blood: A, B, AB, and O. Because the A and B antigens are commonly found on microbes present in and on our bodies, a person's plasma contains antibodies to the antigens that are *not* present on the red blood cells. These antibodies are called anti-A and anti-B. The antibodies present in the plasma of each blood type are shown here:

Blood Type	Antigen on Red Blood Cells	Antibody in Plasma
A	A	Anti-B
B	B	Anti-A
AB	A, B	None
O	None	Anti-A and anti-B

Because type A blood has anti-B and not anti-A antibodies in the plasma, a donor with type A blood can give blood to a recipient with type A blood (Fig. 13.13*a*). But giving type A blood to a type B recipient causes agglutination (Fig. 13.13*b*). *Agglutination,* or clumping of red blood cells, can stop blood from circulating in small blood vessels, leading to organ damage. It may be followed by *hemolysis,* or bursting of red blood cells, which if extensive can be fatal.

Theoretically, which blood type would be accepted by all other blood types? Type O blood has no A or B antigens on the red blood cells and is sometimes called the universal donor. Which blood type could receive blood from any other blood type? Type AB blood has no anti-A or anti-B antibodies in the plasma and is sometimes called the universal recipient. In practice, however, it is not safe to rely solely on the ABO system when matching blood, and instead, samples of the two types of blood are physically mixed, and the result is microscopically examined before blood transfusions are done.

Today, blood transfusions are a matter of concern not only because blood types should match, but also because each person wants to receive blood that is free of infectious agents. Blood is tested for the more serious agents, such as those that cause AIDS, hepatitis, and syphilis.

Rh System

Another important antigen in matching blood types is the Rh factor. Eighty-five percent of the U.S. population have this particular antigen on their red blood cells and are Rh-positive. Fifteen percent do not have this antigen and are Rh-negative. Rh-negative individuals normally do not have antibodies to the Rh factor, but they may make them when exposed to the Rh factor. The designation of blood type usually also includes whether the person has or does not have the Rh factor on the red blood cells (e.g., type A–positive).

During pregnancy, if the mother is Rh-negative and the father is Rh-positive, the fetus may be Rh-positive. The fetal Rh-positive red blood cells may leak across the placenta into the mother's cardiovascular system, especially as placental tissues normally break down during birth (Fig. 13.14*a*). Now, the mother produces anti-Rh antibodies. In a subsequent pregnancy with another Rh-positive child, these antibodies may cross the placenta and destroy the fetal red blood cells (Fig. 13.14*b*). The resulting condition, called hemolytic disease of the newborn, can lead to the death of the fetus.

type A blood of donor + anti-B antibody of type A recipient → no binding 500×

a. No agglutination

type A blood of donor + anti-A antibody of type B recipient → binding 500×

b. Agglutination

Figure 13.13 Blood transfusions. No agglutination (**a**) versus agglutination (**b**) is determined by whether the recipient has antibodies in the plasma that can combine with antigens on the donor's red blood cells. The photos on the right show how the red blood cells appear under the microscope. (both photos): ©J. C. Revy/Medical Images

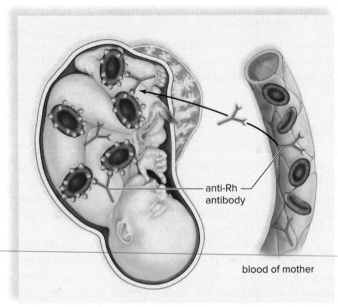

a. Fetal Rh-positive red blood cells leak across placenta into mother's bloodstream.

b. Mother forms anti-Rh antibodies that cross the placenta and attack fetal Rh-positive red blood cells.

Figure 13.14 Hemolytic disease of the newborn. **a.** Due to a pregnancy in which the child is Rh-positive, an Rh-negative mother can begin to produce antibodies against Rh-positive red blood cells. **b.** In the same, but more likely a subsequent, pregnancy these antibodies can cross the placenta and cause hemolysis of an Rh-positive child's red blood cells.

The Rh problem is prevented by giving Rh-negative women an Rh immunoglobulin injection toward the end of the pregnancy, and within 72 hours after giving birth to an Rh-positive child. This injection contains relatively low levels of anti-Rh antibodies that help destroy any baby's red blood cells in the mother's blood before her immune system produces high levels of anti-Rh antibodies.

Tissue Rejection

Certain organs, such as the skin, the heart, and the kidneys, would be relatively easy to transplant from one person to another if the recipient's body did not attempt to reject them. Unfortunately, in addition to their role in antigen presentation to T cells, MHC proteins can also serve as antigens when organs of a different MHC type are transplanted into a recipient. Ideally, the transplanted organ would have exactly the same type of MHC proteins as those of the recipient. Because it is hard to find a perfect MHC match, the chance of organ rejection can also be diminished by administering *immunosuppressive* drugs. Two well-known immunosuppressive drugs, cyclosporine and tacrolimus, both act by inhibiting the production of certain cytokines by T cells.

Xenotransplantation is the use of animal organs instead of human organs in transplant patients. Scientists have chosen to use the pig because pigs are raised as a meat source and are prolific. Genetic engineering can make pig organs less antigenic by removing the MHC antigens. The ultimate goal is to make pig organs as widely accepted as type O blood. Although advances in xenotransplantation are encouraging, many researchers hope that tissue engineering, including the production of human organs that lack MHC antigens, will one day do away with the problem of rejection. Alternatively, scientists are learning how to grow organs from a patient's own tissues. In July 2011, surgeons in Sweden replaced

a cancer patient's diseased trachea with a lab-grown version, produced by seeding the patient's own stem cells onto a trachea-shaped glass scaffold.

Check Your Progress 13.5

1. Explain the immunological reason why immediate allergic responses can occur within seconds after being reexposed to an allergen, whereas delayed-type responses may take days.
2. Discuss why Rh incompatibility is only a problem when a fetus is Rh-positive and the mother is Rh-negative.
3. Discuss the potential side effect of the drugs cyclosporine and tacrolimus, which function by inhibiting cytokine production (rather than inhibiting specific responses to transplanted organs).

13.6 Disorders of the Immune System

Learning Outcomes

Upon completion of this section, you should be able to

1. Define *autoimmune disease*, and list several specific examples of these diseases.
2. Differentiate between acquired and congenital immunodeficiencies.

When a person has an **autoimmune disease,** cytotoxic T cells or antibodies mistakenly recognize the body's own cells or molecules as if they are foreign antigens. Exactly what causes autoimmune diseases is not known. However, sometimes they occur after an individual has recovered from an infection.

SCIENCE IN YOUR LIFE ▶ HEALTH

Opportunistic Infections and HIV

AIDS (acquired immunodeficiency syndrome) is caused by the destruction of the immune system, especially the helper T cells, following an HIV (human immunodeficiency virus) infection. Then, the individual succumbs to many unusual types of infections that would not cause disease in a person with a healthy immune system. Such infections are known as opportunistic infections (OIs).

HIV not only kills helper T cells by directly infecting them, it also causes many uninfected T cells to die by a variety of mechanisms, including apoptosis. Many helper T cells are also killed by the person's own immune system as it tries to overcome the HIV infection. After initial infection with HIV, it may take up to 10 years for an individual's helper T cells to become so depleted that the immune system can no longer organize a specific response to OIs (Fig. 13B). In a healthy individual, the number of helper T cells typically ranges from 800 to 1,000 cells per cubic millimeter (mm³) of blood. The appearance of the following specific OIs can be associated with the helper T-cell count:

- Shingles. Painful infection with varicella-zoster (chicken pox) virus. Helper T-cell count of less than 500/mm³.
- Candidiasis. Yeast infection of the mouth, throat, or vagina. Helper T-cell count of about 350/mm³.
- *Pneumocystis* pneumonia. Fungal infection causing the lungs to fill with fluid and debris. Helper T-cell count of less than 200/mm³.
- Kaposi sarcoma. Cancer of blood vessels due to human herpesvirus 8; gives rise to reddish-purple, coin-sized spots and lesions on the skin. Helper T-cell count of less than 200/mm³.
- Toxoplasmic encephalitis. Protozoan infection characterized by severe headaches, fever, seizures, and coma. Helper T-cell count of less than 100/mm³.
- *Mycobacterium avium* complex (MAC). Bacterial infection resulting in persistent fever, night sweats, fatigue, weight loss, and anemia. Helper T-cell count of less than 75/mm³.
- Cytomegalovirus. Viral infection that leads to blindness, inflammation of the brain, throat ulcerations. Helper T-cell count of less than 50/mm³.

Due to development of powerful drug therapies that slow the progression of AIDS, people infected with HIV in the United States are suffering a lower incidence of OIs than was common in the 1980s and 1990s.

Questions to Consider

1. People who don't have HIV also sometimes get OIs, such as shingles or candidiasis. How would you expect the outcome of, for example, a yeast infection to be different in a healthy person compared to an HIV-infected person?
2. Kaposi sarcoma occurs relatively rarely in people who are infected with human herpesvirus 8, but are HIV-negative. How might an HIV infection increase the odds of getting Kaposi sarcoma?
3. Why do you suppose that the amount of HIV in the blood increases rapidly during the later stages of AIDS, as shown in Figure 13B?

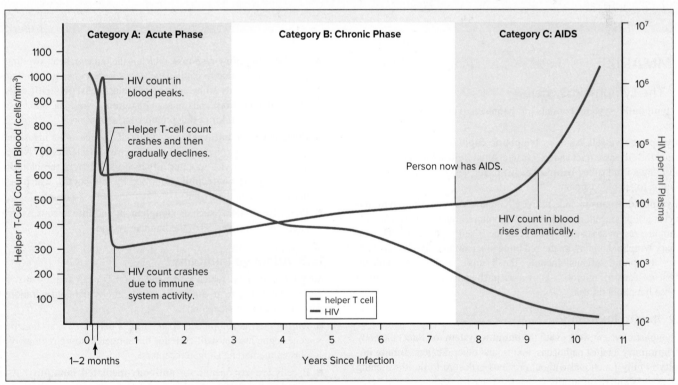

Figure 13B Progression of HIV infection during its three stages.   In category A, the individual may have no symptoms or very mild symptoms associated with the infection. By category B, opportunistic infections have begun to occur, such as candidiasis, shingles, and diarrhea. Category C is characterized by more severe opportunistic infections and is clinically described as AIDS.

In the autoimmune disease *myasthenia gravis* (*MG*), antibodies attach to and interfere with neuromuscular junctions, and muscular weakness results. In *multiple sclerosis* (*MS*), T cells attack the myelin sheath of nerve fibers, and this causes various neuromuscular disorders. A person with *systemic lupus erythematosus* (*SLE*) has various symptoms prior to death, due to kidney damage caused by the deposition of excessive antigen/antibody complexes. In *rheumatoid arthritis,* the joints are affected. As yet, there are no cures for autoimmune diseases, but they can sometimes be controlled with immunosuppressive drugs.

When a person has an **immunodeficiency disease,** the immune system is unable to protect the body against disease. *Acquired immunodeficiency syndrome* (*AIDS*) is an example of an acquired immunodeficiency. Immunodeficiencies may also be primary or congenital (i.e., inherited). This allows opportunistic infections, or those that otherwise would not normally cause harm to the body, to develop in the individual. The Health feature "Opportunistic Infections and HIV" explores this in more detail.

Occasionally, children may be born lacking the ability to produce antibodies. Less commonly, a child may be born with an impaired immune system caused by a defect in T cells. In *severe combined immunodeficiency* (*SCID*), both antibody- and cell-mediated immunity are lacking or inadequate. Without treatment, even common infections can be fatal. Gene therapy has been successful in some SCID patients.

Check Your Progress 13.6

1. Describe four examples of autoimmune diseases.
2. List two immunodeficiency diseases, and identify them as either congenital or acquired.

Conclusion

Developing vaccines for emerging diseases, such as the Zika virus or SARS, is an important focus of the medical community. While our immune systems are remarkably responsive to pathogens, the rapidness of the response is dependent upon our exposure to similar pathogens in our past. The presence of a new virus can present real challenges for our immune system, and the delay in the response can allow the virus to create significant health problems.

Fortunately for us, our understanding not only of how the immune system works, but also the structure of pathogens such as viruses, means that we can develop vaccines at a much faster rate than in the past. In the case of Zika, trials of vaccines have already begun, just a few years after it was recognized as an emerging health concern.

SUMMARIZE

13.1 The Lymphatic System

The **lymphatic system** consists of lymphatic vessels and lymphoid organs:

- **Lymphatic vessels** begin as **lymphatic capillaries** that absorb fats from the digestive tract and excess interstitial fluid at blood capillaries, forming a fluid called **lymph.** The lymphatic vessels eventually return lymph to the bloodstream.
- **Lymphoid organs** include primary lymphoid organs, where **lymphocytes** are produced, and secondary lymphoid organs, where adaptive immune responses can be initiated by the B cells and T cells. The primary lymphoid organs are the **red bone marrow** and **thymus.** Secondary lymphoid organs include the **lymph nodes** and **spleen.** Disease-causing microbes known as **pathogens** may be present in these lymphoid tissues.

13.2 Innate Immunity

The lymphatic system works with the **immune system** to protect the body with **immunity** against pathogens, toxins, and other dangers. **Innate immunity** is fully functional without previous exposure to the threatening substance. **Adaptive immunity** acts against specific antigens.

Mechanisms of innate immunity include:

- Physical and chemical barriers, such as the ciliated cells and mucus of the respiratory tract, and stomach acid.

- The **inflammatory response,** which is the redness, heat, swelling, and pain that occur due to capillary changes in response to chemicals released by tissue-dwelling cells such as **mast cells, dendritic cells,** and **macrophages.** Mast cells release **histamine,** while the latter two cell types release **cytokines** that promote inflammation.
- **Phagocytes** and **natural killer** (**NK**) **cells,** which either ingest and kill microorganisms (phagocytes), or kill virus-infected cells (NK cells). The first phagocytes to enter inflamed tissues from the bloodstream are usually the **neutrophils,** followed by **monocytes,** which become macrophages in the tissues.
- Protective proteins, such as **complement** and **interferons,** which enhance the effectiveness of the immune system.

13.3 Adaptive Immunity

Adaptive immunity, which is dependent on B cells and T cells, is enhanced after exposure to **antigens;** specific molecular configurations that can bind to antigen receptors:

- During their development, B and T cells undergo a rearrangement of gene segments that will code for the antigen-binding regions of their plasma membrane antigen receptors.
- B cells are responsible for **antibody-mediated immunity.** As predicted by the **clonal selection** model, upon exposure to an antigen that binds to its **B-cell receptors,** a B cell replicates and differentiates, producing plasma cells and memory B cells. **Plasma cells** secrete antibodies and eventually undergo apoptosis. **Memory B cells** are

long-lived and produce antibodies if the same antigen enters the body at a later date. **Antibodies,** also called **immunoglobulins (Igs),** are typically Y-shaped proteins with at least two binding sites for a specific antigen. The main classes of human antibodies are IgG, IgM, IgA, IgE, and IgD.

■ T cells are responsible for **cell-mediated immunity.** Each T cell bears **T-cell receptors.** However, for a T cell to recognize an antigen, the antigen must be processed and presented on a cell surface by an **MHC (major histocompatibility complex) protein.** The two main types of T cells are helper T (T_H) cells and cytotoxic T (T_C) cells. **Helper T cells** respond to antigens presented by MHC class II proteins found on **antigen-presenting cells (APCs)** such as dendritic cells or macrophages. The activated T_H cells produce cytokines that stimulate other immune cells. **Cytotoxic T cells** kill virus-infected or cancer cells that bear a "nonself" protein on their MHC class I proteins. Thereafter, the activated T cell undergoes clonal expansion until the illness has been stemmed. Then, most of the activated T cells undergo apoptosis. A few cells remain, however, as **memory T cells.**

13.4 Active Versus Passive Immunity

Immunity is the condition where the body is protected from various threats—usually by the adaptive immune system. Active and passive immunity are induced in different ways:

■ **Active immunity** can be induced through infection or by **immunization.** A person can be immunized with a **vaccine,** which can be a killed or modified form of a pathogen, a subunit, or even pure DNA.

■ **Passive immunity** is acquired from another individual, whether naturally as antibodies passed from a mother to child, or artificially via an injection.

Immune therapies may include cytokines and monoclonal antibodies:

■ Cytokines are chemical messengers used to stimulate blood cell production, and to treat cancer and other diseases.

■ **Monoclonal antibodies,** which are produced by cells derived from the same plasma cell, are used for various purposes, from detecting infections to treating cancer.

13.5 Adverse Effects of Immune Responses

Immune responses to harmless compounds (called allergens) are known as **allergies** or hypersensitivities:

■ **Immediate allergic responses** occur due to an inappropriate production of IgE antibody, which results in an inflammatory response to subsequent exposures to the allergen. The most extreme form of this is **anaphylactic shock,** which can be fatal. **Asthma** also has an immediate allergic component.

■ **Delayed allergic responses,** such as a positive TB skin test, are due to the activity of T cells.

■ The reactions that occur following improperly matched blood types may involve the ABO system, the Rh system, or other blood group antigens. Also, if an Rh-negative woman becomes pregnant with an Rh-positive fetus, hemolytic disease of the newborn may result.

■ The rejection of transplanted tissues is usually due to an imperfect match of MHC proteins between the donor and recipient.

13.6 Disorders of the Immune System

In **autoimmune diseases,** a person's own T cells or antibodies target the body's own cells or molecules. Many types of immunodeficiencies also occur:

■ Examples of autoimmune diseases include myasthenia gravis (MG), multiple sclerosis (MS), systemic lupus erythematosus (SLE), and rheumatoid arthritis.

■ **Immunodeficiency diseases** can be acquired, as occurs with acquired immunodeficiency syndrome (AIDS), or inherited, as is seen in children with severe combined immunodeficiency (SCID), who are born unable to produce their own antibody- and cell-mediated immunity. Depending on the severity, immunodeficiencies can be fatal, usually as a result of other, common opportunistic infections.

ASSESS

Testing Yourself

Choose the best answer for each question.

13.1 The Lymphatic System

1. What is the term for localized swelling caused by fluid accumulation?
   a. lymph   c. edema
   b. lupus   d. complement

2. Which of the following is a function of the spleen?
   a. produces T cells
   b. removes worn-out red blood cells
   c. produces immunoglobulins
   d. produces macrophages
   e. regulates the immune system

3. Which of the following is not a function of the lymphatic system?
   a. produces red blood cells
   b. returns excess fluid to the blood
   c. transports lipids absorbed from the digestive system
   d. defends the body against pathogens

13.2 Innate Immunity

4. Which of the following is most directly responsible for the increase in capillary permeability during the inflammatory reaction?
   a. pain   c. histamine
   b. white blood cells   d. tissue damage

5. Which of the following is not a goal of the inflammatory reaction?
   a. bring more oxygen to damaged tissues
   b. decrease blood loss from a wound
   c. decrease the number of white blood cells in damaged tissues
   d. prevent entry of pathogens into damaged tissues

6. Which of the following is not correct concerning interferon?
   a. Interferon is a protective protein.
   b. Virus-infected cells produce interferon.
   c. Interferon has no effect on viruses.
   d. Interferon can be used to treat certain viral infections.

13.3 Adaptive Immunity

7. The adaptive immune defenses respond to which of the following?
   a. specific antigens
   b. general pathogens
   c. interferon
   d. histamine
   e. All of these are correct.

8. Which of the following does not pertain to B cells?
   a. have passed through the thymus
   b. have specific receptors
   c. are responsible for antibody-mediated immunity
   d. synthesize antibodies

9. Which of the following characteristics pertains to T cells?
   a. have specific receptors
   b. are of more than one type
   c. are responsible for cell-mediated immunity
   d. stimulate antibody production by B cells
   e. All of these are correct.

10. Label a–c on this IgG molecule using these terms: *antigen-binding sites, light chain, heavy chain.* What do *V* and *C* stand for in the diagram?

13.4 Active Versus Passive Immunity

11. Active immunity can be produced by
    a. having a disease.
    b. receiving a vaccine.
    c. receiving gamma globulin injections.
    d. Both a and b are correct.
    e. Both b and c are correct.

12. Which of the following does not occur during a secondary immune response?
    a. Antibodies are made quickly and in great amounts.
    b. Antibody production lasts longer than in a primary response.
    c. Clonal selection occurs for B cells.
    d. All of these occur during a secondary immune response.

13.5 Adverse Effects of Immune Responses

13. A sudden drop in blood pressure in response to an antigen in the body is a characteristic of which of the following?
    a. inflammatory response
    b. passive immunity
    c. cell-mediated immunity
    d. anaphylactic shock
    e. None of these are correct.

14. Which of the following conditions occurs when antibodies attack the myelin sheath covering nerve fibers?
    a. lupus   **c.** multiple sclerosis
    b. rheumatoid arthritis   **d.** myasthenia gravis

13.6 Disorders of the Immune System

15. Which of the following is not an example of an autoimmune disease?
    a. multiple sclerosis
    b. rheumatic fever
    c. hemolytic disease of the newborn
    d. systemic lupus erythematosus

16. Which of the following causes damage to the cells of the immune system and allows for opportunistic infections to develop?
    a. inflammatory response
    b. AIDS
    c. complement
    d. edema
    e. All of these are correct.

ENGAGE

Thinking Critically

1. Some primitive organisms, such as invertebrates, have no lymphocytes and thus lack an adaptive immune system, but they have some components of an innate immune system, including phagocytes and certain protective proteins. What are some general features of innate immunity that make it very valuable to organisms lacking more specific antibody- and cell-mediated responses? What are some disadvantages to having only an innate immune system?

2. Someone bitten by a poisonous snake should be given some antivenom (antibodies) to prevent death. If the person is bitten by the same type of snake 3 years after the initial bite, will he or she have immunity to the venom, or should the person get another shot of antivenom? Justify your response with an explanation of the type of immunity someone gains from a shot of antibodies.

3. Why is it that Rh incompatibility can be a serious problem when an Rh-negative mother is carrying an Rh-positive fetus, but ABO incompatibility between mother and fetus is usually no problem? That is, a type A mother can usually safely carry a type B fetus. (*Hint:* The antibodies produced by an Rh-negative mother against the Rh antigen are usually IgG, whereas the antibodies produced against the A or B antigens are IgM.) Because the Rh antigen obviously serves no vital function (most humans lack it), why do you think it hasn't been completely eliminated during human evolution?

©Ricochet Creative Productions LLC/Synde Mass, photographer

14

Digestive System and Nutrition

Gluten and Celiac Disease

On any trip to the grocery store, you may notice a wide abundance of foods labeled "gluten-free," and almost every restaurant has a gluten-free section of the menu. So what exactly is gluten?

Contrary to what most people believe, gluten is not a carbohydrate. Instead, it is a type of protein commonly found in wheat, rye, and barley. For most of us, gluten is processed by the digestive system in the same manner as any other protein, meaning it is broken down into amino acids and absorbed into the circulatory system.

However, for about one out of every 100 individuals, the body misidentifies gluten as a foreign pathogen. The result is celiac disease, a disorder of the lining of the small intestine. In celiac disease, an example of an autoimmune response, the body's reaction to gluten causes inflammation and loss of the intestinal linings. The damage to the intestines causes malnutrition and other health problems. Individuals with celiac disease must eliminate gluten from their diet. Interestingly, a gluten-free diet does not seem to have health benefits for individuals who do not suffer from celiac disease.

In this chapter, you will learn about the function of the digestive system, as well as the basic principles of nutrients and nutrition.

As you read through this chapter, think about the following questions:

1. How are carbohydrates and proteins processed by the digestive system?
2. What is the role of the small intestine in digestion?

CHAPTER OUTLINE

BEFORE YOU BEGIN

Before beginning this chapter, take a few moments to review the following discussions:

Chapter 2 What are the structural features and major functions of carbohydrates, lipids, proteins, and nucleic acids?

Chapter 7 By what specific biochemical processes does the breakdown of foods result in the production of ATP?

Section 11.5 With what other body systems does the digestive system work, in order to maintain homeostasis?

14.1 The Digestive Tract

The **digestive system** contributes to homeostasis by ingesting food, separating it into chemical nutrients that cells can use, absorbing those nutrients, and eliminating indigestible remains. Digestion takes place within a tube called the *digestive tract,* also called the *gastrointestinal tract,* which begins with the mouth and ends with the anus (Fig. 14.1).

Digestion involves mechanical digestion and chemical digestion. *Mechanical digestion* begins with the chewing of food in the mouth and continues with the churning and mixing of food in the stomach. During *chemical digestion,* many different enzymes break down macromolecules to small organic molecules that can be absorbed. The organs of the digestive tract, and their general functions, are listed in Table 14.1.

The Mouth

The **mouth,** also called the *oral cavity,* is the site where food is ingested. The mouth is bounded externally by the lips and cheeks. The red portion of the lips is poorly keratinized, and this allows blood to show through.

Most people enjoy eating food largely because they like its texture and taste. Sensory receptors called *taste buds* occur primarily on the tongue, and when these are activated by the presence of food, nerve impulses travel by way of cranial nerves to the brain

Figure 14.1 **The human digestive tract.**
The upper part of the tract includes the mouth, pharynx, esophagus, stomach, and small intestine. The large intestine consists of the cecum, the colon (ascending, transverse, descending, and sigmoid colons), the rectum, and the anus. Note also the location of the accessory organs of digestion: the pancreas, liver, salivary glands, and gallbladder.

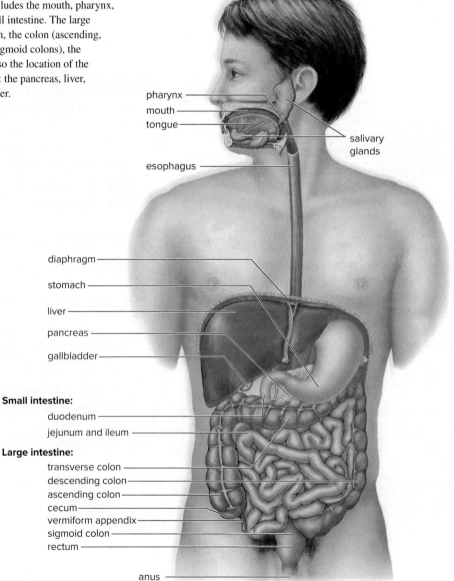

TABLE 14.1 Organs of Digestion

Organ	General Function	Type of Digestion	Accessory Glands
Mouth	Receives food and breaks it into smaller pieces; starts digestion of starch	Mechanical; chemical	Salivary glands
Pharynx	Connects mouth to esophagus; connects respiratory and digestive systems	None	—
Esophagus	Connects pharynx and stomach	None	—
Stomach	Stores food; acidity kills bacteria; starts digestion of protein	Mechanical; chemical	—
Small intestine	Digests all foods; absorbs nutrients	Chemical	Liver, pancreas, gallbladder
Large intestine	Absorbs water; stores indigestible remains	Small amount of chemical by bacteria	—

(see Section 18.3). The *tongue* is composed of skeletal muscle that contracts to change the shape of the tongue. A fold of mucous membrane on the underside of the tongue attaches it to the floor of the mouth.

The roof of the mouth separates the *nasal cavity* from the mouth, and prevents ingested food from entering the nasal cavity. The roof has two parts: an anterior (toward the front) *hard palate* and a posterior (toward the back) *soft palate* (Fig. 14.2*a*). The hard palate contains several bones, but the soft palate is composed of muscle and connective tissue. The soft palate ends in a finger-shaped projection called the *uvula.*

In about one out of 850 newborns, the bones of the hard palate have not fused during fetal development, leaving a gap, or *cleft palate.* If not surgically repaired, such a defect can result in problems with feeding, speech, and hearing (due to ear infections).

The *tonsils* are in the back of the mouth, on either side of the tongue, and in the nasopharynx, where they are called *adenoids.* The tonsils contain lymphoid tissue that helps protect the body against infections (see Section 13.1). If the tonsils become inflamed, the person has *tonsillitis.* If tonsillitis recurs repeatedly, the tonsils may be surgically removed in a procedure called a *tonsillectomy.*

Three pairs of **salivary glands** (see Fig. 14.1) produce saliva. Saliva keeps the mouth moist and contains the enzyme *salivary amylase,* which begins the process of digesting starch. One pair of salivary glands lies on either side of the face, immediately below and in front of the ears. These glands swell when a person has the *mumps,* a disease caused by a viral infection. Another pair of salivary glands lies beneath the tongue, and still another pair lies beneath the floor of the mouth. The ducts from these salivary glands open under the tongue. You may be able to locate the openings if you use your tongue to feel for small flaps on the inside of your cheek and under your tongue.

The Teeth

Teeth are the primary means of mechanical digestion in the digestive tract. During the first 2 years of life, the smaller 20 *deciduous,* or baby, teeth appear. These are eventually replaced by 32 adult teeth (Fig. 14.2*a*). The third pair of molars, called the *wisdom teeth,* sometimes fail to erupt. If they push on the other teeth and/or cause pain, they can be removed by a dentist or oral surgeon.

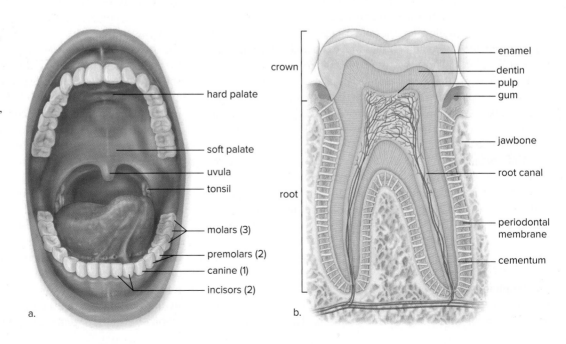

Figure 14.2 Adult mouth and teeth. a. The chisel-shaped incisors bite; the pointed canines tear; the fairly flat premolars grind; and the flattened molars crush food. The last molars, called wisdom teeth, may fail to erupt, or if they do, they are sometimes crooked and useless. **b.** Longitudinal section of a tooth. The crown is the portion that projects above the gum line and can be replaced by a dentist if damaged. When a "root canal" is done, the nerves are removed. When the periodontal membrane is inflamed, the teeth can loosen.

hard palate
soft palate
uvula
tonsil
molars (3)
premolars (2)
canine (1)
incisors (2)

a.

crown
root

enamel
dentin
pulp
gum
jawbone
root canal
periodontal membrane
cementum

b.

Each tooth has two main divisions, a crown and a root (Fig. 14.2*b*). The *crown* has an outer layer of *enamel,* an extremely hard covering made of calcium compounds; *dentin,* a thick layer of bonelike material; and an inner *pulp,* which contains the nerves and the blood vessels. Dentin and pulp are also found in the root. The gum tissue, or *gingiva,* surrounds the teeth and normally forms a tight seal around them.

Bacteria that adhere to the teeth metabolize sugar and give off acids, which may erode the enamel. This can result in *dental caries,* or cavities, in the teeth. The enamel decays very slowly, but once the dentin layer is reached, the damage spreads more rapidly. Three measures can help to prevent tooth decay: limiting sugar intake, daily brushing and flossing, and visiting the dentist regularly. Fluoride treatments, particularly in children, can make the enamel stronger and decay-resistant. Gum disease is more apt to occur with aging. Inflammation of the gums, called *gingivitis,* can spread to the periodontal membrane, which lines the tooth socket. A person then has *periodontitis,* characterized by loss of bone and loosening of the teeth so that extensive dental work may be required. Stimulation of the gums in a manner advised by your dentist is helpful in controlling this condition.

Over time, and especially with exposure to cigarette smoke, soft drinks, or coffee, the enamel layer can become stained. There are a variety of tooth-whitening treatments, most of which contain a bleaching agent such as hydrogen peroxide, that are available over-the-counter to remove these stains.

The Pharynx

The **pharynx** is a region that receives air from the nasal cavities (nasopharynx) and food from the mouth (see Fig. 14.1). From the mouth, food (the bolus) passes through the pharynx and esophagus to the stomach, small intestine, and large intestine. The food passage and air passage cross in the pharynx because the trachea (windpipe) is ventral to (in front of) the esophagus, which takes food to the stomach.

Swallowing, a process that occurs in the pharynx (Fig. 14.3), is a *reflex action* performed automatically, without conscious thought. During swallowing, the soft palate moves back to close off the *nasopharynx,* and the trachea moves up under the *epiglottis* to cover the glottis. The *glottis* is the opening to the larynx (voice box), and therefore the air passage. The up-and-down movement of the Adam's apple, the front part of the larynx, is easy to observe when a person swallows. During swallowing, food normally enters the esophagus because the air passages are blocked. Therefore, we do not breathe when we swallow.

Unfortunately, we have all had the unpleasant experience of having food "go the wrong way." The wrong way may be either into the nasal cavities or into the trachea. If it is the latter, coughing will most likely force the food up out of the trachea and into the pharynx again.

The Esophagus

The **esophagus** is a long muscular tube that passes from the pharynx, through the thoracic cavity and diaphragm, and into the abdominal cavity, where it joins the stomach. The esophagus is

Figure 14.3 Swallowing.   When food is swallowed, the soft palate closes off the nasopharynx, and the epiglottis covers the glottis, forcing the bolus to pass down the esophagus. Therefore, a person does not breathe while swallowing.

ordinarily collapsed, but it opens and receives the bolus when swallowing occurs.

Rhythmic muscular contractions, collectively called **peristalsis,** push the food along the digestive tract. Peristalsis begins in the esophagus and continues in all the organs of the digestive tract. Occasionally, peristalsis begins even though there is no food in the esophagus. This produces the sensation of a lump in the throat.

The esophagus plays no role in the chemical digestion of food. Its sole purpose is to conduct the food bolus from the mouth to the stomach. *Sphincters* are muscles that encircle tubes in the body, acting as valves. The tubes close when sphincters contract, and they open when sphincters relax. The entrance of the esophagus to the stomach is marked by a constriction that is often called a sphincter, although the muscle is not as developed as it would be in a true sphincter. Relaxation of the sphincter allows the bolus to pass into the stomach, while contraction prevents the acidic contents of the stomach from backing up into the esophagus.

Heartburn, which feels like a burning pain rising into the throat, occurs when some of the stomach contents escape into the esophagus. Heartburn is a very common condition, and there are a number of over-the-counter medications and prescription medicines that can alleviate either the symptoms or causes. One of these is Nexium (generic name, esomeprazole), which belongs to a group of drugs called proton pump inhibitors that reduce acid production in the stomach. A more serious form of heartburn is *GERD* (*gastroesophageal reflux disease*), which is frequent or persistent reflux that may lead to more serious problems such as ulcers, difficulty swallowing, or even esophageal cancer due to chronic irritation of the esophageal wall by stomach acid.

Vomiting, which is the forceful expulsion of stomach contents out of the body through the mouth, can help protect against the ingestion of potentially harmful agents. In this scenario, certain cells in the intestinal tract, as well as in the brainstem, trigger the contraction of the abdominal muscles and diaphragm to propel the contents of the stomach upward through the esophagus.

The Wall of the Digestive Tract

The structure of the esophageal wall in the abdominal cavity is representative of that found in the stomach, small intestine, and large intestine. All are composed of the following four layers (Fig 14.4):

- *Mucosa* (mucous membrane layer). A layer of epithelium supported by connective tissue and smooth muscle lines the **lumen** (central cavity) and contains glandular epithelial cells that secrete digestive enzymes and goblet cells that secrete mucus.
- *Submucosa* (submucosal layer). Beneath the mucosa lies the submucosa, a broad band of loose connective tissue that contains blood vessels. Lymph nodules, including some called Peyer's patches, are in the submucosa. Like the tonsils, they help protect us from disease.
- *Muscularis* (smooth muscle layer). Two layers of smooth muscle make up this section. The inner, circular layer encircles the gut; the outer, longitudinal layer lies perpendicular to the circular layer. (The stomach also has oblique muscles.)
- *Serosa* (serous membrane layer). Most of the digestive tract has a serosa, a very thin, outermost layer of squamous epithelium that secretes a serous fluid that keeps the outer surface of the intestines moist so that the organs of the abdominal cavity slide against one another.

The Stomach

The **stomach** (Fig. 14.5a) is a thick-walled, J-shaped organ that lies on the left side of the abdominal cavity, below the liver and diaphragm. It is continuous with the esophagus above and the duodenum of the small intestine below. The stomach receives food from the esophagus, starts the digestion of proteins, and moves food into the small intestine. The human stomach is about 25 cm (10 in.) long, regardless of the amount of food it holds, but the diameter varies, depending on how full it is. As the stomach expands, deep folds in its wall, called *rugae,* gradually disappear. When full, it can hold about 4 liters (1 gal).

The columnar epithelium lining the stomach has millions of gastric pits, which lead into *gastric glands* (Fig. 14.5b). The term *gastric* always refers to the stomach. The gastric glands produce gastric juice, which contains pepsinogen, hydrochloric acid (HCl), and mucus. As discussed in Section 14.3, pepsinogen becomes the enzyme pepsin when exposed to HCl. The high acidity of the stomach (pH of about 2) is also beneficial because it kills most of the bacteria and other microbes present in food.

The stomach acts both physically and chemically on food. Its wall contains three muscle layers: One layer is longitudinal and another is circular, as shown in Figure 14.4a; the third is obliquely arranged. This muscular wall not only moves the food along, but it also churns, mixing the food with gastric juice and breaking it down into small pieces.

Alcohol and other liquids are absorbed in the stomach, but most solid food substances are not. Normally, the stomach empties in 2 to 6 hours. When food leaves the stomach, it is a thick, soupy liquid called **chyme.** Chyme enters the small intestine in squirts by way of the pyloric sphincter, which acts like a valve, repeatedly opening and closing.

Figure 14.4 Wall of the digestive tract. **a.** Several different types of tissues are found in the wall of the digestive tract. Note the placement of circular muscle inside longitudinal muscle. **b.** Micrograph of the wall of the esophagus. (b): ©Biophoto Associates/Science Source

gastric pit

a.

cells that secrete mucus

gastric gland

cells that secrete HCl and enzymes

b.   400×

Figure 14.5 **Anatomy of the stomach.**   **a.** The stomach has a thick wall with folds that allow it to expand and fill with food. **b.** The mucosa contains gastric glands, which secrete gastric juice containing mucus and digestive enzymes. (b): ©Ed Reschke/Getty Images

The Small Intestine

The **small intestine** is named for its small diameter (compared to that of the large intestine), but perhaps it should be called the long intestine. It averages about 6 meters (almost 20 ft) in length, compared to the large intestine, which is about 1.5 meters (5 ft) in length.

The first 25 cm (1 ft) of the small intestine is called the **duodenum.** Ducts from the liver and pancreas join to form one common bile duct that enters the duodenum (see Fig. 14.10). The small intestine receives bile from the liver and pancreatic juice from the pancreas via this duct. The intestine has a slightly basic pH because pancreatic juice contains sodium bicarbonate ($NaHCO_3$), which neutralizes the acid in chyme. The enzymes in pancreatic juice and enzymes produced by the intestinal wall complete the process of food digestion.

The middle part of the small intestine is called the **jejunum,** and the remainder is the **ileum.** The submucosal layer of the ileum contains aggregations of lymphoid tissues called *Peyer's patches,* which are involved in generating immune responses to intestinal pathogens.

The surface area of the small intestine has been compared to that of a tennis court. What factors contribute to such a large surface area? The wall of the small intestine contains fingerlike projections called **villi** (*sing.,* villus), which give the intestinal wall a soft, velvety appearance (Fig. 14.6). A villus has an outer layer of columnar epithelial cells, and each of these cells has thousands of microscopic extensions called microvilli. Microvilli greatly increase the surface area of the villus for the absorption of nutrients.

Nutrients are absorbed into the vessels of a villus. Each villus contains blood capillaries and a small lymphatic capillary, called a **lacteal.** Glycerol and fatty acids (digested from fats) enter the epithelial cells of the villi, where they are joined and packaged as lipoprotein droplets that enter a lacteal. Lacteals are a part of the

Small intestine

Section of intestinal wall

lumen

villus

lacteal

blood capillaries

mucus-producing cell

lymph nodule

venule

lymphatic vessel

arteriole

villus   microvilli

Villi   100×

Figure 14.6 **Anatomy of the small intestine.**   The wall of the small intestine has folds that bear fingerlike projections called villi. Microvilli, which project from the villi, absorb the products of digestion into the blood capillaries and the lacteals of the villi. (photos) (*left*): ©Kage Mikrofotografie/Medical Images; (*right*): ©Science Photo Library /Getty Images lymphatic system (see Section 13.1). The lymphatic system is an assistant to the cardiovascular system; that is, its vessels carry a fluid called lymph to the cardiovascular veins. Sugars and amino acids directly enter the blood capillaries of a villus. After nutrients are absorbed, the bloodstream eventually carries them to all the cells of the body.

Regulation of Digestive Secretions

The secretion of digestive juices is promoted by the nervous system and by hormones. A *hormone* is a substance produced by one set of cells that affects a different set of cells (for more on hormones, see Section 20.1). Hormones are usually transported by the bloodstream. For example, when a person has eaten a meal particularly rich in protein, the stomach produces the hormone *gastrin.* Gastrin enters the bloodstream, and soon the stomach is churning, and the secretory activity of gastric glands is increasing. A hormone produced by the duodenal wall, *GIP (gastric inhibitory peptide),* works opposite to gastrin: It inhibits gastric gland secretion.

Cells of the duodenal wall produce two other hormones that are of particular interest—secretin and CCK (cholecystokinin). Acid, especially hydrochloric acid (HCl) present in chyme, stimulates the release of secretin, while partially digested protein and fat stimulate the release of CCK. Soon after these hormones enter the bloodstream, the pancreas increases its output of pancreatic juice, which helps digest food; the liver increases its output of bile; and the gallbladder contracts to release bile. Figure 14.7 summarizes the actions of gastrin, secretin, and CCK.

The Large Intestine

The **large intestine** includes the cecum, colon, rectum, and anal canal. It is larger in diameter than the small intestine (6.5 cm compared to 2.5 cm), but shorter in length. The large intestine absorbs water, salts, and some vitamins. It also stores indigestible material until it is eliminated as feces.

The **cecum,** which lies below the junction with the small intestine, is a small pouch (6 cm long) that forms the first part of the large intestine. The human cecum has a small projection called the **appendix or vermiform appendix** (*vermiform* means wormlike) (Fig. 14.8). Once thought to be a **vestigial structure,** the appendix is now known to play an important role as a reservoir of beneficial bacteria that help keep the large intestine healthy, thus helping to prevent infection.

This organ is subject to inflammation, a condition called *appendicitis.* If inflamed, the appendix should be removed before the appendix bursts, which could cause *peritonitis,* an inflammation of the lining of the abdominal cavity. Peritonitis can lead to death.

The *colon* includes the ascending colon, which goes up the right side of the body to the level of the liver; the transverse colon, which crosses the abdominal cavity just below the liver and the stomach; the descending colon, which passes down the left side of the body; and the sigmoid colon, which enters the *rectum,* the last 20 cm of the large intestine. The rectum opens at the **anus,** where *defecation,* the expulsion of feces, occurs. When feces are forced into the rectum by peristalsis, a defecation reflex occurs. The stretching of the rectal wall initiates nerve impulses to the spinal cord, and shortly thereafter, the rectal muscles contract and the anal sphincters relax (Fig. 14.9).

Ridding the body of indigestible remains is another way the digestive system helps maintain homeostasis. Feces are normally

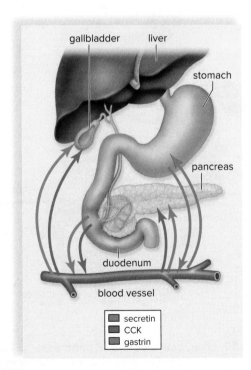

Figure 14.7 Hormonal control of digestive gland secretions. Gastrin (blue), produced by the lower part of the stomach, enters the bloodstream and thereafter stimulates the stomach to produce more digestive juices. Secretin (green) and CCK (purple), produced by the duodenal wall, stimulate the pancreas to secrete its digestive juices and the gallbladder to release bile.

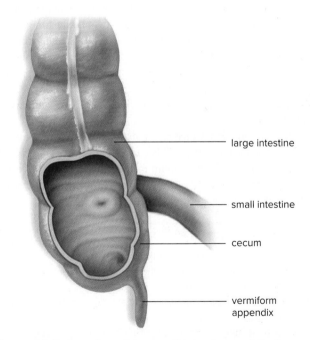

Figure 14.8 Junction of the small intestine and the large intestine.   The cecum is the blind end of the large intestine. The appendix is attached to the cecum.

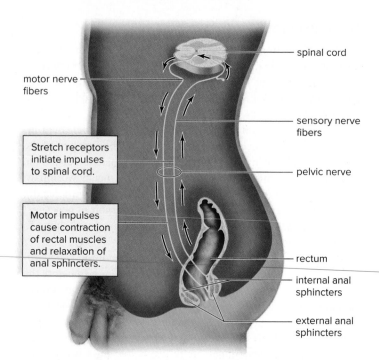

Figure 14.9 Defecation reflex. The accumulation of feces in the rectum causes it to stretch, which initiates a reflex action resulting in rectal contraction and expulsion of the fecal material.

about three-quarters water and one-quarter solids. Bacteria, **fiber** (indigestible plant material), and other indigestible materials are in the solid portion. Bacterial action on undigested materials causes the odor of feces and also accounts for the presence of gas. A breakdown product of bilirubin (described in Section 14.2) and the presence of oxidized iron cause the brown color of feces.

About 40–50% of the fecal mass consists of bacteria and other microbes; in fact, there are about 100 billion bacteria per gram of feces. These include facultative bacteria (which can live with or without oxygen) like *Escherichia coli,* as well as many obligate anaerobes (which die in the presence of oxygen). These bacteria break down some indigestible material (such as fiber), and produce some vitamins that our bodies can absorb and use. In this way, they perform a service for us, and we provide a good environment for them to live in.

Water is considered unsafe for drinking when the coliform (intestinal) bacterial count exceeds a threshold level. A high count indicates that a significant amount of feces has entered the water. The more feces present, the greater the possibility that disease-causing bacteria are also present.

Check Your Progress 14.1

1. List the parts of the human digestive tract that food passes through, and note whether chemical and/or mechanical digestion occurs in each.
2. Compare the major functions of the small intestine and the large intestine.
3. Discuss strategies that might help obese people lose weight by inhibiting the secretion and/or activity of certain hormones.

14.2 Accessory Organs of Digestion

Learning Outcomes

Upon completion of this section, you should be able to
1. Summarize the major functions of the pancreas, the liver, and the gallbladder.
2. Describe the structure and function of the hepatic portal system.

Several other organs play an important role in digestion. These are called accessory organs because they do not process food directly, but rather contribute enzymes and other substances to the digestive tract via ducts. The pancreas, liver, gallbladder, and salivary glands are accessory digestive organs. The salivary glands (see Section 14.1) contribute saliva and enzymes to the mouth. Figure 14.10*a* shows

a.

b.

Figure 14.10 Liver, gallbladder, and pancreas. a. The liver makes bile, which is stored in the gallbladder and sent (black arrow) to the small intestine by way of the common bile duct. The pancreas produces digestive enzymes that are sent (black arrows) to the small intestine by way of the pancreatic duct. **b.** A hepatic lobule. The liver contains over 100,000 lobules, each lobule composed of many cells that perform the various functions of the liver. They remove and add materials to the blood and deposit bile in a duct.

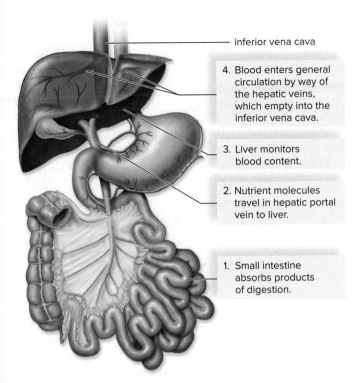

inferior vena cava

4. Blood enters general circulation by way of the hepatic veins, which empty into the inferior vena cava.

3. Liver monitors blood content.

2. Nutrient molecules travel in hepatic portal vein to liver.

1. Small intestine absorbs products of digestion.

Figure 14.11 Hepatic portal system. The hepatic portal vein takes the products of digestion from the digestive system to the liver, where they are processed before entering a hepatic vein.

the location of the pancreas, liver, and gallbladder, and the ducts that deliver the contents of these organs to the duodenum.

The Liver

The **liver,** which is the largest gland in the body, lies mainly in the upper right section of the abdominal cavity, under the diaphragm (see Fig. 14.1). The liver contains approximately 100,000 lobules that serve as its structural and functional units (Fig. 14.10b). Three structures are located between the lobules: a bile duct that takes bile away from the liver; a branch of the hepatic artery that brings O_2-rich blood to the liver; and a branch of the hepatic portal vein that transports nutrients from the intestines. Each lobule has a central vein that enters a hepatic vein. In Figure 14.11, trace the path of blood from the intestines to the liver via the hepatic portal vein and from the liver to the inferior vena cava via the hepatic veins.

In some ways, the liver acts as the gatekeeper to the blood (Table 14.2). As blood from the hepatic portal vein passes through the liver, it removes poisonous substances and detoxifies them. The liver also removes and stores iron and vitamins A, D, E, K, and B_{12}. The liver makes many of the plasma proteins and helps regulate the quantity of cholesterol in the blood.

The liver maintains the blood glucose level at 50–80 mg/100 ml, even though a person eats intermittently. When insulin is present, any excess glucose in the blood is removed and stored by the liver (and muscles) as glycogen. Between meals, glycogen is broken down to glucose, which enters the hepatic veins, and in this way, the blood glucose level remains relatively constant.

If the supply of glycogen is depleted, the liver converts glycerol (from fats) and amino acids to glucose molecules. The

TABLE 14.2 Functions of the Liver
Detoxifies blood by removing and metabolizing poisonous substances
Stores iron (Fe^{2+}) and vitamins A, D, E, K, and B_{12}
Makes many plasma proteins, such as albumins and fibrinogen, from amino acids
Stores glucose as glycogen after a meal, and breaks down glycogen to glucose to maintain the glucose concentration of blood between eating periods
Produces urea after breaking down amino acids
Removes bilirubin, a breakdown product of hemoglobin from the blood, and excretes it in bile, a liver product
Helps regulate blood cholesterol level, converting some to bile salts

conversion of amino acids to glucose necessitates deamination, which involves the removal of amino groups. By a complex metabolic pathway, the liver then combines ammonia with carbon dioxide to form *urea.* Urea is the usual nitrogenous waste product from amino acid breakdown in humans (see Section 16.1).

The liver produces **bile,** which is stored in the gallbladder. Bile has a yellowish-green color because it contains the bile pigment *bilirubin,* derived from the breakdown of hemoglobin, the red pigment in red blood cells. Bile also contains bile salts. Bile salts are derived from cholesterol, and they emulsify fat in the small intestine. When fat is emulsified, it breaks up into droplets, providing a much larger surface area that can be acted upon by pancreatic lipase (see Section 14.3).

The Pancreas

The **pancreas** lies deep in the abdominal cavity, resting on the posterior abdominal wall. It is an elongated and somewhat flattened organ that has both endocrine and exocrine functions. As an exocrine gland, most pancreatic cells produce pancreatic juice, which contains sodium bicarbonate ($NaHCO_3$) to help neutralize the stomach acid, and digestive enzymes for all types of food (see Section 14.3). In its function as an endocrine gland, the pancreas secretes insulin and glucagon, hormones that help keep the blood glucose level within normal limits (see Section 20.5).

The Gallbladder

The **gallbladder** is a pear-shaped, muscular sac attached to the surface of the liver (see Fig. 14.1). The liver produces about 400–800 ml of bile each day, and any excess is stored in the gallbladder. Water is reabsorbed by the gallbladder so that bile becomes a thick, mucuslike material. When needed, bile leaves the gallbladder and proceeds to the duodenum via the common bile duct.

Check Your Progress 14.2

1. Review what it means to say that the pancreas has both endocrine and exocrine functions.
2. List what each accessory gland contributes to the process of digestion.
3. Explain why the liver is necessary for life, whereas the gallbladder can be removed with few consequences.

14.3 Digestive Enzymes

Digestive enzymes, like other enzymes, are proteins that speed up specific chemical reactions. Table 14.3 lists some of the major enzymes produced by the digestive tract. These enzymes help break down carbohydrates, proteins, nucleic acids, and fats, the major components of food.

Starch is a carbohydrate, and its digestion begins in the mouth. Saliva from the salivary glands has a neutral pH and contains **salivary amylase**, the first enzyme to act on starch:

$$\text{starch} + H_2O \xrightarrow{\text{salivary amylase}} \text{maltose}$$

In this equation, salivary amylase is written above the arrow to indicate it is neither a reactant nor a product in the reaction. It merely speeds the reaction in which its substrate, starch, is digested to many molecules of maltose, a disaccharide. Maltose molecules cannot be absorbed by the intestine. Additional digestive action in the small intestine converts maltose to glucose, a monosaccharide, which then can be absorbed.

Protein digestion begins in the stomach. Gastric juice secreted by gastric glands is very acidic (pH of about 2) because it contains hydrochloric acid (HCl). Pepsinogen, a precursor that is converted to the enzyme **pepsin** when exposed to HCl, is also present in gastric juice. Pepsin acts on proteins, which are polymers of amino acids, to produce peptides:

$$\text{protein} + H_2O \xrightarrow{\text{pepsin}} \text{peptides}$$

Peptides vary in length, but they always consist of a number of linked amino acids. Peptides from the stomach are usually too large to be absorbed by the intestinal lining, but later in the small intestine, they are broken down to amino acids, the monomers of a protein molecule.

Pancreatic juice, which enters the duodenum, has a basic pH because it contains sodium bicarbonate ($NaHCO_3$). Sodium bicarbonate neutralizes the acid in chyme, producing the slightly basic pH that is optimum for pancreatic enzymes, including **pancreatic amylase,** which digests starch (Fig. 14.12*a*), and **trypsin,** which digests protein (Fig. 14.12*b*). Trypsin is secreted as trypsinogen, which is converted to trypsin in the duodenum.

Lipase, a third pancreatic enzyme, digests fat molecules in the fat droplets after they have been emulsified by bile salts (Fig 14.12*c*):

$$\text{fat} \xrightarrow{\text{bile salts}} \text{fat droplets}$$
$$\text{fat droplets} + H_2O \xrightarrow{\text{lipase}} \text{glycerol} + 3 \text{ fatty acids}$$

Fats are triglycerides, which means each one is composed of glycerol and three fatty acids. The end products of lipase digestion—glycerol and fatty acid molecules—are small enough to cross the cells of the intestinal villi, where absorption takes place. As mentioned in Section 14.1, glycerol and fatty acids enter the cells of the villi, and within these cells, they are rejoined and packaged as lipoprotein droplets before entering the lacteals (see Fig. 14.6).

Peptidases and **maltase,** enzymes secreted by the surface cells of the small intestinal villi, complete the digestion of protein to amino acids and starch to glucose, respectively. These small molecules can then be absorbed by the cells of the villi. Peptides, which result from the first step in protein digestion, are digested to amino acids by peptidases:

$$\text{peptides} + H_2O \xrightarrow{\text{peptidases}} \text{amino acids}$$

Maltose, a disaccharide that results from the first step in starch digestion, is digested to glucose by maltase:

$$\text{maltose} + H_2O \xrightarrow{\text{maltase}} \text{glucose} + \text{glucose}$$

TABLE 14.3   Major Digestive Enzymes

Enzyme	Produced By	Site of Action	Optimum pH	Digestion
Salivary amylase	Salivary glands	Mouth	Neutral	Starch + H_2O ⟶ maltose
Pancreatic amylase	Pancreas	Small intestine	Basic	Starch + H_2O ⟶ maltose
Maltase	Small intestine	Small intestine	Basic	Maltose + H_2O ⟶ glucose + glucose
Pepsin	Gastric glands	Stomach	Acidic	Protein + H_2O ⟶ peptides
Trypsin	Pancreas	Small intestine	Basic	Protein + H_2O ⟶ peptides
Peptidases	Small intestine	Small intestine	Basic	Peptide + H_2O ⟶ amino acids
Nuclease	Pancreas	Small intestine	Basic	RNA and DNA + H_2O ⟶ nucleotides
Nucleosidases	Small intestine	Small intestine	Basic	Nucleotide + H_2O ⟶ base + sugar + phosphate
Lipase	Pancreas	Small intestine	Basic	Fat droplet + H_2O ⟶ glycerol + fatty acids

Figure 14.12 Digestion and absorption of nutrients. **a.** The breakdown of carbohydrates, such as starch, involves amylase enzymes. **b.** Protein digestion involves the action of protease enzymes. **c.** For fat digestion, bile salts emulsify the fats so lipase enzymes can digest the particles.

a. Carbohydrate digestion

b. Protein digestion

c. Fat digestion

Other disaccharides, each of which has its own enzyme, are digested in the small intestine. The absence of any one of these enzymes can cause illness. The most common situation involves a deficiency of lactase, the enzyme that digests lactose, the sugar found in milk. Interestingly, the rate of this enzyme deficiency varies between human races. For example, only 25% of adult Caucasians, but 75–90% of adult African Americans, are lactase-deficient. Consuming dairy products often gives these individuals the symptoms of *lactose intolerance* (diarrhea, gas, cramps), caused by the fermentation of nondigested lactose by intestinal bacteria. However, lactose-reduced products are available for most kinds of dairy products, including milk, cheese, and ice cream. It is important to include foods that are high in calcium in the diet. Otherwise, osteoporosis, a condition characterized by weak and fragile bones, may develop (see Section 19.6).

As mentioned in Section 6.3, enzymes function best at an optimum temperature and pH, which helps them maintain the proper shape to fit their substrate. The compartmentalization of the digestive system assists in the regulation of enzymatic activity. Because the entire digestive system of humans is maintained at a constant temperature of 37°C (98.6°F), enzymatic activity is largely controlled by pH. As discussed in Section 14.1, the pH of the stomach is between 1 and 2, allowing the activation of pepsinogen enzyme. However, shortly after chyme proceeds into the duodenum (small intestine), the sodium bicarbonate released from the pancreas returns the pH to around 7.4 to 7.8. This "switch" serves to activate digestive enzymes released by the pancreas, thus allowing the digestion of carbohydrates, fats, and proteins.

Check Your Progress 14.3

1. Describe where in the digestive tract the chemical digestion of each of the following types of nutrients occurs: carbohydrates, proteins, and fats.
2. Identify the final molecule (monomer) resulting from the digestion of carbohydrates, proteins, and fats.
3. Explain how the structure of the digestive system assists in regulating the digestive enzymes.

14.4 Human Nutrition

Learning Outcomes

Upon completion of this section, you should be able to

1. List the six major classes of nutrients and their major functions in the body.
2. Describe the relationship between diet and obesity, type 2 diabetes, and cardiovascular disease.
3. Distinguish between vitamins, coenzymes, and minerals.

Every aspect of your body's functioning depends on proper *nutrition,* the intake of nutrients. A *nutrient* is a component of food that is utilized by the body. The six major classes of nutrients are carbohydrates, proteins, lipids, vitamins, minerals, and water. All of these serve essential roles in the body, but overconsumption of almost any nutrient can also lead to health problems.

Food choices to fulfill nutrient needs constitute the *diet.* In June 2011, the U.S. Department of Agriculture (USDA) replaced its 20-year-old symbol of a healthy diet, the food pyramid, with the MyPlate symbol (Fig. 14.13). The MyPlate diagram replaced numerical recommendations of how much of different types of foods (grains, vegetables, fruits, milk, meat, and beans) with a simpler image of a plate symbolizing the need to eat a well-balanced diet.

Obesity

The most common definition of **obesity** in use today is having a body mass index (BMI) of 30 or greater. The body mass index uses a person's height and weight to calculate a general approximation of their percent body fat. This calculation is done by dividing weight in pounds (lb) by height in inches (in.) squared and multiplying by a conversion factor of 703.

$$\text{weight (1bs)/height (in.)}^2 \times 703$$

Figure 14.13 The MyPlate icon.   The U.S. Department of
Agriculture (USDA) released this symbol in 2011 as a guide to better health.
It replaces the more complicated food pyramid models, and simply suggests
eating more balanced portions. Additional information and recommendations
are available at ChooseMyPlate.gov. *Source:* USDA, ChooseMyPlate.gov website

Most people find it is easier to use the table (Fig. 14.14), or
one of the variety of online calculators.

As a general rule:

> A healthy BMI = 18.5 to 24.9;
> An overweight BMI = 25 to 29.9;
> An obese BMI = 30 to 39.9;
> A morbidly obese BMI = 40 or higher.

Keep in mind, however, the BMI is only a general guide. In
general, the taller you are, the more you can weigh without being
overweight or obese. The BMI value doesn't take into account factors like fitness, bone structure, or gender. For example, a weight-lifter might have a high BMI, not because of excess body fat, but
because of increased bone and muscle weight.

According to the Centers for Disease Control and Prevention,
obesity rates in the United States have doubled in the last few decades. As of 2016, about 38% of U.S. adults, and 17% of children
and adolescents aged 2 to 19 years, were obese.

Obesity is most likely caused by a combination of behavioral,
genetic, metabolic, and social factors. Recent studies have also indicated that the composition of bacteria in your large intestine,
called the *bacterial flora* or *microbiota,* may contribute to obesity.
Some people are obese because they suffer from an eating disorder
(see Section 14.5); others may simply have a long-term pattern of

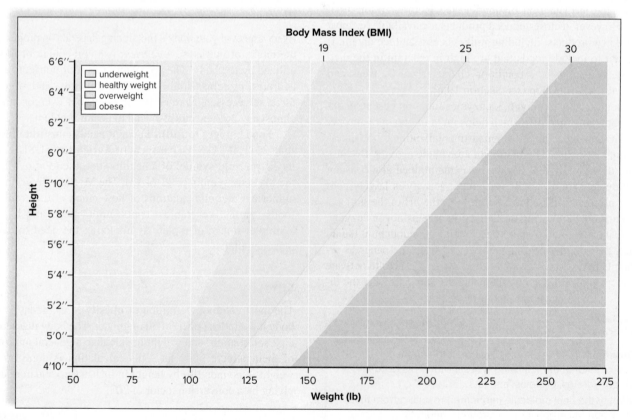

Figure 14.14 Body mass index (BMI).   To calculate your BMI, match your weight with your height, then determine your body mass index (BMI).
Healthy BMI = 18.5 to 24.9; overweight BMI = 25 to 29.9; obese BMI = 30 to 39.9; morbidly obese BMI = 40 or more. *Source:* U.S. Department of Agriculture,
Dietary Guidelines for Americans, 2005.

overeating. People are certainly more sedentary than they used to be and they eat more processed foods, which can contain more calories than meals prepared from fresh foods at home. It is known that obese individuals have more fat cells than normal, and when they lose weight, the fat cells simply get smaller; they don't disappear. An inability to produce sufficient amounts of a satiety hormone, called leptin, might also be involved.

Regardless of the cause, obesity is associated with a higher risk of type 2 diabetes, hypertension, cardiovascular disease, respiratory dysfunction, osteoarthritis, and other conditions. Obesity is one of our nation's most critical health problems and is responsible for an estimated 300,000 deaths each year.

The treatment depends on the degree of obesity. For those who are severely obese, possible treatments include surgery to remove body fat, or even to bypass part of the stomach. But for most people, a lifelong commitment to a healthy diet and regular exercise is the best way to prevent obesity.

Major Classes of Nutrients

Carbohydrates

Carbohydrates are present in food in the form of complex polysaccharides like starch and fiber, and simple sugars like glucose and sucrose (table sugar). Glucose is the quickest, most readily available source of energy for the body. As you will recall from Section 7.1, cells use glucose for the production of ATP by cellular respiration.

As mentioned in Section 14.2, excess glucose is stored by the liver and muscles in the form of glycogen. However, the human liver and muscles can store a total of only about 600 grams of glucose in the form of glycogen; excess glucose is converted into fat and stored in adipose tissues. Between meals, the blood glucose level is maintained at about 50–80 mg/100 ml of blood by the breakdown of glycogen or by the conversion of fat or amino acids to glucose. While most body cells can utilize fatty acids as an energy source, brain cells require a steady supply of glucose.

If you, or someone you know, have lost weight by following a "low-carb" diet, you may think that carbohydrates are unhealthy and should always be avoided. However, according to the American Dietetic Association, some low-carbohydrate, high-fat diets have no benefits over well-balanced diets that include the same number of calories. In fact, a study of over 440 Canadian adults found the lowest risk of obesity in people who consumed about half of their calories from carbohydrates.[1] On the other hand, evidence suggests that many Americans are not eating the right kinds of carbohydrates.

The typical American obtains about one-sixth of his or her daily caloric intake from simple sugars found in foods like candy, cookies, and ice cream. High fructose corn syrup, the most commonly used sweetening agent, is found in soft drinks and a huge variety of other foods. These sugars immediately enter the bloodstream, as do those derived from the digestion of the starch in white bread and potatoes. These foods are said to have a high **glycemic index,** because the blood glucose response to these foods is high. When the blood glucose level rises rapidly, the pancreas

TABLE 14.4	Reducing High-Glycemic Index Carbohydrates

To reduce dietary sugar:
Eat fewer sweets, such as candy, soft drinks, ice cream, and pastry.
Eat fresh fruits or fruits canned without heavy syrup.
Use less sugar—white, brown, or raw—and less honey and syrups.
Avoid sweetened breakfast cereals.
Eat less jelly, jam, and preserves.
Avoid potatoes and processed foods made from refined carbohydrates, such as white bread, rice, and pasta.

produces an overload of insulin to bring the level under control. Cells take up glucose in response to insulin. Hunger quickly returns after eating high-glycemic index foods. A chronically high insulin level may have many harmful effects, such as increased fat deposition and a high blood fatty acid level. Over the years, the body's cells can become insulin-resistant, resulting in type 2 diabetes and other problems.

Table 14.4 gives suggestions on how to reduce your intake of high-glycemic-index carbohydrates.

These recommendations are reinforced by a research study published in June 2011,[2] which combined data from three separate studies of 120,877 U.S. women and men over a 20-year period. The researchers assessed the effects of various dietary and lifestyle choices on body weight. Instead of following people who were already obese and trying to lose weight, this study focused on the weight gain that often occurs gradually over decades. After adjusting for lifestyle factors like physical activity, alcohol use, and smoking, those dietary factors that had the greatest effect on body weight over time are shown in Figure 14.15. The biggest cause of weight gain was potatoes—especially potato chips and french fries. Specifically, adding just a single serving of potato chips a day was associated with a weight gain of 1.69 pounds over four years. Meats, whether processed or not, were also associated with weight gain, as were sweets and refined grains. Vegetables, nuts, whole grains, fruits, and even yogurt were all associated with weight loss.

Complex carbohydrates, like those found in whole-grain foods, are recommended because they are slowly digested to sugars and they contain fiber. **Fiber** includes various indigestible carbohydrates derived from plants. Food sources rich in fiber include beans, peas, nuts, fruits, and vegetables. The typical American consumes only about 15 g of fiber each day; the recommended daily intake is 25 g for women and 38 g for men.

Technically, fiber is not a nutrient for humans because it cannot be digested into smaller molecules and absorbed into the blood. However, insoluble fiber, such as that found in wheat bran, stimulates the movement of food through the large intestine, and may possibly guard against colon cancer by limiting the amount of time cancer-causing substances are in contact with the intestinal wall.

[1] Merchant, A. T., et al. 2009. Carbohydrate Intake and Overweight-Obesity Among Healthy Adults. *Journal of the American Dietetic Association* 109:1165–72.

[2] Mozaffaria, D., et al. 2011. Changes in Diet and Lifestyle and Long-Term Weight Gain in Women and Men. *N Engl J Med.* 364:2392–2404.

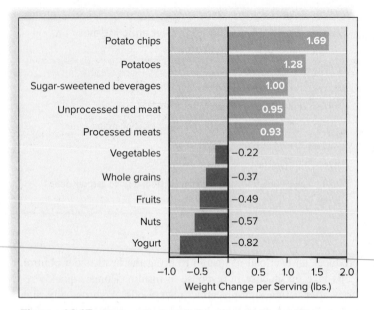

Figure 14.15 Dietary factors influencing body weight. Values indicate the average weight gain, or loss, over time by adding a single daily serving of each item.

Soluble fiber, such as that found in oat bran, combines with cholesterol in the small intestine and prevents it from being absorbed. While the benefits of fiber are clear, some evidence suggests that a diet too high in fiber can be detrimental, possibly impairing the body's ability to absorb iron, zinc, and calcium.

Proteins

Dietary proteins are digested to amino acids, which cells use to synthesize many different types of proteins. Foods rich in protein include red meat, fish, poultry, dairy products, legumes (i.e., peas and beans), nuts, and cereals. Following digestion of protein, amino acids enter the bloodstream and are transported to the tissues. Ordinarily, amino acids are not used as an energy source. Most are incorporated into structural proteins found in muscles, skin, hair, and nails. Others are used to synthesize such proteins as hemoglobin, plasma proteins, enzymes, and hormones.

Adequate protein formation requires 20 different types of amino acids. Of these, eight are required from the diet in adults (nine in children) because the body is unable to produce them. These are termed the **essential amino acids.** The body is capable of producing the other amino acids by simply transforming one type into another type.

Some protein sources, such as meat, milk, and eggs, are complete; they provide all 20 types of amino acids in the approximate amounts needed by the human body. The amino acid composition of other dietary protein sources varies considerably. For example, corn is a poor source of the amino acids lysine and tryptophan. However, high-lysine corn can now be produced through genetic modification of the plant (see Section 26.2). Legumes (beans and peas), seeds, nuts, and grains can be good protein sources but some contain insufficient amounts of at least one essential amino acid. Therefore, vegetarians may be advised to combine two or more incomplete types of plant products to acquire all the essential amino acids.

Amino acids are not stored in the body; thus, a regular supply is needed. However, it does not take very much protein to meet the daily requirement. Two servings of meat a day (one serving is equal in size to a deck of cards) are usually plenty. While the body is harmed if the amount of protein in the diet is severely limited, it is also likely to be harmed by an overabundance of protein. Deamination of excess amino acids in the liver results in urea, our main nitrogenous excretion product. The water needed for excretion of urea can cause dehydration when a person is exercising and losing water by sweating. High-protein diets can also increase calcium loss in urine, which may increase the risk of kidney stones and osteoporosis.

Certain types of meat, especially red meat, are known to be high in saturated fats, while other sources, such as chicken, fish, nuts, and plant proteins, are more likely to be low in saturated fats. As discussed next, a high intake of saturated fats is associated with a greater risk of cardiovascular disease.

Lipids

Lipids include fats, oils, and cholesterol. Fats are a source of energy for the body; 1 gram (g) of fat contains nine calories, which is more than twice the amount found in carbohydrates or protein. *Saturated fats,* which are solid at room temperature, usually have an animal origin. Two well-known exceptions are palm oil and coconut oil, vegetable oils that contain mostly saturated fats. Butter and fats associated with meats (like the fat on steak and bacon) are also high in saturated fats. Saturated fats are particularly associated with cardiovascular disease.

Oils contain *unsaturated fatty acids,* which can be further characterized as monounsaturated and polyunsaturated fatty acids. Corn oil and safflower oil are high in polyunsaturated fatty acids. Polyunsaturated oils are nutritionally essential because they are the only type of fat that contains *linoleic acid* and *linolenic acid,* two fatty acids the body cannot make. Because these fatty acids must be supplied by diet, they are called **essential fatty acids.**

Olive or canola oils contain a larger percentage of monounsaturated fatty acids than other types of cooking oils. In particular, the omega-3 fatty acids are believed to be especially protective against heart disease. Some cold-water fishes, such as salmon, sardines, and trout, are rich sources of omega-3, but they contain only about half as much as flaxseed oil, the best source, from plants.

Perhaps the unhealthiest type of saturated fats are the *trans fatty acids* (often called *trans fats*), which arise when unsaturated fatty acids are hydrogenated to produce a solid fat. Found largely in commercially produced products, trans fats may reduce the function of the plasma membrane receptors that clear cholesterol from the bloodstream. Trans fatty acids are commonly found in cookies and crackers, commercially fried foods, such as french fries from some fast-food chains, and packaged snacks, such as microwave popcorn, as well as in vegetable shortening and some margarines. Any packaged goods that list "partially hydrogenated vegetable oils" or "shortening" on their label likely contain trans fats.

As noted in Section 12.5, cardiovascular disease is often due to blockage of arteries by *plaque,* which contains saturated fats and cholesterol. Cholesterol is carried in the blood by two types of

SCIENCE IN YOUR LIFE ▶ HEALTH

New Dietary Guidelines

Dietary guidelines are typically revised by the U.S. government every 5 years to reflect changes in nutrition science. The latest guidelines were released in 2015 by the Departments of Agriculture and Health and Human Services. The overall purposes of these guidelines were to

1. Promote health.
2. Prevent chronic (long-term) disease.
3. Assist people in reaching and maintaining a healthy weight.

The new guidelines focus less on prescribing quantitative levels for nutrients, and more on establishing healthy eating patterns. These patterns include the following foods:

1. A variety of vegetables, including leafy vegetables, beans, red and yellow vegetables, and starches.
2. Fruits.
3. Grains. At least half of all grains should be whole grains.
4. Fat-free or low-fat dairy products (including soy).
5. Proteins in the form of seafood, lean meats, poultry, eggs, legumes, nuts, and soy products.
6. Oils.

To establish these healthy eating patterns, specific recommendations were made to limit certain nutrients recognized as raising health concerns:

1. Consume less than 10% of calories per day from added sugar.
2. Consume less than 10% of calories per day from saturated fats.
3. Consume less than 2,300 milligrams (mg) per day of sodium.

4. Alcohol should be consumed only in moderation. A maximum of one drink per day for women and up to two drinks per day for men (and only by adults of legal drinking age).

For more information on these guidelines, including the science of how they were determined, visit health.gov/dietaryguidelines/2015/guidelines/.

Questions to Consider

1. Why do you think quantitative values were removed for many of the nutrients?
2. Reflect on your diet over the past 24 hours. How does your diet relate to these general guidelines?

---

lipoproteins: low-density lipoprotein (LDL) and high-density lipoprotein (HDL). LDL is thought of as "bad" because it carries cholesterol from the liver to the cells, while HDL is thought of as "good" because it carries cholesterol to the liver, which takes it up and converts it to bile salts. Saturated fats tend to raise LDL cholesterol levels, whereas unsaturated fats lower LDL cholesterol levels. Figure 14.16 illustrates the amount of saturated fat,

monounsaturated fat, and polyunsaturated fat found in some common cooking items.

Table 14.5 suggests ways to reduce dietary saturated fat and cholesterol. It is not a good idea to rely on commercially produced low-fat foods to reduce total fat intake. In some low-fat products, sugars have replaced the fat, meaning the calorie content can still be high. And even if the caloric content is low, consumers may

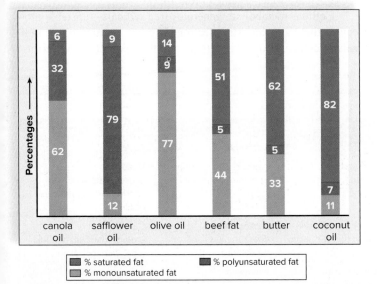

Figure 14.16 Lipid content of selected oils and fats. The consumption of a diet high in saturated fats and cholesterol has been associated with disorders of the cardiovascular and other systems.

TABLE 14.5	Reducing Certain Lipids
To reduce saturated fats and trans fats in the diet:
Choose poultry, fish, or dry beans and peas as a protein source.
Remove skin from poultry and trim fat from red meats before cooking.
Broil, boil, or bake rather than fry.
Limit your intake of butter, cream, trans fats, shortenings, and tropical oils (coconut and palm oils).*
Use herbs and spices to season foods instead of butter or margarine.
Drink skim milk instead of whole milk, and use skim milk in cooking and baking.
To reduce dietary cholesterol:
Limit cheese, liver, and certain shellfish (shrimp and lobster). Preferably, eat white fish and poultry.
Substitute egg whites for egg yolks in both cooking and eating.
Include soluble fiber in the diet. Good sources are oat bran, oatmeal, beans, corn, and fruits such as apples, citrus fruits, and cranberries.

* Although coconut and palm oils are from plant sources, they are mostly saturated fats.

c. Scurvy

a. Rickets

b. Pellagra

Figure 14.17 **Illnesses due to vitamin deficiency.** **a.** Bowing of bones (rickets) due to vitamin D deficiency. **b.** Dermatitis (pellagra) of areas exposed to light due to niacin deficiency. **c.** Bleeding of gums (scurvy) due to vitamin C deficiency. (a, c): ©Biophoto Associates/Science Source; (b): ©Dr. M.A. Ansary/ Science Source

believe they can eat more of a product that is low fat. Manufacturers have also developed fake fats (e.g., olestra), which are not absorbed by the intestine. Nutritionists also express concern about these fake fats, however, because high levels of intake can lead to abdominal cramping, diarrhea, and reduced absorption of fat-soluble vitamins. In addition, recent research suggests that fat substitutes can interfere with the body's ability to regulate the intake of fat, which can result in overeating.

Vitamins

Vitamins are organic compounds (other than carbohydrates, fats, and proteins) that the body needs for metabolic purposes but is unable to produce. Many vitamins serve as coenzymes, which are enzyme helpers. For example, niacin is part of the coenzyme NAD^+, and riboflavin is part of the coenzyme FAD. Our bodies need coenzymes in only small amounts because each can be used over and over again. Not all vitamins are coenzymes; vitamin A, for example, is a precursor for the visual pigment that prevents night blindness. A variety of diseases may occur if vitamins are lacking in the diet (Fig. 14.17). Altogether, there are 13 vitamins, which are divided into those that are fat-soluble (Table 14.6) and those that are water-soluble (Table 14.7).

Cellular metabolism generates free radicals—unstable molecules that carry an extra electron. The most common free radicals in cells are superoxide (O_2^-) and hydroxide (OH^-). Because

TABLE 14.6 Fat-Soluble Vitamins

Vitamin	Functions	Food Sources	Conditions With	
			Deficiency	*Excess*
Vitamin A	Antioxidant synthesized from beta-carotene; needed for healthy eyes, skin, hair, and mucous membranes, and for proper bone growth	Deep yellow/orange and dark, leafy green vegetables, fruits, cheese, whole milk, butter, eggs	Night blindness, impaired growth of bones and teeth	Headache, dizziness, nausea, hair loss, abnormal development of fetus
Vitamin D	A group of steroids needed for development and maintenance of bones and teeth	Milk fortified with vitamin D, fish liver oil; also made in the skin when exposed to sunlight	Rickets, bone decalcification and weakening	Calcification of soft tissues, diarrhea, possible renal damage
Vitamin E	Antioxidant that prevents oxidation of vitamin A and polyunsaturated fatty acids	Leafy green vegetables, fruits, vegetable oils, nuts, whole-grain breads and cereals	Unknown	Diarrhea, nausea, headaches, fatigue, muscle weakness
Vitamin K	Needed for synthesis of substances active in clotting of blood	Leafy green vegetables, cabbage, cauliflower	Easy bruising and bleeding	Can interfere with anticoagulant medication

TABLE 14.7   Water-Soluble Vitamins

Vitamin	Functions	Food Sources	Conditions With Deficiency	Conditions With Excess
Vitamin C	Antioxidant; needed for forming collagen; helps maintain capillaries, bones, and teeth	Citrus fruits, leafy green vegetables, tomatoes, potatoes, cabbage	Scurvy, delayed wound healing, infections	Gout, kidney stones, diarrhea, decreased copper
Thiamine (vitamin B_1)	Part of coenzyme needed for cellular respiration; also promotes activity of the nervous system	Whole-grain cereals, dried beans and peas, sunflower seeds, nuts	Beriberi, muscular weakness, enlarged heart	Can interfere with absorption of other vitamins
Riboflavin (vitamin B_2)	Part of coenzymes, such as FAD; aids cellular respiration, including oxidation of protein and fat	Nuts, dairy products, whole-grain cereals, poultry, leafy green vegetables	Dermatitis, blurred vision, growth failure	Unknown
Niacin (nicotinic acid)	Part of coenzymes NAD^+ and NADP; needed for cellular respiration, including oxidation of protein and fat	Peanuts, poultry, whole-grain cereals, leafy green vegetables, beans	Pellagra, diarrhea, mental disorders	High blood sugar and uric acid, vasodilation, etc.
Folacin (folic acid)	Coenzyme needed for production of hemoglobin and formation of DNA	Dark, leafy green vegetables, nuts, beans, whole-grain cereals	Megaloblastic anemia, spina bifida	May mask B_{12} deficiency
Vitamin B_6	Coenzyme needed for synthesis of hormones and hemoglobin; CNS control	Whole-grain cereals, bananas, beans, poultry, nuts, leafy green vegetables	Rarely, convulsions, vomiting, seborrhea, muscular weakness	Insomnia, neuropathy
Pantothenic acid	Part of coenzyme A needed for oxidation of carbohydrates and fats; aids in the formation of hormones and certain neurotransmitters	Nuts, beans, dark, leafy green vegetables, poultry, fruits, milk	Rarely, loss of appetite, mental depression, numbness	Unknown
Vitamin B_{12}	Complex, cobalt-containing compound; part of the coenzyme needed for synthesis of nucleic acids and myelin	Dairy products, fish, poultry, eggs, fortified cereals	Pernicious anemia	Unknown
Biotin	Coenzyme needed for metabolism of amino acids and fatty acids	Generally in foods, especially eggs	Skin rash, nausea, fatigue	Unknown

they are very reactive, free radicals combine with and damage DNA, proteins (including enzymes), or lipids, which are found in plasma membranes. Vitamins C, E, and A are believed to defend the body against free radicals, and therefore, they are termed *antioxidants.* These vitamins are especially abundant in fruits and vegetables.

One vitamin that has received a lot of attention in the popular media is vitamin D (cholecalciferol). Skin cells contain a precursor cholesterol molecule that is converted to vitamin D after UV exposure. Vitamin D leaves the skin and is modified first in the kidneys and then in the liver until finally it becomes calcitriol. Calcitriol promotes the absorption of calcium by the intestines. The lack of vitamin D leads to rickets in children (Fig. 14.17a). Rickets, characterized by bowing of the legs, is caused by defective mineralization of the skeleton. Since the 1930s, most milk has been fortified with vitamin D, which helps prevent the occurrence of rickets.

Minerals

In addition to vitamins, the body requires various inorganic chemical elements known as **minerals.** Minerals required as nutrients are divided into major minerals and trace minerals. The body contains more than 5 g of each major mineral and less than 5 g of each trace mineral (Fig. 14.18). The major minerals are constituents of cells and body fluids and are structural components of tissues. For example, calcium (present as Ca^{2+}) is needed for the construction

a. Major minerals

b. Trace minerals

Figure 14.18 Minerals in the body. These charts show the usual amount of certain minerals in a 60-kg (135-lb) person. The functions of minerals are given in Table 14.8. **a.** The major minerals are present in amounts larger than 5 g (about a teaspoon). **b.** Trace minerals are present in lesser amounts (notice the smaller scale).

of bones and teeth and for nerve conduction and muscle contraction. Phosphorus (present as PO_4^{3-}) is stored in the bones and teeth and is a part of phospholipids, ATP, and the nucleic acids. Potassium (K^+) is the major positive ion inside cells and is important in nerve conduction and muscle contraction, as is sodium (Na^+). Sodium also plays a major role in regulating the body's water balance, as does chloride (Cl^-). Magnesium (Mg^{2+}) is critical to the functioning of hundreds of enzymes.

The trace minerals are parts of larger molecules. For example, iron (Fe^{2+}) is present in hemoglobin, and iodine (I^-) is a part of thyroxine and triiodothyronine, hormones produced by the thyroid gland. Zinc (Zn^{2+}), copper (Cu^{2+}), and manganese (Mn^{2+}) are present in enzymes that catalyze a variety of reactions. Proteins called zinc-finger proteins, some of which contain zinc, bind to DNA when a particular gene is to be activated. As research continues, more and more elements are added to the list of trace minerals considered essential. During the past 3 decades, for example, very small amounts of selenium, molybdenum, chromium, nickel, vanadium, silicon, and even arsenic have been found to be essential to good health. Table 14.8 lists the functions of various minerals and gives their food sources and signs of deficiency and toxicity.

Occasionally, individuals do not receive enough iron, calcium, magnesium, or zinc in their diets. Adult females need more iron in the diet than males (18 mg compared to 10 mg) because they lose hemoglobin each month during menstruation. Due to its high-fiber content, a vegetarian diet may make zinc less available to the body. However, a varied and complete diet usually supplies enough of each type of mineral.

Diet and Osteoporosis

Many people take calcium supplements to counteract **osteoporosis,** a degenerative bone disease that afflicts an estimated one-fourth of older men and one-half of older women in the United States. Osteoporosis develops because bone-eating cells called osteoclasts become more active than bone-forming cells called osteoblasts. Therefore, the bones become porous, and they break easily because they lack sufficient calcium. Due to recent studies that show consuming more calcium does slow bone loss in elderly people, the guidelines have been revised. Recommended calcium intake varies by age, but in general 1,000 mg a day is recommended for men and women. After age 50 in women, and 70 in men, this increases to 1,200 mg a day. For many people, calcium supplements are needed to obtain these levels.

Because it promotes calcium absorption, vitamin D is an essential companion to calcium in preventing osteoporosis. Other

TABLE 14.8   Minerals

Mineral	Functions	Food Sources	Conditions With	
			Deficiency	*Excess*
Major minerals (more than 100 mg/day needed)				
Calcium (Ca^{2+})	Strong bones and teeth, nerve conduction, muscle contraction	Dairy products, leafy green vegetables	Stunted growth in children, low bone density in adults	Kidney stones; interferes with iron and zinc absorption
Phosphorus (PO_4^{3-})	Bone and soft tissue growth; part of phospholipids, ATP, and nucleic acids	Meat, dairy products, sunflower seeds, food additives	Weakness, confusion, pain in bones and joints	Low blood and bone calcium levels
Potassium (K^+)	Nerve conduction, muscle contraction	Many fruits and vegetables, bran	Paralysis, irregular heartbeat, eventual death	Vomiting, heart attack, death
Sodium (Na^+)	Nerve conduction, pH and water balance	Table salt	Lethargy, muscle cramps, loss of appetite	Edema, high blood pressure
Chloride (Cl^-)	Water balance	Table salt	Not likely	Vomiting, dehydration
Magnesium (Mg^{2+})	Part of various enzymes for nerve and muscle contraction, protein synthesis	Whole grains, leafy green vegetables	Muscle spasm, irregular heartbeat, convulsions, confusion, personality changes	Diarrhea
Trace minerals (less than 20 mg/day needed)				
Zinc (Zn^{2+})	Protein synthesis, wound healing, fetal development and growth, immune function	Meats, legumes, whole grains	Delayed wound healing, night blindness, diarrhea, mental lethargy	Anemia, diarrhea, vomiting, renal failure, abnormal cholesterol levels
Iron (Fe^{2+})	Hemoglobin synthesis	Whole grains, meats, prune juice	Anemia, physical and mental sluggishness	Iron toxicity disease, organ failure, eventual death
Copper (Cu^{2+})	Hemoglobin synthesis	Meat, nuts, legumes	Anemia, stunted growth in children	Damage to internal organs if not excreted
Iodine (I^-)	Thyroid hormone synthesis	Iodized table salt, seafood	Thyroid deficiency	Depressed thyroid function, anxiety
Selenium (SeO_4^{2-})	Part of antioxidant enzyme	Seafood, meats, eggs	Vascular collapse, possible cancer development	Hair and fingernail loss, discolored skin

TABLE 14.9 Reducing Dietary Sodium

To reduce dietary sodium:

1. Use spices instead of salt to flavor foods.
2. Add little or no salt to foods at the table, and add only small amounts of salt when you cook.
3. Eat unsalted crackers, pretzels, potato chips, nuts, and popcorn.
4. Avoid hot dogs, ham, bacon, luncheon meats, smoked salmon, sardines, and anchovies.
5. Avoid processed cheese and canned or dehydrated soups.
6. Avoid brine-soaked foods, such as pickles or olives.
7. Read labels to avoid high-salt products.

nutrients may also be helpful. For example, magnesium has been found to suppress the cycle that leads to bone loss. Along with adequate calcium and vitamin intake, regular exercise also helps to prevent osteoporosis. Factors that increase the risk of osteoporosis include smoking and a high daily caffeine intake. Medications are also available that slow bone loss while increasing skeletal mass. These are still being studied for their effectiveness and possible side effects.

Sodium and Hypertension

The recommended amount of sodium intake per day is less than 2,400 mg, although the average American takes in 4,000–4,700 mg every day. High sodium intake has been linked to hypertension (high blood pressure) in some people. About one-third of the sodium we consume occurs naturally in foods; another one-third is added during commercial processing; and we add the last one-third either during home cooking or at the table in the form of table salt. Table 14.9 gives recommendations for reducing sodium intake.

Dietary Supplements

Dietary supplements are nutrients and plant products (such as herbal teas) that are used to enhance health. The U.S. government does not require dietary supplements to undergo the same safety and effectiveness testing that new prescription drugs must complete before they are approved. Therefore, many of these products have not been tested scientifically to determine their benefits. People often think herbal products are safe because they are "natural," but many plants, such as lobelia, comfrey, and kava kava, can be poisonous.

Supplements can be useful to correct a deficiency, but most people in the United States take them to greatly increase the dietary intake of a particular nutrient. However, ingesting high levels of certain nutrients can cause harm. As mentioned earlier in this section, ingesting too much protein can be harmful, so most nutritionists do not recommend taking protein or amino acid supplements. Most fat-soluble vitamins are stored in the body and can accumulate to toxic levels, particularly vitamins A and D. Certain minerals can also be harmful, even deadly, when ingested in high doses.

Food Additives

Food additives are substances that are added to foods as preservatives, or to enhance flavor or appearance. There are hundreds of food additives, but among the most interesting and controversial are the dyes added to make food look more appealing.

Studies have shown that color is an important psychological part of tasting food. If tasteless yellow coloring is added to vanilla pudding, most consumers report it tastes like banana or lemon. Those who study food marketing know that gray pickles, colorless candy, or tan gelatin desserts would not sell as well. For example, when offered cheese-flavored snacks lacking orange-colored dye, most people rated their taste as bland.

Nine synthetic dyes (mostly derived from petroleum) are currently approved by the FDA for use in food. The most common of these is "Allura Red," also called Red No. 40. It is added to all sorts of candies, flavored drinks, frostings, and jellies. "Brilliant Blue" (Blue No. 1) is found in frozen treats, baked goods, sweets, drinks, and even soaps, mouthwash, and cosmetics.

While there is no doubt about the psychological effect of food coloring, the influence on health is less certain. In March 2011, an FDA advisory panel met to determine whether food dyes may contribute to hyperactivity in children. Many European countries have banned certain dyes, including Blue No. 1, because of this possibility. However, after reviewing the available studies, the panel concluded that the link between food color additives and hyperactivity has not been established. But other experts, have reached a different conclusion. For example, in 2008 the American Academy of Pediatrics suggested that a low-additive diet is a valid intervention for children with ADHD.

Check Your Progress 14.4

1. Discuss what it means to say that a particular food has a high glycemic index.
2. Predict some specific consequences of consuming a diet that is either deficient in protein, or too high in protein.
3. Explain the difference between a vitamin and a mineral in the diet.
4. Describe how a dietary supplement may be harmful for some vitamins or minerals.

14.5 Eating Disorders

Learning Outcomes

Upon completion of this section, you should be able to

1. Describe the three main categories of eating disorders.
2. Discuss some of the psychological and social factors that can lead to eating disorders.
3. Distinguish between compulsive eating, binge eating, and pica.

People with *eating disorders* have attitudes and behaviors toward food that are outside the norm, which causes a serious disturbance to their diet. Anyone can develop an eating disorder, regardless of ethnicity, socioeconomic status, or intelligence level. Eating

a. Anorexia nervosa   b. Bulimia nervosa   c. Pica

Figure 14.19 Characteristics of some eating disorders.   Some common types of eating disorders. **a.** People with anorexia nervosa often think they are overweight, even if they are quite thin. **b.** Those with bulimia nervosa overeat and then purge their bodies of the food they have eaten. **c.** People suffering from pica eat unusual, sometimes bizarre nonfood items. (a): ©Ted Foxx/Alamy; (b): ©Donna Day/Stone/Getty Images; (c): ©McGraw-Hill Education

disorders are usually treated with a combination of medication and psychological counseling. If untreated, the most severe cases can be fatal.

According to the National Institute of Mental Health (NIMH), types of eating disorders include anorexia nervosa, bulimia nervosa, and eating disorders not otherwise specified (EDNOS). The latter category, which includes all eating disorders not meeting the criteria for anorexia nervosa or bulimia nervosa, is the most common diagnosis among people who seek treatment.

Anorexia Nervosa

Anorexia nervosa is a severe psychological disorder characterized by an irrational fear of getting fat. No matter how thin they become, people with this condition think they are overweight. This may be a form of body dysmorphic disorder, in which a person becomes preoccupied with a perceived body defect (Fig. 14.19a). A self-imposed starvation diet is often accompanied by purging episodes involving self-induced vomiting and laxative abuse. Anorexics may also engage in extreme physical activity to avoid weight gain. About 90% of people suffering from anorexia nervosa are young women; an estimated 1 in 200 teenage girls is affected. Athletes such as distance runners, wrestlers, and dancers are at an increased risk because they believe that being thin gives them a competitive edge.

A person with anorexia nervosa has symptoms of starvation, including low blood pressure, irregular heartbeat, constipation, and constantly feeling cold. Decreased bone density, with associated stress fractures, may occur. The body begins to shut down; menstruation ceases in females; the internal organs, including the brain, don't function well; and the skin becomes dry. Impairment of the pancreas and digestive tract may interfere with proper

utilization of any food that is consumed. Death may be imminent. If so, the only recourse may be hospitalization and force-feeding. Psychological counseling may be necessary to convince the person to eat properly.

Bulimia Nervosa

A person with **bulimia nervosa** binge-eats and then purges (vomits) to avoid gaining weight (Fig. 14.19b). Bulimia can coexist with either obesity or anorexia nervosa. According to the NIMH, about 4% of young women, and 1% of males, suffer from bulimia. The desire for a slender body in females or a manly physique in males may lead to excessive dieting, which can be a precursor to bulimia nervosa and anorexia nervosa.

As with anorexia, bulimic individuals are overly concerned about their body shape and weight, and therefore, they may be on a very restrictive diet. In bulimia, the restrictive diet brings on the desire to binge, typically on high-calorie foods like cakes, cookies, and ice cream. Then, a feeling of guilt most likely brings on the next phase, a purging of all the calories that have been taken in.

Bulimia is dangerous to a person's health. The chronic vomiting can alter the blood, leading to an abnormal heart rhythm, and damage to the kidneys can even result in death. At the very least, repeated vomiting leads to inflammation of the pharynx and esophagus, and stomach acids can cause the teeth to erode. The esophagus and stomach may even rupture and tear due to strong contractions during repeated vomiting.

The most important aspect of treatment is to get the patient on a sensible and consistent diet. As with anorexia, psychotherapy can help the patient understand the emotional causes of the behavior. Medications, including antidepressants, have sometimes been effective in reducing the bulimic cycle and restoring normal appetite.

Binge Eating Disorder

Affecting an estimated 3% of adults in the United States, binge eating disorder is the most common eating disorder. People with this disorder frequently eat large amounts of food while feeling a loss of control over their eating. However, they usually do not purge afterward by vomiting or using laxatives. Therefore, although they are not as likely to suffer from immediate medical problems, people with this disorder are often overweight or obese.

The cause of binge eating disorder is unknown. Studies suggest that people with binge eating disorder may have trouble handling some of their emotions. Many people who are binge eaters say that being angry, sad, bored, worried, or stressed can cause them to binge eat. As many as half of all people with binge eating disorder are depressed or have been depressed in the past. Whether depression causes binge eating disorder, or whether binge eating disorder causes depression, is not known. Other research suggests that genes may be involved in binge eating, because the disorder often occurs in several members of the same family. This research is still in the early stages.

Pica

Pica is defined as repeated ingestion of nonfood items, such as ash, charcoal, cotton, soap, soil, wool, or many other items (Fig. 14.19*c*). In most cases, pica affects young children, and stops before adolescence. Pica is often associated with intellectual disability, lack of proper parental care, and neglect. Children with pica are at greater risk of various poisoning (especially lead), as well as parasitic infections.

Adults can also suffer from pica, as documented on the TV show *My Strange Addiction.* On one episode called "Plastic Snacks," an 18-year-old female describes eating plastic forks, cocktail swords, water bottles, CD cases, and other items. She has consumed over 150 pounds of plastic, and prefers it over food. Other episodes have described people who eat chalk, rocks, dirt, their own hair, or even glass. In adults, pica may be a result of nutritional deficiencies or a behavioral response to stress.

Check Your Progress 14.5

1. Compare and contrast anorexia nervosa, bulimia nervosa, binge eating disorder, and pica.
2. Predict some physiological factors that could contribute to an eating disorder.

14.6 Disorders of the Digestive System

Learning Outcomes

Upon completion of this section, you should be able to

1. Describe a common disorder that affects each part of the digestive system.
2. Classify the digestive disorders by the type of cause (e.g., infectious, cancer, inflammatory, etc.).

Disorders of the digestive system can be grouped into two categories: disorders of the tract itself, and disorders of the accessory organs.

Disorders of the Digestive Tract

Stomach Ulcers

A thick layer of mucus normally protects the wall of the stomach. If this protective layer is broken down, the stomach wall can be damaged by the acidic pH of the stomach, resulting in a **stomach ulcer,** or open sore in the wall, caused by a gradual disintegration of the tissue. It now appears that most stomach ulcers are initiated by infection of the stomach by a bacterium called *Helicobacter pylori* (see the Scientific Inquiry feature "A Bacterial Culprit for Stomach Ulcers"), which can impair the ability of the mucous cells to produce protective mucus. Thus, treatment of stomach ulcers now typically includes antibiotics to kill these bacteria, along with medication to reduce the amount of acid in the stomach. Other potential causes of stomach ulcers include certain viral infections, as well as the overuse of anti-inflammatory medications, which can have the side effect of damaging the stomach lining. Duodenal ulcers are also common because the duodenum receives acidic chyme from the stomach.

Intestinal Disorders

The most common problems associated with the small and large intestines are diarrhea and constipation. **Diarrhea,** or loose, watery feces, can be acute or chronic. Most cases of acute (sudden-onset) diarrhea are due to infections of the small or large intestines with any of several bacterial, viral, or protozoal agents. Collectively known as food poisoning, symptoms of these infections can range from a few hours of stomach and bowel discomfort, to life-threatening infections such as cholera, salmonellosis, and the more toxic forms of *E. coli.* In all cases, the intestinal wall becomes irritated by the infection, or by toxins produced by the organism, and peristalsis increases. Less water is absorbed, and the resulting diarrhea helps rid the body of the infectious organisms. Severe diarrhea can lead to dehydration because of water loss and to disturbances in the heart's contraction due to an imbalance of salts in the blood.

As opposed to intestinal infections that come and go, some people are afflicted with chronic diarrhea. About 7 in 100,000 people in the United States suffer from *Crohn's disease,* which is a persistent inflammation of the intestine that results in recurrent bouts of abdominal cramping (which may be severe) and bloody diarrhea. Usually diagnosed between ages 15 and 30, it seems to be caused by a misdirected immune response against one's own intestinal tissues, or perhaps against some of the normal bacteria that live in the gut. The incidence of Crohn's disease is higher among Caucasians compared to other races, and among wealthier, urban populations compared to poorer, rural ones. Genetic predisposition is certainly a factor, as are several environmental triggers, such as smoking and diet.

The opposite of diarrhea is **constipation.** When a person is constipated, the feces are dry and hard. One reason for this condition is that some people have learned to inhibit the defecation reflex to the point that the urge to defecate is ignored. Chronic constipation can lead to the development of *hemorrhoids,* which are enlarged and inflamed blood vessels of the anus. Increasing the amount of water and fiber in the diet can help prevent constipation. Laxatives and enemas can also be used, but frequent use is discouraged because they can be irritating to the colon.

SCIENCE IN YOUR LIFE ▶ SCIENTIFIC INQUIRY

A Bacterial Culprit for Stomach Ulcers

A stomach (gastric) ulcer is a gaping, sometimes bloody, indentation in the stomach lining (Fig. 14A). Australian physician Barry Marshall didn't believe the accepted explanation that the major causes of stomach ulcers were stress or spicy foods. Instead, he and his colleague, Dr. Robin Warren, thought most stomach ulcers were due to infection with a bacterial species called *Helicobacter pylori,* which they had discovered in 1982. At that time, most physicians disagreed with their hypothesis, because they didn't believe any bacteria could survive in the very acidic conditions of the stomach. However, Marshall and Warren, a pathologist, had observed that nearly every biopsy of stomach ulcer tissue contained the corkscrew-shaped *H. pylori* bacteria.

To get the data they needed, Dr. Marshall decided to use himself as a guinea pig. One day in July 1984, he drank a culture containing *H. pylori.* About a week later, he started vomiting and suffering the painful symptoms of an ulcer. Medical tests confirmed that his stomach lining was teeming with bacteria. In a few weeks, his ulcer had disappeared, but the experiment had served its purpose. In recognition of his efforts, in 2005, Marshall and Warren were awarded the Nobel Prize in Physiology or

healthy gastric mucosa

gastric ulcer

Figure 14A   Stomach ulcer.   Drs. Barry Marshall and Robin Warren won the 2005 Nobel Prize in Physiology or Medicine for discovering that most stomach ulcers are caused by a bacterial infection. ©J. James/Science Source

Medicine for their discovery of *H. pylori* and its role in ulcers and other stomach ailments.

Prior to Dr. Marshall's bold experiment, ulcers were treated mainly with drugs to reduce the acidity of the stomach. Patients got temporary relief, but the ulcers tended to return, and serious complications could arise. Some patients had bleeding ulcers and underwent an operation to remove a portion of the stomach. Thanks to the work of Marshall and Warren, in 1994 the National Institutes of Health endorsed antibiotics as a standard treatment for stomach ulcers. An international effort is underway among scientists to develop a vaccine.

Questions to Consider

1. What kinds of adaptations would *H. pylori* need in order to survive in the acid pH of the stomach?
2. Why do you suppose that even after Dr. Marshall's experiment in 1984, it took 10 years for his findings to be widely accepted by the medical community?
3. Do you think Dr. Marshall's experiment was ethical? If so, would it always be okay for scientists to experiment on themselves? If not, what would be a more ethical way to prove that *H. pylori* could cause ulcers?

Polyps and Colon Cancer

The colon is subject to the development of **polyps,** which are small growths arising from the epithelial lining. They may be benign or cancerous. Polyps are usually detected using a procedure known as colonoscopy, during which a long, flexible endoscope is inserted into the colon to visualize the wall of the large intestine. If colon cancer is detected and surgically removed while it is still confined to a polyp, a complete cure is expected.

Some investigators believe that dietary fat increases the likelihood of colon cancer because dietary fat causes an increase in bile secretion. Certain intestinal bacteria may convert bile salts to substances that promote the development of cancer. On the other hand, fiber in the diet seems to inhibit the development of colon cancer, possibly by diluting the concentration of bile salts and facilitating the movement of cancer-inducing substances through the intestine.

Disorders of the Accessory Organs

Disorders of the Pancreas

Pancreatitis is an inflammation of the pancreas. It can be caused by drinking too much alcohol, by gallstones that block the pancreatic duct, or by other unknown factors. In chronic pancreatitis, the

digestive enzymes normally secreted by the pancreas can damage the pancreas and surrounding tissues. Eventually, chronic pancreatitis can decrease the ability of the pancreas to secrete insulin, potentially leading to diabetes.

Unfortunately, **pancreatic cancer** is almost always fatal. Only about 20% of patients are alive 1 year after this diagnosis. Reasons for this include the essential functions of the pancreas, the resistance of many forms of the disease to treatment, and the tendency of the cancer cells to spread to other organs before any symptoms appear. Steve Jobs, Apple CEO, died in 2011 at age 56 after a 7-year battle with pancreatic cancer.

Disorders of the Liver and Gallbladder

The liver performs many functions that are essential to life; therefore, diseases that affect the liver can be life-threatening. A person who has a liver ailment may develop *jaundice,* which is a yellowish coloring in the whites of the eyes, as well as in the skin of light-pigmented persons. Jaundice can be caused by an abnormally large amount of bilirubin in the blood. **Hepatitis,** or inflammation of the liver, is most commonly caused by one of several viruses. Hepatitis A virus is usually acquired by consuming water or food that has been contaminated by sewage. Hepatitis B virus is usually spread by sexual contact, but can also be acquired through blood transfusions or contaminated needles. The hepatitis B virus is more contagious than the AIDS virus, which is spread in similar ways. Thankfully, however, a vaccine is available to prevent hepatitis B. Hepatitis C virus is spread by the same routes as hepatitis B, and it can lead to chronic hepatitis, liver cancer, and death. Unfortunately, no vaccine is available for hepatitis C, although antiviral drugs can be effective.

Cirrhosis is another chronic disease of the liver. It is often seen in alcoholics due to malnutrition and to the toxic effects of consuming excessive amounts of alcohol, which acts as a toxin on the liver. In cirrhosis, the liver first becomes infiltrated with fat, and then the fatty liver tissue is replaced by fibrous scar tissue. Although the liver has amazing regenerative powers, if the rate of damage exceeds the rate of regeneration, there may not be enough time for the liver to repair itself. The preferred treatment in most cases of liver failure is liver transplantation, but the supply of livers currently cannot meet the demand. Artificial livers have been developed, but with very limited success so far.

In some individuals, the cholesterol present in bile can come out of solution and form crystals. If the crystals grow in size, they form **gallstones.** The passage of the stones from the gallbladder may block the common bile duct and cause obstructive jaundice. If this happens, the obstruction can sometimes be relieved; otherwise, the gallbladder must be removed. Interestingly, gallstone formation is particularly common in individuals who have lost a significant amount of weight in a short time, such as those who have undergone gastric bypass procedures.

Check Your Progress 14.6

1. Identify the specific cause of most stomach ulcers.
2. Explain why the pancreas and liver are such important organs to the health of an individual.

Conclusion

The structure of the digestive system is well-suited for its role in the processing of nutrients that our body needs for growth, repair, and development. However, like any organ system in the body, it is susceptible to disease. Such is the case of celiac disease and the small intestine.

Celiac disease occurs when the body mistakenly identifies a nutrient, in this case the protein gluten, as a foreign pathogen. As a result, the body begins a defense response against the gluten protein, causing inflammation and damage to the villi and microvilli lining the walls of the small intestine. In this chapter, we have seen how the small intestine plays an important role in the processing of nutrients, and how any disruption in that function can cause problems for the entire body.

SUMMARIZE

14.1 The Digestive Tract

The **digestive system** is involved in the ingestion and digestion of food and elimination of indigestible material. It includes the following parts:

- The **mouth** takes food into the body. **Salivary glands** send saliva into the mouth, where the teeth chew the food and the tongue forms a bolus for swallowing.
- The **pharynx** is where the air passage and food passage cross. When a person swallows, the air passage is usually blocked off by the epiglottis.
- The **esophagus** is a muscular tube that conducts food from the pharynx to the stomach.
- Beginning with the esophagus, each component of the digestive tract surrounding the central cavity, or **lumen,** has four layers. Muscular contractions called **peristalsis** move food through the tract.
- The **stomach** expands and stores food. When food enters the stomach, the stomach churns, mixing food with the acidic gastric juice. At this point, the food is known as **chyme.**
- The **small intestine** is made up of the **duodenum,** the **jejunum,** and the **ileum.** The duodenum receives bile from the liver and pancreatic juice from the pancreas. The walls of the small intestine have finger-like projections called **villi,** where small nutrient molecules are absorbed. Amino acids and glucose enter the blood vessels of a villus. Glycerol and fatty acids are joined and packaged as lipoproteins before entering lymphatic vessels called **lacteals** in the villi.

- The **large intestine** begins at the **cecum,** which has a pouch called the **appendix (vermiform appendix)** attached. The large intestine continues as the colon and rectum, and ends at the **anus.** The large intestine absorbs water, salts, and some vitamins.

14.2 Accessory Organs of Digestion

Three accessory organs of digestion send secretions to the duodenum via ducts:

- The **pancreas** produces pancreatic juice, which contains sodium bicarbonate and enzymes that chemically digest starch, protein, and fat.
- The **liver** has many important functions. It receives blood from the small intestine by way of the hepatic portal vein and helps to detoxify any harmful substances present. It monitors blood glucose concentrations and helps eliminate excess nitrogen from the body. It also produces many plasma proteins, as well as **bile,** which helps to emulsify (break apart) fats in the intestine.
- The **gallbladder** stores bile produced by the liver, until it is secreted into the duodenum through the common bile duct. Bile emulsifies fat and readies it for digestion by lipase.

14.3 Digestive Enzymes

Like all enzymes, digestive enzymes are specific to their substrate and speed up specific reactions at optimum body temperature and pH. They break down food into smaller molecules like glucose, amino acids, fatty acids, and glycerol, which can be absorbed. As food passes through the digestive tract:

- In the mouth, **salivary amylase** begins digestion of starch.
- In the stomach, **pepsin** begins digesting protein to peptides.

- In the small intestine, **pancreatic amylase, trypsin,** and **lipase** digest starch, protein, and fat. **Peptidases** and **maltase** finish the digestion of protein and starch, respectively.

14.4 Human Nutrition

The nutrients released by the digestive process should provide us with an adequate amount of energy, essential amino acids and fatty acids, and all necessary vitamins and minerals. However, some recommendations can be given, based on scientific studies:

- Minimize your intake of carbohydrates with a high **glycemic index** (simple sugars and refined starches) because they cause a rapid release of insulin that can lead to health problems. An adequate intake of indigestible carbohydrates in the form of **fiber** also promotes health.
- Proteins supply us with **essential amino acids,** but it is wise to avoid meats that are fatty because fats from animal sources are saturated. While unsaturated fatty acids, particularly the omega-3 fatty acids, are protective against cardiovascular disease, the saturated fatty acids lead to plaque, which occludes blood vessels. The body cannot synthesize **essential fatty acids,** which must be supplied by the diet.
- Aside from carbohydrates, proteins, and fats, the body requires vitamins and minerals. **Vitamins** A, E, and C are antioxidants that protect cell contents from damage due to free radicals. The **mineral** calcium is needed to prevent the bone disease **osteoporosis.**

14.5 Eating Disorders

- An eating disorder is an abnormal behavior that causes significant changes in the diet. These changes often have severe health consequences for the individual. **Anorexia nervosa** and **bulimia nervosa** are examples of common eating disorders.

14.6 Disorders of the Digestive System

Various conditions can affect the function of the digestive tract, as well as the accessory organs.

- Most **stomach ulcers** are now thought to be caused by a bacterial infection and thus can be cured with antibiotics.
- **Diarrhea** is an increased frequency and looseness of the feces. Acute diarrhea is most frequently caused by intestinal infections. Chronic (longer-lasting) diarrhea may be due to an immune-mediated condition such as Crohn's disease. Conversely, with **constipation,** the feces become dry and hard.
- Disorders that affect the accessory digestive organs include **pancreatitis, pancreatic cancer, hepatitis,** and **cirrhosis.**

ASSESS

Testing Yourself

Choose the best answer for each question.

14.1 The Digestive Tract

1. The digestive system consists of a series of tissues surrounding a hollow space called the
   - **a.** serosa.
   - **b.** mucosa.
   - **c.** lumen.
   - **d.** muscularis.
2. The large intestine
   - **a.** digests all types of food.
   - **b.** is the longest part of the intestinal tract.
   - **c.** absorbs water.
   - **d.** is connected to the stomach.
   - **e.** is subject to hepatitis.

3. This organ may be part of the immune system and helps prevent infections in the large intestine.
   - **a.** duodenum
   - **b.** rectum
   - **c.** cecum
   - **d.** appendix
4. This structure blocks the opening to the respiratory system during swallowing.
   - **a.** esophagus
   - **b.** pharynx
   - **c.** epiglottis
   - **d.** salivary gland
5. Which of the following is incorrect regarding gastric juice?
   - **a.** It is provided by the pancreas.
   - **b.** It contains pepsin to digest protein.
   - **c.** It has a pH of 2 due to hydrochloric acid.
   - **d.** It is produced by the gastric glands.

14.2 Accessory Organs of Digestion

6. Bile
   - **a.** is an important enzyme for the digestion of fats.
   - **b.** cannot be stored.
   - **c.** is made by the gallbladder.
   - **d.** emulsifies fat.
   - **e.** All of these are correct.
7. Which of the following is not a function of the liver in adults?
   - **a.** produces bile
   - **b.** detoxifies alcohol
   - **c.** stores glucose
   - **d.** produces urea
   - **e.** makes red blood cells
8. This hormone causes the gallbladder to contract and the liver to increase its production of bile.
   - **a.** secretin
   - **b.** pepsin
   - **c.** CCK
   - **d.** gastrin

14.3 Digestive Enzymes

9. Which of these could be absorbed directly without need of digestion?
   - **a.** glucose
   - **b.** nucleic acid
   - **c.** fat
   - **d.** protein
   - **e.** All of these are correct.
10. Which association is incorrect?
    - **a.** protein—trypsin
    - **b.** fat—lipase
    - **c.** maltose—pepsin
    - **d.** starch—amylase
    - **e.** protein—pepsin
11. Which of the following nutrients are loaded into chylomicrons and transported by the lymphatic system following digestion?
    - **a.** carbohydrates
    - **b.** fats
    - **c.** proteins
    - **d.** None of these are correct.

14.4 Human Nutrition

12. Lipids
    - **a.** are never essential in the diet.
    - **b.** include fats and oils.
    - **c.** contain less energy than carbs.
    - **d.** are found only in animals, not in plants.
13. A BMI of 26.7 indicates that a person is
    - **a.** healthy.
    - **b.** obese.
    - **c.** morbidly obese.
    - **d.** underweight.
    - **e.** overweight.

14. The glycemic index is associated with which of the following classes of nutrients?
a. fats and lipids
b. carbohydrates
c. minerals
d. vitamins

14.5 Eating Disorders

15. Compared to anorexia nervosa, bulimia nervosa is *less* likely to be characterized by
a. a restrictive diet, often with excessive exercise.
b. binge eating followed by purging.
c. an obsession with body shape and weight.
d. a distorted body image, so that the person feels fat even when emaciated.
e. a health risk due to this complex.

14.6 Disorders of the Digestive System

16. Pancreatic cancer is one of the most deadly forms of cancer because
a. the pancreas is essential for life.
b. the cancer usually spreads before it is detected.
c. it is often resistant to treatment.
d. All of these are correct.

17. Which of the following may be caused by bacteria?
a. hepatitis   c. constipation
b. stomach ulcers   d. polyps

ENGAGE

Thinking Critically

1. Trace *all* of the types and locations of enzymes that might be involved in digesting a molecule of a complex carbohydrate such as starch, all the way through to the utilization of a glucose molecule by a body cell.

2. Suppose you are taking large doses of creatine, an amino acid supplement that is advertised for its ability to enhance muscle growth. Because your muscles can grow only at a limited rate, what do you suppose happens to the excess creatine that is not used for synthesis of new muscle?

3. Predict and explain the expected digestive results per test tube for this experiment.

C H A P T E R

15

Respiratory System

©wavebreakmedia/Shutterstock

BEFORE YOU BEGIN

Before beginning this chapter, take a few moments to review the following discussions:

Section 7.1 During which stage(s) of cellular respiration are oxygen used and carbon dioxide produced?

Section 11.5 How does the respiratory system contribute to homeostasis in the body?

Section 12.4 What is the path of blood through the pulmonary and systemic circuits?

What Is Asthma?

Asthma is a disease in which the airways become constricted (narrowed) and inflamed (swollen), both of which can result in difficulty breathing. The symptoms often include wheezing and shortness of breath, a frequent cough (often at night), and a feeling of being very tired or weak, especially when exercising. Over 34 million children and adults in the United States have asthma, and the incidence seems to be increasing. Experts offer various explanations for this. One hypothesis is that we may be "too clean," in the sense that we are not exposed to enough common bacteria, viruses, and parasites as children. As a result, our immune systems may react to harmless material we inhale. There are many harmful forms of air pollution, but of increasing concern are tiny, "ultrafine" particles, those less than 0.1 micrometer (µm) across, which are produced at high levels by diesel engines. These particles can bypass the normal defenses of the upper respiratory tract and end up lodging deep in the lungs, with damaging effects.

Recently, there have been breakthroughs in asthma research. In addition to environmental factors, there appear to be a number of genes that increase susceptibility to asthma. For example, one set of genes on chromosome 5 has been linked to asthma and is associated with the body's inflammatory response. Inflammation can constrict the airways and increase fluid flow to the lungs. Variations in a gene on chromosome 17 have also been shown to increase susceptibility, especially in individuals who have been exposed to respiratory viruses. Although there is not currently a cure, the identification of these genes provides hope for those with asthma.

As you read through this chapter, think about the following questions:

1. How would narrowing and swelling of the airways affect the respiratory volumes?

2. How would asthma indirectly affect the other systems of the body?

15.1 The Respiratory System

The primary function of the **respiratory system** is to allow oxygen from the air to enter the blood and let carbon dioxide from the blood exit into the air. During inspiration, or inhalation (breathing in), and expiration, or exhalation (breathing out), air is conducted toward or away from the lungs by a series of cavities, tubes, and openings (Fig. 15.1). **Ventilation,** another term for breathing, involves both inspiration and expiration.

The respiratory system works with the cardiovascular system to accomplish these homeostatic functions:

- External respiration, the exchange of gases (oxygen and carbon dioxide) between air and the blood
- Transport of gases to and from the lungs and the tissues
- Internal respiration, the exchange of gases between the blood and tissue fluid

The oxygen gas is used at the cellular level during the process of cellular respiration (Section 7.1) to produce ATP for cellular activities. The waste product, carbon dioxide, is transported by the circulatory system to the respiratory system in order for gas exchange to occur with the external environment.

The Respiratory Tract

Table 15.1 lists the structures of the respiratory system that are in the path of air as it moves from the nose to the lungs.

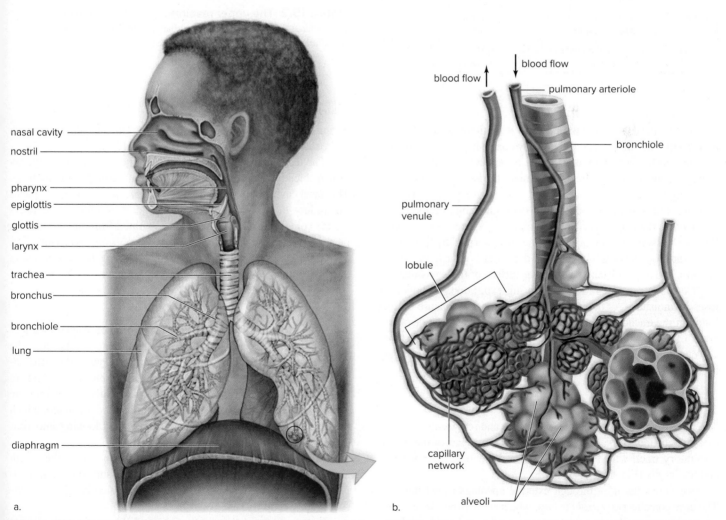

Figure 15.1 The respiratory tract.   **a.** The respiratory tract extends from the nose to the lungs, which are composed of air sacs called alveoli (arrow). **b.** Gas exchange occurs between air in the alveoli and blood within a capillary network that surrounds the alveoli. Notice that the pulmonary arteriole is colored blue—it carries O_2-poor blood away from the heart to the alveoli. Carbon dioxide then leaves the blood, and oxygen enters the blood. The pulmonary venule is colored red—it carries O_2-rich blood from the alveoli toward the heart.

TABLE 15.1	Path of Air	
Structure	**Description**	**Function**
Upper Respiratory Tract		
Nasal cavities	Hollow spaces in nose	Filter, warm, and moisten air
Pharynx	Chamber posterior to oral cavity; lies between nasal cavity and larynx	Connection to surrounding regions
Glottis	Opening into larynx	Passage of air into larynx
Larynx	Cartilaginous organ that houses the vocal cords; voice box	Sound production
Lower Respiratory Tract		
Trachea	Flexible tube that connects larynx with bronchi	Passage of air to bronchi
Bronchi	Paired tubes inferior to the trachea that enter the lungs	Passage of air to lungs
Bronchioles	Branched tubes that lead from bronchi to alveoli	Passage of air to each alveolus
Lungs	Soft, cone-shaped organs of thoracic cavity	Contain alveoli and blood vessels
Alveoli	Thin-walled microscopic air sacs in lungs	Gas exchange between air and blood

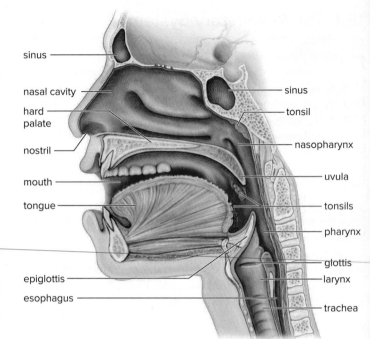

Figure 15.2 The upper respiratory tract.   The upper respiratory tract includes the nose, nasal cavities, pharynx, and larynx. Air passes through these structures, and sometimes the mouth, as it travels to and from the lower respiratory tract via the trachea.

As air moves in along the airways, it is filtered, warmed, and moistened. Filtering is accomplished by coarse hairs just inside the nostrils and by cilia and mucus in the nasal cavities and the other airways of the respiratory tract. In the nose, the hairs and cilia act as screening devices. In the trachea and other airways, the cilia beat upward, carrying mucus, dust, and occasional bits of food that "went down the wrong way" into the pharynx, where the accumulation can be swallowed or expelled. The air is warmed by heat given off by the blood vessels lying close to the surface of the lining of the airways, and it is moistened by the wet surface of these passages.

Conversely, as air moves out during expiration, it cools and loses its moisture. As the air cools, it deposits its moisture on the lining of the trachea and the nose, and the nose may even drip as a result of this condensation. The air still retains so much moisture, however, that upon expiration on a cold day, it condenses and can be seen as a small cloud.

The Nose

The **nose** is a prominent feature of the face and represents the only external portion of the respiratory system. It is a part of the upper respiratory tract, which contains the nasal cavities, the pharynx, and the larynx (Fig. 15.2).

Air enters the nose through external openings called nostrils. The nose contains two *nasal cavities,* which are narrow canals separated from one another by a septum composed of bone and cartilage. Mucous membranes line the nasal cavities. Bony ridges that project laterally into the nasal cavity increase the surface area for moistening and warming air during inhalation, and for trapping water droplets

during exhalation. Odor receptors are on the cilia of cells located high in the recesses of the nasal cavities (see Fig. 18.5).

The tear (lacrimal) glands drain into the nasal cavities by way of tear ducts. For this reason, crying often produces a runny nose. The nasal cavities also communicate with *sinuses,* air-filled spaces that reduce the weight of the skull and act as resonating chambers for the voice. The nasal cavities are separated from the mouth by a partition called the *palate,* which has two portions. Anteriorly, the *hard palate* is supported by bone, while posteriorly the *soft palate* is made of muscle tissue surrounded by a mucous membrane.

The Pharynx

The **pharynx,** also called the throat, is a funnel-shaped passageway that connects the nasal and oral cavities to the larynx. The pharynx consists of three regions: the *nasopharynx,* where the nasal cavities open posterior to the soft palate; the *oropharynx,* where the mouth opens; and the *laryngopharynx,* which opens into the larynx. The soft palate has a soft extension called the *uvula,* which projects into the oropharynx. You can see it by looking into your throat using a mirror.

The *tonsils* form a protective ring at the junction of the mouth and the pharynx. The tonsils are lymphatic tissue containing lymphocytes that protect against invasion by inhaled bacteria and viruses (see Section 13.1).

In the pharynx, the air passage and the food passage cross because the larynx, which receives air, lies above and in front of the esophagus, which receives food. The larynx leads to the trachea. Both the larynx and the trachea are normally open, allowing air to pass, but the esophagus is normally closed and opens only when a person swallows.

The Larynx

The **larynx** is a cartilaginous structure that serves as a passageway for air between the pharynx and the trachea. The larynx can be pictured as a triangular box whose apex, the Adam's apple, is at the front of the neck. The larynx is called the voice box because it houses the vocal cords. The **vocal cords** are mucosal folds supported by elastic ligaments (Fig. 15.3), and the slit between the vocal cords is an opening called the **glottis.**

When air is expelled past the vocal cords through the glottis, the vocal cords vibrate, producing sound. At the time of puberty, the growth of the larynx and the vocal cords is much more rapid and accentuated in the male than in the female, causing the male to develop a more prominent Adam's apple and a deeper voice. The voice "breaks" in the young male due to his inability to control the longer vocal cords.

The high or low pitch of the voice is regulated when speaking and singing by changing the tension on the vocal cords. The greater the tension, as when the glottis becomes more narrow, the higher the pitch. The loudness, or intensity, of the voice depends upon the amplitude of the vibrations—that is, the degree to which the vocal cords vibrate.

When food is swallowed, the larynx moves upward against the **epiglottis,** a flap of tissue that prevents food from passing into the larynx (Fig. 15.3). You can detect this movement by placing your hand gently on your larynx and swallowing.

The Trachea

The **trachea,** commonly called the *windpipe,* is a tube connecting the larynx to the primary bronchi. The trachea lies in front of the esophagus and is held open by C-shaped cartilaginous rings. The open part of the C-shaped rings faces the esophagus, and this allows the esophagus to expand when swallowing. The mucosa that lines the trachea has a layer of pseudostratified ciliated columnar epithelium (see Fig. 11.1). The cilia that project from the epithelium keep the lungs clean by sweeping mucus (produced by goblet cells) and debris toward the pharynx (Fig. 15.4). Smoking (see Section 15.4) is known to destroy these cilia, and consequently, the toxins in cigarette smoke collect in the lungs.

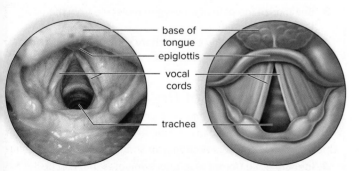

Figure 15.3 Placement of the vocal cords.   Viewed from above, the vocal cords stretch across the glottis, the opening to the trachea. When air is expelled through the glottis, the vocal cords vibrate, producing sound. The glottis is narrow when we make a high-pitched sound, and it widens as the pitch deepens. (photo) ©CNRI/Medical Images

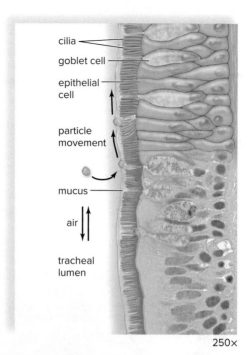

250×

Figure 15.4 The cells lining the trachea.   The lining of the trachea consists of ciliated epithelium with mucus-producing goblet cells. The mucus traps particles, and the cilia help move the mucus toward the throat. (photo): ©Ed Reschke

The Bronchial Tree

The trachea divides into right and left primary **bronchi** (*sing.,* bronchus), which lead into the right and left lungs (see Fig. 15.1). The bronchi branch into a great number of secondary bronchi that eventually lead to **bronchioles.** The bronchi resemble the trachea in structure, but as the bronchial tubes divide and subdivide, their walls become thinner, and the small rings of cartilage are no longer present. Each bronchiole leads to an elongated space enclosed by a multitude of air pockets, or sacs, called **alveoli** (*sing.,* alveolus). The components of the bronchial tree beyond the primary bronchi compose the lungs.

The Lungs

The **lungs** are paired, cone-shaped organs that occupy the thoracic cavity except for a central area that contains the trachea, the thymus, the heart, and the esophagus. The right lung has three lobes, and the left lung has two lobes, allowing room for the heart, whose apex points left. A lobe is further divided into lobules, and each lobule has a bronchiole serving many alveoli. The apex of a lung is narrow, while the base is broad and curves to fit the dome-shaped **diaphragm,** the muscle that separates the thoracic cavity from the abdominal cavity.

Each lung is covered by a very thin serous membrane called a *pleura* (see Fig. 15.8a). Another pleura covers the internal chest wall and diaphragm. Both membranes produce a lubricating serous fluid that helps the pleurae slide freely against each other during inspiration and expiration. Surface tension is the tendency for water molecules to cling to each other due to hydrogen bonding between the molecules. Surface tension holds the two pleural layers together when the lungs recoil during expiration.

Figure 15.5 Gas exchange in the lungs. The lungs consist of portions of the bronchial tree leading to the alveoli. Each alveolus is surrounded by an extensive capillary network. Notice that the pulmonary artery and arteriole carry O_2-poor blood (colored blue) and the pulmonary vein and venule carry O_2-rich blood (colored red).

Blood supply of alveoli

Capillary network of one alveolus

With each inhalation, air passes by way of the respiratory passageways to the alveoli. Each alveolus is made up of simple squamous epithelium surrounded by blood capillaries. Gas exchange occurs between the air in an alveolus and the blood in the capillaries (Fig. 15.5). Oxygen diffuses across the alveolar and capillary walls to enter the bloodstream, while carbon dioxide diffuses from the blood across these walls to enter the alveoli.

If gas exchange is to occur, the alveoli must stay open to receive the inhaled air. The surface tension of fluid coating the alveoli is capable of causing them to close up. To prevent this, the alveoli are coated with *pulmonary surfactant,* a film of lipoprotein that lowers the surface tension and prevents them from closing. The lungs collapse in some newborn babies, especially premature infants, who lack this film. The condition, called *infant respiratory distress syndrome,* is the leading cause of death in babies born prematurely. It is usually treated by inserting a plastic tube into the trachea and administering oxygen as well as surfactant that is either synthetic or extracted from cows' lungs. Severely ill patients may also receive *extracorporeal membrane oxygenation,* during which a cannula is installed in a large blood vessel, through which blood flows into a machine that adds oxygen and removes carbon dioxide prior to being returned to the patient.

Check Your Progress 15.1

1. Identify the structures that comprise the upper and lower respiratory tracts.
2. Discuss the arrangement of the trachea, esophagus, larynx, and epiglottis, which normally prevents food from entering the trachea.
3. Summarize the path of an oxygen molecule from the nose to the bloodstream.
4. Describe the function of pulmonary surfactant.

15.2 Mechanism of Breathing

Learning Outcomes

Upon completion of this section, you should be able to

1. Identify the different respiratory volumes that can be measured with a spirometer.
2. Contrast the mechanisms of inspiration and expiration.
3. Indicate how ventilation is controlled by nervous and chemical mechanisms.

During ventilation (breathing), a free flow of air is vitally important. Medical professionals use a device called a *spirometer* to record the volume of air exchanged during both normal and deep breathing (Fig 15.6*a*). Examining these breathing patterns is also a useful way to increase our understanding of normal inspiration and expiration.

Respiratory Volumes

As shown in Figure 15.6*b,* the amount of air inhaled and exhaled at rest (normally about 0.5 liters) is called the *tidal volume.* It is possible to increase the amount of air inhaled, and therefore the amount exhaled, by deep breathing. The maximum volume of air that can be moved in and out during a single breath is called the *vital capacity,* because breathing is essential to life. A number of the respiratory disorders discussed in Section 15.4 can decrease vital capacity.

When taking a very deep breath, a healthy person can increase the volume of inhaled air beyond the tidal volume by about 3.0 liters, an amount called the *inspiratory reserve volume.* Similarly, one can forcefully exhale well beyond the normal tidal volume. This *expiratory reserve volume* is usually about 1.5 liters of air. Note from Figure 15.6*b* that vital capacity is the sum of the tidal, inspiratory reserve, and expiratory reserve volumes.

These respiratory volumes depend on various factors, such as age (decreasing after age 30), gender (10–20% lower in women), and physical activity (20–30% higher in conditioned athletes). During his bicycling career, Lance Armstrong's vital capacity was over 7 liters, which is 2–3 liters higher than the average adult male's.

During normal breathing, only about 70% of the tidal volume actually reaches the alveoli; 30% remains in the airways. Also, note from Figure 15.6*b* that even after a very deep exhalation, some air (about 1,000 ml) remains in the lungs; this is called the *residual volume.* This air is not as useful for gas exchange because it has been depleted of oxygen. In some lung diseases, the residual volume increases because the individual has difficulty emptying the lungs.

Inspiration and Expiration

To understand ventilation, the manner in which air enters and exits the lungs, it is necessary to remember the following facts:

1. Normally, there is a continuous column of air from the pharynx to the alveoli of the lungs.
2. The lungs lie within the sealed-off thoracic cavity. The rib cage, consisting of the ribs joined to the vertebral column posteriorly and to the sternum anteriorly, forms the top and sides of the thoracic cavity. The intercostal muscles lie between the ribs. The diaphragm and connective tissue form the floor of the thoracic cavity.
3. The lungs adhere to the thoracic wall by way of the pleura. Normally, any space between the two pleurae is minimal due to the surface tension of the fluid between them.
4. Ventilation is governed by Boyle's Law, which states that at a constant temperature the pressure of a given quantity of gas is inversely proportional to its volume. This relationship controls the processes of inhalation and exhalation (Fig. 15.7).

Inspiration

Inspiration is the active phase of ventilation because this is the phase in which the diaphragm and the external intercostal muscles contract (Fig. 15.8*a*). In its relaxed state, the diaphragm is dome-shaped. During deep inspiration, it contracts and lowers. Also, the external intercostal muscles contract, and the rib cage moves upward and outward.

Following this contraction, the volume of the thoracic cavity is larger than it was before. As the thoracic volume increases, the lungs expand. Now the air pressure within the alveoli decreases, creating a partial vacuum. Because alveolar pressure is now less than

a.

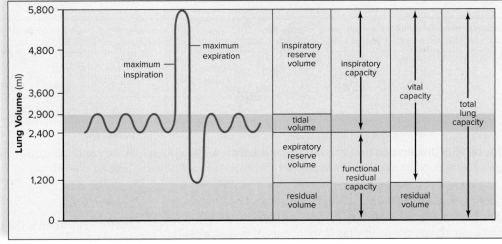

b.

Figure 15.6 Measuring ventilation.   a. A spirometer measures the amount of air inhaled and exhaled with each breath. **b.** During inspiration, the pen moves up, and during expiration, the pen moves down. Vital capacity (red) is the maximum amount of air a person can exhale after taking the deepest inhalation possible. (a): ©Yoav Levy/Medical Images

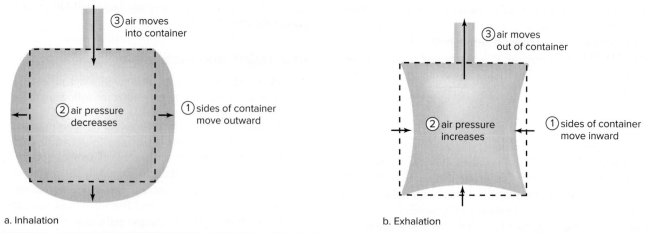

a. Inhalation

b. Exhalation

Figure 15.7 The relationship between air pressure and volume. **a.** When the sides of the container move outward, the volume in the container increases and the air pressure decreases. **b.** When the sides of the container collapse, the volume decreases and air pressure increases.

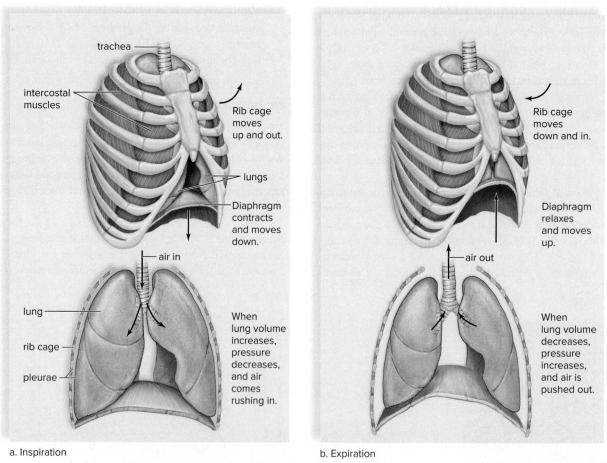

a. Inspiration

b. Expiration

Figure 15.8 Inspiration and expiration compared. **a.** During inspiration, the thoracic cavity and lungs expand so that air is drawn in. **b.** During expiration, the thoracic cavity and lungs resume their original positions and pressures. Now, air is forced out.

atmospheric pressure outside the lungs, air naturally flows from outside the body into the respiratory passages and into the alveoli.

Notice that air comes into the lungs because they have already opened up. Air does not force the lungs open. This is why it is sometimes said that humans inhale by negative pressure. While inspiration is the active phase of breathing, the actual flow of air into the alveoli is passive.

Expiration

Usually, **expiration** is the passive phase of breathing, and no effort is required to bring it about. During expiration, the elastic properties of the thoracic wall and lungs cause them to recoil. In addition, the surface tension of the fluid lining the alveoli tends to draw them closed. During expiration, the abdominal organs press up against the diaphragm, and the rib cage moves down and inward (Fig. 15.8*b*).

The diaphragm and external intercostal muscles are usually relaxed as expiration occurs. However, when breathing is deeper and/or more rapid, expiration can be active. Contraction of the internal intercostal muscles can force the rib cage to move downward and inward, and the increased pressure in the thoracic cavity helps to expel air.

Control of Breathing

Normally, adults have a resting breathing rate of 12 to 20 ventilations per minute. The rhythm of ventilation is controlled by a **respiratory center** located in the medulla oblongata of the brain.

The respiratory center stimulates inspiration by automatically sending impulses to the diaphragm by way of the phrenic nerve, and to the intercostal muscles by way of the intercostal nerves (Fig. 15.9). When the respiratory center temporarily stops sending neuronal signals to the diaphragm and the rib cage, the diaphragm relaxes and resumes its dome shape. Now expiration occurs.

Although the respiratory center automatically controls the rate and depth of breathing, its activity can also be influenced by nervous input and chemical input. Following forced inhalation, *stretch receptors* in the alveolar walls send inhibitory nerve impulses via the vagus nerve to the respiratory center, which temporarily inhibits the respiratory center from sending out nerve impulses.

Chemical Input   As explained in more detail in Section 15.3, most of the carbon dioxide (CO_2) that enters the blood combines with water, forming an acid, which breaks down and gives off hydrogen ions (H^+). The respiratory center contains cells that are sensitive to the levels of both CO_2 and H^+ in the blood. When either rises, the respiratory center increases the rate and depth of breathing. Respiration rate is also influenced by a group of cells called the *carotid bodies,* located in the carotid arteries, and the *aortic bodies,* located in the aorta, which are sensitive mainly to blood oxygen levels. When the concentration of oxygen decreases, these bodies communicate with the respiratory center, and the rate and depth of breathing increase.

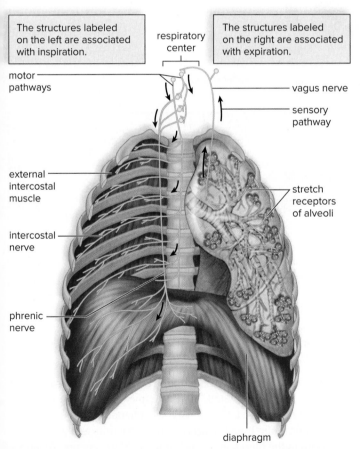

Figure 15.9 Nervous control of breathing.   The respiratory center automatically stimulates the external intercostal (rib) muscles and diaphragm to contract via the phrenic nerve. After forced inhalation, stretch receptors send inhibitory nerve impulses to the respiratory center via the vagus nerve. Usually, expiration automatically occurs due to lack of stimulation from the respiratory center to the diaphragm and intercostal muscles.

Check Your Progress 15.2

1. Compare and contrast tidal volume, vital capacity, expiratory reserve volume, and residual volume.
2. Explain why inspiration is considered the active phase of ventilation, and expiration the passive phase.
3. Discuss the roles of the following in controlling respiration: respiratory center, vagus nerve, and carotid bodies.

15.3 Gas Exchanges in the Body

Learning Outcomes

Upon completion of this section, you should be able to

1. Distinguish between external and internal respiration.
2. Interpret the chemical reaction discussed in this section to explain the effect of hyperventilation or hypoventilation on blood pH.
3. Review how the differences in P_{O_2} and P_{CO_2} in arterial and venous blood determine how gases are exchanged in the lungs versus the tissues.

Respiration includes both the exchange of gases in the lungs (external respiration), and the exchange of gases in the tissues (internal respiration) (Fig. 15.10). Most of the O_2 carried in the blood is attached to the iron-containing *heme* portion of the protein **hemoglobin** (Hb), found in red blood cells.

External Respiration

External respiration refers to the exchange of gases between air in the alveoli and blood in the pulmonary capillaries (see Fig. 15.5). Gases exert pressure, and the amount of pressure each gas exerts is called its partial pressure, symbolized as P_{O_2} and P_{CO_2}. Blood in the pulmonary capillaries has a higher P_{CO_2} than atmospheric air does. Therefore, *CO_2 diffuses out of the plasma into the lungs.* Most of the CO_2 is carried in the plasma as **bicarbonate ions** (HCO_3^-). As free CO_2 begins to diffuse out of the plasma, the enzyme **carbonic anhydrase,** present in red blood cells, speeds

Figure 15.10 External and internal respiration. During external respiration in the lungs, carbon dioxide (CO_2) leaves the blood, and oxygen (O_2) enters the blood. During internal respiration in the tissues, oxygen leaves the blood, and carbon dioxide enters the blood.

the breakdown of carbonic acid (H_2CO_3), driving this reaction to the right:

What happens if you hyperventilate (breathe at a high rate), removing more CO_2 and therefore pushing this reaction further to the right? The blood will have fewer hydrogen ions (H^+), and

respiratory alkalosis, a high blood pH, results. In that case, your breathing will be inhibited, but in the meantime, you might feel dizzy and disoriented. Conversely, what happens if you hypoventilate (hold your breath), driving this reaction to the left? H^+ builds up in the blood, and *respiratory acidosis* occurs. As previously mentioned, increased H^+ and CO_2 levels will stimulate your respiratory center, and your breathing rate will most likely increase. If not, you might become confused, sleepy, and even comatose.

The pressure pattern for O_2 during external respiration is the reverse of that for CO_2. Blood returning from the systemic capillaries

Artificial Lungs

Some organs, such as the kidneys, are relatively easy to transplant from one well-matched individual to another, with a high rate of success. Lungs are more difficult to transplant, however, with a 5-year survival rate of around 51%. However, the number of donor organs does not meet the demand, and thus a number of organizations have been actively researching the development of artificial lungs (Fig. 15A).

One type of artificial lung, called extracorporeal membrane oxygenation (ECMO), is frequently used in premature infants whose lung functions have been compromised. During ECMO, the infant's blood is pumped from the body to a device that contains a series of membranes. These membranes exchange carbon dioxide and oxygen, thus mimicking the function of a natural lung. The device acts as both a heart and a lung, allowing the infant's cardiovascular system time to respond to medical treatment.

For adults, an artificial lung called the BioLung, manufactured by MC3Corp, can help provide lung functions while the individual is awaiting a transplant or is being treated for a serious lung infection. The BioLung is attached to the right ventricle of the heart. Blood leaving the ventricle passes over a series of microfibers that exchange carbon dioxide and oxygen. The oxygenated blood is then returned to the left atrium, thus bypassing the lungs. Unlike the ECMO device, the BioLung is small (about the size of a soda can) and is powered by the contractions of the heart.

Questions to Consider

1. Why do you think the use of artificial lungs is not considered to be a long-term solution for the treatment of lung disease?
2. Besides increased availability, what are other potential advantages of laboratory-grown lungs (or other tissues) compared to regular donor tissues?

Figure 15A An artificial lung.   Artificial lungs can prolong the life of an individual until a lung transplant is available. ©Xinhua/Avalon has a lower P_{O_2} than alveolar air does. Therefore, *in the lungs O_2 diffuses into plasma and then into red blood cells.* Hemoglobin takes up this oxygen and becomes **oxyhemoglobin** (HbO_2):

$$Hb \quad + \quad O_2 \quad \longrightarrow \quad HbO_2$$

hemoglobin   **oxygen**   **oxyhemoglobin**

Interestingly, only a relatively small percentage of the oxygen present in atmospheric air is utilized during normal external respiration. At sea level, air contains about 21% O_2, and exhaled air still contains 16–17% O_2. While this system may not seem very efficient, it does explain how a person whose heart and lungs have failed can sometimes be revived by mouth-to-mouth resuscitation, which involves filling the person's lungs with exhaled air.

Internal Respiration

Internal respiration refers to the exchange of gases between the blood in systemic capillaries and the interstitial fluid. Therefore, internal respiration services tissue cells, and without internal respiration, cells could not continue to produce the ATP that allows them to exist. Blood in the systemic capillaries is a bright red color because red blood cells contain oxyhemoglobin. Oxyhemoglobin gives up O_2, which diffuses out of the blood into the tissues:

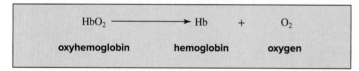

$$HbO_2 \quad \longrightarrow \quad Hb \quad + \quad O_2$$

oxyhemoglobin   **hemoglobin**   **oxygen**

Oxygen diffuses out of the blood into the tissues because the P_{O_2} of interstitial fluid is lower than that of blood. The lower P_{O_2} is due to cells continuously using up oxygen in cellular respiration. *Carbon dioxide diffuses into the blood from the tissues* because the P_{CO_2} of interstitial fluid is higher than that of blood. Carbon dioxide, produced continuously by cells, collects in interstitial fluid.

After CO_2 diffuses into the blood, it enters the red blood cells, where about 10% is taken up by hemoglobin, forming **carbaminohemoglobin** ($HbCO_2$). Most of the remaining CO_2 combines with water in the blood plasma, forming carbonic acid (H_2CO_3), which dissociates to hydrogen ions (H^+) and bicarbonate ions (HCO_3^-). The increased concentration of CO_2 in the blood drives the reaction to the right:

carbonic
anhydrase

$$CO_2 \quad + \quad H_2O \quad \longrightarrow \quad H_2CO_3 \quad \longrightarrow \quad H^+ \quad + \quad HCO_3^-$$

**carbon
dioxide**   **water**   **carbonic
acid**   **hydrogen
ion**   **bicarbonate
ion**

The enzyme carbonic anhydrase, mentioned previously, speeds up the reaction. Bicarbonate ions diffuse out of red blood cells and are carried in the plasma. The globin portion of hemoglobin combines with excess hydrogen ions produced by the overall reaction, and becomes **reduced hemoglobin** (HHb). In this way, the pH of blood remains fairly constant. Blood that leaves the

systemic capillaries is a dark maroon color because red blood cells contain reduced hemoglobin (though veins may actually look blue when seen under the skin, due to the bluish color of carbaminohemoglobin).

In either case, when blood reaches the lungs, the carbon dioxide readily diffuses out of the blood and is exhaled. In contrast, carbon monoxide (CO), a gas that comes primarily from the incomplete combustion of natural gas and gasoline, has a much greater affinity for hemoglobin than oxygen does. Therefore, it stays combined for several hours, making hemoglobin unavailable for oxygen transport. Carbon monoxide detectors in homes can help prevent accidental death due to a malfunctioning furnace or inadequate exhaust system.

Historically, diseases that caused an inability of the lungs to exchange gases (see Section 15.4) could only be treated by a lung transplant. However, as the Scientific Inquiry feature "Artificial Lungs" explains, it is becoming possible to develop artificial lungs that can assist diseased lungs, or replace them for a limited time.

Check Your Progress 15.3

1. Explain the role of hemoglobin in carrying O_2, CO_2, and hydrogen ions. Which is hemoglobin's most essential function?
2. Discuss why arterial blood is bright red in color, but venous blood is a darker blue.
3. Describe why the process of respiration is said to be a passive process.

15.4 Disorders of the Respiratory System

Learning Outcomes

Upon completion of this section, you should be able to

1. Summarize the characteristics of the disorders that affect the upper and lower respiratory tracts.
2. Classify disorders of the respiratory system according to whether they are caused by allergies, infections, a genetic defect, or exposure to toxins.

The respiratory tract is constantly exposed to the air in our environment, and thus it is susceptible to various infectious agents, as well as to pollution and, in some individuals, tobacco smoke. Because of its role in gas exchange, smoking represents one of the greatest health hazards to the respiratory system. As the Health feature "Are E-cigs Safe?" explores, the development of electronic cigarettes (e-cigs) reduces the risk, but questions still remain about the long-term effects of these devices on the respiratory system.

Disorders of the Upper Respiratory Tract

The upper respiratory tract consists of the nose and nasal cavities, the pharynx, and the larynx. Because it is responsible for filtering out many of the pathogens and other materials that may be present

Are E-cigs Safe?

For at least 50 years, the many health risks of smoking have been clear. Despite this, about 36.5 million adult Americans are smokers. Once a person starts smoking, the addictive power of nicotine is strong. But why do young people start smoking? Some may want to look mature or "cool," to be accepted by friends, or to rebel against authority. Some smokers believe the habit helps them control their weight; others admit they simply enjoy the "buzz" that nicotine can provide. Because of the unhealthy side effects of smoking, people often look for alternatives, which explains the growing popularity of electronic cigarettes (e-cigs).

E-cigs are often designed to look like real cigarettes (Fig. 15B), but instead of tobacco, they contain a cartridge filled with an "e-liquid" that consists mainly of nicotine plus propylene glycol or vegetable glycerin. When the device is used, a battery heats the liquid, turning it into a vapor that can be inhaled (explaining the popular term *vaping* for this practice). Often, an LED light at the tip glows, mimicking a lit cigarette. There's no cigarette smell, though, because no tobacco is burning.

Manufacturers claim that the vapor from an e-cig is much safer than cigarette smoke. In contrast, many health experts are concerned that the effects of inhaling pure nicotine vapor haven't yet been sufficiently studied. The nicotine content of e-liquids can be quite variable, and a variety of contaminants—including metals and trace levels of certain carcinogens—have been detected. A recent study suggests that the use of e-cigs may lower the effectiveness of the immune system in fighting disease. And while some smokers say using e-cigs helped them quit cigarettes, health authorities

worry that people who first become hooked on vaping might "graduate" to smoking.

Another concern is that although companies claim they aren't marketing to children, some nicotine solutions contain flavors—such as butterscotch, chocolate, and even cotton candy—that could appeal to the very young. An April 2014 *New York Times* article titled "Selling a Poison by the Barrel" characterized e-liquids as powerful neurotoxins, which when ingested or even absorbed through the skin can be lethal, and poisonings due to contact with or ingestion of e-liquids are on the rise.

So is vaping safer than smoking? Considering all the health risks associated with smoking, it's unlikely that vaping could be worse.

Still, until many issues with e-cigs can be sorted out, it is clearly best to follow the FDA's advice and avoid vaping as well.

Questions to Consider

1. Compared to most drugs, nicotine is unusual, because at low doses it is mainly a stimulant, but at high doses it has more sedative effects. How might these properties contribute to nicotine's addictive potential?
2. Given what you know about the role of the respiratory system in the body, what aspects of homeostasis might using an e-cig interfere with?

Figure 15B   E-cigs.   While e-cigarettes have definite advantages over regular cigarettes, research is under way to determine any health risks associated with their use. ©Martina Paraninfi/Moment/ Getty Images

---

in the air, the upper respiratory tract is susceptible to a variety of viral and bacterial infections. Upper respiratory infections can also spread from these areas to the middle ear or the sinuses.

The Common Cold

Most "colds" are relatively mild viral infections of the upper respiratory tract characterized by sneezing, a runny nose, and perhaps a mild fever. Many different viruses can cause colds, the most common being a group called the rhinoviruses (*rhin* is Greek for "nose"). Most colds last from a few days to a week, when the immune response is able to eliminate the virus. Because colds are caused by viruses, antibiotics do not help, although decongestants and anti-inflammatory medications can ease the symptoms.

Pharyngitis, Tonsillitis, and Laryngitis

Pharyngitis is an inflammation of the throat, usually because of an infection. If the tonsils have not been removed previously, they often become involved as well. What we call "strep throat" is a pharyngitis caused by the bacterium *Streptococcus pyogenes* that can lead to a generalized upper respiratory infection and even a systemic (affecting the body as a whole) infection. The symptoms of strep throat are severe sore throat, high fever, and white patches on a dark-red pharyngeal or tonsillar area (Fig. 15.11). Most cases of strep throat can be successfully treated with antibiotics.

Tonsillitis occurs when the *tonsils,* aggregates of lymphoid tissue in the pharynx, become inflamed and enlarged. The tonsils in the posterior wall of the nasopharynx are often called adenoids. If

Figure 15.11 Strep throat. Pharyngitis, often caused by the bacterium *S. pyogenes,* can lead to swollen, inflamed tonsils, sometimes with whitish patches. ©Dr. P. Marazzi/Science Source

Figure 15.12 A common treatment for middle ear infections. Tympanostomy tubes may be surgically implanted in the eardrums to drain fluid from the middle ears of children with chronic otitis media. ©Clinica Claros/Medical Images

tonsillitis occurs frequently and the enlarged tonsils make breathing difficult, the tonsils can be removed surgically in a *tonsillectomy.* Fewer tonsillectomies are performed today than in the past because we now know that the tonsils help initiate immune responses to many of the pathogens that enter the pharynx. Therefore, they are an important component of the body's immune system.

Laryngitis is an inflammation of the larynx with accompanying hoarseness, often leading to the inability to talk in an audible voice. Usually, laryngitis disappears after resting the vocal cords and treating any infection present. Sometimes benign growths, or polyps, can develop on the vocal cords. Laryngeal polyps are more likely to occur in people who put their vocal cords through excessive wear and tear, such as professional singers. Steven Tyler, lead singer of the rock band Aerosmith, as well as pop music superstar Adele, are two recent examples of singers who had to have surgery to remove polyps that were interfering with their ability to control their vocal cords.

Sinusitis

Sinusitis is an inflammation of the cranial sinuses, the cavities within the facial skeleton that drain into the nasal cavities. Sinusitis develops when nasal congestion blocks the tiny openings leading to the sinuses. Up to 10% of upper respiratory infections are accompanied by sinusitis, and allergies may also play a role. Symptoms may include postnasal discharge, headache, and facial pain that worsens when the patient bends forward. Successful treatment depends on addressing the cause of the inflammation, and restoring proper drainage of the sinuses. Rinsing the sinuses by instilling a warm saline solution into one nostril, and out the other, though somewhat uncomfortable, may help remove irritants and rinse out mucus.

Otitis Media

Otitis media is an inflammation of the middle ear. The middle ear is not a part of the respiratory tract, but this condition is considered here because, especially in children, nasal infections can spread to the ear by way of the *auditory (eustachian) tubes* that lead from the nasopharynx to the middle ear. Pain is the primary symptom of otitis media. A sense of fullness, hearing loss, vertigo (dizziness),

and fever may also be present. If the cause is bacterial, antibiotic therapy is usually very effective. However, if viruses or allergies are the culprit, antibiotics are not effective. Tubes (called tympanostomy tubes) are sometimes surgically placed in the eardrums of children with multiple recurrences to help prevent the buildup of pressure in the middle ear and the possibility of hearing loss (Fig. 15.12). Normally, these tubes fall out with time.

Disorders of the Lower Respiratory Tract

Several disorders of the lower respiratory tract cause problems by obstructing normal airflow. Their causes range from a foreign object lodged in the trachea to excessive mucus in the bronchi and bronchioles. Other conditions tend to restrict the normal elasticity of the lung tissue itself (Fig. 15.13).

Disorders of the Trachea and Bronchi

One of the simplest—but most life-threatening—disorders that affects the trachea is choking. The best way for a person without extensive medical training to help someone who is choking is to perform the Heimlich maneuver, which involves grabbing the choking person around the waist from behind, and forcefully pulling both hands into their upper abdomen to expel whatever is lodged. Trained medical personnel may also insert a breathing tube by way of an incision made in the trachea. The operation is called a *tracheotomy,* and the opening is a *tracheostomy.* Sometimes people whose larynx or trachea has been damaged or destroyed, usually as a result of smoking, must have a permanent tracheostomy tube installed.

Acute bronchitis is an inflammation of the primary and secondary bronchi. Usually it is preceded by a viral infection that has led to a secondary bacterial infection. Most likely, a nonproductive cough has become a deep cough that produces more mucus and perhaps pus (Fig. 15.13*a*). Typically, the bacterial infection can be successfully treated with antibiotics.

In **chronic bronchitis,** the airways are inflamed and filled with mucus. A cough that brings up mucus is common. The bronchi have

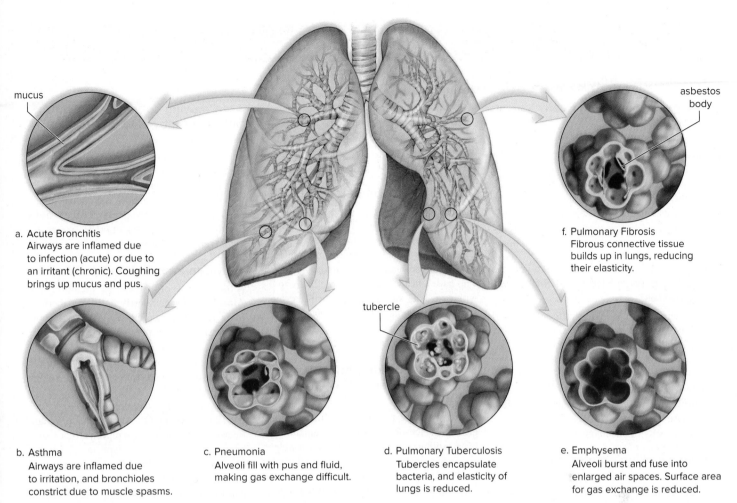

Figure 15.13 Common bronchial and pulmonary diseases. Exposure to infectious pathogens and/or polluted air, including tobacco smoke, causes various diseases and disorders.

mucus

a. Acute Bronchitis
Airways are inflamed due to infection (acute) or due to an irritant (chronic). Coughing brings up mucus and pus.

b. Asthma
Airways are inflamed due to irritation, and bronchioles constrict due to muscle spasms.

c. Pneumonia
Alveoli fill with pus and fluid, making gas exchange difficult.

tubercle

d. Pulmonary Tuberculosis
Tubercles encapsulate bacteria, and elasticity of lungs is reduced.

e. Emphysema
Alveoli burst and fuse into enlarged air spaces. Surface area for gas exchange is reduced.

asbestos body

f. Pulmonary Fibrosis
Fibrous connective tissue builds up in lungs, reducing their elasticity.

undergone degenerative changes, including the loss of cilia and their normal cleansing action. Under these conditions, an infection is more likely to occur. The most frequent cause of chronic bronchitis is smoking, although exposure to environmental pollutants can also be a contributing factor.

Asthma is a disease of the bronchi and bronchioles that is marked by wheezing, breathlessness, and sometimes a cough and expectoration of mucus. The airways are unusually sensitive to specific irritants, which can include a wide range of allergens such as pollen, animal dander, dust, cigarette smoke, and industrial fumes. Even cold air can be an irritant. When exposed to an irritant, the smooth muscle in the bronchioles undergoes spasms (Fig. 15.13*b*). Most asthma patients have some degree of bronchial inflammation that reduces the diameter of the airways and contributes to the seriousness of an attack. Asthma is not curable, but several types of drugs can prevent or treat asthma attacks. One group of drugs, called the beta-agonists, works by dilating the bronchioles. These drugs are usually administered using an asthma inhaler. Corticosteroids can help control the inflammation and hopefully prevent an attack.

Diseases of the Lungs

Pneumonia is an infection of the lungs in which the bronchi or alveoli fill with thick fluid (Fig. 15.13*c*). High fever and chills,

with headache and chest pain, are symptoms of pneumonia. Rather than being a generalized lung infection, pneumonia may be localized in specific lobules of the lungs. Obviously, the more lobules involved, the more serious is the infection. Pneumonia can be caused by several types of bacteria, viruses, and other infectious agents. Certain types of pneumonia mainly strike individuals with reduced immunity. For example, AIDS patients are subject to a particularly rare form of pneumonia caused by the fungus *Pneumocystis carinii*.

Pulmonary tuberculosis is caused by the bacterium *Mycobacterium tuberculosis*. When *M. tuberculosis* bacteria invade the lung tissue, the cells accumulate around the invading bacteria, isolating them from the rest of the body. This accumulation of cells is called a tubercle (Fig. 15.13*d*). If the body's resistance is high, the imprisoned organisms die, but if the resistance is low, the organisms eventually escape and spread. If a chest X-ray detects active tubercles, the individual is put on appropriate drug therapy to ensure the localization of the disease and the eventual destruction of any live bacteria. It is possible to tell if a person has ever been infected with tuberculosis bacteria via a TB skin test, in which a highly diluted extract of the bacteria is injected into the skin of the patient. A person who has never been exposed to *M. tuberculosis* shows no reaction, but one who has had or is fighting an infection shows an area of inflammation that peaks in about 48 hours (see Fig. 13.12).

SCIENCE IN YOUR LIFE ▶ HEALTH

Questions About Smoking and Health

Is Cigarette Smoking Really Addictive?

Yes. Nicotine is an addictive drug (just like heroin and cocaine). Small amounts make the smoker want to smoke more. Also, nicotine not only affects the mood and nature of the smoker but also may cause withdrawal symptoms if the smoker attempts to stop. The younger a person is when he or she begins to smoke, the more likely he or she is to develop an addiction to nicotine.

What Are Some of the Short-Term and Long-Term Effects of Smoking Cigarettes?

Short-term effects include shortness of breath and nagging coughs, diminished ability to smell and taste, premature aging of the skin, and increased risk of sexual impotence in men. Smokers tend to tire easily during physical activity. Long-term effects include many types of cancer, heart disease, aneurysms, bronchitis, emphysema, and stroke. Smoking contributes to the severity of pneumonia and asthma.

Does Smoking Cause Cancer?

Yes. Tobacco use accounts for about one-third of cancer deaths in the United States. Smoking causes almost 90% of lung cancers. Smoking also causes cancers of the oral cavity, pharynx, larynx, and esophagus. It contributes to the development of cancers of the bladder, pancreas, cervix, kidney, and stomach. Smoking is also linked to the development of some leukemias.

Why Do Smokers Have "Smoker's Cough"?

Cigarette smoke contains chemicals that irritate the air passages and lungs. When a smoker inhales these substances, the body tries to protect itself by producing mucus and coughing. The nicotine in smoke decreases the sweeping action of cilia, so some of the poisons in the smoke remain in the lungs.

If You Smoke but Do Not Inhale, Is There Any Danger?

Yes. Wherever smoke touches living cells, it does harm. Even if smokers don't inhale, they are breathing the smoke as secondhand smoke and are still at risk for lung cancer. Pipe and cigar smokers, who often do not inhale, are at an increased risk for lip, mouth, tongue, and several other cancers.

Does Cigarette Smoking Affect the Heart?

Yes. Smoking increases the risk of heart disease, the number one cause of death in the United States. Cigarette smoking is the biggest risk factor for sudden heart attacks. Smokers who have a heart attack are more likely to die within an hour of the heart attack than nonsmokers. Cigarette smoke at very low levels (much lower than the levels that cause lung disease) can cause damage to the heart.

How Does Smoking Affect Pregnant Women and Their Babies?

Smoking during pregnancy is linked with a greater chance of miscarriage, premature delivery, stillbirth, infant death, low birth weight, and sudden infant death syndrome (SIDS). Up to 10% of infant deaths would be prevented if pregnant women did not smoke. When a pregnant woman smokes, the nicotine, carbon monoxide, and other dangerous chemicals in smoke enter her bloodstream and then pass into the baby's body. This prevents the baby from getting essential nutrients and oxygen for growth.

What Are the Dangers of Environmental Tobacco Smoke?

Environmental tobacco smoke, also called secondhand smoke, increases the rate of heart disease by over 25% in nonsmokers and causes an estimated 46,000 deaths from heart disease each year. Children whose parents smoke are more likely to suffer from asthma, pneumonia or bronchitis, ear infections, coughing, wheezing, and increased mucus production in the first 2 years of life.

Are Chewing Tobacco and Snuff Safe Alternatives to Cigarette Smoking?

No. The juice from smokeless tobacco is absorbed directly through the lining of the mouth. This creates sores and white patches that often lead to cancer of the mouth and damage to the teeth and gums. Smokeless tobacco users greatly increase their risk of other cancers, including those of the pharynx.

How Can People Stop Smoking?

A number of organizations, including the American Cancer Society and the American Lung Association, offer suggestions for how to quit smoking. Both organizations also offer support groups to people interested in quitting. There is even advice about how people can help their friends quit smoking. Nicotine Anonymous offers a 12-step program for kicking a nicotine addiction, modeled after the 12-step Alcoholics Anonymous program. Recovering smokers have reported that a support system is very important. Several smoking-cessation drugs have been approved by the FDA, and others are in development phases. One such drug, varenicline (Chantix), affects areas in the brain stimulated by nicotine. Varenicline may lessen withdrawal symptoms by mimicking the effects of nicotine. In addition, several over-the-counter products are available to replace nicotine. These include nicotine patches, gum, and lozenges. These products are designed to supply enough nicotine to alleviate the withdrawal symptoms people experience during attempts to stop smoking.

Questions to Consider

1. What can you do to reduce your exposure to secondhand smoke?
2. Why does cigarette smoke have a negative effect on organ systems that are not in direct contact with the actual smoke?

Emphysema is a chronic and incurable disorder in which the alveoli are distended and their walls damaged so that the surface area available for gas exchange is reduced (Fig. 15.13*e*). Emphysema is often preceded by chronic bronchitis. Air trapped in the lungs leads to alveolar damage and a noticeable ballooning of the chest. The elastic recoil of the lungs is reduced, so the driving force behind expiration is also reduced. The victim often feels out of breath and may have a cough. Because the surface area for gas exchange is reduced, less oxygen reaches the heart and the brain. Because they are often seen in the same patient and tend to recur, emphysema, chronic bronchitis, and asthma are collectively called chronic obstructive pulmonary disease (COPD). COPD is the fourth leading cause of death in the United States, and is usually associated with smoking.

Cystic fibrosis (CF) is an example of a lung disease that is genetic, rather than infectious, although infections do play a role in the disease. One in 31 Americans carries the defective gene, but a child must inherit two copies of the faulty gene to have the disease. Still, CF is the most common inherited disease in the U.S. white population. The gene that is defective in CF codes for a protein called cystic fibrosis conductance transmembrane regulator (CFTR), which is needed for proper transport of Cl^- ions out of the epithelial cells of the lung. Because this also reduces the amount of water transported out of the lung cells, the mucus secretions become very sticky and can form plugs that interfere with breathing. Symptoms of CF include coughing and shortness of breath, and part of the treatment involves clearing mucus from the airways by vigorously slapping the patient on the back, as well as by administering mucus-thinning drugs. None of these treatments is curative, however, and because the lungs can be severely affected, the median survival age for people with CF is only 30 years. Researchers are attempting to develop gene therapy strategies to replace the faulty *CFTR* gene.

Another common lung disease is **pulmonary fibrosis,** in which fibrous connective tissue builds up in the lungs, causing a loss of elasticity (Fig. 15.13*f*). This restricts the ability of the lungs to expand during inhalation, and so reduces the vital capacity and other lung volumes. Pulmonary fibrosis most commonly occurs in elderly persons, and the risk is increased by environmental exposure to silica (sand), various types of dust, and asbestos.

Lung cancer is more prevalent in men than in women, but it is the leading cause of cancer death in both genders. About 87% of lung cancers are associated with cigarette smoking. Autopsies on smokers have revealed the progressive steps by which the most common form of lung cancer develops. The first step appears to be thickening of the cells lining the bronchi. Cilia are then lost, making it impossible to prevent dust and dirt from settling in the lungs. Following this, cells with atypical nuclei appear, followed by a tumor consisting of disordered cells with atypical nuclei. A normal lung and a lung with cancerous tumors are shown in Figure 15.14. A final step occurs when some of these cells break loose and penetrate other tissues, a process called metastasis. Now the cancer has spread. The original tumor may grow until a bronchus is blocked, cutting off the supply of air to that lung. The entire lung then collapses, the secretions trapped in the lung spaces become infected, and pneumonia or a lung abscess (localized area of pus)

a. Normal lung   b. Lung cancer

Figure 15.14 Normal and cancerous lungs compared.
a. Normal lung. Note the healthy red color in both lobes. **b.** Lung of a heavy smoker. Notice that the lung is black except where cancerous tumors have formed. (a): ©Matt Meadows/Getty Images; (b): ©Biophoto Associates/Science Source

results. The only treatment that offers a possibility of cure is *pneumonectomy,* in which a lobe or the whole lung is removed before metastasis has had time to occur. If the cancer has spread, chemotherapy and radiation are also required.

The Health feature "Questions About Smoking and Health" discusses the various illnesses, including cancer, that are apt to occur when a person smokes. If a person stops voluntary smoking and avoids passive smoking, and if the body tissues are not already cancerous, they may return to normal over time.

Check Your Progress 15.4

1. What is the difference between tonsillitis and laryngitis?
2. Explain how, in addition to the directly harmful effects of cigarette smoke, smoking might predispose the lung to damage by other harmful substances.
3. Which of the diseases of the lungs has a hereditary (genetic) component?

Conclusion

Our respiratory system plays a critical role in maintaining homeostasis in the body. The exchange of oxygen and carbon dioxide not only allows our cells to carry out the process of cellular respiration, and thus provide us with the energy we need for our metabolic activities, but this exchange also helps to regulate the pH of our body, allowing enzymes to function at their optimal levels.

Diseases such as asthma, lung cancer, and cystic fibrosis interfere with the ability of the respiratory system to function correctly. Inflammation and destruction of the aveoli in the lungs reduces the surface area over which gas exchange may occur, and in the process, the ability of almost every other organ system to carry out its metabolic functions in the body.

SUMMARIZE

15.1 The Respiratory System

The major functions of the **respiratory system** are to allow O_2 from the air to enter the blood, and CO_2 from the blood to exit the body. This occurs during **ventilation,** which includes inspiration and expiration. The respiratory tract can be divided into two major components:

- The upper respiratory tract, consisting of the **nose** (with nasal cavities), **pharynx, glottis, epiglottis,** and **larynx** (which contains the **vocal cords**).

- The lower respiratory tract, made up of the **trachea, bronchi, bronchioles,** and **lungs.** The lungs contain many **alveoli,** which are air sacs surrounded by a capillary network. The base of each lung contacts the **diaphragm,** a muscle that separates the thoracic and abdominal cavities.

15.2 Mechanism of Breathing

Ventilation includes both inspiration and expiration:

- The volumes of air that move in and out of the respiratory system during ventilation can be studied using a spirometer. The maximum volume of air that can be moved in and out during ventilation is called the vital capacity, which can be affected by many factors, including disease.

- **Inspiration** begins when the **respiratory center** in the brain sends excitatory nerve impulses to the diaphragm and the muscles of the rib cage. As they contract, the diaphragm lowers, and the rib cage moves upward and outward; the lungs expand, creating a partial vacuum, which causes air to rush in. The activity of the respiratory center is influenced by stretch receptors in the alveoli, as well as by the aortic bodies and carotid bodies, which are mainly sensitive to blood oxygen levels.

- **Expiration** occurs as the respiratory center temporarily stops sending impulses to the diaphragm and the muscles of the rib cage. As the diaphragm relaxes, it resumes its dome shape, and as the rib cage retracts, air is pushed out of the lungs.

15.3 Gas Exchanges in the Body

Oxygen and carbon dioxide are exchanged between the blood and tissues during external respiration and internal respiration:

- **External respiration** occurs in the lungs when CO_2 leaves the blood and O_2 enters the blood. In the lungs, the P_{CO_2} is higher in the blood than in alveoli, so CO_2 diffuses from the blood into the alveoli. Because carbon dioxide is present in the blood mainly as **bicarbonate ion** (HCO_3^-), carbonic acid forms first, which is then broken down by the enzyme **carbonic anhydrase** to carbon dioxide and water. Because the P_{O_2} is higher in the alveoli than in blood, O_2 diffuses from the alveoli into the blood where it combines with **hemoglobin** (Hb), forming **oxyhemoglobin** (HbO_2).

- **Internal respiration** occurs in the tissues, when O_2 leaves and CO_2 enters the blood. This occurs because the P_{O_2} is lower in the tissues compared to arterial blood, and the P_{CO_2} is higher. About 10% of this CO_2 enters red blood cells and binds to hemoglobin, forming **carbaminohemoglobin** ($HbCO_2$). Most of the CO_2 combines with water in the plasma, forming carbonic acid that is broken down to bicarbonate and hydrogen ions. Hemoglobin can also combine with hydrogen ions to form **reduced hemoglobin** (HHb).

15.4 Disorders of the Respiratory System

Disorders of the respiratory system can be divided into those that affect the upper respiratory tract and those that affect the lower respiratory tract:

- In the upper respiratory tract, the common cold, **pharyngitis, tonsillitis,** and **laryngitis** are all conditions that most people experience at some time. In addition, these infections can spread to the sinuses (causing **sinusitis**) and middle ear (causing **otitis media**).

- The lower respiratory tract is subject to foreign bodies that can block the trachea and bronchi. In addition, infections can cause **acute bronchitis, chronic bronchitis, pneumonia,** and **pulmonary tuberculosis.** Other common disorders include **asthma, emphysema, cystic fibrosis** (CF), **pulmonary fibrosis,** and **lung cancer.** Smoking is a major risk factor associated with many of these.

ASSESS

Testing Yourself

Choose the best answer for each question.

15.1 The Respiratory System

1. Label this diagram of the human respiratory system.

a. _____
b. _____
c. _____
d. _____
e. _____
f. _____
g. _____
h. _____
i. _____

2. Label the following diagram of the upper respiratory system.

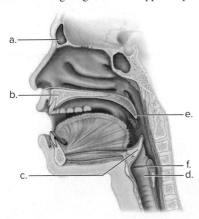

a.
b.
c.
d.
e.
f.

15.2 Mechanism of Breathing

3. The amount of air moved into and out of the lungs during regular breathing is known as the
   a. vital capacity.
   b. tidal volume.
   c. residual volume.
   d. None of these are correct.

4. Which of these statements correctly goes with expiration rather than inspiration?
   a. Rib cage moves up and out.
   b. Diaphragm relaxes and moves up.
   c. Pressure in lungs decreases, and air comes rushing out.
   d. Diaphragm contracts and lowers.

5. Air enters the human lungs because
   a. atmospheric pressure is lower than the pressure inside the lungs.
   b. atmospheric pressure is greater than the pressure inside the lungs.
   c. although the pressures are the same inside and outside, the partial pressure of oxygen is lower in the lungs.
   d. the residual air in the lungs causes the partial pressure of oxygen to be lower than it is outside.

15.3 Gas Exchanges in the Body

6. The enzyme carbonic anhydrase
   a. causes the blood to be more basic in the tissues.
   b. forms carbonic acid in the blood.
   c. actively transports carbon dioxide out of capillaries.
   d. is active only at high altitudes.
   e. All of these are correct.

7. Hemoglobin assists in the transport of gases primarily by
   a. combining with oxygen.
   b. combining with CO_2.
   c. combining with H^+.
   d. being present in red blood cells.
   e. All of these are correct.

8. The binding of small amounts of carbon dioxide to hemoglobin produces
   a. oxyhemoglobin.
   b. carbonic anhydrase.
   c. deoxyhemoglobin.
   d. carbaminohemoglobin.
   e. bicarbonate hemoglobin.

15.4 Disorders of the Respiratory System

9. Which of the following is not true of obstructive pulmonary disorders?
   a. Air does not flow freely in the airways.
   b. Vital capacity is reduced due to loss of lung elasticity.
   c. Disorders include chronic bronchitis, emphysema, and asthma.
   d. Ventilation takes longer to occur.

10. Which of the following is not an infection of the upper respiratory tract?
    a. tonsillitis
    b. sinusitis
    c. otitis media
    d. bronchitis
    e. None of these are correct.

11. Smoking is associated with an increased risk of
    a. emphysema.
    b. several types of cancer.
    c. heart disease.
    d. All of these are correct.

ENGAGE

Thinking Critically

1. The respiratory system of birds is quite different from that of mammals. Unlike mammalian lungs, the lungs of birds are relatively rigid and do not expand much during ventilation. Most birds also have thin-walled air sacs that fill most of the body cavity not occupied by other organs. Inspired air passes through the bird's lungs, into these air sacs, and back through the lungs again on expiration. In what ways might this system be more efficient than the mammalian lung?

2. Children may also suffer from obstructive sleep apnea (OSA). Symptoms experienced during the daytime include breathing through the mouth and difficulty focusing. Children don't often have the excessive sleepiness during the day that adults may have. Children with OSA often have enlarged tonsils or adenoids. Removal of the tonsils and adenoids often alleviates the problem. Continuous positive airway pressure (CPAP) treatment might be necessary if the OSA continues after surgery. Why would enlarged tonsils or adenoids cause OSA?

3. Carbon monoxide (CO) binds to hemoglobin 230 to 270 times more strongly than oxygen does, which means less O_2 is being delivered to tissues. What would be the specific cause of death if too much hemoglobin became bound to CO instead of to O_2?

4. As mentioned in the chapter, sometimes it is necessary to install a permanent tracheostomy—for example, in smokers who develop laryngeal cancer. Besides interfering with the ability to speak, what sorts of health problems would you expect to see in an individual with a permanent tracheostomy, and why?

CHAPTER
16

Urinary System and Excretion

©remik44992/Shutterstock

BEFORE YOU BEGIN

Before beginning this chapter, take a few moments to review the following discussions:

Section 2.3 What determines whether a solution is acidic or basic?

Figure 4.3 What types of molecules can diffuse across the plasma membrane?

Section 11.5 What are the major homeostatic functions of the urinary system?

Kidney Stones

It may start with a pain in your right side, just below the rib cage and extending toward your back. While at first you may think it is just an overworked muscle, maybe due to a new exercise routine, over the next few days you start to feel a constant need to urinate, and notice that the pain is often worse during urination. Then one day, you notice your urine has developed a light pink color. At this point, you decide it is time to see a doctor.

At the clinic, the nurse asks you a number of questions, including how much water and other fluids you drink each day, your diet, and any medications you are currently taking. Your symptoms make the doctor suspect you are experiencing kidney stones (also known as renal lithiasis). Kidney stones are small deposits, usually of calcium or uric acid, that accumulate in the kidney. Often, either an X-ray and/or an ultrasound is used to confirm the diagnosis. In addition, a blood and urine test would help determine the cause of the stones. If these stones are large enough, it may be difficult to pass them in the urine, and as a result, over time they may have a negative effect on the operation of the urinary system.

In this chapter, we will be exploring the function of the urinary system and the role of the kidney as one of the major organs of homeostasis in the body.

As you read through this chapter, think about the following questions:

1. What is the role of the kidneys in the body?

2. How might dehydration cause the development of kidney stones?

3. How would the analysis of the type of kidney stone help determine the cause of the kidney stone?

16.1 The Urinary System

The **urinary system** is involved in **excretion,** the removal of metabolic wastes from the body. People sometimes confuse the terms *excretion* and *defecation*, but they do not refer to the same process. Defecation, the elimination of feces from the body, is a function of the digestive system (see Section 14.1). The undigested food and bacteria that make up feces have never been a part of the functioning of the body, while most substances excreted in urine were once part of the metabolic processes in the body.

Functions of the Urinary System

In humans, the urinary system produces **urine,** a fluid that contains dissolved waste materials, and conducts it to outside the body. As the kidneys produce urine, they carry out the following four functions that contribute to homeostasis:

1. *Excretion of Metabolic Wastes.* The kidneys excrete metabolic wastes, notably nitrogenous wastes. Urea is the primary nitrogenous end product of metabolism in humans, but we also excrete some ammonium, creatinine, and uric acid.

   Urea is a by-product of amino acid metabolism. The breakdown of amino acids in the liver releases **ammonia,** which the liver rapidly combines with carbon dioxide to produce urea. Ammonia is very toxic to cells, but urea is much less toxic. Some ammonia (NH_3) is also excreted as ammonium ion (NH_4^+).

   Creatine phosphate is a high-energy phosphate reserve molecule in muscles. The metabolic breakdown of creatine phosphate results in *creatinine,* which is excreted.

   The breakdown of nucleotides, such as those containing adenine and thymine, produces **uric acid.** Uric acid is rather insoluble. If too much uric acid is present in blood, crystals may form and precipitate out in the joints, producing a painful ailment called **gout.**

2. *Osmoregulation.* A principal function of the kidneys is **osmoregulation,** the maintenance of the appropriate balance of water and salt in the blood. Blood volume is intimately associated with the salt balance of the body. Salts, such as NaCl, have the ability to cause osmosis, the diffusion of water—in this case, into the blood. The more salts there are in the blood, the greater the blood volume and the greater the blood pressure. In this way, the kidneys are also involved in regulating blood pressure.

   The kidneys also help maintain the appropriate level of other ions, such as potassium (K^+), bicarbonate (HCO_3^-), and calcium (Ca^{2+}), in the blood.

3. *Regulation of Acid-Base Balance.* Along with the respiratory system, the kidneys regulate the acid-base balance of the blood. The kidneys monitor and help keep the blood pH at about 7.4, mainly by excreting hydrogen ions (H^+) and reabsorbing the bicarbonate ions (HCO_3^-) as needed. Human urine usually has a pH of 6 or lower because our diet often contains acidic foods.

4. *Secretion of Hormones.* The kidneys assist the endocrine system in hormone secretion. The kidneys release renin, an enzyme that leads to the secretion of the hormone aldosterone from the adrenal cortex, the outer portion of the adrenal glands, which lie atop the kidneys. As described in Section 16.3, aldosterone promotes the reabsorption of sodium ions (Na^+) by the kidneys.

   Whenever the oxygen-carrying capacity of the blood is reduced, or oxygen demand increases, the kidneys secrete the hormone **erythropoietin (EPO),** which stimulates red blood cell production.

   The kidneys also help activate vitamin D from the skin. Vitamin D acts like a hormone in the body because it promotes calcium (Ca^{2+}) absorption from the digestive tract.

Organs of the Urinary System

The urinary system consists of the kidneys, ureters, urinary bladder, and urethra. Figure 16.1 shows these organs and also traces the path of urine.

Kidneys

The **kidneys** are paired, bean-shaped, reddish-brown organs located near the small of the back in the lumbar region, on either side of the vertebral column. About the size of a fist, they lie in depressions dorsal to (behind) the peritoneum, where they receive some protection from the lower rib cage.

Each kidney is covered by a tough fibrous connective tissue layer called a *renal capsule.* The concave side of each kidney has a depression called the hilum, where a *renal artery* enters, and a *renal vein* and a ureter exit, the kidney.

Ureters

The **ureters** conduct urine from the kidneys to the bladder. They are small, muscular tubes about 25 cm long and 5 mm in diameter. The wall of each ureter has three layers: an inner mucosa; a smooth muscle layer; and an outer, fibrous coat of connective tissue. Peristaltic contractions cause urine to enter the bladder in spurts, about one to five times per minute. Because a healthy adult human produces 1–2 liters (l) of urine a day, an average of 40–80 milliliters (ml) per hour enters the bladder.

Urinary Bladder

The **urinary bladder** stores urine until it is expelled from the body. The bladder is located in the pelvic cavity. The urinary bladder has three openings (orifices)—two for the ureters and one for the urethra, which drains the bladder (Fig. 16.2).

The bladder wall is expandable because it contains a middle layer of circular fiber and two layers of longitudinal muscle. The transitional epithelium of the mucosa becomes thinner, and folds in the mucosa called *rugae* disappear as the bladder enlarges.

After urine enters the bladder from a ureter, small folds of bladder mucosa act like a valve to prevent backward flow. Two

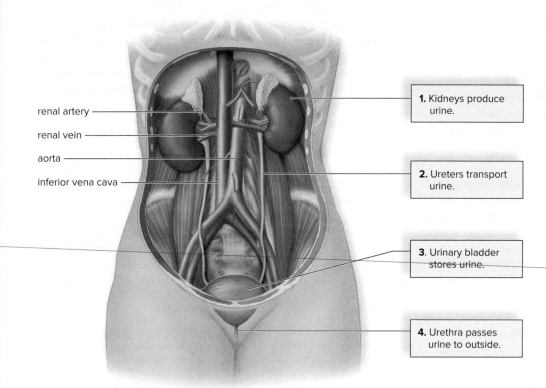

renal artery

renal vein

aorta

inferior vena cava

Figure 16.1 The urinary system.
Urine is found only within the kidneys, the ureters, the urinary bladder, and the urethra. The kidneys are important organs of homeostasis because they excrete metabolic wastes and adjust both the water-salt and acid-base balance of the blood.

1. Kidneys produce urine.

2. Ureters transport urine.

3. Urinary bladder stores urine.

4. Urethra passes urine to outside.

sphincters lie in close proximity where the urethra exits the bladder. The internal sphincter occurs around the opening to the urethra. An external sphincter is composed of skeletal muscle that can be voluntarily controlled. Many people have some degree of *incontinence,* the involuntary loss of urine. At least one in ten people over age 65 have bladder control problems due to an age-related loss of muscle tone in the urinary bladder. Other causes of incontinence include enlarged prostate in men, pregnancy, and nervous system diseases that affect the control of urination.

Urethra

The **urethra** is a small tube that extends from the urinary bladder to an external opening. Therefore, its function is to remove urine from the body. In females, the urethra is only about 4 cm long, whereas in males, the urethra averages 20 cm when the penis is flaccid (nonerect). In females, the reproductive and urinary systems are not connected. In males, the urethra carries urine during urination and semen during ejaculation. As the urethra leaves the male urinary bladder, it is encircled by the prostate gland. In men over age 40, the prostate gland is subject to enlargement, as well as to prostate cancer. Either of these conditions can restrict urination, and may require treatment (see Fig. 21.18).

Urination

When the urinary bladder fills to about 250 ml with urine, stretch receptors send sensory nerve impulses to the spinal cord. Subsequently, motor nerve impulses from the spinal cord cause the urinary bladder to contract and the sphincters to relax so that urination, also called *micturition,* is possible (Fig. 16.2). In older children and adults, the brain controls this reflex, delaying urination until a suitable time.

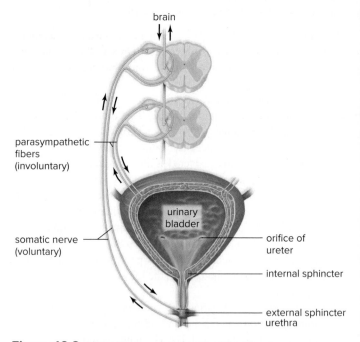

brain

parasympathetic fibers (involuntary)

somatic nerve (voluntary)

urinary bladder

orifice of ureter

internal sphincter

external sphincter
urethra

Figure 16.2 Urination. As the bladder fills with urine, sensory impulses go to the spinal cord and then to the brain. When urination occurs, motor nerve impulses cause the bladder to contract and an internal sphincter to relax. Motor nerve impulses also cause an external sphincter to relax.

Check Your Progress 16.1

1. Summarize the roles of the urinary system in the body.
2. List and briefly describe the functions of the specific components of the urinary system.
3. Explain why urinary system diseases may cause problems with the overall homeostasis of the body.

16.2 Anatomy of the Kidney and Excretion

Learning Outcomes

Upon completion of this section, you should be able to

1. Identify the structures of a human kidney.
2. Identify the parts of a nephron and state the function of each.
3. Distinguish between glomerular filtration, tubular reabsorption, and tubular secretion.

A lengthwise section of a kidney shows that many branches of the renal artery and renal vein reach inside a kidney (Fig. 16.3*a*). A kidney has three regions (Fig. 16.3*c*). The *renal cortex* is an outer, granulated layer that dips down in between a radially striated inner layer called the renal medulla. The *renal medulla* contains cone-shaped tissue masses called renal pyramids. The *renal pelvis* is a central space, or cavity, that is continuous with the ureter.

Anatomy of a Nephron

Under higher magnification, the kidney is composed of over a million **nephrons,** sometimes called renal or kidney tubules

(Fig. 16.3*b*). Each nephron has its own blood supply, including two capillary regions (Fig. 16.4). From the renal artery, an afferent arteriole leads to the **glomerulus,** a knot of capillaries inside the glomerular capsule. Blood leaving the glomerulus enters the efferent arteriole (the term *afferent* means "going toward," while *efferent* means "going away"). The afferent arteriole has a greater diameter than the efferent arteriole, to increase blood pressure into the glomerular capsule. The efferent arteriole takes blood to the *peritubular capillary network,* which surrounds the rest of the nephron. From there, the blood goes into a venule that joins the renal vein.

Parts of a Nephron

Each nephron is made up of several parts. First, the closed end of the nephron is pushed in on itself to form a cuplike structure called the **glomerular capsule** (Bowman capsule). The inner layer of the glomerular capsule is composed of *podocytes* (see Fig. 16.6) that have long cytoplasmic extensions. The podocytes cling to the capillary walls of the glomerulus and leave pores that allow easy passage of small molecules from the glomerulus to the inside of the glomerular capsule.

The glomerular capsule connects to a **proximal convoluted tubule.** The cuboidal epithelial cells lining this part of the

a. Blood vessels

b. Nephrons

c. Renal pelvis, medulla, and cortex

d. Kidney section

Figure 16.3 Anatomy of the kidney.   **a.** A lengthwise section of the kidney showing the blood supply. Note that the renal artery divides into smaller arteries, and these divide into arterioles. Venules join to form small veins, which form the renal vein. **b.** An enlargement showing the placement of nephrons. **c.** The same section as (**a**) without the blood supply. Now it is easier to distinguish the renal pelvis, medulla, and cortex, which connects with the ureter. **d.** A lengthwise section of an actual kidney. (d): ©Kage Mikrofotografie/Medical Images

Figure 16.4 Nephron anatomy. **a.** You can trace the path of blood through a nephron by following the black arrows. **b, c, d.** A nephron is made up of a glomerular capsule, the proximal convoluted tubule, the loop of the nephron, the distal convoluted tubule, and the collecting duct. (b): ©Science Photo Library/Getty Images; (c–d): ©Ed Reschke/Getty Images

nephron have numerous tightly packed microvilli that form a brush border, increasing the surface area for reabsorption. The tube then narrows and makes a U-turn called the **loop of the nephron** (loop of Henle), which is lined with simple squamous epithelium.

Following the loop, the tube becomes the **distal convoluted tubule,** which is composed of cuboidal epithelial cells that have numerous mitochondria but lack microvilli. Because of this, the distal convoluted tubule is not as specialized for reabsorption and is more likely to help move molecules from the blood into the tubule, a process called tubular secretion. The distal convoluted tubules of several nephrons enter one collecting duct. Many **collecting ducts** carry urine to the renal pelvis.

As shown in Figures 16.3*b* and 16.4*a,* the glomerular capsule and the convoluted tubules always lie within the renal cortex. The loop of the nephron dips down into the renal medulla; a few nephrons have a very long loop of the nephron, which penetrates deep into the renal medulla. Collecting ducts are also located in the renal medulla, and they help give the renal pyramids their lined appearance (see Fig. 16.3*d*).

Urine Formation

Figure 16.5 gives an overview of urine formation, which is divided into the following processes: glomerular filtration, tubular reabsorption, and tubular secretion.

Glomerular Filtration

Glomerular filtration occurs when whole blood enters the afferent arteriole and the glomerulus. Due to glomerular blood pressure, water and small molecules move from the glomerulus to the inside of the glomerular capsule. This is a filtration process because large molecules and formed elements are unable to pass through the capillary wall. In effect, then, blood in the glomerulus has two portions—the filterable components and the nonfilterable components:

Filterable Blood Components	Nonfilterable Blood Components
Water	Formed elements (blood cells and
Nitrogenous wastes	platelets)
Nutrients	Plasma proteins
Salts (ions)	

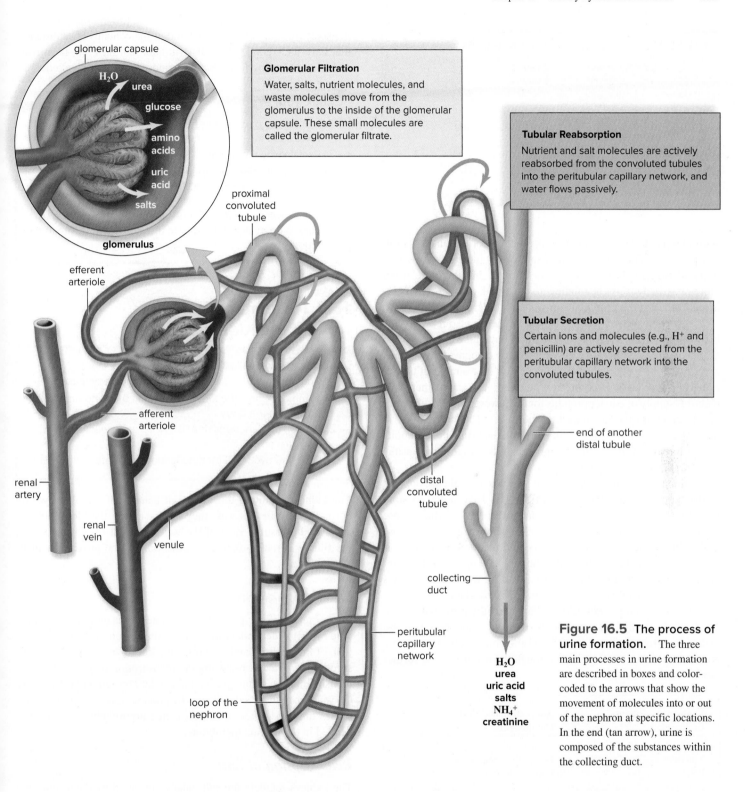

glomerular capsule

H_2O
urea
glucose
amino acids
uric acid
salts

glomerulus

Glomerular Filtration
Water, salts, nutrient molecules, and waste molecules move from the glomerulus to the inside of the glomerular capsule. These small molecules are called the glomerular filtrate.

Tubular Reabsorption
Nutrient and salt molecules are actively reabsorbed from the convoluted tubules into the peritubular capillary network, and water flows passively.

Tubular Secretion
Certain ions and molecules (e.g., H^+ and penicillin) are actively secreted from the peritubular capillary network into the convoluted tubules.

proximal convoluted tubule

efferent arteriole

afferent arteriole

renal artery

renal vein

venule

distal convoluted tubule

end of another distal tubule

collecting duct

peritubular capillary network

loop of the nephron

H_2O
urea
uric acid
salts
NH_4^+
creatinine

Figure 16.5 The process of urine formation.   The three main processes in urine formation are described in boxes and color-coded to the arrows that show the movement of molecules into or out of the nephron at specific locations. In the end (tan arrow), urine is composed of the substances within the collecting duct.

The *glomerular filtrate* contains small dissolved molecules in approximately the same concentration as plasma. Small molecules that escape being filtered and the nonfilterable components leave the glomerulus by way of the efferent arteriole.

As indicated in Table 16.1, nephrons in the kidneys filter 180 liters of water per day, along with a considerable amount of small molecules (such as glucose) and ions (such as sodium). If the composition of urine were the same as that of the glomerular filtrate, the body would continually lose water, salts, and nutrients.

Therefore, the composition of the filtrate is altered as this fluid passes through the remainder of the tubule.

Tubular Reabsorption

Tubular reabsorption occurs as molecules and ions are both passively and actively reabsorbed from the nephron into the blood of the peritubular capillary network. The osmolarity of the blood is maintained by the presence of both plasma proteins and salt. When

TABLE 16.1 Reabsorption from Nephrons

Substance	Amount Filtered (per day)	Amount Excreted (per day)	Reabsorption (%)
Water (l)	180	1.8	99.0
Sodium (g)	630	3.2	99.5
Glucose (g)	180	0.0	100.0
Urea (g)	54	30.0	44.0

l = liters, g = grams

sodium ions (Na^+) are actively reabsorbed, chloride ions (Cl^-) follow passively. The reabsorption of salt (NaCl) increases the osmolarity of the blood compared to the filtrate, and therefore water moves passively from the tubule into the blood. About 65% of Na^+ is reabsorbed at the proximal convoluted tubule.

Nutrients, such as glucose and amino acids, also return to the blood almost exclusively at the proximal convoluted tubule. This is a selective process because only molecules recognized by carrier proteins are actively reabsorbed. Glucose is an example of a molecule that ordinarily is completely reabsorbed because there is a plentiful supply of carrier proteins for it. However, every substance has a maximum rate of transport, and after all its carriers are in use, any excess in the filtrate will appear in the urine. For example, as reabsorbed levels of glucose approach 1.8–2 mg/ml of plasma, the rest appears in the urine.

In diabetes mellitus (see Section 16.4), excess glucose appears in the blood because the liver and muscles have failed to store glucose as glycogen. The kidneys cannot reabsorb all the glucose in the filtrate, and it appears in the urine. The presence of glucose in the filtrate increases its osmolarity, and therefore, less water is reabsorbed into the peritubular capillary network. The frequent urination and increased thirst experienced by untreated diabetics are due to the fact that less water is being reabsorbed.

We have seen that the filtrate that enters the proximal convoluted tubule is divided into two portions—components that are reabsorbed from the tubule into blood and components that are not reabsorbed and continue to pass through the nephron to be further processed into urine:

Reabsorbed Filtrate Components	Nonreabsorbed Filtrate Components
Most water	Some water
Nutrients	Most nitrogenous waste
Required salts (ions)	Excess salts (ions)

The substances not reabsorbed become the tubular fluid, which enters the loop of the nephron.

Tubular Secretion

Tubular secretion is a second way by which substances are removed from blood in the peritubular network and added to the tubular fluid. Hydrogen ions, potassium ions, creatinine, and drugs, such as penicillin, are some of the substances that are removed by active

transport from the blood. In the end, urine contains (1) substances that have undergone glomerular filtration but have not been reabsorbed and (2) substances that have undergone tubular secretion.

Check Your Progress 16.2

1. Identify the parts of a nephron that are found in the renal cortex and the renal medulla.
2. Explain why the proximal convoluted tubule is more specialized for reabsorption than the distal convoluted tubule.
3. Summarize how the kidney uses the processes of diffusion, passive transport, and active transport.
4. Explain where in the nephron, and by what process, glucose is normally returned from the glomerular filtrate to the blood.

16.3 Regulatory Functions of the Kidneys

Learning Outcomes

Upon completion of this section, you should be able to

1. Summarize how the kidneys maintain the water-salt balance in the body.
2. Explain the functions of ADH, ANH, and aldosterone in homeostasis.
3. Describe two ways that the kidneys regulate blood pH.

The kidneys maintain the water-salt balance of the blood within normal limits. In this way, they also maintain the blood volume and blood pressure. Most of the water and salt (NaCl) present in the filtrate is reabsorbed across the wall of the proximal convoluted tubule.

Osmoregulation

The excretion of a hypertonic urine (one that is more concentrated than blood) is dependent upon the reabsorption of water from the loop of the nephron and the collecting duct. We can think of reabsorption of water as requiring (1) reabsorption of salt and (2) establishment of a solute gradient dependent on salt and urea before (3) water is reabsorbed. During the process of reabsorption, water passes through protein channels called **aquaporins** that are embedded in the plasma membrane.

Reabsorption of Salt

The kidneys regulate the salt balance in blood by controlling the excretion and the reabsorption of various ions. Sodium (Na^+) is an important ion in plasma that must be regulated, but the kidneys also excrete or reabsorb other ions, such as potassium ions (K^+), bicarbonate ions (HCO_3^-), and magnesium ions (Mg^{2+}), as needed.

Usually, more than 99% of sodium (Na^+) filtered at the glomerulus is returned to the blood. Most of this sodium (67%) is reabsorbed at the proximal tubule, and a sizable amount (25%) is reabsorbed by the ascending limb of the loop of the nephron. The rest is reabsorbed from the distal convoluted tubule and collecting duct.

Hormones regulate the reabsorption of sodium at the distal convoluted tubule. **Aldosterone,** a hormone secreted by the adrenal cortex, promotes the excretion of potassium ions (K^+) and the reabsorption of sodium ions (Na^+). The release of aldosterone is set in motion by the kidneys themselves. The *juxtaglomerular apparatus* is a region of contact between the afferent arteriole and the distal convoluted tubule (Fig. 16.6). When blood volume, and therefore blood pressure, is not sufficient to promote glomerular filtration, the juxtaglomerular apparatus secretes renin. **Renin** is an enzyme that changes angiotensinogen (a large plasma protein produced by the liver) into angiotensin I. Later, angiotensin I is converted to angiotensin II, a powerful vasoconstrictor that also stimulates the adrenal cortex to release aldosterone. The reabsorption of sodium ions is followed by the reabsorption of water. Therefore, blood volume and blood pressure increase.

Atrial natriuretic hormone (ANH) is a hormone secreted by the atria of the heart when cardiac cells are stretched due to increased blood volume. ANH inhibits the secretion of renin by the juxtaglomerular apparatus and the secretion of aldosterone by the adrenal cortex. Its effect, therefore, is to promote the excretion of Na^+, called natriuresis.

Establishment of a Solute Gradient

The long loop of a nephron, which typically penetrates deep into the renal medulla, is made up of a descending limb and an ascending limb. Salt (NaCl) passively diffuses out of the lower portion of the ascending limb, but the upper, thick portion of the limb actively extrudes salt out into the tissue of the outer renal medulla (Fig. 16.7). Less and less salt is available for transport as fluid moves up the thick portion of the ascending limb. Because of these circumstances, there is an osmotic gradient within the tissues of the renal me-

Figure 16.6 Juxtaglomerular apparatus. The afferent arteriole and the distal convoluted tubule usually lie next to each other. The juxtaglomerular apparatus occurs where they touch. The juxtaglomerular apparatus secretes renin, a substance that leads to the release of aldosterone by the adrenal cortex. Reabsorption of sodium ions and water then occurs. Thereafter, blood volume and blood pressure increase.

proximal convoluted tubule

podocyte

glomerulus

afferent arteriole

juxtaglomerular apparatus

distal convoluted tubule

glomerular capsule

efferent arteriole

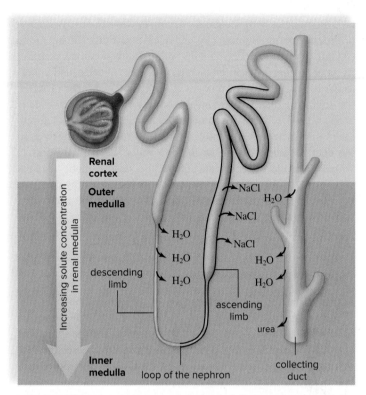

Figure 16.7 Reabsorption of water at the loop of the nephron and the collecting duct. Salt (NaCl) diffuses and is actively transported out of the ascending limb of the loop of the nephron into the renal medulla. Also, urea is believed to leak from the collecting duct and to enter the tissues of the renal medulla. This creates a hypertonic environment, which draws water out of the descending limb and the collecting duct. This water is returned to the cardiovascular system. The thick black outline of the ascending limb indicates it is impermeable to water.

dulla: The concentration of salt is greater in the direction of the inner medulla. Water cannot leave the ascending limb because the limb is impermeable to water.

The large arrow at the far left in Figure 16.7 indicates that the innermost portion of the inner medulla has the highest concentration of solutes. This cannot be due to salt because active transport of salt does not start until fluid reaches the thick portion of the ascending limb. Urea is believed to leak from the lower portion of the collecting duct, and it is this molecule that contributes to the high solute concentration of the inner medulla.

Reabsorption of Water

Because of the osmotic gradient within the renal medulla, water leaves the descending limb along its entire length. This is a countercurrent mechanism: As water diffuses out of the descending limb, the remaining fluid within the limb encounters an even greater osmotic concentration of solute. Therefore, water continues to leave the descending limb from the top to the bottom.

Fluid enters the collecting duct from the distal convoluted tubule. This fluid is hypotonic to the renal cortex, so to this point, the active transport of salt out of the ascending limb has decreased the osmolarity of the filtrate (Fig. 16.7). This allows the production of urine that is hypotonic to the general body fluids—for example, when excess water needs to be excreted. When the urine needs to be hypertonic to body fluids (for example, when the body is dehydrated), the posterior pituitary gland releases **antidiuretic hormone (ADH).** In order to understand the action of this hormone, consider its name. *Diuresis* means an increased amount of urine, and *antidiuresis* means a decreased amount of urine. In the absence of ADH, the collecting duct is impermeable to water, and a dilute urine is produced. ADH increases the number of aquaporins inserted into the plasma membranes of the cells located in the collecting duct, so the duct becomes more permeable to water. Because the filtrate within the collecting duct encounters the same osmotic gradient mentioned earlier, water can diffuse out of the collecting duct into the renal medulla. Usually, more ADH is produced at night, presumably so we don't have to interrupt our sleep as often to urinate. This also explains why the first urine of the day is often more concentrated.

Diuretics

Diuretics are chemicals that increase the production of urine. Drinking alcohol causes diuresis because it inhibits the secretion of ADH. The dehydration that follows contributes to the symptoms of a hangover. Caffeine is a diuretic because it increases the glomerular filtration rate and decreases the tubular reabsorption of Na^+. Diuretic drugs developed to counteract high blood pressure inhibit active transport of Na^+ at the loop of the nephron or at the distal convoluted tubule. A decrease in water reabsorption and a decrease in blood volume follow.

Besides their medical uses, diuretics can also be abused. Bodybuilders and other athletes may take them for a more "cut" look, or to lose weight quickly, in the form of water. Also, because it may be more difficult to detect drugs in dilute urine, drug users who are attempting to pass a urine drug test may use furosemide (Lasix) or other diuretics. While this may be effective, it is also risky. Electrolyte imbalances resulting from inappropriate use of furosemide or other diuretic drugs can result in dehydration, irregular heartbeat, and even death.

Acid-Base Balance

The normal pH for body fluids is about 7.4. This is the pH at which our proteins, such as cellular enzymes, function best. If the blood pH rises above 7.4, a person is said to have *alkalosis,* and if the blood pH falls below 7.4, a person is said to have *acidosis.* Alkalosis and acidosis are abnormal conditions that may need medical attention.

The foods we eat add basic or acidic substances to the blood (such as amino acids), and so do metabolic activities of cells. For example, cellular respiration adds carbon dioxide that combines with water to form carbonic acid, and fermentation adds lactic acid. The pH of body fluids stays at just about 7.4 via several mechanisms, primarily acid-base buffer systems, the respiratory center, and the kidneys.

Acid-Base Buffer Systems

The pH of the blood stays near 7.4 because the blood is buffered. A *buffer* is a chemical or a combination of chemicals that can take up excess hydrogen ions (H^+) or excess hydroxide ions (OH^-). One of the most important buffers in the blood is a combination of carbonic acid (H_2CO_3) and bicarbonate ions (HCO_3^-). When hydrogen ions (H^+) are added to blood, the following reaction occurs:

$$H^+ + HCO_3^- \longrightarrow H_2CO_3$$

When hydroxide ions (OH^-) are added to blood, this reaction occurs:

$$OH^- + H_2CO_3 \longrightarrow HCO_3^- + H_2O$$

These reactions temporarily prevent any significant change in blood pH. A blood buffer, however, can be overwhelmed unless some more permanent adjustment is made. The next adjustment to keep the pH of the blood constant occurs at pulmonary capillaries.

Respiratory Center

As discussed in Section 15.2, the respiratory center in the medulla oblongata increases the breathing rate if the hydrogen ion concentration of the blood rises. Increasing the breathing rate rids the body of hydrogen ions because the following reaction takes place in pulmonary capillaries:

$$H^+ + HCO_3^- \rightleftharpoons H_2CO_3 \rightleftharpoons H_2O + CO_2$$

In other words, the formation of carbon dioxide serves to raise the pH of the blood, and make it less acidic.

It is important to have the correct proportion of carbonic acid and bicarbonate ions in the blood. Breathing readjusts this proportion so that this particular acid-base buffer system can continue to absorb both H^+ and OH^- as needed.

The Kidneys

As powerful as the acid-base buffer and the respiratory center mechanisms are, only the kidneys can rid the body of a wide range of acidic and basic substances and otherwise adjust the pH. The kidneys are slower acting than the other two mechanisms, but they have a more powerful effect on pH. For the sake of simplicity, we can think of the kidneys as reabsorbing bicarbonate ions and excreting hydrogen ions as needed to maintain the normal pH of the blood (Fig. 16.8).

If the blood is acidic, hydrogen ions are excreted, and bicarbonate ions are reabsorbed. If the blood is basic, hydrogen ions are not excreted, and bicarbonate ions are not reabsorbed. Because the urine is usually acidic, it follows that an excess of hydrogen ions is usually excreted. Ammonia (NH_3) provides another means of buffering and removing the hydrogen ions in urine: ($NH_3 + H^+ \rightarrow NH_4^+$). Ammonia (whose presence is quite obvious in the diaper pail or kitty litter box) is produced in tubule cells by the deamination of amino acids. Phosphate provides another means of buffering hydrogen ions in urine.

The importance of the kidneys' ultimate control over the pH of the blood cannot be overemphasized. As mentioned, the enzymes of cells cannot continue to function if the internal environment does not have near-normal pH.

capillary

kidney tubule

Figure 16.8 Acid-base balance. In the kidneys, bicarbonate ions (HCO_3^-) are reabsorbed and hydrogen ions (H^+) are excreted as needed to maintain the pH of the blood. Excess hydrogen ions are buffered, for example, by ammonia (NH_3).

Check Your Progress 16.3

1. Describe the three processes required for the excretion of a hypertonic urine.
2. Summarize the relationship between aldosterone, the juxtaglomerular apparatus, and renin.
3. Describe how a diuretic interacts with the urinary system.
4. Explain the ions that the kidneys control to maintain blood pH homeostasis.

16.4 Disorders of the Urinary System

Learning Outcomes

Upon completion of this section, you should be able to

1. Summarize how damage to the glomeruli/nephrons can lead to uremia and edema.
2. Summarize the different methods of treating kidney failure.
3. Describe three common disorders of the bladder and urethra.

Various disease conditions can affect the kidneys, ureters, urinary bladder, and urethra. The kidneys are essential for life, so if they are damaged sufficiently by disease, their function must somehow be restored if the patient is to survive.

Disorders of the Kidneys

Many major illnesses that affect other parts of the body, especially diabetes, hypertension, and certain autoimmune diseases, can also cause serious kidney disease. Most of these conditions tend to damage the glomeruli, resulting in a decreased glomerular filtration rate, and eventually kidney failure.

Infection of the kidneys is called **pyelonephritis.** Kidney infections usually result from infections of the urinary bladder that spread via the ureter(s) to the kidney(s). Kidney infections are usually curable with antibiotics if diagnosed in time. However, an infection called hemolytic uremic syndrome, caused by various bacteria, namely a strain of *Escherichia coli* known as O157:H7,

can cause serious kidney damage in children even when antibiotics are administered.

As was discussed in the chapter opener, **kidney stones** are hard granules that can form in the renal pelvis. They may be composed of various materials, such as calcium, phosphate, uric acid, and protein. An estimated 5–15% of people will develop kidney stones at some time in their lives. Factors that contribute to the formation of these stones include too much animal protein in the diet, an imbalance in urinary pH, and/or urinary tract infection. Kidney stones may pass unnoticed in the urine flow. However, when a large kidney stone passes, strong contractions within a ureter can be excruciatingly painful. If a kidney stone grows large enough to block the renal pelvis or ureter, a reverse pressure builds up that may destroy the nephrons. To prevent this, it is sometimes necessary to break up a stone using a technique called lithotripsy, in which powerful ultrasound waves shatter the stone into smaller pieces that can be passed.

One of the first signs of kidney damage is the presence of albumin protein, white blood cells, and/or red blood cells in the urine. Albumin is produced by the liver to regulate blood volume. Normally, albumin, or blood cells, should not be present in the urine. Once more than two-thirds of the nephrons have been destroyed by a disease process, urea and other waste products accumulate in the blood, a condition known as *uremia*. Although nitrogenous wastes can cause serious damage, the retention of water and salts by diseased kidneys is of even greater concern. The latter causes *edema*, the accumulation of fluid in body tissues. Imbalances in the ionic composition of body fluids can also affect the operation of the nervous system and circulatory system, which require a stable range of ions (electrolytes) to function correctly. Imbalances have the potential to lead to loss of consciousness and possibly heart failure.

Treatment Options for Kidney Failure

Patients whose kidneys are failing can undergo **hemodialysis,** continuous ambulatory peritoneal dialysis, or kidney transplantation. *Dialysis* is the diffusion of dissolved molecules through a semipermeable natural or synthetic membrane having pore sizes that allow only small molecules to pass through. In hemodialysis, the patient's blood is passed through an artificial kidney machine (Fig. 16.9). This machine contains a membranous tube, which is in contact with a dialysis solution, or *dialysate.* Substances more concentrated in the blood diffuse into the dialysate, and substances more common in the dialysate diffuse into the blood. The dialysate is continuously replaced to maintain favorable concentration gradients. In this way, the artificial kidney can be utilized either to extract substances from the blood, including waste products or toxic chemicals and drugs, or to add substances to the blood—for example, bicarbonate ions (HCO_3^-) if the blood is acidic. In the course of a 3- to 6-hour hemodialysis, 50–250 g of urea can be removed from a patient, which greatly exceeds the amount excreted by normal kidneys. Therefore, a patient needs to undergo treatment only about two or three times a week. Although about 98% of patients opt to be treated at dialysis centers, a growing number are purchasing portable units that can be used in their own homes.

Hemodialysis centers may not be available in all areas, where continuous ambulatory peritoneal dialysis, or CAPD, is more

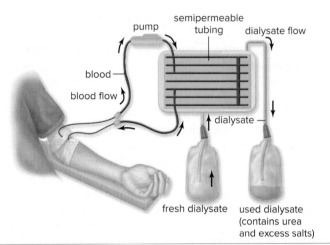

Figure 16.9 An artificial kidney machine. As the patient's blood is pumped through dialysis tubing, the tubing is exposed to a dialysate (dialysis solution). Wastes exit the blood into the solution because of a difference in the concentration gradient between the dialysate and the blood. In this way, blood is not only cleansed, but its water-salt and acid-base balances are also adjusted.

Figure 16.10 Female and male urinary tracts compared. Females have a short urinary tract compared to that of males. This makes it easier for bacteria to invade the urethra and helps explain why females are more likely than males to get a urinary tract infection.

commonly used. In CAPD, the peritoneum serves as the dialysis membrane. A fresh amount of dialysate is introduced directly into the abdominal cavity from a bag that is temporarily attached to a permanently implanted plastic tube. The dialysate flows into the peritoneal cavity by gravity. Waste and salt molecules pass from blood vessels in the abdominal wall into the dialysate before the fluid is collected several hours later. The solution is drained by gravity from the abdominal cavity into a bag, which is then discarded. One advantage of CAPD over an artificial kidney machine is that the individual can go about many of his or her normal activities during CAPD. A disadvantage, however, is an increased likelihood of infections.

Patients with renal failure may undergo a kidney transplant, which is the surgical replacement of a defective kidney with a functioning kidney from a donor. As with all organ transplants, organ rejection is a possibility (see the Scientific Inquiry feature "Matching Organs for Transplantation"). Receiving a kidney from a close relative decreases the chances of rejection. The current 1-year survival rate following a kidney transplant is 95–98%, but donor organs are in short supply. In the future, it may be possible to use kidneys from pigs or kidneys created in the laboratory.

Disorders of the Urinary Bladder and Urethra

Infections are probably the most common cause of problems in the urinary bladder and urethra. Urine leaving the kidneys is normally free of bacteria. The distal parts of the urethra, however, are normally colonized with bacteria. Most of the time, the flow of urine should be one-way from the bladder to the urethra. However, sometimes harmful bacteria from the urethra are able to gain access to the bladder, especially in females, whose urethra is shorter and broader than that of males. This helps explain why females are relatively more prone to bladder infections (Fig. 16.10). In the United States, approximately 25–40% of women between the ages of 20 and 40 years have had a urinary tract infection. Infections of the bladder usually result in inflammation of the bladder, or **cystitis,**

and infection of the urethra may result in *urethritis*. Almost all urinary tract infections can be cured with appropriate antibiotics.

Bladder stones (Fig. 16.11) can form in people of any age. They most commonly occur as a result of another condition that interferes with normal urine flow, such as a bladder infection with associated inflammation, or prostate enlargement in men. They can also form from kidney stones that pass into the urinary bladder and become larger. The most common symptoms of bladder stones are pain, difficulty urinating or increased frequency of urination, and blood in the urine. Similar to kidney stones, bladder stones may be removed surgically or, preferably, by lithotripsy.

The most common type of cancer affecting the urinary system is cancer of the bladder. In the United States, bladder cancer

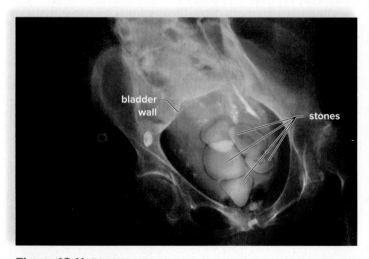

Figure 16.11 Bladder stones. Hard granules such as those seen nearly filling the bladder in this X-ray can form and lodge virtually anywhere in the urinary tract, including the kidneys, ureters, and urethra. ©SPL/Science Source

Matching Organs for Transplantation

As we have seen in this chapter, the kidneys are essential for life. Because humans normally have two kidneys, but can function quite normally with only one, the kidney was an obvious choice as the first complex organ to be successfully transplanted (Fig. 16A). This occurred in 1954 at Peter Bent Brigham Hospital in Boston, where one of 23-year-old Ronald Herrick's kidneys was transplanted into his twin, Richard, who was dying of kidney disease. The operation was successful enough to allow Richard to live another 8 years, while his brother lived with one kidney for over 50 years.

It was no accident, however, that the first successful transplant was performed between twins. Most cells and tissues of the body have molecules on their surface that can vary between individuals; that is, they are polymorphic. The major blood groups are a familiar example. Some individuals have only type A antigen on their red blood cells, some have only type B, some have both (blood type AB), and some have neither (blood type O). Because an individual naturally has antibodies against the A or B antigens their own cells lack, it is essential to make sure donor blood is compatible before giving someone a blood transfusion. Moreover, because other cells besides red blood cells also have these antigens, the first requirement when attempting to match a potential organ donor with a recipient is to make sure the blood types match.

Another step in matching organs for transplantation involves determining whether the donor and the recipient share the same proteins, called human leukocyte antigens (HLAs) or major histocompatibility complex (MHC) proteins. As the latter name implies, these are the major molecules (antigens) that are recognized by the recipient's immune system when transplanted tissue is rejected. Various types of tests can be done to determine what types of

Figure 16A Preparing for transplantation. A donor kidney, kept moist and sterile inside a plastic bag, arrives in an operating room. ©Will & Deni McIntyre/Science Source

MHC molecules the donor cells and recipient cells have, but the ideal situation would be a perfect MHC match between the donor organ and the recipient. This occurs 25% of the time between siblings who have the same mother and father, and also may occur by chance between unrelated individuals.

A final test that is usually performed prior to kidney transplantation is a crossmatching procedure in which the potential donor's cells are mixed with the blood (serum) of the recipient. This is to detect whether the recipient has any antibodies that could damage the cells of the donor kidney. Such antibodies could be present in patients who have been exposed to mismatched MHC molecules through a previous transplantation, blood transfusion, or even pregnancy. Even if the donor and recipient seem closely matched, if the recipient has preformed antibodies that could react with the donor organ, the organ would probably go to another recipient.

Because a perfect match (such as between two identical twins) is rare, transplantations would not have become as commonplace as they are today without the development of immunosuppressive drugs that help prevent transplant rejection. Generally, the drugs used for

this purpose cause an overall suppression of the immune system, leaving transplant patients more susceptible to infections. Because transplant patients usually need to take these drugs for the rest of their lives, there is a need for new drugs that specifically target the cells responsible for the transplant rejection reaction, but leave the rest of the immune system intact.

In order to increase the odds of finding matching kidney donors, the National Kidney Registry has begun using a computerized matching system to find matches. Often, a patient who needs a transplant has one or more potential donors who are willing, but incompatible. The "donor chain" concept allows one of that patient's donors to provide a kidney to a compatible stranger, who in turn has potential donors who could donate to other patients. In an example from 2010, a father in New Jersey donated a kidney to a man in Los Angeles who was part of a chain of five transplants that resulted in a successful kidney transplant for the New Jersey man's 22-month-old child.

Questions to Consider

1. How might the ability to produce organs in a lab from a patient's own stem cells change the process described in this article? What new challenges would it present?
2. Even if two individuals are tested and found to have a perfect match of all of their MHC-I and MHC-II molecules, a kidney transplanted between them still has a higher chance of failing compared to a transplant between identical twins. Can you explain why?
3. Besides the matching criteria described here, what other medical issues should be taken into consideration before a decision is made to transplant an organ into a recipient?

---

is the fourth most common type of cancer in men and the tenth most common in women. Because several toxic by-products of tobacco are secreted in the urine, it is estimated that smoking quintuples the risk of bladder cancer. Certain types of bladder cancer are very malignant. In these cases, the bladder must be removed. With no bladder, the ureters must either be diverted into the large intestine or attached to an opening (or stoma) that is created in the skin near the navel. In the latter case, the patient typically must wear a removable plastic bag to collect the urine. Recently, however, researchers have had success in growing entire human bladders in the lab, and implanting them into a limited

number of patients. Clinical trials of such replacement bladders are ongoing (Fig. 16.12).

Check Your Progress 16.4

1. Explain the specific mechanisms by which kidney failure may cause death.
2. Discuss how smoking may affect the urinary bladder.
3. Summarize the anatomical differences that make females more prone to cystitis. What unique structure may cause difficult urination in males?

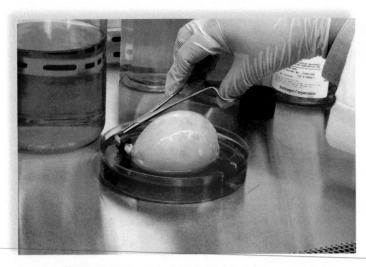

Figure 16.12 Organs grown in the lab. Researchers at Children's Hospital, Boston, have cultured bladder cells on biodegradable bladder-shaped molds to produce an organ that can be implanted in individuals with diseased or defective bladders. ©AP Images/Brian Walker

Conclusion

In this chapter, we explored the role of the kidneys in filtering the blood. Often, kidney stones are caused by an overabundance of uric acid, which may be the result of a combination of a protein-rich diet and dehydration. However, sometimes genetic factors also contribute to the formation of these types of kidney stones. Normally, kidney stones do not present a health problem. However, larger stones can sometimes block the urinary tract and cause problems.

The treatment for this condition involves first breaking the stones into smaller pieces so they can pass through the urinary tract. This is done using a procedure called extracorporeal shock wave lithotripsy (ESWL), which uses sound waves to break up the kidney stones. In addition, a doctor may prescribe a drug called allopurinol to reduce the uric acid levels in the blood, along with recommending a moderation of protein-rich foods in the diet and an increase in water consumption.

SUMMARIZE

16.1 The Urinary System

The **urinary system** has several components, including the kidneys, ureters, urinary bladder, and urethra:

■ The **kidneys** produce **urine.**
■ The **ureters** take urine to the bladder.
■ The **urinary bladder** stores urine.
■ The **urethra** releases urine to the outside.

During urine formation, the kidney serves three homeostatic functions:

■ **Excretion** of metabolic waste, especially nitrogenous waste from the breakdown of amino acids. This process produces **ammonia,** which is excreted mainly as **urea.** Nucleic acids are metabolized to **uric acid,** which may accumulate, causing the disease **gout.**
■ **Osmoregulation,** the maintenance of normal water-salt balance.
■ Helping to regulate the acid-base balance of the blood.

A fourth major function of the kidneys, which is not directly related to urine formation, is the production of the following hormones:

■ **Erythropoietin (EPO),** which stimulates red blood cell production.
■ Renin, an enzyme that leads to the secretion of aldosterone by the adrenal cortex.
■ The kidneys also convert vitamin D to its biologically active form.

16.2 Anatomy of the Kidney and Excretion

Macroscopically, the kidneys are divided into the renal cortex, renal medulla, and renal pelvis. Microscopically, they contain **nephrons.**

■ Each nephron has its own blood supply. The afferent arteriole approaches the **glomerular capsule** and divides to become the **glomerulus,** a capillary tuft. The efferent arteriole leaves the capsule and immediately branches into the peritubular capillary network.
■ In addition to the glomerular capsule, a nephron consists of the **proximal convoluted tubule,** the **loop of the nephron,** the **distal convoluted tubule,** and the **collecting duct.** The spaces between the

podocytes of the glomerular capsule allow small molecules to enter the capsule from the glomerulus. The cuboidal epithelial cells of the proximal convoluted tubule have many mitochondria and microvilli to carry out active transport (and allow passive transport) from the tubule to the blood. The cuboidal epithelial cells of the distal convoluted tubule have numerous mitochondria but lack microvilli. They carry out active transport from the blood to the tubule.

Urine is composed primarily of nitrogenous waste products and salts in water. The following steps occur during urine formation:

■ **Glomerular filtration,** during which water and other small molecules consisting of both nutrients and wastes move from the blood to the inside of the glomerular capsule.
■ **Tubular reabsorption,** during which water and nutrients move back into the blood, mostly at the proximal convoluted tubule.
■ **Tubular secretion,** where certain substances (e.g., hydrogen ions) are transported from the blood into the distal convoluted tubule.

16.3 Regulatory Functions of the Kidneys

Reabsorption of water and the production of a hypertonic urine involves three steps:

■ The reabsorption of salt increases blood volume and pressure because more water is also reabsorbed. Two hormones control the kidneys' reabsorption of sodium (Na^+). Following the secretion of **renin** by the kidneys, **aldosterone,** a hormone secreted by the adrenal cortex, causes reabsorption of sodium ions, and excretion of potassium ions, by the distal convoluted tubules. **Atrial natriuretic hormone (ANH),** secreted by the heart, inhibits the secretion of renin by the kidney, and this inhibits aldosterone. Na^+ is actively transported out of the ascending limb of the loop of the nephron, establishing a solute gradient that increases toward the inner medulla.
■ This gradient draws water from the descending limb of the loop of the nephron and also from the collecting duct. The permeability of the collecting duct is controlled by **antidiuretic hormone (ADH)** from the posterior pituitary gland, which increases the number of plasma membrane channels, called **aquaporins,** in the collecting ducts.

The kidneys also keep blood pH within normal limits:

■ They reabsorb HCO_3^- and excrete H^+ as needed, to maintain the pH at about 7.4. Ammonia also buffers H^+ in the urine.

16.4 Disorders of the Urinary System

These disorders can be divided into those that affect the kidney itself and those that affect the rest of the urinary tract:

■ Disorders specifically affecting the kidneys include **pyelonephritis** (kidney infection) and **kidney stones.**

■ Other major medical conditions, such as diabetes, hypertension, and autoimmune diseases, can lead to kidney failure, which necessitates undergoing **hemodialysis,** continuous ambulatory peritoneal dialysis, or kidney transplantation.

■ The urinary bladder is susceptible to infections (called **cystitis**), **bladder stones,** and cancer.

ASSESS

Testing Yourself

Choose the best answer for each question.

16.1 The Urinary System

1. Label the following diagram of the urinary system and nearby blood vessels.

2. _____ is the elimination of metabolic waste, and _____ is the elimination of undigested food.
   a. Excretion, absorption
   b. Excretion, defecation
   c. Defecation, excretion
   d. None of these are correct.

3. Urea is a by-product of _____ metabolism.
   a. glucose
   b. salt
   c. fat
   d. amino acid

16.2 Anatomy of the Kidney and Excretion

4. Which of the following materials would not be filtered from the blood at the glomerulus?
   a. water
   b. urea
   c. protein
   d. glucose
   e. sodium ions

5. _____ reabsorption from the nephrons is ordinarily 100%.
   a. Sodium
   b. Glucose
   c. Water
   d. Urea

6. The purpose of the loop of the nephron in the process of urine formation is
   a. reabsorption of water.
   b. reabsorption of solutes.
   c. production of filtrate.
   d. secretion of solutes.

7. Label the following diagram of the kidneys.

16.3 Regulatory Functions of the Kidneys

8. Kidneys control blood pH by excreting
   a. HCO_3^-.
   b. H^+.
   c. NH_3.
   d. CO_2.

9. Which of these associations is mismatched?
   a. atrial natriuretic hormone (ANH)—secreted by the heart
   b. renin—secreted by the anterior pituitary
   c. aldosterone—secreted by the adrenal cortex
   d. antidiuretic hormone (ADH)—secreted by the posterior pituitary

10. Which hormone causes a rise in blood pressure?
    a. aldosterone
    b. oxytocin
    c. erythropoietin
    d. atrial natriuretic hormone

16.4 Disorders of the Urinary System

11. Which of these conditions is associated with an infection of the kidney?
    a. urethritis
    b. pyelonephritis
    c. cystitis
    d. All of these are correct.

12. Hemodialysis is done to reduce the effects of which of the following?
    a. uremia
    b. excess EPO in the body
    c. diuretics
    d. cystitis

ENGAGE

Thinking Critically

1. High blood pressure often is accompanied by kidney damage. In some people, the kidney damage is subsequent to the high blood pressure, but in others the kidney damage is what caused the high blood pressure. Explain how a low-salt diet would enable you to determine whether the high blood pressure or the kidney damage came first?

2. The renin-angiotensin-aldosterone system can be inhibited in order to reduce high blood pressure. Usually, the angiotensin-converting enzyme, which converts angiotensin I to angiotensin II, is inhibited by drug therapy. Why would this enzyme be an effective point to disrupt the system?

3. A form of diabetes that may be unfamiliar to most people is diabetes insipidus. When the word *diabetes* is used, everyone tends to think of the type related to insulin. That form of diabetes is diabetes mellitus, which can be type 1 or type 2. In diabetes insipidus, the kidneys are unable to perform their function of water and salt homeostasis due to a lack of ADH in the body. In both types of diabetes, kidney function is affected. Why would frequent urination be a symptom of both diabetes mellitus and diabetes insipidus?

C H A P T E R

17

Nervous System

©Paul Morigi/Getty Images Entertainment/Getty Images

CHAPTER OUTLINE

BEFORE YOU BEGIN

Before beginning this chapter, take a few moments to review the following discussions:

Figure 4.9 How does the sodium–potassium pump move ions across the cell membrane?

Section 11.1 What are the major components of nervous tissue?

Section 11.5 The nervous system interacts with which other body systems?

An Autoimmune Attack

Multiple sclerosis, or MS, is a disease that affects a major control system of the body, the nervous system. The first symptoms of MS tend to be weakness or tingling in the arms and legs, fatigue, a loss of coordination, and blurred vision. As the disease progresses, the individual may experience problems with speech and vision, tremors that make coordinated movement difficult, and numbness in the extremities. We now know that MS is an inflammatory disease that affects the myelin sheaths, which wrap parts of some nerve cells like insulation around an electrical cord. As these sheaths deteriorate, the nerves no longer conduct impulses normally. For unknown reasons, MS often attacks the optic nerves first, before spreading to other areas of the brain. Most researchers think MS results from a misdirected attack on myelin by the body's immune system, although other factors may be involved.

Almost 400,000 people in the United States have MS, and it is the most common disease of the nervous system in young adults. Typically, the first symptoms occur between the ages of 20 and 40. The disease is not contagious, is rarely fatal, and does not appear to be inherited, although some studies suggest a genetic component associated with susceptibility to MS. Like many diseases that affect the nervous system, there is no cure, so affected individuals must deal with their condition for the rest of their lives. The good news is that the disease is not severe in almost 45% of cases, and its symptoms can be controlled with medication.

As you read through this chapter, think about the following questions:

1. What is the function of myelin? What specific type of neurological process will be affected if myelin is damaged?

2. MS affects the CNS only, not the PNS. What cells synthesize myelin in these two systems, and how might this help to explain why MS affects myelin in the CNS only?

3. Would MS cause more of a problem with the white matter or gray matter of the brain?

17.1 Nervous Tissue

Learning Outcomes

Upon completion of this section, you should be able to

1. Compare and contrast the location and function of the central nervous system and the peripheral nervous system.
2. Describe the basic structure of a neuron, and compare the functions of the three classes of neurons.
3. Explain the structure, function, and locations of myelin.

The **nervous system** is responsible for the reception and processing of sensory information from both the external and internal environments. In most animals, including humans, the nervous system has two major anatomical divisions. The **central nervous system (CNS)** consists of the brain and spinal cord, which are located in the midline of the body. The **peripheral nervous system (PNS)** consists of nerves that carry sensory messages to the CNS and motor commands from the CNS to the muscles and glands (Fig. 17.1). Although the CNS and PNS are anatomically and functionally distinct, the two systems work together and are connected to one another.

The nervous system contains two types of cells: neurons and neuroglia (neuroglial cells). **Neurons** are the cells that transmit nerve impulses between parts of the nervous system. **Neuroglia** (Section 11.1) support and nourish neurons, maintain homeostasis, form myelin, and may aid in signal transmission. This section discusses the structure and function of neurons and the myelin sheath that surrounds cell extensions called axons.

Types of Neurons and Neuron Structure

Although the structure of the nervous system is extremely complex, the principles of its operation are simple. There are three classes of neurons: sensory neurons, interneurons, and motor neurons (Fig. 17.2). Their functions are best described in relation to the CNS. A **sensory neuron** takes messages to the CNS. Sensory neurons may be equipped with specialized endings called sensory receptors that detect changes in the environment. An **interneuron** lies entirely within the CNS. Interneurons can receive input from

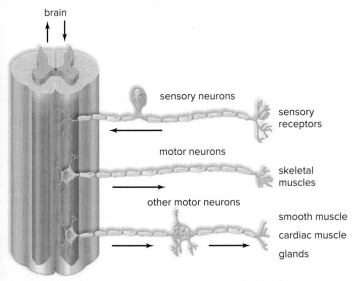

Figure 17.1 Organization of the nervous system. The sensory neurons of the peripheral nervous system take nerve impulses from sensory receptors to the central nervous system (CNS), and motor neurons take nerve impulses from the CNS to muscles and glands.

Figure 17.2 Types of neurons. Sensory neurons, interneurons, and motor neurons have different arrangements in the body. How does each neuron's arrangement correlate with its function? (photos) (axon of neuron): ©Steve Gschmeissner/Science Photo Library/Alamy; (cell body): ©David M. Phillips/ Science Source

sensory neurons and also from other interneurons in the CNS. Thereafter, they sum up all the messages received from these neurons before they communicate with motor neurons. A **motor neuron** takes messages away from the CNS to an effector (an organ, muscle fiber, or gland). Effectors carry out our responses to environmental changes.

Neurons vary in appearance, but they generally consist of three regions: a cell body, dendrites, and an axon. The **cell body** contains the nucleus, as well as other organelles. **Dendrites** are extensions leading toward the cell body that receive signals from other neurons and send them on to the cell body. An **axon** conducts nerve impulses away from the cell body toward other neurons or effectors.

Myelin Sheath

Some axons are covered by a protective **myelin sheath** (Fig. 17.3). In the PNS, this covering is formed by a type of neuroglia called **Schwann cells,** which contain the lipid substance myelin in their plasma membranes. The myelin sheath develops when Schwann cells wrap themselves around an axon as many as 100 times, and in this way lay down many layers of plasma membrane. Because each Schwann cell myelinates only part of an axon, the myelin sheath is interrupted. The gaps where there is no myelin sheath are called **nodes of Ranvier.** Each Schwann cell covers only about 1 millimeter (mm) of an axon. Because a single axon may be several feet long (if it runs from the spinal cord to the foot, for example), several hundred or more Schwann cells may be required to myelinate a single axon.

In the PNS, myelin gives nerve fibers their white, glistening appearance. The myelin sheath also plays an important role in nerve

a.

node of Ranvier

axon

myelin sheath

Schwann cell

b.

Figure 17.3 Myelin sheath.   **a.** In the PNS, a myelin sheath forms when Schwann cells wrap themselves around an axon. **b.** Electron micrograph of a cross-section of an axon surrounded by a myelin sheath.

(b): ©McGraw-Hill Education/Dr. Dennis Emery, Iowa State University, photographer

regeneration within the PNS. If an axon is accidentally severed, the myelin sheath may remain and serve as a passageway for new fiber growth. In the CNS, myelin is produced by a type of neuroglia called **oligodendrocytes.** Unlike in the PNS, nerve regeneration does not seem to occur to any significant degree in the CNS.

The CNS is composed of two types of nervous tissue—gray matter and white matter. **Gray matter** is gray because it contains neurons with short, nonmyelinated axons. **White matter** is white because it contains myelinated axons that run together in bundles called **tracts.** In the CNS, neurons with short axons make up the gray matter because no part of these neurons has a myelin sheath. However, some neurons in the CNS have long axons with a myelin sheath. These long axons, which make up the white matter of the brain, are carrying messages from one part of the CNS to another. The surface layer of the brain is gray matter, and the white matter lies deep within the gray matter. The central part of the spinal cord consists of gray matter, and the white matter surrounds the gray matter.

Check Your Progress 17.1

1. Identify the three classes of neurons and describe their relationship to each other.
2. Describe the three parts of a neuron.
3. Distinguish the cell types that form the myelin in the PNS versus the CNS.
4. Review the structure of gray matter and white matter, and describe where each is found in the CNS and the PNS.

17.2 Transmission of Nerve Impulses

Learning Outcomes

Upon completion of this section, you should be able to

1. Summarize the activities that generate and propagate an action potential.
2. Describe the process of saltatory conduction.
3. Define synaptic integration.
4. Explain the role of neurotransmitters in the transmission of a nerve impulse across a synapse.

The nervous system uses the **nerve impulse** to convey information. The nature of a nerve impulse has been studied by using excised axons and a voltmeter called an oscilloscope. Voltage, which is expressed in millivolts (mV), is a measure of the electrical potential difference between two points. In the case of a neuron, specialized instruments can measure the difference in potentials between the inside and the outside of the axon. The changes in these voltages can be recorded over time in order to visualize a nerve impulse.

Resting Potential

The experimental setup shown in Figure 17.4 illustrates the generation of a nerve impulse. When the axon is not conducting an impulse, the voltmeter records a potential difference across an axonal membrane (membrane of axon) equal to about −70 mV. This reading indicates that the inside of the axon is negative compared to the outside. This is called the **resting potential** because the axon is not conducting an impulse.

The existence of the polarity (charge difference) correlates with a difference in ion distribution on either side of the axonal membrane. As Figure 17.4a shows, the concentration of sodium ions (Na^+) is greater outside the axon than inside, and the concentration of potassium ions (K^+) is greater inside the axon than outside. The unequal distribution of these ions is maintained by the action of carrier proteins in the membrane called **sodium–potassium pumps** (see Fig. 4.9), which actively transport Na^+ out of and K^+ into the axon.

These pumps are always working because the membrane is somewhat permeable to these ions, and they tend to diffuse toward their lesser concentration. Because the membrane is more permeable to K^+ than to Na^+, there are always more positive ions outside the membrane than inside. This accounts for the polarity recorded by the voltmeter. Large, negatively charged organic ions in the axoplasm also contribute to the polarity across a resting axonal membrane.

Action Potential

An **action potential** is a rapid change in polarity across the membrane of an axon as the nerve impulse occurs. An action potential

a. Resting potential: Na^+ outside the axon, K^+ and large anions inside the axon. Separation of charges polarizes the cell and causes the resting potential.

b. Stimulus causes the axon to reach its threshold; the axon potential increases from −70 to −55. The action potential has begun.

Figure 17.4 Resting and action potentials of the axonal membrane. a. Resting potential. A voltmeter indicates the axonal membrane has a resting potential of −70 mV. There is a preponderance of Na^+ outside the axon and a preponderance of K^+ inside the axon. The permeability of the membrane to K^+ compared to Na^+ causes the inside to be negative compared to the outside. **b.** During an action potential, depolarization occurs when Na^+ gates open and Na^+ begins to move inside the axon. **c.** Depolarization continues until a potential of +35 mV is reached. **d.** Repolarization occurs when K^+ gates open and K^+ moves outside the axon. **e.** Graph of an action potential.

c. Depolarization continues as Na^+ gates open and Na^+ moves inside the axon.

d. Action potential ends: repolarization occurs when K^+ gates open and K^+ moves to outside the axon. The sodium–potassium pump returns the ions to their resting positions.

e. An action potential can be visualized if voltage changes are graphed over time.

is an all-or-none phenomenon. If a stimulus causes the axon membrane to de-polarize to a certain level, called *threshold,* an action potential occurs. The strength of an action potential does not change, but an intense stimulus can cause an axon to fire (start an axon potential) more often in a given time interval than a weak stimulus.

The action potential requires two types of gated channel proteins in the membrane (Fig. 17.4*a*). One gated channel protein opens to allow Na$^+$ to pass through the membrane to inside the cell, and the second gated channel opens to allow K$^+$ to pass through the membrane to outside the cell.

Sodium Gates Open   When an action potential begins, the gates of sodium channels open first, and Na$^+$ flows down its concentration gradient into the axon (Fig 17.4*b*). As Na$^+$ moves inside the axon, the membrane potential changes from −70 mV to +35 mV. This is called a *depolarization* because the charge inside the axon changes from negative to positive (Fig. 17.4*c*).

Potassium Gates Open   Second, the gates of potassium channels open, and K$^+$ flows down its concentration gradient to outside the axon. As K$^+$ moves outside the axon, the action potential changes from +35 mV back to −70 mV. This is a *repolarization* because the inside of the axon resumes a negative charge as K$^+$ exits the axon (Fig. 17.4*d*). An action potential takes only 2 milliseconds (msec). In order to visualize such rapid fluctuations in voltage across the axon membrane, researchers generally find it useful to plot the voltage changes over time.

Conduction of an Action Potential   In nonmyelinated axons, the action potential travels down an axon one small section at a time. As soon as an action potential has moved on, the previous section undergoes a **refractory period,** during which the sodium gates are unable to open. Notice, therefore, that the action potential cannot move backward and instead always moves down an axon toward its terminals. When the refractory period is over, the sodium–potassium pump has restored the previous ion distribution by pumping Na$^+$ to outside the axon and K$^+$ to inside the axon.

In myelinated axons, the gated ion channels that produce an action potential are concentrated at the nodes of Ranvier. Because ion exchange occurs only at the nodes, the action potential travels faster than in nonmyelinated axons. This is called **saltatory conduction,** meaning the action potential "jumps" from node to node. Speeds of 200 m per second (450 mph) have been recorded.

Transmission Across a Synapse

Every axon branches into many fine endings, each tipped by a small swelling called an *axon terminal.* Each terminal lies very close to either the dendrite or the cell body of another neuron (or a muscle cell). This region of close proximity is called a **synapse** or chemical synapse (Fig. 17.5). It is important to note that the two neurons at a synapse don't ever physically touch each other.

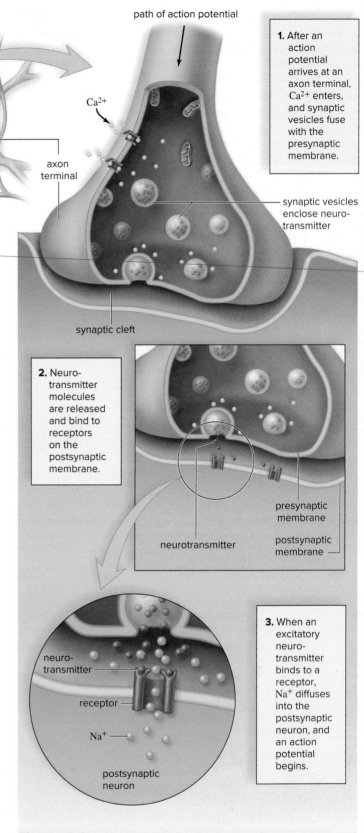

Figure 17.5 Structure and function of a synapse.
Transmission across a synapse from one neuron to another occurs when an action potential causes a neurotransmitter to be released at the presynaptic membrane. The neurotransmitter diffuses across a synaptic cleft and binds to a receptor in the postsynaptic membrane. An action potential may begin in the postsynaptic membrane if the concentration of incoming Na$^+$ reaches threshold.

Instead, they are separated by a tiny gap called the **synaptic cleft.** At a synapse, the membrane of the first neuron is called the *pre*synaptic membrane, and the membrane of the next neuron is called the *post*synaptic membrane.

An action potential cannot cross a synapse. Communication between the two neurons at a chemical synapse is carried out by molecules called **neurotransmitters,** which are stored in synaptic vesicles in the axon terminals. When nerve impulses traveling along an axon reach an axon terminal, gated channels for calcium ions (Ca^{2+}) open, and calcium enters the terminal. This sudden rise in Ca^{2+} stimulates synaptic vesicles to merge with the presynaptic membrane, and neurotransmitter molecules are released into the synaptic cleft. They diffuse across the cleft to the postsynaptic membrane, where they bind with specific receptor proteins.

Depending on the type of neurotransmitter and receptor, the response of the postsynaptic neuron can be toward excitation (causing an action potential to happen) or toward inhibition (stopping an action potential from happening).

Synaptic Integration

The dendrites and cell body of a neuron can have synapses with many other neurons, thus a single neuron may receive many excitatory and inhibitory signals. Excitatory signals have a depolarizing effect, causing the charge across the neuron membrane to move closer to the threshold needed to trigger an action potential. Inhibitory signals usually have a hyperpolarizing effect; that is, they increase the potential difference across the axonal membrane. **Integration** is the summing up of excitatory and inhibitory signals by a neuron (Fig. 17.6). If a neuron receives many excitatory signals (either from different synapses or at a rapid rate from one synapse), the chances are that the neuron's axon will transmit a nerve impulse. On the other hand, if a neuron receives both inhibitory and excitatory signals, the summing up of these signals may prohibit the axon from firing.

Neurotransmitters

More than 100 substances are known or suspected to be neurotransmitters. Two of the more common ones are *acetylcholine* (*ACh*) and *norepinephrine* (*NE*).

Once a neurotransmitter has been released into a synaptic cleft and has initiated a response, it is removed from the cleft. In some synapses, the postsynaptic membrane contains enzymes that rapidly inactivate the neurotransmitter. For example, the enzyme *acetylcholinesterase* (*AChE*) breaks down acetylcholine. In other synapses, the presynaptic membrane rapidly reabsorbs the neurotransmitter, possibly for repackaging in synaptic vesicles. The short existence of neurotransmitters at a synapse prevents continuous stimulation (or inhibition) of postsynaptic membranes.

It is of interest to note here that many drugs that affect the nervous system act by either interfering with or potentiating (enhancing) the action of neurotransmitters. Drugs can enhance or block the release of a neurotransmitter, mimic the action of a neurotransmitter or block the receptor, or interfere with the removal of a neurotransmitter from a synaptic cleft (see Section 17.6). In addition, several naturally occurring venoms and poisons, organophosphate insecticides, and nerve agents such as sarin gas, all

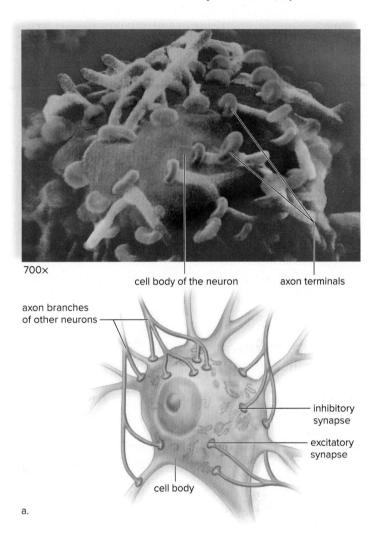

700×

cell body of the neuron   axon terminals

a.

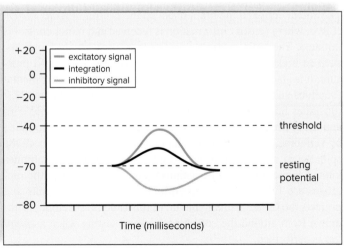

b.

Figure 17.6 Synaptic integration.   a. Inhibitory signals and excitatory signals are summed up in the dendrites and cell body of the postsynaptic neuron. Only if the combined signals cause the membrane potential to rise above threshold does an action potential occur. **b.** In this example, threshold was not reached. (photo) (a): ©Science Source inhibit the AChE enzyme, thus prolonging the activity of ACh. In contrast, the bacterium responsible for botulism produces a toxin that inhibits ACh release, which can paralyze muscles. Botox, a highly diluted preparation of this toxin, is now being used to treat a variety of conditions, from eyelid spasms and back pain to the wrinkles that appear on our faces with age.

Check Your Progress 17.2

1. Describe the activity of the sodium–potassium pump present in neurons.
2. Explain how the changes in Na⁺ and K⁺ ion concentrations that occur during an action potential are associated with depolarization and repolarization.
3. Explain how the nerve impulse is transmitted across a synapse.
4. Define *refractory period, saltatory conduction,* and *synaptic integration.*

17.3 The Central Nervous System

Learning Outcomes

Upon completion of this section, you should be able to

1. Understand the relationship between the CNS and PNS in the nervous system.
2. Describe the structure and function of the spinal cord.
3. Identify the major regions of the brain and provide the main functions of each region.
4. Compare the functions of the primary motor and somato-sensory areas of the brain versus the association areas and processing centers.

The spinal cord and the brain make up the *central nervous system* (*CNS*), where sensory information is received and motor control is initiated. The brain controls or influences many bodily functions, such as breathing, heart rate, body temperature, and blood pressure. It is also the source of our emotions, as well as higher mental functions such as reasoning, memory, and creativity. Figure 17.7 illustrates how the CNS relates to the PNS. Both the spinal cord and the brain are protected by bone. The spinal cord is surrounded by vertebrae, and the brain is enclosed by the skull. Also, both the spinal cord and the brain are wrapped in protective membranes known as **meninges** (*sing.,* meninx). The spaces between the meninges are filled with **cerebrospinal fluid,** which cushions and protects the CNS. A small amount of this fluid is sometimes withdrawn from around the cord for laboratory testing when a *spinal tap* is performed.

The brain has hollow interconnecting cavities termed **ventricles,** which also connect with the hollow *central canal* of the spinal cord. The ventricles produce and serve as a reservoir for cerebrospinal fluid, as does the central canal. Normally, any excess cerebrospinal fluid drains away into the cardiovascular system. However, blockages can occur. In an infant, the brain and skull can enlarge due to cerebrospinal fluid accumulation, resulting in a condition called hydrocephalus ("water on the brain"). If cerebrospinal fluid collects in an adult, the brain cannot enlarge and instead

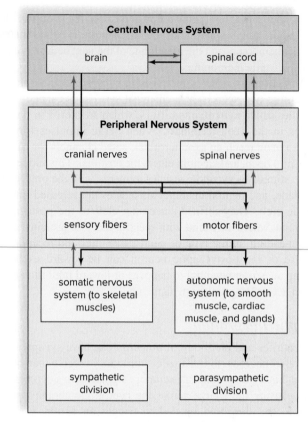

Figure 17.7 Organization of the nervous system.   The CNS is composed of the spinal cord and brain. The PNS is composed of the cranial and spinal nerves. Nerves contain both sensory and motor fibers. In the somatic system, nerves conduct impulses from sensory receptors to the CNS, and motor impulses from the CNS to the skeletal muscles. In the autonomic system, consisting of the sympathetic and parasympathetic divisions, motor impulses travel to smooth muscle, cardiac muscle, and glands.

is pushed against the skull. Such situations can cause brain damage if not quickly corrected.

The Spinal Cord

The **spinal cord** extends from the base of the brain through a large opening in the skull called the foramen magnum and into the vertebral canal formed by openings in the vertebrae.

Structure of the Spinal Cord

Figure 17.8*a* shows how an individual vertebra protects the spinal cord. The spinal nerves project from the cord between the vertebrae that make up the vertebral column. Fluid-filled *intervertebral discs* cushion and separate the vertebrae. If a disc ruptures, the vertebrae press on the spinal cord and spinal nerves, causing pain and loss of motor function.

A cross-section of the spinal cord shows a central canal, gray matter, and white matter (Fig. 17.8*b, c*). The central canal contains cerebrospinal fluid, as do the meninges that protect the spinal cord. The gray matter is centrally located and shaped like the letter *H.* Portions of sensory neurons and motor neurons are found there, as are interneurons that communicate with these two types of neurons.

a.

dorsal

dorsal root

dorsal root
ganglion

ventral root

spinal
nerve

white matter

gray matter

central canal

meninges

b.   ventral

5x

white   central   gray
matter   canal   matter
c.

Figure 17.8 Spinal cord. **a.** The spinal cord passes through the vertebral canal formed by the vertebrae. **b.** The spinal cord has a central canal filled with cerebrospinal fluid, gray matter in an H-shaped configuration, and white matter. The white matter contains tracts that take nerve impulses to and from the brain. **c.** Micrograph of a cross-section of the spinal cord. (c): ©Karl E. Deckart/Medical Images

The dorsal root of a spinal nerve contains sensory fibers entering the gray matter, and the ventral root of a spinal nerve contains motor fibers exiting the gray matter. The dorsal and ventral roots join before the spinal nerve leaves the vertebral canal. Spinal nerves are a part of the PNS.

The white matter of the spinal cord surrounds the gray matter. The white matter contains ascending tracts taking information to the brain (primarily located dorsally) and descending tracts taking information from the brain (primarily located ventrally). Because the tracts cross after they enter and exit the CNS, the left side of the brain controls the right side of the body, and the right side of the brain controls the left side of the body.

Functions of the Spinal Cord

The spinal cord provides a means of communication between the brain and the peripheral nerves that leave the cord. When someone touches your hand, sensory receptors generate nerve impulses that pass through sensory fibers to the spinal cord and up ascending tracts to the brain. When we voluntarily move our limbs, motor impulses originating in the brain pass down descending tracts to the spinal cord and out to our muscles by way of motor fibers. Therefore, if the spinal cord is severed, we suffer a loss of sensation and a loss of voluntary control—that is, paralysis (see Section 17.7).

In Section 17.5, we will see that the spinal cord is also the center for thousands of reflex arcs, which allow the nerves and muscles to respond very quickly to potentially dangerous stimuli. A stimulus causes sensory receptors to generate nerve impulses that travel in sensory axons to the spinal cord. Interneurons integrate the incoming data and relay signals to motor neurons. The reflex is complete when motor axons cause skeletal muscles to contract or a gland or an organ to respond. Each interneuron in the spinal cord has synapses with many other neurons and, therefore, they send signals to several other interneurons and motor neurons.

The spinal cord plays a similar role for the internal organs. For example, when blood pressure falls, sensory receptors in the carotid arteries and aorta generate nerve impulses that pass through sensory fibers to the cord and then up an ascending tract to a cardiovascular center in the brain. The brain then sends nerve impulses down a descending tract to the spinal cord. Motor impulses then travel via spinal nerves to smooth muscle in the walls of arteries and arterioles, causing their constriction so that the blood pressure rises.

The Brain

The human **brain** has been called the last great frontier of biology. The goal of modern neuroscience is to understand the structure and function of the brain's various parts well enough that it will be possible to prevent or correct the many disorders that can afflict it, robbing many people of a normal life. This section provides only a glimpse of what is known about the brain and the modern avenues of research.

We will discuss the brain with reference to its four major parts: the cerebrum, the diencephalon, the cerebellum, and the brain stem. It may be helpful for you to associate these four regions with the brain's ventricles (fluid-filled spaces): the cerebrum with the two lateral ventricles, the diencephalon with the third ventricle,

Figure 17.9 The human brain.   a. The cerebrum, seen here in longitudinal section, is the largest part of the brain in humans. The right cerebral hemisphere is shown here. **b.** The cerebrum has left and right cerebral hemispheres, which are connected by the corpus callosum. (b): ©Colin Chumbley/Science Source

and the brain stem and the cerebellum with the fourth ventricle (Fig. 17.9*a*).

The Cerebrum

The **cerebrum** is the largest portion of the brain in mammals, such as humans. The cerebrum is the last center to receive sensory input and carry out integration before commanding voluntary motor responses. It communicates with and coordinates the activities of the other parts of the brain.

The Cerebral Hemispheres

Just as the human body has two halves, so does the cerebrum. These halves are called the left and right **cerebral hemispheres** (Fig. 17.9*b*). A deep groove, the *longitudinal fissure,* divides the left and right cerebral hemispheres. The two cerebral hemispheres communicate via the *corpus callosum,* an extensive bridge of nerve tracts.

Shallow grooves called *sulci (sing.,* sulcus) divide each hemisphere into lobes (Fig. 17.10). The *frontal lobe* is the most ventral of the lobes (directly behind the forehead). The *parietal lobe* is posterior to the frontal lobe. The *occipital lobe* is posterior to the parietal lobe (at the rear of the head). The *temporal lobe* lies inferior to the frontal and parietal lobes (at the temple and the ear). Each lobe is associated with particular functions, as indicated in Figure 17.10.

The **cerebral cortex** is a thin, highly convoluted outer layer of gray matter that covers the cerebral hemispheres. Each fold, or convolution, in the cortex is called a *gyrus.* The cerebral cortex contains over 1 billion cell bodies and is the region of the brain that accounts for sensation, voluntary movement, and all the thought processes we associate with consciousness.

Primary Motor and Sensory Areas of the Cortex   The cerebral cortex contains motor areas, sensory areas, and association areas. The *primary motor area* is in the frontal lobe just anterior to the central sulcus (Fig. 17.10). Voluntary commands to skeletal muscles begin in the primary motor area, and each part of the body is controlled by a certain section (Fig. 17.11*a*). Large areas of cerebral cortex are devoted to controlling structures that carry out very fine, precise movements, such as those associated with the movements of the hand. The *primary somatosensory area* is just posterior to the central sulcus in the parietal lobe. Sensory information from the skin and skeletal muscles arrives here, where each part of the body is sequentially represented (Fig. 17.11*b*). As with the primary motor area, large parts of the primary visual area are dedicated to receiving information from body areas generating the most sensory information, such as the face and hands.

Association Areas   *Association areas* are places where integration occurs and where memories are stored. Anterior to the primary motor area is a premotor area. The premotor area organizes motor functions for skilled motor activities, such as walking and talking at the same time. Next, the primary motor area sends signals to the cerebellum, which integrates them.

The *somatosensory association area,* located just posterior to the primary somatosensory area, processes and analyzes sensory information from the skin and muscles. The *visual association area* associates new visual information with previously received visual information. It might "decide," for example, whether we have seen this face or scene or symbol in the past before. The *auditory association area* performs the same functions with regard to sounds.

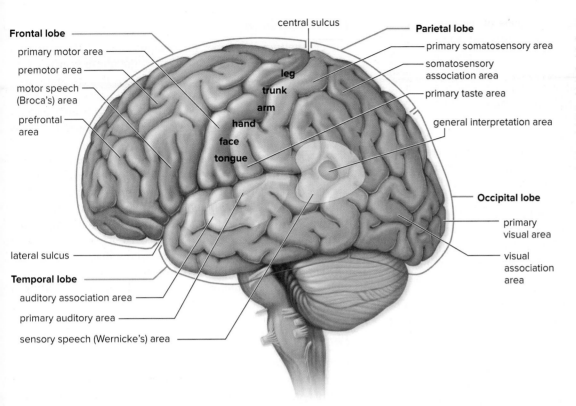

Figure 17.10 The lobes of a cerebral hemisphere. Each cerebral hemisphere is divided into four lobes: frontal, parietal, temporal, and occipital. The frontal lobe contains centers for reasoning and movement, the parietal lobe for somatic sensing and taste, the temporal lobe for hearing, and the occipital lobe for vision.

a. Primary motor area

b. Primary somatosensory area

Figure 17.11 The primary motor and somatosensory areas. In these drawings, the size of the body part reflects the amount of cerebral cortex devoted to that body part. For example, the amount of primary motor cortex (**a**) and somatosensory cortex (**b**) devoted to the thumb, fingers, and hand is greater than that for the foot and toes.

Processing Centers   Processing centers of the cortex receive information from the other association areas and perform higher-level analytical functions. The *prefrontal area,* a processing center in the frontal lobe, receives information from the other association areas and uses it to reason and plan our actions. Integration in this center accounts for our most cherished human abilities (e.g., to think critically and to formulate appropriate behaviors).

The unique ability of humans to speak is partially dependent upon two processing centers found only in the left cerebral cortex: *Wernicke's area* in the posterior part of the left temporal lobe and *Broca's area* in the left frontal lobe. Broca's area is located just anterior to the portion of the primary motor area for speech musculature (lips, tongue, larynx) (see Fig. 17.14). Wernicke's area helps us understand both the written and spoken word and sends the information to Broca's area. Broca's area adds grammatical refinements and directs the primary motor area to stimulate the appropriate muscles for speaking.

Central White Matter   Much of the rest of the cerebrum beneath the cerebral cortex is composed of white matter. Tracts within the cerebrum take information between the different sensory, motor, and association areas pictured in Figures 17.10 and 17.11. The corpus callosum contains tracts that join the two cerebral hemispheres. Descending tracts from the primary motor area communicate with various parts of the brain, and ascending tracts from lower brain centers send sensory information up to the primary somatosensory area.

Basal Nuclei   Although the majority of the cerebrum is composed of tracts, there are masses of gray matter located deep within the white matter. These **basal nuclei** integrate motor commands, ensuring that proper muscle groups are activated or inhibited.

The Diencephalon

The hypothalamus and the thalamus are in the *diencephalon,* a region that encircles the third ventricle (see Fig. 17.9a). The **hypothalamus** forms the floor of the third ventricle. The hypothalamus is an integrating center that helps maintain homeostasis. It regulates hunger, sleep, thirst, body temperature, and water balance. The hypothalamus controls the pituitary gland and thereby serves as a link between the nervous and endocrine systems.

The **thalamus** consists of two masses of gray matter that form the sides and roof of the third ventricle. The thalamus receives all sensory input except for smell. Visual, auditory, taste, and somatosensory information arrives at the thalamus via the cranial nerves and tracts from the spinal cord. The thalamus integrates this information and sends it on to the appropriate portions of the cerebrum. The thalamus is also involved in arousal of the cerebrum, and it participates in higher mental functions, such as memory and emotions.

The **pineal gland,** which secretes the hormone melatonin, is also located in the diencephalon (see Fig. 17.9a). *Melatonin* is a hormone that is involved in maintaining our normal sleep-wake cycle. It is sometimes recommended for people suffering from insomnia, jet lag, or that have nontraditional work schedules, such as night or rotating shifts. It is also used to treat *seasonal affective disorder* (*SAD*), a form of depression that occurs at specific times of the year, usually in the winter months.

The Cerebellum

The **cerebellum** lies under the occipital lobe of the cerebrum and is separated from the brain stem by the fourth ventricle (see Fig. 17.9a). The cerebellum has two portions that are joined by a narrow, median portion. Each portion is primarily composed of white matter, which, in longitudinal section, has a treelike pattern. Overlying the white matter is a thin layer of gray matter that forms a series of complex folds.

The cerebellum receives sensory input from the joints, muscles, and other sensory pathways about the present position of body parts. It also receives motor output from the cerebral cortex about where these parts should be located. After integrating this information, the cerebellum sends motor signals by way of the brain stem to the skeletal muscles. In this way, the cerebellum maintains posture and balance. It ensures that muscles work together to produce smooth, coordinated voluntary movements, such as those required when playing the piano or hitting a baseball.

The Brain Stem

The **brain stem** contains the midbrain, the pons, and the medulla oblongata (see Fig. 17.9a). The **midbrain** acts as a relay station for tracts passing between the cerebrum and the spinal cord or cerebellum. It also has reflex centers for visual, auditory, and tactile responses. The word *pons* means "bridge" in Latin, and true to its name, the **pons** contains bundles of axons traveling between the cerebellum and the rest of the CNS. In addition, the pons functions with the medulla oblongata to regulate the breathing rate.

The **medulla oblongata** regulates vital functions like heartbeat, breathing, and blood pressure. It also contains reflex centers for vomiting, coughing, sneezing, hiccuping, and swallowing. The medulla oblongata lies just superior to the spinal cord. It contains tracts that ascend or descend between the spinal cord and higher brain centers.

The **reticular activating system (RAS)** contains the *reticular formation,* a complex network of nuclei and nerve fibers that extend the length of the brain stem. The reticular formation receives sensory signals that it sends up to higher centers, and motor signals that it sends to the spinal cord.

The RAS arouses the cerebrum via the thalamus and causes a person to be alert. Apparently, the RAS can filter out unnecessary sensory stimuli, explaining why you can study with the TV on. If you want to awaken the RAS, surprise it with a sudden stimulus, like an alarm clock, smelling salts, or splashing your face with cold water; if you want to deactivate it, remove visual and auditory stimuli. A severe injury to the RAS can cause a person to be comatose, from which recovery may be impossible.

Electroencephalograms

The electrical activity of the brain can be recorded in the form of an *electroencephalogram* (*EEG*). Electrodes are taped to different parts of the scalp, and an instrument records the so-called brain waves. The EEG is a diagnostic tool. For example, an irregular pattern can signify epilepsy or a brain tumor. A flat EEG is often used as a clinical and legal criterion for brain death.

17.4 The Limbic System and Higher Mental Functions

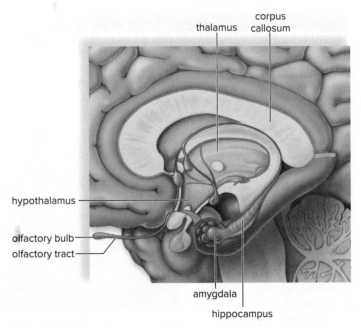

Figure 17.12 The limbic system.   Structures deep within the cerebral hemispheres and surrounding the diencephalon join higher mental functions such as reasoning with more primitive feelings such as fear and pleasure.

Emotions and higher mental functions are associated with the limbic system in the brain. The limbic system blends primitive emotions (rage, fear, joy, sadness) and higher mental functions (reason, memory) into a united whole. It accounts for why activities such as sexual behavior and eating seem pleasurable and also for why, say, mental stress can cause high blood pressure.

Anatomy of the Limbic System

The **limbic system** is a complex network of tracts and nuclei that includes portions of the cerebral lobes, the basal nuclei, and the diencephalon. Two significant structures within the limbic system are the hippocampus and the amygdala, which are essential for learning and memory (Fig. 17.12). The **hippocampus** is a seahorse-shaped structure deep in the temporal lobe that is well situated to communicate with the prefrontal area of the brain, which is also involved in learning and memory. The **amygdala** is an almond-shaped structure in the limbic system that allows us to respond to and display anger, avoidance, defensiveness, and fear. The amygdala prompts release of adrenaline and other hormones into the bloodstream.

A case report published in 2010 described a woman who lacked an amygdala due to a rare genetic disease. Not only did she seem to exhibit no response to typically fearful stimuli like snakes, tarantulas, and haunted houses, but she also could not recognize fearful facial expressions of others. She retained her ability to feel other emotions like surprise, happiness, and disgust, however. While acknowledging they need to study more subjects, the researchers suggest this case shows that the amygdala is a critical brain region for triggering a state of fear.[1]

Higher Mental Functions

The well-developed human cerebrum is responsible for higher mental functions such as memory and learning, as well as language and speech. As in other areas of biological study, brain research has progressed due to technological breakthroughs. Neuroscientists now have a wide range of techniques at their disposal for studying the human brain, including modern technologies that allow us to record its functioning.

Memory and Learning

Just as the connecting tracts of the corpus callosum are evidence that the two cerebral hemispheres work together, so the limbic system indicates that cortical areas may work with lower centers to produce memory and learning. **Memory** is the ability to hold a thought in mind or to recall events from the past, ranging from a word we learned only yesterday to an early emotional experience that has shaped our lives. **Learning** takes place when we retain and utilize past memories.

Types of Memory   The last time you decided to try a new restaurant, you may have looked up the number and tried to remember it for a short period of time. If you said you were trying to keep it in the forefront of your brain, you were exactly correct. The prefrontal area, which is active during *short-term memory,* lies just behind our forehead! On the other hand, there are probably some telephone numbers you have memorized. These have gone into *long-term memory.* Think of the phone number of someone close to you whom you call frequently. Can you bring the number to mind without also thinking about the place or person associated with that number? Most likely you cannot, because long-term memory is typically a mixture of *semantic memory* (of ideas, concepts, and meanings) and *episodic memory* (of specific facts,

---

[1] Feinstein, J. S. et al. "The Human Amygdala and the Induction and Experience of Fear," *Current Biology* 21(1):34–38 (2011). Epub 2010 December 16.

Why Do We Sleep?

When was the last time you went without enough sleep? How did you feel the next day? What do you think would happen if you slept only 3 or 4 hours a night for a month? But if you sleep 8 hours a night, that's about 122 days a year. Couldn't you do something productive or fun with all that extra time?

Our bodies need sleep as much as they need food. Sleep deprivation affects us both mentally and physically. Without enough sleep, we become irritable and have trouble concentrating. A 2008 study showed that adults deprived of sleep for 35 hours had increased activity in the amygdala, the brain's emotional center. Sleep-deprived bodies lose coordination and agility. Our reaction times are slower. Studies are also showing links between too little sleep and obesity, heart disease, and diabetes.

Virtually all mammals and birds sleep. Most fish and insects do, too, or at least they enter a sleeplike state. Even dolphins, which need to keep surfacing to breathe, seem to sleep in one-half of their brain at a time! And yet, scientists are still debating exactly why we need to sleep at all. Considering that we spend around a third of our lives in this inactive, seemingly unproductive state, the essential function of sleep, as one sleep researcher noted, "may be the biggest open question in biology."

Most people probably believe that sleep is simply a time for the body to rest. However, no matter how quietly you may sit or lie down, watching TV or just meditating all day, you will still need to sleep. Another idea is that sleeping serves to protect various creatures by keeping them still and quiet at times when predators are more common. However, this does not explain the severe effects of sleep deprivation—and besides, a sleeping animal cannot run away or defend itself as well as an awake one.

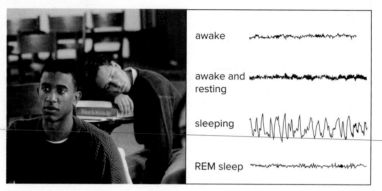

Figure 17A The sleeping brain.   In addition to an increased need for daytime naps, chronic sleep deprivation may lead to serious health problems. When we are awake, attentive, or in different stages of sleep, the brain produces characteristic types of electrical waves that can be measured with an EEG.
(students): ©Doug Menuez/Getty Images; (EEG): ©BSIP SA/Alamy

Prior to 1953, most scientists believed that during sleep the brain was completely inactive. That was the year that a graduate student at the University of Chicago studied his sleeping son and noticed there were periods when his eyes darted rapidly back and forth. Subsequent research has shown that during this rapid eye movement (REM) sleep the brain is very active. REM sleep may be important for brain development; infants spend about 50% of their sleeping time in this state compared to 15–20% for adults. REM sleep may also be a time when the brain is building neurological connections to consolidate memories learned from experiences during that day. A 2004 study showed that subjects who were asked to work on a math problem that contained a "hidden solution" were about twice as likely to realize that solution when they returned after a night's sleep compared to subjects who spent an equal amount of time awake.

If a person is hooked up to an EEG during REM sleep (Fig. 17A), the brain appears to be awake, but the muscles are very still. If people are awakened during REM sleep, they are very likely to report that they have been dreaming,

and dreams may also play a role in sorting through our experiences, saving some as memories and deleting others.

Another, often overlooked, task the brain must perform is forgetting. During the activities of an average day, the brain is taking in mountains of information by building new connections between neurons. Because maintaining these new connections requires energy, if some were not broken down, the brain would soon require more energy than the body could supply. Studies show that this reduction of connections in the brain occurs during non-REM sleep. Perhaps sleep allows the brain time to "clean out" a large percentage of these memories, while solidifying others.

So if you have trouble sleeping, also known as insomnia, how can you get a better night's rest? Experts note that caffeine, nicotine, and even alcohol can disrupt sleep patterns. Exercising at night can raise metabolism when it needs to be slowing down for sleep. Taking naps longer than 15 minutes can make it harder to fall asleep later, as can exercising within 3 hours of bedtime. Regardless of the true purpose of sleep, all sleep researchers would agree that we can't rob ourselves of sleep for long without serious consequences.

Questions to Consider

1. Briefly, how would you define what sleep is?
2. Do you agree that the necessity of sleep for mammals and birds is one of the biggest mysteries in biology? Why do you think that the answer has eluded scientists for so long?
3. Of the several ideas presented here regarding the purpose of sleep, which one makes the most sense to you?

persons, events, etc.). Due to brain damage, some people lose one type of memory but not the other. For example, without a working episodic memory, they can carry on a conversation but have no recollection of recent events. If you are talking to them and then leave the room, they don't remember you when you come back!

Skill memory is another type of memory that can exist independent of episodic memory. Skill memory is involved in performing motor activities, such as riding a bike, playing ice hockey, or typing a letter using a computer keyboard. When a person first learns a skill, more areas of the cerebral cortex are involved than after the skill is perfected. In other words, you have to think about what you are doing when you learn a skill, but later the actions become automatic. Skill memory involves all the motor areas of the cerebrum below the level of consciousness.

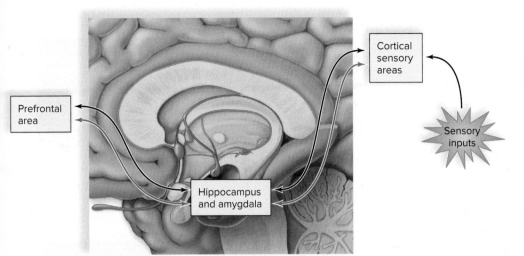

Figure 17.13 Long-term memory circuits. The hippocampus and the amygdala are believed to be involved in the storage and retrieval of memories. Semantic memory (red arrows) and episodic memory (black arrows) are stored separately. Therefore, you can lose one without losing the other.

Long-Term Memory Storage and Retrieval The first step toward curing memory disorders is to know what parts of the brain are functioning when we remember something. Our long-term memories are stored in bits and pieces throughout the sensory association areas of the cerebral cortex. Visions are stored in the visual association area, sounds are stored in the auditory association area, and so forth. The hippocampus serves as a bridge between the sensory association areas and the prefrontal area. A classic case that demonstrated this involved a physicist identified as S.S., whose hippocampus was selectively destroyed by a herpesvirus infection. Even though he still had a high IQ and could remember childhood events and physics equations, S.S. forgot recent experiences within a few minutes. By observing S.S. and patients like him, we can surmise that the hippocampus must be involved in the conversion of short-term memories to long-term memories. As discussed in the Health feature "Why Do We Sleep?," different stages of sleep may be involved in the formation of memory connections in the brain.

So why are some memories so emotionally charged? The amygdala is responsible for fear conditioning and associating danger with sensory stimuli received from both the diencephalon and the cortical sensory areas. Figure 17.13 diagrammatically illustrates the long-term memory circuits of the brain.

In addition to studying what regions of the brain are involved in memory formation, neurobiologists want to know what exactly neurons are doing when we store memories and bring them back. They have discovered a chemical process called *long-term potentiation (LTP)*, which is an enhanced response at synapses within the hippocampus. LTP seems to mainly involve glutamate, an amino acid that can function as a neurotransmitter. To prove this, investigators at the Massachusetts Institute of Technology produced mice that lacked the receptor for glutamate only in the hippocampus. Unlike control mice, the defective mice could not learn to run the maze, demonstrating an important role for this chemical in memory.

Language and Speech

Language is obviously dependent upon semantic memory. Therefore, we would expect some of the same areas in the brain to be involved in both memory and language. Seeing words and hearing words depend on the primary visual cortex in the occipital lobe and the primary auditory cortex in the temporal lobe, respectively (Fig. 17.14*a*). And speaking words depends on motor centers in the frontal lobe. Functional imaging by a technique known as PET (positron emission tomography) supports these suppositions (Fig. 17.14*b*).

From studies of patients with speech disorders, it has been known for some time that damage to the motor speech (Broca's) area results in an inability to speak, while damage to the sensory speech (Wernicke's) area results in the inability to comprehend speech. Actually, any disruption of pathways between various centers of the brain can contribute to an inability to comprehend our environment and use speech correctly.

An additional observation pertaining to language and speech is the recognition that the left brain and right brain have different functions. Only the left hemisphere, not the right, contains a Broca's area and Wernicke's area. In an attempt to cure epilepsy in the early 1940s, the corpus callosum was surgically severed in some patients. Later studies showed that these split-brain patients could only name objects seen by the left hemisphere. If objects were viewed only by the right hemisphere, a split-brain patient could choose the proper object for a particular use but was unable to name it. Based on these and various other studies, the popular idea developed that the left brain can be contrasted with the right brain along these lines:

Left Hemisphere	Right Hemisphere
Verbal	Nonverbal, visual, spatial
Logical, analytical	Intuitive
Rational	Creative

Further, researchers generally came to believe that one hemisphere was dominant in each person, accounting in part for personality traits. However, recent studies suggest that the hemispheres simply process the same information differently. The right hemisphere is more global in its approach, whereas the left hemisphere is more specific. For example, if only the left cerebral cortex is functional, a person has difficulty communicating the emotions involved with language. Therefore, both sides of the brain play a role in language use.

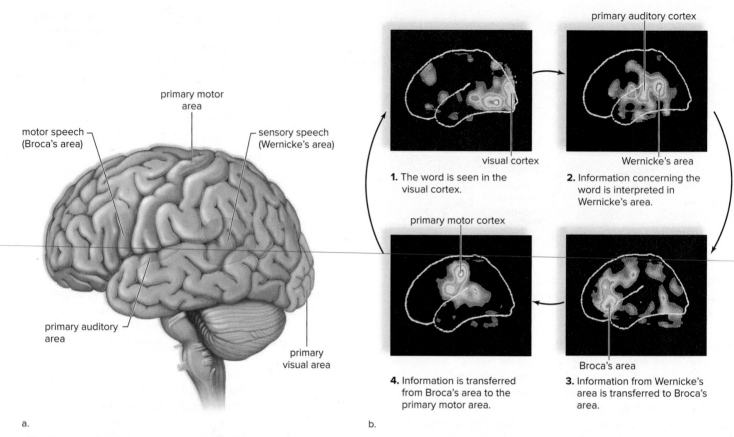

Figure 17.14 Brain regions involved in language and speech. **a.** The labeled areas are thought to be involved in speech comprehension and use. **b.** These PET images show the cortical pathway for reading words and then speaking them. Red indicates the most active areas of the brain, and blue indicates the least active areas. (b, all): ©Marcus Raichle

---

Check Your Progress 17.4

1. Summarize the role of the major structure of the limbic system in memory.
2. Discuss why two forms of memory are needed by the body.
3. Summarize the evidence from the mouse study that showed the role of glutamate in memory formation.
4. Describe the specific functions of Broca's and Wernicke's areas.

17.5 The Peripheral Nervous System

Learning Outcomes

Upon completion of this section, you should be able to

1. Describe the overall anatomy of the peripheral nervous system, including the cranial nerves and spinal nerves.
2. Distinguish between the somatic and autonomic divisions of the peripheral nervous system.
3. Contrast the overall functions of the sympathetic and parasympathetic divisions of the autonomic nervous system.

The *peripheral nervous system* (*PNS*) lies outside the central nervous system and is composed of nerves and ganglia. In the PNS, **nerves** are bundles of axons. The axons that occur in nerves are called **nerve fibers.**

Sensory fibers carry information to the CNS, and motor fibers carry information away from the CNS. **Ganglia** (*sing.,* ganglion) are swellings associated with nerves that contain collections of cell bodies.

Humans have 12 pairs of **cranial nerves** attached to the brain. By convention, the pairs of cranial nerves are referred to by Roman numerals (Fig. 17.15a). Some of the cranial nerves contain only sensory input fibers, while others contain only motor output fibers. The remaining are mixed nerves that contain both sensory and motor fibers. Cranial nerves are largely concerned with the head, neck, and facial regions of the body. However, the vagus nerve (X) has branches not only to the pharynx and larynx, but also to most of the internal organs.

The **spinal nerves** of humans emerge in 31 pairs from between openings in the vertebral column of the spinal cord. Each spinal nerve originates when two short branches, or roots, join together (Fig. 17.15b). The dorsal root (at the back) contains sensory fibers that conduct impulses inward (toward the spinal cord) from sensory receptors. The cell body of a sensory neuron is in a

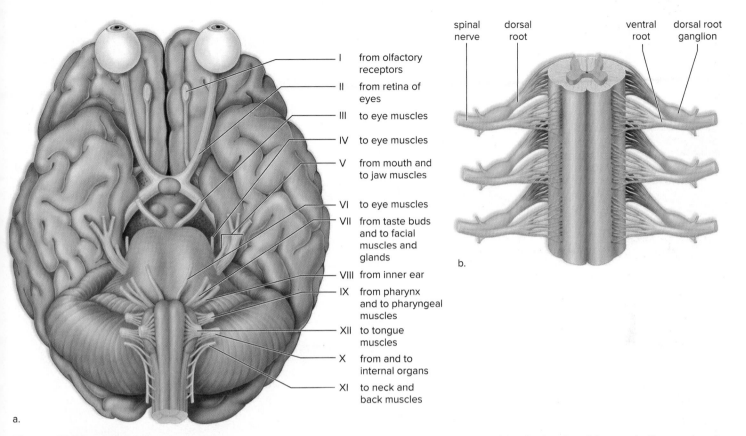

	from olfactory receptors
I	from olfactory receptors
II	from retina of eyes
III	to eye muscles
IV	to eye muscles
V	from mouth and to jaw muscles
VI	to eye muscles
VII	from taste buds and to facial muscles and glands
VIII	from inner ear
IX	from pharynx and to pharyngeal muscles
XII	to tongue muscles
X	from and to internal organs
XI	to neck and back muscles

Figure 17.15 Cranial and spinal nerves.   **a.** Ventral surface of the brain showing the attachments of the 12 pairs of cranial nerves. **b.** Cross-section of the spinal cord showing three pairs of spinal nerves. The human body has 31 pairs of spinal nerves altogether, and each spinal nerve has a dorsal root and a ventral root attached to the spinal cord.

dorsal root ganglion. The ventral root (at the front) contains motor fibers that conduct impulses outward (away from the cord) to effectors. Notice, then, that all spinal nerves are mixed nerves that contain many sensory and motor fibers. Each spinal nerve serves the particular region of the body in which it is located. For example, the intercostal muscles of the rib cage are innervated by thoracic nerves.

Somatic System

The PNS is subdivided into the somatic system and the autonomic system. The **somatic system** serves the skin, skeletal muscles, and tendons. It includes nerves that take sensory information from external sensory receptors to the CNS and motor commands away from the CNS to the skeletal muscles. Some actions in the somatic system are due to **reflex actions,** which are automatic responses to a stimulus. A reflex occurs quickly, without our even having to think about it. Other actions are voluntary, and these always originate in the cerebral cortex, as when we decide to move a limb.

The Reflex Arc

A reflex arc is a nerve pathway that carries out a reflex. Reflexes are programmed, built-in circuits that allow for protection and

survival. They enable the body to react swiftly to stimuli that could disrupt homeostasis. They are present at birth and require no conscious thought to take place. Figure 17.16 illustrates the path of a withdrawal reflex that involves only the spinal cord. If your hand touches a sharp pin, sensory receptors in the skin generate nerve impulses that move along sensory fibers through the dorsal root ganglia toward the spinal cord. Sensory neurons that enter the cord dorsally pass signals on to many interneurons. Some of these interneurons synapse with motor neurons whose short dendrites and cell bodies are in the spinal cord. Nerve impulses travel along these motor fibers to an effector, which brings about a response to the stimulus. In this case, the effector is a muscle, which contracts so that you withdraw your hand from the pin. Various other reactions are also possible—you will most likely look at the pin, wince, and cry out in pain. This whole series of responses occurs because some of the interneurons involved carry nerve impulses to the brain. The brain makes you aware of the stimulus and directs these other reactions to it. You do not feel the pain until your brain receives the information and interprets it.

Autonomic System

The **autonomic system** of the PNS regulates the activity of cardiac and smooth muscle and glands. The system is composed of the

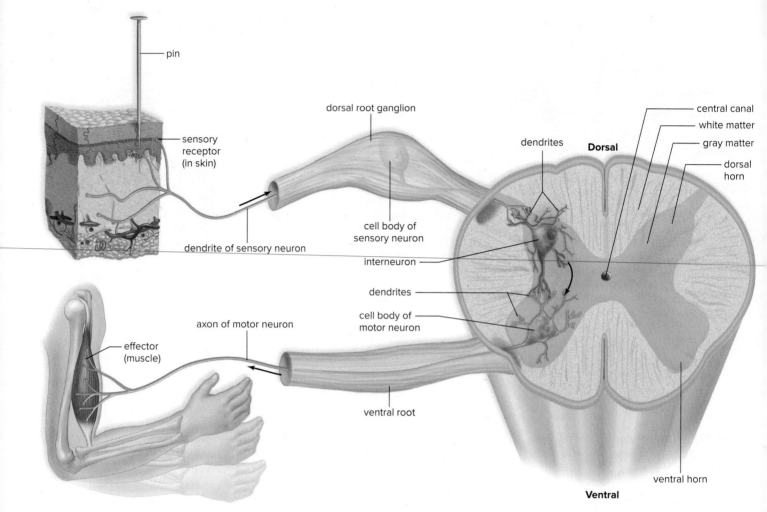

Figure 17.16 A somatic reflex arc showing the path of a spinal reflex.   A stimulus (e.g., sharp pin) causes sensory receptors in the skin to generate nerve impulses that travel in sensory axons to the spinal cord. Interneurons integrate data from sensory neurons and then relay signals to motor axons. Motor axons convey nerve impulses from the spinal cord to a skeletal muscle, which contracts. Movement of the hand away from the pin is the response to the stimulus.

sympathetic and parasympathetic divisions (Fig. 17.17). These two divisions have several features in common: (1) they function automatically and usually in an involuntary manner; (2) they innervate all internal organs; and (3) for each signal, they utilize two motor neurons that synapse at a ganglion. The first neuron has a cell body within the CNS, and its axon is called the preganglionic fiber. The second neuron has a cell body within the ganglion, and its axon is termed the postganglionic fiber.

Reflex actions, such as those that regulate the blood pressure and breathing rate, are especially important to the maintenance of homeostasis. These reflexes begin when the sensory neurons in contact with internal organs send messages to the CNS. They are completed by motor neurons within the autonomic system.

Sympathetic Division

The cell bodies of preganglionic fibers in the **sympathetic division** are located in the middle, or thoracolumbar, portion of the spinal cord. The sympathetic division is especially important during emergency situations when you might be required to fight back or

run away. It accelerates the heartbeat and dilates the bronchi. Active muscles, after all, require a ready supply of glucose and oxygen. On the other hand, the sympathetic division inhibits the digestive tract—digestion is not an immediate necessity if you are under attack.

One of the functions of the sympathetic division is to activate the adrenal medulla to secrete the hormones epinephrine (adrenaline) and norepinephrine (NE) into the blood (see Section 20.4). In fact, one special preganglionic sympathetic neuron goes right to the adrenal gland without stopping at a ganglion. The adrenaline and NE released by the adrenal medulla bind to receptors on various cell types, adding to the "fight or flight" response. Interestingly, the neurotransmitter released by sympathetic postganglionic axons is also NE. A small amount of the NE released by these axons spills over into the blood. During times of high sympathetic nerve activity, the amount of NE entering the blood from these neurons increases significantly.

Because they cause the heart to beat stronger and faster and constrict certain blood vessels, adrenaline and NE tend to increase blood pressure. Certain drugs, called beta blockers, can be used in

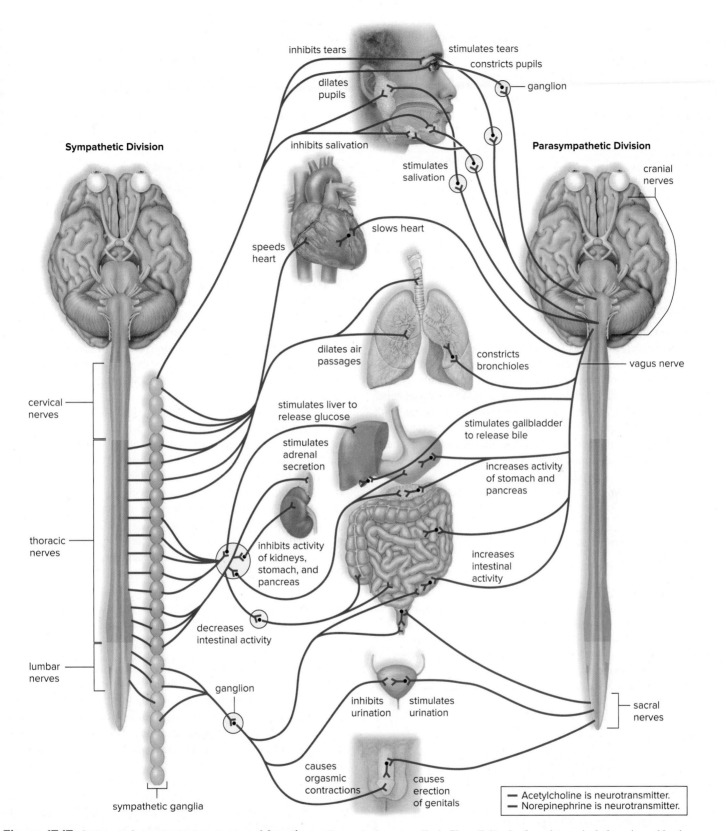

Figure 17.17 Autonomic system structure and function. Sympathetic preganglionic fibers (*left*) arise from the cervical, thoracic, and lumbar portions of the spinal cord. Parasympathetic preganglionic fibers (*right*) arise from the cranial and sacral portions of the spinal cord. Each system innervates the same organs but has contrary effects.

TABLE 17.1   Comparison of Somatic Motor and Autonomic Motor Pathways

| | Somatic Motor Pathway | Autonomic Motor Pathways | |
		Sympathetic	Parasympathetic
Type of control	Voluntary/involuntary	Involuntary	Involuntary
Number of neurons per message	One	Two (preganglionic shorter than postganglionic)	Two (preganglionic longer than postganglionic)
Location of motor fiber	Most cranial nerves and all spinal nerves	Thoracolumbar spinal nerves	Cranial (e.g., vagus) and sacral spinal nerves
Neurotransmitter	Acetylcholine	Norepinephrine	Acetylcholine
Effectors	Skeletal muscles	Smooth and cardiac muscle, glands	Smooth and cardiac muscle, glands

patients with hypertension to block the activity of adrenaline and NE, which slows the heart and decreases blood pressure.

Parasympathetic Division

The **parasympathetic division** includes a few cranial nerves (e.g., the vagus nerve), as well as fibers that arise from the sacral (bottom) portion of the spinal cord. Therefore, this division is often referred to as the craniosacral portion of the autonomic system. The parasympathetic division, sometimes called the housekeeper division, promotes all the internal responses we associate with "rest and digest." Opposite to the sympathetic system, it causes the pupil of the eye to contract, promotes digestion of food, and retards the heartbeat. The major neurotransmitter utilized by the parasympathetic division is acetylcholine (ACh).

Table 17.1 compares and contrasts the characteristics of the somatic and autonomic systems.

Check Your Progress 17.5

1. Contrast cranial and spinal nerves.
2. Explain why, if you touch a hot stove, you usually withdraw your hand before feeling the pain.
3. Summarize the features of the autonomic system that are different from the somatic system.
4. Provide the neurological explanation for the following: You eat a big lunch, then go for a jog, during which your stomach starts to ache.

17.6 Drug Abuse

Learning Outcomes

Upon completion of this section, you should be able to

1. Explain the specific mechanisms whereby the most common illicit drugs affect the brain.
2. Classify illicit drugs as to whether they have a depressant, stimulant, or psychoactive effect on the body.
3. List the long-term effects of drug use on the body.

A wide variety of drugs affect the nervous system and can alter the mood and/or emotional state. The first time someone tries a drug

such as nicotine, alcohol, or cocaine, the person may experience strong feelings of pleasure. However, with continued use, larger amounts of the drug may be needed to create the same sensation. At some point, the users may become addicted, meaning they develop a physiological dependence on the drug and can't feel normal without it. At the extreme, drug addicts may prioritize getting "high" over food, hygiene, and interpersonal relationships. Even drug abusers who want to quit may face an enormous challenge from intense cravings and withdrawal symptoms when they stop taking a drug.

Most illicit drugs affect the action of a particular neurotransmitter at synapses in the brain (Fig. 17.18). Stimulants are drugs that increase the likelihood of neuron excitation, and depressants decrease the likelihood of excitation. Increasingly, researchers believe that dopamine is one of the main neurotransmitters in the brain that is responsible for mood. Cocaine is known to potentiate the effects of dopamine by interfering with its uptake from synaptic clefts (Fig. 17.19). Many of the new medications developed to counter drug dependence and mental illness affect the release, reception, or breakdown of dopamine.

Some Specific Drugs of Abuse

Nicotine

Although the numbers have been decreasing since 2011 according to the CDC, in 2015, 25.3% of high school students and 7.4% of middle school students reported using a tobacco product. When a person smokes a cigarette or cigar, or chews tobacco, nicotine is quickly distributed throughout the body. This causes a release of epinephrine from the adrenal cortex, which increases blood sugar levels and causes an initial feeling of stimulation. Then, as blood sugar falls, depression and fatigue set in, leading the smoker to seek more nicotine. In the CNS, nicotine causes neurons to release excess dopamine, which has a reinforcing effect that leads to dependence on the drug. Nicotine induces both physiological and psychological dependence, and once they start, many tobacco users find it extremely difficult to give up the habit. Withdrawal symptoms include headache, stomach pain, irritability, and insomnia. Pregnant women who smoke are more likely to have stillborn or premature infants, or babies born with other health problems.

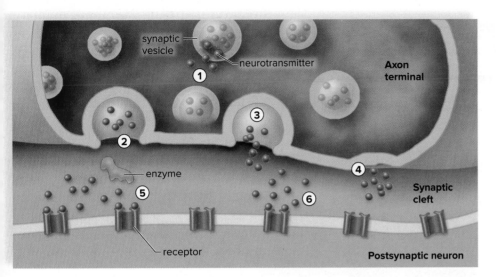

Figure 17.18 Drug actions at a synapse.
A drug can ① cause the neurotransmitter (NT) to leak out of a synaptic vesicle into the axon terminal; ② prevent release of NT into the synaptic cleft; ③ promote release of NT into the synaptic cleft; ④ prevent reuptake of NT by the presynaptic membrane; ⑤ block the enzyme that causes breakdown of the NT; or ⑥ bind to a receptor, mimicking the action of an NT.

Alcohol (Ethanol)

With the possible exception of caffeine (see the Health feature "Caffeine: Good or Bad for You?"), alcohol is the most used and abused drug among America's teenagers. According to a 2015 national survey, 26.9% of high school students reported drinking some alcohol (down from 37.4% in 2014), and 7% binge drank (five-plus drinks in one setting) during the 30 days preceding the survey. Among adults, 86.4% reported that they had consumed alcohol during their lifetime, with 56% stating that they had used alcohol in the past month.

When ingested, ethanol acts as a drug in the brain by influencing the action of gamma aminobutyrate (GABA), an inhibitory neurotransmitter, and other neurotransmitters like serotonin and dopamine. Alcohol is primarily metabolized in the liver, where it disrupts the normal workings of this organ so that fats cannot be broken down. Fat accumulation, the first stage of liver deterioration, begins after only a single night of heavy drinking. If heavy drinking continues, fibrous scar tissue is deposited during the next stage of liver damage. If heavy drinking stops at this point, the liver may still recover and become normal again. If not, the final and irrevocable stage, cirrhosis of the liver, occurs. Liver cells die, harden, and turn orange (cirrhosis means orange).

Alcohol is a carbohydrate and as such can be converted into energy for the body to use. However, alcoholic beverages lack the vitamins, minerals, essential amino acids, and fatty acids the body needs. For this reason, many alcoholics are vitamin-deficient, undernourished, and prone to illness.

Based upon extensive research, the Surgeon General of the United States recommends that pregnant women drink no alcohol at all. Alcohol crosses the placenta freely and causes fetal alcohol syndrome in newborns, which is characterized by intellectual disability and various physical defects.

Marijuana

Marijuana is the most commonly used illegal drug in the United States. Surveys vary, but in 2015 about 52% of young adults reported using marijuana in their lifetime, and 46% of the U.S. population had tried it at least once. Marijuana, or "pot," refers to the dried flowering tops, leaves, and stems of the Indian hemp plant *Cannabis sativa,* which contain and are covered by a resin that is rich in THC (tetrahydrocannabinol). Researchers have found that THC binds to a receptor in the brain for anandamide, a naturally occurring neurotransmitter that is important for short-term memory processing, and perhaps for feelings of contentment. Usually, marijuana is smoked in a cigarette form called a "joint," or in pipes or other devices, but it can also be consumed. The occasional user experiences a mild euphoria along with alterations in vision and judgment, which result in distortions of space and time. Motor incoordination, including the inability to speak coherently, takes place. Heavy use can result in hallucinations, anxiety, depression,

brain activity ⟶

Before chronic cocaine use, brain is more active (red areas).

After chronic cocaine use, brain is less active.

Figure 17.19 Drug abuse. PET scans show that the usual activity of the brain is reduced and many areas become inactive after a person uses cocaine. ©Science Source

Caffeine: Good or Bad for You?

Red Bull. Rockstar. Monster. Full Throttle. Walk down the beverage aisle of most U.S. grocery stores, and the colorful labels of these and other "energy drinks" compete for your attention (Fig. 17B). That's no surprise—the United States is the world's largest consumer of energy drinks, totaling roughly 1 billion units in 2016.

Around 90% of American adults report using caffeine every day. Caffeine, or 1,3,7-trimethylxanthine, occurs naturally in over 60 plant species worldwide. In plants, it is thought to function as a natural pesticide, killing certain insects that attempt to feed on the plant. It is naturally present in coffee, tea, and chocolate. It is also added to soft drinks and energy drinks, possibly to add flavor, though studies seem to prove that most people cannot taste the difference when caffeine is added to a caffeine-free beverage.

For humans, caffeine is a central nervous system stimulant. It readily crosses the blood-brain barrier, where it seems to block the effects of adenosine, which it resembles chemically. In the brain, adenosine normally seems to suppress brain activity, and by blocking this effect, caffeine can increase alertness. However, in response to continuing doses of caffeine, brain cells adjust by producing more adenosine receptors, so that increased doses of caffeine are needed for the same effect. If a regular user suddenly stops ingesting caffeine, withdrawal symptoms—such as headache, inability to concentrate, and insomnia—may ensue within 12 to 24 hours and last from 1 to 5 days.

Americans still get most of their caffeine from that tried-and-true source, the cup of coffee, which averages about 100 mg of caffeine. By comparison, a 12-oz can of Diet Coke contains about 38 mg (about 10 mg more than regular Coca-Cola), Mountain Dew has 55 mg, and Diet Pepsi Max has 69 mg. Energy drinks

Figure 17B A sampling of energy drinks found in most U.S. grocery stores.   The caffeine content of these beverages varies widely. ©McGraw-Hill Education

vary even more, with Red Bull at 80 mg, and Full Throttle and Monster doubling that to around 160. At the extreme end of the spectrum, a 20-oz bottle of Fixx Energy Drink contains 500 mg of caffeine!

Increasingly, scientists and physicians have become concerned that the combination of high doses of caffeine in energy drinks, coupled with a high intake of drinks by some individuals, may place drinkers at possible risk for caffeine intoxication. As a stimulant, high doses of caffeine can have serious consequences. Adverse effects seen with high doses of caffeine include nervousness, irritability, tremors, insomnia, headaches, and high blood pressure. High levels of caffeine can interfere with the normal operation of the heart, causing palpitations and irregular heartbeat. Caffeine reduces absorption of iron from the intestine, and because it acts as a diuretic (see Section 16.3), it increases calcium loss in the urine. Some

individuals, perhaps 20%, are more sensitive to lower levels of caffeine, so even a few milligrams may cause problems for them. Pregnant and nursing women should also be careful to limit their caffeine intake. Fortunately, caffeine-induced deaths are rare. The lethal dose for adults has been estimated at 5,000 mg.

So there is definitely a downside to caffeine use (or abuse). But might there be an upside? In 2008, researchers at Harvard published results of an 18-year study of thousands of men and women. They found that women who drank two to three cups of caffeinated coffee per day had a 25% lower risk of death from heart disease, and women who drank five to seven cups per week had a 7% lower risk of death from all causes than nondrinkers. For men, although the benefits were not statistically significant, the scientists found no health risk to drinking up to seven cups of coffee per day. Other studies have suggested a link between caffeine and increased athletic and academic performance, as well as a reduced risk of Parkinson disease, migraine headaches, and even diabetes. So although the debate over caffeine's risks versus benefits may never be completely resolved, for many people, it is possible that consumption of a moderate amount of caffeine may end up doing more good than harm.

Questions to Consider

1. Estimate how much caffeine you consume per day. What effects can you notice after consuming caffeine?
2. Would you agree that manufacturers should be required to display the caffeine content of soft drinks and energy drinks on the label? Why or why not?
3. Do you believe there should be any restrictions on selling, for example, an energy drink containing 500 mg of caffeine? If so, what should the minimum age be?

body image distortions, paranoid reactions, and similar psychotic symptoms. Some researchers believe that long-term marijuana use leads to brain impairment. Fetal cannabis syndrome, which resembles fetal alcohol syndrome, has also been reported.

Cocaine and Crack

Cocaine is an alkaloid derived from the shrub *Erythroxylon coca.* It is sold in powder form and as crack, a more potent extract. Crack is cocaine that comes in a rocklike crystal that can be heated and its vapors inhaled. The term *crack* refers to the crackling sound heard when it is heated. Cocaine prevents the synaptic uptake of

dopamine, and this causes the user to experience a rush sensation. The epinephrine-like effects of dopamine account for the state of arousal that lasts for several minutes after the rush experience.

A cocaine binge can go on for days, after which the user suffers a crash. During the binge period, the user is hyperactive and has little desire for food or sleep but has an increased sex drive. During the crash period, the user is fatigued, depressed, and irritable, has memory and concentration problems, and displays no interest in sex. Indeed, men who use "coke" often become impotent.

Cocaine causes extreme physical dependence. With continued use, the body begins to make less dopamine to compensate for a seemingly excess supply. The user subsequently experiences tolerance, withdrawal symptoms, and an intense craving for cocaine. Overdosing on cocaine can cause seizures and cardiac and respiratory arrest. It is possible that long-term cocaine abuse causes brain damage, and babies born to addicts suffer withdrawal symptoms and may have neurological and behavioral problems.

Heroin

Heroin is derived from morphine, an alkaloid of opium. Once heroin is injected into a vein, a feeling of euphoria, along with relief of any pain, occurs within 3 to 6 minutes. Heroin binds to receptors meant for the endorphins, naturally occurring neurotransmitters that kill pain and produce a feeling of tranquility. With repeated heroin use, the body's production of endorphins decreases. Tolerance develops so that the user needs to take more of the drug just to prevent withdrawal symptoms, and the original euphoria is no longer felt. Heroin withdrawal symptoms include perspiration, dilation of pupils, tremors, restlessness, abdominal cramps, vomiting, and increased blood pressure and respiratory rate. People who are excessively dependent may experience convulsions, respiratory failure, and death. Infants born to women who are physically dependent also experience those withdrawal symptoms.

"Club" Drugs

Ecstasy (MDMA [3,4-methylenedioxymethamphetamine]), Rohypnol (flunitrazepam), and ketamine are examples of drugs that are abused by many teens and young adults who attend night-long dances called raves or trances. Of course, these drugs may also be abused by others as well. MDMA is chemically similar to methamphetamine. Many users say that "XTC" increases their feelings of well-being and friendliness, even love, for other people. However, it can cause many of the same side effects as other stimulants, such as an increase in heart rate and blood pressure, muscle tension, and blurred vision. In high doses, MDMA can interfere with temperature regulation, leading to hyperthermia, followed by damage to the liver, kidneys, and heart. Chronic MDMA use can also damage memory and lead to depression. Perhaps because of these side effects, several recent surveys have reported that MDMA use has been decreasing among youths aged 12 to 17.

Rohypnol is in the same class of drugs as diazepam (Valium), a popular sedative. When mixed with alcohol, Rohypnol, or "roofies," can render victims incapable of resisting, for example, sexual assault. It can also produce anterograde amnesia, which means victims may not remember events they experienced while under the influence of the drug. Ketamine, or "Special K," is an anesthetic that veterinarians use to perform surgery on animals. When administered to humans, it typically induces a dreamlike state, sometimes with hallucinations. Because it can render the user unable to move, ketamine has also been used as a date rape drug. At higher doses, ketamine can cause dangerous reductions in heart and respiratory functions.

Methamphetamine

Methamphetamine, also called "meth" or "crank," is a powerful CNS stimulant. Meth is often synthesized or "cooked" in makeshift home laboratories, usually starting with either ephedrine or pseudoephedrine, common ingredients in many cold and asthma medicines. Many states have passed laws making these medications more difficult to purchase. Meth is usually produced as a powder (speed) that can be snorted, or as crystals (crystal meth) that can be smoked. The most immediate effect is an initial "rush" of euphoria, due to high amounts of dopamine released in the brain. The user may also experience an increased energy level, alertness, and elevated mood. After the initial rush, a state of high agitation typically occurs that, in some individuals, leads to violent behavior. Chronic use can lead to paranoia, irritability, hallucinations, insomnia, tremors, hyperthermia, cardiovascular collapse, and death. Similar to cocaine, long-term meth abuse renders the brain less able to produce normal levels of dopamine. The user may have strong cravings for more meth, without being able to reach a satisfactory high.

"Bath Salts"

One of the newer illicit drugs to become available are "bath salts," synthetic powders sold in colorful containers under innocent-sounding names like "Blue Wave" and "Vanilla Sky." However, unlike the innocuous sodium salts and fragrances that have been added to bathwater for centuries, these new products often contain synthetic amphetamine-like chemicals such as methylenedioxypyrovalerone and mephedrone. Although they are often labeled "Not For Human Consumption," when ingested, injected, or snorted, these chemicals inhibit the reuptake of several neurotransmitters, producing a euphoric sensation. Unfortunately, a long list of side effects includes severe chest pain, hallucinations, extreme paranoia, and violent episodes often leading to suicide. In response to dramatic increases in bath salts–related emergency-room visits and deaths, in 2012 a law was passed that makes it illegal to possess or sell many of the chemicals required to make bath salts. Unfortunately, illicit drugmakers are presumably working to develop new "designer drugs" to circumvent these bans.

Check Your Progress 17.6

1. Examine Figure 17.18 and hypothesize why, because illicit drugs affect the activity of neurotransmitters, it might take increasing amounts of a drug to obtain the same effect over time.
2. Detail the specific modes of action of cocaine, heroin, methamphetamine, and bath salts.

17.7 Disorders of the Nervous System

Learning Outcomes

Upon completion of this section, you should be able to

1. Describe two abnormalities seen in the brain of Alzheimer disease patients.
2. Compare the mechanism causing the following diseases: Parkinson disease, MS, stroke, meningitis, and prion disease.
3. Describe the symptoms seen in several disorders of the brain and spinal cord and discuss why it is difficult to find cures for each.

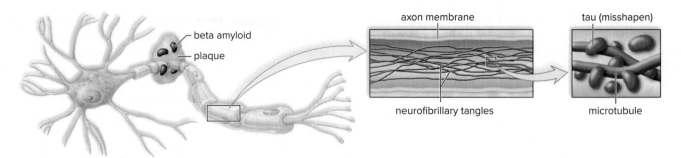

Figure 17.20 Alzheimer disease.   Some of the neurons of patients with Alzheimer disease have beta amyloid plaques and neurofibrillary tangles. These neurons are present throughout the brain, but concentrated in the hippocampus and amygdala.

Myriad abnormal conditions can affect the nervous system. We will consider only a few of the most serious, common, and/or interesting ones here.

Disorders of the Brain

Alzheimer disease is the most common cause of dementia, an impairment of brain function significant enough to interfere with the patient's ability to carry on daily activities. Signs of Alzheimer disease may appear before the age of 50, but the condition is usually seen in individuals past the age of 65. It is estimated that at least 40% of individuals older than 80 develop Alzheimer disease. One of the early symptoms is loss of memory, particularly of recent events. Characteristically, the person asks the same question repeatedly and becomes disoriented even in familiar places. Gradually, the person loses the ability to perform any type of daily activity and becomes bedridden. Death is often due to pneumonia resulting from the patient's debilitated state.

In Alzheimer patients, abnormal neurons are present throughout the brain but especially in the hippocampus and amygdala. These neurons have two abnormalities: (1) plaques, containing a protein called beta amyloid, envelop the axons; and (2) neurofibrillary tangles are in the axons (Fig. 17.20) and may extend upward to surround the nucleus. The neurofibrillary tangles arise because a protein called tau no longer has the correct shape. Tau holds microtubules in place so that they support the structure of neurons. When tau changes shape, it grabs onto other tau molecules, and the tangles result.

Researchers are working to discover the cause of beta amyloid plaques and neurofibrillary tangles. Several genes that predispose a person to Alzheimer disease have been identified, but it is unknown why inheritance of these genes leads to Alzheimer disease. Moreover, although several treatments have been tried to date, none has been shown to prevent or cure Alzheimer disease. A variety of medications can help ease symptoms such as memory loss, agitation, and depression, but brain cells still die. Most drugs currently available for treating symptoms of Alzheimer disease are acetylcholinesterase inhibitors, which increase the amount of time that acetylcholine can reside in synapses to transmit nerve impulses.

Parkinson disease is characterized by a gradual loss of motor control, typically beginning between the ages of 50 and 60. Eventually, the person develops a wide-eyed, unblinking expression, involuntary tremors of the fingers and thumbs, muscular rigidity, and a shuffling gait. Speaking and performing ordinary daily tasks becomes laborious. In Parkinson patients, the basal nuclei (see Section 17.3) function improperly because of a degeneration of the dopamine-releasing neurons in the brain. Without dopamine, the excessive excitatory signals from the motor cortex result in the symptoms of Parkinson disease.

Unfortunately, it is not possible to give Parkinson disease patients dopamine directly because of the impermeability of the capillaries serving the brain. However, symptoms can be alleviated by giving patients L-dopa, a chemical that can be changed to dopamine in the body, until too few cells are left to do the job. Then, patients must turn to a number of controversial surgical procedures. In 1998, the actor Michael J. Fox, who had been diagnosed with Parkinson disease 7 years earlier at the age of 30, had a surgical procedure called a thalamotomy, which has been successful in reducing the symptoms of some PD patients. However, in his biography *Lucky Man: A Memoir,* Fox admits that after a temporary improvement, his symptoms returned in full force. In 2007, the first report was published on treating Parkinson disease with gene therapy (see Section 26.3), in which a virus engineered to produce the required neurotransmitter was injected into the brains of 12 Parkinson disease patients. In 1 year, motor function had improved in these patients from 25% to 65% of normal.

Multiple sclerosis (MS) is the most common neurological disease that afflicts young adults. Nearly 400,000 people in the United States have MS, with approximately 10,000 new cases diagnosed each year. MS affects the myelin sheath of neurons in the white matter of the brain. Although the initiating cause remains unknown, MS is considered an autoimmune disease in which the patient's own white blood cells attack the myelin, oligodendrocytes, and eventually, neurons in the CNS. Specifically, MS is characterized by an infiltration of lymphocytes and macrophages into the brain, brain stem, optic nerves, and spinal cord. The most common symptoms of MS include fatigue, vision problems, limb weakness and/or pain, and abnormal sensations such as numbness or tingling. These symptoms, however, can be similar to those of several other brain disorders, so a technique called magnetic resonance imaging (MRI) can be very helpful in both diagnosing and monitoring the disease (Fig. 17.21). MS can also take several clinical forms, from the milder relapsing-remitting form to more relentlessly progressive forms. Although MS is not generally considered a fatal disease, people with MS die an average of 7 years earlier than the general population. No treatment has been

Figure 17.21 Multiple sclerosis. This MRI scan shows the brain of an MS patient. The white areas, also known as MS plaques, are sites of active inflammation and myelin degradation. ©Sovereign/Medical Images

shown to reverse the damage to the myelin of MS patients, but several medications are effective in slowing the progression of the disease, especially in the early stages.

A **stroke** results in disruption of the blood supply to the brain. There are two major forms of stroke: hemorrhagic and ischemic. In hemorrhagic stroke, bleeding occurs into the brain due to leakage from small arteries, often after years of damage by high blood pressure. Ischemic stroke occurs when there is a sudden loss of blood supply to an area of the brain, usually due to a thrombus or thromboembolism (see Section 12.5). In both types of stroke, the symptoms depend on the amount and specific area of brain tissue affected. However, some degree of paralysis and aphasia (loss of speech) are common signs. The risk of stroke increases with age, and African Americans have a higher incidence than other races in the United States. Physicians treating stroke patients must be careful to determine which type they are dealing with. For example, tissue plasminogen activator (tPA) can be used to help dissolve clots when treating ischemic stroke, but could have disastrous consequences if administered to a hemorrhagic stroke patient.

Meningitis is an infection of the meninges that surround the brain and spinal cord. Most cases of meningitis are caused by bacteria or viruses that have usually gained access to the meninges after first infecting the blood, middle ear, or sinuses. Certain viruses are also capable of infecting the brain and meninges by infecting peripheral nerve cells and being transported back to the brain. Because cells of the immune system have somewhat limited access to the brain, the infection may spread into the brain tissue. Bacterial meningitis is especially serious, even life-threatening if untreated. Physicians may diagnose meningitis based on clinical signs such as a stiff neck, fever, and headache, and may confirm the diagnosis by examining a sample of cerebrospinal fluid, usually obtained by a lumbar puncture. Most cases of bacterial meningitis are caused by organisms called *Haemophilus influenzae* or *Neisseria meningitidis.* Outbreaks of *N. meningitidis,* also called meningococcus, have occurred recently among college students. Fortunately, vaccines are available to prevent infection with both organisms.

Several brain diseases of humans and animals are caused by **prions,** which are infectious agents that most researchers believe are composed of protein only (see Section 28.5). Examples of human diseases include kuru, Creutzfeldt–Jakob disease, and fatal familial insomnia. Prion diseases of animals include scrapie in sheep, mad cow disease in cattle, and chronic wasting disease in deer. Mad cow disease seems able to cross the species barrier and infect humans who consume infected tissues; about 200 human cases have been confirmed, mostly in the United Kingdom. All prion diseases are characterized by long incubation periods between initial exposure to the agent and onset of clinical signs, followed by rapid progression of the disease and, uniformly, death. Prion diseases are thought to result from the accumulation in the tissues of an abnormal form of a normal host (prion) protein. When this occurs in the brain, nerve cells die, forming areas that look like holes, which pathologists refer to as spongiform changes (Fig. 17.22). These diseases are also unique in that they can be inherited or can be spread by either eating infected tissues (as in mad cow disease) or through transplantation of certain organs or tissues (such as corneas) from an infected individual. Therefore, prion diseases are also called transmissible spongiform encephalopathies (TSEs). No treatment has yet been shown to delay the inevitably fatal progression of these diseases.

Disorders of the Spinal Cord

Spinal cord injuries may result from car accidents or other trauma. Because little or no nerve regeneration is possible in the CNS, any resulting disability is usually permanent. The cord may be completely cut across (called a transection) or only partially severed (a partial section). The location and extent of the damage produce a variety of effects, depending on the partial or complete stoppage of impulses passing up and down the spinal cord. If the spinal cord is completely transected, no sensations or somatic motor impulses are able to pass the point of injury. If the injury is between the first thoracic vertebra (T1) and the second lumbar vertebra (L2), paralysis of the lower body and legs occurs, a condition known as *paraplegia* (see Fig. 19.6 for the specific locations of these vertebrae). If the injury is in the neck region, the entire body below the neck is usually affected, a condition called *quadriplegia.* If the

Figure 17.22 Effect of prions on the brain. In this stained section of brain tissue from a patient with a prion disease, the spongiform changes (holes) in the tissue are due to accumulation of the abnormal prion protein. ©Ralph C. Eagle, Jr./Science Source

injury is above the second or third cervical vertebrae (C2 or C3), nervous control of respiration is also affected. In 1995, actor Christopher Reeve, best known for his movie role as "Superman," was thrown from his horse, crushing his top two cervical vertebrae. He immediately lost control of his arms and legs, and had to be placed on a respirator. Through an intense exercise and physical therapy program, Reeve regained the ability to move slightly (e.g., taking tiny steps while being held upright in a pool). Unfortunately, he died of cardiac arrest in 2004.

Amyotrophic lateral sclerosis (ALS), also known as Lou Gehrig's disease, is a rare but devastating condition that affects the motor nerve cells of the spinal cord. ALS is incurable, and most people die within 5 years of being diagnosed, mainly due to failure of the respiratory muscles. However, about 10% of ALS patients survive for 10 or more years. Typically, the first signs of ALS occur between the ages of 40 and 60 in the form of mild weakness or tremors in one or more limbs, starting in the fingers or toes and gradually involving the larger muscles nearer to the trunk of the body. The cause(s) of nerve cell death in ALS is (are) unclear. About 5–10% of all ALS cases are genetic in nature, but the majority of victims have no known risk factors. The cerebrospinal fluid of people with ALS often contains abnormally high levels of the neurotransmitter *glutamate,* which can be toxic to neurons. Autoimmune reactions may also play a role, as may environmental toxins.

Other than Lou Gehrig, the most famous person with ALS was probably physicist Stephen Hawking, who lived for roughly 50 years with the disease (Fig. 17.23) before passing away in 2018. The fact that Hawking remained an active scientist and author during those decades shows that ALS has very little effect on brain function.

Disorders of the Peripheral Nerves

Guillain–Barré syndrome (GBS) is an inflammatory disease that causes demyelination of peripheral nerve axons. It is thought to result from an abnormal immune reaction to one of several types of infectious agents. In GBS patients, antibodies formed against these bacteria or viruses seem to cross-react with the myelin. Symptoms usually begin with weakness or unsteadiness 2 to 4 weeks after an infection or immunization. The muscle weakness typically starts in the lower limbs and ascends to the upper limbs over a period of days to weeks. The disease process may eventually affect the respiratory muscles, so that mechanical ventilation is required. In most cases, however, the paralysis begins to improve in a few weeks, and most patients make a full recovery within 6 to 12 months.

Myasthenia gravis (MG) is an autoimmune disorder in which antibodies are formed that react against the acetylcholine receptor (AChR) at the neuromuscular junction of the skeletal muscles. Normally, when the action potential arrives at the presynaptic terminal of peripheral motor nerves, acetylcholine (ACh) is released into the synaptic cleft and binds to the AChR on the muscle cell, stimulating contraction. In MG, antibodies bind to the AChR and block binding of ACh, preventing muscle stimulation.

Interestingly, the most common symptom of MG is weakness of the muscles of the eye or eyelid, although over time the weakness usually spreads to the face, trunk, and then the limbs. In severe cases, the respiratory muscles can be affected, leading to death if untreated. Even though there is no cure, MG patients often respond very well to treatment with immunosuppressive drugs and inhibitors of acetylcholinesterase (AChE), an enzyme that normally destroys the ACh after it is released. A technique called *plasmapheresis* can also be used in both GBS and MG to remove the harmful antibodies from the patient's blood.

Check Your Progress 17.7

1. Summarize the major symptoms of Alzheimer disease and explain the age at which these symptoms may appear.
2. Distinguish between Alzheimer disease and Parkinson disease.
3. Explain why MS is an autoimmune disease.
4. Discuss the following: One cause of bacterial meningitis, *Haemophilus influenzae,* can also be found in the throat (pharynx) of healthy people. What factors might determine when *H. influenzae* causes disease?

Figure 17.23 Physicist Stephen Hawking. Dr. Hawking was a productive scientist and author for over half a century, even though he had amyotrophic lateral sclerosis (ALS). ©Jason Bye/Alamy

Conclusion

Multiple sclerosis (MS) is an autoimmune disease affecting the myelin. The initiating cause of MS is unknown, but most researchers agree that there are most likely many contributing factors, including infections and other environmental influences. Studies have shown that the risk of contracting MS depends in part on where in the world a person lives, although the specific factor related to geographic location has not been identified.

Although there is no cure, many individuals with MS are able to control their symptoms with the help of immunosuppressive medications such as corticosteroids and beta interferon. However, in most cases, the disease does progress slowly, often over a period of decades.

SUMMARIZE

17.1 Nervous Tissue

The **nervous system** coordinates and regulates the body's other systems:

■ In most animals, the nervous system is divided into a **central nervous system (CNS)** and **peripheral nervous system (PNS).**

■ Nervous tissue contains **neurons** and **neuroglia.** There are three types of neurons. **Sensory neurons** take information from sensory receptors to the CNS; **interneurons** occur within the CNS; and **motor neurons** take information from the CNS to muscles or glands.

■ A neuron is composed of **dendrites,** a **cell body,** and an **axon.** Some axons are covered by a **myelin sheath.** In the CNS, myelin is synthesized by neuroglia called **oligodendrocytes;** in the PNS, myelin is produced by **Schwann cells.** Gaps in the myelin covering of axons are called **nodes of Ranvier.**

■ In the CNS, areas called **gray matter** contain short, nonmyelinated nerves, while **white matter** has neurons with longer, myelinated axons that run in **tracts.**

17.2 Transmission of Nerve Impulses

The nervous system transmits information using **nerve impulses:**

■ In order to transmit a nerve impulse, an **action potential** must be generated. When an axon is not conducting a nerve impulse, the inside of the axon is negative (-70 mV) compared to the outside. **Sodium–potassium pumps** actively transport Na^+ out of an axon and K^+ to the inside of an axon. The **resting potential** is due to the leakage of K^+ to the outside of the neuron. When an axon is conducting a nerve impulse, Na^+ first moves into the axoplasm, causing membrane depolarization, until the inside of the axon is positive ($+35$ mV) compared to the outside. K^+ then moves out of the axoplasm, causing repolarization. Immediately following an action potential, there is a **refractory period,** during which the sodium gates are unable to open.

■ In myelinated axons, action potentials are generated only at the unmyelinated nodes of Ranvier. Therefore, the action potentials travel much faster as they "jump" from node to node, a phenomenon known as **saltatory conduction.**

■ At the junction of an axon with another nerve cell, or a muscle cell, a **synapse** forms, where the cell membranes are very close but don't touch. In order to send messages, chemicals called **neurotransmitters** are released into this gap, or **synaptic cleft.**

■ After being released by the presynaptic neuron, binding of the neurotransmitter to receptors in the postsynaptic membrane can cause excitation or inhibition.

■ **Integration** is the summing of excitatory and inhibitory signals.

17.3 The Central Nervous System

The central nervous system (CNS), consisting of the spinal cord and brain, receives and integrates sensory input and formulates motor output:

■ The spinal cord and brain are protected by membranes called **meninges.** Spaces between the meninges are filled with **cerebrospinal fluid,** which also fills hollow cavities in the brain called **ventricles.**

■ The **spinal cord** sends sensory information to the brain, receives motor output from the brain, and carries out reflex actions.

■ The **brain** can be divided into several functional and anatomical regions. Sensation, reasoning, learning and memory, and language and speech take place in the **cerebrum.** The **cerebral cortex** is a thin layer of gray matter covering the cerebrum. The cerebral cortex of each **cerebral hemisphere** has four lobes: a frontal, parietal, occipital, and temporal lobe. The primary motor area, in the frontal lobe, sends out motor commands to lower brain centers, which pass them on to motor neurons. The primary somatosensory area, in the parietal lobe, receives sensory information from lower brain centers in communication with sensory neurons. The cortex also contains processing centers for reasoning and speech. Deep within the cerebrum, the **basal nuclei** integrate motor activity.

■ Other regions of the brain include the **hypothalamus,** which controls homeostasis, and the **thalamus,** which specializes in sending sensory input on to the cerebrum. The **cerebellum** primarily coordinates skeletal muscle contractions. The **brain stem** includes the **midbrain, medulla oblongata,** and **pons,** which relay messages to and from the cerebrum, and have centers for vital functions, such as breathing and the heartbeat. Also found in the brain stem is the **reticular activating system (RAS),** a relay center that is also involved in alertness.

17.4 The Limbic System and Higher Mental Functions

The **limbic system** is associated with emotions, as well as higher brain functions like **memory** and **learning:**

■ The limbic system contains the **hippocampus,** which acts as a conduit for sending messages to long-term memory and retrieving them, and the **amygdala,** which adds emotional overtones to memories. In addition, it has various connections between the cerebral cortex and the hypothalamus, the thalamus, and the basal nuclei.

■ Two basic types of memory are short-term and long-term memory. The hippocampus is essential for formation of long-term memories. At the chemical level, this process involves glutamate and is known as long-term potentiation.

■ Language and speech involve many areas of the brain, but Broca's area and Wernicke's area are indispensable. The right and left hemispheres play different, though overlapping, roles.

17.5 The Peripheral Nervous System

The peripheral nervous system (PNS) lies outside of the CNS, where it conducts nerve impulses to and from the CNS:

■ The PNS contains only nerves and ganglia. **Nerves** are bundles of axons; **ganglia** are collections of cell bodies. Each axon in a nerve is called a **nerve fiber.**

■ Humans have 12 paired **cranial nerves,** and 31 paired **spinal nerves,** that emerge from the CNS. These may carry sensory or motor impulses, or both. The cell bodies of spinal nerves carrying sensory information lie in the **dorsal root ganglia.**

■ The PNS can be subdivided into the somatic system and the autonomic system. The **somatic system** takes information from external sensory receptors to the CNS, and voluntary motor commands from the cerebrum to skeletal muscles. It also carries out **reflex actions** in which interneurons in the spine quickly send messages to motor nerves without CNS input.

■ The **autonomic system** involuntarily controls the smooth muscle of the internal organs and glands. The **sympathetic division** is associated with responses that occur during times of stress, and the **parasympathetic division** is associated with responses that occur during times of relaxation.

17.6 Drug Abuse

Most illicit drugs affect neurotransmitters in the brain, either promoting or inhibiting the action of a particular neurotransmitter. Examples include the following:

- Nicotine, which increases dopamine release in the brain.
- Cocaine, which inhibits uptake of dopamine.
- Heroin, which binds to receptors for natural painkillers called endorphins.

17.7 Disorders of the Nervous System

The precise cause of many of the disorders affecting the CNS remains unclear, and many are untreatable. Examples of brain disorders include the following:

- **Alzheimer disease** seems to be due to an accumulation of abnormal proteins in the brain.
- **Parkinson disease** results from a degeneration of dopamine-secreting neurons.
- **Multiple sclerosis (MS)** is the result of autoimmune destruction of myelin and myelin-producing cells.
- **Stroke,** whether hemorrhagic or ischemic, occurs when the blood supply to the brain is disrupted.
- Infections of the brain include **meningitis** and **prion** diseases.
- Similar to brain disorders, disorders of the spine and peripheral nerves can be serious and difficult to treat. Injuries to the spinal cord can cause a spectrum of problems, from weakness in one limb to complete paralysis.
- **Amyotrophic lateral sclerosis (ALS)** is a fatal disease resulting from a loss of motor neurons in the spinal cord.
- **Guillain–Barré syndrome** and **myasthenia gravis** are both autoimmune disorders that attack the peripheral nerves.

ASSESS

Testing Yourself

Choose the best answer for each question.

17.1 Nervous Tissue

1. Which of the following neuron parts receive(s) signals from sensory receptors of other neurons?
   a. cell body
   b. axon
   c. dendrites
   d. Both a and c are correct.

2. The neuroglia cells that form myelin sheaths in the CNS are called
   a. oligodendrocytes.
   b. ganglionic cells.
   c. Schwann cells.
   d. astrocytes.
   e. microglia.

17.2 Transmission of Nerve Impulses

3. Which of these correctly describes the distribution of ions on either side of an axon when it is not conducting a nerve signal?
   a. more sodium ions (Na^+) outside and more potassium ions (K^+) inside
   b. more K^+ outside and more Na^+ inside
   c. charged protein outside and Na^+ and K^+ inside
   d. Na^+ and K^+ outside and water only inside
   e. chloride ions (Cl^-) outside and K^+ and Na^+ inside

4. When the action potential begins, sodium gates open, allowing Na^+ to cross the membrane. This causes the charge on the inside of the neuron to become
   a. more negative.
   b. more positive.
   c. neutral.
   d. None of these are correct.

5. Repolarization of an axon during an action potential is produced by
   a. inward diffusion of Na^+.
   b. outward diffusion of K^+.
   c. inward active transport of Na^+.
   d. active extrusion of K^+.

6. Transmission of the nerve signal across a synapse is accomplished by the
   a. movement of Na^+ and K^+.
   b. release of a neurotransmitter by a dendrite.
   c. release of a neurotransmitter by an axon.
   d. release of a neurotransmitter by a cell body.
   e. All of these are correct.

17.3 The Central Nervous System

7. Which of the following cerebral areas is not correctly matched with its function?
   a. occipital lobe—vision
   b. parietal lobe—somatosensory area
   c. temporal lobe—primary motor area
   d. frontal lobe—Broca's motor speech area

8. This part of the brain forms the link between the nervous system and the endocrine system.
   a. corpus callosum
   b. amygdala
   c. reticular formation
   d. hypothalamus

9. The cerebellum
   a. coordinates skeletal muscle movements.
   b. receives sensory input from the joints and muscles.
   c. receives motor output from the cerebral cortex.
   d. All of these are correct.

17.4 The Limbic System and Higher Mental Functions

10. The regulation of the information that is to be relayed to memory is the function of the
    a. reticular formation.
    b. hippocampus.
    c. hypothalamus.
    d. cerebellum.
    e. pons.

11. Memories are stored in the sensory association areas of the
    a. cerebral cortex.
    b. spinal cord.
    c. brain stem.
    d. hypothalamus.

17.5 The Peripheral Nervous System

12. Label the following diagram.

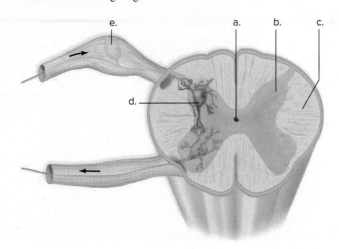

13. Which of the following is correct regarding the autonomic nervous system?
   a. It uses both cranial and spinal nerves.
   b. It is divided into sympathetic and parasympathetic divisions.
   c. Most organs under ANS control are under dual innervation—that is, by both sympathetic and parasympathetic divisions.
   d. Major responsibilities are regulation and repair of cardiac muscle, smooth muscle, organs, and glands.
   e. All of these are correct.

17.6 Drug Abuse

14. Effects of drugs can include
   a. prevention of neurotransmitter release.
   b. prevention of reuptake by the presynaptic membrane.
   c. blockages to a receptor.
   d. All of these are correct.
   e. None of these are correct.

15. Which of the following does not act as a stimulant to the CNS?
   a. cocaine
   b. methamphetamine
   c. ecstasy
   d. alcohol

17.7 Disorders of the Nervous System

16. A stroke results from
   a. hemorrhage in the brain.
   b. a blockage of a blood vessel in the brain.
   c. a deficiency of tissue plasminogen activator.
   d. Both a and b are correct.

17. What do Creutzfeldt–Jakob and mad cow disease have in common?
   a. Both are fatal diseases.
   b. Both are caused by prions.
   c. Both cause spongiform changes in the brain.
   d. All of these are correct.

ENGAGE

Thinking Critically

1. How do you suppose scientists and physicians have determined which areas of the brain are responsible for different activities, as shown in Figures 17.10, 17.11, and 17.14?

2. Dopamine is a key neurotransmitter in the brain. Patients with Parkinson disease suffer because of a lack of this chemical, while abusers of drugs such as nicotine, cocaine, and methamphetamine enjoy an enhancement of dopamine activity. Would these drugs be a possible treatment, or even a cure, for Parkinson disease? Why or why not?

3. What are some factors that make brain diseases such as Alzheimer disease, Parkinson disease, and MS so difficult to treat?

4. Explain the following: How does the bite of a black widow spider, which injects a powerful AChE inhibitor, cause muscle cramps, salivation, fast heart rate, and high blood pressure?

C H A P T E R

18
Senses

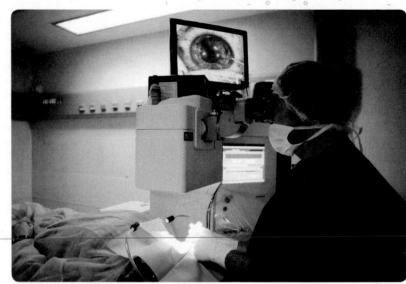

©Amelie Benoist/Science Source

BEFORE YOU BEGIN

Before beginning this chapter, take a few moments to review the following discussions:

Section 11.4 What specific types of receptors in the skin provide information about the external environment?

Section 17.1 What are the roles of the central and peripheral nervous systems in the body?

Section 17.3 What are the functions of the primary somatosensory area and somatosensory association area of the cerebral cortex?

Improving Your Eyesight

John had always worn glasses. As a computer programmer, he often spent 6–10 hours a day looking at a computer screen. Initially, he tried using bifocal and progressive lenses, but they never seemed to adjust his vision correctly. After a visit to his eye doctor, John decided he would like to permanently correct his vision, and remove his need for glasses, by undergoing LASIK surgery.

LASIK, which stands for laser-assisted in situ keratomileusis, is a quick and relatively painless procedure that involves the use of a laser to permanently change the shape of the cornea. During the procedure, a small flap of tissue (the conjunctiva) is cut away from the front of the eye. The flap is folded back, exposing the cornea, allowing the surgeon to remove a defined amount of tissue from the cornea. Each pulse of the laser removes a small amount of corneal tissue, allowing the surgeon to flatten or increase the steepness of the curve of the cornea. After the procedure, the flap of tissue is put back into place and allowed to heal on its own.

John was an ideal candidate for this type of surgery—he was over 18 years of age and his last two eye exams had indicated his vision was relatively stable. He knew there would be a small amount of discomfort from the procedure, but his doctor informed him that the procedure was generally safe and that almost 80% of patients improved their vision to the point that they no longer needed glasses.

In this chapter, we will explore not only the structure and function of the eye, but also of the other sense organs that contribute information to our central nervous system for interpretation.

As you read through this chapter, think about the following questions:

1. What is the role of the cornea in vision?
2. How does the eye normally adjust to looking at objects at different distances?

18.1 Sensory Receptors and Sensations

Sensory receptors are specialized cells that detect certain types of *stimuli* (*sing.,* stimulus). Based on the source of the stimulus, sensory receptors can be broadly classified as interoceptors or exteroceptors.

Interoceptors receive stimuli from inside the body, such as changes in blood pressure, blood volume, and the pH of the blood. Interoceptors are directly involved in homeostasis and are regulated by a negative feedback mechanism. For example, when blood pressure rises, pressoreceptors signal a regulatory center in the brain, which then sends out nerve impulses to the arterial walls, causing them to relax so that the blood pressure falls. Once the pressoreceptors are no longer stimulated, the feedback mechanism shuts down.

Exteroceptors are sensory receptors that detect stimuli from outside the body, such as those related to touch, taste, smell, vision, hearing, and equilibrium (Table 18.1). Their function is to send messages to the central nervous system to report changes in environmental conditions.

Types of Sensory Receptors

Based on the nature of the stimulus, interoceptors and exteroceptors can be further classified into four categories: chemoreceptors, photoreceptors, mechanoreceptors, and thermoreceptors.

Chemoreceptors respond to chemical substances in the immediate vicinity. As Table 18.1 indicates, taste and smell are dependent on this type of sensory receptor, but certain chemoreceptors in various other organs are sensitive to internal conditions. For example, chemoreceptors that monitor blood pH are located in the carotid arteries and aorta. If the pH lowers (becomes acidic), the breathing rate increases. As more carbon dioxide is exhaled, the blood pH rises (becomes more basic).

Photoreceptors respond to light energy. Our eyes contain photoreceptors that are sensitive to a specific range of wavelengths of light, and thereby provide us with the sense of vision. Stimulation of the photoreceptors known as rod cells results in black-and-white vision, whereas stimulation of the photoreceptors known as cone cells results in color vision (see Section 18.4).

Mechanoreceptors are stimulated by mechanical forces, which most often result in pressure of some sort. For example, when we hear, airborne sound waves are converted to fluid-borne pressure waves that can be detected by mechanoreceptors in the inner ear. Similarly, mechanoreceptors are responding to fluid-borne pressure waves when we detect changes in gravity and motion, helping us keep our balance. These receptors are in the vestibule and semicircular canals of the inner ear, respectively (see Section 18.6).

The sense of touch is dependent on pressure receptors in the skin and other organs that are sensitive to strong or slight pressures. Pressoreceptors located in certain arteries detect changes in blood pressure, and stretch receptors in the lungs detect the degree of lung inflation. Proprioceptors, which respond to the stretching of muscle fibers, tendons, joints, and ligaments, make us aware of the position of our limbs.

Thermoreceptors, located in the hypothalamus and skin, are stimulated by changes in temperature. Those that respond when temperatures rise are called warmth receptors, and those that respond when temperatures fall are called cold receptors.

How Sensation Occurs

Sensory receptors respond to environmental stimuli by generating nerve impulses. The nervous system detects these stimuli when environmental changes, such as pressure to the fingertips or light to the eye, stimulate sensory receptors. **Sensation** occurs when nerve impulses arrive at the cerebral cortex of the brain. **Perception** occurs when the brain interprets the meaning of stimuli.

As we discussed in Section 17.5, sensory receptors are the first element in a reflex arc. We are aware of a reflex action only when sensory information reaches the brain. At that time, the brain integrates this information with other information received from other sensory receptors. After all, if you burn yourself and quickly remove your hand from a hot stove, the brain receives information not only from your skin, but also from your eyes, nose, and all sorts of sensory receptors.

TABLE 18.1   Examples of Exteroceptors

Sensory Receptor	Stimulus	Category	Sense	Sensory Organ
Taste cells	Chemicals	Chemoreceptor	Taste	Taste buds
Olfactory cells	Chemicals	Chemoreceptor	Smell	Olfactory epithelium
Rod cells and cone cells in retina	Light rays	Photoreceptor	Vision	Eye
Hair cells in spiral organ	Sound waves	Mechanoreceptor	Hearing	Ear
Hair cells in semicircular canals	Motion	Mechanoreceptor	Rotational equilibrium	Ear
Hair cells in vestibule	Gravity	Mechanoreceptor	Gravitational equilibrium	Ear

Some sensory receptors are free nerve endings or encapsulated nerve endings, while others are specialized cells closely associated with neurons. The plasma membrane of a sensory receptor contains proteins that react to the stimulus. For example, the receptor proteins in the plasma membrane of chemoreceptors bind to certain molecules. When this happens, ion channels open, and ions flow across the plasma membrane. If the stimulus is sufficient, nerve impulses begin and are carried by a sensory nerve fiber within the PNS to the CNS (Fig. 18.1). In this process, known as **sensory transduction,** energy from the chemical or physical stimulus is transduced by the sensory receptors into an electrical signal. This electrical signal is in the form of nerve impulses (action potentials) that travel along the sensory nerve fiber. The stronger the stimulus, the greater the frequency of nerve impulses. Nerve impulses that reach the spinal cord first are conveyed to the brain by ascending tracts. If nerve impulses finally reach the cerebral cortex, sensation and perception occur.

All sensory receptors initiate nerve impulses; the sensation that results depends on the part of the brain receiving the nerve impulses. Nerve impulses that begin in the optic nerve reach the visual cortex, and thereafter, we see objects. Nerve impulses that begin in the auditory nerve reach the auditory cortex, and thereafter, we hear sounds.

Before sensory receptors initiate nerve impulses, they carry out **integration,** the summing up of signals. One type of integration is called **sensory adaptation**, a decrease in the response to a stimulus. For example, as you sit reading this, you are probably not aware of the feeling of the chair beneath you, until you were reminded by reading this. When sensory adaptation occurs, the sensory receptors send fewer impulses to the brain. The thalamus (see Section 17.3) also has a role in sensory adaptation. The majority of the sensory information is conveyed from the brain stem through the thalamus to the cerebral cortex. The thalamus acts as a gatekeeper and only passes on information of immediate importance. Just as we gradually become unaware of particular environmental stimuli, we can suddenly become aware of stimuli that may have been present for some time. This can be attributed to the workings of the thalamus, which has synapses with all the major ascending sensory tracts.

The functioning of our sensory receptors makes a significant contribution to homeostasis. Without sensory input, we would not receive information about our internal and external environment. This information leads to appropriate reflex and voluntary actions to keep the internal environment constant.

Check Your Progress 18.1

1. Describe how interoceptors are often involved in homeostasis.
2. List a specific example of each of the four types of sensory receptors.
3. Explain why sensory adaptation is important to an organism.

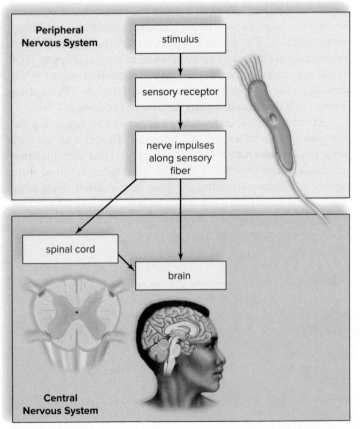

Figure 18.1 Sensation.   The stimulus is received by a sensory receptor, which generates nerve impulses (action potentials). Nerve impulses are conducted to the CNS by sensory nerve fibers within the PNS, and only those impulses that reach the cerebral cortex result in sensation and perception.

18.2 Somatic Senses

Learning Outcomes

Upon completion of this section, you should be able to

1. Compare and contrast the functions of proprioceptors, cutaneous receptors, and pain receptors.
2. List specific types of cutaneous receptors that are sensitive to fine touch, pressure, and temperature.
3. Explain the importance of pain receptors to the survival of an organism.

Senses whose receptors are associated with the skin, muscles, joints, and viscera (organs located in the body cavities) are called the *somatic senses.* Somatic sensory receptors can be categorized into three types: proprioceptors, cutaneous receptors, and pain receptors. Some of these are interoceptors, while others are exteroceptors. Regardless, these receptors send nerve impulses via the spinal cord to the somatosensory areas of the cerebral cortex (see Fig. 17.11).

Proprioceptors

Proprioceptors are mechanoreceptors involved in reflex actions that maintain muscle tone, and thereby the body's equilibrium and posture. They help us know the position of our limbs in space by detecting the degree of muscle relaxation, the stretch of tendons,

Figure 18.2 Muscle spindles and Golgi tendon organs. **1.** When a muscle is stretched, muscle spindles send sensory nerve impulses to the spinal cord. **2.** Motor nerve impulses from the spinal cord result in muscle fiber contraction so that muscle tone is maintained. **3.** When tendons are stretched excessively, Golgi tendon organs cause muscle relaxation.

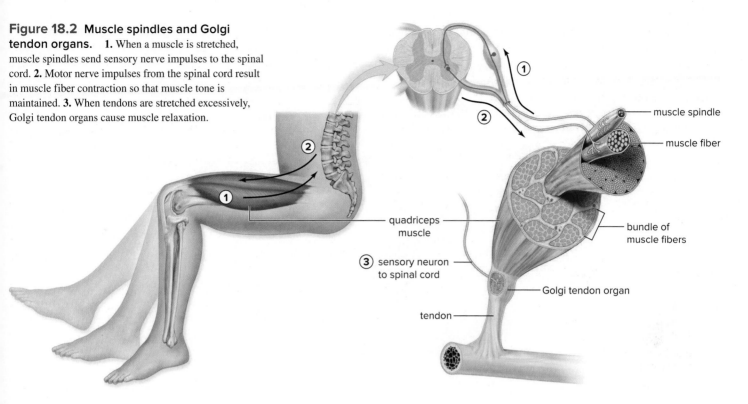

muscle spindle

muscle fiber

quadriceps muscle

bundle of muscle fibers

sensory neuron to spinal cord

Golgi tendon organ

tendon

and the movement of ligaments. Two main types of proprioceptors help to maintain the amount of tension, or tone, in skeletal muscles. Muscle spindles act to increase the degree of muscle contraction, and Golgi tendon organs act to decrease it.

Figure 18.2 illustrates the activity of muscle spindles and Golgi tendon organs. In a muscle spindle, sensory nerve endings are wrapped around thin muscle cells within a connective tissue sheath. When the muscle relaxes and stretching of the muscle spindle occurs, nerve impulses are generated and travel to the spinal cord. The rapidity of these nerve impulses is proportional to the stretching of a muscle. A reflex action then occurs, which results in contraction of muscle fibers adjoining the muscle spindle. The knee-jerk reflex, which involves muscle spindles, offers an opportunity for physicians to test a reflex action.

Golgi tendon organs are sensory receptors buried in the collagen fibers of tendons. When skeletal muscles, and therefore tendons, are stretched excessively, these proprioceptors generate nerve impulses that travel via sensory neurons to the spinal cord, inhibiting further muscle contraction. In addition to maintaining proper muscle tone, Golgi tendon organs also protect muscles and tendons from injury due to overstretching.

Cutaneous Receptors

As was described in Section 11.4, the skin is composed of two layers: the epidermis and the dermis (Fig. 18.3). The epidermis is stratified squamous epithelium in which cells become keratinized as they rise to the surface, where they are sloughed off. The dermis is a thick connective tissue layer. Both layers of the skin contain **cutaneous receptors,** which make the skin sensitive to touch, pressure, pain, and temperature (warmth and cold).

At least four types of cutaneous receptors are sensitive to fine touch. *Meissner corpuscles* and *Krause end bulbs* are concentrated in the fingertips, the palms, the lips, the tongue, the nipples, the penis, and the clitoris. *Merkel disks* are found where the epidermis meets the dermis. A free nerve ending called a *root hair plexus* winds around the base of a hair follicle and fires if the hair is touched.

At least two different types of cutaneous receptors are sensitive to pressure. *Pacinian corpuscles* are onion-shaped sensory receptors that lie deep inside the dermis. *Ruffini endings* are encapsulated by sheaths of connective tissue and contain lacy networks of nerve fibers.

Temperature receptors are simply free nerve endings in the epidermis. Some free nerve endings are responsive to cold; others are responsive to warmth. Cold receptors are far more numerous than warmth receptors, but the two types have no known structural differences.

Although the existence of different types of cutaneous receptors has been recognized for many years, researchers are just beginning to understand how the brain collects and processes sensory information from the skin. In one study, researchers at Johns Hopkins University genetically engineered mice so that different types of neurons that receive sensory information from the skin could be visualized by their expression of different fluorescent proteins. They determined that each neuron was receiving information from between 30 and 150 cutaneous receptors, and that different types of neurons receive information from different types of receptors. Moreover, they found that nerves bringing sensory information from small areas of skin line up in small columns in the spinal cord (about 3,000 to 5,000 columns per mouse). The researchers admit they don't fully understand how this large amount of information is

free nerve endings
(pain, heat, cold)

Merkel disks (touch)

Krause end
bulbs (touch)

root hair
plexus (touch)

epidermis

Meissner
corpuscles (touch)

Pacinian corpuscles
(pressure)

Ruffini endings
(pressure)

dermis

Figure 18.3 Cutaneous
receptors in human skin.
The general function of each sensory
receptor is shown here. However,
receptors are not always this
specialized. For example, microscopic
examination of the skin of the ear
shows only free nerve endings (pain
receptors), yet the skin of the ear is
sensitive to all sensations.

processed by the brain, but they suspect that the way these columns are arranged in the spinal cord is a key factor.

Pain Receptors

The skin and many internal organs and tissues have **pain receptors** (free nerve endings, also called *nociceptors*), which are sensitive to chemicals released by damaged cells. When damage occurs due to mechanical, thermal, or electrical stimulation, or a toxic substance, cells release chemicals that stimulate pain receptors. Aspirin and ibuprofen reduce pain by inhibiting the synthesis of one class of these chemicals.

Sometimes, stimulation of internal pain receptors is felt as pain from the skin, as well as the internal organs. This is called *referred pain.* Some internal organs have a referred pain relationship with areas located in the skin. For example, pain from the heart is typically felt in the left shoulder and arm. This most likely happens when nerve impulses from the pain receptors of internal organs travel to the spinal cord and synapse with neurons also receiving impulses from the skin. Pain receptors are protective because they alert us to possible danger. For example, without the pain of appendicitis, we might never seek the medical help needed to avoid a ruptured appendix.

Check Your Progress 18.2

1. Classify which specific types of somatic sensory receptors are interoceptors and which are exteroceptors.
2. Classify the cutaneous receptors by their location (i.e., in the dermis or epidermis).
3. Predict the likely outcome if an individual lacked muscle spindles; or pain receptors.

18.3 Senses of Taste and Smell

Learning Outcomes

Upon completion of this section, you should be able to
1. Identify the structures that are involved in taste, including the five types of taste receptors.
2. Describe the sensory receptors involved in smell.
3. Compare and contrast how the brain receives information about taste versus smell.

Taste and smell are called *chemical senses* because their receptors are sensitive to molecules in the food we eat and the air we breathe. Taste cells and olfactory cells are classified as chemoreceptors.

Sense of Taste

The sensory receptors for the sense of taste, the taste cells, are located in about 10,000 **taste buds** located primarily on the tongue (Fig. 18.4). Many lie along the walls of the *papillae,* the small elevations on the tongue that are visible to the naked eye. Isolated taste buds are also present on the hard palate, the pharynx, and the epiglottis. Different taste cells can detect at least five primary types of taste: salty, sour, bitter, sweet, and umami (Japanese; savory, delicious). Umami receptors are capable of detecting certain amino acids. In particular, this receptor detects the amino acid glutamate, which is part of the flavor enhancer monosodium glutamate (MSG). Taste buds containing sensory receptors for each type of taste are located throughout the tongue.

How the Brain Receives Taste Information

Taste buds open at a taste pore. They have supporting cells and a number of elongated taste cells that end in microvilli. When molecules bind to receptor proteins of the microvilli, nerve impulses are generated

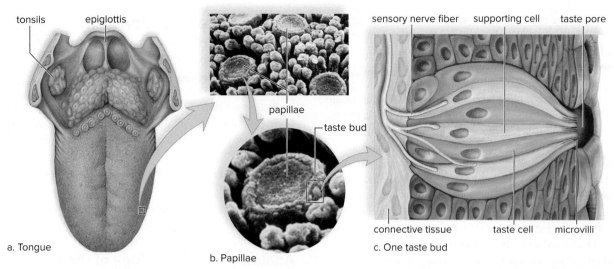

a. Tongue

b. Papillae

c. One taste bud

Figure 18.4 Taste buds in humans. **a.** Papillae on the tongue contain taste buds that are sensitive to sweet, sour, salty, bitter, and perhaps umami. **b.** Taste buds occur mainly along the walls of the papillae. **c.** Taste cells end in microvilli that bear receptor proteins for certain molecules. When molecules bind to the receptor proteins, nerve impulses are generated and go to the brain, where the sensation of taste occurs. (b, both): ©Omikron/SPL/Science Source

in sensory nerve fibers that go to the brain. When they reach the gustatory (taste) cortex, they are interpreted as particular tastes.

Sense of Smell

Approximately 80–90% of what we perceive as "taste" is actually due to the sense of smell, which explains why food tastes dull when we have a head cold or a stuffed-up nose. Our sense of smell is dependent on **olfactory cells** located within olfactory epithelium high in the roof of the nasal cavity (Fig. 18.5). Olfactory cells are modified neurons. Each cell ends in a tuft of about five olfactory cilia, which bear receptor proteins for odor molecules.

How the Brain Receives Odor Information

Each olfactory cell has only one out of about 1,000 different types of receptor proteins. Nerve fibers from like olfactory cells lead to

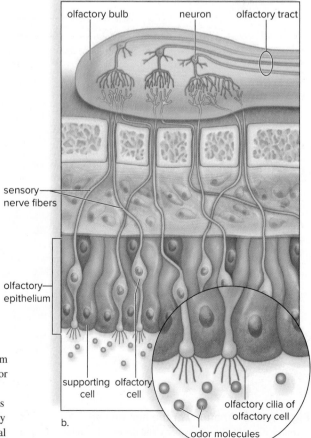

Figure 18.5 Olfactory cell location and anatomy. **a.** The olfactory epithelium in humans is located high in the nasal cavity. **b.** Olfactory cells end in cilia that bear receptor proteins for specific odor molecules. The cilia of each olfactory cell can bind to only one type of odor molecule (signified here by color). For example, if a rose causes olfactory cells sensitive to "blue" and "green" odor molecules to be stimulated, then neurons designated by blue and green in the olfactory bulb are activated. The primary olfactory area of the cerebral cortex interprets the pattern of stimulation as the scent of a rose.

the same neuron in the olfactory bulb, an extension of the brain. An odor contains many odor molecules, which activate a characteristic combination of receptor proteins. For example, a rose might stimulate olfactory cells, designated by blue and green in Figure 18.5b, while a hyacinth might stimulate a different combination. An odor's signature in the olfactory bulb is determined by which neurons are stimulated. The neurons communicate this information via the olfactory tract to the brain. When the information reaches the olfactory cortex, we know we have smelled a rose or a hyacinth.

Have you ever noticed that a certain aroma vividly brings to mind a certain person or place? A person's perfume may remind you of someone else, or the smell of boxwood may remind you of your grandfather's farm. The olfactory bulbs (see Fig. 17.12) have direct connections to the limbic system and its centers for emotions and memory. One investigator showed that when subjects smelled an orange while viewing a painting, they not only remembered the painting when asked about it later, but they also had many deep feelings about it.

Check Your Progress 18.3

1. Compare and contrast the functions of the chemoreceptors on the tongue and in the nasal cavity.
2. Predict some structural variations that could provide certain animals, such as dogs, with a sense of smell thousands of times more sensitive than that of humans.
3. Describe the anatomical connections that exist in the brain between smells and memories.

18.4 Sense of Vision

Learning Outcomes

Upon completion of this section, you should be able to

1. Identify the structures of the human eye.
2. Explain how the eye focuses on near and far objects.
3. Discuss the function and distribution of rods and cones.
4. Trace the path of a nerve impulse from the retina to the visual cortex.

Vision requires the work of the eyes and the brain. Researchers estimate that at least a third of the cerebral cortex takes part in processing visual information.

Anatomy and Physiology of the Eye

The eye, an elongated sphere about 2.5 cm in diameter, has three layers, or coats: the sclera, the choroid, and the retina (Fig. 18.6). Only the retina contains photoreceptors for light energy. Table 18.2 lists the functions of the parts of the eye.

The outer layer, called the **sclera,** is white and fibrous except for the **cornea,** which is made of transparent collagen fibers. The cornea is the window of the eye. Beginning at the edge of the cornea, a thin layer of epithelial cells forms a mucous membrane called the **conjunctiva** that covers the sclera and inner eyelids, keeping them moist. An inflammation of this membrane, called

sclera
choroid
retina
retinal blood vessels
optic nerve
fovea centralis
posterior compartment filled with vitreous humor
retina
choroid
sclera
ciliary body
lens
iris
pupil
cornea
anterior compartment filled with aqueous humor
suspensory ligament

Figure 18.6 Anatomy of the human eye.   Notice that the sclera (the outer layer of the eye) becomes the cornea and that the choroid (the middle layer) is continuous with the ciliary body and the iris. The retina (the inner layer) contains the photoreceptors for vision. The fovea centralis is the region where vision is most acute.

TABLE 18.2 Functions of the Parts of the Eye

Part	Function
Sclera	Protects and supports the eye
Cornea	Refracts light rays
Choroid	Absorbs stray light
Ciliary body	Holds lens in place; aids accommodation
Iris	Regulates light entrance
Pupil	Admits light
Retina	Contains sensory receptors for sight
Rod cells	Make black-and-white vision possible
Cone cells	Make color vision possible
Fovea centralis	Makes acute vision possible
Other	
Lens	Refracts and focuses light rays
Humors	Transmit light rays and support the eye
Optic nerve	Transmits impulse to brain

pink eye or *conjunctivitis,* can be caused by viruses, bacteria, irritants, or allergic reactions. The middle, thin, darkly pigmented layer, the **choroid,** is vascular and absorbs stray light rays that photoreceptors have not absorbed. Toward the front, the choroid becomes the donut-shaped **iris.** The iris contains smooth muscle that regulates the size of the **pupil,** a hole in the center of the iris through which light enters the eye. The color of the iris (and therefore, the color of your eyes) correlates with its pigmentation. Heavily pigmented eyes are brown, while lightly pigmented eyes are green or blue. Behind the iris, the choroid thickens and forms the circular ciliary body. The *ciliary body* contains the **ciliary muscle,** which controls the shape of the lens for near and far vision.

The **lens,** attached to the ciliary body by ligaments, divides the eye into two compartments. The one in front of the lens is the anterior compartment, and the one behind the lens is the posterior compartment. The anterior compartment is filled with a clear, watery fluid called the *aqueous humor.* A small amount of aqueous humor is continually produced each day. Normally, it leaves the anterior compartment by way of tiny ducts. When a person has *glaucoma,* these drainage ducts are blocked, and aqueous humor builds up, causing an increase in pressure (see Section 18.7).

The third layer of the eye, the **retina,** is located in the posterior compartment, which is filled with a clear, gelatinous material called the *vitreous humor.* The retina contains photoreceptors called rod cells and cone cells. The rods are very sensitive to light, but they do not respond to color. Therefore, at night or in a darkened room, we see only shades of gray. The cones, which require bright light, are sensitive to different wavelengths of light, which give us the ability to distinguish colors. The retina has a special region called the **fovea centralis,** where cone cells are densely packed. Light is normally focused on the fovea when we look directly at an object. This is helpful because

vision is most acute in the fovea centralis. Sensory nerve fibers from the retina form the *optic nerve,* which takes nerve impulses to the visual cortex.

Function of the Lens

The lens, assisted by the cornea and the humors, focuses images on the retina (Fig. 18.7). Focusing starts with the cornea and continues as the rays pass through the lens and the humors. The image produced is much smaller than the object because light rays are bent (refracted) when they are brought into focus.

Visual accommodation occurs for close vision. During this process, the lens rounds up in order to bring the image to focus on the retina. The shape of the lens is controlled by the ciliary muscle within the ciliary body. When you view a distant object, the ciliary muscle is relaxed, causing the suspensory ligaments attached to the ciliary body to be taut. Therefore, the lens remains relatively flat (Fig. 18.7*a*). When you view a near object, the ciliary muscle contracts, releasing the tension on the suspensory ligaments, and the lens rounds up due to its natural elasticity (Fig. 18.7*b*). Now the image is focused on the retina. Because close work requires contraction of the ciliary muscle, it very often causes muscle fatigue, known as *eyestrain.* Usually after the age of 40, the lens loses some of its elasticity and is unable to accommodate, a condition known as *presbyopia.* The use of reading glasses may then be necessary, or bifocal lenses for those who already have corrective lenses.

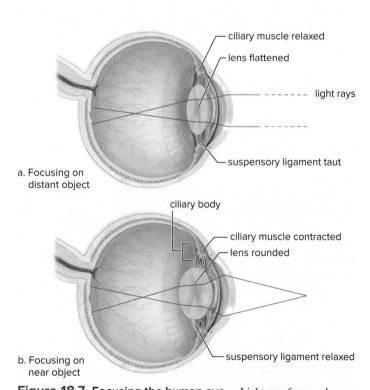

a. Focusing on distant object

b. Focusing on near object

Figure 18.7 Focusing the human eye.   Light rays from each point on an object are bent by the cornea and the lens in such a way that an inverted and reversed image forms on the retina. **a.** When focusing on a distant object, the lens is flat because the ciliary muscle is relaxed and the suspensory ligament is taut. **b.** When focusing on a near object, the lens accommodates; that is, it becomes rounded because the ciliary muscle contracts, causing the suspensory ligament to relax.

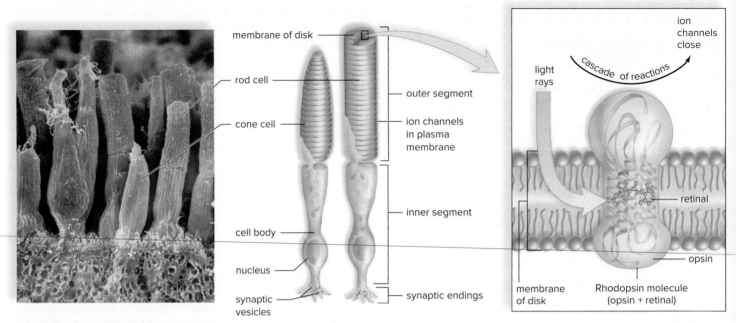

Figure 18.8 Photoreceptors in the eye. The outer segment of rods and cones contains stacks of membranous disks that contain visual pigments. In rods, the membrane of each disk contains rhodopsin, a complex molecule composed of the protein opsin and the pigment retinal. When rhodopsin absorbs light energy, it splits, releasing opsin, which sets in motion a cascade of reactions that cause ion channels in the plasma membrane to close. Thereafter, nerve impulses go to the brain. (photo): ©Science Source

Visual Pathway to the Brain

The pathway for vision begins once light has been focused on the photoreceptors in the retina. Some integration occurs in the retina, where nerve impulses begin before the optic nerve transmits them to the brain.

Function of Photoreceptors   The photoreceptors in the eye are of two types—**rod cells** and **cone cells.** Figure 18.8 illustrates their structure. Both rods and cones have an outer segment joined to an inner segment by a stalk. Pigment molecules are embedded in the membrane of the many disks present in the outer segment. Synaptic vesicles are located at the synaptic endings of the inner segment.

The visual pigment in rods is a deep purple pigment called rhodopsin. **Rhodopsin** is a complex molecule made up of the protein opsin and a light-absorbing molecule called **retinal,** which is a derivative of vitamin A. When a rod absorbs light, rhodopsin splits into opsin and retinal, leading to a cascade of reactions and the closure of ion channels in the rod cell's plasma membrane. The release of inhibitory transmitter molecules from the rod's synaptic vesicles ceases. Thereafter, signals go to other neurons in the retina. Rods are very sensitive to light, and are therefore suited to night vision. Because carrots are rich in vitamin A, it is true that eating carrots can improve your night vision. With the exception of the fovea centralis, rod cells are plentiful throughout the entire retina. Therefore, they also provide us with peripheral vision and perception of motion.

The cones, on the other hand, are located primarily in the fovea centralis and are activated by bright light. They allow us to detect the fine detail and the color of an object. **Color vision** depends on three different kinds of cones, which contain pigments

called the B (blue), G (green), and R (red) pigments. Each pigment is an iodopsin made up of retinal and an opsin. There is a slight difference in the opsin structure of each, which accounts for the ability of the pigments to absorb different wavelengths (colors) of visible light. Various combinations of cones are believed to be stimulated by in-between shades of color.

Function of the Retina   The retina has three layers of neurons (Fig. 18.9). The layer closest to the choroid (at the back of the eye) contains the rod cells and cone cells; the middle layer contains bipolar cells; and the innermost layer contains ganglion cells, whose sensory fibers become the optic nerve. Only the rod cells and the cone cells are sensitive to light, and therefore, light must penetrate to the back of the retina before they are stimulated.

The rod cells and the cone cells synapse with the bipolar cells, which in turn synapse with ganglion cells whose axons become the optic nerve. Notice in Figure 18.9 that there are many more rod cells and cone cells than ganglion cells. In fact, the retina has as many as 150 million rod cells and 6.5 million cone cells, but only 1 million ganglion cells. The sensitivity of cones versus rods is due in part to how directly they connect to ganglion cells. As many as 150 rods, covering about one square millimeter of retina (about the size of a thumbtack hole) may synapse on the same ganglion cell. No wonder stimulation of rods results in vision that is blurred and indistinct. In contrast, some cone cells in the fovea centralis activate only one ganglion cell. This explains why cones, especially in the fovea centralis, provide us with a sharper, more detailed image of an object.

Blind Spot   Notice in Figure 18.9*a* that there are no rods and cones where the optic nerve exits the retina. Therefore, no vision is

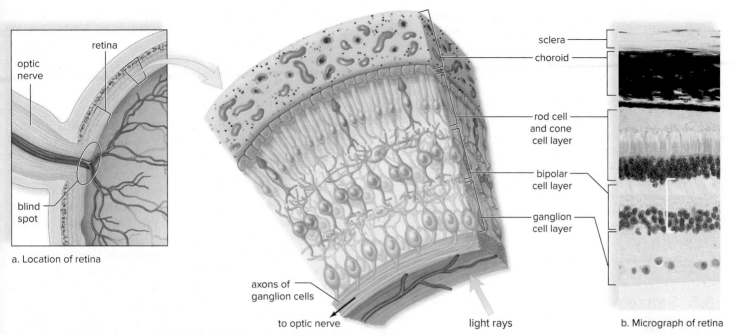

a. Location of retina

b. Micrograph of retina

to optic nerve

light rays

Figure 18.9 Structure and function of the retina. **a.** The retina is the inner layer of the eye. **b.** Rod cells and cone cells, located at the back of the retina nearest the choroid, synapse with bipolar cells, which synapse with ganglion cells. Integration of signals occurs at these synapses. Therefore, much processing occurs in bipolar and ganglion cells. Notice also that many rod cells share one bipolar cell, but cone cells do not. Certain cone cells synapse with only one ganglion cell. Cone cells, in general, distinguish more detail than do rod cells. **c.** Micrograph shows that the sclera and choroid are relatively thin compared to the retina, which has several layers of cells. (b): ©Biophoto Associates/Science Source

possible in this area. You can prove this to yourself by putting a dot to the right of center on a piece of paper. Use your right hand to move the paper slowly toward your right eye, while you look straight ahead. The dot will disappear at one point—this is your right eye's **blind spot.**

From the Retina to the Visual Cortex The axons of ganglion cells in the retina assemble to form the optic nerves. The optic nerves carry nerve impulses from the eyes to the *optic chiasma.* The *optic chiasma* has an X shape, formed by a crossing-over of optic nerve fibers. Fibers from the right half of each retina converge and continue on together in the *right optic tract,* and fibers from the left half of each retina converge and continue on together in the *left optic tract* (Fig. 18.10).

The optic tracts sweep around the hypothalamus, and most fibers synapse with neurons in nuclei (masses of neuron cell bodies) within the thalamus. Axons from the thalamic nuclei form optic radiations that take nerve impulses to the *visual area* within the occipital lobe. Notice in Figure 18.10 that the image arriving at the thalamus, and therefore the visual area, has been split because the left optic tract carries information about the right portion of the visual field (shown in green) and the right optic tract carries information about the left portion of the visual field (shown in red). Therefore, the right and left visual areas must communicate with each other for us to see the entire visual field. Also, because the image is inverted and reversed, it must be righted in the brain for us to correctly perceive the visual field.

The most surprising finding has been that the primary visual area acts like a post office, parceling out information regarding

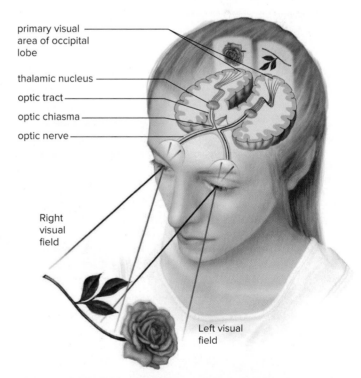

Figure 18.10 Optic chiasma. Both eyes "see" the entire visual field. Because of the optic chiasma, data from the right half of each retina (red lines) go to the right visual cortex, and data from the left half of each retina (green lines) go to the left visual cortex. These data are then combined to allow us to see the entire visual field. Note that the visual pathway to the brain includes the thalamus, which has the ability to filter sensory stimuli.

color, form, motion, and possibly other attributes to different portions of the adjoining visual association area. Therefore, the brain has taken the visual field apart, even though we see a unified visual field. The visual association areas are believed to rebuild the field and give us an understanding of it at the same time.

Check Your Progress 18.4

1. List all tissues that light passes through from when it enters the eye until it is converted into a nerve impulse.
2. Describe how the ciliary muscle controls the shape of the lens.
3. Explain how the number of rod or cone cells synapsing on a single ganglion cell allows cones to produce a clearer image.
4. Describe how the anatomy of the optic chiasma implies that the brain interprets visual information that has been "split" in half.

18.5 Sense of Hearing

Learning Outcomes

Upon completion of this section, you should be able to

1. List the parts of the ear that make up the outer ear, middle ear, and inner ear.
2. Describe the mechanism by which sound waves in the outer ear are converted into nerve impulses in the inner ear.
3. Explain how the inner ear is able to distinguish sounds of different pitches and volumes.

The ear functions in hearing as well as balance (equilibrium). The sensory receptors for both senses are mechanoreceptors located in the inner ear, each consisting of **hair cells** with stereocilia (long microvilli) that are sensitive to mechanical stimulation.

Anatomy of the Ear

Figure 18.11 shows that the ear has three divisions: outer, middle, and inner. The **outer ear** consists of the **pinna** (external flap), which collects and funnels sound into the **auditory canal.** The opening of the auditory canal is lined with fine hairs and sweat glands. Modified sweat glands are located in the upper wall of the canal. They secrete *earwax,* a substance that helps guard the ear against the entrance of foreign materials, such as air pollutants and microorganisms.

The **middle ear** begins at the **tympanic membrane** (eardrum) and ends at a bony wall containing two small openings covered by membranes. These openings are called the **oval window** and the **round window.** Between the tympanic membrane and the oval window are three small bones, collectively called the **ossicles.** Individually, they are the **malleus** (hammer), the **incus** (anvil), and the **stapes** (stirrup) because their shapes resemble these objects. The malleus adheres to the tympanic membrane, and the stapes touches the oval window. An **auditory tube,** also called the eustachian tube, which extends from the middle ear to the nasopharynx, permits equalization of air pressure. Chewing gum, yawning, and swallowing in elevators and airplanes help move air through the auditory tubes upon ascent and descent. As this occurs, we often hear the ears "pop."

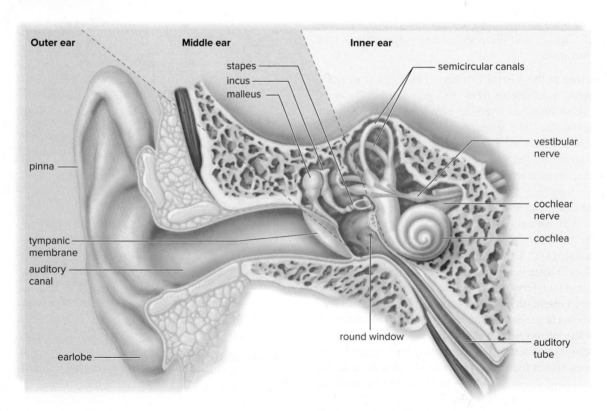

Figure 18.11 Anatomy of the human ear.   In the middle ear, the malleus (hammer), the incus (anvil), and the stapes (stirrup) amplify sound waves. In the inner ear, the mechanoreceptors for equilibrium are in the semicircular canals and the vestibule, and the mechanoreceptors for hearing are in the cochlea.

Whereas the outer ear and the middle ear contain air, the inner ear is filled with fluid. Anatomically speaking, the **inner ear** has three areas: the **semicircular canals** and the **vestibule** function in equilibrium; the **cochlea** functions in hearing. The cochlea resembles the shell of a snail because it spirals.

Auditory Pathway to the Brain

Hearing requires the ear, the cochlear nerve, and the auditory cortex of the brain. The sound pathway begins with the auditory canal.

Through the Auditory Canal and Middle Ear   The process of hearing begins when sound waves enter the auditory canal. Just as ripples travel across the surface of a pond, sound waves travel by the successive vibrations of molecules. When these vibrations impinge upon the tympanic membrane, it begins to vibrate at the same frequency as the sound waves. The malleus then takes the pressure from the inner surface of the tympanic membrane and passes it, by means of the incus, to the stapes in such a way that the pressure is multiplied about 20 times as it moves. The stapes strikes the membrane of the oval window, causing it to vibrate, and in this way, the pressure is passed to the fluid within the cochlea.

From the Cochlea to the Auditory Cortex   Examining part of the cochlea in cross-section reveals that it has three canals: the *vestibular canal,* the *cochlear canal,* and the *tympanic canal* (Fig. 18.12). Each of these canals is filled with fluid. The fluid within the cochlear canal is called *endolymph;* that within the vestibular and tympanic canals is called *perilymph.* The sense organ for hearing, called the **organ of Corti** (also called the *spiral organ*), is located in the cochlear canal. The spiral organ consists of little hair cells and a gelatinous material called the *tectorial membrane.* The hair cells sit on the basilar membrane, and their stereocilia are embedded in the tectorial membrane.

When the stapes strikes the membrane of the oval window, pressure waves move from the vestibular canal to the tympanic canal across the basilar membrane. The basilar membrane moves up and down, and the stereocilia of the hair cells embedded in the tectorial membrane bend. Then, nerve impulses begin in the *cochlear nerve* (also called the *auditory nerve*) and travel to the brain. When they reach the auditory cortex in the temporal lobe, they are interpreted as a sound.

Each part of the organ of Corti is sensitive to different wave frequencies, or pitch. Near the tip, it responds to low pitches, such as a tuba, and near the base, it responds to higher pitches, such as a bell or a whistle. The nerve fibers from each region along the length of the organ of Corti lead to slightly different areas in the auditory cortex. The pitch sensation we experience depends upon which region of the basilar membrane vibrates and which area of the auditory cortex is stimulated.

Volume is a function of the amplitude of sound waves. Loud noises cause the fluid within the vestibular canal to exert more pressure and the basilar membrane to vibrate to a greater extent. The brain interprets the resulting increased stimulation as volume. Researchers believe that the brain interprets the tone of a sound based on the distribution of the hair cells stimulated. Exposure to loud noises over a long period of time is one of the causes of

Cochlea cross-section

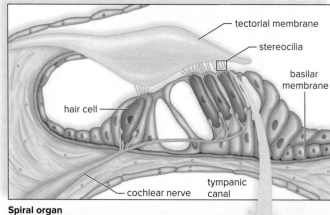

Spiral organ

Stereocilia   3,900×

Figure 18.12 Mechanoreceptors for hearing.   The spiral organ is located within the cochlea. In the uncoiled cochlea, note that the spiral organ consists of hair cells resting on the basilar membrane, with the tectorial membrane above. Pressure waves move from the vestibular canal to the tympanic canal, causing the basilar membrane to vibrate. This causes the stereocilia (or at least a portion of the more than 20,000 hair cells) embedded in the tectorial membrane to bend. Nerve impulses traveling in the cochlear nerve to the auditory cortex result in hearing. (photo): ©P. Motta/SPL/Science Source hearing loss. For more on hearing loss, see the Health feature "Preventing Hearing Loss" in Section 18.7.

18.6 Sense of Equilibrium

In addition to detecting sound waves, mechanoreceptors in the inner ear detect rotational and/or angular movement of the head, or **rotational equilibrium,** as well as movement of the head in the vertical or horizontal planes, or **gravitational equilibrium** (Fig. 18.13).

a. Rotational equilibrium: receptors in ampullae of semicircular canal

b. Gravitational equilibrium: receptors in utricle and saccule of vestibule

Figure 18.13 Mechanoreceptors for equilibrium.   **a.** Rotational equilibrium. The ampullae of the semicircular canals contain hair cells with stereocilia embedded in a cupula. When the head rotates, the cupula is displaced, bending the stereocilia. This sends nerve impulses via the vestibular nerve to the brain. **b.** Gravitational equilibrium. The utricle and the saccule contain hair cells with stereocilia embedded in an otolithic membrane. When the head bends, otoliths are displaced, causing the membrane to sag and the stereocilia to bend. The direction the stereocilia bend tells the brain in which direction the head moved.

Through their communication with the brain, these mechanoreceptors help us maintain balance and equilibrium, but other structures are also involved. For example, proprioceptors provide information on the positions of muscles and limbs. Vision, if available, also provides extremely helpful information the brain can act upon to keep balance.

Rotational Equilibrium Pathway

Rotational equilibrium (Fig 18.13*a*) involves the three semicircular canals, which are arranged so that there is one in each dimension of space. The base of each of the three canals, called an ampulla, is slightly enlarged. Little hair cells, whose stereocilia are embedded within a gelatinous material called a cupula, are found within the ampullae. Because of the way the semicircular canals are arranged, each ampulla responds to head rotation in a different plane of space. As fluid within a semicircular canal flows over and displaces a cupula, the stereocilia of the hair cells bend, and the pattern of impulses carried by the vestibular nerve to the brain stem and cerebellum changes. The brain uses information from the hair cells within the ampullae of the semicircular canals to maintain rotational equilibrium through appropriate motor output to various skeletal muscles that can right our present position in space as need be.

Gravitational Equilibrium Pathway

Gravitational equilibrium (Fig. 18.13*b*) depends on the **utricle** and the **saccule,** two membranous sacs located in the vestibule. Both of these sacs contain little hair cells, whose stereocilia are embedded within a gelatinous material called an otolithic membrane. Calcium carbonate ($CaCO_3$) granules, or **otoliths,** rest on this membrane. The utricle is especially sensitive to horizontal (back-and-forth) movements and the bending of the head, while the saccule responds best to vertical (up-and-down) movements.

When the body is still, the otoliths in the utricle and the saccule rest on the otolithic membrane, above the hair cells. When the head bends or the body moves in the horizontal and vertical planes, the otoliths are displaced and the otolithic membrane sags, bending the stereocilia of the hair cells beneath. If the stereocilia move toward the largest stereocilium, called the kinocilium, nerve impulses increase in the vestibular nerve. If the stereocilia move away from the kinocilium, nerve impulses decrease in the vestibular nerve. If you are upside down, nerve impulses in the vestibular nerve cease. These data reach the vestibular cortex, which uses them to determine the direction of the movement of the head at the moment. The brain uses this information to maintain gravitational equilibrium through appropriate motor output to various skeletal muscles that can right our present position in space as need be.

Check Your Progress 18.6

1. Explain the role, and specific location, of mechanoreceptors involved in maintaining balance and equilibrium.
2. Classify each of the following as being associated with either the semicircular canals or the vestibule: ampulla, cupula, otoliths, saccule, and utricle.
3. Explain how having two separate systems for equilibrium provides more complete information than one system could.

18.7 Disorders That Affect the Senses

Learning Outcomes

Upon completion of this section, you should be able to
1. List three common disorders that can affect the senses of taste and smell.
2. Explain the anatomical abnormalities that cause color blindness, nearsightedness, farsightedness, and astigmatism.
3. Characterize the three most common causes of blindness in the United States.
4. Describe some common disorders of hearing and equilibrium.

Disorders of Taste and Smell

It is hard for most of us to imagine our world without the ability to taste and smell. Disorders that affect these senses may not sound very serious, but these senses contribute substantially to our enjoyment of life. In addition, unpleasant smells and tastes can warn us about dangers such as fire, poisonous fumes, and spoiled food.

In most people, the sense of smell begins to decline after age 60, and a large proportion of elderly people lose some of their olfactory sense. Some people are born with a poor sense of smell or taste, and others completely lack a sense of smell, a condition known as *anosmia.* However, the most common conditions that affect our sense of taste and smell are upper respiratory infections, allergies, exposure to certain drugs and chemicals, and certain types of brain tumors. Brain injury due to trauma can also cause smell or taste problems, and smoking impairs the ability to identify odors and diminishes the sense of taste.

The extent of a loss of smell or taste can be tested using the lowest concentration of a chemical that a person can detect and recognize. Scientists have developed an easily administered "scratch-and-sniff" test to evaluate the sense of smell. For taste, patients are asked to evaluate different chemical concentrations, sometimes applied directly to specific areas of the tongue.

Disorders of the Eye

Color blindness and problems with visual focus are two common abnormalities of the eye. More serious disorders can result in blindness.

Color Blindness

There are many different forms of color blindness. Complete **color blindness** is extremely rare. Instead, in most instances a particular type of cone is lacking or deficient in number. The most common mutation is a lack of either red or green cones, resulting in various types of color blindness (Fig. 18.14). If the eye lacks red cones, the green colors are accentuated, and vice versa. Most genetic mutations affecting color vision are X-linked, recessive traits. Because females have two copies of the X chromosome, they are more likely to have a normal copy of the gene. Therefore, color blindness is much more common in males. Approximately 5–8% of males and 0.5% of females have some type of deficiency in color vision. See Section 24.1 for additional discussion of the genetics of color blindness.

Figure 18.14 Testing for color blindness.   This is one type of test plate used to diagnose color blindness. There are many different types of color blindness, and thus many different tests are used. In this case, you should be able to see the number "74" in green among the dots. ©Steve Allen/Getty Images

a. Nearsightedness

b. Farsightedness

c. Astigmatism

Figure 18.15 Common abnormalities of the eye and possible corrective lenses.   **a.** A concave lens in nearsighted persons focuses light rays on the retina. **b.** A convex lens in farsighted persons focuses light rays on the retina. **c.** An uneven lens in persons with astigmatism focuses light rays on the retina.

Visual Focus

The majority of people can see what is designated as a size 20 letter 20 feet away, and so are said to have 20/20 vision. Persons who can see close objects but cannot see the letter from this distance are said to be **nearsighted.** These individuals usually have an elongated shape to the eye, so that when they attempt to look at a distant object, the lens of the eye brings the image into focus in front of the retina (Fig. 18.15a). They can see close objects because the lens can compensate for the longer shape of the eye. To see distant objects, these people can wear concave lenses, which diverge the light rays so that the image focuses on the retina.

Persons who can easily see the optometrist's chart but cannot see close objects well are **farsighted.** These individuals can see distant objects better than they can see close objects. Their eyes have a shortened shape, and when they try to see close objects, the image is focused behind the retina. When the object is distant, the lens can compensate for the shorter structure of the eye. When the object is close, these persons must wear convex lenses to increase the bending of light rays so that the image can be focused on the retina (Fig. 18.15b). People over the age of 40 usually begin to have trouble seeing objects close up due to a condition called presbyopia, in which the lens of the eye begins to lose its elasticity and thus cannot focus on close objects.

When the cornea or lens is uneven, the image is fuzzy. The light rays cannot be evenly focused on the retina. This condition, known as **astigmatism,** can be corrected by wearing an unevenly ground lens to compensate for the uneven cornea (Fig. 18.15c). Rather than wearing glasses or contact lenses, many people with problems of visual focus are now choosing to undergo a laser surgery procedure called LASIK (laser-assisted in situ keratomileusis; see the chapter opener).

Common Causes of Blindness

The most frequent causes of blindness in adults are retinal disorders, glaucoma, and cataracts, in that order. Retinal disorders are varied. In **diabetic retinopathy,** the leading cause of blindness in the United States for people under the age of 65, capillaries to the retina become damaged, and hemorrhages and blocked vessels can occur. Surgical lasers can be used to seal the leaking blood vessels, but eventually scarring can lead to blindness. A common condition in people over 50 years of age is **macular degeneration,** in which the cones are destroyed because thickened choroid vessels no longer function as they should. With **retinal detachment,** the retina becomes torn or separated from the choroid, often following trauma to the eye or head. If diagnosed quickly enough, the retina can usually be reattached with a laser.

Glaucoma occurs when the drainage system of the eyes fails, so fluid builds up and the resulting increase in pressure destroys the nerve fibers responsible for peripheral vision. If glaucoma is untreated, total blindness can result. Eye doctors usually measure the pressure inside the eye during routine exams, but it is advisable for everyone to be aware of the disorder because it can come on quickly, and damage to the nerves can be permanent. Those who have experienced acute glaucoma report that the eye feels as heavy as a stone. Rarely, the abnormalities that lead to glaucoma can be present at birth, even though the disease symptoms may not appear for a few months or years. Regular visits to an eye-care specialist,

Figure 18.16 Cataract in a human eye. A cataract is a lens that has become opaque or cloudy. Risk factors that increase the chances of cataract formation include UV exposure, diabetes, and smoking. Cataracts can usually be surgically removed and replaced with an artificial lens. ©Sue Ford/Science Source

especially by the elderly, are a necessity in order to catch conditions such as glaucoma early enough to allow effective treatment.

With aging, the eye is increasingly subject to **cataracts,** cloudy spots on the lens of the eye. Eventually these pervade the whole lens, which takes on a milky-white to yellowish color and becomes incapable of transmitting light (Fig. 18.16). Factors that increase the risk of cataracts include exposure to ultraviolet or other types of radiation, diabetes, and heavy alcohol consumption. In addition, smoking can double the risk of cataracts, possibly by reducing the delivery of blood, and therefore nutrients, to the lens. In most cases, vision can be restored by surgical removal of the unhealthy lens and replacement with a clear plastic artificial lens. An estimated 22.3 million Americans over the age of 40 have cataracts. Cataract surgery is the most frequently performed surgery in the United States. Surgeons can now replace cataracts with multifocal lenses that allow patients to focus at different distances, even eliminating the need for glasses in some patients.

Several published studies have shown that a certain form of hereditary blindness could be successfully treated with gene therapy. These studies involved injecting a genetically modified virus into one eye of patients with Leber's congenital amaurosis (LCA), a rare condition in which retinal degeneration begins in childhood and eventually causes total blindness. Starting 2 weeks after the injections, vision in the injected eyes improved from detecting only hand movements to reading lines on an eye chart. Equally significant, the injected eyes showed no signs of inflammation in the retina or other toxic side effects. This treatment has recently been approved by the Food and Drug Administration (FDA).

Although it may sound like science fiction, researchers are currently testing an artificial eye of sorts, which may allow blind people to see. For blind patients who still have intact connections between their retinas and brain, an array of electrodes can be surgically implanted in the retina. These electrodes receive input from a tiny digital camera mounted on eyeglass frames worn by the patient. This input is converted into a nerve impulse, which the individual's brain interprets as sight. Clinical trials have shown that by using such a device, some formerly blind patients can detect light or even distinguish between objects such as a cup or plate, and more advanced devices may become available for testing soon.

Because several disorders of the eye have been associated with exposure to ultraviolet radiation, it is recommended that everyone, especially those who live in sunny climates or work outdoors, wear sunglasses that absorb ultraviolet light. The Sunglass Association of America has adopted the following system for categorizing sunglasses:

- Cosmetic lenses, which absorb at least 70% of UVB (the more harmful type of ultraviolet radiation), 20% of UVA, and 60% of visible light. Such lenses are worn for comfort or appearance rather than protection.
- General-purpose lenses that absorb at least 95% of UVB, 60% of UVA, and 60–92% of visible light. These lenses are good for outdoor activities in temperate regions.
- Special-purpose lenses, which block at least 99% of UVB, 60% of UVA, and 60–92% of visible light. They are good for bright sun combined with sand, snow, or water.

Disorders of Hearing and Equilibrium

Hearing loss can develop gradually or suddenly and has many potential causes. Especially when we are young, the middle ear is subject to infections that can lead to hearing impairment if not treated promptly by a physician (see Section 15.4). Hearing loss is also one of the most common conditions of older adults, affecting one in three people over age 60. Age-associated hearing loss usually develops gradually. The first sign may be problems understanding conversations when there is noise in the background. Actually, in most individuals, age-related hearing loss probably begins at around age 20, with the ability to hear high-pitched sounds affected first. With age, the mobility of ossicles decreases, and in otosclerosis, new filamentous bone grows over the stirrup, impeding its movement. Surgical treatment is the only remedy for this type of conduction deafness. Some types of hearing loss seem to run in families, but most can be attributed to the effect of years of frequent exposure to loud noise (see the Health feature "Preventing Hearing Loss"). Some types of deafness can be treated with cochlear implants.

Sudden deafness, which may occur all at once or over a period of up to 3 days, most commonly affects people between the ages of 30 and 50, usually in only one ear. Potential causes include infections, head trauma, and the side effects of certain drugs. Depending on the cause, sudden deafness often resolves itself within a few days.

Deafness can also be present at birth. Several genetic disorders are associated with deafness at birth, as are infections with German measles (rubella) and mumps virus when contracted by a woman during her pregnancy. For this reason, every female should be immunized against these viruses before she reaches childbearing age.

Disorders of equilibrium often manifest as **vertigo,** the feeling that a person or the environment is moving when no motion is occurring. It is possible to simulate a feeling of vertigo by spinning

SCIENCE IN YOUR LIFE ▶ HEALTH

Preventing Hearing Loss

Some degree of age-associated hearing loss is very common, but deafness due to stereocilia damage from exposure to loud noises is preventable. Hospitals are now aware that even the ears of newborns need to be protected from noise and are taking steps to make sure neonatal intensive care units and nurseries are as quiet as possible.

In today's society, exposure to the types of noises listed in Table 18A is common. Noise is measured in decibels (db), and any noise above a level of 80 db could result in damage to the hair cells of the spiral organ. Hearing loss occurs once 25–30% of these cells are damaged. Eventually, the stereocilia and then the hair cells disappear completely (Fig. 18A).

Legendary rock musician Pete Townsend of The Who recently admitted that long-term exposure to loud music has caused him to experience hearing loss. Digital music players may pose an additional risk because they can produce music as loud as 120 db, plus hold thousands of songs and play for hours without recharging. Like all the sensory systems, the ears can become desensitized to loud music, so that it seems less loud. The response of many music enthusiasts, of course, is to turn up the volume. In 2006, an iPod user filed a lawsuit against Apple, claiming the company did not warn users sufficiently about the dangers of loud music, and asking the company to offer a software update to limit the device's output to 100 db, as it has already done in France.

The first hint of danger could be temporary hearing loss, a "full" feeling in the ears, muffled hearing, or tinnitus (e.g., ringing in the ears). If you have any of these symptoms, modify your listening habits immediately to prevent further damage. If exposure to noise is unavoidable, specially designed noise-reduction earmuffs are available, and it is also possible to purchase earplugs made from a compressible, spongelike material at the drugstore or sporting-goods store. These earplugs are not the same as those worn for swimming, and they should not be used interchangeably.

In addition to loud music, noisy indoor or outdoor equipment can also be harmful to hearing if hearing protection is not used. Even motorcycles and recreational vehicles such as snowmobiles can contribute to a gradual loss of hearing. Exposure to intense sounds of short duration, such as a burst of gunfire, can result in an immediate hearing loss. Hunters may have a significant hearing reduction in the ear opposite the shoulder where the gun is held.

The butt of the rifle offers some protection to the ear nearest the gun when it is shot.

Finally, people need to be aware that some medicines are ototoxic. Anticancer drugs, most notably cisplatin, and certain antibiotics (e.g., streptomycin, kanamycin, and gentamicin) make ears especially susceptible to a hearing loss. Anyone taking such medications needs to be especially careful to protect the ears from loud noises.

Questions to Consider

1. Do you think that manufacturers of digital music players should be required to restrict the output of these devices to 100 db in the United States?
2. If you listen to loud music or are exposed to other noise, are you concerned that the hair cells in the spiral organs of your inner ears may become damaged? Why or why not?

a.   b.

Figure 18A
Hearing loss.
a. Normal hair cells in the spiral organ of a guinea pig. **b.** Damaged hair cells in the spiral organ of a guinea pig. This damage occurred after 24-hour exposure to a noise level typical of a rock concert. (both): ©Kresge Hearing Research Institute, University of Michigan

TABLE 18A Noises That Affect Hearing

Type of Noise	Sound Level (decibels)	Effect
Jet engine, shotgun, rock concert	Over 125	Beyond threshold of pain; potential for hearing loss high
Music on ear buds (maximum), thunderclap	Over 120	Hearing loss likely
Chain saw, pneumatic drill, jackhammer, snowmobile, cement mixer	100–200	Regular exposure of more than 1 minute risks permanent hearing loss
Farm tractor, newspaper press, subway, motorcycle	90–100	Fifteen minutes of unprotected exposure potentially harmful
Lawn mower, food blender	85–90	Continuous daily exposure for more than 8 hours can cause hearing damage
Diesel truck, average city traffic noise	80–85	Annoying; constant exposure may cause hearing damage

Source: National Institute on Deafness and Other Communication Disorders, National Institutes of Health.

your body rapidly and then stopping suddenly. Vertigo can be caused by problems in the brain, as well as the inner ear. An estimated 20% of those who experience symptoms of vertigo have benign positional vertigo (BPV), which may result from the formation of particles in the semicircular canals. When individuals with BPV move their heads suddenly, especially when lying down, these particles shift like pebbles inside a tire. After the head movement, there is an initial delay, but when the movement stops, the particles tumble down with gravity, stimulating the stereocilia and resulting in the sensation of movement. Interestingly, most cases of BPV either resolve on their own or can be cured using a series of precise head movements designed to cause the particles to settle into a position where they can no longer stimulate the stereocilia.

Because the senses of hearing and equilibrium are anatomically linked, certain disorders can affect both. An example of this phenomenon is *Meniere's disease,* which is usually characterized by vertigo, a feeling of fullness in the affected ear(s), tinnitus (ringing in the ears), and hearing loss. In about 80% of patients, only one ear is affected, and the disease usually strikes between the ages of 20 and 50. The exact cause of Meniere's disease is unknown, but it seems related to an increased volume of fluid in the semicircular canals, vestibule, and/or cochlea. For this reason, it has been called "glaucoma of the ear." There is no cure for Meniere's disease, but the vertigo and feeling of pressure can often be managed by adhering to a low-salt diet, which is thought to decrease the volume of fluid produced. Any hearing loss that occurs, however, is usually permanent.

> ### Check Your Progress 18.7
> 1. Describe how a lack of smell can be life-threatening.
> 2. Explain how the shape of the eyeball can affect vision.
> 3. Distinguish how blindness can occur in macular degeneration, glaucoma, and cataracts.
> 4. List some common causes of hearing loss.

Conclusion

Following the surgery, John received eyedrops and some pain medication to help relieve the discomfort associated with the procedure. He was also informed not to expect immediate changes in his vision. While some patients see improvements in their vision as early as the day after the surgery, it takes 2 to 3 months for their vision to stabilize.

John's doctor scheduled several appointments to assess how his vision was progressing. At his first appointment, he complained of some redness and dryness in his eyes, but was informed by his doctor that this was a normal outcome of the surgery.

His doctor was pleased with John's progress, and informed him that most patients will have vision close to 20/20, but even with LASIK it was possible that his vision could still change over time.

SUMMARIZE

18.1 Sensory Receptors and Sensations

Sensory receptors are cells capable of converting many types of stimuli into nerve impulses, a process called **sensory transduction.** They can be classified in several ways:

- According to whether they receive stimuli from inside or outside the body—that is, as **interoceptors** or **exteroceptors.**
- According to the nature of the stimulus they receive—that is, as **chemoreceptors, photoreceptors, mechanoreceptors,** or **thermoreceptors.**

The conscious perception of stimuli is the result of several processes:

- Detection occurs when environmental changes stimulate sensory receptors, generating nerve impulses.
- **Sensation** occurs when these nerve impulses reach the cerebral cortex.
- **Perception** occurs when the brain interprets the meaning of sensations.
- **Integration** occurs when the incoming signals are summed together.
- **Sensory adaptation** occurs when a receptor reduces sensitivity to a stimulus.

18.2 Somatic Senses

The somatic sensory receptors include proprioceptors, cutaneous receptors, and pain receptors:

- **Proprioceptors** help maintain equilibrium and posture. Proprioception is illustrated by the action of muscle spindles that are stimulated when muscle fibers stretch, and Golgi tendon organs that are stimulated when muscles contract.
- **Cutaneous receptors** in the skin detect touch, pressure, pain, and temperature (warmth and cold).
- **Pain receptors** are free nerve endings that respond to chemicals released by damaged cells. The pain of internal organs is sometimes felt in the skin and is called referred pain.

18.3 Senses of Taste and Smell

The senses of taste and smell involve the detection of molecules in food and in the air by chemoreceptors:

- Located mainly on the tongue, **taste buds** contain taste cells. After food molecules bind to plasma membrane receptor proteins on the microvilli of taste cells, as well as on the cilia of **olfactory cells** in the nasal cavity, nerve impulses eventually reach the cerebral cortex, which determines the taste and odor according to the pattern of stimulation. The gustatory (taste) cortex responds to a range of salty, sour, bitter, sweet, and perhaps "umami" tastes.

18.4 Sense of Vision

Vision requires the eye, the optic nerves, and the visual cortex in the occipital lobe:

- The eye has three layers. The outer layer, called the **sclera,** can be seen as the white of the eye. It also forms the transparent portion of the front of the eye called the **cornea.** The exposed surface of the sclera is covered by the **conjunctiva.** The middle pigmented layer, called the **choroid,** absorbs stray light rays. At the front of the eye the choroid becomes the **iris,** which forms the **pupil,** or opening through which light passes. As light passes through the eye, it is focused by the **lens** on its way to the **retina,** the inner layer of the eyeball. The choroid also forms the **ciliary muscle,** which controls the shape of the lens. To see a close object, **visual accommodation** occurs as the lens rounds up.

■ Light is normally focused on the **fovea centralis,** an area of densely packed cone cells. Photoreceptors called **rod cells** (sensory receptors for dim light) and **cone cells** (sensory receptors for bright light and color) are located in the retina. The visual pathway begins when light strikes **rhodopsin** within the membranous disks of rod cells. A cascade of reactions leads to the closing of ion channels in a rod cell's plasma membrane, and signals are passed to other neurons in the retina. From the rod and cone cell layer, impulses pass to the bipolar cell layer and the ganglion cell layer. Many rods may synapse on the same ganglion cell, but some cone cells may activate only one ganglion cell. There are no photoreceptors in the area of the retina known as the **blind spot,** where the optic nerve exits.

■ The axons of ganglion cells become the optic nerve, which takes nerve impulses to the visual cortex in the occipital lobe. The primary visual area parcels out signals for color, form, and motion to the visual association area. Then, the visual cortex rebuilds the field.

18.5 Sense of Hearing

Hearing is dependent on the ear, the cochlear nerve, and the auditory cortex of the brain:

■ The ear is divided into three parts. The **outer ear** consists of the **pinna** and the **auditory canal,** which directs waves to the middle ear. The **middle ear** begins with the **tympanic membrane** and contains the **ossicles** (**malleus, incus,** and **stapes**). The malleus is attached to the **tympanic membrane,** and the stapes is attached to the **oval window,** which is covered by a membrane. Within the **inner ear,** the **cochlea** functions in hearing, while the **vestibule** and **semicircular canals** function in equilibrium.

■ The auditory pathway begins when the outer ear receives and the middle ear amplifies the sound waves, which then strike the oval window membrane. Its vibrations set up pressure waves across the cochlear canal, which contains the **organ of Corti** (also called the spiral organ), consisting of hair cells whose stereocilia are embedded within the tectorial membrane. When the basilar membrane vibrates, the stereocilia of the hair cells bend. This movement is transduced into nerve impulses that are carried by the cochlear nerve to the primary auditory area in the temporal lobe of the cerebral cortex.

18.6 Sense of Equilibrium

The ear also contains mechanoreceptors for our sense of equilibrium, or balance:

■ **Rotational equilibrium,** which is the sensing of head movements, is dependent on the stimulation of hair cells within the ampullae of the semicircular canals.

■ **Gravitational equilibrium,** which is the sensing of the position of the head in space, relies on the stimulation of hair cells within the **utricle** and the **saccule** by small granules called **otoliths.**

18.7 Disorders That Affect the Senses

Most conditions that affect the sensory organs are not life-threatening, but can affect the quality of life:

■ Disorders of taste and smell may result from infections, allergies, certain drugs, or tumors.

■ Disorders of the eye include **color blindness** and problems with visual focus such as **astigmatism** and being **nearsighted** or **farsighted.** The most common causes of blindness are retinal disorders like **diabetic retinopathy, macular degeneration,** and **retinal detachment,** as well as **glaucoma** (increased intraocular pressure) and **cataracts** (cloudy lens).

■ Disorders of hearing are very common, especially in the aging population. Possible causes include excessive noise exposure, infections, trauma, and certain drugs. **Vertigo** is an equilibrium disorder characterized by a feeling of motion or dizziness.

ASSESS

Testing Yourself

Choose the best answer for each question.

18.1 Sensory Receptors and Sensations

1. These receptors detect stimuli within the body.
   - a. interoceptors
   - b. exteroceptors
   - c. homeoreceptors
   - d. reflex receptors

2. A receptor that detects changes in pH, or specific molecules in the environment, would be classified as a
   - a. mechanoreceptor.
   - b. photoreceptor.
   - c. chemoreceptor.
   - d. thermoreceptor.
   - e. None of these are correct.

3. Where does the process of sensation occur in the body?
   - a. at the sensory receptor
   - b. in the spinal cord
   - c. within the synapses between neurons of the PNS
   - d. in the cerebral cortex
   - e. All of these are correct.

18.2 Somatic Senses

4. Which type of receptor detects the chemicals released by damaged tissues?
   - a. nociceptors
   - b. proprioceptors
   - c. Meissner corpuscles
   - d. Ruffini endings
   - e. None of these are correct.

5. Which of the following types of receptors assists in the maintenance of muscle tone?
   - a. nociceptors
   - b. proprioceptors
   - c. Pacinian corpuscles
   - d. Krause end bulbs
   - e. All of these are correct.

18.3 Senses of Taste and Smell

6. The senses of taste and smell rely primarily on what type of receptors?
   - a. mechanoreceptors
   - b. nociceptors
   - c. protoreceptors
   - d. proprioceptors
   - e. chemoreceptors

7. Olfactory bulbs are located
   - a. on the tongue.
   - b. in the nasal cavity.
   - c. in the brain stem.
   - d. in the aorta.
   - e. None of these are correct.

18.4 Sense of Vision

8. Label the following diagram of the human eye.

9. Which structure of the eye is incorrectly matched with its function?
   a. lens—focusing
   b. cones—color vision
   c. iris—regulation of amount of light
   d. choroid—location of cones
   e. sclera—protection

10. Adjustment of the lens to focus on objects close to the viewer is called
   a. convergence.
   b. focusing.
   c. visual accommodation.
   d. constriction.

18.5 Sense of Hearing

11. Label the following diagram of the human ear.

12. Which of the following is (are) not involved in the sense of hearing?
   a. auditory canal
   b. tympanic membrane
   c. ossicles
   d. semicircular canals
   e. cochlea

13. Which one of these correctly describes the location of the spiral organ?
   a. between the tympanic membrane and the oval window in the inner ear
   b. in the utricle and saccule within the vestibule
   c. between the tectorial membrane and the basilar membrane in the cochlear canal
   d. between the nasal cavities and the throat
   e. between the outer and inner ear within the semicircular canals

18.6 Sense of Equilibrium

14. Which of the following structures would allow you to know that you were upside down, even if you were in total darkness?
   a. utricle and saccule
   b. cochlea
   c. semicircular canals
   d. tectorial membrane

15. Rotational equilibrium is established by which of the following?
   a. the semicircular canals
   b. the utricle and saccule
   c. the cochlea
   d. the auditory canal
   e. None of these are correct.

18.7 Disorders That Affect the Senses

16. A color-blind person has (an) abnormal
   a. rods.
   b. cochlea.
   c. cornea.
   d. cones.
   e. None of these are correct.

17. Which pair of terms is properly matched?
   a. nearsightedness—light rays focus in front of the retina
   b. nearsightedness—light rays focus behind the retina
   c. farsightedness—light rays focus in front of the retina
   d. None of these are correct.

ENGAGE

Thinking Critically

1. Some sensory receptors, such as those for taste, smell, and pressure, readily undergo the process of sensory adaptation, or decreased response to a stimulus. In contrast, receptors for pain are less prone to adaptation. Why does this make good biological sense? What do you think happens to children who are born without the ability to feel pain normally?

2. Trace the path of a photon of light through all the layers of the retina discussed in this chapter. Then, trace the nerve impulse from a photoreceptor located on the right side of the left eyeball through the visual cortex.

3. The density of taste buds on the tongue can vary. Some obese individuals have a lower density of taste buds than usual. Assume that taste perception is related to taste bud density. If so, what hypothesis would you test to see if there is a relationship between taste bud density and obesity?

4. List the types of senses and receptors that are active as you walk into a restaurant and sit down at a table.

CHAPTER

19

Musculoskeletal System

Knee with osteoarthritis

Normal knee

©Scott Camazine/Science Source

CHAPTER OUTLINE

BEFORE YOU BEGIN

Before beginning this chapter, take a few moments to review the following discussions:

Section 7.1 What biochemical pathways do cells use to produce ATP?

Section 11.1 What is the role of connective tissue in the body?

Section 11.5 How does the musculoskeletal system contribute to homeostasis?

Replacing Joints

Jackie was an outstanding athlete in high school, and even as she reached her early fifties, she tried to stay in shape. But during her customary 3-mile jogs, she was having an increasingly hard time ignoring the pain in her left knee. She had torn some ligaments there playing intramural softball in college, and her knee had never quite felt the same. In her forties, she was able to control the pain by taking over-the-counter medications, but 2 years ago she had arthroscopic surgery to remove some torn cartilage and calcium deposits from the knee joint. Now that the pain was getting worse, she knew her best option might be a total knee replacement.

Although it sounds drastic, replacing old, arthritic joints with shiny new artificial ones is becoming increasingly routine. Over 4.5 million Americans have an artificial knee, and the numbers are increasing as the population ages. During the procedure, a surgeon removes bone from the bottom of the femur and the top of the tibia, and replaces each with caps made of metal or ceramic, held in place with bone cement. A plastic plate is installed to allow the femur and tibia to move smoothly against each other, and a smaller plate is attached to the kneecap (patella) so it can function properly. The average cost for knee replacement surgery is about $58,000. The knee is the most common joint that is replaced, followed by the hip. And although it is more difficult, each year about 10,000 artificial ankles are being installed worldwide. In this chapter, we will examine the anatomy and physiology of bones, joints, and muscles.

As you read through this chapter, think about the following questions:

1. Why does Jackie's physical condition make her an ideal candidate for a knee replacement?
2. What is the role of cartilage in the knee joint?
3. Why would it be more difficult to replace an ankle than a knee or hip, or even an elbow or shoulder?

19.1 Overview of Bone and Cartilage

The skeleton provides attachment sites for the muscles, whose contraction makes the bones move so that we can walk, play sports, send a text message, and perform all sorts of other activities. Together, the bones and muscles make up the **musculoskeletal system.** In this chapter, we discuss the structures and functions of first the skeleton and then the muscles. Bones and the tissues associated with them are connective tissues, whereas muscles are, of course, composed of muscular tissue.

Organization of Tissues in the Skeleton

A **bone** is classified by its shape: long, short, flat, or irregular. Long bones do not have to be terribly long—the bones in your fingers are classified as long. But a long bone must be longer than it is wide. The bones of the arms and legs are long bones, except for the kneecap and the wrist and ankle bones.

A long bone of the arm or leg can be used to illustrate the organization of tissues of the skeleton (Fig. 19.1a). A bone is enclosed by a tough, fibrous, connective tissue covering called the **periosteum,** which is continuous with the ligaments that go across a joint. The periosteum contains blood vessels that enter the bone and give off branches that service various types of cells found in bone. Each expanded end of a long bone is termed an epiphysis; the longest portion, between the epiphyses, is called the diaphysis. The epiphysis and diaphysis are separated by a small region called the metaphysis. Where the diaphysis of a long bone contacts (articulates with) another bone at a **joint,** it will be covered by a layer of hyaline cartilage (see below), which is also called articular cartilage.

Structure of Bone and Associated Tissues

The primary connective tissues of the skeleton are bone, cartilage, and dense fibrous connective tissue. All connective tissues contain cells separated by a matrix that contains fibers. Bone tissue is particularly strong because the matrix contains collagen fibers plus mineral salts, notably calcium phosphate.

Bone

There are two types of bone tissue, compact bone and spongy bone. **Compact bone** is highly organized and composed of tubular units called *osteons.* Each bone cell, or **osteocyte,** lies in a *lacuna,* or tiny chamber. Lacunae are arranged in concentric circles around a *central canal* (Fig. 19.1b). Central canals contain blood vessels, lymphatic vessels, and nerves. Tiny canals called *canaliculi* run through the matrix, connecting the lacunae with each other and with the central canal. In this way, the canaliculi bring nutrients from the blood vessel in the central canal to the cells in the lacunae.

Compared to compact bone, **spongy bone** has an unorganized appearance (Fig. 19.1b). Cells called osteocytes are found in *trabeculae,* numerous thin plates separated by unequal spaces. Although the spaces make spongy bone lighter than compact bone, spongy bone is still very strong. Just as braces are used for support in buildings, the plates of spongy bone follow lines of stress. The *epiphyses* (ends) of a long bone are composed largely of spongy bone.

The long part of the bone is called the shaft, or *diaphysis* (Fig. 19.1a). The diaphysis is separated from the epiphyses by the *metaphysis,* which contains the *epiphyseal plate* (also called the growth plate), a region of cartilage that allows for bone growth.

If you split the diaphysis open, you will see it is not solid but instead has a medullary cavity containing *yellow bone marrow.* Yellow bone marrow contains a large amount of fat. The spaces of spongy bone are often filled with **red bone marrow,** a specialized tissue that produces all types of blood cells. In infants, red bone marrow is found in the cavities of most bones. In adults, red bone marrow occurs in a more limited number of bones (see Section 19.2).

Cartilage

Cartilage is not as strong as bone, but it is more flexible because the matrix is gel-like and contains many collagenous and elastic fibers. The cells lie within lacunae, which are irregularly grouped. Cartilage has no blood vessels, and therefore, injured cartilage is slow to heal.

All three types of cartilage—hyaline cartilage, fibrocartilage, and elastic cartilage—are associated with bones. **Hyaline cartilage** is firm and somewhat flexible. The matrix appears uniform and glassy (Fig. 19.1), but actually it contains a generous supply of collagen fibers. Hyaline cartilage is found in the joints at the ends of long bones, in the nose, at the ends of the ribs, and in the larynx and trachea.

Fibrocartilage is stronger than hyaline cartilage because the matrix contains wide rows of thick collagen fibers. Fibrocartilage is able to withstand both tension and pressure. This type of cartilage is found where support is of prime importance—in the discs between the vertebrae and in the cartilaginous discs between the bones of the knee.

Elastic cartilage is more flexible than hyaline cartilage because the matrix contains mostly elastin fibers. This type of cartilage is found in the ear flaps and epiglottis.

Dense Fibrous Connective Tissue

Dense fibrous connective tissue contains rows of cells called fibroblasts separated by bundles of collagen fibers. Dense fibrous connective tissue forms the flared sides of the nose. **Ligaments,** which bind bone to bone, are dense fibrous connective tissue, and so are **tendons,** which connect muscle to bone at joints.

Bone Growth and Remodeling

Bones are composed of living tissues, as exemplified by their ability to undergo growth and remodeling.

Figure 19.1 Anatomy of a long bone. **a.** A long bone is encased by fibrous membrane (periosteum) except where it is covered at the ends by hyaline cartilage, shown on the micrograph (*right, top*). **b.** Spongy bone, located beneath the cartilage, may contain red bone marrow. The central shaft contains yellow bone marrow and is bordered by compact bone, which is shown in the enlargement and micrograph (*right, center*). (photos) (hyaline cartilage, compact bone): ©Ed Reschke; (osteocyte): ©Biophoto Associates/Science Source

Bone Development and Growth

The bones of the human skeleton, except those of the skull, first appear during embryonic development as hyaline cartilage. These cartilaginous structures are then gradually replaced by bone, a process called *endochondral ossification* (Fig. 19.2).

During endochondral ossification, the cartilage begins to break down in the center of a long bone, which is now covered by a periosteum. **Osteoblasts** invade the region and begin to lay down spongy bone in what is called a primary ossification center. Other osteoblasts lay down compact bone beneath the periosteum. As the compact bone thickens, **osteoclasts,** which are large, multinucleated, macrophage-like cells, break down the spongy bone, and the cavity created becomes the medullary cavity.

The ends of developing bone continue to grow, but soon, secondary ossification centers appear in these regions. Here, spongy bone forms and does not break down. Also, the growth plate remains between the primary ossification center and each secondary center. The limbs keep increasing in length as long as the growth plates are present. The rate of growth is mainly controlled by growth hormone and sex hormones. Eventually, the growth plates become ossified, and the bone stops growing.

Remodeling of Bones

In the adult, bone is continually being broken down and built up again, a process called remodeling. Osteoclasts break down bone, resulting in increased calcium levels in the blood. Simultaneously, the bone is repaired by the work of osteoblasts. As they form new bone, osteoblasts take calcium from the blood. Eventually, some of these cells get caught in the matrix they secrete and are converted to osteocytes, the cells found within the lacunae of osteons. Because of continual remodeling, the thickness of bones can change due to excessive use or disuse. Hormones such as calcitonin and parathyroid hormone (see Section 20.3) can also affect the thickness of bones. When women reach menopause, the likelihood of a bone disease called osteoporosis increases, in part due to reduced estrogen levels (see Section 19.6).

Bone remodeling also occurs when a fractured bone is healing. Bone healing can be divided into three major phases: (1) the *reactive phase,* which includes the body's inflammatory response to the injury; (2) the *reparative phase,* when the body generates a bone callus, which is a temporary connective tissue bridge that is gradually replaced in the last phase; and (3) the *remodeling phase,* during which osteoclasts and osteoblasts gradually replace the temporary bone with compact bone. Depending on factors such as the type of fracture, the age of the patient, and medical intervention, some fractures are healed in as little as 3 or 4 weeks, while others may not be properly healed after a year or more.

Check Your Progress **19.1**

1. Describe the typical anatomy of compact and spongy bone.
2. List the three types of cartilage and where they are found in the body.
3. Compare and contrast ligaments and tendons.
4. Describe the role of osteoclasts and osteoblasts in remodeling bone.

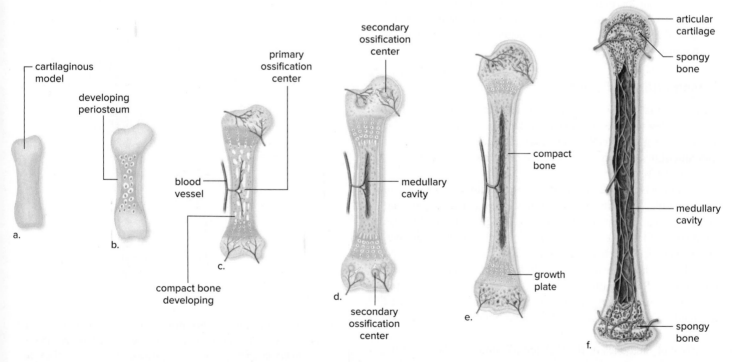

Figure 19.2 Endochondral ossification of a long bone. **a.** A cartilaginous model develops during embryonic development. **b.** A periosteum develops. **c.** A primary ossification center contains spongy bone surrounded by compact bone. **d.** The medullary cavity forms in the shaft, and secondary ossification centers develop at the ends of the long bone. **e.** Growth is still possible as long as cartilage remains at the growth plate. **f.** When the bone is fully formed, the growth plate disappears.

19.2 Bones of the Skeleton

Learning Outcomes

Upon completion of this section, you should be able to

1. Summarize the major functions of the skeleton.
2. List the major bones that comprise the axial and appendicular skeletons.
3. Compare the structure and function of fibrous joints, cartilaginous joints, and synovial joints.

The skeletal system has a wide array of functions in the body. These include:

■ *Support.* The bones of the lower limbs (the femur in particular and also the tibia) support the entire body when we are standing, and the coxal bones of the pelvic girdle support the abdominal cavity.

■ *Protection.* The bones of the skull protect the brain. The rib cage, composed of the ribs, thoracic vertebrae, and sternum, protects the heart and lungs.

■ *Producing blood cells.* All bones in the fetus have spongy bone containing red bone marrow that produces blood cells. In the adult, the flat bones of the skull, ribs, sternum, clavicles, and also the vertebrae and pelvis produce blood cells.

■ *Storage.* All bones have an extracellular substance that contains calcium phosphate. When bones are remodeled, osteoclasts break down bone and return calcium ions and phosphate ions to the bloodstream. Fat is stored in the yellow bone marrow (see Fig. 19.1).

■ *Assisting with movement.* While articulations (joints) occur between all the bones, we associate body movement, in particular, with the bones of the lower limbs (especially the femur and tibia) and the feet (tarsals, metatarsals, and phalanges) because we use them when walking.

Classification of the Bones

The approximately 206 bones of the skeleton are classified according to whether they occur in the axial skeleton or the appendicular skeleton. The **axial skeleton** is in the midline of the body, and the **appendicular skeleton** consists of the limbs along with their girdles (Fig. 19.3).

Long bones, such as the humerus and femur, are longer than they are wide. In contrast, short bones, such as the carpals and tarsals, are cube-shaped—that is, their lengths and widths are about equal. Flat bones, such as those of the skull, are platelike, with broad surfaces. Irregular bones, such as the vertebrae and facial bones, have varied shapes that permit connections with other bones. Some authorities recognize another category, the round bones, which are circular in shape and usually embedded in a tendon. A good example of a round bone is the patella, or kneecap.

None of the bones of the skeleton are smooth. They have articulating depressions and protuberances at various joints. They also have projections, often called *processes,* where the muscles attach, as well as openings for nerves and/or blood vessels to pass through.

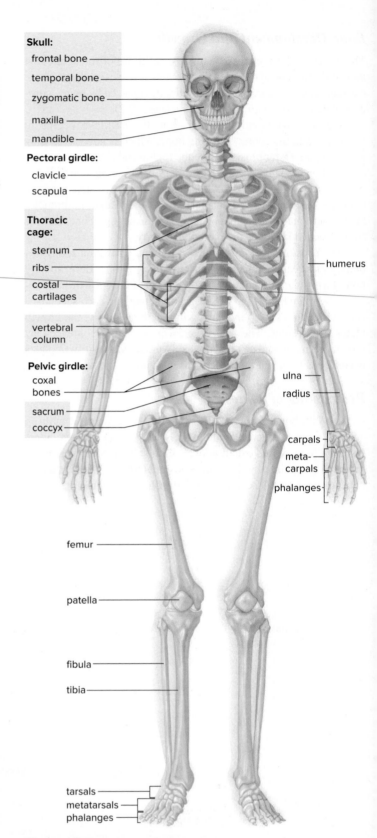

Figure 19.3 The human skeleton.   The skeleton of a human adult contains bones that belong to the axial skeleton (shaded in blue) and those that belong to the appendicular skeleton (unshaded). The bones of the axial skeleton are located along the body's axis, and the bones of the appendicular skeleton are located in the girdles and appendages. Not all bones are labeled.

The Axial Skeleton

The axial skeleton lies in the midline of the body and consists of the skull, hyoid bone, vertebral column, rib cage, and ossicles.

The Skull

The **skull** is formed by the cranium (braincase) and the facial bones. It should be noted, however, that some cranial bones also help to form the face.

The Cranium The *cranium* protects the brain. In adults, it is composed of eight bones fitted tightly together. In newborns, certain cranial bones are joined by membranous regions called **fontanels,** which not only allow the cranium to be compressed during birth, but also allow for expansion of the brain during the rapid growth spurt that follows birth. The fontanels usually close by the age of 24 months by the process of intramembranous ossification.

Some of the bones of the cranium contain **sinuses,** air spaces lined by mucous membrane. The sinuses reduce the weight of the skull and give a resonant sound to the voice. Two sinuses, called the *mastoid sinuses,* drain into the middle ear. Infections of the upper respiratory tract can spread into various sinuses, sometimes resulting in chronic sinusitis that is difficult to treat (see Section 15.4).

The major bones of the cranium have the same names as the lobes of the brain: frontal, parietal, occipital, and temporal. On the top of the cranium (Fig. 19.4a), the *frontal bone* forms the forehead, the *parietal bones* extend to the sides, and the *occipital bone*

curves to form the base of the skull. At the base, there is a large opening, the **foramen magnum** (Fig. 19.4b), through which the spinal cord passes and connects with the brain stem. Below the much larger parietal bones, each *temporal bone* has an opening (external auditory canal) that leads to the middle ear.

The *sphenoid bone,* which is shaped like a bat with wings outstretched, extends across the floor of the cranium from one side to the other. The sphenoid is the keystone of the cranial bones because all the other bones articulate with it. The sphenoid completes the sides of the skull and also helps form the eye sockets. The eye sockets are called *orbits* because of our ability to rotate our eyes. The *ethmoid bone,* which lies in front of the sphenoid, also helps form the orbits and the nasal septum. The orbits are completed by various facial bones.

The Facial Bones The most prominent facial bones are the mandible, the maxillae (maxillary bones), the zygomatic bones, and the nasal bones.

The *mandible,* or lower jaw, is the only movable portion of the skull (Fig. 19.5a), and its action permits us to chew our food. It also forms the chin. Tooth sockets are located on the mandible and on the two *maxillae,* which form the upper jaw and also the anterior portion of the hard palate. The *palatine bones* make up the posterior portion of the hard palate and the floor of the nose (see Fig. 19.4b).

The lips and cheeks have a core of skeletal muscle. The *zygomatic bones* are the cheekbone prominences, and the nasal bones form the bridge of the nose. Other bones (e.g., ethmoid and vomer)

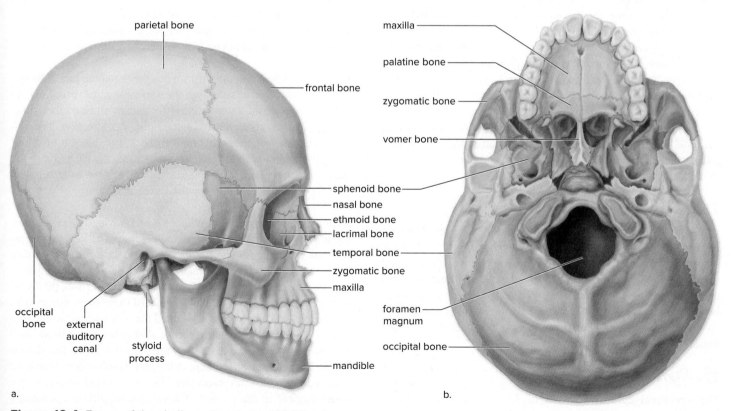

Figure 19.4 Bones of the skull. **a.** Lateral view. **b.** Inferior view.

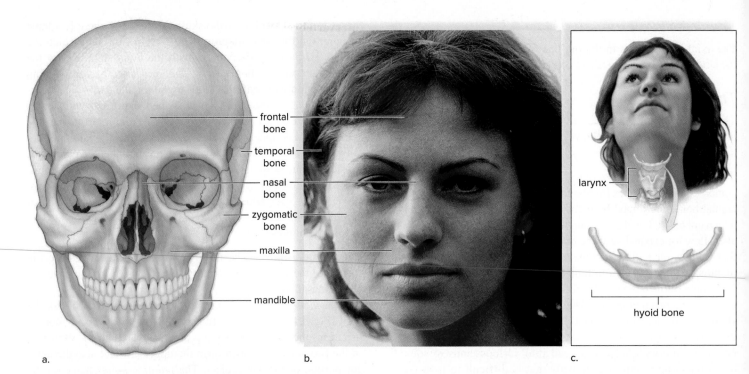

Figure 19.5 Bones of the face and the hyoid bone.   a. The frontal bone forms the forehead and eyebrow ridges, and the zygomatic bones form the cheekbones. The maxillae have numerous functions: assisting in the formation of the eye sockets and the nasal cavity; forming the upper jaw; and containing sockets for the upper teeth. The mandible is the lower jaw with sockets for the lower teeth. The mandible has a projection we call the chin. **b.** The maxillae, frontal bones, and nasal bones help form the external nose. **c.** The hyoid bone attaches to the larynx as shown. (b) ©Corbis VCG/Getty Images

are a part of the nasal septum, which divides the interior of the nose into two nasal cavities. The *lacrimal bone* (see Fig. 19.4*a*) contains the opening for the *nasolacrimal canal,* which brings tears from the eyes to the nose.

The temporal and frontal bones are cranial bones that contribute to the face. The temporal bones account for the flattened areas we call the temples. The frontal bone forms the forehead and has supraorbital ridges where the eyebrows are located. Glasses sit where the frontal bone joins the nasal bones (Fig. 19.5*b*).

While the ears are formed only by cartilage and not by bone, the nose is a mixture of bones, cartilages, and connective tissues. The cartilages complete the tip of the nose, and fibrous connective tissue forms the flared sides of the nose.

The Hyoid Bone

Although the *hyoid bone* is not part of the skull, we include it here because it is part of the axial skeleton. The hyoid is the only bone in the body that does not articulate with another bone. It is attached to processes of the temporal bones by muscles and ligaments and to the larynx by a membrane (Fig. 19.5*c*). The larynx is the voice box at the top of the trachea in the neck region. The hyoid bone anchors the tongue and serves as the site of attachment for the muscles associated with swallowing.

The Vertebral Column

The **vertebral column** consists of 33 vertebrae (Fig. 19.6). Normally, the vertebral column has four curvatures that

provide resilience and strength for an upright posture. *Scoliosis* is an abnormal lateral (sideways) curvature of the spine. Two other well-known abnormal curvatures are *kyphosis,* an abnormal posterior curvature that often results in a hunchback, and *lordosis,* an abnormal anterior curvature resulting in a swayback.

The vertebral column forms when the vertebrae join. The spinal cord, which passes through the vertebral canal, gives off the spinal nerves at the intervertebral foramina. Among other functions, spinal nerves control skeletal muscle contraction. The *spinous processes* of the vertebrae can be felt as bony projections along the midline of the back. The spinous processes and also the *transverse processes,* which extend laterally, serve as attachment sites for the muscles that move the vertebral column.

Types of Vertebrae   The various vertebrae are named according to their location in the vertebral column. The *cervical vertebrae* are in the neck. The first cervical vertebra, called the *atlas,* holds up the head. It is so named because Atlas, of Greek mythology, held up the world. Movement of the atlas permits the "yes" motion of the head. It also allows the head to tilt from side to side. The second cervical vertebra is called the *axis* because it allows a degree of rotation, as when someone shakes his or her head "no." The *thoracic vertebrae* have long, thin, spinous processes and articular facets for the attachment of the ribs. *Lumbar vertebrae* have a large body and thick processes. The five *sacral vertebrae* are fused together in the *sacrum.* The *coccyx,* or tailbone, is usually composed of four fused vertebrae.

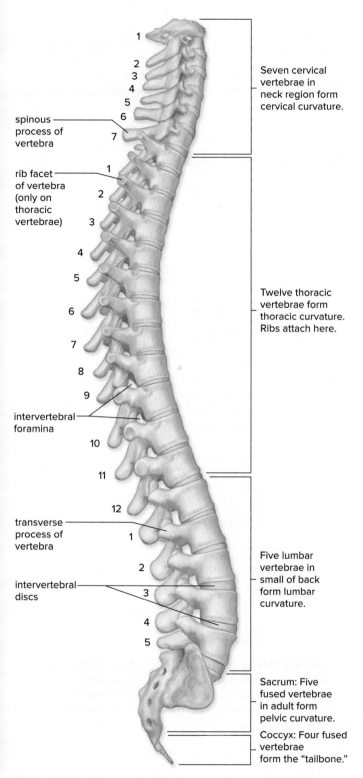

spinous process of vertebra

rib facet of vertebra (only on thoracic vertebrae)

intervertebral foramina

transverse process of vertebra

intervertebral discs

Seven cervical vertebrae in neck region form cervical curvature.

Twelve thoracic vertebrae form thoracic curvature. Ribs attach here.

Five lumbar vertebrae in small of back form lumbar curvature.

Sacrum: Five fused vertebrae in adult form pelvic curvature.

Coccyx: Four fused vertebrae form the "tailbone."

Figure 19.6 The vertebral column.   The vertebral column is flexible because the vertebrae are separated by intervertebral discs. The vertebrae are named for their location in the vertebral column. For example, the thoracic vertebrae are in the thorax. Note that humans have a coccyx, which is also called a tailbone.

Intervertebral Discs   Between the vertebrae are **intervertebral discs** composed of fibrocartilage that act as a kind of padding. They prevent the vertebrae from grinding against one another and absorb shock caused by movements such as running, jumping, and even walking. Unfortunately, these discs become weakened with age and can even herniate and rupture. A herniated or "slipped" disc may press against the spinal cord and/or spinal nerves, resulting in pain, numbness, or even loss of function in the areas innervated by the spinal nerves at or below that region. Herniated discs are most common in the lumbar region; about 25% of people with lower back pain actually have a herniated disc. Although rest and treatment of symptoms usually results in much improvement, surgical treatment may be necessary, especially if there is chronic pain or loss of nerve function.

The Rib Cage

The *rib cage,* also called the thoracic cage, is composed of the thoracic vertebrae, the ribs and their associated cartilages, and the sternum (Fig. 19.7). The rib cage is part of the axial skeleton.

The rib cage demonstrates how the skeleton is protective but also flexible. The rib cage protects the heart and lungs. Yet, it swings outward and upward upon inspiration, and then downward and inward upon expiration. Because of their proximity to vital organs, however, the ribs also have the potential to cause harm. A fractured rib can puncture a lung.

The Ribs

There are 12 pairs of ribs, and all 12 connect directly to the thoracic vertebrae in the back. Each rib is a flattened bone that originates at a particular thoracic vertebra and proceeds toward the ventral thoracic wall. A rib articulates with the body and transverse process of its corresponding thoracic vertebra. It curves outward and then forward and downward.

The upper seven pairs of ribs connect directly to the sternum by means of costal cartilages. These are called the "true

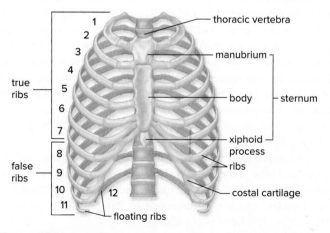

true ribs

false ribs

thoracic vertebra

manubrium

body

sternum

xiphoid process

ribs

costal cartilage

floating ribs

Figure 19.7 The rib cage.   The rib cage consists of the 12 thoracic vertebrae, the 12 pairs of ribs, the costal cartilages, and the sternum. The rib cage protects the lungs and the heart.

ribs." The remaining pairs of ribs do not connect directly to the sternum, and they are called the "false ribs." They attach to the sternum by means of a common cartilage. The last two pairs are also called "floating ribs" because they do not attach to the sternum at all.

The Sternum

The *sternum,* or breastbone, lies in the midline of the body. Along with the ribs, it helps protect the heart and lungs. The sternum is a flat bone that is shaped like a knife.

The sternum is composed of three bones that fuse during fetal development. These bones are the *manubrium* (the handle), the *body* (the blade), and the *xiphoid process* (the point of the blade). The manubrium articulates with the clavicles of the appendicular skeleton and the first pair of ribs. The manubrium joins with the body of the sternum at an angle. This is an important anatomical landmark because it occurs at the level of the second rib, and therefore allows the ribs to be counted. Medical personnel sometimes count the ribs to locate the apex of a patient's heart, which is usually between the fifth and sixth ribs.

The xiphoid process is the third part of the sternum. Composed of hyaline cartilage in the child, it becomes ossified in the adult. The variably shaped xiphoid process serves as an attachment site for the diaphragm, which divides the thoracic cavity from the abdominal cavity.

The Appendicular Skeleton

The appendicular skeleton consists of the bones within the pectoral and pelvic girdles and their attached limbs. The two (left and right) pectoral girdles and upper limbs are specialized for flexibility. The pelvic girdle and lower limbs are specialized for strength.

The Pectoral Girdles and Upper Limbs

Each **pectoral girdle** (shoulder girdle) consists of a scapula (shoulder blade) and a clavicle (collarbone) (Fig. 19.8). The *clavicle* extends across the top of the thorax. It articulates with (joins with) the sternum and the *acromion process* of the *scapula,* a bone external to the ribs in the back. The muscles of the arm and chest attach to the *coracoid process* of the scapula. The *glenoid cavity* of the scapula articulates with the head of the humerus, the single long bone in the arm. This joint allows the arm to move in almost any direction, but has reduced stability and thus is the joint most apt to be dislocated. Tendons that encircle and help form a socket for the humerus are collectively called the *rotator cuff.* Vigorous circular movements of the arm (such as pitching a baseball) can lead to rotator cuff injuries.

The upper limb consists of the *humerus* in the arm and the *radius* and *ulna* in the forearm. The shaft of the humerus has a tuberosity (protuberance) where the deltoid, a prominent muscle of the shoulder, attaches. Enlargement of this tuberosity occurs in people who do a lot of heavy lifting.

The far end of the humerus has two protuberances, called the *capitulum* and the *trochlea,* which articulate respectively with the radius and the ulna at the elbow. The bump at the back of the elbow

Figure 19.8 Bones of the pectoral girdle and upper limb. The pectoral girdle consists of the clavicle (collarbone) and the scapula (shoulder blade). The humerus is the single bone of the arm. The forearm is formed by the radius and ulna. A hand contains carpals, metacarpals, and phalanges.

is the *olecranon process* of the ulna. This area is sometimes called the "funny bone" because the ulnar nerve passes behind it with only skin for protection—thus it is susceptible to being bumped, resulting in pain or numbness.

When the arm is held with the palm turned forward, the radius and ulna are about parallel to one another. When the arm is turned so that the palm faces backward, the radius crosses in front of the ulna, a feature that contributes to the easy twisting motion of the forearm.

The hand has many bones, and this increases its flexibility. The wrist has eight *carpal bones,* which look like small pebbles. From these, five *metacarpal bones* fan out to form a framework for the palm. The metacarpal bone that leads to the thumb is opposable to the other digits. The term *digits* refers to either fingers or toes. The knuckles are the enlarged distal ends of the metacarpals. Beyond the metacarpals are the *phalanges,* the bones of the fingers and the thumb. The phalanges of the hand are long, slender, and lightweight.

The Pelvic Girdle and Lower Limb

Figure 19.9 shows how the lower limb is attached to the pelvic girdle. The **pelvic girdle** consists of two heavy, large coxal bones (hipbones). The pelvis is a basin composed of the pelvic girdle, sacrum, and coccyx. The pelvis bears the weight of the body, protects the organs within the pelvic cavity, and serves as the site of attachment for the lower limbs.

Each *coxal bone* has three parts: the ilium, the ischium, and the pubis, which are fused in the adult (Fig. 19.9). The hip socket, called the acetabulum, occurs where these three bones meet. We sit on the *ischium,* which has a posterior spine called the ischial spine. The *pubis,* from which the term *pubic hair* is derived, is the anterior part of a coxal bone. The two pubic bones are joined together by a fibrocartilage disc at the pubic symphysis.

The male and female pelves differ from one another. In the female, the ilia are more flared, the pelvic cavity is broader, and the outlet is wider. These adaptations facilitate giving birth.

The *femur* (thighbone) is the longest and strongest bone in the body. The head of the femur articulates with the coxal bones at the *acetabulum,* and the short neck better positions the legs for walking. The femur has two large processes, the greater and lesser trochanters, which are places of attachment for the thigh muscles and the muscles of the buttocks. At its distal end, the femur has medial and lateral condyles that articulate with the *tibia.* This is the region of the knee and the patella, or kneecap. The *patella* is held in place by the quadriceps tendon, which continues as a ligament that attaches to the tibial tuberosity. At the distal end, the medial malleolus of the tibia causes the inner bulge of the ankle. The *fibula* is the more slender bone in the leg. The fibula has a head that articulates with the tibia and a prominence (lateral malleolus) that forms the outer bulge of the ankle.

Each foot has an ankle, an instep, and five toes. The many bones of the foot give it considerable flexibility, especially on rough surfaces. The ankle contains seven *tarsal* bones, one of which (the talus) can move freely where it joins the tibia and fibula. Strange to say, the calcaneus, or heel bone, is also considered part of the ankle. The *talus* and *calcaneus* support the weight of the body.

The instep has five elongated *metatarsal bones.* The distal end of the metatarsals forms the ball of the foot. If the ligaments that bind the metatarsals together become weakened, flat feet are apt to result. The bones of the toes are called phalanges, just like those of the fingers, but in the foot, the phalanges are stout and extremely sturdy.

Joints

Bones are joined at the joints, which are classified as fibrous, cartilaginous, or synovial based on their structure and their ability to move. Most **fibrous joints,** such as the **sutures** between the cranial bones, are immovable. **Cartilaginous joints** are connected by

Figure 19.9 Bones of the pelvic girdle and lower limb. The ilium, ischium, and pubis join at the acetabulum (hip socket) to form a coxal bone. The pelvis is completed by the addition of the sacrum and coccyx. The femur (thighbone) and tibia and fibula (shinbones) form the leg. Tarsals, metatarsals, and phalanges construct the foot.

Identifying Skeletal Remains

Regardless of how, when, and where human bones are found unexpectedly, many questions must be answered. How old was this person at the time of death? Are these the bones of a male or female? What was the ethnicity? Are there any signs that this person was murdered?

Clues about the identity and history of the deceased person are available throughout the skeleton (Fig. 19A). Age is approximated by *dentition,* or the structure of the teeth in the upper jaw (maxilla) and lower jaw (mandible). For example, infants between 0 and 4 months of age will have no teeth present; children approximately 6 to 10 years of age will have missing deciduous, or "baby," teeth; young adults acquire their last molars, or "wisdom teeth," around age 20. The age of older adults can be approximated by the number and location of missing or broken teeth. Studying areas of bone ossification also gives clues to the age of the deceased at the time of death. In older adults, signs of joint breakdown provide additional information about age. Hyaline cartilage becomes worn, yellowed, and brittle with age, and the hyaline cartilages covering bone ends wear down over time. The amount of yellowed,

brittle, or missing cartilage helps scientists guess the person's age.

Figure 19A Forensic investigators uncover a skeleton.
A knowledge of bone structure and how bones age will help identify these remains. ©Michael Donne/Science Source

If skeletal remains include the individual's pelvic bones, these provide the best method for determining an adult's gender. The pelvis is shallower and wider in the female than in the male. The long bones, particularly the humerus and femur, give information about gender as well. Long bones are thicker and denser in males, and points of muscle attachment are bigger and more prominent. The skull of a

male has a square chin and more prominent ridges above the eye sockets, or orbits.

Determining ethnic origin of skeletal remains can be difficult, because so many people have a mixed racial heritage. Forensic anatomists rely on observed racial characteristics of the skull. In general, individuals of African or African American descent have a greater distance between the eyes, eye sockets that are roughly rectangular, and a jaw that is large and prominent. Skulls of Native Americans typically have round eye sockets, prominent cheek (zygomatic) bones, and a rounded palate. Caucasian skulls usually have a U-shaped palate, and a suture line between the frontal bones is often visible. Additionally, the external ear canals in Caucasians are long and straight, so that the auditory ossicles can be seen.

Once the identity of the individual has been determined, the skeletal remains can be returned to the family for proper burial.

Questions to Consider

1. Why do you think compact bones are most often found by forensic investigators?
2. How does an examination of bone ossification provide an indication of age?

---

hyaline cartilage, as in the costal cartilages that join the ribs to the sternum, or by fibrocartilage, as in the intervertebral discs. Cartilaginous joints tend to be slightly movable.

In **synovial joints,** which are freely movable, the two bones are separated by a cavity (Fig. 19.10). Ligaments hold the two bones in place and help form a capsule around the joint. Tendons also help stabilize the joint. The joint capsule is lined by a **synovial membrane,** which produces a lubricating *synovial fluid.* Major synovial joints include the shoulder, elbow, hip, and knee. Aside from articular cartilage, the knee contains *menisci* (*sing.,* meniscus), crescent-shaped pieces of hyaline cartilage between the bones (Fig. 19.10*b*). These give added stability and act as shock absorbers. Unfortunately, athletes often suffer injury to the menisci, known as torn cartilage. The anterior and posterior cruciate ligaments, which help stabilize the knee by attaching the femur to the tibia, are also prone to being torn (sometimes called "ruptured"). The knee joint also contains 13 fluid-filled sacs called **bursae** (*sing.,* bursa), which ease friction between the ten-

dons and ligaments. Inflammation of the bursae is called **bursitis.** *Tennis elbow* is a form of bursitis.

There are different types of synovial joints. The shoulder is a *ball-and-socket joint,* which allows movement in all planes, even rotational movement (Fig. 19.10*c*). The knee and elbow joints are *hinge joints* because, like a hinged door, they largely permit movement in one direction only (Fig. 19.10*d*). The joint between the radius and ulna is a *pivot joint* in which only rotation is possible. Some of the movements at synovial joints are shown in Figure 19.11.

Check Your Progress 19.2

1. Explain the difference between the axial and appendicular skeletons.
2. Define the difference between true ribs, false ribs, and floating ribs.
3. List the three major types of joints and give a specific example of each.

bursae

joint cavity filled with synovial fluid

articular cartilage

meniscus

meniscus

ligament

ligament

b. Generalized synovial joint

head of humerus

scapula

c. Ball-and-socket joint

ulna

humerus

a. A gymnast depends on flexible joints.

d. Hinge joint

Figure 19.10 The structure of a synovial joint. a. Synovial joints like the hips and shoulders are the most flexible type of joint. **b.** The ends of bones in synovial joints are capped by articular cartilage. The joint capsule is lined with synovial membrane, which gives off synovial fluid as a lubricant. Menisci (*sing.*, meniscus) cushion and stabilize the joint, and bursae are fluid-filled sacs that reduce friction. **c.** Ball-and-socket joints form the shoulder and hip. **d.** Hinge joints construct the knee and elbow.

(a): ©Gerard Vandystadt/Science Source

Figure 19.11 Types of movement allowed by synovial joints. a. Flexion and extension. **b.** Adduction and abduction. **c.** Rotation and circumduction. **d.** Inversion and eversion. Red dots indicate pivot points.

extension

flexion

flexion

flexion

extension

extension

abduction

abduction

adduction

adduction

abduction

adduction

Rotation: Body part moves around its own axis.

Supination: Hand faces anterior or downward.
Pronation: Hand faces posterior or downward.

Circumduction: Body part moves so that a cone shape is outlined.

Inversion: Sole of foot turns inward.

Eversion: Sole of foot turns outward.

Flexion: Joint angle decreases.

Extension: Joint angle increases.

Adduction: Body part moves toward midline.

Abduction: Body part moves away from midline.

a.

b.

c.

d.

367

19.3 Skeletal Muscles

Learning Outcomes

Upon completion of this section, you should be able to

1. Explain how antagonistic muscle groups can move bones in different directions.
2. Identify the major muscles in the human body.
3. Describe six characteristics used to name skeletal muscles.

Humans have three types of muscle tissue: smooth, cardiac, and skeletal (see Fig. 11.4). **Skeletal muscle** makes up the greatest percentage of muscle tissue in the body. Skeletal muscle is voluntary because its contraction can be consciously stimulated and controlled by the nervous system.

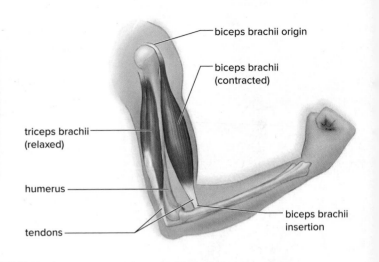

Skeletal Muscles Work in Pairs

Skeletal muscles are covered by several layers of fibrous connective tissue called fascia, which extends beyond the muscle to become its tendon. Tendons are attached to the skeleton, and the contraction of skeletal muscle causes the bones at a joint to move. When skeletal muscles contract, one bone remains fairly stationary, and the other one moves. The **origin** of a muscle is on the stationary bone, and the **insertion** of a muscle is on the bone that moves. Figure 19.11 illustrates some terms used to describe these movements, and gives examples. *Flexion* means bending bones at a joint so that the angle between the bones decreases, and the involved bones move closer together, while *extension* increases that angle and moves the parts farther apart. *Abduction* at a joint moves the involved part (usually a limb) away from the middle of the body, while *adduction* brings it back toward the midline. *Rotation* moves the part around an axis, due to a twisting motion at the joint. Note that not all of these movements can occur at all joints.

Most muscles have antagonists, and antagonistic pairs bring about movement in opposite directions. For example, the biceps brachii and the triceps brachii are antagonists; one flexes the forearm, and the other extends the forearm (Fig. 19.12).

Nomenclature

Skeletal muscles are named based on the following characteristics:

- *Size*. The gluteus maximus that makes up the buttocks is the largest muscle (*maximus* means "greatest"). Other terms used to indicate size are *vastus* ("huge") and *longus* ("long").
- *Shape*. The deltoid is shaped like a triangle. (The Greek letter delta has this appearance: Δ.) The trapezius is shaped like a trapezoid. Another term used to indicate shape is *latissimus* ("wide").
- *Location*. The frontalis overlies the frontal bone. The external obliques are located outside the internal obliques. Another term used to indicate location is *pectoralis* ("chest").
- *Direction of muscle fibers*. The rectus abdominis is a longitudinal muscle of the abdomen (*rectus* means "straight"). The orbicularis oculi is a circular muscle around the eye.
- *Number of attachments*. The biceps brachii has two attachments, or origins (*bi-* means "two").

Figure 19.12 Antagonistic muscles.   Muscles can exert force only by shortening; therefore, they often work as antagonistic pairs. The biceps and triceps brachii exemplify an antagonistic pair of muscles that act opposite to one another. The biceps brachii raises the forearm, and the triceps brachii lowers the forearm.

- *Action*. The extensor digitorum extends the fingers (digits). The adductor longus is a long muscle that adducts the thigh. Another term used to indicate action is *masseter* ("to chew").

Major Skeletal Muscles

There are approximately 600 muscles in the human body. The major ones are shown in Figure 19.13. As listed in Tables 19.1 and 19.2, the muscles of the head are involved in facial expression and chewing; those of the neck participate in head movements; the muscles of the trunk and upper limb move the arm, the forearm, and the fingers; and the muscles of the buttocks and lower limb move the thigh, the leg, and the toes.

orbicularis oculi

frontalis

zygomaticus

orbicularis oris

masseter

sternocleidomastoid

trapezius

deltoid

pectoralis major

latissimus dorsi

biceps brachii

rectus abdominis

external oblique

flexor carpi group

flexor digitorum

iliopsoas

adductor longus

sartorius

quadriceps femoris group

peroneus longus

tibialis anterior

gastrocnemius

extensor digitorum longus

occipitalis

sternocleido-mastoid

trapezius

deltoid

triceps brachii

gluteus medius

extensor carpi group

extensor digitorum

latissimus dorsi

external oblique

gluteus maximus

hamstring group

peroneus longus

a. Anterior view

b. Posterior view

Figure 19.13 Major muscles of the human body. Skeletal muscles located near the surface in (**a**) anterior and (**b**) posterior views.

TABLE 19.1   Major Human Muscles (Fig. 19.13a: Anterior View)

Name	Action
Head and neck	
Frontalis	Wrinkles forehead and lifts eyebrows
Orbicularis oculi	Closes eye (winking)
Zygomaticus	Raises corner of mouth (smiling)
Masseter	Closes jaw (chewing)
Orbicularis oris	Closes and protrudes lips (kissing)
Upper limb and trunk	
External oblique	Compresses abdomen; rotates trunk
Rectus abdominis	Flexes spine; compresses abdomen
Pectoralis major	Flexes and adducts arm ventrally (pulls arm across chest)
Deltoid	Abducts and moves arm up and down in front
Biceps brachii	Flexes forearm and rotates hand outward
Lower limb	
Adductor longus	Adducts and flexes thigh
Iliopsoas	Flexes thigh at hip joint
Sartorius	Raises and rotates thigh and leg
Quadriceps femoris group	Extends leg at knee; flexes thigh
Peroneus longus	Everts foot
Tibialis anterior	Dorsiflexes and inverts foot
Flexor digitorum longus	Flexes toes
Extensor digitorum longus	Extends toes

TABLE 19.2   Major Human Muscles (Fig, 19.13b: Posterior View)

Name	Action
Head and neck	
Occipitaiis	Moves scalp backward
Sternocleidomastoid	Turns head to side; flexes neck and head
Trapezius	Extends head; raises scapula as when shrugging shoulders
Upper limb and trunk	
Latissimus dorsi	Extends and adducts arm dorsally (pulls arm across back)
Deltoid	Abducts and moves arm up and down in front
External oblique	Rotates trunk
Triceps brachii	Extends forearm
Flexor carpi group	Flexes hand
Extensor carpi group	Extends hand
Flexor digitorum	Flexes fingers
Extensor digitorum	Extends fingers
Buttocks and lower limb	
Gluteus medius	Abducts thigh
Gluteus maximus	Extends thigh back (forms buttocks)
Hamstring group	Flexes leg and extends thigh
Gastrocnemius	Plantar flexes foot (tiptoeing)

Check Your Progress 19.3

1. State the three types of muscle tissue in the human body.
2. Describe the difference between a muscle's origin and its insertion.
3. Identify the major skeletal muscles of the body.
4. Define *flexion, extension, abduction, adduction*, and *rotation*.

19.4 Mechanism of Muscle Fiber Contraction

Learning Outcomes

Upon completion of this section, you should be able to

1. Describe the microscopic structure of a muscle fiber.
2. Review the sliding filament model of muscle contraction.
3. Summarize how activities within the neuromuscular junction control muscle fiber contraction.
4. Indicate three ways that muscle cells can generate ATP.

In Section 11.1, we examined the structure of skeletal muscle. Recall that skeletal muscle tissue has alternating light and dark bands, giving it a striated appearance (see Fig. 11.4). These bands are due to the arrangement of myofilaments in a muscle fiber.

Structure of a Muscle Fiber

A muscle fiber is a cell containing the usual cellular components, but special names have been assigned to some of these components (Table 19.3). Figure 19.14 shows the structure of a muscle fiber. The plasma membrane is called the **sarcolemma;** the cytoplasm is the sarcoplasm; and the endoplasmic reticulum is the **sarcoplasmic reticulum.** A muscle fiber also has some unique anatomical characteristics. One feature is its T (for transverse) system; the sarcolemma forms **T (transverse) tubules** that penetrate, or dip down, into the cell so that they come in contact—but do not fuse—with expanded portions of the sarcoplasmic reticulum. The expanded portions of the sarcoplasmic reticulum are calcium storage sites. Calcium ions (Ca^{2+}), as we shall see, are essential for muscle contraction.

The sarcoplasmic reticulum encases hundreds and sometimes even thousands of **myofibrils,** each about 1 μm in diameter, which are the contractile portions of the muscle fibers. Any other organelles, such as mitochondria, are located in the sarcoplasm between the myofibrils. The sarcoplasm also contains glycogen, which provides stored energy for muscle contraction, and the red pigment myoglobin, which binds oxygen until it is needed for muscle contraction.

Myofibrils and Sarcomeres

Myofibrils are cylindrical in shape and run the length of the muscle fiber. The light microscope shows that skeletal muscle fibers have light and dark bands called *striations*. The electron microscope shows that the striations of skeletal muscle fibers are formed by the placement of myofilaments within units of myofibrils called

sarcomeres. A sarcomere extends between two dark lines, called the *Z lines*. A sarcomere contains two types of protein myofilaments. The thick filaments are made up of a protein called **myosin,** and the thin filaments are made up of a protein called **actin.** Other proteins are also present. The I band is light colored because it contains only actin filaments attached to a Z line. The dark regions of the A band contain overlapping actin and myosin filaments, and its H band has only myosin filaments.

Myofilaments

The thick and thin filaments differ in the following ways:

Thick Filaments   A thick filament is composed of several hundred molecules of myosin. Each myosin molecule is shaped like a golf club, with the straight portion of the molecule ending in a double globular head. The myosin heads occur on each side of a sarcomere but not in the middle.

Thin Filaments   A thin filament primarily consists of two intertwining strands of actin. Two other proteins, called tropomyosin and troponin, also play a role, as we will discuss later in this section.

Sliding Filament Model   We will also see that when muscles are stimulated, impulses travel down a T tubule, and calcium is released from the sarcoplasmic reticulum. Calcium triggers muscle fiber contraction as the sarcomeres within the myofibrils shorten. When a sarcomere shortens, the actin (thin) filaments slide past the myosin (thick) filaments and approach one another. This causes the I band to shorten and the H band to almost or completely disappear. The movement of actin filaments in relation to myosin filaments is called the **sliding filament model** of muscle contraction. During the sliding process, the sarcomere shortens, even though the filaments themselves remain the same length. ATP supplies the energy for muscle contraction. Although the actin filaments slide past the myosin filaments, it is the myosin filaments that do the work. Myosin filaments break down ATP and form cross-bridges that pull the actin filaments toward the center of the sarcomere.

Skeletal Muscle Contraction

Muscle fibers are stimulated to contract by motor neurons whose axons are in nerves (Fig. 19.15). The axon of one motor neuron can stimulate from a few to several muscle fibers of a muscle because each axon has several branches. A branch of an axon ends in an axon terminal that lies in close proximity to the sarcolemma of a muscle fiber. A small gap, called a synaptic cleft, separates the axon terminal from the sarcolemma. This entire region is called a **neuromuscular junction** (Fig. 19.15).

Axon terminals contain synaptic vesicles filled with the neurotransmitter acetylcholine (ACh). When nerve impulses traveling down a motor neuron arrive at an axon terminal, the synaptic vesicles release ACh into the synaptic cleft. ACh quickly diffuses across the cleft and binds to receptors in the sarcolemma. Now the sarcolemma generates impulses that spread over itself and down T tubules to the sarcoplasmic reticulum. The release of Ca^{2+} ions from the sarcoplasmic reticulum causes the filaments within the sarcomeres to slide past one another. Sarcomere contraction causes

TABLE 19.3	Microscopic Anatomy of a Muscle
Name	**Function**
Sarcolemma	Plasma membrane of a muscle fiber that forms T tubules
Sarcoplasm	Cytoplasm of a muscle fiber that contains the organelles, including myofibrils
Glycogen	A polysaccharide that stores energy for muscle contraction
Myoglobin	The pigment that stores oxygen for muscle contraction
T tubule	Extension of the sarcolemma that extends into the muscle fiber and conveys impulses that cause Ca^{2+} to be released from the sarcoplasmic reticulum
Sarcoplasmic reticulum	The smooth ER of a muscle fiber that stores Ca^{2+}
Myofibril	A bundle of myofilaments that contracts
Myofilament	Actin filaments and myosin filaments whose structure and functions account for muscle striations and contractions

Figure 19.14 Skeletal muscle fiber structure and function. A muscle fiber contains many myofibrils, divided into sarcomeres, which are contractile. When the myofibrils of a muscle fiber contract, the sarcomeres shorten. The actin (thin) filaments slide past the myosin (thick) filaments toward the center so that the H band gets smaller, to the point of disappearing. (photo): ©Biology Media/Science Source

a. One motor axon causes several muscle fibers to contract.

skeletal muscle fiber

axon branch

axon terminal

myofibril

neuromuscular junction

100×

muscle fiber

axon branch

plasma membrane of axon

axon terminal

synaptic vesicle

synaptic cleft

sarcolemma

mitochondrion

myofibril

nucleus

synaptic vesicle

synaptic cleft

acetylcholine (ACh)

Na+

folded sarcolemma

ACh receptor

b. A neuromuscular junction is the juxtaposition of an axon terminal and the sarcolemma of a muscle fiber.

c. The release of a neurotransmitter (ACh) causes receptors to open and Na+ to enter a muscle fiber.

Figure 19.15 Neuromuscular junction. **a.** The branch of a motor nerve fiber ends in an axon terminal that meets but does not touch a muscle fiber. **b.** A synaptic cleft separates the axon terminal from the sarcolemma of the muscle fiber. **c.** Nerve impulses traveling down a motor fiber cause synaptic vesicles to discharge a neurotransmitter (ACh) that diffuses across the synaptic cleft. When the neurotransmitter is received by the sarcolemma of a muscle fiber, impulses begin that lead to muscle fiber contraction. (photo) (a): ©Ed Reschke/Oxford Scientific/Getty Images

myofibril contraction, which in turn results in the contraction of a muscle fiber and, eventually, a whole muscle.

Chemicals produced by some organisms can interfere with the operation of the neuromuscular junction. For example, tetanus toxin, produced by a different bacterial species, interferes with muscle relaxation. Thus, it causes excessive contraction, which can be fatal if it affects the respiratory muscles. Fortunately, the tetanus vaccine is very effective at preventing this. The Scientific Inquiry feature "The Discovery of Botox" explores how a drug called Botox, the trade name for a bacterial neurotoxin called botulinum toxin A, was discovered as a treatment for several medical conditions, including prevention of wrinkling on the face.

The Molecular Mechanism of Contraction

As shown in Figure 19.16, two other proteins are associated with actin filaments. Threads of *tropomyosin* wind about an actin filament, and *troponin* occurs at intervals along the threads. When Ca^{2+} ions are released from the sarcoplasmic reticulum, they combine with troponin, and this causes the tropomyosin threads to shift their position.

The double globular heads of a myosin filament have ATP binding sites, where an ATPase splits ATP into ADP and $\circledP$. The ADP and $\circledP$ remain on the myosin heads until the heads attach to an actin filament, forming cross-bridges. Now, ADP and $\circledP$ are released, and the cross-bridges change their positions. This is the power stroke that pulls the actin filament toward the center of the sarcomere. When ATP molecules again bind to the myosin heads, the cross-bridges are broken, and heads detach from the actin filament. Actin filaments move nearer the center of the sarcomere each time the cycle is repeated. When nerve impulses cease, the sarcoplasmic reticulum actively transports Ca^{2+} ions back into the sarcoplasmic reticulum, and the muscle relaxes.

Energy for Muscle Contraction

ATP produced previous to strenuous exercise lasts a few seconds, and then muscles acquire new ATP in three different ways: creatine

The Discovery of Botox

Several of the most important bacterial pathogens that cause human diseases—including cholera, diphtheria, tetanus, and botulism—do so by secreting potent toxins capable of sickening or killing their victims. The botulinum toxin, produced by the bacterium *Clostridium botulinum,* is one of the most lethal substances known. Less than a microgram (μg) of the purified toxin can kill an average-size person, and 4 kilograms (kg) (8.8 pounds [lb]) would be enough to kill all humans on Earth! Given this scary fact, it seems that the scientists who discovered the lethal activity of this bacterial toxin nearly 200 years ago could never have anticipated that the intentional injection of a very dilute form of botulinum toxin (now known as Botox) would become the most common nonsurgical cosmetic procedure performed by many physicians. As with many breakthroughs in science and medicine, the pathway from thinking about botulism as a deadly disease to using botulinum toxin as a beneficial treatment involved the hard work of many scientists, mixed with a considerable amount of luck.

In the 1820s, a German scientist, Justinus Kerner, was able to prove that the deaths of several people had been caused by their consumption of spoiled sausage (in fact, botulism is named for the Latin word for "sausage," *botulus*). A few decades later, a Belgian researcher named Emile Pierre van Ermengem identified the specific bacterium responsible for producing the botulinum toxin, which can cause symptoms ranging from droopy eyelids to paralysis and respiratory failure.

By the 1920s, medical scientists at the University of California had obtained the toxin in pure form, which allowed them to determine that it acts by preventing nerves from communicating with muscles, specifically by interfering with the release of acetylcholine from the axon terminals of motor nerves.

Scientists soon began testing very dilute concentrations of the toxin as a treatment for conditions in which the muscles contract too much, such as crossed eyes or spasms of the facial muscles or vocal cords. In 1989, the FDA first approved diluted botulinum toxin (Botox) for treating specific eye conditions

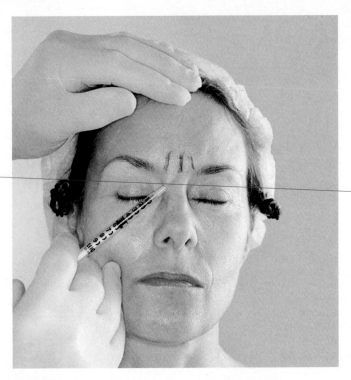

Figure 19B **Treating facial wrinkles with Botox.**
©Thinkstock/Getty Images

called blepharospasm (eyelid spasm) and strabismus (crossing of the eyes).

Right around this time, a lucky break occurred that eventually would open the medical community's eyes to the greater potential of the diluted toxin. A Canadian ophthalmologist, Jean Carruthers, had been using it to treat her patients' eye conditions when she noticed that some of their wrinkles had also subsided. One night at a family dinner, Dr. Carruthers shared this information with her husband, a dermatologist, who decided to investigate whether he could reduce the deep wrinkles of some of his patients by injecting the dilute toxin into their skin. The treatment worked well, and after trying it on several more patients (as well as on themselves!), the Canadian doctors spent several years presenting their findings at scientific meetings and in research journals. Although they were initially considered "crazy," the Carruthers eventually were able to convince the scientific community that diluted botulinum

toxin was effective in treating wrinkles; however, they never patented it for that use, so they missed out on much of the $3.67 billion in annual sales the drug now earns for the company that did patent it.

The uses of diluted botulinum toxin seem to be growing since it was FDA approved for the treatment of frown lines in 2002 (Fig. 19B). In March 2010, it was approved for the treatment of muscle stiffness in people with upper limb spasticity, and the company has approval for as many as 90 uses of diluted botulinum toxin, including treatment of migraine headaches.

Questions to Consider

1. Considering that botulism is caused by a preformed toxin, how do you suppose it can be treated?
2. Do you think companies should be allowed to patent a naturally occurring molecule such as botulinum toxin? Why or why not?

a. Function of Ca²⁺ ions in muscle contraction

b. Function of cross-bridges in muscle contraction

Figure 19.16 The role of calcium and ATP in muscle contraction. **a.** Upon release, calcium binds to troponin, exposing myosin binding sites. **b.** After breaking down ATP (1), myosin heads bind to an actin filament (2), and later, a power stroke causes the actin filament to move (3). When another ATP binds to myosin, the head detaches from actin (4), and the cycle begins again. Although only one myosin head is featured, many heads are active at the same time.

phosphate breakdown, cellular respiration, and fermentation (Fig. 19.17). Creatine phosphate breakdown and fermentation are anaerobic, meaning they do not require oxygen. Creatine phosphate breakdown, used first, is a way to acquire ATP before oxygen starts entering mitochondria. Cellular respiration is aerobic and takes place only when oxygen is available. If exercise is vigorous to the point that oxygen cannot be delivered fast enough to working muscles, fermentation occurs. Fermentation brings on oxygen debt.

Creatine Phosphate Breakdown

Creatine phosphate is a high-energy compound built up when a muscle is resting. Creatine phosphate cannot participate directly in muscle contraction. Figure 19.17a illustrates how this pathway generates ATP.

This reaction occurs in the midst of sliding filaments, and therefore is the speediest way to make ATP available to muscles.

Creatine phosphate provides enough energy for only about 8 seconds of intense activity, and then it is spent. Creatine phosphate is rebuilt when a muscle is resting by transferring a phosphate group from ATP to creatine. Creatine phosphate is also a very popular dietary supplement for training athletes, and there is evidence that consuming creatine in moderate doses is safe and may increase some types of athletic performance.

Fermentation

Like creatine phosphate breakdown, fermentation supplies ATP without consuming oxygen. During fermentation, glucose is broken down to lactate (Fig. 19.17b), which in the water environment of the cell forms lactic acid (also see Section 7.3).

The accumulation of lactate in a muscle fiber makes the cytoplasm more acidic, and eventually enzymes cease to function well. If fermentation continues longer than 2 or 3 minutes, cramping and

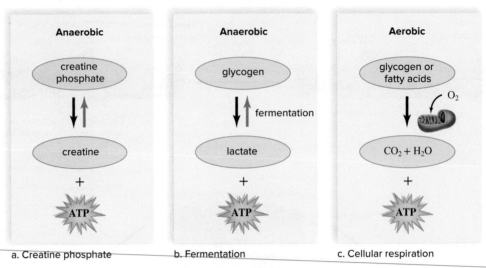

Figure 19.17 The three pathways by which muscle cells produce the ATP energy needed for contraction.   The (**a**) creatine phosphate pathway and (**b**) fermentation pathway are both anaerobic. **c.** The process of cellular respiration is an aerobic pathway and produces the most amount of ATP.

fatigue set in. Cramping seems to be caused by lack of the ATP needed to pump calcium ions back into the sarcoplasmic reticulum and to break the linkages between the actin and myosin filaments so that muscle fibers can relax.

Cellular Respiration

Cellular respiration, completed in mitochondria, usually provides most of a muscle's ATP (see Section 7.1). Glycogen and fat are stored in muscle cells. Therefore, a muscle cell can use glucose (from glycogen) and fatty acids (from fat) as fuel to produce ATP if oxygen is available (Fig. 19.17*c*).

Myoglobin, an oxygen carrier similar to hemoglobin, is synthesized in muscle cells, and its presence accounts for the reddish-brown color of skeletal muscle fibers. Myoglobin has a higher affinity for oxygen than does hemoglobin. Therefore, myoglobin can pull oxygen out of blood and make it available to muscle mitochondria that are carrying on cellular respiration. Then, too, the ability of myoglobin to temporarily store oxygen reduces a muscle's immediate need for oxygen when cellular respiration begins. The end products (carbon dioxide and water) are usually no problem. Carbon dioxide leaves the body at the lungs, and water simply enters the extracellular space. The by-product, heat, keeps the entire body warm.

Oxygen Debt

Because both the creatine phosphate and fermentation pathways are anaerobic processes, when a muscle uses these pathways to supply its energy needs, it incurs an **oxygen debt.** Oxygen debt is obvious when a person continues to breathe heavily after exercising. The ability to run up an oxygen debt is one of muscle tissue's greatest assets. Without oxygen, brain tissue cannot last nearly as long as muscles can.

In people who exercise regularly, the number of muscle mitochondria increases, so fermentation is not needed to produce ATP. Their bodies' mitochondria can start consuming oxygen as soon as the ADP concentration begins to rise during muscle contraction.

Because mitochondria can break down fatty acids instead of glucose, blood glucose is spared for the activity of the brain. Glucose is the preferred energy source of the brain. Because less lactate is produced in people who train, the pH of the blood remains steady, and there is less oxygen debt.

Repaying an oxygen debt requires replenishing creatine phosphate supplies and disposing of lactate. Lactate can be changed back to pyruvate and metabolized completely in mitochondria, or pyruvate can be sent to the liver to reconstruct glycogen. A marathon runner who has just crossed the finish line is not usually exhausted due to oxygen debt. Instead, the runner has used up all the muscles', and probably the liver's, glycogen supply. It takes about 2 days to replace glycogen stores on a high-carbohydrate diet.

Check Your Progress 19.4

1. Identify three microscopic components of muscle cells not found in other types of cells.
2. Explain how the thin and thick filaments interact in the sliding filament model.
3. Describe how neurons specifically control muscle contraction.
4. List one specific advantage of each mechanism in generating ATP for muscle contraction.

19.5 Whole Muscle Contraction

Learning Outcomes

Upon completion of this section, you should be able to

1. Explain what is happening to a muscle fiber during summation, tetanus, and fatigue.
2. Describe how muscles generate different levels of contractive force.
3. Explain the types of exercise that require more fast-twitch versus slow-twitch muscle fibers.

Researchers sometimes study muscles in the laboratory in an effort to understand whole muscle contraction in the body.

In the Laboratory

When a muscle fiber is isolated, placed on a microscope slide, and provided with ATP plus various salts, it contracts completely along its entire length. This observation has resulted in the *all-or-none law,* which states that a muscle fiber contracts completely or not at all. In contrast, a whole muscle shows degrees of contraction. To study whole muscle contraction in the laboratory, an isolated muscle is stimulated electrically, and the mechanical force of contraction is recorded as a visual pattern called a *myogram.* When the strength of the stimulus is above a threshold level, the muscle contracts and then relaxes. This action—a single contraction that lasts only a fraction of a second—is called a **muscle twitch.**

Figure 19.18*a* is a myogram of a muscle twitch, which is customarily divided into three stages: the *latent period,* or the period of time between stimulation and initiation of contraction; the *contraction period,* when the muscle shortens; and the *relaxation period,* when the muscle returns to its former length. It's interesting to use our knowledge of muscle fiber contraction to understand these events. From our study thus far, we know that a muscle fiber

in an intact muscle contracts when calcium leaves storage sacs and relaxes when calcium returns to storage sacs.

But unlike the contraction of a muscle fiber, a whole muscle has degrees of contraction, and a twitch can vary in height (strength) depending on the degree of stimulation. Why? Obviously, a stronger stimulation causes more individual fibers to contract than before.

If a whole muscle is given a rapid series of stimuli, it can respond to the next stimulus without relaxing completely. Summation is increased muscle contraction until maximal sustained contraction, called **tetanus,** is achieved (Fig. 19.18*b*). The myogram no longer shows individual twitches; rather, the twitches are fused and blended completely into a straight line. Tetanus continues until the muscle becomes fatigued due to depletion of energy reserves. Fatigue is apparent when a muscle relaxes even though stimulation continues.

In the Body

In the body, nerves cause muscles to contract. Each axon within a nerve stimulates a number of muscle fibers. A nerve fiber, together with all of the muscle fibers it innervates, is called a **motor unit.** A motor unit obeys the all-or-none law. Why? Because all the muscle fibers in a motor unit are stimulated at once, and they all either contract or do not contract. A variable of interest is the number of muscle fibers within a motor unit. For example, in the ocular muscles that move the eyes, the innervation ratio is one motor axon per 23 muscle fibers, while in the gastrocnemius muscle of the leg, the ratio is about one motor axon per 1,000 muscle fibers. Thus, moving the eyes allows finer control than moving the legs.

Tetanic contractions ordinarily occur in the body because, as the intensity of nervous stimulation increases, more and more motor units are activated. This phenomenon, known as recruitment, results in stronger and stronger muscle contractions. But while some muscle fibers are contracting, others are relaxing. Because of this, intact muscles rarely fatigue completely. Even when muscles appear to be at rest, they exhibit **muscle tone,** in which some of their fibers are always contracting. Muscle tone is particularly important in maintaining posture. If all the fibers within the muscles of the neck, trunk, and legs were to suddenly relax, the body would collapse.

Medical professionals can measure the electrical activity of muscles using a test called an *electromyogram* (*EMG*). Most commonly, needles are inserted at various points along a muscle suspected of functioning abnormally, and the patient is asked to smoothly contract the muscles. Because different conditions have varying effects on the electrical patterns generated by muscles, the EMG test can help in the diagnosis of many neuromuscular conditions.

Athletics and Muscle Contraction

Athletes who excel in a particular sport, and much of the general public as well, are interested in staying fit by exercising.

Exercise and the Size of Muscles Muscles that are not used or that are used for only very weak contractions decrease in size, or

a.

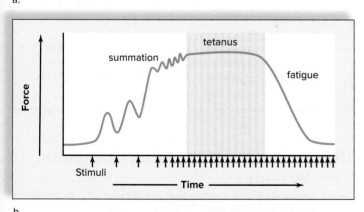

b.

Figure 19.18 Physiology of skeletal muscle contraction.
Stimulation of a muscle dissected from a frog resulted in these myograms.
a. A simple muscle twitch has three periods: latent, contraction, and relaxation. **b.** Summation and tetanus. When the muscle is not permitted to relax completely between stimuli, the contraction gradually increases in intensity. The muscle becomes maximally contracted until it fatigues.

Fast-twitch muscle fiber
• is anaerobic
• has explosive power
• fatigues easily

Slow-twitch muscle fiber
• is aerobic
• has steady power
• has endurance

Figure 19.19   Slow- and fast-twitch muscle fibers.   If your muscles contain many slow-twitch fibers (dark color), you probably do better at a sport such as cross-country running. But if your muscles contain many fast-twitch fibers (light color), you probably do better at a sport such as weightlifting.
(weightlifter): ©Image Source; (muscle fibers): ©Cultura RM/Alamy; (runner): ©John Giustina/Getty Images

atrophy. Atrophy can occur when a limb is placed in a cast or when the nerve serving a muscle is damaged. If nerve stimulation is not restored, muscle fibers are gradually replaced by fat and fibrous tissue. Unfortunately, atrophy can cause muscle fibers to shorten progressively, leaving body parts contracted in contorted positions.

Forceful muscular activity over a prolonged period causes muscle to increase in size as the number of myofibrils within the muscle fibers increases. Increase in muscle size, called *hypertrophy,* occurs only if the muscle contracts to at least 75% of its maximum tension.

Slow-Twitch and Fast-Twitch Muscle Fibers   We have seen that all muscle fibers metabolize both aerobically and anaerobically. Some muscle fibers, however, utilize one method more than the other to provide myofibrils with ATP. Slow-twitch fibers tend to be aerobic, and fast-twitch fibers tend to be anaerobic (Fig. 19.19).

Slow-twitch fibers have a steadier tug and more endurance, despite having motor units with a smaller number of fibers. These muscle fibers are most helpful in sports such as long-distance running, biking, jogging, and swimming. Because they produce most of their energy aerobically, they tire only when their fuel supply is gone. Slow-twitch fibers have many mitochondria and are dark in color because they contain myoglobin, the respiratory pigment found in muscles. They are also surrounded by dense capillary beds and draw more blood and oxygen than fast-twitch fibers. Slow-twitch fibers have a low maximum tension, which develops slowly, but these muscle fibers are highly resistant to fatigue. Because slow-twitch fibers have a substantial reserve of glycogen and fat, their abundant mitochondria can maintain a steady, prolonged production of ATP when oxygen is available.

Fast-twitch fibers tend to be anaerobic and seem designed for strength because their motor units contain many fibers. They

provide explosions of energy and are most helpful in sports activities such as sprinting, weightlifting, swinging a golf club, or throwing a shot. Fast-twitch fibers are light in color because they have fewer mitochondria, little or no myoglobin, and fewer blood vessels than slow-twitch fibers. Fast-twitch fibers can develop maximum tension more rapidly than slow-twitch fibers, and their maximum tension is greater. However, their dependence on anaerobic energy leaves them vulnerable to an accumulation of lactate that causes them to fatigue quickly.

Check Your Progress 19.5

1. List the three typical stages of a muscle twitch, as seen in a myogram.
2. Estimate which would have a greater ratio of motor nerve axons per muscle fibers: a muscle in your index finger or one in your back.
3. Explain why some domesticated birds, like turkeys and chickens, have mostly fast-twitch fibers in their breast (wing) muscles, whereas wild birds like geese and ducks have mostly slow-twitch fibers in those muscles.

19.6 Disorders of the Musculoskeletal System

Learning Outcomes

Upon completion of this section, you should be able to

1. List three common disorders that affect the bones and joints and three that affect the muscles.
2. Identify some of the most common steps that can be taken to help prevent osteoporosis.

Disorders of the Skeleton and Joints

Because bones are rigid, they are susceptible to being broken, or fractured (Fig. 19.20). The most common cause of bone fractures is trauma, although pathological fractures can also occur during normal everyday activities if the bones are diseased. The naming of fractures tells you what type of break has occurred. A fracture is complete if the bone is broken clear through, and is incomplete if the bone is not separated into two parts. A fracture is simple if it does not pierce the skin, and is compound if it does pierce the skin. Impacted means that the broken ends are wedged into each other. A spiral fracture occurs when the break is ragged due to twisting of a bone.

Stress fractures are small cracks in the bone that occur most commonly on the lower leg when muscles become fatigued and unable to absorb the stress of athletic activities. These small fractures usually heal with rest—that is, by avoiding the activity that produced them. Other types of uncomplicated fractures may be treated with casts or splints. However, some fractures require surgery to realign bones or replace bone fragments. In children, the growth plate is the last part of the bone to ossify, leaving it more susceptible to fracture. Fracture of the growth plate may cause the affected bone to stop growing, resulting in limbs that are crooked or uneven in length.

Osteoporosis is a condition in which bone loses mass and mineral content. This leads to an increased risk of fractures, especially of the pelvis, vertebrae, and wrist. As the bone softens, affected individuals often develop a characteristic curvature of the spine (Fig. 19.21). Eighty percent of those affected by osteoporosis are women. About one in four women will have an osteoporosis-related fracture in their lifetime. Women generally have about 30% less bone mass than men to begin with, and after menopause women typically lose about 2% of their bone mass per year. Estrogen replacement therapy (ERT) has been shown to increase bone mass and reduce fractures, but because it also can increase the risk of cardiovascular disease and certain types of cancer, patients must

Figure 19.21 Osteoporosis.   Osteoporosis can eventually result in the slumped posture and loss of height characteristically seen in some elderly people. (skeletons): ©Alfred Pasieka/Science Source; (woman): ©Yoav Levy/Medical Images

be well informed before beginning ERT. Other factors that decrease the risk of osteoporosis for everyone include consuming 1,000–1,200 mg of calcium per day (depending on age) and engaging in regular physical exercise.

There are many different types of **arthritis,** or inflammation of the joints. The Arthritis Foundation estimates that nearly one in three adult Americans has some degree of chronic joint pain. Arthritis is second only to heart disease as a cause of work disability in the United States. The most common form of arthritis is **osteoarthritis,** or degenerative joint disease, which results from the deterioration of the cartilage in one or more synovial joints. As the cartilage continues to soften, crack, and wear away, the resulting bone-on-bone contact causes pain and a decreased range of motion. Symptoms of OA, such as pain or stiffness in the joints after periods of inactivity, typically begin after age 40 and progress slowly. The hands, hips, knees, lower back, and neck are most often affected (Fig. 19.22). **Rheumatoid arthritis (RA)** is considered an autoimmune disease, in which the body's immune system attacks the joints, as well as other tissues. RA affects more women than men, and often first strikes people in their twenties and thirties. The initiating cause of RA remains to be discovered, but genetic factors may play a major role. Over-the-counter anti-inflammatory drugs such as ibuprofen, or more powerful prescription drugs such as corticosteroids, can reduce the pain of most

Figure 19.20 A fracture.   This X-ray shows a tibia (*right*) that has fractured into several pieces. The fibula (*left*) is also fractured. ©Scott Camazine/Medical Images

Figure 19.22 The effects of arthritis.   Arthritis often affects the hands, resulting in swelling, pain, and loss of the ability to perform normal functions. ©Dr. P. Marazzi/Science Source

types of arthritis. Some studies have shown that dietary supplements containing glucosamine and chondroitin sulfate are also beneficial. The damaged joint can also be surgically replaced with a prosthesis (artificial joint).

Disorders of the Muscles

Muscle cramps and twitches don't usually rise to the level of a "disorder," but almost everyone has experienced them. A true muscle cramp doesn't originate in the muscle itself, but rather is caused by abnormal nerve activity. This type of cramp is more common at night or after exercise, and can usually be ended by stretching the muscle. Most of us have also experienced occasional muscle twitches, or *fasciculations*. These also result from abnormal nerve activity, but involve only a small part of a muscle. Twitches most commonly occur in the eyelids, between the thumb and forefinger, and in the feet. They may occur more frequently due to stress, caffeine, or lack of sleep.

Fibromyalgia is a disorder that causes chronic pain in the muscles and ligaments. It affects about 4 million Americans, mostly women in their mid-30s to late-50s. The pain is usually most severe in the neck, shoulders, back, and hips. Although the cause of fibromyalgia is unknown, it may be due to low levels of serotonin or other neurotransmitters involved in pain perception and sleep regulation. As a result, patients with fibromyalgia may be more sensitive to pain, and they usually are unable to get enough deep sleep. As a result, chronic fatigue is also a major symptom. There are no specific treatments, but most people can reduce their symptoms by combining medications to reduce pain and improve sleep with regular exercise and a healthy diet.

Muscular dystrophy (MD) refers to a group of genetic diseases that affect the muscles. Most people are probably familiar with these diseases from the annual Labor Day Telethon sponsored by the Muscular Dystrophy Association. These disorders vary considerably in their degree of severity. The more severe forms may cause problems beginning at a very young age, while other forms may not cause problems until middle age or later. The most common type of MD, called Duchenne MD, is due to an abnormality in the gene coding for a muscle-associated protein called dystrophin. Because it is an X-linked recessive disorder, Duchenne MD

mainly affects boys. Muscle weakness is usually noticeable by age 3 to 5. By age 12, most affected children are unable to walk, and will eventually need a respirator to breathe. Corticosteroids have been shown to slow the degenerative process of Duchenne MD, but these drugs often have side effects. Gene therapy, in which the correct form of the faulty gene is inserted into the patient's cells, is being investigated as a potential cure for MD.

Check Your Progress 19.6

1. Describe the likely outcome if a bone fracture is not treated.
2. Predict why astronauts working in space would be likely to lose bone density.
3. Differentiate between the cause of rheumatoid arthritis and the cause of osteoarthritis.

Conclusion

For the first painful month or so after having her knee replaced, Jackie wondered if she had made the right decision. Just walking down the hall or up stairs was excruciating at first. Within 2 months, however, she was walking and swimming. Her physical therapist attributed her rapid recovery to her previous habit of staying in shape.

Jackie knows that without twenty-first-century medicine, she might have had a difficult time walking by the time she was 60. And she is reminded of the new metallic part of her every time she sets off an airport metal detector. Still, her doctor has reminded her that even in adults, bones are dynamic structures. Whereas her natural bone could be replaced by bone remodeling, the plastic and ceramic parts of her artificial knee would eventually wear out. So there was a very good chance she would have to undergo a repeat replacement of her knee within 10 to 20 years. For Jackie, the ability to lead an active lifestyle would be worth going through the whole procedure again.

SUMMARIZE

19.1 Overview of Bone and Cartilage

The **musculoskeletal system** includes the bones and muscles. The skeleton is made up of bone, cartilage, and dense fibrous connective tissue:

■ **Bone** is a connective tissue that contains mineral salts, which increase its strength. In a long bone, the wall of the shaft is **compact bone,** which is made up of units called osteons, consisting of a central canal surrounded by tiny canals called canaliculi and chambers called lacunae. Within the lacunae are **osteocytes.** The outermost layer of a bone is a tough connective tissue layer called **periosteum.** The ends of a

long bone contain **spongy bone,** which is less organized, although still very strong. The spaces in spongy bones often contain **red bone marrow,** which produces all types of blood cells.

■ Bone is a living tissue that is continuously being remodeled by **osteoblasts,** which build bone, and **osteoclasts,** which break it down. In children, the growth of long bones occurs at a band of cartilage called the growth plate.

■ **Cartilage** is not as strong as bone, but is more flexible because it contains many collagenous and elastic fibers. **Hyaline cartilage** covers the ends of long bones, whereas **fibrocartilage** is mainly used for support. **Elastic cartilage** is the most flexible type.

- **Dense fibrous connective tissue,** which contains fibroblasts and bundles of collagen fibers, forms the ligaments and tendons. **Ligaments** connect bone to bone, and **tendons** connect muscle to bone at **joints.**

19.2 Bones of the Skeleton

The skeleton supports and protects the body, produces blood cells, serves as a storehouse for fats and mineral salts, and permits flexible movement. It can be divided into two major parts, plus the associated joints:

- The **axial skeleton** lies in the midline of the body and consists of the skull, hyoid bone, vertebral column, and rib cage. The **skull** contains the cranium, which protects the brain, and the facial bones. In newborns, bones of the cranium are joined by membranous **fontanels.** Certain cranial bones contain air spaces called **sinuses.** The spinal cord exits the base of the skull through a large opening called the **foramen magnum.** The **vertebral column** protects the spinal cord. Individual vertebrae are separated by **intervertebral discs** made of fibrocartilage.
- The **appendicular skeleton** consists of the bones of the **pectoral girdles,** upper limbs, **pelvic girdle,** and lower limbs.
- There are three types of joints. **Fibrous joints** tend to be immobile, like the **sutures** between the cranial bones. **Cartilaginous joints,** which are slightly movable, include the joints between the ribs and sternum. Flexible joints like the shoulder and knees are the **synovial joints,** in which two bones are separated by a cavity. This cavity, or joint capsule, is lined with a **synovial membrane** that is lubricated by synovial fluid. Many synovial joints contain fluid-filled sacs, or **bursae;** inflammation of these is called **bursitis.**

19.3 Skeletal Muscles

Skeletal muscles are striated and under voluntary control. Most work in antagonistic pairs, causing bones to move at a joint:

- The **origin** of a muscle attaches to a stationary bone; the **insertion** is on the bone that moves.
- Terms used to describe the types of movements due to muscle contraction include *flexion, extension, abduction,* and *adduction.*
- Muscles are grouped by location and named for characteristics such as size, shape, location, and action.

19.4 Mechanism of Muscle Fiber Contraction

Examining the microscopic anatomy of a muscle fiber helps explain the molecular mechanism of muscle contraction:

- The **sarcolemma** of a muscle fiber forms **T (transverse) tubules** that extend into the fiber and almost touch the **sarcoplasmic reticulum,** which stores calcium ions. Each muscle fiber contains many **myofibrils,** which contain myofilaments made of **actin** and **myosin.** These are arranged in subunits called **sarcomeres,** which are the contractile units of a muscle fiber.
- At a **neuromuscular junction,** synaptic vesicles release acetylcholine (ACh), which binds to protein receptors on the sarcolemma, causing impulses to travel down T tubules and calcium to leave the sarcoplasmic reticulum. Calcium ions bind to troponin and cause tropomyosin threads to shift their position, revealing myosin binding sites on actin. Myosin is an ATPase, and once it breaks down ATP, the myosin head is ready to attach to actin. The release of ADP + ⓟ causes the head to change its position. This is the power stroke that causes the actin filament to slide toward the center of a sarcomere. This is the **sliding filament model** of muscle contraction.

A muscle fiber has three ways to acquire ATP after muscle contraction begins:

- **Creatine phosphate,** built up when a muscle is resting, can donate phosphates to ADP, forming ATP.
- Oxygen-dependent cellular respiration can occur within the mitochondria of the muscle cell. Myoglobin, an oxygen carrier that has a higher affinity for oxygen than does hemoglobin, provides oxygen for this process.
- Fermentation of glucose, with the concomitant production of lactate, quickly produces ATP. Fermentation can result in **oxygen debt,** and lactate accumulation may cause fatigue and cramping.

19.5 Whole Muscle Contraction

The contraction of whole muscles can be studied in the laboratory, as well as in the body:

- In the laboratory, a single all-or-none contraction of a muscle fiber is called a **muscle twitch.** In whole muscle contraction, summation is followed by **tetanus,** or maximum sustained contractions, until the muscle must relax from fatigue.
- A **motor unit** consists of a single nerve fiber and all the muscle fibers it innervates. Muscles exhibit **muscle tone** because some muscle fibers are contracting at any time. Two major types of muscle fibers can be identified, with respect to their predominant method of ATP production. Slow-twitch fibers, specialized for endurance, are darker in color and have more mitochondria and myoglobin for aerobic cellular respiration. Fast-twitch fibers have more muscle fibers in their motor units and rely more on an anaerobic means of acquiring ATP. These fibers provide quick bursts of power, but they fatigue quickly.

19.6 Disorders of the Musculoskeletal System

Disorders affecting the skeletal system can be grouped into those affecting the bones and joints, and those affecting the muscles:

- Disorders of the skeleton and joints include fractures, osteoporosis, and arthritis. **Osteoporosis** is a softening of bone due to loss of mass and mineral content, which affects women more commonly than men. Among the many types of **arthritis,** or joint inflammation, are **osteoarthritis,** and **rheumatoid arthritis,** which is an autoimmune disease.
- Muscles can be affected by cramps and twitches, which are very common but usually just an annoyance. More serious muscle disorders include **fibromyalgia,** which causes intense pain and may be due to low levels of serotonin, and **muscular dystrophy (MD),** a group of genetic disorders that affect muscle development.

ASSESS

Testing Yourself

Choose the best answer for each question.

19.1 Overview of Bone and Cartilage

1. Spongy bone
   a. contains osteons.
   b. contains red bone marrow, where blood cells are formed.
   c. weakens bones.
   d. takes up most of a leg bone.
   e. All of these are correct.

2. _____ are involved in the breakdown of bone.
   a. Monocytes
   b. Lymphocytes
   c. Osteoblasts
   d. Osteoclasts

3. _____ connect bone to bone, and _____ connect muscle to bone.
   a. Ligaments, ligaments
   b. Ligaments, tendons
   c. Tendons, ligaments
   d. None of these are correct.

19.2 Bones of the Skeleton

4. Which of the following is not a function of the skeletal system?
   a. production of blood cells
   b. movement
   c. storage of minerals and fat
   d. production of body heat

5. Label the indicated bones of the human skeleton.

19.3 Skeletal Muscles

6. The _____ of the muscle attaches to the bone that is stationary during movement.
   a. origin
   b. diaphysis
   c. insertion
   d. None of these are correct.

7. The biceps and triceps are considered
   a. synergists.
   b. protagonists.
   c. antagonists.
   d. None of these are correct.

8. Muscles can be named based on
   a. size.
   b. action.
   c. shape.
   d. location.
   e. All of these are correct.

19.4 Mechanism of Muscle Fiber Contraction

9. Label the indicated areas in the diagram below.

10. Nervous stimulation of muscles
    a. occurs at a neuromuscular junction.
    b. results in an action potential that travels down the T system.
    c. causes calcium to be released from expanded regions of the sarcoplasmic reticulum.
    d. All of these are correct.

11. As ADP and Ⓟ are released from a myosin head,
    a. actin filaments move toward the H band.
    b. a sarcomere shortens.
    c. myosin filaments pull toward the thin filaments.
    d. All of these are correct.

19.5 Whole Muscle Contraction

12. Which of the following represents the maximal amount of muscle contraction?
    a. tetanus
    b. tone
    c. twitch
    d. fermentation

13. Fast-twitch muscle fibers have which of the following characteristics?
    a. fewer number of mitochondria
    b. rely on fermentation and creatine phosphate pathways
    c. are primarily anaerobic
    d. provide for a quick release of energy
    e. All of these are correct.

19.6 Disorders of the Musculoskeletal System

14. A _____ is an injury to muscle due to twisting or stretching.
   a. sprain
   b. strain
   c. sarcoma
   d. myalgia

15. Which of the following is a form of degeneration found in synovial joints?
   a. muscular dystrophy
   b. osteoporosis
   c. osteoarthritis
   d. fibromylagia

ENGAGE

Thinking Critically

1. The dystrophin protein is located between the sarcolemma and the outer myofilaments of the muscle. It is responsible for conducting the force of the muscle contraction from the myofilaments to the connective tissue of the muscle. In many cases, a mutation in this protein causes the symptoms of muscular dystrophy.
   a. Why would a loss of dystrophin protein cause weakness and a loss of coordination?
   b. Muscular dystrophy is frequently referred to as a muscle-wasting disease, in which the muscles lose mass over time. How would a defect in dystrophin contribute to the wasting of muscle tissue?

2. Similar to the activity in our muscle cells, certain bacteria can utilize sugars by aerobic cellular respiration (which requires oxygen) or by fermentation (which does not). However, like muscle cells, these bacteria use fermentation only when oxygen is not available. Why is aerobic respiration preferable? What is the specific role of oxygen in this process? (You may want to refer back to Section 7.1.)

3. Most authorities emphasize the importance of calcium consumption for the prevention of osteoporosis. However, surveys have shown that osteoporosis is rare in some countries, such as China, where calcium intake is lower than in the United States. These studies often point to the higher level of animal protein consumed in the United States as having a detrimental effect on bone mass. If this were true, what physiological mechanism might explain it? Can you think of any other differences between the lifestyles in China and the United States that also might explain the difference?

C H A P T E R

20

Endocrine System

©Jim Varney/Science Source

BEFORE YOU BEGIN

Before beginning this chapter, take a few moments to review the following discussions:

Figure 2.22 What is the chemical structure of a steroid?

Section 11.5 How do negative feedback mechanisms function in homeostasis?

Section 17.3 Where is the hypothalamus located, and what are its functions?

Diabetes in Young Adults

For some time, Ethan had been feeling very sluggish and had been losing weight. However, he was beginning to notice that he was always thirsty and was urinating much more frequently than usual. Concerned about his health, Ethan visited a local health clinic to meet with a physician. The doctor mentioned that his symptoms were consistent with many disorders, including both viral infections and diabetes. As a quick test, the doctor ordered a urinalysis test to see if there was any glucose in his urine, which would indicate that Ethan's symptoms were caused by diabetes mellitus, a disease that affects over 29.1 million Americans. The results of the urinalysis indicated that there were small amounts of glucose in Ethan's urine, a sign that Ethan's body may not be adequately maintaining its blood glucose levels. The doctor scheduled Ethan for a blood glucose test the following morning and instructed him to not eat or drink anything for 8 hours prior to the test.

During a blood glucose test, a small vial of blood is removed and the amount of glucose measured. Normally, after 8 hours of fasting, the blood glucose level should be between 70 and 100 mg per deciliter (mg/dl) of blood. Ethan's value was slightly above this, but it was not high enough for the doctor to conclude that diabetes was the cause of Ethan's symptoms. The next test was an oral glucose tolerance test (OGTT). In this test, Ethan drank a solution containing 100 grams (g) of glucose. Then, over the next 3 hours, five additional vials of blood were drawn and tested for glucose levels. In a normal individual participating in this test, blood glucose levels rise rapidly and then fall to below 140 mg/dl within 2 hours. In Ethan's case, the response was much slower, and his 2-hour blood glucose level was 150 mg/dl. The physician told Ethan that the cause of his symptoms was most likely type 2 diabetes mellitus, a disease of the endocrine system, the organ system that is responsible for the long-term homeostasis of the body.

As you read through this chapter, think about the following questions:

1. Which hormones control the level of glucose in the blood?

2. How do feedback mechanisms help control blood glucose levels?

20.1 Overview of the Endocrine System

The **endocrine system** consists of glands and tissues that secrete hormones. **Hormones** are chemicals that affect the behavior of other glands or tissues, which can be located far away from the sites of hormone production. The major endocrine glands are shown in Figure 20.1.

The endocrine system is rivaled only by the nervous system with respect to the degree that both influence the body's other systems. Many hormones help regulate the function of other organ systems in the body. Hormones help maintain homeostasis by influencing cellular metabolism, as well as the growth and development of body parts. Hormones are an essential component of the body's response to stress, and some (especially the reproductive hormones) act on the brain to influence behavior. Some hormones, called **growth factors,** promote cell division and mitosis (see Section 5.2).

Endocrine glands have no ducts. In contrast, *exocrine glands* secrete their products through ducts. For example, the salivary glands send saliva into the mouth by way of the salivary ducts. Instead, endocrine glands usually secrete their hormones into the interstitial fluid. From there, hormones diffuse into the bloodstream for distribution throughout the body. However, not all hormones are carried in the bloodstream. Some, called *local hormones,* affect neighboring cells. As we shall see, prostaglandins, which mediate pain and inflammation, are good examples of local hormones.

Table 20.1 lists the major hormones produced by the endocrine system.

Hormones and Homeostasis

Like the nervous system, the endocrine system is intimately involved in homeostasis. However, while the nervous system is organized to respond rapidly to stimuli by transmitting nerve impulses,

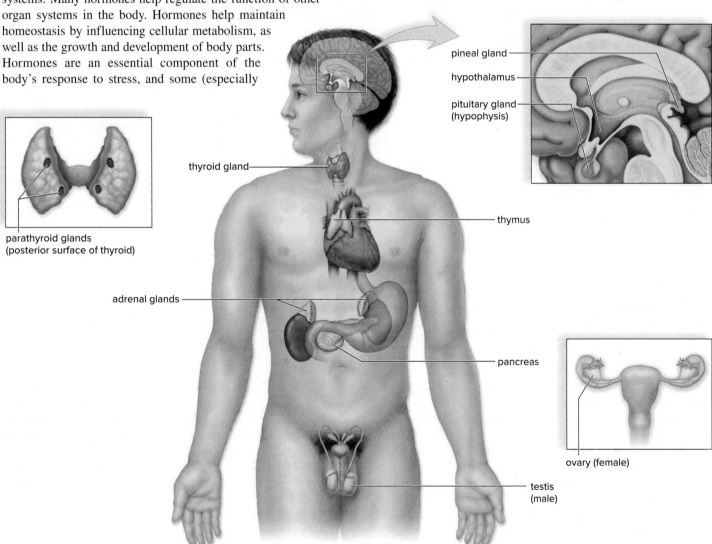

parathyroid glands (posterior surface of thyroid)

thyroid gland

pineal gland

hypothalamus

pituitary gland (hypophysis)

thymus

adrenal glands

pancreas

ovary (female)

testis (male)

Figure 20.1 The endocrine system.   Location of major endocrine glands in the body. Other organs, such as the kidneys and heart, produce hormones. However, hormone production is not a primary function of these organs.

TABLE 20.1   Principal Endocrine Glands and Hormones

Endocrine Gland	Hormone Released	Chemical Class	Target Tissues/Organs	Chief Functions of Hormone
Hypothalamus	Hypothalamic-releasing and hypothalamic-inhibiting	Peptide	Anterior pituitary	Regulate anterior pituitary hormones
Pituitary gland				
Posterior pituitary	Antidiuretic (ADH)	Peptide	Kidneys	Stimulates water reabsorption by kidneys
	Oxytocin	Peptide	Uterus, mammary glands	Stimulates uterine muscle contraction, release of milk by mammary glands
Anterior pituitary	Thyroid-stimulating (TSH)	Glycoprotein	Thyroid	Stimulates thyroid
	Adrenocorticotropic (ACTH)	Peptide	Adrenal cortex	Stimulates adrenal cortex
	Gonadotropic (FSH, LH)	Glycoprotein	Gonads	Egg and sperm production; sex hormone production
	Prolactin (PRL)	Protein	Mammary glands	Milk production
	Growth (GH)	Protein	Soft tissues, bones	Cell division, protein synthesis, and bone growth
	Melanocyte-stimulating (MSH)	Peptide	Melanocytes in skin	Unknown function in humans; regulates skin color in lower vertebrates
Thyroid	Thyroxine (T_4) and triiodothyronine (T_3)	Iodinated amino acid	All tissues	Increases metabolic rate; regulates growth and development
	Calcitonin	Peptide	Bones, kidneys, intestine	Lowers blood calcium level
Parathyroids	Parathyroid (PTH)	Peptide	Bones, kidneys, intestine	Raises blood calcium level
Adrenal gland				
Adrenal cortex	Glucocorticoids (cortisol)	Steroid	All tissues	Raise blood glucose level; stimulate breakdown of protein
	Mineralocorticoids (aldosterone)	Steroid	Kidneys	Reabsorb sodium and excrete potassium
	Sex hormones	Steroid	Gonads, skin, muscles, bones	Stimulate reproductive organs and bring about sex characteristics
Adrenal medulla	Epinephrine and norepinephrine	Modified amino acid	Cardiac and other muscles	Released in emergency situations; raise blood glucose level
Pancreas	Insulin	Protein	Liver, muscles, adipose tissue	Lowers blood glucose level; promotes formation of glycogen
	Glucagon	Protein	Liver, muscles, adipose tissue	Raises blood glucose level
Gonads				
Testes	Androgens (testosterone)	Steroid	Gonads, skin, muscles, bones	Stimulate male sex characteristics
Ovaries	Estrogens and progesterone	Steroid	Gonads, skin, muscles, bones	Stimulate female sex characteristics
Thymus	Thymosins	Peptide	T lymphocytes	Stimulate production and maturation of T lymphocytes
Pineal gland	Melatonin	Modified amino acid	Brain	Controls circadian and circannual rhythms; possibly involved in maturation of sexual organs

hormones secreted by the endocrine system must reach their target organs via the blood, resulting in a slower but often a more prolonged response. The blood concentration of a substance often prompts an endocrine gland to secrete its hormone. For example, the parathyroid glands secrete a hormone when the blood calcium level falls below normal. This hormone causes osteoclasts to slowly release calcium from bone. It takes time for the cells to respond, but the effect is longer lasting.

The production of most hormones is controlled in two ways: by negative feedback, and by the action of other hormones. When controlled by **negative feedback,** an endocrine gland can be sensitive to either the condition it is regulating or the blood level of the hormone it is producing. For example, when the blood glucose

level rises, the pancreas produces insulin. Insulin causes the liver to store glucose, and glucose is removed from the blood for liver storage. The stimulus for the production of insulin is, thereby, inhibited, and the pancreas stops producing insulin. Or, the anterior pituitary produces a thyroid-stimulating hormone that causes the thyroid to produce two hormones. When these hormones rise, the anterior pituitary stops producing thyroid-stimulating hormone.

The effect of a hormone also can be controlled by the release of an *antagonistic* hormone. The effect of insulin, for example, is offset by the production of glucagon by the pancreas. Insulin lowers the blood sugar level, while glucagon raises it. Similarly, calcitonin produced by the thyroid gland lowers the blood calcium level, but parathyroid hormone increases blood calcium.

The Action of Hormones

Hormones are a type of chemical signal. Chemical signals are a means of communication between cells, between body parts, or even between individuals. Hormones have a wide range of effects on cells. Some of these effects induce a target cell to increase its uptake of particular substances (e.g., glucose) or ions (e.g., calcium). Other hormones bring about an alteration of the target cell's structure in some way. They typically affect the metabolism of target cells that have receptors to receive them (Fig. 20.2).

Most hormones fall into two basic chemical classes: (1) **peptide hormones** are either peptides, proteins, glycoproteins, or modified amino acids; (2) **steroid hormones** always have the same complex of 4-carbon rings, but each specific hormone has different side chains.

Peptide Hormones

The majority of the hormones produced by the endocrine system are peptide hormones. The receptors for peptide hormones are found on the cell surface. Because of this, peptide hormones can target specific cells of the body, and the sensitivity can be regulated by the number of receptors.

In our example, we will use epinephrine, a typical peptide hormone, that binds to a receptor on muscle cells (Fig. 20.3). This leads to a conversion of ATP to cyclic AMP (adenosine monophosphate), or cAMP, inside the cell. In this chemical signaling process, the peptide hormone is called the **first messenger,** and cAMP is called the **second messenger.**

The second messenger typically sets in motion an enzymatic pathway sometimes termed an enzyme cascade, because each enzyme, in turn, activates another. Because enzymes work over and over, every step in an enzyme cascade leads to more reactions—the binding of even a single peptide hormone molecule can result in a thousandfold response. In muscle cells, the activation of these

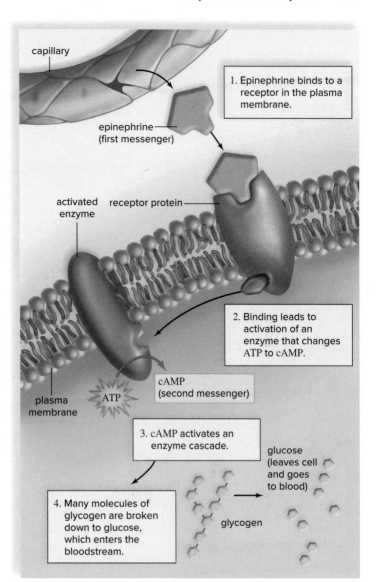

Figure 20.3 Action of epinephrine, a peptide hormone.   Epinephrine (first messenger) binds to a receptor in the plasma membrane of a muscle cell. Thereafter, a second messenger (cyclic AMP, in this case) forms and activates an enzyme cascade.

cellular signaling pathways by epinephrine leads to the breakdown of glycogen to glucose (Fig. 20.3).

To better understand the terms *first messenger* and *second messenger,* imagine that the adrenal medulla, which produces epinephrine, is the home office that sends out a courier (i.e., the first messenger, epinephrine) to a factory (the cell). The courier doesn't have clearance to enter the factory, so when he arrives there, he tells a supervisor through a screen door that the home office wants the factory to produce a particular product. The supervisor (i.e., cAMP, the second messenger) walks over and flips a switch (the enzymatic pathway), and a product is made.

Steroid Hormones

Steroid hormones act in a somewhat different way. After diffusing through the plasma membrane, these hormones typically bind to receptors in the cytoplasm or nucleus (Fig. 20.4). Within the cell,

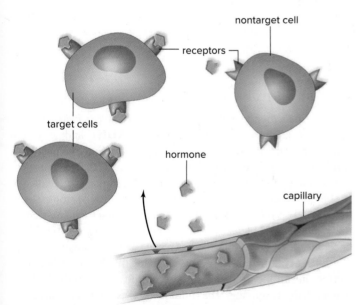

Figure 20.2 Target cell concept.   Most hormones are distributed by the bloodstream to target cells. Target cells have receptors for the hormone, and the hormone combines with the receptor as a key fits a lock.

1. Hormone diffuses through plasma membrane because it is lipid soluble.

steroid hormone

plasma membrane

cytoplasm

nucleus

2. Hormone binds to receptor inside nucleus.

DNA

receptor protein

mRNA

protein

ribosome

mRNA

3. Hormone-receptor complex activates gene and synthesis of a specific mRNA molecule.

4. mRNA moves to ribosomes, and protein synthesis occurs.

Figure 20.4 Steroid hormone action. A steroid hormone, such as the sex hormones testosterone and estrogen, passes directly through the target cell's plasma membrane before binding to a receptor in the nucleus or cytoplasm. The hormone-receptor complex binds to DNA, and gene expression follows.

the hormone-receptor complex usually binds to DNA and activates transcription of certain genes. Translation of messenger RNA (mRNA) transcripts results in enzymes and other proteins that can carry out a response to the hormonal signal (see Section 25.3). Because it takes longer to synthesize new proteins than to activate enzymes already present in cells, steroid hormones generally act more slowly than peptide hormones, but their actions usually last longer.

Pheromones

Chemical signals that act between individuals of the same species are called **pheromones.** The effects of pheromones are better documented in other animals than in humans. For example, female moths release a sex attractant that is received by male moth antennae even several miles away. Humans do produce a rich supply of airborne chemicals from a variety of areas, including the scalp,

oral cavity, armpit (axilla), genital areas, and feet. Studies suggest that women prefer the axillary odors of men who are a different major histocompatibility complex (MHC) type than themselves. Choosing a mate of a different MHC type could conceivably improve the immune response of offspring (see Section 13.3 for an explanation of MHC functions). Several studies indicate that human pheromones can affect the menstrual cycle. Women who live in the same household often have menstrual cycles in synchrony.

Check Your Progress 20.1

1. List six major human endocrine organs.
2. Explain what it means to say that secretion of a hormone is regulated by negative feedback.
3. Explain why second messengers are needed for peptide hormones.
4. Provide a specific example that suggests human behavior may be influenced by pheromones.

20.2 Hypothalamus and Pituitary Gland

Learning Outcomes

Upon completion of this section, you should be able to

1. Describe the relationship between the hypothalamus and the pituitary gland.
2. List two hormones released by the posterior pituitary and six produced by the anterior pituitary.
3. Recognize which three pituitary hormones act on other endocrine glands.

The **hypothalamus** helps regulate the internal environment in two ways. Through the autonomic system, it modulates heartbeat, blood pressure, hunger and appetite, body temperature, and water balance. It also controls the glandular secretions of the **pituitary gland** (*hypophysis*). The pituitary, a small gland about 1 cm in diameter, is connected to the hypothalamus by a stalklike structure. The pituitary has two portions: the posterior pituitary and the anterior pituitary.

Posterior Pituitary

Neurons in the hypothalamus, called *neurosecretory cells,* produce the hormones **antidiuretic hormone (ADH)** and oxytocin (Fig. 20.5). These hormones pass through axons into the **posterior pituitary,** where they are stored in axon terminals. Certain neurons in the hypothalamus are sensitive to the water-salt balance of the blood. When these cells determine that the blood is too concentrated, ADH is released from the posterior pituitary. Upon reaching the kidneys, ADH causes water to be reabsorbed. As the blood becomes diluted by reabsorbed water, ADH is no longer released.

Oxytocin, the other hormone made in the hypothalamus, causes uterine contraction during childbirth and milk letdown when a baby is nursing. The more the uterus contracts during labor, the more nerve impulses reach the hypothalamus, causing oxytocin to be released. Similarly, the more a baby suckles, the more oxytocin is released. In both instances, the release of

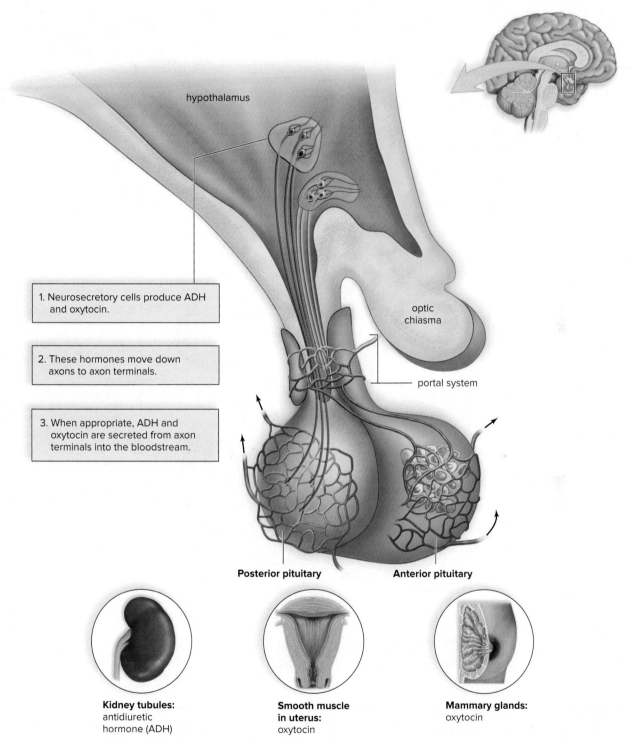

1. Neurosecretory cells produce ADH and oxytocin.

2. These hormones move down axons to axon terminals.

3. When appropriate, ADH and oxytocin are secreted from axon terminals into the bloodstream.

hypothalamus

optic chiasma

portal system

Posterior pituitary

Anterior pituitary

Kidney tubules:
antidiuretic
hormone (ADH)

**Smooth muscle
in uterus:**
oxytocin

Mammary glands:
oxytocin

Figure 20.5 Hormones produced by the hypothalamus and posterior pituitary.   The hypothalamus produces two hormones, ADH and oxytocin, stored and secreted by the posterior pituitary.

oxytocin from the posterior pituitary is controlled by **positive feedback**—that is, the stimulus continues to bring about an effect that ever increases in intensity.

Anterior Pituitary

A *portal system,* consisting of two capillary networks connected by a vein, lies between the hypothalamus and the anterior pituitary (Fig. 20.6). The hypothalamus controls the anterior pituitary by producing **hypothalamic-releasing hormones** and, in some instances, **hypothalamic-inhibiting hormones.** For example, one hypothalamic-releasing hormone stimulates the anterior pituitary to secrete a thyroid-stimulating hormone, and a particular hypothalamic-inhibitory hormone prevents the anterior pituitary from secreting prolactin.

Four of the six hormones produced by the **anterior pituitary** have an effect on other glands: **thyroid-stimulating hormone (TSH)**

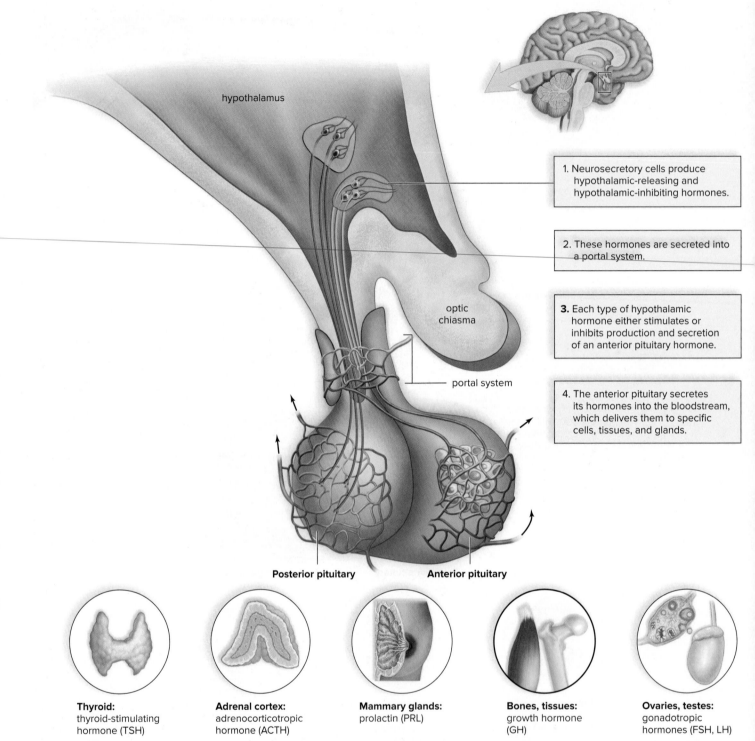

hypothalamus

1. Neurosecretory cells produce hypothalamic-releasing and hypothalamic-inhibiting hormones.

2. These hormones are secreted into a portal system.

optic chiasma

3. Each type of hypothalamic hormone either stimulates or inhibits production and secretion of an anterior pituitary hormone.

portal system

4. The anterior pituitary secretes its hormones into the bloodstream, which delivers them to specific cells, tissues, and glands.

Posterior pituitary   **Anterior pituitary**

Thyroid:
thyroid-stimulating hormone (TSH)

Adrenal cortex:
adrenocorticotropic hormone (ACTH)

Mammary glands:
prolactin (PRL)

Bones, tissues:
growth hormone (GH)

Ovaries, testes:
gonadotropic hormones (FSH, LH)

Figure 20.6 Hormones produced by the anterior pituitary. The hypothalamus controls the secretions of the anterior pituitary, and the anterior pituitary controls the secretions of the thyroid, adrenal cortex, and gonads, which are also endocrine glands.

stimulates the thyroid to produce the thyroid hormones. **Adrenocorticotropic hormone (ACTH)** stimulates the adrenal cortex to produce cortisol. **Follicle-stimulating hormone (FSH)** and **luteinizing hormone (LH)** are both examples of **gonadotropic hormones.** They stimulate the gonads—the testes in males and the ovaries in females—to produce gametes and sex

hormones. In each instance, the blood level of the last hormone in the hypothalamus-anterior pituitary-target gland control system exerts negative feedback over the secretions of the first two structures (Fig. 20.7).

The other three hormones produced by the anterior pituitary do not affect other endocrine glands. **Prolactin** is produced in

Human Growth Hormone: Does Rambo Know Something We Don't?

In some individuals, abnormally low levels of growth hormone production by the anterior pituitary can result in short stature. Suppose you have a son or daughter who is considerably shorter than most kids his or her age. The average adult American male is 5 ft 9 in. tall. If you knew your son would reach only 5 ft, do you think it would be ethical to treat him with human growth hormone (HGH), which could easily add 3 or 4 in. to his height? And because HGH injections can cost between $10,000 and $60,000 per year, should that be covered by health insurance, even if it raises insurance rates for other people?

Now suppose you are 55 years old. You are finding that even though you run and lift weights 3 days a week, you are slowly losing muscle mass, and you don't feel nearly as energetic as you once felt, even in your forties. Then you hear about how some professional athletes and movie stars are swearing by HGH as a way to recapture their youth. You find a website touting "legal" HGH at affordable prices, and claiming HGH will help you lose weight, tone your muscles, and give you younger-looking skin in a matter of weeks. Assuming it was legal and you could afford it, would you give it a try?

Prior to the development of recombinant DNA techniques, treatment of children who failed to grow due to insufficient HGH production required isolation of HGH from the pituitary glands of human cadavers. In 1985, the FDA approved the first recombinant HGH,

produced by inserting the *HGH* gene into *E. coli* bacteria. In 2003, the FDA approved a new use for HGH, in the treatment of children of short stature due to unknown causes. However, HGH must still be administered by a physician, and it is illegal to use it for fitness purposes.

Figure 20A At age 61, actor Sylvester Stallone admitted to using human growth hormone and anabolic steroids to help him get in shape for filming his latest *Rambo* movie.
©Ramey Photo Agency

The production of recombinant HGH has been very beneficial in specific medical cases, but its use for other purposes remains controversial. Prior to filming his fourth *Rambo* movie in 2007, actor Sylvester Stallone admitted to using HGH, as well as anabolic steroids, to help tone his body (Fig. 20A). That same year, Stallone was fined over $10,000 for attempting to smuggle 48 vials of HGH and 4 vials of testosterone into Australia. When he made that film, at age 61, he seemed to be in incredible condition, but it is hard to say how much of that was due to hard workouts, genetics, or pharmaceuticals. Studies show that HGH can cut down on body fat and increase muscle mass, but no antiaging effects have been proven, and the list of scary side effects is long. Still, it's worth asking: Would you be willing to take the risk?

Questions to Consider

1. Would you ever consider using HGH under any circumstances? Why or why not?
2. Do you believe that HGH should be illegal for personal use? If so, do you believe that taking HGH is worse than taking currently legal athletic supplements, such as creatine? Would it make a difference if someone was taking it to grow taller instead of for other perceived benefits?
3. Do you think that celebrities have an increased responsibility to avoid promoting behaviors that could be harmful to others?

Figure 20.7 Negative feedback mechanisms in the endocrine system. Feedback mechanisms (red arrows) provide means of controlling the amount of hormones produced (blue arrows) by the hypothalamus and pituitary glands.

quantity only during pregnancy and after childbirth. It causes the mammary glands in the breasts to develop and produce milk and suppress ovulation for a time in the nursing mother. It also plays a role in carbohydrate and fat metabolism.

Melanocyte-stimulating hormone (MSH) causes skin-color changes in many fishes, amphibians, and reptiles that have melanophores, special skin cells that produce color variations. The concentration of this hormone in humans is very low.

Growth hormone (GH) or somatotropic hormone, promotes skeletal and muscular growth. It increases the rate at which amino acids enter cells and causes increased protein synthesis. It also promotes fat metabolism as opposed to glucose metabolism.

The anterior pituitary normally produces higher levels of GH during childhood and adolescence, when most body growth is occurring. The amount of growth hormone produced during childhood is clearly one factor that influences height (see Section 20.7), although other factors, such as the genetic predisposition to attain a certain height, are also important.

20.3 Thyroid and Parathyroid Glands

Learning Outcomes

Upon completion of this section, you should be able to

1. Identify the anatomical location of the thyroid and parathyroid glands.
2. Distinguish between the functions of T_3, T_4, calcitonin, and parathyroid hormone.

The **thyroid gland** is a large gland located in the neck, where it is attached to the trachea just below the larynx (see Fig. 20.1). The **parathyroid glands** are embedded in the posterior surface of the thyroid gland.

Thyroid Gland

The thyroid gland contains a large number of *follicles,* each a small spherical structure made of thyroid cells that produce *triiodothyronine* (T_3), which contains three iodine atoms, and **thyroxine (T_4),** which contains four iodine atoms. To produce T_3 and T_4, the thyroid gland actively acquires iodine from the bloodstream. The concentration of iodine in the thyroid gland is approximately 25 times that found in the blood. The primary source for iodine in our diets is iodized salt.

The thyroid hormones T_3 and T_4 increase the metabolic rate. They do not have a single target organ. Instead, they stimulate most cells of the body to metabolize more glucose and utilize more energy. Although the thyroid releases about ten times more T_4 than T_3, T_3 has a much more potent activity. Fortunately, the liver is able to convert most of the T_4 produced by the thyroid gland to T_3. Interestingly, even though T_3 and T_4 are considered peptide hormones because they are synthesized from the amino acid tyrosine, their receptor is actually located inside cells, more like a steroid hormone receptor.

The thyroid gland also produces a hormone called **calcitonin,** which helps control blood calcium levels (Fig. 20.8). Calcium (Ca^{2+}) plays a significant role in many processes, including nerve conduction, muscle contraction, and blood clotting. The thyroid gland secretes calcitonin when the blood calcium level rises. The primary effect of calcitonin is to bring about the deposition of calcium in the bones. It does this by temporarily reducing the activity

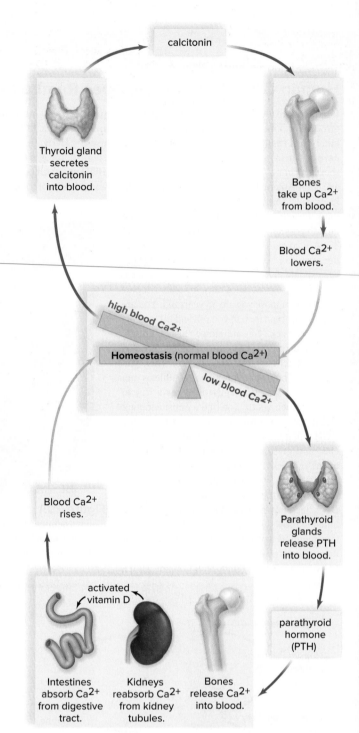

Figure 20.8 Regulation of blood calcium level.   *Top:* When the blood calcium (Ca^{2+}) level is high, the thyroid gland secretes calcitonin. Calcitonin promotes the uptake of Ca^{2+} by the bones, and therefore the blood Ca^{2+} level returns to normal. *Bottom:* When the blood Ca^{2+} level is low, the parathyroid glands release parathyroid hormone (PTH). PTH causes the bones to release Ca^{2+} and the kidneys to reabsorb Ca^{2+} and activate vitamin D. Thereafter, the intestines absorb Ca^{2+}. Therefore, the blood Ca^{2+} level returns to normal.

and number of osteoclasts, the cells that break down bone (see Section 19.1). When the blood calcium level decreases to normal, the thyroid stops releasing calcitonin.

Parathyroid Glands

Many years ago, the four parathyroid glands were sometimes mistakenly removed during thyroid surgery because they are so small. **Parathyroid hormone (PTH),** the hormone produced by the parathyroid glands, causes the blood phosphate (HPO_4^{2-}) level to decrease and the blood calcium level to increase.

PTH promotes the activity of osteoclasts and the release of calcium from the bones. PTH also promotes the reabsorption of calcium by the kidneys, where it helps activate vitamin D. Vitamin D, in turn, stimulates the absorption of calcium from the intestine. These effects bring the blood calcium level back to the normal range. As soon as blood calcium is back to normal, the parathyroid glands no longer secrete PTH.

Check Your Progress 20.3

1. Explain how the amount of T_3 and T_4 secreted by the thyroid gland is controlled.
2. Describe the symptoms that might be expected in a person with a thyroid tumor that was producing excess calcitonin.
3. Predict the expected effect(s) of removal of the parathyroid glands.

20.4 Adrenal Glands

Learning Outcomes

Upon completion of this section, you should be able to
1. Describe the location of the adrenal glands in the body.
2. List the hormones produced by the adrenal medulla and adrenal cortex and provide a function for each.
3. Distinguish between mineralocorticoid and glucocorticoid hormones.

The **adrenal glands** sit atop the kidneys (see Fig. 20.1). Each adrenal gland consists of an inner portion called the **adrenal medulla** and an outer portion called the **adrenal cortex.** These portions, like the anterior pituitary and the posterior pituitary, are two functionally distinct endocrine glands. The adrenal medulla is under nervous control, and a portion of the adrenal cortex is under the control of adrenocorticotropic hormone (ACTH), from the anterior pituitary. Stress of all types, including emotional and physical trauma, prompts the hypothalamus to stimulate a portion of the adrenal glands (Fig. 20.9).

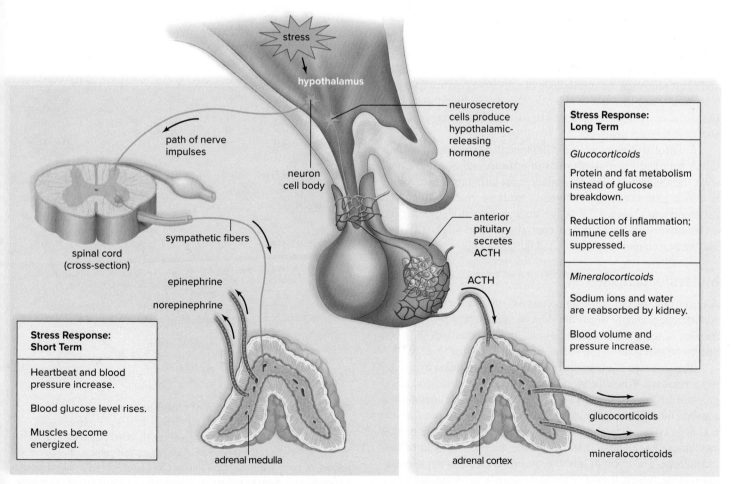

Figure 20.9 The stress response.   Both the adrenal cortex and the adrenal medulla are under the control of the hypothalamus when they help us respond to stress. *Left:* Nervous stimulation causes the adrenal medulla to provide a rapid, but short-term, stress response. *Right:* The adrenal cortex provides a slower, but long-term, stress response. ACTH causes the adrenal cortex to release glucocorticoids. Independently, the adrenal cortex releases mineralocorticoids.

Adrenal Medulla

The hypothalamus initiates nerve impulses that travel by way of the brain stem, spinal cord, and sympathetic nerve fibers to the adrenal medulla, which then secretes its hormones.

Epinephrine (adrenaline) and **norepinephrine** (noradrenaline) produced by the adrenal medulla rapidly bring about all the body changes that occur when an individual reacts to an emergency situation in a fight-or-flight manner (Fig. 20.9, *left*). The effects of these hormones provide a short-term response to stress.

Adrenal Cortex

The hormones produced by the adrenal cortex provide a longer-term response to stress (Fig. 20.9, *right*). The two major types of hormones produced by the adrenal cortex are the mineralocorticoids and the glucocorticoids. **Mineralocorticoids** regulate salt and water balance, leading to increases in blood volume and blood pressure. **Glucocorticoids** regulate carbohydrate, protein, and fat metabolism, leading to an increase in the blood glucose level. The adrenal cortex also secretes small amounts of male and female sex hormones in both sexes.

Glucocorticoids

ACTH stimulates the portion of the adrenal cortex that secretes the glucocorticoids, of which **cortisol** is the most significant. Cortisol raises the blood glucose level in at least two ways: (1) it promotes the breakdown of muscle proteins to amino acids, which are taken up from the bloodstream and converted into glucose by the liver; and (2) it promotes the metabolism of fatty acids, rather than carbohydrates, and this spares glucose. The rise in blood glucose is beneficial to a person under stress because glucose is the preferred energy source for neurons.

Cortisol also counteracts the inflammatory response that leads to the pain and swelling of joints in arthritis and bursitis. *Cortisone,* the medication often administered for inflammation of joints, is also a glucocorticoid. Very high levels of glucocorticoids in the blood can suppress the body's defense system, including the inflammatory response that occurs at infection sites. This can make a person more susceptible to injury and infection.

Mineralocorticoids

Aldosterone is the most important of the mineralocorticoids. Aldosterone primarily targets the kidneys, where it promotes renal absorption of sodium (Na^+) and renal excretion of potassium (K^+), and thereby helps regulate blood volume and blood pressure (Fig. 20.10).

The secretion of mineralocorticoids is not controlled by the anterior pituitary. When the blood sodium level and, therefore, blood pressure are low, the kidneys secrete **renin.** Renin is an enzyme that converts the plasma protein *angiotensinogen* to *angiotensin I,* which is changed to *angiotensin II* by a converting enzyme found in lung capillaries. Angiotensin II stimulates the adrenal cortex to release aldosterone. The effect of this system, called the renin-angiotensin-aldosterone system, is to raise blood pressure in two ways: angiotensin II constricts the arterioles, and aldosterone causes the kidneys to reabsorb sodium. When the blood sodium level rises, water is

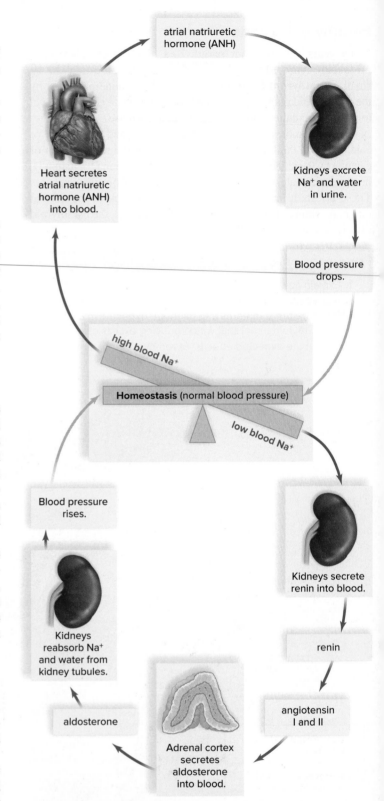

Figure 20.10 Regulation of blood pressure and volume.   *Top:* When the blood Na^+ is high, high blood volume causes the heart to secrete atrial natriuretic hormone (ANH). ANH causes the kidneys to excrete Na^+, and water follows. The blood volume and pressure then return to normal. *Bottom:* When the blood sodium (Na^+) level is low, low blood pressure causes the kidneys to secrete renin. Renin leads to the secretion of aldosterone from the adrenal cortex. Aldosterone causes the kidneys to reabsorb Na^+, and water follows, so that blood volume and pressure return to normal.

reabsorbed, in part because the hypothalamus secretes ADH (see Section 20.2). Blood pressure then increases to normal.

When the atria of the heart are stretched due to a great increase in blood volume, cardiac cells release a hormone called **atrial natriuretic hormone (ANH)** which inhibits the secretion of aldosterone from the adrenal cortex. The effect of this hormone is to cause the excretion of sodium—that is, *natriuresis*. When sodium is excreted, so is water, and thus blood pressure lowers to normal.

Check Your Progress 20.4

1. Explain how the secretion of hormones by the adrenal medulla and cortex is controlled.
2. Summarize how the hormones produced by the adrenal glands help the body respond to stress.
3. Describe the specific effect of aldosterone on the kidney.

20.5 Pancreas

Learning Outcomes

Upon completion of this section, you should be able to

1. Explain how the pancreas has both endocrine and exocrine functions.
2. Describe the functions of insulin, glucagon, and somatostatin.

The **pancreas** is a long organ that lies in the abdomen, between the kidneys and near the duodenum of the small intestine (see Fig. 20.1). It is composed of two types of tissue. Exocrine tissue produces and secretes digestive juices by way of ducts to the small intestine (see Section 14.2). The **pancreatic islets** (islets of Langerhans) are clusters of at least three types of endocrine cells: (1) *alpha cells* that produce glucagon, (2) *beta cells* that produce insulin, and (3) *delta cells* that produce somatostatin. Insulin and glucagon are important in regulating the blood glucose level (Fig. 20.11).

Insulin is secreted when the blood glucose level is high, which usually occurs just after eating. Insulin stimulates the uptake of glucose by cells, especially liver cells, muscle cells, and adipose tissue cells. In liver and muscle cells, glucose is then stored as glycogen. In muscle cells, the breakdown of glucose supplies energy for protein metabolism, and in fat cells, the breakdown of glucose supplies glycerol for the formation of fat. In these various ways, insulin lowers the blood glucose level.

Glucagon is secreted from the pancreas, usually before eating, when the blood glucose level is low. The major target tissues of glucagon are the liver and adipose tissue. Glucagon stimulates the liver to break down glycogen to glucose. Fat and protein are also used as energy sources by the liver, thus sparing glucose. Adipose tissue cells break down fat to glycerol and fatty acids. The liver takes these up and uses them as substrates for glucose formation. In these various ways, glucagon raises the blood glucose level.

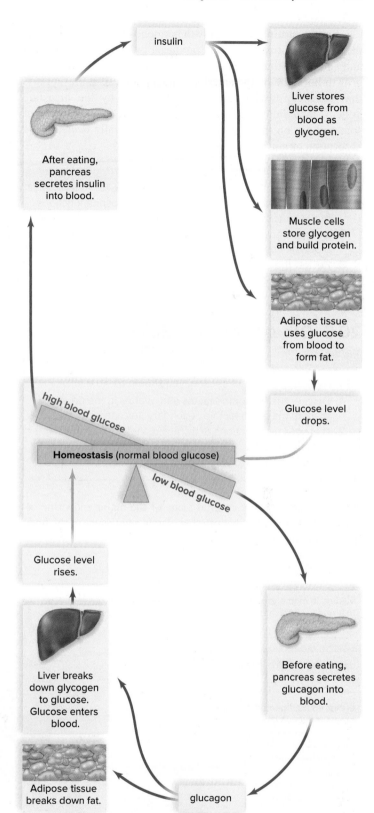

Figure 20.11 Regulation of blood glucose level. *Top:* When the blood glucose level is high, the pancreas secretes insulin. Insulin promotes the storage of glucose as glycogen and the synthesis of proteins and fats (as opposed to their use as energy sources). Therefore, insulin lowers the blood glucose level to normal. *Bottom:* When the blood glucose level is low, the pancreas secretes glucagon. Glucagon acts opposite to insulin. Therefore, glucagon raises the blood glucose level to normal.

Somatostatin is also a growth hormone inhibitor. Besides the pancreas, somatostatin is also produced by cells in the stomach and small intestine. Its main effects are to inhibit the release of growth hormone by the anterior pituitary, as well as to suppress the release of various hormones produced by the digestive system, including insulin and glucagon. The overall effect of somatostatin on the digestive tract is to generally decrease the absorption of nutrients.

Check Your Progress 20.5

1. Distinguish between the endocrine and exocrine functions of the pancreas.
2. Describe how pancreatic hormones interact to regulate blood glucose levels.
3. Review the effect of somatostatin on other hormones, such as insulin and glucagon.

20.6 Other Endocrine Glands

Learning Outcomes

Upon completion of this section, you should be able to

1. Discuss the role of the testes and ovaries in the development of male and female secondary sex characteristics.
2. Describe the functions of thymosin, melatonin, leptin, and prostaglandins.
3. Define and list two examples of growth factors.

While we have already covered many of the major endocrine organs of the body, a number of other glands and tissues of the body have endocrine functions.

Testes and Ovaries

The **gonads** are the testes in males and the ovaries in females. The testes are located in the scrotum, and the ovaries are located in the pelvic cavity. Under the influence of the gonadotropic hormones secreted by the anterior pituitary, the **testes** produce sperm and **androgens** (e.g., **testosterone**), which are the male sex hormones. Under the influence of the gonadotropic hormones, the **ovaries** produce eggs, as well as estrogen and progesterone, the female sex hormones. The hypothalamus and the pituitary gland control the hormonal secretions of these organs in a similar manner as previously described for the thyroid gland.

Greatly increased testosterone secretion at the time of puberty stimulates the growth of the penis and the testes. Testosterone also brings about and maintains the male secondary sex characteristics that develop during puberty, such as the growth of a beard, axillary (underarm) hair, and pubic hair. Testosterone also prompts the larynx and the vocal cords to enlarge, causing the voice to change. It is partially responsible for the muscular strength of males. This is the reason some athletes take supplemental amounts of illegal **anabolic steroids,** which are either testosterone or related chemicals. The side effects of taking anabolic steroids are listed in Figure 20.12. Testosterone also stimulates oil and sweat glands in the skin. Therefore, it is largely responsible for acne and body odor. Another side effect of testosterone is baldness. Genes for baldness are probably inherited by both sexes, but baldness occurs more often in males because of the presence of testosterone.

Estrogen and **progesterone** have many effects on the body. In particular, estrogen secreted at the time of puberty stimulates the growth of the uterus and the vagina. Estrogen is necessary for egg maturation and is largely responsible for the secondary sex characteristics in females, including female body hair and fat distribution. In general, females have a more rounded appearance than males

balding in men and women; hair on face and chest in women

deepening of voice in women

breast enlargement in men and breast reduction in women

liver dysfunction and cancer

kidney disease and retention of fluids, called "steroid bloat"

reduced testicular size, low sperm count, and impotency

'roid mania—delusions and hallucinations; depression upon withdrawal

severe acne

high blood cholesterol and atherosclerosis; high blood pressure and damage to heart

in women, increased size of ovaries; cessation of ovulation and menstruation

stunted growth in adolescents by causing premature ossification of growth plates

Figure 20.12 The side effects of anabolic steroid use.   The use of steroid hormones has an effect on many of the organ systems of the body.

because of a greater accumulation of fat beneath the skin. Also, the pelvic girdle is wider in females than in males, resulting in a larger pelvic cavity. Both estrogen and progesterone are required for breast development and regulation of the uterine cycle, which includes monthly menstruation (see Section 21.3).

Thymus

The **thymus** is a lobular organ that lies just beneath the sternum (see Fig. 20.1). This organ reaches its largest size and is most active during childhood. With aging, the organ gets smaller and becomes fatty. Lymphocytes that originate in the bone marrow and then pass through the thymus differentiate into T lymphocytes (T cells). The lobules of the thymus are lined by epithelial cells that secrete hormones called *thymosins*. These hormones aid in the differentiation of T cells packed inside the lobules.

Pineal Gland

The **pineal gland,** which is located in the brain (see Fig. 20.1), produces the hormone **melatonin,** primarily at night. Melatonin is involved in our daily sleep-wake cycle. Normally, we grow sleepy at night when melatonin levels increase and awaken once daylight returns and melatonin levels are low. Daily 24-hour cycles such as this are called **circadian rhythms,** and they are controlled by a biological clock located in the hypothalamus, as discussed in the Health feature "Melatonin As a Nutritional Supplement."

Hormones from Other Tissues

Some organs that are usually not considered endocrine glands can secrete hormones. We have already mentioned in this chapter that the heart produces atrial natriuretic hormone. Figure 14.7 showed that the stomach and small intestine produce peptide hormones that regulate digestive secretions. Other tissues that are usually not considered endocrine glands also secrete hormones.

Leptin

Leptin is a protein hormone produced by adipose tissue. Leptin acts on the hypothalamus, where it signals satiety—that the individual has had enough to eat. Paradoxically, the blood of obese individuals may be rich in leptin. It is possible that the leptin they produce is ineffective because of a genetic mutation, or else their hypothalamic cells lack a suitable number of receptors for leptin.

Growth Factors

A number of different types of organs and cells produce **growth factors,** which stimulate cell division and mitosis. Like hormones, they act on cell types with specific receptors to receive them. Some, such as lymphokines, are released into the blood. Others diffuse to nearby cells. Growth factors of particular interest include:

Granulocyte-macrophage colony-stimulating factor is secreted by many different tissues. It causes bone marrow stem cells to form various types of white blood cells.
Platelet-derived growth factor is released from platelets and from many other cell types. It helps in wound healing and causes an increase in the number of fibroblasts, smooth muscle cells, and certain cells of the nervous system.

Epidermal growth factor and *nerve growth factor* stimulate the cells indicated by their names, as well as many others. These growth factors are also important in wound healing.

Prostaglandins

Prostaglandins are potent chemical signals produced within cells from *arachidonate,* a fatty acid. Prostaglandins are generally not distributed in the blood. Instead, they act locally, quite close to where they were produced. In the uterus, prostaglandins cause muscles to contract. Therefore, they are implicated in the pain and discomfort of menstruation in some women. Also, prostaglandins mediate the effects of *pyrogens,* chemicals believed to reset the temperature regulatory center in the brain. Aspirin reduces body temperature and controls pain because it inhibits the synthesis of prostaglandins.

Check Your Progress 20.6

1. Summarize the role of testosterone and estrogen in the body.
2. List some of the undesirable side effects of anabolic steroid use.
3. Describe the normal function of leptin.
4. Explain why prostaglandins are considered to be local hormones.

20.7 Disorders of the Endocrine System

Learning Outcomes

Upon completion of this section, you should be able to

1. Identify the cause of each of the following conditions: diabetes insipidus, pituitary dwarfism, gigantism, acromegaly, Cushing syndrome, and Addison disease.
2. Discuss the specific causes and likely outcomes of hypothyroidism and hyperthyroidism.
3. Compare and contrast type 1 versus type 2 diabetes mellitus.

The endocrine glands play a major role in regulating the development and function of many body systems. Therefore, an increase or decrease in the production of most hormones can cause significant disease. Increased production of a hormone is often due to cancer affecting the endocrine gland, while decreased production can be due to various conditions that result in destruction of the gland.

Disorders of the Pituitary Gland

Because it controls the secretions of several other endocrine glands, the pituitary has been called the "master gland" of the endocrine system. As such, disorders of the pituitary gland can have dramatic effects on the body. As discussed in Section 20.2, the posterior pituitary produces antidiuretic hormone (ADH), which reduces the amount of urine formed by increasing water reabsorption by the collecting ducts of the kidney (see Section 16.3). If too little ADH is secreted, a condition known as **diabetes insipidus** results. Patients with DI are usually very thirsty, produce large

SCIENCE IN YOUR LIFE ▶ HEALTH

Melatonin As a Nutritional Supplement

The hormone melatonin is now sold as a nutritional supplement. The popular press promotes its use in pill form for sleep, aging, cancer treatment, sexuality, and more. At best, melatonin may have some benefits in certain sleep disorders. But most physicians do not yet recommend it for that use because so little is known about its dosage requirements and possible side effects.

Melatonin is produced by the pineal gland in greatest quantity at night and smallest quantity during the day. Notice in Figure 20B that melatonin's production cycle accompanies our natural sleep-wake cycle. Rhythms with a period of about 24 hours are called circadian ("around the day") rhythms. All circadian rhythms seem to be controlled by an internal biological clock because they are free-running—that is, they have a regular cycle even in the absence of environmental cues. In scientific experiments, humans have lived in underground bunkers where they never see the light of day. In a few people, the sleep-wake cycle drifts badly, but in most, the daily activity schedule is just about 25 hours.

An individual's internal biological clock is reset each day by the environmental day-night cycle. Characteristically, biological clocks that control circadian rhythms are reset by environmental cues, or else they drift out of phase with the environmental day-night cycle.

Recent research suggests that our biological clock lies in a cluster of neurons within the hypothalamus, called the suprachiasmatic nucleus (SCN). The SCN undergoes spontaneous cyclical changes in activity, and therefore, it can act as a pacemaker for circadian rhythms. Neural connections between the retina and the SCN indicate that reception of light by the eyes most likely resets the SCN and keeps our biological rhythms on a 24-hour cycle. Some people suffer from seasonal affective disorder, or SAD. As the days get darker and darker during the fall and winter, they become depressed, sometimes severely. They find it difficult to keep up because their biological clock has fallen behind without early morning light to reset it. If so, a half-hour dose of simulated daylight from a portable light box first thing in the morning makes them feel operational again. The SCN also controls the secretion of melatonin by the pineal gland, and, in turn, melatonin may quiet the operation of the neurons in the SCN.

Research is still going forward to see if melatonin will be effective for circadian rhythm disorders such as SAD, jet lag, sleep phase problems, recurrent insomnia in the totally blind, and some other less common disorders. So-called jet lag occurs when you travel across several time zones and your biological clock is out of phase with local time. Jet-lag symptoms gradually disappear as your biological clock adjusts to the environmental signals of the local time zone.

Many young people have a sleep phase problem because their circadian cycle lasts 25 to 26 hours. As they lengthen their day, they get out of sync with normal times for sleep and activity. Medical researchers have speculated that melatonin could be used to adjust biological clocks; however, there is still much concern about possible side effects of melatonin supplementation.

An issue that highlights this concern is the widespread availability of melatonin-laced products like "relaxation brownies," as well as beverages with names like "Slowtivate" and "Marley's Mellow Mood." Although many such products bear label warnings such as "For adults only; not suitable for children," there is very little regulation of who can purchase these products, or even the

Figure 20C Packaging of products containing melatonin. ©Marcus Funston/ Ricochet Creative Productions LLC

levels of melatonin and other chemicals that are added.

An increasing number of states are moving to ban these products. In Arizona, after a 2-year-old was rushed to a hospital in 2011 after eating part of a brownie and falling into a deep sleep, the state department of health issued a mass recall of the product. Parents in other states have also complained about the type of packaging used for these products, which may use cartoon characters or brightly colored designs that some say are trying to target children in a not-so-subtle way (Fig. 20C).

Questions to Consider

1. Have you or someone you know ever experienced seasonal affective disorder? If so, do you think it was at least partly psychological, or did you feel it was a "purely" chemical disorder over which you had no control?
2. Are you more of a "morning person" or a "night person"? Can you think of any way that variations in melatonin secretion could account for how different people feel about, for example, getting up early in the morning?
3. Currently, melatonin is sold without a prescription in health food stores and drugstores in the United States, but a typical recommended dose is two to three times lower than that found in most "relaxation brownies." Do you see a difference between selling melatonin over the counter in health food stores versus selling melatonin-laced products in convenience stores? Should both be banned? Why or why not?

| 6 P.M.   6 A.M. | 6 P.M.   6 A.M. | 6 P.M.   6 A.M. |
| a. Experimental | b. Winter | c. Summer |

Figure 20B Melatonin production.
Melatonin production is greatest at night, when we are sleeping. Light suppresses melatonin production (a) so its duration is longer in the winter (b) than in the summer (c). (photo): ©Evelyn Jo Johnson

Figure 20.13 Growth hormone influences height.
Irregularities in growth hormone can lead to gigantism. ©Xinhua News/ Associated Press

Normal hand

Acromegaly hand

Figure 20.14 Acromegaly.   Acromegaly is caused by overproduction of GH in the adult. It is characterized by enlargement of the bones in the face, the hands, and the feet as a person ages. (normal hand): ©Bart's Medical Library/Medical Images; (acromegaly hand): ©Bart's Medical Library/ Medical Images; (man): ©Yasser Al-Zayyat/AFP/Getty Images

volumes of urine, and can become severely dehydrated if the condition is untreated. Another type of DI occurs when the kidneys are unable to respond to ADH produced by the pituitary. Consumption of alcohol also inhibits ADH production, such that the symptoms of a hangover are, in part, caused by dehydration.

The anterior pituitary produces several important hormones, including growth hormone (GH) and adrenocorticotropic hormone (ACTH). If too little GH is produced during childhood, the individual will have **pituitary dwarfism,** characterized by normal proportions but small stature. Conversely, if too much GH is secreted during childhood, the result may be **gigantism** (Fig. 20.13). People with gigantism usually have poor health, primarily because GH has a secondary effect on the blood glucose level, promoting diabetes mellitus (discussed later in this section).

On occasion, GH is overproduced in adulthood, resulting in a condition known as **acromegaly.** Because long bone growth is no longer possible in adults, only the feet, hands, and face (particularly the chin, nose, and eyebrow ridges) can grow, and these bones become overly large (Fig. 20.14).

Some pituitary tumors produce large amounts of ACTH, which is a common cause of **Cushing syndrome.** The excess ACTH stimulates the adrenal cortex to overproduce cortisol. The increased amount of cortisol causes muscle protein to be metabolized and subcutaneous fat to be deposited in the face and midsection (Fig. 20.15). Depending on the cause and duration of Cushing syndrome, some people may experience more dramatic changes, including masculinization, high blood pressure, and weight gain. Cushing

a.   b.

Figure 20.15 Cushing syndrome.   a. The tumors associated with adrenal hyperplasia often contribute to Cushing syndrome. **b.** A man showing some of the common characteristics of Cushing syndrome. (adrenal gland): ©McGraw-Hill Education; (man): ©Biophoto Associates/Science Source

syndrome can also be caused by tumors of the adrenal gland or by taking high or prolonged doses of cortisone or similar drugs.

Disorders of the Thyroid, Parathyroid, and Adrenal Glands

If the thyroid gland does not produce enough thyroid hormones for any reason, **hypothyroidism** occurs. Failure of the thyroid to function properly in infancy or childhood results in *congenital*

a. Congenital hypothyroidism

b. Simple goiter

affected eye
c. Exophthalmic goiter

Figure 20.16 Abnormalities of the thyroid gland. **a.** Individuals who develop hypothyroidism during infancy or childhood do not grow and develop as others do. **b.** An enlarged thyroid gland is often caused by a lack of iodine in the diet. Without iodine, the thyroid is unable to produce its hormones, and increased release of TSH from the anterior pituitary stimulation causes the gland to enlarge and form a simple goiter. **c.** In hyperthyroidism, accumulation of fluid behind the eyes can cause the eyes to protrude (exophthalmia). This individual's left eye was affected. (a): ©Medical-on-Line/Alamy; (b): ©Biophoto Associates/ Science Source; (c): ©Dr. P. Marazzi/SPL/Science Source

hypothyroidism (Fig. 20.16*a*). Individuals with this condition are short and stocky. Thyroid hormone therapy can initiate growth, but unless treatment is begun relatively quickly, stunted growth and intellectual disability may result. Hypothyroidism can also occur in adults, most often when the immune system produces antibodies that destroy the gland, a condition known as *Hashimoto thyroiditis.* Untreated hypothyroidism in adults may produce a group of clinical symptoms called *myxedema,* which is characterized by lethargy, weight gain, hair loss, constipation, slow heart rate, and thickened or puffy skin. The administration of adequate doses of thyroid hormones restores normal body functions and appearance.

If iodine is lacking in the diet, the thyroid gland is unable to produce sufficient amounts of T_3 and T_4. In response to constant stimulation by TSH released by the anterior pituitary, the thyroid gland enlarges, resulting in a *simple goiter* (Fig. 20.16*b*). The incidence of simple goiters was much higher in the United States prior to the addition of iodine to most salt.

Hyperthyroidism results from the oversecretion of thyroid hormones. In **Graves disease,** antibodies are produced that react with the TSH receptor on thyroid follicular cells, mimicking the effect of TSH and causing the production of too much T_3 and T_4. One typical sign of Graves disease is *exophthalmia,* or excessive protrusion of the eyes (Fig. 20.16*c*). One or both eyes protrude because of the edema in eye socket tissues and swelling of the muscles that move the eyes. The patient with Graves disease usually becomes hyperactive, nervous, and irritable, and may suffer from insomnia, diarrhea, and abnormal heart rhythms. Drugs that inhibit the synthesis or release of thyroid hormones, or their effects on body tissues, are used to manage Graves disease. Hyperthyroidism can also be caused by a thyroid tumor, which is usually detected as a lump during physical examination. Removal or destruction of a portion of the thyroid gland, either by surgery or by administration of radioactive iodine, is often curative.

When insufficient parathyroid hormone production leads to a dramatic drop in the blood calcium level, tetany results. In *tetany,* the body shakes from continuous muscle contraction, brought about by increased excitability of the nerves.

As mentioned previously, an increased production of cortisol by the adrenal glands is known as Cushing syndrome. In contrast, adrenal gland insufficiency is called **Addison disease.** Perhaps the most famous person to suffer from Addison disease was President John F. Kennedy. The most common cause of Addison disease in the United States is destruction of the adrenal cortex by the immune system. Clinical symptoms do not appear until about 90% of both adrenal cortexes have been destroyed. These symptoms are fairly nonspecific and may include weakness, weight loss, abdominal pain, mood disturbances, and hyperpigmentation of the skin (Fig. 20.17). Because the decreased production of mineralocorticoids can affect the balance of sodium and potassium, which in turn can have dramatic effects on the heart, Addison disease can be fatal unless properly treated by replacing the missing hormones.

Diabetes Mellitus

The National Diabetes Information Clearinghouse estimates that about 30.3 million Americans, or 9.4% of the population, have **diabetes mellitus,** often referred to simply as diabetes, a condition that affects their ability to regulate their glucose metabolism. People with diabetes either do not produce enough insulin (type 1) or cannot properly use the insulin they produce (type 2). In either case, although blood glucose levels rise, cellular famine exists in the midst of plenty. Some of the excess glucose in the blood is excreted into the urine, and water follows, causing the volume of urine to increase. Because the glucose in the blood cannot be used, the body turns to the metabolism of fat, which leads to the buildup of ketones in the blood. In turn, these ketones are metabolized to

Figure 20.17 Addison disease.
Addison disease is characterized by a peculiar bronzing of the skin, particularly noticeable in light-skinned individuals. Note the color of the hands compared to the hand of an individual without the disease. ©BSIP/Science Source

form various acids, which can build up in the blood (acidosis) and lead to coma and death.

The glucose tolerance test is often used to diagnose diabetes mellitus (see the chapter opener). After a person ingests a known amount of glucose, the blood glucose concentration is measured at intervals. In a diabetic, the blood glucose level usually rises greatly and remains elevated for hours (Fig. 20.18). In the meantime, glucose appears in the urine. In a nondiabetic, the blood glucose level rises somewhat, and then returns to normal after about 2 hours.

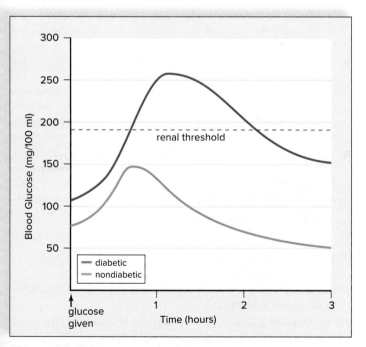

Figure 20.18 Glucose tolerance test.   Following the administration of 100 g of glucose, the blood glucose level rises dramatically in the diabetic and glucose appears in the urine. Also, the blood glucose level at 2 hours is equal to or more than 200 mg/100 dl.

About 10% of diabetics in the United States have *type 1 diabetes.* This condition usually begins in childhood or adolescence, and thus it is sometimes called juvenile-onset diabetes. However, type 1 diabetes can also occur later in life, following a viral infection, an autoimmune reaction, or an environmental agent that leads to destruction of the pancreatic islets. Because individuals with type 1 diabetes are suffering from an insulin shortage, treatment simply consists of providing the needed insulin through daily injections. These injections usually control the diabetic symptoms, but if too much insulin is administered, or a meal is missed, the result can be hypoglycemia (low blood sugar). Symptoms of hypoglycemia include perspiration, pale skin, shallow breathing, and anxiety. Because the brain requires a constant supply of glucose, unconsciousness can result. The treatment is quite simple: Immediate ingestion of a sugary snack or fruit juice can very quickly counteract hypoglycemia. Better control of glucose levels can sometimes be achieved with an insulin pump, a small device worn outside the body that is connected to a plastic catheter inserted under the skin. It is also possible to transplant a working pancreas, or even fetal pancreatic islet cells, into patients with type 1 diabetes. Some researchers believe that some form of artificial pancreas will become available in the next decade.

Most diabetics in the United States have *type 2 diabetes.* Often, the patient is overweight or obese, and adipose tissue may produce a substance that impairs insulin receptor function. Because type 2 diabetes usually doesn't occur before age 40, it was formerly called adult-onset diabetes. However, due to the increasing prevalence of obesity in children, type 2 diabetes is occurring at younger ages. Normally, the binding of insulin to its receptor on cell surfaces causes the number of glucose transporters to increase in the plasma membrane, but not in the type 2 diabetic. Treatment usually involves weight loss, which can sometimes control symptoms in the type 2 diabetic. However, many type 2 diabetics also have low insulin levels, so they may require injections of insulin, along with medications to increase the effectiveness of the insulin they produce.

Regardless of the type of diabetes they have, diabetics usually need to monitor their blood glucose levels several times a day. This is usually done by pricking a finger to obtain a drop of blood, which is tested with an external device. A U.S. company is currently testing a special tattoo ink that changes colors in response to glucose levels in the skin. The FDA has already approved a watch-like device that measures glucose levels in tiny amounts of fluid it painlessly extracts from the skin every 20 minutes. For insulin administration, insulin pumps are now replacing the need for injecting insulin with a needle and syringe.

Long-term complications of both types of diabetes are blindness, kidney disease, and cardiovascular disorders, including reduced circulation. The latter can lead to gangrene in the arms and legs. Diabetic women who become pregnant have an increased risk of diabetic coma, and the child of a diabetic is more likely to be stillborn or to die shortly after birth. These complications are uncommon, however, if the mother's blood glucose level is carefully regulated during pregnancy.

Check Your Progress 20.7

1. Explain how ADH is related to an increased urge to urinate when one drinks alcohol.
2. Describe why a goiter forms in iodine deficiency.
3. Differentiate the role of antibodies in causing Graves disease versus Addison disease.
4. Describe the causes and treatment of type 1 and type 2 diabetes.

Conclusion

Over time, Ethan has learned to manage his condition better, even though he feels a sense of resigned frustration that he may always have to deal with type 1 diabetes. He recently began using an insulin pump, which eliminated the need for daily insulin injections, although he has to insert his infusion catheter (through which the pump dispenses insulin as needed) at a different site every couple of days. He also must remember to input his daily intake of carbohydrates into the machine, and he still must check his glucose levels in a drop of blood extracted from a finger-stick before eating, exercising, or going to bed at night.

Continuous glucose monitors have been available for almost a decade. Most of these devices include a probe that must be inserted under the skin, and a receiver that can display blood glucose readings taken hundreds of times a day. However, most of these machines still require checking blood from a finger-stick several times a day. Newer technologies, such as a remote glucose monitor, are allowing individuals to monitor and record blood glucose readings from a distance. Research is also underway to develop an artificial pancreas that has the ability to not only monitor blood glucose levels, but deliver the precise doses of insulin without the diabetic being aware. These technologies will help diabetics, like Ethan, to manage their disease more effectively.

SUMMARIZE

20.1 Overview of the Endocrine System

The **endocrine system** consists of **endocrine glands** that produce hormones. **Hormones** are chemical signals that affect other tissues, often at a distance. **Negative feedback** and antagonistic hormonal actions usually control the secretion of hormones. **Growth factors** are chemical signals that affect cell growth.

■ Hormones are either peptides or steroids. Binding of a **peptide hormone** to a receptor at the plasma membrane typically activates an enzyme cascade inside the cell. A **steroid hormone** combines with a receptor in the cell, and that complex activates transcription of certain genes.

■ A peptide hormone typically acts as a **first messenger.** This means the hormone binds to a cellular receptor, triggering activation of a **second messenger** inside the cell.

■ A **pheromone** is a chemical signal that affects other individuals. There is some evidence for pheromone activity in humans.

20.2 Hypothalamus and Pituitary Gland

The **hypothalamus** controls hormone secretion by the **pituitary gland** through two distinct mechanisms:

■ The hypothalamus itself synthesizes **antidiuretic hormone (ADH)** and **oxytocin,** which are stored in the **posterior pituitary** until they are released. In contrast to most hormones, the release of oxytocin is regulated by **positive feedback.**

■ The hypothalamus communicates with the **anterior pituitary** via a portal system, through which **hypothalamic-releasing hormones** and **hypothalamic-inhibiting hormones** pass.

■ Several hormones produced by the anterior pituitary stimulate other hormonal glands: **thyroid-stimulating hormone (TSH)** acts on the thyroid gland; **adrenocorticotropic hormone (ACTH)** stimulates the adrenal cortex; and the **gonadotropic hormones, follicle-stimulating hormone (FSH)** and **luteinizing hormone (LH)** stimulate the gonads.

■ Three other hormones are produced by the anterior pituitary. **Prolactin (PRL)** causes mammary gland development after childbirth; **melanocyte-stimulating hormone (MSH)** affects skin color in many animals; and **growth hormone (GH)** promotes skeletal and muscular growth.

20.3 Thyroid and Parathyroid Glands

The **thyroid gland** produces three hormones:

■ **Thyroxine** (T_4) and triiodothyronine (T_3) increase the metabolic rate of most cells in the body.

■ **Calcitonin** helps lower the blood calcium level by reducing the activity of osteoclasts in the bone.

The **parathyroid glands** secrete **parathyroid hormone (PTH)**, which raises the blood calcium, and decreases the blood phosphate levels.

20.4 Adrenal Glands

The paired **adrenal glands** respond to stress. The adrenal medulla is controlled directly by the nervous system, allowing an almost immediate response. The adrenal cortex responds on a longer-term basis to ACTH released by the anterior pituitary:

- The **adrenal medulla** secretes **epinephrine** and **norepinephrine,** which bring about fight-or-flight responses we associate with emergency situations.
- The **adrenal cortex** produces the **glucocorticoids** (e.g., cortisol) and the **mineralocorticoids** (e.g., aldosterone). **Cortisol** stimulates the conversion of amino acids to glucose, raising the blood glucose level. It can also suppress inflammatory responses. **Aldosterone** causes the kidneys to reabsorb sodium ions (Na^+) and to excrete potassium ions (K^+), which raises the blood pressure. The release of aldosterone is inhibited by **atrial natriuretic hormone (ANH)**, released by cells in the atrium of the heart.

20.5 Pancreas

The **pancreas** has both exocrine and endocrine functions. The **pancreatic islets** secrete three hormones:

- **Insulin** lowers the blood glucose level by stimulating the uptake of glucose by various types of cells.
- **Glucagon** raises the blood glucose level.
- **Somatostatin** inhibits growth hormone release and decreases absorption of nutrients from the digestive tract.

20.6 Other Endocrine Glands

Other organs and tissues produce hormones:

- The male **gonads (testes)** produce **androgens** (male sex hormones) like **testosterone,** while in the female the **ovaries** produce **estrogen** and **progesterone.** Athletes who abuse **anabolic steroids,** due to their influence on muscle-building, risk serious side effects.
- The **thymus** secretes thymosins, which stimulate T-lymphocyte production and maturation.
- The **pineal gland** produces **melatonin,** which may be involved in **circadian rhythms** and reproductive organ development.
- Tissues also produce hormones. Adipose tissue produces **leptin,** which acts on the hypothalamus, and various tissues produce **growth factors** that stimulate cell division. **Prostaglandins** act locally, near where they are produced.

20.7 Disorders of the Endocrine System

Increases or decreases in the amounts of hormones synthesized by endocrine glands cause several common disorders:

- Disorders caused by pituitary gland dysfunction include **diabetes insipidus** (insufficient ADH), **pituitary dwarfism** (GH deficiency), **gigantism** (too much GH in child), **acromegaly** (too much GH in adult), and **Cushing syndrome** (too much ACTH).
- **Hypothyroidism** is the result of insufficient thyroid hormones, while **hyperthyroidism** is caused by an overproduction of thyroid hormones. **Graves disease** is a type of hyperthyroidism.
- **Addison disease** is a serious condition in which the adrenal cortexes have been damaged, often by an autoimmune process.
- **Diabetes mellitus** is a common condition affecting the body's ability to metabolize glucose. In type 1 diabetes, the pancreas doesn't produce enough insulin. In type 2 diabetes, insulin receptors do not function properly. Insulin injections may be required for both types of diabetes, but type 2 diabetes can be more difficult to treat, resulting in long-term complications.

ASSESS

Testing Yourself

Choose the best answer for each question.

20.1 Overview of the Endocrine System

1. Name the indicated endocrine glands in the image below:

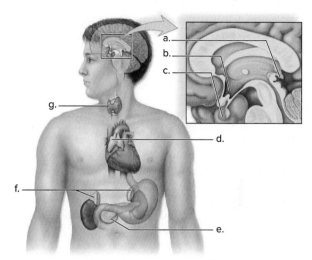

2. Peptide hormones interact with which structures on the surface of a cell?
   a. receptor proteins
   b. second messenger systems
   c. pheromones
   d. neurotransmitters
   e. digestive enzymes

3. A _____ requires the use of a second messenger system to enter a cell.
   a. steroid hormone
   b. peptide hormone
   c. pheromone
   d. All of these are correct.

20.2 Hypothalamus and Pituitary Gland

4. Which of the following acts as the link between the nervous system and the endocrine system?
   a. posterior pituitary gland
   b. hypothalamus
   c. anterior pituitary gland
   d. parathyroid

5. Growth hormone (GH) is released by which endocrine gland?
   a. posterior pituitary gland
   b. hypothalamus
   c. anterior pituitary gland
   d. adrenal gland

6. Which of the following hormones is regulated by positive feedback mechanisms?
   a. thyroid-stimulating hormone (TSH)
   b. gonadotropic hormone
   c. oxytocin
   d. growth hormone
   e. None of these are correct.

20.3 Thyroid and Parathyroid Glands

7. Thyroid hormones directly regulate which aspect of human physiology?
   a. circadian rhythm
   b. stress response
   c. sex hormone production
   d. metabolic rate
   e. None of these are correct.

8. Which of the following hormones increases blood calcium levels?
   a. calcitonin
   b. thyroxine (T_4)
   c. parathyroid hormone
   d. mineralocorticoid

20.4 Adrenal Glands

9. Which of the following hormones is (are) not produced by the adrenal cortex?
   a. glucocorticoids
   b. mineralocorticoids
   c. gonadocorticoids
   d. norepinephrine

10. Which of the following is not correct regarding aldosterone?
    a. It is produced by the adrenal cortex.
    b. It is inhibited by the action of epinephrine.
    c. Its release is regulated by renin from the kidneys.
    d. It causes the kidneys to reabsorb sodium (Na+) ions.
    e. All of these are correct.

20.5 Pancreas

11. Which of the following correctly describes the hormone insulin?
    a. It is produced by B cells in the pancreas.
    b. It increases glucose uptake by liver and muscle cells.
    c. It is a peptide hormone.
    d. It lowers blood glucose levels.
    e. All of these are correct.

12. The hormone antagonistic to insulin is
    a. epinephrine.
    b. parathyroid hormone.
    c. glucagon.
    d. cortisol.
    e. progesterone.

20.6 Other Endocrine Glands

13. The hormone produced by the pineal gland to regulate the circadian rhythm is called
    a. estradiol.
    b. melatonin.
    c. renin.
    d. aldosterone.
    e. None of these are correct.

20.7 Disorders of the Endocrine System

14. The disease that is believed to be caused by an autoimmune response that destroys the pancreatic islets is called
    a. Addison disease.
    b. diabetes insipidus.
    c. type 1 diabetes mellitus.
    d. type 2 diabetes mellitus.
    e. gestational diabetes.

15. Which hormone and condition are mismatched?
    a. growth hormone—acromegaly
    b. thyroxine—goiter
    c. parathyroid hormone—tetany
    d. cortisol—myxedema
    e. insulin—diabetes

ENGAGE

BioNOW

Want to know how this science is relevant to your life? Check out the BioNOW video below:

■ Quail Hormones

From your understanding of the endocrine system in humans, which endocrine glands and hormones are most likely involved in this response of the quails to the changes in their environment?

Thinking Critically

1. Because some of their functions overlap, why is it necessary to have both a nervous system and an endocrine system?

2. Certain endocrine disorders, such as Cushing syndrome, can be caused by excessive secretion of a hormone (in this case, ACTH) by the pituitary gland, or by a problem with the adrenal gland itself. If you were able to measure the ACTH levels of a Cushing patient, how could you tell the difference between a pituitary problem and a primary adrenal gland problem?

3. In animals, pheromones can affect many different behaviors. Because humans do produce a number of airborne chemicals, what potential human behaviors could be affected by these chemicals? How would these pheromones be perceived, and would we necessarily be conscious of their effects on us?

©MedicalRF.com

C H A P T E R

21

Reproductive System

BEFORE YOU BEGIN

Before beginning this chapter, take a few moments to review the following discussions:

Section 5.4 How are human gametes produced?

Section 20.1 What is the difference between a steroid hormone and a peptide hormone?

Section 20.6 Where are the male and female sex hormones produced, and what are their functions?

Three Parents—One Baby

In vitro fertilization (IVF) technology has been used for many years to help couples conceive a child. The process involves taking an egg from the mother and fertilizing it with paternal sperm to create embryos that are then implanted into the mother (or a surrogate) to carry the child to term. A new technique has been proposed that allows for a donor egg to be used in cases where the mother's cells contain defective genes in the mitochondria.

Mitochondria are organelles inside a cell that generate ATP energy the cell requires for its metabolism. These organelles contain genetic material (DNA) that codes for 37 genes, 14 of which are related to specific proteins related to efficient ATP production. Any defect in these genes leads to serious conditions like Alpers disease and various mitochondrial myopathies (muscle disorders). While there are numerous other genes found in the nucleus related to mitochondria, it is this smaller set of genes found in the mitochondria themselves that have become a matter of controversy. At issue is an innovative therapy to prevent genetic diseases directly tied to these mitochondrial genes.

A new form of in vitro fertilization creates what is sometimes called a "three-parent baby" because a third party donates an egg to be fertilized with a maternal nucleus and paternal sperm. The donor's nuclear genes are removed but the genes inside the mitochondria remain behind. The resulting baby has all of the nuclear genes of the mother and father but the 37 genes found in the mitochondria belong to the donor of the original egg cell.

In this chapter, we will explore the structure and function of the reproductive system in humans, as well as some of the diseases that may cause individuals to need processes such as IVF.

As you read through this chapter, think about the following questions:

1. Which sex normally contributes the mitochondrial genes to the offspring?

2. How does IVF differ from other assisted reproductive technologies?

Used with permission from Ricochet Creative Productions ©2015.

21.1 Male Reproductive System

Learning Outcomes

Upon completion of this section, you should be able to
1. Describe the path of sperm from the testes to the urethra.
2. List the organs that produce components of seminal fluid.
3. Analyze the role of gonadotropin-releasing hormone, luteinizing hormone, follicle-stimulating hormone, and testosterone in male sexual reproduction.

TABLE 21.1   Male Reproductive Structures

Organ	Function
Testes	Produce sperm and sex hormones
Epididymides	Ducts where sperm mature and are stored
Vasa deferentia	Conduct and also store sperm
Seminal vesicles	Contribute nutrients and fluid to semen
Prostate gland	Contributes basic fluid to semen
Urethra	Conducts sperm
Bulbourethral glands	Contribute viscous fluid to semen
Penis	Organ of sexual intercourse

Sexual reproduction involves the production of **gametes,** haploid sex cells that become united during fertilization to produce a diploid organism (see Section 5.4). In humans, the production of gametes is the role of the **reproductive system.** For humans, these systems are different between males and females. In this section, we will explore the structure and function of the male reproductive system, which includes the organs depicted in Figure 21.1 and listed in Table 21.1.

Genital Tract

The male gonads are paired **testes** (*sing.,* testis). Sperm produced by the testes mature within the epididymides (*sing.,* **epididymis**), which are tightly coiled ducts lying outside the testes. Maturation seems to be required in order for sperm to swim to the egg. When sperm leave an epididymis, they enter a **vas deferens** (*pl.,* vasa deferentia), where they may also be stored for a time. Each vas deferens passes into the abdominal cavity, where it curves around the urinary bladder and empties into an ejaculatory duct. The ejaculatory ducts connect to the **urethra.**

At the time of ejaculation, sperm leave the penis in a fluid called **semen (seminal fluid).** Three glands add secretions to semen. The paired **seminal vesicles** lie at the base of the bladder, and each has a duct that joins with a vas deferens. The **prostate gland** is a single, donut-shaped gland that surrounds the upper portion of the urethra just below the urinary bladder. **Bulbourethral glands** (Cowper glands) are pea-sized organs that lie underneath the prostate on either side of the urethra.

Each component of semen seems to have a particular function. Sperm are more viable in a basic solution, and semen, which is

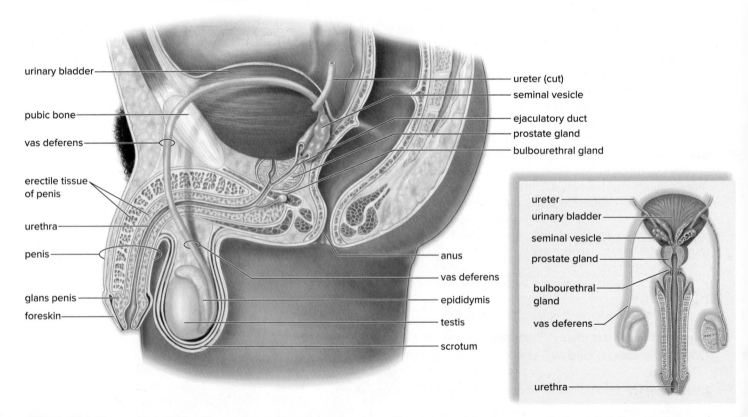

Figure 21.1 The male reproductive system.   The testes produce sperm. The seminal vesicles, the prostate gland, and the bulbourethral glands provide a fluid medium for the sperm, which move from the vasa deferens through the ejaculatory ducts to the urethra in the penis.

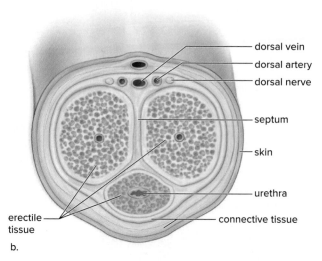

Figure 21.2 Penis anatomy. **a.** Beneath the skin and the connective tissue lies the urethra, surrounded by erectile tissue. This tissue expands to form the glans penis, which in uncircumcised males is partially covered by the foreskin (prepuce). **b.** Two other columns of erectile tissue in the penis are located dorsally.

milky in appearance, has a slightly basic pH (about 7.5). Swimming sperm require energy, and semen contains the sugar fructose, which serves as an energy source. Semen also contains *prostaglandins,* local hormones that cause the uterus to contract. These contractions help propel the sperm toward the egg.

The **penis** (Fig. 21.2) is the male organ of sexual intercourse. The penis has a long shaft and an enlarged tip called the *glans penis.* At birth, the glans penis is covered by a layer of skin called the *foreskin.* **Circumcision,** the surgical removal of the foreskin, is usually done for religious reasons or perceived health benefits, including easier hygiene, decreased urinary tract infections, and reduced risk of penile cancer. A study of 3,000 Ugandan men showed that compared to uncircumcised men, circumcised men were 25–35% less likely to contract HIV, human papillomavirus, or genital herpes. However, opponents of the procedure focus on the risks, including pain and potential loss of sensitivity in the penis, and the American Academy of Pediatrics' position is that the benefits are not sufficient to recommend routine neonatal circumcision.

Erection and Orgasm in Males

Spongy, erectile tissue containing distensible blood spaces extends through the shaft of the penis. During sexual arousal, autonomic nerves release nitric oxide (NO). This stimulus leads to the production of cGMP (cyclic guanosine monophosphate), a chemical signal that causes the smooth muscle of incoming arterial walls to relax and the erectile tissue to fill with blood. The veins that take blood away from the penis are compressed, and the penis becomes erect. **Erectile dysfunction (ED)** (also sometimes called impotency) is the inability to achieve or maintain an erection suitable for sexual intercourse. ED may be caused by a number of factors, including poor blood flow, certain medications, and many illnesses. Medications for the treatment of erectile dysfunction inhibit the enzyme that breaks down cGMP, ensuring that a full erection will take place. Some

of these medications can cause vision problems, because the same enzyme occurs in the retina.

When sexual stimulation intensifies, sperm enter the urethra from the vasa deferentia, and the accessory glands contribute secretions to the semen. Once semen is in the urethra, rhythmic muscle contractions cause it to be ejaculated from the penis in spurts. During *ejaculation,* a sphincter normally closes off the urinary bladder so that no urine enters the urethra, and no semen enters the bladder. (Notice that the urethra carries urine or semen at different times.)

The contractions that expel semen from the penis are a part of male *orgasm,* the physiological and psychological sensations that occur at the climax of sexual stimulation. The psychological sensation of pleasure is centered in the brain, but the physiological reactions involve the genital (reproductive) organs and associated muscles, as well as the entire body.

Following ejaculation and/or loss of sexual arousal, the penis returns to its normal flaccid state. After ejaculation, a male typically experiences a period of time, called the *refractory period,* during which stimulation does not bring about an erection. The length of the refractory period may be from minutes to hours, although it tends to increase with age.

There may be in excess of 400 million sperm in the approximately 3.5 ml of semen expelled during ejaculation. However, the sperm count can be much lower than this, and fertilization of the egg by a sperm still can take place.

Male Gonads, the Testes

The testes, which produce sperm as well as the male sex hormones, lie outside the abdominal cavity of the human male, within the saclike **scrotum.** The testes begin their development inside the abdominal cavity but descend into the scrotal sacs during the last 2 months of fetal development. If the testes fail to descend properly, and thus remain in the abdomen, male infertility may result. This is because the internal temperature inactivates some of the

c. Spermatogenesis

Figure 21.3 Testis and spermatogenesis. **a.** The lobules of a testis contain seminiferous tubules. **b.** Light micrograph of a cross-section of the seminiferous tubules, where spermatogenesis occurs. Note the location of interstitial cells in clumps among the seminiferous tubules. **c.** Diagrammatic representation of spermatogenesis, which occurs in the walls of the tubules. **d.** A sperm has a head, a middle piece, a tail, and an end piece. The nucleus is in the head, which is capped by the enzyme-containing acrosome. (b): ©Ed Reschke

enzymes necessary to produce viable sperm. This condition can usually be surgically corrected, however. The scrotum helps regulate the temperature of the testes by holding them closer to or farther away from the body. In fact, any activity that increases testicular temperature, such as taking hot baths, using a laptop computer, or even sitting for extended periods of time, may decrease sperm production.

Seminiferous Tubules

A longitudinal section of a testis shows that it is composed of compartments called *lobules,* each of which contains one to three tightly coiled **seminiferous tubules** (Fig. 21.3a). These tubules have a combined length of approximately 250 meters. A microscopic

cross-section of a seminiferous tubule reveals that it is packed with cells undergoing **spermatogenesis** (Fig. 21.3b, c), the production of sperm. Newly formed *spermatogonia* (*sing.,* spermatogonium) move away from the outer wall and become *primary spermatocytes* that undergo meiosis I to produce *secondary spermatocytes* with 23 chromosomes. Secondary spermatocytes undergo meiosis II to produce four *spermatids* that are also haploid. Spermatids then differentiate into sperm. Also present are **Sertoli cells** (sustentacular cells), which support, nourish, and regulate the spermatogenic cells.

Sperm, or spermatozoa, have three main parts: a head, a middle piece, and a tail (Fig. 21.3d). Mitochondria in the middle piece provide energy for the movement of the tail, which has the structure of a flagellum (see Fig. 3.15). The head contains a nucleus covered by a cap called the *acrosome,* which stores enzymes needed to penetrate the egg. The ejaculated semen of a normal human male contains several hundred million sperm, but only one sperm normally enters an egg. Sperm may live between 3 and 5 days in the female genital tract.

The male sex hormones, the androgens, are secreted by cells that lie between the seminiferous tubules. Therefore, they are called **interstitial cells** (Fig. 21.3*b*). The most important of the androgens is testosterone, whose functions are discussed next.

Hormonal Regulation in Males

The hypothalamus has ultimate control of the testes' sexual function because it secretes a hormone called **gonadotropin-releasing hormone (GnRH)** that stimulates the anterior pituitary to secrete the gonadotropic hormones. There are two gonadotropic hormones—**follicle-stimulating hormone (FSH)** and **luteinizing hormone (LH)**—in both males and females. In males, FSH promotes the production of sperm in the seminiferous tubules, which also release the hormone *inhibin*. Inhibin, as its name suggests, inhibits further FSH synthesis (Fig. 21.4).

Testosterone, the main sex hormone in males, is essential for the normal development and functioning of the organs listed in Table 21.1. Testosterone also brings about and maintains the male

secondary sex characteristics that develop at the time of puberty. Males are generally taller than females and have broader shoulders and longer legs relative to trunk length. The deeper voices of males compared to those of females are due to a larger larynx with longer vocal cords. Because the Adam's apple (laryngeal prominence) is a part of the larynx, it is usually more prominent in males than in females. Testosterone causes males to develop noticeable hair on the face, chest, and occasionally other regions of the body, such as the back. Testosterone also leads to the receding hairline and pattern baldness that occur in aging males.

Testosterone is responsible for the greater muscular development of males. Knowing this, both males and females sometimes take *anabolic steroids,* which are either testosterone or related steroid hormones resembling testosterone. Health problems involving the kidneys and the cardiovascular system, as well as hormonal imbalances, can arise from such use. In males, the testes shrink in size, and feminization of other male traits occurs (see Fig. 20.12).

Check Your Progress 21.1

1. Compare the functions of the seminiferous tubules, interstitial cells, epididymis, vasa deferentia, and urethra.
2. Describe the process of spermatogenesis.
3. Explain the role of the following hormones in male reproduction: GnRH, FSH, LH, and testosterone.

21.2 Female Reproductive System

Learning Outcomes

Upon completion of this section, you should be able to

1. Identify where an oocyte is produced and how it is transported to the uterus.
2. Describe the major components of the female external genitalia.
3. Describe the events that occur during a female orgasm.

The female reproductive system includes the organs depicted in Figure 21.5 and listed in Table 21.2. The female gonads are paired **ovaries** that lie in shallow depressions, one on each side of the upper pelvic cavity. **Oogenesis** is the production of an egg, or **oocyte,** the female gamete (see Section 21.3). The ovaries usually

Figure 21.4 Hormonal control of testes.   GnRH stimulates the anterior pituitary to produce FSH and LH. FSH stimulates the testes to produce sperm, and LH stimulates the testes to produce testosterone. Testosterone from interstitial cells and inhibin from the seminiferous tubules exert negative feedback control over the hypothalamus and the anterior pituitary, and this ultimately regulates the level of testosterone in the blood.

TABLE 21.2	Female Reproductive Structures
Organ	**Function**
Ovaries	Produce oocyte and sex hormones
Uterine tubes (oviducts or fallopian tubes)	Conduct oocyte; location of fertilization; transport early zygote
Fimbriae	Movement of oocyte within the uterine tubes
Uterus (womb)	Houses developing fetus
Cervix	Contains opening to uterus
Vagina	Receives penis during sexual intercourse; serves as birth canal and as the exit for menstrual flow

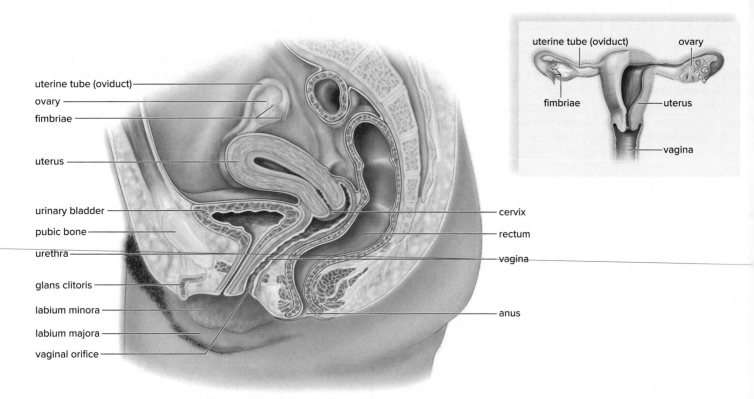

Figure 21.5 The female reproductive system.   The ovaries release one egg a month; fertilization occurs in the uterine tube, and development occurs in the uterus. The vagina is the birth canal, as well as the organ of sexual intercourse and the outlet for menstrual flow.

alternate in producing one oocyte per month. **Ovulation** is the process by which an oocyte bursts from an ovary and usually enters a uterine tube.

The Genital Tract

The **uterine tubes,** also called oviducts or fallopian tubes, extend from the uterus to the ovaries. However, the uterine tubes are not attached to the ovaries. Instead, they have fingerlike projections called **fimbriae** (*sing.,* fimbria) that sweep over the ovaries. When an oocyte bursts from an ovary during ovulation, it usually is swept into a uterine tube by the combined action of the fimbriae and the beating of cilia that line the uterine tubes.

Once in the uterine tube, the oocyte is propelled slowly by ciliary movement and tubular muscle contraction toward the uterus. An oocyte lives only approximately 6 to 24 hours unless fertilization occurs. Fertilization, and therefore formation of a **zygote,** usually takes place in the uterine tube. The developing embryo normally arrives at the uterus after several days and then embeds, or *implants,* itself in the uterine lining, which has been prepared to receive it. Occasionally however, the embryo may implant in the uterine tube or elsewhere, resulting in an *ectopic pregnancy.*

The **uterus** is a thick-walled, muscular organ about the size and shape of an inverted pear. The uterine tubes join the uterus at its upper end, while at its lower end, the **cervix** connects with the vagina nearly at a right angle (Fig. 21.5).

Development of the embryo normally takes place in the uterus. This organ, sometimes called the womb, is approximately

5 cm wide in its usual state but is capable of stretching to over 30 cm wide to accommodate a growing baby. The lining of the uterus, called the **endometrium,** participates in the formation of the placenta (see Section 21.3), which supplies nutrients needed for embryonic and fetal development. In the nonpregnant female, the functional layer of the endometrium varies in thickness according to a monthly cycle of events called the uterine cycle.

A small opening in the cervix leads to the vaginal canal. The **vagina** is a tube that lies at a 45-degree angle to the small of the back. The mucosal lining of the vagina lies in folds and can extend. This is especially important when the vagina serves as the birth canal, and it facilitates sexual intercourse when the vagina receives the penis. The vagina also acts as the exit for menstrual flow.

External Genitals

The external genital organs of the female are known collectively as the **vulva** (Fig. 21.6). The vulva includes two large, hair-covered folds of skin called the *labia majora* (*sing.,* labium major). The labia majora extend backward from the *mons pubis,* a fatty prominence underlying the pubic hair. The *labia minora* are two small folds lying just inside the labia majora. The *glans clitoris* is the external portion of the **clitoris,** the organ of sexual arousal in females. Like the penis, the clitoris contains a shaft of erectile tissue that becomes engorged with blood during sexual stimulation.

The cleft between the labia minora contains the openings of the urethra and the vagina. The vagina may be partially closed by a ring of tissue called the *hymen.* The hymen is ordinarily ruptured by sexual intercourse or by other types of physical activity.

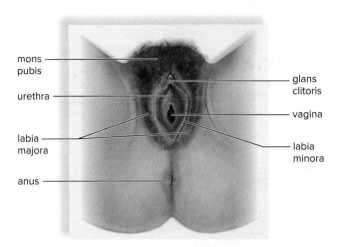

Figure 21.6 External genitals of the female. The opening of the vagina is partially blocked by a membrane called the hymen. Physical activities and sexual intercourse disrupt the hymen.

If remnants of the hymen persist after sexual intercourse, they can be surgically removed.

Notice that the urinary and reproductive systems in the female are entirely separate. The urethra carries only urine, and the vagina serves only as the birth canal and the organ for sexual intercourse.

Orgasm in Females

Upon sexual stimulation, the labia minora, the vaginal wall, and the clitoris become engorged with blood. The breasts also swell, and the nipples become erect. The labia majora enlarge, redden, and spread away from the vaginal opening.

The vagina expands and elongates. Blood vessels in the vaginal wall release small droplets of fluid that seep into the vagina and lubricate it. Mucus-secreting glands beneath the labia minora on either side of the vagina also provide lubrication for entry of the penis into the vagina. Although the vagina is the organ of sexual intercourse in females, the extremely sensitive clitoris plays a significant role in the female sexual response. The thrusting of the penis and the pressure of the pubic symphyses of the partners acts to stimulate the clitoris, which may swell to two or three times its usual size.

Orgasm occurs at the height of the sexual response. Blood pressure and pulse rate rise, breathing quickens, and the walls of the vagina, uterus, and uterine tubes contract rhythmically. A sensation of intense pleasure is followed by relaxation when organs return to their normal size. Females have little or no refractory period between orgasms, and multiple orgasms can occur during a single sexual experience.

Check Your Progress 21.2

1. Explain how a tubal ligation, or cutting and sealing off the uterine tubes, prevents pregnancy.
2. Explain how males and females differ in the specialization of their genital tract for reproduction versus urination.
3. Describe the event(s) that occur(s) during a female orgasm that may increase the chances of fertilization.

21.3 Ovarian and Uterine Cycles

Learning Outcomes

Upon completion of this section, you should be able to

1. List the stages of the ovarian cycle and explain what is occurring in each stage.
2. Describe the process of oogenesis.
3. Summarize how estrogen and progesterone affect the uterus.
4. Describe changes that occur in the uterus during menstruation, pregnancy, and menopause.

Hormone levels cycle in the female on a monthly basis, and the ovarian cycle drives the uterine cycle.

The Ovarian Cycle

A longitudinal section through an ovary shows it is made up of an outer *cortex* and an inner *medulla* (Fig. 21.7a). In the cortex are many **follicles,** each one containing an immature oocyte. A female is born with all the ovarian follicles she will ever have, an estimated 700,000. However, only about 400 of these follicles will ever mature because a female usually produces only one oocyte per month during her reproductive years. Because oocytes are present at birth, they age as the woman ages. This may be one reason older women are more likely to produce children with genetic defects.

The **ovarian cycle** occurs as a follicle changes from a primary to a secondary to a vesicular (Graafian) follicle (Fig. 21.7a). Epithelial cells of a *primary follicle* surround a primary oocyte. Pools of follicular fluid surround the oocyte in a *secondary follicle.* In a *vesicular follicle,* a fluid-filled cavity increases to the point that the follicle wall balloons out on the surface of the ovary.

As a follicle matures, oogenesis, depicted in Figure 21.7b, is initiated and continues. The *primary oocyte* divides, producing two haploid cells. One cell is a *secondary oocyte,* and the other is a *polar body.* The vesicular follicle bursts, releasing the secondary oocyte. This process is referred to as ovulation. Once a vesicular follicle has lost the secondary oocyte, it develops into a **corpus luteum,** a glandlike structure that produces progesterone and some estrogen.

The secondary oocyte enters a uterine tube. If a sperm enters the secondary oocyte, fertilization occurs, and the secondary oocyte completes meiosis. An egg with 23 chromosomes and a second polar body results. When the sperm nucleus unites with the egg nucleus, a zygote with 46 chromosomes is produced. If zygote formation and pregnancy do not occur, the corpus luteum begins to degenerate after about 10 days.

Phases of the Ovarian Cycle

The ovarian cycle is commonly divided into two phases. The first half of the cycle is called the follicular phase, and the second half is the luteal phase. During the *follicular phase,* follicle-stimulating hormone (FSH), produced by the anterior pituitary, promotes the development of a follicle in the ovary, which secretes **estrogen** and some **progesterone** (Fig. 21.8). As the estrogen level in the blood rises, it exerts negative feedback control over the anterior pituitary secretion of FSH so that the follicular phase comes to an end.

2. The secondary follicle contains a primary oocyte and produces the sex hormones estrogen and some progesterone.

3. Vesicular (Graafian) follicle develops.

50×

secondary follicle   oocyte   vesicular (Graafian) follicle

primary follicles

1. A primary follicle contains an oocyte and begins producing the sex hormone estrogen.

6. Corpus luteum degenerates.

a.

corpus luteum

5. Corpus luteum produces the sex hormones progesterone and some estrogen.

secondary oocyte

4. Ovulation: The secondary oocyte is released.

First polar body

sperm

Second polar body

Sperm nucleus and egg nucleus fuse; zygote with 46 chromosomes results.

primary oocyte (46 chromosomes)

→ meiosis I →

secondary oocyte (23 chromosomes)

→ meiosis II →

zygote

b.

Figure 21.7 Ovarian cycle. **a.** A single follicle goes through six stages in one place within the ovary. As a follicle matures, layers of follicle cells surround the developing oocyte. Eventually, the mature follicle ruptures, and the secondary oocyte is released. The follicle then becomes the corpus luteum, which eventually disintegrates. **b.** During oogenesis, the chromosome number is reduced from 46 to 23. Fertilization restores the full number of chromosomes.
(photo) (a): ©Ed Reschke/Getty Images

Presumably, an estrogen spike causes a sudden secretion of a large amount of GnRH from the hypothalamus. This leads to a surge of luteinizing hormone (LH) from the anterior pituitary, which causes ovulation at about the 14th day of a 28-day cycle.

Now, the *luteal phase* begins. During this phase, LH promotes the development of the corpus luteum, which secretes progesterone and some estrogen. As the blood level of progesterone rises, it exerts negative feedback control over the anterior pituitary secretion of LH so that the corpus luteum in the ovary begins to degenerate. As the luteal phase comes to an end, the low levels of progesterone and estrogen in the body cause menstruation to begin, as discussed next.

The Uterine Cycle

The female sex hormones estrogen and progesterone have numerous functions. One of their functions is to affect the endometrium, causing the uterus to undergo a cyclical series of events known as the **uterine cycle** (Fig. 21.9, *bottom*). Twenty-eight-day cycles are divided as follows:

During *days 1–5,* a low level of female sex hormones in the body causes the endometrium to disintegrate and its blood vessels to rupture. On days 1 to 5 of the cycle, a flow of blood and tissues passes out of the vagina during menstruation, also called the menstrual period.

Figure 21.8 Hormonal control of ovaries. The hypothalamus produces GnRH, which stimulates the anterior pituitary to produce FSH and LH. FSH stimulates the follicle to produce primarily estrogen, and LH stimulates the corpus luteum to produce primarily progesterone. Estrogen and progesterone maintain the sex organs (e.g., uterus) and the secondary sex characteristics, and they exert feedback control over the hypothalamus and the anterior pituitary. Feedback control regulates the relative amounts of estrogen and progesterone in the blood.

During *days 6–13,* increased production of estrogen by a new ovarian follicle in the ovary causes the endometrium to thicken and become vascular and glandular. This is called the *proliferative phase* of the uterine cycle.

On *day 14* of a 28-day cycle, ovulation usually occurs.

During *days 15–28,* increased production of progesterone by the corpus luteum in the ovary causes the endometrium of the uterus to double or triple in thickness (from 1 mm to 2–3 mm) and the uterine glands to mature, producing a thick mucous secretion. This is called the *secretory phase* of the uterine cycle. The endometrium is now prepared to receive the developing embryo. If this does not occur, the corpus luteum in the ovary degenerates, and the low level of sex hormones in the female body results in the endometrium breaking down during menstruation.

Table 21.3 and Figure 21.9 compare the stages of the uterine cycle with those of the ovarian cycle.

Estrogen and progesterone affect not only the uterus but other parts of the body as well. Estrogen is largely responsible for the secondary sex characteristics in females, including body hair and fat distribution. In general, females have a more rounded appearance

TABLE 21.3	Ovarian and Uterine Cycles
Ovarian Cycle	**Events**
Follicular phase—Days 1–13*	FSH secretion begins.
	Follicle maturation occurs. Estrogen secretion is prominent.
Ovulation—Day 14*	LH spike occurs.
Luteal phase—Days 15–28*	LH secretion continues. Corpus luteum forms and secretes progesterone.
Uterine Cycle	**Events**
Menstruation—Days 1–5	Endometrium breaks down.
Proliferative phase—Days 6–13	Endometrium rebuilds.
Secretory phase—Days 15–28	Endometrium thickens, and glands are secretory.

* Assuming a 28-day cycle.

than males because of a greater accumulation of fat beneath the skin. The pelvic girdle becomes wider and deeper in females, so the pelvic cavity usually has a larger relative size in women compared to men. Like males, females develop axillary and pubic hair during puberty. Both estrogen and progesterone are also required for breast development.

Menstruation

During **menstruation,** arteries that supply the uterine lining constrict and the capillaries weaken. Blood spilling from the damaged vessels detaches layers of the lining, not all at once, but in random patches. Endometrium, mucus, and blood descend from the uterus, and through the vagina, creating menstrual flow. *Fibrinolysin,* an enzyme released by dying cells, prevents the blood from clotting. Normal menstruation lasts from 3 to 10 days. Some abdominal cramping, breast tenderness, and even moodiness are considered to be normal during the menstrual period. See Section 21.6 for a discussion of abnormal menstruation.

Fertilization and Pregnancy

If fertilization does occur, an embryo begins development even as it travels down the uterine tube to the uterus. The endometrium is now prepared to receive the developing embryo, which becomes implanted in the lining several days following fertilization (Fig. 21.10). Pregnancy has now begun.

The **placenta,** which sustains the developing embryo and later fetus, originates from both maternal and fetal tissues. It is the region of exchange of molecules between fetal and maternal blood, although the two rarely mix. At first, the placenta produces **human chorionic gonadotropin (HCG),** which maintains the corpus luteum in the ovary until the placenta begins its own production of progesterone and estrogen. A pregnancy test can usually detect the presence of HCG in the blood or urine around 9 days after fertilization. Recently, however, more sensitive tests claim to be able to detect HCG as early as 4 days following fertilization, but this is

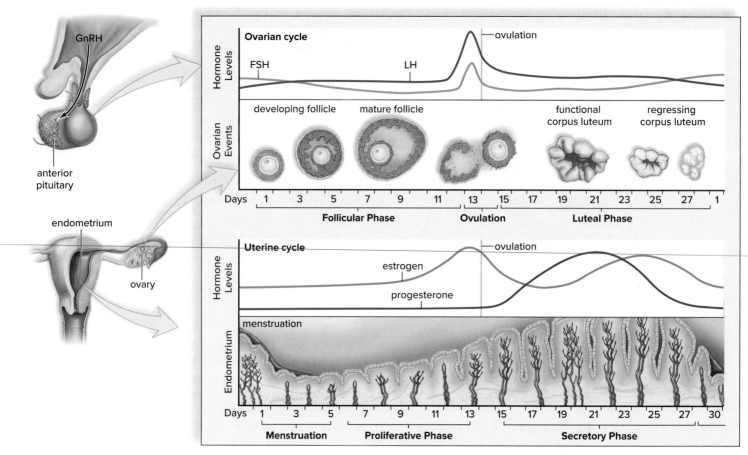

Figure 21.9 Female hormone levels during the ovarian and uterine cycles.   During the follicular phase of the ovarian cycle (*top*), FSH released by the anterior pituitary promotes the maturation of a follicle in the ovary. The ovarian follicle produces increasing levels of estrogen, which causes the endometrium to thicken during the proliferative phase of the uterine cycle (*bottom*). After ovulation and during the luteal phase of the ovarian cycle, LH promotes the development of the corpus luteum. This structure produces increasing levels of progesterone, which causes the endometrium to become secretory. Menstruation and the proliferative phase begin when progesterone production declines to a low level.

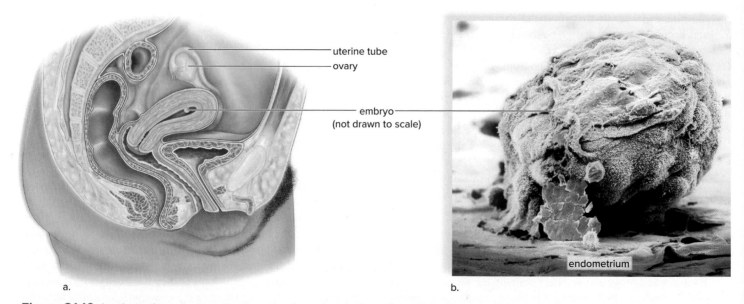

Figure 21.10 Implantation.   a. Site of implantation of an embryo in the uterine wall. **b.** A scanning electron micrograph showing an embryo implanted in the endometrium on day 12 following fertilization. (b): ©Bettman/Contributor/Getty Images

highly dependent on the amount of HCG being produced by the female.

Progesterone and estrogen produced by the placenta have two effects: They shut down the anterior pituitary so that no new follicle in the ovaries matures, and they maintain the endometrium so that the corpus luteum in the ovary is no longer needed. Usually, no menstruation occurs during pregnancy.

Menopause

Menopause, the period in a woman's life during which the ovarian and uterine cycles cease, usually occurs between ages 45 and 55. The ovaries become unresponsive to the gonadotropic hormones produced by the anterior pituitary, and they no longer secrete estrogen or progesterone. At the onset of menopause, the uterine cycle becomes irregular. However, it is still possible for a woman to conceive during this time of transition. Therefore, a woman is usually not considered to have completed menopause (and thus be infertile) until menstruation has been absent for a year.

The hormonal changes during menopause often produce physical symptoms, such as "hot flashes" (caused by circulatory irregularities), dizziness, headaches, insomnia, sleepiness, and depression. Until recently, many women took hormonal replacement therapies (HRTs) that consisted of combined estrogen-progestin drugs, to ease menopausal symptoms. Some studies have shown that HRT therapies may have some benefits for younger women. However, accumulating data show long-term use of the combined drugs by postmenopausal women may cause increases in breast cancer, heart attacks, strokes, and blood clots, and are therefore generally not recommended.

Check Your Progress 21.3

1. Summarize the events that occur during the two phases of the ovarian cycle.
2. Distinguish between the proliferative phase and the secretory phase of the uterine cycle and the hormones that promote each.
3. Describe the changes that occur in the ovarian and uterine cycles during menstruation, pregnancy, and menopause.

21.4 Control of Reproduction

Learning Outcomes

Upon completion of this section, you should be able to

1. Distinguish several birth control methods that are designed to prevent ovulation from those that prevent fertilization.
2. Summarize the birth control methods that are effective at preventing the spread of sexually transmitted diseases.
3. Describe two medications used as emergency contraceptives.

Birth control methods are used to regulate the number of children an individual or couple will have.

Birth Control Methods

The most reliable method of birth control is *abstinence*—that is, not engaging in sexual contact. This form of birth control has the added advantage of preventing the spread of sexually transmitted diseases (STDs). Nearly as effective in preventing pregnancy are the surgical methods of sterilization: vasectomy in males and tubal ligation in females. A **vasectomy** involves cutting and sealing off each vas deferens, so that sperm are unable to reach the seminal fluid. **Tubal ligation** consists of cutting and sealing the uterine tubes.

Table 21.4 lists other means of birth control, and rates their effectiveness when used correctly. For example, the birth control pill is nearly 100% effective, meaning that very few sexually active women will get pregnant while taking the pill. On the other hand, with natural family planning, we expect that 70% will not get pregnant and 30% will get pregnant within the year.

Figure 21.11 shows some of the most commonly used means of birth control. **Contraceptives** are medications and devices that reduce the chance of pregnancy. Oral contraception involves taking **birth control pills** on a daily basis (Fig. 21.11a). Most of these contain a combination of estrogen and progesterone that mimics the hormonal condition of pregnancy, thus shutting down the pituitary production of both FSH and LH so no follicle begins to develop in the ovary. Because ovulation does not occur, pregnancy cannot take place. A male birth control pill, or *male hormonal contraception,* will likely be available within a few years. Most approaches use a combination of a progesterone-like hormone to inhibit sperm production, along with testosterone to combat side effects. Concerns persist, however, about whether most men would be willing to consistently take any contraceptive, whether in the form of a pill, an injection, or an implant.

An **intrauterine device (IUD)** (Fig. 21.11b) is a small device made of molded plastic or copper that is inserted into the uterus by a physician. IUDs are believed to alter the environment of the uterus and oviducts so that fertilization probably will not occur—but if it does, implantation cannot take place. Another type of IUD has a hormone cylinder that releases a synthetic progesterone.

The **diaphragm** is a soft latex cup with a flexible rim that lodges behind the pubic bone and fits over the cervix (Fig. 21.11c). Each woman must be properly fitted by a physician, and the diaphragm can be inserted into the vagina no more than 2 hours before sexual relations. Also, it must be used with spermicidal jelly or cream and should be left in place for at least 6 hours after sexual relations. The *cervical cap* is a mini-diaphragm.

In addition to helping prevent pregnancy, both male and female condoms provide some degree of protection against sexually transmitted diseases (Fig. 21.11d). However, condoms are not 100% effective in preventing pregnancy or STDs. A **male condom** is most often a latex sheath that fits over the erect penis. The ejaculate is trapped inside the sheath, and thus does not enter the vagina. A **female condom** consists of a large polyurethane tube with a flexible ring that fits onto the cervix. The open end of the tube has a ring that covers the external genitals. Better protection is achieved when a condom is used in conjunction with a spermicide.

Contraceptive implants (Fig. 21.11e) use synthetic progesterone to prevent ovulation by disrupting the ovarian cycle. The

TABLE 21.4 Common Birth Control Methods*

Name	Procedure	Methodology	Effectiveness	Risk
Abstinence	Refrain from sexual intercourse	No sperm in vagina	100%	None
Vasectomy	Vasa deferentia cut and tied	No sperm in semen	Almost 100%	Irreversible sterility
Tubal ligation	Uterine tubes cut and tied	No oocytes in uterine tube	Almost 100%	Irreversible sterility
Oral contraception	Hormone medication taken daily	Anterior pituitary does not release FSH and LH	About 91%	Increased risk of heart attack, stroke, and thromboembolism
Contraceptive implants	Tubes of progestin (form of progesterone) implanted under skin	Anterior pituitary does not release FSH and LH	More than 99%	Similar to other hormone treatments; rare problems with insertion/removal
Contraceptive injections/patch	Injections of hormones	Anterior pituitary does not release FSH and LH	About 91%	Similar to other hormone treatments
Vaginal contraceptive ring	Ring inserted into vagina each month	Anterior pituitary does not release FSH and LH	About 91%	May be somewhat lower than other hormone treatments
Intrauterine device (IUD)	Plastic coil inserted into uterus by physician	Prevents implantation	About 99%	PID†, rare problems with insertion/removal
Diaphragm	Latex cup inserted into vagina to cover cervix before intercourse	Blocks entrance of sperm to uterus	With spermicide, about 90%	Latex, spermicide allergy
Cervical cap	Latex cap held by suction over cervix	Delivers spermicide near cervix	About 70–86%	UTI‡, latex or spermicide allergy
Male condom	Latex sheath fitted over erect penis	Traps sperm and prevents STDs	About 85%	Latex allergy
Female condom	Polyurethane liner fitted inside vagina	Blocks entrance of sperm to uterus and prevents STDs	About 85%	Latex allergy
Coitus interruptus	Penis withdrawn before ejaculation	Prevents sperm from entering	About 75%	None
Spermicidal jellies, creams, or foams	Spermicidal products inserted before intercourse	Kill a large number of sperm	About 75%	UTI‡, vaginitis, allergy
Natural family planning	Day of ovulation determined by record-keeping, various methods of testing	Intercourse avoided on certain days of the month	About 75–80%	None
Vaginal film	Thin, waferlike substance inserted into vagina, near cervix	Kills a large number of sperm	74–94%	UTI‡, vaginitis, allergy

*Assuming a 28-day cycle; † = pelvic inflammatory disease; ‡ = urinary tract infection.

Figure 21.11 Various methods of birth control. **a.** Oral contraception (birth control pills). **b.** Intrauterine device. **c.** Spermicidal jelly and diaphragm. **d.** Male and female condoms. **e.** Contraceptive implant and patch. **f.** A vaginal contraceptive ring. (a): ©McGraw-Hill Education/Bob Coyle, photographer; (b): ©imagebroker/Alamy; (c): ©McGraw-Hill Education/Jill Braaten, photographer; (d): ©Scott Camazine/Science Source; (e) (implant): ©Population Council, Karen Tweedy-Holmes; (patch): ©Bob Pardue/Alamy; (f): ©Michaela Begsteiger Image Broker/Newscom

a.

b.

c.

d.

e.

f.

newest version consists of a single capsule that remains effective for about 3 years. A **contraceptive patch** is also available. Known by the brand name Ortho Evra, the patch is applied to the skin once a week for 3 weeks, then no patch is worn during the menstrual period. **Contraceptive injections** are available as progesterone only or as a combination of estrogen and progesterone. The time between injections can vary from 1 to several months.

One of the newer developments in hormonal contraceptives is the **vaginal contraceptive ring** (Fig. 21.11*f*). Known by the brand name NuvaRing, this device is a flexible vinyl ring about 2 inches in diameter, that contains estrogen and progesterone. Currently available by prescription only, it is inserted into the vagina on about the fifth day after menstruation ceases, and left in place for 21 days. Used properly, the NuvaRing prevents ovulation for that month's cycle. It also thickens the cervical mucus, preventing sperm from entering the uterus.

Contraceptive vaccines are now being developed. For example, a vaccine intended to immunize women against HCG, the hormone so necessary to maintaining the implantation of the embryo, was successful in a limited clinical trial. Because HCG is not normally present in the body, no autoimmune reaction is expected, but the immunization does wear off with time. Similar approaches are being investigated to develop a safe antisperm vaccine for use in men or women.

Emergency Contraception

Emergency contraception, or "morning-after pills," consists of medications that can prevent pregnancy after unprotected intercourse. The expression "morning-after" is a misnomer, in that some treatments can be started up to 5 days after unprotected intercourse.

The first FDA-approved medication produced for emergency contraception was a kit called Preven. Preven included four synthetic progesterone pills; two were taken up to 72 hours after unprotected intercourse, and two more were taken 12 hours later. The hormone disrupted the normal uterine cycle, making it difficult for an embryo to implant in the endometrium. One study estimated that Preven was 85% effective in preventing unintended pregnancies. The Preven kit also included a pregnancy test; women were instructed to take the test first before using the hormone, because the medication was not effective on an established pregnancy.

In 2006, the FDA approved another drug, called Plan B One-Step, which is up to 89% effective in preventing pregnancy if taken within 72 hours after unprotected sex. It is available without a prescription to women age 17 and older. In August 2010, ulipristal acetate (also known as ella) was also approved for emergency contraception. It can be taken up to 5 days after unprotected sex, and studies indicate it is somewhat more effective than Plan B One-Step. Unlike Plan B One-Step, however, a prescription is required.

Mifepristone, also known as Mifeprex, RU-486, or the "abortion pill," can cause the loss of an implanted embryo by blocking the progesterone receptors of endometrial cells. This causes the endometrium to slough off, carrying the embryo with it. When taken in conjunction with a prostaglandin to induce uterine contractions, RU-486 is 95% effective at inducing an abortion up to the 49th day of gestation. Because of its mechanism of action, the use of RU-486 is more controversial compared to other medications, and while it is currently available in the United States for early medical abortion, it is not approved for emergency contraception.

Check Your Progress 21.4

1. Classify which birth control methods in Table 21.4 prevent ovulation, and which prevent fertilization.
2. List the birth control methods discussed in this section that physically block sperm from entering the uterus.
3. Summarize which birth control methods reduce the risk of STDs.
4. Define *emergency contraception.*

21.5 Sexually Transmitted Diseases

Learning Outcomes

Upon completion of this section, you should be able to

1. List the specific cause and common symptoms of three STDs that are caused by viruses, as well as three caused by bacteria.
2. Describe how HIV specifically affects the immune system and how this explains the three categories of HIV infection.
3. Discuss four types of antiretroviral drugs.
4. Explain why antibiotics should not be used to treat viral infections.

Many infectious diseases are transmitted by sexual contact and are therefore called sexually transmitted diseases (STDs). Note that these diseases may or may not affect the reproductive organs. The Health feature "Preventing Transmission of STDs" provides general guidelines on how to protect yourself from these infections.

STDs Caused by Viruses

AIDS, genital herpes, genital warts, and hepatitis B are common STDs caused by viruses. Therefore, they do not respond to traditional antibiotics. Antiviral drugs have been developed to treat some of these, but in many cases infection is lifelong.

AIDS

Acquired immunodeficiency syndrome (AIDS) is caused by the **human immunodeficiency virus (HIV).** AIDS is considered a pandemic because the disease is prevalent in the entire human population around the globe. The statistics on HIV and AIDS are updated every 2 to 3 years. As of the end of 2016, the World Health Organization reports that AIDS has killed nearly 35 million people worldwide, and about 36.7 million more are currently estimated to be infected with HIV. It is estimated that over 3.2 million children are infected with the HIV virus. The epidemic has been particularly devastating in sub-Saharan Africa, where 30–40% of the population is infected in some countries. In the United States, an estimated 1.1 million people are infected, and over 600,000 have

SCIENCE IN YOUR LIFE ▶ HEALTH

Preventing Transmission of STDs

Being aware of how STDs are spread (Fig. 21A) and then observing the following guidelines will greatly help prevent the transmission of STDs.

Sexual Activities Transmit STDs

Abstain from sexual intercourse or develop a long-term monogamous (always the same partner) *sexual relationship* with a partner who is free of STDs.

Refrain from multiple sex partners or having relations with someone who has multiple sex partners. If you have sex with two other people and each of these has sex with two people and so forth, the number of people having sexual contact is quite large.

Remember that, although the prevalence of AIDS is presently higher among homosexuals and bisexuals, the highest rate of increase is now occurring among heterosexuals. The lining of the uterus is only one cell thick, and thus it can allow infected cells from a sexual partner to enter the blood.

Be aware that having relations with an intravenous drug user is risky because the behavior of this group risks hepatitis and an HIV infection. Be aware that anyone who already has another sexually transmitted disease is more susceptible to an HIV infection.

Avoid anal-rectal intercourse (in which the penis is inserted into the rectum) because the lining of the rectum is thin and cells infected with HIV can easily enter the body there.

Unsafe Sexual Practices Transmit STDs

Always use a latex condom during sexual intercourse if you do not know for certain that your partner has been free of STDs for some time. Be sure to follow the directions supplied by the manufacturer.

Avoid fellatio (kissing and insertion of the penis into a partner's mouth) *and cunnilingus* (kissing and insertion of the tongue into the vagina) because they may be a means of transmission. The mouth and gums often have cuts and sores that facilitate the entrance of infected cells.

Be cautious about using alcohol or any drug that may prevent you from being able to control your behavior.

Drug Use Transmits Hepatitis and HIV

Stop, if necessary, or do not start the habit of injecting drugs into your veins. Be aware that hepatitis and HIV can be spread by blood-to-blood contact.

Always use a new, sterile needle for injection or use one that has been cleaned in bleach if you are a drug user and cannot stop your behavior.

Questions to Consider

1. How can you be certain your partner doesn't have any STDs? Do people necessarily know they have an STD?
2. What does it mean if a person tells you he or she just had an AIDS test, and it was negative? What does this test detect in the blood?
3. Which methods of birth control are most effective in preventing transmission of HIV? Which are least effective?

a.   b.

Figure 21A Transmission of STDs.   a. Sexual activities transmit STDs. **b.** Sharing needles transmits STDs. In the United States, more than 65 million people are living with an incurable STD. An additional 15 million people become infected with one or more STDs each year. Approximately one-fourth of these new infections occur in teenagers. Despite the fact that STDs are quite widespread, many people remain unaware of the risks and the sometimes deadly consequences of becoming infected. (a) (man and woman): ©Igormakarov/Shutterstock; (male couple): ©Image Source/Getty Images; (b): ©Don Mason/Getty Images

died. While homosexual men make up the largest proportion of people with AIDS in the United States, the fastest rate of increase is now seen among heterosexuals, and over half of all new HIV infections are occurring in people under the age of 25. One in four new infections now occur in the 13–24 age bracket.

HIV is transmitted by sexual contact with an infected person, including vaginal or anal intercourse and oral/genital contact.

Also, needle-sharing among intravenous drug users is a very efficient way to transmit HIV. Babies born to HIV-infected mothers may become infected before or during birth or through breastfeeding after birth. Cultural factors may also play a very important role in the transmission of HIV. For example, polygamy is common in many African countries, as are misconceptions about the effectiveness of birth control, and, indeed, about the fact that HIV is the cause of AIDS. Such practices and beliefs may significantly increase the risk of HIV infection.

Stages of an HIV Infection   When a person initially becomes infected with HIV, the virus replicates at a high level in the macrophages and CD4+ helper T cells, and spreads throughout the lymphoid tissues of the body (Fig. 21.12). Even though the person's blood contains a high "viral load" at this point, most experience only mild symptoms such as fever, chills, aches, and swollen lymph nodes, and soon these symptoms disappear. Even if a person had an HIV test at this time, it would probably be negative because it tests for the presence of anti-HIV antibodies, which are not detectable for an average of 25 days. This means that early in the infection, a person can be highly infectious, even though the HIV antibody test is negative.

Usually, no other HIV-related symptoms occur for several years. Even so, the virus is beginning to cause a loss of CD4+ T cells, which is one of the hallmarks of an HIV infection that is progressing to AIDS. HIV is thought to kill these cells directly by infecting them and also indirectly by inducing apoptosis, even of uninfected cells. Cytotoxic T cells are also thought to kill large numbers of HIV-infected cells. During this stage of infection, the body is able to maintain a reasonably healthy immune system by producing as many as one to two billion new CD4+ helper T cells per day. As long as the body's production of new CD4+ T cells is able to keep pace with the loss of CD4+ T cells (much like leaving the water running in a sink with the drain open), the person stays relatively healthy. However, even during this asymptomatic stage, evidence indicates that the virus continues to replicate at a high rate. Eventually, the number of CD4+ T cells falls so low that the patient begins to suffer from **opportunistic infections,** which have the *opportunity* to occur because the immune system is weakened.

In order to help physicians assess the prognosis of individuals at different stages of HIV infection, the U.S. Centers for Disease Control and Prevention (CDC) has defined three categories of HIV infection, based mainly on the type of clinical disease seen in the patient (Fig. 21.13). Within each category, the number of CD4+

a. AIDS patient Tom Moran, at diagnosis

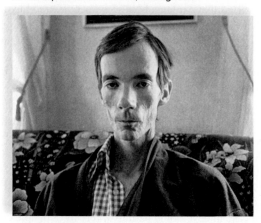

b. AIDS patient Tom Moran, 6 months later

c. AIDS patient Tom Moran, in category C of AIDS

Figure 21.13 The course of an AIDS infection.   These photos show the effect of an HIV infection in one individual who progressed through all the phases of HIV infection. At first, the body can fight off the effects of the virus, but as the infection takes over, the number of CD4+ T cells drops from thousands to hundreds. Eventually, an AIDS patient will die from the opportunistic diseases that flourish because the immune system has been destroyed. (all): ©Nicholas Nixon

Figure 21.12 HIV, the AIDS virus.   False-colored micrographs show HIV particles budding from an infected helper T cell. These viruses can infect helper T cells and also macrophages, which work with helper T cells to stem the infection. ©Scott Camazine/Science Source

T cells is also monitored. Patients in *category A* usually have CD4⁺ T-cell counts of 500 or greater cells per cubic millimeter (cells/mm³) of blood, and either lack symptoms completely or have persistently enlarged lymph nodes. Patients in *category B* have symptoms that are indicative of an unhealthy immune system. Their CD4⁺ T-cell count is usually between 200 and 499 cells/mm³ of blood. Examples of category B conditions include thrush, a yeast infection of the oral cavity, and repeated episodes of shingles, a reactivation of a childhood chickenpox virus infection. Patients in *category C* usually have CD4⁺ T-cell counts less than 200 cells/mm³ of blood, and have one or more AIDS-indicator conditions, such as recurrent bacterial pneumonia, various fungal infections, Kaposi sarcoma (caused by a human herpesvirus), mycobacterial infections, and toxoplasmosis of the brain (see the Health feature "Opportunistic Infections and HIV" in Section 13.6).

Although not composing an official category of HIV infection, another small, yet very interesting group of HIV-infected people remain healthy for at least 10 years, even without any antiviral treatments. Often called *long-term nonprogressors,* these people have been very interesting for HIV researchers to study because their viral loads remain low and their CD4⁺ T-cell counts usually remain normal. As of this writing, no single factor has been identified that explains how some people are able to remain relatively healthy in the face of HIV infection, but their existence continues to provide hope that effective host resistance to HIV is possible.

Treatments for HIV Infection Prior to the development of antiviral drugs, most HIV-infected individuals eventually progressed to category C, or full-blown AIDS, which was almost inevitably fatal. There is still no cure for HIV infection, but since 1995 an increasing number of drugs have become available to help control the replication and spread of HIV, and the life span of many HIV-infected patients has increased to the point where HIV is no longer considered an automatic death sentence. Unfortunately, this is not the case in developing nations where most people cannot afford these drugs.

In order to understand how antiretroviral drugs work, a brief review of the life cycle of HIV is necessary (Fig. 21.14). HIV is a type of RNA virus called a retrovirus. HIV has a protein on its outer surface called gp120, which attaches to a CD4 molecule, plus another molecule called CCR5 or CXCR4 on a host T cell, macrophage, or dendritic cell. This attachment causes *fusion* of HIV with the host plasma membrane, releasing HIV's inner core inside the cell. This core contains the viral RNA, plus several viral enzymes. One such enzyme, called *reverse transcriptase,* makes a DNA copy of the viral RNA, called cDNA. Duplication occurs and the result is double-stranded DNA (dsDNA). During *integration,* the viral enzyme integrase inserts the viral dsDNA into host DNA, where it will remain for the life of the cell. Following integration, viral DNA serves as a template for production of more viral RNA. Some of the viral RNA proceeds to the ribosomes, where it brings about the synthesis of long polypeptides that need to be cleaved into smaller pieces before they can be assembled into new viruses. This cleavage process depends on a third viral enzyme, called *protease.* Finally, the new virus particles are assembled and then released from the host plasma membrane by a process called budding.

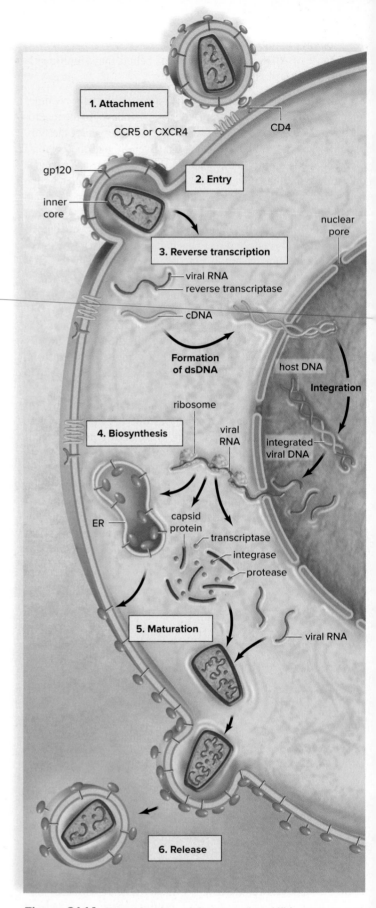

Figure 21.14 Reproduction of the retrovirus HIV. HIV attaches to CD4 plus CCR5 or CXCR4, and then uses reverse transcription to produce a double-stranded DNA copy of viral RNA. The viral dsDNA integrates into the cell's chromosomes before the virus reproduces and buds from the cell.

a.   b.

Figure 21.15 **Two STDs caused by viruses.** **a.** There are several types of herpes viruses, and usually the type called herpes simplex 2 causes genital herpes. Symptoms of genital herpes are due to an outbreak of blisters, which can be present on the labia of females (seen here) or on the penis of males. **b.** Several types of human papillomaviruses (HPVs) can cause genital warts, seen here on the penis. In the female, genital warts may occur externally on the vulva, or internally in the vagina or cervix. Certain HPV types can also cause cancer. (a): ©Bart's Medical Library/Medical Images; (b) Source: Dr. M.F. Rein/CDC

The National Institute of Allergy and Infectious Disease lists five classes of currently available antiretroviral drugs: (1) *reverse transcriptase inhibitors,* which block the production of viral DNA from RNA; (2) *protease inhibitors,* which block the viral enzyme that cleaves long viral polypeptides prior to assembly; (3) *fusion/ entry inhibitors,* which interfere with HIV's ability to fuse with the host plasma membrane during initial infection; (4) *integrase inhibitors,* which inhibit the insertion of viral DNA into host DNA; and (5) *multidrug combination products* that incorporate drugs from more than one class into a single product.

When the first antiretroviral drug, known as AZT, became available in 1995, it was initially very effective at reducing the viral load in many patients, but the virus tended to become resistant to the drug very quickly. This is because the HIV reverse transcriptase enzyme is very error-prone, which means the virus mutates at a very high rate, resulting in drug-resistant mutants. However, today, HIV-infected patients are typically treated with three or four antiretroviral drugs at the same time, a combination called *highly active antiretroviral therapy (HAART).* Because it is unlikely that a particular mutant virus will be resistant to several drugs all at once, HAART is usually able to inhibit HIV replication to the point that no virus is detectable in the blood. It is important to note, however, that these patients presumably still have large numbers of HIV-infected cells, and thus they can still transmit the virus. Other potential problems with HAART include various side effects, which can be severe, the high cost, and the complexity of dosing schedules for the different medications.

Investigators have found that when HAART is discontinued, the virus rebounds. Therefore, therapy is usually continued indefinitely. An HIV-positive pregnant woman who takes reverse transcriptase inhibitors during her pregnancy reduces the chances of HIV transmission to her newborn. If possible, drug therapy should be delayed until the 10th to 12th weeks of pregnancy to minimize any adverse effects of antiretroviral drugs on fetal development.

Genital Herpes

Genital herpes is caused by the *herpes simplex virus,* of which there are two types: type 1, which usually causes cold sores and fever blisters, and type 2, which more often causes genital herpes. In the United States, it is estimated that approximately one in four women, and one in five men, have had a genital herpes infection, but 20% may never have symptoms. Even without symptoms, however, the virus can be transmitted. After becoming infected, most individuals experience a tingling or itching sensation before blisters appear at the infected site within 2 to 20 days (Fig. 21.15*a*). Once the blisters rupture, they leave painful ulcers, which take from 5 days to 3 weeks to heal. These symptoms can be accompanied by fever, pain upon urination, and swollen lymph nodes. Several antiviral medications are available to treat herpesvirus infections. These can be used topically in ointment form, taken orally, or in severe cases, injected. All of these medications inhibit the replication of viral DNA.

After the blisters heal, the virus becomes latent (dormant), and blisters can recur repeatedly at variable intervals. Sunlight, sexual intercourse, menstruation, and stress seem to cause the symptoms of genital herpes to recur. While the virus is latent, it primarily resides in the ganglia of the sensory nerves associated with the affected skin. The ability of the virus to "hide" in this manner has made it more difficult to develop an effective vaccine.

Infection of the newborn can occur if the child comes in contact with a lesion in the birth canal. Within 1 to 3 weeks, the infant may be gravely ill and can become blind, have neurological disorders, including brain damage, or die. Birth by cesarean section prevents these adverse developments.

Human Papillomavirus

An estimated 79 million Americans are currently infected with the *human papillomavirus (HPV).* There are over 100 types of HPV.

The Challenges of Developing a Vaccine Against HIV

An ideal vaccine against the HIV virus would be inexpensive and able to provide lifelong protection against all strains of HIV (Fig. 21B). Is this possible? Effective vaccines have been developed against diseases such as hepatitis B, smallpox, polio, tetanus, influenza, and the measles. But what about AIDS? The top obstacles in the development of an AIDS vaccine include the following:

1. The ideal AIDS vaccine would prevent HIV entry into human cells and prevent the progression and transmission of the disease. However, no vaccine has ever proved to be 100% effective at blocking a virus from entry into cells. Instead, most vaccines prevent, modify, or weaken the disease caused by the infection.
2. Due to HIV's high rate of mutation, there are several genetically different types and subtypes of HIV. HIV strains may differ by 10% within one person and by 35% in people across the globe. Viruses that are genetically different may have different surface proteins, and HIV surface proteins are the focus of many AIDS vaccines. The question is, will scientists need a vaccine for each HIV subtype, or will one vaccine provide protection for all HIV variants?
3. If a vaccine produces only short-term protection, people would need to continue to get booster shots similar to shots given yearly for the flu.
4. There are concerns that an AIDS vaccine would make people more vulnerable to HIV infection. It is believed that in some diseases, such as yellow fever and Rift Valley fever, the antibodies produced after receiving the vaccine helped the virus infect more cells.
5. HIV can be transmitted as a free virus and in infected cells, so perhaps a vaccine would need to stimulate both cellular and antibody-mediated responses. This is complicated by the fact that HIV infects and destroys immune cells, particularly

Figure 21B **A vaccine against HIV.** Different strains of HIV and a high viral mutation rate are just two of the reasons it has been difficult to develop a vaccine against HIV.
©McGraw-Hill Education/Jill Braaten, photographer

T cells. In addition, most successful vaccines stimulate only antibody production.
6. Most vaccines in use today against other diseases are prepared from live-attenuated (weakened) forms of the infectious virus. There are concerns that an AIDS vaccine made using weakened forms of the live virus will cause HIV disease (AIDS).
7. HIV inserts its genetic material into human cells, where it can hide from the immune system.

Although there are many difficulties in vaccine development, AIDS vaccine trials are underway. The process can take many years. After a vaccine has been tested in animals, it must pass through three phases of clinical trials before it is marketed or administered to the public. In Phases I and II, the vaccine is tested from 1 to 2 years in a small number of HIV-uninfected volunteers. The most effective vaccines move into Phase III. In Phase III, the vaccine is tested 3 to 4 years in thousands of HIV-uninfected people. A Phase III trial of the RV144 HIV vaccine was conducted from 2003 to 2009 in Thailand and involved over 16,000 volunteers. The vaccine was reported to reduce the rate of HIV infection by 32%. In February 2015, a second trial commenced in South Africa.

Although the results from the RV144 trial yielded a lower percentage than will be needed for a vaccine to be considered effective, most scientists have many reasons to be optimistic that an AIDS vaccine can and will be developed. The RV144 results, and other successes with vaccinating monkeys, suggest that the research is beginning to yield results. But the most compelling reason for optimism is the human body's ability to suppress the infection. The immune system is able to successfully and effectively decrease the HIV viral load in the body, helping delay the onset of AIDS an average of 10 years in 60% of people who are HIV-infected in the United States. Studies have shown that a small number of people remain HIV uninfected after repeated exposure to the virus, and a few HIV-infected individuals maintain a healthy immune system for over 15 years. It is these stories of the human body's ability to fight HIV infection that keep scientists hopeful that there is a way to help the body fight HIV infection. Most scientists agree that—despite the obstacles or the setbacks—an AIDS vaccine is possible in our future.

Questions to Consider

1. Why might an HIV/AIDS vaccine actually give some people a false sense of security?
2. What cultural factors may be inhibiting the development of a vaccine?

Most cause warts, and about 30 types cause **genital warts,** which are sexually transmitted. Genital warts may appear as flat or raised warts on the penis and/or foreskin of males (Fig. 21.15*b*), and on the vulva, vagina, and/or cervix of females. Note that if the wart(s) is (are) only on the cervix, there may be no outward signs or symptoms of the disease. Newborns can also be infected with HPV during passage through the birth canal.

About ten types of HPV can cause **cervical cancer,** which affects around 15,000 women each year in the United States. These HPV types produce a viral protein that inactivates a host protein called p53, which normally acts as a "brake" on cell division. Once p53 has been inactivated in a particular cell, that cell is more prone to the uncontrolled cell division characteristic of cancer. Early detection of cervical cancer is possible by means of a **Pap test,** in which a few cells are removed from the region of the cervix for microscopic examination. If the cells are cancerous, a hysterectomy (removal of the uterus) may be recommended. In males, HPV can cause cancers of the penis, anus, and other areas. According to a 2008 study, HPV now causes as many cancers of the mouth and throat in U.S. males as does tobacco, perhaps due to an increase in oral sex, as well as a decline in tobacco use.

Currently, there is no cure for an HPV infection, but the warts can be treated effectively by surgery, freezing, application of an acid, or laser burning. However, even after treatment, the virus can sometimes be transmitted. Therefore, once someone has been diagnosed with genital warts, abstinence or the use of a condom is recommended to help prevent transmission of the virus.

In June 2006, the U.S. Food and Drug Administration licensed Gardasil, an HPV vaccine that is effective against the four most common types of HPV found in the United States, including the two types that cause about 70% of cervical cancers. Because the vaccine doesn't protect those who are already infected, ideally children should be vaccinated before they become sexually active. This vaccine has been controversial, however. For example, in February 2007, Texas governor Rick Perry issued an executive order requiring that all girls entering the sixth grade be vaccinated for HPV. Within 2 months, however, the Texas legislature passed a bill blocking public officials from carrying out the mandatory vaccination.

In 2009, the U.S. Food and Drug Administration approved Gardasil for use in males. The CDC recommends that 11- to 12-year-old girls and boys receive three doses of the vaccine. Nonpregnant females between the ages of 13 and 26, and males from age 13 to 21, can also be vaccinated if they did not receive any or all of the three recommended doses when they were younger. Older individuals should speak with their doctor to find out if getting vaccinated is right for them.

Hepatitis B

Of the several forms of viral **hepatitis,** or inflammation of the liver, *hepatitis B* is the most likely to be transmitted sexually. Hepatitis B virus (HBV) is more contagious than HIV, and HBV is transmitted by similar routes. About 50% of persons infected with HBV develop flulike symptoms, including fatigue, fever, headache, nausea, vomiting, and muscle aches. As the virus replicates in the liver, a dull pain in the upper right half of the abdomen may occur, as may jaundice, a yellowish cast to the skin. Some persons

have an acute infection that lasts only a few weeks, while others develop a chronic form of the disease that leads to liver failure and the need for a liver transplant. A safe and effective vaccine is available for the prevention of HBV infection. This vaccine is now on the list of recommended immunizations for children.

STDs Caused by Bacteria

Chlamydia, gonorrhea, and syphilis are bacterial diseases that are usually curable with appropriate antibiotic therapy if diagnosed early enough. However, an increasing number of these bacteria are becoming resistant to antibiotics.

Chlamydia

Chlamydia is named for the tiny bacterium that causes it, *Chlamydia trachomatis.* New chlamydial infections are more numerous than any other sexually transmitted disease. Some estimate that the actual incidence could be as high as 6 million new cases per year. For every reported case in men, more than five cases are detected in women. The low rates in men suggest that many of the sex partners of women with chlamydia are not diagnosed or reported.

Chlamydial infections of the lower reproductive tract are usually mild or asymptomatic. About 8 to 21 days after infection, men experience *urethritis,* an inflammation of the urethra, resulting in a mild burning sensation on urination and a mucous discharge. The infection can spread to the prostate gland and epididymides in males.

Women may have a vaginal discharge, along with the symptoms of a urinary tract infection. If not properly treated, the infection can eventually cause **pelvic inflammatory disease (PID).** This condition is characterized by inflamed uterine tubes that become partially or completely blocked by scar tissue. As a result, the woman may become sterile or subject to ectopic pregnancy, which can be a medical emergency. Chlamydial infections may also increase the possibility of premature and stillborn births. If a newborn is exposed to chlamydia during delivery, pneumonia or infection of the eyes can result (Fig. 21.16).

New and faster laboratory tests are now available for detecting a chlamydial infection. Their expense sometimes prevents public

Figure 21.16 Chlamydial eye infection.   This newborn's eyes were infected after passing through the birth canal of an infected mother.
©Dr. Allan Harris/Medical Images

clinics from testing for chlamydia. Thus, the following criteria have been suggested to help physicians decide which women should be tested: no more than 24 years old; a new sex partner within the preceding 2 months; cervical discharge; bleeding during parts of the vaginal exam; and use of a nonbarrier method of contraception. Some doctors, however, are routinely prescribing additional antibiotics appropriate to treating chlamydia for anyone who has gonorrhea, because 40% of females and 20% of males with gonorrhea also have chlamydia.

Gonorrhea

Gonorrhea is caused by the bacterium *Neisseria gonorrhoeae.* Although as many as 20% of infected males are asymptomatic, most have urethritis, and experience pain at urination with a thick, greenish-yellow urethral discharge 3 to 5 days after contact with an infected partner. In contrast, 60–80% of infected females are asymptomatic until they develop severe PID. Similarly, scarring of each vas deferens may occur in untreated males. There is also strong evidence that infection with *N. gonorrhoeae* increases the risk of transmitting HIV.

Gonorrhea proctitis, or infection of the anus, with symptoms that may include pain in the anus and blood or pus in the feces, is spread by anal intercourse. Oral sex can cause infection of the throat and the tonsils. Gonorrhea can also spread to other parts of the body, causing heart damage or arthritis. If, by chance, the person touches infected genitals and then his or her eyes, a severe eye infection can result.

Eye infection leading to blindness can occur as a baby passes through the birth canal. Because of this, all vaginally delivered infants receive eyedrops containing antibacterial agents, such as silver nitrate, tetracycline, or penicillin, as a protective measure.

Reported gonorrhea rates declined steadily until the late 1990s, and since then, they have stayed fairly constant. Annually, approximately 330,000 cases of gonorrhea are reported in the United States, but the CDC estimates that there are probably over 820,000 new cases each year. About 75% of all reported cases are found in people between 15 and 29 years of age. Gonorrhea rates among African Americans are about 20 times greater than among Caucasians. Antibiotic-resistant strains of *N. gonorrhoeae* are also becoming more common. As with AIDS, condoms can help protect against gonorrheal and chlamydial infections.

Syphilis

Syphilis is caused by the bacterium *Treponema pallidum,* an actively motile corkscrewlike organism (Fig. 21.17a). Rates of infection have mostly been decreasing since reporting began in the 1940s, largely because of the development of effective diagnostic and treatment methods. In 2013, approximately 17,000 cases were reported in the United States. However, syphilis remains a serious health problem in certain areas, mainly in the South and in large metropolitan areas. About 67% of new cases occur in men who have sex with men, with rates rising fastest among males aged 20 to 29 years. Worldwide, an estimated 10 million new infections occur each year.

Syphilis has three stages, which can be separated by intervening latent periods during which the bacteria are not multiplying. During the *primary stage,* a hard chancre (ulcerated sore with hard edges) indicates the site of infection (Fig. 21.17b). The chancre can go unnoticed, especially because it usually heals spontaneously, leaving little scarring. During the *secondary stage,* the individual breaks out in a rash, coinciding with the replication and spread of bacteria throughout the body. Curiously, the rash does not itch and is even seen on the palms of the hands and the soles of

a.   c.   d.

Figure 21.17 Syphilis.   **a.** Scanning electron micrograph of *Treponema pallidum,* the cause of syphilis. **b.** The primary stage of syphilis is a chancre at the site where the bacterium enters the body. **c.** The secondary stage is a body rash that occurs even on the palms of the hands and soles of the feet. **d.** In the tertiary stage, gummas may appear on the skin or internal organs. (a): ©Melba Photo Agency/Alamy; (b): Source: Dr. N.J. Fiumara, Dr. Gavin Hart/CDC; (c): ©Collection CNRI/Medical Images; (d): Source: Susan Lindsley/CDC the feet (Fig. 21.17c). Hair loss can occur, and infectious gray patches may develop on the mucous membranes, including the mouth. These symptoms disappear of their own accord.

Not all cases of secondary syphilis go on to the *tertiary stage.* During this stage, which lasts until the patient dies, syphilis may affect the cardiovascular system by weakening arterial walls (particularly in the aorta), allowing aneurysms to develop. The disease may also affect the nervous system, causing psychological problems. Sometimes during the tertiary stage, gummas, large destructive ulcers, may develop on the skin or within the internal organs (Fig. 21.17d).

Congenital syphilis, which is present at birth, is caused by *T. pallidum* bacteria crossing the placenta. The child is born blind and/or with numerous anatomical malformations.

As with most bacterial diseases, antibiotics are effective in treating syphilis, especially in the early stages. Therefore, it is crucial for all sexual contacts to be identified and treated.

Check Your Progress 21.5

1. List three STDs that are caused by bacteria and three that are caused by viruses.
2. Describe three ways that HIV may destroy CD4$^+$ T cells.
3. List four classes of antiretroviral drugs and the specific stage of the HIV life cycle that each class targets.
4. Explain why viral diseases, in general, are more difficult to treat than bacterial diseases.

21.6 Reproductive Disorders and Assisted Reproductive Technologies

Learning Outcomes

Upon completion of this section, you should be able to

1. Describe some common disorders of the male reproductive system.
2. Describe some common disorders of the female reproductive system.
3. Discuss four assisted reproductive technologies that can help infertile couples conceive a child.

Several common disorders affect the reproductive system. While some of these affect the ability of young adult males or females to reproduce, others tend to occur in older individuals, and may affect their overall health.

Common Conditions Affecting the Male Reproductive System

Erectile Dysfunction

It is estimated that about 50% of men aged 40 to 75 have experienced some degree of erectile dysfunction (ED), or impotence, which is the inability to produce or maintain an erection sufficient to perform sexual intercourse. Since the late 1990s, a variety of new drugs have been developed to treat ED. All of these drugs—like Viagra, Levitra, and Cialis—work by increasing blood flow to the penis during sexual arousal. Specifically, they inhibit an enzyme called PDE-5, found mainly in the penis, which normally breaks down some of the chemicals responsible for an erection. Surveys have shown these medications to be helpful in about 65% of men who are experiencing ED.

ED can have many causes. It is very likely that changes in the cardiovascular system associated with aging alone can also produce some degree of ED. A normal erection is dependent on blood flow, so almost any condition that affects the cardiovascular system, such as atherosclerosis, high blood pressure, or diabetes mellitus, can lead to ED. Certain medications can cause ED, as can smoking, alcohol, and some illicit drugs. Psychological factors, especially depression, can also be very important contributors to ED. Depending on the cause, other therapies besides the Viagra-type drugs may be indicated. For example, chemicals that dilate blood vessels can be injected directly into the penis, applied as a gel or a patch, or inserted into the urethra as a suppository. If chemical treatments do not work, it is possible to surgically implant a device into the penis that either causes it to remain semirigid all the time or that must be inflated with a small pump, which is usually implanted in the scrotum. Obviously, such devices are usually a last resort in men with ED for whom other approaches have been unsuccessful.

Disorders of the Prostate

Beginning around age 40, most men have some enlargement of the prostate gland, which may eventually grow from its normal size of a walnut to that of a lime or even a lemon. This condition is called **benign prostatic hyperplasia (BPH).** Because the prostate gland wraps around the urethra, this enlargement initially may force the bladder muscle to work harder to empty the urethra, leading to irritation of the bladder and an urge to urinate more frequently (Fig. 21.18). As the prostate grows larger, this increased pressure may result in bladder or kidney damage, and may even completely block urination, a condition that is fatal if untreated.

Early signs of an enlarged prostate include a more frequent urge to urinate, a weak stream of urine, and/or a sense of not fully emptying the bladder after urination. The presence of an enlarged prostate can usually be confirmed by a simple digital exam of the prostate, which a physician performs by inserting the finger of a gloved hand into the rectum. Depending on the severity of the symptoms, treatments for BPH range from avoiding liquids in the evening (to reduce the need to urinate during the night), to taking medications that shrink the prostate and/or improve urine flow, to having surgery.

Prostate enlargement is often due to an enzyme (5-alpha reductase) that acts on the male sex hormone testosterone, converting it to a substance that promotes prostate growth. That growth is needed during puberty, but continued growth in the adult is undesirable. Several chemical substances have been shown to interfere with the action of this enzyme. Saw palmetto, which is sold in tablet form as an over-the-counter nutritional supplement, has been shown to be effective. Prescription drugs such as finasteride (Proscar) are more powerful inhibitors of the enzyme, but men taking these drugs may experience erectile dysfunction and loss of libido.

If drugs don't control symptoms or the symptoms are severe, surgery may be necessary to reduce the pressure on the urethra

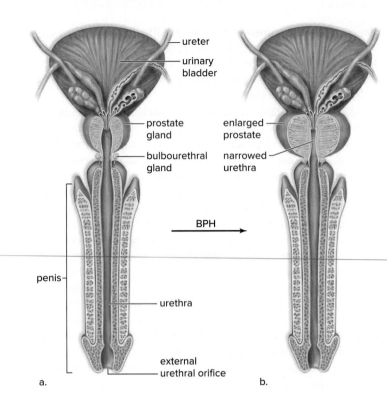

Figure 21.18 Benign prostatic hyperplasia (BPH). **a.** This longitudinal section shows how the prostate gland normally surrounds the urethra. **b.** When the prostate gland enlarges, it can constrict the urethra as it exits the bladder, making it difficult to fully empty the bladder.

and/or reduce the size of the prostate. In the most common operation, called *transurethral resection* of the prostate, the surgeon scrapes away the innermost core of the gland using a small instrument inserted into the urethra. In a more limited operation, called transurethral incision of the prostate, the doctor simply makes one or two small cuts in the prostate to relieve pressure. Alternatively, a newer procedure called *transurethral microwave thermotherapy* uses microwaves to produce heat that destroys, and hopefully shrinks, the enlarged gland.

Prostate cancer is the most commonly diagnosed cancer in American males, with one in ten men expected to develop the disease in their lifetime. After lung cancer, prostate cancer is the second leading cancer-related cause of death in men in the United States, because it has a tendency to metastasize and spread throughout the body. Many men are concerned that BPH may be associated with prostate cancer, but the two conditions are not necessarily related. If prostate cancer is detected early, usually by a blood test for prostate-specific antigen (PSA), it can usually be successfully treated. However, some experts now recommend against the routine screening of PSA levels, mainly because this test often detects prostate cancers that are small or slow growing and thus may not require treatment.

Testicular Cancer

Testicular cancer is the most common type of cancer in males aged 15 to 35. There are several types of testicular cancer, and some of them have a greater tendency to metastasize to other tissues. If discovered and treated early, testicular cancer is one of the most curable of all cancers. Therefore, experts recommend that men perform a monthly testicular self-examination after a hot shower or bath, when the scrotal skin is looser. Men should examine each testicle for any lumps or abnormal tenderness and for any change in the size of one testicle compared to the other. Any such findings should be reported to a doctor as soon as possible. If testicular cancer is confirmed, treatment usually includes surgery to remove the affected testicle, plus radiation or chemotherapy for types of cancer that have a high risk of spreading. However, if diagnosed and treated before it has spread, the cure rate for testicular cancer approaches 100%.

Common Conditions Affecting the Female Reproductive System

Endometriosis

Endometriosis, the presence of endometrial-like tissue at locations outside the uterine cavity, is the most common cause of chronic pelvic pain in women. Estimates indicate that between 10% and 33% of women aged 25 to 35 years will have some degree of this condition. The abnormal tissue often grows on the ovaries, the uterine tubes, the outside of the uterus, or on virtually any other abdominal organ. Endometriosis can induce a chronic inflammatory reaction, which often results in pain, abnormal menstrual cycles, or infertility. The exact cause of endometriosis is unclear, but various theories have been suggested. During menstruation, it is possible that live cells shed from the endometrium actually flow backward, through the oviducts and into the abdomen, where they are able to form endometrial tissue. Another hypothesis suggests that under certain hormonal conditions, the cells that normally line the abdominal organs can become endometrial tissue.

Endometriosis is often diagnosed when a surgeon sees the abnormal tissue during a surgical procedure to investigate the cause of abdominal pain (Fig. 21.19). The initial treatment usually focuses on controlling the pain, but may also involve drugs, such as oral contraceptives, which inhibit ovulation, and thus reduce the normal hormonally induced changes in the endometrial tissue. If these approaches do not work, surgery may be performed to either remove the abnormal endometrial tissue or to remove the uterus and ovaries in women who do not wish to have children.

Ovarian Cancer

Ovarian cancer accounts for about 3% to 4% of all women's cancers in the United States (around 22,000 new cases), causing about 14,000 deaths a year. Unlike testicular cancer, ovarian cancer typically occurs in individuals over the age of 40. However, it also has a worse prognosis than testicular cancer, with only 50% of women surviving 5 years after diagnosis. The main reason for this may be that ovarian cancer is often "silent," showing no obvious signs or symptoms until late in its development. The most common symptoms include abdominal pain, feeling bloated or full, and a frequent urge to urinate.

Women over age 40 should have a cancer-related checkup every year. Early detection requires periodic, thorough pelvic examinations. The Pap test, useful in detecting cervical cancer, does not reveal ovarian cancer. Testing for the level of tumor marker

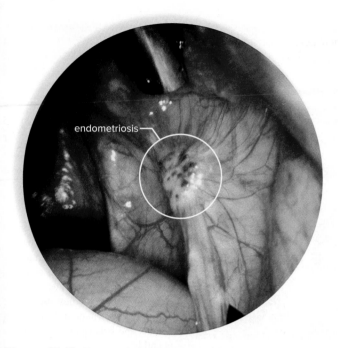

Figure 21.19 Endometriosis. Endometriosis is a common cause of pelvic pain, abnormal menstruation, and infertility in women. The condition occurs when endometrial tissue, which is often darkly pigmented, grows outside the uterus (usually in the abdominal cavity). ©Laparoscopic Appearance of Endometriosis (copyright 1991), www.memfert.com, Dan Martin, M.D.

CA-125, a protein antigen, is helpful. Surgery, radiation therapy, and drug therapy are treatment options. Surgery usually involves removing one or both ovaries, the uterus, and the uterine tubes.

Women who have never had children are twice as likely to develop ovarian cancer as those who have. Early age at first pregnancy, early menopause, and the use of oral contraceptives, which reduces ovulation frequency, can help protect against ovarian cancer. If a woman has had breast cancer, her chances of developing ovarian cancer double.

Disorders of Menstruation

Most menstrual cycles occur every 21 to 35 days, with 3 to 10 days of bleeding during which 30–40 ml of blood are lost. Abnormalities include too little bleeding (or a complete lack of menstruation) or too much bleeding. The former usually indicates a problem with the control of menstruation by the hypothalamus or pituitary gland. This may be seen, for example, in young women who are undergoing strenuous athletic training, and/or who have very low levels of body fat. Excessive bleeding may be due to a large number of other disorders, including various STDs, certain medications, or problems with blood clotting.

The most common disorder of menstruation is *dysmenorrhea,* or painful menstruation. About half of all women experience this condition at some point, which includes such symptoms as agonizing cramps, headaches, backaches, and nausea. Painful menstruation is the leading cause of lost time from school and work among young women. There are many potential causes of dysmenorrhea. One common factor is a high level of prostaglandins released by the endometrium during the luteal and menstrual phases of the cycle. Research has shown that prostaglandin levels are four or five

times higher than normal in women who experience painful menstruation. Because prostaglandins are known to cause uterine contractions, investigators suspect that these chemicals are responsible for much of the pain these women experience.

Other potential causes of painful menstruation include endometriosis (discussed previously) and various types of *ovarian cysts.* These fluid-filled sacs can develop on the ovaries of females of all ages, but are more common in the childbearing years. Two major types of ovarian cysts are follicular cysts, usually caused by the failure of a developing follicle to burst, and corpus luteum cysts, which occur when the corpus luteum doesn't break down normally. On rare occasions, surgery may be needed to remove ovarian cysts.

Premature ovarian failure may occur in some women under the age of 40. In this condition, the ovaries either do not release oocytes on a regular basis, or do not produce a sufficient amount of estrogen. Women with premature ovarian failure have irregular periods, but may still conceive.

Premenstrual syndrome (*PMS*) is a group of symptoms related to the menstrual cycle. Anywhere from 2 weeks to a few days before menstruation, a significant number of women experience such symptoms as unexplained changes in mood, headaches, joint pain, digestive upsets, and sore breasts. The mood swings can be particularly troublesome if they lead to behaviors uncharacteristic of the individual.

Self-help measures for painful menstruation include applying a heating pad to the lower abdomen, drinking warm beverages, and taking warm showers or baths. Over-the-counter anti-inflammatory medications may be helpful; if not, a physician can prescribe other medications. For PMS, avoiding salt, sugar, caffeine, and alcohol may also help, especially when symptoms are present.

Infertility

Infertility is the failure of a couple to achieve pregnancy after one year of regular, unprotected intercourse. The American Medical Association estimates that 15% of all couples are infertile. The cause of infertility can be attributed to the male (40%), the female (40%), or both (20%).

Causes of Infertility

The most frequent cause of infertility in males is low sperm count and/or a large proportion of abnormal sperm, which can have many causes. The public is particularly concerned about endocrine-disrupting contaminants, which are discussed in the Ecology feature "Endocrine-Disrupting Contaminants." Other possible causes of male infertility include various STDs, abnormal hormone levels, smoking, and alcohol. As noted in Section 21.1, activities that increase testicular temperature can lower sperm count.

Common causes of infertility in females include blocked oviducts due to pelvic inflammatory disease and endometriosis, as discussed earlier in this section. Infertility problems also begin to increase sharply after age 35 in women, more so than in men. Several studies have also linked obesity and infertility in both genders, but the exact mechanism(s) involved remain unclear.

Sometimes the cause of infertility can be corrected by medical intervention so that couples can have children. It is possible to give

Endocrine-Disrupting Contaminants

Rachel Carson's book *Silent Spring,* published in 1962, predicted that pesticides would have a deleterious effect on animal life. Soon thereafter, it was found that pesticides caused the eggshells of bald eagles to become so thin that their eggs broke and the chicks died. Additionally, populations of terns, gulls, cormorants, and lake trout declined after they ate fish contaminated by high levels of environmental toxicants. The concern was so great that the U.S. Environmental Protection Agency (EPA) came into existence. This agency, together with civilian environmental groups, has brought about a reduction in pollution release and a cleaning up of emissions. Even so, we are now aware of more subtle effects of pollutants.

Hormones influence nearly all aspects of physiology and behavior in animals. Therefore, when wildlife in contaminated areas began to exhibit certain abnormalities, researchers began to think that certain pollutants could affect the endocrine system. In England, male fish exposed to sewage developed ovarian tissue and produced a metabolite normally found only in females during egg formation. In California, western gulls displayed abnormalities in gonad structure and nesting behaviors. Hatchling alligators in Florida possessed abnormal gonads and hormone concentrations linked to nesting.

At first, such effects seemed to involve only the female hormone estrogen, and researchers therefore called the contaminants ecoestrogens. Many of these chemicals interact with hormone receptors, and in that way cause developmental effects. Others bind directly with sex hormones, such as testosterone and estradiol (a potent estrogen). Still others alter the physiology of growth hormones and neurotransmitters responsible for brain development and behavior. Therefore, the preferred term today for these pollutants is *endocrine-disrupting contaminants (EDCs).*

Figure 21C Exposure to endocrine-disrupting contaminants. Various types of wildlife, as well as humans, are exposed to endocrine-disrupting contaminants that can seriously affect their health and reproductive abilities. (owl): ©Art Wolfe/Iconica/Getty Images; (fish): ©Vilainecrevette/123RF

Many EDCs are chemicals used as pesticides and herbicides in agriculture, and some are associated with the manufacture of various other organic molecules, such as PCBs (polychlorinated biphenyls). PCBs have been banned in the United States since 1979, but because they are chemically stable, they are still present in the environment. Other chemicals shown to influence hormones are found in plastics, food additives, and personal hygiene products. In mice, phthalate esters, which are plastic components, affect neonatal development when present in the part-per-trillion range. It is of great concern that EDCs have been found at levels comparable to functional hormone levels in the human body.

Therefore, it is not surprising that EDCs are affecting the endocrine systems of a wide range of organisms (Fig. 21C).

Scientists and representatives of industrial manufacturers continue to debate whether EDCs pose a health risk to humans. Some suspect that EDCs lower sperm counts, reduce male and female fertility, and increase rates of certain cancers (breast, ovarian, testicular, and prostate). Additionally, some studies suggest that EDCs contribute to learning deficits and behavioral problems in children.

Laboratory and field research continues to identify chemicals that have the ability to influence the endocrine system. Millions of tons of potential EDCs are produced annually in the United States, and the EPA is under pressure to certify these compounds as safe. The European Economic Community has already restricted the use of certain EDCs and has banned the production of specific plastic components intended for use by children. Only through continued scientific research and the cooperation of industry can we identify the risks the EDCs pose to the environment, wildlife, and humans.

Questions to Consider

1. There are undeniable benefits to the use of pesticides, one of the major ones being more attractive, insect-free food. How would you personally weigh these benefits compared to the risks of ingesting endocrine-disrupting contaminants?

2. How might endocrine-disrupting chemicals specifically affect cells? (Review how peptide and steroid hormones work in Section 20.1.)

3. What would be some implications for agricultural and manufacturing industries if the use of all potential endocrine-disrupting chemicals were banned? How much extra would you be willing to pay for your monthly food (and other products) if their costs increased?

females fertility drugs, which are gonadotropic hormones that stimulate the ovaries and bring about ovulation, although this may cause multiple births.

When reproduction does not occur in the usual manner, many couples adopt a child. Others sometimes try one of the assisted reproductive technologies discussed later in this section.

Assisted Reproductive Technologies

Assisted reproductive technologies (*ART*) are techniques that increase the chances of pregnancy.

Artificial Insemination by Donor (AID)   During artificial insemination, sperm are placed in the vagina by a physician. Sometimes a woman is artificially inseminated by her partner's sperm. This is especially helpful if the partner has a low sperm count, because the sperm can be collected over a period of time and concentrated so that the sperm count is sufficient to result in fertilization. Often, however, a woman is inseminated by sperm acquired from a donor who is a complete stranger to her. At times, a combination of partner and donor sperm is used.

A variation of AID is *intrauterine insemination* (*IUI*). In IUI, fertility drugs are given to stimulate the ovaries, and then the donor's sperm is placed in the uterus, rather than in the vagina.

If the prospective parents wish, sperm can be sorted into those believed to be X chromosome–bearing or Y chromosome–bearing to increase the chances of having a child of the desired sex.

In Vitro Fertilization (IVF)   During IVF, conception occurs in laboratory glassware. Ultrasound machines can now spot follicles in the ovaries that hold immature oocytes. Therefore, the latest method is to forgo the administration of fertility drugs and retrieve immature oocytes by using a needle. The immature oocytes are then brought to maturity in glassware before concentrated sperm are added. After about 2 to 4 days, the embryos are ready to be transferred to the uterus of the woman, who is now in the secretory phase of her uterine cycle. If desired, the embryos can be tested for a genetic disease, so that only those free of disease are used. If implantation is successful, development is normal and continues to term.

Gamete Intrafallopian Transfer (GIFT)   The term *gamete* refers to a sex cell, either a sperm or an oocyte. GIFT was devised to overcome the low success rate (15–20%) of in vitro fertilization. The method is exactly the same as in vitro fertilization, except the oocytes and the sperm are placed in the uterine tubes immediately after they have been brought together. GIFT has the advantage of being a one-step procedure for the woman—the oocytes are removed and reintroduced all in the same time period. A variation on this procedure is to fertilize the oocytes in the laboratory and then place the zygotes in the uterine tubes.

Surrogate Mothers   In some instances, women are contracted and paid to have babies. These women are called surrogate mothers. The sperm and even the oocyte can be contributed by the contracting parents.

Intracytoplasmic Sperm Injection (ICSI)   In this highly sophisticated procedure, a single sperm is injected into an oocyte (Fig. 21.20). It is used effectively when a man has severe infertility problems.

Figure 21.20 Intracytoplasmic sperm injection.   This procedure uses a microscope connected to a monitor to indicate the injection of sperm into the egg using a microscopic needle. ©CC Studio/SPL/Science Source

Check Your Progress 21.6

1. Explain how sildenafil (Viagra) and similar drugs work to enhance erections.
2. Describe how endometriosis causes infertility.
3. List some factors that explain why ovarian cancer is more frequently fatal compared to testicular cancer.
4. Compare the degree of medical intervention required for artificial insemination by donor (AID), in vitro fertilization (IVF), gamete intrafallopian transfer (GIFT), and intracytoplasmic sperm injection (ICSI).

Conclusion

The three-parent in vitro fertilization procedure is up for review by the FDA to be approved for clinical studies. However, in 2015 the United Kingdom became the first country to approve three-parent IVF to prevent the inheritance of genetic diseases. Despite the acceptance in the UK, this procedure has been a source of controversy by many parties interested in reproductive issues.

While popularized in the media as three-parent babies, the genes in question amount to just 37 out of the estimated 23,000 to 25,000 total genes in a human cell. Therefore, many argue that this does not justify calling it a third-parent baby because the donated mitochondrial genes are insignificant in number compared to the rest of the genome derived from the parents.

But this is just the start; additional technologies are being developed that may be even more controversial. In 2015, researchers announced that it may be possible to fertilize an egg using another egg, thus eliminating the need for sperm. While the deployment of these technologies may be years away, there is already opposition developing to changing the one egg–one sperm traditional view of reproduction.

SUMMARIZE

21.1 Male Reproductive System

The **reproductive system** produces **gametes,** haploid sex cells that join during fertilization. The male genital tract consists of **testes** that produce **sperm,** accessory organs that add secretions to the **semen (seminal fluid),** and the external genitalia:

■ **Spermatogenesis** occurs in the **seminiferous tubules** of the testes, producing sperm that mature and are stored in each **epididymis.** As they develop, the sperm are supported by **Sertoli cells.** Sperm enter each **vas deferens** before entering the **urethra** along with seminal fluid produced by the **seminal vesicles, prostate gland,** and **bulbourethral glands.**

■ The external genitals of males are the **penis,** the organ of sexual intercourse, and the testes, which are within the **scrotum.** Erection of the penis occurs when expandable tissue in the penis fills with blood. **Erectile dysfunction (ED)** occurs when an erection cannot be achieved.

■ Orgasm in males is a physical and emotional climax that results in ejaculation of semen from the penis. **Circumcision** is the surgical removal of the foreskin of the penis.

The hypothalamus controls the male reproductive system through the secretion of hormones:

■ **Gonadotropin-releasing hormone (GnRH)** from the hypothalamus causes the anterior pituitary to secrete **follicle-stimulating hormone (FSH)** and **luteinizing hormone (LH).** FSH promotes spermatogenesis in the seminiferous tubules of the testes. LH promotes testosterone production by **interstitial cells. Testosterone** is required for proper functioning of the male sex organs, as well as for the development of male secondary sex characteristics.

21.2 Female Reproductive System

The female genital tract includes the internal organs that produce the female sex cells and allow fertilization to occur, plus the external genitalia:

■ **Oogenesis** is the production of an egg, or **oocyte,** by an **ovary.** After **ovulation,** the release of an oocyte, it is transported by **fimbriae** through a **uterine tube** (oviduct) to the **uterus,** where a fertilized oocyte, or **zygote,** can implant in the **endometrium** and develop. The lower part of the uterus is called the **cervix.**

■ The external genitals of the female include the **vagina,** the **clitoris,** and the **vulva** (labia majora and labia minora). The vagina is the organ of sexual intercourse, the birth canal, and the exit for menstrual fluids. The external genitals, especially the clitoris, play an active role in orgasm, which culminates in vaginal, uterine, and oviduct contractions.

21.3 Ovarian and Uterine Cycles

The **ovarian cycle** is under the control of the hypothalamus and anterior pituitary:

■ During the follicular phase, FSH causes maturation of a **follicle** that secretes estrogen and some progesterone. This phase ends with ovulation.

■ During the luteal phase, which occurs during the second half of the cycle, LH converts the follicle into the **corpus luteum,** which secretes **progesterone** and **estrogen.**

The **uterine cycle** varies depending on whether fertilization occurs:

■ After ovulation, estrogen produced by a new ovarian follicle causes the endometrium to thicken. Progesterone from the corpus luteum causes the endometrium to become secretory, in preparation to receive a fertilized egg.

■ If no pregnancy occurs, the low level of hormones causes the endometrium to break down, resulting in **menstruation.** If a fertilized embryo implants itself in the thickened endometrium, pregnancy occurs. Initially, production of **human chorionic gonadotropin (HCG)** by the **placenta** maintains the corpus luteum. As the placenta develops, it takes over production of estrogen and progesterone.

■ During **menopause,** the ovaries become unresponsive to gonadotropic hormones, and the ovarian and uterine cycles cease.

21.4 Control of Reproduction

Only abstinence provides 100% protection against STDs, although condoms are somewhat effective. Common methods of preventing pregnancy are summarized here:

■ Abstaining from sex is the most reliable birth control method.

■ Surgical approaches to birth control—that is, **vasectomy** and **tubal ligation**—are virtually 100% effective.

■ Oral **contraceptives,** or **birth control pills,** are nearly 100% effective in preventing ovulation.

■ An **intrauterine device (IUD)** is inserted into the uterus to prevent either fertilization or implantation of a fertilized embryo.

■ A **diaphragm** is inserted to cover the cervix, to prevent sperm from entering the uterus.

■ **Male condoms** and **female condoms** both form a barrier that prevents sperm from entering the uterus, while providing some protection against STDs.

■ **Contraceptive implants, contraceptive patches,** and **contraceptive injections** all utilize estrogen and/or progesterone to prevent ovulation.

■ A **vaginal contraceptive ring** can be inserted into the vagina every month to prevent ovulation.

■ **Emergency contraception** can be taken after sexual activity (and fertilization) has occurred. It works by preventing implantation of the embryo.

21.5 Sexually Transmitted Diseases

Many infectious diseases can be transmitted sexually. Several common STDs are caused by viruses:

■ **Acquired immunodeficiency syndrome (AIDS),** caused by the **human immunodeficiency virus (HIV),** devastates the immune system, resulting in **opportunistic infections.** Inflammation of the liver, or **hepatitis,** can be caused by several viruses, including hepatitis B virus, which can be transmitted sexually.

■ **Genital herpes** may cause blisters that flare up repeatedly.

■ **Cervical cancer, genital warts,** and cancers of the anus and other tissues are caused by human papillomaviruses. Cervical cancer is often detected using the **Pap test.**

■ Hepatitis B can lead to liver failure.

Several common STDs are caused by bacteria:

- **Chlamydia** and **gonorrhea** can cause urethritis, and possibly **pelvic inflammatory disease (PID)** in females.
- **Syphilis** may cause cardiovascular and neurological complications if untreated.

21.6 Reproductive Disorders and Assisted Reproductive Technologies

The reproductive system is affected by several other common conditions:

- Common disorders of the male reproductive tract include erectile dysfunction (ED), **benign prostatic hyperplasia (BPH), prostate cancer,** and **testicular cancer.**
- Common disorders of the female reproductive tract include **endometriosis, ovarian cancer,** and abnormal menstruation.
- Couples experiencing **infertility** may use assisted reproductive technologies in order to have a child. Such technologies include artificial insemination, in vitro fertilization, gamete intrafallopian transfer, and intracytoplasmic sperm injection.

ASSESS

Testing Yourself

Choose the best answer for each question.

21.1 Male Reproductive System

1. Label the indicated structures of the male reproductive system.

2. Testosterone is produced and secreted by
   a. spermatogonia.
   b. sustentacular cells.
   c. seminiferous tubules.
   d. interstitial cells.

3. Spermatogenesis occurs in this structure of the male reproductive system.
   a. prostate
   b. penis
   c. Sertoli cells
   d. seminiferous tubules
   e. None of these are correct.

21.2 Female Reproductive System

4. Label the indicated structures of the female reproductive system.

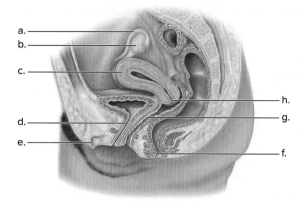

5. Implantation occurs in which structure of the female reproductive system?
   a. uterine tubes
   b. uterus
   c. vagina
   d. cervix
   e. None of these are correct.

6. This structure connects the ovary with the uterus.
   a. cervix
   b. uterine tube
   c. vagina
   d. endometrium

21.3 Ovarian and Uterine Cycles

7. The release of the oocyte from the follicle is caused by
   a. a decreasing level of estrogen.
   b. a surge in the level of follicle-stimulating hormone.
   c. a surge in the level of luteinizing hormone.
   d. progesterone released from the corpus luteum.

8. Which of the following is not an event of the ovarian cycle?
   a. FSH promotes the development of a follicle.
   b. The endometrium thickens.
   c. The corpus luteum secretes progesterone.
   d. Ovulation of an egg occurs.

9. During pregnancy,
   a. the ovarian and uterine cycles occur more quickly than before.
   b. GnRH is produced at a higher level than before.
   c. the ovarian and uterine cycles do not occur.
   d. the female secondary sex characteristics are not maintained.

21.4 Control of Reproduction

In questions 10–12, match each method of protection with a means of birth control in the key.

Key:

a. vasectomy   d. diaphragm
b. oral contraception   e. male condom
c. intrauterine device (IUD)

10. Blocks entrance of sperm to uterus
11. Traps sperm and prevents STDs
12. Prevents implantation of an embryo

21.5 Sexually Transmitted Diseases

13. Which of the following can be treated using antibiotics?
   a. genital herpes
   b. hepatitis
   c. chlamydia
   d. HIV
   e. None of these are correct.

14. The bacterium *Treponema pallidum* causes which of the following?
   a. gonorrhea
   b. syphilis
   c. hepatitis
   d. chlamydia

21.6 Reproductive Disorders and Assisted Reproductive Technologies

15. Which of the following would not occur in a male?
   a. endometriosis
   b. benign prostatic hyperplasia
   c. prostate cancer
   d. infertility
   e. All of these are correct.

16. In which of the following does fertilization occur outside the body?
   a. in vitro fertilization
   b. gamete intrafallopian transfer
   c. surrogate mothers
   d. All of these are correct.

ENGAGE

Thinking Critically

1. The average sperm count in males is now lower than it was several decades ago. The reasons for the lower sperm count usually seen today are not known. What data might be helpful in order to formulate a testable hypothesis?

2. In the animal kingdom, only primates menstruate. Other mammals come into season, or "heat," during certain times of the year, while still others ovulate only after having sex. What would be some possible advantages of a monthly menstrual cycle?

3. The condition called benign prostatic hyperplasia is usually not life-threatening, but prostate cancer can be. Because the prostate gland is typically enlarged in both conditions, why is one condition benign and the other potentially life-threatening?

4. Female athletes who train intensively often stop menstruating. The important factor appears to be the reduction of body fat below a certain level. Give a possible evolutionary explanation for a relationship between body fat in females and reproductive cycles.

©Image Source/Getty Images

CHAPTER

22

Development and Aging

How Long Will You Live?

From the time we are born, our bodies begin to age. Due to advances in medicine and nutrition, our life span expectancy has increased from about 45 years of age in 1900, to just under 80 years of age in the United States today. In some countries, the average life expectancy is over 83 years of age. Increasingly, more individuals are living past the age of 100 (called centenarians). In the United States, there are over 55,000 individuals, or around 0.02% of the population, who are centenarians. The rate is almost triple this in Japan and Okinawa.

In the past several years, researchers have begun to use new molecular techniques to unravel some of the mysteries as to why some individuals live past the age of 100. A variation of one gene, *FOX03A*, has been identified to be more prevalent in centenarians than in the general population. This gene regulates the insulin pathways of the body and appears to control the genetic mechanisms that protect cells against free radicals. Both insulin regulation and protection against free radicals have been known for some time to enhance longevity in a variety of organisms. Interestingly, *FOX03A* is also believed to be involved in the process of apoptosis and may help protect the body against cancer. Although researchers do not think that *FOX03A* is the sole gene for longevity, it is providing a starting point for larger genomic studies on human aging.

In this chapter, we will explore what happens to the human body from the time we are conceived through the development of old age.

As you read through this chapter, think about the following questions:

1. What genetic factors control the process of development in humans?
2. What are the major causes of aging at the cellular and physiological levels?

CHAPTER OUTLINE

22.1 Fertilization and Early Stages of Development

22.2 Processes of Development

22.3 Human Embryonic and Fetal Development

22.4 Human Pregnancy, Birth, and Lactation

22.5 Aging

BEFORE YOU BEGIN

Before beginning this chapter, take a few moments to review the following discussions:

Section 5.4 How does meiosis in animals result in the production of haploid gametes?

Figures 21.3 and 21.7 What are the differences between spermatogenesis and oogenesis?

Sections 21.2 and 21.3 Where does fertilization usually occur, and what hormonal changes happen after fertilization?

433

22.1 Fertilization and Early Stages of Development

Animal development begins with a single cell that multiplies and changes to form a complete organism.

Fertilization

Fertilization is the union of a sperm and an oocyte, resulting in a **zygote**. A sperm has three distinct parts: a *head*, a *middle piece*, and a *tail*. The head contains a nucleus and is capped by a membrane-bound *acrosome*.

In humans, the plasma membrane of the oocyte is surrounded by an extracellular matrix termed the *zona pellucida*. In turn, the zona pellucida is surrounded by a few layers of adhering follicle cells called the *corona radiata*. These cells nourished the oocyte when it was in a follicle of the ovary. During fertilization, which usually occurs in a uterine tube, a sperm cell must make its way to the inside of the oocyte. As illustrated in Figure 22.1, the stages of fertilization are (1) the sperm squeezes through the corona radiata; (2) the sperm releases acrosomal enzymes that allow it to penetrate the zona pellucida; (3) the plasma membrane of the sperm head fuses with the plasma membrane of the oocyte; (4) the sperm nucleus enters the oocyte and releases its chromatin; (5) oocyte *cortical granules* secrete enzymes that turn the zona pellucida into a *fertilization membrane;* and (6) the sperm chromatin reorganizes into a sperm pronucleus that can fuse with the egg pronucleus, forming a single nucleus.

Normally, out of perhaps hundreds of sperm that reach an oocyte, only one is able to penetrate the egg. If more than one sperm enters an oocyte, an event called *polyspermy,* the zygote would have too many chromosomes, and development would be abnormal. Prevention of polyspermy depends on changes in the oocyte's plasma membrane and in the zona pellucida. As soon as a sperm touches an oocyte, the oocyte's plasma membrane depolarizes (from –65 mV to 10 mV), and this prevents the binding of any other sperm. In addition, the enzymes secreted by cortical granules

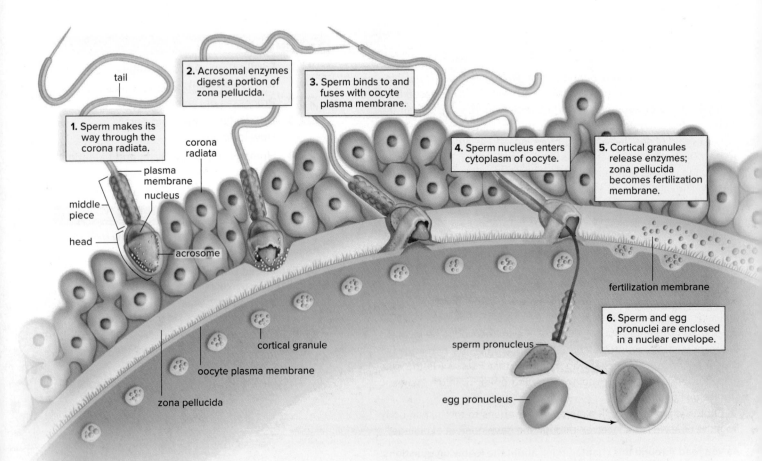

Figure 22.1 Fertilization.   During fertilization, a single sperm enters the oocyte. A sperm makes its way through the corona radiata surrounding the oocyte. The head of a sperm has a membrane-bound acrosome filled with enzymes. When released, these enzymes digest a pathway for the sperm through the zona pellucida. After it binds to the plasma membrane of the oocyte, a sperm enters the oocyte. When the sperm pronucleus joins with the oocyte pronucleus, fertilization is complete.

during formation of the fertilization membrane cause the zona pellucida to lift away from the surface of the oocyte. Now, other sperm cannot bind to the zona pellucida either.

Early Stages of Animal Development

The early stages of animal development occur at the cellular, tissue, and organ levels of organization.

Cellular Stages of Development

The cellular stages of development are (1) cleavage resulting in a multicellular embryo, and (2) formation of the blastula. **Cleavage** is cell division without growth. DNA replication and mitotic cell division occur repeatedly, and the cells get smaller with each division. In other words, cleavage increases only the number of cells. It does not change the original volume of the egg cytoplasm.

As shown in Figure 22.2, cleavage in a lancelet (Fig. 22.2*a*), a primitive chordate, is *equal* (cells of uniform size) and results in a **morula,** which is a ball of cells. The next cellular stage in lancelet development is formation of a **blastula,** which is a hollow ball of cells having a fluid-filled cavity called a **blastocoel.** The blastocoel

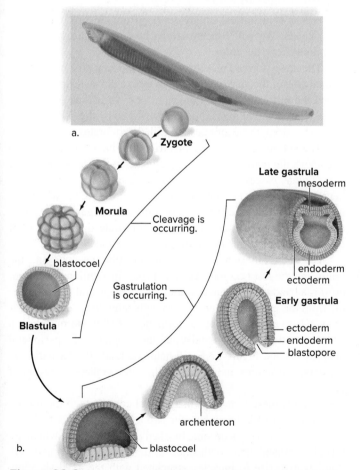

a.

b. **Figure 22.2 Lancelet early development.** **a.** A lancelet. **b.** The early stages of development are exemplified in the lancelet. Cleavage produces a number of cells that form a ball called a morula. The blastula then develops a cavity, the blastocoel. Invagination during gastrulation produces the germ layers ectoderm and endoderm. The mesoderm then arises.

(a): ©Patrick J. Lynch/Science Source

forms when the cells of the morula pump Na$^+$ into extracellular spaces and water follows by osmosis. The water collects in the center, and the result is a hollow ball of cells.

The zygotes of other animals also undergo cleavage and form a morula. In frogs, cleavage is *not equal* because of the presence of yolk, a dense nutrient material. When yolk is present, the zygote and embryo are said to exhibit polarity because the embryo has an *animal pole* (smaller cells) and a *vegetal pole* (larger cells containing yolk). A chicken develops on land and lays a hard-shelled egg containing a large amount of yolk. In the developing chick, the yolk does not participate in cleavage. All vertebrates have a blastula stage, but the appearance of the blastula can be different from that of a lancelet. In the chick, the blastula forms as a layer of cells that spreads out over the yolk. The blastocoel is a space that separates these cells from the yolk:

(chick): ©Small_frog/Getty Images

The blastula of humans resembles that of the chick embryo, but this resemblance cannot be related to the amount of yolk because the human egg contains little yolk. Rather, the evolutionary history of these two animals can explain this similarity. Both birds and mammals are related to reptiles, and all three groups develop similarly, despite a difference in the amount of yolk in their eggs.

Tissue Stages of Development

The tissue stages of development are (1) the early gastrula and (2) the late gastrula. The early gastrula stage begins when certain cells begin to push, or invaginate, into the blastocoel, creating a double layer of cells (Fig. 22.2*b*). Cells migrate during this and other stages of development, sometimes traveling quite a distance before reaching a destination, where they continue developing.

Gastrulation involves the formation of three layers of cells that will develop into adult organs. An early gastrula has two layers of cells. The outer layer of cells is called the **ectoderm,** and the inner layer is called the **endoderm.** The endoderm borders the gut, which at this point is termed either the *archenteron* or the primitive gut. The pore, or hole, created by invagination is the **blastopore,** and in a lancelet, the blastopore eventually becomes the anus. In addition to ectoderm and endoderm, the late gastrula has a middle layer of cells called the **mesoderm.**

Figure 22.2 illustrates gastrulation in a lancelet, and Figure 22.3 compares the lancelet, frog, and chick late gastrula stages. In the lancelet, mesoderm formation begins as outpocketings from the archenteron (Fig. 22.3*a*). These outpocketings will grow in size until they meet and fuse, forming two layers of mesoderm. The space between them is the coelom. The *coelom* is a body cavity lined by mesoderm that contains internal organs. In humans, the coelom becomes the thoracic and abdominal cavities.

In the frog, the cells containing yolk do not participate in gastrulation, and therefore do not invaginate. Instead, a slitlike blastopore

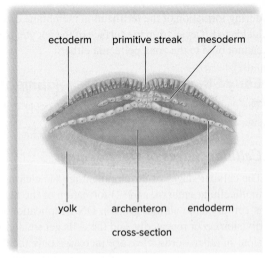

a. Lancelet late gastrula   b. Frog late gastrula   c. Chick late gastrula

Figure 22.3 Comparative development of mesoderm.   a. In the lancelet, mesoderm forms by an outpocketing of the archenteron. **b.** In the frog, mesoderm forms by migration of cells between the ectoderm and endoderm. **c.** In the chick, mesoderm also forms by invagination of cells.

is formed when the animal pole cells begin to invaginate from above, forming endoderm. Animal pole cells also move down over the yolk, to invaginate from below. Some yolk cells, which remain temporarily in the region of the blastopore, are called the *yolk plug.* Mesoderm forms when cells migrate between the ectoderm and endoderm (Fig. 22.3*b*). Later, a splitting of the mesoderm creates the coelom.

The chicken egg contains so much yolk that endoderm formation does not occur by invagination. Instead, an upper layer of cells becomes ectoderm, and a lower layer becomes endoderm. Mesoderm arises by an invagination of cells along the edges of a longitudinal furrow in the midline of the embryo. Because of its appearance, this furrow is called the *primitive streak* (Fig. 22.3*c*). Later, the newly formed mesoderm splits to produce a coelomic cavity.

Ectoderm, mesoderm, and endoderm are called the embryonic **germ layers.** No matter how gastrulation takes place, the result is the same: three germ layers are formed. It is possible to relate the development of future organs to these germ layers (Table 22.1).

TABLE 22.1   Embryonic Germ Layers

Embryonic Germ Layer	Vertebrate Adult Structures
Ectoderm (outer layer)	Nervous system; epidermis of skin; epithelial lining of oral cavity and rectum
Mesoderm (middle layer)	Musculoskeletal system; dermis of skin; cardiovascular system; urinary system; reproductive system—including most epithelial linings; outer layers of respiratory and digestive systems
Endoderm (inner layer)	Epithelial lining of digestive tract and respiratory tract, associated glands of these systems, epithelial lining of urinary bladder

Organ Stages of Development

The organs of an animal's body develop from the three embryonic germ layers. We will concentrate on how one organ system, the nervous system, develops.

The newly formed mesoderm cells lying along the main longitudinal axis of the animal coalesce to form a dorsal supporting rod called the **notochord.** The notochord persists in lancelets, but in frogs, chicks, and humans, it is later replaced by the vertebral column. Therefore, these animals are called vertebrates.

The nervous system develops from midline ectoderm located just above the notochord. At first, a thickening of cells, called the *neural plate,* is seen along the dorsal surface of the embryo. Then, neural folds develop on either side of a neural groove, which becomes the *neural tube* when these folds fuse. Figure 22.4 shows cross-sections of frog development to illustrate the formation of the neural tube. At this point, the embryo is called a *neurula.*

Later, the anterior end of the neural tube develops into the brain, and the rest becomes the spinal cord. In addition, the *neural crest* is a band of cells that develops where the neural tube pinches off from the ectoderm. Neural crest cells migrate to various locations, where they contribute to formation of skin and muscles, in addition to the adrenal medulla and the ganglia of the peripheral nervous system. Figure 22.5 will help you relate the formation of vertebrate structures and organs to the three embryonic layers of cells.

Midline mesoderm cells that did not contribute to the formation of the notochord now become two longitudinal masses of tissue. These two masses become blocked off into *somites,* which are serially arranged along both sides for the length of the notochord. Somites give rise to muscles associated with the axial skeleton, and to the vertebrae. The serial origin of axial muscles and the vertebrae testifies that vertebrates are segmented animals.

A primitive gut tube is formed by endoderm as the body itself folds into a tube. The heart, too, begins as a simple tubular pump. Organ formation continues until the germ layers have given rise to the specific organs listed in Table 22.1.

Figure 22.4 Development of the neural tube and coelom in a frog embryo. **a.** Ectodermal cells that lie above the future notochord (called presumptive notochord) thicken to form a neural plate. **b.** Splitting of the mesoderm produces a coelom, which is completely lined by mesoderm. **c.** A neural tube and a coelom have now developed.

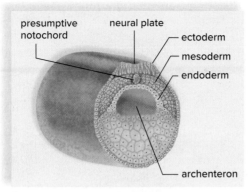

presumptive notochord — neural plate — ectoderm — mesoderm — endoderm — archenteron

a.

notochord — coelom — gut — yolk

b.

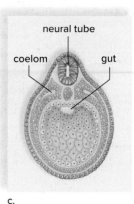

neural tube — coelom — gut

c.

neural tube — somite — notochord

ectoderm
mesoderm
endoderm

gut — coelom

Figure 22.5 Vertebrate embryo, cross-section. At the neurula stage, each of the germ layers, indicated by color (see key), can be associated with the later development of particular parts (see Table 22.1). The somites give rise to the muscles of each segment and to the vertebrae, which replace the notochord in vertebrates.

Check Your Progress 22.1

1. Outline the steps of fertilization, beginning with sperm in the uterine tube.
2. Describe two ways that an oocyte avoids polyspermy.
3. Compare and contrast the cellular and tissue stages of development in the lancelet, frog, and chick.
4. Identify the organ systems formed from each of the primary germ layers.

22.2 Processes of Development

Learning Outcomes

Upon completion of this section, you should be able to

1. Explain how cellular differentiation and morphogenesis play a role in development.
2. Explain how cytoplasmic segregation and induction are able to influence cellular differentiation.
3. Describe how morphogen genes, homeotic genes, and apoptosis help determine pattern formation during development.

Aside from growth, the process of development requires (1) cellular differentiation and (2) morphogenesis. **Cellular differentiation** occurs when cells become specialized in structure and function. For example, a muscle cell looks different and acts differently than a nerve cell. **Morphogenesis** produces the shape and form of the body. One of the earliest indications of morphogenesis is cell migration to form the germ layers. Later, morphogenesis includes the process of **pattern formation,** which directs how tissues and organs are arranged in the body. As we will see, apoptosis, or programmed cell death (see Section 5.1), is an important part of pattern formation.

Cellular Differentiation

At one time, investigators mistakenly believed that irreversible genetic changes must account for cellular differentiation. Perhaps the genes are parceled out as development occurs, they thought, and that's why cells of the body have different structures and functions. However, the process of somatic cell cloning shows that every cell in the body contains all the genes necessary to develop into an organism. In other words, all an organism's genes are in each cell of the body.

The process by which different tissues come about becomes clearer when we consider that specialized cells produce only certain proteins. In other words, we now know that specialization is not due to the parceling out of genes; rather, it is due to differential gene expression. Certain genes and not others are turned on in differentiated cells.

Therefore, in recent years, investigators have turned their attention to discovering the mechanisms that lead to differential gene expression during development. Differentiation must begin long before we can recognize specialized types of cells. Ectodermal, endodermal, and mesodermal cells in the gastrula look quite similar, but they must be different because they develop into different organs.

Investigations by early researchers showed that the cytoplasm of a frog's egg is not uniform. It is polar and has both an anterior/posterior axis and a dorsal/ventral axis, which can be correlated with the **gray crescent,** a gray area that appears after the sperm fertilizes the egg (Fig. 22.6a). When researchers in the lab divided the fertilized egg such that each half contained part of the gray crescent, each experimentally separated daughter cell developed into a complete embryo (Fig. 22.6b). However, if the egg was divided so that only one daughter cell received the gray crescent, only that cell became a complete embryo (Fig. 22.6c). This experiment allowed investigators to hypothesize that the gray crescent contains particular chemical signals needed for development to proceed normally.

a. Frog's egg is polar and has axes.

b. Each cell receives a part of the gray crescent.

c. Only the cell on the left receives the gray crescent.

Figure 22.6 Experimental determination of cytoplasmic influence on development. **a.** A frog's egg has anterior/posterior and dorsal/ventral axes that correlate with the position of the gray crescent. **b.** When researchers cut the fertilized egg such that the gray crescent is divided in half, each daughter cell is capable of developing into a complete tadpole. **c.** If the fertilized egg is cut such that only one daughter cell receives the gray crescent, then only that cell can become a complete embryo.

Cytoplasmic Segregation

The oocyte is now known to contain substances called *maternal determinants* that influence the course of development. These determinants consist of RNAs and proteins synthesized from the maternal genome and stored in the egg. **Cytoplasmic segregation** is the parceling out of maternal determinants as mitosis occurs:

Cytoplasmic segregation helps determine how the various cells of the morula will later develop. As development proceeds, specialization of cells is influenced by maternal determinants and the signals given off by neighboring cells.

Induction

Induction is the ability of one embryonic tissue to influence the development of another tissue by the use of signals called inducers. Inducers are chemical signals that alter the metabolism of the receiving cell and activate particular genes.

Induction and Frog Experiments A frog embryo's gray crescent becomes the dorsal lip of the blastopore, where gastrulation

begins. Because this region is necessary for complete development, early investigators called the dorsal lip of the blastopore the *primary organizer.* The cells closest to the primary organizer become endoderm, those farther away become mesoderm, and those farthest away become ectoderm. This suggests that a molecular concentration gradient may act as a chemical signal (inducer) to cause germ layer differentiation.

The gray crescent of a frog's egg marks the dorsal side of the embryo where the mesoderm becomes notochord and ectoderm becomes nervous system. Experiments have shown that presumptive (potential) notochord tissue induces the formation of the nervous system. If presumptive nervous system tissue, located just above the presumptive notochord, is cut out and transplanted to another region of an embryo, nervous tissue does not develop in that region. The presumptive nervous system is missing some sort of signal to differentiate. On the other hand, if presumptive notochord tissue is cut out and transplanted beneath what would be ectoderm in another region, this other ectoderm does differentiate into nervous tissue. This experiment shows that signals from the presumptive notochord system cause ectoderm to differentiate into nervous tissue.

A well-known series of inductions accounts for the development of the vertebrate eye. An optic vesicle, which is a lateral outgrowth of a developing brain, induces the overlying ectoderm to thicken and become the lens of the eye. The developing lens, in turn, induces an optic vesicle to form an optic cup, where the retina forms.

Induction and Roundworm Experiments Research with the roundworm *Caenorhabditis elegans* has also shown that induction is necessary to the process of differentiation. *C. elegans* is only 1 mm long, and vast numbers can be raised in the laboratory in either petri dishes or a liquid medium. The worm is hermaphroditic, and self-fertilization is the rule. Development of *C. elegans* takes 3 days, and the adult worm contains only 959 cells. Investigators have been able to watch the developmental process from beginning to end, especially because the worm is transparent. Its entire genome has been sequenced, and many modern genetic studies have been done utilizing this worm.

Fate maps, diagrams that trace the differentiation of developing cells, have been developed that show the destiny of each cell of *C. elegans* as it arises following successive cell divisions (Fig. 22.7). Some investigators have studied in detail the development of the vulva, a pore through which eggs are laid. A cell called the *anchor cell* induces the vulva to form. The cell closest to the anchor cell receives the most inducer and becomes the inner vulva. This cell, in turn, produces another inducer, which acts on its two neighboring cells, and they become the outer vulva. Additional work with *C. elegans* has shown that a series of inductions occurs as development proceeds.

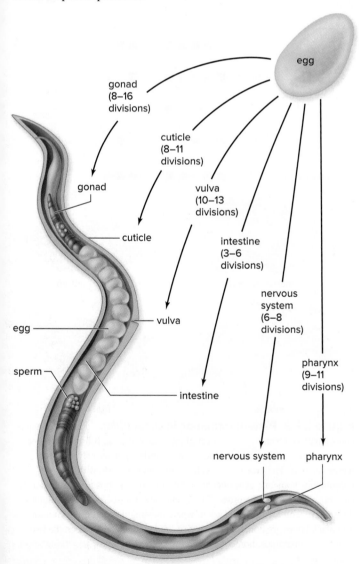

Figure 22.7 Development of *C. elegans,* a nematode. A fate map of the worm, showing that as cells arise by cell division, they are destined to become particular structures.

Morphogenesis

Pattern formation is the ultimate in morphogenesis. To understand the concept of pattern formation, think about the pattern of your own body. Your vertebral column runs along the main axis, and your arms and legs occur in certain locations. Pattern formation refers to how this arrangement of body parts comes about during development. Fruit fly experiments, in particular, have contributed to our knowledge of pattern formation. A few pairs of fruit flies can produce hundreds of offspring in a couple of weeks, all within a small bottle kept on a laboratory bench.

Morphogen Genes

In fruit flies, investigators have discovered certain genes, now called **morphogen genes,** that determine the relationship of individual parts. For example, some genes control which end of the animal will be the head and which the tail, and others determine how many segments the animal will have.

Each of these genes codes for a protein that is present in a gradient—cells at the start of the gradient contain high levels of the protein, and cells at the end of the gradient contain low levels of the protein. These gradients are called morphogen gradients because they determine the shape of the organism (Fig. 22.8). A morphogen gradient is efficient because it has a range of effects, depending on the particular concentration in a portion of the animal. Sequential sets of master genes code for morphogen gradients that activate the next set of master genes in turn. This is one reason development turns out to be so orderly.

Homeotic Genes

Homeotic genes, such as the *Hox* family of genes, act by controlling the identity of each segment. They encode master regulatory proteins that control the expression of other genes, which in turn are responsible for the development of segment-specific structures. Mutations in homeotic genes cause particular segments to develop as if they were located elsewhere in the body. For example, the second thoracic segment of the fruit fly should develop a pair of wings. However, a mutation in the homeotic gene that controls the development of the third thoracic segment allows that segment to develop wings when it should not, resulting in a four-winged fly (Fig. 22.9*a*).

Homeotic genes have now been found in many other organisms, and surprisingly, they all contain the same particular sequence of nucleotides, called a *homeobox*. The homeobox codes for a particular sequence of 60 amino acids called a **homeodomain.** A homeodomain protein binds to DNA and helps determine which genes are turned on:

In *Drosophila,* homeotic genes are located on a single chromosome. In mice and also humans, the same four clusters of homeotic genes are located on four different chromosomes (Fig. 22.9*b*). In all three types of animals, homeotic genes are expressed from anterior

a.

b.

Figure 22.8 Morphogen gradients in the fruit fly. Different morphogen gradients appear as development proceeds. **a.** This early gradient determines which end is the head and which is the tail. The different colors show the concentration gradients of two different proteins. **b.** Another gradient determines the number of segments. (both): ©Steve Paddock, Howard Hughes Medical Research Institute

to posterior in the same order. The first clusters determine the final development of anterior segments of the embryo, while those later in the sequence determine the final development of posterior segments of the embryo.

Because the homeotic genes of so many different organisms contain the same homeodomain, we know that this nucleotide sequence arose early in the history of life and that it has been largely conserved as evolution occurred. In general, it has been intriguing to learn that developmental processes are similar in organisms ranging from yeasts to plants to a wide variety of animals, including humans.

Apoptosis

We have already discussed the importance of **apoptosis** (programmed cell death) in the normal day-to-day operation of the body (see Fig. 5.3). Apoptosis is also an important part of pattern formation in all organisms. In humans, for example, we know that apoptosis is necessary for the shaping of the hands and feet. If it does not occur, the child is born with webbing between the fingers and toes.

The fate maps of *C. elegans* indicate that apoptosis occurs in 131 cells as development takes place. When a cell-death signal is received, an inhibiting protein becomes inactive, allowing a cell-death cascade to proceed, which ends in enzymes destroying the cell.

a.

SEM 50×

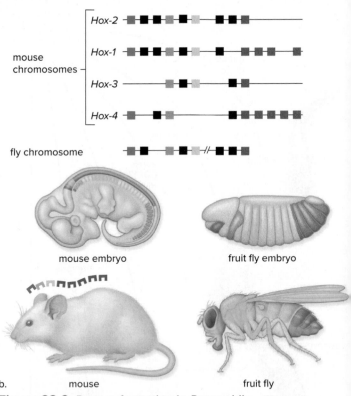

b.

Figure 22.9 Pattern formation in *Drosophila*. Homeotic genes control pattern formation, an aspect of morphogenesis. **a.** If homeotic genes are activated at inappropriate times, abnormalities such as a fly with four wings occur. **b.** The green, blue, yellow, and red colors show that homologous homeotic genes occur on four mouse chromosomes and on a fly chromosome in the same order. These genes are color-coded to the region of the embryo, and therefore the adult, where they regulate pattern formation. The black boxes are homeotic genes that are not identical between the two animals. In mammals, homeotic genes are called *Hox* genes. (a): ©David Scharf/ Science Source

22.3 Human Embryonic and Fetal Development

Learning Outcomes

Upon completion of this section, you should be able to

1. Identify the extraembryonic membranes and provide a function for each.

2. Summarize, in chronological order, the major events that occur during fetal development from months three to nine.

3. Describe the flow of blood in a fetus and explain the role of the placenta.

In humans, the length of time from *conception* (fertilization followed by implantation) to birth is approximately 9 months. It is customary to calculate a baby's "due date" by adding 280 days to the start of the last menstrual period because this date is usually known, whereas the day of fertilization is usually unknown. Because the time of birth is influenced by so many variables, only about 5% of babies actually arrive on the forecasted date.

Human development before birth is often divided into embryonic development (months one and two) and fetal development (months three to nine). **Embryonic development** consists of early formation of the major organs, and fetal development is the refinement of these structures.

Extraembryonic Membranes

The **extraembryonic membranes** lie outside of the embryo. The names of these membranes are derived from their function in reptiles and birds. In reptiles, these membranes made development on land possible. If an embryo develops in the water, the water supplies oxygen for the embryo and takes away waste products. The surrounding water prevents desiccation, or drying out, and provides a protective cushion. For animals that live on land, all these functions are performed by the extraembryonic membranes.

In the chick, the extraembryonic membranes develop from extensions of the germ layers, which spread out over the yolk. Figure 22.10*a* shows the chick surrounded by the membranes. The **chorion** lies next to the shell and carries on gas exchange. The **amnion** contains the protective amniotic fluid, which bathes the developing embryo. The **allantois** collects nitrogenous wastes, and the **yolk sac** surrounds the remaining yolk, which provides nourishment.

Humans (and other mammals) also have these extraembryonic membranes (Fig. 22.10*b*). As will be discussed further in this section, the functions of the extraembryonic membranes in humans have been modified to suit internal development. However, their very presence indicates our relationship to birds and to reptiles. It is interesting to note that all chordate animals develop in water, either in bodies of water or within amniotic fluid.

Embryonic Development

Embryonic development encompasses the first 2 months of development following fertilization.

The First Week

Fertilization usually occurs in the distal third (closest to the ovary) of a uterine tube (Fig. 22.11), and cleavage begins even as the

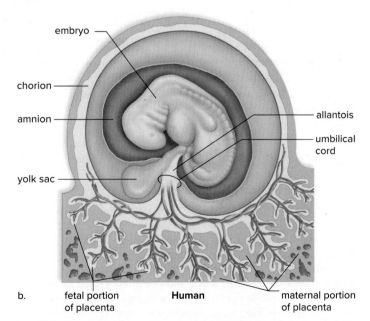

Figure 22.10 Extraembryonic membranes. Extraembryonic membranes, which are not part of the embryo, are found during the development of (**a**) chicks and (**b**) humans. Each has a specific function.

embryo passes down this duct to the uterus. By the time the embryo reaches the uterus on the third day, it is called a *morula*. By about the fifth day, the morula is transformed into the blastocyst. The **blastocyst** has a fluid-filled cavity, a single layer of outer cells called the **trophoblast,** and an inner cell mass. Later, the trophoblast, reinforced by a layer of mesoderm, gives rise to the chorion, one of the extraembryonic membranes (see Fig. 22.10). The chorion develops into the fetal half of the placenta. The inner cell mass eventually becomes the embryo, which develops into a fetus.

The Second Week

At the end of the first week, the embryo begins the process of implanting in the wall of the uterus. The trophoblast secretes enzymes to digest away some of the tissue and blood vessels of the endometrium of the uterus (Fig. 22.11). The embryo is now about the size of the period at the end of this sentence. The trophoblast begins to secrete the hormone **human chorionic gonadotropin (HCG),** which is the basis for the pregnancy test and serves to maintain the corpus luteum past the time it normally disintegrates (see Section 21.3). Because of this, the endometrium is maintained and menstruation does not occur.

As the week progresses, the inner cell mass detaches itself from the trophoblast, and two more extraembryonic membranes form (Fig. 22.12*a*). The yolk sac, which forms below the embryonic disk, has no nutritive function as it does in chicks, but it is the first site of blood cell formation. The amnion has a cavity where the embryo (and then the fetus) develops. In humans, amniotic fluid acts as an insulator against cold and heat and also absorbs shock, such as that caused by the mother exercising.

Gastrulation occurs during the second week. The inner cell mass now has flattened into the **embryonic disk,** composed of two layers of

cells: ectoderm above and endoderm below. Once the embryonic disk elongates to form the primitive streak, the third germ layer, mesoderm, forms by invagination of cells along the streak. The trophoblast is reinforced by mesoderm and becomes the chorion (Fig. 22.12*b*). It is possible to relate the development of future organs to these germ layers (see Table 22.1).

The Third Week

Two important organ systems appear during the third week. The nervous system is the first organ system to be visually evident. At first, a thickening appears along the entire dorsal length of the embryo, and then invagination occurs as neural folds appear. When the neural folds meet at the midline, the neural tube, which later develops into the brain and the nerve cord, is formed (see Fig. 22.4). After the notochord is replaced by the vertebral column, the nerve cord is called the spinal cord.

Development of the heart begins in the third week and continues into the fourth week. At first, there are right and left heart tubes. When these fuse, the heart begins pumping blood, even though its chambers are not fully formed.

Figure 22.11 Human development before implantation. Structures and events proceed counterclockwise. (1) At ovulation, the secondary oocyte leaves the ovary. A single sperm nucleus enters the oocyte and (2) fertilization occurs in the uterine tube. As the zygote moves along the uterine tube, it undergoes (3) cleavage to produce (4) a morula. (5) The blastocyst forms and (6) implants itself in the uterine lining.

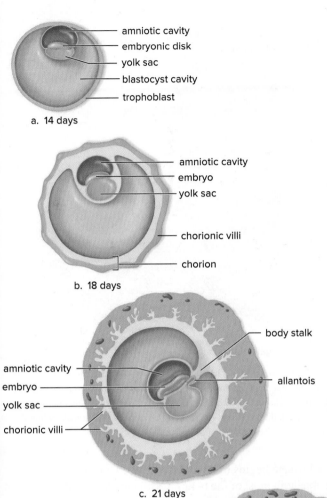

a. 14 days

- amniotic cavity
- embryonic disk
- yolk sac
- blastocyst cavity
- trophoblast

b. 18 days

- amniotic cavity
- embryo
- yolk sac
- chorionic villi
- chorion

c. 21 days

- body stalk
- allantois
- amniotic cavity
- embryo
- yolk sac
- chorionic villi

The Fourth and Fifth Weeks

At four weeks, the embryo is barely larger than the height of this print. A bridge of mesoderm called the *body stalk* connects the caudal (tail) end of the embryo with the chorion, which has treelike projections called *chorionic villi* (Fig. 22.12*c, d*). The fourth extra-embryonic membrane, the allantois, is contained within this stalk, and its blood vessels become the umbilical blood vessels. The head and the tail then lift up, and the body stalk moves anteriorly by constriction. Once this process is complete, the **umbilical cord,** which connects the developing embryo to the placenta, is fully formed (Fig. 22.12*e*).

Little flippers called *limb buds* appear (Fig. 22.13). Later, the arms and the legs develop from the limb buds, and even the hands and the feet become apparent. At the same time—during the fifth week—the head enlarges, and the sense organs become more prominent. It is possible to make out the developing eyes, ears, and even the nose.

The Sixth Through Eighth Weeks

During the sixth through eighth weeks of development, the embryo becomes more recognizable as human. Concurrent with brain development, the head achieves its normal relationship with the body as a neck region develops. The nervous system is developed well enough to permit reflex actions, such as a startle response to touch. At the end of this period, the embryo is about 38 mm (1.5 inches) long and weighs less than 1 gram (g), even though all organ systems have been established.

Fetal Development

Fetal development includes the third through ninth months of development. At this time, the fetus looks human (Fig. 22.14).

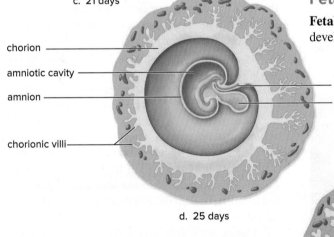

d. 25 days

- chorion
- amniotic cavity
- amnion
- chorionic villi
- allantois
- yolk sac

e. 35+ days

- chorion
- digestive tract
- chorionic villi
- amnion
- amniotic cavity
- umbilical cord

Figure 22.12 Human embryonic development. **a.** At first, the embryo contains no organs, only tissues. The amniotic cavity is above the embryonic disk, and the yolk sac is below. **b.** The chorion develops villi, the structures so important to the exchange between mother and child. **c, d.** The allantois and yolk sac, two more extraembryonic membranes, are positioned inside the body stalk as it becomes the umbilical cord. **e.** At 35+ days, the embryo has a head region and a tail region. The umbilical cord takes blood vessels between the embryo and the chorion (placenta).

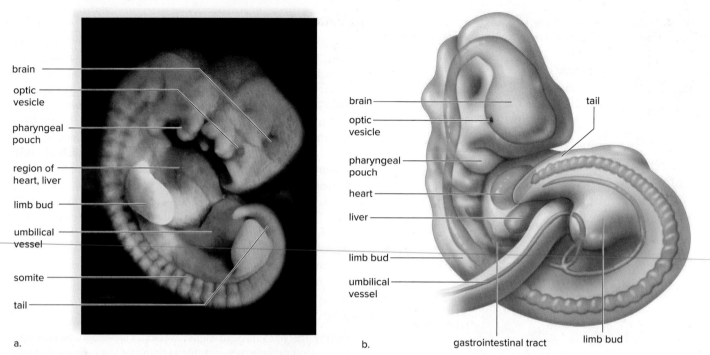

brain

optic
vesicle

pharyngeal
pouch

region of
heart, liver

limb bud

umbilical
vessel

somite

tail

a.

brain

optic
vesicle

pharyngeal
pouch

heart

liver

limb bud

umbilical
vessel

tail

gastrointestinal tract   limb bud

b.

Figure 22.13 Human embryo at beginning of fifth week.   **a.** Scanning electron micrograph. **b.** The embryo is curled so that the head touches the heart and liver, the two organs whose development is farther along than the rest of the body. The organs of the gastrointestinal tract are forming, and the arms and the legs develop from the limb buds. The bones of the tail regress and become those of the coccyx (tailbone). The pharyngeal pouches become functioning gills only in fishes and amphibian larvae. In humans, the first pair of pharyngeal pouches becomes the auditory tubes. The second pair becomes the tonsils, while the third and fourth pairs become the thymus and the parathyroid glands. (a): ©Anatomical Travelogue/Science Source

The Third and Fourth Months

At the beginning of the third month, the fetal head is still relatively large, the nose is flat, the eyes are far apart, and the ears are well formed. Head growth now begins to slow down as the rest of the body increases in length. Epidermal refinements, such as eyelashes, eyebrows, hair on head, fingernails, and nipples, appear.

Figure 22.14 Three- to four-month-old fetus.   At this stage, the fetus looks human, with its face, hands, and fingers well defined.
©John Watney/Science Source

Cartilage begins to be replaced by bone as ossification centers appear in most of the bones. Cartilage remains at the ends of the long bones, and ossification is not complete until age 18 or 20 years. The skull has six large membranous areas called **fontanels,** which permit a certain amount of flexibility as the head passes through the birth canal and allow rapid growth of the brain during infancy.

Sometime during the third month, it is possible to distinguish males from females. The presence or absence of the *SRY* gene on the Y chromosome determines whether the gonads will differentiate into testes or ovaries. Once these have differentiated, they produce the sex hormones (see Section 20.6) that influence the differentiation of the genital tract. At this time, either testes or ovaries are located within the abdominal cavity, but later, in the last trimester of fetal development, the testes descend into the scrotal sacs (see Section 21.1).

During the fourth month, the fetal heartbeat is loud enough to be heard when a physician applies a stethoscope to the mother's abdomen. By the end of this month, the fetus is about 152 mm (6 in.) in length and weighs about 170 g (6 oz).

The Fifth Through Seventh Months

During the fifth through seventh months, the mother begins to feel movement. At first, there is only a fluttering sensation, but as the fetal legs grow and develop, kicks and jabs are felt. The fetus, though, is in the fetal position, with the head bent down and in contact with the flexed knees.

The wrinkled, translucent skin is covered by a fine down called **lanugo.** This, in turn, is coated with a white, greasy, cheeselike substance called **vernix caseosa,** which probably protects the delicate skin from the amniotic fluid. The eyelids are now fully open.

At the end of this period, the length has increased to about 300 mm (12 in.), and the weight is about 1,380 g (3 lb). If born now, it is likely that the baby will survive.

Fetal Circulation

The fetus has circulatory features that are not present in the adult circulation (Fig. 22.15). All of these features are necessary because the fetus does not use its lungs for gas exchange. For example, much of the blood entering the right atrium is shunted into the

Figure 22.15 Fetal circulation and the placenta.   a. The lungs are not functional in the fetus, and the blood passes directly from the right atrium to the left atrium or from the right ventricle to the aorta. The umbilical arteries take fetal blood to the placenta, and the umbilical vein returns fetal blood from the placenta. **b.** At the placenta, an exchange of molecules between fetal and maternal blood takes place across the walls of the chorionic villi. Oxygen and nutrient molecules diffuse into the fetal blood, and carbon dioxide and urea diffuse out of the fetal blood.

SCIENCE IN YOUR LIFE ▶ HEALTH

Preventing and Testing for Birth Defects

Birth defects, or congenital disorders, are abnormal conditions that are present at birth. According to the Centers for Disease Control and Prevention (CDC), about 1 in 33 babies born in the United States has a birth defect. Those due to genetics can sometimes be detected before birth by one of the methods shown in Figure 22A; however, these methods are not without risk. Some birth defects are not serious, and not all can be prevented. But women can take steps to increase their chances of delivering a healthy baby.

Eat a Healthy Diet

Certain birth defects occur because the developing embryo does not receive sufficient nutrition. For example, women of childbearing age are urged to make sure they consume adequate amounts of the vitamin folic acid in order to prevent neural tube defects such as spina bifida and anencephaly. In spina bifida, part of the vertebral column fails to develop properly and cannot adequately protect the delicate spinal cord. With anencephaly, most of the fetal brain fails to develop. Anencephalic infants are stillborn or survive for only a few days after birth. Fortunately, folic acid is plentiful in leafy green vegetables, nuts, and citrus fruits. It is also present in multivitamins, and many breads and cereals are fortified with it. The CDC recommends that all women of childbearing age get at least 400 micrograms (μg) of folic acid every day through supplements, in addition to eating a healthy, folic acid–rich diet. Unfortunately, neural tube birth defects can occur just a few weeks after conception, when many women are still unaware that they are pregnant—especially if the pregnancy is unplanned.

Avoid Alcohol, Smoking, and Drug Abuse

As shown in Figure 22.15, the placenta is the region for exchange between the fetal and maternal circulatory systems. Alcohol consumption during pregnancy is a leading cause of birth defects. In severe instances, the baby is born with fetal alcohol syndrome (FAS), estimated to occur in 0.2 to 1.5 of every 1,000 live births in the United States. Children with FAS are frequently underweight, have an abnormally small head, abnormal facial development, and intellectual disability. As they mature, children with FAS often exhibit short attention span, impulsiveness, and poor judgment, as well as serious difficulties with learning and memory. Heavy alcohol use also reduces a woman's folic acid level, increasing the risk of the neural tube defects just described.

Cigarette smoking causes many birth defects. Babies born to smoking mothers typically have low birth weight, and are more likely to have defects of the face, heart, and brain than those born to nonsmokers. Illegal drugs should also be avoided. For example, cocaine causes blood pressure fluctuations that deprive the fetus of oxygen. Cocaine-exposed babies may have problems with vision and coordination, and may be intellectually disabled.

Alert Medical Personnel If You Are or May Be Pregnant

Several medications that are safe for healthy adults may pose a risk to a developing fetus. If pregnant women need to be immunized, they are usually given killed or inactivated forms of the vaccine, because live forms often present a danger to the fetus. Because the rapidly dividing cells of a developing embryo or fetus are very susceptible to damage from radiation, pregnant women should avoid unnecessary X-rays. If a medical X-ray is unavoidable, the woman should notify the X-ray technician so her fetus can be protected as much as possible, for example, by draping a lead apron over her abdomen if this area is not targeted for X-ray examination.

Avoid Infections That Cause Birth Defects

Certain pathogens, such as rubella, toxoplasmosis, herpes simplex, and cytomegalovirus, may cause birth defects.

Question to Consider

1. Besides tobacco, alcohol, and illegal drugs, what are some other potential risks a pregnant woman should avoid? Why?

a. Amniocentesis

b. Chorionic villus sampling

c. Preimplantation genetic diagnosis

Figure 22A Three methods for genetic defect testing before birth. About 20% of all birth defects are due to genetic or chromosomal abnormalities, which may be detected before birth. **a.** Amniocentesis is usually performed from the fifteenth to the seventeenth week of pregnancy. **b.** Chorionic villus sampling is usually performed from the eighth to the twelfth week of pregnancy. **c.** Preimplantation genetic diagnosis can be performed prior to in vitro fertilization, either on oocytes that have been collected from the woman, or on the early embryo.

left atrium through the **foramen ovale** between the two atria. Also, any blood that does enter the right ventricle and is pumped into the pulmonary trunk is shunted into the aorta by way of the **ductus arteriosus.**

Blood within the aorta travels to the various branches, including the iliac arteries, which connect to the *umbilical arteries* leading to the placenta. Exchange of gases and nutrients between maternal blood and fetal blood takes place at the placenta. The *umbilical vein* carries blood rich in nutrients and oxygen to the fetus. The umbilical vein enters the liver and then joins the *ductus venosus,* which merges with the inferior vena cava, a vessel that returns blood to the heart. The umbilical arteries and vein run alongside one another in the umbilical cord, which is tied off and cut at birth, leaving only the umbilicus (navel).

The most common of all cardiac defects in the newborn is the persistence of the foramen ovale. Once the umbilical cord has been severed and the lungs have expanded, blood enters the lungs in quantity. The return of this blood to the left side of the heart usually causes a flap to cover the opening. Incomplete closure occurs in nearly one out of four individuals, but even so, passage of the blood from the right atrium to the left atrium rarely occurs because either the opening is small or it closes when the atria contract. In a small number of cases, the passage of impure blood from the right side to the left side of the heart is sufficient to cause a "blue baby." Such a condition can usually be corrected by threading a catheter into the heart, and sealing the defect.

The ductus arteriosus normally closes because endothelial cells divide and block off this duct. Remains of the ductus arteriosus and parts of the umbilical arteries and vein are later transformed into connective tissue.

Structure and Function of the Placenta Humans belong to the group of mammals called placental mammals. The **placenta** is firmly attached to the uterine wall by the allantois and by fingerlike projections called the chorionic villi. The placenta is a structure that functions only before birth. When the child is born, the placenta becomes part of the *afterbirth.*

The placenta functions in gas, nutrient, and waste exchange between the embryonic (later fetal) and maternal circulatory systems. The placenta begins to form once the embryo is fully implanted. At first, the entire chorion has chorionic villi that project into the endometrium. Later, these disappear in all areas except where the placenta develops. By the tenth week, the placenta is fully formed and is producing progesterone and estrogen. These hormones have two effects: (1) due to their negative feedback control of the hypothalamus and the anterior pituitary, they prevent any new follicles from maturing, and (2) they maintain the lining of the uterus so that the corpus luteum is no longer needed. Usually no menstruation occurs during pregnancy.

The placenta has a fetal side contributed by the chorion and a maternal side consisting of uterine tissues. Notice in Figure 22.15 how the chorionic villi are surrounded by maternal blood. Normally, maternal and fetal blood never mix, because exchange always takes place across the walls of the chorionic villi.

The umbilical cord stretches between the placenta and the fetus. The umbilical cord is the lifeline of the fetus because it contains the umbilical arteries and vein, which transport waste molecules (carbon dioxide and urea) to the placenta for disposal and take oxygen and nutrient molecules from the placenta to the rest of the fetal circulatory system.

As discussed in the Health feature "Preventing and Testing for Birth Defects," harmful chemicals, such as alcohol, can also cross the placenta. This is of particular concern during the embryonic period, when various structures are first forming. Each organ or part seems to have a sensitive period during which a substance can alter its normal development. Even over-the-counter drugs should not be taken during pregnancy without permission from your physician.

Check Your Progress 22.3

1. Name the extraembryonic membrane that gives rise to each of the following: umbilical blood vessels, the first blood cells, and the fetal half of the placenta.
2. Summarize the major events by month during fetal development.
3. Describe the path of blood flow in the fetus starting with the placenta, and name the structures unique to fetal circulation.
4. Explain the function of the placenta.

22.4 Human Pregnancy, Birth, and Lactation

Learning Outcomes

Upon completion of this section, you should be able to
1. Describe changes that occur in female physiology during pregnancy.
2. Outline the stages of birth.
3. Summarize the advantages of breast-feeding.

Pregnancy

Many changes that take place in the mother's body during pregnancy are due to the hormones progesterone and estrogen. Others are due to the increasing size of the uterus, from a nonpregnant weight of 60–80 g to 900–1,200 g at term.

Morning Sickness and Energy Level

About 6 weeks into her pregnancy, the mother may experience nausea and vomiting, loss of appetite, and fatigue. These symptoms usually subside by around the 12th week. Subsequently, some mothers then report increased energy levels and a general sense of well-being despite gaining weight. The weight gain of pregnancy is due to breast and uterine enlargement; weight of the fetus; amount of amniotic fluid; size of the placenta; the mother's own increase in total body fluid; and an increase in storage of proteins, fats, and minerals.

Effects on Smooth Muscle

Progesterone decreases uterine motility by relaxing smooth muscle, including the smooth muscle in the walls of arteries. The arteries expand, and this leads to a low blood pressure that sets in motion the renin-angiotensin-aldosterone mechanism (see Section 20.4). Aldosterone promotes sodium and water retention, and blood volume increases until it reaches its peak sometime during weeks 28 to 32 of pregnancy. Altogether, blood volume increases from the normal 5 liters up to 7 liters—a 40% rise. An increase in the number of red blood cells follows. With the rise in blood volume, cardiac output increases by 20–30%. Blood flow to the kidneys, placenta, skin, and breasts rises significantly. Smooth muscle relaxation also explains the common gastrointestinal effects of pregnancy. The heartburn experienced by many is due to relaxation of the esophageal sphincter and reflux of stomach contents into the esophagus. Constipation can be a result of decreased intestinal tract motility.

Other Effects

Compression of the ureters and urinary bladder by an enlarged uterus can result in incontinence, or the involuntary leakage of urine from the bladder. Compression of the inferior vena cava, especially when lying down, decreases venous return, and the result is edema (fluid accumulation) and varicose veins.

In addition to the steroid hormones progesterone and estrogen, the placenta also produces some peptide hormones. Several of these may make cells resistant (unresponsive) to insulin, and the result can be gestational diabetes. Hormones released by the placenta, rather than stretching of the skin, may also be the cause of *striae gravidarum*, commonly called "stretch marks," which form over the abdomen and lower breasts during pregnancy. Melanocyte activity also increases during pregnancy. Darkening of certain areas of the skin, including the face, neck, and breast areolae, is common.

Birth

The uterus has contractions during the last trimester of the pregnancy. At first, these are light, lasting about 20 to 30 seconds and occurring every 15 to 20 minutes. Near the end of pregnancy, the contractions may become stronger and more frequent. However, the onset of true labor is marked by uterine contractions that occur regularly every 10 to 15 minutes and last for 40 seconds or longer.

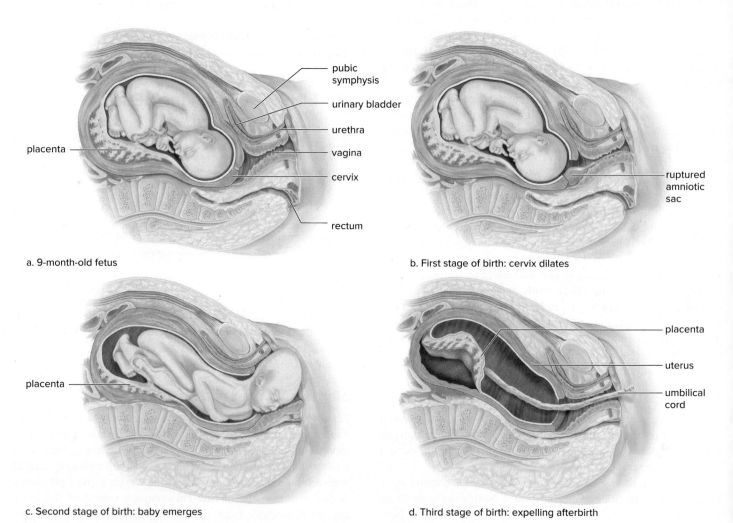

a. 9-month-old fetus

b. First stage of birth: cervix dilates

c. Second stage of birth: baby emerges

d. Third stage of birth: expelling afterbirth

Figure 22.16 Three stages of parturition (birth).   **a.** Position of fetus just before birth begins. **b.** Dilation of cervix. **c.** Birth of baby. **d.** Expulsion of afterbirth.

These contractions are generally called "labor pains," although when they first appear, they are not usually painful and only become painful at a later stage.

A positive feedback mechanism regulates the onset and continuation of labor. Uterine contractions are induced by stretching of the cervix, which also brings about the release of oxytocin from the posterior pituitary. Oxytocin stimulates uterine contractions, which push the fetus downward, and the cervix stretches even more. This cycle keeps repeating itself until the baby is born. Three events, in any order, indicate that delivery will soon occur: (1) The uterine contractions are occurring about every 5 minutes and becoming stronger. (2) The amnion, which contains amniotic fluid, ruptures, causing water to flow out of the vagina. This event is sometimes referred to as "breaking water." (3) A plug of mucus from the cervical canal leaves the vagina. This plug prevents bacteria and sperm from entering the uterus during pregnancy. Its expulsion may be the least noticeable of the events, unless it is mixed with blood.

Stage 1

Parturition, the process of giving birth to an offspring, can be divided into three stages (Fig. 22.16). Prior to the first stage, there can be a "bloody show" caused by expulsion of the mucus plug. At first, the uterine contractions of labor occur in such a way that the cervical canal slowly disappears as the lower part of the uterus is pulled upward toward the baby's head (Fig. 22.16*b*). This process is called effacement, or "taking up the cervix." With further contractions, the baby's head acts as a wedge to assist cervical dilation. If the amniotic membrane has not already ruptured, it is apt to do so during this stage, releasing the amniotic fluid, which leaks out the vagina. The first stage of parturition ends once the cervix is dilated completely.

Stage 2

During the second stage of parturition, the uterine contractions occur every 1 to 2 minutes and last about 1 minute each. They are accompanied by a desire to push, or bear down. As the baby's head gradually descends into the vagina, the desire to push becomes greater. When the baby's head reaches the exterior, it turns so that the back of the head is uppermost (Fig. 22.16*c*). If the vaginal orifice does not expand enough to allow passage of the head, an **episiotomy** may be performed. This incision, which enlarges the opening, is sewn together later. As soon as the head is delivered, the baby's shoulders rotate so that the baby faces either to the right or the left. At this time, the physician may hold the head and guide it downward, while one shoulder and then the other emerges. The rest of the baby follows easily.

Once the baby is breathing normally, the umbilical cord is cut and tied, severing the child from the placenta. The stump of the cord shrivels and leaves a scar, which is the umbilicus.

Stage 3

The placenta, or **afterbirth,** is delivered during the third stage of parturition (Fig. 22.16*d*). About 15 minutes after delivery of the baby, uterine muscular contractions shrink the uterus and dislodge

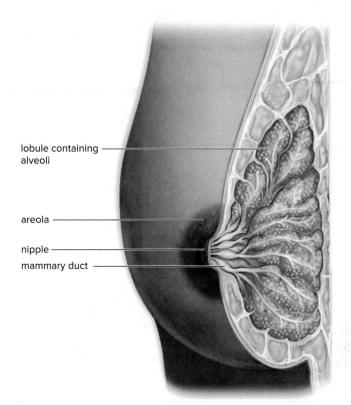

lobule containing alveoli

areola

nipple

mammary duct

Figure 22.17 Female breast anatomy.   The female breast contains lobules consisting of ducts and alveoli. The alveoli are lined by milk-producing cells in the lactating (milk-producing) breast. Milk droplets congregate in the ducts that enter sinuses just behind the nipple. When a baby suckles, the sinuses are squeezed.

the placenta. The placenta is then expelled into the vagina. As soon as the placenta and its membranes are delivered, the third stage of parturition is complete.

Female Breast and Lactation

A female breast contains 15 to 25 lobules, each with a milk duct. Each milk duct begins at the nipple and divides into numerous other ducts that end in blind sacs called alveoli (Fig. 22.17).

During pregnancy, the breasts enlarge as the ducts and alveoli increase in number and size. The same hormones that affect the mother's breasts can also affect the child's. Some newborns, including males, even secrete a small amount of milk for a few days.

Usually, no milk is produced during pregnancy. The hormone prolactin is needed for lactation to begin, and the production of this hormone is suppressed because of the feedback control that the increased amount of estrogen and progesterone during pregnancy has on the pituitary. Once the baby is delivered, however, the pituitary begins secreting prolactin. It takes a couple of days for milk production to begin, and in the meantime, the breasts produce **colostrum,** a thin, yellow, milky fluid rich in protein, including antibodies.

The continued production of milk requires a suckling child. When a breast is suckled, the nerve endings in the areola are stimulated, and a nerve impulse travels along neural pathways from the nipples to the hypothalamus, which directs the pituitary gland to release the hormone oxytocin. When this hormone arrives at the breast, it causes contraction of the lobules so that milk flows into the ducts (called milk letdown), where it may be drawn out of the nipple by the suckling child.

Whether to breast-feed or not is a private decision based in part on a woman's particular circumstances. However, it is well known that breast milk contains antibodies produced by the mother that can help a baby survive. Babies have immature immune systems, less stomach acid to destroy foreign antigens, and also unsanitary habits. Breast-fed babies are less likely to develop stomach and intestinal illnesses, including diarrhea, during the first 13 weeks of life. Breast-feeding also has physiological benefits for the mother. Suckling causes uterine contractions that can help the uterus return to its normal size, and breast-feeding uses up calories and thus can help a woman return to her normal weight.

Check Your Progress 22.4

1. Describe the physiological changes that occur in a pregnant woman's circulatory system.
2. Identify the stage of parturition in which the baby is delivered.
3. Define colostrum and list three advantages of breast-feeding.

22.5 Aging

Learning Outcomes

Upon completion of this section, you should be able to
1. Describe the factors that contribute to aging.
2. Discuss the effects of aging on various body systems.

Development does not cease once birth has occurred but continues throughout the stages of life: infancy, childhood, adolescence, and adulthood. Infancy, the toddler years, and preschool years are times of remarkable growth. During the birth to 5-year-old stage, humans acquire gross motor and fine motor skills. These include the ability to sit up and then to walk, as well as being able to hold a spoon and manipulate small objects. Language usage begins during this time and will become increasingly sophisticated throughout childhood. As infants and toddlers explore their environment, their senses—vision, taste, hearing, smell, and touch—mature dramatically. Socialization is very important, as a child forms emotional ties with its caregivers and learns to separate self from others. Babies do not all develop at the same rate, and there is a large variation in what is considered normal.

The preadolescent years, from 6 to 12 years of age, are a time of continued rapid growth and learning. Preadolescents form identities apart from parents, and peer approval becomes very important. Adolescence begins with the onset of *puberty,* as the young person achieves sexual maturity. For girls, puberty begins between 10 and 14 years of age, whereas for boys it generally occurs between ages 12 and 16. During this time, the sex-specific hormones cause the secondary sexual characteristics to appear. Profound

social and psychological changes are also associated with the transition from childhood to adulthood.

Aging encompasses these progressive changes, which contribute to an increased risk of infirmity, disease, and death. Today, there is great interest in **gerontology,** the study of aging, because our society includes more older individuals than ever before. In the next half-century, the number of people over age 65 will increase by 147%. The human *life span* is judged to be a maximum of 120 to 125 years. The goal of gerontology is not necessarily to increase the life span, but to increase the *health span,* the number of years an individual enjoys the full functions of all body parts and processes.

Aging and death are as much a part of biology as are embryonic development and birth. If animals—and indeed, all forms of life—did not age and eventually die, their exponentially increasing numbers would presumably overwhelm the ability of the Earth to support life. Moreover, aging and death are a critical part of the evolutionary process, because animals that age and die are continuously replaced by individuals with new genetic combinations that can adapt to environmental changes that occur.

Before discussing the effects of aging on the various body systems, it is useful to look at some of the reasons why our bodies age.

Factors Influencing Aging

Aging is a complex process affected by multiple factors. Most scientists who study gerontology believe that aging is partly genetically preprogrammed. This idea is supported by the observation that longevity runs in families—that is, the children of long-lived parents tend to live longer than those of short-lived parents. As would also be expected, studies show that identical twins have a more similar life span than nonidentical twins.

Hormones

Many laboratory studies of aging have been performed in the nematode *Caenorhabditis elegans,* in which single-gene mutations have been shown to influence the life span. For example, mutations that decrease the activity of a hormone receptor similar to the insulin receptor more than double the life span of the worms, which also behave and look like much younger worms throughout their prolonged lives. Interestingly, small breed dogs, such as poodles and terriers, which may live 15 to 20 years, have lower levels of an analogous receptor compared to large breed dogs that live 6 to 8 years.

Telomeres

Studies of the behavior of cells grown in the lab also suggest a genetic influence on aging. Most types of differentiated cells can divide only a limited number of times. One factor that may control the number of cell divisions is the length of the telomeres, sequences of DNA, at the ends of chromosomes. Telomeres protect the ends of chromosomes from deteriorating or fusing with other chromosomes. Each time a cell divides, the telomeres normally shorten, and cells with shorter telomeres tend to undergo fewer divisions (see Section 5.1). Some cells, such as stem cells, possess an enzyme called telomerase, which replenishes the length of the telomeres, effectively making stem cells immortal. Cancer cells, which behave in a similar manner to stem cells, frequently possess an active telomerase enzyme, which allows them to replicate continuously (see Section 25.5). Studies using stem cells and cancer cells have begun to close in on the genetic factors that cause cellular aging. For example, in 2012, researchers used gene therapy (see Section 26.3) to introduce an active telomerase enzyme into mice, thus slowing the aging process.

Mitochondria and Diet

The mitochondria are the powerhouses of the cell (see Section 3.3). As the mitochondria harvest the energy contained in carbohydrates, fats, and proteins, they generate free radicals. Free radicals are unstable molecules that carry an extra electron. To become stable, free radicals donate an electron to another molecule, such as DNA, proteins (e.g., enzymes), or lipids found in plasma membranes. Eventually, these molecules are unable to function, and the cell loses internal functions. This may lead to cell death. Scientists have determined that high-calorie diets increase the levels of free radicals, thus accelerating cellular aging. Multiple studies on model organisms, including mice, have supported that a low-calorie diet can extend the life span. It is also possible to reduce the negative effects of free radicals by increasing one's consumption of natural antioxidants, such as those present in brightly colored and dark-green vegetables and fruits. Chemicals in nuts, fish, shellfish, and red wine have also been shown to reduce our exposure to free radicals and slow the aging process.

Damage Accumulation

Another set of hypotheses proposes that aging involves the accumulation of damage over time. In 1900, the average human life expectancy was around 45 years. A baby born in the United States this year has a life expectancy of about 78 years. Because human genes have presumably not changed much in such a short time, most of this gain in life span is due to better medical care, along with the application of scientific knowledge about how to prolong our lives.

Two basic types of cellular damage can accumulate over time. The first type can be thought of as agents that are unavoidable—for example, the accumulation of harmful DNA mutations or the buildup of harmful metabolites. A second form involves the fact that proteins—such as the collagen fibers present in many support tissues—may become increasingly cross-linked as people age. This cross-linking may account for the inability of such organs as the blood vessels, heart, and lungs to function as they once did. Some researchers have now found that glucose has the tendency to attach to any type of protein, which is the first step in a cross-linking process. They are currently experimenting with drugs that can prevent cross-linking. However, other sources of cellular damage may be avoidable, such as a poor diet or exposure to the sun.

The Effects of Aging on Organ Systems

Integumentary System

As aging occurs, the skin becomes thinner and less elastic because the number of elastic fibers decreases and the collagen fibers become increasingly cross-linked to each other, reducing their flexibility. There is also less adipose tissue in the subcutaneous layer;

therefore, older people are more likely to feel cold. Together these changes typically result in sagging and wrinkling of the skin.

As people age, the sweat glands also become less active, resulting in decreased tolerance to high temperatures. There are fewer hair follicles, so the hair on the scalp and the limbs thins out. Older people also experience a decrease in the number of melanocytes, making their hair gray and their skin pale. In contrast, some of the remaining pigment cells are larger, and pigmented blotches ("age spots") appear on the skin.

Cardiovascular System

Common problems with cardiovascular function are usually related to diseases, especially atherosclerosis (see Section 12.5). However, even with normal aging, the heart muscle weakens somewhat, and may increase slightly in size as it compensates for its decreasing strength. The maximum heart rate decreases even in the fittest older athlete, and it takes longer for the heart rate and blood pressure to return to normal resting levels following stress. Some part of this decrease in heart function may also be due to blood leaking back through heart valves that have become less flexible.

Aging also affects the blood vessels. The middle layer of arteries contains elastic fibers, which, like collagen fibers in the skin, become more rigid with time. These changes, plus a frequent decrease in the internal diameter of arteries due to atherosclerosis, contribute to a gradual increase in blood pressure with age. Indeed, nearly 50% of older adults have chronic hypertension. Such changes are common in individuals living in Western industrialized countries but not in agricultural societies. This indicates that a diet low in cholesterol and saturated fatty acids, along with a sensible exercise program, may help to prevent age-related cardiovascular disease.

Immune System

As people age, many of their immune system functions become compromised. Because a healthy immune system normally protects the entire body from infections, toxins, and at least some types of cancer, some investigators believe that decreasing immune function can play a major role in the aging process.

As noted in Section 13.1, the thymus is an important site for T-cell maturation. Beginning in adolescence, the thymus begins to involute, gradually decreasing in size, and eventually becoming replaced by fat and connective tissue. The thymus of a 60-year-old adult is about 5% of the size of the thymus of a newborn, resulting in a decrease in the ability of older people to generate T-cell responses to new antigens. The evolutionary rationale for this may be that the thymus is energetically expensive for an organism to maintain, and that compared to younger animals that must respond to a high number of new infections and other antigens, older animals have already responded to most of the antigens to which they will be exposed in their life.

Aging also affects other immune functions. Because most B-cell responses are dependent on T cells, antibody responses also begin to decline. This, in turn, may explain why the elderly do not respond as well to vaccinations as young people do. This presents challenges in protecting older people against diseases like influenza and pneumonia, which can otherwise be prevented by annual vaccination.

Not all immune functions decrease in older animals. The activity of *natural killer cells,* which are a part of the innate immune system, seems to change very little with age. Perhaps by investigating how these cells remain active throughout a normal human life span, researchers can learn to preserve other aspects of immunity in the elderly.

Digestive System

The digestive system is perhaps less affected by the aging process than other systems. Because secretion of saliva decreases, more bacteria tend to adhere to the teeth, causing more decay and periodontal disease. Blood flow to the liver is reduced, resulting in less efficient metabolism of drugs or toxins. This means that, as a person gets older, less medication is needed to maintain the same level in the bloodstream.

Respiratory System

Cardiovascular problems are often accompanied by respiratory disorders, and vice versa. Decreasing elasticity of lung tissues means that ventilation is reduced. Because we rarely use the entire vital capacity, these effects may not be noticed unless the demand for oxygen increases, such as during exercise.

Excretory System

Blood supply to the kidneys is also reduced. The kidneys become smaller and less efficient at filtering wastes. Salt and water balance are difficult to maintain, and the elderly dehydrate faster than young people. Urinary incontinence (lack of bladder control) increases with age, especially in women. In men, an enlarged prostate gland may reduce the diameter of the urethra, causing frequent or difficult urination.

Nervous System

Between the ages of 20 and 90, the brain loses about 20% of its weight and volume. Neurons are extremely sensitive to oxygen deficiency, and neuron death may be due not to aging itself but to reduced blood flow in narrowed blood vessels. However, contrary to previous scientific opinion, recent studies using advanced imaging techniques show that most age-related loss in brain function is not due to whatever loss of neurons is occurring. Instead, decreased function may occur due to alterations in complex chemical reactions, or increased inflammation in the brain. For example, an age-associated decline in levels of dopamine can affect brain regions involved in complex thinking.

Perhaps more important than the molecular details, recent studies have confirmed that lifestyle factors can affect the aging brain. For example, animals on restricted calorie diets developed fewer Alzheimer-like changes in their brains. Other positive factors that may help maintain a healthy brain include continuing your education (the "use it or lose it" idea), regular exercise, and sufficient sleep.

Sensory Systems

In general, with aging, more stimulation is needed for taste, smell, and hearing receptors to function as before. A majority of people over age 80 have a significant decline in their sense of smell, and about 15% suffer from anosmia, or a total inability to smell. The latter condition can be a serious health hazard, due to the inability to detect smoke, gas leaks, or spoiled food. After age 50, most people gradually begin to lose the ability to hear tones at higher frequencies, and this can make it difficult to identify individual voices and to understand conversation in a group.

Starting at about age 40, the lens of the eye does not accommodate as well, resulting in *presbyopia,* or difficulty focusing on near objects, which causes many people to require reading glasses as they reach middle age. Finally, as noted in Section 18.7, cataracts and other eye disorders become much more common in the aged.

Musculoskeletal System

For the average person, muscle mass peaks between age 16 to 19 for females and 18 to 24 for males. Beginning in the twenties or thirties, but accelerating with increasing age, muscle mass generally decreases, due to decreases in both the size and number of muscle fibers. Most people who reach age 90 have 50% less muscle mass compared to when they were 20. Although some of this loss may be inevitable, regular exercise can slow this decline.

Like muscles, bones tend to shrink in size and density with age. Due to compression of the vertebrae, along with changes in posture, most of us lose height as we age. Those who reach age 80 will be about 2 inches shorter than they were in their twenties. Women lose bone mass more rapidly than men do, especially after menopause. As discussed in Section 19.6, osteoporosis is a common disease in the elderly. Although some decline in bone mass is a normal result of aging, certain extrinsic factors are also important. A proper diet and moderate exercise program has been found to slow the progressive loss of bone mass.

Endocrine System

As with the immune system, aging of the hormonal system can affect many different organs of the body. These changes are complex, however, with some hormone levels tending to decrease with age, while others increase. The activity of the thyroid gland generally declines, resulting in a lower basal metabolic rate. The production of insulin by the pancreas may remain stable, but cells become less sensitive to its effects, resulting in a rise in fasting glucose levels of about 10 mg/dl each decade after age 50.

Human growth hormone (HGH) levels also decline with age, but it is very unlikely that taking HGH injections will "cure" aging, despite Internet claims. In fact, one study found that people with lower levels of HGH actually lived longer than those with higher levels.

Reproductive System

Testosterone levels are highest in men in their twenties. After age 30, testosterone levels decrease by about 1% per year. Extremely low testosterone levels have been linked to a decreased sex drive, excessive weight gain, loss of muscle mass, osteoporosis, general fatigue, and depression. However, the levels below which testosterone treatment should be initiated remain controversial. Testosterone

SCIENCE IN YOUR LIFE ▶ HEALTH

Alzheimer Disease

Alzheimer disease is characterized by the presence of abnormally structured neurons and a reduced amount of acetylcholine (ACh), primarily in an area of the brain called the hippocampus (see Section 17.4). A neuron that has been affected by Alzheimer disease has two characteristic features. Bundles of fibrous protein, called neurofibrillary tangles, extend from the axon to surround the nucleus of the neuron. Tangles form when the supporting protein, called tau, becomes malformed (Fig. 22B) and twists the neurofibrils, which are normally straight. In addition, protein-rich accumulations called amyloid plaques envelop branches of the axon. The plaques grow so dense that they trigger an inflammatory reaction that causes neuron death.

Treatment involves using one of two categories of drugs. Cholinesterase inhibitors work at neuron synapses in the brain, slowing the activity of acetylcholinesterase, the enzyme that breaks down ACh. Allowing ACh to accumulate in synapses keeps memory pathways in the brain functional for a longer period. The second drug, memantine, blocks excitotoxicity, the tendency of diseased neurons to self-destruct. This recently approved medication is used only in moderately to severely affected patients. Using the drug allows neurons involved in memory pathways to survive longer in affected patients. Successes with these medications indicates that treatment for AD patients should begin as soon as possible after diagnosis and continue indefinitely. However, neither type of medication cures the disease; both merely slow the progress of symptoms, allowing the patient to function independently for a longer time. Additional research is currently under way to test the effectiveness of anticholesterol statin drugs, as well as anti-inflammatory medications, in slowing the progress of the disease.

Much of the current research on Alzheimer disease focuses on the prevention and cure of the disease. Scientists believe that a cure will require an early diagnosis, because it is thought that the disease may begin in the brain 15 to 20 years before symptoms develop. Currently, diagnosis can't be made with absolute certainty until the brain is examined at autopsy. A new test on the cerebrospinal fluid may allow early detection of amyloid proteins and a much earlier diagnosis. Researchers are also testing vaccines that would target the patient's immune system to destroy amyloid plaques.

Early findings have shown that risk factors for cardiovascular disease also contribute to an increased incidence of Alzheimer disease. Risk factors for cardiovascular disease include elevated blood cholesterol and blood pressure, smoking, obesity, sedentary lifestyle, and diabetes mellitus. Thus, evidence suggests that a lifestyle tailored for good cardiovascular health may also prevent Alzheimer disease.

Questions to Consider

1. What is the role of the hippocampus, and how does this relate to the symptoms of Alzheimer disease?
2. How might anti-inflammatory drugs slow the progression of Alzheimer disease?

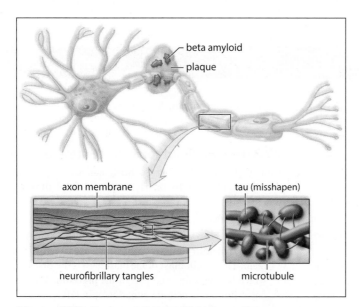

Figure 22B Tau protein and Alzheimer disease. Some of the neurons of Alzheimer disease patients have beta amyloid plaques and neurofibrillary tangles. These neurons are present throughout the brain but concentrated in the hippocampus and amygdala.

replacement therapy, whether through injection, patches, or gels, is associated with side effects like enlargement of the prostate, acne or other skin reactions, and the production of too many red blood cells.

Menopause, the period in a woman's life during which the ovarian and uterine cycles cease, usually occurs between ages 45 and 55. The ovaries become unresponsive to the gonadotropic hormones produced by the anterior pituitary, and they no longer secrete estrogen or progesterone. At the onset of menopause, the uterine cycle becomes irregular, but as long as menstruation occurs, it is still possible for a woman to conceive. Therefore, a woman is usually not considered to have completed menopause (and thus be infertile) until menstruation has been absent for a year.

The hormonal changes during menopause often produce physical symptoms such as "hot flashes" (caused by circulatory irregularities), dizziness, headaches, insomnia, sleepiness, and depression. To ease these symptoms, female hormone replacement therapy (HRT) was routinely prescribed until 2002, when a large clinical study showed that in the long term, HRT caused more health problems than it prevented. For this reason, most doctors no longer routinely recommend long-term HRT for the prevention of postmenopausal conditions.

It is also of interest that, as a group, females live longer than males. It is likely that estrogen offers women some protection against cardiovascular disorders when they are younger. Males suffer a marked increase in heart disease in their forties, but an increase is not noted in females until after menopause, when women lead men in the incidence of stroke. Men remain more likely than women to have a heart attack at any age, however.

Check Your Progress 22.5

1. Explain how telomeres influence cellular aging.
2. Summarize how diet may contribute to aging.
3. Summarize the diseases associated with the aging process and explain the body systems each is associated with.

Conclusion

Aging is a complex process that is influenced by factors such as genetics and our environments. The genetics of longevity are not associated with a single gene, such as *FOXO3A*. Rather, minor variations in a number of genes, and gene combinations, make contributions to our life span. By examining the genetic basis of diseases such as Alzheimer disease, scientists are better able to determine how our genes affect aging at the cellular and physiological levels.

However, it is important to note that the environment also plays a major role. Exposure to the sun, the types of food we eat, and our behaviors all can affect our life span. Often, environmental factors interact with variations in genes, meaning that for some people, specific behaviors, such as drinking alcohol, may have a different effect on the aging process than for another individual with another combination of genes.

SUMMARIZE

22.1 Fertilization and Early Stages of Development

Animal development begins with a single cell that develops into an embryo. Development proceeds through cellular, organ, and tissue stages:

- Development begins at **fertilization.** Only one sperm enters the oocyte, and its nucleus fuses with the oocyte nucleus, forming a **zygote.**
- During the cellular stages, **cleavage** (cell division) occurs, without overall growth. The result is a **morula,** which becomes the **blastula** when an internal cavity (the **blastocoel**) appears.
- During the tissue stages, **gastrulation** results in formation of the **germ layers: ectoderm, mesoderm,** and **endoderm.** Invagination of the gastrula forms a **blastopore.** Both the cellular and tissue stages can be affected by the amount of yolk.
- Organ formation can be related to germ layers. For example, the nervous system develops from midline ectoderm, just above the **notochord.**

22.2 Processes of Development

Development involves an increasing specialization of cells, as well as **morphogenesis,** the formation of the shape of the body. Morphogenesis progresses to **pattern formation,** the positioning of tissues and organs.

- **Cellular differentiation** begins with **cytoplasmic segregation** in the egg. An example of this is the **gray crescent** of a frog's egg.
- Some cellular differentiation is due to **induction,** an ongoing process in which one tissue regulates the development of another tissue through chemical signals. In simple animals, **fate maps** have been constructed to map the destiny of individual cells.
- During morphogenesis, sequential sets of **morphogen genes** code for morphogen gradients that, in turn, activate the next set of morphogen genes. **Homeotic genes** code for proteins that contain a **homeodomain,** a sequence of 60 amino acids that is highly conserved among a wide

variety of organisms. **Apoptosis,** or programmed cell death, also plays an important role in development.

22.3 Human Embryonic and Fetal Development

Human development before birth can be divided into **embryonic development** (months one to two) and **fetal development** (months three to nine).

- Shortly after fertilization, the embryo forms a ball of cells called the morula. This develops into a hollow sphere of cells called the **blastocyst.**
- The **extraembryonic membranes** appear early in human development. The **trophoblast** of the blastocyst is the first sign of the **chorion,** which becomes the fetal part of the **placenta.** This structure also produces **human chorionic gonadotropin (HCG),** the basis of the pregnancy test. The **amnion** contains amniotic fluid, which cushions and protects the embryo. The **yolk sac** serves to produce blood cells. The **allantois** gives rise to the **umbilical cord.**
- Gastrulation is marked by the appearance of the **embryonic disk,** followed by a steady progression of organ formation during embryonic development. During fetal development, features are refined, and the fetus adds weight. The skin is covered by fine hairs called **lanugo,** as well as by **vernix caseosa.**
- Because the fetus receives oxygen and nutrients from the placenta, the fetal heart contains a **foramen ovale** and a **ductus arteriosus.**

22.4 Human Pregnancy, Birth, and Lactation

Many physiological changes occur in a woman's body during pregnancy. Birth, or **parturition,** requires three stages, and occurs about 280 days after the start of the mother's last menstruation.

- Many pregnant women experience morning sickness early in their pregnancy. Due to a general relaxation of smooth muscles, a pregnant woman's body generally takes on fluid, and blood flow to many organs increases. Stretch marks, or striae gravidarum, may appear on the skin.

■ During stage one of parturition, uterine contractions begin to push the baby's head against the cervix. Uterine contractions increase during stage two, as the baby's head enters the birth canal and the rest of the body usually follows. An **episiotomy** may be required to enlarge the vaginal opening. The placenta, or **afterbirth,** is delivered during stage three.

■ After a baby is born, prolactin from the pituitary gland stimulates milk production. For the first day or two after birth, the breasts produce **colostrum,** an antibody- and nutrient-rich fluid. Breast-feeding is usually beneficial to the newborn and the mother.

22.5 Aging

The aging process begins as soon as we are born. **Gerontology** is the study of **aging.** A number of factors influence the aging process, including hormones, telomere length, mitochondria and diet, and the accumulation of cellular damage. Aging affects all of the body systems and is evident in diseases such as Alzheimer disease and osteoporosis.

ASSESS

Testing Yourself

Choose the best answer for each question.

22.1 Fertilization and Early Stages of Development

1. Only one sperm enters an egg because
   a. sperm have an acrosome.
   b. the corona radiata gets larger.
   c. changes occur in the zona pellucida.
   d. the cytoplasm hardens.
   e. All of these are correct.

2. The fusion of a sperm and an egg cell initally produces which of the following?
   a. blastocyst   c. zygote
   b. gamete   d. stem cell

3. Which primary germ layer is not correctly matched to an organ system or organ that develops from it?
   a. ectoderm—nervous system
   b. endoderm—lining of the digestive tract
   c. mesoderm—skeletal system
   d. endoderm—cardiovascular system

22.2 Processes of Development

4. Which process is responsible for the shape and form of the body?
   a. cleavage   c. induction
   b. growth   d. morphogenesis

5. The ability of one embryonic tissue to influence the development of another tissue is called
   a. induction.
   b. cellular differentiation.
   c. apoptosis.
   d. pattern formation.
   e. morphogenesis.

6. The identity of each segment of a developing embryo is controlled by
   a. fate maps.
   b. the gray crescent.
   c. homeotic genes.
   d. the blastocyst.
   e. None of these are correct.

22.3 Human Embryonic and Fetal Development

7. Label each stage of the diagram below.

8. Which developmental stage is mismatched?
   a. second week—implantation
   b. third week—heart development begins
   c. fourth and fifth week—limb buds appear
   d. sixth through eighth week—all organ systems present
   e. None of these stages are mismatched.

9. Which of these statements is correct?
   a. Fetal circulation, like adult circulation, takes blood equally to a pulmonary circuit and a systemic circuit.
   b. Fetal circulation shunts blood away from the lungs but makes full use of the systemic circuit.
   c. Fetal circulation includes an exchange of substances between fetal blood and maternal blood at the placenta.
   d. Unlike adult circulation, fetal blood always carries oxygen-rich blood and therefore has no need for the pulmonary circuit.
   e. Both b and c are correct.

22.4 Human Pregnancy, Birth, and Lactation

10. Which of the following hormones is not matched correctly to its effect on the mother?
    a. peptide hormones—increased resistance to insulin
    b. progesterone—reduced chances of the mother having an immune reaction to the fetus
    c. estrogen—increased uterine blood flow
    d. progesterone—relaxation of smooth muscle and decrease in blood pressure
    e. All of these are matched correctly.

11. Which of the following is increased in the mother during pregnancy?
    a. metabolic rate
    b. blood volume
    c. cardiac output
    d. All of these are correct.

22.5 Aging

12. After each cell division, these chromosomal structures shorten, thus regulating the life span of a cell.
    a. mitochondria
    b. free radicals
    c. telomeres
    d. plasma membranes

13. Which of the following is associated with the aging process?
    a. reduction in the production of sex hormones, such as testosterone
    b. increased chances of dehydration
    c. decrease in bone density
    d. decrease in the ability of the senses to detect stimuli
    e. All of these are correct.

ENGAGE

Thinking Critically

1. Mitochondria contain their own genetic material. Mitochondrial mutations are inherited from mother to child, without any contribution from the father. Why?

2. At-home pregnancy tests check for the presence of HCG in a female's urine. Where does HCG come from? Why is HCG found in a pregnant woman's urine?

3. A variety of fertility tests are available for women to purchase over the counter—for example, at drugstores. Because more women now delay their first pregnancy until they are older, these tests are becoming more popular. Many of these tests incorporate antibodies that specifically react with hormones found in a woman's urine. What specific hormones could be measured in these tests, and how might these tests actually work?

©McGraw-Hill Education/Jill Braaten, photographer

CHAPTER

23

Patterns of Gene Inheritance

BEFORE YOU BEGIN

Before beginning this chapter, take a few moments to review the following discussions:

Chapter 2 What are the roles of DNA and proteins in a cell?

Section 5.4 How are chromosomes segregated during the process of meiosis?

Section 5.6 How does the process of meiosis relate to the human life cycle?

Genetics of Phenylketonuria

Have you ever looked at the back of a can of diet soda and wondered what the warning "Phenylketonurics: Contains Phenylalanine" means? Phenylalanine is an amino acid found in many foods and in the artificial sweetener aspartame. For most people, phenylalanine does not present any problems, because any excess is broken down by enzymes in the body. However, some people, called phenylketonurics, lack a functional copy of this enzyme and thus are unable to break down the phenylalanine. The excess may accumulate in the body, causing a variety of nervous system disorders. An estimated 1 in 10,000 people in the United States are phenylketonurics.

Like the rest of us, you are the product of your family tree. The DNA you inherit from your parents directly affects the proteins that enable your body to function properly. Rare genetic disorders, such as phenylketonuria, pique our curiosity about how traits are inherited from one generation to the next. The process of meiosis can be used to predict the inheritance of a trait, and the genetic diversity produced through meiosis can sometimes lead to phenylketonuria.

Through the patterns of inheritance first described by Mendel, you will learn that certain traits, such as phenylketonuria, are recessive and that it takes two nonfunctional copies of that gene before you are affected. This chapter will introduce you to observable patterns of inheritance, including some human genetic disorders, such as phenylketonuria.

As you read through this chapter, think about the following questions:

1. How does both the environment and genetics influence genetic traits such as phenylketonuria?

2. What is the difference between an autosomal dominant and autosomal recessive pattern of inheritance?

3. How does meiosis help predict the probability of producing gametes and inheriting a trait?

23.1 Mendel's Laws

The science of genetics explains the process of inheritance (why you are human, as are your parents) and also why there are variations between offspring from one generation to the next (why you have a different combination of traits than your parents). Virtually every culture in history has attempted to explain observed inheritance patterns. An understanding of these patterns has always been important to agriculture, animal husbandry (the science of breeding animals), and medicine.

Gregor Mendel

Our modern understanding of genetics is based largely on the work of Gregor Mendel (Fig. 23.1), an Austrian monk who developed the basic principles of inheritance after performing a series of ingenious experiments in the 1860s. Mendel's parents were farmers, which provided him with the practical experience he needed to grow his experimental organism, the pea plant.

Mendel was also a mathematician. He kept careful and complete records, even though he crossed and catalogued some 24,034 plants through several generations. He concluded that the plants transmitted distinct factors to offspring. We know now that the factors that control traits are called genes, and that genes are found on chromosomes.

In pea plants and in humans, the chromosomes are found in pairs, called *homologous chromosomes.* One member of the pair is inherited from the mother, while the other member is inherited from the father. Homologous chromosomes have the same length and centromere location and are similar in the types of genes they contain. For example, one trait in humans is finger length. One member of the homologous pair of chromosomes could contain the gene for long fingers, whereas the other member could contain the gene for short fingers. These alternate forms of a gene for the same trait are called **alleles.** Alleles are always at the same spot, or **locus,** on each member of the homologous pair. In Figure 23.2, the letters on the homologous chromosomes stand for alleles. *G* is an allele of *g,* and vice versa; *R* is an allele of *r,* and vice versa. *G* could never be an allele of *R* because *G* and *R* are at different loci and govern different traits.

Mendel knew nothing about homologous pairs of chromosomes, but from his work with pea plants, as described in the Scientific Inquiry feature "The Investigations of Gregor Mendel," he concluded that pea plants have two factors for every trait, such as stem length. He observed that one of the factors controlling the same trait can be dominant over the other, which is recessive. Therefore, a tall pea plant could have one factor for long stem and another for short stem. While most of the offspring from two such tall pea plants were tall, some were short. Mendel reasoned that this would be possible if the gametes (i.e., sperm and oocyte) contained only one factor for each trait. Only in that way could an offspring receive two factors for short stem. On the

Figure 23.1 Gregor Mendel.   Mendel was a skilled mathematician with excellent skills of observation. His work established the early principles of inheritance that serve as the basis for modern genetic studies. ©Ned M. Seidler/National Geographic Creative

Figure 23.2 Homologous chromosomes.   a. The letters represent alleles—that is, alternate forms of a gene. Each allelic pair, such as *Gg* or *Tt,* is located on homologous pairs of chromosomes at a particular gene locus. **b.** Sister chromatids carry the same alleles in the same order. Proteins made from each allele determine the observable traits.

basis of such studies, Mendel formulated his first law, the law of segregation.

> The **law of segregation** states the following:
> - Each individual has two factors for each trait.
> - The factors segregate (separate) during the formation of the gametes.
> - Each gamete contains only one factor from each pair of factors.
> - Fertilization gives each new individual two factors for each trait.

The Inheritance of a Single Trait

The **phenotype** of an organism refers to the individual's actual appearance. Phenotype may describe either an individual's physical characteristics, or microscopic and metabolic characteristics. In this chapter, we will use the traits of freckles and finger length in our examples (Fig. 23.3). Both of these traits have variations in the phenotype, which are determined by the combination of alleles that the individual inherited. The **genotype** refers to the alleles the chromosomes carry that are

responsible for that trait. In diploid organisms, such as humans, a pair of homologous chromosomes contains two alleles for each trait, one on each member of the homologous pair. Mendel suggested using letters to indicate these factors, which we now call alleles. A capital letter indicates a **dominant allele** and a lowercase letter indicates a **recessive allele.** The word *dominant* is not meant to imply that the dominant allele is better or stronger than the recessive allele. Dominant means that this allele will mask the expression of the recessive allele when they are together in the same organism.

Many factors influence freckles, but a dominant allele (*F*) causes widespread freckling patterns on the body. There are also many different alleles that control finger length. However, a dominant allele (*S*) produces a condition called brachydactyly, or shortened fingers.

In the case of a single trait, such as freckles, there are three possible combinations, or genotypes, of the two alleles: *FF, Ff,* and *ff*. If the two members of the allelic pair are the same (*homo*), the organism is said to be **homozygous.** Both *FF* and *ff* are homozygous, with *FF* homozygous dominant, and *ff* homozygous recessive. If the two members of the allelic pair are different (*hetero*), the organism is said to be **heterozygous.** Only *Ff* is a heterozygous genotype. But what about the phenotype? When an organism has an *F,* it exhibits the dominant phenotype regardless of whether the other allele is *F* or *f.* Both *FF* and *Ff* exhibit the dominant phenotype, widespread freckles. The recessive allele *f* is masked by the expression of the dominant allele *F* in the heterozygous genotype. The only time the recessive phenotype, no freckles, is expressed, is when the genotype is homozygous recessive, or *ff*. The phenotypes and genotypes for this trait are summarized in Table 23.1.

Gamete Formation

The genotype has two alleles for each trait, whereas the gametes (i.e., sperm and oocyte) have only one allele for each trait in accordance with Mendel's law of segregation. During meiosis, the homologous chromosomes separate (see Section 5.4), resulting in only one member of each pair of chromosomes in each haploid cell. In humans, these haploid cells go on to form gametes. Therefore, there is only one allele for each trait, such as the widespread presence of freckles, in each gamete. When working problems in genetics, *no two letters in a gamete can be the same letter of the alphabet.* If the genotype of an individual is *Ff,* the gametes for this individual will contain either an *F* or an *f,* but not both. If the genotype is *FfSs,* each gamete receives an allele for both freckles and finger length.

a. Freckles: *FF* or *Ff*

b. No freckles: *ff*

c. Short fingers: *SS* or *Ss*

d. Long fingers: *ss*

Figure 23.3 Dominance of human traits. The allele for freckles (**a**) is dominant to (**b**) no freckles. The allele for (**c**) short finger length (brachydactyly) is dominant over (**d**) long fingers. (a): ©BananaStock/age fotostock; (b): ©BDLM/Getty Images; (c): ©McGraw-Hill Education/Bob Coyle, photographer; (d): ©McGraw-Hill Education/Bob Coyle, photographer

TABLE 23.1	Genotype Related to Phenotype	
Genotype	**Genotype**	**Phenotype**
FF	Homozygous dominant	Freckles
Ff	Heterozygous	Freckles
ff	Homozygous recessive	No freckles

The Investigations of Gregor Mendel

When Gregor Mendel began his work, most plant and animal breeders acknowledged that both sexes contribute equally to a new individual. They thought that parents of contrasting appearance always produced offspring of intermediate appearance. This concept, called the *blending concept of inheritance,* meant that a cross between plants with red flowers and plants with white flowers would yield only plants with pink flowers. When red and white flowers reappeared in future generations, the breeders mistakenly attributed this to instability in the genetic material.

Mendel chose to work with the garden pea, *Pisum sativum* (Fig. 23A). The garden pea was a good choice for several reasons, including the fact that they were easy to cultivate and had a short generation time. Although peas normally self-pollinate (pollen only goes to the same flower), they could be cross-pollinated by hand by transferring pollen from an anther to a stigma. Many varieties of peas were available, and Mendel chose 22 for his experiments. When these varieties self-pollinated, they were *true-breeding.* A true-breeding line always produced individuals of the same phenotype. In contrast to his predecessors, Mendel studied the inheritance of relatively simple and discrete traits, such as seed shape, seed color, and flower color.

Mendel utilized both one-trait and two-trait crosses in his research. He cross-pollinated the plants for the F_1 (first filial) generation. None of the offspring of this generation were intermediate between the parents. All of the offspring exhibited the trait of one or the other parent, which Mendel termed the dominant trait. The trait that disappeared in the offspring he called the recessive trait. He then allowed the F_1 plants to self-pollinate to obtain an F_2 generation. He counted many plants and always found approximately the same ratios of plants exhibiting dominant and recessive traits in this generation (Fig. 23B).

As Mendel followed the inheritance of individual traits, he kept careful records, and he used his understanding of the mathematical laws of probability and statistics to interpret his results. Mendel recognized that discrete units of heredity, which he called "elementens," were being passed from generation to generation. We know that elementens are actually genes. His results became the basis of the *particulate theory of inheritance.* Inheritance involves reshuffling the same genes from generation to generation.

Mendel achieved his success in genetics by studying large numbers of offspring,

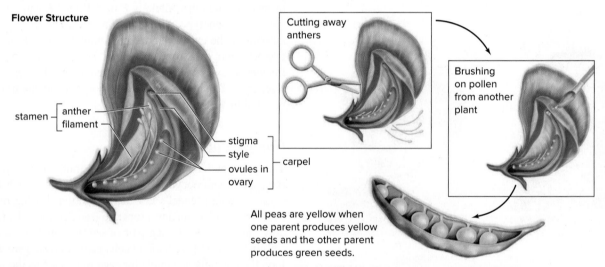

Figure 23A Garden pea anatomy and a few traits.   In the garden pea, *Pisum sativum,* pollen grains produced in the anther contain sperm, and ovules in the ovary contain oocytes. Mendel brushed pollen from one plant onto the stigma of another plant. After fertilization, the ovules developed into seeds (peas). The open pod shows the results of a cross between plants with yellow seeds and plants with green seeds.

One-Trait Crosses

It is now possible for us to consider a particular cross. If a homozygous man with freckles (Fig. 23.3*a*) reproduces with a woman without freckles (Fig. 23.3*b*), will their children have freckles? To solve this problem, (1) use Table 23.1 to determine the genotype of each parent; (2) list the possible gametes from each parent; (3) combine all possible gametes; and (4) determine the genotypes and the phenotypes of all the offspring. When writing a heterozygous genotype (*Ff*), always put the capital letter first to avoid confusion.

In the following diagram, the letters in the first row are the genotypes of the parents. Each parent has only one type of gamete

in regard to the freckle phenotype, and therefore all the offspring have similar genotypes and phenotypes. The children are heterozygous (*Ff*) and would have freckles:

keeping careful records, and treating his data quantitatively. However, a certain amount of luck was involved in the traits Mendel chose. None of the traits he analyzed exhibited the more complicated patterns of inheritance.

Questions to Consider

1. From a statistical perspective, why are large numbers of offspring needed?
2. Besides the reasons listed in this feature, what other reasons would make the pea plant a better object for genetic study than humans?
3. What "more complicated patterns of inheritance" could have confused Mendel's results?

Trait	Characteristics				F_2 Results		
	Dominant		Recessive		Dominant	Recessive	Ratio
Stem length	Tall		Short		787	277	2.84:1
Pod shape	Inflated		Constricted		882	299	2.95:1
Seed shape	Round		Wrinkled		5,474	1,850	2.96:1
Seed color	Yellow		Green		6,022	2,001	3.01:1
Flower position	Axial		Terminal		651	207	3.14:1
Flower color	Purple		White		705	224	3.15:1
Pod color	Green		Yellow		428	152	2.82:1
				Totals:	14,949	5,010	2.98:1

Figure 23B The traits Mendel selected for study.   The parent (P generation) plants were true-breeding and were cross-pollinated to produce an F_1 (first filial) generation. These plants always resembled the parent with the dominant characteristic. The F_1 plants were allowed to self-pollinate. In the F_2 (second filial) generation, he always achieved a 3:1 (dominant to recessive) phenotypic ratio. The results of all of Mendel's crosses are shown here.

Because the children are heterozygous for this single pair of alleles, we can say they are monohybrids (a hybrid is a combination of two different traits). What would happen if two monohybrid individuals reproduced? This is called a **monohybrid cross** ($Ff \times Ff$). Notice that each parent has two possible types of gametes (F or f), and we must ensure that all types of sperm have an equal chance to fertilize all possible types of eggs. One way to visualize this is to use a **Punnett square** (Fig. 23.4), in which all possible types of sperm are lined up vertically, and all possible types of eggs are lined up horizontally (or vice versa), and every possible combination of gametes occurs within the squares.

After we determine the genotypes and the phenotypes of the offspring, we can determine the genotypic and the phenotypic ratios. The genotypic ratio is 1 *FF*: 2 *Ff*: 1 *ff*, or simply 1:2:1, but the phenotypic ratio is 3:1. Why? Because three individuals have freckles and one does not have freckles. This 3:1 phenotypic ratio is always expected for a monohybrid cross when one allele is completely dominant over the other. The exact ratio is more likely to be observed if a large number of matings take place and if a large number of offspring result. Only then do all possible kinds of sperm have an equal chance of fertilizing all possible kinds of oocytes. Naturally, in humans, we do not routinely observe hundreds of offspring from

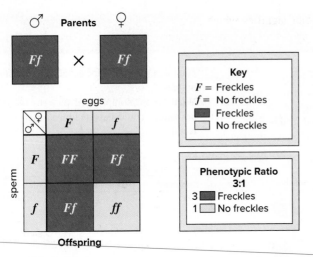

Figure 23.4 Monohybrid cross.   A Punnett square diagrams the results of a cross. When the parents are heterozygous, each child has a 75% chance of having the dominant phenotype and a 25% chance of having the recessive phenotype.

a single type of cross. The best interpretation of Figure 23.4 in humans is to say that each child has three chances out of four to have freckles, or one chance out of four to lack freckles. It is important to realize that *chance has no memory;* for example, if two heterozygous parents already have three children with freckles and are expecting a fourth child, this child still has a 75% chance of having freckles and a 25% chance of not having freckles.

One-Trait Crosses and Probability

Another method of calculating the expected ratios uses the rules of probability, or chance. First, we must know the *product rule of probability,* which states that the chance of two or more independent events occurring together is the product (multiplication) of their chance of occurring separately. For example, if you flip a coin, the chance of getting "heads" is $1/2$ (one head out of two possibilities, or $1/2$). If you flip two coins, the chance of getting two

heads is $1/2 \times 1/2$, or $1/4$. You must multiply together the probability of each head occurring separately ($1/2$).

In the cross just considered ($Ff \times Ff$), what is the chance of obtaining either an F or an f from a parent? Because the parents are heterozygous, for each parent the chance of getting an F allele is $1/2$, and the chance of getting an f allele is $1/2$.

Therefore, the probability of having these genotypes is as follows:

- The chance of getting $FF = 1/2 \times 1/2 = 1/4$
- The chance of getting $Ff = 1/2 \times 1/2 = 1/4$
- The chance of getting $fF = 1/2 \times 1/2 = 1/4$
- The chance of getting $ff = 1/2 \times 1/2 = 1/4$

Now we must consider the *sum rule of probability,* which states that the chance of an event that can occur in more than one way is the sum (addition) of the individual chances. Therefore, to calculate the chance of an offspring having freckles, add the chances of *FF, Ff,* or *fF* from the preceding list to arrive at $3/4$, or 75%. The chance of offspring without freckles (only the *ff* genotype) is $1/4$, or 25%.

Notice that the chances of an individual being heterozygous from this cross is $1/2$. This is because it is possible to have either the *fF* or *Ff* genotypes and be heterozygous. Because each of these occurs $1/4$ of the time, we add them together to get $1/2$.

The One-Trait Testcross

It is not possible to tell by observation if an individual expressing a dominant allele is homozygous dominant or heterozygous. However, this may be determined by carefully analyzing their offspring, or by performing a testcross. In a **testcross,** an organism showing the dominant phenotype, but unknown genotype, is crossed with a homozygous recessive individual. Why use a homozygous recessive individual? Because this is the one case where the phenotype indicates a known genotype. The results of a testcross should indicate the genotype of an organism with the dominant phenotype (Fig. 23.5).

a.

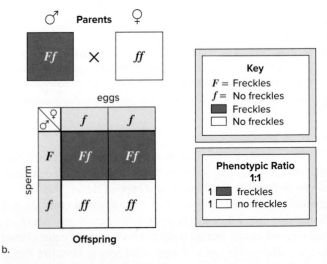

b.

Figure 23.5 Determining if an individual is homozygous dominant or heterozygous for a single trait.   This cross determines if an individual with the dominant phenotype is homozygous or heterozygous. **a.** Because all offspring show the dominant characteristic, the individual is most likely homozygous, as shown. **b.** Because the offspring show a 1:1 phenotypic ratio, the individual is heterozygous, as shown.

The results of crosses between people also tell us the genotype of the parents. For example, Figure 23.5 shows two possible results when a man with freckles reproduces with a woman who lacks freckles. If the man is homozygous dominant, all his children will have freckles. If the man is heterozygous, each child has a 50% chance of having freckles. The birth of just one child without widespread freckles indicates that the man is heterozygous.

Historically, testcrosses were used extensively by animal breeders and agricultural geneticists to determine the genotype of an individual with a dominant phenotype. Obviously, testcrosses are not appropriate for humans, but an analysis of families with large numbers of children can shed light on the genotype of the parents using the principles of a testcross. Generally, testcrosses have been largely replaced by molecular techniques that are able to more rapidly assess the alleles that are present in an individual (see Section 26.1).

The Inheritance of Two Traits

Figure 23.6 represents two pairs of homologues undergoing meiosis. The homologues are distinguished by length—that is, one pair of homologues is short and the other is long. In Figure 23.6, the color signifies which parent the chromosome is inherited from: One homologue of each pair is the "paternal" chromosome, and the other is the "maternal" chromosome. When the homologues separate (segregate) during meiosis, each gamete receives one member from each pair of homologues. The homologues separate independently. It does not matter which member of a pair goes into which gamete. In the simplest terms, a gamete in Figure 23.6 will *receive one short and one long chromosome of either color.* Therefore, all possible combinations of chromosomes are in the gametes.

We will assume that the alleles for two genes are on these homologues. The alleles *S* and *s* are on one pair of homologues, and the alleles *F* and *f* are on the other pair of homologues. Because there are no restrictions as to which homologue goes into which gamete, a gamete can receive either an *S* or an *s* and either an *F* or an *f* in any combination. In the end, the gametes will collectively have all possible combinations of alleles.

Independent Assortment

Mendel had no knowledge of meiosis, but he could see that his results were attainable only if the sperm and eggs contain every possible combination of factors. This caused him to formulate his

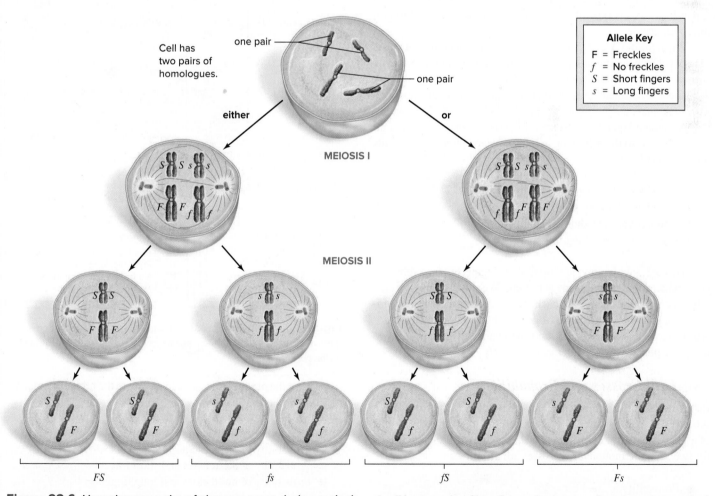

Figure 23.6 Homologous pairs of chromosomes during meiosis.   A cell has two pairs of homologous chromosomes (homologues), recognized by length, not color. The long pair of homologues carries alleles for freckles and the short pair of homologues carries alleles for finger length. The homologues, and the alleles they carry, align independently during meiosis. Therefore, all possible combinations of chromosomes and alleles occur in the gametes, as shown in the last row of cells.

second law of inheritance, also called the law of independent assortment.

> The **law of independent assortment** states the following:
> - Each pair of factors separates independently (without regard to how the others separate).
> - All possible combinations of factors can occur in the gametes.

Figure 23.6 illustrates the law of segregation and the law of independent assortment. Because it does not matter which homologue of a pair faces which spindle pole, a daughter cell can receive either homologue and either allele.

Two-Trait Crosses

In the two-trait cross depicted in Figure 23.7, a person homozygous for freckles and short fingers (*FFSS*) reproduces with one who has no freckles and long fingers (*ffss*). The law of segregation tells us that the gametes for the *FFSS* parent have to be *FS* and the gametes for the *ffss* parent have to be *fs*. Therefore, their offspring will all have the genotype *FfSs* and the same phenotype (freckles with short fingers). This genotype is called a dihybrid because the individual is heterozygous in two regards: presence of freckles and finger length. This type of genetic analysis is also referred to as a **dihybrid cross.**

When the dihybrid *FfSs* reproduces with another dihybrid that is *FfSs*, what gametes are possible? The law of segregation tells us that each gamete can have only one letter of each kind, and the law of independent assortment tells us that all combinations are possible. Therefore, these are the gametes for both dihybrids: *FS, Fs, fS,* and *fs*.

A Punnett square represents all possible sperm fertilizing all possible eggs, so the following are the expected phenotypic results for a dihybrid cross:

> - 9 freckles and short fingers
> - 3 freckles and long fingers
> - 3 no freckles and short fingers
> - 1 no freckles and long fingers

This 9:3:3:1 phenotypic ratio is always expected for a dihybrid cross when simple dominance is present. We can use this expected ratio to predict the chances of each child receiving a certain phenotype. For example, the chance of getting the two dominant phenotypes together is 9 out of 16, and the chance of getting the two recessive phenotypes together is 1 out of 16.

Two-Trait Crosses and Probability

Instead of using a Punnett square, it is also possible to use the product rule and the sum rule of probability to predict the results of a dihybrid cross. For example, we know that the probable results for two separate monohybrid crosses are as follows:

> - Probability of freckles = $^3/_4$
> - Probability of no freckles = $^1/_4$
> - Probability of short fingers = $^3/_4$
> - Probability of long fingers = $^1/_4$

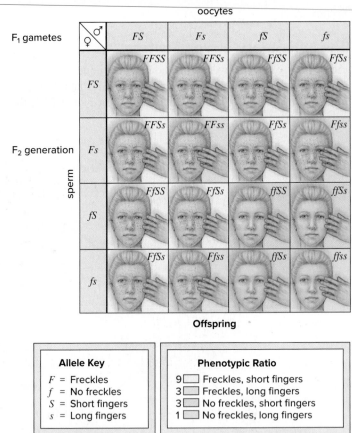

Allele Key

F = Freckles
f = No freckles
S = Short fingers
s = Long fingers

Phenotypic Ratio

9 ☐ Freckles, short fingers
3 ☐ Freckles, long fingers
3 ☐ No freckles, short fingers
1 ☐ No freckles, long fingers

Figure 23.7 Dihybrid cross. Each dihybrid can form four possible types of gametes, so four different phenotypes occur among the offspring in the proportions shown.

Using the product rule, we can calculate the probable outcome of a dihybrid cross as follows:

> - Probability of freckles and short fingers = $^3/_4 \times ^3/_4 = ^9/_{16}$
> - Probability of freckles and long fingers = $^3/_4 \times ^1/_4 = ^3/_{16}$
> - Probability of no freckles and short fingers = $^1/_4 \times ^3/_4 = ^3/_{16}$
> - Probability of no freckles and long fingers = $^1/_4 \times ^1/_4 = ^1/_{16}$

In this way, the rules of probability tell us that the expected phenotypic ratio when all possible sperm fertilize all possible eggs is 9:3:3:1.

The Two-Trait Testcross

It is impossible to tell by inspection whether an individual expressing the dominant allele for two traits is homozygous dominant or heterozygous in regard to these traits. In this case, to determine the genotype, the individual in question is crossed with an individual that is homozygous recessive for both traits. The recessive phenotype is used because it has a *known* phenotype.

For example, if a man who is homozygous dominant for freckles and short fingers reproduces with a woman who is homozygous recessive for both traits, then all his children will have the dominant phenotypes. However, if a man is heterozygous for both traits, then each child has a 25% chance of showing either one or both recessive traits. A Punnett square (Fig. 23.8) shows that the expected ratio is 1 freckles with short fingers: 1 freckles with long fingers: 1 without freckles with short fingers: 1 without freckles with long fingers, or 1:1:1:1.

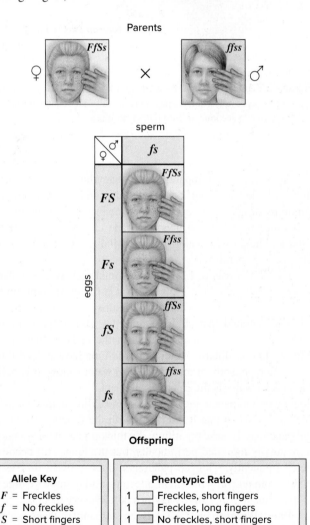

Figure 23.8 Determining if an individual is homozygous dominant or heterozygous for two traits.   The results of this cross indicate that the individual with the dominant phenotypes is heterozygous for both traits, because some of the children are homozygous recessive for one or both traits. The chance of receiving any possible phenotype is 25%.

Check Your Progress 23.1

1. Distinguish between the terms *genotype* and *phenotype*.
2. Explain the purpose of a testcross.
3. Summarize Mendel's law of segregation and law of independent assortment.
4. Describe the types of crosses that result in a 3:1 and a 9:3:3:1 ratio.

23.2 Pedigree Analysis and Genetic Disorders

Learning Outcomes

Upon completion of this section, you should be able to

1. Recognize autosomal dominant and autosomal recessive patterns of inheritance when examining a pedigree.
2. Analyze pedigrees to determine the probability of a genetic disorder.

Many traits and disorders in humans, and other organisms also, are genetic in origin and follow Mendel's laws. These traits are often controlled by a single pair of alleles on the autosomal chromosomes. An autosome is any chromosome other than those associated with determining sex (X or Y in humans). The patterns of inheritance associated with the sex chromosomes are discussed in Section 24.2.

Patterns of Inheritance

In humans, it is unethical or impossible to carry out some of the crosses that can be done in plants, such as a testcross or self-pollination. Instead, human inheritance is often diagrammed as a pedigree. A **pedigree** is a chart of a family's history with regard to a particular genetic trait. Pedigrees can be done for humans as well as other animals, such as dogs and livestock. In the chart, males are designated by squares and females by circles. A line between a square and a circle represents a mating. A vertical line going downward leads directly to a single child. If there are more children, they are placed off a horizontal line. If the purpose of the pedigree is to follow the inheritance of a particular trait, those individuals with the trait, called the affected individuals, are usually shaded. A pedigree is often used to determine the inheritance of a genetic disease.

Pedigrees for Autosomal Disorders

It is possible to decide if an inherited condition is due to an autosomal dominant or an autosomal recessive allele by studying a pedigree. Does the following pattern of inheritance represent an autosomal dominant or an autosomal recessive characteristic?

Pedigree

Generations

Key
aa = affected
Aa = carrier (unaffected)
AA = unaffected
$A\_$ = unaffected
(one allele unknown)

Autosomal recessive disorders
- Most affected children have unaffected parents.
- Heterozygotes (Aa) have an unaffected phenotype.
- Two affected parents will always have affected children.
- Close relatives who reproduce are more likely to have affected children.
- Both males and females are affected with equal frequency.

Figure 23.9 Autosomal recessive pedigree. The list gives ways to recognize an autosomal recessive disorder.

Figure 23.10 Neuron affected by Tay-Sachs disease. In Tay-Sachs disease, a lysosomal enzyme is missing. This causes the substrate of that enzyme to accumulate within the lysosomes.

In this pattern, the child is affected, but neither parent is. This can happen only if the disorder is recessive and the parents are heterozygotes. Notice that the parents are carriers because they are unaffected but are capable of having a child with the genetic disorder. If the family pedigree suggests that the parents are carriers for an autosomal recessive disorder, it would be possible to do prenatal testing for the genetic disorder.

Notice that in the pedigree in Figure 23.9, cousins (connected by the double line in generation III) are the parents of three children, two of whom have the disorder. Aside from illustrating that reproduction between cousins is more likely to bring out recessive traits, this pedigree also shows that "chance has no memory." Each child born to heterozygous parents has a 25% chance of having the disorder. In other words, it is possible that if a heterozygous couple has four children, each child might have the condition.

Autosomal Recessive Disorders

Inheritance of two recessive alleles is required before an autosomal recessive disorder will appear.

Tay-Sachs disease is a well-known autosomal recessive disorder that usually occurs among Jewish people in the United States, most of whom are of central and eastern European descent. Tay-Sachs disease results from a lack of the enzyme hexosaminidase A (Hex A) and the subsequent storage of its substrate, a glycosphingolipid, in lysosomes. Lysosomes build up in many body cells, but the primary sites of storage are the cells of the brain, which accounts for the onset of symptoms and the progressive deterioration of psychomotor functions (Fig. 23.10).

At first, it is not apparent that a baby has Tay-Sachs disease. However, development begins to slow down between 4 and 8 months of age, and neurological impairment and psychomotor difficulties then become apparent. The child gradually becomes blind and helpless, develops uncontrollable seizures, and eventually becomes paralyzed.

Cystic fibrosis (CF) is an autosomal recessive disorder that occurs among all ethnic groups, but it is the most common lethal genetic disorder among Caucasians in the United States. Research has demonstrated that chloride ions (Cl⁻) fail to pass through a plasma membrane channel protein in the cells of these patients (Fig. 23.11). Ordinarily, after chloride ions have passed through the membrane, sodium ions (Na⁺) and water follow. It is believed that lack of water is the cause of abnormally thick mucus in bronchial tubes and pancreatic ducts. In these children, this mucus is so thick and viscous that it interferes with the function of the lungs and pancreas. To ease breathing, the thick mucus in the lungs must be manually loosened periodically, but the lungs still become infected frequently. Clogged pancreatic ducts prevent digestive enzymes from reaching the small intestine, and to improve digestion, CF patients take digestive enzymes before every meal.

Phenylketonuria (PKU) is an autosomal recessive metabolic disorder that affects nervous system development. Affected individuals lack the enzyme needed for the normal metabolism of the amino acid phenylalanine, and so it appears in the urine and blood. Newborns are routinely tested in the hospital for elevated levels of phenylalanine in the blood. If an elevated level is detected, the newborn will develop normally if placed on a diet low in phenylalanine, which must be continued until the brain is fully developed, around the age of 7, or else severe intellectual

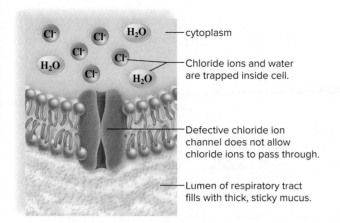

Figure 23.11 **Cystic fibrosis disease.** Cystic fibrosis is due to a faulty protein that is supposed to regulate the flow of chloride ions into and out of cells through a channel protein.

disabilities will develop. Some doctors recommend that the diet continue for life, but in any case, a pregnant woman with phenylketonuria must be on the diet in order to protect her unborn child from harm.

Sickle-cell disease is an autosomal recessive disorder in which the red blood cells are not biconcave disks like normal red blood cells; rather, they are irregular, and often sickle-shaped (Fig. 23.12). This defect is caused by an abnormal type of hemoglobin that differs from normal hemoglobin by one amino acid in the protein globin. The single amino acid change causes hemoglobin molecules to stack up and form insoluble rods, and the red blood cells become sickle-shaped.

Because sickle-shaped cells can't pass along narrow capillary passageways as disk-shaped cells can, they clog the vessels and break down. This is why persons with sickle-cell disease experience poor circulation, anemia, and low resistance to infection. Internal hemorrhaging leads to further complications, such as jaundice, episodic pain in the abdomen and joints, and damage to internal organs. Interestingly, sickle-cell disease does provide some protection against malaria, which is why the allele still

persists in the human population. This is covered in greater detail in Section 27.3.

Variations in the sickle-cell gene enable heterozygous individuals to express variations of the recessive phenotype. Sickle-cell heterozygotes have sickle-cell traits in which the blood cells are normal unless they experience dehydration or mild oxygen deprivation. Intense exertion may cause sickling of some red blood cells for a short period of time. These patients can experience episodes and symptoms very similar to those of patients with the autosomal recessive genotype.

Autosomal Dominant Disorders

Inheritance of only one dominant allele is necessary for an autosomal dominant genetic disorder to be displayed.

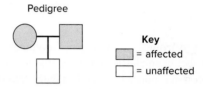

In this pattern, the child is unaffected, but the parents are both affected. This can happen if the condition is autosomal dominant and the parents are heterozygotes. Figure 23.13 lists other ways to recognize an autosomal dominant pattern of inheritance. This pedigree also illustrates that when both parents are unaffected, all their children are unaffected. Why? Because neither parent has a dominant gene that passes the condition on. Inheritance of only one dominant allele is necessary for an autosomal dominant genetic disorder to appear. The following describe some examples of autosomal dominant disorders in humans.

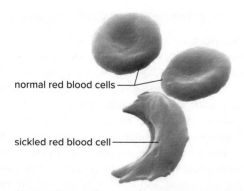

normal red blood cells

sickled red blood cell

Abnormally shaped red blood cells in sickle-cell disease

Figure 23.12 **Sickle-cell disease.** An autosomal recessive mutation in the gene that codes for the globin protein causes normal red blood cells to develop a sickled shape. ©Eye of Science/Science Source

Figure 23.13 **Autosomal dominant pedigree.** The list gives ways to recognize an autosomal dominant disorder.

Many neurons in normal brain

Loss of neurons in Huntington brain

Figure 23.14 Autosomal dominant genetic disorder.   Huntington disease is characterized by increasingly serious psychomotor and mental disturbances because of the loss of nerve cells. (both): ©Dr. Hemachandra Reddy, The Neurological Science Institute, Oregon Health & Science University

Marfan syndrome, an autosomal dominant disorder, is caused by a defect in an elastic connective tissue protein called fibrillin. This protein is normally abundant in the lens of the eye, the bones of the limbs, fingers, and ribs, and also in the wall of the aorta. The affected person often has a dislocated lens, long limbs and fingers, and a caved-in chest. The aorta wall is weak and may burst without warning. A tissue graft can strengthen the aorta, but people with Marfan syndrome still should not overexert themselves.

Huntington disease is a neurological disorder that leads to progressive degeneration of brain cells (Fig. 23.14). The disease is caused by a mutated copy of the gene for a protein called huntingtin. Most patients appear normal until they are of middle age and have already had children, who may later also be stricken. Occasionally, the first sign of the disease appears during the teen years or even earlier. There is no effective treatment, and death comes 10 to 15 years after the onset of symptoms.

Several years ago, researchers found that the gene for Huntington disease was located on chromosome 4. They developed a test to detect the presence of the gene. However, few people want to know they have inherited the gene because there is no cure. At least now we know that the disease stems from a mutation that causes the huntingtin protein to have too many copies of the amino acid glutamine. The normal version of huntingtin has stretches of between 10 and 25 glutamines. If huntingtin has more than 36 glutamines, it changes shape and forms large clumps inside neurons. Even worse, it attracts and causes other proteins to clump with it. One of these proteins, called CBP, which helps nerve cells survive, is inactivated when it clumps with huntingtin. Researchers hope to combat the disease by boosting CBP levels.

Check Your Progress 23.2

1. Describe the characteristics of a pedigree for an autosomal dominant and an autosomal recessive trait.

2. Distinguish between autosomal recessive and autosomal dominant genetic disorders.

23.3 Beyond Simple Inheritance Patterns

Learning Outcomes

Upon completion of this section, you should be able to

1. Explain how the phenotypes for incompletely dominant and codominant traits differ from dominant/recessive trait phenotypes.

2. Explain how ABO blood types are an example of multiple allele inheritance.

3. Describe polygenic inheritance.

Since the time of Mendel, scientists have discovered that for most traits, the patterns of inheritance do not precisely follow what is predicted by Mendelian genetics. Living organisms are complex biological systems, with many genes and environmental factors interacting to maintain life. However, the principles established by Mendel, including the laws of segregation and independent assortment (see Section 23.1), can be used to help us understand the basis of the complex patterns of inheritance we will explore in this section.

Incomplete Dominance and Codominance

Incomplete dominance occurs when the heterozygote is intermediate between the two homozygotes. An example is hair structure in Caucasians (Fig 23.15a). When a curly-haired Caucasian reproduces with a straight-haired Caucasian, their children have wavy hair, an intermediate phenotype. When two wavy-haired persons reproduce, the expected phenotypic ratio among the offspring is 1:2:1—that is, one curly-haired child to two with wavy hair to one with straight hair (Fig. 23.15b). We can explain incomplete dominance by assuming that only one allele codes for a product, and the single dose of the product gives the intermediate result. Notice in Figure 23.15 that we do not use the same capital and lowercase letters as we used with dominant/recessive traits.

a. A person with straight hair (H₁H₁)

A person with naturally curly hair (H₂H₂)

Heterozygous Parents (H₁H₂)

♂ H_1H_2 × ♀ H_1H_2

oocytes

Key
- Straight hair
- Wavy hair
- Curly hair

♂ \ ♀	H_1	H_2
H_1	H_1H_1	H_1H_2
H_2	H_1H_2	H_2H_2

sperm

Phenotypic Ratio
1 ▢ Straight hair
2 ▢ Wavy hair
1 ▢ Curly hair

b.   **Offspring**

Figure 23.15 Incomplete dominance.   Among Caucasians, neither (**a**) straight nor curly hair is dominant. (**b**) When two wavy-haired individuals reproduce, each offspring has a 25% chance of having either straight or curly hair and a 50% chance of having wavy hair, the intermediate phenotype. (a, man): ©Umbar Shakir/Getty Images; (a, woman): ©Larry Williams/Corbis/Getty Images

Figure 23.16 Incompletely dominant genetic disorder.
Familial hypercholesterolemia is incompletely dominant. Persons with one mutated allele (heterozygotes) have an abnormally high level of cholesterol in the blood, and those with two mutated alleles (homozygotes) have a higher level still. In homozygotes, the elevated cholesterol level may sometimes form visible deposits in the skin. (photo): ©Medical-On-Line/Alamy

Codominance occurs when alleles are equally expressed in a heterozygote. A familiar example is the human blood type AB, in which the red blood cells have the characteristics of both type A and type B blood. We can explain codominance by assuming that both genes code for a product, and we observe the results of both products being present. As we will see later in this section, blood type inheritance is also said to be an example of multiple alleles.

Incompletely Dominant Disorders

The severity of symptoms exhibited in **familial hypercholesterolemia** parallels the number of LDL-cholesterol receptor proteins in the plasma membrane. A person with two mutated alleles lacks LDL-cholesterol receptors; a person with only one mutated allele has half the normal number of receptors; and a person with two normal alleles has the usual number of receptors. The presence of excessive cholesterol in the blood causes cardiovascular disease. Therefore, people with no receptors die of cardiovascular disease before 30 years of age. These individuals develop cholesterol deposits in the skin, tendons, and cornea. Individuals with one-half the number of receptors may die while still young or after

they have reached middle age. People with the full number of cholesterol receptors do not have familial hypercholesterolemia (Fig. 23.16).

Multiple Allele Inheritance

When a trait is controlled by **multiple alleles,** the gene exists in several allelic forms. But as is the case with any trait, each person has only two of the possible alleles.

ABO Blood Types

Three alleles for the same gene control the inheritance of ABO blood types. These alleles determine the presence or absence of antigens on red blood cells:

I^A = A antigen on red blood cells
I^B = B antigen on red blood cells
i = Neither A nor B antigen on red blood cells

Each person has only two of the three possible alleles, and both I^A and I^B are dominant over i. Therefore, there are two possible genotypes for type A blood and two possible genotypes for type B blood. On the other hand, I^A and I^B are fully expressed in the presence of the other. Therefore, if a person inherits one of each of these alleles, that person will have type AB blood. Type O blood can result only from the inheritance of two i alleles.

The possible genotypes and phenotypes for blood type are as follows:

Phenotype	Genotype
A	I^AI^A, I^Ai
B	I^BI^B, I^Bi
AB	I^AI^B
O	ii

SCIENCE IN YOUR LIFE ▶ SCIENTIFIC INQUIRY

Genetics of Eye Color

As we have already noted, genes rarely act alone to produce a phenotype. In many cases, multiple genes may be part of a metabolic pathway consisting of the interaction of multiple enzymes and proteins.

An example of a trait determined by gene interaction is human eye color (Fig. 23C). Multiple pigments are involved in producing eye color, including melanin. While a number of factors (including structure of the eye) control minor variations in eye color (such as green or hazel eyes), a gene called *OCA2* is involved in producing the melanin that establishes the brown/blue basis of eye color. Individuals who lack *OCA2* completely (as occurs in albinism) have red eyes. A recessive form of *OCA2* results in blue eyes in homozygous recessive individuals, whereas a single dominant *OCA2* allele produces brown eyes (the most common eye color).

The *HERC2* gene overrides the instructions of the *OCA2* gene. If individuals are homozygous for a recessive allele of *HERC2*, they will have blue eyes regardless of the genotype associated with *OCA2*. Geneticists call this type of interaction, where one gene can override another, an *epistatic interaction*.

Questions to Consider

1. What would be the potential consequences if one of the genes involved in a multiple gene interaction were to develop an error?
2. How difficult would it be to use genetic engineering to modify a trait that is determined by a multiple gene interaction?

Figure 23C Genetics of eye color.   Individuals with red eyes (*left*) lack the ability to produce eye pigments. The presence of a dominant *OCA2* allele produces the common brown eye color (*center*). Blue eyes (*right*) may occur either from being homozygous recessive for an *OCA2* allele, or due to an interaction with a second gene, *HERC2*. (red eye): ©Mediscan/ Alamy; (brown eye): ©Stylephotographs/123RF; (blue eye): ©Lightpoet/123RF

In the past, blood typing was used in paternity suits. However, the results of a blood test can only suggest that the tested individual *might* be the father, not that he definitely *is* the father. As an example, see Figure 23.17. It is possible, but not definite, that a man with type B blood (genotype I^Ai) is the father of a child with type O blood, depending on the genotype of the mother. On the other hand, a blood test can sometimes definitely prove that a man is *not* the father. For example, a man with type AB blood cannot possibly be the father of a child with type O blood. Therefore, blood tests can be used in legal cases only to exclude a man from possible paternity.

As a point of interest, the Rh factor is inherited separately from A, B, AB, or O blood types. When you are Rh-positive, your red blood cells have a particular antigen, and when you are Rh-negative, that antigen is absent. There are multiple recessive alleles for Rh-negative, but they are all recessive to Rh-positive. More information on the interactions of the ABO and Rh blood types with the immune system is provided in Section 13.4.

Polygenic Inheritance

Many traits, such as size or height, shape, weight, skin color, metabolic rate, and behavior, are governed by several genes (each with sets of alleles). **Polygenic inheritance** occurs when a trait is

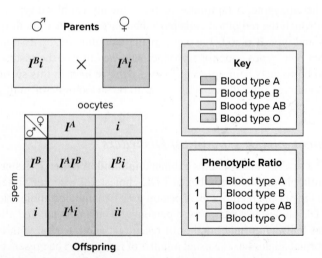

Figure 23.17 Inheritance of blood types.   Blood type exemplifies multiple allele inheritance. A mating between individuals with type A blood and type B blood can result in any one of the four blood types. Why? Because the parents are I^Ai and I^Bi. If both parents had type AB blood, predict the blood types of their potential offspring.

Figure 23.18 Polygenic inheritance.   Skin color is controlled by many pairs of alleles, which results in a range of phenotypes. The vast majority of people have skin colors in the middle range, while fewer people have skin colors in the extreme range.

governed by two or more genes (sets of alleles). The individual has a copy of all allelic pairs, possibly located on many different pairs of chromosomes. Each dominant allele codes for a product, and therefore the dominant alleles have a quantitative effect on the phenotype, and these effects are additive. The result is a *continuous variation* of phenotypes, resulting in a distribution of these phenotypes that resembles a bell-shaped curve. The more genes involved, the more continuous are the variations and distribution of the phenotypes.

Skin Color

Skin color is the result of pigmentation produced by cells called melanocytes in the skin (see Section 11.4). We now know there are over 100 different genes that influence skin color. For simplicity, we will assume that skin has only three pairs of alleles (*Aa, Bb,* and *Cc*) and that each capital letter contributes pigment to the skin. When a very dark person reproduces with a very light person, the children have medium-brown skin. When two people with the genotype *AaBbCc* reproduce with one another, the individuals may range in skin color from very dark to very light. The distribution of these phenotypes typically follows a bell-shaped curve, meaning that few people have the extreme phenotypes and most people have the phenotype that lies in the middle. A bell-shaped curve is a common identifying characteristic of a polygenic trait (Fig. 23.18).

However, skin color is also influenced by the sunlight in the environment. Notice how a range of phenotypes exists for each genotype. For example, individuals who are *AaBbCc* may vary in their skin color, even though they possess the same genotype, and several possible phenotypes fall between the two extremes. The interaction of the environment with polygenic traits is discussed in Section 23.4.

> **Check Your Progress 23.3**
>
> 1. Interpret the genotype of the heterozygote if the inheritance pattern for a genetic disorder is shown to be incompletely dominant.
> 2. Determine the genotype of the child, the mother, and the possible genotypes of the father for a child with type O blood who is born to a mother with type A blood.

23.4 Environmental Influences

> **Learning Outcomes**
>
> Upon completion of this section, you should be able to
> 1. Describe ways in which the environment can influence genetic traits.
> 2. Understand how scientists determine the effect of nature versus nurture.

Environmental factors, such as nutrition or temperature, can also influence the expression of genetic traits. Polygenic traits seem to be particularly influenced by the environment. For example, in the case of height, both genetics and the environment, namely nutrition, are involved. Polygenic traits that have environmental influences are often identified by a bell-shaped curve of phenotypes (Fig. 23.19).

Temperature can also affect the phenotypes of plants and animals. For example, primroses have white flowers when grown above 32°C and red flowers when grown at 24°C. The coats of Siamese cats and Himalayan rabbits are darker in color at the ears, nose, paws, and tail. Himalayan rabbits are known to be homozygous for the allele *ch,* which is involved in the production of melanin. Experimental evidence suggests that the enzyme coded for by this gene is active only at low temperatures, and therefore black fur occurs only at the extremities, where body heat is lost to the environment (Fig. 23.20).

Investigators try to determine what percentage of various human traits is due to nature (inheritance) and what percentage is due to nurture (the environment). Some studies use identical and fraternal twins who have been separated since birth and thus have been raised in different environments. The supposition is that if identical twins in different environments share the same trait, that trait is most likely inherited. These studies have found that identical twins are more similar in their intellectual talents, personality traits, and levels of lifelong happiness than are fraternal twins when both groups have been separated since birth. Biologists conclude that all behavioral traits are partly inheritable and that genes exert their effects by acting together in complex combinations susceptible to environmental influences.

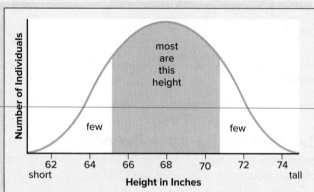

Figure 23.19 Polygenic inheritance and environmental effects. A person's height is determined by both environmental and genetic factors. This results in a characteristic bell-shaped distribution curve.
(photo): ©McGraw-Hill Education/David Hyde and Wayne Falda, photographers

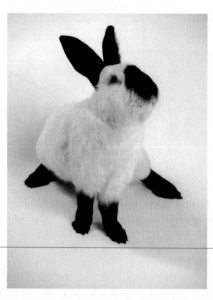

Figure 23.20 Coat color in Himalayan rabbits. The influence of the environment on the phenotype has been demonstrated by plucking out the fur from one area and applying an ice pack. The new fur that grew in was black instead of white, showing that the enzyme that produces melanin (a dark pigment) in the rabbit is active only at low temperatures. ©Neil Twigg/Alamy

Scientists have also learned how to manipulate the environment in order to prevent the damaging effects of some genetic disorders. For example, the bodies of people who have the genetic disorder phenylketonuria (PKU) lack an enzyme that breaks down the amino acid phenylalanine. The accumulation of phenylalanine leads to neurological damage in these patients. However, scientists have discovered that if these patients modify their diet so that they consume very little protein containing phenylalanine, they can avoid many of the damaging effects of this disease.

Check Your Progress 23.4

1. Summarize the environmental factors that may influence polygenic traits such as height and skin color.
2. Discuss how studies using identical twins may help determine how much of a phenotype is based on genetics and how much is environmental.

Conclusion

Phenylketonuria is a genetic disorder that displays an autosomal recessive pattern of inheritance. However, it is also an example of a trait that has an environmental influence, because the disorder is expressed only in the presence of an environmental factor, which in this case is the phenylalanine in the artificial sweetener. Because individuals who are homozygous recessive for this trait lack the ability to break down this compound, they may experience the symptoms of phenylketonuria, such as seizures.

Geneticists recognize that very few traits exhibit a simple recessive or dominant pattern of inheritance. A variety of other factors, including the sex of the individual (covered in the next chapter), may influence the expression of a trait. Still, by understanding these simple patterns of inheritance, it is often easier to identify outside factors that are causing the trait to deviate from a typical Mendelian pattern.

SUMMARIZE

23.1 Mendel's Laws

■ Genes carry instructions for a trait. Alternative forms of a gene (producing differences in the trait) are called **alleles.** The location of a gene on a chromosome is called its **locus.** In genetics, **phenotype** refers to the physical appearance of the individual, whereas the **genotype** lists the individual's combination of alleles. For a trait, **dominant alleles** mask the expression of **recessive alleles.** It is customary to use letters

to represent the genotypes of individuals. **Homozygous** dominant is indicated by two capital letters, and homozygous recessive is indicated by two lowercase letters. **Heterozygous** is indicated by a capital letter and a lowercase letter.

■ An individual has two alleles for each trait, whereas the gametes have one allele for each trait. Therefore, in keeping with Mendel's **law of segregation,** each heterozygous individual can form two types of gametes.

■ When performing an actual cross, it is assumed that all possible types of sperm fertilize all possible types of oocytes. The Punnett square is based on this assumption. The results may be expressed as a probable phenotypic ratio. It is also possible to state the chance of an offspring having a particular phenotype. When a heterozygote reproduces with a heterozygote (called a **monohybrid cross**), the phenotypic ratio is 3:1; that is, each child has a 75% chance of having the dominant phenotype and a 25% chance of having the recessive phenotype. **Punnett squares** are often used to indicate the potential offspring of a genetic cross.

- A **testcross** is sometimes used to determine the genotype of an individual with a dominant phenotype.

- In keeping with Mendel's **law of independent assortment,** an individual heterozygous for two traits (**dihybrid**) can form four types of gametes. Therefore, the phenotypic ratio for a dihybrid cross is 9:3:3:1; in other words, $9/16$ of the offspring have the two dominant traits, $3/16$ have one of the dominant traits with one of the recessive traits, $3/16$ have the other dominant trait with the other recessive trait, and $1/16$ have both recessive traits.

23.2 Pedigree Analysis and Genetic Disorders

- A **pedigree** is a chart of a family's history with regard to a particular genetic trait. It may be possible to decide if an inherited condition is due to an autosomal dominant or an autosomal recessive allele by studying a pedigree.

- **Tay-Sachs disease** (a lysosomal storage disease), **cystic fibrosis (CF)** (faulty regulator of chloride channel), **phenylketonuria (PKU)** (inability to metabolize phenylalanine), and **sickle-cell disease** (sickle-shaped red blood cells) are all autosomal recessive disorders.

- **Marfan syndrome** (defective elastic connective tissue) and **Huntington disease** (abnormal huntingtin protein) are both autosomal dominant disorders.

23.3 Beyond Simple Inheritance Patterns

- There are many exceptions to Mendel's laws, including **incomplete dominance** (curly hair), **codominance,** and **multiple alleles** (ABO blood types).

- **Familial hypercholesterolemia** (liver cells lack cholesterol receptors) is an incompletely dominant autosomal disorder.

- In **polygenic inheritance,** such as height and skin color in humans, several genes each contribute to the overall phenotype in equal, small degrees. If the environment is involved, the result is a bell-shaped curve of phenotypes.

23.4 Environmental Influences

- Environmental factors, such as nutrition and temperature, affect the expression of certain traits. In humans, nutrition helps determine height. In Siamese cats and Himalayan rabbits, the effect of temperature is apparent in the distribution of fur color.

- Humans can manipulate the environment to lessen the damaging effects of genetic defects.

ASSESS

Testing Yourself

Choose the best answer for each question.

23.1 Mendel's Laws

1. The law of segregation states all of the following except
   a. factors separate during formation of the gametes.
   b. each individual has two factors for each trait.
   c. gametes contain a single factor for each trait.
   d. factors assort independently of each other by meiosis.

2. If two parents with short fingers (dominant) have a child with long fingers, what is the chance their next child will have long fingers?
   a. There is no chance.
   b. $1/16$
   c. $1/2$
   d. $3/16$
   e. $1/4$

3. Using the product rule, determine the probability that an Aa individual will be produced from an Aa × Aa cross.
   a. 25%
   b. 50%
   c. 75%
   d. 100%
   e. 0%

4. According to the law of independent assortment,
   a. all possible combinations of factors can occur in the gametes.
   b. only the parental combinations of gametes can occur in the gametes.
   c. only the nonparental combinations of gametes can occur in the gametes.

5. The genotype of an individual with the dominant phenotype can be determined best by reproduction with
   a. a homozygous recessive individual.
   b. an individual with the dominant phenotype.
   c. a heterozygote.
   d. None of these are correct.

6. In a dihybrid cross of AaBb × AaBb, what is the probability that offspring will have the dominant phenotype for trait A but the recessive phenotype for trait B?
   a. $1/16$
   b. $3/16$
   c. $7/16$
   d. $9/16$
   e. None of these are correct.

23.2 Pedigree Analysis and Genetic Disorders

7. Which of the following does not have an autosomal recessive pattern of inheritance?
   a. cystic fibrosis
   b. Huntington disease
   c. phenylketonuria
   d. sickle-cell disease
   e. None of these are correct.

8. Which of the following does not describe the pedigree analysis of an autosomal dominant disorder?
   a. heterozygotes are affected
   b. two affected parents will always have affected children
   c. affected children usually have an affected parent
   d. two unaffected parents will not have affected children

23.3 Beyond Simple Inheritance Patterns

9. If a child has type O blood and the mother is type A, then which of the following might be the blood type of the child's father?
   a. A only
   b. A or O
   c. B only
   d. A, B, or O
   e. O only

10. A trait in which the heterozygotes have a phenotype that is intermediate between the dominant and recessive phenotypes is an indication of
    a. codominance.
    b. pleiotropy.
    c. a sex-linked trait.
    d. incomplete dominance.
    e. None of these are correct.

11. In the analysis of a genetic disorder, you notice that the phenotypic distribution follows a bell-shaped curve, and that multiple genes are probably involved. This suggests what pattern of inheritance?
    a. autosomal dominant
    b. autosomal recessive
    c. incomplete dominance
    d. polygenic inheritance
    e. None of these are correct.

23.4 Environmental Influences

12. Which of the following does not have an environmental factor that affects the distribution of the phenotype?
    a. phenylketonuria
    b. skin color
    c. height
    d. Marfan syndrome

ENGAGE

BioNOW

Want to know how this science is relevant to your life? Check out the BioNOW video below:

■ Glowing Fish Genetics

How might an experiment be designed to test whether the pattern of inheritance in the glowfish is dominant or codominant?

Thinking Critically

1. You are working with the fruit fly *Drosophila* and want to determine whether a newly found characteristic is dominant or recessive. Design an experiment that will help you determine the dominance of the trait.

2. How would you determine whether a disease in humans is simply polygenic or has an environmental influence?

3. If identical twins have exactly the same genetic makeup, why are there any differences at all between them?

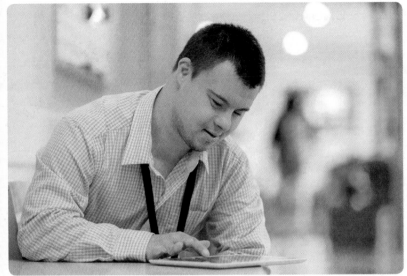
©Disability Images/Blend Images LLC

Chromosomal Basis of Inheritance

Down Syndrome

Sarah's youngest son, Brandon, acted like any other 6-year-old boy. He loved to eat and play, and enjoyed spending time with his family. However, Brandon was born with Down syndrome, a genetic disorder that affects over 400,000 Americans. Down syndrome occurs when an individual inherits an extra copy of chromosome 21. It is the most common chromosomal anomaly in humans and the incidence of the condition is slowly rising in Western cultures. From 1983 to 2003, the total number of cases of Down syndrome increased by almost 30%.

People with Down syndrome have similar characteristics, including an upward slant to the eyes, low muscle tone, a deep crease across the center of the palm, and an increased risk of cardiac defects. There is some degree of intellectual disability, but most individuals with Down syndrome have IQs in the mild to moderate disability range. Despite these health concerns, the life expectancy of Down syndrome individuals has increased tremendously in recent years, from 25 years in 1983 to over 60 years today.

The National Down Syndrome Society tells us that individuals with Down syndrome can be integrated into all aspects of society. They attend school, work, and participate in community activities. Some individuals graduate from high school and attend college. Some socialize, date, and even marry.

In this chapter, we will examine genetic conditions that are caused by changes in chromosomal number and structure.

As you read through this chapter, think about the following questions:

1. What causes a change in the number of chromosomes?

2. What is the genetic basis that causes the symptoms of Down syndrome?

3. How can the chromosome number of an individual be tested before he or she is born?

BEFORE YOU BEGIN

Before beginning this chapter, take a few moments to review the following discussions:

Section 5.4 What is meant by the term *homologous chromosomes*?

Section 23.1 What does Mendel's law of segregation tell us about chromosomal inheritance?

Section 23.2 How does a Punnett square help predict the genotypes and phenotypes of a genetic cross?

24.1 Gene Linkage

Not long after Mendel's study of inheritance (see Section 23.1), investigators began to realize that many different types of genes are on a single chromosome. A chromosome doesn't contain just one or two genes; it contains a long series of genes in a definite sequence. While genes may have multiple variations, or alleles, the sequence of the genes is fixed because each gene has its own particular locus on a chromosome.

All the genes on one chromosome form a **linkage group** because they tend to be inherited together. Figure 24.1 shows some of the traits in the chromosome 19 linkage group of humans. When we do two-trait crosses (see Fig. 23.7), we are assuming that the genes for the different traits are on nonhomologous chromosomes and therefore are not linked. The genes that are linked do not show independent assortment. Figure 24.2a shows the results of a cross when linkage is complete: A dihybrid produces only two types of gametes in equal proportion.

Muscular dystrophy

Atherosclerosis susceptibility

Insulin-resistant diabetes mellitus

Ichthyosis

Eye color (green/blue)

Hair color (brown)

Acute myeloid leukemia

Alzheimer disease (late onset)

Deafness

Hemolytic anemia

Ovarian carcinoma

Colorectal cancer

Cardiomyopathy

Figure 24.1 Human chromosome 19.   A selection of traits located on human chromosome 19 (not to scale). Each of these traits is associated with a gene. Collectively, these genes form a single linkage group. In human genetics, genes are sometimes named after the disease they are associated with, not their function in the body.

crossing-over during meiosis

no crossing-over during meiosis

crossing-over during meiosis

50%   50%

97%

3% recombinant gametes

2 types of gametes in equal proportions

a. Genes closely linked

4 types of gametes in unequal proportions

b. Genes farther apart on chromosome

Figure 24.2 Recombination and linkage groups.   In this example, alleles *A* and *B* are on one member of a homologous pair, and alleles *a* and *b* are on the other member. **a.** When the genes for *A* and *B* are located closely together, recombination between the alleles is rare and only two types of gametes are produced in equal proportion. **b.** When the genes are farther apart, crossing-over occurs more frequently between the alleles, producing four different types of gametes. The recombinant gametes occur in reduced proportion because crossing-over occurs infrequently.

During meiosis, crossing-over sometimes occurs between the nonsister chromatids of a tetrad (see Fig. 5.13). During crossing-over, the chromatids exchange genetic material, and therefore genes. If the genes contain different alleles, we can observe the genetic content of one chromatid moving to the sister chromatid. If crossing-over occurs between the two genes of interest, a dihybrid produces four types of gametes instead of two (Fig. 24.2*b*). However, it is important to note that there are fewer recombinant gametes because crossing-over is infrequent, and will not always occur between two genes.

The occurrence of crossing-over can help tell the sequence of genes on a chromosome because crossing-over occurs more often between distant genes (Fig. 24.2*b*) than between genes that are close together on a chromosome (Fig. 24.2*a*). Historically, maps of chromosomes were based on the recombination frequencies between genes, with each *map unit* equal to an observed 1% of crossing-over. Thus, in Figure 24.2*b*, the distance between the *A* and *B* genes would be three map units.

Check Your Progress 24.1

1. Explain why linked alleles do not show independent assortment.
2. Explain what a distance of 10 map units tells you about the distance between two genes.

24.2 X-Linked Inheritance

Learning Outcomes

Upon completion of this section, you should be able to

1. Explain X-linked genetic inheritance.
2. Solve crosses involving sex-linked traits.
3. Describe four X-linked recessive disorders.

Normally, both males and females have 23 pairs of chromosomes; 22 pairs are called **autosomes,** and one pair is the sex chromosomes. The **sex chromosomes** are named because they differ between the sexes. In humans, males have the sex chromosomes X and Y, and females have two X chromosomes.

Traits controlled by genes on the sex chromosomes are said to be **sex-linked;** a gene on an X chromosome is **X-linked,** and a gene on the Y chromosome is *Y-linked.* The X and Y chromosomes are not homologous, and contain different combinations of genes. The X chromosome contains over 2,000 genes, while only around 200 genes have been identified on the Y chromosome. For that reason, we will focus our discussion on X-linked traits.

It would be logical to suppose that a sex-linked trait is passed from father to son or from mother to daughter, but this is not always the case. For X-linked traits, a male always receives an X-linked allele from his mother, from whom he inherited an X chromosome. Because the X and Y chromosomes are not homologous, the Y chromosome from the father does not carry an allele for the trait. Most sex-linked genetic disorders are recessive. Therefore, a female must receive two alleles, one from each parent, before she has the condition.

Sex-Linked Alleles

When considering X-linked traits, the allele on the X chromosome is shown as a letter attached to the X chromosome. For example, following is the key for red-green color blindness, a well-known X-linked recessive disorder:

$$X^B = \text{normal vision} \quad X^b = \text{color blindness}$$

The possible genotypes and phenotypes in both males and females are:

Genotypes	Phenotypes
$X^B X^B$	Female who has normal color vision
$X^B X^b$	Carrier female with normal color vision
$X^b X^b$	Female who is color blind
$X^B Y$	Male who has normal vision
$X^b Y$	Male who is color blind

The heterozygous female (second genotype in the list) is sometimes referred to as a *carrier* female because, although a female with this genotype has normal color vision, she is capable of passing on an allele for color blindness. Color-blind females are rare because they must receive the allele from both parents. Color-blind males are more common because they need only one recessive allele to be color blind. The allele for color blindness has to be inherited from their mother because it is on the X chromosome. Males inherit the Y chromosome only from their father.

Now let us consider a mating between a man with normal vision and a heterozygous woman (Fig. 24.3). What is the chance this couple will have a color-blind daughter? A color-blind son? All daughters will have normal color vision because they all receive an X^B from their father. The sons, however, have a 50% chance of being color blind, depending on whether they receive an X^B or an X^b from their mother. The inheritance of a Y chromosome

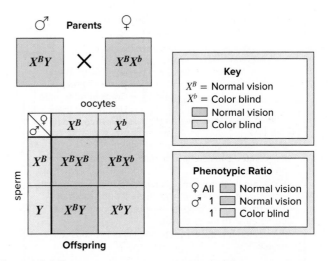

Figure 24.3 Cross involving an X-linked allele.   The male parent is normal, but the female parent is a carrier—that is, an allele for color blindness is located on one of her X chromosomes. Therefore, each son has a 50% chance of being color blind. The daughters will have normal color vision, but each one has a 50% chance of being a carrier.

from their father cannot offset the inheritance of an X^b from their mother. Because the Y chromosome doesn't have an allele for the trait, it can't possibly prevent color blindness in a son. Note in Figure 24.3 that the phenotypic results for sex-linked traits are given separately for males and females.

Pedigree for X-Linked Disorders

Like color blindness, most sex-linked disorders are carried on the X chromosome. Figure 24.4 gives a pedigree for an *X-linked recessive disorder*. More males than females have the disorder because recessive alleles on the X chromosome are always expressed in males. The Y chromosome lacks an allele for the disorder. X-linked recessive conditions often pass from grandfather to grandson because the daughters of a male with the disorder are carriers. Figure 24.4 lists various ways to recognize a recessive X-linked disorder.

Only a few known traits are *X-linked dominant*. One example is incontinentia pigmenti, which causes discoloration of the skin. If a disorder is X-linked dominant, affected males pass the trait *only* to daughters, who have a 100% chance of having the condition. Females can pass an X-linked dominant allele to both sons and daughters. If a female is heterozygous and her partner is normal, each child has a 50% chance of escaping an X-linked dominant disorder. This depends on which maternal X chromosome is inherited.

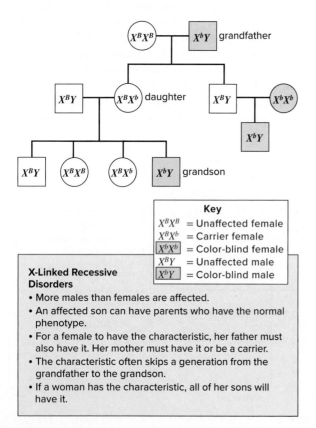

Figure 24.4 X-linked recessive pedigree. This pedigree for color blindness exemplifies the inheritance pattern of an X-linked recessive disorder. The list gives various ways of recognizing the X-linked recessive pattern of inheritance.

X-Linked Recessive Disorders of Interest

Some forms of **color blindness** are associated with recessive alleles on the X chromosome. The most common of these are red-green color blindness. Between 7% and 10% of men have red-green color blindness, depending on the ethnic background of the individual (higher in Caucasians, lower in some Native American populations). Most of them see brighter greens as tans, olive greens as browns, and reds as reddish browns. A few cannot tell reds from greens at all. They see only yellows, blues, blacks, whites, and grays.

Duchenne muscular dystrophy is an X-linked recessive disorder characterized by wasting away of the muscles. Symptoms, which include a waddling gait, toe walking, frequent falls, and difficulty in rising, may appear as soon as the child starts to walk. Muscle weakness intensifies until the individual is confined to a wheelchair. Death usually occurs by age 20; therefore, affected males are rarely fathers. The recessive allele remains in the population by passing from carrier mother to carrier daughter.

The absence of a protein, now called dystrophin, is the cause of Duchenne muscular dystrophy. Much investigative work determined that the lack of dystrophin causes calcium to leak into the cell, which promotes the action of an enzyme that dissolves muscle fibers. When the body attempts to repair the tissue, fibrous tissue forms (Fig. 24.5), and this cuts off the blood supply so that more and more cells die. Immature muscle cells can be injected into muscles, but it takes 100,000 cells for dystrophin production to increase by 30–40%.

Fragile X syndrome is caused by an abnormal number of repeat sequences in the genome and is the most common cause of inherited mental impairment. These impairments can range from mild learning disabilities to more severe intellectual disabilities. It is also associated with some forms of *autism,* a class of social, behavioral, and communication disorders. Males with full symptoms of the condition have characteristic physical abnormalities. A long face, prominent jaw, and large ears are facial features

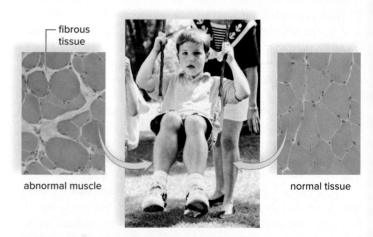

Figure 24.5 Duchenne muscular dystrophy.   In Duchenne muscular dystrophy, an X-linked recessive disorder, the calves of the legs enlarge because fibrous tissue develops as muscles waste away due to lack of the protein dystrophin. (left & right): ©Dr. Rabi Tawil, Director, Neuromuscular Pathology Laboratory, University of Rochester Medical Center; (center): ©Muscular Dystrophy Association

SCIENCE IN YOUR LIFE ► HEALTH

Trinucleotide Repeat Expansion Disorders (TREDs)

Several of the genetic diseases we have studied result from a single change in the DNA that results in a single change in an amino acid in a particular protein. For example, the most common form of sickle-cell disease results from a one-nucleotide change in the DNA that causes a one-amino-acid difference in the protein hemoglobin. However, in at least 15 human diseases, including Huntington disease and fragile X syndrome, the problem lies in a DNA sequence that is repeated too many times, called a *trinucleotide repeat.*

As described in Section 23.3, Huntington disease is a neurological disorder in which the huntingtin protein is altered. The trinucleotide sequence CAG normally appears several times at the beginning of the gene that encodes the huntingtin protein. The presence of multiple CAGs results in multiple glutamines in the protein. People with 10 to 35 copies of the CAG display no signs of Huntington disease.

However, people with more than 40 copies have the disease.

Fragile X syndrome is one of the most common genetic causes of mental impairment. As children, fragile X individuals may be hyperactive or autistic. Their speech is delayed in development and often repetitive in nature. As adults, males have large testes and big, usually protruding ears. They are short in stature, but the jaw is prominent, and the face is long and narrow (Fig. 24Aa). The term *fragile X syndrome* originated because diagnosis used to be dependent upon observation of an X chromosome whose tip is attached to the rest of the chromosome by only a thin thread (Fig. 24Ab). This disease results from repeats of the trinucleotide CGG. People with 6 to 50 copies appear normal, whereas those with over 230 copies have the disease. Those with between 50 and 230 copies are carriers of fragile X, with either no symptoms or very mild symptoms.

People who have the intermediate number of repeats are said to have a pre-mutation. Pre-mutations can lead to full-blown mutations in future generations by expansion of the number of repeats. This is sometimes called genetic anticipation. The number of repeats increases with each generation, as does the severity of the disease (Fig. 24Ac). It is not known why the repeats expand in number. Using the current molecular technology, it is possible to accurately determine the number of trinucleotide repeats present in an individual.

Questions to Consider

1. The repeat expansion appears to occur during meiosis but not mitosis. Why might this be true?
2. If a male has fragile X syndrome, is his mother or father more likely to be a carrier for the disease? Why?
3. Why is fragile X, but not Huntington disease, considered a syndrome?

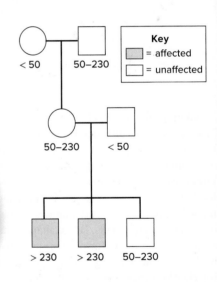

a. Young male with fragile X syndrome   Same individual when mature   b. Fragile X chromosome   c. Inheritance pattern for fragile X syndrome

Figure 24A TRED in fragile X syndrome.   a. A young male with fragile X syndrome appears normal, but with age develops an elongated face with a prominent jaw and ears that noticeably protrude. **b.** An arrow points out the fragile site of this fragile X chromosome. **c.** The number of repeats at the fragile site are given for each person, showing that the incidence and severity increase with each succeeding generation. A grandfather who has a premutation with 50 to 230 repeats has no symptoms but transmits the condition to his grandsons through his daughter. Two grandsons have full-blown mutations with more than 230 repeats. (a, both): ©R. Simensen; (photo, b): ©Monica Schroeder/Science Source

often seen in fragile X males. Excessively flexible joints and genital abnormalities are common as well. Females with the condition have variable symptoms. Most of the same traits seen in males with fragile X have been reported in females as well, but often the symptoms are milder in females and present with lower frequency. The Health feature "Trinucleotide Repeat Expansion Disorders (TREDs)" provides additional information on fragile X syndrome.

Figure 24.6 A pedigree showing X-linked inheritance of hemophilia in European royal families. Because Queen Victoria was a carrier, each of her sons had a 50% chance of having the disorder, and each of her daughters had a 50% chance of being a carrier. This pedigree shows only a portion of the European royal families.

1. Victoria
2. Edward VII
3. Irene
4. George V
5. George VI
6. Margaret
7. Victoria
8. Alfonso XIII
9. Juan

10. Alexandra
11. Charles
12. Diana
13. Andrew
14. Edward
15. Anne
16. Sarah

Hemophilia, of which there are two common types, is another X-linked recessive disorder. Hemophilia A is due to the absence or minimal presence of a clotting factor known as factor VIII, and hemophilia B is due to the absence of clotting factor IX. Hemophilia is called the bleeder's disease because the affected person's blood either does not clot or clots very slowly. Although hemophiliacs bleed externally after an injury, they also bleed internally, particularly around joints. Hemorrhages can be stopped with transfusions of fresh blood (or plasma) or concentrates of the missing clotting protein. Factors VIII and IX are also now available as biotechnology products.

At the turn of the century, hemophilia was prevalent among the royal families of Europe, and all of the affected males could trace their ancestry to Queen Victoria of England (Fig. 24.6). Of Queen Victoria's 26 grandchildren, four grandsons had hemophilia, and four granddaughters were carriers. Because none of Queen Victoria's ancestors were affected, it seems that the faulty allele she carried arose by mutation, either in Victoria or in one of her parents. Her carrier daughters, Alice and Beatrice, introduced the allele into the ruling houses of Russia and Spain, respectively. Alexei, the last heir to the Russian throne before the Russian Revolution, was a hemophiliac. There are no hemophiliacs in the present British royal family because Victoria's eldest son, King Edward VII, did not receive the allele.

Check Your Progress 24.2

1. Explain how sex-linked patterns of inheritance differ from autosomal inheritance.
2. Describe the phenotypic ratio that is expected for a cross in which both parents have one X-linked recessive allele.
3. Summarize the causes of the X-linked recessive disorders discussed in this chapter.

24.3 Changes in Chromosome Number

Learning Outcomes

Upon completion of this section, you should be able to

1. Describe how chromosome number disorders arise.
2. Give several examples of chromosome number disorders.

The Karyotype

Physicians and prospective parents sometimes want to view an unborn child's chromosomes to determine whether a chromosomal abnormality exists. Any cell in the body except red blood cells, which lack a nucleus, can be a source of chromosomes for examination. In adults, it is easiest to obtain and use white blood cells separated from a blood sample for the purpose of looking at the chromosomes. For an examination of fetal chromosomes, a physician may recommend procedures such as chorionic villus sampling or amniocentesis (see Section 22.3).

After a cell sample has been obtained, the cells are stimulated to divide in a culture medium. When a cell divides, chromatin condenses to form chromosomes. The nuclear envelope fragments, liberating the chromosomes. Next, a chemical is used to stop the division process when the chromosomes are most compacted and visible microscopically. Stains are applied to the slides, and the cells can be photographed with a camera attached to a microscope. Staining causes the chromosomes to have dark and light cross-bands of varying widths, and these can be used—in addition to size and shape—to help pair up the chromosomes. A computer is used to arrange the chromosomes in pairs (Fig. 24.7). The display is called a **karyotype.**

Changing the Number of Chromosomes

Normally, a human receives 22 pairs of autosomes and two sex chromosomes. Sometimes individuals are born with either too many or too few autosomes or sex chromosomes. This occurs as a result of **nondisjunction,** which is a failure of the chromosomes or sister chromatids to separate during meiosis. As shown in Figure 24.8, nondisjunction may occur during meiosis I, when

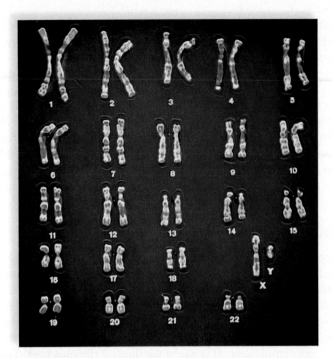

Figure 24.7 Karyotype of human chromosomes. In body cells, the chromosomes occur in pairs. In a karyotype, the pairs have been numbered and arranged by size from largest to smallest. These chromosomes are duplicated, and each one is composed of two sister chromatids. This karyotype illustrates the 46 chromosomes of a male. ©CNRI/SPL/Science Source

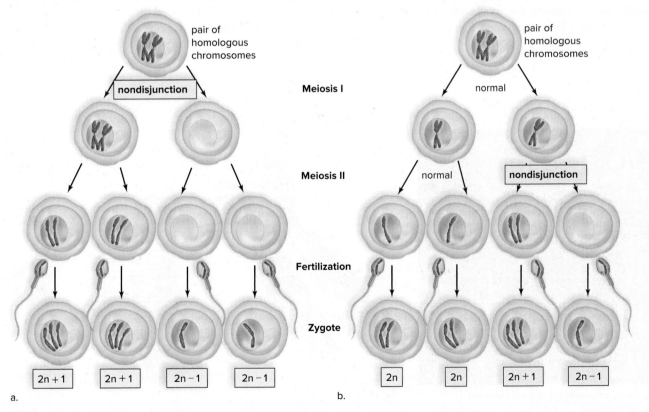

Figure 24.8 Nondisjunction of chromosomes during oogenesis. **a.** Nondisjunction can occur during meiosis I and results in abnormal eggs that also have one more or one less than the normal number of chromosomes. Fertilization of these abnormal eggs with normal sperm results in a zygote with abnormal chromosome numbers. **b.** Nondisjunction can also occur during meiosis II if the sister chromatids separate but the resulting daughter chromosomes go into the same daughter cell. The egg will then have one more or one less than the usual number of chromosomes. Fertilization of these abnormal eggs with normal sperm produces a zygote with abnormal chromosome numbers.

both members of a homologous pair go into the same daughter cell, or during meiosis II, when the sister chromatids fail to separate and both daughter chromosomes go into the same gamete.

Notice in Figure 24.8 that nondisjunction causes some of the egg cells to be missing a chromosome, while others have two copies of the same chromosome. When these eggs are fertilized by a sperm cell, the result is a zygote with an incorrect complement of chromosomes. If the zygote has an extra chromosome (2n + 1), the result is called a **trisomy,** because one type of chromosome is present in three copies. If the zygote is missing a chromosome (2n − 1), the result is called a **monosomy,** because one type of chromosome is present in a single copy.

Changes in Autosome Chromosome Number

Normal development depends on the presence of exactly two of each kind of chromosome. Too many chromosomes are tolerated better than a deficiency of chromosomes, and several autosome trisomies are known to occur in humans (see Table 24.1). Among autosomal trisomies, only trisomy 21 (Down syndrome) has a reasonable chance of survival after birth. This is most likely due to the fact that chromosome 21 is one of the smallest chromosomes. Trisomies of chromosome 13 (Patau syndrome) and 18 (Edwards syndrome) also occur, but the individuals rarely survive past the first few months of life.

Down Syndrome

The most common autosomal trisomy seen among humans is trisomy 21, also called Down syndrome (Fig. 24.9*a*). This syndrome

TABLE 24.1		Syndromes from Abnormal Chromosome Numbers	
Syndrome	**Sex**	**Chromosomes**	**Chromosome Number**
Patau	M or F	Trisomy 13	47
Edwards	M or F	Trisomy 18	47
Down	M or F	Trisomy 21	47
Poly-X	F	XXX (or XXXX)	47 or 48
Klinefelter	M	XXY (or XXXY)	47 or 48
Jacobs	M	XYY	47
Turner	F	X	45

is easily recognized by the following characteristics: short stature; an eyelid fold; a flat face; stubby fingers; a wide gap between the first and second toes; a large, fissured tongue; a round head; a palm crease (called the simian line); and mental impairment, which may range from mild to severe.

Persons with Down syndrome usually have three copies of chromosome 21 because the egg had two copies instead of one. However, in around 23% of the cases, the extra chromosome 21 came from the sperm. The chances of a woman having a Down syndrome child increase rapidly with age, starting at about age 40, and the reasons for this are still being determined.

Although an older woman is more likely to have a Down syndrome child, most babies with Down syndrome are born to

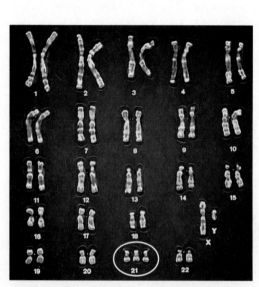

a. Down syndrome karyotype with an extra chromosome 21

b.

Figure 24.9 Trisomy 21.   a. Karyotype of an individual with Down syndrome shows an extra chromosome 21. **b.** More sophisticated technologies allow investigators to pinpoint the location of specific genes associated with the syndrome. An extra copy of the *Gart* gene, which leads to a high level of purines in the blood, may account for the mental impairment seen in persons with Down syndrome. (a): ©CNRI/SPL/Science Source women younger than age 40 because this is the age group having the most babies. A karyotype can detect changes in chromosome number such as occurs with Down syndrome. However, young women are not routinely encouraged to undergo the procedures necessary to get a sample of fetal cells (i.e., amniocentesis or chorionic villus sampling) because the risk of complications is greater than the risk of having a Down syndrome child. Fortunately, a test based on substances in maternal blood can help identify fetuses that may need to be karyotyped. The Health feature "Preventing and Testing for Birth Defects" (see Section 22.3) describes the processes of amniocentesis and chorionic villus sampling.

The genes that cause Down syndrome are located on the long arm of chromosome 21 (Fig. 24.9b), and extensive investigative work has been directed toward discovering the specific genes responsible for the characteristics of the syndrome. Thus far, investigators have discovered several genes that may account for various conditions seen in persons with Down syndrome. For example, they have located genes most likely responsible for the increased tendency toward leukemia, cataracts, and accelerated rate of aging. Researchers have also discovered that an extra copy of the *Gart* gene causes an increased level of purines in the blood, a finding associated with mental impairment. One day, it may be possible to control the expression of the *Gart* gene even before birth so that at least this symptom of Down syndrome does not appear.

Changes in Sex Chromosome Number

An abnormal sex chromosome number is the result of inheriting too many or too few X or Y chromosomes. Figure 24.8 can be used to illustrate nondisjunction of the sex chromosomes during oogenesis, if you assume that the chromosomes shown represent X chromosomes. Nondisjunction during oogenesis or spermatogenesis can result in gametes that have too few or too many X or Y chromosomes. After fertilization, the syndromes associated with the sex chromosomes in Table 24.1 are possibilities.

The chances of survival are greater when trisomy or monosomy involves the sex chromosomes. In normal XX females, one of the X chromosomes becomes a darkly stained mass of chromatin called a **Barr body.** A Barr body is an inactive X chromosome. Therefore, we now know that the cells of females function with a single X chromosome just as those of males do. This is most likely the reason that a zygote with one X chromosome (Turner syndrome) can survive. Then, too, all extra X chromosomes beyond a single one become Barr bodies, and this explains why poly-X females and XXY males are seen fairly frequently. An extra Y chromosome, called Jacobs syndrome, is tolerated in humans, most likely because the Y chromosome carries few genes.

A person with only one sex chromosome, an X (Turner syndrome), is a female, and a person with more than one X chromosome plus a Y (Klinefelter syndrome) is a male. This shows that in humans, the presence of a Y chromosome, not the number of X chromosomes, determines maleness. The *SRY* (*sex-determining region of Y*) gene, on the short arm of the Y chromosome, produces a hormone called testis-determining factor, which plays a critical role in the development of male genitals.

Turner Syndrome

Females with Turner syndrome have only one sex chromosome, an X. They are usually short and may have malformed features, such as a webbed neck, high palate, and small jaw. Many have congenital heart and kidney defects. Most have ovarian failure and do not undergo puberty or menstruate without sex hormone replacement therapy. However, pregnancy has been achieved through in vitro fertilization using donor eggs. Women with Turner syndrome have a normal range of intelligence but often have a nonverbal learning disability. They can lead very successful, fulfilling lives if they receive appropriate care.

Klinefelter Syndrome

About 1 in 650 males is born with two X chromosomes and one Y chromosome. The symptoms of this condition (referred to as "47, XXY") are often so subtle that only 25% are ever diagnosed, and those are usually not diagnosed until after age 15. Earlier diagnosis opens the possibility for educational accommodations and other interventions that can help mitigate common symptoms, which include speech and language delays. Those 47, XXY males who develop more severe symptoms as adults are referred to as having "Klinefelter syndrome." All 47, XXY adults will require assisted reproduction in order to father children. Affected individuals commonly receive testosterone supplementation beginning at puberty.

Poly-X Females

A poly-X female has more than two X chromosomes and thus extra Barr bodies in the nucleus. Females with three X chromosomes (sometimes called triplo-X) have no distinctive phenotype, aside from a tendency to be tall and thin. Although some have delayed motor and language development, most poly-X females are not mentally impaired. Some may have menstrual difficulties, but many menstruate regularly and are fertile. Their children usually have a normal karyotype.

Females with more than three X chromosomes occur rarely. Unlike XXX females, XXXX females are tall and more likely to be mentally impaired. They exhibit various physical abnormalities, but may menstruate normally.

Jacobs Syndrome

XYY males, who have Jacobs syndrome, can only result from nondisjunction during spermatogenesis, specifically due to a nondisjunction event during meiosis II of spermatogenesis. We know this because two Ys are present only during meiosis II in males. Affected males are usually taller than average, suffer from persistent acne, and tend to have speech and reading problems. Despite the extra Y chromosome, there is no difference in behavior between XYY males and XY males.

24.4 Changes in Chromosome Structure

Learning Outcomes

Upon completion of this section, you should be able to

1. Describe a chromosomal deletion, duplication, translocation, and inversion.
2. Summarize the disorders caused by changes in chromosome structure.

Chromosomal mutations occur when breaks occur in the backbone of the DNA molecules within a chromosome (see Section 25.1). Various environmental agents—radiation, certain organic chemicals, or even viruses—can cause chromosomes to break apart. Ordinarily, when breaks occur, the segments reunite to give the same sequence of genes. But their failure to reunite correctly can result in one of several types of mutations: deletion, duplication, translocation, or inversion. Chromosomal mutations can occur during meiosis, and if the offspring inherits the abnormal chromosome, a syndrome may develop.

Deletions and Duplications

A **deletion** occurs when a single break causes a chromosome to lose an end piece or when two simultaneous breaks lead to the loss of an internal chromosomal segment. An individual who inherits a normal chromosome from one parent and a chromosome with a deletion from the other parent no longer has a pair of alleles for each trait, and a syndrome can result.

Williams syndrome occurs when chromosome 7 loses a tiny end piece (Fig. 24.10). Children who have this syndrome look like pixies because they have turned-up noses, wide mouths, small chins, and large ears. Although their academic skills are poor, they exhibit excellent verbal and musical abilities. The gene that governs the production of the protein elastin is missing, which affects the health of the cardiovascular system and causes their skin to age prematurely. Such individuals are very friendly but need an ordered life, perhaps because of the loss of a gene for a protein that is normally active in the brain.

Cri du chat (cat's cry) syndrome is seen when chromosome 5 is missing an end piece. The affected individual has a small head, is mentally impaired, and has facial abnormalities. Abnormal development of the glottis and larynx results in the most characteristic symptom—the infant's cry resembles that of a cat.

In a **duplication,** a chromosomal segment is repeated in the same chromosome or in a nonhomologous chromosome. In any case, the individual has more than two alleles for certain traits. An inverted duplication is known to occur in chromosome 15. **Inversion** means that a segment joins in the direction opposite from normal. Children with this syndrome, called inv dup 15 syndrome, have poor muscle tone, mental impairment, seizures, a curved spine, and autistic characteristics, including poor speech, hand flapping, and lack of eye contact (Fig. 24.11).

The Health feature "Trinucleotide Repeat Expansion Disorders (TREDs)" (Section 24.2) describes a type of duplication

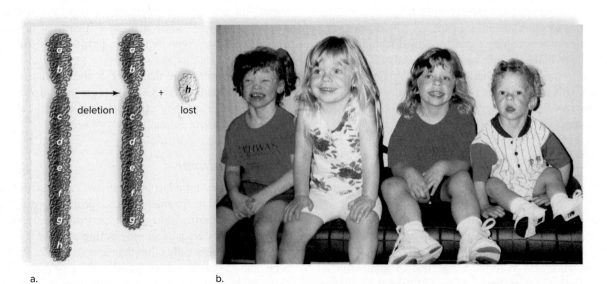

a.   b.

Figure 24.10 Deletion.   **a.** When chromosome 7 loses an end piece, the result is Williams syndrome. **b.** These children, although unrelated, have similar appearance, health, and behavioral problems. (b): ©The Williams Syndrome Association

a.   b.

Figure 24.11 Duplication. **a.** When a piece of chromosome 15 is duplicated and inverted, (**b**) a syndrome results (called *inv dup 15*) that is characterized by poor muscle tone and autistic characteristics. (b): ©Kathy Wise

called a trinucleotide repeat expansion. In the case of Huntington disease, the trinucleotide repeat consists of the sequence CAG repeated over and over again. In the case of fragile X syndrome, the trinucleotide repeat is CGG. Persons with duplications over a particular number, which is specific for each disease, exhibit symptoms of the disease. In the case of Huntington disease, sufferers have over 40 copies of the CAG sequence within the huntingtin protein gene.

Translocation

A **translocation** is the exchange of chromosomal segments between two nonhomologous chromosomes. A person who has both of the involved chromosomes has the normal amount of genetic material and is healthy, unless the chromosome exchange breaks an allele into two pieces. The person who inherits only one of the translocated chromosomes will have only one copy of certain alleles and three copies of certain other alleles.

In 5% of cases, a translocation that occurred in a previous generation between chromosomes 21 and 14 is the cause of one type of Down syndrome. The affected person inherits two normal chromosomes 21 and an abnormal chromosome 14 that contains a segment of chromosome 21. In these cases, Down syndrome is not related to the age of the mother, but instead tends to run in the family of either the father or the mother.

Figure 24.12 shows the phenotype of an individual with a translocation between chromosomes 2 and 20. Although they have the normal amount of genetic material, the rearrangement of the material produces the phenotypic changes associated with Alagille syndrome. People with this syndrome ordinarily have a deletion on chromosome 20. Therefore, it can be deduced that the translocation disrupted an allele on chromosome 20. The symptoms of Alagille syndrome range from mild to severe, so some people may not be aware they have the syndrome.

Inversions

An inversion occurs when a segment of a chromosome is turned 180 degrees. You might think this is not a problem because the same genes are present, but the reverse sequence of alleles can lead to altered gene activity.

Crossing-over between an inverted chromosome and the noninverted homologue can lead to recombinant chromosomes that have both duplicated and deleted segments. This happens because alignment between the two homologues is possible only when the inverted chromosome forms a loop (Fig. 24.13).

a.

b.

Figure 24.12 Translocation. **a.** When chromosomes 2 and 20 exchange segments, (**b**) Alagille syndrome may result in both cyanosis (blue coloration) and clubbing (widening of the fingers), because the translocation disrupts an allele on chromosome 20. (b): ©LISSAC/BSIP/age fotostock inverted segment   region of crossing-over

homologous chromosomes

duplication and deletion in both

Figure 24.13 Inversion. *Left:* A segment of one homologue is inverted. Notice that, in the inverted segment, *edc* occurs instead of *cde*. *Middle:* The two homologues can pair only when the inverted sequence forms an internal loop. After crossing-over, a duplication and a deletion can occur. *Right:* The homologue on the left has *AB* and *ba* sequences and neither *gf* nor *FG* genes. The homologue on the right has *gf* and *FG* sequences, and neither *AB* nor *ba* genes.

Check Your Progress 24.4

1. Explain why a duplication on one chromosome is usually associated with a deletion on the corresponding homologous chromosome.
2. Explain how it is possible for a person with a translocation or an inversion to be phenotypically normal.

Conclusion

Research on the causes of Down syndrome has indicated that 95% of Down syndrome cases are a result of nondisjunction of chromosome 21, resulting in three copies of this chromosome. However, we now also know that some cases are caused by translocations. Because of this research, scientists have been able to identify specific genes associated with the symptoms of the syndrome. For Down syndrome individuals, such as Brandon, these medical advances have ensured he can lead a long, productive life.

SUMMARIZE

24.1 Gene Linkage

■ All the genes on one chromosome form a **linkage group,** which is broken only when crossing-over occurs. Genes that are linked tend to go together into the same gamete.

■ If crossing-over occurs, a dihybrid cross yields all possible phenotypes among the offspring, but the expected ratio is greatly changed because recombinant phenotypes are reduced in number.

24.2 X-Linked Inheritance

In humans, there are 22 pairs of **autosomes** (1 to 22), and one pair of **sex chromosomes** (X and Y). Some traits are **sex-linked,** meaning that although they do not determine gender, they are carried on the sex chromosomes. Most of the alleles for these traits are carried on the X chromosome (**X-linked**), whereas the Y does not bear alleles for those same traits.

■ **Color blindness, Duchenne muscular dystrophy, fragile X syndrome,** and **hemophilia** are all X-linked recessive disorders.

24.3 Changes in Chromosome Number

■ **Nondisjunction** during meiosis can result in an abnormal number of autosomes or sex chromosomes in the gametes. In a **trisomy,** individuals have an extra chromosome, whereas in a **monosomy** they are missing a chromosome. If nondisjunction occurs with the X chromosome, extra X chromosomes may be inactivated, forming **Barr bodies.** Monosomies and trisomies are often detected by **karyotype** analysis.

■ Down syndrome results when an individual inherits three copies of chromosome 21.

■ Females who possess a single X have Turner syndrome, and those who are XXX are poly-X females. Males with Klinefelter syndrome are XXY. Males who are XYY have Jacobs syndrome.

24.4 Changes in Chromosome Structure

■ Changes in chromosome structure also affect the phenotype. **Chromosomal mutations** include **deletions, duplications, translocations, and inversions.**

■ Translocations do not necessarily cause any difficulties if the person has inherited both translocated chromosomes. However, the translocation can disrupt a particular allele, and then a syndrome will follow.

■ An inversion can lead to chromosomes that have a deletion and a duplication when the inverted piece loops back to align with the noninverted homologue, and crossing-over follows between the nonsister chromatids.

ASSESS

Testing Yourself

Choose the best answer for each question.

24.1 Gene Linkage

1. All the genes on one chromosome are said to form a
   a. chromosomal group.
   b. recombination group.
   c. linkage group.
   d. crossing-over group.

2. If alleles *R* and *s* are linked on one chromosome and *r* and *S* are linked on the homologous chromosome, what gametes will be produced? (Assume that no crossing-over occurs.)
   a. *RS, Rs, rS, rs*
   b. *RS, rs*
   c. *Rs, rS*
   d. *R, S, r, s*

24.2 X-Linked Inheritance

3. The following pedigree pertains to color blindness. Using the letter *B* for the normal allele, what is the genotype of the individual with the asterisk?

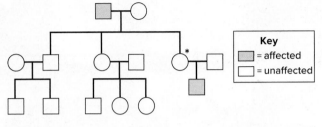

Key:
■ = affected
□ = unaffected

a. $X^b X^b$
b. $X^B X^b$
c. $X^B X^B$

4. Assume that two parents with normal vision have a son who has red-green color blindness. Which parent is responsible for the son's color blindness?
   a. the mother
   b. the father
   c. either parent
   d. None of these are correct.

24.3 Changes in Chromosome Number

For questions 5–9, match the chromosome disorder to its description in the key.

Key:
   a. female with undeveloped ovaries and uterus, unable to undergo puberty, normal intelligence, can live normally with hormone replacement
   b. XXY male, can inherit more than two X chromosomes
   c. male or female, mentally impaired, short stature, flat face, stubby fingers, large tongue, simian palm crease
   d. XXX or XXXX female
   e. caused by nondisjunction during spermatogenesis

5. Klinefelter syndrome
6. Poly-X female
7. Down syndrome
8. Turner syndrome
9. Jacobs syndrome

24.4 Changes in Chromosome Structure

For questions 10–13, match the chromosomal mutation to its description in the key.

Key:
   a. turned-up nose, wide mouth, small chin, large ears, poor academic skills, excellent verbal and musical abilities, prematurely aging cardiovascular system
   b. deletion in chromosome 5
   c. poor muscle tone, mental impairment, seizures, curved spine, autistic characteristics, poor speech, hand flapping, lack of eye contact
   d. translocation between chromosomes 2 and 20

10. Alagille syndrome
11. inv dup 15 syndrome
12. Williams syndrome
13. cri du chat syndrome

ENGAGE

Thinking Critically

1. Although most men with Klinefelter syndrome are infertile, some are able to father children. It was found that most fertile individuals with Klinefelter syndrome exhibit mosaicism, in which some cells are normal (46, XY), but others contain the extra chromosome (47, XXY). How might this mosaicism come about? What effects might result? Why might expectant parents want to undergo fetal genetic testing if they know that, regardless of the findings, they plan to have the baby?

2. Why do you think there are no viable trisomies of chromosome 1?

3. Why can a person carrying a translocation be normal except for the inability to have children?

4. Both the husband and wife have normal vision. The wife gives birth to a color-blind daughter. Is it more likely the father has normal vision or is color blind? What does this lead you to deduce about the girl's parentage?

25

DNA Structure and Gene Expression

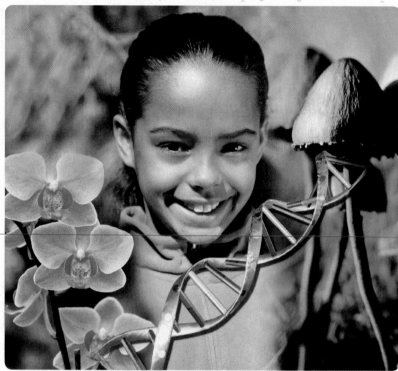

(girl): ©Jodi Matthews/Getty Images; (flower): ©Sombat Muycheen/Shutterstock; (DNA): ©Radius Images/Alamy; (mushroom): ©IT Stock/age fotostock

CHAPTER OUTLINE

BEFORE YOU BEGIN

Before beginning this chapter, take a few moments to review the following discussions:

Section 2.8 What are the three components of a nucleotide?

Section 2.8 How does the structure of DNA differ from that of RNA?

Section 23.1 What is the relationship between the genotype and phenotype?

Synthetic DNA

Deoxyribonucleic acid, or DNA, is the genetic material of all life on Earth. Each of the organisms in the photo above, and all of the estimated 2.3 million species currently identified on the planet, use a combination of four nucleotides (A, C, G, and T) to code for 20 different amino acids. The combination of these amino acids produces the tremendous variation that gives us the diversity of life.

However, a biotechnology company named Synthorx has recently expanded this genetic alphabet by creating two new forms of nucleotides, called X and Y. When these nucleotides are incorporated into the DNA of *E. coli* bacteria, the number of potential amino acids that can be coded for increases from 20 to 172.

What does this mean for us? Scientists believe that with this innovation they will be able to develop new drugs (including antibiotics) and vaccines that can help protect us from some diseases. It may also be possible to develop genetically modified organisms that resist pathogens, grow faster, or produce foods with higher nutritional content.

In this chapter, we will explore the structure and function of the genetic material and examine how the instructions contained within DNA are expressed to form the proteins that create the characteristics of life.

As you read through this chapter, think about the following questions:

1. How is information contained within a molecule of DNA?

2. How does the flow of genetic information from DNA to protein to trait work?

3. What mechanisms are in place to ensure that genetic information is accurately expressed?

25.1 DNA Structure

The middle of the twentieth century was an exciting period of scientific discovery. On one hand, geneticists were busy determining that **DNA (deoxyribonucleic acid)** is the genetic material of life. On the other hand, biochemists were in a frantic race to describe the structure of DNA. The classic experiments performed during this era set the stage for an explosion in our knowledge of modern molecular biology.

When researchers began their work, they knew that the genetic material must be (1) able to store information that pertains to the development, structure, and metabolic activities of the cell or organism; and (2) stable so it can be replicated with high accuracy during cell division and be transmitted from generation to generation.

In this section, we will explore how the structure of a DNA molecule satisfies both of these requirements.

The Nature of the Genetic Material

During the late 1920s, the bacteriologist Frederick Griffith was attempting to develop a vaccine against a form of bacteria (*Streptococcus pneumoniae*) that causes pneumonia in mammals. In 1931, he performed a classic experiment with the bacterium. He noticed that when these bacteria are grown on culture plates, some, called S strain bacteria, produce shiny, smooth colonies, and others, called R strain bacteria, produce colonies that have a rough appearance. Under the microscope, S strain bacteria have a capsule (mucous coat) that makes them smooth, but R strain bacteria do not.

When Griffith injected mice with the S strain of bacteria, the mice died (Fig. 25.1a), and when he injected mice with the R strain, the mice did not die (Fig. 25.1b). In an effort to determine whether the capsule alone was responsible for the virulence (ability to kill) of the S strain bacteria, he injected mice with heat-killed S strain bacteria. The mice did not die (Fig. 25.1c).

Finally, Griffith injected the mice with a mixture of heat-killed S strain and live R strain bacteria. Most unexpectedly, the mice died—and living S strain bacteria were recovered from the bodies! Griffith concluded that some substance necessary for the bacteria to produce a capsule and be virulent must have passed from the dead S strain bacteria to the living R strain bacteria, thus transforming the R strain bacteria (Fig. 25.1d). This change in the phenotype of the R strain bacteria must be due to a corresponding change in their genotype. Thus, the transforming substance was potentially the genetic material. Reasoning such as this prompted investigators at the time to begin looking for the transforming substance to determine the chemical nature of the genetic material.

By the 1940s, scientists recognized that genes are on chromosomes, and that chromosomes contain both proteins and nucleic acids. However, there was some debate about whether protein or DNA was the genetic material. Many thought the protein component of chromosomes must be the genetic material because proteins contain up to 20 different amino acids that can be sequenced in any particular way. On the other hand, nucleic acids—DNA and RNA—contain only four types of nucleotides as basic building blocks. Some argued that DNA did not have enough variability to be able to store information and be the genetic material.

In 1944, after 16 years of research, Oswald Avery and his coinvestigators, Colin MacLeod and Maclyn McCarty, published a paper demonstrating that the transforming substance that allows

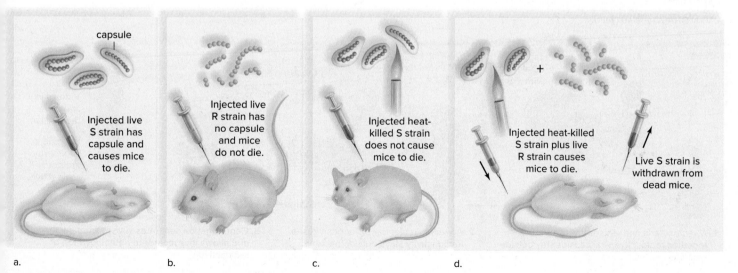

a.   b.   c.   d.

Figure 25.1 Griffith's experiment.   **a.** Encapsulated S strain is virulent and kills mice. **b.** Nonencapsulated R strain is not virulent and does not kill mice. **c.** Heat-killed S strain bacteria do not kill mice. **d.** If heat-killed S strain and R strain are both injected into mice, they die because the R strain bacteria have been transformed into the virulent S strain.

S. pneumoniae to produce a capsule and be virulent is DNA. This meant that DNA is the genetic material. Here is what they found out:

- DNA from S strain bacteria causes R strain bacteria to be transformed so that they can produce a capsule and be virulent.
- The addition of DNase, an enzyme that digests DNA, prevents transformation from occurring. This supports the hypothesis that DNA is the genetic material.
- The molecular weight of the transforming substance is large. This suggests the possibility of genetic variability.
- The addition of enzymes that degrade proteins has no effect on the transforming substance, nor does RNase, an enzyme that digests RNA. This shows that neither protein nor RNA is the genetic material.

These experiments showed that DNA is the transforming substance and, therefore, the genetic material. Although some scientists remained skeptical, many felt that the evidence for DNA being the genetic material was overwhelming.

An experiment by Alfred Hershey and Martha Chase in the early 1950s helped to firmly establish DNA as the genetic material. Hershey and Chase used a virus called a T phage, composed of radioactively labeled DNA and capsid coat proteins, to infect *Escherichia coli* bacteria (Fig 25.2). They discovered that the radioactive tracers for DNA, but not protein, ended up inside the bacterial cells, causing them to become transformed. As only the genetic material could have caused this transformation, Hershey and Chase determined that DNA must be the genetic material.

Structure of DNA

The structure of DNA was determined by James Watson and Francis Crick in the early 1950s. The data they used and how they interpreted the data to deduce DNA's structure are reviewed in the Scientific Inquiry feature "Finding the Structure of DNA."

DNA is a chain of nucleotides. Each nucleotide is a complex of three subunits—phosphoric acid (phosphate), a pentose sugar (deoxyribose), and a nitrogen-containing base. There are four possible bases: two are **purines** with a double ring, and two are **pyrimidines** with a single ring. Adenine (A) and guanine (G) are purines; thymine (T) and cytosine (C) are pyrimidines.

A DNA polynucleotide strand has a backbone made up of alternating phosphate and sugar molecules. The bases are attached to

a. Viral DNA is labeled (yellow).

b. Viral capsid is labeled (yellow).

Figure 25.2 Hershey–Chase experiments. These experiments concluded that (**a**) viral DNA, not (**b**) protein, was responsible for directing the production of new viruses.

Finding the Structure of DNA

In 1953, James Watson, an American biologist, and Francis Crick, a British physicist, set out not only to determine the structure of DNA, but also to build a model that would explain how DNA, the genetic material, can vary from species to species and even from individual to individual. In the process, they also proposed how DNA replicates (makes a copy of itself) so that daughter cells receive an identical copy.

The bits and pieces of data available to Watson and Crick were like puzzle pieces they had to fit together. This is what they knew from the research of others:

1. DNA is a polymer of nucleotides, each one having a phosphate group, the sugar deoxyribose, and a nitrogen-containing base. There are four types of nucleotides because there are four different bases: adenine (A) and guanine (G) are purines, while cytosine (C) and thymine (T) are pyrimidines.
2. A chemist, Erwin Chargaff, had determined in the late 1940s that regardless of the species under consideration, the number of purines in DNA always equals the number of pyrimidines. Further, the amount of adenine equals the amount of thymine (A = T), and the amount of guanine equals the amount of cytosine (G = C). These findings came to be known as Chargaff's rules.
3. Rosalind Franklin (Fig. 25A*a*), working with Maurice Wilkins at King's College, London, had just prepared an X-ray diffraction photograph of DNA (Fig. 25A*b, c*). It showed that DNA is a double helix of constant diameter and that the bases are regularly stacked on top of one another.

Using these data, Watson and Crick deduced that DNA has a twisted, ladderlike structure. The sugar-phosphate molecules make up the sides of the ladder, and the bases make up the rungs. The double helices of the DNA molecule are antiparallel, meaning they are oriented in opposite directions. Further, they determined that if A is normally hydrogen-bonded with T, and G is normally hydrogen-bonded with C (Chargaff's rules), then the rungs always have a constant width, consistent with the X-ray photograph.

Watson and Crick built an actual model of DNA out of wire and tin. This double-helix model does indeed allow for differences in DNA structure between species because the base pairs can be in any order. Also, the model suggests that complementary base pairing plays a role in the replication of DNA. As Watson and Crick pointed out in their original paper, "It has not escaped our notice that the specific pairing we have postulated immediately suggests a possible copying mechanism for the genetic material."

When the Nobel Prize in Physiology or Medicine for the discovery of the double helix was awarded in 1962, the honor went to James Watson, Francis Crick, and Maurice Wilkins. Rosalind Franklin was not listed as one of the recipients. Tragically, Franklin developed ovarian cancer in 1956, and she died in 1958 at the age of 37. Franklin had not been nominated for the award, and according to the rules at that time, was ineligible for the Nobel Prize.

Questions to Consider

1. Based on Chargaff's rules, if a segment of DNA is composed of 20% adenine (A) bases, what is the percentage of guanine (G)?
2. Watson and Crick's discovery of DNA is clearly one of the most important biological discoveries in the last century. What advances in medicine and science can you think of that are built on knowing the structure of DNA?
3. Describe why the structure of DNA led Watson and Crick to point out "a possible copying mechanism for the genetic material."

Figure 25A X-ray diffraction of DNA.   a. Rosalind Franklin. **b.** The X-ray diffraction system used by Franklin. **c.** The diffraction pattern of DNA produced by Rosalind Franklin. The crossed (X) pattern in the center told investigators that DNA is a helix, and the dark portions at the top and the bottom told them that some feature is repeated over and over. Watson and Crick determined that this feature was the result of the hydrogen-bonded bases.

(a, c): ©Science Source

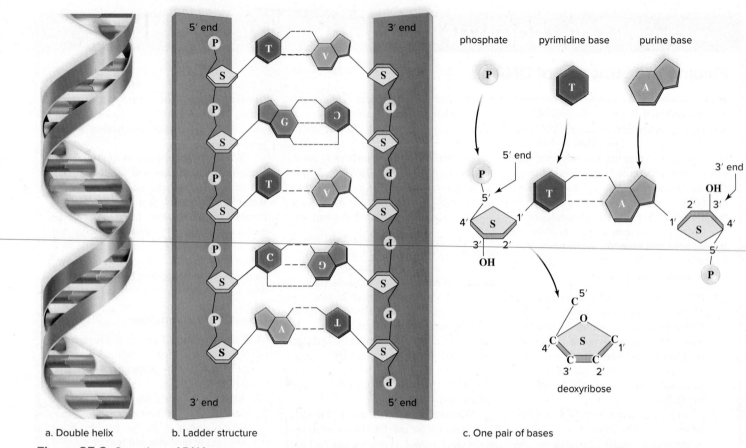

Figure 25.3 Overview of DNA structure.   **a.** DNA double helix. **b.** Unwinding the helix reveals a ladder configuration in which the uprights are composed of sugar and phosphate molecules, and the rungs are complementary bases. The bases in DNA pair in such a way that the sugar-phosphate backbones are oriented in different directions. **c.** Notice that 3′ and 5′ are part of the system for numbering the carbon atoms that make up the sugar. (DNA): ©Radius Images/Alamy RF

a. Double helix   b. Ladder structure   c. One pair of bases

the sugar but project to one side. DNA has two such strands, and the two strands twist about one another in the form of a **double helix** (Fig. 25.3*a*). The strands are held together by hydrogen bonding between the bases: A always pairs with T by forming two hydrogen bonds, and G always pairs with C by forming three hydrogen bonds. Notice that a purine is always bonded to a pyrimidine. This is called **complementary base pairing.** When the DNA helix unwinds, it resembles a ladder (Fig. 25.3*b*). The sides of the ladder are the sugar-phosphate backbones, and the rungs of the ladder are the complementary paired bases.

The two DNA strands are *antiparallel,* meaning they are oriented in opposite directions, which you can verify by noticing that the sugar molecules are oriented differently. The carbon atoms in a sugar molecule are numbered, and the fifth carbon atom (5′) is uppermost in the strand on the left, while the third carbon atom (3′) is uppermost in the strand on the right (Fig. 25.3*c*).

Check Your Progress 25.1

1. Summarize the significance of the Griffith and Avery experiments.
2. Explain how, at the completion of the Hershey–Chase experiment, the results suggested that DNA was the genetic material.
3. Describe the structure of the DNA molecule.

25.2 DNA Replication

Learning Outcomes

Upon completion of this section, you should be able to
1. Explain why DNA replication is semiconservative.
2. Summarize the events that occur during the process of DNA replication.

Cell division occurs whenever the body grows or heals itself. During cell division, each new cell requires an exact copy of the DNA contained in the chromosomes. The process of copying one DNA double helix into two identical double helices is called **DNA replication.** In the S phase of interphase, during mitosis when DNA is replicated (see Section 5.1), the double-stranded structure of DNA allows each original strand to serve as a template for the formation of a complementary new strand (Fig. 25.4*a*). As a result, DNA replication is termed *semiconservative* because a new double helix has one conserved old strand and one new strand (Fig. 25.4*b*). Replication results in two DNA helices that are identical to each other and to the original molecule.

At the molecular level, several enzymes and proteins participate in the synthesis of the new DNA strands. This process is summarized in Figure 25.5:

1. The enzyme *DNA helicase* unwinds and "unzips" the double-stranded DNA by breaking the weak hydrogen bonds between the paired bases.
2. New complementary DNA nucleotides fit into place by the process of complementary base pairing. These are positioned and joined by the enzyme *DNA polymerase*. The DNA polymerase uses each original strand as a template for the formation of a complementary new strand.
3. Because the strands of DNA are oriented in an antiparallel configuration, and the DNA polymerase may add new nucleotides only to one end of the chain, DNA synthesis occurs in opposite directions. The *leading strand* follows the helicase enzyme, while synthesis on the *lagging strand* results in the formation of short segments of DNA called *Okazaki fragments*.
4. To complete replication, the enzyme *DNA ligase* connects the Okazaki fragments and seals any breaks in the sugar-phosphate backbone.
5. The two double-helix molecules are identical to each other and to the original DNA molecule.

A number of chemotherapeutic drugs for cancer treatment stop replication and, therefore, cell division. Some chemotherapeutic drugs are analogs that have a similar, but not identical, structure to one of the four nucleotides in DNA. When these are mistakenly used by the cancer cells to synthesize DNA, replication stops, and the cancer cells die off.

a. The mechanism of DNA replication

b. The products of replication

Figure 25.4 Overview of DNA replication. **a.** After the DNA double helix unwinds, each parental strand serves as a template for the formation of the new daughter strands. Complementary free nucleotides hydrogen bond to a matching base (e.g., A with T; G with C) in each parental strand and are joined to form a complete daughter strand. **b.** Two helices, each with a daughter and parental strand (semiconservative), are produced during replication.

Check Your Progress 25.2

1. Explain why DNA replication is said to be semiconservative.
2. Summarize the sequence of events that occur during DNA replication.

Figure 25.5 Molecular mechanisms of DNA replication. A general diagram of the process of DNA replication, showing the major enzymes and proteins involved in the process.

25.3 Gene Expression

The process of using the information within a gene to synthesize a protein is called gene expression. Gene expression relies on the participation of several different forms of **RNA (ribonucleic acid)** molecules (see Section 2.8), most important of which are messenger RNA (mRNA), transfer RNA (tRNA), and ribosomal RNA (rRNA). We will take a closer look at each one of these molecules in the content below.

Overall, gene expression requires two processes, called transcription and translation (Figure 25.6). In eukaryotes, transcription takes place in the nucleus and translation takes place in the cytoplasm. During **transcription,** a portion of DNA serves as a template for mRNA formation. During **translation,** the sequence of mRNA bases (which are complementary to those in the template DNA) determines the sequence of amino acids in a polypeptide. So, in effect, genetic information lies in the sequence of the bases in DNA, which through mRNA determines the sequence of amino acids in a protein. Transfer RNA assists mRNA during protein

synthesis by bringing amino acids to the ribosomes. Proteins differ from one another by the sequence of their amino acids, and proteins determine the structure and function of cells and the phenotype of the organism.

Transcription

During transcription, a segment of the DNA, called a **gene,** serves as a template for the production of an RNA molecule. Historically, a gene was considered to be a nucleic acid sequence that coded for the sequence of amino acids in a protein. We now know that not all genes contain instructions for protein formation. Some genes include instructions for the formation of RNA molecules, such as mRNA, tRNA, and rRNA. We also know that protein-coding regions can be interrupted by regions that do not code for a protein. Thus, the definition of a gene is changing. Mark Gerstein (a bioinformatic professor at Yale University) has suggested a gene be defined as follows: "A gene is a genomic sequence (either DNA or RNA) directly encoding functional products, either RNA or protein."[1] Although all three classes of RNA are formed by transcription, we will focus on transcription to form messenger RNA (mRNA), the first step in protein synthesis.

Messenger RNA

The purpose of **messenger RNA (mRNA)** is to carry genetic information from the DNA to the ribosomes for protein synthesis. Messenger RNA is formed by the process of transcription which, in eukaryotes, occurs in the nucleus.

Transcription begins when the enzyme **RNA polymerase** binds tightly to a **promoter,** a region of DNA that contains a special sequence of nucleotides. This enzyme opens up the DNA helix just in front of it so that complementary base pairing can occur in the same way as in DNA replication. Then, RNA polymerase inserts the RNA nucleotides, and an mRNA molecule results. When mRNA forms, it has a sequence of bases complementary to that of the DNA; wherever A, T, G, or C are present in the DNA template, U, A, C, or G, respectively, are incorporated into the mRNA molecule (Fig. 25.7). Now, mRNA is a faithful copy of the sequence of bases in DNA.

Processing of mRNA   In eukaryotic cells, the mRNA molecule that was produced by transcription, called pre-mRNA, is typically processed before exiting the nucleus and becoming a mature mRNA molecule.

Not all of the DNA in the pre-mRNA molecule contains information needed for the final product. The information needed to produce the functional product, such as a protein, is contained within DNA sequences called **exons.** The regions between the exons are called **introns,** or intergenic sequences (Fig. 25.8). While the DNA within the intron does not contain information that contributes to the final structure of the protein, we now know these regions often have important regulatory functions.

During the processing of the pre-mRNA molecule:

1. The 5′ end of the mRNA is modified by the addition of a cap, composed of an altered guanine nucleotide.
2. The 3′ end is modified by the addition of a poly-A tail, a series of adenosine nucleotides.

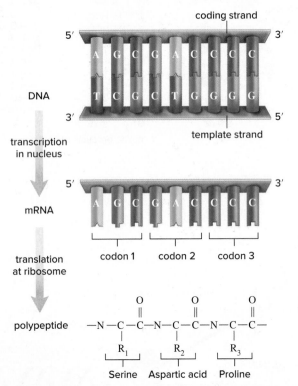

Figure 25.6 Overview of gene expression.   One strand of DNA acts as a template for mRNA synthesis, and the sequence of bases in mRNA determines the sequence of amino acids in a polypeptide.

[1]Gerstein, M. B., Bruce, C., and Rozowsky, J. S. et al. "What is a gene, post-ENCODE? History and updated definition," *Genome Research* 17:669–681 (2007).

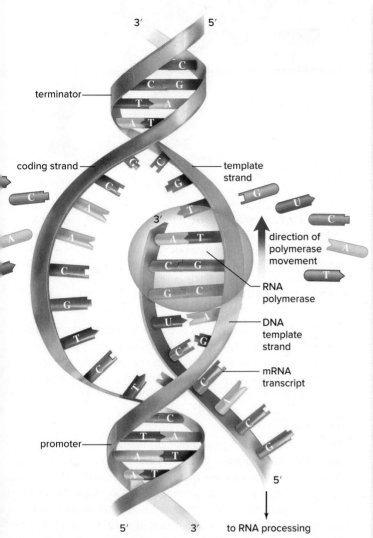

Figure 25.7 Transcription of DNA to form mRNA. During transcription, complementary RNA is made from a DNA template. At the point of attachment of RNA polymerase, the DNA helix unwinds and unzips, and complementary RNA nucleotides are joined together. After RNA polymerase has passed by, the DNA strands rejoin and the mRNA transcript is released.

3. The introns are removed, and the exons are joined to form a mature mRNA molecule consisting of continuous exons.

Ordinarily, processing brings together all the exons of a gene. In some instances, cells use only certain exons to form a mature RNA transcript. The result can be a different protein product in each cell. Alternate mRNA splicing accounts for the ability of white blood cells to produce a specific antibody for each type of bacteria and virus we encounter on a daily basis (see Section 13.3).

Translation

Translation is the second process by which gene expression leads to protein synthesis. Translation occurs at the ribosome and requires several enzymes, and several different types of RNA molecules, including mRNA, tRNA, and rRNA.

Figure 25.8 mRNA processing. During processing, a cap and tail are added to the pre-mRNA molecule, and the introns are removed so that only exons remain.

The Genetic Code

The sequence of bases in DNA is transcribed into mRNA, which ultimately codes for a particular sequence of amino acids to form a polypeptide. Can four mRNA bases (A, C, G, U) provide enough combinations to code for 20 amino acids? If only one base stood for an amino acid (i.e., a "singlet code"), then only four amino acids would be possible. If two bases stood for one amino acid, there would only be 16 possible combinations (4 × 4). If the code is a triplet, then there are 64 possible triplets of the four bases (4 × 4 × 4). Each triplet of nucleotides is called a **codon.** The code is also *degenerate,* meaning that most amino acids are coded for by more than one codon (Fig. 25.9). For example, leucine has six codons and serine has four codons. This degeneracy offers some protection against possibly harmful mutations that change the sequence of bases. Of the 64 codons, 61 code for amino acids, whereas the remaining three are *stop codons* (UAA, UGA, UAG), codons that do not code for amino acids but instead signal polypeptide termination.

The genetic code is just about universal in all living organisms. This means that a codon in a fruit fly codes for the same

a.   b.

Figure 25.10 Transfer RNA: amino acid carrier.   a. A tRNA is a polynucleotide that folds into a bootlike shape because of complementary base pairing. At one end of the molecule is its specific anticodon—in this case, GCU (which hybridizes to the codon CGA). At the other end, an amino acid attaches that corresponds to this anticodon—in this case, arginine. **b.** tRNA is represented like this in the illustrations in this text.

amino acid as in a bird, a fern, or a human. There are some minor variations in some of the oldest forms of prokaryotes (see Section 28.3). The universal nature of the genetic code suggests that it dates back to the very first organisms on Earth and that all living organisms have a common evolutionary history.

Transfer RNA

Transfer RNA (tRNA) molecules bring amino acids to the ribosomes, the site of protein synthesis. Each tRNA molecule is a single-stranded polynucleotide that doubles back on itself such that complementary base pairing creates a bootlike shape. On one end is an amino acid, and on the other end is an **anticodon,** a triplet of three bases complementary to a codon of mRNA (Fig. 25.10). Although there are 64 possible codons, there are only 40 different tRNA molecules. This is because of the *wobble effect,* which states that for some tRNAs, the third nucleotide in the mRNA codon may vary. This is believed to provide additional degeneracy to the genetic code, and help protect against mutations that may alter the amino acid sequence of a protein.

When a tRNA–amino acid complex comes to the ribosome, its anticodon pairs with an mRNA codon. For example, if the codon is CGG, what is the anticodon, and what amino acid will be attached to the tRNA molecule? Based on Figure 25.9, the answer to this question is as follows:

Codon (mRNA)	Anticodon (tRNA)	Amino Acid (protein)
CGG	GCC	Arginine

The order of the codons of the mRNA determines the order that tRNA–amino acids come to a ribosome and, therefore, the final sequence of amino acids in a protein. During translation, tRNAs bring amino acids to the ribosomes in the order dictated by the base sequence of mRNA.

Ribosomes and Ribosomal RNA

Ribosomes are the site of translation. They are small structural bodies that may be found both in the cytoplasm and on the rough endoplasmic reticulum. Ribosomes are composed of many proteins and several **ribosomal RNAs (rRNAs).** In eukaryotic cells,

First Base	Second Base				Third Base
	U	**C**	**A**	**G**	
U	UUU phenylalanine	UCU serine	UAU tyrosine	UGU cysteine	U
	UUC phenylalanine	UCC serine	UAC tyrosine	UGC cysteine	C
	UUA leucine	UCA serine	UAA *stop*	UGA *stop*	A
	UUG leucine	UCG serine	UAG *stop*	UGG tryptophan	G
C	CUU leucine	CCU proline	CAU histidine	CGU arginine	U
	CUC leucine	CCC proline	CAC histidine	CGC arginine	C
	CUA leucine	CCA proline	CAA glutamine	CGA arginine	A
	CUG leucine	CCG proline	CAG glutamine	CGG arginine	G
A	AUU isoleucine	ACU threonine	AAU asparagine	AGU serine	U
	AUC isoleucine	ACC threonine	AAC asparagine	AGC serine	C
	AUA isoleucine	ACA threonine	AAA lysine	AGA arginine	A
	AUG *(start)* methionine	ACG threonine	AAG lysine	AGG arginine	G
G	GUU valine	GCU alanine	GAU aspartic acid	GGU glycine	U
	GUC valine	GCC alanine	GAC aspartic acid	GGC glycine	C
	GUA valine	GCA alanine	GAA glutamic acid	GGA glycine	A
	GUG valine	GCG alanine	GAG glutamic acid	GGG glycine	G

Figure 25.9 The genetic code.   Notice that in this chart, each of the codons (in boxes) is composed of three letters representing the first base, second base, and third base of a codon on the mRNA molecule. For example, find the box where C for the first base and A for the second base intersect. You will see that U, C, A, or G can be the third base. The bases CAU and CAC are codons for histidine.

a. Structure of a ribosome

large subunit

small subunit

b. Binding sites of ribosome

5′

mRNA

E   P   A

3′

tRNA binding sites

polypeptide

incoming tRNA

mRNA

outgoing tRNA

c. Function of ribosomes

d. Polyribosome

Figure 25.11 Polyribosome structure and function. **a.** Structure of a ribosome. **b.** Internal view of a ribosome showing the tRNA binding sites. **c.** Several ribosomes, collectively called a polyribosome, move along an mRNA at one time. Therefore, several polypeptides can be made at the same time. **d.** Electron micrograph of a polyribosome. (d): ©Science Source

rRNA is produced in a nucleolus within the nucleus. The rRNA then joins with proteins manufactured in and imported from the cytoplasm to form two ribosomal subunits, one large and one small (Fig. 25.11a). The subunits leave the nucleus and join together in the cytoplasm to form a ribosome just as protein synthesis begins.

A ribosome has a binding site for mRNA, as well as binding sites for three tRNA molecules (Fig. 25.11b). These binding sites facilitate complementary base pairing between tRNA anticodons and mRNA codons. As the ribosome moves down the mRNA molecule, new tRNAs arrive, and a polypeptide forms and grows longer (Fig. 25.11c). Translation terminates once the polypeptide is fully formed and an mRNA stop codon is reached. The ribosome then dissociates into its two subunits and falls off the mRNA molecule.

As soon as the initial portion of mRNA has been translated by one ribosome and the ribosome has begun to move down the mRNA, another ribosome attaches to the same mRNA. Therefore, several ribosomes are often attached to and translating a single mRNA, thus forming several copies of a polypeptide simultaneously. The entire complex is called a *polyribosome* (Fig. 25.11d).

Translation Requires Three Steps

During translation, the codons of an mRNA base pair with the anticodons of tRNA molecules carrying specific amino acids. The order of the codons determines the order of the tRNA molecules at a ribosome and the sequence of amino acids in a polypeptide. The

process of translation must be extremely orderly so the amino acids of a polypeptide are sequenced correctly.

Protein synthesis involves three steps: initiation, elongation, and termination. Enzymes are required for each of the three steps to function properly. The first two steps, initiation and elongation, require energy. In our discussion of the process, we will use an example from a prokaryotic cell, such as a bacteria.

Initiation

Initiation is the step that brings all the components of the translation machinery together. Proteins called initiation factors are required to assemble the small ribosomal subunit, mRNA, initiator tRNA, and the large ribosomal subunit for the start of protein synthesis.

Initiation is shown in Figure 25.12. To begin, a small ribosomal subunit attaches to the mRNA in the vicinity of the *start codon* (AUG). The first or initiator tRNA pairs with this codon because its anticodon is UAC. Then, a large ribosomal subunit joins to the small subunit (Fig. 25.12). Although similar in many ways, initiation in eukaryotes is much more complicated and will not be discussed here.

A ribosome has three binding sites for tRNAs. One of these is called the P (for peptide) site, and the other is the A (for amino acid) site. The tRNA exits at the E site. The initiator tRNA happens to be capable of binding to the P site, even though it carries only the amino acid methionine (see Fig. 25.9). The A site is for tRNA carrying the next amino acid.

A small ribosomal subunit binds to mRNA; an initiator tRNA pairs with the mRNA start codon AUG.

The large ribosomal subunit completes the ribosome. Initiator tRNA occupies the P site. The A site is ready for the next tRNA.

Initiation

Figure 25.12 Initiation. During initiation, participants in the translation process assemble as shown. The start codon, AUG, also codes for the first amino acid, methionine.

Elongation

Elongation is the protein synthesis step in which a polypeptide increases in length one amino acid at a time (Fig. 25.13). In addition to the participation of tRNAs, elongation requires elongation factors, which facilitate the binding of tRNA anticodons to mRNA codons at a ribosome.

Elongation consists of a series of four steps (Fig. 25.13):

1. A tRNA with an attached peptide is already at the P site, and a tRNA carrying the next amino acid in the chain is just arriving at the A site.
2. Once the next tRNA is in place at the A site, the peptide chain will be transferred to this tRNA.
3. Energy and part of the ribosomal subunit are needed to bring about this transfer. The energy contributes to peptide bond formation, which makes the peptide one amino acid longer by adding the peptide from the A site.
4. Next, translocation occurs—the mRNA moves forward one codon length, and the peptide-bearing tRNA is now at the ribosome P site. The "spent" tRNA now exits. The new codon is at the A site and is ready to receive the next complementary tRNA.

The complete cycle in steps 1–4 is repeated at a rapid rate (e.g., about 15 times each second in the bacteria *E. coli*).

Termination

Termination is the final step in protein synthesis. During termination, as shown in Figure 25.14, the polypeptide and the assembled components that carried out protein synthesis are separated from one another.

Termination of polypeptide synthesis occurs at a stop codon. Termination requires a protein called a release factor, which cleaves the polypeptide from the last tRNA. After this occurs, the polypeptide is set free and begins to take on its

1. A tRNA–amino acid approaches the ribosome and binds at the A site.

2. Two tRNAs can be at a ribosome at one time; the anticodons are paired to the codons.

3. Peptide bond formation attaches the peptide chain to the newly arrived amino acid.

4. The ribosome moves forward; the "empty" tRNA exits from the E site; the next amino acid–tRNA complex is approaching the ribosome.

Elongation

Figure 25.13 Elongation. Note that a polypeptide is already at the P site. During elongation, polypeptide synthesis occurs as amino acids are added one at a time to the growing chain.

The ribosome comes to a stop codon on the mRNA. A release factor binds to the site.

release factor

stop codon

The release factor hydrolyzes the bond between the last tRNA at the P site and the polypeptide, releasing them. The ribosomal subunits dissociate.

Termination

Figure 25.14 Termination.   During termination, the finished polypeptide is released, as are the mRNA and the last tRNA.

three-dimensional shape. The ribosome dissociates into its two subunits.

Properly functioning proteins are of paramount importance to the cell and to the organism. For example, if an organism inherits a faulty gene, the result can be a genetic disorder (such as Huntington disease) caused by a malfunctioning protein or a propensity toward cancer. Proteins are the link between genotype and phenotype. The DNA sequence underlying these proteins distinguishes different types of organisms. In addition to accounting for the difference between cell types, proteins account for the differences between organisms.

Review of Gene Expression

A gene is expressed when product, in our case a protein, has been synthesized. Protein synthesis requires the process of transcription and translation (Fig. 25.15). During transcription, a segment of a DNA strand serves as a template for the formation of messenger RNA (mRNA). The bases in mRNA are complementary to those in DNA. Every three mRNA bases is a *codon* (a triplet code) for a certain amino acid. Messenger RNA is processed before it leaves the nucleus, during which time the introns are removed and the ends are modified. Messenger RNA carries a sequence of codons to the *ribosomes*.

During translation, tRNAs bring attached amino acids to the ribosomes. Because tRNA anticodons pair with codons, the amino acids become sequenced in the order originally specified by DNA. The genes we receive from our parents determine the proteins in our cells and these proteins are responsible for our inherited traits!

Check Your Progress 25.3

1. Explain the role of mRNA, tRNA, and rRNA in gene expression.
2. Describe the movement of information from the nucleus to the formation of a functional protein.
3. Discuss why the genetic code is said to be degenerate.

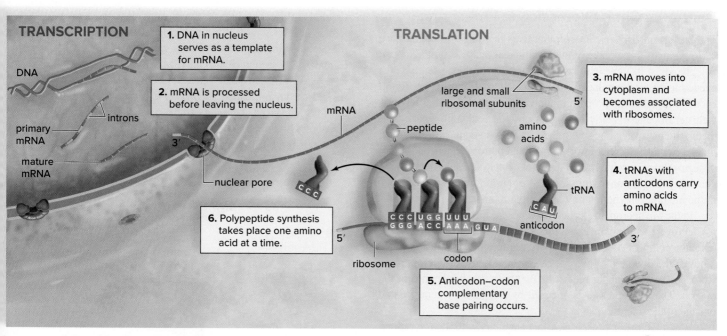

TRANSCRIPTION

1. DNA in nucleus serves as a template for mRNA.

DNA

2. mRNA is processed before leaving the nucleus.

primary mRNA

introns

mature mRNA

nuclear pore

6. Polypeptide synthesis takes place one amino acid at a time.

ribosome

TRANSLATION

large and small ribosomal subunits

mRNA

peptide

amino acids

3. mRNA moves into cytoplasm and becomes associated with ribosomes.

4. tRNAs with anticodons carry amino acids to mRNA.

tRNA

anticodon

codon

5. Anticodon–codon complementary base pairing occurs.

Figure 25.15 Review of gene expression.   Messenger RNA is produced and processed in the nucleus during transcription, and protein synthesis occurs at the ribosomes (in cytoplasm and rough ER) during translation.

25.4 Control of Gene Expression

The human body contains many types of cells that differ in structure and function. Each cell type must contain its own mix of proteins that make it different from all other cell types. Therefore, only certain genes are active in cells that perform specialized functions, such as nerve, muscle, gland, and blood cells.

Some of these active genes are called housekeeping genes because they govern functions that are common to many types of cells, such as glucose metabolism. But otherwise, the activity of selected genes accounts for the specialization of cells:

Cell type	Red blood	Muscle	Pancreatic
Gene type			
Housekeeping	▭	▭	▭
Hemoglobin	▭	▭	▭
Insulin	▭	▭	▭
Myosin	▭	▭	▭

In other words, gene expression is controlled in a cell, and this control accounts for its specialization. Let's begin by examining a simpler system—the control of transcription in prokaryotes.

Control of Gene Expression in Prokaryotes

The bacterium *Escherichia coli* that lives in your intestine can use various sugars as a source of energy and carbon. This organism can quickly adjust its gene expression to match your diet. The enzymes needed to break down lactose, the sugar present in milk, are found encoded together in an operon in the bacterial DNA. An **operon** is a cluster of genes usually coding for proteins related to a particular metabolic pathway, along with the short DNA sequences that coordinately control their transcription. The control sequences consist of a *promoter,* a sequence of DNA where RNA polymerase first attaches to begin transcription, and an *operator,* a sequence of DNA in the *lac* operon where a repressor protein binds (Fig. 25.16).

In the *lac* operon, the structural genes for three enzymes needed for lactose metabolism are under the control of one promoter/operator complex. If lactose is absent, a protein called a *repressor* binds to the operator. When the repressor is bound to the operator, RNA polymerase cannot transcribe the three structural genes of the operon. The *lac* repressor is encoded by a regulatory gene located outside of the operon.

When lactose is present, the lactose binds with the *lac* repressor so the repressor is unable to bind to the operator. Then, RNA polymerase is able to transcribe the structural genes into a single mRNA, which is then translated into the three different enzymes. In this way, the enzymes needed to break down lactose are synthesized only when lactose is present.

The *lac* operon is considered an inducible operon. It is activated only when lactose induces its expression. Other bacterial operons are repressible operons. They are usually active until a repressor turns them off.

Control of Gene Expression in Eukaryotes

In prokaryotes, a single promoter serves several genes that make up a transcription unit or operon. In eukaryotes, each gene has its own promoter where RNA polymerase binds. In contrast to the prokaryotes, eukaryotes employ a variety of mechanisms to regulate gene expression. These mechanisms affect whether the gene is expressed, the speed with which it is expressed, and how long it is expressed.

Levels of Gene Control

Eukaryotic genes exhibit control of gene expression at five different levels: (1) pretranscriptional control, (2) transcriptional control, (3) posttranscriptional control, (4) translational control, and (5) posttranslational control.

Pretranscriptional Control   Eukaryotes use DNA methylation and chromatin packing as a way to keep genes turned off. Genes within darkly staining, highly condensed portions of chromatin, called *heterochromatin,* are inactive. A dramatic example of this occurs with the X chromosome in mammalian females. Females have two X chromosomes, while males have only one X chromosome. This inequality is balanced by the inactivation of one X chromosome in every cell of the female body. Each inactivated X chromosome, called a Barr body in honor of its discoverer, can be seen as a small, darkly staining mass of condensed chromatin along the inner edge of the nuclear envelope. During early prenatal development, one or the other X chromosome in a cell is randomly shut off by chromatin condensation into heterochromatin. All of the cells that form from division of that cell will have the same X chromosome inactivated. Therefore, females have patches of tissue that differ in which X chromosome is being expressed. If that female is heterozygous for a particular X-linked gene, she will be a mosaic, containing patches of cells expressing different alleles. This can be clearly seen in calico cats (Fig. 25.17).

Active genes in eukaryotic cells are associated with more loosely packed chromatin, called *euchromatin.* However, even euchromatin must be "unpacked" before it can be transcribed. The presence of nucleosomes limits access to the DNA by the transcription machinery. A chromatin remodeling complex pushes the nucleosomes aside to open up sections of DNA for expression:

a. **Lactose absent.** Enzymes needed to metabolize lactose are not produced.

b. **Lactose present.** Enzymes needed to metabolize lactose are produced only when lactose is present.

Figure 25.16 The *lac* operon.   a. When lactose is absent, the regulator gene codes for a repressor that is normally active. When it binds to the operator, RNA polymerase cannot attach to the promoter, and structural genes are not expressed. **b.** When lactose is present, it binds to the repressor, changing its shape so it is inactive and cannot bind to the operator. Now, RNA polymerase binds to the promoter, and the structural genes are expressed.

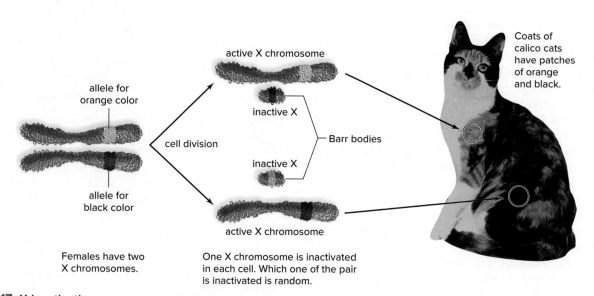

Figure 25.17 X-inactivation.   In cats, the alleles for black or orange coat color are carried on the X chromosome. Random X-inactivation occurs in females. Therefore, in heterozygous females, some of the cells express the allele for black coat color, while other cells express the allele for orange coat color. The white color on calico cats is provided by another gene. (photo): ©Linn Currie/Shutterstock

Transcriptional Control As in prokaryotes, eukaryotic transcriptional control is dependent on the interaction of proteins with particular DNA sequences. The proteins are called transcription factors and activators, while the DNA sequences they are associated with are called *enhancers* and *promoters*. In eukaryotes, each gene has its own promoter. **Transcription factors** are proteins that help RNA polymerase bind to a promoter. Several transcription factors per gene form a *transcription initiation complex* that also helps pull double-stranded DNA apart and even acts to release RNA polymerase so that transcription can begin. The same transcription factors are used again at other promoters, so it is easy to imagine that, if one malfunctions, the result could be disastrous to the cell.

In eukaryotes, *transcription activators* are proteins that speed transcription dramatically. In general, transcription activators are themselves activated in response to a signal received and transmitted by a signal transduction pathway. They bind to enhancer regions in the DNA and stimulate transcription.

Posttranscriptional Control Following transcription, messenger RNA (mRNA) is processed before it leaves the nucleus and passes into the cytoplasm. The primary mRNA is converted to the mature mRNA by the addition of a poly-A tail and a guanine cap, and by the removal of the introns and splicing back together of the exons. The same mRNA can be spliced in different ways to make slightly different products in different tissues. For example, both the hypothalamus and the thyroid gland produce the hormone calcitonin, but the calcitonin mRNA that exits the nucleus contains different combinations of exons in the two tissues.

The speed of transport of mRNA from the nucleus into the cytoplasm can ultimately affect the amount of gene product following transcription. There is a difference in the length of time it takes various mRNA molecules to pass through a nuclear pore.

Translational Control The longer an mRNA remains in the cytoplasm before it is broken down, the more gene product can be translated. Differences in the poly-A tail or the guanine cap can determine how long a particular transcript remains active before it is destroyed by a ribonuclease associated with ribosomes. Hormones can cause the stabilization of certain mRNA transcripts. For example, the mRNA for vitelline, an egg membrane protein, can persist for 3 weeks if it is exposed to estrogen, as opposed to 15 hours without estrogen.

Posttranslational Control Some proteins are not active immediately after synthesis. After translation, insulin is folded into a three-dimensional structure that is inactive. Then, a sequence of about 30 amino acids is enzymatically removed from the middle of the molecule, leaving two polypeptide chains that are bonded together by disulfide (S—S) bonds. This activates the protein. Other modifications, such as phosphorylation, also affect the activity of a protein. Many proteins only function a short time before they are degraded or destroyed by the cell.

The various levels of gene control are summarized in Figure 25.18.

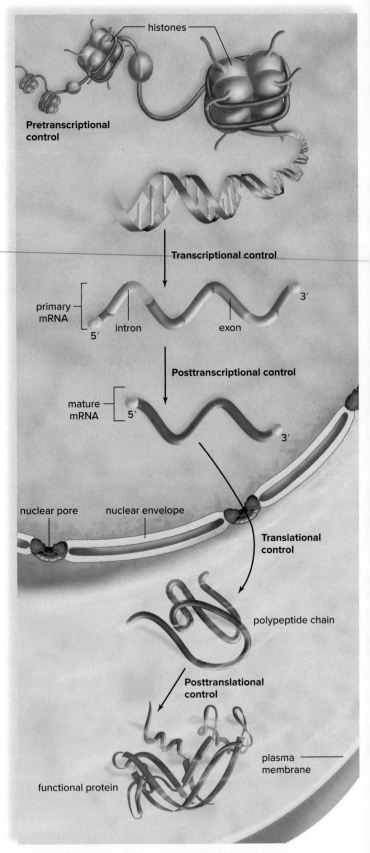

Figure 25.18 Levels at which control of gene expression occurs in eukaryotic cells. The five levels of control are (1) pretranscriptional control, (2) transcriptional control, and (3) posttranscriptional control, all of which occur in the nucleus, and (4) translational and (5) posttranslational control, which occur in the cytoplasm.

25.5 Gene Mutations and Cancer

Learning Outcomes

Upon completion of this section, you should be able to
1. Summarize the causes of gene mutations.
2. Describe why cancer is a failure of genetic control.
3. Describe the characteristics of cancer cells.

A **gene mutation** is a permanent change in the sequence of bases in DNA. Gene mutations may have a variety of effects on a cell. For example, it may change how the gene is expressed in the cell. A mutation may also change the structure of a protein so its function is altered, or completely inactivated.

Mutations that occur in the sex cells, also called *germ cells,* may be passed on from generation to generation. This is the basis of heredity, and may be responsible for susceptibility to some forms of cancer. Mutation in the remainder of the cells of the body, called *somatic cells,* are passed on to daughter cells by cell division. These are not passed on to future generations, but may sometimes lead to the development of cancer.

Causes of Mutations

Some mutations are spontaneous—they happen for no apparent reason—whereas others are induced by environmental influences. In most cases, *spontaneous mutations* arise as a result of abnormalities in normal biological processes. *Induced mutations* may result from exposure to toxic chemicals or radiation, which induce (cause) changes in the base sequence of DNA.

Errors in DNA Replication   DNA replication errors are a rare source of mutations. DNA polymerase, the enzyme that carries out replication, proofreads the new strand against the old strand. Usually, mismatched pairs are then replaced with the correct nucleotides. In the end, there is typically only one mistake for every 1 billion nucleotide pairs replicated.

Mutagens   Environmental influences called **mutagens** cause mutations in humans. Mutagens include radiation (e.g., radioactive elements, X-rays, ultraviolet [UV] radiation) and certain organic chemicals (e.g., chemicals in cigarette smoke and certain pesticides). The rate of mutations resulting from mutagens is generally low because DNA repair enzymes constantly monitor and repair any irregularities.

Transposons   Transposons are specific DNA sequences that have the remarkable ability to move within and between chromosomes. Their movement to a new location sometimes alters neighboring genes, particularly by increasing or decreasing their expression. Although "movable elements" in corn were described over 50 years ago by Barbara McClintock, their significance was only realized recently. Also called jumping genes, transposons have now been discovered in almost every species, including bacteria, fruit flies, and humans. McClintock described how the presence of white kernels in corn is due to a transposon located within a gene coding for a pigment-producing enzyme (Fig. 25.19*a, b*). "Indian corn" displays a variety of colors and patterns because of transposons (Fig. 25.19*c*). In a rare human neurological disorder called Charcot-Marie-Tooth disease, a transposon called Mariner causes the muscles and nerves of the legs and feet to gradually wither away.

Effect of Mutations on Protein Activity

Point mutations involve a change in a single DNA nucleotide. That change alters transcription and possibly changes the specific amino acid. One type of point mutation is a *base substitution* resulting in one DNA nucleotide being replaced with another incorrect nucleotide. Notice the base difference in the second row of Figure 25.20 and how it changes the resultant amino acid sequence.

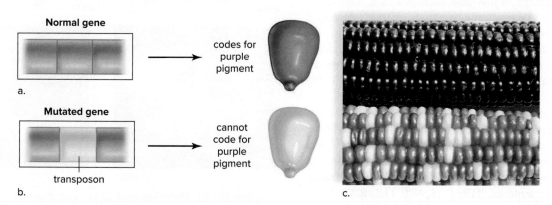

Figure 25.19 Transposon.   **a.** A purple coding gene ordinarily codes for a purple pigment. **b.** A transposon "jumps" into the purple-coding gene. This mutated gene is unable to code for purple pigment and so a white kernel results. **c.** Indian corn displays a variety of colors and patterns due to transposon activity.
(c): ©Mondae Leigh Baker

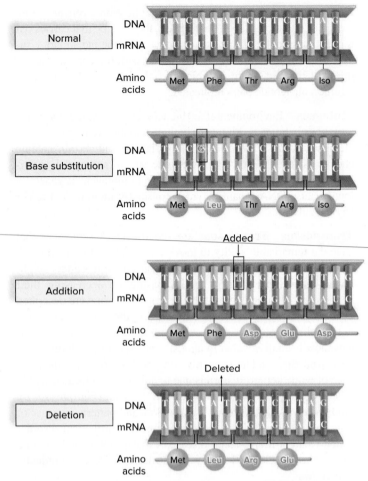

Figure 25.20 Point mutations.   The effect of a point mutation can vary. Starting at the *top:* Normal sequence of bases results in a normal sequence of amino acids. Next, a base substitution can result in the wrong amino acid. In the final two rows, an addition or deletion can result in a frameshift mutation, altering all the codons downstream of the point mutation.

Sometimes a base substitution has little or no effect on the final protein produced, but in some cases early stop codons can be introduced, or coding for the wrong amino acid can severely alter the protein shape. Such is the case with the genetic disorder sickle-cell disease (Fig. 25.21). In this gene, there is a base substitution that alters the mRNA codon for glutamic acid. Instead, the codon for valine is present, altering the final shape of hemoglobin—the protein that carries oxygen in the blood. The abnormal hemoglobin molecules form semirigid rods, and the red blood cells become sickle-shaped, resulting in decreased blood flow through tiny blood vessels.

Frameshift mutations (see Fig. 25.20) occur most often because one or more nucleotides are either inserted or deleted from DNA. The result of a frameshift mutation can be a completely new sequence of codons and nonfunctional protein. Here is how this occurs: The sequence of codons is read from a specific starting point, as in this sentence, THE CAT ATE THE RAT. If the letter *C* is deleted from this sentence and the reading frame is shifted, we read THE ATA TET HER AT—something that doesn't make sense.

Nonfunctional Proteins

A single nonfunctioning protein can have a dramatic effect on the phenotype, because enzymes are often a part of metabolic pathways. One particular metabolic pathway in cells is as follows:

If a faulty code for enzyme E_A is inherited, a person is unable to convert the molecule A to B. Phenylalanine builds up in the system, and the excess causes mental impairment and the other symptoms of the genetic disorder phenylketonuria (PKU). In the same pathway, if a person inherits a faulty code for enzyme E_B, then B cannot be converted to C, and the individual is an albino.

A rare condition called androgen insensitivity is due to a faulty receptor for androgens, which are male sex hormones such as testosterone. Although there is plenty of testosterone in the blood, the cells are unable to respond to it. Female instead of male external genitals form, and female instead of male secondary sex characteristics occur. The individual, who appears to be a normal female, may be prompted to seek medical advice when menstruation never occurs. The karyotype is that of a male rather than a female, and the individual does not have the internal sexual organs of a female.

Mutations Can Cause Cancer

It is estimated that one-third of us will develop cancer at some time in our lives. While the 5-year survival rate for individuals with cancer has increased from 50% in 1977 to 68% today, around 1,600 people each day will die of the disease. In the United States, the three deadliest forms of cancer are lung cancer, colorectal cancer, and breast cancer.

Figure 25.21 Mutation and sickle-cell disease.   Due to a base substitution in the hemoglobin gene, the DNA now codes for valine instead of glutamic acid, and the result is that normal red blood cells become sickle-shaped. (photos): ©Eye of Science/Science Source

The development of cancer involves a series of accumulating mutations that can be different for each type of cancer. As discussed in Section 5.2, tumor suppressor genes ordinarily act as brakes on cell division, especially when it begins to occur abnormally. Proto-oncogenes stimulate cell division but are usually turned off in fully differentiated nondividing cells. When proto-oncogenes mutate, they become oncogenes that are active all the time. Carcinogenesis begins with the loss of tumor suppressor gene activity and/or the gain of oncogene activity.

It often happens that tumor suppressor genes and proto-oncogenes code for transcription factors or proteins that control transcription factors. As we have seen, transcription factors are a part of the rich and diverse types of mechanisms that control gene expression in cells. They are of fundamental importance to DNA replication and repair, cell growth and division, control of apoptosis (programmed cell death), and cellular differentiation. Therefore, it is not surprising that inherited or acquired defects in transcription factor structure and function contribute to the development of cancer.

To take an example, a major tumor suppressor gene called *p53* is more frequently mutated in human cancers than any other known gene. It has been found that the p53 protein acts as a transcription factor, and as such is involved in turning on the expression of genes whose products are cell cycle inhibitors. The p53 protein also promotes apoptosis when it is needed. The retinoblastoma (RB) protein controls the activity of a transcription factor for cyclin D and other genes whose products promote entry into the S stage of the cell cycle. When the tumor suppressor gene *p16* mutates, it is unable to control the cell cycle, the RB protein is continuously expressed, and the result is too much active cyclin D in the cell.

Mutations in many other genes also contribute to the development of cancer. Several proto-oncogenes code for ras proteins, which are needed for cells to grow, to make new DNA, and to not grow out of control. A point mutation is sufficient to turn a normally functioning *ras* proto-oncogene into an oncogene. Abnormal growth results.

Although cancers vary greatly, they usually follow a common multistep progression (Fig. 25.22). Most cancers begin as an abnormal cell growth that is **benign,** or not cancerous, and usually does not grow larger. However, additional mutations may occur, causing the abnormal cells to fail to respond to inhibiting signals that control the cell cycle. When this occurs, the growth becomes **malignant,** meaning it is cancerous and possesses the ability to spread. The Health feature "Benign Versus Malignant Tumors" explores these differences in more detail.

Characteristics of Cancer Cells

The primary characteristics of cancer cells are as follows:

- *Cancer cells are genetically unstable.* Generation of cancer cells appears to be linked to mutagenesis. A cell acquires a mutation that allows it to continue to divide. Eventually, one of the progeny cells will acquire another mutation and gain the ability to form a tumor. Further mutations occur, and the most

New mutations arise, and one cell (brown) has the ability to start a tumor.

Cancer in situ. The tumor is at its place of origin. One cell (purple) mutates further.

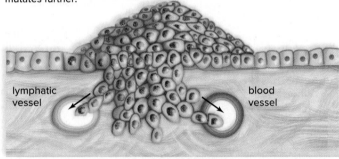

Cancer cells now have the ability to invade lymphatic and blood vessels and travel throughout the body.

New metastatic tumors are found some distance from the primary tumor.

Figure 25.22 Progression of cancer. A single abnormal cell begins the process, and the most aggressive cell, thereafter, becomes the one that divides the most and forms the tumor. Eventually, cancer cells gain the ability to invade underlying tissue and travel to other parts of the body, where they develop new tumors.

aggressive cell becomes the dominant cell of the tumor. Tumor cells undergo multiple mutations and also tend to have chromosomal aberrations and rearrangements.

- *Cancer cells do not correctly regulate the cell cycle.* Cancer cells continue to cycle through the cell cycle. The normal controls of the cell cycle do not operate to stop the cycle and allow the cells to differentiate. Because of that, cancer cells tend to be nonspecialized. Both the rate of cell division and the number of cells increase.

- *Cancer cells escape the signals for cell death.* A cell that has genetic damage or problems with the cell cycle will initiate apoptosis, or programmed cell death. However, cancer cells do not respond to internal signals to die, and they continue to divide even with genetic damage. Cells from the

SCIENCE IN YOUR LIFE ▶ HEALTH

Benign Versus Malignant Tumors

Cancer is a disease characterized by unrestricted cell growth. As a cancer cell continues its unregulated division, it may form a population of cells called a tumor. There are two different types of tumors: benign and malignant.

A benign tumor (Fig. 25A) is usually surrounded by a fibrous capsule (usually made of connective tissue). Because of this capsule, a benign tumor does not invade the adjacent tissue. The cells of a benign tumor resemble normal cells fairly closely. Many people have benign tumors. For example, a mole (or nevus) is a benign tumor of skin cells called melanocytes. However, not all benign tumors are harmless. Sometimes these tumors restrict a normal tissue's blood supply, or cause normal tissue to not function correctly. If this is in a critical area of the body, such as areas of the brain that control the heartbeat or breathing, the results may be fatal.

The cells of malignant tumors do not resemble normal cells, and are able to invade surrounding tissues (Fig 25B). As the cells of the tumor invade nearby tissues, they may come in contact with either blood or lymph vessels, allowing the tumor to spread throughout the body. The tumors associated with lung cancer are often malignant, and may easily spread to other organs, such as the brain or liver. The more deformed a cancer cell, the greater the chance it will be malignant, and thus dangerous to the individual.

Figure 25A Benign Tumor. A mole is an example of a benign tumor. ©Dr. P. Marazzi/Science Source

Figure 25B A malignant tumor. Melanoma is a form of skin cancer that is an example of a malignant tumor. ©Mediscan/Alamy

Questions to Consider

1. Why might oncologists (cancer physicians) look for abnormal cells in the blood of a suspected cancer patient?
2. What differences in the cell cycle might occur between the cells of a benign and malignant tumor?

immune system, when they detect an abnormal cell, will send signals to that cell, inducing apoptosis. Cancer cells also ignore these signals. Most normal cells have a built-in limit to the number of times they can divide before they die. One of the reasons normal cells stop entering the cell cycle is that the telomeres become shortened. **Telomeres** are sequences at the ends of the chromosomes that keep them from fusing with each other. With each cell division, the telomeres shorten, eventually becoming short enough to signal apoptosis. Cancer cells turn on the gene that encodes the enzyme telomerase, which is capable of rebuilding and lengthening the telomeres. Cancer cells thus show characteristics of "immortality" in that they can enter the cell cycle repeatedly.

■ *Cancer cells can survive and proliferate elsewhere in the body.* Many of the changes that must occur for cancer cells to form tumors elsewhere in the body are not understood. The cells apparently disrupt the normal adhesive mechanism and move to another place within the body. They travel through the blood and lymphatic vessels and then invade new tissues, where they form tumors. This process is known as **metastasis.** As a tumor grows, it must increase its blood supply by forming new blood vessels, a process called **angiogenesis.** Tumor cells switch on genes that code for the production of growth factors, encouraging blood vessel formation. These new blood vessels supply the tumor cells with the nutrients and oxygen they require for rapid growth, but they also rob normal tissues of nutrients and oxygen.

Check Your Progress 25.5

1. Explain how gene mutations occur.
2. Distinguish between a point mutation and a frameshift mutation.
3. Explain the characteristics of cancer cells.

Conclusion

DNA is the genetic information that contains the instructions required for the production of the proteins (and enzymes) involved in the majority of body functions. The development of synthetic DNA, with its unique collection of genetic information, may result in the capability to produce proteins that have a number of benefits to humans.

Prevention of Cancer

Be tested for cancer.

Do a shower test for breast cancer or testicular cancer. Women should get a Pap smear for cervical cancer annually if they are over 21 or are sexually active. Have a friend or physician check your skin annually for any unusual moles. Have other exams done regularly by a physician.

Be aware of occupational hazards.

Exposure to several different industrial agents (nickel, chromate, asbestos, vinyl chloride, etc.) and/or radiation increases the risk of various cancers. Your employer should notify you of potential hazards in your workplace. The risk of several of these cancers is increased when combined with cigarette smoking.

Use sunscreen.

Almost all cases of basal cell and squamous cell skin cancers are sun-related. Use a sunscreen of at least SPF 30 (Fig. 25C) and wear protective clothing if you are going to be out during the brightest part of the day. Don't sunbathe on the beach or in a tanning salon.

Check your home for radon.

Excessive radon exposure in homes increases the risk of lung cancer, especially in cigarette smokers. It is best to test your home and take the proper remedial actions.

Avoid unnecessary X-rays.

Even though most medical and dental X-rays are adjusted to deliver the lowest dose possible, unnecessary X-rays should be avoided. Sensitive areas of the body that are not being X-rayed should be protected with lead screens.

Practice safe sex.

Human papillomaviruses (HPVs) cause genital warts and are associated with cervical cancer, as well as tumors of the vulva, vagina, anus, penis, and mouth. HPVs may be involved in 90–95% of all cases of cervical cancer, and 20 million people in the United States have an infection that can be transmitted to others.

Carefully consider hormone therapy.

Estrogen therapy to control menopausal symptoms increases the risk of endometrial cancer and heart attack. However, including

Figure 25C UV protection. Use of sunscreen with SPF of 30 or higher can help prevent skin cancer, as can limiting midday exposure. ©moodboard/Superstock

progesterone in estrogen replacement therapy helps minimize this risk.

Maintain a healthy weight.

The risk of cancer (especially colon, breast, and uterine cancers) is 55% greater among obese women and 33% greater among obese men, compared to people of normal weight. Eating a low-fat, healthy diet helps maintain your weight, as well as reducing your risk of cancer.

Exercise regularly.

Regular activity can reduce the incidence of certain cancers, especially colon and breast cancer.

Increase consumption of foods that are rich in vitamins A and C.

Beta-carotene, a precursor of vitamin A, is found in dark-green, leafy vegetables, carrots, and various fruits. Vitamin C is present in citrus fruits. These vitamins are called antioxidants because they prevent the formation of

chemicals called free radicals that can damage the DNA in the cell. Vitamin C also prevents the conversion of nitrates and nitrites into carcinogenic nitrosamines in the digestive tract.

Include vegetables from the cabbage family in your diet.

The cabbage family includes cabbage, broccoli, Brussels sprouts, kohlrabi, and cauliflower. These vegetables may reduce the risk of gastrointestinal and respiratory tract cancers.

Limit consumption of salt-cured, smoked, or nitrite-cured foods.

Salt-cured or pickled foods may increase the risk of stomach and esophageal cancer. Smoked foods, such as ham and sausage, contain chemical carcinogens similar to those in tobacco smoke. Nitrites are sometimes added to processed meats (e.g., hot dogs and cold cuts) and other foods to protect them from spoilage. These are converted to carcinogenic nitrosamines in the digestive tract.

Be moderate in the consumption of alcohol.

Cancers of the mouth, throat, esophagus, larynx, and liver occur more frequently among heavy drinkers, especially when accompanied by smoking cigarettes or chewing tobacco.

Don't smoke.

Cigarette smoking accounts for about 30% of all cancer deaths. Smoking is responsible for 90% of lung cancer cases among men and 79% among women—about 87% altogether. People who smoke two or more packs of cigarettes a day have lung cancer mortality rates 13 to 23 times greater than those of nonsmokers. Cigars and smokeless tobacco (chewing tobacco or snuff) increase the risk of cancers of the mouth, larynx, throat, and esophagus.

Questions to Consider

1. Which of these cancer-preventive measures are you currently practicing?
2. Why do you think tobacco use increases the risk of other types of cancer besides lung cancer?
3. Why does the use of tanning beds also increase the incidence of skin cancer?

SUMMARIZE

25.1 DNA Structure

- In a series of experiments, researchers were able to determine that **DNA (deoxyribonucleic acid)** was the genetic material.
- DNA is structured as a **double helix,** with two sugar-phosphate backbones and paired nitrogen-containing bases (which may be either **purines** or **pyrimidines**). **Complementary base pairing** of A (adenine) with T (thymine), and G (guanine) with C (cytosine) occurs between the strands. The DNA strands of the double helix are oriented in an antiparallel configuration.

25.2 DNA Replication

- **DNA replication** is semiconservative, meaning that, following replication, each double helix contains one old strand and one new strand. Enzymes, including DNA polymerase, helicase, and DNA ligase are involved in the replication process.

25.3 Gene Expression

Genes contain the information for the synthesis of proteins or **RNA (ribonucleic acid)** molecules. RNA plays an active role in all stages of gene expression. Gene expression involves two processes: transcription and translation.

- **Transcription** produces an RNA sequence complementary to one of the DNA strands. Transcription uses an **RNA polymerase** enzyme, which binds to a region of the DNA called a **promoter. Messenger RNA (mRNA)** produced in eukaryotes must be processed by the addition of a cap and tail, and removal of **introns,** before being transported to the cytoplasm where the information in the **exons** can be translated.
- DNA specifies the synthesis of proteins using a **codon** consisting of three bases of RNA. Each codon codes for one amino acid.
- During **translation, transfer RNA (tRNA)** molecules, attached to their own particular amino acid, travel to a ribosome, and through complementary base pairing between **anticodons** of tRNA and codons of mRNA, the tRNAs, and the amino acids they carry, are sequenced in a predetermined way to form a polypeptide chain.
- Translation consists of three stages: **initiation, elongation,** and **termination.** A ribosome, which contains **ribosomal RNA (rRNA)**, has a binding site for two tRNAs at a time. The tRNA at the P site passes a peptide to a newly arrived tRNA–amino acid at the A site. Then, translocation occurs: The ribosome moves, and the tRNA with the peptide is at the P site. In this way, an amino acid grows one amino acid at a time until a polypeptide is formed.

25.4 Control of Gene Expression

- Each cell type in the body contains its own mix of proteins that make it different from all other cell types. Only certain genes are active in cells that perform specialized functions. In prokaryotes, genes are clustered into operons. An **operon** contains a group of genes, usually coding for proteins related to a particular metabolic pathway, along with the promoter, a DNA sequence where RNA polymerase binds to begin transcription, and an operator, where a repressor protein binds. In the case of the *lac* operon, the structural genes code for enzymes that metabolize lactose. When lactose is absent, the repressor binds to the operator and shuts off the operon. When lactose is present, it binds to the repressor so it can no longer bind to the operator, and this allows RNA polymerase to transcribe the structural genes.

- Control of gene expression is more complicated in eukaryotes. Eukaryotes have five levels of control: pretranscriptional control, transcriptional control, posttranscriptional control, translational control, and posttranslational control.
- Genes found within highly condensed heterochromatin are inactive. A dramatic example of this is inactivated X chromosomes, called Barr bodies. Even genes found in looser-packed DNA, called euchromatin, must have the nucleosomes shifted in order to be expressed.
- The transcriptional level of control includes **transcription factors** that assist the RNA polymerase and transcriptional activators that greatly increase the rate of transcription.
- In posttranscriptional control, differences in mRNA processing affect gene expression.
- Translational controls occur in the cytoplasm and involve the length of time the mRNA is functional.
- Posttranslational controls also occur in the cytoplasm and involve the length of time the protein is functional. Proteins can be activated by a variety of methods.

25.5 Gene Mutations and Cancer

- A **gene mutation** is a permanent change in the sequence of bases in DNA. The effect of a DNA base sequence change on protein activity can range from no effect to complete inactivity.
- Gene mutations result in a change in the sequence of nucleotides in the DNA and can be caused by errors in replication, **mutagens,** and **transposons.**
- Point mutations involve a change in a single DNA nucleotide and, therefore, a possible change in a specific amino acid. Frameshift mutations occur most often because one or more nucleotides are either inserted or deleted from DNA.
- The development of cancer involves a series of accumulating mutations that can be different for each type of cancer. Carcinogenesis begins with the loss of tumor suppressor gene activity and/or the gain of oncogene activity. Mutations in many other genes also contribute to the development of cancer. Although cancers vary greatly, they usually follow a common multistep progression. Cancer cells defy the normal regulation of the cell cycle and can invade and colonize other areas of the body. Cancer cells do not exhibit contact inhibition and thus form **benign** tumors. When tumor cells gain the ability to invade surrounding tissues, they are said to be **malignant.**
- In terms of their primary characteristics, cancer cells are genetically unstable, do not correctly regulate the cell cycle, cause **angiogenesis,** escape the signals for cell death, and can survive and proliferate elsewhere in the body, a process called **metastasis.** Multiple mutations and chromosomal aberrations are present in cancer cells. They continue to cycle through the cell cycle and tend to be nonspecialized. They do not respond to either internal or external signals for apoptosis, and they induce the expression of telomerase to repair their **telomeres.**

ASSESS

Testing Yourself

Choose the best answer for each question.

25.1 DNA Structure

1. The double-helix model of DNA resembles a twisted ladder in which the rungs of the ladder are
   a. a purine paired with a pyrimidine.
   b. A paired with G, and C paired with T.
   c. sugar-phosphate paired with sugar-phosphate.
   d. a 5′ end paired with a 3′ end.
   e. Both a and b are correct.

2. If 30% of an organism's DNA is thymine, then
   a. 70% is purine.
   b. 20% is guanine.
   c. 30% is adenine.
   d. 70% is pyrimidine.
   e. Both b and c are correct.

3. If the sequence of bases in one strand of DNA is 5′ TAGCCT 3′, then the sequence of bases in the other strand is
   a. 3′ TCCGAT 5′.
   b. 3′ TAGCCT 5′.
   c. 3′ ATCGGA 5′.
   d. 3′ AACGGUA 5′.
   e. None of these are correct.

25.2 DNA Replication

4. DNA replication is said to be semiconservative because
   a. one of the new molecules conserves both of the original DNA strands.
   b. the new DNA molecule contains two new DNA strands.
   c. both of the new molecules contain one new strand and one old strand.
   d. DNA polymerase conserves both of the old strands.

5. The enzyme responsible for separating double-stranded DNA into single-stranded DNA is
   a. DNA helicase.
   b. DNA primase.
   c. DNA polymerase.
   d. DNA ligase.

6. The enzyme responsible for adding new nucleotides to a growing DNA chain during DNA replication is
   a. helicase.   c. DNA polymerase.
   b. RNA polymerase.   d. DNA ligase.

25.3 Gene Expression

7. Which of the following processes occurs in the nucleus and forms a complementary copy of one strand of the DNA molecule for gene expression?
   a. DNA replication   d. RNA processing
   b. translation   e. None of these are correct.
   c. transcription

8. RNA processing
   a. removes the exons, leaving only the introns.
   b. is the same as transcription.
   c. is an event that occurs after RNA is transcribed.
   d. is the rejection of old, worn-out RNA.
   e. All of these are correct.

9. If the sequence of bases in the coding strand of a DNA molecule is TAGC, then the sequence of bases in the mRNA will be
   a. AUCG.
   b. TAGC.
   c. UAGC.
   d. ATCG.

25.4 Control of Gene Expression

10. An operon is a short sequence of DNA
    a. that prevents RNA polymerase from binding to the promoter.
    b. that prevents transcription from occurring.
    c. and the sequences that control its transcription.
    d. that codes for the repressor protein.
    e. that functions to prevent the repressor from binding to the operator.

11. How is transcription directly controlled in eukaryotic cells?
    a. through the use of phosphorylation
    b. by means of apoptosis
    c. using transcription factors and activators
    d. when chromatin is packed to keep genes turned on
    e. None of these are correct.

25.5 Gene Mutations and Cancer

12. Mutation rates are typically low due to
    a. proofreading by DNA polymerase.
    b. DNA repair enzymes.
    c. silent mutations.
    d. None of these are correct.
    e. All of these are correct.

13. Which of these is a characteristic of cancer cells?
    a. They do not exhibit contact inhibition.
    b. They lack specialization.
    c. They have abnormal chromosomes.
    d. They fail to undergo apoptosis.
    e. All of these are correct.

ENGAGE

Thinking Critically

1. What kind of genes do you think would be included in the category of "housekeeping" genes?

2. When the enzyme telomerase was first discovered, some people thought it might be "the fountain of youth" because it could immortalize cells. Why do you think this may not be the case?

3. Why is it only the *risk* for cancer that is inherited?

26

Biotechnology and Genomics

©ESB Professional/Shutterstock

BEFORE YOU BEGIN

Before beginning this chapter, take a few moments to review the following discussions:

Section 25.1 What are the structural characteristics of a DNA molecule?

Section 25.2 What is the role of the DNA polymerase in DNA replication?

Section 25.3 What are the stages in gene expression?

Genome Editing and CRISPR

Biotechnology has come a long way since it first began in the early 1950s. Historically, biotech has focused on inserting modified genes into bacteria and plants to yield genetically modified organisms that can withstand changes in the climate, gain enhanced nutritional value, or produce new drugs to target human diseases.

Recently, a new form of genome editing called CRISPR is revolutionizing biotechnology. Based on a series of enzymes that protect bacteria from viruses, CRISPR acts as a "molecular scalpel," allowing researchers to target specific sequences of a genome for editing. Early experiments were able to correct defects in mice, and experiments have already been conducted on human embryos. Researchers are actively examining how to modify the pig genome to produce more hearts and organs for human transplants. Almost monthly, new applications of this technology are revealed in scientific journals.

These advances present tremendous opportunities to address human disease, but they also give rise to ethical questions. For example, to what extent is it ethical to improve the human genome, and should society allow couples to produce "designer babies"? In this chapter, we will explore some of these techniques and areas of study, so that you may be better informed to make decisions.

As you read through this chapter, think about the following questions:

1. What is the process by which a gene from one species is inserted into the genome of another species?

2. What is the difference between a genetically modified and transgenic organism?

26.1 DNA Technology

Knowledge of DNA biology has led to an ability to manipulate the genes of organisms. We can clone genes and then use them to alter the **genome** (the complete genetic makeup of an organism) of viruses and cells, whether bacterial, plant, or animal cells. This practice, called **genetic engineering,** has innumerable uses, from producing a product to treating cancer and genetic disorders.

Forms of Cloning

We often think of **cloning** as the production of identical copies of an organism through some asexual means. The members of a bacterial colony on a petri dish are clones because they all came from the division of the same original cell. Human identical twins are also considered clones, because the first two cells of the embryo separated and each became a complete individual. The Scientific Inquiry feature "Forms of Cloning" explores some of the different ways this term is used in biology.

Cloning a Gene

Another major biological application of cloning is **gene cloning,** which is the production of many identical copies of a single gene. Biologists clone genes for a number of reasons. They might want to produce large quantities of the gene's protein product, such as human insulin, learn how a cloned gene codes for a particular protein, or use the genes to alter the phenotypes of other organisms in a beneficial way. When cloned genes are used to modify a human, the process is called gene therapy (see Section 26.3). Otherwise, organisms with foreign DNA or genes inserted into them are called **transgenic organisms,** which are frequently used to produce a product desired by humans. While a variety of techniques now exist to produce cloned DNA, most processes rely on recombinant DNA technology and the polymerase chain reaction (PCR).

Recombinant DNA Technology

Recombinant DNA (rDNA) contains DNA from two or more different sources, such as the human cell and the bacterial cell in Figure 26.1. To make rDNA, a researcher needs a **vector,** a piece of DNA that can be manipulated such that foreign DNA can be added to it. One common vector is a plasmid. **Plasmids** are small accessory rings of DNA from bacteria that are not part of the bacterial chromosome and are capable of self-replicating. Plasmids

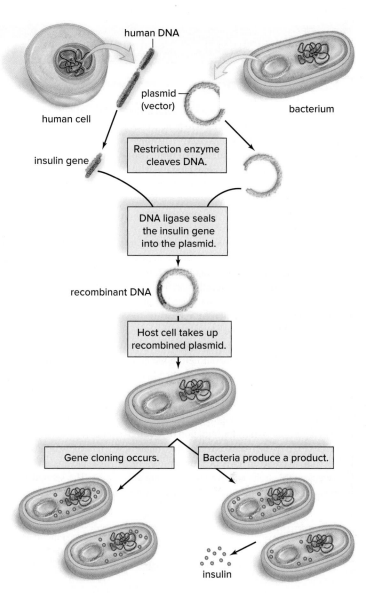

Figure 26.1 Cloning a human gene. Human DNA and plasmid DNA are cleaved by a specific type of restriction enzyme. For example, human DNA containing the insulin gene is spliced into a plasmid by the enzyme DNA ligase. Gene cloning is achieved after a bacterium takes up the plasmid. If the gene functions normally as expected, the product (e.g., insulin) may also be retrieved.

were discovered by investigators studying the bacterium *Escherichia coli.*

Two enzymes are needed to introduce foreign DNA into vector DNA: (1) a **restriction enzyme** to cleave the vector DNA and (2) **DNA ligase** to seal DNA into an opening created by the restriction enzyme. Hundreds of restriction enzymes occur naturally in bacteria, where they act as a primitive immune system by cutting up any viral DNA that enters the cell. They are called restriction enzymes because they *restrict* the growth of viruses, but they can also be used as molecular scissors to cut double-stranded DNA at a specific site. For example, the restriction enzyme called *Eco*RI always recognizes and cuts double-stranded DNA

SCIENCE IN YOUR LIFE ▶ SCIENTIFIC INQUIRY

Forms of Cloning

The term *cloning* may be used in several different ways in biology. In addition to gene cloning (see Fig. 26.1), there is reproductive cloning—the ability to clone an adult animal from a normal body cell—and therapeutic cloning, which allows the rapid production of mature cells of a specific type. Both types of cloning are a direct result of recent discoveries about how the cell cycle is controlled.

Reproductive cloning, or the cloning of adult animals, was once thought to be impossible, because investigators found it difficult to have the nucleus of an adult cell "start over" with the cell cycle, even when it was placed in an egg cell that had had its own nucleus removed.

In 1997, Dolly the sheep demonstrated that reproductive cloning is indeed possible. The donor cells were starved before the cell's nucleus was placed in an egg that had had its nucleus removed (called an enucleated egg). This caused the cells to stop dividing and go into a G_0 (resting) stage, making the nuclei amenable to cytoplasmic signals for initiation of development (Fig. 26A*a*). This advance has made it possible to clone all sorts of farm animals that have desirable traits and even to clone rare animals that might otherwise become extinct.

In 2018, Chinese scientists announced that they have cloned the first primates, a pair of monkeys. This development offers new opportunities in the use of cloned animals in research programs.

In *therapeutic cloning,* however, the objective is to produce mature cells of various cell types rather than an individual organism. The purposes of therapeutic cloning are (1) to learn more about how specialization of cells occurs and (2) to provide cells and tissues that could be used to treat human illnesses, such as diabetes, or major injuries, such as strokes or spinal cord injuries.

There are two possible ways to carry out therapeutic cloning. The first way is to use the same procedure as reproductive cloning, except that embryonic stem cells (ESCs) are separated and each is subjected to a treatment that causes it to develop into a particular type of cell, such as red blood cells, muscle cells, or nerve cells (Fig. 26A*b*). Some have ethical concerns about this type of therapeutic cloning, because if the embryo were allowed to continue development it would become an individual.

The second way to carry out therapeutic cloning is to use adult stem cells. Stem cells are found in many organs of an adult's body; for example, the bone marrow has stem cells that produce new blood cells. However, adult stem cells are limited in the number of cell types they may become. Nevertheless, scientists are beginning to overcome this obstacle. In 2006, by adding just four genes to adult skin stem cells, called fibroblasts, into becoming induced pluripotent stem cells (iPS), a type of stem cell similar to an ESC. The researchers were then able to create heart and brain cells from the adult stem cells. Other researchers have used this technique to reverse Parkinson-like symptoms in rats.

Although questions exist on the benefits of iPS cells, these advances demonstrate that scientists are actively investigating methods of overcoming the current limitations and ethical concerns of using embryonic stem cells.

Questions to Consider

1. How might the study of therapeutic cloning benefit scientific studies of reproductive cloning?
2. What types of diseases might not be treatable using therapeutic cloning?

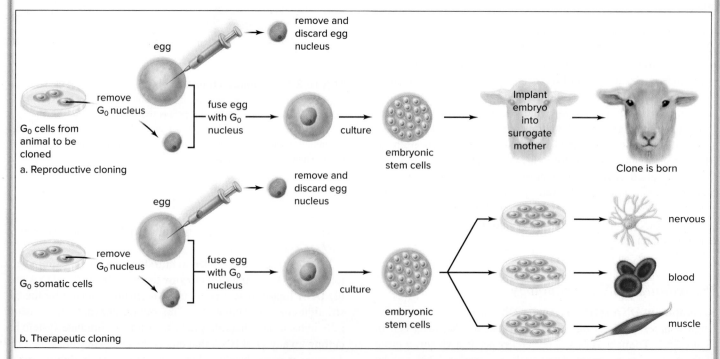

Figure 26A Reproductive and therapeutic cloning. **a.** The purpose of reproductive cloning is to produce an individual that is genetically identical to the one that donated a nucleus. The nucleus is placed in an enucleated egg, and after several mitotic divisions, the embryo is implanted into a surrogate mother for further development. **b.** The purpose of therapeutic cloning is to produce specialized tissue cells. A nucleus is placed in an enucleated egg, and after several mitotic divisions, the embryonic cells (called embryonic stem cells) are separated and treated to become specialized cells. ©Science Source in the following manner when DNA has the sequence of bases GAATTC:

Notice that a gap now exists into which a piece of foreign DNA can be placed if it ends in bases complementary to those exposed by the restriction enzyme. To ensure this, it is necessary only to cleave the foreign DNA with the same type of restriction enzyme. The single-stranded, but complementary, ends of the two DNA molecules are called "sticky ends" because they can bind a piece of foreign DNA by complementary base pairing. Sticky ends facilitate the insertion of foreign DNA into vector DNA.

DNA ligase, an enzyme that functions in DNA replication, is then used to seal the foreign piece of DNA into the vector. Bacterial cells take up recombinant plasmids, especially if the cells are treated to make them more permeable. Thereafter, as the plasmid replicates, so does the foreign DNA, and thus the gene is cloned.

DNA Sequencing

DNA sequencing is a procedure that determines the order of nucleotides in a segment of DNA, often within a specific gene. DNA sequencing allows researchers to identify specific alleles associated with a disease and thus facilitate the development of medicines or treatments. This information also serves as the foundation for the study of forensic biology and even contributes to our understanding of our evolutionary history (see Section 32.5).

When DNA technology was in its inception in the early 1970s, this technique was performed manually using dye-terminator substances or radioactive tracer elements attached to each of the four nucleotides during DNA replication, with results being deciphered from their pattern on a gel plate. Modern-day sequencing involves dyes attached to the nucleotides and use of a laser to detect the different dyes by an automated sequencing machine, which shows the order of nucleotides on a grid computer screen. To begin sequencing a segment of DNA, many copies of the segment are made, or replicated, using a procedure called the polymerase chain reaction.

The Polymerase Chain Reaction

The **polymerase chain reaction (PCR)** can create billions of copies of a segment of DNA in a test tube in a matter of hours. PCR is very specific—it *amplifies* (makes copies of) a targeted DNA sequence, usually a few hundred bases in length. PCR requires the use of DNA polymerase, the enzyme that carries out DNA replication, and a supply of nucleotides for the new DNA strands. PCR involves three basic steps that occur repeatedly (Fig. 26.2), usually for about 35 to 40 cycles: (1) a denaturation step at 95°C, where DNA is heated to become single-stranded; (2) an annealing step at a temperature usually between 50° and 60°C, where an oligonucleotide primer hybridizes to each of the single DNA strands; and (3) an extension step at 72°C, where an engineered DNA polymerase adds complementary bases to each of the single DNA strands, creating double-stranded DNA.

PCR is a chain reaction because the targeted DNA is repeatedly replicated, much in the same way natural DNA replication occurs, as long as the process continues. Notice in Figure 26.2 that the amount of DNA doubles with each replication cycle. Thus, assuming you start with only one copy of DNA, after one cycle, you will have two copies, after two cycles four copies, and so on. PCR

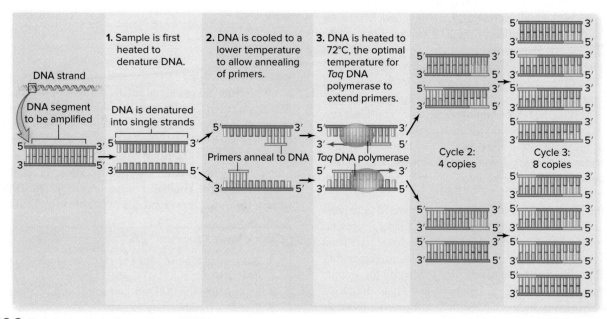

Figure 26.2 Polymerase chain reaction (PCR).   PCR allows the production of many identical copies of DNA in a laboratory setting. Assuming you start with only one copy, you can create millions of copies with only 20 to 25 cycles. For this reason, only a tiny DNA sample is necessary for forensic genetics.

has been in use since its development in 1985 by Kary Banks Mullis, and now almost every laboratory has automated PCR machines to carry out the procedure. Automation became possible after a temperature-insensitive (thermostable) DNA polymerase was extracted from the bacterium *Thermus aquaticus,* which lives in hot springs. The enzyme can withstand the high temperature used to denature double-stranded DNA. Therefore, replication does not have to be interrupted by the need to add more enzyme.

DNA amplified by PCR is often analyzed for various purposes. For example, mitochondrial DNA base sequences have been used to decipher the evolutionary history of human populations. Because so little DNA is required for PCR to be effective, it is commonly used as a forensic method for analyzing DNA found at crime scenes—only a drop of semen, a flake of skin, or the root of a single hair is necessary!

DNA Analysis

Analysis of DNA following PCR has undergone improvements over the years. At first, the entire genome was treated with restriction enzymes, and because all individuals have their own restriction enzyme sites, they would have a unique collection of DNA fragment sizes. During a process called *gel electrophoresis,* whereby an electrical current is used to force DNA through a porous gel material, these fragments are separated according to their size. Smaller fragments move farther through the gel than larger fragments, and result in a pattern of distinctive bands, called a DNA profile or **DNA fingerprint.**

Now, **short tandem repeat (STR) profiling** is the method of choice. STRs are the same short sequence of DNA bases that recur several times, as in GATAGATAGATA. STR profiling is advantageous because it doesn't require the use of restriction enzymes. Instead, PCR is used to amplify target sequences of DNA, which are fluorescently labeled. The PCR products are placed in an automated DNA sequencer. As the sequences move through the sequencer, the fluorescent labels are picked up by a laser. A detector then records the length of each DNA fragment. The fragments are different lengths because all individuals have their own number of repeats at the particular location of the STR on the chromosome (i.e., each STR locus). That is, the greater the number of STRs at a locus, the longer the DNA fragment amplified by PCR. If individuals are homozygotes, they will have a single fragment, and heterozygotes will have two fragments of different lengths (Fig. 26.3). The more STR loci employed, the more confident scientists can be of distinctive results for each person.

The FBI has collected a large database of human STR profiles for current and future crime scene analyses, and genetic profiles may someday be on people's driver's licenses in the United States.

When the DNA matches that of a virus or mutated gene, it is known that a viral infection, genetic disorder, or cancer is present. DNA fingerprints from blood or tissues at a crime scene have been successfully used in convicting criminals. DNA fingerprinting through STR profiling was extensively used to identify the victims of the tsunamis in recent years in Indonesia and Japan. Relatives can be found, paternity suits can be settled, and genetic disorders can be detected. PCR has also shed new light on evolutionary

Figure 26.3 DNA fingerprinting.   To establish a DNA fingerprint, short segments of DNA from samples are amplified by a PCR reaction. These fragments are then separated using gel electrophoresis (or a detector) to look for small variations in the length of the fragments.

studies by comparing DNA extracted from human mummies thousands of years old or animal fossils millions of years old.

Applications of DNA technology are limited only by our imaginations. For example, the process of genome editing using processes such as CRISPR (see the Scientific Inquiry feature "Genome Editing"), unheard of just 20 years ago, is now allowing scientists to edit specific sequences of DNA. In another application, the National Football League uses synthetic DNA to mark each of the Super Bowl footballs to be able to authenticate them.

Check Your Progress 26.1

1. Summarize the two required steps for producing recombinant DNA.
2. Explain how the PCR reaction amplifies a segment of DNA.
3. Explain why STRs may be used for identification.

Genome Editing

A relatively new advance in DNA technology is genome editing, the targeting of specific sequences in the DNA for removal or replacement. There are several methods by which editing may be done; the most widely used is called CRISPR (*clustered regularly interspaced short palindromic repeats*).

CRISPR was first discovered in prokaryotes, where it acts as a form of immune defense against invading viruses. Viruses function by inserting their DNA into host cells, causing those cells to form new viruses (see Section 28.5). The CRISPR system is based on an endonuclease enzyme called Cas9, which is capable of identifying specific sequences of nucleotides in the genomic DNA of the invading virus and breaking both of the DNA strands, thus inactivating the virus.

Cas9 identifies the specific nucleotides to be cut using a guide RNA molecule that complementary base-pairs to the genomic DNA sequence (Fig. 26B). To protect the bacteria from Cas9 activity against its own DNA, a sequence called PAM (which is not found in bacterial cells) must be adjacent to the target DNA sequence.

The CRISPR system can be used by researchers to target a specific sequence of nucleotides, in almost any organism, for editing. If the genomic sequence of the target is known, a complementary RNA strand can be used by Cas9 to produce a break in the DNA.

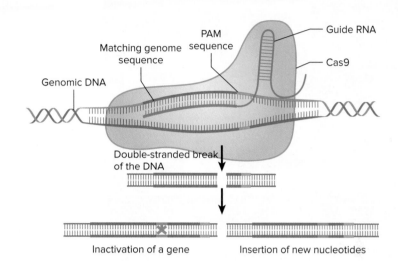

Figure 26B CRISPR and genome editing. Genome editing involves using a nuclease—in this case, Cas9—to target specific sequences of DNA. The double-stranded break in the DNA created by Cas9 may be used to inactivate a gene to study its function, or to insert new sequences of nucleotides.

This break can be used to inactivate the gene and thus study the role of the gene in the cell, or Cas9 can act as a form of molecular scissors to insert new nucleotides at specific DNA locations. CRISPR and other genome editing technologies continue to develop. Scientists are investigating ways of making the processes more efficient, as well as new applications for genome editing in humans and other organisms.

Questions to Consider

1. Recently, research has shown that CRISPR may cause unintentional breaks in DNA sequences other than the target sequence. What may be the consequences of this?
2. The chapter opener introduced the concept of designer babies. In what way may the altering of an embryo's DNA be beneficial? Can you think of any concerns about this process?

26.2 Biotechnology Products

Learning Outcomes

Upon completion of this section, you should be able to

1. Identify ways in which human society benefits from genetically modified bacteria, plants, and animals.
2. Describe the steps involved in the production of a transgenic animal.

Today, bacteria, plants, and animals are genetically engineered to produce biotechnology products. Recall that a *genetically modified organism (GMO)* is one whose genome has been modified in some way, usually by using recombinant DNA technology. Organisms that have had a foreign gene inserted into their genome are called *transgenic organisms.*

The use of the biotechnology processes outlined in Section 26.1, and specifically the development of the CRISPR genome editing process, has greatly facilitated the ability of scientists to generate GMOs. Almost every day, new advances are being reported by scientists. In this section, we will take a look at some of the ways bacteria, plants, and animals have been genetically modified to benefit humans.

Transgenic Bacteria

Recombinant DNA technology is used to produce transgenic bacteria, which may be grown commercially in huge vats called bioreactors. The bacteria express the cloned gene, and the gene product is usually collected from the medium in which the bacteria are grown. Biotechnology products produced by bacteria include insulin, human growth hormone, tPA (tissue plasminogen activator), and hepatitis B vaccine.

Transgenic bacteria have many other uses as well. Some have been produced to promote the health of plants. For example, bacteria that normally live on plants and encourage the formation of ice crystals have been changed from frost-plus to frost-minus bacteria. As a result, new crops such as frost-resistant strawberries are

Figure 26.4 Bioremediation. Bacteria capable of decomposing oil have been engineered and patented by researchers such as Dr. Chakrabarty. (bacteria): ©Medical-on-Line/Alamy; (scientist): ©CEK/AP Images

Figure 26.5 Example of a genetically modified plant. Genetic modification of these apples (*bottom row*) allows them to resist browning, compared to other apples (*top row*). ©Okanagan Specialty Fruits Inc.

being developed. Also, a bacterium that normally colonizes the roots of corn plants has now been endowed with genes (from another bacterium) that code for an insect toxin. The toxin protects the roots from insects.

Bacteria can be selected for their ability to degrade a particular substance, and this ability can then be enhanced by bioengineering. For instance, naturally occurring bacteria that eat oil can be genetically engineered to do an even better job of cleaning up beaches after oil spills (Fig. 26.4), such as the 2010 *Deepwater Horizon* spill in the Gulf of Mexico. Bacteria can also remove sulfur from coal before it is burned, resulting in cleaner emissions. One bacterial strain was given genes that allowed it to clean up levels of toxins that would have killed other bacterial strains. Further, these bacteria were given "suicide" genes that caused them to self-destruct when their job was done.

Organic chemicals are often synthesized by having catalysts act on precursor molecules or by using bacteria to carry out the synthesis. Today, it is possible to go one step further and manipulate the genes that code for these enzymes. For instance, biochemists discovered a strain of bacteria that is especially good at producing phenylalanine, an organic chemical needed to make aspartame, better known as NutraSweet. They isolated, altered, and cloned the appropriate genes so various bacteria could be genetically engineered to produce phenylalanine.

Transgenic Plants

Techniques have been developed to introduce foreign genes into immature plant embryos or into plant cells called *protoplasts* that have had their cell wall removed. It is possible to treat protoplasts with an electric current while they are suspended in a liquid containing foreign DNA. The electric current makes tiny, self-sealing holes in the plasma membrane through which the desired genetic

material can enter. Protoplasts go on to develop into mature plants containing and expressing the foreign DNA.

In a more recent method, foreign DNA is inserted into the plasmid of the bacterium *Agrobacterium,* which normally infects plant cells. The plasmid then contains recombinant DNA because it has genes from different sources—namely, those of the plasmid and also the isolated genes of interest. When a bacterium with a recombinant plasmid infects a plant, the recombinant plasmid enters the cells of the plant and can naturally incorporate into the plant's DNA.

Foreign genes transferred to strains of cotton, corn, potato, and even bananas have made these plants resistant to pests such as fungi and insects, because their cells now produce a chemical that is toxic to the pest species. Similarly, soybeans have been made resistant to a common herbicide. Some corn and cotton plants are both pest- and herbicide-resistant. A strain of rice called Golden Rice has been engineered to have a higher vitamin A content.

Another focus of genetic engineering in plants has been the development of crops with improved qualities, especially improvements that reduce waste from food spoilage. For example, by knocking out a gene that causes browning in apples, a company called Okanagan Specialty Fruits produced the Arctic Apple, a genetically modified apple with an increased shelf life (Fig. 26.5). "Innate" is a genetically modified potato in which a process called RNA interference turns off the expression of genes associated with bruising.

Some progress has also been made in increasing the food quality of crops. Soybeans have been developed that mainly produce monounsaturated fatty acid, a change that may improve human health. These altered plants also produce acids that can be used as hardeners in paints and plastics. The necessary genes were derived from *Vernonia* and castor bean seeds and were transferred into the soybean DNA.

Other types of genetically engineered plants are also expected to increase productivity. Stomata might be altered to take in more carbon dioxide or lose less water. A team of Japanese scientists is working on introducing the C_4 photosynthetic cycle into rice (see the Ecology feature "The New Rice" in Section 8.4).

Are Genetically Engineered Foods Safe?

A series of focus groups conducted by the Food and Drug Administration (FDA) in 2000 showed that although most participants believed that genetically engineered foods (also called GMOs) might offer benefits, they feared possible unknown long-term health consequences. The discovery by activists that a type of genetically engineered corn called StarLink had inadvertently made it into the food supply triggered the recall of taco shells, tortillas, and many other corn-based foodstuffs from supermarkets. Further, the makers of StarLink were forced to buy back StarLink from farmers and to compensate food producers at an estimated cost of several hundred million dollars in late 2000.

StarLink is a type of "Bt" corn. It contains a foreign gene taken from a common soil organism, *Bacillus thuringiensis,* which makes a protein that is toxic to many insect pests. About a dozen Bt varieties, including corn, potato, and even a tomato, have now been approved for human consumption. These strains contain a gene for an insecticidal protein called CryIA. Instead, StarLink contained a gene for a related protein called Cry9C, which researchers thought might slow down the chances of pest resistance to Bt corn. In order to get FDA approval for use in foods, the makers of StarLink performed the required tests. Like the other now-approved strains, StarLink wasn't poisonous to rodents, and its biochemical structure is not similar to those of most chemicals in food that commonly cause allergic reactions in humans (called allergens). But the Cry9C protein resisted digestion longer than the other Bt proteins when it was put in simulated stomach acid and subjected to heat. Because most food allergens resist

digestion in a similar fashion, StarLink was not approved for human consumption.

The scientific community is now trying to devise more tests for allergens because it has not been possible to determine conclusively whether Cry9C is or is not an allergen. Also, at this point it is unclear how resistant to digestion a protein must be in order to be an allergen, and it is also unclear what degree of amino acid sequence similarity a potential allergen must have to a known allergen to raise concern. Dean D. Metcalfe, chief of the Laboratory of Allergic Diseases at the National Institute of Allergy and Infectious Diseases, said "We need to understand thresholds for sensitization to food allergens and thresholds for elicitation of a reaction with food allergens."

Other scientists are concerned about the following potential drawbacks to the planting of Bt corn: (1) resistance among populations of the target pest, (2) exchange of genetic material between the transgenic crop and related plant species, and (3) Bt crops' impact on nontarget species. They feel that many more studies are needed before stating for certain that Bt corn has no ecological drawbacks.

Despite controversies, the planting of genetically engineered corn has increased annually. The USDA reports that U.S. farmers planted genetically engineered corn on between 80% to 89% of all corn acres in 2014, up from 26% in 2001 (Fig. 26C). Today, almost 94% of soybean acreage, and 84% to 91% of cotton acreage is genetically modified. Some groups advocate that GMOs should be labeled as such, but this may not be easy to accomplish because, for example, most cornmeal is derived from

both conventional and genetically engineered corn. So far, there has been no attempt to sort out one type of food product from the other. However, at many health food stores, foods that *do not* contain GMOs are now labeled.

Questions to Consider

1. Do you think genetically modified organisms (GMOs) should be labeled? Construct an argument both for and against labeling GMOs.

2. Some people are strongly advocating the complete removal of GMOs from the market. A few people, called ecoterrorists, are taking such drastic actions as burning crops or even setting biotechnology labs on fire. Do you think GMOs should be removed from the market? Why or why not? What further information would you need to make your decision?

3. Rice is a staple in the diet of millions of people worldwide, many of them living in less-developed countries. In some of those same countries, vitamin A deficiency is a major cause of blindness in small, malnourished children. Scientists have developed a new form of rice, called "golden rice," which is genetically modified to assist with the metabolism of vitamin A and might prevent millions of cases of blindness. The scientists who created golden rice claim it is safe for consumption, although critics say the health effects are not yet fully understood. Given its nutritional potential, should golden rice be planted and distributed on a wide scale? What information would you need to make your decision?

Figure 26C Genetically engineered crops. Genetically engineered **(a)** corn, **(b)** soybeans, and **(c)** cotton crops are increasingly being planted by today's farmers. (a): Source: USDA/Doug Wilson, photographer; (b): ©Norm Thomas/Science Source; (c): ©Pixtal/age fotostock

a.

b.

c.

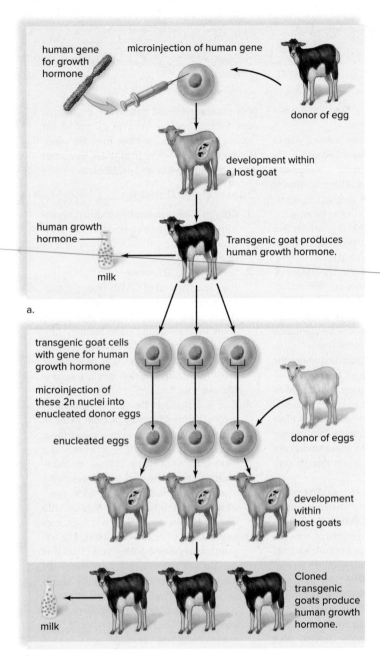

a.

b.

Figure 26.6 Production of transgenic animals. a. A genetically engineered egg develops in a host to create a transgenic goat that produces a biotechnology product in its milk. **b.** Nuclei from the transgenic goat are transferred into donor eggs, which develop into cloned transgenic goats.

Unlike C₃ plants, C₄ plants do well in hot, dry weather. These modifications would require a more complete reengineering of plant cells than the single-gene transfers that have been done so far.

However, as the Health feature "Are Genetically Engineered Foods Safe?" discusses, the public is concerned about the possible effect of genetically modified organisms, also called GMOs, on human health, as well as the environment.

Like bacteria, plants are also being engineered to produce human proteins, such as hormones, clotting factors, and antibod-

ies, in their seeds. One type of antibody made by corn can deliver radioisotopes to tumor cells, and another made by soybeans can be used to treat genital herpes. Recently, antibodies against the Ebola virus are being produced by genetically engineered tobacco plants.

Transgenic Animals

Techniques have been developed to insert genes into the eggs of animals. It is possible to microinject foreign genes into eggs by hand, but another method uses vortex mixing. The eggs are placed in an agitator with DNA and silicon-carbide needles. The needles make tiny holes in the eggs through which the DNA can enter. When these eggs are fertilized, the resulting offspring are transgenic animals. Using this technique, many types of animal eggs have acquired the gene for bovine growth hormone (BGH). The procedure has been used to produce larger fishes, cows, pigs, rabbits, and sheep.

Gene pharming, the use of transgenic farm animals to produce pharmaceuticals, is being pursued by a number of firms. Genes that code for therapeutic and diagnostic proteins are incorporated into an animal's DNA, and the proteins appear in the animal's milk. Plans are underway to produce drugs for the treatment of cystic fibrosis, cancer, blood diseases, and other disorders by this method. Figure 26.6 outlines the procedure for producing transgenic animals: DNA containing the gene of interest is injected into donor eggs. Following in vitro fertilization, the zygotes are placed in host females, where they develop. After female offspring mature, the product is secreted in their milk.

Eliminating a gene is another way to study a gene's function. A knockout mouse has had both alleles of a gene removed or made nonfunctional. For example, scientists have constructed a knockout mouse lacking the *CFTR* gene, the same gene mutated in cystic fibrosis patients. The mutant mouse has a phenotype similar to a human with cystic fibrosis and can be used to test new drugs for the treatment of the disease.

Check Your Progress 26.2

1. List some of the beneficial applications of transgenic bacteria, plants, and animals.
2. Distinguish between a transgenic organism and a cloned organism.

26.3 Gene Therapy

Learning Outcome

Upon completion of this section, you should be able to

1. Compare and contrast in vivo and ex vivo gene therapy.
2. Explain how RNA interference represents a form of gene therapy.

Once a genetic disorder is detected, gene therapy is a potential course of treatment. **Gene therapy** is the insertion of genetic

Brain
(gene transfer by injection)*
• Huntington disease
• Alzheimer disease
• Parkinson disease
• brain tumors

Skin
(gene transfer by modified
blood cells)**
• skin cancer

Lungs
(gene transfer by aerosol spray)*
• cystic fibrosis
• hereditary emphysema

Liver
(gene transfer by modified
implants)**
• familial hypercholesterolemia

Blood
(gene transfer by bone
marrow transplant)*
• sickle-cell disease

Endothelium
(blood vessel lining)
(gene transfer by
implantation of
modified implants)**
• hemophilia
• diabetes mellitus

Muscle
(gene transfer by injection)*
• Duchenne muscular dystrophy

Bone marrow
(gene transfer by implantation
of modified stem cells)**
• SCID
• sickle-cell disease

* in vivo
** ex vivo

Figure 26.7 Gene therapy. Some of the ways that in vivo and ex vivo gene therapy are being used to treat human diseases.

material into human cells for the treatment of genetic disorders and various other human illnesses, such as cardiovascular disease and cancer.

Figure 26.7 shows regions of the body that have received copies of normal genes by various methods of gene transfer. Viruses genetically modified to be safe can be used to ferry a normal gene into the body, and so can liposomes, which are microscopic globules of lipids specially prepared to enclose the normal gene (i.e., ex vivo gene therapy). On the other hand, sometimes the gene is injected directly into a particular region of the body (i.e., in vivo gene therapy).

Note that despite its promise for treating disorders, gene therapy may have detrimental side effects for some patients, such as causing leukemia or invoking an immune response. Nonetheless, some strides are being made for a number of human diseases, some of which are described in the following sections.

Ex Vivo Gene Therapy

Figure 26.8 describes an ex vivo methodology for treating children who have SCID (severe combined immunodeficiency). These children lack the enzyme ADA (adenosine deaminase), which is involved in the maturation of T and B cells. Therefore, these children are prone to constant infections and may die without treatment. To carry out gene therapy, bone marrow stem cells are removed from the bone marrow of the patient and infected with a virus that carries a normal gene for the enzyme into their DNA. Then, the cells are returned to the patient, where it is hoped they will divide to produce more blood cells with the same genes. Patients who have undergone this procedure show significantly improved immune function associated with a sustained rise in the level of ADA enzyme activity in the blood.

Another example of ex vivo gene therapy has been used to treat familial hypercholesterolemia, a condition that develops when liver cells lack a receptor protein for removing cholesterol from the blood. The high levels of blood cholesterol make the patient subject to fatal heart attacks at a young age. A small portion of the liver is surgically excised and then infected with a virus containing a normal gene for the receptor before being returned to the patient. Patients are expected to experience lowered serum cholesterol levels following this procedure.

In Vivo Gene Therapy

Cystic fibrosis patients lack a gene that codes for the transmembrane carrier of the chloride ion (for genetics of this disease, see Section 23.2). They often die due to numerous infections of the respiratory tract because a thick mucus forms in the lungs and attracts bacteria and other antigens. In gene therapy trials, the gene needed to cure cystic fibrosis is sprayed into the nose or delivered to the lower respiratory tract by an adenovirus vector or by using liposomes. So far, these treatments have met with limited success, but investigators are trying to improve uptake by using a combination of different vectors.

Gene therapy is increasingly relied upon as a part of cancer treatment. Genes are being used to make healthy cells more tolerant of chemotherapy, while making tumor cells more sensitive. Knowing that the tumor suppressor gene *p53* brings about apoptosis (cell death), researchers are interested in finding a way to selectively introduce *p53* into cancer cells, and in that way, kill them.

RNA Interference

RNA interference, or RNAi, is a procedure in which small pieces of RNA are used to "silence" the expression of specific alleles. These RNA sequences are designed to be complementary to the mRNA transcribed by a gene of interest. Once the complementary RNA sequences enter the cell, they bind with the target RNA, producing double-stranded RNA molecules. These double-stranded RNA molecules are then broken down by a series of enzymes within the cell. First discovered in worms, RNAi is believed to have evolved in eukaryotic organisms as a protection against certain types of viruses. Research into developing RNAi treatments for a number of human diseases, including cancer and hepatitis, is currently underway.

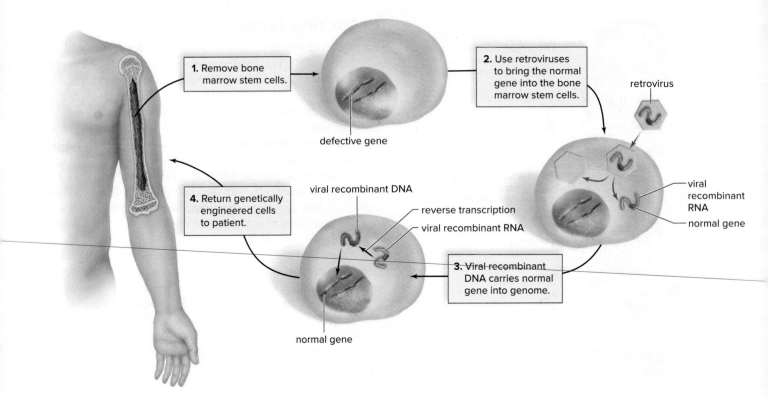

1. Remove bone marrow stem cells.

defective gene

2. Use retroviruses to bring the normal gene into the bone marrow stem cells.

retrovirus

viral recombinant RNA

normal gene

viral recombinant DNA

reverse transcription

viral recombinant RNA

4. Return genetically engineered cells to patient.

3. Viral recombinant DNA carries normal gene into genome.

normal gene

Figure 26.8 Ex vivo gene therapy in humans. Bone marrow stem cells are withdrawn from the body, an RNA retrovirus is used to insert a normal gene into them, and they are then returned to the body.

Check Your Progress 26.3

1. Summarize the methods being used to introduce genes into humans for gene therapy.
2. Explain the difference between RNA interference and in vivo gene therapy.
3. Provide examples of both ex vivo and in vivo gene therapy.

26.4 Genomics, Proteomics, and Bioinformatics

Learning Outcomes

Upon completion of this section, you should be able to

1. Discuss the implications of knowing the human genome sequence.
2. Describe a major insight from comparative genomics.
3. Compare and contrast functional genomics and proteomics.

In the preceding century, researchers discovered the structure of DNA, how DNA replicates, and how DNA and RNA are involved in the process of protein synthesis. Genetics in the twenty-first century largely concerns **genomics,** the study of the complete genetic sequences of humans and other organisms. Knowing the sequence of bases in genomes is the first step, and mapping their location on the chromosomes is the next step. The enormity of the task can be appreciated by knowing not only that we have approximately 19,000 genes that code for proteins (the actual number has yet to be determined), but also that nearly 98% of the approximately 3.2 billion bases of our genome is noncoding and contains many repetitive sequences of unidentified function. Many other organisms have an even larger number of protein-coding genes, but fewer noncoding regions when compared to the human genome.

Sequencing the Genome

We now know the sequence of the roughly 3.2 billion pairs of DNA bases in our genome. Stretched out, the DNA in each of our cells is about 5 feet long, and the nucleotides would make a book over a half million pages if printed as text. This feat, which has been compared to completion of the periodic table of the elements in chemistry, was accomplished by the **Human Genome Project (HGP),** a 13-year effort that involved both university and private laboratories around the world. How did they do it? First, investigators developed a laboratory procedure that would allow them to decipher a short sequence of base pairs, and then instruments became available that could carry out sequencing automatically. Over the 13-year span, DNA sequencers were constantly improved, and today's instruments can automatically analyze up to 120 million base pairs of DNA in a 24-hour period. So, new genomes are being sequenced all the time, and at a much faster rate than the human genome. For example, the genome of the African clawed frog, *Xenopus laevis,* which is roughly the same size as the human genome, was sequenced in under a year.

Completion of the human genome sequence has opened up great possibilities for biomedical research and treatment. These methods are widely being used to screen individuals for risk associated with specific diseases (cancer, diabetes, etc). The HGP also led to the discovery of many small regions of DNA that vary among individuals (polymorphisms). Most of these are *single nucleotide polymorphisms* (*SNPs*), meaning they have a difference of only one nucleotide. Many SNPs have no effect. Others may contribute to protein-coding differences affecting the phenotype. It's possible that certain SNP patterns change an individual's susceptibility to disease and alter his or her response to medical treatments. These discoveries have now led pharmaceutical companies to consider producing "designer drugs," which are tailored for an individual's genotype.

Determining that humans have approximately 23,000 genes required a number of techniques, many of which relied on identifying RNAs in cells and then working backward to find the DNA that can pair with that RNA. Structural genomics—knowing the sequence of the bases and how many genes we have—is now being followed by functional genomics. Most of the known human genes are expected to code for proteins. However, most of the human genome is noncoding because it does not specify the order of amino acids in a polypeptide. This noncoding DNA, once dismissed as "junk DNA," is now known to serve many important functions, including the regulation of genes that code for proteins.

Genome Architecture

Initially, researchers were somewhat surprised to discover that nearly 98% of the human genome is DNA that does not directly code for amino acid sequences. Some of the DNA that does not specify polypeptides is transcribed into ribosomal RNA and transfer RNA, both structural molecules involved in protein assembly. The rest of the genome consists of transposable elements (or transposons), repetitive DNA elements, and sequences with unknown function. Transposable elements make up approximately 44% of the human genome. Transposable elements, originally discovered by Barbara McClintock in 1950 (who in 1983 won a Nobel Prize in Physiology or Medicine for her work), are short sequences of DNA that are able to jump from one location on a chromosome to another. Their movement to a new location sometimes alters neighboring genes, particularly decreasing their expression. In other words, a transposon sometimes acts like a regulator gene. The movement of transposons throughout the genome is thought to be a driving force in the evolution of life.

Around 44% of the human genome is made up of repetitive elements, which occur when the same sequence of two or more nucleotides (e.g., CACACA) are repeated many times along the length of one or more chromosomes. Although many scientists still dismiss them as having no function, others point out that the centromeres and telomeres of chromosomes are composed of repetitive elements and, therefore, repetitive DNA elements may not be as useless as once thought. Telomeres are repetitive DNA sequences found near the ends of chromosomes and are thought to help maintain their structural stability. In addition, repetitive sequences in centromeres may perhaps help with segregating sister chromatids during cell division.

Redefining the Gene

Knowledge of the human genome sequences has changed the way researchers think about the concept of a "gene." Historically, a gene was thought of as a particular location (locus) of a chromosome.

While prokaryotes typically possess a single circular chromosome with genes that are tightly packed together, eukaryotic chromosomes are much more complex. The genes are seemingly randomly distributed along the length of a chromosome and are fragmented into exons, with intervening sequences called introns scattered throughout the length of the gene. In fact, 95% or more of most human genes is composed of introns. Recall that after transcription, introns need to be spliced out and exons joined together to form a functional mRNA transcript that will next be translated into a protein. Once regarded as merely intervening sequences, introns are now attracting attention as regulators of gene expression. The presence of introns allows exons to be put together in various sequences so that different mRNAs and proteins can result from a single gene. It could also be that introns function to regulate gene expression and help determine which genes are to be expressed and how they are to be spliced. In fact, entire genes have been found embedded within the introns of other genes.

Thus, perhaps the modern definition of a gene should take the emphasis away from the chromosome and place it on the results of transcription. Previously, molecular genetics considered a gene to be a nucleic acid sequence that codes for the sequence of amino acids in a protein. In contrast to this definition, we have known for some time that all three types of RNA (rRNA, mRNA, and tRNA) are transcribed from DNA and that these RNAs are useful products. We also know that protein-coding regions can be interrupted by regions that do not code for a protein but do produce RNAs with various functions. In light of these new findings, Mark Gerstein and associates suggested a new definition in 2007: "A gene is a genomic sequence (either DNA or RNA) directly encoding functional products, either RNA or protein." This definition takes into account three new things we have learned by investigating genomic sequences: (1) a gene product may not necessarily be a protein; (2) a gene may not be found at a particular locus on a chromosome; and (3) the genetic material need not be only DNA—some prokaryotes have RNA genes.

Functional and Comparative Genomics

In addition to the human genome, the genomes of many other organisms, including a common bacterium, a yeast, and a mouse, are also complete (Table 26.1). Because we now know the nucleotide sequence of many genomes, we can focus on comparative and functional genomics.

Using **comparative genomics,** researchers have identified many similarities between the sequence of human bases and those of other organisms. Model organisms (e.g., those found in Table 26.1) can be used in these types of genetic analyses because they share many mechanisms and cellular pathways with other organisms, including humans. For example, scientists inserted a human gene associated with Parkinson disease into the fruit fly, *Drosophila melanogaster,* and the flies showed symptoms similar to those seen in humans with the disorder. These studies confirmed

TABLE 26.1 Comparison of Sequenced Genomes

Organism	Estimated Size	Approximate Number of Genes	Chromosome Number
Homo sapiens (human)	3.2 billion bases	~19,000	46
Mus musculus (mouse)	2.5 billion bases	~20,000	40
Drosophila melanogaster (fruit fly)	180 million bases	~14,000	8
Arabidopsis thaliana (flowering plant)	120 million bases	~26,000	10
Caenorhabditis elegans (roundworm)	100 million bases	~19,000	12
Saccharomyces cerevisiae (yeast)	12.1 million bases	~6,300	32

(human): ©Image Source/Jupiterimages; (mouse): ©imageBroker/Superstock; (fruit fly): ©Hermann Eisenbeiss/Science Source; (plant): Source: USDA/Peggy Greb, photographer; (roundworm): ©Sinclair Stammers/Science Source; (yeast): ©G. Wanner/Getty Images

Comparing genomes will also help us understand the evolutionary relationships among organisms. One surprising discovery is that the genomes of all vertebrates are similar. Researchers were not surprised to find that the genomes of humans and chimpanzees were approximately 98% alike, but they did not expect to find that the human and mouse sequences were 85% similar. Genomic comparisons will likely yield improved ability to reconstruct evolutionary relationships among organisms, as discussed in Chapter 27. Comparative genomics also allows researchers to infer function of unknown genes in one species through amino acid similarity of those genes in another species.

The aim of **functional genomics** is to understand the function of the various genes discovered within each genomic sequence and how these genes interact. In fact, functional genomics has utilized comparative genomics to assess similarities between human genes and genes of other organisms to help deduce the probable function of many of our estimated 19,000 genes. Functional genomics also uses DNA microarrays to monitor the expression of thousands of genes simultaneously. The use of a microarray can tell what genes are turned on in a specific cell or tissue type in a particular organism at a particular point in time and under certain environmental circumstances. For example, we could compare gene expression of a patient in different stages of cancer growth to assist with treatment. As discussed in Section 26.3, DNA microarrays can also be used to identify various mutations in a human's genome. This is called the person's **genetic profile,** which can be used to determine if various genetic diseases are likely, as well as to suggest which drug therapy may be most appropriate based on the individual's genotype.

Proteomics

Now that entire genomes are being published for different species, there is a race to sequence their proteomes, or a species' entire collection of proteins. **Proteomics** is the study of the structure, function, and interaction of cellular proteins, which differ depending on each cell type. Each cell produces hundreds of different proteins that can vary between cells and within the same cell, depending on conditions. Therefore, the goal of proteomics is an overwhelming endeavor.

Computer modeling of the three-dimensional shape of these proteins is an important part of proteomics. The study of protein shape and function is essential to the discovery of better drugs so that their chemical structure can match that of different protein shapes. One day, it may be possible to correlate drug treatment to the particular genome of the individual to increase efficiency and decrease side effects.

Bioinformatics

Bioinformatics is the application of computer technologies, specially developed software, and statistical techniques to the study of biological information, particularly databases that contain much genomic and proteomic information (Fig. 26.9). The new data produced by structural genomics and proteomics have produced literally terabytes of raw data stored in databases that are readily available to research scientists. They are called raw data because billions of base pairs of DNA nucleotide sequence have little

that the suspected gene is involved in Parkinson disease and suggested we might be able to use fruit flies to test potential therapies.

Comparative genomics also offers a way to study changes in the genome through time because some diseases, as well as model organisms, have a shorter generation time than humans. In this way, we have been able to track the evolution of the human immunodeficiency virus (HIV), the virus that causes AIDS, in individual patients. Tracking the genome sequence of the virus through time has allowed scientists and doctors to understand how the virus responds to different drug therapy regimens and, in some cases, modify treatment to improve a patient's life span.

Figure 26.9 Bioinformatics. New computer programs are being developed to make sense out of the raw data generated by genomics and proteomics. Bioinformatics allows researchers to study both functional and comparative genomics in a meaningful way. ©Geno Zhou/FeatureChina/Newscom

meaning by themselves. Functional genomics and proteomics are dependent on computer analysis to find significant patterns in the raw data. For example, BLAST, which stands for *b*asic *l*ocal *a*lignment *s*earch *t*ool, is a computer program that can identify homologous genes among the genomic sequences of model organisms. Homologous genes are genes that code for the same proteins, although the base sequence may be slightly different. Finding these differences can help identify the putative function of genes as new organisms' genomes are sequenced, and also help to trace the history of evolution among a group of organisms. For example, researchers found the function of the protein that causes cystic fibrosis by using the computer to search for genes in model organisms that have the same sequence. Because they knew the function of this same gene in model organisms, they could deduce the function in humans. This was a necessary step toward possibly developing specific treatments for cystic fibrosis.

Bioinformatics has various applications in human genetics. The human genome has 3.2 billion known base pairs, and without the computer it would be almost impossible to make sense of these

data. For example, it is now known that an individual's genome often contains multiple copies of a gene. But individuals may differ as to the number of copies—called copy number variations. Now it seems that the number of copies of a gene in a genome can be associated with specific diseases. The computer can help make correlations between genomic differences among large numbers of people and certain diseases.

It is safe to say that without bioinformatics, our progress in assembling DNA sequences into genomes; determining the function of DNA sequences; mapping genes on chromosomes; comparing our genome to model organisms; knowing how genes and proteins interact in cells; and so forth would be extremely slow. Instead, with the help of bioinformatics, progress should proceed rapidly in these and other areas.

Check Your Progress 26.4

1. Explain the difference between genomics and proteomics.
2. Explain how comparative genomics can provide insights into gene function.
3. Discuss the importance of bioinformatics to the study of genomics and proteomics.

Conclusion

Through reading this chapter, it should be clear that advances in genomics and biotechnology will allow us to better understand the causes of human diseases, as well as to develop potential treatments and cures. One of the great challenges of the twenty-first century will be to learn how to assemble and interpret the massive quantities of data generated through genome sequencing projects, comparative genomics, and proteomics. An even greater challenge to society is to use these data and the technologies that result in an ethically responsible manner. There are many ethical dilemmas facing the use of biotechnology, and thus the need for everyone to understand the principles of the material presented in this chapter.

SUMMARIZE

26.1 DNA Technology

■ Knowledge of the **genome** has enabled scientists to conduct **genetic engineering** and several types of **cloning. Gene cloning** can be used to isolate a gene and produce many copies of it. The gene can be studied in the laboratory or inserted into a bacterium, plant, or animal, producing a **transgenic organism.** Common methods of studying DNA molecules include **recombinant DNA (rDNA)** technology and the **polymerase chain reaction (PCR).** Recombinant DNA contains DNA from two different sources. A **restriction enzyme** cleaves both

vector (often a **plasmid**) DNA and foreign DNA. The resulting "sticky ends" facilitate insertion of foreign DNA into vector DNA. The foreign gene is sealed into the vector DNA by **DNA ligase.**

■ PCR uses a heat-resistant DNA polymerase to quickly make multiple copies of a specific piece (target) of DNA. PCR is a chain reaction because the targeted DNA is replicated over and over again. Analysis of DNA segments following PCR may generate a **DNA fingerprint** (or profile). These have multiple uses, from assisting genomic research to DNA forensics studies. Increasingly, **short tandem repeat (STR)** profiling is being used to identify individuals or for forensic studies.

26.2 Biotechnology Products

- Transgenic organisms, also called genetically modified organisms (GMOs), have had a foreign gene inserted into them.
- Genetically modified bacteria, agricultural plants, and farm animals now produce biotechnology products of interest to humans, such as hormones and vaccines. Bacteria usually secrete the product, but the seeds of plants and the milk of animals contain the product.
- Transgenic bacteria have also been engineered to promote the health of plants, extract minerals, and produce medically important chemicals.
- Transgenic crops, engineered to resist herbicides and pests, are commercially available.
- Transgenic animals have been given various genes, in particular the one for bovine growth hormone (BGH). Cloning of whole animals is now possible.

26.3 Gene Therapy

- **Gene therapy,** by either ex vivo or in vivo methods, is used to correct the genotype of humans and to cure various human ills by giving the patient a foreign gene.
- During ex vivo therapy, cells are removed from the patient, treated, and returned to the patient. Ex vivo gene therapy has apparently helped children with SCID lead normal lives.
- In vivo therapy consists of directly giving the patient a foreign gene that will improve his or her health. Although it has limited success for treating cystic fibrosis, a number of in vivo therapies are being employed in the war against cancer and other human illnesses, such as cardiovascular disease.

26.4 Genomics, Proteomics, and Bioinformatics

- **Genomics** is the study of the complete genetic sequences of a species. Because of the **Human Genome Project (HGP),** researchers now know the sequence of all the base pairs of the human genome. So far, approximately 19,000 genes that code for proteins have been found; the rest of our DNA consists of noncoding regions.
- Noncoding regions include introns, transposable elements, and repetitive DNA. Noncoding DNA, once dismissed as "junk," is now thought to have important functions that may include maintaining structural integrity of chromosomes and gene regulation.
- Currently, researchers are placing an emphasis on functional and comparative genomics. **Functional genomics** aims to understand the function of protein-coding regions and noncoding regions of our genome. To that end, researchers are utilizing new tools such as DNA microarrays to generate **genetic profiles.**
- **Comparative genomics** has revealed little difference between the DNA sequence of our bases and those of many other organisms. Genome comparisons have revolutionized our understanding of evolutionary relations by revealing previously unknown relationships between organisms.
- **Proteomics** is the study of which genes are active in producing proteins in which cells and under which circumstances.
- **Bioinformatics** is the use of computers to assist with analysis of data from proteomics and functional and comparative genomics.

ASSESS

Testing Yourself

Choose the best answer for each question.

26.1 DNA Technology

1. Which of the following enzymes are needed to introduce foreign DNA into a vector?
   a. DNA gyrase and DNA ligase
   b. DNA ligase and DNA polymerase
   c. DNA gyrase and DNA polymerase
   d. restriction enzyme and DNA gyrase
   e. restriction enzyme and DNA ligase

2. In this process, a gene of interest is inserted into the chromosome of a bacterium, allowing the gene to be expressed within the cell.
   a. polymerase chain reaction
   b. DNA sequencing
   c. DNA cloning
   d. microarray technology
   e. DNA replication

3. The polymerase chain reaction
   a. uses RNA polymerase.
   b. takes place in huge bioreactors.
   c. uses a temperature-insensitive enzyme.
   d. makes lots of nonidentical copies of DNA.
   e. All of these are correct.

26.2 Biotechnology Products

4. Which of the following statements is incorrect?
   a. Bacteria usually secrete the biotechnology product into the medium.
   b. Plants are being engineered to have human proteins in their seeds.
   c. Animals are engineered to have a human protein in their milk.
   d. Animals can be cloned, but plants and bacteria cannot.

5. Bacteria are able to successfully transcribe and translate human genes because
   a. both bacteria and humans contain plasmid vectors.
   b. bacteria can replicate their DNA, but humans cannot.
   c. human and bacterial ribosomes are vastly different.
   d. the genetic code is nearly universal.

26.3 Gene Therapy

6. In this process, cells are removed from the body and a vector is used to insert a gene of interest.
   a. DNA fingerprinting
   b. in vivo gene therapy
   c. ex vivo gene therapy
   d. DNA sequencing

7. When a cloned gene is used to modify a human disease, the process is called
   a. bioremediation.
   b. gene therapy.
   c. genetic profiling.
   d. gene pharming.
   e. DNA fingerprinting.

26.4 Genomics, Proteomics, and Bioinformatics

8. This field of study examines the function and interaction of proteins within a cell or an organism.

    a. genomics   **d.** gene therapy

    b. bioengineering   **e.** None of these are correct.

    c. proteomics

9. Comparative genomics

    a. is the application of computer technologies to the study of the genome.

    b. is the study of the structure, function, and interaction of cellular proteins.

    c. can be used to understand human gene function by investigating genes in other species.

    d. involves studying all the genes that occur in a cell.

    e. is the study of a person's complete genotype, or genetic profile.

10. Bioinformatics can

    a. assist genomics and proteomics.

    b. compare our genome to that of a monkey.

    c. depend on computer technology.

    d. match up genes with proteins.

    e. All of these are correct.

ENGAGE

BioNOW

Want to know how this science is relevant to your life? Check out the BioNOW video below:

■ Glowing Fish Genetics

Are the glowing fish examples of a genetically modified organism? What DNA technologies might have been used to produce these fish?

Thinking Critically

1. We can use transgenic viruses to infect humans and help treat genetic disorders. That is, the viruses are genetically modified to contain "normal" human genes to try to replace nonfunctional human genes. Using this type of gene therapy, a person is infected with a particular virus, which then delivers the "normal" human gene to cells by infecting them. What are some pros and cons of viral gene therapy?

2. In a genomic comparison between humans and yeast, what genes would you expect to be similar?

C H A P T E R

27

Evolution of Life

©Mediscan/Alamy

CHAPTER OUTLINE

BEFORE YOU BEGIN

Before beginning this chapter, take a few moments to review the following discussions:

Sections 3.2 and 3.3 What are the differences between eukaryotic and prokaryotic cells?

Section 23.1 What is the genetic basis of inheritance?

Section 25.1 How does the structure of DNA relate to its role as the genetic material?

Evolution of Antibiotic Resistance

You have probably heard of antibiotic-resistant bacteria such as MRSA (methicillin-resistant *Staphylococcus aureus*), a concern of hospitals and medical facilities. Unfortunately, examples of antibiotic-resistant bacteria are becoming more common. One of these is *Shigella sonnei* (shown above), a bacteria that is commonly found in human feces, but when it is ingested, can cause diarrhea and intestinal problems. Globally, around 100 million people per year are infected with *Shigella*, and over 600,000 die from it. Normally, regular hand washing with soap and warm water prevents most *Shigella* infections. Historically, antibiotics have been used to treat *Shigella*, but an antibiotic-resistant strain of *Shigella* has developed that is becoming a concern of the health industry.

Use and overuse of antibiotics have resulted in evolution of resistant bacterial strains. Although we tend to think of evolution as happening over long timescales, human activities can accelerate the process of evolution quite rapidly. In fact, evolution of resistance to the antibiotic methicillin occurred in just one year! Antibiotic-resistant strains of bacteria are generally hard to treat, and treatment of infected patients can cost thousands of dollars.

Some scientists believe that "superbugs," or bacteria that have evolved antibiotic resistance, will be a far bigger threat to human health than emerging diseases such as H1N1 flu and AIDS. The good news is that our understanding of evolutionary biology has helped change human behavior to deal with superbugs. For example, doctors no longer prescribe antibiotics unless they are relatively certain a patient has a bacterial infection. Antibiotic resistance is an example of why evolution is important in people's everyday lives. In this chapter, you will learn about evidence that indicates evolution has occurred and about how the evolutionary process works.

As you read through this chapter, think about the following questions:

1. In the case of evolution of antibiotic resistance, what type of selection is operating?

2. Why would resistance to antibiotics evolve more quickly in some species than others?

27.1 Theory of Evolution

Learning Outcomes

Upon completion of this section, you should be able to

1. List two observations Darwin made on his voyage that influenced the development of his theory.
2. Identify how Darwin's and Lamarck's theories of evolution are different.
3. Summarize the four main observations that make up Darwin's theory of natural selection.

In December 1831, a new chapter in the history of biology had its humble origins. A 22-year-old naturalist, Charles Darwin, set sail on a journey of a lifetime aboard the British naval vessel HMS *Beagle* (Fig. 27.1). Darwin's primary mission on this voyage was to serve as the ship's naturalist—to collect and record the geological and biological diversity he saw during the voyage.

Initially, Darwin was a supporter of the long-held idea that species had remained unchanged since the time of creation. During the 5-year voyage of the *Beagle,* Darwin's observations challenged his own belief that species do not change over time. The idea that species adapt and change in response to their environment, what is now called **evolution,** was not a new concept in Darwin's time. What Darwin was able to format was the mechanism by which this change occurs—the concept of **natural selection.**

This new view was not readily accepted by Darwin's peers, but it gained gradual credibility as a result of a scientific and intellectual revolution that began in Europe in the late 1800s. Today, over 150 years since Darwin first published his idea of natural selection, the principle he proposed has been subjected to rigorous scientific tests, so much so that it is now considered one of the unifying theories of biology. Darwin's theory of evolution by natural selection explains both the unity and diversity of life on Earth, how all living organisms share a common ancestor, and how species adapt to various habitats and ways of life.

Although many have believed that Darwin forged this change in worldview by himself, several biologists during the preceding century and some of Darwin's contemporaries had a large influence

Figure 27.1 Voyage of the HMS *Beagle*. The map shows the journey of the HMS *Beagle* around the world. As Darwin traveled along the east coast of South America, he noted that a bird called a rhea looked like the African ostrich. On the Galápagos Islands, marine iguanas, found no other place on Earth, use their large claws to cling to rocks and their blunt snouts for eating marine algae. (photos) (Darwin): Source: National Library of Medicine; (rhea): ©Nicole Duplaix/National Geographic/Getty Images; (HMS *Beagle*): ©Mary Evans Picture Library/The Image Works; (iguana): ©FAN Travelstock/Alamy

Charles Darwin

HMS *Beagle*

marine iguana

Galápagos Islands

rhea

on Darwin as he developed his theory. The European scientists of the eighteenth and nineteenth centuries were keenly interested in understanding the nature of biological diversity. This was a time of exploration and discovery as the natural history of new lands was mapped and documented. Shipments of new plants and animals from newly explored regions were arriving in England to be identified and described by biologists—it was a time of rapid expansion of our understanding of the Earth's biological diversity. In this atmosphere of discovery, Darwin's theory first took root and grew.

One scientist that influenced Darwin as he developed his theory was Jean-Baptiste de Lamarck. Lamarck, a predecessor of Darwin, proposed one of the first hypotheses to explain how species evolve and adapt to the environment. Lamarck developed the theory of **inheritance of acquired characteristics**—that the environment can bring about inherited change. One example he gave, and the one for which he is most famous, is that the long neck of a giraffe developed over time because animals stretched their necks to reach food high in trees and then passed gradually longer necks to their offspring (Fig. 27.2*a*). This hypothesis for the inheritance of acquired characteristics has never been verified. The reason is due to the molecular mechanism of inheritance. Changes to an organism's visible characteristics, or *phenotype* (see Section 23.1), acquired during an organism's lifetime do not result in genetic changes that are heritable. As an example, consider tail cropping in Doberman pinschers. All Doberman puppies are born with tails, even though their parents' tails are most often cropped. That is, tail cropping is a phenotypic change that is not inherited in the DNA.

While Lamarck's inheritance of acquired characteristics does not adequately explain the overall process of evolution, his work was important in advancing the discussion about evolutionary change over time. Recent findings in genetics also suggest that **epigenetic inheritance,** where chemical modifications to the DNA that occur in one generation may be passed on to later generations, actually represents a form of Lamarckian inheritance.

In contrast to Lamarck, Darwin proposed that traits that provided an advantage must be passed on to the next generation—that is, must be heritable for adaptation to occur (Fig. 27.2*b*). In the example of the giraffe, Darwin's theory would predict that giraffe populations, which have variation in the length of necks, would contain some individuals with longer necks than others. Those with the longer necks would be able to reach more leaves, acquire more energy, and potentially produce more offspring. These offspring would inherit the long neck trait from their parents. Over time, the longer neck trait would become more common in a population, resulting in an adaptation to the environment.

Darwin called this process of adaptation evolution by natural selection. Darwin's theory developed over a long period of time. During his voyage on the HMS *Beagle,* he made many observations that shaped the development of his theory of natural selection. Darwin observed that some animals, such as the African ostrich and the South American rhea, looked very similar, but could not be related because they lived on different continents. He reasoned that the similarity between these two large, flightless

Figure 27.2 A comparison of Lamarck's and Darwin's theories of evolution.   a. Jean-Baptiste de Lamarck's proposal of the inheritance of acquired characteristics. **b.** Charles Darwin's ideas developed into the theory of natural selection.

birds was due to adaptation to similar environments. While visiting the Galápagos Islands, Darwin observed that the finches on each island were similar in overall appearance, but had differences in beak size and shape (Fig. 27.3). To explain these observations, Darwin speculated whether these different species of finches could have descended from a mainland finch species. In other words, he wondered if a finch from South America was the common ancestor to all the types on the Galápagos Islands. Perhaps new species had arisen because the geographic distance between the islands isolated populations of birds long enough for them to evolve independently. And perhaps the present-day species had resulted from accumulated changes that occurred within each of these isolated populations.

With this and other types of evidence, Darwin returned home from his voyage and spent the next 20 years developing and

a. Large, ground-dwelling finch   b. Warbler-finch   c. Cactus-finch

Figure 27.3 Three Galápagos finches. Each of the 13 species of finches has a beak adapted to a particular way of life. For example: (**a**) the heavy beak of the large ground-dwelling finch (*Geospiza magnirostris*) is suited to a diet of large seeds; (**b**) the beak of the warbler-finch (*Certhidea olivacea*) is suited to feeding on insects found among ground vegetation or caught in the air; and (**c**) the longer beak, somewhat decurved, and the split tongue of the cactus-finch (*Cactornis scandens*) are suited to extracting the flesh of cactus fruit. (a): ©Miguel Castro/Science Source; (b): ©Celia Mannings/Alamy; (c): ©Michael Stubblefield/Alamy

fine-tuning his theory of natural selection. His theory can be summarized as follows (see Fig. 27.2*b*):

- *Individual organisms within a species exhibit variation that can be passed from one generation to the next—that is, they have heritable variation.* Darwin emphasized that the members of a population vary in their functional, physical, and behavioral traits. He emphasized that variation, abundant in natural populations, must be heritable for the process of natural selection to operate. We now know that genes, along with the environment, determine the phenotype of individuals and trait variation within populations.

- *Organisms compete for available resources.* Darwin read an essay by Thomas Malthus in which he proposed that human population growth is kept in check because an environment can support only a limited number of people. Darwin realized that if all offspring born to a population were to survive, there would not be sufficient resources to support it indefinitely. He calculated the reproductive potential of elephants, assuming an average life span of 100 years and that each female bears no fewer than six young over her lifetime. If all these young were to survive and reproduce, then after only 750 years, each breeding pair would have 19 million descendants! Obviously, no environment has the resources to support an elephant population of this magnitude, and no such elephant population has ever existed.

- *Individual organisms within a population differ in terms of their reproductive success.* Darwin observed that there are some individuals in a population that have favorable traits that enable them to better compete for limited resources, and thus devote more energy to reproduction. Darwin called this ability to have more offspring *differential reproductive success*. Natural selection occurs because certain members of a population happen to have an advantageous trait that allows them to survive and reproduce to a greater extent than do other members. For example, a mutation in a wild dog that increases its sense of smell may help it find prey.

- *Organisms become adapted to conditions as the environment changes.* Darwin's theory proposes that natural selection

through differential reproductive success shapes traits, inherited from predecessors, in response to the environment. The result of this *descent with modification* is adaptation. An **adaptation** is any evolved trait that helps an organism be more suited to its environment. Adaptations are especially recognizable when unrelated organisms living in a particular environment display similar characteristics. For example, manatees, penguins, and sea turtles all have flippers, which help them move through the water.

Such adaptations to specific environments result from natural selection. Over time, selection can cause adaptive traits to be increasingly represented in the population. Evolution includes other processes in addition to natural selection (see Sections 27.3 and 27.4), but natural selection is the only process that results in adaptation to the environment.

> **Check Your Progress 27.1**
>
> **1.** Describe the four critical elements of Darwin's theory.
> **2.** Explain why Lamarck's "inheritance of acquired characteristics" does not explain evolutionary change.
> **3.** Summarize the process of natural selection.

27.2 Evidence of Evolution

> **Learning Outcomes**
>
> Upon completion of this section, you should be able to
> **1.** Define *biological evolution*.
> **2.** Describe, with examples, four lines of evidence for evolution.

Biological evolution, or simply evolution, is all the changes that have occurred, due to differential reproductive success, in living organisms over geological time. As we discussed in Section 27.1, differential reproductive success indicates that some individuals reproduce more than others because they are better suited to their environment. The Earth is about 4.6 billion years old, and prokaryotes, probably the first living organisms, evolved about 3.5 billion years ago. The eukaryotic cell arose about 2.1 billion years ago, but multicellularity did not begin until perhaps 700 million years ago. This means that only single-celled organisms were present for 80% of the time that life has existed on Earth. Consequently, most evolutionary events we will be discussing in the next few chapters occurred in less than 20% of the history of life!

Because of descent with modification, all living organisms share the same fundamental characteristics: They are made of cells, take chemicals and energy from the environment, respond to external stimuli, and reproduce (see Section 1.1). Life is diverse because living organisms are adapted to different environments and the features that enable them to survive in those environments vary tremendously.

Many fields of biology provide evidence that evolution through descent with modification occurred in the past and is still occurring. Let us look at the various types of evidence for evolution.

Fossil Evidence

Fossils are the remains and traces of past life or any other direct evidence of past life. Most fossils consist only of hard parts of organisms, such as shells, bones, or teeth, because these are usually preserved after death. The soft parts of a dead organism are often consumed by scavengers or decomposed by bacteria.

Occasionally, however, an organism is buried quickly and in such a way that decomposition is never completed or is completed so slowly that the soft parts leave an imprint of their structure. Traces include trails, footprints, burrows, worm casts, or even preserved droppings.

The great majority of fossils are found embedded in sedimentary rock. Sedimentation, a process that has been going on since Earth formed, can take place on land or in bodies of water. The weathering and erosion of rocks produces particles that vary in size and are called sediment. As such particles accumulate, the sediments form strata (*sing.*, stratum), or recognizable layers of rock. Any given stratum is older than the one above it and younger than the one immediately below it, so that the relative age of fossils can be determined based on their depth.

Paleontologists are biologists who study the fossil record and from it draw conclusions about the history of life. When fossils are arranged from oldest to youngest, they can provide evidence of evolutionary change through time. A particularly good example of this is the horse, which evolved from a dog-sized, forest-dwelling tree browser with forward-looking eyes and teeth geared toward chewing leaves, to the animal we recognize as a horse today. Modern horses are adapted for an open field–type habitat, with eye sockets more to the sides of their head to allow for better peripheral vision and thereby detection of possible predators, as well as teeth geared toward grinding grasses.

Particularly interesting are the fossils that serve as **transitional links** between groups. A famous example are the fossils of *Archaeopteryx* that lived about 165 million years ago (Fig. 27.4). The fossil clearly seems to be an intermediate form between dinosaurs and birds. *Archaeopteryx* had dinosaur-like features including jaws with teeth and a long, jointed tail, but it also had feathers and wings similar to those of modern birds. The fossils of *Archaeopteryx* are similar to other transitional fossils in that they have some traits like their ancestors and others like their descendants, rather than expressing intermediate traits.

Fossils have been discovered that support the hypothesis that whales had terrestrial ancestors. *Ambulocetus natans* (meaning the walking whale that swims) was the size of a large sea lion, with broad, webbed feet on its forelimbs and hindlimbs that enabled it to both walk and swim. It also had tiny hooves on its toes and the primitive skull and teeth of early whales (Fig. 27.5). Modern whales still have remnants of a hindlimb consisting of only a few bones that are very reduced in size. As the ancestors of whales adopted an increasingly aquatic lifestyle, the location of the nasal opening underwent a transition, from the tip of the snout as in *Ambulocetus,* to midway between the tip of the snout and the skull in *Basilosaurus,* to the very top of the head in modern whales (Fig. 27.5). Other transitional links among fossil vertebrates suggest that fishes evolved before amphibians, which evolved before reptiles, which evolved before mammals in the history of life.

Figure 27.4 Transitional fossils. **a.** A fossil of *Archaeopteryx,* now considered the first bird, has features of both birds and dinosaurs. Fossils indicate it had feathers and wing claws. Most likely, it was a poor flier. Perhaps it ran over the ground on strong legs and climbed into trees with the assistance of these claws. **b.** *Archaeopteryx* also had a feather-covered, reptilelike tail that shows up well in this artist's representation.
(a): ©Jason Edwards/Getty Images; (b): ©Joe Tucciarone

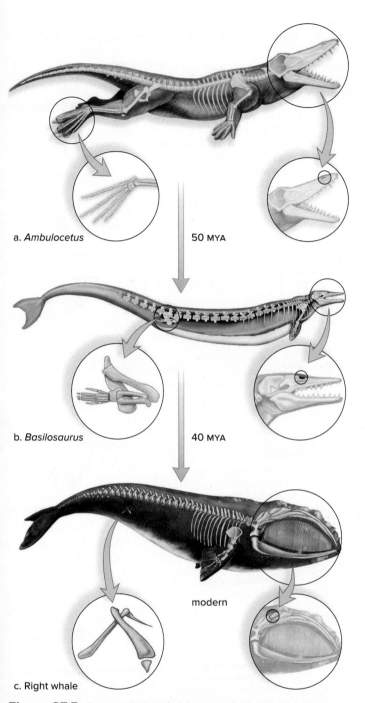

a. *Ambulocetus*

50 MYA

b. *Basilosaurus*

40 MYA

modern

c. Right whale

Figure 27.5 Anatomical transitions during the evolution of whales. Fossil evidence suggests that modern whales evolved from terrestrial ancestors that walked on four limbs. Transitional fossils show a gradual reduction in the hindlimb and a movement of the nasal openings from the tip of the nose to the top of the head—both are adaptations to living in water. **a.** *Ambulocetus,* an ancestor to whales (50 million years ago [MYA]), had a distinct hindlimb and a nose at the tip of its snout. **b.** *Basilosaurus,* a more recent ancestor of whales (40 MYA), had a greatly reduced hindlimb and a nasal opening mid-snout. **c.** The modern Right whale has a vestigial hindlimb that is inside the body. Its nasal opening is far back on the top of the head, from which it breathes as it surfaces from under the water.

Geological Timescale

As a result of studying strata, scientists have divided Earth's history into eras, and then periods and epochs (Table 27.1). The fossil record has helped determine the dates given in the table. There are two ways to age fossils. The relative dating method determines the relative order of fossils and strata depending on the layer of rock in which they were found, but it does not determine the actual date they were formed.

The absolute dating method relies on radioactive dating techniques to assign an actual date to a fossil. All radioactive isotopes have a particular *half-life,* the length of time it takes for half of the radioactive isotope to change into another stable element. Carbon 14 (^{14}C) is the only radioactive isotope in organic matter. Assuming a fossil contains organic matter, half of the ^{14}C will have changed to nitrogen 14 (^{14}N) in 5,730 years. For example, if a fossil has one-fourth the amount of radioactive ^{14}C as a modern sample, then the fossil is approximately 11,460 years old (two half-lives).

Biogeographical Evidence

The field of **biogeography** is the study of the range and distribution of plants and animals in different places throughout the world. Such distributions are consistent with the hypothesis that, when forms are related, they evolved in one locale and then spread to accessible regions. Therefore, a different mix of plants and animals would be expected whenever geography separates continents, islands, seas, and so on. Darwin's observations of biogeography while on the HMS *Beagle* helped to form his theory of evolution. One observation in particular was that of the variation among the finches on the Galápagos Islands (see Section 27.1).

Many of the barriers between landmasses arose through a process called **continental drift.** That is, the position of the continents has never been fixed. Rather, their positions and the positions of the oceans have changed over time (Fig. 27.6). During the Permian period, all the present landmasses belonged to one continent and then later drifted apart. As evidence of this, fossils of one species of seed fern (*Glossopteris*) have been found on all the southern continents separated by oceans. This species' presence on Antarctica is evidence that this continent was not always frozen. In contrast, many Australian species are restricted to that continent, including the majority of marsupials (pouched mammals such as the kangaroo). What is the explanation for these distributions? Some organisms must have evolved and spread out before the continents broke up. Then they became extinct.

The world's six major biogeographical regions each have their own distinctive mix of living organisms. Darwin noted that South America lacked rabbits, even though the environment was quite suitable to them. He concluded there are no rabbits in South America because rabbits evolved somewhere else and had no means of reaching South America. To take another example, both cacti and spurges (*Euphorbia*) are plants adapted to a hot, dry environment—both are succulent, spiny, flowering plants. Why do cacti grow in the American deserts, while most *Euphorbia* grow in African deserts, when each would do well on the other continent? It seems obvious that they just happened to evolve on their respective continents. What is the best explanation for this phenomenon? Different mammals and flowering plants evolved separately in

TABLE 27.1 The Geological Timescale: Major Divisions of Geological Time and Some of the Major Evolutionary Events of Each Time Period

Era	Period	Epoch	Million Years Ago (MYA)	Plant Life	Animal Life	
Cenozoic	Quaternary	Holocene	current	Humans influence plant life.	Age of *Homo sapiens*	
			Significant Extinction Event Underway			
	Quaternary	Pleistocene	0.01	Herbaceous plants spread and diversify.	Presence of ice age mammals. Modern humans appear.	
	Neogene	Pliocene	2.6	Herbaceous angiosperms flourish.	First hominids appear.	
	Neogene	Miocene	5.3	Grasslands spread as forests contract.	Apelike mammals and grazing mammals flourish; insects flourish.	
	Neogene	Oligocene	23.0	Many modern families of flowering plants evolve; appearance of grasses.	Browsing mammals and monkeylike primates appear.	
	Paleogene	Eocene	33.9	Subtropical forests with heavy rainfall thrive.	All modern orders of mammals are represented.	
	Paleogene	Paleocene	55.8	Flowering plants continue to diversify.	Ancestral primates, herbivores, carnivores, and insectivores appear.	
			Mass Extinction: 50% of all species, dinosaurs and most reptiles			
Mesozoic	Cretaceous		65.5	Flowering plants spread; conifers persist.	Placental mammals appear; modern insect groups appear.	
	Jurassic		145.5	Flowering plants appear.	Dinosaurs flourish; birds appear.	
			Mass Extinction: 48% of all species, including corals and ferns			
	Triassic		199.6	Forests of conifers and cycads dominate.	First mammals appear; first dinosaurs appear; corals and molluscs dominate seas.	
			Mass Extinction ("The Great Dying"): 83% of all species on land and sea			
Paleozoic	Permian		251.0	Gymnosperms diversify.	Reptiles diversify; amphibians decline.	
	Carboniferous		299.0	Age of great coal-forming forests: ferns, club mosses, and horsetails flourish.	Amphibians diversify; first reptiles appear; first great radiation of insects.	
			Mass Extinction: Over 50% of coastal marine species, corals			
	Devonian		359.2	First seed plants appear. Seedless vascular plants diversify.	First insects and first amphibians appear on land.	
	Silurian		416.0	Seedless vascular plants appear.	Jawed fishes diversify and dominate the seas.	
			Mass Extinction: Over 57% of marine species			
	Ordovician		443.7	Nonvascular land plants appear.	Invertebrates spread and diversify; first jawless and then jawed fishes appear.	
	Cambrian		488.3	Marine algae flourish.	All invertebrate phyla present; first chordates appear.	
				630		First soft-bodied invertebrates evolve.	
			1,000	Protists diversify.		
			2,100	First eukaryotic cells evolve.		
			2,700	O_2 accumulates in atmosphere.		
			3,500	First prokaryotic cells evolve.		
			4,570	Earth forms.		

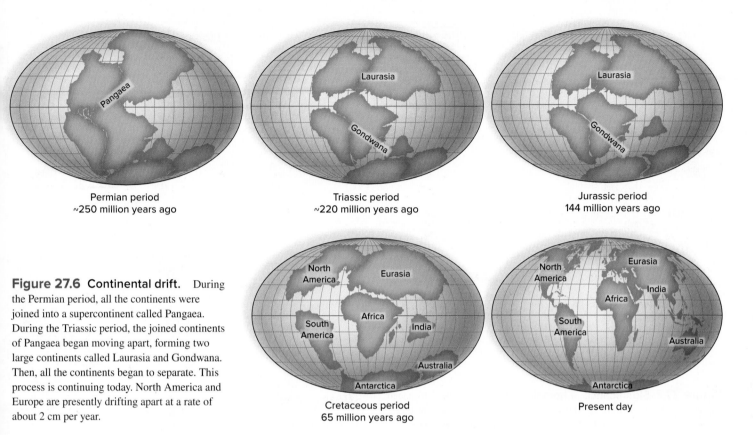

Figure 27.6 Continental drift. During the Permian period, all the continents were joined into a supercontinent called Pangaea. During the Triassic period, the joined continents of Pangaea began moving apart, forming two large continents called Laurasia and Gondwana. Then, all the continents began to separate. This process is continuing today. North America and Europe are presently drifting apart at a rate of about 2 cm per year.

each biogeographical region, and barriers such as mountain ranges and oceans prevented them from migrating to other regions.

Mass Extinctions

Extinction is the death of every member of a species. During mass extinctions, a large percentage of species become extinct within a relatively short period of time. So far, there have been five major mass extinctions. These occurred at the ends of the Ordovician, Devonian, Permian, Triassic, and Cretaceous periods (see Table 27.1), and a sixth is likely occurring now, probably as a result of human activities (discussed in Section 37.3). Following mass extinctions, the remaining groups of organisms are likely to spread out and fill the habitats vacated by those that have become extinct.

Evidence suggests that the Cretaceous extinction (or "Cretaceous crisis") was due to an asteroid that exploded, producing meteorites that fell to Earth. A large meteorite striking Earth could have produced a cloud of dust that mushroomed into the atmosphere, blocking out the sun and causing plants to freeze and die. In support of this hypothesis, a huge crater that could have been caused by a meteorite involved in the Cretaceous extinction was found in the Caribbean–Gulf of Mexico region on the Yucatán Peninsula. During the Cretaceous period, great herds of dinosaurs roamed the plains, but all dinosaur species went extinct near the end of the Cretaceous period.

Certainly, continental drift contributed to the Ordovician extinction. This extinction occurred after Gondwana arrived at the South Pole. Immense glaciers, which drew water from the oceans, chilled even once-tropical land. Marine invertebrates and coral reefs, which were especially hard hit, didn't recover until Gondwana drifted away from the pole and warmth returned. The mass extinction at the end of the Devonian period saw an end to 70% of marine invertebrates. Other scientists believe that this mass extinction could have been due to movement of Gondwana back to the South Pole.

Anatomical Evidence

The concept of common descent offers a plausible explanation for anatomical similarities among organisms. Vertebrate forelimbs are used for flight (birds and bats), orientation during swimming (whales and seals), running (horses), climbing (arboreal lizards), or swinging from tree branches (monkeys). Yet all vertebrate forelimbs contain the same sets of bones organized in similar ways, despite their dissimilar functions. The most plausible explanation for this unity is that the basic forelimb plan belonged to a common ancestor, and then the plan was modified in the succeeding groups as each continued along its own evolutionary pathway. Structures that are anatomically similar because they are inherited from a common ancestor are called **homologous structures** (Fig. 27.7). In contrast, **analogous structures** serve the same function, but are not constructed similarly, nor do they share a common ancestry. The wings of birds and insects and the eyes of octopi and humans are analogous structures and are similar due to a common environment, not common ancestry. The presence of homology, not analogy, is evidence that organisms are related.

Vestigial structures are anatomical features that are fully developed in one group of organisms but that are reduced and may

Figure 27.7 Significance of homologous structures. Although the specific design details of vertebrate forelimbs are different, the same bones are present (note color-coding). Homologous structures provide evidence of a common ancestor.

have no function in similar groups. Most birds, for example, have well-developed wings for flight. However, some bird species (e.g., ostrich) have greatly reduced wings and do not fly. Similarly, snakes have no use for hindlimbs, and yet some have remnants of hindlimbs in a pelvic girdle and legs. The presence of vestigial structures can be explained by common descent. Vestigial structures occur because organisms inherit their anatomy from their ancestors. They are traces of an organism's evolutionary history.

The homology shared by vertebrates extends to their embryological development (Fig. 27.8). At some time during development, all vertebrates have a postanal tail and exhibit paired pharyngeal pouches. In fishes and amphibian larvae, these pouches develop into functioning gills. In humans, the first pair of pouches becomes the cavity of the middle ear and the auditory tube. The second pair becomes the tonsils, while the third and fourth pairs become the thymus and parathyroid glands. Why should terrestrial vertebrates develop and then modify structures like pharyngeal pouches that have lost their original function? The most likely explanation is that fishes are ancestral to other vertebrate groups.

In 1859, Charles Darwin speculated that whales evolved from a land mammal. His hypothesis has now been substantiated. In recent years, the fossil record has yielded an incredible parade of fossils that link modern whales and dolphins to land ancestors. The presence of a vestigial pelvic girdle and legs in modern whales is also significant evidence that ancestors of these creatures once walked on land. However, that these structures are severely reduced in modern whales is consistent with the fact that they are exclusively aquatic today (see Fig. 27.5).

Figure 27.8 Significance of developmental similarities. At these comparable developmental stages, vertebrate embryos have many features in common, which suggests they evolved from a common ancestor. (These embryos are not drawn to scale.)

Biochemical Evidence

Almost all living organisms use the same basic biochemical molecules, including DNA, ATP (adenosine triphosphate), and many identical or nearly identical enzymes. Further, organisms use the same DNA triplet code for the same 20 amino acids in their proteins. Because the sequences of DNA bases in the genomes of many organisms are now known, it has become clear that humans share a large number of genes with much simpler organisms. It appears that life's vast diversity has come about by only a slight difference in many of the same genes and regulatory genes often found in introns and other regions of the genome. The result has been widely divergent types of bodies.

When the degree of similarity in DNA nucleotide sequences or the degree of similarity in amino acid sequences of proteins is

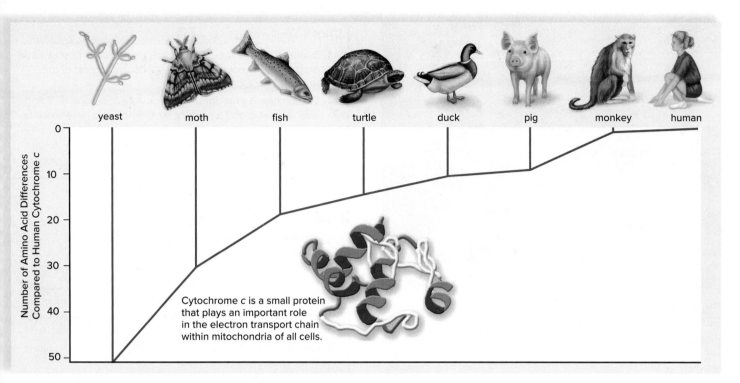

Figure 27.9 **Significance of biochemical differences.**   The branch points in this diagram indicate the number of amino acids that differ between human cytochrome *c* and the organisms depicted. These biochemical data are consistent with those provided by a study of the fossil record and comparative anatomy.

examined, the more similar the DNA sequences are, generally the more closely related the organisms are. For example, humans and chimpanzees are about 97% similar! Cytochrome *c* is a molecule used in the electron transport chain of all the organisms appearing in Figure 27.9. Data regarding differences in the amino acid sequence of cytochrome *c* show that the sequence in a human differs from that in a monkey by only one amino acid, from that in a duck by 11 amino acids, and from that in a yeast by 51 amino acids. These data are consistent with other data regarding the anatomical similarities of these organisms and, therefore, how closely they are related. In addition, we now know that chromosomal changes, such as translocations and inversions (see Section 24.4), play an important role in evolutionary change.

Evolution is no longer considered a hypothesis. It is one of the great unifying theories of biology. In science, the word *theory* is reserved for those conceptual schemes supported by a large number of observations and scientific experiments (see Section 1.3). The theory of evolution has the same status in biology that the theory of gravity has in physics.

We Can Observe Selection at Work

Table 27.1 outlines major events in the evolution of life over a timescale of millions and billions of years. But evolution does not necessarily happen over long periods of time. Researchers have recorded the evolution of traits in natural populations over decades. Over the course of history, humans have used the power of evolution to shape traits of livestock, agricultural plants, and even our animal companions, over relatively short periods of time.

Humans as Agents of Evolution

Darwin noted that humans could artificially modify desired traits in plants and animals by selecting to breed individuals with preferred traits. For example, the diversity of domestic dogs has resulted from prehistoric humans selectively breeding wolves with particular traits, such as hair length, height, and guarding behavior. This type of human-controlled breeding to increase the frequency of desired traits is called **artificial selection.** Artificial selection, like natural selection, is possible only because the original population exhibits a variety of characteristics, allowing humans to select traits they prefer. For dogs, the result of artificial selection is the existence of many breeds of dogs all descended from the wolf (Fig. 27.10). Darwin surmised that if humans could create such a wide variety of organisms by artificial selection, then natural selection could also produce diversity. But in this case, the environment, and not humans, is the force selecting for particular traits.

Evolution in Natural Populations

The Galápagos finches have beaks adapted to the food they eat, with different species of finches on each island (see Fig. 27.3). Today, many investigators, including Peter and Rosemary Grant of Princeton University, are documenting natural selection as it occurs on the Galápagos Islands. In 1973, the Grants began a study of the various finches on Daphne Major, an island near the center of the Galápagos Islands. The rainfall on this island varies widely from wet years to dry years. The Grants measured the beaks of each finch on the island and found that the beak depth of the medium ground finch, *Geospiza fortis,* changed between wet and dry years (Fig. 27.11). Medium ground finches like to eat small, tender

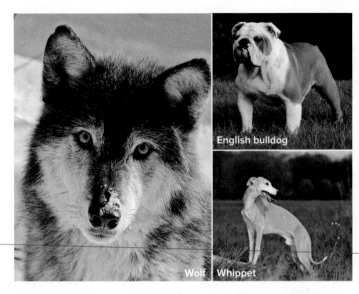

Figure 27.10 Artificial selection. All dogs, *Canis lupus familiaris,* are descended from the gray wolf, *Canis lupus,* which humans started to domesticate about 14,000 years ago. The process of selective breeding by humans has led to extreme differences among breeds. (wolf): ©Nicholas James McCollum/Moment Open/Getty Images; (bulldog): ©Willee Cole/Getty Images; (whippet): ©Daniela Jakob/123RF

seeds, but when the weather turns dry, they must eat larger, drier seeds that are harder to crush. During dry years, the majority of finches died from starvation because their beaks were not well equipped to feed on the harder seeds.

But the few finches with deeper beaks were better able to crush larger, harder seeds, and these finches survived to reproduce. Therefore, among the next generation of *G. fortis* birds, the mean, or average, beak was deeper than the previous generation. The Grants' research demonstrates that evolutionary change can sometimes be observed within the time frame of a human life span.

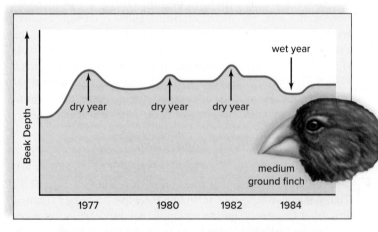

Figure 27.11 Evolution in action. The average beak depth of ground finches varies from generation to generation, according to the amount of rainfall. The amount of rainfall affects the hardness and size of seeds on the islands, and different beak depths were better suited to eating different types of seeds. Average beak features were observed to change many times over a period of a decade. This is one way in which evolution by natural selection has been observable over short periods of time.

27.3 Microevolution

Many traits can change temporarily in response to a varying environment. For example, the color change in the fur of an Arctic fox from brown to white in winter, the increased thickness of your dog's fur in cold weather, or the bronzing of your skin when exposed to the sun last only for a season.

These are not evolutionary changes. Changes to traits over an individual's lifetime are not evidence that an individual has evolved, because these traits are not heritable (see Fig. 27.2). In order for traits to evolve, they must have the ability to be passed on to subsequent generations. Evolution is about change in a heritable trait within a population, not within an individual, over many generations.

Darwin observed that populations, not individuals, evolve, but he could not explain how traits change over time. Now we know that genes interact with the environment to determine traits. Because genes and traits are linked, evolution is really about genetic change—or more specifically, evolution is the *change in allele frequencies in a population over time.* This type of evolution is called **microevolution.** In this chapter, we use the example of the peppered moth to examine how populations evolve.

Microevolution in the Peppered Moth

Population genetics, as its name implies, is the field of biology that studies the diversity of populations at the level of the gene. A **population** is all the members of a single species that occupy a particular area at the same time and that interbreed and exchange genes. Population geneticists are interested in how genetic diversity in populations changes over generations, and in the forces that cause populations to evolve. Population geneticists study microevolution by measuring the diversity of a population in terms of allele and genotype frequencies over generations.

You may recall from Section 23.1 that diploid organisms, such as moths, carry two copies of each chromosome, with one copy of each gene on each chromosome. A single gene can come in many forms, or alleles, that encode variations of a single trait. In the peppered moth, a single gene for body color has two alleles, *D* (dark color) and *d* (light color), with *D* dominant to *d* (Fig. 27.12).

We know that with only two alleles there are three possible genotypes (the combinations of alleles in an individual) for the color gene in the peppered moth: *DD* (homozygous dominant), *Dd* (heterozygous), or *dd* (homozygous recessive). *DD* or *Dd* genotypes produce dark moths, and the *dd* genotype produces light moths (Fig. 27.12).

Allele Frequencies

Suppose a population geneticist collected a population of 25 peppered moths, some dark and some light, from a forest outside London (Fig. 27.13). In this population, you would expect to find a mixture of *D* and *d* alleles in the **gene pool**—the alleles of all genes in all individuals in a population. The population geneticist ran tests to determine the alleles present in each moth. Of the 50 alleles in the peppered moth gene pool (2 alleles × 25 moths), 10 were *D* and 40 were *d*. Thus the frequency of the *D* and the *d*

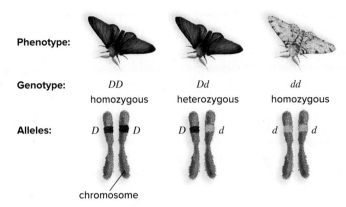

Figure 27.12 The genetic basis of body color in the peppered moth.   Light or dark body color in the peppered moth is determined by a gene with two alleles: *D* and *d*. Genotypes *DD* and *Dd* produce dark body color, and *dd* produces light body color. The *D* and *d* alleles are variants of a gene at a particular locus on a chromosome.

Figure 27.13 How Hardy-Weinberg equilibrium is estimated.   **a.** A population of 25 moths contains a gene pool of *D* and *d* alleles. **b.** The frequencies of the *D* and *d* alleles can be estimated from the gene pool. Under Hardy-Weinberg equilibrium, the frequencies of *D* and *d* alleles should produce in the next generation a predictable frequency of genotypes that can be calculated with the Hardy-Weinberg principle

TABLE 27.2   Hardy-Weinberg Proportions Can Be Used to Determine If Evolution Has Occurred

Hardy-Weinberg Condition	Deviation from Condition	Effect of Deviation on Population	Expected Deviation from HWE	Evolution Occurred?
Random mating	Nonrandom mating	Allelles do not assort randomly	Change in genotype frequencies	No
No selection	Selection	Certain alleles are selected for or against	Change in allele frequencies	Yes
No mutation	Mutation	Addition of new alleles	Change in allele frequencies	Yes
No migration	Immigration or emigration	Individuals carry alleles into, or out of, the population	Change in allele frequencies	Yes
Large population (no genetic drift)	Small population (genetic drift) 1. bottleneck effect 2. founder effect	Loss of allele diversity; some allelles may disappear	Change in allele frequencies	Yes

alleles would be 10/50, or 0.20, and 40/50, or 0.80. The allele frequency, as illustrated in this example, is the proportion of each allele in a population's gene pool (Fig. 27.13).

Notice that the frequencies of D and d add up to 1. This relationship is true of the sum of allele frequencies in a population for any gene of any diploid organism. This relationship is described by the expression $p + q = 1$, where p is the frequency of one allele, in this case D, and q is the frequency of the other allele, d (Fig. 27.13).

For the next three seasons, samples of 25 moths were collected from the same forest, and the allele frequencies were always the same: 0.20 D and 0.80 d. Because allele frequencies in this population did not change over generations, we could conclude that this population has not evolved.

Hardy-Weinberg Equilibrium

A population in which allele frequencies do not change over time, such as in the moth population just described, is said to be in **genetic equilibrium**, or **Hardy-Weinberg equilibrium**—a stable, nonevolving state. Hardy-Weinberg equilibrium is derived from the work of British mathematician Godfrey H. Hardy and German physician Wilhelm Weinberg, who in 1908 developed a mathematical model to estimate genotype frequencies of a population that is in genetic equilibrium. Their mathematical model proposes that the genotype frequencies of a nonevolving population can be described by the equation $p^2 + 2pq + q^2$, again with p and q representing the frequency of alleles D and d. Recall that in our moth population, D = p and d = q, so that D^2 is the frequency of the DD genotype, 2Dd is the frequency of the Dd genotype, and d^2 is the frequency of the dd genotype (Fig. 27.13).

Conditions of Hardy-Weinberg Equilibrium

The principles of the Hardy-Weinberg equation indicate that allele frequencies in a gene pool will remain at equilibrium, and thus constant, after one generation of random mating in a large, sexually reproducing population as long as five conditions are met:

1. *No mutations.* Genetic mutations are an alteration in an allele, due to a change in DNA composition. Under Hardy-Weinberg assumptions, allele changes do not occur, or changes in one direction are balanced by changes in the opposite direction.
2. *No genetic drift.* Genetic drift is random changes in allele frequencies by chance. If a population is very large, changes in allele frequencies due to chance alone are insignificant.

3. *No gene flow.* Gene flow is the sharing of alleles between two populations through interbreeding. If there is no gene flow, then migration of individuals, and therefore their genes, into or out of the population does not occur.
4. *Random mating.* Random mating occurs when individuals pair by chance, not according to their genotypes or phenotypes.
5. *No selection.* Often, the environment selects certain phenotypes to reproduce and have more offspring than other phenotypes. If selection does not occur, no phenotype is favored over another to reproduce.

Although possible in theory, Hardy-Weinberg equilibrium is never achieved in wild populations because all of the five required conditions are never met in the real world. The peppered moth population we are using as an example obeys all five of the conditions for genetic equilibrium, but in reality, populations are constantly evolving from generation to generation. Hardy-Weinberg equilibrium does not typically occur in natural populations, but it is an important tool for population geneticists because the violation of one or more of the five conditions causes the allele and/or genotype frequencies of a population to change in predictable ways (Table 27.2). For example, change in genotype frequencies without a change in allele frequencies over generations suggests that a population is not mating randomly.

Check Your Progress 27.3

1. List the five conditions necessary to maintain a Hardy-Weinberg equilibrium in a population.
2. Explain what deviations from Hardy-Weinberg equilibrium tell us about a population.

27.4 Processes of Evolution

Learning Outcomes

Upon completion of this section, you should be able to
1. Define the five agents of evolutionary change.
2. List the four requirements of evolution by natural selection.
3. Identify examples of the three types of natural selection.

Five Agents of Evolutionary Change

In Section 27.3, we outlined the five conditions for a population to be in genetic equilibrium. The opposite of these conditions can cause evolutionary change—namely mutations, genetic drift, gene flow, nonrandom mating, and natural selection. In this section, we define each of these five agents of evolutionary change.

Mutations

The principles of Hardy-Weinberg equilibrium recognize that new mutations can cause the allele frequencies in a population to change. **Mutations,** which are genetic changes, are the only source of new variation in a population. Without mutations, there could be no new heritable genetic diversity to be shaped by the forces of natural selection. It is important to realize that mutations are random events, they do not arise because the organism "needs" one. For example, a mutation that gives bacteria resistance to antibiotics appears by chance. It is because the mutation provides a survival advantage that it becomes increasingly common in the bacteria population over generations, not because the bacteria needed to evolve resistance to antibiotics. Not all mutations provide an advantage. Some mutations are harmful. But in reality, most mutations have little to no effect on an organism's fitness. For example, one type of mutation, called a *point mutation,* is a change in a single nucleotide in a gene. A point mutation is called "silent" if it does not result in a change to the function of the protein encoded by the gene. Mutations that cause a malfunctioning protein can be harmful to an organism.

Genetic Drift

Genetic drift refers to changes in the allele frequencies of a gene pool due to the random meeting of gametes during fertilization. Each time an organism reproduces, one in millions of sperm will fertilize one of many potential eggs. The random selection and assortment of gametes in a population causes allele frequencies to shift each generation. Thus, populations are always undergoing genetic drift.

As you can imagine, genetic drift has greater effects in smaller populations because there are fewer gametes to assort. Removal of gametes from a population due to random events can have an effect on allele frequencies in the next generation. For example, the chance death of one individual in a population of a million will not have an appreciable effect on allele frequencies, but the chance death of one individual in a population of ten could change the frequency of an allele by 10%, or even cause its loss altogether (if that individual was the only one with that allele) (Fig. 27.14).

In nature, the *founder effect* and *bottleneck effect* occur when populations are drastically reduced in size. For example, a **bottleneck effect** can occur following a natural disaster that kills a large proportion of a population (Fig. 27.15). Both of these effects result in a severe reduction in the gene pool that can drastically affect allele frequencies.

The **founder effect** occurs when a few individuals form a new colony, and only a fraction of the total genetic diversity of the original

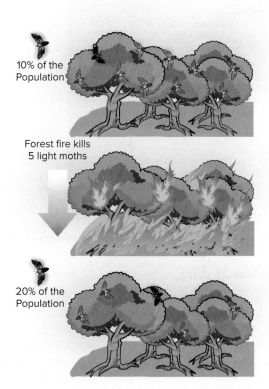

Figure 27.14 Genetic drift.   Genetic drift occurs when, by chance, only certain members of a population (in this case, peppered moths) reproduce and pass on their alleles to the next generation. A natural disaster can cause the allele frequencies of the next generation's gene pool to be markedly different from those of the previous generation.

Figure 27.15 Bottleneck and founder effects change allele frequencies.   **a.** The gene pool of a large population contains four different alleles, represented by colored marbles in a bottle, each with a different frequency. **b.** A population bottleneck occurs. The marbles, or alleles, that exit the bottle must pass through the narrow neck into the cup. The new gene pool will have a fraction of the alleles from the original population. **c.** The gene pool of the new population has changed from the original. Some alleles are in high frequency, while some are not present. **d.** A founder event is the same as a bottleneck, except that in the founder event, the original population still exists.

SCIENCE IN YOUR LIFE ▶ SCIENTIFIC INQUIRY

Inbreeding in the Pingelapese

One of the requirements of a population in Hardy-Weinberg equilibrium is that mates are chosen at random—that is, without preference for a particular trait. Most populations, however, do not meet this requirement because some traits are more attractive in a mate than others. Humans, for example, select mates based on a set of traits that we find appealing in a mate. This type of assortative mating based on trait preference is common in many species. Inbreeding is a unique form of nonrandom, or assortative, mating where individuals mate with close relatives, such as cousins.

One consequence of inbreeding is an increase in the frequency of homozygous genotypes in a population. For the most part, an increase in homozygotes does not have a large detrimental effect, especially if the population is very large. But when populations are small, inbreeding can have a large impact on the health of a population. Many human diseases are caused by the inheritance of two recessive alleles, such that the disease appears only in persons who are homozygous

recessive for the disease-causing allele. In very small populations, the probability that individuals carrying the recessive allele will mate increases because the mating pool is very small. In this case, inbreeding significantly increases the frequency of homozygous recessive genotypes, and thus those afflicted with a disease.

Human cultures tend to have social rules that discourage inbreeding, but in very small populations, such as those following a bottleneck or founder event, inbreeding is sometimes unavoidable. One example of the effects of inbreeding on human populations is the occurrence of a rare form of non-sex-linked color blindness, achromatopsia, on the island of Pingelap, a small coral atoll in the Pacific Ocean. People who are achromatic are completely color blind (Fig. 27A). Normal human color vision is possible because of cells in the back of the eye, called cones. Those with achromatopsia do not have any cone cells because they are homozygous for a rare recessive allele that prevents cone cells from developing.

Complete color blindness is so common on Pingelap that it is considered part of everyday life for most families.

Experts propose that the high frequency of achromatopsia appeared following a severe population bottleneck in 1775 when a typhoon killed 90% of the inhabitants. Approximately 20 people survived the typhoon. Four generations after the typhoon struck, achromatopsia began to appear frequently in the population. How could this happen? Figure 27B shows how inbreeding in a population can produce homozygous recessive genotypes. Geneticists explain that the high frequency of this rare genetic disorder on Pingelap is consistent with a large degree of intermarriage among relatives, or inbreeding, following the typhoon.

Mwanenised, a male survivor of the typhoon, was a carrier, or a heterozygote, for the achromatopsia allele. He had ten children, which was a large proportion of the first generation of Pingelapese after the typhoon. Geneticists would predict that, on average, 50% of his children would have been homozy-

Figure 27A Achromatopsia.   Complete achromatopsia is a recessive genetic disorder that causes complete color blindness. People with complete achromatopsia see no color, only shades of gray. This image shows how a person with this form of color blindness would see the world. (both) ©Ingram Publishing/Superstock

gene pool is represented in these individuals (Fig. 27.15). The particular alleles carried by the founders are dictated by chance alone.

Gene Flow

Gene flow is the movement of alleles among populations when individuals or their gametes migrate from one population to another and breed in that new population. For example, adult plants are not able to migrate, but their pollen is often blown by the wind or carried by insects into different populations.

Gene flow mixes genetic diversity and keeps the gene pools of populations similar.

Nonrandom Mating

Nonrandom mating occurs when individuals are selective about choosing a mate with a preferred trait. Because most sexually reproducing organisms select their mates based on some trait, random mating is never observed in natural populations. *Inbreeding,* or mating between close relatives, is an example of nonrandom mating. Inbreeding changes genotype frequencies by decreasing the proportion of heterozygotes and increasing the proportion of homozygotes for all genes. This increase in the frequency of homozygotes also increases the frequency of recessive homozygous disorders in a population. For example, the Amish population of gous for the normal allele, and 50% would have been heterozygous carriers of the color-blind allele. Thus, none of Mwanenised's children would have been color blind (Fig. 27B). However, intermarriage of the next generations would unavoidably have brought together men and women who were carriers for the achromatopsia allele. Thus, in subsequent generations the homozygous recessive genotype, and complete color blindness, did appear due to inbreeding within a small population (Fig. 27B). On Pingelap, an increase in color blindness was observed by the fourth generation.[1]

The effect of inbreeding can be long-term, especially if a population remains relatively small and isolated. Today, 1 in 12 Pingelapese suffers from achromatopsia, more than 3,000 times more frequently than in the United States, where it occurs in approximately 1 in 40,000 births.

[1] *See* Sacks, Oliver (1998). *The Island of the Colorblind* (Vintage/Anchor Books, New York). An interesting, nontechnical account of the Pingelapese.

Questions to Consider

1. What might have happened to the color-blind allele in the Pingelapese population following the typhoon if Mwanenised had not had any children? Only five children?
2. Would you predict that the Pingelapese population is in Hardy-Weinberg equilibrium? How would you measure this?
3. Has the Pingelapese population evolved? Explain. (*Hint:* Does inbreeding cause a change in allele frequencies?)

Figure 27B Inbreeding in small populations increases the frequency of recessive genetic disorders.   On the island of Pingelap, a high level of incidental inbreeding in a small population has resulted in a higher than expected occurrence of complete achromatopsia.

Lancaster, Pennsylvania, is an isolated religious sect descended from a few German founders. Intermarriage has increased the frequency of Ellis–van Creveld syndrome, a rare form of dwarfism, that occurs in individuals who are homozygous recessive for a mutation to the *EVC* gene (for more information on the effects of inbreeding in human populations, see the Scientific Inquiry feature "Inbreeding in the Pingelapese").

Natural Selection

In Section 27.1, we introduced natural selection as the process that allows some individuals, with an advantage over others, to produce more offspring. Here, we restate these steps in the context of modern evolutionary theory. Evolution by natural selection requires the following:

1. *Individual variation.* The members of a population differ from one another.
2. *Inheritance.* Many of these differences are heritable genetic differences.
3. *Overproduction.* Individuals in a population are engaged in a struggle for existence because breeding individuals in a population tend to produce more offspring than the environment can support.

Evolution of Antibiotic Resistance

The antibiotic penicillin was discovered by Alexander Fleming in 1928. On November 25, 1930, penicillin successfully cured a bacterial infection for the first time. Since the discovery of penicillin, modern science has produced an array of antibiotics effective against most bacterial infections. Prior to the discovery of penicillin, bacterial infections were a major cause of death worldwide, so much so that today it is difficult to imagine dying from a bacterial infection. Yet we are now coming full circle; many of the antibiotics that have saved millions of lives have evolved resistance to antibiotics (Fig. 27C). In 1945, the first resistance to penicillin was discovered in a strain of *Staphylococcus aureus*. Today, over 90% of *S. aureus* strains are resistant to penicillin.

Through Darwin's theory of natural selection, we have come to understand how bacteria become resistant to antibiotics. The process by which bacteria evolve antibiotic resistance is the same as natural selection—antibiotics kill bacteria that are susceptible, but bacterial populations are usually so large that some resistant individuals are likely to survive and reproduce. Through time, the frequency of resistant individuals increases in the population to the point that certain antibiotics are no longer effective. Overall, antibiotic resistance is evolving at an alarming rate, to the point where bacterial infections that just decades ago were treatable with antibiotics can now be fatal. In fact, resistance often evolves soon after the introduction of new antibiotics. This is the type of "accelerated evolution" that creates "superbugs," such as methicillin-resistant *Staphylococcus aureus*, or MRSA, which you learned about in the beginning of this chapter (Fig. 27Ca). Resistant bacteria, such as MRSA, can cause abscessed lesions on the skin that are difficult to treat. In

a. *Staphylococcus aureus*   SEM 9,560×

b. *Pseudomonas aeruginosa*   SEM 4,590×

c. *Mycobacterium tuberculosis*   SEM 6,200×

Figure 27C   Bacteria with antibiotic-resistant strains.   a. Methicillin-resistant *Staphylococcus aureus* (MRSA). **b.** *Pseudomonas aeruginosa*. **c.** *Mycobacterium tuberculosis*.

(a): ©Science Source; (b): ©Steve Gschmeissner/Science Source; (c): ©SPL/Science Source

some cases, resistant bacteria cause a systemic infection of the blood called *bacterial septicemia*. *Pseudomonas aeruginosa* (Fig. 27Cb) produces a toxin that causes abscesses to erupt on the skin in patients who have *P. aeruginosa*

septicemia. Bacterial resistance creates the need for even newer antibiotics. The development of a single new antibiotic is estimated to cost between $400 and $500 million. Antibiotic resistance adds over $30 billion to annual medical costs in the United States alone!

Our understanding of evolutionary biology has helped doctors treat patients appropriately. Doctors generally no longer prescribe antibiotics for colds or flu because they are viral infections. Antibiotics kill only bacteria, not viruses. In the case of tuberculosis (Fig. 27Cc), the strain infecting a patient is tested for antibiotic resistance so that an appropriate antibiotic treatment can be administered. The antibiotic isoniazid is used to treat patients with nonresistant strains, whereas a four-drug regimen is recommended for treatment of resistant strains.

Questions to Consider

1. In New York City, state health officials have the power to quarantine tuberculosis patients who do not take their medicine. That is, they can essentially lock them up for as long as needed (often up to 18 months) to treat their illness. Note also that some of the medications can have serious side effects. What do you think about this policy?
2. Many people in less-developed countries die from tuberculosis, not because their disease is incurable, but simply because they do not have health insurance and cannot afford the medications. Why might it be advantageous for the United States to pay more for our medications so that pharmaceutical companies can provide them to lower-income people at a reduced cost or for free?

---

4. *Differential reproductive success.* Individuals that are better adapted to their environment produce more offspring than those that are not as well adapted, and consequently, their fertile offspring will make up a greater proportion of the next generation.

In biology, the **fitness** of an individual is measured by the number of fertile offspring produced throughout its lifetime. Gene mutations are the ultimate source of variation because they provide new alleles. Sometimes, these mutations result in positive fitness consequences for an organism, perhaps providing variation for adaptation to a new environmental change. However, in sexually reproducing organisms, genetic variation can also result from crossing-over and

independent assortment of chromosomes during meiosis, and also fertilization, when gametes are combined. A different combination of alleles can lead to a new and different phenotype.

In this context, consider that most of the traits on which natural selection acts are polygenic, and thus, controlled by more than one gene. Such traits have a range of phenotypes that often follow a bell-shaped curve.

The three main types of natural selection are stabilizing selection, directional selection, and disruptive selection.

Stabilizing Selection   With **stabilizing selection,** extreme phenotypes are selected against, and individuals near the average are favored. Stabilizing selection can improve adaptation of the population to those aspects of the environment that remain constant. Human birth weight is an example of stabilizing selection. Over many years, hospital data have shown that human infants born with an intermediate birth weight (3–4 kg) have a better chance of survival than those at either extreme (either much less or much greater than average). When a baby is small, its systems may not be fully functional, and when a baby is large, it may have experienced a difficult delivery. Stabilizing selection reduces the variability in birth weight in human populations (Fig. 27.16).

Directional Selection **Directional selection** occurs when an extreme phenotype is favored and the distribution curve shifts in that direction. This changes the average phenotype in a population. Such a shift can occur when a population is adapting to a changing environment. For example, directional selection can be seen in the changes of a bacterial population (Fig. 27.17). Initially, most of the bacteria are susceptible (nonresistant) to the antibiotic drug. When the bacteria in the petri dish are exposed to the antibiotic, the non-resistant bacteria die or decline in numbers, mostly before they reproduce. Subsequent generations result in bacteria that are antibiotic resistant and thus have a greater ability to survive and reproduce.

This antibiotic resistance allele becomes prevalent in the population, and so a shift in the distribution curve occurs. As discussed in the chapter opener, and the Health Feature "Evolution of Antibiotic Resistance," new species of bacteria are becoming resistant to antibiotics every year as a result of directional selection.

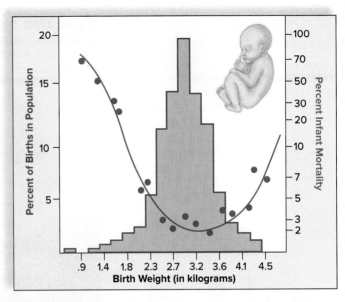

Figure 27.16 Human birth weight. The birth weight (blue) is influenced by the mortality rate (red). Babies that are born weighing 3–4 kg are more likely to survive.

Disruptive Selection In **disruptive selection,** two or more extreme phenotypes are favored over any intermediate phenotype (Fig. 27.18). For example, British land snails (*Cepaea nemoralis*) have a wide habitat range that includes grass fields, hedgerows,

a.

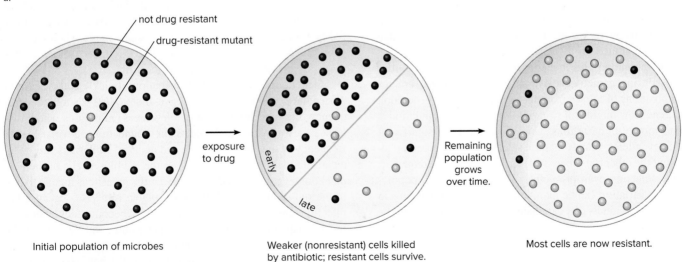

b.

Figure 27.17 Directional selection in a population of bacteria. Directional selection occurs when natural selection favors one extreme phenotype (in this case, antibiotic resistance), resulting in a shift in the distribution curve.

Figure 27.18 Disruptive selection. **a.** Disruptive selection favors two or more extreme phenotypes. **b.** Today, British land snails primarily possess one of two different phenotypes, each adapted to a different habitat. Snails with dark shells are more prevalent in forested areas, and light-banded snails are more prevalent in areas with low-lying vegetation. (b) (left): ©NHPA/Superstock; (right): ©W. Layer/Blickwinkel/age fotostock

a.

b.

and forested areas. In areas with low-lying vegetation, thrushes feed mainly on snails with dark shells that lack light bands, and in forested areas, they feed mainly on snails with light-banded shells. Therefore, the two different habitats have resulted in two different phenotypes in the population.

Maintenance of Variation

A population nearly always shows some genotypic variation. The maintenance of variation is beneficial because populations with limited variation may not be able to adapt to new conditions should the environment change, and may become extinct. How can variation be maintained in spite of selection constantly working to reduce it? First, we must remember that the forces that promote variation are still at work: Mutation still generates new alleles, recombination and independent assortment still shuffle the alleles during gametogenesis, and fertilization still creates new combinations of alleles from those present in the gene pool. Second, gene flow might still occur. If the receiving population is small and mostly homozygous, gene flow can be a significant source of new alleles. Finally, natural selection favors certain phenotypes, but the other types may still remain in reduced frequency. Disruptive selection even promotes polymorphism in a population. Genetic variation can be maintained in diploid organisms in heterozygote individuals that can maintain recessive alleles in the gene pool.

The Heterozygote Advantage

Only alleles that are expressed (cause a phenotypic difference) are subject to natural selection. In diploid organisms, this fact makes the heterozygote a potential protector of recessive alleles that might otherwise be weeded out of the gene pool by natural selection. Because the heterozygote remains in a population, so does the possibility of the recessive phenotype, which might have greater fitness in a changing environment. When natural selection favors the ratio of two or more phenotypes in generation after generation, it is called balanced polymorphism. Sickle-cell disease offers an example of a *balanced polymorphism* (Fig. 27.19).

Individuals with sickle-cell disease have the genotype $Hb^S Hb^S$ (Hb = hemoglobin, the oxygen-carrying protein in red blood cells; S = sickle cell) and tend to die at an early age due to hemorrhaging and organ destruction. Those who are heterozygous and have sickle-cell trait ($Hb^A Hb^S$; A = normal) are better off because their red blood cells usually become sickle-shaped only when the oxygen content of the environment is low. Ordinarily, those with a normal genotype ($Hb^A Hb^A$) are the most fit.

Geneticists studying the distribution of sickle-cell disease in Africa have found that the recessive allele (Hb^S) has a higher frequency (0.2 to as high as 0.4 in a few areas) in regions where malaria is also prevalent. What is the connection between higher frequency of the recessive allele and malaria? Malaria is caused by a parasite that lives in and destroys the red blood cells of the normal homozygote ($Hb^A Hb^A$). However, the parasite is unable to live in the red blood cells of the heterozygote ($Hb^A Hb^S$) because the infection causes the red blood cells to become sickle-shaped. Sickle-shaped red blood cells lose potassium, and this causes the parasite to die. Thus, in an environment where malaria is prevalent, the heterozygote is favored. Each of the homozygotes is selected against but is maintained because the heterozygote is favored in parts of Africa subject to malaria. Figure 27.19 summarizes the effects of the three possible genotypes.

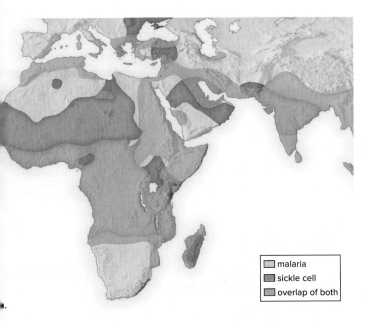

Genotype	Phenotype	Result
$Hb^A Hb^A$	Normal	Dies due to malarial infection
$Hb^A Hb^S$	Sickle-cell trait	Lives due to protection from both
$Hb^S Hb^S$	Sickle-cell disease	Dies due to sickle-cell disease

Figure 27.19 Sickle-cell disease.   a. Sickle-cell disease is more prevalent in areas of Africa where malaria is more common. **b.** Whether or not a person dies from the sickle-cell disease, malaria, or survives depends upon the genotype. Heterozygotes for the sickle-cell trait are protected against both diseases.

Check Your Progress 27.4

1. Summarize the five agents of evolutionary change.
2. Explain how the agents of evolutionary change relate to the process of natural selection.
3. Describe how malaria makes having a sickle-cell allele advantageous, and explain if having a sickle-cell allele would be advantageous in all parts of the world.

27.5 Macroevolution and Speciation

Learning Outcomes

Upon completion of this section, you should be able to

1. Distinguish between prezygotic and postzygotic isolation mechanisms.
2. Explain how adaptive radiation can lead to speciation.
3. Compare and contrast gradualism with punctuated equilibrium.

In this section, we turn our attention to evolution on a large scale—that is, **macroevolution.** The history of life on Earth is a part of macroevolution. Macroevolution involves speciation, or the splitting of one species into two or more species. The same microevolutionary agents that are at play within populations—genetic drift, natural selection, mutation, and migration—also shape the evolution of new species. Thus, microevolution and macroevolution are the result of the same agents, differing only in the scale at which they occur. Macroevolution is the result of the accumulation of microevolutionary change that results in the formation of new species.

There are several different ways of defining a species. In this section, we will focus on the biological species concept, which states that a **species** is a group of organisms capable of interbreeding that are isolated reproductively from other organisms. Under the biological species concept, if organisms cannot mate and produce offspring in nature, or if their offspring are sterile, they are defined as different species.

Reproductive isolation of similar species is accomplished by the isolating mechanisms illustrated in Figure 27.20. **Prezygotic isolating mechanisms** are in place before fertilization, and thus reproduction is never attempted. **Postzygotic isolating mechanisms** are in place after fertilization, so reproduction may take place, but it does not produce fertile offspring.

The Process of Speciation

Speciation has occurred when one species gives rise to two species, each of which continues on its own evolutionary pathway. How can we recognize speciation? Whenever reproductive isolation develops between two formerly interbreeding groups or populations, speciation has occurred. One type of speciation, called **allopatric speciation,** usually occurs when populations become separated by a geographic barrier and gene flow is no longer possible. Figure 27.21 illustrates an example of allopatric speciation that has been extensively studied in California. Apparently, members of an ancestral population of *Ensatina* salamanders existing in the Pacific Northwest migrated southward, establishing a series of populations. Each population was exposed to its own selective pressures along the coastal and Sierra Nevada Mountains. Due to the presence of the Central Valley of California, which is largely dry and thus an unsuitable habitat for amphibians, gene flow rarely occurs between eastern and western populations of *Ensatina*. Genetic differences also increased from north to south, resulting in two distinct forms of *Ensatina* salamanders in Southern California that differ dramatically in color.

It is also possible that a single population could suddenly divide into two reproductively isolated groups without being geographically isolated. The best evidence for this type of speciation, called **sympatric speciation,** is found among plants, where multiplication of the chromosome number in one plant prevents it from successfully reproducing with others of its kind. Self-reproduction can maintain such a new plant species.

Adaptive Radiation

One of the best examples of speciation is provided by the finches on the Galápagos Islands, located 600 miles west of Ecuador in South America. As Darwin proposed, these 13 species of finches

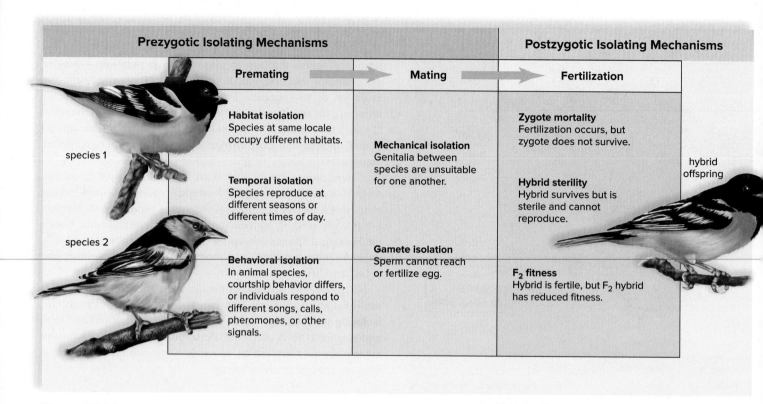

Figure 27.20 Reproductive barriers. Prezygotic isolating mechanisms prevent mating attempts, or a successful outcome should mating take place. No zygote ever forms. Postzygotic isolating mechanisms prevent the zygote from developing—or should an offspring result, it is not fertile.

Ensatina ring species

- *E. eschscholtzii oregonensis*
- *E. eschscholtzii picta*
- central valley mountains
- central valley barrier
- *E. eschscholtzii xanthoptica*
- coastal mountains
- *E. eschscholtzii platensis*
- Sierra Nevada mountains
- *E. eschscholtzii croceater*
- *E. eschscholtzii eschscholtzii*
- *E. eschscholtzii klauberi*

Figure 27.21 Allopatric speciation. In this example of allopatric speciation, the Central Valley of California is reproductively separating a range of populations of *Ensatina eschscholtzii* that are all descended from the same northern ancestral species.

are descended from a species of mainland finch that migrated to one of the islands. It is likely that after the original species colonized a single island, some individuals dispersed to other islands where they adapted, via natural selection, to feed in various habitats on ecologically different islands. Remarkably, the shape of the beak of each finch species is well matched to the abundant food type on each island.

Today, there are seed-eating ground finches, cactus-eating ground finches, insect-eating tree finches, all with different-sized beaks; and a warbler-type tree finch, with a beak adapted to eating insects and gathering nectar. Among the tree finches is a woodpecker type, which lacks the long tongue of a true woodpecker, but makes up for this by using a cactus spine or a twig to ferret out insects. Darwin's finches are an example of **adaptive radiation,** or the proliferation of a species by adaptation to different ways of life.

The Pace of Speciation

Currently, there are two hypotheses about the pace of speciation and, therefore, evolution. One hypothesis is called the phyletic gradualism model, and the other is called the punctuated equilibrium model. Each model gives a different answer to the question of why so few transitional links are found in the fossil record.

a.

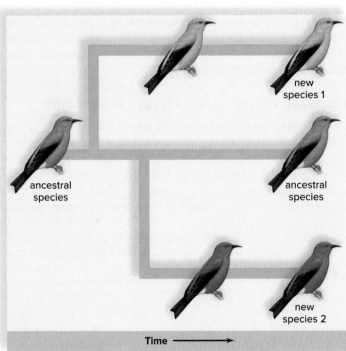

b.

Figure 27.22 Gradualism compared to punctuated equilibrium.   **a.** Supporters of the phyletic gradualism model hypothesize that speciation takes place gradually and many transitional links occur. **b.** Supporters of the more recent, punctuated equilibrium model hypothesize that speciation occurs rapidly, with no transitional links.

Traditionally, evolutionary biologists, including Darwin, have supported a process called the **gradualistic model,** which states that change is very slow but steady within a lineage before and after a divergence (splitting of the line of descent) (Fig. 27.22*a*). Therefore, it is not surprising that few transitional links such as

Archaeopteryx (see Fig. 27.4) have been found. Indeed, the fossil record, even if it were complete, might be unable to show when speciation has occurred. Why? Because a new species comes about after reproductive isolation, and reproductive isolation cannot be detected in the fossil record! Only when a new species evolves and displaces the existing species is the new species likely to show up in the fossil record.

A model of evolution called **punctuated equilibrium** has also been proposed (Fig. 27.22*b*). It says that long periods of stasis, or no visible change, are followed by rapid periods of speciation. With reference to the length of the fossil record (about 3.5 billion years), speciation occurs relatively rapidly, and this can explain why few transitional links are found. Mass extinction events are often followed by rapid (relative to the age of the Earth) periods of speciation.

> **Check Your Progress 27.5**
>
> **1.** Distinguish between a prezygotic isolation mechanism and a postzygotic isolation mechanism.
> **2.** Explain the following: You find evidence of an adaptive radiation in the fossil record, followed by a long period of little change. Which model of evolution would this support?

27.6 Systematics

> **Learning Outcomes**
>
> Upon completion of this section, you should be able to
> **1.** Recognize how phylogenetics, the theory of evolution, and classification are interrelated.
> **2.** Define the goal of systematic biology and its two branches of study: taxonomy and phylogenetics.
> **3.** Interpret the information provided in both a cladogram and a phylogenetic tree.

In Section 27.5, we examined macroevolution, or evolutionary change that results in the formation of new species. Macroevolution is the source of past and present biodiversity. **Systematic biology** (or **systematics**) is the study of the evolutionary history of biodiversity. Systematic biology is a quantitative science that uses characteristics of living and fossil organisms, or traits, to infer the evolutionary relationships among organisms, and to then organize biodiversity based upon these relationships.

Classification

Taxonomy is the branch of systematic biology concerned with identifying, naming, and classifying organisms. A **taxon** (*pl.,* **taxa**) is the general name for organisms that exhibit a set of shared traits. **Classification** is the process of naming and assigning organisms or groups of organisms to a taxon. For example, the taxon Vertebrata contains organisms with a bony spinal column. As another example, the taxon Canidae contains the wolf (*Canis lupus*) and its close relative, the domestic dog (*Canis familiaris*). The binomial system of nomenclature is used to assign a two-part name (*genus* and *species*) to each taxon. For example, the scientific name for modern humans is *Homo sapiens.* Notice that the scientific name is in italics and only the genus is capitalized. The name of an organism usually tells you something about the organism. In this instance, the species name, *sapiens,* refers to a large brain.

Today, **taxonomists,** scientists that study taxonomy, use several categories of classification created by Swedish biologist Carl Linnaeus in the 1700s to show varying levels of similarity: species, genus, family, order, class, phylum, and kingdom. Recently, the domain, a higher taxonomic category, has been added to this list. There can be several species within a genus, several genera within a family, and so forth. In this hierarchy, the higher the category, the more inclusive it is. Therefore, species in the same domain have general traits in common, whereas those in the same genus have quite specific traits in common.

Phylogeny

Phylogenetics is the branch of systematic biology that studies the evolutionary relatedness among groups of organisms. Phylogeneticists use characters from the fossil record, comparative anatomy and development, and the sequence, structure, and function of RNA and DNA molecules to construct a **phylogeny,** a hypothesis of evolutionary relatedness among taxa represented by a "family tree." In essence, systematic biology is the study of the evolutionary history of biodiversity, and a phylogeny is the estimation, and visual representation, of this history. Taxonomists can then use a phylogeny to classify organisms into taxa based upon shared evolutionary history.

Cladistics is one method phylogeneticists use to construct a phylogeny called a **cladogram.** Branches in a cladogram are called *clades,* and each clade contains a most recent common ancestor and all its descendants. Cladistics uses only those traits shared among all individuals to define a clade. These **shared derived traits** are not found in organisms outside of the clade, called "outgroups." Figure 27.23 depicts a cladogram for seven groups of vertebrates. Only the lamprey, the "outgroup," lacks jaws, but the other six groups of vertebrates are in the same clade because they all have jaws, a shared derived trait. In the same way, the vertebrates in the clade without the lamprey and the shark all have the shared derived trait of lungs, and so forth. Figure 27.23 is somewhat

misleading because, although single traits are noted on the tree, cladistics uses much more data to arrange groups of organisms into clades. Cladistics is based on the premise of **parsimony,** which considers the simplest explanation—that is, the cladogram, which requires the fewest number of evolutionary changes—to be the best hypothesis of evolutionary history.

It is important to note that a phylogeny (or in the case of cladistics, a cladogram) represents evolutionary relatedness only if homologous traits, or those that share a common ancestry (see Fig. 27.7), are used. Deciphering homology is sometimes difficult because of convergent evolution. **Convergent evolution** is the independent evolution of analogous traits, or the same or similar traits, in distantly related lines of descent (see Section 27.2). Analogous structures have the same function in different groups, but organisms with these structures do not share a recent common ancestor. Instead, analogous structures arise because of adaptations to similar environments. DNA nucleotide sequences are traits that are commonly used to construct a phylogeny. Modern computer systems and laboratory techniques have made acquiring DNA sequences easy and economical. Homology of nucleotides in a DNA sequence is determined by alignment, or matching sections of sequence from the same section of the genome among all the taxa under study. Figure 27.24 is a phylogeny of primates based on DNA data, if we assume that DNA differences, or mutations, occur at a relatively constant rate. This is called a *molecular clock,* because the number of differences can be used to estimate the absolute age of each branch in the phylogeny.

Linnaean Classification Versus Phylogenetics

The Linnaean classification system was developed long before Darwin proposed his theory of evolution, and even longer before the field of phylogenetics was born. The Linnaean classification system still carries historical taxa that were assembled without an understanding of evolutionary relatedness. This can cause problems for the taxonomist, because in some cases analogous, and not homologous, traits were used to classify organisms. Figure 27.25 illustrates the types of problems that arise when trying to reconcile Linnaean classification with the principles of phylogenetics. The phylogeny in this figure proposes that birds are in a clade with

Figure 27.23 Cladogram.
A cladogram gives comparative information about relationships. Organisms in the same clade share the same derived traits. Humans and all the other vertebrates shown, except lampreys, are in the same clade as sharks because they all have jaws. However, humans are alone on a branch because only they are bipedal. This cladogram is simplified; phylogeneticists actually use a great deal more data to construct cladograms.

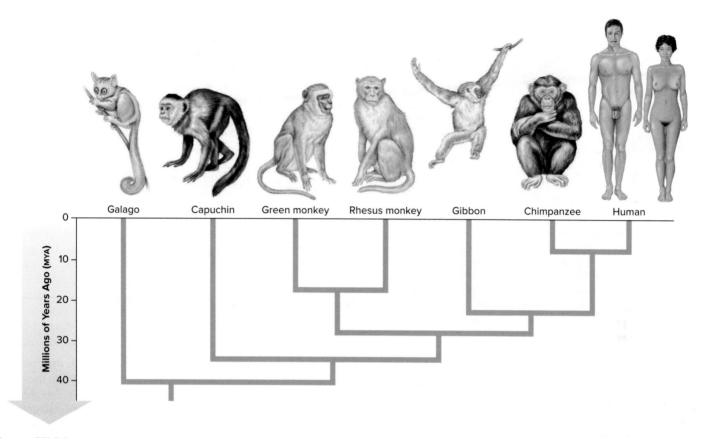

Figure 27.24 Molecular data.   The relationships of certain primate species are based on a study of their genomes. The length of the branches indicates the relative number of DNA base-pair differences between the groups. These data make it possible to suggest a date for the origin of other branches in the tree by using a molecular clock.

crocodiles because they share a recent common ancestor. This ancestor had a gizzard, among other shared derived traits of the skull. Yet Linnaean classification places birds in their own group, separate from crocodiles and separate from reptiles in general. In many other instances, also, Linnaean classification is not consistent with new understandings about phylogenetic relationships.

Therefore, some phylogeneticists have proposed a different system of classification called the International Code of Phylogenetic Nomenclature, or PhyloCode, which sets forth rules to follow in naming of clades. Other biologists are hoping to modify Linnaean classification to be consistent with the principles of modern systematic biology.

Three-Domain Classification System

Classification systems change over time. Historically, most biologists have utilized the five-kingdom system of classification: plants, animals, fungi, protists, and monerans. Organisms were classified into these kingdoms based on type of cell (prokaryotic or eukaryotic), level of organization (unicellular or multicellular), and type of nutrition. In the five-kingdom system, the monerans were distinguished by their structure—they were prokaryotic (lack a membrane-bound nucleus)—whereas the organisms in the other kingdoms were eukaryotic (have a membrane-bound nucleus). The kingdom Monera contained all prokaryotes, which according to

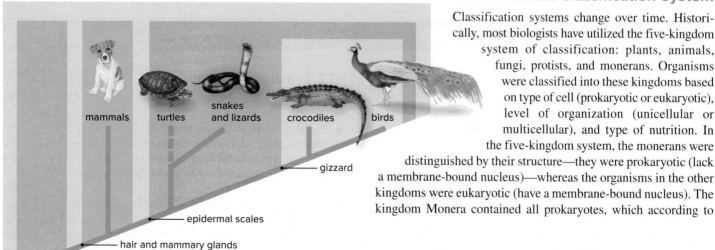

Figure 27.25 Cladistic classification.   Taxonomic designations are based upon evolutionary history. Each taxon includes a common ancestor and all of its descendants.

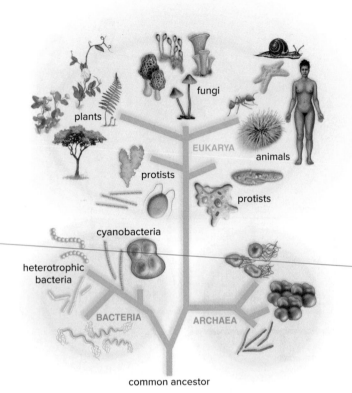

Figure 27.26 The three-domain system of classification.
Representatives of each domain are depicted. The phylogenetic tree shows that domain Archaea is more closely related to domain Eukarya than either is to domain Bacteria.

the fossil record evolved first (see Section 28.1 for discussion about the origin of the first cell).

Ribosomal RNA (rRNA) genes and new data about cells challenged the five-kingdom system of classification. These data support that there are two, not one, groups of prokaryotes. Based on this information, a **three-domain system** of classification was developed (Fig. 27.26). A domain is a classification category

higher than the kingdom. The two prokaryote groups were classified into the **domain Bacteria** and the **domain Archaea** because they are so fundamentally different from each other that they warrant being assigned to separate domains. A third domain, the eukaryotes, were classified into the **domain Eukarya**, which contains kingdoms for all eukaryotes, including protists, animals, fungi, and plants. Interestingly, the Archaea and Eukarya are more closely related to each other than either is to domain Bacteria.

Check Your Progress 27.6

1. Explain the differences between the study of taxonomy and phylogenetics. How do they relate to the classification of organisms?
2. Describe the differences between a cladogram and a phylogenetic tree.
3. List the three domains of life.

Conclusion

In this chapter, you have learned about both macroevolution and microevolution. Although people tend to think of evolution occurring over long timescales, it can occur quickly, particularly under human influence. The evolution of antibiotic resistance and pesticide resistance are examples of how such rapid evolution can be important in your everyday life. The evolution of antibiotic-resistant strains of bacteria, such as *Shigella sonnei,* is an increasing health problem, costing billions in increased medical costs and even resulting in deaths of some affected patients. By the same token, pesticide-resistant insects are costing billions in crop damage and decreasing yields. An evolutionary approach to treating bacterial infections involves careful use of antibiotics, thereby slowing the rate at which newly resistant strains evolve.

SUMMARIZE

27.1 Theory of Evolution

Charles Darwin served as the naturalist aboard the HMS *Beagle* during its five-year voyage. Darwin's observations of geology and species variation led him to his theory of **evolution:** how species change over time.

- Jean-Baptiste de Lamarck proposed, before Darwin, a theory of evolution called **inheritance of acquired characteristics,** which proposed that changes to the phenotype acquired during a lifetime could be passed on to subsequent generations.
- Darwin proposed the theory of **natural selection,** which proposes that **adaptation** to the environment is due to heritable changes that, if advantageous, are passed to future generations differentially because of high reproductive success.

27.2 Evidence of Evolution

Darwin's theory of evolution by natural selection is supported by the following:

- The fossil record and biogeography, as well as studies of comparative anatomy and development, and biochemistry, all provide evidence for evolution. **Transitional links** are **fossils** that have traits of both ancestor and descendant, and represent evolutionary shifts from one trait to another.
- **Biogeography** shows that the distribution of organisms on Earth can be influenced by a combination of evolutionary and geological processes. **Continental drift** is one geological process that affected the biogeographical distribution of organisms.
- Comparing the anatomy and the development of organisms distinguishes **homologous structures** from **analogous structures.** Only

homologous structures provide information about common ancestry. **Vestigial structures** are remnants of past adaptations, but which are no longer used in the current environment. **Artificial selection** by humans for preferred traits has led to evolution of many plants and animals. In natural populations, evolution can be observed to occur over periods as short as decades, such as the beak size of the Galápagos medium ground finch.

■ All organisms have certain biochemical molecules in common, and these chemical similarities indicate the degree of relatedness.

27.3 Microevolution

■ **Microevolution** is a process that involves a change in allele frequencies within the gene pool of a sexually reproducing population. **Population genetics** is the study of the change in allele frequencies in the **gene pool** of a **population.** The work of Hardy-Weinberg indicates that gene pool frequencies arrive at a **genetic equilibrium,** or **Hardy-Weinberg equilibrium,** that is maintained generation after generation unless disrupted by mutations, genetic drift, gene flow, nonrandom mating, or natural selection. Any change from the initial allele frequencies in the gene pool of a population signifies that evolution has occurred.

27.4 Processes of Evolution

■ The principles of the Hardy-Weinberg equation state that gene pool frequencies arrive at an equilibrium that is maintained generation after generation unless disrupted by **mutations, genetic drift, gene flow, nonrandom mating,** or natural selection.

■ Mutations may influence the **fitness,** or reproductive success, of the individual. The **founder effect** and **bottleneck effect** are special types of genetic drift due to severe reduction in population size.

■ **Stabilizing selection, directional selection,** and **disruptive selection** are different ways in which the distribution of a trait changes in response to natural selection.

27.5 Macroevolution and Speciation

■ **Macroevolution** is the origin of new **species,** or **speciation.** A new species forms when the two populations can no longer interbreed because of either (or both) **prezygotic** or **postzygotic isolating mechanisms.**

■ **Allopatric speciation,** the most common form of speciation, occurs when populations become reproductively isolated because of a geographic barrier.

■ **Sympatric speciation** occurs when organisms overlap in their distributions.

■ The evolution of several species of finches on the Galápagos Islands is an example of speciation caused by **adaptive radiation,** because each one has a different way of life.

■ Currently, there are two models about the pace of speciation. The **gradualistic model** represents a slow, steady change over time, while during **punctuated equilibrium,** long periods of no change are interrupted by periods of rapid change.

27.6 Systematics

■ **Systematic biology,** or **systematics,** involves **phylogenetics,** the study of evolutionary relatedness (or **phylogeny**) among **taxa** (*sing.,* **taxon**), and **taxonomy,** the **classification** of taxa into a hierarchy of categories: domain, kingdom, phylum, class, order, family, genus, and species. **Taxonomists** are scientists who classify biodiversity.

■ **Cladistics** is one method of phylogenetics that applies the principle of **parsimony** to **shared derived traits** to develop a **cladogram,** or a

treelike hypothesis of evolutionary history. Cladistics strives to classify organisms based upon evolutionary relatedness. This may be complicated by **convergent evolution** and analogous traits.

■ Each clade, or branch, contains the most recent common ancestor and all its descendants. Modern phylogenetic methods most commonly analyze similarities and differences among nucleotide sequences, aided by computers, to reconstruct phylogenies.

■ The **three-domain system (domain Bacteria, domain Archaea,** and **domain Eukarya),** based on molecular data, is currently preferred to the formerly used five-kingdom system (Monera, Protista, Fungi, Plantae, Animalia). Both bacteria and archaea are prokaryotes. Members of the kingdoms Protista, Fungi, Plantae, and Animalia are eukaryotes.

ASSESS

Testing Yourself

Choose the best answer for each question.

27.1 Theory of Evolution

1. Inheritance of acquired characteristics is associated with
   a. Darwin.
   b. Lamarck.
   c. Malthus.
   d. None of these are correct.

2. A(n) _____ makes a species better suited to its environment.
   a. homologous trait
   b. adaptation
   c. analogous trait
   d. mutation

27.2 Evidence of Evolution

3. Fossils that serve as transitional links allow scientists to
   a. determine how prehistoric animals interacted with each other.
   b. deduce the order in which various groups of animals arose.
   c. relate climate change to evolutionary trends.
   d. determine why evolutionary changes occur.

4. The flipper of a dolphin and the fin of a tuna are
   a. homologous structures.
   b. homogeneous structures.
   c. analogous structures.
   d. reciprocal structures.

27.3 Microevolution

5. Assuming a Hardy-Weinberg equilibrium, 21% of a population is homozygous dominant, 50% is heterozygous, and 29% is homozygous recessive. What percentage of the next generation is predicted to be homozygous recessive?
   a. 21%
   b. 50%
   c. 29%
   d. 42%
   e. 58%

6. Which of the following would not be true of a population that is exhibiting Hardy-Weinberg equilibrium for a trait?
   a. There is no genetic drift.
   b. There is random mating.
   c. There is no evidence of selection.
   d. There is no gene flow with another population.
   e. It is evolving with regard to that trait.

27.4 Processes of Evolution

For questions 7–11, match the description with the appropriate term in the key.

Key:

    a. mutation
    b. natural selection
    c. founder effect
    d. bottleneck
    e. gene flow
    f. nonrandom mating

7. The northern elephant seal went through a severe population decline as a result of hunting in the late 1800s. As a result of a hunting ban, the population has rebounded but is now homozygous for nearly every gene studied.

8. A small, reproductively isolated religious sect called the Dunkers was established by 27 families that came to the United States from Germany over 200 years ago. The frequencies for blood group alleles in this population differ significantly from those in the general U.S. population.

9. Turtles on a small island tend to mate with relatives more often than turtles on the mainland.

10. Within a population, plants that produce an insect toxin are more likely to survive and reproduce than plants that do not produce the toxin.

11. The gene pool of a population of bighorn sheep in the southwestern United States is altered when several animals cross over a mountain pass and join the population.

12. People who are heterozygous for the cystic fibrosis gene are more likely than others to survive a cholera epidemic. This heterozygote advantage seems to explain why homozygotes are maintained in the human population, and is an example of
    a. disruptive selection.
    b. balanced polymorphism.
    c. high mutation rate.
    d. nonrandom mating.

27.5 Macroevolution and Speciation

13. The formation of new species due to geographic barriers is called
    a. isolation speciation.
    b. allopatric speciation.
    c. allelomorphic speciation.
    d. sympatric speciation.
    e. symbiotic speciation.

14. Which of the following models is supported by the observation that few transitional links are found in the fossil record?
    a. gradualism
    b. punctuated equilibrium
    c. Hardy-Weinberg
    d. None of these are correct.

27.6 Systematics

15. The three-domain classification system has recently been developed based on
    a. mitochondrial biochemistry and plasma membrane structure.
    b. cellular and rRNA sequence data.
    c. plasma membrane and cell wall structure.
    d. rRNA sequence data and plasma membrane structure.
    e. nuclear and mitochondrial biochemistry.

16. The branch of systematic biology that assigns organisms to specific classifications (domain, phyla, class, etc.) is called
    a. phylogenetics.
    b. taxonomy.
    c. systematics.
    d. cladistics.

ENGAGE

Thinking Critically

1. The frequency of a rare disorder expressed as an autosomal recessive trait is 0.0064. Using the Hardy-Weinberg equation, determine the frequency of carriers for the disease in this population.

2. Viruses such as HIV are rapidly replicated and have very high mutation rates. Thus, evolution of the virus can be observed in a single infected person. Using what you have learned in this chapter, explain why HIV is so hard to treat, even though multiple drugs to treat HIV have been developed.

3. You observe a wasting disease in cattle that you know is genetically caused and thus heritable. The disease is fatal in young cattle. The allele frequency for the gene that causes the disease is 0.05 in the United States, but 0.35 in South America. Explain why such a difference in allele frequencies might exist.

4. If $p^2 = 0.36$, what percentage of the population has the recessive phenotype, assuming a Hardy-Weinberg equilibrium?

5. In a population of snails, 10 had no antennae (*aa*), 180 were heterozygous with antennae (*Aa*), and 810 were homozygous with antennae (*AA*). What is the frequency of the *a* allele in the population?

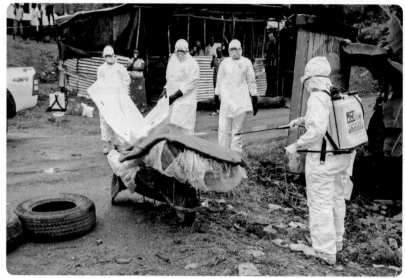
©epa european pressphoto Agency b.v./Alamy

The West Africa Ebola Outbreak

In 2013, an outbreak of Ebola, one of the most feared viruses on the planet, began in the West African nation of Guinea. It is believed that a one-year-old boy contracted the disease while playing near a tree that housed a species of bat that is known to carry the virus. By early 2014, the disease had become widespread in the neighboring countries of Sierra Leone and Liberia, with cases in Nigeria, Mali, and Senegal. According to the CDC, there were over 28,000 confirmed cases of Ebola in West Africa, and over 11,000 confirmed deaths. But most agencies believe that this is an underestimate and that the complete toll of this outbreak may never be known.

What makes Ebola so feared is that it belongs to a family of viruses that cause hemorrhagic fever, a disease that targets several different cell types of the body, including macrophages of the immune system, and the endothelial cells in the circulatory system and liver. While Ebola is frequently described as a disease that causes widespread bleeding, most deaths are actually due to fluid loss, organ failure (such as the liver), or an overall failure of the immune system. It is transmitted through direct contact with the body fluids of an infected person.

Like many viruses, there are many misconceptions regarding the Ebola virus. These include that the disease is airborne, you can get the virus from contact with cats and dogs, and that antibiotics are an effective treatment. In fact, in many ways Ebola is similar to any virus—it must invade specific cells of the body in order to hijack the cell's metabolic machinery to make more copies of itself.

In this chapter, we will not only examine the interaction of viruses with living organisms, but also other members of the microbial world, such as bacteria and prions.

As you read through this chapter, think about the following questions:

1. How do viruses, such as Ebola, infect the cells of the body?

2. Are viruses classified as living organisms?

3. What are some other human diseases caused by viruses?

CHAPTER OUTLINE

28.1 The Microbial World

28.2 Origin of Microbial Life

28.3 Archaea

28.4 Bacteria

28.5 Viruses, Viroids, and Prions

BEFORE YOU BEGIN

Before beginning this chapter, take a few moments to review the following discussions:

Section 1.2 To which two domains do the prokaryotes belong?

Section 3.2 What are some basic structural features of prokaryotic cells?

Table 27.1 When did prokaryotic cells first appear on Earth?

28.1 The Microbial World

Antonie van Leeuwenhoek was an early eighteenth-century Dutch tradesman and scientist. He was apparently very skilled at working with glass lenses, which enabled him to greatly improve the microscopes that already existed in the late 1600s. Using these instruments, he was among the first to view microscopic life-forms in a drop of water, which he called "animalcules" and described in this way:

> [They] were incredibly small, nay so small, in my sight, that I judged that even if 100 of these very wee animals lay stretched out one against another, they could not reach to the length of a grain of coarse sand.[1]

Leeuwenhoek and others after him believed that the "wee animals" he had observed could arise spontaneously from inanimate matter. For about 200 years, scientists carried out various experiments to determine the origin of microscopic organisms in laboratory cultures. Finally, in about 1859, Louis Pasteur devised the experiment shown in Figure 28.1.

---

[1] Antonie van Leeuwenhoek

In the first experiment, when flasks containing sterilized broth were exposed to either outdoor or indoor air, they often became contaminated with microbial growth. In the second experiment, however, if the neck of the flask was curved so that microbes could not enter the broth from the air, no growth occurred. Thus, the bacteria were not arising spontaneously in the broth, but rather were coming from the air. In 1884, Pasteur also suggested that something even smaller than a bacterium was the cause of rabies, and it was he who chose the word *virus* from a Latin word meaning "poison." Pasteur went on to develop a vaccine for rabies, which he used to save the life of a young French boy who had been bitten by a rabid dog.

Microbiology is the study of *microbes,* a term that includes bacteria, archaea, protists, fungi, viruses, viroids, and prions. Most, but not all, of these organisms are so small that they require a microscope to be seen. Today we know that bacteria and other microbes are incredibly numerous in air, water, soil, and on objects. A single spoonful of soil can contain 10^{10} bacteria, and the total number of bacteria on Earth has been estimated at around 10^{30}, which clearly exceeds the numbers of any other type of living organism on Earth. It is a good thing bacteria are microscopic—if they were the size of beetles, the Earth would be covered in a layer of bacteria several miles deep!

We need only visit almost any news site to hear about diseases caused by microbes. Viral diseases such as avian flu and Ebola, and new strains of antibiotic-resistant bacteria such as *Staphylococcus aureus* and *Mycobacterium tuberculosis* will make you think that all microbes cause disease. However, many microbes have beneficial, and sometimes essential, roles to play in human health, as well as in the proper functioning of the biosphere. For example, for every cell in your body, there are 10 bacterial cells living on or within you. While you are generally unaware of them, most of these

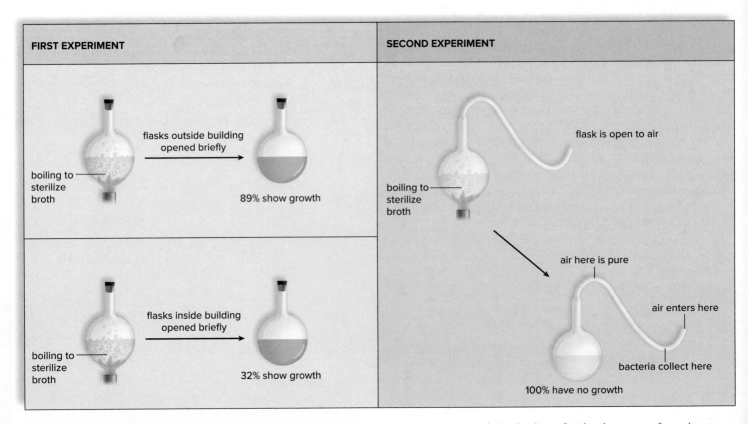

Figure 28.1 Pasteur's experiments. Pasteur disproved the theory of spontaneous generation of microbes by performing these types of experiments.

microbes, also known as the **microbiota,** have beneficial effects. Bacteria that live on your skin help crowd out harmful microbes that might grow in those areas, and bacteria in the intestines aid in digestion and synthesize vitamin K and vitamin B_{12}. Our bodies are actually a finely balanced ecosystem for microbes.

Microbes play an important role in many biological processes:

- *Nutrient cycling.* Both bacteria and fungi play essential roles in various nutrient cycles on Earth by breaking down organic and inorganic materials that can be reused by plants and animals.
- *Photosynthesis.* Photosynthetic algae and other protists are primary producers that capture energy from the sun or inorganic material, providing nutrients for more complex organisms. Some bacteria also perform photosynthesis, often in ways very different from algae or plants. These organisms, therefore, all occupy an important position in the food chain.
- *Bioremediation.* Microbes, specifically the bacteria, are also able to consume virtually any substance found on Earth. Soil contaminated by oil spills or just about any other toxic compounds can be cleaned by encouraging the growth of bacteria that eat these compounds. Wastewater that leaves your bathroom and kitchen is probably cleaned at a treatment plant by processes that rely on the activity of bacteria.

Not only do many microbes play essential roles in the environment, but they are also valuable for industrial processes, particularly food processing. Also, most antibiotics known today were first discovered in soil bacteria or fungi. Genetic techniques can be used to alter the products generated by bacterial cultures. A variety of valuable products can be made in this way, including insulin and vaccines. Although it might not seem so to an individual suffering from tuberculosis, AIDS, or another serious infectious disease, most microbes are far more beneficial than harmful—in fact, we could not live without them!

In this chapter, we will explore prokaryotic bacteria, viruses, viroids, and prions, and then we will take a closer look at fungi and protists in Chapter 29.

Check Your Progress 28.1

1. Describe the experiments Pasteur used to disprove the idea of spontaneous generation.
2. List the major groups of microbes studied in microbiology.
3. Explain what is meant by the term *microbiota.*

28.2 Origin of Microbial Life

Learning Outcomes

Upon completion of this section, you should be able to

1. Distinguish between chemical and biological evolution.
2. List and briefly describe the four stages thought to have led to the formation of the first cells.
3. Explain how the Miller—Urey experiment (among others) provided support for the "primordial soup" hypothesis.
4. Compare the implications of the protein-first, RNA-first, and membrane-first hypotheses for the evolution of the first living cell.

As noted in Section 27.1, Darwin's theory of evolution by natural selection is guided by the principle of *common ancestry,* that all life on Earth can be traced back to a single ancestor. This ancestor, also called the **last universal common ancestor (LUCA),** is common to all organisms that live, and have lived, on Earth since life began (Fig. 28.2). What were the properties of this common ancestor? How did life first begin on Earth? Darwin mused that perhaps the first living organism arose in a "warm little pond with all sorts of ammonia and phosphoric salts, light, heat, electricity, etc." With typical insight, Darwin was describing some of the actual conditions of ancient Earth that gave rise to the first forms of life.

In Section 1.1, we considered the characteristics shared by all organisms. Organisms acquire energy through *metabolism,* or the chemical reactions that occur within cells. They also respond and interact with their environment, self-replicate, and are subject to the forces of natural selection that drive adaptation to the environment. The molecules of organisms, called **biomolecules,** are organic molecules (see Section 2.4). The first organisms on Earth would have had all of these characteristics. Yet early Earth was very different from the Earth we know today, and consisted mainly of inorganic substances. How, then, did life get started in this inorganic "warm little pond"?

Advances in chemistry, evolutionary biology, paleontology, microbiology, and other branches of science have helped to develop new, and test old, hypotheses about the origin of life. These studies contribute to an ever-growing body of scientific evidence that life originated 3.5–4 billion years ago (BYA) from nonliving matter in a series of four stages:

Stage 1. *Organic monomers.* Simple organic molecules, called monomers, evolved from inorganic compounds prior to the existence of cells. Amino acids, the basis of proteins, and nucleotides, the building blocks of DNA and RNA, are examples of organic monomers.

Stage 2. *Organic polymers.* Organic monomers were joined, or polymerized, to form organic polymers, such as DNA, RNA, and proteins.

Stage 3. *Protobionts.* Organic polymers became enclosed in a membrane to form the first cell precursors, called protobionts (protocells).

Stage 4. *Living cells.* Protobionts acquired the ability to self-replicate, as well as other cellular properties.

Scientists have performed experiments to test hypotheses at each stage of the origin of life. Stages 1–2 involve the processes of a "chemical evolution," before the origin of life (see Fig. 28.2). Stages 3 and 4 are when life first evolved through the processes of "biological evolution."

Evolution of Monomers

In the 1920s, Russian biochemist Alexander Oparin and British geneticist J. B. S. Haldane independently hypothesized that the first stage in the origin of life was the evolution of simple organic molecules from the inorganic compounds present in the Earth's early atmosphere. The Oparin-Haldane hypothesis, sometimes called the *"primordial soup" hypothesis,* proposes that early Earth had very little oxygen (O_2), but instead was made up of water vapor (H_2O), hydrogen gas (H_2), methane (CH_4), and ammonia (NH_3).

Figure 28.2 Stages of the origin of life. The first organic molecules (*bottom*) originated from chemically altered inorganic molecules present on early Earth (Stage 1). More complex organic macromolecules were synthesized to create polymers (Stage 2) that were then enclosed in a plasma membrane to form the protobiont, or protocell (Stage 3). The protobiont underwent biological evolution to produce the first true, self-replicating, living cell (Stage 4). This first living cell underwent continued biological evolution (*top*), with a single surviving lineage, the LUCA, which became the common ancestor of all life on Earth.

Methane and ammonia are reducing agents because they readily donate their electrons. In the absence of oxygen, their reducing capability is powerful. Thus, the early Earth had a reducing atmosphere in which oxidation-reduction (redox) reactions could have driven the chemical evolution, or abiotic synthesis, of organic monomers from inorganic molecules in the presence of strong energy sources. Note that *oxidation* used in this context refers to a chemical redox reaction (see Section 6.4), and not to oxygen gas.

In 1953, the American chemist Stanley Miller, working under Harold Urey, performed a famous experiment to test the primordial soup hypothesis of early chemical evolution (Fig. 28.3). He reasoned that the energy sources on early Earth included heat from volcanoes and meteorites, radioactivity from isotopes, powerful electric discharges in lightning, and solar radiation. For his experiment, Miller placed a mixture of methane (CH_4), ammonia (NH_3), hydrogen (H_2), and water (H_2O) in a closed system, heated the mixture, and circulated it past an electric spark (simulating lightning). After a week's run, Miller discovered that a variety of amino acids and other organic acids had been produced.

The *Miller–Urey experiment* has been repeatedly tested over the decades since it was first performed. In 2008, a group of scientists examined 11 vials of compounds produced from variations of the Miller–Urey experiment, and found a greater variety of organic molecules than Miller reported, including all 22 amino acids. If early atmospheric gases did react with one another to produce small organic compounds, neither oxidation (no free oxygen was present) nor decay (no decomposers existed) would have destroyed these molecules, and rainfall would have washed them into the ocean, where they would have accumulated for hundreds of millions of years. Therefore, the oceans would have been a thick, warm organic soup—much like Darwin's "warm little pond."

Figure 28.3 Stanley Miller's experiment. Gases that were thought to be present in the early Earth's atmosphere were admitted to the apparatus, circulated past an energy source (electric spark), and cooled to produce a liquid that could be withdrawn. Upon chemical analysis, the liquid was found to contain various small organic molecules, which could serve as monomers for large cellular polymers.

The *Oparin-Haldane hypothesis* was an important contribution to our understanding of the early stages of life's origins, but other hypotheses have also been proposed and tested. In the late 1980s, German chemist Günter Wächtershäuser proposed that thermal vents at the bottom of the Earth's oceans (Fig. 28.4) provided all the elements and conditions necessary to synthesize organic monomers. According to his *"iron-sulfur world" hypothesis,* dissolved gases emitted from thermal vents, such as carbon monoxide (CO), ammonia, and hydrogen sulfide, pass over iron and nickel sulfide minerals, also present at thermal vents. The iron and nickel sulfide molecules act as catalysts that drive the chemical evolution from inorganic to organic molecules.

A very different line of thinking involves the comets and meteorites that have pelted the Earth throughout history. In recent years, scientists have confirmed the presence of organic molecules in some meteorites. Many scientists feel that these organic molecules could have seeded the chemical origin of life on early Earth.

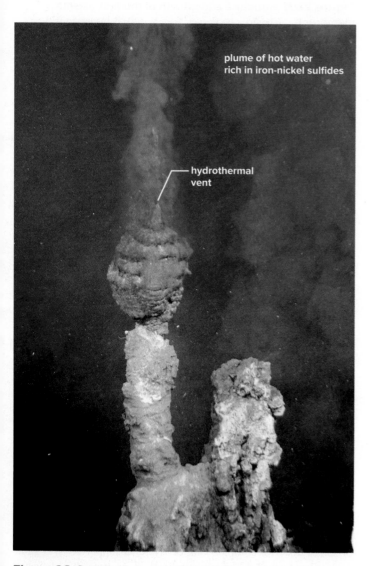

Figure 28.4 Chemical evolution at hydrothermal vents. Minerals that form at deep-sea hydrothermal vents like this one can catalyze the formation of ammonia and even monomers of larger organic molecules that occur in cells. Source: New Zealand American Submarine Ring of Fire 2007 Exploration, NOAA Vents Program

Others even hypothesize that bacterium-like cells evolved first on another planet and then were carried to Earth, an idea known as *panspermia.* The fact that Mars possessed liquid water in its past, and that life may have evolved there and then was carried to Earth, is one hypothesis that is being examined. Another event that lends evidence to panspermia is the discovery of glycerine and phosphorus found in the Churyumov–Gerasimenko comet in 2015. Glycerine is important because it is found in fat and is necessary for metabolic processes. The discovery of phosphorus is significant because it is an important element in so many biological molecules like ATP, DNA, RNA, and phospholipids.

Evolution of Polymers

Within a cell's cytoplasm, organic monomers join to form polymers in the presence of enzymes—such as the synthesis of protein polymers from amino acids by ribosomes. Enzymes themselves are proteins, but presumably there were no proteins present on early Earth. How did the first organic polymers form if no enzymes were present?

Building on the Miller–Urey experiment, American biochemist Sidney Fox suggested that once amino acids were present in the oceans, they could have collected in shallow puddles along the rocky shore. Then, the heat of the sun could have caused them to form *proteinoids,* small polypeptides that have some catalytic properties.

The formation of proteinoids has been simulated in the laboratory. When placed in water, proteinoids form microspheres, structures composed only of protein that have many properties of a cell. According to the **protein-first hypothesis,** some of these newly formed polypeptides had enzymatic properties. Moreover, if a certain level of enzyme activity provided an advantage over others, this would have set the stage for natural selection to shape the evolution of these first organic polymers. Those that evolved to be part of the first cell would have had a selective advantage over those that did not become part of a cell.

Fox's protein-first hypothesis assumes that protein enzymes arose prior to the first DNA molecule. Thus, the genes that encode proteins followed the evolution of the first polypeptides.

In contrast, the **RNA-first hypothesis** suggests that only the macromolecule RNA was needed to progress toward formation of the first cell or cells. Thomas Cech and Sidney Altman shared the 1989 Nobel Prize in Chemistry for their discovery that RNA can be both a substrate and an enzyme. Some viruses today have RNA genes; therefore, the first genes could have been RNA. It would seem, then, that RNA could have carried out the processes of life commonly associated today with DNA and proteins. Those who support this hypothesis say it was an "RNA world" some 4 BYA.

Evolution of Protobionts

Before the first true cell arose, a **protobiont** (also called a protocell) would have emerged. Protobionts are characterized as having an outer membrane. This membrane would have provided a boundary between the inside of the cell and its outside world, which was critical to the proper regulation and maintenance of cellular activities. Therefore, the evolution of a membrane was a critical step in the origin of life.

The plasma membrane of modern cells is made up of phospholipids assembled in a bilayer. The first plasma membranes were

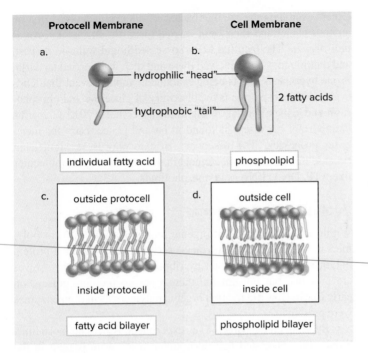

| Protocell Membrane | Cell Membrane |

a. hydrophilic "head"
hydrophobic "tail"

b. hydrophilic "head"
hydrophobic "tail"
2 fatty acids

individual fatty acid

phospholipid

c. outside protocell
inside protocell
fatty acid bilayer

d. outside cell
inside cell
phospholipid bilayer

Figure 28.5 A comparison of protobiont and modern cell plasma membranes.   The first lipid membrane was likely made of a single layer of fatty acids. These first protobiont (protocell) membranes and the modern cell membrane have features in common. **a.** Individual fatty acids have a single fatty acid chain with a hydrophilic head and hydrophobic tail. **b.** Phospholipids of modern cell membranes are made of two hydrophobic fatty acid chains (the "tails") attached to a hydrophilic head. **c.** Protobiont membranes were likely made up of a bilayer of fatty acids, with hydrophilic heads pointing outward, and hydrophobic tails pointing inward. **d.** Modern cells are organized in a similar fashion. Both bilayers create a semipermeable barrier between the inside and outside of a cell.

likely made up of fatty acids, which are smaller than phospholipids but like phospholipids have a hydrophobic "tail" and hydrophilic "head" (Fig. 28.5). Fatty acids are one of the organic polymers that could have formed from chemical reactions at deep-water thermal vents early in the history of life. In water, fatty acids assemble into small spheres called **micelles,** consisting of a single layer of fatty acids organized with their heads pointing out and tails pointing toward the center of the sphere.

Under appropriate conditions, micelles can merge to form vesicles. **Vesicles** are larger than micelles and are surrounded by a bilayer (two layers) of fatty acids (Fig. 28.6), similar to the phospholipid bilayer of modern cell membranes. An important feature of a vesicle lipid bilayer is that the individual fatty acids can flip between the two layers, which helps to move select molecules, such as amino acids, from outside the vesicle to its insides. The first protobiont would likely have been a type of vesicle with this type of fatty acid bilayer membrane. Interestingly, if lipids are made available to protein microspheres, lipids tend to become associated with microspheres, producing a lipid-protein membrane. Lipid-protein microspheres share some interesting properties with modern cells: they resemble bacteria; they have an electrical potential difference; and they divide and perhaps are subject to selection.

In the early 1960s, British biophysicist Alec Bangham discovered that when he extracted lipids from egg yolks and placed them

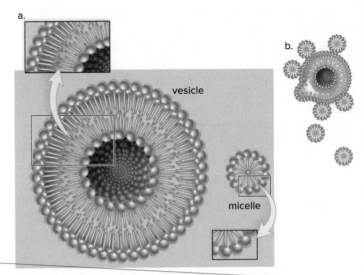

a.
vesicle

b.
micelle

Figure 28.6 Structure and growth of the first plasma membrane.   The first plasma membrane was likely made of a fatty acid bilayer, similar to that seen in vesicles. Protobionts (protocells), the ancestor of modern cells, are thought to have had this type of membrane. **a.** A cross-section of a vesicle reveals the fatty acid bilayer of a vesicle membrane. Micelles are spherical droplets formed by a single layer of fatty acids, and are much smaller than vesicles. **b.** Under proper conditions, micelles can merge to form vesicles. As micelles are added to the growing vesicle, individual fatty acids flip their hydrophilic heads toward the inside and outside of the vesicle, and their hydrophobic tails toward each other. This process forms a bilayer of fatty acids.

in water, the lipids would naturally organize themselves into double-layered bubbles roughly the size of a cell. Bangham's bubbles soon became known as **liposomes.** Bangham and others soon realized that liposomes might have provided life's first membranous boundary. Perhaps liposomes with a phospholipid membrane engulfed early molecules that had enzymatic, even replicative, abilities. The liposomes would have protected the molecules from their surroundings and concentrated them so they could react (and evolve) quickly and efficiently. These investigators called this the **membrane-first hypothesis,** meaning that the first cell had to have a plasma membrane before any of its other parts.

A protobiont would have had to acquire nutrition—that is, other molecules—so it could grow. One hypothesis suggests that protobionts were **heterotrophs,** organisms that consume preformed organic molecules. However, if the protobiont evolved at hydrothermal vents, it may have carried out *chemosynthesis*—the synthesis of organic molecules from inorganic molecules and nutrients. Indeed, many modern bacteria are **chemoautotrophs** that obtain energy by oxidizing inorganic compounds, such as hydrogen sulfide (H_2S), a molecule that is abundant at thermal vents. When hydrothermal vents in the deep and extremely dark ocean were first discovered in the 1970s, investigators were surprised to discover complex vent ecosystems supported by organic molecules formed by chemosynthesis, a process that does not require the energy of the sun.

ATP (adenosine triphosphate) is the most important energy-carrying molecule in living organisms. Because all life on Earth uses ATP to fuel cellular metabolism, the evolution of a means to synthesize ATP must have occurred very early in the history of life. The first protobionts evolved in the oxygen-poor environment of early Earth. Thus, it is probable that ATP was likely synthesized first by fermentation (see Section 7.3). The subsequent evolution of oxidative phosphorylation would have provided an advantage because it greatly increased the amount of ATP synthesized per unit of energy. Mitochondria share a common ancestor with a group of bacteria that synthesize ATP via an electron transport chain. Oxidative phosphorylation is possible in eukaryotes because mitochondria provide an electron transport chain ATP factory. At first, the protobiont must have had limited ability to break down organic molecules, and scientists speculate that it took millions of years for complex biochemical pathways to evolve completely.

Evolution of Living Cells

Today's cell is able to carry on protein synthesis in order to produce the enzymes that allow DNA to replicate. The **central dogma** of biology states that DNA directs protein synthesis and that information flows from DNA to RNA to protein. It is possible that this sequence developed in stages.

According to the RNA-first hypothesis, RNA would have been the first to evolve, and the first true cell would have had RNA genes. These genes would have directed and enzymatically carried out protein synthesis. As noted, today we know a number of viruses have RNA genes. These viruses have a protein enzyme called *reverse transcriptase* that uses RNA as a template to form DNA. Perhaps with time, reverse transcription occurred within the protobiont, and this is how DNA-encoded genes arose. If so, RNA was responsible for both DNA and protein formation.

According to the protein-first hypothesis, proteins, or at least polypeptides, were the first of the three (DNA, RNA, and protein) to arise. Only after the protobiont developed a plasma membrane and sophisticated enzymes did it have the ability to synthesize DNA and RNA from small molecules provided by the ocean. Because a nucleic acid is a complicated molecule, it is unlikely that RNA arose spontaneously from simple chemicals. It seems more likely that enzymes were needed to guide the synthesis of nucleotides and then nucleic acids. Once there were DNA genes, protein synthesis would have been carried out in the manner dictated by the central dogma of genetics.

The Scottish chemist Alexander Cairns-Smith proposed that polypeptides and RNA evolved simultaneously. Therefore, the first true cell would have contained RNA genes that could have replicated because of the presence of proteins. This eliminates the baffling chicken-or-egg paradox: Assuming a plasma membrane, which came first, proteins or RNA? It means, however, that two unlikely events would have had to happen at the same time.

After DNA formed, the genetic code had to evolve before DNA could store genetic information. The present genetic code is subject to fewer errors than a million other possible codes. Also, the present code is among the best at minimizing the effect of mutations. A single-base change in a present codon is likely to result in the substitution of a chemically similar amino acid and, therefore, minimal changes in the final protein. This evidence suggests that the genetic code did undergo a natural selection process before finalizing into today's code.

Check Your Progress 28.2

1. Explain the role of biomolecules in chemical and biological evolution.
2. List the four stages of the evolution of life, and explain why they likely occurred in a particular order.
3. Compare and contrast the "primordial soup" and the "iron-sulfur world" hypotheses.
4. Explain why the protein-first versus RNA-first debate can be compared to a chicken-or-egg paradox.

28.3 Archaea

Learning Outcomes

Upon completion of this section, you should be able to
1. List some of the major criteria used to distinguish the domains Archaea, Bacteria, and Eukarya.
2. Review the structural features of the plasma membranes and cell walls of archaea.
3. Distinguish between halophiles, thermoacidophiles, and methanogens.

As noted in Section 1.2, the bacteria (domain Bacteria) and archaea (domain Archaea) are prokaryotes, but each is placed in its own domain because of molecular and cellular differences. **Prokaryotes** are single-celled organisms that lack a nucleus and membrane-bound cytoplasmic organelles found in eukaryotic cells. Archaea and bacteria are not close relatives, even though both are prokaryotes. Based on a number of criteria, including nucleic acid similarities, the eukarya are more closely related to archaea than to bacteria (Table 28.1).

Archaeal Size and Structure

Archaea usually range from 0.1 to 15 μm in size. Their genome is a single, closed, circular DNA molecule, often smaller than a bacterial genome. Like bacteria, archaea reproduce asexually by *binary fission*.

The plasma membrane of archaea differs markedly from those of bacteria and eukaryotes. Rather than a lipid bilayer, archaea often have a monolayer of lipids with branched side chains. These chemical characteristics help many archaea tolerate acid and heat.

TABLE 28.1 Comparison of Domains Archaea and Eukarya

Feature	Domain	
	Archaea	Eukarya
Nucleus	No	Yes
Organelles	No	Yes
Introns	Sometimes	Yes
Histones	Yes	Yes
RNA polymerase	Several types	Several types
Methionine is at start of protein synthesis	Yes	Yes

The cell walls of archaea may also facilitate their survival in extreme environments. The cell walls of archaea lack peptidoglycan, which is a distinguishing feature of bacterial cell walls (see Section 28.4). In some archaea, the cell wall is largely composed of polysaccharide, and in others, it is pure protein. A few have no cell wall.

Types of Archaea

Three main types of archaea are distinguished based on their unique habitats and metabolic activities: halophiles, thermoacidophiles, and methanogens (Fig. 28.7). Other archaea are found in moderate environments, such as lake sediments and soil, where they are likely involved in nutrient cycling. Several archaea have been found living in symbiotic relationships with animals, including sponges, sea cucumbers, and in the digestive tracts of humans and other animals (see the content below under "Methanogens"). However, no parasitic archaea have yet been found—that is, they are not known to cause any infectious disease.

Halophiles

The **halophiles** (salt lovers) usually thrive at salt concentrations of around 12–15%; in contrast, ocean water is about 3.5% salt. Halophiles have been isolated from environments such as the Great Salt Lake in Utah, the Dead Sea, and hypersaline soils, where most other organisms cannot survive.

These archaea have evolved a number of mechanisms enabling them to survive in very salty environments. Their plasma membranes often have chloride pumps that contain a light-powered protein called *halorhodopsin* (related to the rhodopsin pigment found in the human retina). This unique protein pumps chloride and water into the cell, thereby preventing excessive dehydration.

Some halophiles also perform a type of photosynthesis that uses a pigment called *bacteriorhodopsin* instead of chlorophyll. When the pigment absorbs solar energy, it moves a hydrogen ion to outside the plasma membrane. This establishes an H⁺ gradient that promotes ATP synthesis.

Thermoacidophiles

Another major type of archaea, the **thermoacidophiles,** are found in extremely hot, acidic, aquatic environments such as hot springs, geysers, and underwater volcanoes. Unlike the great majority of life-forms on Earth, the lipid membranes and proteins of these archaea have evolved to withstand and function optimally at temperatures as high as 80°C. Some can even grow at 105°C (remember that water boils at 100°C)!

One thermoacidophilic species, *Picrophilus torridus*, has been particularly well studied for its ability to survive at a pH of less than 1, similar to that found in a 1.2-molar sulfuric acid solution (car battery acid is around 4.5 molar). Sequencing of the genome of this organism has shown that a relatively high number (12%) of its genes are involved in transport functions of its plasma membrane. Many of these transporter proteins are presumably involved in pumping excess H⁺ ions out of the cell, as well as in taking advantage of the high density of H⁺ outside the cell to transport other desired molecules into the cell. As one of the most acid-tolerant organisms known, *P. torridus* will likely provide enzymes for various biotechnology processes that require acidic conditions.

a. Halophiles

b. Thermoacidophiles

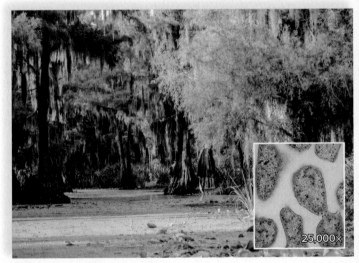

c. Methanogens

Figure 28.7 Archaea in extreme habitats.   Many archaea thrive in unusual environmental conditions. **a.** Halophilic archaea can live in salt lakes. **b.** Thermophilic archaea can live in the hot springs of Yellowstone National Park. **c.** Methanogens live in anaerobic swamps and in the guts of animals. (a) (Salt Lake): ©Marco Regalia Sell/Alamy; (halophiles): ©Eye of Science/Science Source; (b) (geyser): ©Alfredo Mancia/Getty Images; (thermoacidophile): ©Eye of Science/Science Source; (c) (swamp): ©iStockphoto/Getty Images; (methanogens): ©Dr. M. Rohde, GBF/Science Source

Methanogens

The **methanogens** (methane-makers) mostly use carbon dioxide and hydrogen as energy sources, producing methane as a by-product. Methanogens are found in anaerobic environments such as swamps, lake sediments, rice paddies, and the intestines of animals. Cows, which have large populations of methanogens in their digestive tracts, release a significant amount of methane into the environment. Because methane is a greenhouse gas that may contribute to global warming and climate change, some scientists are suggesting that the amount of methane produced by cattle should be reduced by changing the animals' diets, adding substances to their feed to alter the microbial populations, or developing a vaccine to specifically inhibit the growth of methanogens. However, the relative contribution of cattle, compared to other potential sources of methane, remains controversial.

Check Your Progress 28.3

1. List some key characteristics of archaea, bacteria, and eukarya.
2. Explain how the plasma membranes of archaea differ from those of bacteria and eukaryotes.
3. Describe adaptations that allow archaea to survive in extremely hot or salty environments.
4. Discuss why a dedicated environmentalist might give up eating beef.

28.4 Bacteria

Learning Outcomes

Upon completion of this section, you should be able to

1. Identify the major structural features of bacteria.
2. Describe bacterial reproduction, including three mechanisms by which bacteria increase variation.
3. Explain why bacteria are considered to be metabolically diverse.
4. List several major bacterial diseases of humans and describe how they are treated.

Bacteria (domain Bacteria) are the most common type of prokaryote on Earth. The amount of bacteria on our planet is amazing: Even though we can't see them with our naked eye, the biomass of bacteria on Earth exceeds that of plants and animals combined. They are found in practically every kind of environment on Earth. In this section, we consider the bacteria—their characteristics, metabolism, and lifestyle.

Bacterial Size and Structure

Most bacteria are between 0.2 and 10 μm in size. A few, however, are quite large, including one species that is about the same size as the period at the end of this sentence. Bacteria have three basic shapes: rod (bacillus, *pl.,* bacilli); spherical (coccus, *pl.,* cocci); and spiral-shaped or helical (spirillum, *pl.,* spirilla) (Fig. 28.8). A bacillus or coccus can occur singly or may occur in particular arrangements. For example, when cocci form a cluster, they are called staphylococci, whereas cocci that form chains are called streptococci.

The typical structure of a bacterium is shown in Figure 28.9. All bacterial cells have a plasma membrane, which is a lipid bilayer, similar to the plasma membrane in plant and animal cells. Most bacterial cells are further protected by a cell wall that contains the unique molecule peptidoglycan.

Bacteria can be classified by differences in their cell walls, which are detected using a staining procedure devised more than 100 years ago by Hans Christian Gram. When you go to the doctor for a bacterial infection, one of the most common tests performed is the **Gram stain,** because different antibiotics tend to be more effective against Gram-positive or Gram-negative bacteria. Cell walls that have a thick layer of peptidoglycan outside the plasma membrane stain purple with the Gram stain procedure, and are called Gram-positive bacteria. If the peptidoglycan layer is either thin or lacking altogether, the cells stain pink and are considered Gram-negative. In addition to their plasma membrane, Gram-negative bacteria have an outer membrane that contains **lipopolysaccharide** molecules. When these Gram-negative cells are killed by your immune system, these molecules are released, stimulating inflammation and fever. Beyond the cell wall, some bacteria have a slimy polysaccharide capsule that can protect the cell from dehydration and the immune system.

a. Bacilli (rod): SEM 13,300×
 Escherichia coli

b. Cocci: SEM 6,250×
 Streptococcus thermophilus

c. Spirillum: LM 400×
 Spirillum volutans

Figure 28.8 Typical shapes of bacteria. **a.** Bacilli, rod-shaped bacteria. **b.** Cocci (spherical). **c.** Spirillum (curved). (a): ©Eye of Science/Science Source; (b): ©SciMAT/Science Source; (c): ©Ed Reschke/Getty Images

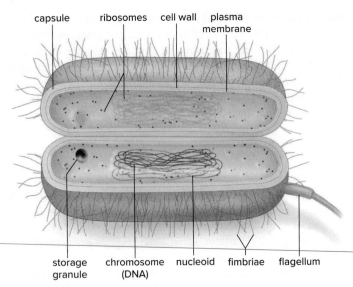

Figure 28.9 Typical bacterial cell.   Bacteria are prokaryotes and thus have no membrane-bound nuclei or organelles.

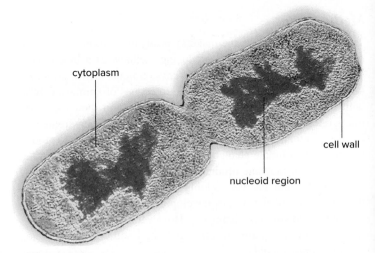

Figure 28.10 Binary fission.   When conditions are favorable for growth, prokaryotes divide to reproduce. Binary fission is a form of asexual reproduction because the daughter cells have exactly the same genetic material as the parent. ©CNRI/SPL/Science Source

Motile bacteria have **flagella** for locomotion, but never cilia. The flagellum is a stiff, curved filament that rotates like a propeller. Bacterial flagella are structurally distinct from eukaryotic flagella. Some bacteria have *fimbriae* that bind to various surfaces. For example, bacteria that cause urinary tract infections can bind to urinary tract cells. Drinking cranberry juice inhibits this binding.

Most bacteria have a single circular chromosome, which is located in a **nucleoid** region of the cytoplasm, rather than in a membrane-bound nucleus. Many bacteria also harbor accessory rings of DNA called *plasmids* that can carry genes for antibiotic resistance, among other things, and are commonly used to carry foreign DNA into other bacteria during genetic engineering (see Section 26.1). Bacterial cells also contain an abundance of ribosomes, as well as various types of granules that store nutrients such as glycogen and lipids. Notice that bacteria do not have an endoplasmic reticulum, a Golgi apparatus, mitochondria, or chloroplasts. This diagram summarizes the major structural features of bacteria:

Bacterial Reproduction and Gene Transfer

Bacteria reproduce asexually. After a period of sufficient growth, the bacterial cell simply divides into two new cells, with each cell getting a copy of the genome and approximately half of the cytoplasm. This process is known as **binary fission** (Fig. 28.10). Each daughter cell is a *clone,* or exact copy, of the parent cell. When bacteria are spread out and grown on agar plates, each individual

cell can give rise to a colony of millions of cells. In the absence of genetic mutations, each of these cells is a clone of the initial cell. Some bacteria need only 20 minutes to reproduce, whereas others grow more slowly, with generation times of a day or more.

Under unfavorable conditions, some bacteria, such as *Clostridium tetani,* the cause of tetanus, can produce resistant structures called **endospores,** thick-walled, dehydrated structures capable of surviving the harshest conditions, perhaps even for thousands of years. Endospores are not for reproduction, but are simply a way to survive unfavorable conditions.

Sexual reproduction does not occur among prokaryotes, but at least three means of *horizontal gene transfer* are known. **Conjugation** takes place when a donor cell passes DNA to a recipient cell by way of a *sex pilus.* **Transformation** occurs when a bacterium takes up DNA released into the medium by dead bacteria. During **transduction,** viruses carry portions of bacterial DNA from one bacterium to another. As discussed in Section 26.2, scientists are using these processes to build genetically modified organisms (GMOs).

Bacterial Metabolism

Although most bacteria are structurally similar to each other, they demonstrate a remarkable range of metabolic abilities. Most bacteria are heterotrophs that require an outside source of organic compounds in the same way animals do. Some heterotrophic bacteria are anaerobic and cannot use oxygen to capture electrons at the end of their electron transport chain (which is physically located in their plasma membrane, as all prokaryotes lack mitochondria). Instead, they use a variety of substances as final electron acceptors. Sulfate-reducing bacteria transfer the electrons to sulfate, producing hydrogen sulfide, which smells like rotten eggs. Denitrifying bacteria use nitrate, whereas others use minerals, such as iron or manganese, as electron receivers.

Other bacteria are chemoautotrophs. They reduce carbon dioxide to an organic compound by using energetic electrons derived from chemicals, such as ammonia, hydrogen gas, and hydrogen sulfide. Electrons can also be extracted from certain minerals, such as iron.

Some bacteria are photosynthesizers that use solar energy to produce their own food. **Cyanobacteria** are believed to have arisen some 3.8 billion years ago and to have produced much of the oxygen in the atmosphere at that time. Sometimes erroneously called *blue-green algae*, these organisms contain green chlorophyll and have other pigments that give them a bluish-green color (Fig. 28.11). Other bacterial photosynthesizers split hydrogen sulfide, instead of water, in anaerobic environments. Therefore, they produce sulfur rather than oxygen as a by-product of photosynthesis.

Aside from producing oxygen, cyanobacteria are often the first colonizers of rocks. Many are capable of both carbon fixation and nitrogen fixation, needing only minerals, air, sunlight, and water for growth. Cyanobacteria are also notorious for forming toxic blooms in waters enriched by nutrients, which may require that some lakes be closed to human use. Cyanobacteria (and in some cases eukaryotic algae) form a symbiotic relationship with fungi in a *lichen*. The fungi provide a place for the cyanobacteria to grow and obtain water and mineral nutrients. Some cyanobacteria even appear to live in a symbiotic relationship in the fur of sloths, providing the animals with a form of camouflage in their dense green forest home!

a. *Chroococcus*

b. *Oscillatoria*

Figure 28.11 Cyanobacteria. Cyanobacteria are photosynthetic bacteria that contain chlorophyll. **a.** In *Chroococcus,* single cells are grouped in a common gelatinous sheath. **b.** Filaments of cells occur in *Oscillatoria,* a common inhabitant in freshwater environments. (a): ©Biology Pix/Science Source; (b): ©M. I. Walker/Alamy

Bacterial Diseases of Humans

Most types of bacteria do not cause disease, but a significant number do. Why does one microbe cause disease, whereas another, closely related species is completely harmless? Pathogenic microbes often carry genes that code for specific virulence factors that determine the type and extent of illness they are capable of causing. For example, right now you have billions of *Escherichia coli* living in your large intestine, without causing any problems. However, some strains of *E. coli* have acquired genes that make them dangerous invasive pathogens. The strain of *E. coli* called O157:H7 has the ability to generate a toxin that damages the lining of the intestine, resulting in a bloody diarrhea. It has also obtained other virulence factors that help it stick to the lining of the gut more efficiently. In 2011, around 60 people in ten states became ill from *E. coli* O157:H7 contamination in Romaine lettuce. That same year, an outbreak of pathogenic *E. coli* in Germany sickened 852 people and caused 32 deaths.

By acquiring genes coding for virulence factors (i.e., via conjugation, transformation, and/or transduction), harmless bacteria may be converted into pathogens. Antibiotic resistance genes may pass between organisms in the same ways, creating antibiotic-resistant pathogens. Figure 27.17 and the Health feature "Evolution of Antibiotic Resistance" in Section 27.4 describe how antibiotic resistance evolves in bacteria.

Some bacterial diseases that are transmitted sexually were discussed in Section 21.5. Next, we discuss a few more common types of bacterial diseases in humans.

Streptococcal Infections

More different types of human disease are caused by bacteria from the genus *Streptococcus* than by any other type of bacterium. *Streptococcus pneumoniae* can cause pneumonia, meningitis, and middle ear infections. Up to 70% of apparently healthy adults are carriers of this potentially harmful bacterium, which causes disease mainly in children and the elderly. *S. mutans* is found on the teeth and contributes to tooth decay and the formation of dental caries.

The most common illness caused by streptococci is pharyngitis, commonly called strep throat, which is usually due to infection by *S. pyogenes* (Fig. 28.12*a*). *S. pyogenes* also causes relatively mild skin diseases, such as impetigo in infants (Fig. 28.12*b*). A number of more serious and invasive diseases can occur after or during infections with *S. pyogenes*. The nature and severity of these diseases depend on which virulence factors the pathogen carries. Scarlet fever is a strep infection caused by a *Streptococcus* strain that produces a toxin causing a red rash. An intense immune response to the infection can lead to rheumatic fever, which is characterized by a high temperature, swollen joints that form nodules, and heart damage. Although rarely seen in the United States today, rheumatic fever killed more school-age children than all other diseases combined in the early twentieth century.

By releasing enzymes that destroy connective and muscle tissues and kill cells, *S. pyogenes* infection can lead to large-scale cell lysis and tissue destruction. "Flesh-eating" bacteria cause necrotizing fasciitis (Fig. 28.12*c*), in which rapid tissue damage can require amputation of infected limbs or, in 40% of cases, rapid death.

a. *Streptococcus pyogenes*   20,000×   b. Impetigo   c. Flesh-eating disease

Figure 28.12 *Streptococcus pyogenes*. **a.** *Streptococcus pyogenes,* shown here in an electron micrograph, often forms chains of cells, as seen here. **b.** Impetigo is a common mild skin infection. **c.** Flesh-eating disease is rare but life-threatening. (a): ©Dr. Kari Lounatmaa/Science Source; (b): ©SPL/Science Source; (c): ©ZUMA Press, Inc./Alamy

Staphylococcus aureus and MRSA

Staphylococcus aureus is perhaps the bacterial species that has received the most media attention recently. About 20% of people are carriers of this bacterium (mostly on their skin or in their nostrils) without any symptoms. When *S. aureus* does cause disease, it is usually limited to skin infections. However, in people who are very young, very old, or immunocompromised for any other reason, it can invade the body and cause life-threatening disease. Moreover, a strain of *S. aureus* that is resistant to methicillin, called MRSA, is killing an increasing number of young, otherwise healthy individuals. Besides being resistant to many antibiotics, MRSA strains often possess genes coding for toxins not found in other *S. aureus* strains. Some of these toxins can be very damaging to tissues. Since the peak in MRSA infections between 2006 and 2008, the number of annual cases have been declining, mostly because health-care facilities have initiated infection-control procedures and increased awareness among health professionals.

Tuberculosis

Tuberculosis (TB) is one of the leading worldwide causes of death due to infectious disease. When first described in 1882 by the father of medical microbiology, Robert Koch, TB caused one of every seven deaths in Europe, and one-third of all deaths of young adults. Today, it is estimated that one-third of the world's population is infected with the TB bacterium, causing approximately 2 million deaths each year. Tuberculosis is a chronic disease caused by *Mycobacterium tuberculosis,* a pathogen closely related to *M. leprae,* the causative agent of leprosy. Generally, *M. tuberculosis* infection occurs in the lungs, but it can occur elsewhere as well.

M. tuberculosis is very slow growing, and the extent of the disease is determined by host susceptibility. An immune response results in inflammation of the lungs. Active lesions can produce dense structures called *tubercles* (Fig. 28.13). These can persist for years, causing symptoms such as coughing, and spread the bacteria to other areas of the lungs and body, as well as to other individuals. Eventually, the damaged lung tissue hardens and calcifies, leaving characteristic spots that can be observed with chest X-rays. The number of TB cases in the United States surged in the late 1980s, mainly due to increased rates of infection with HIV, which damages

tubercle

a.   b.   4×

Figure 28.13 Tuberculosis. Tuberculosis, an infection caused by the bacterium *Mycobacterium tuberculosis,* usually settles in the lungs. **a.** Lung tissue with a large cavity surrounded by tubercles, which are hard, calcified nodules where the bacteria are trapped. **b.** Photomicrograph shows a cross-section of one large tubercle, with a diseased center. (a): ©McGraw-Hill Education; (b): ©Carolina Biological Supply Company/Medical Images

helper T cells needed to fight the TB bacterium. Since 1992, however, the prevalence of TB in the United States has steadily declined to a rate of three cases per 100,000 people in 2015. This reduction is mainly attributed to better testing and treatment of infected persons. This trend is also occurring worldwide, although numbers of TB cases continue to increase in Africa and Southeast Asia.

The 2007 story of a 31-year-old Atlanta lawyer with TB shows how easily an infectious agent can spread around the world. Despite knowing he was infected with *M. tuberculosis,* he flew from Atlanta to Paris to get married, traveling through five countries in the process. Making matters worse, he was infected with a strain of *M. tuberculosis* called XDR, which is resistant to nearly all drugs available to treat TB. Upon returning to the United States, he was taken into forced quarantine for 4 weeks, the first time an American had been forcibly isolated by the Centers for Disease Control and Prevention (CDC) in several decades. He was released after it was determined that he was no longer infectious. It should be noted, however, that there is no way to know how many people infected with *M. tuberculosis* are traveling on airplanes every day.

Antibiotics and Probiotics

In 1928 at a London lab, Scottish scientist Alexander Fleming was researching substances that might inhibit bacterial growth when he took a 2-week vacation. Upon returning, he noticed some petri dishes that had bacteria growing on them were contaminated with *Penicillium* fungus. In one of the most famous moments in medical history, Fleming realized that the fungus was producing a chemical that inhibited the growth of the bacteria (Fig. 28A*a*). Of course that substance turned out to be penicillin, and the practice of medicine was changed forever. Many bacterial infections that had previously meant serious illness or even death were now treatable.

As often happens, however, the widespread use of Fleming's new drug, and many others like it, had unintended consequences. Especially in patients who take antibiotics frequently or for a prolonged period, normal microbial populations in the body can be disrupted, resulting in diarrhea, yeast infections, or worse. A bacterium called *Clostridium difficile*, which is normally present in the intestines at relatively low levels, tends to overgrow in these patients, and may cause a severe, or even fatal, inflammation of the colon. Treatment for this disease is usually more antibiotics, but some authorities are suggesting that probiotics may be helpful in preventing this and many similar diseases.

Grocery store shelves and Internet health sites are full of products claiming to contain "probiotics," usually containing bacteria of the genera *Lactobacillus* or *Bifidobacterium*. According to the U.N. Food and Agricultural Organization, probiotics are "live microorganisms, which, when administered in adequate amounts, confer a health benefit on the host."[2] By this definition, foods like yogurt, cheeses, and other fermented dairy products, are probiotics. But because of the increased interest in the health benefits of "friendly" bacteria, live bacteria are being added to an increasing variety of products (Fig. 28A*b*). Americans spent over $1 billion on probiotic products in 2011, and are expected to spend twice as much on probiotics by 2016.

How do probiotics work? As already mentioned in this chapter, the normal microflora of the body can produce vitamins and aid digestion. They also inhibit harmful microbes by simply taking up space and nutrients, and in some cases, by secreting chemicals that directly kill pathogens. But the beneficial effects of normal microflora go beyond competing with other microbes. Body surfaces such as the intestinal wall are fortified with large populations of immune cells that guard against invasion of the tissues by pathogens. An ongoing interaction between these cells and the normal microflora is necessary to maintain a healthy immune system. Experimentally, animals raised under gnotobiotic (germ-free) conditions have poorly developed immune systems, and replenishing their bodies with "good" bacteria restores healthy immune functions, sort of like priming a pump. That is one reason why probiotics are also being tested for treatment of inflammatory conditions such as irritable bowel syndrome, ulcerative colitis, and Crohn's disease. Future applications may also include treating urinary tract and vaginal infections, inhibiting food allergies, preventing tooth decay, and improving the effectiveness of vaccines!

Despite their potential, however, one aspect of probiotics to be aware of is quality control. Currently, there is very little regulation of which bacterial strains, or even the number of beneficial organisms, a product must contain to be labeled as probiotic. And of course, it is extremely unlikely that all the potential benefits of probiotics will pan out. In a period of less than 100 years, however, we have gone from discovering a new way to inhibit or kill harmful bacteria that invade our bodies, to realizing how closely our health is linked to our relationship with harmless microflora.

Questions to Consider

1. If no antibiotics existed, how might your life be different?
2. Besides taking antibiotics, what other factors might influence the numbers and types of microflora in your body?
3. High numbers of "good" bacteria are found in the intestine and on the skin. The immune system needs to protect these areas from invading microbes, but cannot respond as strongly to the normal microflora without causing problems. What are some possible ways that immune cells could distinguish "good" from "bad" bacteria?

---

[2] U.N. Food and Agricultural Organization

Figure 28A **Changing attitudes about microbes.**
a. When Fleming discovered the inhibition of bacterial growth by *Penicillium* (a fungal contaminant) in 1928, few would have predicted the (**b**) benefits of bacteria less than 100 years later.
(a): ©Bettmann/Contributor/ Getty Images;
(b): ©McGraw-Hill Education

a.   b.

Food Poisoning

Whether the source was mishandled food at a salad bar, or potato salad that sat in the sun too long at a picnic, all of us have probably experienced the intestinal discomfort known as food poisoning. Two basic types of bacteria cause food poisoning: those that produce toxins while they are growing in food, and those that cause infections once they are in the intestine.

Several species of bacteria can produce toxins in foods, especially those containing dairy, eggs, or meat products. The symptoms, which consist mainly of vomiting and diarrhea, tend to appear suddenly within a few hours of ingestion, and are usually self-limiting. In contrast, *Clostridium botulinum,* the causative agent of *botulism,* produces one of the most toxic substances on Earth. When people are canning and don't heat foods to a high enough temperature, *Clostridium* can produce endospores that survive the canning process. These endospores then germinate in the airless environment of the can or bottle and become toxin-producing cells. If untreated, about 25% of people who ingest botulism toxin die from respiratory paralysis.

Salmonella is a classic example of a bacterial food poisoning agent that does not produce gastroenteritis until it has reproduced in the intestines for several days. A recent large outbreak of salmonellosis in the United States occurred between late summer of 2008 and January 2009, prompting the FDA to recall every product made from peanuts processed by a Georgia factory. Over 700 people became ill, and at least 9 died after ingesting contaminated products.

Drug Control of Bacterial Diseases

Most **antibiotics** kill or inhibit bacteria by interfering with their unique metabolic pathways. Therefore, they are not expected to harm human cells. For example, erythromycin and tetracyclines inhibit bacterial protein synthesis by binding to bacterial ribosomes, whereas penicillins and cephalosporins inhibit bacterial cell wall synthesis.

There are problems associated with antibiotic therapy. Some individuals are allergic to certain antibiotics, and the reaction may even be fatal. Antibiotics not only kill off disease-causing bacteria, they may also reduce the number of beneficial bacteria in the intestinal tract and vagina. This may allow the overgrowth of certain harmful bacteria or yeast, especially in the intestine, vagina, or mouth. See the Health feature "Antibiotics and Probiotics" for a discussion of how probiotics can be used to replenish these beneficial organisms.

Most important perhaps is the growing resistance of bacteria to antibiotics, as discussed in the story that opened Chapter 27. Antibiotics were introduced in the 1940s, and for several decades they worked so well it appeared that infectious diseases had been brought under control. However, we now know that bacterial strains can mutate and become resistant to a particular antibiotic. Worse yet, when bacteria exchange genetic material, resistance can pass between different bacteria. Penicillin and tetracycline now have a failure rate of more than 22% against *Neisseria gonorrhoeae,* which causes gonorrhea. Bacteria with multiple drug resistances, such as MRSA and multidrug-resistant *M. tuberculosis,* are an especially serious threat in hospitals, prisons, and long-term care facilities. With the rise of antibiotic resistance, many companies are now working to develop innovative kinds of antibacterial therapies.

Recent advances in the study of genomics (see Section 26.4) has enabled scientists to rapidly screen the genomes of bacteria for the genes that produce novel compounds that may be used to produce new classes of antibiotics. There has been some recent success in these efforts, but it still takes years of research and development to develop these new drugs for widespread use.

Check Your Progress 28.4

1. Describe the three basic shapes of bacteria.
2. Explain how bacterial conjugation differs from transformation and transduction.
3. Compare the nutritional strategy of a heterotrophic bacterium with that of a chemoautotroph.
4. Consider how antibiotics work, and then describe two specific mechanisms bacteria could use to become resistant.

28.5 Viruses, Viroids, and Prions

Learning Outcomes

Upon completion of this section, you should be able to

1. Identify the major structural features of viruses.
2. Describe the steps in a typical viral reproductive cycle and explain how some viruses become latent.
3. Describe several viral diseases of humans and explain why it is difficult to produce vaccines against some viruses.
4. Distinguish among viruses, viroids, and prions.

As we learned in Section 1.1, all living organisms are composed of cells. **Viruses** are acellular, meaning they are not composed of cells. Also, viruses are obligate parasites, meaning they can reproduce only inside a living cell (called the host cell) by utilizing at least some of the machinery (ribosomes, certain enzymes, etc.) of that cell. Therefore, the question arises, "Are viruses alive?"

Scientists and philosophers have long argued this question. Some say that viruses are not alive. After all, not only are they acellular, but also some have been synthesized in the lab from chemicals! Moreover, when viruses are outside a host cell, they are totally quiescent, exhibiting no metabolic activity. Others argue that viruses are alive because they have a genome that directs their reproduction when they are inside the host cell.

If viruses are not considered alive, then certainly the simpler viroids and prions are not alive either. Viroids are strands of RNA that can reproduce inside a cell, and prions are protein molecules that cause other proteins to become prions.

Viral Size and Structure

Most viruses are much smaller than bacteria. Viruses typically measure between 0.03 and 0.2 μm, whereas most bacteria measure at least 0.5 μm. Interestingly, the largest virus was recently discovered in Siberia. Called Pithovirus, it is about 1.5 μm in size, which is larger than the smallest known bacterium.

Viruses come in a variety of shapes, including helices, spheres, polyhedrons, and more complex forms. A virus always has at least two parts—an outer **capsid** composed of protein subunits, which protects an inner core of nucleic acid (Fig. 28.14). The viral genome can be single- or double-stranded DNA, or single- or double-stranded RNA. This diversity in genetic material is different from all cellular organisms, which always have a double-stranded DNA genome. Special viral enzymes that help a virus reproduce can also be inside the capsid.

In some viruses, especially those that infect animals, the capsid is surrounded by a membrane called an **envelope.** This envelope is made of lipid, and is usually derived from the host cell's plasma membrane when the virus buds from the host cell. Viral glycoproteins called spikes often extend from the envelope. These spikes are critical to viral infection because they help the virus bind to the surface of the host cell before entering it.

Viral Reproduction

Viruses infect almost every type of organism on Earth. Viruses called *bacteriophages* infect bacterial cells, usually just a particular group or species of bacteria. Some viruses infect only plants, whereas others infect only animals. Some viruses can infect and cause disease only in humans. The specificity of a virus for a host occurs because the spike of a virus and a receptor molecule on a cell's plasma membrane fit together like a hand fits a glove, which triggers events allowing viral entry into the cell.

Figure 28.15 illustrates the reproductive cycle of a typical enveloped animal RNA virus. This reproductive cycle has six steps:

1. During *attachment,* the spikes of the virus bind to a specific receptor molecule on the surface of a host cell. The host cell normally uses this receptor for another purpose, but it is effectively "hijacked" by the virus for its own purposes.
2. During *entry,* also called penetration, the viral envelope fuses with the host's plasma membrane, and the rest of the virus (capsid and viral genome) enters the cell. The genome is freed when cellular enzymes remove the capsid, a process called *uncoating.*
3. *Replication* occurs when a viral enzyme makes complementary copies of the genome, which is RNA in this case.
4. During *biosynthesis,* some of these RNA molecules serve as mRNA for the production of more capsid and spike proteins, using host ribosomes.
5. At *assembly,* a mature capsid forms around a copy of the viral genome.
6. During *budding,* new viruses are released from the cell surface. During this process, they acquire a portion of the host cell's plasma membrane and spikes, which were specified by viral genes during biosynthesis. The enveloped viruses are now free to spread the infection to other cells.

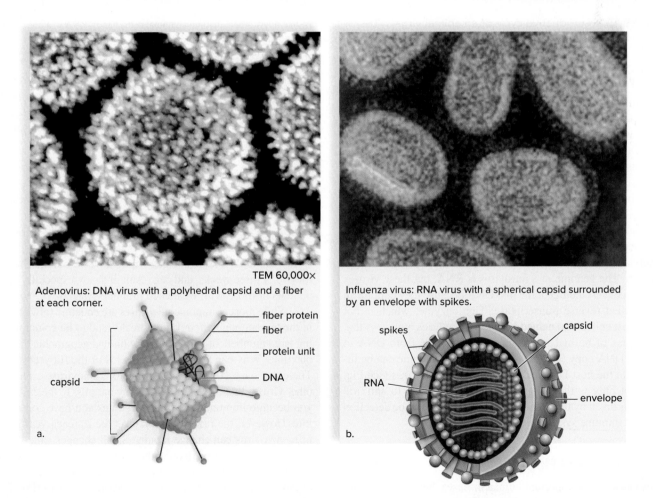

Figure 28.14 Viruses.   a. Despite their diversity, all viruses have an outer capsid composed of protein subunits and a nucleic acid core that is composed of either DNA or RNA. **b.** Some types of viruses also have a membranous envelope. (photos) (a): ©Biophoto Associates/Science Source; (b): Source: Cynthia Goldsmith/CDC

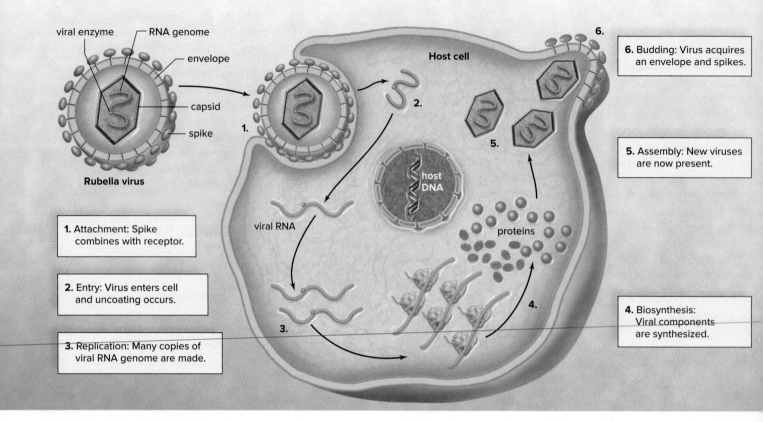

viral enzyme — RNA genome

— envelope

— capsid

— spike

Rubella virus

Host cell

6. Budding: Virus acquires an envelope and spikes.

5. Assembly: New viruses are now present.

host DNA

proteins

viral RNA

1. Attachment: Spike combines with receptor.

2. Entry: Virus enters cell and uncoating occurs.

3. Replication: Many copies of viral RNA genome are made.

4. Biosynthesis: Viral components are synthesized.

Figure 28.15 Reproductive cycle of an animal virus.   The rubella virus, like many others that infect animal cells, has an RNA genome. Notice also that the mode of entry requires uncoating and that the virus acquires an envelope and spikes when it buds from the host cell.

Latency

Some animal viruses can become latent (hidden) inside the host cell. Herpesviruses and retroviruses are well known for using this strategy, which helps them avoid detection by the host immune system. During latency, new viruses are not produced, but the viral genome is reproduced along with the host cell. Environmental stresses, such as ultraviolet radiation, can induce the latent virus to enter the biosynthesis stage, leading to the production of new virus particles. Borrowing terms that were first defined in bacterio-phages, latent viruses are said to be "lysogenic," whereas viruses that are actively reproducing are known as "lytic."

Retroviruses add an interesting twist to the story of viral re-production. The genome of a retrovirus is RNA, but these viruses are able to convert their genome into DNA because they contain an enzyme called reverse transcriptase. This enzyme, which is not found in host cells, is so named because it catalyzes a process that is the reverse of normal transcription, which goes from DNA to RNA. The DNA copy (cDNA) of the retroviral genome can be in-tegrated into the host DNA, where it is called a *provirus* (see Fig. 21.14). Not only is the integrated provirus resistant to antiviral medications taken by the host, but it is also able to escape detection by the host immune system.

Viral Diseases of Humans

Viruses cause many important human diseases, but we have room to discuss only a few of them: colds, influenza, measles, and four herpes viral infections. Some viral diseases that are transmitted sexually, including HIV/AIDS, were discussed in Section 21.5.

The best protection against most viral diseases is immuniza-tion utilizing a vaccine. Of the viral diseases we will be discussing, vaccines are currently available for influenza, measles, chicken-pox, and shingles, but not for the common cold, herpes simplex, or Epstein–Barr viruses.

The Common Cold and Influenza

Colds are most commonly caused by rhinoviruses, and the symp-toms usually include a runny nose, mild fever, and fatigue. The flu, caused by the influenza virus, is characterized by more severe symptoms, such as a high fever, body aches, and severe fatigue. Cold symptoms tend to subside within a week, but the flu may last for two or three weeks, and can result in death, especially in elderly patients or those with weakened immune systems.

While most common cold viruses are endemic (always present) in the human population, flu outbreaks tend to be epidemic, affect-ing large numbers of people in more limited geographic areas at any one time. Why can you get several colds or the flu year after year? There are over 100 different strains of rhinoviruses, plus several other viruses that cause very similar symptoms. In the case of colds, you become immune only to those strains you have contracted be-fore. However, the reason you may get the flu each year is that the influenza virus can change rapidly. Small changes called *antigenic drift*, especially those affecting the surface spikes, may be enough to make the virus capable of temporarily evading the immune response of individuals who were immune to the original virus (Fig. 28.16a). Therefore, manufacturers of influenza vaccines try to incorporate the strains that are predicted to cause the highest number of cases each year. Still, antigenic drift can lead to local epidemics.

Figure 28.16 Antigenic drift and shift. **a.** In antigenic drift, small mutations gradually change surface antigens so that antibodies to the original virus become less effective. As time goes by, people are more likely to become ill upon exposure to the virus. **b.** In antigenic shift, even more serious major changes take place on surface antigens as genome segments are reassorted between two influenza viruses that infect the same cell. Now, antibodies are more likely to be ineffective, and most people will become ill when they are exposed to the virus.

The genome of the influenza virus is composed of eight segments of RNA. When two different influenza viruses infect the same cell, these RNA segments can get mixed up as the viruses reproduce. When new virus particles are assembled that contain RNA from both original viruses, this reassortment event, called an *antigenic shift,* may lead to new combinations of surface spikes (Fig. 28.16*b*). Because the human population has not previously been exposed to this combination of antigens, the result can be a pandemic, or worldwide epidemic. The "avian flu" virus, which goes by the names H5N1 and H7N9, and which originated in Southeast Asia, contains genes from a bird virus, as well as a human virus, and thus may have resulted from a reassortment event. Fortunately, neither of these viruses currently seem to be efficiently transmitted to humans. However, the most recent "swine flu" outbreak of 2009 serves to remind us it is impossible to predict what future reassortment events may occur, so vaccine manufacturers must play catch-up whenever a new virus evolves.

Measles

Measles is one of the most contagious human diseases. An unvaccinated person can become infected simply by breathing viral particles in the air of a room where an infected person has been, even 2 hours later. After exposure, there is a 7- to 12-day incubation period before the onset of flulike symptoms, including fever. A red rash develops on the face and moves to the trunk and limbs. The rash lasts for several days. In more-developed nations, the fatality rate is about 1 in 3,000, while in less-developed countries, the fatality rate is 10–15%. The number of deaths is decreasing in many areas, however, due to efforts to vaccinate children, who still make

up the majority of measles fatalities. In the United States, the number of measles cases dropped from 5 million each year to a few thousand after introduction of the measles vaccine in 1963. Between 2000 and 2013, measles deaths worldwide fell by 75%, from an estimated 750,000 to 146,000. The measles vaccine is usually given to infants in combination with vaccines against mumps and rubella, collectively called the MMR vaccine. However, recently there has been an increase in the rates of measles in the United States, which is largely attributed to the decision by some parents not to vaccinate their children with the MMR vaccine due to religious or personal choice.

Herpesviruses

Herpesviruses cause chronic infections that remain latent for much of the time. Eight herpesviruses capable of infecting humans have been described. Some of these infect epithelial cells and neurons, whereas others infect blood cells. While some herpesviruses do not seem to cause any problems, we will explore a few that are known to cause disease.

Herpes simplex virus type 1 (HSV-1) is usually associated with cold sores and fever blisters around the mouth. It is estimated that 70–90% of the adult population is infected with HSV-1. Herpes infections of the genitals are generally caused by HSV-2 and are transmitted through sexual contact. Painful blisters fill with clear fluid that is very infectious (see Fig. 21.15). Genital herpes infections are widespread in the United States, affecting 20–25% of the adult population. Most individuals infected with genital herpes will experience recurrences of symptoms induced by various stresses.

Another herpesvirus that infects humans is the varicella-zoster virus, which causes chickenpox. Later in life, usually after age 60, the latent virus can reemerge as a related disease called shingles. Painful blisters form in an area innervated by a single sensory neuron, often on the upper chest or face. The symptoms may last for several weeks or, in some cases, months. The FDA approved a vaccine for the prevention of chickenpox in 1995, and many states are now requiring that children receive this vaccine. In May 2006, the FDA approved the first vaccine for prevention of shingles. It is for use in people over age 60.

Epstein–Barr virus (EBV) is the herpesvirus associated with infectious mononucleosis, or "mono." Like HSV-1, EBV is very common—as many as 95% of adults aged 35 to 40 have been infected. When infected with EBV as children, most people have no symptoms. In contrast, those infected as teenagers and young adults often develop infectious mononucleosis. The disease is named after the tendency of the cells infected by the virus, which are also known as mononuclear cells, to become more numerous in the blood. The symptoms of mono usually include fever, fatigue, and swollen lymph nodes. Like all herpesviruses, EBV establishes a lifelong latent infection. EBV has also been associated with other conditions, such as chronic fatigue syndrome and certain rare cancers.

Antiviral Drugs

Because viruses use the machinery of host cells for viral replication, it is difficult to develop drugs that affect viral replication without harming host cells. As noted earlier, however, many viruses use their own enzymes to copy their genetic material, and many antiviral drugs inhibit these enzymes. As discussed in Section 21.5, a variety of antiretroviral agents have been developed to inhibit the reverse transcriptase and protease enzymes used by HIV, which can delay the onset of AIDS. Other drugs, such as acyclovir and valacyclovir (or Valtrex), are commonly used to control recurrences of HSV-1 and HSV-2 infections. These compounds mimic nucleotides, and thus can inhibit viral genome synthesis. Other antiviral drugs may affect virus attachment, entry, or assembly. If no effective antiviral drugs are available for a particular viral infection, patients are often told to simply let the virus run its course, because antibiotics are not effective against viral infections.

Viroids and Prions

Viroids and prions are also acellular pathogens. A **viroid** consists of a circular piece of naked RNA. The viroid RNA is ten times shorter than that of most viral genomes, and apparently does not code for any proteins. Viroid replication causes diseases in plants, the only known hosts. The mechanism of viroid diseases, such as potato spindle tuber and apple scar skin, are not known.

Prions are proteinaceous infectious particles that cause degenerative diseases of the nervous system in humans and other animals. They are derived from normal proteins of unknown function in the brains of healthy individuals. Disease occurs when the normal proteins change into the abnormal, prion shape. This forms more prion proteins, which go on to convert other normal proteins into the wrong shape. These abnormal proteins seem to build up in the brain, causing a loss of neurons and what appear to be holes in the tissue (see Fig. 17.22).

The first prion disease to be described was *scrapie,* which occurs in sheep. It is called scrapie because, as the disease affects the animal's brain, an affected sheep begins to scrape off much of its wool. Scrapie is not capable of causing disease in humans. However, another prion disease in cattle, called *bovine spongiform encephalopathy* (BSE), or mad cow disease, does appear capable of infecting humans and has caused over 150 human deaths in Great Britain and a few other countries. This disease is now called *variant Creutzfeldt–Jakob disease* (vCJD), to distinguish it from the "classic" form of CJD, which occurs spontaneously in a very low percentage of people. In April 2012, a California dairy cow was diagnosed with BSE, which was the fourth case of BSE detected in U.S. cattle since 2003. Three human cases of vCJD have been documented in the United States, but all three victims were thought to have acquired the disease in other countries prior to moving to the United States.

Chronic wasting disease is another prion disease that occurs in deer, elk, and moose, but there have been no documented cases of transmission to humans. One prion disease that passes directly from human to human is kuru, which was transmitted among a cannibalistic tribe in New Guinea due to their traditional practice of consuming their dead relatives. Even though prion diseases are as frightening as they are fascinating, their incidence in humans remains very low.

Check Your Progress 28.5

1. Describe the basic structure of an enveloped virus, including capsid, nucleic acid, envelope, and spikes.
2. Explain the six stages of a typical animal virus reproductive cycle.
3. List three viral diseases for which a vaccine is available, and three for which there is no vaccine.
4. Compare how viruses differ from viroids and prions.

Conclusion

Historically, Ebola epidemics and outbreaks have been very short-lived and have affected only a relatively small number of individuals. The duration and extent of the 2014–2015 Ebola outbreak in West Africa caused immediate concern throughout the globe. The fact that, for the first time, cases were reported in both Europe and the United States, increased public awareness and concern about Ebola.

The attention to this disease has focused both governments and private agencies on the need to develop methods of managing outbreaks of highly contagious diseases, as well as investing in research on how to treat and prevent these diseases.

Advances in medical research, specifically in the field of genomics, has already had an impact on how outbreaks of infectious diseases will be addressed in the future. Even before the disease had peaked in West Africa, researchers were testing new vaccines and using the data from these trials to possibly prevent future widespread outbreaks of Ebola and other hemorrhagic fevers.

SUMMARIZE

28.1 The Microbial World

Microbiology is primarily the study of bacteria, archaea, protists, fungi, and viruses. Most microbes are microscopic, and some are extremely numerous.

■ Two of the most significant contributors to the early discipline of microbiology were Antonie van Leeuwenhoek, who was the first to visualize microbes using his microscopes, and Louis Pasteur, whose many accomplishments include disproving the idea that microbes could arise spontaneously.

■ Microbes are perhaps best known for causing infectious diseases, which can range from the merely annoying to the swiftly fatal. However, most microorganisms are far more beneficial than harmful, not only to humans, but to Earth's ecosystem. These include the **microbiota** that reside on and in the human body, as well as microbes that perform the functions of decomposition, photosynthesis, and bioremediation in ecosystems.

28.2 Origin of Microbial Life

A central principle of evolutionary theory is that all life on Earth has arisen from a **last universal common ancestor (LUCA).** Chemical reactions are hypothesized to have led to the formation of **biomolecules,** as well as the first true cells in the following stages:

■ In stage 1, simple organic molecules, or monomers, arise from inorganic chemicals. The "primordial soup" hypothesis suggested by Oparin and Haldane is an influential idea about this stage, which is supported by the work of Miller, Urey, and others.

■ In stage 2, monomers join to form polymers such as proteins, RNA, or membranes. The **protein-first hypothesis** suggests that some early proteins were enzymes that could synthesize other molecules. The **RNA-first hypothesis** holds that RNA was the first macromolecule.

■ In stage 3, the evolution of a **protobiont** (protocell) very likely involved the formation of **micelles,** which would have fused to form larger **vesicles.** The **membrane-first hypothesis** focuses on the need for an outer membrane to contain other biochemical reactions. The first membranes may have been similar to **liposomes,** which form spontaneously from lipids in water. The protobiont may have been a **heterotroph** that consumed preformed organic molecules, or a **chemoautotroph** that required only inorganic nutrients.

■ Stage 4 produced living cells that were able to reproduce. According to the **central dogma** of genetics, information flows from DNA to RNA to protein. However, this may not have been the case in the earliest cells.

28.3 Archaea

Archaea and bacteria are **prokaryotes,** single-celled microbes that lack a nucleus or membrane-bound organelles. However, the two groups are classified into different domains of life due to nucleic acid sequence and biochemical differences.

■ Structurally, archaeal plasma membranes have unique lipids that help some to survive in extreme environments.

■ Archaea are often associated with extreme habitats, but actually are widespread in the environment. Three major types of archaea are the **halophiles, thermoacidophiles,** and **methanogens.**

28.4 Bacteria

Bacteria are the most widespread, and in many ways successful, type of organism on Earth.

■ Most bacteria have a cell wall that contains peptidoglycan, and most bacteria can be classified as either Gram-positive or Gram-negative using the **Gram stain** procedure. The outer membrane of Gram-negative bacteria contains a toxic substance called **lipopolysaccharide.** Many bacteria have **flagella** for motility.

■ All bacteria have a DNA genome, usually found on a single, circular chromosome in the **nucleoid** region. Bacteria reproduce by **binary fission,** producing identical daughter cells. Horizontal gene transfer also occurs, by **conjugation, transformation,** or **transduction.** Some bacteria can survive for long periods by producing **endospores.**

■ Most bacteria are heterotrophs, requiring carbon in the form of organic molecules. Some are chemoautotrophs, which acquire carbon from carbon dioxide and energy from chemicals. **Cyanobacteria** are an important group of photosynthetic bacteria.

■ Bacterial species that cause significant diseases in humans include members of the genera *Staphylococcus, Streptococcus, Mycobacterium,* and *Salmonella.* **Antibiotics** are chemicals that either kill or inhibit the growth of bacteria.

28.5 Viruses, Viroids, and Prions

Viruses are minute, acellular pathogens that reproduce as obligate intracellular parasites.

■ Structural features shared by all viruses include an outer **capsid** composed of protein and an inner core of nucleic acid. The viral genome can be single- or double-stranded DNA or RNA. Many animal viruses also have an **envelope** with spikes.

■ The reproductive cycle of an animal virus typically has six steps: attachment, entry, genome replication, biosynthesis, assembly, and budding. During budding, enveloped viruses usually acquire a portion of the host cell's plasma membrane. **Retroviruses** (e.g., HIV) have RNA genomes that are converted to DNA by reverse transcriptase, before integrating into the host cell's chromosome as a provirus.

■ Viruses infect almost all types of organisms and cause many important diseases. Viral diseases of humans include the common cold, influenza, measles, and chickenpox. Antiviral drugs usually inhibit viral enzymes.

Like viruses, viroids and prions are acellular pathogens.

■ **Viroids** are obligate, intracellular plant pathogens that are autonomously replicating short RNA molecules. They contain no protein.

■ **Prions** contain no nucleic acid. They are proteinaceous, infectious particles that convert a normal cellular protein into the abnormal form, which accumulates and damages brain tissue, inducing dementia. Diseases caused by prions include scrapie, kuru, and mad cow disease.

572 UNIT 6 Evolution and Diversity

ASSESS

Testing Yourself

Choose the best answer for each question.

28.1 The Microbial World

1. Decomposers
   a. break down dead organic matter in the environment by secreting digestive enzymes.
   b. break down living organic matter by secreting digestive enzymes.
   c. destroy living cells and then break them down with digestive enzymes.
   d. live in close association with another species.

2. Which of the following best describes the term *microbiota*?
   a. microbes that are smaller than a eukaryotic cell
   b. microbes that cause disease
   c. microbes that are decomposers
   d. microbes that are naturally found on or within our bodies

28.2 Origin of Microbial Life

3. The RNA-first hypothesis for the origin of cells is supported by the discovery of
   a. ribozymes.
   b. proteinoids.
   c. polypeptides.
   d. nucleic acid polymerization.

4. Which of these steps probably occurred first in the chemical and biological evolution of life?
   a. The cell membrane of a protobiont formed.
   b. The first organic molecules formed.
   c. The original genetic material was RNA.
   d. Simple organic monomers formed from inorganic materials.

28.3 Archaea

5. Archaea differ from bacteria in that they
   a. have a nucleus.
   b. have membrane-bound organelles.
   c. have peptidoglycan in their cell walls.
   d. are often photosynthetic.
   e. None of these are correct.

6. While studying an ancient lake, you discover that the water in the lake had an exceptionally high concentration of salt. What type of bacteria would you have expected to live in this lake?
   a. cyanobacteria
   b. thermoacidophiles
   c. methanogens
   d. halophiles

28.4 Bacteria

7. Which of the following describes a bacterium that is spherical in shape?
   a. bacillus
   b. coccus
   c. spirillum

8. In this process, bacteria pick up DNA from the environment.
   a. binary fission
   b. transduction
   c. conjugation
   d. transformation

9. The DNA of a bacteria is contained where in the cell?
   a. in the nucleus
   b. in a region of the cytoplasm called the nucleoid
   c. at the ribosomes
   d. within the lipopolysaccharide layer

28.5 Viruses, Viroids, and Prions

9. Prions contain
   a. DNA only.
   b. protein only.
   c. RNA only.
   d. DNA, RNA, and protein.

10. Small changes in influenza surface antigens lead to
    a. antigenic drift.
    b. antigenic shift.
    c. antigenic recombination.
    d. antigenic schism.

11. The envelope of an animal virus is usually derived from the _____ of its host cell.
    a. cell wall
    b. plasma membrane
    c. capsule
    d. receptors

12. Capsid proteins are synthesized during which phase of viral replication?
    a. replication
    b. biosynthesis
    c. assembly
    d. proteination
    e. All of these are correct.

ENGAGE

Thinking Critically

1. Model organisms are those widely used by researchers who wish to understand basic processes that are common to many species. Bacteria such as *Escherichia coli* are model organisms for modern geneticists. Give three reasons why bacteria would be useful in genetic experiments.

2. Many viruses contain their own enzymes for replicating their genetic material, whereas others use the host cell's enzymes. Would DNA viruses or RNA viruses be more likely to produce their own enzyme(s) for this purpose, and why?

3. Explain a method by which an antiviral drug could interfere with the replication cycle of a virus.

4. Antibiotic medications work by targeting specific structures and functions in bacterial cells. Side effects on the patient are usually minimal, because their eukaryotic cells do not possess the same structures and characteristics as the prokaryotic pathogens. What structures or functions of the prokaryotic cell would serve as good targets for new antibiotics?

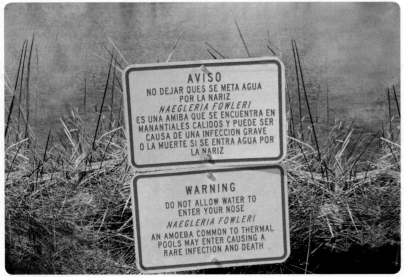

©FLPA/SuperStock

Dangerous Protists

Protected from the heat wave in an air-conditioned lab, a pathologist pulls a petri plate covered in *E. coli* bacteria from an incubator. A boy has suddenly passed away, and a sample of his brain tissue is in the center of the plate. Radiating from the tissue are tiny pathways cut through the film of bacteria, pathways characteristic of the "brain-eating amoeba" *Naegleria fowleri*. The tiny protists must have been the cause of death, and they were feeding their way through the *E. coli*, just as they had fed through the boy's brain. Five days ago, the boy suddenly experienced a high fever, stiff neck, and vomiting. Doctors suspected bacterial meningitis, but antibiotics didn't help, and he died before more tests could be performed.

But how was the boy infected? At the start of the oppressive heat wave, he and his family had gone to the lake to cool off. Stirred up from the bottom by the boy's boisterous play, the amoeba likely entered his nose during a handstand attempt and burrowed toward his brain, virtually guaranteeing his death. Fortunately, *N. fowleri* is just one rare species among thousands of other beneficial protists that live around us, on us, and in us. In the same lake where the boy was infected, there are critical species producing oxygen, recycling nutrients, and forming the base of the aquatic food chain.

In this chapter, we will not only explore the diversity of life found in Kingdom Protista, but also that of Kingdom Fungi. While we often hear about the negative aspects of organisms in these kingdoms, we will learn that many species of protistans and fungi benefit humans.

As you read through this chapter, think about the following questions:

1. Are all protistans parasites?
2. How are protists related to other eukaryotes, such as humans?
3. How do protists and fungi impact human health and welfare?

CHAPTER OUTLINE

29.1 Protists
29.2 Fungi

BEFORE YOU BEGIN

Before beginning this chapter, take a few moments to review the following discussions:

Section 1.2 Protists and fungi belong to which domain of life?

Section 3.5 How does the endosymbiotic theory explain the origin of energy-producing organelles in the eukaryotic cell?

Section 5.4 What is the difference between a haploid and a diploid eukaryotic cell?

573

29.1 Protists

The **domain Eukarya** is divided into four kingdoms: Protista, Fungi, Plantae, and Animalia. For the most part, **protists are single-celled eukaryotic** microbes, although some exist as colonies of cells or are multicellular. They are such a diverse group, however, that perhaps the best definition of a protist is any eukaryotic organism that is not a plant, animal, or fungus.

Protists offer us a glimpse into the past because they are most likely related to the first eukaryotic cell to have evolved. The first eukaryotic cell is thought to have arisen from a prokaryotic cell around 1.7 billion years ago. According to the **endosymbiotic theory,** mitochondria may have resulted when a nucleated cell engulfed aerobic bacteria, and chloroplasts may have originated when a nucleated cell with mitochondria engulfed cyanobacteria (see Fig. 3.16). Protists also bridge the gap between the first eukaryotic cells and multicellular organisms—the other three types of eukaryotes (fungi, plants, and animals) all trace their ancestry to a protist.

Characteristics of Protists

Protists vary in size from microscopic algae and protozoans to kelp that can exceed 200 m in length. Kelp, a brown alga, is multicellular; *Volvox,* a green alga, is colonial; *Spirogyra,* also a green alga, is filamentous. Most protists are single-celled, but despite their small size they have attained a high level of complexity. The amoeboids and ciliates possess unique organelles—their contractile vacuole is an organelle that assists in water regulation.

Protists are sometimes grouped according to how they acquire organic nutrients. The algae are a diverse group of photoautotrophic protists that synthesize organic compounds via photosynthesis. Protozoans are a group of heterotrophic protists that obtain organic compounds from the environment. Some protozoans, such as *Euglena,* are able to combine autotrophic and heterotrophic nutritional modes.

Protists reproduce sexually and asexually. Asexual reproduction by mitosis is the norm in protists. Sexual reproduction generally occurs only when environmental conditions are unfavorable. Protists can form **spores** or **cysts,** which are dormant phases of the protist life cycle, that can survive until favorable conditions return. Parasitic protists form cysts for the transfer to a new host.

Many protists cause diseases in humans, but many others have significant ecological importance. Aquatic photoautotrophic protists produce oxygen and are the foundation of the food chain in both freshwater and saltwater ecosystems. They are a part of **plankton,** organisms suspended in the water that serve as food for heterotrophic protists and animals. Interestingly, whales, the largest animals in the sea, feed on plankton, one of the smallest.

Diversity of Protists

Protists were once classified together as a single kingdom. However, DNA evidence has suggested that protists do not all belong to the same evolutionary lineage, meaning they did not evolve from a single common ancestor. In fact, protists and other eukaryotes, including the plants, fungi, and animals, are currently classified into five supergroups (Table 29.1). A **supergroup** is a high-level taxonomic group below domain and above kingdom. Each supergroup represents a separate evolutionary lineage.

The DNA evidence supports these multiple protist lineages, but the relationships among the lineages are difficult to decipher. Protist lineages are very long and old, dating back to the origin of the first eukaryotes. As lineages stretch back in time, we can be less and less certain about how they are related to each other, just as the history of humans is less complete the further back in time we look. New research in the evolution of protists has helped clarify some of the evolutionary relationships among eukaryote lineages (Fig. 29.1), but much research still needs to be done.

In the following sections, we examine the various supergroups into which protists and other eukaryotes have been placed based on our current understanding.

Supergroup Archaeplastids

The archaeplastids include land plants and other photosynthetic organisms, such as green and red algae, that have chloroplasts (also called *plastids*) derived from endosymbiotic cyanobacteria (see Fig. 3.16).

The **green algae** are protists that contain both chlorophylls *a* and *b*. They inhabit a variety of environments, including oceans, fresh water, snowbanks, the bark of trees, and the backs of turtles. Green algae occur in many different forms. The majority are single-celled; however, filamentous and colonial forms exist. Seaweeds are multicellular green algae that resemble lettuce leaves. Some of the 8,000 species of green algae also form symbiotic relationships with plants, animals, and fungi in lichens (see Section 29.2). Despite the name, green algae are not always green; some have additional pigments that give them an orange, red, or rust color.

Biologists propose that land plants are closely related to the green algae, because both land plants and green algae have chlorophylls *a* and *b,* a cell wall that contains cellulose, and food reserves made of starch. Molecular data suggest that the green algae are subdivided into two groups, the *chlorophytes* and the *charophytes* (Fig. 29.1). Charophytes are thought to be the green algae group most closely related to land plants.

Chlamydomonas is an example of a charophyte that inhabits still, freshwater pools. Its fossil ancestors date back over a billion years. The anatomy of *Chlamydomonas* is best seen using an

TABLE 29.1   Eukaryotic Supergroups

Supergroup	Types of Organisms
Archaeplastids	Land plants, as well as green and red algae
SAR Supergroup	Stramenopiles, alveolates, and rhizaria
Excavates	Euglenids and certain other flagellates
Amoebozoans	Amoeboids, as well as plasmodial and cellular slime molds
Opisthokonts	Animals, fungi, and certain flagellates

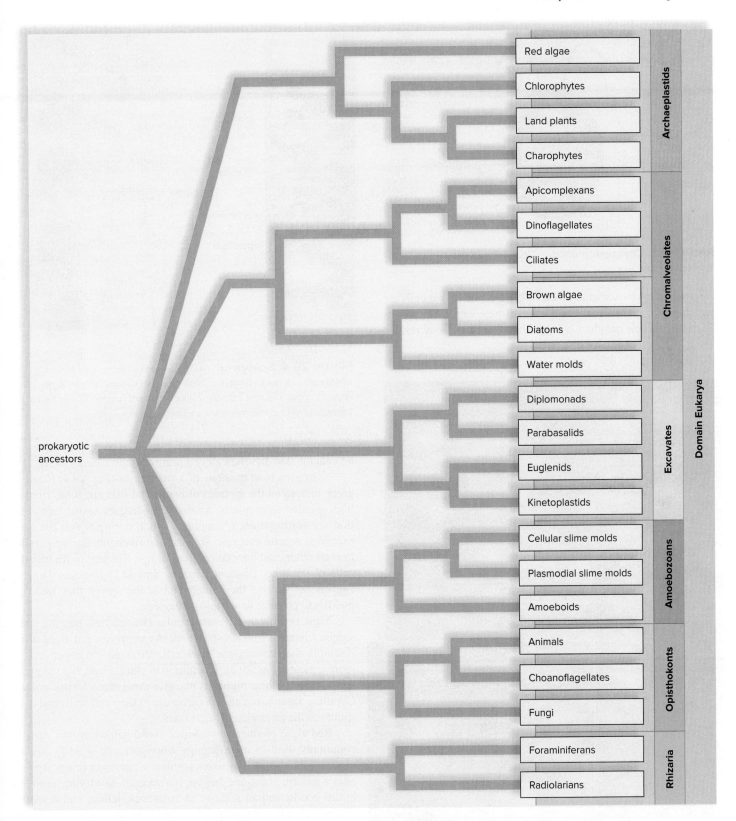

Figure 29.1 Evolutionary relationship of the eukaryotic supergroups.   Molecular data are used to determine the relatedness of the supergroups and their constituents. This is a simplified tree that does not include all members of each supergroup.

electron microscope, because it is less than 25 μm long (Fig. 29.2). It has a defined cell wall and a single, large, cup-shaped chloroplast that contains a *pyrenoid,* a dense body where starch is synthesized. In many species, a bright red, light-sensitive eyespot helps guide individuals toward light for photosynthesis.

A number of colonial forms occur among the chlorophytes. *Volvox* is a well-known colonial green alga (Fig 29.3). A colony is a loose association of independent cells. A *Volvox* colony is a hollow sphere with thousands of cells arranged in a single layer surrounding a watery interior. *Volvox* cells move the colony by

Figure 29.2 *Chlamydomonas*. *Chlamydomonas* is a motile, single-celled green alga.

coordinating the movement of their flagella. Some *Volvox* cells are specialized for reproduction, and each of these can divide asexually to form a new daughter colony. This daughter colony resides

a. Cell anatomy
b. Conjugation
50×

Figure 29.4 *Spirogyra*. **a.** *Spirogyra* is a filamentous green alga in which each cell has a ribbonlike chloroplast. **b.** During conjugation, the cell contents of one filament enter the cells of another filament. Zygote formation follows. (b): ©M. I. (Spike) Walker/Science Source

for a time within the parent colony, but then it escapes by releasing an enzyme that dissolves away a portion of the parent colony.

Spirogyra is an example of a charophyte (Fig. 29.4) found in green masses on the surfaces of ponds and streams. It has ribbon-like, spiraled chloroplasts. *Spirogyra* undergoes sexual reproduction via **conjugation,** a temporary union during which the cells exchange genetic material. Two haploid filaments line up parallel to each other, and the cell contents of one filament move into the cells of the other filament, forming diploid zygospores. Diploid zygospores survive the winter, and in the spring they undergo meiosis to produce new haploid filaments.

Most **red algae** are multicellular charophytes, ranging from simple filaments to leafy structures. As *seaweeds*, red algae often resemble the brown algae seaweeds, although red algae can be more delicate (Fig. 29.5). In addition to chlorophyll, the red algae contain red and blue pigments that give them characteristic colors. Coralline algae have calcium carbonate in their cell walls and contribute to the formation of coral reefs.

Red algae produce a number of useful gelling agents. *Agar* is commonly used in microbiology laboratories to solidify culture media and commercially to encapsulate vitamins or drugs. Agar is also a gelatin substitute for vegetarians, an antidrying agent in baked goods, and an additive in cosmetics, jellies, and desserts. *Carrageenan* is a related product used in cosmetics and chocolate manufacture. The species *Porphyra,* a red seaweed, is popular as a sushi wrap in Japan.

The SAR Supergroup

The SAR supergroup includes three large subgroups: the stramenopiles, the alveolates, and the rhizaria. Until recently, the rhizarians were classified as a separate supergroup. Also, stramenopiles and alveolates were in another supergroup called Chromalveolata. However,

17×

daughter colony

vegetative cells

Figure 29.3 *Volvox*. *Volvox* is a colonial chlorophyte. The adult *Volvox* colony often contains daughter colonies, which are asexually produced by special cells. (both): ©Manfred Kage/Science Source

Figure 29.5 Red algae.   Red algae contain accessory pigments in addition to chlorophyll. Represented here by *Chondrus crispus,* they are smaller and more delicate than brown algae. ©Dan Ippolito

Figure 29.6 Brown algae.   Brown algae have accessory pigments that allow them to obtain their light energy while living in deeper waters compared to other algae. This is a type of brown algae known as bull kelp, *Nereocystis luetkeana.* The photo on the upper right shows flotation bladders that brown algae form to keep their blades close to the surface. (both): ©Dan Ippolito

recent genetic studies demonstrated closer relationships among all three and led to the formation of the new SAR supergroup.

Stramenopiles

The stramenopiles include the brown algae, diatoms, golden brown algae, and water molds. Some members of this group are photosynthetic, but they have a different evolutionary lineage from the green and red algae.

Brown algae are the conspicuous multicellular seaweeds that dominate rocky shores along cold and temperate coasts. The color of brown algae is due to accessory pigments that actually range from pale beige to yellow-brown to almost black. These pigments allow the brown algae to extend their range down into deeper waters because the pigments are more efficient than green chlorophyll in absorbing the sunlight away from the ocean surface. The alga produces a slimy matrix that retains water when the tide is out and the seaweed is exposed. This gelatinous material, algin, is used in ice cream, cream cheese, and some cosmetics.

The Sargasso Sea in the North Atlantic Ocean commonly has large floating mats of brown algae called *sargassum.* The most familiar brown algae are kelp (genus *Laminaria*) in coastal regions and giant marine kelp (genus *Macrocystis*) in deeper waters (Fig. 29.6). Tissue differentiation in kelp results in blades, stalks, and holdfasts, which are analogous to the leaves, stems, and roots of plants.

Diatoms are tiny, single-celled chromalveolates with an ornate silica shell. The shell is made up of upper and lower shelves, called valves, that fit together. Diatoms have a photosynthetic accessory pigment that gives them an orange-yellow color. Diatoms make up a significant part of plankton, which serves as a source of oxygen and food for heterotrophs in both freshwater and marine ecosystems.

Diatoms reproduce asexually and sexually. Asexual reproduction occurs by diploid parents undergoing mitosis to produce two diploid daughter cells. Each time a diatom reproduces asexually, the size of the daughter cells decreases until diatoms are about 30% of their original size. At this point, they begin to reproduce sexually. The diploid cell produces gametes by meiosis. Gametes fuse to produce a diploid zygote, which grows and then divides via mitosis to produce new diploid diatoms of normal size.

Diatoms are easy to recognize under the microscope because they have a wide variety of elaborate shells made of silica (Fig. 29.7). The two overlapping shells have intricately shaped depressions, pores, and passageways that bring the diatom's plasma membrane in contact with the environment. The fossilized remains of diatoms, also called *diatomaceous earth,* accumulate on the ocean floor and can be mined for use as filtering agents and abrasives.

Most **water molds** are **saprotrophs,** meaning they feed on dead organic matter. They usually live in water, where they decompose remains and form furry growths when they parasitize fish (Fig. 29.8). In spite of their common name, some water molds live on land and parasitize insects and plants. The water mold *Phytophthora* was responsible for the 1840s potato famine in Ireland.

Water molds, also called oomycetes ("egg fungi"), used to be grouped with fungi, because they are similar to fungi in many ways. The funguslike water molds have a filamentous body similar to the hyphae of fungi, but their cell walls are composed of cellulose instead of chitin. The life cycle of water molds also differs from that of the fungi. Water molds have a filamentous body, as do

Figure 29.7 Assorted fossilized diatoms.   Diatoms have overlapping shells made of silica. Scientists use their delicate markings to identify the particular species. ©Darlyne A. Murawski/Getty Images

Figure 29.8 Water mold. *Saprolegnia*, a water mold, feeding on a dead goldfish. Water molds are not fungi. ©Noble Proctor/Science Source

Figure 29.9 Dinoflagellates. Dinoflagellates have cellulose plates. Here we see *Gonyaulax*, a dinoflagellate that contains a reddish-brown pigment and is responsible for occasional "red tides." (photo): ©Kevin Schafer/Alamy

fungi, but the cell walls of water molds are largely composed of cellulose, whereas fungi have cell walls made of chitin. During asexual reproduction, water molds produce flagellated spores. During sexual reproduction, they produce eggs and sperm.

Alveolates

Alveolates have alveoli (small air sacs) lying just beneath their plasma membranes that are thought to lend support to the cell surface or aid in membrane transport. Alveolates are single-celled organisms.

Dinoflagellates are best known for the **red tide** they cause when they greatly increase in number, an event called an algal bloom. *Gonyaulax*, one species implicated in red tides, produces a very potent toxin (Fig. 29.9). This can be harmful by itself, but it also accumulates in shellfish. The shellfish are not damaged, but people can become quite ill when they eat them.

Despite some harmful effects, dinoflagellates are important members of the phytoplankton in marine and freshwater ecosystems. Generally, dinoflagellates are photosynthetic, although colorless heterotrophic forms are known to live as symbionts inside other organisms. Corals, which build coral reefs, contain large numbers of symbiotic dinoflagellates.

Many dinoflagellates have protective cellulose plates that become encrusted with silica, turning them into hard shells. Dinoflagellates have two flagella. One is located in a groove that encircles the protist, and the other is in a longitudinal groove and has a free end. The arrangement of the flagella makes a dinoflagellate whirl as it moves. In fact, the name dinoflagellate is derived from the Greek *dino-*, for "whirling." It is interesting to note that luminous dinoflagellates produce a twinkling light that can give seas a phosphorescent glow at night, particularly in the tropics.

Ciliates are the largest group of animal-like protists, also called **protozoans.** All of them have cilia, hairlike structures that rhythmically beat, moving the cell forward or in reverse. Typically, the cell rotates as it moves. Cilia also help capture prey and particles and then move them toward the mouthparts. After phagocytic vacuoles engulf food, they combine with lysosomes, which supply the enzymes needed for digestion. A **contractile vacuole** helps to maintain water balance with the surrounding environment.

Some ciliates are up to 3 mm long, which is large enough to be seen with the naked eye. Most are freely motile, but some can be anchored to a surface using a stalk. *Paramecium* is the most widely

known ciliate, and it is commonly used for research and teaching. It is shaped like a slipper and has visible contractile vacuoles (Fig. 29.10*a*). Members of the genus *Paramecium* have a large *macronucleus* and a small *micronucleus*. The macronucleus produces mRNA and directs metabolic functions. The micronucleus is important during sexual reproduction. *Paramecium* has been important for studying ciliate sexual reproduction, which involves *conjugation*, with interactions between micronuclei and macronuclei (Fig. 29.10*b*).

The apicomplexa are commonly called **sporozoans** because they produce spores. All phases of the life cycle are generally nonmotile, with the exception of male gametes and zygotes. All sporozoans are either intercellular or extracellular parasites. An apical complex composed of fibrils, microtubules, and organelles is found at one end of the single cell. Enzymes for attacking host cells are secreted through a pore at the end of the apical complex. Specific diseases of humans caused by sporozoans are discussed at the end of this section.

Rhizaria

The rhizarians consist of the **foraminiferans** and the radiolarians, organisms with fine, threadlike pseudopods. Although rhizarians were once classified with amoebozoans, they are now assigned to a different supergroup. Foraminiferans and radiolarians both have a skeleton, called a *test*. Because each geological time period has a distinctive form of foraminiferan, they can be used to date sedimentary rock. Depositions over millions of years followed by geological upheaval formed the White Cliffs of Dover along the southern coast of England (Fig. 29.11*a*). Also, the great Egyptian pyramids are built of foraminiferous limestone. In radiolarians, the test is internal (Fig. 29.11*b*). The tests of dead foraminiferans and radiolarians form a deep layer of sediment on the ocean floor. Their presence is used as an indicator of oil deposits on land or sea.

a. *Paramecium*

b. During conjugation, two paramecia first unite at oral areas.   100×

Figure 29.10 Ciliates. Ciliates are the most complex of the protists. **a.** Diagram of the structure of a *paramecium*, next to an electron micrograph. Note the oral groove, the gullet, and the anal pore. **b.** A form of sexual reproduction called conjugation occurs periodically. (photos) (a): ©Carolina Biological Supply/Medical Images; (b): ©Ed Reschke/Photolibrary/Getty Images

a. Foraminiferan, *Globigerina,* and the White Cliffs of Dover, England

b. Radiolarian tests   SEM 150×

Figure 29.11 Foraminiferans and radiolarians.   a. Pseudopods of a live foraminiferan project through holes in the calcium carbonate shell. Fossilized shells were so numerous they became a large part of the White Cliffs of Dover when a geological upheaval occurred. **b.** Skeletal tests of radiolarians. (a) (foraminiferan): ©NHPA/Superstock; (cliffs): ©Stockbyte/Getty Images; (b): ©Eye of Science/Science Source

The Excavates

The excavates have atypical or absent mitochondria and distinctive flagella and/or deep (excavated) oral grooves. They are sometimes referred to as **flagellates** due to the fact that they propel themselves using one or more flagella.

Euglenids are freshwater single-celled organisms that typify the problem of classifying protists. Many euglenids have chloroplasts, but some do not. Those that lack chloroplasts ingest or absorb their food. Those that have chloroplasts are believed to have originally acquired them by ingestion and subsequent endosymbiosis of a green algal cell. Three, rather than two, membranes surround these chloroplasts. The outermost membrane is believed to represent the plasma membrane of an original host cell that engulfed a green alga.

Euglenids have two flagella, one of which is typically much longer than the other and projects out of the anterior, vase-shaped invagination (Fig. 29.12). Near the base of this flagellum is an **eyespot apparatus,** which is a photoreceptive organelle for detecting light. Because euglenids are bounded by a flexible **pellicle** composed of protein strips lying side by side, they can assume different shapes as the underlying cytoplasm undulates and contracts.

Parabasalids and diplomonads are single-celled, flagellated excavates that are endosymbionts of animals. They are able to survive in anaerobic, or low-oxygen, environments. These protozoans lack mitochondria; instead, they rely on fermentation for the production of ATP. A variety of forms are found in the guts of termites, where they assist with the breakdown of cellulose.

Parabasalids have a unique, fibrous connection between the Golgi apparatus and flagella. The most common sexually transmitted disease, trichomoniasis, is caused by the parabasalid *Trichomonas vaginalis.* Infection causes vaginitis in women. The parasite may also infect the male genital tract; however, the male may have no symptoms.

A diplomonad cell has two nuclei and two sets of flagella. The diplomonad *Giardia lamblia* forms cysts, which are transmitted by contaminated water. *Giardia* attaches to the human intestinal wall, causing severe diarrhea (Fig. 29.13). This protozoan lives in the digestive tracts of a variety of other mammals as well. Beavers are

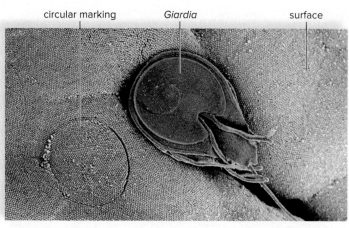

Figure 29.13 *Giardia lamblia.* This protist adheres to any surface, including epithelial cells, by means of a sucking disk. Characteristic markings can be seen on the host cell after the disk detaches. ©Cultura RM/Alamy

Figure 29.12 *Euglena.* Euglenoids have a flexible pellicle and a long flagellum that propels the body. (photo): ©Kage Mikrofotografie/Medical Images

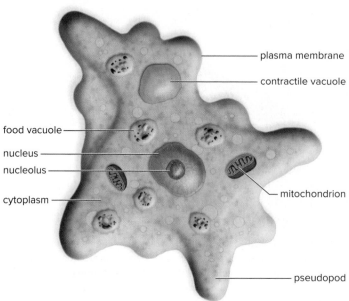

Figure 29.14 Amoeba. This amoeboid is common in freshwater ponds. Bacteria and other microorganisms are digested in food vacuoles, and contractile vacuoles rid the body of excess water.

known to be a reservoir of *Giardia* infection in the mountains of the western United States, and many cases of infection have been acquired by hikers who filled their canteens at a beaver pond.

The Amoebozoa

Amoeboids move by *pseudopods,* processes that form when cytoplasm streams forward in a particular direction (Fig. 29.14). They usually live in aquatic environments, such as oceans and freshwater lakes and ponds, where they are a part of the zooplankton. When amoeboids feed, their pseudopods surround and engulf their prey, which may be algae, bacteria, or other protists. Digestion then occurs within a food vacuole.

In forests and woodlands, slime molds feed on dead plant material. They also feed on bacteria, keeping their population under control. The vegetative cells of slime molds are mobile and amoeboid. They ingest their food by phagocytosis. **Plasmodial slime molds** exist as a *plasmodium,* a diploid, multinucleated, cytoplasmic mass enveloped by a slime sheath, that creeps along, phagocytizing decaying plant material in a forest or agricultural field (Fig. 29.15). At times unfavorable to growth, such as during a drought, the plasmodium develops many *sporangia,* reproductive structures that produce spores. The spores produced by a sporangium can survive until moisture is sufficient for them to germinate. In plasmodial slime molds, spores release a haploid flagellated cell or an amoeboid cell. Eventually, two of these fuse to form a zygote that feeds and grows, producing a multinucleated plasmodium once again.

Cellular slime molds can exist as individual amoeboid cells. They are common in soil, where they feed on bacteria and yeasts. *Dictyostelium discoideum* is the commonly researched example. As food supplies dwindle, the cells release a chemical that causes them to aggregate into a sluglike *pseudoplasmodium,* which eventually gives rise to a fruiting body that produces spores. When favorable conditions return, the spores germinate, releasing haploid amoeboid cells, and this cycle, which is asexual, begins again. A sexual cycle is known to occur under very moist conditions.

Genetic analyses of several slime mold species has revealed that these may have been some of the earliest eukaryotes to adapt

Plasmodium, *Physarum*

Sporangia, *Hemitrichia*

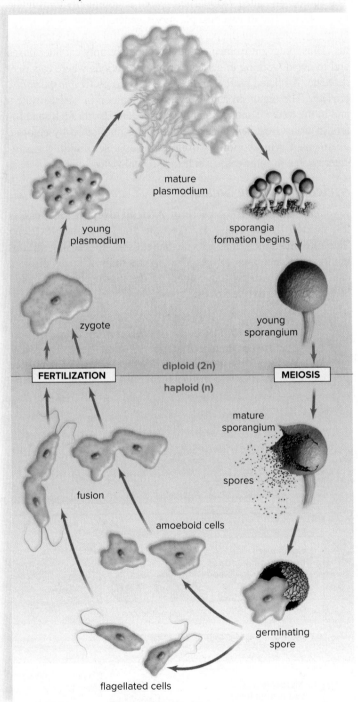

mature plasmodium

young plasmodium

sporangia formation begins

zygote

young sporangium

FERTILIZATION

diploid (2n)

haploid (n)

MEIOSIS

mature sporangium

spores

fusion

amoeboid cells

germinating spore

flagellated cells

Figure 29.15 Plasmodial slime molds.   The diploid adult forms sporangia during sexual reproduction, when conditions are unfavorable to growth. Haploid spores germinate, releasing haploid amoeboid or flagellated cells that fuse. (photos): top left: ©NHPA/Superstock; top right: ©NaturePL/Superstock

to life on land, perhaps a billion years ago. Recent research has also shown that slime mold behavior can be surprisingly complex. Japanese researchers showed that when placed into a "maze" constructed on an agar plate, slime molds could find the shortest path to food. When presented with different types of food, the slime molds chose the most nutritious type.

The Opisthokonta

Animals and fungi are opisthokonts, along with several closely related protists. This supergroup includes both single-celled and multicellular protozoans. Among the opisthokonts are the **choanoflagellates,** animal-like protozoans that are closely related to sponges. The choanoflagellates, including single-celled as well as colonial forms, are filter-feeders with cells that bear a striking resemblance to the choanocytes that line the inside of sponges (see Section 31.2). Each choanoflagellate has a single posterior flagellum surrounded by a collar of slender microvilli (Fig 29.16). Beating of the flagellum creates a water current that flows through the collar, where food particles are taken in by phagocytosis.

Protistans and Human Disease

Compared to the number of diseases caused by bacteria and viruses, a relatively few protists cause human diseases, and all of these are caused by heterotrophic protozoans. However, some cause significant sickness and mortality.

Malaria

The most widespread and dangerous protozoan disease is **malaria.** According to the World Health Organization (WHO), about half of the world's population lives in areas where malaria is endemic. In 2016, an estimated 216 million people were infected with malaria, resulting in over 445,000 deaths. As significant as those numbers are, they are actually decreasing, at least in part due to the concerted efforts of many governments and private organizations.

Malaria can be caused by several sporozoan parasites in the genus *Plasmodium*. All of these have a complex life cycle that involves transmission by a mosquito vector (Fig. 29.17). The sexual reproduction phase in the mosquito ends with the migration of a developmental stage called sporozoites into the salivary glands. These sporozoites are then transmitted by the mosquito's bite to a human, where they reproduce asexually in the liver and red blood

Figure 29.16 Choanoflagellates.   The choanoflagellates are most closely related to the animals.

cells, forming merozoites. The infected red blood cells frequently rupture, releasing merozoites and toxins, and causing the person to experience the chills and fever characteristic of malaria. Some of these merozoites become gametocytes, which, if ingested by another mosquito, can start the sexual reproduction phase again. People who survive malaria have only a limited immunity to reinfection.

Interestingly, the sickle-cell trait found in up to one-third of sub-Saharan Africans provides some protection against malaria. Efforts to control mosquito populations have been successful in the United States, where malaria is rarely seen. The control of mosquitoes worldwide has now been hampered by the rise of resistance to pesticides like DDT. Scientists are working to find new ways to inhibit mosquitoes, such as genetically engineering them to be resistant to *Plasmodium* infection. A Seattle company is even designing a laser system that could protect villages by zapping mosquitoes in flight!

An intense research effort is underway to develop a malaria vaccine, funded largely by the Bill and Melinda Gates Foundation, which has contributed about $1.2 billion to malaria research since 2000. A 2011 study published in the *New England Journal of Medicine* reported the results of a malaria vaccine trial conducted in seven countries and involving 15,460 children. Overall, the vaccine reduced the incidence of malaria in children by approximately 50% during the 12 months after vaccination.

Toxoplasmosis

The sporozoan called *Toxoplasma gondii* is commonly transmitted by cat feces. The disease toxoplasmosis generally causes no appreciable symptoms, but the parasite can be harmful to a developing fetus. Thus, pregnant women are advised not to empty cat litter boxes and to avoid working in gardens or other locations where cats may defecate. A related protozoal disease is caused by *Cryptosporidium parvum*. The organism and its cysts are common in surface waters and in the feces of animals and birds, and can pass through sand filters in water treatment plants, while being unaffected by chlorine treatment. Cryptosporidiosis usually causes a self-limiting gastroenteritis, but in some cases can lead to a fatal watery diarrhea.

African Sleeping Sickness

Trypanosoma brucei, the cause of **African sleeping sickness,** is transmitted by the tsetse fly (see the Health feature "African Sleeping Sickness"). It attacks the patient's blood, causing inflammation that decreases oxygen flow to the brain. Chagas disease, caused by *T. cruzi*, is transmitted by the kissing bug, so called because it tends to bite lips.

Figure 29.17 Life cycle of *Plasmodium vivax*, a species that causes malaria. Asexual reproduction of this sporozoan occurs in humans, while sexual reproduction takes place within the *Anopheles* mosquito.

SCIENCE IN YOUR LIFE ▶ HEALTH

African Sleeping Sickness

The World Health Organization (WHO) recognizes several neglected tropical diseases (NTDs) that affect more than 1 billion people exclusively in the most impoverished communities. Several of these NTDs, including African sleeping sickness, are caused by parasitic protists that are transmitted to humans from the bite of an infected insect.

African sleeping sickness, also called African human trypanosomiasis, is the deadliest of all of the NTDs. It is caused by a type of parasitic, single-celled protist, *Trypanosoma brucei,* which is transmitted to humans in the bite of the blood-sucking tsetse fly (*Glossina*). The tsetse fly is the transmission vector of the disease because it carries the trypanosomes and transmits the long, blade-shaped protists (Fig. 29A) into the human bloodstream while feeding. The flies first become infected with the parasite when they bite either an infected human or other mammal, such as domestic cattle and pigs, which carry *T. brucei*.

Figure 29B illustrates the relationship between humans, the tsetse fly, and domestic livestock. Unlike humans, livestock can carry *T. brucei* without getting sick. Therefore, they act as a reservoir for the disease. Humans also can serve as a reservoir for new tsetse fly infections because the parasite can be picked up from humans by a fly bite and transmitted to others.

It is estimated that as many as 50,000 people are plagued by this disease. In the later stages of infection, *T. brucei* attacks the brain, causing behavioral changes and a shift in sleeping patterns. If caught early enough, it can be cured with medication, but because it is most common in very poor nations, medication and treatment are difficult to come by. Without treatment, the disease is fatal.

The tsetse fly prefers shaded, cool, moist habitats, such as the sandy edges of pools and rivers in central Africa. Regions that are plagued by sleeping sickness have no running water, so daily trips to local rivers and watering holes are necessary to collect water. The people who live and work in tsetse habitat are the most vulnerable to the disease. As a result, many communities have abandoned the areas infested with the tsetse fly. The abandonment of these fertile valleys has had a huge impact on the ecology of surrounding areas and the economy of communities and nations on a large scale.

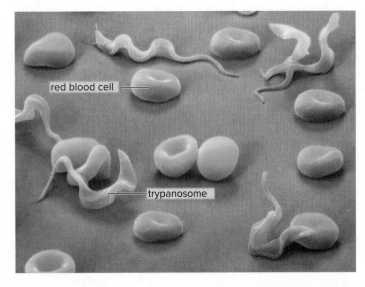

Figure 29A
Trypanosoma brucei.
A micrograph showing *T. brucei*, a causal agent of African sleeping sickness, among red blood cells. ©Eye of Science/Science Source

Recently, an effort to eliminate the tsetse fly from these lands has been successful. The WHO reports that the number of new cases of sleeping sickness has dropped to the lowest levels in 50 years. The treatment and control of sleeping sickness has allowed the resettlement of more than 25 million hectares of prime agricultural land.

Questions to Consider

1. How does poverty reinforce a high occurrence of African sleeping sickness?
2. How are the economy, ecology, and disease biology of African sleeping sickness interdependent?

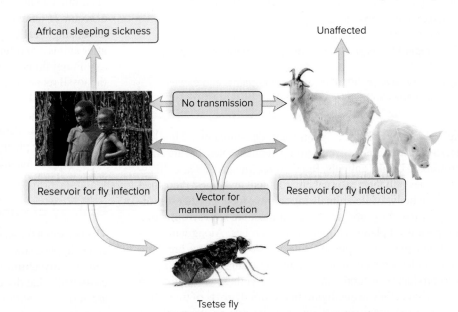

Figure 29B **The transmission pattern of *Trypanosoma brucei*.**   Tsetse flies can become infected with *T. brucei* by feeding on infected humans or livestock. The flies can then transmit the parasitic protist to other humans, causing African sleeping sickness. (photos) (children): ©DLILLC/Corbis/VCG/Corbis Documentary/Getty Images; (tsetse fly): ©Frank Greenaway/Getty Images; (goat): ©Eureka/Alamy; (pig): ©Anatolii Tsekhmister/123RF

Amoeboid-Related Diseases

Parasitic amoeboids in the genus *Entamoeba* cause **amoebic dysentery.** Complications arise when the parasite invades the intestinal lining and reproduces there. If the parasite enters the body proper, liver and brain involvement can be fatal. Another example is an amoeboid in the genus *Acanthamoeba* which can cause corneal inflammation as well as serious infections. Even more foreboding, the amoeboid protist *Naegleria fowleri* can invade and attack the human nervous system, nearly always resulting in the death of the victim. It is usually acquired by swimming in warm bodies of fresh water, such as ponds, lakes, rivers, and unchlorinated swimming pools. Fortunately, this scenario is extremely rare.

Check Your Progress 29.1

1. Describe the endosymbiotic theory.
2. Describe the circumstances by which some protists produce spores or cysts, and explain the benefit this provides to the organism.
3. Compare the type of nutrition that is typically required by an alga, a protozoan, a water mold, and a cellular slime mold.
4. Identify the type of protist that causes each of the following diseases: malaria, African sleeping sickness, and amoebic dysentery.

29.2 Fungi

Learning Outcomes

Upon completion of this section, you should be able to

1. Explain how most fungi obtain their nutrients.
2. Describe the structural components of a fungal body.
3. Compare the reproductive strategies of the major groups of fungi.
4. Identify several fungal diseases of humans, and explain why many can be difficult to treat.

Fungi (domain Eukarya, supergroup Opisthokonts, kingdom Fungi) are a structurally diverse group of eukaryotes that are strict heterotrophs. Unlike animals, fungi (*sing.,* fungus) release digestive enzymes into their external environment and digest their food outside the body, while animals ingest their food and digest it internally. A few fungi are parasitic, but most are **saprotrophs** that decompose dead plants, animals, and microbes. Along with bacteria, fungi play an important role in ecosystems by breaking down complex organic molecules and returning inorganic nutrients to producers of food—that is, photosynthesizers. Fungi can degrade even cellulose and lignin in the woody parts of trees. It is common to see fungi (brown rot or white rot) on the trunks of fallen trees. The body of a fungus can become large enough to cover acres of land.

Evolution of Fungi

Figure 29.18 illustrates our current understanding of the evolutionary relationships among the fungi. The evolutionary tree is a

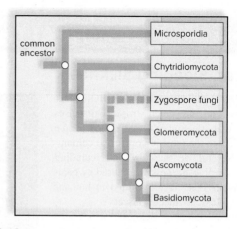

Figure 29.18 Evolutionary relationships of the fungi. While DNA evidence is still unraveling the relationships among the fungi, this diagram presents the most accepted hypothesis on the evolutionary history of the fungi.

hypothesis about how these groups are related. The Microsporidia and chytrids are different from all other fungi, because they are single-celled. The chytrids are aquatic and have flagellated spores and gametes. Our description of fungal structure applies best to the zygospore fungi, sac fungi, and club fungi. The AM fungi are interesting because they exist only as mycorrhizae in symbiotic association with plant roots.

Protists evolved some 1.5 BYA (billion years ago). Plants, animals, and fungi can all trace their ancestry to protists, but molecular data tell us that animals and fungi shared a more recent common ancestor than animals and plants. Therefore, animals and fungi, both in the supergroup Opisthokonta, are more closely related to each other than either is to plants (see Fig. 29.1). The common ancestor of animals and fungi was most likely an aquatic, flagellated, single-celled protist. Multicellular forms evolved sometime after animals and fungi split into two different lineages.

Fungi do not fossilize well, so it is difficult to estimate from the fossil record when they first evolved. The earliest known fossil fungi are dated at 460 MYA (million years ago), but fungi probably evolved a lot earlier. Mycorrhizae are evident in plant fossils, also some 460 MYA. Perhaps fungi were instrumental in the colonization of land by plants. Much of the fungal diversity most likely had its origin in an adaptive radiation when organisms began to colonize land.

Biology of Fungi

The body of a fungus is composed of a mass of individual filaments called hyphae (*sing.,* **hypha**). Collectively, the mass of filaments is called a **mycelium** (*pl.,* mycelia) (Fig. 29.19). Some fungi have cross-walls that divide a hypha into a chain of cells. These hyphae are termed *septate.* Septa have pores that allow cytoplasm and even organelles to pass from one cell to the other along the length of the hypha. *Nonseptate* fungi have no cross-walls, and their hyphae are multinucleated.

Hyphae give the mycelium quite a large surface area per volume of cytoplasm, which facilitates absorption of nutrients into the body of a fungus. Hyphae grow at their tips, and the mycelium absorbs and then passes nutrients on to the growing tips. Individual

a. Fungal mycelia

b. Individual strands of hyphae   170×

septate

aseptate

c. Septate and aseptate hyphae

Figure 29.19 Fungal mycelia and hyphae.   a. Fungal mycelium made from hyphae, growing as a white mass on strawberries. **b.** Hyphal filaments growing on the surface of a plant. **c.** Hyphae are either septate (do have cross-walls) or aseptate (do not have cross-walls). (a): ©Matt Meadows/Photolibrary/Getty Images; (b): ©Andrew Syred/Science Source

hyphae can grow quite rapidly, as much as 18 ft per day. Altogether, a single mycelium may add as much as a kilometer of new hyphae in a single day!

Fungal cells are quite different from plant cells, not only because they lack chloroplasts, but also because their cell walls contain chitin, not cellulose. **Chitin,** like cellulose, is a polymer of glucose, but in chitin, each glucose molecule has a nitrogen-containing amino group attached to it. Chitin is also the major structural component of the exoskeleton of arthropods, such as insects, lobsters, and crabs. Unlike plants, the energy reserve of fungi is not starch, but glycogen, as in animals. Fungi are nonmotile and most do not have flagella at any stage in their life cycle. They move toward a food source by growing toward it.

Although a few fungal species are aquatic, most have adapted to life on land by producing windblown spores during both asexual and sexual reproduction (Fig. 29.20). A spore is a haploid reproductive cell that develops into a new organism without the need to fuse with another reproductive cell. In fungi, spores germinate into new mycelia. Sexual reproduction in fungi involves conjugation of hyphae from two different mating types (usually designated + and −). Often, the haploid nuclei from the two hyphae do not immediately fuse to form a zygote. The hyphae contain + and − nuclei for long periods of time. Eventually, the nuclei fuse to form a zygote that undergoes meiosis, followed by spore formation.

Diversity of Fungi

Fungi are traditionally classified based on their mode of sexual reproduction. Figure 29.18 is an evolutionary tree showing how the five major groups of fungi that will be discussed in this section are believed to be related. Major fungal groups (phyla) include the microsporidians, chytrids, zygospore fungi, sac fungi, club fungi, and AM fungi.

Microsporidians

The single-celled **Microsporidia** are parasites of animal cells, most often seen in insects but also found in vertebrates, such as fish, rabbits, and humans. Biologists once believed that Microsporidia were an ancient line of protist due in part to their lack of mitochondria. However, genome sequencing of the microsporidian *Encephalitozoon cuniculi* revealed genes that are mitochondrial,

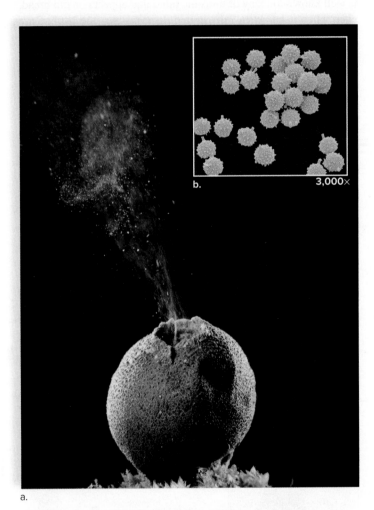
b.   3,000×

a.

Figure 29.20 Fungi reproduce by spore formation.   a. Here a pear-shaped puffball, *Lycoperdon pyriforme,* is releasing perhaps billions of spores. **b.** The spores of a puffball fungus as viewed under an electron microscope. (a): ©Richard Packwood/Getty Images; (b): ©Steve Gschmeissner/Getty Images

leading to the hypothesis that this organism once had a mitochondrion, which later became greatly reduced. In addition, microsporidians have the smallest known eukaryotic genome. New sequence information places *E. cuniculi* and other microsporidia in fungi, rather than with the protists.

Chytrid Fungi

The **chytrid fungi** (phylum Chytridomycota) may have been the first type of fungi to evolve. They are unique among the fungi because they are aquatic, and they produce flagellated reproductive cells. Most chytrids reproduce asexually through the production of *zoospores,* which grow into new chytrids. Many chytrids play a role in the decay and digestion of dead aquatic organisms, but some are parasitic on plants, animals, and protists (Fig. 29.21).

Zygospore Fungi

The **zygospore fungi** (phylum Zygomycota) are mainly saprotrophs, but some are parasites of small soil protists or worms and even insects, such as the housefly. *Rhizopus stolonifer* (Fig. 29.22) is well known to many of us as the mold that appears on old bread, even when it has been refrigerated. In *Rhizopus,* the hyphae are specialized: Some are horizontal and exist on the surface of the bread; others grow into the bread to anchor the mycelium and carry out digestion; and still others are stalks that bear sporangia. A **sporangium** is a capsule that produces spores.

When *Rhizopus* reproduces sexually, the ends of + and − hyphae join, haploid nuclei fuse, and a thick-walled **zygospore** results. The zygospore undergoes a period of dormancy before meiosis and germination take place. Following germination, aerial hyphae, with sporangia at their tips, produce many spores. The spores are dispersed by air currents and give rise to new mycelia.

Sac Fungi

Approximately 75% of all known fungi are **sac fungi** (phylum Ascomycota), named for their characteristic cuplike sexual reproductive structure called an **ascocarp.** Many sac fungi reproduce by

Figure 29.21 Chytrids parasitizing a protist. These aquatic chytrids (*Chytriomyces hyalinus*) have penetrated the cell walls of this algal protist and are absorbing nutrients meant for their host. They will produce flagellated zoospores that will go on to parasitize other protists. Reproduced with permission of the Freshwater Biological Association on behalf of The Estate of Dr Hilda Canter-Lund. ©The Freshwater Biological Association.

producing chains of asexual spores called *conidia* (*sing.,* conidium). Cup fungi, morels, and truffles have conspicuous ascocarps (Fig. 29.23). Truffles, underground symbionts of hazelnut and oak trees, are highly prized as gourmet delights. Pigs and dogs are trained to sniff out truffles in the woods, but they are also cultivated on the roots of seedlings.

Morels (Fig. 29.23c) are often collected by chefs and other devotees because of their unique and delicate flavor. Collectors must learn, however, to avoid "false morels," which can be

Figure 29.22 Black bread mold, *Rhizopus stolonifer.* The mycelium of this mold utilizes sporangia to produce windblown spores. A zygospore forms during the sexual life cycle. (photo): ©Garry DeLong/Getty Images

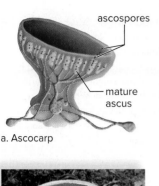

ascospores

mature ascus

a. Ascocarp

ascocarp

b. Cup fungi

ascocarp

c. Morels

Figure 29.23 Sexual reproduction in sac fungi. Some sac fungi are known to us by their ascocarp, the structure that produces spores. **a.** Diagram of an ascocarp containing asci, where spores are produced. **b.** The ascocarps of cup fungi. **c.** The ascocarp of a morel, which is a prized delicacy. (b): ©Carol Wolfe; (c): Bear Dancer Studios/Mark Dierker

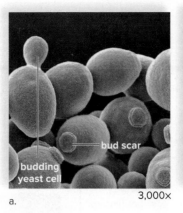

bud scar

budding yeast cell

a. 3,000×

b. 425×

Figure 29.24 Asexual reproduction in sac fungi. **a.** Yeasts, unique among fungi, reproduce by budding. **b.** The sac fungi usually reproduce asexually by producing spores called conidia. (a): ©Science Photo Library/Getty Images; (b): ©Dennis Kunkel Microscopy, Inc./Medical Images

poisonous. Interestingly, even true morels should not be eaten raw. As saprophytes, these organisms secrete digestive enzymes capable of digesting wood or leaves. If consumed prior to being inactivated by cooking, these enzymes can cause digestive problems.

Most fungal plant pathogens are sac fungi. Examples of these pathogenic fungi include powdery mildews that grow on plant leaves, as well as chestnut blight and Dutch elm disease that destroy trees. *Ergot,* a parasitic sac fungus that infects rye, produces hallucinogenic compounds similar to LSD. Psychoses resulting from ingesting grain contaminated with rye ergot may have led to the Salem, Massachusetts, witch trials in 1692.

Some of the most familiar fungi were formerly considered "imperfect fungi," because their means of sexual reproduction is unknown. However, based on DNA sequencing and other recent information, these fungi are now classified as sac fungi. Examples include *Penicillium,* the original source of penicillin, a breakthrough antibiotic that led to the important class of "cillin" antibiotics that have saved millions of lives. Other species of *Penicillium—P. roquefortii* and *P. camemberti*—are necessary to the production of blue cheeses. *Aspergillus* is used widely for its ability to produce citric acid and various enzymes. Certain species also cause serious infections in people with compromised immune systems.

Yeasts The term **yeasts** is generally applied to single-celled fungi, and many of these organisms are sac fungi. *Saccharomyces cerevisiae,* brewer's yeast, is representative of budding yeasts. When unequal binary fission occurs, a small cell gets pinched off and then grows to full size. Asexual reproduction occurs when the food supply runs out, producing spores (Fig. 29.24).

When some yeasts ferment, they produce ethanol and carbon dioxide. In the wild, yeasts grow on fruits, and historically, the yeasts already present on grapes were used to produce wine. Today, selected yeasts are added to relatively sterile grape juice in

order to make wine. Also, yeasts are added to grains to make beer and liquor. Both the ethanol and carbon dioxide are retained in beers and sparkling wines, while carbon dioxide is released from wines. In breadmaking, the carbon dioxide produced by yeasts causes the dough to rise, and the ethanol quickly evaporates. The gas pockets are preserved as the bread bakes.

Club Fungi

Club fungi (phylum Basidiomycota) are named for their characteristic sexual reproductive structure called a **basidium.** Basidia are enclosed within a basidiocarp, which develops after + and − hyphae join. The union of + and − nuclei occurs in the basidium, which produces spores by meiosis. We recognize most club fungi by their basidiocarp (Fig. 29.25).

When you eat a mushroom, you are consuming a basidiocarp. Certain mushrooms are poisonous, especially those of the genus *Amanitas. A. phalloides,* also known as the death cap, causes 90% of fatalities related to mushroom poisoning, due to its production of a toxin that interferes with gene transcription by inhibiting RNA polymerase. Other club fungi, mainly of the genus *Psilocybe,* produce a hallucinogenic chemical called psilocybin that is a structural analog of LSD. These mushrooms have been used in religious ceremonies since ancient times, though their recreational use is currently illegal in the United States. See the Health feature "Deadly Fungi" for a further discussion of fungi that produce toxins dangerous to humans.

Shelf or bracket fungi found on dead trees are also basidiocarps. Less well known are puffballs and stinkhorns. In *puffballs,* spores are produced inside parchmentlike membranes, and the spores are released through a pore or when the membrane breaks down. *Stinkhorns* resemble a mushroom, but they emit a very disagreeable odor. Flies are attracted by the odor. When they linger to feed on the sweet jelly, they pick up spores and later distribute them.

Smuts and *rusts* are club fungi that parasitize cereal crops, such as corn, wheat, oats, and rye. They are of great economic importance because of the crops they destroy each year. The corn smut mycelium grows between corn kernels and secretes substances that cause tumorlike swellings to develop (Fig. 29.26*a*).

nuclei in basidium

fusion meiosis

spores

gill of mushroom

basidiocarp

(−) (+)

a. Sexual reproduction

Figure 29.25 Sexual reproduction in club fungi. Most club fungi are known to us by their basidiocarp, the structure that produces spores as a part of sexual reproduction. **a.** Sexual life cycle of a basidiocarp and a basidium (*pl.*, basidia) where spores are produced. Also shown are basidiocarps of (**b**) scarlet hood mushroom, (**c**) chicken-of-the-woods shelf fungi, and (**d**) a giant puffball. Puffballs contain many spores, and their basidiocarps tend to be rounded. When they are mature, any pressure from outside, such as a raindrop or the kick of a shoe, ejects the spores through a hole as a cloud of dust. (b): ©Biophoto Associates/Science Source; (c): ©Inga Spence/Science Source; (d): ©L. West/Science Source

b. Mushroom

c. Shelf fungi

d. Giant puffball

The life cycle of rusts often requires two different plant host species to complete the cycle, and one way to keep them in check is to eradicate the alternate host. Wheat rust (Fig. 29.26*b*) can also be controlled by producing new and resistant strains of wheat.

Symbiotic Relationships of Fungi

Several instances in which fungi are parasites of plants have already been mentioned. Fungi also commonly form *symbiotic* relationships with other organisms, in which both partners benefit.

Lichens

Lichens are associations between fungi and cyanobacteria or green algae. The different lichen species are identified according to the fungal partner. Lichens are efficient at acquiring nutrients and moisture, and therefore, they can survive in poor soils, as well as on rocks with no

a. Corn smut, *Ustilago*

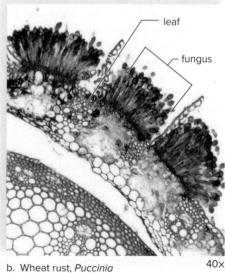

leaf

fungus

b. Wheat rust, *Puccinia*

40×

Figure 29.26 Smuts and rusts. a. Corn smut. **b.** Wheat rust. Both are important fungal diseases affecting agricultural crops in the United States and other countries. (a): ©Marvin Dembinsky Photo Associates/Alamy; (b): ©Patrick Lynch/Science Source

Deadly Fungi

It is unwise for amateurs to collect mushrooms in the wild because certain species are poisonous. The red and yellow *Amanita* mushrooms are especially dangerous. These species are also known as fly agaric because they were once thought to kill flies (the mushrooms were gathered and then sprinkled with sugar to attract flies). Its toxins include muscarine and muscaridine, which produce symptoms similar to those of acute alcoholic intoxication. In 1 to 6 hours, the victim staggers, loses consciousness, and becomes delirious, sometimes suffering from hallucinations, manic conditions, and stupor. Luckily, it also causes vomiting, which rids the system of the poison, so death occurs in less than 1% of cases. The death angel mushroom (*Amanita phalloides,* Fig. 29C) causes 90% of the fatalities attributed to mushroom poisoning. When this mushroom is eaten, symptoms don't begin until 10 to 12 hours later. Abdominal pain, vomiting, delirium, and hallucinations are not the real problem; rather, a poison interferes with RNA transcription by inhibiting RNA polymerase, and the victim dies from liver and kidney damage.

Some hallucinogenic mushrooms are used in religious ceremonies, particularly among Mexican Indians. *Psilocybe mexicana* contains a chemical called psilocybin that is a structural analogue of LSD and mescaline. It produces a dreamlike state in which visions of colorful patterns and objects seem to fill up space and dance past in endless succession. Other senses are also sharpened to produce a feeling of intense reality.

The only reliable way to tell a nonpoisonous mushroom from a poisonous one is to be able to correctly identify the species. Poisonous mushrooms cannot be identified with simple tests, such as whether they peel easily, have a bad odor, or blacken a silver coin during cooking. Only consume mushrooms identified by an expert!

Like club fungi, some sac fungi also contain chemicals that can be dangerous to people. *Claviceps purpurea,* the ergot fungus, infects rye and replaces the grain with ergot—hard, purple-black bodies consisting of tightly cemented hyphae (Fig. 29D). When ground with the rye and made into bread, the fungus releases toxic alkaloids that cause the disease ergotism. In humans, vomiting, feelings of intense heat or cold, muscle pain, a yellow face, and lesions on the hands and feet are accompanied by hysteria and hallucinations. Ergotism was common in Europe during the Middle Ages. During this period, it was known as St. Anthony's Fire and was responsible for 40,000 deaths in an epidemic in A.D. 994. We now know that ergot contains lysergic acid, from which LSD is easily synthesized. Based on recorded symptoms, historians believe that those individuals who were accused of practicing witchcraft in Salem, Massachusetts, during the seventeenth century were actually suffering from ergotism. It is also speculated that ergotism is to blame for supposed demonic possessions throughout the centuries. As recently as 1951, an epidemic of ergotism occurred in Pont-Saint-Esprit, France. Over 150 persons became hysterical, and 4 died.

Because the alkaloids that cause ergotism stimulate smooth muscle and selectively block the sympathetic nervous system, they can be used in medicine to cause uterine contractions and to treat certain circulatory disorders, including migraine headaches. Although the ergot fungus can be cultured in petri dishes, no one has succeeded in inducing it to form ergot in the laboratory. So far, the only way to obtain ergot, even for medical purposes, is to collect it in an infected field of rye.

Figure 29D **Ergot infection of rye, caused by** *Claviceps purpurea.* ©Wildlife GmbH/Alamy

Questions to Consider

1. Why do you think that some species of fungi contain chemicals that are harmful to people?
2. How might an understanding of fungal poisons aid in the development of new drugs for humans?

Figure 29C **Poisonous mushroom,** *Amanita phalloides.* ©De Agostini/Getty Images

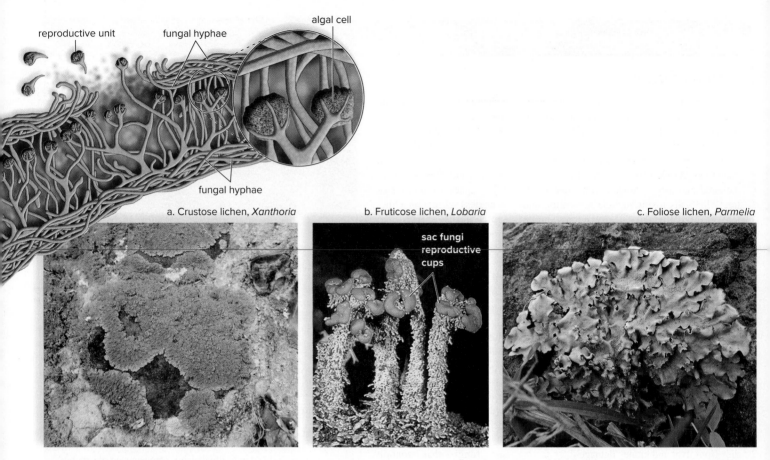

reproductive unit   fungal hyphae   algal cell

fungal hyphae

a. Crustose lichen, *Xanthoria*   b. Fruticose lichen, *Lobaria*   c. Foliose lichen, *Parmelia*

sac fungi reproductive cups

Figure 29.27 Lichen morphology.   a. A section of a compact crustose lichen shows the placement of the algal cells and the fungal hyphae, which encircle and penetrate the algal cells. **b.** Fruticose lichens are shrublike. **c.** Foliose lichens are leaflike. (photos) (a): ©Dr. Jeremy Burgess/SPL/Science Source; (b): ©Steven P. Lynch; (c): ©Yogesh More/Alamy

soil. These primary colonizers produce organic matter and create new soil, allowing plants to invade the area. Lichens exhibit three structures: compact *crustose* lichens, often seen on bare rocks or tree bark; shrublike *fruticose* lichens; and leaflike *foliose* lichens (Fig. 29.27).

The body of a lichen has three layers. The fungus forms a thin, tough upper layer and a loosely packed lower layer. These shield the photosynthetic cells in the middle layer. Specialized fungal hyphae, which penetrate or envelop the photosynthetic cells, transfer organic nutrients directly to the fungus. The fungus protects the algae from predation and desiccation and provides them with minerals and water. Lichens can reproduce asexually by releasing fragments that contain hyphae and an algal cell. As with many symbioses, the relationship between fungi and algae was likely a pathogen-and-host interaction originally, but became mutually beneficial over evolutionary time.

Mycorrhizae

Mycorrhizae are mutualistic relationships between soil fungi and the roots of most plants. Plants whose roots are invaded by mycorrhizae grow more successfully in dry or poor soils, particularly those deficient in phosphates. The relationship is very ancient, because it is seen in early plant fossils. Perhaps it helped plants adapt to life on dry land. Mycorrhizal fungi generally go unnoticed, except for the truffle, which is collected for food.

Mycorrhizal fungi may live on the outside of roots, enter the cortex of roots, or penetrate root cells. The fungus and plant cells can easily exchange nutrients, with the plant providing organic nutrients to the fungus and the fungus bringing water and minerals to the plant. The fungal hyphae greatly increase the surface area from which the plant can absorb water and nutrients.

AM fungi (phylum Glomeromycota) are a group whose name stands for *a*rbuscular *m*ycorrhizal fungi. Arbuscules are branching invaginations the fungus makes when it invades plant roots. AM fungi are one type of mycorrhiza, fungi that form mutually beneficial relationships with the roots of plants.

Fungal Diseases of Humans

For reasons that aren't completely understood, fungi tend to cause disease mainly in people whose immune system isn't working properly. **Mycoses,** fungal diseases of animals, vary in levels of seriousness.

Superficial Mycoses

Most *superficial mycoses* are confined to the outer layers of the skin, hair, or nails. Fungi called *dermatophytes* cause infections of the skin called *tineas*. Athlete's foot is a tinea characterized by itching and peeling of the skin between the toes (Fig. 29.28*a*). In **ringworm,**

a.   b.   c.

Figure 29.28 Human fungal diseases.   a. Athlete's foot and (**b**) ringworm are caused by fungi called *dermatophytes*. **c.** Thrush, or oral candidiasis, is characterized by the formation of white patches on the tongue. (a): ©P. Marazzi/SPL/Science Source; (b): ©John Hadfield/SPL/Science Source; (c): ©Dr. M.A. Ansary/Science Source

which is not caused by a worm but instead by several different fungi, the fungus releases enzymes that degrade keratin and collagen in skin. The area of infection becomes red and inflamed, and the fungal colony grows outward, forming a ring of inflammation. As the center of the lesion begins to heal, ringworm acquires its characteristic appearance, a red ring surrounding an area of healed skin (Fig. 29.28*b*). Some people can harbor and spread the fungus, but show no signs of disease. At any one time, an estimated 3–8% of the U.S. population is infected with scalp ringworm.

Candida albicans causes a wide variety of fungal infections. Superficial *C. albicans* infections ("yeast infections") tend to occur when the balance of the normal microbiota in an area of the body is disturbed, or the immune system is suppressed. For example, resident bacteria of the genus *Lactobacillus* normally produce organic acids that lower the pH of the vagina, and this inhibits *C. albicans,* a normal vaginal inhabitant, from proliferating. When lactobacilli are killed off by antibiotics, the yeast proliferates, resulting in inflammation, itching, and discharge. **Thrush,** a *Candida* infection of the mouth, is common in newborns and AIDS patients (Fig. 29.28*c*). In immunosuppressed individuals, *Candida* may cause an invasive infection that can damage the heart, brain, and other organs.

Systemic Mycoses

Systemic mycoses refer to fungal infections affecting the internal organs, mainly the lungs. Especially in immunocompromised individuals, the fungi can then spread through the bloodstream to multiple organs, eventually resulting in the death of the patient. This is well known in HIV infection leading to AIDS, and several fungal diseases are considered to be AIDS-defining illnesses (see the Health feature "Opportunistic Infections and HIV" in Section 13.6).

Occasionally, systemic fungal infections occur in otherwise healthy patients. Due to an effective immune response, most of these infections are either asymptomatic, or cause only mild disease. In rare situations, however, an invasive and progressive infection occurs. Most serious systemic fungal infections start with respiratory symptoms such as coughing, chest pain, hoarseness, and perhaps blood in the sputum. Depending on which fungal species is involved, different organ systems like the skin, central nervous system, bones and joints, and even the heart can be affected.

The majority of people living in the Midwest and eastern United States have been infected with *Histoplasma capsulatum,* a common soil fungus often associated with bird droppings. Only about 5% of infected individuals notice any symptoms, but about 3,000 a year develop a lung disease called *histoplasmosis,* which resembles tuberculosis. Similarly, *Cryptococcus neoformans* occurs throughout the world, mainly in soils contaminated with the droppings of pigeons and chickens. *Coccidioidomycosis,* also called valley fever, can range from benign to severe infections. The causative agent, *Coccidioides immitis,* is found mainly in desert areas, such as in the southwestern United States. Research has shown that 80% of people become infected with *C. immitis* within 4 years of moving to an area where the organism is prevalent.

Also known as black mold, *Stachybotrys chartarum* grows well on building materials—especially those containing cellulose—that become moist. It grows best in dark, unventilated locations. *S. chartarum* is thought to play a role in "sick building syndrome," in which individuals exposed to toxins produced by the fungus may experience allergies, flulike symptoms, headaches, fatigue, and dermatitis.

Antifungal Drugs

As eukaryotes, fungal cells contain a nucleus, organelles, and ribosomes that are like those of human cells. These similarities make it a challenge to design antimicrobials against fungi that do not also harm humans. It is generally safer to treat fungal skin infections with a topical medication, which is not absorbed. For systemic fungal infections, medication must be taken into the body, and thus side effects are more of a problem. To minimize this, researchers exploit any biochemical differences they can discover. The biosynthesis of membrane sterols differs somewhat between fungi and humans. Therefore, a variety of fungicides are directed against sterol biosynthesis.

Check Your Progress 29.2

1. Explain how the type of nutrition required by fungal cells differs from that required by algae or amoeboids.
2. Define *hypha, mycelium,* and *chitin*.
3. Compare the modes of reproduction of the zygospore fungi, sac fungi, yeasts, and club fungi.
4. List two types of superficial mycoses, and two systemic mycoses.

Conclusion

The *N. fowleri* amoeba that infected the boy has certain ecological requirements, most notably a lake or river with high water temperature and low water level. Climate change has made those conditions more common. For example, in 2010 and 2012, the first-ever cases of *N. fowleri* infection were documented in Minnesota, over 500 miles north of any other recorded infections. Higher than normal summer temperatures allowed the pathogen to survive in a swimming lake and take the lives of two children.

Changing climate influences not only the pathogen but also the human behavior that makes infection more likely. During the summers of 2010 and 2012, in the same heat that allowed *N. fowleri* to survive, large numbers of people visited area lakes and rivers to cool down. More people in the water can create a higher chance of infection, especially for a protist, such as *N. fowleri,* that is stirred up from the bottom by splashing and jumping.

SUMMARIZE

29.1 Protists

The **domain Eukarya** contains the **protists,** a diverse group of mostly single-celled **eukaryotes** that are widespread in aquatic and moist environments. It is likely that the first eukaryotic cells to evolve were protists, and as stated by the **endosymbiotic theory,** that mitochondria and chloroplasts first arose when a nucleated cell engulfed an aerobic or photosynthetic bacterium, respectively. The biology of protists can be quite complex.

Protists may carry out both sexual and asexual reproduction. **Spores** or **cysts** may be formed when reproduction is not possible.

Plankton are protists found suspended in water, which make up an important part of many food chains.

The protists are currently classified based on molecular data into **supergroups** that include all eukaryotes. The following summarizes these supergroups:

The Archaeplastids

- **Green algae** are photosynthetic protists that derive energy from the sun. There are two forms of green algae—the charophytes and the chlorophytes. All green algae possess chlorophyll, store reserve food as starch, and have cell walls of cellulose, as do plants. Green algae can be single-celled, form colonies, or live as multicellular organisms. Some green algae reproduce by **conjugation.**

- **Red algae** are charophytes that contain red and blue photosynthetic pigments. Red algae are used in a variety of consumer goods.

The SAR Supergroup

- **Brown algae,** also called seaweeds, generally live at deeper depths than other algae.

- **Diatoms,** which have an outer layer of silica, are extremely numerous in both marine and freshwater ecosystems.

- **Water molds,** which ordinarily are **saprotrophs,** are sometimes parasitic.

- **Dinoflagellates** usually have cellulose plates and two flagella. They are extremely numerous in the ocean and can produce a toxin when they form **red tides.**

- **Ciliates** are animal-like protistans that are commonly referred to as **protozoans.** They move by means of their many cilia. They are remarkably diverse in form—and as exemplified by *Paramecium,* they show how complex a protist can be, despite being a single cell. Many contain a **contractile vacuole** to regulate water in their cells.

- Apicomplexans are spore-producing organisms (**sporozoans**) that are commonly parasites of other organisms.

- The Rhizaria consist of the **foraminiferans,** and the radiolarians are organisms with fine, threadlike pseudopods.

The Excavates

- **Euglenids** are flagellated protists, or **flagellates,** with a **pellicle** instead of a cell wall. Their chloroplasts are most likely derived from a green alga, through endosymbiosis. They have an **eyespot apparatus** that acts as a photoreceptor. Euglenids are both autotrophic and heterotrophic.

- Parabasalids and diplomonads are single-celled, flagellated excavates that are endosymbionts of animals. *Trichomonas vaginalis* and *Giardia lamblia* are examples.

The Amoebozoa

- **Amoeboids** move and feed by forming pseudopods. Parasitic amoeboids, in the genus *Entamoeba,* cause amoebic dysentery.

- **Plasmodial slime molds** are amoeboid and ingest food by phagocytosis. **Cellular slime molds** exist as individual amoeboid cells until nutrients become limited. The cells then fuse into a pseudoplasmodium that produces spores.

The Opisthokonta

- This supergroup includes the multicellular fungi and animals.

- **Choanoflagellates** are animal-like protistans closely related to the sponges.

Protistans and Human Disease

A few protists cause some important human diseases, as follows:

- Sporozoans cause **malaria,** one of the most significant diseases in the world. Toxoplasmosis and cryptosporidiosis are other diseases caused by sporozoans.

- The most important diseases caused by flagellates are **African sleeping sickness** and Chagas disease. Infections with *Giardia lamblia* and *Trichomonas vaginalis* are also very common.

- The most common human disease caused by an amoeboid is **amoebic dysentery.** Rarely, amoeboids can cause a severe keratitis, and even a fatal brain infection.

29.2 Fungi

Fungi are single-celled or multicellular eukaryotes that are strict heterotrophs. Many are **saprotrophs,** meaning they act as decomposers.

- After external digestion, fungi absorb the resulting nutrient molecules. Most fungi are saprotrophs that aid the cycling of chemicals in ecosystems by decomposing dead remains.

- The body of a fungus is composed of thin filaments, each of which is called a **hypha.** A mass of hyphae is termed a **mycelium.** Nonseptate hyphae have no cross-walls, forming multinucleate cells. Septate hyphae have cross-walls, but pores allow for exchange of cytoplasm. The cell wall contains **chitin.**

- Most fungi produce nonmotile, and often windblown, spores during both asexual and sexual reproduction. During sexual reproduction, hyphae tips fuse so that hyphae containing both + and − nuclei usually result. Following nuclear fusion, meiosis occurs during the production of the sexual spores.

Fungi can be classified into the following major groups:

- The single-celled **Microsporidia** are parasites of animal cells.
- **Chytrid fungi** are aquatic and have flagella at some stages.
- **Zygospore fungi,** of which the common bread mold *Rhizopus stolonifer* is a member, reproduce by forming a capsule called a **sporangium** that produces thick-walled **zygospores.**
- **Sac fungi** are named for their production of a cuplike **ascocarp,** which produces spores called conidia. Most known fungi are sac fungi, including morels, truffles, *Penicillium, Aspergillus,* and plant pathogens like Dutch Elm disease and ergot. Some sac fungi, like the brewer's yeast *Saccharomyces cerevisiae,* produce single-celled forms called **yeasts.**
- **Club fungi** are named for their production of a club-shaped reproductive structure called a **basidium.** Some club fungi produce toxins; those in the genus *Psilocybe* produce an LSD-like substance.

Fungi often form the following symbiotic relationships with other organisms:

- **Lichens** are symbiotic associations between fungi and cyanobacteria or green algae. The algae provide organic nutrients to the fungi, while the fungi provide protection and enhanced absorption of water and minerals.
- **Mycorrhizae** enjoy a mutualistic relationship with plant roots, such that the fungus helps the plant absorb minerals and water, while the plant supplies the fungus with organic nutrients. **AM fungi** are a type of mycorrhiza that invades plant roots.

A few fungi are parasites of animals, causing fungal infections or **mycoses.** Most of these fungi are opportunists that mainly infect those with a compromised immune system.

- Superficial mycoses affect the outermost layers of skin, or sometimes the nails or hair. **Ringworm** is a type of tinea, or fungal infection of the skin. *Candida albicans,* a normal inhabitant of several body areas, can overgrow and cause "yeast infections" of the vagina, or **thrush** in the oral cavity.
- Systemic mycoses affect other body systems, especially the lungs.

ASSESS

Testing Yourself

Choose the best answer for each question.

29.1 Protists

1. Which of these is a green alga?
   a. *Volvox*
   b. *Gelidium*
   c. *Euglena*
   d. *Paramecium*
   e. *Plasmodium*

2. Dinoflagellates
   a. usually reproduce sexually.
   b. have protective cellulose plates.
   c. are insignificant producers of food and oxygen.
   d. have cilia instead of flagella.
   e. tend to be larger than brown algae.

3. Ciliates
   a. move by pseudopods.
   b. are not as varied as other protists.
   c. feed and move using cilia.
   d. do not divide by binary fission.
   e. are closely related to the radiolarians.

4. Which of the following is most closely related to the land plants?
   a. dinoflagellates   c. chlorophytes
   b. charophytes   d. euglenids

5. Protists that may lack mitochondria and possess deep oral grooves are known as
   a. excavates.   d. amoeboids.
   b. apicomplexans.   e. ciliates.
   c. alveolates.

6. Plasmodial and cellular _____ are important decomposers that have an amoeboid vegetative state.
   a. brown algae   d. slime molds
   b. water molds   e. red algae
   c. filamentous algae

7. Through what process did eukaryotic cells gain chloroplasts and mitochondria?
   a. engulfment
   b. endosymbiosis   d. phagocytosis
   c. endocytosis   e. conjugation

8. All protists are
   a. photosynthetic.   d. eukaryotes.
   b. parasites.   e. All of these are correct.
   c. single-celled.

29.2 Fungi

9. Which of the following represents an organism that uses decomposition as its mode of nutrition?
   a. parasite
   b. saprotroph
   c. autotroph
   d. All of these are correct.

10. Which of the following is not a characteristic of Microsporidia?
    a. parasitic
    b. lives in host cells
    c. multicellular
    d. can infect invertebrates like insects
    e. can infect vertebrates like humans

11. Which of the following is not a characteristic of chytrids?
    a. They have flagellated spores.
    b. They have flagellated gametes.
    c. They can live on very dry land.
    d. They can be single cells.
    e. They can be animal parasites.

12. The collective mass of hyphae is called a (n)
    a. sporophyte.
    b. mycelium.
    c. ascocarp.
    d. lichen.

13. Mycorrhizae
    a. are a type of lichen.
    b. are mutualistic relationships.
    c. help plants gather solar energy.
    d. help plants gather inorganic nutrients.
    e. Both b and d are correct.

14. Fungi are classified according to
    a. sexual reproductive structures.
    b. the shape of their hyphae.
    c. the mode of nutrition.
    d. the type of cell wall.
    e. All of these are correct.

15. Lichens
    a. cannot reproduce.
    b. need a nitrogen source to live.
    c. are parasitic on trees.
    d. are able to live in extreme environments.

ENGAGE

BioNOW

Want to know how this science is relevant to your life? Check out the BioNOW video below:

■ Nut Fungus

Fungi have an important role in the ecosystem as decomposers. How does their unique feeding method allow for this role?

Thinking Critically

1. Sickle-cell disease occurs mostly in people of African ancestry. When a person inherits two copies of the sickle-cell gene, his or her hemoglobin is abnormal, causing red blood cells to be more fragile and sticky than usual. This leads to anemia and circulatory problems. About 100,000 Americans have sickle-cell disease, and about 2 million carry one copy of the gene. Even though it can cause a serious disease, the sickle-cell gene is thought to have been favored by natural selection, because it gives some protection against malaria. Considering the type of cells infected by *Plasmodium* species, how might this protection work? What do you think may happen to the prevalence of the sickle-cell gene if an effective vaccine against malaria is ever developed and widely used?

2. You are trying to develop a new anti-termite chemical that will not harm environmentally beneficial insects. Because termites are adapted to eat only wood, they will starve if they cannot digest this food source. Termites have two symbiotic partners: the protozoan *Trichonympha collaris* and the bacteria it harbors that actually produce the enzyme that digests the wood. Knowing this, how might you prevent termite infestations without targeting the termites directly?

3. Many fungal infections of humans are considered to be opportunistic, meaning that fungi that are normally free-living (usually in soil) can sometimes survive, and even thrive, on or inside the human body. From the fungal "point of view," what unique challenges would be encountered when trying to survive on human skin? What about inside human lungs?

©Monty Rakusen/Getty Images

The Source of Coal

Our industrial society runs on fossil fuels, such as coal. The term *fossil fuel* refers to the remains of organic material from ancient times. During the Carboniferous period more than 300 million years ago, a great swamp forest encompassed what is now northern Europe, the Ukraine, and the Appalachian Mountains in the United States. The weather was warm and humid, and the plants grew very tall. These were not plants that would be familiar to us; instead, they were related to various groups of plants that are less well known—the lycophytes, horsetails, and ferns.

The amount of biomass in a Carboniferous swamp forest was enormous, and occasionally the swampy water rose, covering the plants that had died. Dead plant material under water does not decompose well. The partially decayed remains became covered by sediment, which changed them into sedimentary rock. Exposed to significant amounts of pressure and over long periods of time, the organic material became coal. This process continued for millions of years, resulting in immense deposits of coal. Geological upheavals raised the deposits to the levels where they can be mined today.

With a change of climate, many of the plants of the Carboniferous period became extinct. However, as we will see in this chapter, some of their smaller herbaceous relatives have evolved and survived to our time. We owe the industrialization of today's society to these ancient forests.

As you read through this chapter, think about the following questions:

1. Where do lycophytes, horsetails, and ferns fit in the evolutionary history of plants?

2. What adaptations in plants are associated with angiosperms?

CHAPTER OUTLINE

BEFORE YOU BEGIN

Before beginning this chapter, take a few moments to review the following discussions:

Section 3.3 What cellular structures are unique to plants?

Section 5.4 What is the role of meiosis in a cell?

Section 9.1 How does vascular tissue contribute to the success of the various groups of plants?

30.1 Evolutionary History of Plants

Plants (domain Eukarya, supergroup Archaeplastids, kingdom Plantae) are vital to human survival. As explained by the Ecology feature, "Plants and Humans," our dependence on them is nothing less than absolute.

Although a land environment does offer advantages such as plentiful light, it also has certain challenges such as the constant threat of desiccation (drying out). Most important, all stages of reproduction—gametes, zygote, and embryo—must be protected from the drying effects of air. To keep the internal environment of cells moist, a land plant must acquire water and transport it to all parts of its body. The evolution of an internal vascular system has made certain groups of plants incredibly successful. The evolutionary history of plants begins in the water. Most likely, land plants evolved from a type of freshwater green algae some 450 MYA (million years ago): both green algae and plants (1) contain chlorophylls

a and *b* and various accessory pigments, (2) store excess carbohydrates as starch, and (3) have cellulose in their cell walls. A comparison of DNA and RNA base sequences suggests that land plants are most closely related to a group of freshwater green algae known as charophytes. Although the common ancestor of modern charophytes and land plants no longer exists, if it did, it would have features resembling those of the *Chara* and *Coleochaete* (Fig. 30.1).

Figure 30.1 Charophytes. Charophytes (represented here by *Chara* and *Coleochaete*) are the green algae most closely related to land plants. (*Chara*): ©Heather Angel/Natural Visions; (*Coleochaete*): ©Chloe Shaut/Ricochet Creative Productions LLC

SCIENCE IN YOUR LIFE ▶ ECOLOGY

Plants and Humans

Humans derive most of their nourishment from three flowering plants: wheat, corn, and rice. All three of these plants are members of the grass family and are collectively called grains (Fig. 30A). Wheat is commonly used to produce flour and bread. It was first cultivated in the Middle East about 8000 BCE, and is thought to be one of the earliest cultivated plants. Early settlers brought wheat to North America, where many varieties of corn, more properly called maize, were already in existence. Rice originated in southeastern Asia several thousand years ago, where it grew in swamps. Many foods of both plant and animal origin are considered bland or tasteless without seasonings or spices. The Americas were discovered when Columbus was seeking a new route to the Far East to acquire spices, such as nutmeg, oregano, rosemary, and sage. In addition, our primary sweetener, sugar, comes almost exclusively from two plants—sugarcane (grown in South America, Africa, Asia, and the Caribbean) and sugar beets (grown mostly in Europe and North America). Our most popular drinks—coffee, tea, and cola—also come from flowering plants. Coffee originated in Ethiopia,

whereas tea is thought to have been first used somewhere in central Asia.

Cotton and rubber are two plants that still have many uses today (Fig. 30B). Until a few decades ago, cotton and other natural fibers were our only source of clothing. Interestingly, when Levi Strauss wanted to make a tough pair of jeans, he needed a stronger fiber than cotton, so he used hemp. Hemp comes from a plant closely related to marijuana, but unlike marijuana, it has extremely small amounts of the chemical tetrahydrocannabinol (THC), which causes the hallucinogenic effect when marijuana is smoked. Today, hemp is increasingly used to make clothing due to its toughness and wearability. Rubber had its origin in Brazil from the thick, white sap of the rubber tree. To produce a stronger rubber, such as that in tires, sulfur is added, and the sap is heated in a process called vulcanization. This produces a flexible material less sensitive to temperature changes. However, at present, most rubber is synthetically produced.

Wheat plants, *Triticum*

Rice plants, *Oryza*

Figure 30A Plants used for food. (wheat): ©Pixtal/age fotostock; (rice grain head): ©Thanakrit Sathavornmanee/Shutterstock; (rice plants): ©Wesley Hitt/Getty Images

Chara are commonly known as stoneworts because they are encrusted with calcium carbonate deposits. The body consists of a single file of very long cells anchored in mud by thin filaments. Whorls of branches occur at regions called nodes, located between the cells of the main axis. Male and female reproductive structures grow at the nodes. A *Coleochaete* grows on flat, aquatic surfaces, such as rocks in streams. They are being used as a model organism in the study of biotechnology. A key feature in the evolution of plants was the protection of the zygote. Land plants not only protect the zygote, they also protect and nourish the resulting embryo. This may be the first derived feature that separates land plants from green algae.

Figure 30.2 traces the evolutionary history of land plants and will serve as a backdrop as we discuss major groups of plants in the pages that follow.

To better understand the evolution of land plants, it is possible to associate each group with one of the major evolutionary events representing adaptations to a land existence.

- Mosses were the first group to have some form of protection for the embryo to prevent it from drying out.
- Lycophytes were the first group to evolve **vascular tissue,** which transports water and nutrients throughout the plant body.
- Ferns evolved **megaphylls,** large leaves that increase the surface area for photosynthesis.
- Gymnosperms were the first plants with **seeds.** Seeds contain the embryo and supportive nutrients within a protective coat.
- Angiosperms are the only group to have flowers. Flowers attract a variety of pollinators and give rise to fruits.

Alternation of Generations

All plants have a life cycle that includes an **alternation of generations.** In this life cycle, two multicellular individuals alternate, each producing the other (Fig. 30.3). The two individuals are (1) a sporophyte, which represents the diploid (2n) generation, and (2) a gametophyte, which represents the haploid (n) generation.

The **sporophyte** (2n) is the structure that produces spores by meiosis. A spore is a haploid reproductive cell that develops into a new organism without needing to fuse with another reproductive cell. In the plant life cycle, a spore undergoes mitosis and becomes a gametophyte.

The **gametophyte** (n) is named because of its role in the production of gametes. In plants, eggs and sperm are produced by mitotic cell division. A sperm and egg fuse, forming a diploid zygote that undergoes mitosis and becomes the sporophyte.

Two observations are in order. First, meiosis produces haploid spores. This is consistent with the realization that the sporophyte is the diploid generation and spores are haploid reproductive cells. Second, mitosis occurs as a spore becomes a gametophyte, and

Rubber tree, *Hevea*

Cotton, *Gossypium*

Figure 30B Plants used commercially. (rubber tree): ©Underverse/Getty Images; (cotton): ©Pixtal/age fotostock

Plants have also been used for centuries for a number of important household items, including the house itself. We are most familiar with lumber as the major structural portion in buildings. This wood comes mainly from a variety of conifers: pine, fir, and spruce, among others. In the tropics, trees and even herbs provide important components for houses. In rural parts of Central and South America, palm leaves are preferable to tin for roofs, because they last as long as 10 years and are quieter during rainstorms. In the Middle East, numerous houses along rivers are made entirely of reeds.

Today, plants are increasingly researched for their use in medicinal products. Currently, about 50% of all pharmaceutical drugs originate from plants or are derived from plant products. Some cancers are even treatable with medicinal plants, such as the rosy periwinkle, extracts from which have shown some success in treating childhood leukemia. Indeed, the National Cancer Institute and most pharmaceutical companies have spent millions of dollars to send botanists out to collect and test plant samples from around the world. Tribal medicine men, or shamans, of South America and Africa have already been of great importance in developing numerous drugs. However, plant extracts also continue to be misused for their hallucinogenic or other effects on the human body; examples are coca, the source of cocaine and crack, and the opium poppy, the source of heroin.

In addition to all these uses of plants, we should not forget their aesthetic value. Flowers brighten any yard, ornamental plants accent landscaping, and trees provide cooling shade during the summer, as well as shelter from harsh winds during the winter.

Questions to Consider

1. Why do you think plants are such a good source of drugs for human use? What advantage does this give the plants?
2. With over 369,000 species of angiosperms alone, why do you think we get most of our nourishment from only three plants (wheat, corn, and rice)?
3. What are some uses of plants, other than those just discussed?

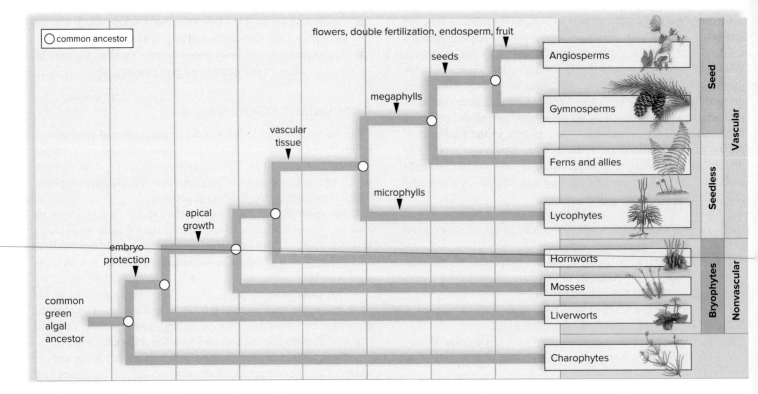

Figure 30.2 Evolutionary history of plants. The evolution of plants involves several significant innovations.

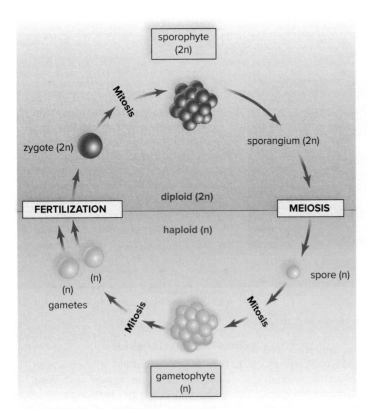

Figure 30.3 Alternation of generations. Plants alternate between a diploid sporophyte stage and a haploid gametophyte stage.

mitosis also occurs as a zygote becomes a sporophyte. Indeed, it is the occurrence of mitosis at these times that results in two generations.

Plants differ as to which generation is the dominant one. In plants, the dominant generation carries out the majority of photosynthesis. In nonvascular plants, the gametophyte is dominant, but in the other three groups of plants, the sporophyte is dominant. In the history of plants, only the sporophyte evolves vascular tissue. Therefore, the shift to sporophyte dominance is an adaptation to life on land. Notice that as the sporophyte gains in dominance, the gametophyte becomes microscopic (Fig. 30.4) and dependent upon the sporophyte.

All plants have an alternation-of-generations life cycle, but the appearance of the generations varies widely. In ferns, the gametophyte is a small, independent, heart-shaped structure. Archegonia are female reproductive organs that produce eggs. Antheridia are male reproductive organs that produce motile sperm. The eggs are fertilized by flagellated sperm, which swim to the archegonia in a film of water. In contrast, the female gametophyte of an angiosperm, called the embryo sac, is retained within the body of the sporophyte plant and consists of seven cells within a structure called an ovule. Following fertilization, the ovule becomes a seed. In seed plants, pollen grains are mature, sperm-bearing male gametophytes. Pollen grains can be transported by wind, insects, bats, or birds, and therefore do not need free water to reach the egg. In seed plants, reproductive cells are protected from drying out in the land environment.

Check Your Progress 30.1

1. Summarize the major adaptations in the evolution of the land plants.
2. Identify the role of each generation in the alternation-of-generations life cycle.

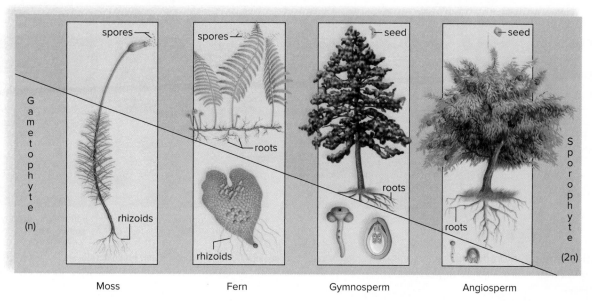

Figure 30.4 Reduction in the size of the gametophyte. As plants became adapted to life on land, the size of the gametophyte decreased, and the size of the sporophyte increased, as evidenced by these representatives of today's plants. In the moss and fern, spores disperse the gametophyte. In gymnosperms and angiosperms, seeds disperse the sporophyte.

30.2 Nonvascular Plants

Learning Outcomes

Upon completion of this section, you should be able to
1. Identify the traits of nonvascular plants that enable them to survive on land.
2. Compare and contrast the features of liverworts and mosses.

The **bryophytes**—the liverworts, hornworts, and mosses—were the first plants to colonize land. They only superficially appear to have roots, stems, and leaves because, by definition, true roots, stems, and leaves must contain vascular tissue. **Vascular tissue** is specialized for the transport of water and organic nutrients throughout the body of a plant. Bryophytes, which lack vascular tissue, are often called the **nonvascular plants.** Vascular tissue also provides support to the plant body, so bryophytes typically are low-lying; some mosses reach a maximum height of only about 20 cm.

Bryophytes do share other traits with the vascular plants. For example, they have an alternation-of-generations life cycle similar to all other plants. Their bodies are covered by a cuticle, which is interrupted in hornworts and mosses by stomata, and they have apical tissue that produces complex tissues. However, bryophytes are the only land plants in which the gametophyte is dominant (see Fig. 30.4). Antheridia produce flagellated sperm, which means they need a film of moisture in order to swim to eggs located inside archegonia. The bryophytes' lack of vascular tissue and the presence of flagellated sperm mean you are apt to find bryophytes in moist locations. Some bryophytes compete well in harsh environments because they can reproduce asexually.

Liverworts

Liverworts (phylum Hepaticophyta) come in two types: those that have a flat, lobed body called a *thallus* (*pl.,* thalli) and those that are leafy. *Marchantia* is a familiar liverwort (Fig. 30.5) that has a smooth upper surface. The lower surface bears numerous *rhizoids* (rootlike hairs) that project into the soil. *Marchantia* reproduces both asexually and sexually. Gemmae cups on the upper surface of the thallus contain gemmae, groups of cells that detach from the thallus and can asexually start a new plant. Sexual reproduction depends on disk-headed stalks that bear **antheridia** (*sing.,* antheridium), where flagellated sperm are produced, and on umbrella-headed stalks that bear **archegonia** (*sing.,* archegonium), where eggs are produced. Following fertilization, tiny sporophytes composed of a foot, a short stalk, and a capsule begin growing within archegonia. Windblown spores are produced within the capsule.

Hornworts

Although some species of hornworts live on trees, most live in moist, well-shaded areas. They photosynthesize but also have a symbiotic relationship with cyanobacteria, which, unlike plants, can fix nitrogen from the air.

The small sporophytes of a hornwort resemble tiny green broom handles rising from a thin gametophyte, usually less than 2 cm in diameter (Fig. 30.6). Like the gametophyte, a sporophyte can photosynthesize, although it has only one chloroplast per cell. A hornwort can bypass alternation of generations by reproducing asexually through fragmentation.

Mosses

Mosses (phylum Bryophyta) can be found from the Arctic through the tropics to parts of the Antarctic. Although most live

gemma cup

thallus

rhizoids

a. Gemma cup

gemma

Thallus with gemmae cups

male gametophyte

female gametophyte

b. Male gametophytes bear antheridia

c. Female gametophytes bear archegonia

Figure 30.5 Liverwort, *Marchantia*.   **a.** A gemma is a group of cells that can detach and start a new plant. **b.** Antheridia are present in disk-shaped stalks. **c.** Archegonia are present in umbrella-shaped stalks. (photos) (a): ©Ed Reschke; (b): ©J.M. Conrader/Nat'l Audubon Society/Science Source; (c): ©Ed Reschke/Oxford Scientific/Getty Images

sporophyte

gametophyte

Figure 30.6 Hornworts.   The "horns" of a hornwort, such as this species of *Anthoceros*, are sporophytes that grow continuously from a base anchored in gametophyte tissue. (photo): ©Steven P. Lynch

in damp, shaded locations in the temperate zone, some survive in deserts, whereas others inhabit bogs and streams. In forests, mosses frequently form a mat that covers the ground or the surfaces of rotting logs. Mosses can store large quantities of water in their cells, but if a dry spell continues for long, they become dormant until it rains.

Most mosses can reproduce asexually by fragmentation. Just about any part of the plant is able to grow and eventually produce leaflike thalli. Figure 30.7 describes the life cycle of a typical temperate-zone moss. The gametophyte of mosses has two stages. First, there is the algalike *protonema*, a branching filament of cells. Then, after about 3 days of favorable growing conditions, upright leafy thalli appear at intervals along the protonema. Rhizoids anchor the thalli, which bear antheridia and archegonia. An antheridium consists of a short stalk, an outer layer of sterile cells, and an inner mass of cells that become the flagellated sperm. An

archegonium, which looks like a vase with a long neck, has a single egg located inside its base.

The dependent sporophyte consists of a *foot*, which grows down into the gametophyte tissue; a *stalk;* and an upper *capsule,* or *sporangium,* where windblown spores are produced. At first, the sporophyte is green and photosynthetic. At maturity, it is brown and nonphotosynthetic. Because the gametophyte is the dominant generation, it seems consistent for spores to be dispersal agents— that is, when the haploid spores germinate, the gametophyte is in a new location.

Adaptations and Uses of Nonvascular Plants

Although mosses are nonvascular, they are better than flowering plants at living on stone walls, fences, and even in the shady cracks of hot, exposed rocks. For these particular microhabitats, being small and simple seems to be a selective advantage. When bryophytes colonize bare rock, they help convert the rocks to soil that can be used for the growth of other organisms.

In areas such as bogs, where the ground is wet and acidic, dead mosses, especially *Sphagnum*, do not decay. The accumulated moss, called peat or bog moss, can be used as fuel. Peat moss is also commercially important because it has special nonliving cells that can absorb moisture. Thus, peat moss is often used in gardens to improve the water-holding capacity of the soil.

Check Your Progress 30.2

1. List two features limiting the adaptation of nonvascular plants.
2. Describe two features of nonvascular plants that indicate they are adapted to the land environment.
3. Describe the differences between a liverwort and a moss.

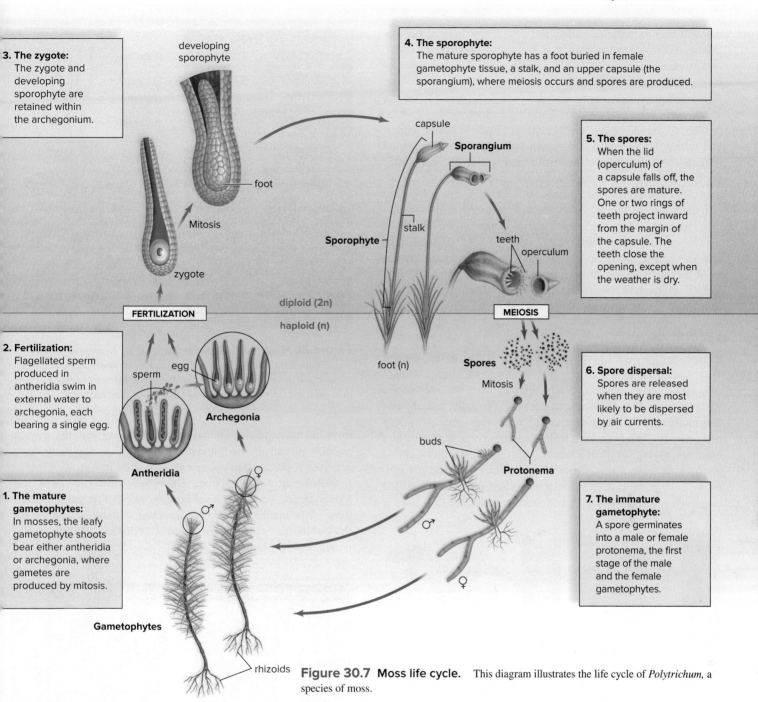

3. The zygote:
The zygote and developing sporophyte are retained within the archegonium.

developing sporophyte

foot

Mitosis

zygote

4. The sporophyte:
The mature sporophyte has a foot buried in female gametophyte tissue, a stalk, and an upper capsule (the sporangium), where meiosis occurs and spores are produced.

capsule

Sporangium

Sporophyte

stalk

teeth

operculum

5. The spores:
When the lid (operculum) of a capsule falls off, the spores are mature. One or two rings of teeth project inward from the margin of the capsule. The teeth close the opening, except when the weather is dry.

diploid (2n)

FERTILIZATION

haploid (n)

MEIOSIS

foot (n)

Spores

Mitosis

2. Fertilization:
Flagellated sperm produced in antheridia swim in external water to archegonia, each bearing a single egg.

sperm

egg

Archegonia

Antheridia

6. Spore dispersal:
Spores are released when they are most likely to be dispersed by air currents.

buds

Protonema

♂

♀

1. The mature gametophytes:
In mosses, the leafy gametophyte shoots bear either antheridia or archegonia, where gametes are produced by mitosis.

♂

♀

7. The immature gametophyte:
A spore germinates into a male or female protonema, the first stage of the male and the female gametophytes.

Gametophytes

rhizoids

Figure 30.7 Moss life cycle. This diagram illustrates the life cycle of *Polytrichum,* a species of moss.

30.3 Seedless Vascular Plants

All the other plants we will study are **vascular plants.** Vascular tissue in these plants consists of **xylem,** which conducts water and minerals up from the soil, and **phloem,** which transports organic nutrients from one part of the plant to another. The vascular plants usually have true roots, stems, and leaves. The roots absorb water from the soil, and the stem conducts water to the leaves. Xylem, with its strong-walled cells, supports the body of the plant against the pull of gravity. The leaves are fully covered by a waxy cuticle except where it is interrupted by stomata, little pores whose size can be regulated to control water loss.

The sporophyte is the dominant generation in vascular plants. Seedless vascular plants include two groups: lycophytes and ferns and their allies. Both of these groups disperse their offspring by producing windblown spores. This is advantageous because the sporophyte is the generation with vascular tissue. Another advantage of having a dominant sporophyte relates to its being diploid. If a faulty gene is present, it can be masked by a functional gene. Then, too, the greater the amount of genetic material, the greater the possibility of mutations that will lead to increased variety and complexity. Indeed, vascular plants are complex, extremely varied, and widely distributed.

Some vascular plants do not produce seeds. Seedless vascular plants include whisk ferns, club mosses, horsetails, and ferns, which disperse their offspring by producing windblown spores. When the spores germinate, they produce a small gametophyte that is independent of the sporophyte for its nutrition. In these plants, antheridia release flagellated sperm, which swim in a film of external water to the archegonia, where fertilization occurs. Because spores are dispersal agents and the nonvascular gametophyte is independent of the sporophyte, these plants cannot wholly benefit from the adaptations of the sporophyte to a terrestrial environment.

The seedless vascular plants formed the great swamp forests of the Carboniferous period around 299 MYA (see Table 27.1). A large number of these plants died but did not decompose completely. Instead, they were compressed to form the coal (see the chapter opener) that we still mine and burn today. Oil has a similar origin, but it typically forms in marine sedimentary rocks and includes animal remains.

Lycophytes

Lycophytes (phylum Lycophyta) are also called **club mosses** and were among the first land plants to have vascular tissue. Unlike true mosses (bryophytes), the lycophytes have well-developed vascular tissue in roots, stems, and leaves. Typically, a fleshy underground and horizontal stem, called a rhizome, sends up upright aerial stems. Tightly packed, scalelike leaves cover the stems and branches, giving the plant a mossy look. The small leaves, called *microphylls,* have a single vein composed of xylem and phloem. As we will see, most other vascular plants have larger leaves, called *megaphylls,* that contain branched vascular tissue.

Microphyll   Megaphyll

single strand of vascular tissue

branched vascular tissue

The sporangia of the lycophytes are borne on terminal clusters of leaves, called strobili, which are club-shaped. The spores are sometimes harvested and sold as lycopodium powder, or vegetable sulfur, for use in pharmaceuticals and in fireworks because it is highly flammable. Members of the genus *Lycopodium* featured in Figure 30.8 are common in moist woodlands in temperate climates, where they are called ground pines; they are also abundant in the tropics and subtropics.

Ferns and Their Allies

In this section, we will discuss the characteristics of the seedless vascular plants called the **pteridophytes,** which include ferns, whisk ferns, and horsetails. Unlike lycophytes, pteridophytes possess megaphylls as leaves.

Ferns

Ferns (phylum Polypodiophyta) are the largest group of plants other than the flowering plants, and they display great diversity in

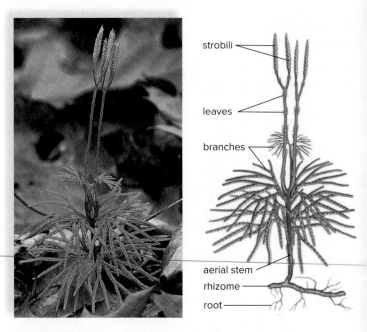

strobili

leaves

branches

aerial stem

rhizome

root

Figure 30.8 Ground pine, *Lycopodium.* Lycophytes such as *Lycopodium* have vascular tissue and thus true roots, stems, and leaves. The *Lycopodium* sporophyte develops an underground rhizome system. A rhizome is an underground stem; this rhizome produces roots along its length. (photo): ©Steve Solum/Avalon

form and habitat. Ferns are most abundant in warm, moist, tropical regions, but can also be found in northern regions and in relatively dry, rocky, places. They range in size from low-growing plants resembling mosses to tall trees. The *fronds* (leaves) that grow from a rhizome can vary. The royal fern has fronds that stand 6 ft tall; those of the maidenhair fern are branched, with broad leaflets; and those of the hart's tongue fern are straplike and leathery. Figure 30.9 provides some examples of ferns.

spores on fertile frond

frond (undivided)

Cinnamon fern, *Osmunda cinnamomea*

Hart's tongue fern, *Campyloneurum scolopendrium*

Figure 30.9 Fern diversity. The many different types of ferns grow in places that offer moisture. These photos show the dominant sporophyte. The separate, delicate, and almost microscopic gametophyte is water dependent and produces sperm that require moisture to swim to the egg. (cinnamon fern): ©Laszlo Podor Photography/Getty Images; (Hart's tongue fern): ©Organics image library/Alamy

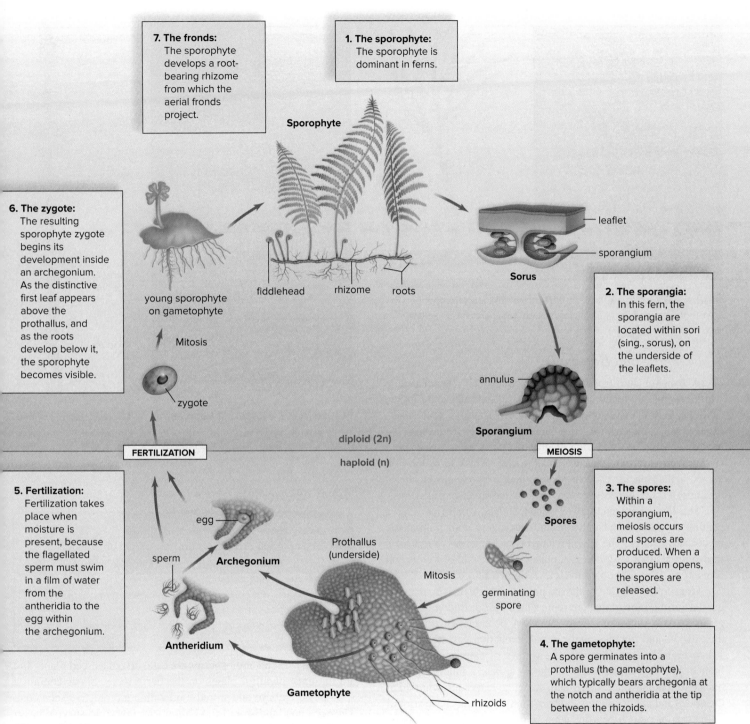

7. The fronds: The sporophyte develops a root-bearing rhizome from which the aerial fronds project.

1. The sporophyte: The sporophyte is dominant in ferns.

Sporophyte

leaflet

sporangium

Sorus

6. The zygote: The resulting sporophyte zygote begins its development inside an archegonium. As the distinctive first leaf appears above the prothallus, and as the roots develop below it, the sporophyte becomes visible.

young sporophyte on gametophyte

fiddlehead   rhizome   roots

Mitosis

zygote

2. The sporangia: In this fern, the sporangia are located within sori (sing., sorus), on the underside of the leaflets.

annulus

Sporangium

diploid (2n)

FERTILIZATION

haploid (n)

MEIOSIS

5. Fertilization: Fertilization takes place when moisture is present, because the flagellated sperm must swim in a film of water from the antheridia to the egg within the archegonium.

egg

sperm

Archegonium

Antheridium

Prothallus (underside)

Mitosis

germinating spore

Spores

Gametophyte

rhizoids

3. The spores: Within a sporangium, meiosis occurs and spores are produced. When a sporangium opens, the spores are released.

4. The gametophyte: A spore germinates into a prothallus (the gametophyte), which typically bears archegonia at the notch and antheridia at the tip between the rhizoids.

Figure 30.10 Fern life cycle.   Stages of the life cycle of a typical fern.

Life Cycle, Adaptations, and Uses of Ferns   Ferns have true roots, stems, and leaves that contain vascular tissue. The well-developed leaves fan out, capture solar energy, and photosynthesize. The life cycle of a typical temperate fern is shown in Figure 30.10. The dominant sporophyte produces windblown spores. When the spores germinate, a tiny green and independent gametophyte develops. The gametophyte is water dependent because it lacks vascular tissue. Also, flagellated sperm produced within antheridia require an outside source of moisture to swim to the eggs in the archegonia. Upon fertilization the zygote develops into a sporophyte. In nearly all ferns, the leaves of the sporophyte first appear in a curled-up form called a fiddlehead, which unrolls as it

grows. Once established, some ferns, such as the bracken fern *Pteridium aquilinum,* can spread into drier areas by means of vegetative (asexual) reproduction. For example, ferns can spread when their rhizomes grow horizontally in the soil, producing fiddleheads that become new fronds.

In terms of economic value, people use ferns in decorative bouquets and as ornamental plants in the home and garden. Wood from tropical tree ferns often serves as a building material because it resists decay, as well as damage by termites. Ferns, especially the ostrich fern, are eaten as food, and in the Northeast, many restaurants feature fiddleheads as a special treat, although studies have shown that some are carcinogenic.

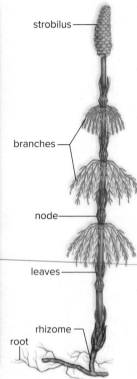

Figure 30.11 Whisk fern, *Psilotum.* Whisk ferns have no roots or leaves—the branches carry on photosynthesis. The sporangia are yellow.
(photo): ©Biophoto Associates/Science Source

Whisk Ferns and Horsetails

Whisk ferns (phylum Psilotophyta), named for their resemblance to whisk brooms, are found in Arizona, Texas, Louisiana, and Florida, as well as Hawaii and Puerto Rico. *Psilotum* (Fig. 30.11) looks like a rhyniophyte, an ancient vascular plant known only from the fossil record. Its aerial stem forks repeatedly and is attached to a *rhizome,* a fleshy horizontal stem that lies underground. Whisk ferns have no leaves, so the branches carry on photosynthesis. Sporangia, located at the ends of short branches, produce spores that are dispersal agents.

The *Psilotum* gametophyte is independent, small (less than a centimeter), and found underground associated with mutualistic mycorrhizal fungi. The resemblance of its life cycle to that of a fern suggests that *Psilotum* is actually a fern devoid of leaves and roots. If so, it could be said to be a vestigial fern.

Horsetails (phylum Equisetophyta), which thrive in moist habitats around the globe, are represented by *Equisetum,* the only genus in existence today (Fig. 30.12). A perennial rhizome produces roots and photosynthetic aerial stems that can stand about 1.3 m high. In some species, the whorls of slender green side branches at the joints (nodes) of the stem bear a fanciful resemblance to a horse's tail. The small, scalelike leaves are whorled at the nodes and are nonphotosynthetic. Many horsetails have strobili at the tips of branch-bearing stems. Others send up special buff-colored, naked stems that bear the strobili.

The stems of horsetails are tough and rigid because of silica deposited in their cell walls. Early Americans, in particular, used horsetails to scour pots and called them "scouring rushes." Today, they are still used as ingredients in a few abrasive powders. *Equisetum* is sometimes considered a weed in the garden, and it is difficult to control because any portion of the rhizome left in the soil will regenerate.

Check Your Progress 30.3

1. Explain what makes lycophytes significant in plant evolution.
2. Describe the difference between a microphyll and a megaphyll.
3. Summarize the life cycle of a fern.

Figure 30.12 Horsetail, *Equisetum.* Whorls of branches and tiny leaves are at the nodes of the stem. Spore-producing sporangia are borne in strobili (cones). (photo): ©Robert P. Carr/Avalon

30.4 Seed Plants

Learning Outcomes

Upon completion of this section, you should be able to

1. List the features of seed plants that make them fully adapted to life on land.
2. Compare and contrast the life cycles of gymnosperms with angiosperms.
3. List the key features of monocots and eudicots.

Gymnosperms and angiosperms are categorized as seed plants and are the most plentiful plants on the Earth today. Seeds contain a sporophyte embryo and stored food within a protective seed coat. The seed coat and stored food allow an embryo to survive harsh conditions during long periods of dormancy (arrested state) until environmental conditions become favorable for growth. Seeds can even remain dormant for hundreds of years. When a seed germinates, the stored food is a source of nutrients for the growing seedling. The survival value of seeds largely accounts for the dominance of seed plants today.

Seed plants, such as the gymnosperms, are *heterosporous,* meaning they have two types of spores and produce a male and female gametophyte during the course of their life cycle (Fig. 30.13). **Pollen grains,** which resist drying out, become the multicellular male gametophyte. **Pollination** occurs when a pollen grain is brought to the vicinity of a female gametophyte by wind or a pollinator. For angiosperms, the **pollinator** is an animal, often an insect, that carries pollen from flower to flower. After a pollen

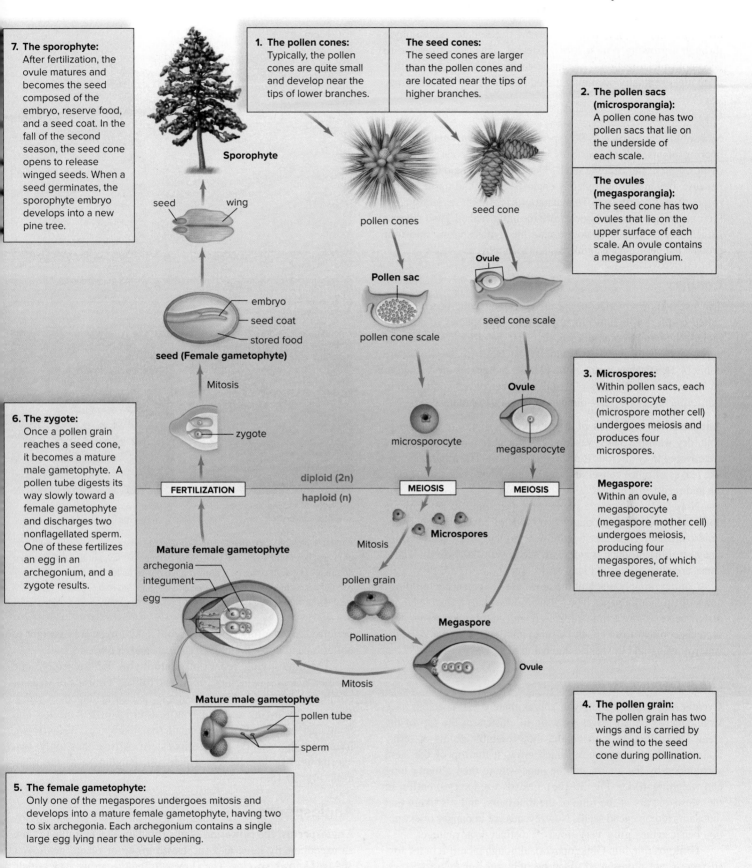

7. The sporophyte:
After fertilization, the ovule matures and becomes the seed composed of the embryo, reserve food, and a seed coat. In the fall of the second season, the seed cone opens to release winged seeds. When a seed germinates, the sporophyte embryo develops into a new pine tree.

1. The pollen cones:
Typically, the pollen cones are quite small and develop near the tips of lower branches.

The seed cones:
The seed cones are larger than the pollen cones and are located near the tips of higher branches.

2. The pollen sacs (microsporangia):
A pollen cone has two pollen sacs that lie on the underside of each scale.

The ovules (megasporangia):
The seed cone has two ovules that lie on the upper surface of each scale. An ovule contains a megasporangium.

3. Microspores:
Within pollen sacs, each microsporocyte (microspore mother cell) undergoes meiosis and produces four microspores.

Megaspore:
Within an ovule, a megasporocyte (megaspore mother cell) undergoes meiosis, producing four megaspores, of which three degenerate.

6. The zygote:
Once a pollen grain reaches a seed cone, it becomes a mature male gametophyte. A pollen tube digests its way slowly toward a female gametophyte and discharges two nonflagellated sperm. One of these fertilizes an egg in an archegonium, and a zygote results.

4. The pollen grain:
The pollen grain has two wings and is carried by the wind to the seed cone during pollination.

5. The female gametophyte:
Only one of the megaspores undergoes mitosis and develops into a mature female gametophyte, having two to six archegonia. Each archegonium contains a single large egg lying near the ovule opening.

Figure 30.13 Pine life cycle. The sporophyte is dominant, and its sporangia are borne in cones. There are two types of cones: pollen cones (microstrobili) and seed cones (megastrobili).

grain germinates, sperm move toward the female gametophyte through a growing **pollen tube.** External water is not needed to accomplish fertilization. The female gametophyte develops within an **ovule,** eventually becoming a seed that disperses the offspring.

Gymnosperms

Most of the approximately 900 species of **gymnosperms** are cone-bearing plants. On the surfaces of their female cone scales are ovules, which later become seeds. These ovules are said to be naked because they are not completely enclosed by diploid tissue (as the ovules of angiosperms are). This characteristic gives the gymnosperms their name; *gymnos* is a Greek word meaning "naked," and *sperma* means "seed." Botanists now divide the gymnosperms into four groups: conifers, cycads, ginkgos, and gnetophytes.

Conifers

The better-known gymnosperms are evergreen, cone-bearing trees called **conifers** (phylum Pinophyta). The 630 species of conifers include the familiar pine, spruce, fir, cedar, hemlock, redwood, and cypress. Perhaps the oldest and largest trees in the world are conifers. Bristlecone pines in the Nevada mountains are known to be more than 4,900 years old, and a number of redwood trees in California are 2,000 years old and more than 100 meters tall.

Adaptations and Uses of Conifers   Conifers are adapted to cold, dry weather. Thus, vast areas of northern temperate regions are covered in coniferous forests (Fig. 30.14*a*). The tough, needle-like leaves of pines conserve water because they have a thick cuticle and recessed stomata.

Note in the life cycle of the pine (see Fig. 30.13) that the dominant sporophyte produces two kinds of cones. **Pollen cones** (microstrobili) produce windblown pollen, and **seed cones** (megastrobili) produce windblown seeds (Fig. 30.14*b, c*). These cones are adaptations to a land environment.

Conifers supply much of the wood used to construct buildings and to manufacture paper. They also produce many valuable chemicals, such as those extracted from resin (pitch), a substance that protects conifers from attack by fungi and insects, and is used for waterproofing and in varnish and paints.

Other Gymnosperms

Cycads (phylum Cycadophyta, approximately 320 species) have large, finely divided leaves growing in clusters at the top of the stem. Depending on their height, they resemble palms or ferns. Pollen cones or seed cones, which grow at the top of the stem surrounded by the leaves, can be huge—more than a meter long and weighing 40 kg (Fig. 30.15*a*). Cycads were very plentiful in the Mesozoic era at the time of the dinosaurs, and it's likely that dinosaurs fed on cycad seeds. Now, cycads are in danger of extinction because they grow very slowly, a distinct disadvantage.

Ginkgos (phylum Ginkgophyta), although plentiful in the fossil record, are represented today by only one surviving species, *Ginkgo biloba,* the maidenhair tree, so named because its leaves resemble those of the southern maidenhair fern, *Adiantum capillus-veneris.* Female trees produce fleshy seeds, which ripen in the fall, and give off such a foul odor that male trees are usually

a. A northern coniferous forest

b. Cones of lodgepole pine, *Pinus contorta*

c. Fleshy seed cones of juniper, *Juniperus*

Figure 30.14 Conifers.   **a.** Conifers are adapted to living in cold climates. **b.** In pine trees, the seed cones become woody, but the pollen cones do not. **c.** The seed cones of a juniper have fleshy scales that fuse into a berrylike structure. (a): ©Pete Ryan/National Geographic/Getty Images; (b): ©Kathy Merrifield/Science Source; (c): ©McGraw-Hill Education/Evelyn Jo Johnson, photographer

preferred for planting (Fig. 30.15*b*). Ginkgo trees are resistant to pollution and do well along city streets and in city parks.

The three living genera of **gnetophytes** don't resemble one another. *Gnetum,* which occurs in the tropics, consists of trees or climbing vines with broad, leathery leaves arranged in pairs. *Ephedra,* occurring only in southwestern North America and Southeast Asia, is a shrub with small, scalelike leaves. *Welwitschia,* living in the deserts of southwestern Africa, has only two enormous, straplike leaves, which split lengthwise as the plant ages (Fig. 30.15*c*).

Angiosperms

Angiosperms, the flowering plants (phylum Magnoliophyta), are an exceptionally large and successful group of plants, with over 369,000 known species. The flowers of angiosperms vary widely in appearance (Fig. 30.16). Angiosperms live in all sorts of habitats, from fresh water to desert, and from the frigid north to the torrid tropics. They range in size from the tiny, almost microscopic duck weed to *Eucalyptus* trees over 100 m tall. It would be pollen cone   seed cone

a.

b.

c.

Figure 30.15 Gymnosperm diversity.   a. Cycad cones. **b.** The
female ginkgo trees are known for their malodorous seeds. **c.** *Welwitschia* is
known for its enormous straplike leaves. (a, pollen cone): ©Steven P. Lynch; (a, seed
cone): ©Steven P. Lynch; (b): ©Kathy Merrifield/Science Source; (c): ©Fiona Norris

Cactus, *Opuntia*   Water lily, *Nymphaea odorata*

Blue flag iris, *Iris versicolor*   Amborella, *Amborella trichopoda*

Snow trillium, *Trillium nivale*   Apple blossom, *Malus domestica*

Butterfly weed, *Asclepias tuberosa*

Figure 30.16 Flower diversity.   Angiosperm flowers vary widely
in their appearance, but their function is similar. (cactus flowers): ©Liane M/
Shutterstock; (water lily): ©McGraw-Hill Education/Dave Moyer, photographer; (iris):
©Steffen Hauser/botanikfoto/Alamy; (*Amborella*): ©Stephen McCabe/Arboretum at
University of California Santa Cruz; (snow trillium): ©Adam Jones/Science Source; (apple
blossoms): ©Inga Spence/Science Source; (butterfly weed): ©Evelyn Jo Johnson

impossible to exaggerate the importance of angiosperms in our
everyday lives. As discussed in the Ecology feature "Plants and
Humans" in Section 30.1, they provide us with clothing, food,
medicines, and other commercially valuable products.

The name angiosperm is derived from the Greek words *angio,*
meaning "vessel," and *sperma,* meaning "seed." The seed develops
from an ovule within an ovary, which becomes a fruit. Therefore,
angiosperms produce covered seeds (in contrast to the exposed
seeds of gymnosperms). Although the oldest fossils of angiosperms
date back no more than about 135 million years, they are most
likely much older. Molecular evidence suggests the possible
ancestors of angiosperms may have originated as far back as
200 million years ago.

Scientists hypothesize that angiosperms coevolved with the
insects that act as pollinators. As angiosperms became diverse,
flying insects also enjoyed an adaptive radiation, increasing in
morphological and ecological diversity.

Monocots and Eudicots

Most flowering plants belong to one of two classes. There are about
60,000 known species of Monocotyledones (Liliopsida), often
referred to as the **monocots.** The remainder of the flowering plants

belong to the Eudicotyledones, commonly called the **eudicots.**
Table 30.1 lists the differences between monocots and eudicots.

Monocots are named because they have only one cotyledon in
their seeds, whereas eudicots have two cotyledons. Cotyledons are
the seed leaves—they contain nutrients that nourish the developing embryo.

TABLE 30.1 Monocots and Eudicots

Monocots	Eudicots
One cotyledon	Two cotyledons
Flower parts in threes or multiples of three	Flower parts in fours or fives or multiples of four or five
Usually herbaceous	Woody or herbaceous
Usually parallel venation	Usually net venation
Scattered bundles in stem	Vascular bundles in a ring
Fibrous root system	Taproot system
Pollen grain with one pore	Pollen grain with three pores

The Flower

The flowers of angiosperms have certain structures in common (Fig. 30.17). The flower stalk expands slightly at the tip into a **receptacle,** which bears the other flower parts. These parts, called **sepals, petals, stamens,** and **carpels,** are attached to the receptacle in whorls (circles). The sepals, collectively called the **calyx,** protect the flower bud before it opens. The sepals may drop off or may be colored like the petals. Usually, however, sepals are green and remain attached to the receptacle. The petals, collectively called the **corolla,** are quite diverse in size, shape, and color. The corolla often attracts a particular pollinator.

Each stamen consists of a saclike **anther,** where pollen is produced, and a stalk called a **filament.** In most flowers, the anther is positioned where the pollen can be carried away by wind or a pollinator. One or more carpels is at the center of a flower. A carpel has three major regions: ovary, style, and stigma. The combination of the ovary, style, stigma, and carpels is sometimes called a *pistil.* The swollen base is the **ovary,** which contains from one to hundreds of ovules. The **style** elevates the **stigma,** which is adapted to receive pollen grains. Glands in the region of the ovary produce nectar, a nutrient that pollinators gather as they go from flower to flower. The flower contributes to the success of the angiosperms.

Life Cycle, Adaptations, and Uses of Flowering Plants

Figure 30.18 depicts the life cycle of a typical flowering plant. Sexual reproduction in flowering plants is dependent on the flower, which produces both pollen and seeds. A gymnosperm, such as a pine tree, depends wholly on the wind to disperse both pollen and seeds. In contrast, among angiosperms, some species have windblown pollen, whereas others rely on a pollinator to carry pollen to another member of the same species. Bees, wasps, flies, butterflies, moths, beetles, and even bats can act as pollinators. The flower provides the pollinator with nectar, a sugary substance that, in the case of bees, will become honey in the hive. Several species of bees (such as honeybees and bumblebees) also have two "pollen baskets," one on each of their hind legs. When bees make a brief stop on a flower, they suck up nectar with specialized mouthparts and gather pollen grains, which are stored in their pollen baskets. Honey and pollen are the main diets of immature bees (called larvae). When a bee visits another flower, some of the pollen grains are rubbed off, and in this way the bee delivers pollen from one flower to another. Other types of pollinators also carry pollen

Figure 30.17 Generalized flower. A flower has four main parts: sepals, petals, stamens, and carpels. Each stamen has an anther and a filament. A carpel has a stigma, a style, and an ovary. The ovary contains ovules.

between flowers of the same species, because pollinators and flowers are adapted to one another.

Flowering plants that have windblown pollen are usually not very showy, whereas insect-pollinated flowers and bird-pollinated flowers are often colorful. Many flowers are adapted to attract specific pollinators. For example, bee-pollinated flowers are usually blue or yellow and have ultraviolet shadings that lead the pollinator to the location of nectar at the base of the flower. Night-blooming flowers are usually aromatic and white or cream-colored, so that their smell alone, rather than color, can attract nocturnal pollinators such as bats.

Fruits, the final products of a flower, aid in the dispersal of seeds (Fig. 30.19). The production of fruit is another advantage angiosperms have over gymnosperms. Dry fruits, such as pods, assist the dispersal of windblown seeds. Mature pods sometimes explode and shoot out the seeds. Other pods, such as those of peas and beans, simply break open and scatter the seeds. The famed dandelion produces a one-seeded fruit with an umbrella-like crown of intricately branched hairs. The slightest gust of wind catches the hairs, raising and propelling the fruit into the air like a parachute.

Fruits also help disperse angiosperm seeds in two other ways. One means is exemplified by the coconut, a fruit that can drift for thousands of kilometers across seas and oceans. A second method of seed dispersal occurs because the seeds of some fruits have a

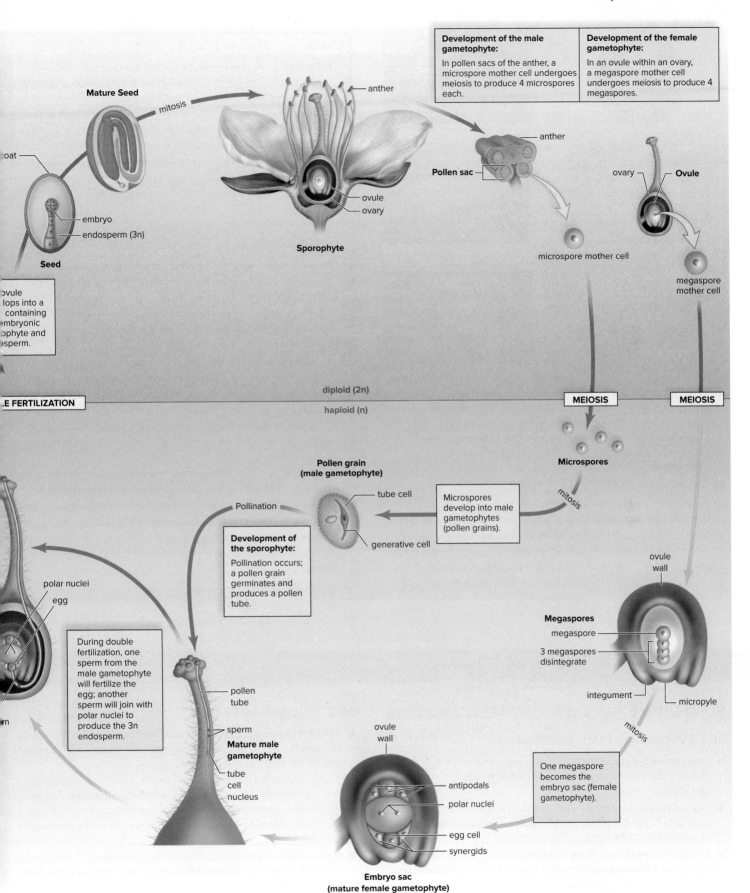

Development of the male gametophyte:

In pollen sacs of the anther, a microspore mother cell undergoes meiosis to produce 4 microspores each.

Development of the female gametophyte:

In an ovule within an ovary, a megaspore mother cell undergoes meiosis to produce 4 megaspores.

Mature Seed

mitosis

anther

coat

embryo

endosperm (3n)

Seed

ovule
lops into a
containing
mbryonic
ophyte and
sperm.

.E FERTILIZATION

anther

Pollen sac

ovule

ovary

Sporophyte

ovary

Ovule

microspore mother cell

megaspore mother cell

diploid (2n)

haploid (n)

MEIOSIS

MEIOSIS

Pollen grain
(male gametophyte)

tube cell

Pollination

generative cell

Microspores
develop into male
gametophytes
(pollen grains).

mitosis

Microspores

Development of
the sporophyte:

Pollination occurs;
a pollen grain
germinates and
produces a pollen
tube.

ovule
wall

polar nuclei

egg

During double
fertilization, one
sperm from the
male gametophyte
will fertilize the
egg; another
sperm will join with
polar nuclei to
produce the 3n
endosperm.

pollen
tube

sperm

Mature male
gametophyte

tube
cell
nucleus

ovule
wall

Megaspores

megaspore

3 megaspores
disintegrate

integument

micropyle

mitosis

antipodals

polar nuclei

egg cell

synergids

Embryo sac
(mature female gametophyte)

One megaspore
becomes the
embryo sac (female
gametophyte).

Figure 30.18 Flowering plant life cycle. The parts of the flower involved in reproduction are the stamens and the carpel. Reproduction has been divided into significant stages: female gametophyte development, male gametophyte development, and sporophyte development.

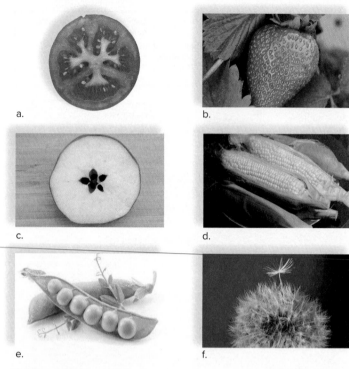

a.
b.
c.
d.
e.
f.

Figure 30.19 Fruits. Angiosperms have diverse fruits. **a.** A tomato is technically a berry. **b.** A strawberry is an aggregate fruit formed from many carpels within a single flower. **c.** Apples have seeds inside a fleshy fruit. **d.** A corn cob carries many fruits (kernels). **e.** Peas have seeds in pods. **f.** Dandelion fruits are carried by wind. (a): ©Stockbyte/PunchStock; (b–c): ©Mondae Leigh Baker; (d): ©Crezalyn Nerona Uratsuji/Getty Images; (e): ©Valentyne Volkov/Shutterstock; (f): ©Steve Allen/Stockbyte/Getty Images

fleshy covering that animals eat as a source of food. Only the fleshy part is digested, so the stones or seeds pass through the animal's digestive system and are deposited perhaps far away from the parent plant. Still other fruits, like those of the cocklebur, have seeds with hooks that catch on the fur of animals and are carried away.

Check Your Progress 30.4

1. Recognize the two types of spores that occur in the seed plant life cycle, and describe where they develop and what role they play in the life cycle.
2. List the features of the flowering plant life cycle that are not found in any other group of plants.
3. Distinguish between a monocot and a eudicot plant.

Conclusion

While we may consider coal an important resource in modern society, it is in fact part of an evolutionary history of one of the most important groups of life on the planet—the plants. Over millions of years, plants—like other organisms—have evolved unique features that gave them an advantage over previous groups.

Coal is a reminder that plants dominated the Earth's surface long before the appearance of humans, or even most animals we would recognize today. When coal deposits were forming over 300 million years ago, the reptiles were just beginning their age of dominance, although it would still be tens of millions of years before dinosaurs walked the Earth.

Coal is just one of the many benefits plants provide for us. Plants, and particularly angiosperms, are responsible for important products such as food, clothing, and medicines. Land plants also provide shelter for many animals, release oxygen to the air, and help buffer noise in cities. Without plants, the majority of life on the planet would quickly become extinct.

SUMMARIZE

30.1 Evolutionary History of Plants

- Plants are multicellular, photosynthetic organisms adapted to a land existence that probably evolved from a multicellular, freshwater green alga about 500 MYA.
- Five significant events associated with plant adaptation to a land existence are (1) embryo protection, (2) **vascular tissue,** (3) **megaphylls,** (4) **seeds,** and (5) the flower.
- The life cycle of plants is characterized by **alternation of generations**—some have a dominant **gametophyte** (haploid) generation and others have a dominant **sporophyte** (diploid) generation.
- Land plants are divided into nonvascular plants, seedless vascular plants, and seed plants, based on the presence/absence of vascular tissue and their dispersal mechanism.

30.2 Nonvascular Plants

- The **bryophytes** are the group of **nonvascular plants** that include the **liverworts** and **mosses.** They lack **vascular tissue**—and therefore, they do not have true roots, stems, or leaves.
- The gametophyte is dominant. **Antheridia** produce swimming sperm that use external water to reach eggs in the **archegonia.** Following fertilization, the dependent moss sporophyte consists of a foot, a stalk, and a capsule within which windblown spores are produced by meiosis.
- Use of windblown spores to disperse the gametophyte is nonvascular plants' chief adaptation for reproducing on land. Each spore germinates to produce a gametophyte.

30.3 Seedless Vascular Plants

- **Vascular plants** have vascular tissue—**xylem** and **phloem.** Xylem transports water and minerals and also supports plants against the pull of gravity. Phloem transports organic nutrients from one part of the plant to another.

- Lycophytes are often called **club mosses.** They contain true roots, stems, and leaves because they possess vascular tissue. Lycophytes have narrow leaves called microphylls.

- The pteridophytes include **ferns,** whisk ferns, and **horsetails.** They have larger leaves, called megaphylls, which have branched vascular tissue.

- The sporophyte is dominant. In the life cycle of seedless vascular plants, windborne spores disperse the gametophyte, and the separate gametophyte produces flagellated sperm within antheridia, and eggs within archegonia.

- Vegetative (asexual) reproduction is sometimes used to disperse ferns in dry habitats.

30.4 Seed Plants

- Seed plants have a life cycle that is fully adapted to existence on land. The spores develop inside the body of the sporophyte and are of two types. The male gametophyte is the **pollen grain,** which produces nonflagellated sperm. **Pollination** occurs when the pollen grain is transferred to the egg by the wind or carried by an animal (**pollinator**). The sperm use a **pollen tube** in order to reach the egg. The egg is located within an ovule and produced by the female gametophyte.

- **Gymnosperms,** which include the **conifers, cycads, ginkgos,** and **gnetophytes,** produce seeds that are uncovered (naked). Male gametophytes develop in **pollen cones.** The female gametophyte develops within an ovule located on the scales of **seed cones.** Following pollination and fertilization, the ovule becomes a winged seed dispersed by wind. Seeds contain the next sporophyte generation.

- **Angiosperms,** also called the flowering plants, produce seeds covered by fruits. The **petals** of flowers attract pollinators, and the ovary develops into a fruit, which aids seed dispersal. Monocotyledones (**monocots**) and Eudicotyledones (**eudicots**) are the two main classes of flowering plants. A flower consists of a **corolla** (petals), **calyx (sepals), stamens (anther** and **filament),** and **carpel (stigma, style, ovary,** and **ovule).** Angiosperms provide most of the food that sustains terrestrial animals, and they are the source of many products used by humans.

ASSESS

Testing Yourself

Choose the best answer for each question.

30.1 Evolutionary History of Plants

1. The spores produced by a plant are
   a. haploid and genetically different from each other.
   b. haploid and genetically identical to each other.
   c. diploid and genetically different from each other.
   d. diploid and genetically identical to each other.

2. The gametophyte is the dominant generation in
   a. ferns.
   b. gymnosperms.
   c. mosses.
   d. angiosperms.

3. Label the stages in the following diagram of the plant life cycle.

30.2 Nonvascular Plants

4. In mosses, meiosis occurs in
   a. antheridia.
   b. archegonia.
   c. capsules.
   d. protonema.

5. In bryophytes, sperm usually move from the antheridium to the archegonium by
   a. swimming.
   b. flying.
   c. insect pollination.
   d. wind pollination.
   e. bird pollination.

30.3 Seedless Vascular Plants

6. Microphylls
   a. have a single strand of vascular tissue.
   b. evolved before megaphylls.
   c. evolved as extensions of the stem.
   d. are found in lycophytes.
   e. All of these are correct.

7. How are ferns different from mosses?
   a. Only ferns produce spores as dispersal agents.
   b. Ferns have vascular tissue.
   c. In the fern life cycle, the gametophyte and sporophyte are both independent.
   d. Ferns do not have flagellated sperm.
   e. Both b and c are correct.

7. Ferns have
   a. a dominant gametophyte generation.
   b. vascular tissue.
   c. seeds.
   d. Both a and b are correct.
   e. All of these are correct.

30.4 Seed Plants

8. Which statement about the conifer life cycle is false?
   a. Meiosis produces spores.
   b. The gametophyte is the dominant generation.
   c. The seed is a dispersal agent.
   d. Male gametophytes are carried by the wind.

 9. Which of these pairs is mismatched?
   a. anther—produces microspores
   b. carpel—produces pollen
   c. ovule—becomes seed
   d. ovary—becomes fruit
   e. flower—is a reproductive structure
10. Unlike eudicots, monocots have
   a. woody tissue.
   b. two seed leaves.
   c. scattered vascular bundles in their stems.
   d. flower parts in multiples of four or five.
11. Label the parts of the flower in the following diagram.

a. _____

b. _____

c. _____

d. _____   e. _____

ENGAGE

Thinking Critically

 1. What types of environmental stresses are placed on plants living on dry land, and how do plants overcome these stresses?

 2. What characteristics of angiosperms have allowed this group of plants to dominate Earth?

 3. Compare and contrast the plant alternation-of-generations life cycle with the human life cycle.

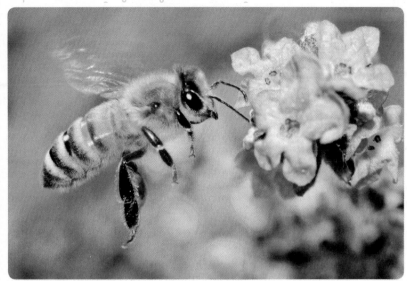
©ElementalImaging/Getty Images

Animals: The Invertebrates

BEFORE YOU BEGIN

Before beginning this chapter, take a few moments to review the following discussions:

Section 1.2 What is the basis for classifying an organism as a member of kingdom Animalia?

Section 3.3 Which cellular features are unique to animals?

Table 27.1 When did animals first appear in the tree of life?

The Vanishing Bees

What is happening to the bees? Honeybees first evolved around 130 million years ago, during the Cretaceous period. Flowering plants present at that time developed nectar that contained sugars to attract bees, encouraging them to feed on the flowers, and thus spread pollen. Humans have been keeping bees for thousands of years, to produce honey and wax. However, over the past decade, bee populations around the world have been decreasing, with as many as one-third of bees dying each winter. The causes of this "colony collapse disorder (CCD)" appear to be coming from many directions. Scientists have un-covered evidence that multiple stressors including disease, pesticides, loss of biodiversity, and climate change add up to diminish the health of a colony. While a colony or individual might overcome one stressor, such as a disease, additional burdens increase the chance of collapse.

Although beekeepers are generally able to replace their bee colo-nies, it has been noted that bee pollination is involved in about one in three bites of food we eat and, because of lost colonies, the cost of bee pollination services has increased by 20%. If bees were to disappear from the planet, it is not likely that humans would run out of food, but many of our favorite fruits and vegetables would become more expen-sive or rare. This story of dying bees illustrates how important inverte-brates can be to the health of humans, as well as the planet.

In this chapter, you will read about the evolution and diversity of the major groups of invertebrates, as well as the traits that have enabled them to successfully adapt to a wide range of habitats.

As you read through this chapter, think about the following questions:

1. Where do invertebrates fit into the evolutionary history of eukaryotes? Of animals?

2. What evolutionary advantage might there be to invertebrates having multiple developmental stages, each with a different body form, habitat, and lifestyle?

31.1 Evolutionary Trends Among Animals

The traditional five-kingdom classification placed animals in the kingdom Animalia. However, the modern three-domain system places animals in the domain Eukarya. Within the Eukarya, animals belong in the supergroup Opisthokonta, along with fungi and certain protistans, notably the choanoflagellates (see Section 29.1).

While animals are extremely diverse, they share some important differences from the other multicellular eukaryotes. Unlike plants, which make their food through photosynthesis, animals are heterotrophs, and must acquire nutrients from an external source. Unlike fungi, which digest their food externally and absorb the breakdown products, animals ingest (eat) whole food and digest it internally. In general, animals share the following characteristics (Fig. 31.1):

- Typically have the power of movement or locomotion by means of muscle fibers
- Multicellular; most have specialized cells that form tissues and organs
- Have a life cycle in which the adult is typically diploid
- Usually undergo sexual reproduction and produce an embryo that goes through developmental stages
- Heterotrophic; usually acquire food by ingestion, followed by digestion

The animal kingdom is currently divided into between 35 and 40 groups, or phyla. The more common of these are presented in Table 31.1. It is believed animals evolved from a protistan ancestor approximately 600 million years ago.

In this chapter, we will explore the phyla of animals called invertebrates. **Invertebrates** lack an internal skeleton, or endoskeleton, of bone or cartilage. The invertebrates evolved first, and far outnumber the **vertebrates,** which are the animals with an endoskeleton. Because the fossil record is not always complete (see Section 27.2), most evolutionary trees (Fig. 31.2) are based on molecular (DNA and RNA) comparisons. For molecular comparisons, it is assumed that the more closely related two organisms are, the more nucleotide sequences in their DNA they will have in common (see Section 26.4).

Anatomical Data

A number of anatomical criteria are used to classify animals (Fig. 31.2). Next, we will focus on the general characteristics of symmetry and embryonic development.

a.

b.

c.

d.

Figure 31.1 General features of animals. **a.** Animals are multicellular, with specialized cells that form tissues and organs. **b.** Animals are heterotrophs, meaning they obtain nutrition from external sources. **c.** Animals are typically motile, due to their well-developed nervous and muscular systems. **d.** Most animals reproduce sexually, beginning life as a 2n zygote, which undergoes development to produce a multicellular organism that has specialized tissues. (a): ©Salvanegra/iStock/Getty Images; (b): ©McGraw-Hill Education/Mark Dierker, photographer; (c): ©JonathanC Photography/Shutterstock; (d): ©Carolina Biological Supply Company/Medical Images

TABLE 31.1 Animal Characteristics

DOMAIN: Eukarya
KINGDOM: Animalia

CHARACTERISTICS
Multicellular, usually with specialized tissues;
ingest or absorb food; adults are diploid

Invertebrates

Sponges (bony, glass, spongin): Asymmetrical, saclike body perforated by pores; internal cavity lined by choanocytes; spicules serve as internal skeleton. *Number of species: 5,150*

Radiata

Comb jellies: Possess eight longitudinal rows of cilia (combs) that aid in locomotion; lack nematocysts. *Number of species: 150*

Cnidarians (hydra, jellyfish, corals, sea anemones): Radially symmetrical with two tissue layers; incomplete digestive tract; tentacles with nematocysts. *Number of species: 10,000*

Protostomia (Lophotrochozoa)

Lophophorates (bryozoans): Filter feeders with a mouth surrounded by a ciliated tentacle-like structure. *Number of species: 5,000*

Flatworms (planarians, tapeworms, flukes): Bilateral symmetry with cephalization; three tissue layers and organ systems; acoelomate with incomplete digestive tract that can be lost in parasites; hermaphroditic. *Number of species: 20,000*

Rotifers (wheel animals): Microscopic animals with a corona (crown of cilia) that looks like a spinning wheel when in motion. *Number of species: 2,200*

Molluscs (chitons, clams, snails, squids): Coelom, all have a foot, mantle, and visceral mass; foot is variously modified; in many, the mantle secretes a calcium carbonate shell as an exoskeleton; true coelom and all organ systems. *Number of species: 110,000*

Annelids (polychaetes, earthworms, leeches): Segmented with body rings and setae; cephalization in some polychaetes; hydroskeleton; closed circulatory system. *Number of species: 16,000*

Protostomia (Ecdysozoa)

Roundworms (roundworms, pinworms, hookworms, filarial worms): Pseudocoelom and hydroskeleton; complete digestive tract; free-living forms in soil and water; parasites common. *Number of species: 25,000*

Arthropods (crustaceans, spiders, scorpions, centipedes, millipedes, insects): Chitinous exoskeleton with jointed appendages undergoes molting; insects—most have wings—are most numerous of all animals. *Number of species: 1,000,000*

Deuterostomes

Echinoderms (sea stars, sea urchins, sand dollars, sea cucumbers): Radial symmetry as adults; unique water-vascular system and tube feet; endoskeleton of calcium plates. *Number of species: 7,000*

Chordates (tunicates, lancelets, vertebrates): All have notochord, dorsal tubular nerve cord, pharyngeal pouches, and postanal tail at some time; contains mostly vertebrates in which notochord is replaced by vertebral column. *Number of species: 56,000*

Vertebrates

Fishes (jawless, cartilaginous, bony): Endoskeleton, jaws, and paired appendages in most; internal gills; single-loop circulation; usually scales. *Number of species: 28,000*

Amphibians (frogs, toads, salamanders): Most have jointed limbs; lungs; three-chambered heart with double-loop circulation; moist, thin skin. *Number of species: 6,900*

Reptiles (snakes, turtles, crocodiles): Amniotic egg; rib cage in addition to lungs; three- or four-chambered heart typical; scaly, dry skin; copulatory organ in males and internal fertilization. *Number of species: 8,000*

Birds (songbirds, waterfowl, parrots, ostriches): Endothermy, feathers, and skeletal modifications for flying; lungs with air sacs; four-chambered heart. *Number of species: 10,000*

Mammals (monotremes, marsupials, placental): Hair and mammary glands. *Number of species: 4,800*

The number of species for each group is based on current estimates.

Type of Symmetry

Animals can be asymmetrical, radially symmetrical, or bilaterally symmetrical. **Asymmetrical** animals have no particular symmetry, such as some species of sponges. **Radial symmetry** means that the animal is organized circularly, similar to a wheel, so that no matter how the animal is sliced longitudinally, mirror images are obtained (Fig. 31.3*a*). **Bilateral symmetry** means that the animal has definite right and left halves; only a longitudinal cut down the center of the animal will produce a mirror image (Fig. 31.3*b*).

Radially symmetrical animals have the advantage of being able to reach out in all directions from one center. They may also be **sessile,** which means they are permanently attached to a substrate, or are free floating, such as jellyfish. Bilaterally symmetrical animals tend to be more active, with movement directed toward

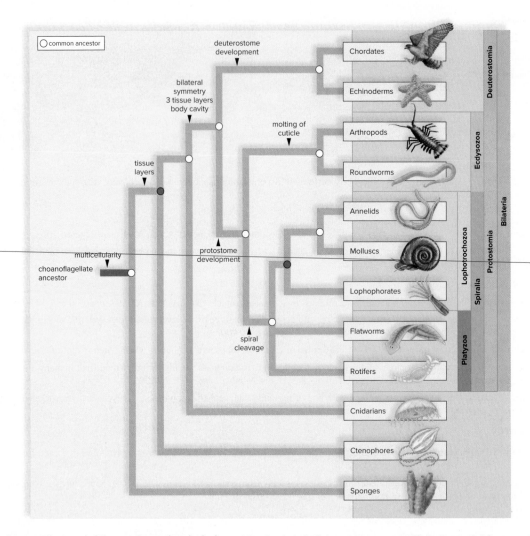

Figure 31.2 Phylogenetic tree of the major animal phyla. All animal phyla living today are most likely descended from a colonial flagellated protist living about 600 MYA. This phylogenetic (evolutionary) tree is a hypothesis and uses morphological and molecular data (rRNA sequencing) to determine which phyla are most closely related to one another. While many relationships are still not clear, placement of Ctenophora and relationships among Spiralia are most uncertain (marked with red dots).

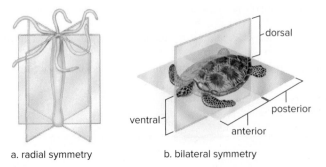

a. radial symmetry   b. bilateral symmetry

Figure 31.3 Symmetry in animals. The body planes of animals can be (**a**) radially symmetrical or (**b**) bilaterally symmetrical.

an anterior end (head). During the evolution of animals, bilateral symmetry is accompanied by **cephalization,** localization of a brain and specialized sensory organs at the anterior end.

Embryonic Development

Sponges are multicellular animals, but unlike other animals, they do not have true tissues. Therefore, sponges have a *cellular level of organization.* Because of this, sponges are often classified as a separate group, called parazoans (see Fig. 31.2), from other animals.

True tissues appear in the other animals (the eumetazoans) as they undergo embryological development. The first three tissue layers are often called germ layers because they give rise to the organs and organ systems of complex animals. Animals such as the cnidarians are diploblastic, meaning they have only two tissue layers (ectoderm and endoderm) as embryos. These animals display a *tissue level of organization* (see Fig. 1.2). Those animals that develop further and have all three tissue layers (ectoderm, mesoderm, and endoderm) as embryos are triploblastic and have an *organ level of organization.* Notice in the phylogenetic tree that the animals with three tissue layers are either **protostomes** (*proto* is Greek for "first"; *stoma* is Greek for "mouth") or **deuterostomes** (*deuter* is Greek for "second").

Differences Between Protostomes and Deuterostomes   The differences between protostome and deuterostome development may be identified by three major events (Fig. 31.4):

1. Cleavage, the first event of development, is cell division without cell growth. In protostomes, spiral cleavage occurs, and daughter cells sit in grooves formed by the previous cleavages. The fate of these cells is fixed and determinate in protostomes; each can contribute to development in only one particular way. In deuterostomes, radial cleavage occurs, and the daughter cells sit right on top of the previous cells. The fate of these cells is indeterminate—that is, if they are separated from one another, each cell can go on to become a complete organism.

2. As development proceeds, a hollow sphere of cells, or blastula, forms and the indentation that follows develops into an opening called the blastopore. In protostomes, the mouth appears at or near the blastopore, hence the origin of their name. In deuterostomes, the anus appears at or near the blastopore, and only later does a second opening form the mouth, hence the origin of their name.

3. Certain members of the protostomes and all deuterostomes have a body cavity completely lined by mesoderm, called a true coelom. However, a true coelom develops differently in the two groups. In protostomes, the mesoderm arises from cells located near the embryonic blastopore, and a splitting occurs that produces the coelom. In deuterostomes, the coelom arises as a pair of mesodermal pouches from the wall of the primitive gut. The pouches enlarge until they meet and fuse.

The protostomes are divided into two groups: the ecdysozoa and the spiralia. The spiralia (see Section 31.3) contain the lophophorans, which have a mouth surrounded by ciliated tentacle-like structures, and the platyzoans, which have bilateral symmetry but no circulatory or respiratory systems. The ecdysozoans include the roundworms and arthropods (see Section 31.4). Both of these types of animals molt, and so shed their outer covering as they grow. Ecdysozoa means molting animals. The deuterostomes include the echinoderms (see Section 31.5) and the chordates (see Section 32.1).

Check Your Progress   31.1

1. Distinguish an animal from a plant or fungi.
2. List the general characteristics that animals have in common.
3. Using Figure 31.2, identify the general characteristics of a mollusc.

31.2 The Simplest Invertebrates

Learning Outcomes

Upon completion of this section, you should be able to

1. Describe the anatomy and life cycle of sponges and cnidarians.
2. Contrast the way of life of a sponge with that of a cnidarian.
3. Identify the basic morphology of a hydra.

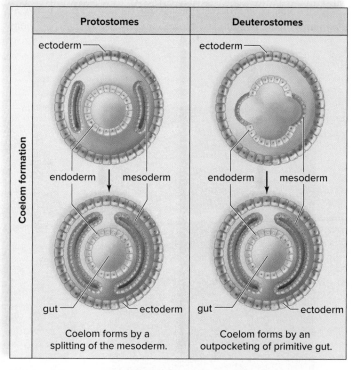

Figure 31.4 Protostomes compared to deuterostomes.
Left: In the embryo of protostomes, cleavage is spiral—new cells are at an angle to old cells—and each cell has limited potential and cannot develop into a complete embryo; the blastopore is associated with the mouth; and the coelom, if present, develops by a splitting of the mesoderm. *Right:* In deuterostomes, cleavage is radial—new cells sit on top of old cells—and each one can develop into a complete embryo; the blastopore is associated with the anus; and the coelom, if present, develops by an outpocketing of the primitive gut.

Porifera

While all animals are multicellular, the members of the phylum **Porifera** are the only animals to lack true tissues and to have only a cellular level of organization. They have only a few cell types, and they lack the nerve and muscle cells seen in more complex animals. Molecular data place them at the base of the evolutionary tree of animals (see Fig. 31.2).

Sponges (Fig. 31.5) are poriferans that are aquatic, largely marine animals that vary greatly in size, shape, and color. They are multicellular, but have few cell types and no nerve or muscle cells, nor any organized tissues. However, despite their apparent simplicity, molecular data show they are at the base of the evolutionary tree of animals.

In sponges, the outer layer of the body wall contains flattened epidermal cells, some of which have contractile fibers; the middle layer is a semifluid matrix with wandering amoeboid cells; and the inner layer is composed of flagellated cells called **collar cells,** or choanocytes (Fig. 31.5). The beating of the flagella produces water currents that flow through the pores into the central cavity and out through the osculum, the upper opening of the body. Even a simple sponge only 10 cm tall is estimated to filter as much as 100 liters of water each day. It takes this much water to supply the needs of the sponge. A sponge is a sessile filter feeder, an organism that filters its food from the water by means of a straining device—in this case, the pores of the walls and the microvilli making up the collar of collar cells. Microscopic food particles that pass between the microvilli are engulfed by the collar cells and digested by them in food vacuoles or are passed to the amoeboid cells for digestion. The amoeboid cells also act as a circulatory device to transport nutrients from cell to cell, and they produce the sex cells (the egg and the sperm) and spicules.

Sponges can reproduce both asexually and sexually. They reproduce asexually by fragmentation followed by regeneration, gemmule formation, or budding. A gemmule is like a spore and is resistant to drying out, freezing, or lack of oxygen. Gemmule formation occurs in freshwater sponges, and when conditions are favorable, a gemmule gives rise to an adult sponge. During sexual reproduction, sperm are released through the osculum and drawn into other sponges through the pores, where they fertilize eggs within the body. Most sponges are **hermaphroditic,** meaning they possess both male and female sex organs. However, they usually do not self-fertilize. After fertilization, the zygote develops into a flagellated larva that may swim to a new location. Sponges are capable of regeneration—if the cells of a sponge are separated, they are capable of reassembling and regenerating into a complete and functioning organism!

Sponges are classified on the basis of their skeleton. Some sponges have an internal skeleton composed of **spicules** (Fig. 31.5), small needle-shaped structures with one to six rays. Chalk sponges have spicules made of calcium carbonate, while glass sponges have spicules that contain silica. Most sponges have a skeleton composed of fibers of spongin, a soft protein. A bath sponge is the dried spongin skeleton. Today, however, commercial "sponges" are usually synthetic. Note that the popular "loofah" is not really a sponge; it is actually a spongy plant related to cucumbers.

Figure 31.5 Simple sponge anatomy.   a. A simple sponge. **b.** The wall of a sponge contains two layers of cells, the outer epidermal cells and the inner collar cells as shown in (**c**). The collar cells (**d,** enlarged) have flagella that beat, moving the water through pores as indicated by the blue arrows in (**b**). Food particles in the water are trapped by the collar cells and digested within their food vacuoles. Amoebocytes transport nutrients from cell to cell. Spicules form an internal skeleton in some sponges. (a): ©Andrew J. Martinez/Science Source

Comb Jellies and Cnidarians

These two groups of animals have true tissues and, as embryos, they have ectoderm and endoderm as their germ layers. They are radially symmetrical as adults, which offers the advantage of being able to reach out in all directions for food.

Comb Jellies

Comb jellies (phylum Ctenophora) are solitary, free-swimming marine invertebrates found primarily in warm waters. Ctenophores propel themselves by beating their cilia (Fig. 31.6) and can range in size from a few centimeters to 1.5 meters (m) in length. Their body is made up of a transparent jellylike substance called **mesoglea.** Most ctenophores capture their prey by using sticky adhesive cells called colloblasts. Some ctenophores are bioluminescent, enabling them to produce their own light.

Cnidarians

Cnidarians (phylum Cnidaria) are tubular or bell-shaped animals that reside mainly in shallow coastal waters. However, there are some freshwater, brackish, and oceanic forms. The term *cnidaria* is derived from the presence of specialized stinging cells called **cnidocytes** (see Fig. 31.9). Each cnidocyte has a toxin-filled capsule called a **nematocyst** that contains a long, spirally coiled hollow thread. When the trigger of the cnidocyte is touched, the nematocyst is discharged. Some threads merely trap prey, while others have spines that penetrate and inject paralyzing venom.

The body of a cnidarian is a two-layered sac. The outer tissue layer is a protective epidermis derived from ectoderm. The inner tissue layer, which is derived from endoderm, secretes digestive juices into the internal cavity, called the **gastrovascular cavity,** which is involved in the digestion of food and circulation of nutrients. The fluid-filled gastrovascular cavity also serves as a supportive hydrostatic skeleton. This type of skeleton offers some resistance to the contraction of muscle but permits flexibility. The two tissue layers are separated by mesoglea.

Two basic body forms are seen among cnidarians (Fig. 31.7). The mouth of a **polyp** is directed upward, while the mouth of a jellyfish, or medusa, is directed downward. The bell-shaped medusa has more mesoglea than a polyp, and the tentacles are concentrated on the margin of the bell. At one time, both body forms may have been a part of the life cycle of all cnidarians. When both are present, the animal is dimorphic: the sessile polyp stage produces a medusa by asexual budding, and the motile **medusa** stage produces egg and sperm. In some cnidarians, one stage is dominant and the other is reduced; in other species, one form is absent altogether.

Cnidarian Diversity

Cnidarians are a very diverse group of aquatic animals. Sea anemones (Fig. 31.8a) are sessile polyps that live attached to a submerged substrate. Most sea anemones range in size from 0.5 to 20 cm in length and 0.5 to 10 cm in diameter and are often colorful. Their upward-turned oral disk that contains the mouth is surrounded by a large number of hollow tentacles harboring nematocysts.

Corals (Fig. 31.8b) resemble sea anemones encased in a calcium carbonate (limestone) house. The coral polyp can extend into the water to feed on microorganisms and retreat into the house for safety. Some corals can be solitary, but the vast majority live in colonies that vary in shape from rounded to branching. Many corals exhibit elaborate geometric designs and stunning colors, and are responsible for the building of coral reefs. The slow accumulation of

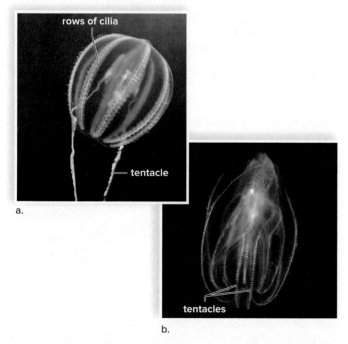

Figure 31.6 Comb jellies. **a**. *Pleurobrachia pileus;* **b**. *Mnemiopsis leidyi.* (a): ©Jeff Rotman/Alamy; (b): ©Andrey Nekrasov/Alamy

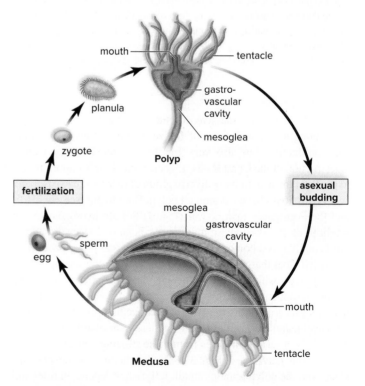

Figure 31.7 Cnidarian life cycle. Some cnidarians have both a polyp stage and a medusa stage. In others, one stage may be dominant or absent altogether.

a. Sea anemone, *Actinia*

b. Cup coral, *Dendrophyllia*

c. Portuguese man-of-war, *Physalia*

d. Jellyfish, *Aurelia*

Figure 31.8 **Cnidarian diversity.** **a.** The anemone, which is sometimes called the flower of the sea, is a solitary polyp. **b.** Corals are colonial polyps residing in a calcium carbonate or proteinaceous skeleton. **c.** The Portuguese man-of-war is a colony of modified polyps and medusae. **d.** True jellyfishes undergo the complete life cycle. This is the medusa stage. The polyp is small. (a): ©AlbyDeTweede/iStock/Getty Images; (b): ©Carson Ganci/ Design Pics; (c): ©Islands in the Sea 2002, NOAA/OER; (d): ©Vladimir Wrangel/ Shutterstock

limestone can result in massive structures, such as the Belize Barrier Reef along the eastern coast of Belize. Coral reef ecosystems are very productive, but also very fragile (see the Ecology feature "Destruction of the Coral Reefs"). Their protection is important due to the fact that an extremely diverse group of marine life exists there.

Hydrozoans have a dominant polyp. Both hydra (see Fig. 31.9) and the Portuguese man-of-war (Fig. 31.8*c*) are hydrozoans. You might think the Portuguese man-of-war is an odd-shaped medusa, but actually it is a colony of polyps. The original polyp becomes a gas-filled float that provides buoyancy, keeping the colony afloat. Other polyps, which bud from this one, are specialized for feeding or for reproduction. A long, single tentacle armed with numerous nematocysts arises from the base of each feeding polyp. Swimmers who accidentally come upon a Portuguese man-of-war can receive painful, even serious, injuries from these stinging tentacles.

In true jellyfishes (Fig. 31.8*d*), the medusa is the primary stage, and the polyp remains small. Jellyfishes depend on tides and currents for their primary means of movement. They feed on a variety of invertebrates and fishes and are themselves food for marine animals.

Hydra

The body of a hydra (Fig. 31.9 *top*) is a small tubular polyp about one-quarter inch in length. Hydras are often studied in biology

Figure 31.9 **Anatomy of *Hydra*.** *Top:* The body of *hydra* is a small, tubular polyp whose wall contains two tissue layers. *Hydras* reproduce asexually by forming outgrowths called buds (see photo) that develop into a complete animal. *Middle:* Various types of cells in the body wall. *Bottom:* Cnidocytes are cells that contain nematocysts. (photo) ©NHPA/M. I. Walker/Avalon

SCIENCE IN YOUR LIFE ▶ ECOLOGY

Destruction of the Coral Reefs

Coral reefs are among the most biologically diverse and productive communities on Earth. They tend to be found in warm, clear, and shallow tropical waters worldwide and are typically formed by reef-building corals, which are cnidarians. Aside from being beautiful and giving shelter to many colorful species of fishes, coral reefs help generate economic income from tourism, protect ocean shores from erosion, and may serve as the source of medicines derived from antimicrobial compounds that reef-dwelling organisms produce.

However, coral reefs around the globe are being destroyed for a variety of reasons, most of them linked to human development. Deforestation, for example, causes tons of soil to settle on the top of coral reefs. This sediment prevents photosynthesis of symbiotic algae that provide food for the corals. When the algae die, so do the corals, which then turn white (Fig. 31A). This is called coral "bleaching," and it has been seen in the Pacific Ocean and the Caribbean Sea. Evidence suggests that climate change is one of the factors leading to coral bleaching because corals can tolerate only a narrow range of temperatures. As global temperatures rise, the corals may expel the symbiotic algae they depend upon for nutrients. This causes the white coloring, or bleaching, of the corals. While the coral is still alive at this point, it is more susceptible to pathogens that can kill corals, similar to pathogens that cause cholera in humans. Increases in aquatic nutrients from fertilizers that wash into the ocean also make corals more susceptible to diseases, which can also kill them.

Scientists estimate that 90% of coral reefs in the Philippines are dead or deteriorating due to human activities such as pollution and, especially, overfishing. Fishing methods that employ dynamite or cyanide to kill or stun the fish for food or the pet trade can easily kill corals. Paleobiologist Jeremy Jackson of the Smithsonian Tropical Research Institute in Panama

Figure 31A   Coral bleaching.   ©Secret Sea Visions/Getty Images

estimates that we may lose 60% of the world's coral reefs by the year 2050.

Questions to Consider

1. What features of coral reefs help explain why they are so biologically diverse?
2. Considering what is causing the loss of coral reefs, is it possible to save them? How?

---

classes and labs as an example of a cnidarian. They are likely to be found attached to underwater plants or rocks in most lakes and ponds. The only opening (the mouth) is in a raised area surrounded by four to six tentacles that contain a large number of nematocysts.

The outer tissue layer is a protective epidermis derived from ectoderm. The inner tissue layer, derived from endoderm, is called a gastrodermis. The two tissue layers are separated by mesoglea. There are both circular and longitudinal muscle fibers. Nerve cells located below the epidermis, near the mesoglea, interconnect and form a nerve net that communicates with sensory cells throughout the body. The nerve net allows transmission of impulses in several directions at once. Because they have both muscle fibers and nerve fibers, cnidarians are capable of directional movement.

The body of a hydra can contract or extend, and the tentacles that ring the mouth can reach out and grasp prey and discharge nematocysts (Fig. 31.9). Digestion begins within the central cavity but is completed within the food vacuoles of gastrodermal cells. Nutrient molecules are passed to the other cells of the body by diffusion. The large gastrovascular cavity allows digestion and gastrodermal cells to exchange gases directly with a watery medium.

Although hydras exist only as polyps (there are no medusae), they can reproduce either sexually or asexually. When sexual reproduction is going to occur, an ovary or a testis develops in the body wall. Like the sponges, cnidarians have great regenerative powers, and hydras can grow an entire organism from a small piece. When conditions are favorable, hydras reproduce asexually by making small outgrowths, or buds, that pinch off and begin to live independently.

Check Your Progress 31.2

1. Describe three main characteristics of sponges.
2. Explain how cnidarians are more anatomically complex than sponges.
3. Explain the basic anatomical characteristics of a hydra.

31.3 The Spiralia

Learning Outcomes

Upon completion of this section, you should be able to

1. Describe the basic features of lophotrochozoa.
2. Identify the unique features of molluscs.
3. Contrast the differences between flatworms and annelids.

Spiralia are a diverse group of protostomes. The precise phylogeny of spiralians is under debate, but common classification includes two large groups: lophotrochozoans and platyzoans. These animals are bilaterally symmetrical during at least one stage of their development and undergo spiral cleavage, compared to radial cleavage in other protostomes. As embryos, they have three germ layers, and as adults they achieve the organ level of organization.

Figure 31.10 Lophotrochozoan characteristics. a. The lophophore feeding apparatus. **b.** A trochophore larva. (a): ©The Natural History Museum/Alamy

Lophotrochozoa

All lophotrochozoa are bilaterally symmetrical, at least in some stage of their development. As embryos, they have three germ layers, and as adults, they have an organ level of organization. Lophotrochozoans are protostomes and include the **lophophorans** (bryozoans, phoronids, and brachiopods) and the **trochozoans** (molluscs and annelids). Lophophorans are aquatic and have a feeding apparatus called the lophophore, which is a mouth surrounded by a ciliated tentacle-like structure (Fig. 31.10). Today, trochophores either have a trochophore larva (molluscs and annelids), or an ancestor had one in the past (flatworms and rotifers).

The evolutionary story of the lophophorans is still being researched, with molecular evidence suggesting that the ancestry of this group is more complex than originally thought. For our purposes, we will focus on the traditional classification, which includes the bryozoans, brachiopods, and the phoronids.

Bryozoans (phylum Bryozoa) are aquatic, colonial lophophorans. Colonies are made up of individuals called zooids. Zooids are not independent animals, but instead are single members of a colony that cooperate as a single organism. Some zooids specialize in feeding and thus filter particles from the water with the lophophore; some specialize in reproduction; and some can perform both functions. Zooids coordinate functions within a colony by communicating through chemical signals. Zooids have protective exoskeletons, which they use to attach to substrates, including the bottoms of ships, where they can be a nuisance, increasing drag and impeding maneuverability.

Brachiopods (phylum Brachiopoda) are a small group of lophophorans that have two hinged shells, much as molluscs do—but instead of having a left and right shell, they have a top and a bottom shell. Brachiopods affix themselves to hard surfaces with a muscular pedicel. Like other lophophorans, brachiopods use their lophophore to feed by filtering particles from the water.

Phoronids (phylum Phoronida) live inside a long tube formed from their own chitinous secretions. The tube is buried in the

ground and their lophophore extends from it, but it can retract very quickly when needed. Only about 15 species of phoronids exist worldwide.

Trochozoans

Molluscs and annelids are classified as trochozoans, the second major group of lophotrochozoans.

Molluscs

The **molluscs** (phylum Mollusca) are the second most numerous group of animals, numbering over 110,000 species. They inhabit a variety of environments, including marine, freshwater, and terrestrial habitats. Although almost everyone enjoys looking at the intricate patterns and beauty of seashells, few people are aware of the tremendous diversity found in the phylum Mollusca (Fig. 31.11). Molluscs include chitons, limpets, slugs, snails, abalones, conchs, nudibranchs, clams, scallops, squid, and octopuses. Molluscs have a true **coelom,** and all coelomates have bilateral symmetry, three germ layers, an organ level of organization, and a complete digestive tract. The advantages of having a coelom include freer body movements, space for development of complex organs, greater surface area for absorption of nutrients, and protection of internal organs from damage.

The Unique Characteristics of Molluscs Despite being a very large and diversified group, all molluscs have a body composed of at least three distinct parts:

1. The **visceral mass** is the soft-bodied portion that contains internal organs, including a highly specialized digestive tract, paired kidneys, and reproductive organs. The nervous system of a mollusc consists of several ganglia connected by nerve cords. The amount of cephalization and sensory organs varies from nonexistent in clams to complex in squid and octopuses.
2. The **foot** is a strong, muscular portion used for locomotion. Molluscan groups can be distinguished by modifications of the foot. Molluscs exhibit varying amounts of mobility. Oysters are sessile; snails are extremely slow moving; and squid are fast-moving, active predators.
3. The **mantle,** a membranous or sometimes muscular covering, envelops but does not completely enclose the visceral mass. The *mantle cavity* is the space between the two folds of the mantle. The mantle may secrete a *shell,* which is an exoskeleton. Some molluscs have a univalve (single shell) while others are bivalves (possessing two shells).

Another feature often present is a rasping, tonguelike *radula,* an organ that bears many rows of teeth and is used to obtain food.

Gastropods The **gastropods** (class Gastropoda) include nudibranchs, conchs, and snails. In gastropods, whose name means "stomach-footed," the foot is ventrally flattened, and the animal moves by muscle contractions that pass along the foot. Many are herbivores that use their radulas to scrape algae from surfaces. Others are carnivores, using their radulas to bore through surfaces, such as bivalve shells, to obtain food.

While nudibranchs, also called sea slugs, lack a shell, conchs and snails have a univalve coiled shell in which the visceral mass

a. Chiton, *Tonicella*

b. Scallop, *Pecten* sp.

c. Nudibranch, *Glossodoris macfarlandi*

d. Common octopus, *Octopus vulgaris*

Figure 31.11 Diversity of molluscs. a. A chiton is believed to be most like the ancestral mollusc that gave rise to all the other types. As with the chiton, recent evidence suggests that ancestral molluscs were segmented, but that this segmentation was later lost throughout much of the phylum. **b.** Like clams, a scallop is a bivalve. **c.** Similar to snails, a nudibranch is a gastropod. **d.** Like squids, an octopus is a cephalopod. (a): ©Jeff Rotman/Photolibrary/Getty Images; (b): ©Larry S. Roberts; (c): ©Kenneth W. Fink/Bruce Coleman/Avalon; (d): ©Juniors Bildarchiv GmbH/Alamy

spirals. Land snails, such as *Helix aspera,* have a head with two pairs of tentacles; one pair bears eyes at the tips (Fig. 31.12). The shell not only offers protection, but also prevents desiccation (drying out). While aquatic gastropods have gills, land snails have a mantle that is richly supplied with blood vessels and functions as a lung. Reproduction is also adapted to a land existence. Land snails are hermaphroditic. When two snails meet, each inserts its penis into the vagina of the other, and following fertilization and the deposit of eggs externally, development proceeds directly without the formation of swimming larvae.

Cephalopods In **cephalopods** (class Cephalopoda, meaning head-footed), which includes octopuses, squid, and nautiluses, the foot has evolved into a funnel or siphon about the head (Fig. 31.13). Aside from the tentacles, which seize prey, cephalopods have a powerful beak and a radula (toothy tongue) to tear prey apart. Cephalization is apparent. The eyes have a lens and a retina with photoreceptors similar to those of vertebrates. However, the eye is constructed so differently from the vertebrate eye that the so-called camera-type eye must have evolved independently in both the molluscs and in the vertebrates. In cephalopods, the brain is formed from a fusion of ganglia, with nerves leaving the brain that supply various parts of the body. An especially large pair of nerves controls the rapid contraction of the mantle, allowing these animals

to move quickly via the jet propulsion of water. Rapid movement and the secretion of a dark ink help cephalopods escape their enemies. Octopuses have no shell, and squid have only a remnant of a shell concealed beneath the skin. Octopuses, like some other species of cephalopods, are thought to be among the most intelligent invertebrates and are even capable of learning!

Bivalves Clams, mussels, oysters, and scallops are called **bivalves** (class Bivalvia) because their shells have two parts. In a clam, the shell, which is secreted by the mantle, is composed of protein and calcium carbonate, with an inner layer of mother-of-pearl. If a foreign body is placed between the mantle and the shell, pearls form as concentric layers of shell are deposited about the particle.

Figure 31.14 shows the internal anatomy of the freshwater clam, *Anodonta*. The adductor muscles hold the valves of the shell together. Within the mantle cavity, the gills, which are the organs for gas exchange in aquatic forms, hang down on either side of the visceral mass, which lies above the foot.

The clam is a filter feeder. Food particles and water enter the mantle cavity by way of the incurrent siphon, a posterior opening between the two valves. Mucous secretions cause smaller particles to adhere to the gills, and ciliary action sweeps them toward the mouth.

Figure 31.12 Gastropod anatomy. Snails have a long muscular foot that allows them to creep along slowly by way of muscular contraction. Note the absence of lungs in this land snail. (photo): ©Rosemary Calvert/Photographer's Choice/Getty Images

Figure 31.13 Cephalopod anatomy. Anatomy of a squid, a cephalopod. Squids are torpedo-shaped and adapted for fast swimming via jet propulsion. (photo): ©Georgette Douwma/Science Source

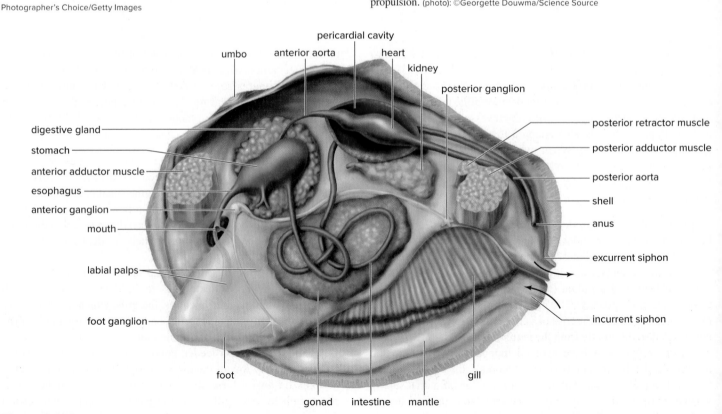

Figure 31.14 Clam, *Anodonta*. In this drawing of a clam, one of the bivalve shells and the mantle have been removed from one side. Follow the path of food from the incurrent siphon to the gills, the mouth, the stomach, the intestine, the anus, and the excurrent siphon. Locate the three ganglia: anterior, foot, and posterior. The heart lies in the reduced coelom.

The heart of a clam lies just below the hump of the shell within the pericardial cavity, the only remains of the reduced coelom. The heart pumps blood into a dorsal aorta that leads to the various organs of the body. Within the organs, however, blood flows through spaces, or sinuses, rather than through vessels. This is called an **open circulatory system** because the blood is not entirely contained within blood vessels. This type of circulatory system functions well in a slow-moving or sessile animal.

The nervous system of a clam is composed of three pairs of ganglia (anterior, foot, and posterior), which are all connected by nerves. Clams lack cephalization. The "hatchet" foot projects anteriorly from the shell, and by expanding the tip of the foot with blood and pulling the body after it, the clam moves forward.

The digestive system of the clam includes a mouth with labial palps, an esophagus, a stomach, and an intestine, which coils about in the visceral mass and then is surrounded by the heart as it extends to the anus. Two excretory kidneys remove waste from the pericardial cavity for excretion into the mantle cavity.

The sexes are usually separate. The gonad (i.e., ovary or testis) is located around the coils of the intestine. All clams have some type of larval stage, and marine clams have a trochophore larva.

Annelids   Annelids (phylum Annelida, about 12,000 species) are *segmented,* as evidenced by the rings encircling the outside of their bodies. Segmentation is also seen in arthropods and chordates, although annelids are the only trochozoan with segmentation and a well-developed coelom. Internally, the segments of annelids are partitioned by septa. Worms, in general, do not have an internal or external skeleton, but most often have a **hydrostatic skeleton,** a fluid-filled interior that supports muscle contraction and enhances flexibility. Along with the partitioning of the fluid-filled coelom, this hydroskeleton permits each body segment to move independently. Locomotion occurs by contraction and expansion of each body segment, propelling the animal forward. Thus, a terrestrial annelid is capable of crawling on the surface in addition to burrowing in the mud. Although the most familiar members of this phylum are leeches and earthworms, the majority of annelids are marine. Annelids vary in size from microscopic to tropical earthworms as long as 4 m.

The body plan in annelids has led to specialization of the digestive tract. The digestive system may include a pharynx, esophagus, crop, gizzard, intestine, and accessory glands. Annelids have an extensive **closed circulatory system** with blood vessels that run the length of the body and branch to every segment. The nervous system consists of a brain connected to a ventral solid **nerve cord,** with ganglia in each segment. The excretory system consists of nephridia in most segments. A **nephridium** (*pl.,* nephridia) is a tubule that collects waste material and excretes it through an opening in the body wall.

Polychaetes   Marine annelids are the Polychaeta, which refers to the presence of many setae. **Setae** are bristles that anchor the worm or help it move. The setae are in bundles on parapodia, which are paddlelike appendages found on most segments. Parapodia are used in swimming, but can also be used as respiratory organs. Clam worms, such as *Nereis* (Fig. 31.15*a*), prey on crustaceans and other small animals, which they capture using a pair of strong, chitinous jaws that extend with the pharynx. Associated with its

sensory projections

parapodia

a. Clam worm, *Nereis*

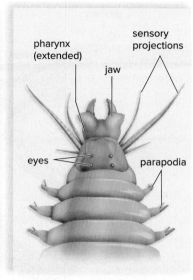

pharynx (extended)

sensory projections

jaw

eyes

parapodia

b. Head region of *Nereis*

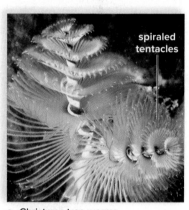

spiraled tentacles

c. Christmas tree worms, *Spirobranchus*

Figure 31.15 Annelid diversity. **a.** Clam worms are predatory polychaetes that have a well-defined head region. **b.** Note also the parapodia, which are used for swimming and as respiratory organs. **c.** Christmas tree worms (a type of tube worm) are sessile filter feeders whose ciliated tentacles spiral.

(a): ©Heather Angel/Natural Visions;
(c): ©Diane R. Nelson

way of life, *Nereis* has a well-defined head region with eyes and other sense organs (Fig. 31.15*b*).

Other polychaetes are sessile tube worms, with tentacles that form a funnel-shaped fan (Fig. 31.15*c*). Water currents, created by the action of cilia, trap food particles that are directed toward the mouth of these filter feeders.

Polychaetes have breeding seasons, and only during these times do they possess functional sex organs. In *Nereis,* many worms concurrently shed a portion of their bodies containing either eggs or sperm, which float to the surface where fertilization then takes place. The zygote rapidly develops into a trochophore larva, just as in marine clams. The existence of this larval form in both the annelids and molluscs is evidence that these two groups of animals are evolutionarily related.

Oligochaetes   The oligochaetes (class Oligochaeta), which include earthworms, have few setae per segment. Earthworms (e.g., *Lumbricus*) (Fig. 31.16) do not have a well-developed head or parapodia. Their setae protrude in pairs directly from the surface of the body. Locomotion, which is accomplished section by section, utilizes muscle contraction and the setae. When longitudinal muscles contract, segments bulge, and their setae protrude into the soil. Then, when circular muscles contract, the setae are withdrawn,

a.

b.

c.

Figure 31.16 Earthworm, *Lumbricus*.   a. Internal anatomy of the anterior part of an earthworm. Each body segment has an internal septa that divides the coelom into compartments. **b.** Cross-section of an earthworm. **c.** When earthworms mate, they are held in place by a mucus secreted by the clitellum. The worms are hermaphroditic, and when mating, sperm pass from the seminal vesicles of each to the seminal receptacles of the other.

(c): ©MikeLane45/iStock360/Getty Images

and these segments move forward in sequence, pushing the whole animal forward.

Earthworms reside in soil where there is adequate moisture to keep the body wall moist for gas exchange. They are scavengers that do not have an obvious head, and feed on leaves or any other organic matter. Food drawn into the mouth by the action of the muscular pharynx is stored in a crop and ground up in a thick, muscular gizzard. Digestion and absorption occur in a long intestine, whose dorsal surface has an expanded region, called a typhlosole, that increases the surface for absorption.

Segmentation   Earthworm segmentation is visible internally by the presence of septa. The long, ventral solid nerve cord leading from the brain has ganglionic swellings and lateral nerves in each segment. The paired nephridia in most segments have two openings: One is a ciliated funnel that collects coelomic fluid, and the other is an exit in the body wall. Between the two openings is a convoluted region where waste material is removed from the blood vessels about the tubule. Red blood moves anteriorly in the dorsal blood vessel, which connects to the ventral blood vessel by five pairs of connectives called "hearts." Pulsations of the dorsal blood vessel and the five pairs of hearts are responsible for blood flow. As the ventral vessel takes the blood toward the posterior regions of the worm's body, it gives off branches in every segment. Altogether, segmentation is evidenced by (1) body rings, (2) coelom divided by septa, (3) setae on most segments, (4) ganglia and lateral nerves in each segment, (5) nephridia in most segments, and (6) branch blood vessels in each segment.

Reproduction   Earthworms are hermaphroditic: the male organs are the testes, the seminal vesicles, and the sperm ducts, and the female organs are the ovaries, the oviducts, and the seminal receptacles. When mating, two worms lie parallel to each other facing in opposite directions. The fused midbody segment, called a clitellum, secretes mucus that protects the sperm from drying out as they pass between the worms. After the worms separate, the clitellum of each produces a slime tube, which is moved along over the anterior end by muscular contractions. As it passes, eggs and the sperm received earlier are deposited, and fertilization occurs. The slime tube then forms a cocoon to protect the worms as they develop. There is no larval stage.

Comparison with Clam Worms   Comparing the anatomy of marine clam worms (class Polychaeta) to that of terrestrial earthworms (class Oligochaeta) highlights the manner in which earthworms are adapted to life on land. A lack of cephalization is seen in the nonpredatory earthworms that extract organic remains from the soil they eat. Their lack of parapodia helps reduce the possibility of water loss and facilitates burrowing in soil. The clam worm makes use of external water, while the earthworm provides a mucous secretion to aid fertilization. It is the aquatic form that has the swimming, or trochophore larva, not the land form.

Leeches   Leeches (class Hirudinea) are usually found in fresh water, but some are marine or even terrestrial. They have the same body plan as other annelids, but they have no setae, and each body ring has several transverse grooves. Most leeches are only 2–6 cm in length, but some, including the medicinal leech, are as long as 20 cm.

Among their modifications are two suckers, a small one around the mouth and a large, posterior one. While some leeches are free-living, feeding on plant material or invertebrates, most are fluid feeders that attach themselves to open wounds. Some bloodsuckers, such as the medicinal leech (Fig. 31.17), can cut through tissue. Leeches are able to keep blood flowing and prevent clotting by means of a substance in their saliva known as hirudin, a powerful anticoagulant.

Platyzoans

The second major group of Spiralia is Platyzoa. It includes organisms such as the flatworms and rotifers. These groups have bilateral symmetry and no circulatory or respiratory system.

Flatworms

Flatworms (phylum Platyhelminthes) are aptly named because they have an extremely flat body. Like the cnidarians, flatworms have an incomplete digestive tract and only one opening, the mouth. When one opening is present, the digestive tract is said to be *incomplete,* and when two openings are present, the digestive tract is said to be *complete.* Also, flatworms have no body cavity, and instead the third germ layer, mesoderm, fills the space between their organs.

Among flatworms, planarians are free-living, whereas flukes and tapeworms are parasitic. Free-living flatworms have muscles and excretory, reproductive, and digestive systems. The worms lack respiratory and circulatory systems, however. Because the body is flat and thin, diffusion alone can pass needed oxygen and other substances from cell to cell.

Free-Living Flatworms Freshwater planarians, shown in Figure 31.18, are small (several millimeters to several centimeters) worms that live in lakes, ponds, streams, and springs. Some tend to be colorless; others have brown or black pigmentation. They feed on small living or dead organisms, such as worms and crustaceans.

Planarians have an excretory system that consists of a network of interconnecting canals extending through much of the

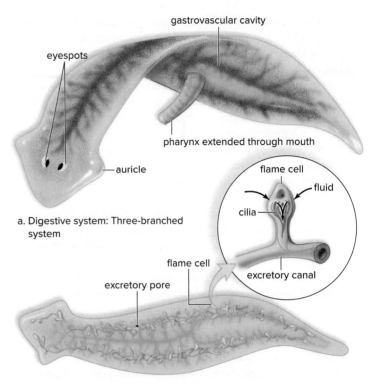

a. Digestive system: Three-branched system

b. Excretory system: Flame-cell system

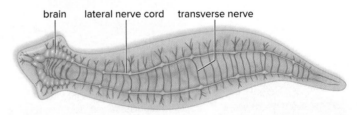

c. Reproductive system: Hermaphroditic system

d. Nervous system: Ladder-style system

e. Photomicrograph of a planarian

Figure 31.18 Planarian anatomy.   a. A planarian extends its pharynx to suck food into a gastrovascular cavity that branches throughout the body. **b.** The excretory system includes flame cells. **c.** The reproductive system (shown in pink and blue) has both male and female organs. **d.** The nervous system has a ladderlike appearance. **e.** A flatworm, such as *Dugesia,* is bilaterally symmetrical and has a head region with eyespots. (e): ©NHPA/M. I. Walker/Avalon

Figure 31.17 Medicinal leech, *Hirudo medicinalis.*   Leeches have been used in medicinal practices for thousands of years.

body. The beating of cilia in the flame cells (named because the beating of the cilia reminded early investigators of the flickering of a flame) keeps the water moving toward the excretory pores.

Planarians have a ladderlike nervous system. They possess a small anterior brain and two lateral nerve cords that are joined by cross-branches. Planarians exhibit cephalization; in addition to a brain, they have light-sensitive organs (eyespots) and chemosensitive organs located on the auricles. Their three muscle layers—outer circular, inner longitudinal, and diagonal—allow for varied movement. A ciliated epidermis allows planarians to glide along a film of mucus.

Planarians capture food by wrapping around the prey, entangling it in slime, and pinning it down. Then, they extend a muscular pharynx, and by a sucking motion, tear up and swallow the food. The pharynx leads into a three-branched gastrovascular cavity within which digestion is completed. The digestive tract is considered incomplete because it has only one opening.

Planarians can reproduce both sexually and asexually. Although *hermaphroditic* (possessing both male and female reproductive organs), they typically practice cross-fertilization, in which the penis of one is inserted into the genital pore of the other (and vice versa), and reciprocal transfer of sperm takes place. Fertilized eggs hatch in 2 to 3 weeks as tiny worms. Asexual reproduction occurs by regeneration—if you slice a planarian in half, two new

planarians will grow. You can even make a two-headed planarian by slicing the head in half!

Parasitic Flatworms   There are several classes of parasitic flatworms, two of which are discussed here: the tapeworms (class Cestoda) and the flukes (class Trematoda).

Tapeworms   As adults, tapeworms are endoparasites (internal parasites) of various vertebrates, including humans. They vary in length from a few millimeters to nearly 20 m.

Tapeworms have a tough integument, a specialized body covering resistant to the host's digestive juices. Their excretory, muscular, and nervous systems are similar to those of other flatworms. Tapeworms have a well-developed anterior region, called the **scolex,** which bears hooks for attachment to the intestinal wall of the host and suckers for feeding. Behind the scolex are **proglottids,** a series of reproductive units with a full set of male and female sex organs. Each proglottid fertilizes its own eggs, which number in the thousands. Immature proglottids are located closer to the scolex, while mature (or gravid) proglottids are farther away. The gravids, which contain fertilized eggs, break away and are eliminated in the host's feces. A secondary host must ingest the eggs for the life cycle to continue. Figure 31.19 illustrates the life cycle of the pork tapeworm, *Taenia solium,* where the human is the primary host and the pig is the secondary host. The larvae burrow through the intestinal wall and travel in the bloodstream to finally

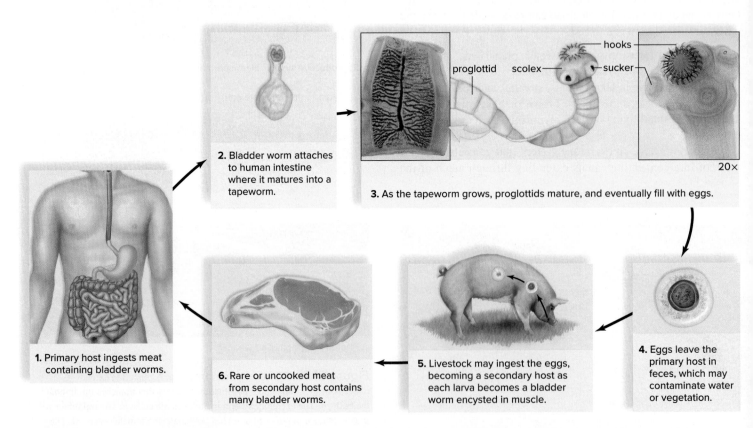

2. Bladder worm attaches to human intestine where it matures into a tapeworm.

proglottid   scolex   hooks   sucker

20×

3. As the tapeworm grows, proglottids mature, and eventually fill with eggs.

1. Primary host ingests meat containing bladder worms.

6. Rare or uncooked meat from secondary host contains many bladder worms.

5. Livestock may ingest the eggs, becoming a secondary host as each larva becomes a bladder worm encysted in muscle.

4. Eggs leave the primary host in feces, which may contaminate water or vegetation.

Figure 31.19 Life cycle of a tapeworm, *Taenia*.   The life cycle includes a human (primary host) and a pig (secondary host). The adult worm is modified for its parasitic way of life. It consists of a scolex and many proglottids, which become bags of eggs. (photos) (proglottid): Source: CDC; (scolex) ©James Webb/Medical Images lodge and encyst in muscle. The cyst is a small, hard-walled structure that contains a larva called a bladder worm. When humans eat infected meat that has not been thoroughly cooked, the bladder worm breaks out of the cyst, attaches itself to the intestinal wall, and grows to adulthood. Then, the life cycle begins again.

Flukes Flukes are endoparasites of various vertebrates. Their flattened and oval-to-elongated body is covered by a nonciliated integument. At the anterior end of these animals, an oral sucker is surrounded by sensory papillae, and there is at least one other sucker used for attachment to the host. Although the digestive system is reduced compared to that of free-living flatworms, the alimentary canal is well developed. The excretory and muscular systems are similar to those of free-living flatworms, but the nervous system is reduced, with poorly developed sense organs. Most flukes are hermaphroditic.

Flukes are usually named for the type of vertebrate organ they inhabit; for example, there are blood, liver, and lung flukes. The blood fluke (*Schistosoma* spp.) occurs predominantly in the Middle East, Asia, and Africa. Each year, over 65 million persons are infected with schistosomiasis. Adult flukes are small (approximately 2.5 cm long) and may live for years in their human hosts. The Chinese liver fluke, *Clonorchis sinensis,* is a major parasite of humans, cats, dogs, and pigs. This 20 mm-long fluke is commonly found in many regions of the Orient. It requires two intermediate hosts: a snail and a fish. Eggs are shed into the water in feces of the human or other mammal host and enter the body of a snail, where they undergo development. Larvae escape into the water and bore into the muscles of a fish. When humans eat infected fish, the juveniles migrate into the bile duct, where they mature. A heavy infection can cause destruction of the liver and death.

Rotifers

Rotifers (phylum Rotifera) are trochozoans related to the flatworms. With sizes ranging between 0.5 mm and 3 mm, rotifers are microscopic animals. Anton van Leeuwenhoek, an early eighteenth-century scientist, was one of the first to view rotifers (which he called "wheeled animalcules") through a microscope. Rotifers have a crown of cilia, known as the corona, on their heads (Fig. 31.20). The corona looks like a spinning wheel when in motion. This disc of beating cilia serves as an organ of locomotion and also directs food into the mouth. The approximately 2,200 species primarily live in fresh water; however, some marine and terrestrial forms exist. The majority of rotifers are transparent, but some are very colorful. Many species of rotifers can desiccate during harsh conditions and remain dormant for lengthy periods of time. This characteristic has earned them the title "resurrection animalcules."

Check Your Progress 31.3

1. List the characteristics that unite the molluscs and annelids.
2. Compare features of the flatworm, mollusc, and annelid body cavity, digestive tract, and circulatory system.
3. Identify three characteristics found in all molluscs.

Figure 31.20 Rotifer. Rotifers are microscopic animals. The beating of cilia on two lobes at the anterior end of the animal gives the impression of a pair of spinning wheels.

31.4 The Ecdysozoa

Learning Outcomes

Upon completion of this section, you should be able to

1. Identify the characteristics unique to ecdysozoans.
2. Describe the characteristics contributing to the success and diversity of arthropods.
3. List the major features that distinguish the crustaceans, insects, and arachnids.

The **ecdysozoans** include roundworms and arthropods. They are one of the major groups of animals, and contain the largest number of species.

Ecdysozoans construct an outer covering called a **cuticle,** which protects and supports the animal and is periodically shed to allow growth. Unlike many other invertebrate species that reproduce by releasing large amounts of eggs and sperm into their aqueous environment, many ecdysozoans have evolved separate sexes, which come together and copulate to deliver sperm into the female body, or to directly fertilize eggs as they are released.

Roundworms

Roundworms (phylum Nematoda, about 25,000 species) are nonsegmented worms that are prevalent in almost any environment. Generally, nematodes are colorless and range in size from microscopic to exceeding 1 m in length. The internal organs, including the tubular reproductive organs, lie within the pseudocoelom. A **pseudocoelom** is a body cavity that is incompletely lined by mesoderm. In other words, mesoderm occurs inside the body

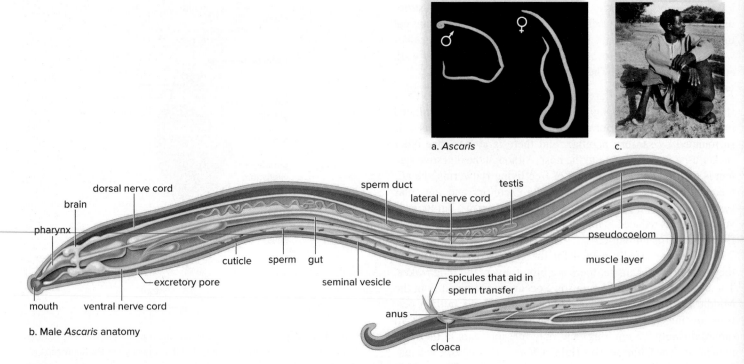

a. *Ascaris*

c.

b. Male *Ascaris* anatomy

Figure 31.21 Roundworm anatomy. **a.** The roundworm *Ascaris*. **b.** Roundworms such as *Ascaris* have a pseudocoelom and a complete digestive tract with a mouth and an anus. **c.** A filarial worm infection causes elephantiasis, which is characterized by a swollen body part when the worms block lymphatic vessels. (a): ©Kim Scott/Ricochet Creative Productions LLC; (c): ©Vanessa Vick/The New York Times/Redux

wall but not around the digestive cavity (gut). The fluid-filled pseudocoelom provides space for the development of organs, substitutes for a circulatory system by allowing easy passage of molecules, and provides a type of skeleton.

Nematodes have developed a variety of lifestyles from free-living to parasitic. One species, *Caenorhabditis elegans,* a free-living nematode, is a model animal used in genetics and developmental biology, because it was one of the first species to have its genome sequenced.

Biology of a Roundworm

In the roundworm, *Ascaris lumbricoides* (Fig. 31.21*a*), females tend to be larger (20–35 cm in length) than males (Fig. 31.21*b*). Both sexes move by a characteristic whiplike motion because they have only longitudinal muscles and no circular muscles next to the body wall. *Ascaris* species are most commonly parasites of humans and pigs.

A female *Ascaris* is very reproductively prolific, producing over 200,000 eggs daily. The eggs are passed with host feces and, under the right conditions, can develop into a new worm within 2 weeks. The eggs need to enter a new host's body via uncooked vegetables, soiled fingers, or ingested fecal material, and hatch in the intestines. The juveniles make their way into the veins and lymphatic vessels and are carried to the heart and lungs. From the lungs, the larvae travel up the trachea, where they are swallowed and eventually reach the intestines. There, the larvae mature and begin feeding on intestinal contents. The symptoms of an *Ascaris* infection depend upon the site and the stage of infection.

Other Roundworms

Trichinosis is a fairly serious infection caused by *Trichinella spiralis,* a roundworm that rarely infects humans in the United States. Humans contract the disease when they eat undercooked pork that contains encysted larvae. After maturation, the female adult burrows into the wall of the host's small intestine, where she deposits live larvae that are then carried by the bloodstream and encyst in the skeletal muscles. When the adults are in the small intestine, digestive disorders, fatigue, and fever occur. When the larvae encyst, the symptoms include aching joints, muscle pain, and itchy skin.

Elephantiasis is caused by a roundworm called the filarial worm (Fig. 31.21*c*), which utilizes the mosquito as an intermediate host. When a mosquito bites an infected person, it transports larvae to a new host. Because the adult worms reside in lymphatic vessels, fluid return is impeded, and the limbs of an infected human can swell to an enormous size, even resembling those of an elephant.

Other roundworm infections are more common in the United States. Children frequently acquire a pinworm infection, and hookworm is seen in the southern states, as well as worldwide. Good hygiene, proper disposal of sewage, and cooking meat thoroughly usually protect people from parasitic roundworms.

Arthropods

Arthropods (phylum Arthropoda) are extremely diverse (Fig. 31.22), ranging in size from less than 0.1 mm (mites) to 4 m (Japanese crab) in length. Well over a million species have been discovered and described, mostly insects, but some experts suggest

a. Flat-backed millipede, *Sigmoria*

b. Tarantula, *Aphonopelma*

c. Dungeness crab, *Cancer*

d. Paper wasp, *Polistes*

e. Indian giant tiger centipede,
Scolopendra hardwickei

Figure 31.22 Arthropod diversity. **a.** A millipede has only one pair of antennae, and the head is followed by a series of segments, each with two pairs of appendages. **b.** The hairy tarantulas of the genus *Aphonopelma* are dark in color and move carefully and steadily. Their bite is harmless to people. **c.** A crab is a crustacean with a calcified exoskeleton, one pair of claws, and four other pairs of walking legs. **d.** A wasp is an insect with two pairs of wings, both used for flying, and three pairs of walking legs. **e.** A centipede has only one pair of antennae, and the head is followed by a series of segments, each with a single pair of appendages. (a): ©John MacGregor/Getty Images; (b): ©G.C. Kelley/Science Source; (c): ©Bob Evans/Getty Images; (d): ©James Robinson/Science Source; (e): ©Matthijs Kuijpers/Alamy

that many more species exist. Arthropods occupy almost every type of habitat and are considered the most successful group of all the animals.

The remarkable success of arthropods is dependent on these characteristics:

1. *Segmentation.* The body of an arthropod is divided into segments. In some arthropods, the individual segments are fused into functional regions called the head, thorax, and abdomen (Fig. 31.23*a*). Segmentation allows the attachment of specialized, jointed appendages.

2. *Jointed appendages.* Jointed appendages allow for specialization of the body segments. The jointed appendages of arthropods are basically hollow tubes moved by muscles (Fig 31.23*b*). Typically, the appendages are highly adapted for a particular function, such as food gathering, reproduction, and locomotion. In addition, many appendages are associated with sensory structures and used for tactile purposes.

3. *An exoskeleton* (Fig. 31.23*c*). The exoskeleton is composed primarily of chitin, a strong, flexible, nitrogenous polysaccharide. The exoskeleton serves many functions, including protection, attachment for muscles, locomotion, and prevention of desiccation. However, because it is hard and nonexpandable, arthropods must molt, or shed, the exoskeleton in order to grow larger.

In addition, arthropods have a well-developed nervous system that consists of a brain and a ventral nerve cord. The head bears various types of sense organs, including eyes of two types—simple and compound. The compound eye is composed of many complete visual units, each of which operates independently (Fig. 31.23*d*). The lens of each visual unit focuses an image on a small number of photoreceptors within that unit. The simple eye, like that of vertebrates, has a single lens that brings the image to focus onto many receptors, each of which receives only a portion of the image. In addition to sight, many arthropods have well-developed touch, smell, taste, balance, and hearing.

Respiration in arthropods depends on the environment in which the organism lives. Marine forms use gills, which are vascularized, highly convoluted, thin-walled tissue specialized for gas exchange. Terrestrial forms have book lungs (e.g., spiders) or air tubes called tracheae. Tracheae serve as a rapid way to transport oxygen directly to the cells.

The majority of arthropods undergo metamorphosis, which is the drastic change in form and physiology that occurs as an immature stage, called a larva, becomes an adult. Among arthropods, the larva eats different food and lives in a different environment than the adult. For example, larval crabs live among and feed on plankton, while adult crabs are bottom dwellers that catch live prey or scavenge dead organic matter. Among insects, such as butterflies, the caterpillar feeds on leafy vegetation, while the adult feeds on nectar.

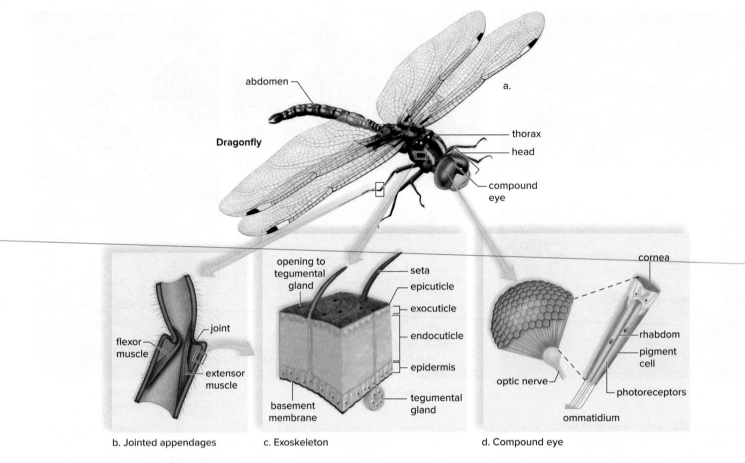

b. Jointed appendages   c. Exoskeleton   d. Compound eye

Figure 31.23 Characteristics of arthropods.   **a.** The body of an arthropod is segmented. These segments are often fused into a head, thorax, and abdomen. **b.** Jointed appendages help specialize the body segments. In the hollow appendages, muscles allow movement. **c.** The exoskeleton is secreted by the epidermis and consists of the endocuticle; the exocuticle, hardened by the deposition of calcium carbonate; and the epicuticle, a waxy layer. Chitin makes up the bulk of the exoskeleton. **d.** Arthropods have a more complex nervous system, which consists of a compound eye that contains many individual units, each with its own lens and photoreceptors.

Crustaceans

Crustaceans (subphylum Crustacea) are a group of largely marine arthropods that include barnacles, shrimps, lobsters, and crabs. There are also some freshwater crustaceans, including the crayfish, and some terrestrial ones, including the sowbug, or pillbug.

Crustaceans are named for their hard shells. The exoskeleton is calcified to a greater degree in some forms than in others. Although crustacean anatomy is extremely diverse, the head usually bears a pair of compound eyes and five pairs of appendages. The first two pairs, called antennae, lie in front of the mouth and have sensory functions. The other three pairs are mouthparts used in feeding.

In crayfish, such as *Cambarus,* the thorax bears five pairs of walking legs. The first walking leg is a pinching claw (Fig. 31.24). The gills are situated above the walking legs. The head and thorax are fused into a cephalothorax, which is covered on the top and sides by a nonsegmented carapace. The abdominal segments, which contain much musculature, are equipped with small paddle-like structures called swimmerets. The last two segments bear the uropods and the telson, which make up a fan-shaped tail to propel the crayfish backward.

The digestive system includes a stomach, which is divided into two main regions. The anterior portion consists of a gastric mill, a special grinding apparatus equipped with chitinous teeth. The posterior region acts as a filter to prevent coarse particles from entering the digestive glands where absorption takes place. *Green glands* lying in the head region, anterior to the esophagus, excrete metabolic wastes through a duct that opens externally at the base of the antennae. The coelom is reduced to a space around the reproductive system. A heart within a pericardial cavity pumps blood containing the respiratory pigment hemocyanin into a **hemocoel** consisting of sinuses (open spaces). Hemocyanin contains copper (compared to hemoglobin—the red, iron-containing pigment of humans), causing the blood to have a blueish color. These sinuses are common in an open circulatory system, where the hemolymph flows freely about the organs and is not typically contained within blood vessels.

The nervous system of the crayfish is very similar to that of the earthworm. The crayfish has a brain and a ventral nerve cord

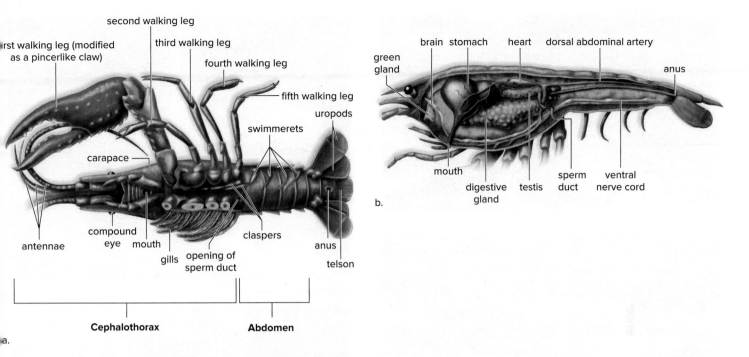

Figure 31.24 Male crayfish, *Cambarus*. **a.** Externally, it is possible to observe the crayfish's jointed appendages, including the swimmerets, and the walking legs, which include claws. These appendages, plus a portion of the carapace, have been removed from the right side so that the gills are visible. **b.** An internal view shows the parts of the digestive and circulatory systems. Note the ventral nerve cord.

that passes posteriorly. Along the length of the nerve cord, segmental ganglia give off 8 to 19 paired lateral nerves.

The sexes are separate in the crayfish, with the gonads located just ventral to the pericardial cavity. In the male, a coiled sperm duct opens to the outside at the base of the fifth walking leg. Sperm transfer is accomplished by the first two pairs of swimmerets, which are enlarged and quite strong. In the female, the ovaries open at the bases of the third walking legs. A stiff fold between the bases of the fourth and fifth pairs of walking legs serves as a seminal receptacle. Following fertilization, the eggs are attached to the swimmerets of the female.

Insects

Insects (subphylum Uniramia) are so numerous (well over 1 million identified species) and so diverse (Fig. 31.25) that the study of this one group is a major specialty in biology called entomology. Some insects show remarkable behavioral adaptations, as exemplified by the social systems of bees, ants, termites, and other colonial insects.

Insects have certain features in common. The body is divided into a head, a thorax, and an abdomen. The head usually bears a pair of sensory antennae, a pair of compound eyes, and several simple eyes. The mouthparts are adapted to the specific insect's way of life. For example, a grasshopper has mouthparts that chew, whereas a butterfly has a long tube for siphoning the nectar of flowers. The abdomen contains most of the internal organs. The thorax bears three pairs of legs and zero to three pairs of wings. Wings, if present, enhance an insect's ability to

survive by providing a way of escaping enemies, finding food, facilitating mating, and dispersing the offspring. The exoskeleton of an insect is lighter and contains less chitin than that of many other arthropods.

Here, we will discuss the features of a grasshopper as a representative example of insect form and function. In the grasshopper (Fig. 31.26), the third pair of legs is suited to jumping. There are two pairs of wings. The forewings are tough and leathery, and when folded back at rest, they protect the broad, thin hindwings. On the lateral surface, the first abdominal segment bears a large tympanum on each side for the reception of sound waves.

Internal Organs The digestive system of a grasshopper is suitable for a herbivorous diet. The mouthparts chew the food, which is temporarily stored in the crop before passing into a gastric mill, where it is finely ground before digestion is completed in the stomach. Nutrients are absorbed into the hemocoel from outpockets called gastric ceca. The stomach leads into an intestine and a rectum, which empties by way of an anus. The excretory system consists of **Malpighian tubules,** which extend into the hemocoel and collect nitrogenous wastes that are excreted into the digestive tract. The formation of a solid nitrogenous waste, namely uric acid, conserves water.

The respiratory system begins with openings in the exoskeleton called spiracles. From here, the air enters small tubules called **tracheae** (Fig. 31.26a). The tracheae branch and rebranch until they end intracellularly, where the actual exchange of gases takes place. The movement of air through this complex of tubules is not passive. Rather, air is actively pumped by alternate

a. Housefly, *Musca*

b. Walking stick, *Diapheromera*

c. Bee, *Apis*

d. Dragonfly, *Aeshna*

e. Birdwing butterfly, *Troides*

Figure 31.25 Insect diversity. **a.** Flies have a single pair of wings and lapping mouthparts. **b.** Walking sticks are herbivorous, with biting and chewing mouthparts. **c.** Bees have four translucent wings and a thorax separated from the abdomen by a narrow waist. **d.** Dragonflies have two pairs of similar wings. They catch and eat other insects while flying. **e.** Butterflies have forewings larger than their hindwings. Their mouthparts form a long tube for siphoning up nectar from flowers. (a): ©McPhoto/SHU/INSADCO Photography/Alamy; (b): ©Creatas/age fotostock; (c): ©MedioImages/Photodisc/Getty Images; (d): ©Jupiterimages/Ablestock.com; (e): ©Darlyne A. Murawski/National Geographic/Getty Images

contraction and relaxation of the body wall. Breathing by tracheae is suitable to the small size of insects (most are less than 60 mm in length), because the tracheae would be crushed by any significant amount of weight.

The circulatory system contains a slender, tubular heart that lies against the dorsal wall of the abdominal exoskeleton and pumps hemolymph into an aorta that leads to a hemocoel, where it circulates before returning to the heart again. The hemolymph is colorless and lacks a respiratory pigment because the tracheal system is responsible for gas exchange.

Reproduction and Development Grasshopper reproduction is adapted to life on land. The male grasshopper has a penis, and sperm passed to the female are stored in a seminal receptacle. Internal fertilization protects both gametes and zygotes from drying out. The female deposits the fertilized eggs in the ground with her ovipositor.

Metamorphosis is a change in form and physiology that occurs as an immature stage, called a larva, becomes an adult. Grasshoppers undergo gradual metamorphosis as they mature. The immature grasshopper, called a nymph, looks like an adult grasshopper, even though it differs somewhat in shape and form. Other insects, such as butterflies, undergo complete metamorphosis, involving drastic changes in form. At first, the animal is a wormlike larva (caterpillar) with chewing mouthparts. It then forms a case, or cocoon, about itself and becomes a pupa. During this stage, the body parts are completely reorganized. The adult then emerges from the cocoon. This life cycle allows the larvae and adults to use different food sources. Most eating by insects occurs during the larval stage. The field of forensic science has found a way to put

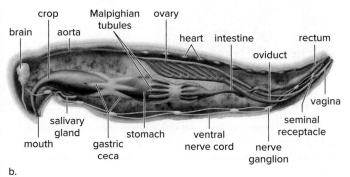

Figure 31.26 Female grasshopper, *Romalea*. a. Externally, the body of a grasshopper is divided into three sections and has three pairs of legs. The tympanum receives sound waves, and the jumping legs and the wings are for locomotion. **b.** Internally, the digestive system is specialized. The Malpighian tubules excrete a solid nitrogenous waste (uric acid). A seminal receptacle receives sperm from the male, which has a penis.

our knowledge of insect life cycle stages to good use, as presented in the Scientific Inquiry feature, "Maggots: A Surprising Tool for Crime Scene Investigation."

Comparison with Crayfish The grasshopper and the crayfish share a common ancestor, but they have diverged in their morphology to adapt to aquatic versus terrestrial environments. In crayfish, gills take up oxygen from water, while in the grasshopper, tracheae allow oxygen-laden air to enter the body. Appropriately, the crayfish has an oxygen-carrying pigment, but a grasshopper has no such pigment in its blood. The crayfish excretes a liquid nitrogenous waste (ammonia), while the grasshopper excretes a solid nitrogenous waste (uric acid). Only grasshoppers have (1) a tympanum for the reception of sound waves, and (2) a penis in males for passing sperm to females without possible desiccation, and an ovipositor in females for laying eggs in soil. Crayfish utilize their uropods when they swim; a grasshopper has legs for hopping and wings for flying.

Arachnids

The **arachnids** (subphylum Chelicerata) include scorpions, spiders, ticks, and mites (Fig. 31.27). In this group, the cephalothorax bears six pairs of appendages: the chelicerae and the pedipalps, and four pairs of walking legs. The cephalothorax is followed by an abdomen that contains internal organs.

Scorpions are the oldest terrestrial arthropods. Today, they are widely distributed in the tropics, subtropics, desert, and temperate regions, with a notable absence from New Zealand. They are nocturnal and spend most of the day hidden under a log or a rock. In scorpions, the pedipalps are large pincers, and the long abdomen ends with a stinger that contains venom.

Ticks and mites are parasites. Ticks suck the blood of vertebrates and sometimes transmit diseases, such as Rocky Mountain spotted fever or Lyme disease. Chiggers, the larvae of certain mites, feed on the skin of vertebrates.

SCIENCE IN YOUR LIFE ▶ SCIENTIFIC INQUIRY

Maggots: A Surprising Tool for Crime Scene Investigation

A fresh human corpse is lying on the floor. Within 10 to 15 minutes, blowflies with blue and green metallic bodies arrive. An animal corpse is a perfect site for egg production. The flies lay eggs in the mouth, nostrils, ears, eyes, and any open wounds. Depending on the temperature, the eggs will hatch in the next 12 to 24 hours. The small hatchling larval flies, more commonly known as maggots, feed on fat and collagen in the corpse in order to grow. Their development occurs in three stages called instars, during which the larvae shed skin and grow. When the third instar grows, it leaves the corpse and forms a pupa that takes about 14 days (depending on the temperature) to emerge as an adult blowfly. The adult has to wait a day or two until it can fly because the wings need to expand and the body has to harden. Once capable of

flight, the adult leaves to find a mate, as well as another corpse in which to lay eggs, thereby completing the life cycle.

What do blowflies have to do with crime scene investigation? If the corpse involved is a human, the timing of the various stages of the life cycle of the flies (e.g., newly hatched eggs, maggot size, maggot weight, or pupa remains) can help forensic scientists determine the postmortem interval (or PMI) since the time of death. Forensic entomology uses insects such as blowflies to determine the time of death. A forensic entomologist can even determine if a human corpse has been moved based on the fact that different species of blowflies prefer to lay eggs in different environments. For example, some species lay their eggs in shade, whereas others lay eggs in sunny habitats.

Similarly, some species inhabit urban areas, whereas others inhabit rural areas. Thus, the combination of life stage(s) and species of blowfly found on a corpse can help a forensic entomologist determine the time and possible location of a person's death.

Questions to Consider

1. Given that the hatching of eggs and the growth rate of blowfly larvae depend on temperature, how sure do you think forensic entomologists can be about time of death?

2. As a juror, how would you feel about forensic entomology evidence entered into court?

3. When might forensic entomology evidence be useful in a murder case?

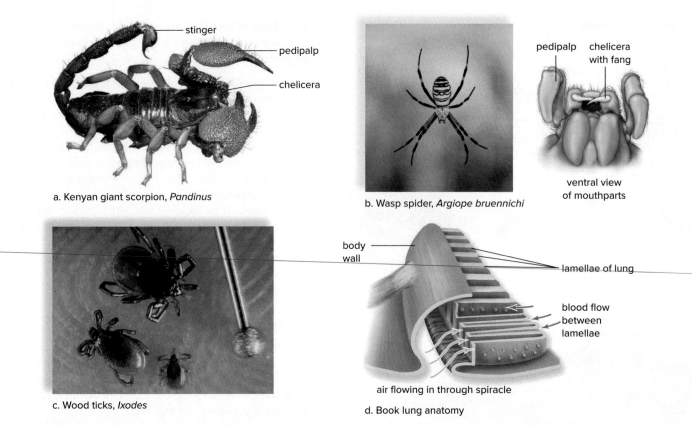

Figure 31.27 Arachnid diversity. **a.** A scorpion has pincerlike chelicerae and pedipalps. Its long abdomen ends with a stinger that contains venom. **b.** The wasp spider is named because of its markings. It is not poisonous to humans. **c.** In the western United States, the wood tick carries a debilitating disease called Rocky Mountain spotted fever. **d.** Arachnids breathe by means of book lungs, in which the "pages" are double sheets of thin tissue (lamellae). (photos) (a): ©Tom McHugh/Science Source; (b): ©imagebroker.net/Superstock; (c): ©Scott Camazine/Science Source

Spiders, the most familiar arachnids, have a narrow waist that separates the cephalothorax from the abdomen. Each chelicera consists of a basal segment and a fang that delivers venom to paralyze or kill prey. Spiders use silk threads for all sorts of purposes, from lining their nests to catching prey. Biologists have used web-building behavior to discover how spider families are related.

The internal organs of spiders also show how they are adapted to a terrestrial way of life. Malpighian tubules work in conjunction with rectal glands to reabsorb ions and water before a relatively dry nitrogenous waste (uric acid) is excreted. Invaginations of the inner body wall form the lamellae ("pages") of spiders' so-called book lungs. Air flowing into the folded lamellae on one side exchanges gases with blood flowing in the opposite direction on the other side.

Check Your Progress 31.4

1. Name two ways in which roundworms are anatomically similar to arthropods.
2. Explain the features that account for the great success of arthropods.
3. Identify the major anatomical characteristics of a crustacean.

31.5 Invertebrate Deuterostomes

Learning Outcomes

Upon completion of this section, you should be able to

1. Recognize the basic morphological characteristics of echinoderms.
2. Describe how sea stars move, feed, and reproduce.

Among animals, chordates are most closely related to the echinoderms (phylum Echinodermata) as witnessed by their similar embryological development, even though echinoderms are invertebrates. Both echinoderms and chordates are deuterostomes, in which the second embryonic opening becomes the mouth, and the coelom forms by outpocketing of the primitive gut (see Fig. 31.4). Protostomes, in contrast, are characterized by a mouth that forms from the first embryonic opening (the blastopore).

Characteristics of Echinoderms

Echinoderms are a diverse group of marine animals. There are no terrestrial echinoderms. They have an endoskeleton (internal skeleton) consisting of spine-bearing, calcium-rich plates. The spines, which stick out through their delicate skin, account for their name.

a. Brittle star, *Ophiothrix*

b. Sea cucumber, *Apostichopus*

c. Sea urchin, *Strongylocentrotus*

d. Feather stars, *Antedon*

Figure 31.28 Echinoderm diversity. **a.** Brittle stars have long, slender arms, which make them the most mobile echinoderms. **b.** Sea cucumbers look like a cucumber. They lack arms but have tentacle-like tube feet with suckers around the mouth. **c.** Sea urchins have large, colored, external spines for protection. **d.** A feather star extends its arms to filter suspended food particles from the sea. (a): ©Diane R. Nelson; (b): ©Andrey Nekrasov/Alamy; (c): ©Fuse/Getty Images; (d): ©Philippe Bourseiller/Getty Images

It may seem surprising that echinoderms, although related to chordates, lack those features we associate with vertebrates such as humans. For example, echinoderms are often radially, not bilaterally, symmetrical. Their larva is a free-swimming filter feeder with bilateral symmetry, but it typically metamorphoses into a radially symmetrical adult. Recall that with radial symmetry, it is possible to obtain two mirror images, no matter how the animal is sliced longitudinally, whereas in bilaterally symmetrical animals, only a longitudinal cut gives two mirror images.

Echinoderm Diversity

Echinoderms are quite diverse (Fig. 31.28). They include sea lilies, motile feather stars, brittle stars, and sea cucumbers. Sea cucumbers actually look like a cucumber, except they have feeding tentacles surrounding their mouth. You may be more familiar with sea urchins and sand dollars in the class Echinoidea, which have spines for locomotion, defense, and burrowing. Sea urchins are food for sea otters off the coast of California.

Sea Stars

Sea stars (starfish) are commonly found along rocky coasts, where they feed on clams, oysters, and other bivalve molluscs.

The five-rayed body has an oral, or mouth, side (the underside) and an aboral, or anus, side (the upper side) (Fig. 31.29). Various structures project through the body wall: (1) spines from the endoskeletal plates offer some protection; (2) pincerlike structures called pedicellariae keep the surface free of small particles; and (3) skin gills, tiny fingerlike extensions of the skin, are used for gas exchange. On the oral surface, each arm has a groove lined by small tube feet.

To feed, a sea star positions itself over a bivalve such as a clam and attaches some of its tube feet to each side of the shell (Fig. 31.29*a*). By working its tube feet in alternation, it pulls the shell open. A very small crack is enough for the sea star to evert its cardiac stomach and push it through the crack to contact the soft parts of the bivalve. The stomach secretes enzymes, and digestion begins even while the bivalve is attempting to close its shell. Later, partly digested food is taken into the sea star's body, where digestion continues in the pyloric stomach, using enzymes from digestive glands found in each arm. A short intestine opens at the anus on the aboral side.

In each arm, the well-developed coelomic cavity contains not only a pair of digestive glands, but also gonads (either male or female), which open on the aboral surface by very small pores. The nervous system consists of a central nerve ring from which radial nerves extend into each arm. A light-sensitive eyespot is at the tip

a. Rainbow star, *Orthasterias koehleri*.

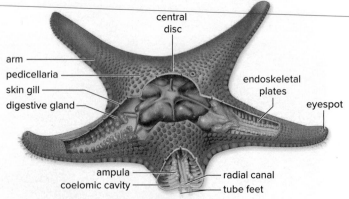

b. Aboral side showing ray cross-section

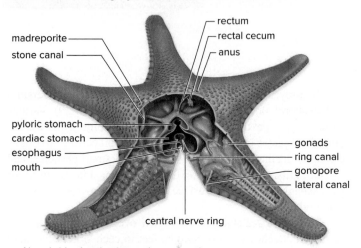

c. Aboral side showing internal cross-section

Figure 31.29 Sea star anatomy and behavior. **a.** A sea star uses the suction of its tube feet to open a bivalve mollusc, its primary source of food. **b.** Each arm of a sea star contains digestive glands, gonads, and portions of the water vascular system. This system (colored orange) terminates in tube feet. **c.** A different cross-section shows the orientation of mouth, digestive system, and anus. (a): ©Mark Conlin/Alamy

of each arm. Sea stars are capable of coordinated but slow responses and body movements.

Locomotion depends on their water vascular system. Water enters this system through a structure on the aboral side called the madreporite, or sieve plate. From there, it passes down a stone canal into a ring canal, which surrounds the mouth. A radial canal in each arm branches off from the central ring canal. For locomotion, water enters the ampullae from the radial canals. Contraction of ampullae forces water into the tube foot, expanding it. When the foot touches a surface, the center is withdrawn, giving it suction so it can adhere to the surface. By alternating the expansion and contraction of the tube feet, a sea star moves slowly along.

Echinoderms don't have a respiratory, excretory, or circulatory system. Fluids within the coelomic cavity and the water vascular system carry out many of these functions. For example, gas exchange occurs across the skin gills and the tube feet. Nitrogenous wastes diffuse through the coelomic fluid and the body wall. Cilia on the peritoneum lining the coelom keep the coelomic fluid moving.

Sea stars reproduce both asexually and sexually. If the body is fragmented, each fragment can regenerate a whole animal as long as the fragment contains part of the central disk. Fishermen who try to get rid of sea stars by cutting them up and tossing them overboard are merely propagating more sea stars! Sea stars also spawn, releasing either eggs or sperm. The larva is bilaterally symmetrical and metamorphoses to become the radially symmetrical adult.

Check Your Progress 31.5

1. Explain why echinoderms are more closely related to chordates than invertebrates.
2. Identify the basic structures of an echinoderm.

Conclusion

Colony collapse disorder not only threatens to affect the production of agriculturally important crops, such as fruits and vegetables, but the loss of pollinating bees also has the potential to disrupt the function of many of the world's ecosystems.

The invertebrates—animals that lack a backbone—are far more diverse than the vertebrates. While various invertebrates are associated with disease and health problems, many are very beneficial to the survival of the human species. Various insects, for example, act as pollinators for a large number of commercially grown crops. Overall, invertebrates possess an enormous variety of traits that have allowed them to successfully adapt to a wide range of lifestyles and habitats.

SUMMARIZE

31.1 Evolutionary Trends Among Animals

Animals are motile, multicellular heterotrophs that ingest their food. They exhibit a diploid life cycle.

- Animals are divided into two categories—**invertebrates,** those that lack an endoskeleton—and **vertebrates,** those that possess an endoskeleton.
- Symmetry varies among animals. Some are **asymmetrical,** whereas others have **radial symmetry** or **bilateral symmetry.** Some animals live portions of their life as **sessile** individuals. **Cephalization** is often associated with bilateral symmetry.
- Animals are further divided into the **protostomes** or **deuterostomes** based upon their developmental features and processes.

31.2 The Simplest Invertebrates

- **Porifera** (sponges) have a cellular level of organization, lack true tissues, and are typically asymmetrical.
- They use **collar cells** to create a water current, enabling them to act as filter feeders. The internal skeleton is composed of **spicules,** which is often used to classify them.
- Most sponges are **hermaphroditic,** meaning they have both male and female sex organs.
- **Comb jellies** and **cnidarians** are radially symmetrical and have two tissue layers derived from the germ layers ectoderm and endoderm. The body of comb jellies is made up of a transparent jellylike substance called mesoglea.
- The **gastrovascular cavity** serves as a location for digestion and circulation of nutrients.
- The defining feature of the cnidarians is the presence of stinging cells called **cnidocytes** that contain a toxin-filled capsule called a **nematocyst.**
- Cnidarians exist as either a **polyp** or a **medusa,** or they can alternate between the two.

31.3 The Spiralia

Spiralia include lophotrochozoans and platyzoans, protostome animals that undergo spiral cleavage.

- The lophophorans include the **bryozoans, brachiopods,** and **phoronids.** All have a ciliated feeding apparatus.
- The **trochozoans** have a stage of development that uses cilia for locomotion. The trochozoans include:
  - **Flatworms** have three tissue layers and no coelom. Planaria have cephalization, muscles, and a ladder-type nervous system and also use flame cells to help propel water toward excretory pores. Tapeworms have a specialized structure called the **scolex** that enables them to attach onto their host, followed by multiple reproductive units called **proglottids.**
  - Rotifers are microscopic and have a corona that resembles a spinning wheel when in motion.
  - The body of a **mollusc** typically contains a **visceral mass,** a **mantle,** and a **foot.** Many molluscs also have a head and a radula.
  - Molluscs often have a reduced coelom and an open circulatory system.

- **Gastropods** include the nudibranchs, conchs, and snails.
- **Cephalopods,** such as squid, display marked cephalization, move rapidly by jet propulsion, and have a **closed circulatory system.**
- **Bivalves** (e.g., clams) are generally filter feeders that possess an **open circulatory system,** a radula, and a hatchet foot.
- **Annelids** are segmented both externally and internally. Their **hydrostatic skeleton** is a fluid-filled interior that supports muscle contraction and enhances flexibility. The nervous system consists of a ventral solid **nerve cord** and brain.
- The **nephridium** functions as a waste collection and excretory organ.
- Polychaetes are marine worms that have parapodia that contain tiny bristlelike hairs called **setae.**
- Earthworms are oligochaetes that are segmented and hermaphroditic scavengers that feed on organic matter in the soil.
- **Leeches** lack the setae found in other annelids. They contain suckers on each end of the body that enable attachment and feeding.
- Platyzoans include flatworms and rotifers. Free-living flatworms (planarians) have three tissue layers and no coelom. They are hermaphroditic, and usually reproduce sexually. Planarians have muscles and a ladder-type nervous system with paired ganglia serving as the brain. They take in food through an extended pharynx, leading to a gastrovascular cavity that extends through the body.
- Flukes and tapeworms are common parasites of vertebrate animals. Flukes have two suckers by which they attach to and feed from their hosts. Tapeworms have a **scolex** with hooks and suckers for attaching to the host's intestinal wall. The body of a tapeworm is made up of **proglottids,** which, when mature, contain thousands of eggs.
- **Rotifers** are microscopic and aquatic and have a corona that resembles a spinning wheel when in motion.

31.4 The Ecdysozoa

The **ecdysozoans** all possess a **cuticle** and undergo molting during their larval-stages. They represent the most abundant form of animal on the planet.

- **Roundworms** are usually small, very diverse, and are present almost everywhere in great numbers. Roundworms have a pseudocoelom. **Trichinosis** and **elephantiasis** are two diseases associated with roundworm infections.
- **Arthropods** are the most diverse group of animals. Their success is largely attributable to a flexible exoskeleton, segmentation, a well-developed nervous system, respiratory organs, and **metamorphosis.** Like many other arthropods, **crustaceans** have a head that bears compound eyes, antennae, and mouthparts. Blood is pumped into the **hemocoel,** where hemolymph will flow about the organs.
- Like many other **insects,** grasshoppers have wings and three pairs of legs attached to the thorax. Grasshoppers also have several adaptations to a terrestrial life, including respiration by **tracheae** and an excretory system consisting of **Malpighian tubules,** which collect and excrete nitrogenous wastes.
- Spiders are **arachnids** with chelicerae, pedipalps, and four pairs of walking legs attached to a cephalothorax. Spiders, too, are adapted to life on land, and they spin silk that is used in various ways.
- The crayfish, a crustacean, also has an open circulatory system, respiration by gills, and a ventral solid nerve cord.

31.5 Invertebrate Deuterostomes

Echinoderms and chordates are deuterostomes, whereby the second embryonic opening becomes the mouth and the coelom forms by outpocketing of the primitive gut. Echinoderms (e.g., sea stars, sea urchins, sea cucumbers, sea lilies) have radial symmetry as adults (not as larvae) and spines from endoskeletal plates.

■ Typical of echinoderms, sea stars have tiny skin gills, a central nerve ring with branches, and a water vascular system for locomotion. Each arm of a sea star contains components of the nervous, digestive, and reproductive systems.

ASSESS

Testing Yourself

Choose the best answer for each question.

31.1 Evolutionary Trends Among Animals

1. Which of these is not a characteristic of animals?
   a. heterotrophic
   b. ingest their food
   c. cells lack cell walls
   d. have chlorophyll
   e. multicellular

2. The phylogenetic tree of animals shows that
   a. rotifers are closely related to flatworms.
   b. both molluscs and annelids are protostomes.
   c. some animals have radial symmetry.
   d. sponges were the first to evolve from an ancestral protist.
   e. All of these are correct.

3. Which of these does not pertain to a protostome?
   a. spiral cleavage
   b. blastopore is associated with the anus
   c. coelom, splitting of mesoderm
   d. annelids, arthropods, and molluscs
   e. mouth is associated with first opening

31.2 The Simplest Invertebrates

4. Comb jellies are most closely related to
   a. cnidarians.
   b. sponges.
   c. flatworms.
   d. roundworms.
   e. rotifers.

5. Corals belong to which of the following?
   a. comb jellies
   b. sponges
   c. cnidarians
   d. rotifers
   e. None of these are correct.

6. Which of the following is responsible for the process of filter feeding in a sponge?
   a. flame cells
   b. collar cells
   c. cnidocytes
   d. mesoglea

31.3 The Spiralia

7. A ciliated feeding structure in the larval stage is a characteristic of a
   a. trochozoan.
   b. parazoan.
   c. lophophoran.
   d. deuterostome.

8. Which of the following has a complete digestive system?
   a. tapeworm
   b. sponge
   c. cnidarian
   d. earthworm

9. Bivalves, cephalopods, and gastropods are all
   a. arthropods.
   b. lophophorans.
   c. cnidarians.
   d. annelids.
   e. molluscs.

10. Which of the following is a characteristic of an annelid?
    a. nephridia
    b. closed circulatory system
    c. a nerve cord
    d. All of these are correct.

31.4 The Ecdysozoa

11. Which characteristic(s) account(s) for the success of arthropods?
    a. jointed appendages
    b. an exoskeleton
    c. segmentation
    d. All of these are correct.

12. Which of the following is not an arthropod?
    a. arachnids
    b. crustaceans
    c. rotifers
    d. insects
    e. All of the above are arthropods.

13. Label the indicated parts of the diagram below:

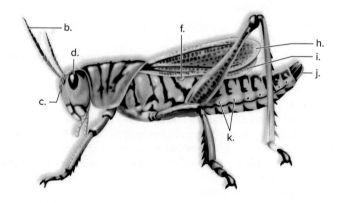

31.5 Invertebrate Deuterostomes

14. The two phyla in the Deuterostomia are
    a. Arthropoda and Nematoda.
    b. Brachiopoda and Platyhelminthes.
    c. Cestoda and Rotifera.
    d. Cnidaria and Ctenophora.
    e. Echinodermata and Chordata.

15. Which is not a characteristic feature of sea stars?
    a. coelom
    b. eyespot
    c. skin gills
    d. trachea
    e. tube feet

16. The water vascular system of sea stars functions in
    a. the circulation of gases and nutrients.
    b. digestion.
    c. locomotion.
    d. reproduction.
    e. Both a and c are correct.

ENGAGE

Thinking Critically

1. Cnidarians are radially symmetrical, whereas flatworms, roundworms, molluscs, annelids, and arthropods are bilaterally symmetrical. How does the type of symmetry relate to the lifestyle of each type of animal?

2. Multicellular animals are thought to have evolved from single-celled protists resembling the choanoflagellates that exist today. Describe some specific advantages to multicellularity that gave animals an advantage over their single-celled relatives.

3. Many species of invertebrates (e.g., flatworms, roundworms, arthropods) are parasites of other animals. List several advantages, as well as challenges, inherent in a parasitic lifestyle.

CHAPTER

32

Animals: Chordates and Vertebrates

©FLPA/Superstock

BEFORE YOU BEGIN

Before beginning this chapter, take a few moments to review the following discussions:

Section 27.2 How does the fossil record support evolutionary theory?

Section 27.6 How does a phylogenetic tree indicate the evolutionary relatedness of species?

Section 31.1 What phyla of animals are most closely related to the chordates?

Evolution of Man's Best Friend

Dogs are probably our most familiar and beloved domesticated animals, and well deserving of the moniker "man's best friend." They are our companions, as well as our working partners, in war, law enforcement, herding, service, rescue, and therapy. But when and where did domestic dogs first evolve? Scientists have long suspected that *Canis familiaris* evolved from *Canis lupus,* the gray wolf, but genetic analyses now indicate that modern dogs evolved from a species of wolf that is now extinct. And while previous studies supported the idea that dogs were first domesticated in China or the Middle East, comparisons of mitochondrial DNA isolated from modern dogs, wolves, and ancient dog fossils now indicate that the first domesticated dogs most likely originated in Europe.

However, the time frame of this event still remains in question. While initially the domestication of the dog is believed to have occurred around 16,000 years ago, newer molecular analytical techniques suggest this may have happened as far back as 130,000 years ago. If this is the case, then humans and canines have been around each other for much longer than anyone thought.

The New Guinea singing dog is named for its habit of howling at different pitches. It is also called Stone Age dog, after the stone tool–using people who took it to New Guinea 6,000 years ago. Living on an island, it has remained geographically and reproductively isolated ever since. Therefore, it might be a "living fossil," an organism with the same genes and characteristics as its original ancestor. If so, the New Guinea singing dog offers an opportunity to study what early domesticated dogs were like. However, like many other vertebrates, which are the topic of this chapter, the New Guinea singing dog is threatened with extinction. Expeditions into the highlands of New Guinea have yielded only a few droppings, tracks, and haunting howls in the distance.

As you read through this chapter, think about the following questions:

1. When during the history of life did vertebrates evolve?
2. Which characteristics set vertebrates apart from other animals?
3. Which other vertebrates might be considered "living fossils"?

32.1 Chordates

Learning Outcomes

Upon completion of this section, you should be able to

1. Identify the four morphological characteristics unique to the chordates.
2. Describe the features of the two groups of nonvertebrate chordates.

To be classified as a **chordate** (phylum Chordata, around 60,000 species), an animal must have, at some time during its life history, the following characteristics (Fig. 32.1):

- A dorsal supporting rod called a **notochord.** The notochord is located just below the nerve cord, toward the back (i.e., dorsal). Vertebrates have an embryonic notochord that is replaced by the vertebral column during development.
- A dorsal tubular **nerve cord.** Tubular means that the cord contains a canal filled with fluid. In vertebrates, the nerve cord is protected by the vertebrae. Therefore, it is called the spinal cord because the vertebrae form the spine.
- **Pharyngeal pouches.** Most vertebrates have pharyngeal pouches only during embryonic development. In the nonvertebrate chordates, the fishes, and some amphibian larvae, the pharyngeal pouches develop into functioning gills. Water passing into the mouth and the pharynx goes through the gill slits, which are supported by gill arches. In terrestrial vertebrates that breathe by lungs, the pouches are modified for various

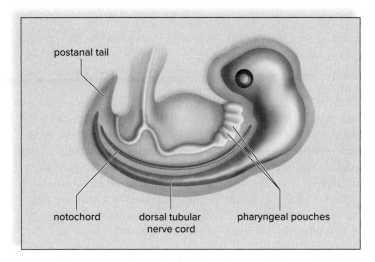

Figure 32.1 Chordate characteristics. All chordates possess these traits at some point in their life cycle.

purposes. In humans, the first pair of pouches becomes the auditory tubes. The second pair becomes the tonsils, while the third and fourth pairs become the thymus and the parathyroids.

- **Postanal tail.** The tail extends beyond the anus, hence the term *postanal.*

Evolutionary Trends Among the Chordates

Figure 32.2 depicts the phylogenetic tree of the chordates, and lists at least one main evolutionary trend that distinguishes each group of animals from the preceding group.

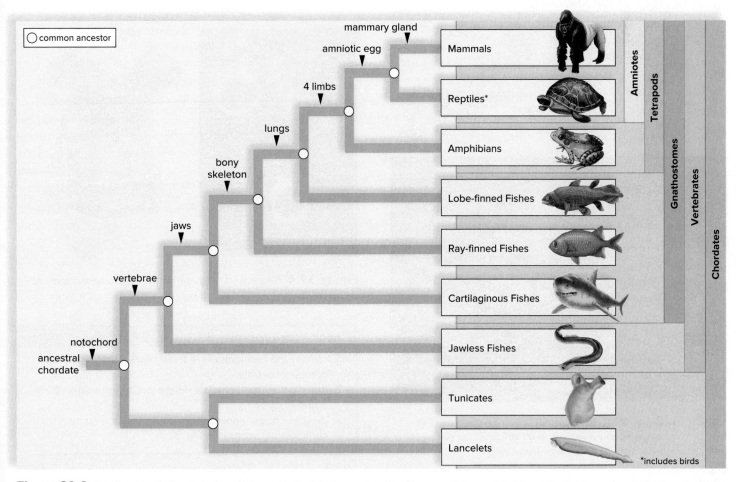

Figure 32.2 Phylogenetic tree of chordates.   Each of the innovations listed is an evolutionary trend shared by the classes beyond the branch point.

The tunicates and lancelets are nonvertebrate chordates, meaning they do not have vertebrae. The vertebrates include the fishes, amphibians, reptiles, birds, and mammals. The cartilaginous fishes were the first to have jaws, and some early bony fishes had lungs. However, not all bony fishes have lungs. Amphibians were the first group to clearly have jointed appendages and to invade land. But, the fleshy appendages of lobe-finned fishes from the Devonian era contained bones homologous to those of terrestrial vertebrates. These fishes are believed to be ancestral to the amphibians. Reptiles and mammals have a means of reproduction suitable for land. During development, an amnion and other extraembryonic membranes are present. The amnion and extraembryonic membranes support the embryo and prevent it from drying out as it develops into a particular species' offspring.

Nonvertebrate Chordates

Nonvertebrate chordates are characterized by the fact that the notochord never becomes a vertebral column.

Lancelets (subphylum Cephalochordata, about 23 species) all belong to one genus, *Branchiostoma* (formally *Amphioxus*).

Figure 32.3 Lancelet, *Branchiostoma.* Lancelets are filter feeders. Water enters the mouth and exits at the atriopore after passing through the gill slits. (photo): ©Heather Angel/Natural Visions

Lancets are marine chordates only a few centimeters long. Lancets are named for their resemblance to a lancet—a small, two-edged surgical knife (Fig. 32.3). Lancelets are found in shallow water along most coasts, where they usually lie partly buried in sandy or muddy substrates with only their anterior mouth and gill apparatus exposed. They feed on microscopic particles filtered out of a constant stream of water that enters the mouth and exits through the gill slits.

Lancelets retain the four chordate characteristics as adults. In addition, segmentation is present, as evidenced by the segmental arrangement of the muscles and the periodic branches of the dorsal tubular nerve cord.

Tunicates (subphylum Urochordata, about 1,250 species) live on the ocean floor and take their name from a tunic that makes the adults look like thick-walled, squat sacs (Fig. 32.4). They are also called sea squirts because they squirt water from one of their siphons when disturbed. The tunicate larva is bilaterally symmetrical and has the four chordate characteristics. Metamorphosis produces the sessile adult, which has incurrent and excurrent siphons.

The pharynx is lined by numerous cilia. Beating of these cilia creates a current of water that moves into the pharynx and out the numerous gill slits, the only chordate characteristic that remains in the adult. Microscopic particles adhere to a mucous secretion and are digested.

Some biologists have suggested that tunicates are directly related to vertebrates. They hypothesize that a larva with the four chordate characteristics may have become sexually mature, and evolution thereafter produced a fishlike vertebrate.

Figure 32.4 Sea squirt, *Clavelina.* Note that the only chordate characteristic remaining in the adult is gill slits. (photo): ©Diane R. Nelson

Check Your Progress 32.1

1. Identify the four features that distinguish chordates from other animal phyla.
2. Identify the morphological differences between a tunicate and a lancelet.
3. Describe three innovations that distinguish mammals from cartilaginous fishes (see Fig. 32.2).

32.2 Vertebrates: Fish and Amphibians

Learning Outcomes

Upon completion of this section, you should be able to

1. Compare and contrast the characteristics of a jawless fish to those of a jawed fish.

2. Describe the major evolutionary innovations that distinguish the fishes from the amphibians.

3. Identify the features of the three groups of living amphibians.

At some time in their life history, all **vertebrates** (subphylum Vertebrata, about 58,000 species) have the characteristics of a chordate. The embryonic notochord, however, is generally replaced by a vertebral column composed of individual vertebrae. The vertebral column, which is part of the flexible but strong jointed endoskeleton, provides evidence that vertebrates are segmented. The skeleton protects the internal organs and serves as a place of attachment for muscles. Together, the skeleton and muscles form a system that permits rapid, and more efficient, movement. Two pairs of appendages are typical. The pectoral and pelvic fins of fishes evolved into the jointed appendages that allowed vertebrates to move onto land.

The main axis of the endoskeleton consists of not only the vertebral column, but also a skull that encloses and protects the brain. The high degree of cephalization is accompanied by complex sense organs. The eyes develop as outgrowths of the brain. The ears are primarily equilibrium devices in aquatic vertebrates, but they also function as sound-wave receivers in land vertebrates.

The evolution of jaws in vertebrates has allowed them to diversify into a variety of biological roles. Vertebrates have a complete digestive tract and a large coelom. Their circulatory system is closed, meaning that the blood is contained entirely within blood vessels. Vertebrates have an efficient means of obtaining oxygen from water or air, as appropriate. The kidneys are important excretory and water-regulating organs. The sexes are generally separate, and reproduction is usually sexual. The evolution of the amnion allowed reproduction to take place on land. Many species of reptiles, and a few species of mammals, lay a shelled egg. In placental mammals, development takes place within the uterus of the female. Although the use of vertebrates, as well as other animals, in laboratory research has been very beneficial in many ways, it also generates controversy for some. The Health feature "Vertebrates and Human Medicine" discusses some ways that products from vertebrates are used in medicine.

A strong, jointed endoskeleton, a vertebral column composed of vertebrae, a closed circulatory system, efficient respiration and excretion, and a high degree of cephalization are characteristics demonstrating vertebrates are adapted to an active lifestyle. Figure 32.5 shows major milestones in the history of vertebrates: the evolution of jaws, limbs, and the amnion, an extraembryonic membrane that is first seen in the shelled **amniotic egg** of reptiles.

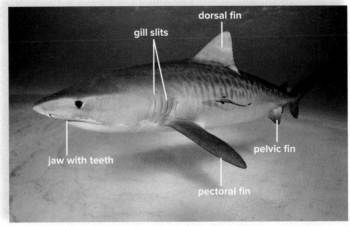

a. Tiger shark, *Galeocerdo cuvier*

b. Blue-spotted salamander, *Ambystoma*

c. Rhinoceros iguanas, *Cyclura*

Figure 32.5 Milestones in vertebrate evolution. a. The evolution of jaws in fishes allows animals to be predators and to feed off other animals. **b.** The evolution of limbs in most amphibians is adaptive for locomotion on land. **c.** The evolution of an amnion and a shelled egg in reptiles is adaptive for reproduction on land. (a): ©Alastair Pollock Photography/ Getty Images; (b): ©Marvin Dembinsky Photo Associates/Alamy; (c): ©DEA Picture Library/Getty Images

Vertebrates and Human Medicine

Hundreds of pharmaceutical products come from other vertebrates, and even those that produce poisons and toxins provide medicines that benefit us.

Natural Products with Medical Applications

The black-and-white spitting cobra of Southeast Asia paralyzes its victims with a potent venom, which eventually leads to respiratory arrest. However, that venom is also the source of the drug Immunokine, which can inhibit some harmful effects of an overactive immune system. It is approved in Thailand for use in combating the side effects of cancer therapy, and it is being studied for use in treating AIDS, autoimmune diseases, and other disorders.

Although snakebites can be very painful, certain components found in venom actually relieve pain (Fig. 32A*a*). The black mamba, found mainly in sub-Saharan Africa, is one of the most lethal snakes on Earth. Compounds in its venom called mambalgins, however, block pain signals by inhibiting the flow of certain ions through nerves that carry pain messages. When tested in mice, these compounds were as effective as morphine, with fewer side effects.

Another compound, known as epibatidine, derived from the skin of an endangered Ecuadorian poison-dart frog, is 50–200 times more powerful than morphine in relieving chronic and acute pain, without the addictive properties. Unfortunately, it can also have serious side effects, so companies have synthesized compounds with a similar structure, hoping to improve its safety profile.

Other venoms mainly affect blood clotting. Eptifibatide is derived from the venom of the pigmy rattlesnake, which lives in the southeastern United States. Because it binds to blood platelets and reduces their tendency to clump together, this drug is used to reduce the risk of clot formation in patients at risk for heart attacks. Alternatively, the venom of several pit vipers, such as the copperhead, contains "clot-busting" (thrombolytic) substances, which can be used to dissolve abnormal clots that have already formed.

Sharks produce a variety of chemicals with potentially medicinal properties. Squalamine is a steroidlike molecule that was first isolated from the liver of dogfish sharks. It has broad antimicrobial properties, and it can inhibit the abnormal growth of new blood vessels, which is a factor in cancer and a variety of other diseases. Squalamine is also safe enough to be used in the eye and is currently being tested as a potential treatment for macular degeneration, an eye disease that will affect 3 million Americans by 2020.

Animal Pharming

Some of the most powerful applications of genetic engineering can be found in the development of drugs and therapies for human diseases. In fact, this technology has led to a new industry: animal pharming, which uses genetically altered vertebrates, such as mice, sheep, goats, cows, pigs, and chickens, to produce medically useful pharmaceutical products.

To accomplish this, the human gene for some useful product is inserted into the embryo of a vertebrate. That embryo is implanted into a foster mother, which gives birth to the transgenic animal, which contains genes from the two sources. An adult transgenic vertebrate produces large quantities of the pharmed product in its blood, eggs, or milk, from which the product can be easily harvested and purified.

An example of a pharmed product used in human medicine is human antithrombin. This medication is important in the treatment of individuals who have a hereditary deficiency of this protein and so are at high risk for life-threatening blood clots. Approved by the FDA in 2009, the bioengineered drug, known by the brand name ATryn, is purified from the milk of transgenic goats (Fig. 32A*b*).

Xenotransplantation

There is an alarming shortage of human donor organs to fill the need for hearts, kidneys, and livers. One solution is xenotransplantation, the transplantation of nonhuman vertebrate tissues and organs into humans. The first such transplant occurred in 1984 when a team of surgeons implanted a baboon heart into an infant, who, unfortunately, lived only 20 days.

Although apes are more closely related to humans, pigs are considered to be the best source for xenotransplants. Pig organs are similar to human organs in size, anatomy, and physiology, and large numbers of pigs can be produced quickly. Most infectious microbes of pigs are unlikely to infect a human recipient. Currently, pig heart valves and skin are routinely used for treatment of humans (Fig. 32A*c*). Miniature pigs, whose heart size is similar to that of humans, are being genetically engineered to make their tissues less foreign to the human immune system, to minimize rejection.

Questions to Consider

1. Is it ethical to change the genetic makeup of vertebrates in order to use them as drug or organ factories?
2. What are some of the health concerns that may arise due to xenotransplantation?

a. Snake

b. Goats

c. Pig heart

Figure 32A Medical uses of animals.   a. Snake venom may be used to create pain medications. **b.** Mammals, such as these goats, may express pharmaceutical compounds in their milk. **c.** Pigs are now being genetically altered to provide a supply of hearts for heart transplant operations. (a): ©Marie Holding/iStock/Getty Images Plus; (b): ©tomophotography/Moment/Getty Images; (c): ©Victoria Dolidovich/Hemera/Getty Images Plus

Fishes: Evolution of the Jaw

The first vertebrates were jawless fishes that wiggled through the water and sucked up food from the ocean floor. Today, there are three living groups of fishes: jawless fishes, cartilaginous fishes, and bony fishes. The two latter groups have *jaws,* tooth-bearing structures in the head. Jaws evolved from the first pair of gill arches, structures that ordinarily support gills (Fig. 32.6). The second pair of gill arches became support structures for the jaws, instead of gills. The presence of jaws permits a predatory way of life in many species of fishes.

Fishes are adapted to life in the water. Usually, they shed their sperm and eggs into the water, where fertilization occurs. The zygote develops into a swimming larva, which must fend for itself until it develops into the adult form.

Jawless Fishes

Living representatives of the **jawless fishes** (about 65 species) are cylindrical and up to a meter long. They have smooth, scale-less skin and no jaws or paired fins. The two groups of living jawless fishes are *hagfishes* (class Myxini) and *lampreys* (class Cephalaspidomorphi). Although both are jawless, there are distinct differences. Hagfishes are an ancient group of fish. They possess a skull, but lack the vertebrae found in the other classes of vertebrates. However, molecular evidence suggests that these were once present in these fish, so they are traditionally classified with the vertebrates. Hagfishes are scavengers, feeding mainly on dead fishes.

Lampreys possess a true vertebral column. Most are parasites that use their round mouth as a sucker to attach themselves to another fish and tap into its circulatory system. Unlike other fishes, the lamprey cannot take in water through its mouth. Instead, water moves in and out through the gill openings.

Cartilaginous Fishes

Cartilaginous fishes (class Chondrichthyes, about 750 species) are the sharks (see Fig. 32.5*a*), the rays (Fig. 32.7*a*), and the skates. This group of fishes is so named because they have skeletons of cartilage instead of bone. The small dogfish shark is often dissected

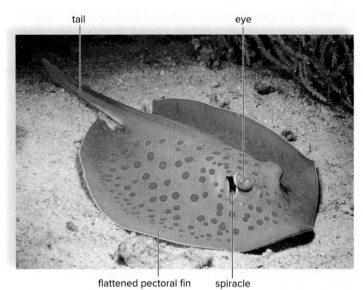

a. Blue-spotted stingray, *Taeniura lymma*

b. A ray-finned fish

c. Coelacanth, *Latimeria chalumnae*

Figure 32.7 **Jawed fishes.** **a.** Rays, such as this blue-spotted stingray, are cartilaginous fishes. **b.** A soldierfish has the typical appearance and anatomy of a ray-finned fish, the most common type of bony fish. **c.** A coelacanth is a lobe-finned fish once thought to be extinct. (a): ©Comstock Images/PictureQuest; (c): ©Peter Scoones/SPL/Science Source

Figure 32.6 **Evolution of the jaw.** The first jaw evolved from the first and second gill arches of fishes.

in biology laboratories. One of the most dangerous sharks, and one that is capable of living in both fresh and salt water, is the bullshark, although, in general, people are rarely attacked by sharks. Each year, more people are injured or killed by dog attacks than shark attacks. The largest sharks, the whale sharks, feed on small fishes and marine invertebrates and do not attack humans. Skates and rays are rather flat fishes that live partly buried in the sand and feed on mussels, clams, and various crustaceans.

Three well-developed senses enable sharks and rays to detect their prey: (1) They have the ability to sense electric currents in water—even those generated by the muscle movements of animals. (2) They have a lateral line system, a series of pressure-sensitive cells along both sides of the body that can sense pressure caused by nearby movement in the water. (3) They have a keen sense of smell. The part of the brain associated with this sense is enlarged relative to the other parts. Sharks can detect some scents from about a third of a mile away in the open ocean.

Bony Fishes

Bony fishes (about 30,000 species) are by far the most numerous and diverse of all the vertebrates.

Most of the bony fishes we eat, such as cod, trout, salmon, and haddock, are **ray-finned fishes** (class Actinopterygii). Their paired fins, which they use to balance and propel the body, are thin and supported by bony rays. Ray-finned fishes have various ways of life. Some, such as herring, are filter feeders; others, such as trout, are opportunists; and still others are predaceous carnivores, such as piranhas and barracudas.

Ray-finned fishes have a swim bladder, which usually regulates buoyancy (Fig. 32.7b). By secreting gases into or absorbing gases from the bladder, these fishes can change their density, allowing them to go up or down in the water. The streamlined shape, fins, and muscle action of ray-finned fishes are all suited for locomotion in the water. Their skin is covered by bony scales that protect the body but do not prevent water loss. The gills of ray-finned fish are kept continuously moist by the passage of water through the mouth and out the gill slits during respiration. As the water passes over the gills, oxygen is absorbed and carbon dioxide is given off. Ray-finned fishes have a single-circuit circulatory system. The heart is a simple pump, and the blood flows through the chambers, including a nondivided atrium and ventricle, to the gills. Oxygenated blood leaves the gills and is circulated throughout the body by the heart (Fig. 32.8a).

Another type of bony fish, called the **lobe-finned fishes** (class Sarcopterygii), are the ancestors of the amphibians. These fishes not only had fleshy appendages that could be adapted to land locomotion, but most also had a **lung** that was used for respiration. One type of lobe-finned fish, the coelacanth, was thought to have gone extinct about 20,000 years ago. However, some coelacanths have been discovered off the coasts of Eastern Africa and Indonesia, making them the only "living fossil" among these fishes (see Fig. 32.7c).

Amphibians: Transition to Land

Amphibians (class Amphibia, about 6,000 species), whose class name means living on both land and in the water, are represented today by frogs, toads, newts, and salamanders. Most members of this group lead a lifestyle that includes a larval stage that lives in the water, and an adult stage that lives on land. Caecilians are also classified as amphibians even though they are fossorial, wormlike amphibians that spend most of their life underground.

Amphibians evolved from the lobe-finned fishes with lungs by way of transitional forms. Two hypotheses have been suggested to account for the evolution of amphibians from lobe-finned fishes. Perhaps lobe-finned fishes had an advantage over others because they could use their lobed fins to move from pond to pond. Or perhaps the supply of food on land in the form of plants and insects—and the absence of predators—promoted further adaptations to the land environment. Paleontologists have recently found a well-preserved transitional fossil from the late Devonian period in Arctic Canada that represents an intermediate between lobe-finned fishes and tetrapods with limbs. This fossil, named *Tiktaalik roseae,* provides unique insights into how the legs of tetrapods arose (Fig. 32.9).

Aside from jointed appendages (which have been lost in the limbless caecilians), amphibians have other features not seen in bony fishes: They usually have four limbs, eyelids for keeping their eyes moist, ears (a tympanum) for picking up sound waves, and a voice-producing larynx. Relative to body size, the brain is larger than that of a fish. Adult amphibians usually have small lungs, but some species respire entirely through their skin. Air enters the mouth by way of nostrils, and when the floor of the mouth is raised,

Figure 32.8 Vertebrate circulatory pathways. a. The single-loop pathway of fishes has a two-chambered heart. **b.** The double-loop pathway of other vertebrates sends blood to the lungs and to the body. In amphibians and most reptiles, limited mixing of oxygen-rich and oxygen-poor blood takes place in the single ventricle of their three-chambered heart. **c.** The four-chambered heart of some reptiles (crocodilians and birds) and mammals sends only oxygen-poor blood to the lungs and oxygen-rich blood to the body.

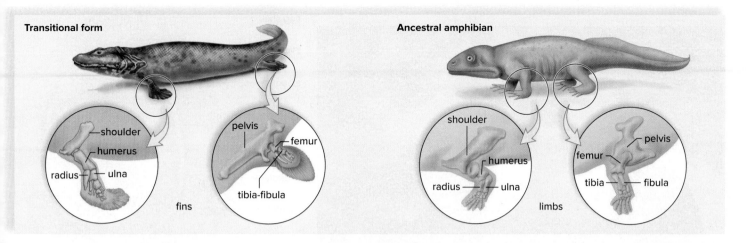

Figure 32.9 Lobe-finned fishes to amphibians. This transitional form links the lobes of lobe-finned fishes to the limbs of ancestral amphibians. Compare the fins of the (**a**) transitional form to the limbs of the (**b**) ancestral amphibian.

air is forced into the relatively small lungs. Respiration is supplemented by gas exchange through the smooth, moist, and glandular skin. The amphibian heart has a divided atrium but only a single ventricle. The right atrium receives deoxygenated blood from the body, and the left atrium receives oxygenated blood from the lungs. These two types of blood are partially mixed in the single ventricle (see Fig. 32.8*b*). Mixed blood is then sent to all parts of the body. Some is sent to the skin, where it is further oxygenated.

Currently, there is widespread concern about the future of amphibians because over 50% of all amphibian species are threatened with extinction. Because of their permeable skin and a life cycle that often depends on both water and land, they are thought to be "indicator species" of environmental quality. That is, they are the first to respond to environmental degradation, and disappearances of populations and species thus generate concern for other species—even humans.

Check Your Progress 32.2

1. Recognize characteristics that distinguish fish from other vertebrates.
2. List the features that distinguish amphibians from fish.
3. Describe the features that allow most amphibians to have life stages in water and on land.

32.3 Vertebrates: Reptiles and Mammals

Learning Outcomes

Upon completion of this section, you should be able to

1. Compare the major evolutionary innovations that distinguish the reptiles and mammals.
2. List the features that distinguish birds from the rest of the reptile lineage.
3. Identify the unique features that define the three living lineages of mammals.

Reptiles: Amniotic Egg

Reptiles (class Reptilia, about 17,000 species, including birds) diversified and were most abundant between 245 and 65 million years ago. These reptiles included the mammal-like reptiles (now extinct), the ancestors of today's living reptiles, and the dinosaurs (now extinct). Some dinosaurs are remembered for their great size. *Brachiosaurus,* an herbivore, was about 23 m (75 ft) long and about 17 m (56 ft) tall. *Tyrannosaurus rex* was 5 m (16 ft) tall when standing on its hind legs. The bipedal stance of some reptiles preceded the evolution of wings in birds. In fact, molecular and morphological evidence suggests that birds are the descendants of dinosaurs.

The reptiles living today are mainly turtles, alligators, snakes, lizards, and birds. The reptile body is covered with hard, keratinized scales, which protect the animal from desiccation and predators. Another adaptation for a land existence is the manner in which snakes typically use their tongue as a sense organ. Reptiles have well-developed lungs enclosed by a protective rib cage. When the rib cage expands, a partial vacuum establishes a negative pressure, which causes air to rush into the lungs. The atrium of the heart is always separated into right and left chambers, but division of the ventricle varies. An interventricular septum is incomplete in certain species. Therefore, some oxygenated and deoxygenated blood is mixed between the ventricles (see Fig. 32.8*c*).

Perhaps the most outstanding evolutionary innovation that first appeared in reptiles is their means of reproduction that is suitable to a land existence. The penis of the male passes sperm directly to the female. Fertilization is internal, and the female typically possesses leathery, flexible, shelled eggs. Some snakes lay eggs, while other snakes actually give birth to live young. This *amniotic egg* made development on land possible and eliminated the need for a swimming-larval stage during development. The egg provides the developing embryo with atmospheric oxygen, food, and water; it removes nitrogenous wastes; and it protects the embryo from drying out and from mechanical injury. This is accomplished by the presence of extraembryonic membranes such as the chorion (Fig. 32.10).

a. Crocodile, *Crocodylus niloticus*

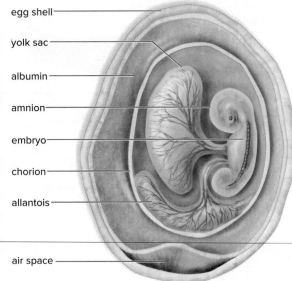

egg shell
yolk sac
albumin
amnion
embryo
chorion
allantois

air space

b.

Figure 32.10 The reptilian egg allows reproduction on land. **a.** Baby American crocodile, *Crocodylus,* hatching out of its shell. Note that the shell is leathery and flexible, not brittle like birds' eggs. **b.** Inside the egg, the embryo is surrounded by extraembryonic membranes. The chorion aids gas exchange, the yolk sac provides nutrients, the allantois stores waste, and the amnion encloses a fluid that prevents drying out and provides protection. (a): ©Heinrich van den Berg/Gallo Images/Getty Images

Fishes, amphibians, and living reptiles other than birds are **ectothermic,** meaning that their body temperature matches the temperature of the external environment. If it is cold externally, they are cold internally; if it is hot externally, they are hot internally. Reptiles try to regulate their body temperatures by basking in the sun if they need warmth or by hiding in the shadows if they need cooling off.

Feathered Reptiles

To many people, the **birds** (class Aves) are the most conspicuous, melodic, beautiful, and fascinating group of vertebrates. Over 9,000 species of birds have been recognized, and they range in size from the tiny "bee" hummingbird at 1.8 g to the ostrich at a maximum weight of 160 kg and a height of 2.7 m. The fossil record of birds, such as *Archaeopteryx* and newly described fossils from Mongolia and Spain, provides evidence that birds evolved from reptiles. Combined with molecular studies, birds are now considered part of the Reptilia. Their legs have scales and their feathers are modified reptilian scales. However, birds typically lay a hard-shelled amniotic egg, rather than the leathery egg of reptiles.

Anatomy and Physiology of Birds

Nearly every anatomical feature of a bird can be related to its ability to fly (Fig. 32.11*a*). The forelimbs are modified as wings. The hollow, very light bones are laced with air cavities. A horny beak has replaced jaws equipped with teeth, and a slender neck connects the head to a rounded, compact torso. The sternum is enlarged and has a keel, to which strong muscles are attached for flying. Respiration is efficient because the lobular lungs form anterior and posterior air sacs. The presence of these sacs means that the air

circulates one way through the lungs and that gases are continuously exchanged across respiratory tissues (Fig. 32.11*b*). Another benefit of air sacs is that they lighten the body for more efficient flying.

Flight requires acute sense organs and an intricate nervous system. Birds have particularly good vision and well-developed brains. Their muscle reflexes are excellent. An enlarged portion of the brain seems to be the area responsible for instinctive behavior. A ritualized courtship often precedes mating. Many newly hatched birds require parental care before they are able to fly away and seek food for themselves. A remarkable aspect of bird behavior is the seasonal migration of many species over very long distances. Birds navigate by day and night, whether it's sunny or cloudy, by using the sun and stars and even Earth's magnetic field to guide them.

Diversity of Birds

The majority of birds, including eagles, geese, and mockingbirds, have the ability to fly. However, some birds, such as emus, penguins, and ostriches, are flightless. Traditionally, birds have been classified based on beak and foot type (Fig. 32.12) and, to some extent, on their habitat and behavior. The various orders include birds of prey with notched beaks and sharp talons; shorebirds with long, slender, probing beaks and long, stiltlike legs; woodpeckers with sharp, chisel-like beaks and grasping feet; waterfowl with broad beaks and webbed toes; penguins with wings modified as paddles; and songbirds with perching feet.

Mammals: Hair and Mammary Glands

Mammals (class Mammalia, about 4,600 species) also evolved from the reptiles. The first mammals were small, about the size

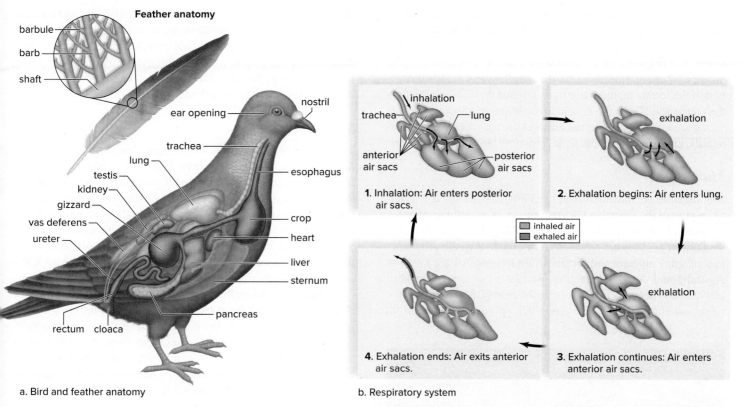

Feather anatomy

barbule
barb
shaft

ear opening
nostril
trachea
lung
testis
kidney
gizzard
vas deferens
ureter
rectum   cloaca
esophagus
crop
heart
liver
sternum
pancreas

a. Bird and feather anatomy

inhalation
trachea
lung
anterior
air sacs
posterior
air sacs

1. Inhalation: Air enters posterior air sacs.

exhalation

2. Exhalation begins: Air enters lung.

inhaled air
exhaled air

exhalation

4. Exhalation ends: Air exits anterior air sacs.

3. Exhalation continues: Air enters anterior air sacs.

b. Respiratory system

Figure 32.11 Bird anatomy and physiology. **a.** The anatomy of a pigeon is representative of bird anatomy. **b.** In birds, air passes one way through the lungs.

a. Cardinal, *Cardinalis cardinalis*

b. Bald eagle, *Haliaetus leucocephalus*

c. Flamingo, *Phoenicopterus ruber*

Figure 32.12 Bird beaks. **a.** A cardinal's beak allows it to crack tough seeds. **b.** A bald eagle's beak allows it to tear prey apart. **c.** A flamingo's beak strains food from the water with bristles that fringe the mandibles. (a): ©Gary W. Carter/Corbis Documentary/Getty Images; (b): ©Dale DeGabriele/Getty Images; (c): ©Medford Taylor/Getty Images

of mice. During the dinosaurs' reign (165–65 MYA), mammals remained small in size and changed little evolutionarily. Some of the earliest mammalian groups are still represented today by monotremes and marsupials. The placental mammals that evolved later went on to live in many habitats, including air, land, and sea.

The chief characteristics of mammals are body hair and milk-producing mammary glands. Almost all mammals are endothermic and maintain a constant internal temperature. Many of the adaptations of mammals are related to temperature control. *Hair,* for example, provides insulation against heat loss and allows mammals to be active, even in cold weather. Like birds, mammals have efficient respiratory and circulatory systems, which ensure a ready supply of oxygen to muscles whose contraction produces body heat. Also, like birds, mammals have double-loop circulation and a four-chambered heart (see Fig. 32.8*c*).

Mammary glands enable females to feed (nurse) their young a nutrient-rich source of food. Nursing also creates a bond between mother and offspring that helps ensure parental care while the young are helpless. In all mammals (except monotremes), the young are born alive after a period of development in the uterus, a part of the female reproductive tract. Internal development shelters the young and allows the female to move actively about while the young are maturing. Mammals are classified as monotremes, marsupials, or placental mammals, according to how they reproduce.

Monotremes

Monotremes are mammals that, like birds, have a *cloaca*, a terminal region of the digestive tract that serves as a common chamber for feces, excretory wastes, and sex cells. They also lay hard-shelled amniotic eggs. They are represented by only five species: four species of spiny anteater and the duckbill platypus. All are found in Australia and New Guinea (Fig. 32.13*a*). The female duckbill platypus lays her eggs in a burrow in the ground. She incubates the eggs, and after hatching, the young lick up milk that seeps from

a. Duckbill platypus, *Ornithorhynchus anatinus*

b. Koala,
   Phascolarctos cinereus

c. Virginia opossum,
   Didelphis virginianus

Figure 32.13 Monotremes and marsupials. a. The duckbill platypus is a monotreme that inhabits Australian streams. **b.** The koala is an Australian marsupial that lives in trees. **c.** The opossum is the only marsupial in the Americas. The Virginia opossum is found in a variety of habitats. (a): ©Worldslifewonders/Shutterstock; (b): ©John White Photos/Getty Images; (c): ©John Macgregor/Photolibrary/Getty Images

modified sweat glands on the abdomens of both males and females. The spiny anteater has a pouch on the belly side formed by swollen mammary glands and longitudinal muscle. The egg moves from the cloaca to this pouch, where hatching takes place and the young remain for about 53 days. They then stay in a burrow, where the mother periodically visits and nurses them.

Marsupials

The young of **marsupials** begin their development inside the female's body, but they are born in a very immature condition. Newborns crawl up into a pouch on their mother's abdomen. Inside the pouch, they attach to the nipples of mammary glands and continue to develop. Frequently, more are born than can be accommodated by the number of nipples, and it's "first come, first served."

Marsupial mammals are found mainly in Australia, where they underwent adaptive radiation for several million years without competition from placental mammals. Among the herbivorous marsupials in Australia, koalas are tree-climbing browsers (Fig. 32.13*b*), and kangaroos are grazers. The Virginia opossum (*Didelphis virginiana*) is the only marsupial that occurs north of Mexico (Fig. 32.13*c*). Other species of marsupials live in Central and South America. The Tasmanian devil, a carnivorous marsupial about the size of a small to medium-sized dog, is now threatened with extinction due to a transmissible tumor disease.

Placental Mammals

The vast majority of living mammals are **placental mammals** (Fig. 32.14). In these mammals, the extraembryonic membranes of the reptilian egg (see Fig. 32.10) have been modified for internal development within the uterus of the female. The chorion contributes to the fetal portion of the placenta, while a part of the uterine wall contributes to the maternal portion. Here, nutrients, oxygen, and waste are exchanged between fetal and maternal blood.

Mammals, with the exception of marine mammals, are adapted to life on land and have limbs that allow them to move rapidly. In fact, an evaluation of mammalian features leads us to the conclusion that they lead active lives. The lungs are expanded not only by the action of the rib cage, but also by the contraction of the diaphragm, a horizontal muscle that separates the thoracic cavity from the abdominal cavity. The heart has four chambers. The internal temperature is constant, and hair, when present, helps insulate the body.

The mammalian brain is well developed and enlarged due to the expansion of the cerebral hemispheres, which control the rest of the brain. The brain is not fully developed until after birth, and the young learn to take care of themselves during a period of dependency on their parents.

Mammals have differentiated teeth. Typically, in the front, the incisors and canine teeth have cutting edges for capturing and killing prey. On the sides, the premolars and molars chew food. The specific shape and size of the teeth may be associated with whether the mammal is an herbivore (eats vegetation), a carnivore (eats meat), or an omnivore (eats both meat and vegetation). For example, mice (order Rodentia) have continuously growing incisors; horses have large, grinding molars; and dogs (order Carnivora) have long canine teeth. Placental mammals are classified based on their methods of obtaining food and their mode of locomotion. For

a. White-tailed deer, *Odocoileus virginianus*

b. African lioness, *Panthera leo*

Figure 32.14
Placental mammals.
Placental mammals have adapted to various ways of life. **a.** Deer are herbivores that live in forests. **b.** Lions are carnivores on the African plain. **c.** Monkeys inhabit tropical forests. **d.** Whales are sea-dwelling placental mammals. (a): ©Paul E. Tessier/Getty Images; (b): ©John Downer/The Image Bank/Getty Images; (c): ©Paul Souders/Corbis/Getty Images; (d): ©2011 Tory Kallman/Getty Images

c. White-faced monkey, *Cebus capucinus*

d. Killer whale, *Orcinus orca*

example, bats (order Chiroptera) have membranous wings supported by digits; horses (order Perissodactyla) have long, hoofed legs; and whales (order Cetacea) have paddlelike forelimbs.

Primates

Primata are members of the order **Primates.** In contrast to the other orders of placental mammals, most primates are adapted to an arboreal life—that is, for living in trees. Primate limbs are mobile, and the hands and feet have five digits each. Many primates have a big toe and a thumb that are both opposable, meaning the big toe or thumb can touch each of the other toes or fingers. Humans don't have an opposable big toe, but the thumb is opposable, and this results in a powerful and precise grip. The opposable thumb allows a primate to easily reach out and bring food to the mouth. When locomoting, primates grasp and release tree limbs freely because they have nails instead of the claws of their evolutionary ancestors.

In primates, the snout is shortened considerably, allowing the eyes to move to the front of the head. The stereoscopic vision (or depth perception) that results permits primates to make accurate judgments about the distance and position of adjoining tree limbs. Some primates that are active during the day, such as humans, have color vision and strong visual acuity because the retina contains cone cells in addition to rod cells. Cone cells require bright light, but the image is sharp and in color. The lens of the eye focuses light directly on the fovea, a region of the retina where cone cells are concentrated.

The evolutionary trend among primates is toward a larger and more complex brain. The brain size is smallest in prosimians and

largest in modern humans. Increases in size of the cerebral cortex, the brain region associated with learning, memory, thought, and awareness, is primarily responsible for the trend toward increased brain size. The portion of the brain devoted to smell gets smaller, and the portions devoted to sight increase in size and complexity during primate evolution. Also, more and more of the brain is involved in controlling and processing information received from the hands and the thumb. The result is good hand-eye coordination in humans.

It is difficult to care for several offspring while moving from limb to limb, so one offspring per birth interval tends to be the norm in most primates. The juvenile period of dependency on parental care is extended, and there is an emphasis on learned behavior and complex social interactions.

The order Primates has two suborders: the **strepsirhini** (lemurs, aye-ayes, bush babies, and lorises) and the **haplorhini** (monkeys, apes, and humans). Thus, humans are more closely related to the monkeys and apes than they are to the strepsirhini.

Check Your Progress 32.3

1. List three features that distinguish reptiles from mammals.
2. Recognize the features that make placental mammals unique.
3. Identify the characteristics that separate primates from other mammals.

32.4 Evolution of the Hominins

The evolutionary tree in Figure 32.15 shows that all primates share a common ancestor and that over time they diverged, producing the various lines of descent, or **lineage,** we see today. Notice that prosimians (also called strepsirhini), represented by lemurs, aye-ayes, bush babies, and lorises, were the first types of primates to diverge from the primate line of descent, and African apes were the last group to diverge. Also note that the tree suggests that humans are most closely related to African apes. One of the most unfortunate misconceptions concerning human evolution is the belief that Darwin and others suggested that humans evolved from apes. On the contrary, humans and apes shared a common apelike ancestor. Today's apes are our distant cousins, and we couldn't have evolved from our cousins because we are contemporaries—living on Earth at the same time.

Molecular data have been used to estimate the date of the split between the human lineage and that of apes. When two lines of descent first diverge from a common ancestor, the genes of the two lineages are nearly identical. But as time goes by, each lineage accumulates genetic changes. Many genetic changes are neutral (not tied to adaptation) and accumulate at a fairly constant rate. Such changes can be used as a kind of **molecular clock** to infer how closely two groups are related and the point at which they diverged. Molecular data suggest that the split between the ape and human lineages occurred about 6 to 8 MYA. Currently, humans and chimpanzees are probably the most closely related, sharing more than 90% of our DNA.

Hominins (the designation that includes humans and species very closely related to humans) first evolved about 5 MYA. Molecular data show that hominins, chimpanzees, and gorillas are all closely related and these groups must have shared a common ancestor sometime during the Miocene. Hominins, chimpanzees, and gorillas are now grouped together as hominines. The hominids include the hominines and the orangutan. The hominoids include the gibbon and the hominids. The hominoid common ancestor first evolved at the beginning of the Miocene about 23 MYA.

The First Hominins

There have been many recent advances in the study of the hominins, and recent discoveries of fossils in Africa are challenging our view of how early hominins evolved. Paleontologists use certain anatomical features when they try to determine if a fossil is a hominin. These features include **bipedalism** (walking on two feet), the shape of the face, and brain size. Today's humans have a flatter face and a more pronounced chin than do the apes, because the human jaw is shorter than that of the apes. Then, too, our teeth are generally smaller and less specialized. We don't have the sharp canines of an ape, for example. Chimpanzees have a brain size of about 400 cubic centimeters (cc), and modern humans have a brain size of about 1,360 cc.

It's hard to decide which fossils are hominins, because human features evolved gradually and at different rates. Most investigators rely first and foremost on bipedalism as the hallmark of a hominin, regardless of the size of the brain.

There have been a number of recent fossil discoveries in Africa that have the potential to reshape our understanding of early hominin evolution, specifically the occurrence of the split between the ape and human lineages. One of the oldest of these fossils, called *Sahelanthropus tchadensis,* dated at 7 MYA, was found in Chad, in central Africa, far from eastern and southern Africa, where other hominid fossils were excavated. The only find, a skull, appears to be that of a hominin, because it has smaller canines and thicker tooth enamel than an ape. The braincase, however, is very apelike. It is impossible to tell if this hominin walked upright. Some suggest this fossil is ancestral to the gorilla.

Orrorin tugenensis, dated at 6 MYA and found in eastern Africa, is thought to be another early hominin, especially because the limb anatomy suggests a bipedal posture. However, the canine teeth are large and pointed, and the arm and finger bones retain adaptations for climbing. Some suggest this fossil is ancestral to the chimpanzee.

Two species of **ardipithecines** have been uncovered: *Ardipithecus kadabba* and *A. ramidus.* Only teeth and a few bone bits have been found for *A. kadabba,* and these have been dated to around 5.6 MYA. A more extensive collection of fossils has been collected for *A. ramidus.* To date, over 100 skeletons, all dated to 4.4 MYA, have been identified from this species; all were collected near a small town in Ethiopia, East Africa. These fossils have been reconstructed to form a female fossil specimen, affectionately called Ardi.

Some of Ardi's features are primitive, like that of an ape, but others are like that of a human. Ardi was about the size of a chimpanzee, standing about 120 cm (4 ft) tall and weighing about 55 kg (110 lb). It appears that males and females were about the same size.

Ardi had a small head compared to the size of her body. The skull had the same features as *Sahelanthropus tchadensis,* but was smaller. Ardi's brain size was around 300 to 350 cc, slightly less than that of a chimpanzee brain (around 400 cc), and much smaller than that of a modern human (1,360 cc). The muzzle (area of the nose and mouth) projects forward, and the forehead is low with heavy eyebrow ridges, a combination that makes the face more primitive than that of the australopithecines (discussed next). However, the projection of the face is less than that of a chimpanzee, because Ardi's teeth were small and like those of an omnivore. She lacked the strong, sharp canines of a chimpanzee, and her diet probably consisted mostly of soft, rather than tough, plant material.

Ardi could walk erect, but she spent a lot of time in trees. Ardi's feet had a bone, missing in apes, that kept her feet squarely on the ground, a sure sign that she was bipedal and not a quadruped like the apes. Nevertheless, like the apes, she had an opposable big toe. Opposable toes allow an animal's feet to grab hold of a tree limb. The wrists of Ardi's hands were flexible, and most likely she moved along tree limbs on all fours, as ancient apes did.

Until recently, it was suggested that bipedalism evolved when a dramatic change in climate caused the forests of East Africa to be replaced by grassland. However, evidence suggests that Ardi lived in the woods, which questions the advantage that walking erect would have afforded her. Bipedalism does provide an advantage in caring for a helpless infant by allowing it to be carried by hand

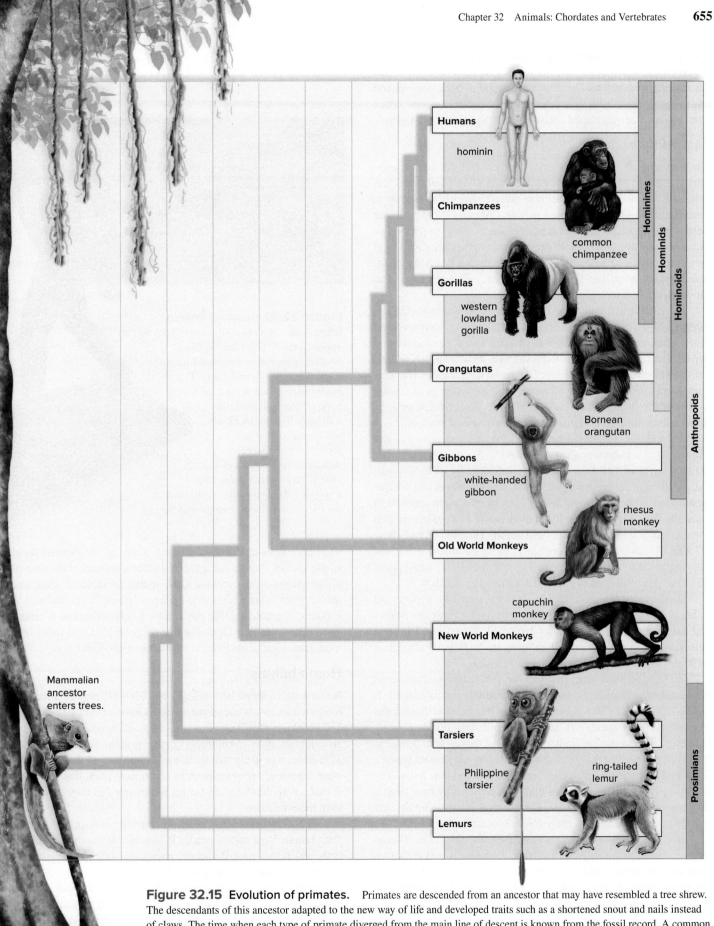

Figure 32.15 Evolution of primates.   Primates are descended from an ancestor that may have resembled a tree shrew. The descendants of this ancestor adapted to the new way of life and developed traits such as a shortened snout and nails instead of claws. The time when each type of primate diverged from the main line of descent is known from the fossil record. A common ancestor was living at each point of divergence—for example, there was a common ancestor for hominines about 7 MYA, for the hominoids about 15 MYA, and one for anthropoids about 45 MYA.

from one location to another. It is also possible that bipedalism benefited the males of the species as they foraged for food on the floor of the forests. More evidence is needed to better understand this mystery, but one thing is clear—the ardipithecines represent a link between our quadruped ancestors and the bipedal hominins.

Australopithecines

The hominin line of descent begins in earnest with the australopithecines, a group of species that evolved and diversified in Africa. Originally, some **australopithecines** were classified according to their frame. Some were termed gracile ("slender") types. Some were robust ("powerful") and tended to have strong upper bodies and especially massive jaws. Recent changes in the classification of these groups has separated the gracile types into genus *Australopithecus* and the robust types into the genus *Paranthropus*. The genus *Australopithecus* gave rise to genus *Homo*.

The first australopithecine to be discovered was unearthed in southern Africa by Raymond Dart in the 1920s. This hominin, named *Australopithecus africanus,* dates to about 2.9 MYA and had a brain size of about 500 cc. Limb anatomy suggests these hominids walked upright. However, the proportions of the limbs were apelike. The fore-limbs were longer than the hindlimbs. Some argue that *A. africanus,* with its relatively large brain, is a possible ancestral candidate for early *Homo,* whose limb proportions are similar to those of this fossil.

In the 1970s, a team led by Donald Johanson unearthed nearly 250 fossils of a hominin called *A. afarensis.* A now-famous female skeleton dated at 3.18 MYA is known worldwide by its field name, Lucy. Although her brain was small (400 cc), the shapes and relative proportions of her limbs indicate that Lucy stood upright and walked bipedally (Fig. 32.16*a*). Even better evidence of bipedal locomotion comes from a trail of footprints in Laetoli, Tanzania, dated about 3.7 MYA. The larger prints are double, as though a smaller being was stepping in the footfalls of another. There are additional small prints off to the side, within hand-holding distance (Fig. 32.16*b*).

The fact that the australopithecines were apelike with regard to brain size and walked erect like more modern humans shows that human characteristics did not evolve all at one time. The term *mosaic evolution* is applied when different body parts change at different rates and, therefore, at different times.

In 2008, the American anthropologist Lee Berger discovered the bones of a 2-million-year-old australopithecine he named *A. sediba* ("wellspring"). The small brain (500 cc) and long arms suggest that this species is an australopith that climbed trees. However, it also had a humanlike pelvis, nose, and dentition (teeth). A series of papers published in 2013 by the journal *Science* described *A. sediba's* mosaic of primitive and modern traits. The discovery has led to hypotheses that this species may have been a direct ancestor of the genus *Homo* or a lineage of austrolopiths that existed at the same time as early *Homo* species.

Australopithecus afarensis is most likely ancestral to the *Paranthropus* genus found in eastern Africa: *P. robustus* and *P. boisei.* *Paranthropus boisei* had a powerful upper body and the largest molars of any hominin. This genus died out; therefore, it is possible that *A. afarensis* is ancestral to both *A. africanus* and early *Homo.*

Evolution of Early *Homo*

The study of our genus, *Homo,* has undergone considerable scrutiny in the past several years. Increasingly, there is evidence that what

b.

Figure 32.16 *Australopithecus afarensis.* **a.** A reconstruction of Lucy on display at the St. Louis Zoo. **b.** These fossilized footprints occur in ash from a volcanic eruption some 3.7 MYA. The larger footprints are double, and a third, smaller individual was walking to the side. (A female holding the hand of a youngster may have been walking in the footprints of a male.) The footprints suggest that *A. afarensis* walked bipedally. (a): ©Dan Dreyfus and Associates; (b): ©John Reader/Science Source

a.

were once considered separate species are, in fact, variations of a single species. Notice in Figure 32.17 that a number of the members of the genus *Homo* include slash marks to indicate areas under debate. Fossils are assigned to the genus *Homo* if (1) the brain size is 600 cc or greater, (2) the jaw and teeth resemble those of humans, and (3) tool use is evident. For our purposes, we will focus on the more traditional classification of the members of our genus.

Homo habilis

Homo habilis, dated between 2.0 and 1.9 MYA, may be ancestral to modern humans. Some of these fossils have a brain size as large as 775 cc, about 45% larger than that of *A. afarensis.* The cheek teeth are smaller than even those of the gracile australopithecines. Therefore, it is likely that these early members of the genus *Homo* were omnivores who ate meat in addition to plant material. Bones at their campsites bear cut marks, indicating that they used tools to strip them of meat.

The stone tools made by *H. habilis,* whose name means "handyman," are rather crude. It is possible these are the cores from which they took flakes sharp enough to scrape away hide, cut tendons, and easily remove meat from bones.

Early *Homo* skulls suggest that the portions of the brain associated with speech were enlarged. We can speculate that the ability to speak may have led to hunting cooperatively. Other members of the group may have remained plant gatherers. If so, both hunters and gatherers most likely ate together and shared their food. In this way, society and culture could have begun.

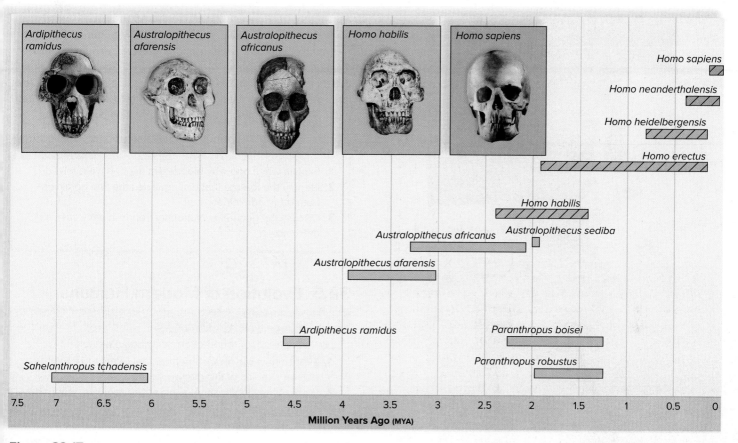

Figure 32.17 Human evolution.   Several groups of extinct hominins preceded the evolution of modern humans. The groups have been divided into the early humanlike hominins (orange), later humanlike hominins (green), early *Homo* species (lavender), and finally the later *Homo* species (blue). The slash marks or similar colors indicate areas where current research is focusing on combining groups into single species. (photos) (*A. ramidus*): ©Richard T. Nowitz/Science Source; (*A. afarensis*): ©Scott Camazine/Alamy; (*A. africanus*): ©Philippe Plailly/Science Source; (*H. habilis*): ©Kike Calvo VWPics/Superstock; (*H. sapiens*): ©Kenneth Garrett/Getty Images

Homo ergaster *and* Homo erectus

Homo erectus fossils are found in Africa, Asia, and Europe and have been dated to between 1.9 and 0.3 MYA. A Dutch anatomist named Eugene Dubois was the first to unearth *H. erectus* bones in Java in 1891. Since that time, many other fossils have been found in the same area. Although all fossils assigned the name *H. erectus* are similar in appearance, enough discrepancy exists to suggest that several different species have been included in this group. In particular, some experts suggest that the Asian form is *Homo erectus* and the African form is *Homo ergaster* (Fig. 32.18).

Compared with *H. habilis*, *H. erectus* had a larger brain (about 1,000 cc) and a flatter face. The nose projected, however. This type of nose is adaptive for a hot, dry climate, because it permits water to be removed before air leaves the body. The recovery of an almost complete skeleton of a 10-year-old boy indicates that this species was much taller than the hominids discussed thus far. Males were 1.8 m tall (about 6 ft), and females were 1.55 m (approaching 5 ft). Indeed, these hominids were erect and most likely had a striding gait like ours. The robust, and most likely heavily muscled, skeleton still retained some australopithecine features. Even so, the size of the birth canal indicates that infants were born in an immature state that required an extended period of care.

Homo erectus may have first appeared in Africa and then migrated into Asia and Europe (Fig. 32.19). At one time, the migration was thought to have occurred about 1 MYA. Recently,

H. erectus fossil remains in Java and the Republic of Georgia have been dated at 1.9 and 1.6 MYA, respectively. These remains push the evolution of *H. erectus* in Africa to an earlier date than has yet been determined. In any case, such an extensive population movement is a first in the history of humankind and a tribute to the intellectual and physical skills of the species.

Homo erectus was the first hominid to use fire, and fashioned more advanced tools than early *Homos*. These hominids used heavy, teardrop-shaped axes and cleavers. Flake tools were probably used for cutting and scraping. It could be that this species was a systematic hunter and brought kills to the same site over and over. In one location, researchers have found over 40,000 bones and 2,647 stones. These sites could have been "home bases," where social interaction occurred and a prolonged childhood allowed time for learning. Perhaps a language evolved and a culture more like our own developed.

Homo floresiensis   In 2004, scientists announced the discovery of the fossil remains of *Homo floresiensis*. The 18,000-year-old fossil of a 1-m tall, 25-kg adult female was discovered on the island of Flores in the South Pacific. This important finding suggests that a species of *Homo* coexisted with modern *Homo sapiens* much more recently than Neandertals, which went extinct about 28,000 years ago. The specimen was the size of a 3-year-old *Homo sapiens sapiens* (a subspecies of *Homo sapiens*) but possessed a braincase only one-third the size of that of a modern human.

Figure 32.18 *Homo ergaster.*   This skeleton of a 10-year-old boy who lived 1.6 MYA in eastern Africa shows femurs that are angled because the neck of the femur is quite long. ©Associated Press

Figure 32.19 **Migration out of Africa.**   The dates indicate the migration of early *Homo erectus* from Africa.[1]

[1]Derived from Anton, S., et al., "Evolution of Early *Homo:* An Integrated Biological Perspective," *Science*, vol. 345, July 4, 2014, 6192.

Apparently, *H. floresiensis* used tools and fire. A 2007 study supports the hypothesis that this diminutive hominin and her peers evolved from normal-sized, island-hopping *H. erectus* populations that reached Flores about 840,000 years ago. Some scientists think that its small size is due to island dwarfing, the reduction in size of large animals when their gene pool is limited to a small environment. This phenomenon is seen in other island populations.

> **Check Your Progress 32.4**
>
> 1. Explain when and why bipedalism might have evolved.
> 2. Identify the feature that distinguishes the first hominids from other hominines.
> 3. Discuss the general evolutionary trends in early species of *Homo*.

32.5 Evolution of Modern Humans

> **Learning Outcomes**
>
> Upon completion of this section, you should be able to
>
> 1. Explain how the replacement model relates to evolution of later species of the genus *Homo*.
> 2. Discuss the significance of increased tool use, language, and agriculture in Cro-Magnons.

Many early *Homo* species in Europe are now classified as *Homo heidelbergensis.* Just as *H. erectus* is believed to have evolved from *H. ergaster* in Asia, so *H. heidelbergensis* is believed to have evolved from *H. ergaster* in Europe. However, this is an active area of research, and new findings are constantly changing our understanding of human evolution during this time period. For the sake of discussion, we will group together the *H. ergaster* in Africa, *H. erectus* in Asia, and *H. heidelbergensis* in Europe as a collective group of early *Homo* species who lived between 1.5 and 0.25 MYA (see Fig. 32.17).

The most widely accepted hypothesis for the evolution of modern humans from earlier *Homo* species is referred to as the **replacement model,** or out-of-Africa hypothesis (Fig. 32.20).

The replacement model proposes that modern humans evolved from earlier *Homo* species only in Africa, and then modern humans migrated to Asia and Europe, where they replaced the existing species about 100,000 years ago. Recent fossil discoveries in northern Africa are challenging this hypothesis, but more research is needed before scientists can modify the model.

Neandertals

The **Neandertals,** which are classified as *Homo neandertalensis,* are a species that lived between 200,000 and 28,000 years ago. Neandertal fossils have been found from the Middle East throughout Europe. Neandertals take their name from Germany's Neander Valley, where one of the first Neandertal skeletons, estimated to be about 200,000 years old, was discovered.

According to the replacement model, Neandertals were eventually supplanted by modern humans. The exact mechanism of how this occurred is still being researched, with some studies suggesting that *Homo sapiens* outcompeted the Neandertals for resources. However, this view is being challenged by studies of the Neandertal

Figure 32.20 The replacement model.   According to the replacement model, modern humans evolved in Africa and then replaced early *Homo* species in Asia and Europe.

genome (completed in 2010), which suggests that not only did Neandertals interbreed with *Homo sapiens,* but between 1% and 4% of the genomes of non-African *Homo sapiens* contain remnants of the Neandertal genome. Some scientists are suggesting that Neandertals were not a separate species, but simply a race of *Homo sapiens* that was eventually absorbed into the larger population. Research continues into these and other hypotheses that explain these similarities.

Physiologically, the Neandertal brain was, on average, slightly larger than that of *H. sapiens* (1,400 cc, compared with 1,360 cc in most modern humans). The Neandertals were heavily muscled, especially in the shoulders and neck. The bones of the limbs were shorter and thicker than those of modern humans. It is hypothesized that a larger brain than that of modern humans was required to control the extra musculature. The Neandertals lived in Europe and Asia during the last Ice Age, and their sturdy build could have helped conserve heat.

The Neandertals give evidence of being culturally advanced. Most lived in caves, but those living in the open may have built houses. They manufactured a variety of stone tools, including spear points, which could have been used for hunting. Scrapers and knives could have helped in food preparation. They most likely successfully hunted bears, woolly mammoths, rhinoceroses, reindeer, and other contemporary animals. They used and could control fire, which probably helped them cook meat and keep themselves warm. They even buried their dead with flowers and tools and may have had a religion. Perhaps they believed in life after death. If so, they were capable of thinking symbolically.

Denisovans

In 2008, a fragment of a finger bone was discovered in Denisova Cave in southern Siberia. Initially, scientists thought that this

might be the remains of a species of early *Homo,* possibly related to *Homo erectus.* However, mitochondrial DNA studies indicated that the fossil belonged to a species that existed around 1 million years ago, around the same time as the Neandertals. The analyses suggest that the **Denisovans** and Neandertals shared a common ancestor but did not interbreed with one another, possibly because of their geographic locations. However, what is interesting is the fact that *Homo sapiens* in the Oceania region (New Guinea and nearby islands) share around 5% of their genomes with the Denisovans. In 2014, researchers reported that an allele that allows for high-elevation living in Tibetans originated with the Denisovans (see the Scientific Inquiry feature "Adapting to Life at High Elevations" in Section 1.1). When coupled with the Neandertal data, this suggests that modern *Homo sapiens* did not simply replace groups of early *Homo* species, but instead may have assimilated them by inbreeding. Scientists are just beginning to unravel the implications of these Denisovan discoveries.

Cro-Magnons

Cro-Magnons are the oldest fossils to be designated *Homo sapiens.* In keeping with the replacement model, the Cro-Magnons are named after a fossil location in France, where modern humans entered Asia and Europe from Africa approximately 100,000 years ago. They probably reached western Europe about 40,000 years ago. Cro-Magnons had a thoroughly modern appearance. They had lighter bones, flat high foreheads, domed skulls housing brains of 1,350 cc, small teeth, and a distinct chin. They made advanced stone tools, including compound tools, such as by fitting stone flakes to a wooden handle. They may have been the first to make knifelike blades and to throw spears, enabling them to kill animals from a distance. They were such accomplished hunters that some researchers hypothesize they may have been responsible for the extinction of many larger mammals, such as the giant sloth, the mammoth, the saber-toothed tiger, and the giant ox, during the late Pleistocene epoch. This event is known as the Pleistocene overkill.

Cro-Magnons hunted cooperatively, and perhaps were the first to have language. They are believed to have lived in small groups, with the men hunting by day while the women remained at home with the children. It's quite possible that this hunting way of life among prehistoric people influences our behavior even today. The Cro-Magnon culture included art. They sculpted small figurines out of reindeer bones and antlers. They also painted beautiful drawings of animals on cave walls in Spain and France.

Cro-Magnons have been widely distributed about the globe ever since they evolved. As with any other species that has a wide geographic distribution, phenotypic and genotypic variations are noticeable between these populations. The Scientific Inquiry feature "Human Ethnic Groups" explores some of these differences.

> **Check Your Progress 32.5**
>
> **1.** Discuss the importance of the replacement model in the evolution of the genus *Homo.*
> **2.** Compare and contrast Neandertals and Cro-Magnons.

SCIENCE IN YOUR LIFE ▶ SCIENTIFIC INQUIRY

Human Ethnic Groups

Evolutionists have hypothesized that human variations evolved as adaptations to local environmental conditions. One obvious difference among people is skin color. A darker skin is protective against the high UV intensity of bright sunlight. On the other hand, a white skin ensures vitamin D production in the skin when the UV intensity is low. Harvard University geneticist Richard Lewontin points out, however, that this hypothesis concerning the survival value of dark and light skin has never been tested.

Two correlations between body shape and environmental conditions have been noted since the nineteenth century. The first, known as Bergmann's rule, states that animals in colder regions of their range have a bulkier body build. The second, known as Allen's rule, states that animals in colder regions of their range have shorter limbs, digits, and ears. Both of these effects help regulate body temperature by increasing the surface-area-to-volume ratio in hot climates and decreasing the ratio in cold climates. For example, Figure 32B, shows that the Maasai of East Africa tend to be slightly built with elongated limbs, while the Eskimos, who live in northern regions, are bulky and have short limbs.

Other anatomical differences among ethnic groups, such as hair texture, a fold on the upper eyelid (common in Asian peoples), and the shape of lips, cannot be explained as adaptations to the environment. Perhaps these features became fixed in different populations due simply to genetic drift. As far as intelligence is concerned, no significant disparities have been found among different ethnic groups.

Genetic Evidence for Common Ancestry

The replacement model for the evolution of humans, discussed earlier in this section, pertains to the origin of ethnic groups. This hypothesis proposes that all modern humans have a relatively recent common ancestor—that is, Cro-Magnon—who evolved in Africa and then spread into other regions. Paleontologists tell us that the variation among modern populations is considerably less than that which existed among archaic human populations some 250,000 years ago. If so, all ethnic groups evolved from the same single ancestral population.

A comparative study of mitochondrial DNA shows that the differences among human populations are consistent with their having a common ancestor no more than a million years ago. Lewontin has also found that the genotypes of different modern populations are extremely similar. He examined variations in 17 genes, including blood groups and various enzymes, among seven major geographic groups: Europeans (Caucasians), black Africans, mongoloids, South Asian Aborigines, Amerinds, Oceanians, and Australian Aborigines. He found that the great majority of genetic variation—85%—occurs within ethnic groups, not between them. In other words, the amount of genetic variation between individuals of the same ethnic group is greater than the variation between any two ethnic groups.

Questions to Consider

1. Why might a person's outward appearance not necessarily be an indication of his or her ethnic background?
2. Some ethnic groups are more highly disposed to certain diseases than others, such as hypertension and cardiovascular disease. Explain what factors may have contributed to this in this ethnic population.

a.   b.

Figure 32B Human ethnic groups.   a. The Maasai live in East Africa. **b.** Eskimos live near the Arctic Circle. (a): ©Sylvia S. Mader; (b): ©B & C Alexander/Science Source

Conclusion

Like all vertebrates, and all life, canines have an evolutionary history that may be traced back not only through the fossil record, but by the record contained within their genomes. The development of new genetic techniques has expanded our ability to analyze the genomes of a wide variety of organisms, from dogs to whales to humans. Evolutionary geneticists are using this information to redefine some of the previously established relationships—such as between humans and Neandertals.

Often, these studies reveal interesting information about our evolutionary past. For example, studies of the canine and human genomes reveal that changes in the neurological processes, diet, and digestion of these two species evolved together over time, suggesting that our interactions with man's best friend may have played an important role not only in our behavior, but in how we evolved as a species.

SUMMARIZE

32.1 Chordates

- Organisms that are classified as **chordates** (tunicates, lancelets, and vertebrates) have a dorsal supporting rod known as a **notochord,** a dorsal tubular **nerve cord, pharyngeal pouches,** and a **postanal tail** at some time in their life history.

- **Lancelets** and **tunicates** are the nonvertebrate chordates. Adult tunicates lack chordate characteristics except gill slits, but adult lancelets have the four chordate characteristics.

32.2 Vertebrates: Fish and Amphibians

- **Vertebrates** have the four chordate characteristics as embryos, but the notochord is replaced by the vertebral column.

- Internal organs are well developed, and cephalization is apparent. The **amniotic egg** was a major evolutionary milestone as well.

- The vertebrate classes trace their evolutionary history as follows:
  - **Jawless fishes** were the first vertebrates and are represented by hagfishes and lampreys that lacked jaws and fins.
  - **Cartilaginous fishes** include the sharks, rays, and skates. Their skeleton is composed of cartilage. They possess jaws and fins.
  - **Bony fishes** have jaws and two pairs of fins. The bony fishes include the **ray-finned fishes** and the **lobe-finned fishes** (some of which have **lungs**).
  - **Amphibians** (e.g., frogs and salamanders) evolved from lobe-finned fishes and have two pairs of limbs. Most amphibian species return to the water to reproduce, and their tadpoles or larvae then metamorphose into terrestrial adults.

32.3 Vertebrates: Reptiles and Mammals

- **Reptiles** most often possess shelled eggs, which contain extraembryonic membranes, including an amnion that allows them to reproduce on land. They are **ectothermic,** meaning their body temperature matches the environment.

- **Birds** are feathered reptiles; feathers help maintain a constant body temperature. They are endothermic, meaning they can maintain an internal temperature. Hollow bones, lungs with air sacs that penetrate bones and allow one-way ventilation, and a keeled breastbone adapt birds for flight.

- **Mammals** have hair to help them maintain a constant body temperature, and mammary glands for nursing of young. **Monotremes** lay eggs; **marsupials** have a pouch in which the newborn matures; and **placental mammals,** which are far more varied and numerous, retain offspring inside the uterus until birth.

- Primates are mammals adapted to living in trees. The order **Primates** includes the strepsirhini (lemurs, aye-ayes, bush babies, and lorises) and the haplorhini (monkeys, apes, and humans).

- Strepsirhini diverged first, followed by the monkeys and then the apes.

32.4 Evolution of the Hominins

- The hominin lineage diverged from the same primate line as the rest of the apes. A **molecular clock** can be used to infer the relatedness between the two groups. Molecular evidence suggests humans are most closely related to African apes, whose ancestry split from ours about 6 to 10 MYA.

- Hominin evolution began in eastern Africa with the rise of **ardipithecines** and the **australopithecines. Bipedalism,** or walking upright, is one of the main features that separates modern humans. The most famous australopithecine is Lucy (3.2 MYA), who walked bipedally and had a small brain.

- *Homo habilis,* present about 2 MYA, is certain to have made tools.

- *Homo erectus,* with a brain capacity of 1,000 cc and a striding gait, was the first to migrate out of Africa.

- The recent discovery of the fossil, *Homo floresiensis,* however, suggests a species of *Homo* coexisted with modern humans as recently as 18,000 years ago.

32.5 Evolution of Modern Humans

- The **replacement model** suggests that modern humans, called **Cro-Magnons,** originated in Africa and, after migrating into Europe and Asia, replaced the other *Homo* species (including **Neandertals** and **Denisovans**) found there. Various lines of evidence support this hypothesis.

- The Neandertals lived approximately 200,000 to 28,000 years ago. They were physically adapted to the last Ice Age. Cro-Magnons first appeared approximately 130,000 years ago. They had advanced culture and technology.

ASSESS

Testing Yourself

Choose the best answer for each question.

32.1 Chordates

1. Which of these is not a chordate characteristic?
   a. dorsal supporting rod, the notochord
   b. dorsal tubular nerve cord
   c. pharyngeal pouches
   d. postanal tail
   e. vertebral column

2. Label the diagram of a chordate below:

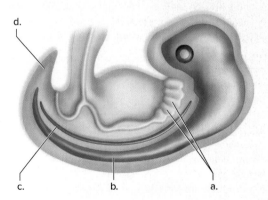

32.2 Vertebrates: Fish and Amphibians

3. Which of these is not characteristic of all vertebrates?
   a. complete digestive system
   b. closed circulatory system
   c. skin with either scales or feathers
   d. endoskeleton made of bone or cartilage
   e. vertebral column

4. The first vertebrates to evolve were
   a. amphibians.
   b. jawed fishes.
   c. jawless fishes.
   d. amniotes.
   e. reptiles.

5. Bony fishes are divided into which two groups?
   a. hagfishes and lampreys
   b. sharks and ray-finned fishes
   c. ray-finned fishes and lobe-finned fishes
   d. jawless fishes and cartilaginous fishes

6. Amphibians evolved from which type of ancestral fish?
   a. sea squirts and lancelets
   b. ray-finned fishes
   c. cartilaginous fishes
   d. lobe-finned fishes
   e. jawless fishes

32.3 Vertebrates: Reptiles and Mammals

7. The amniotes include all but the
   a. birds.   c. mammals.
   b. reptiles.   d. amphibians.

8. Which of the following groups has a heart with four distinct chambers?
   a. birds
   b. lampreys
   c. reptiles other than birds
   d. amphibians

9. The first mammals to evolve were
   a. aquatic.   d. primates.
   b. placental.   e. monotremes.
   c. marsupials.

32.4 Evolution of the Hominins

10. Choose the correct order of primate evolution, from the oldest to the most recent group.
    a. prosimians—anthropoids—hominoids—hominids—hominines
    b. hominines—hominids—hominoids—anthropoids—prosimians
    c. prosimians—anthropoids—hominines—hominids—hominoids
    d. anthropoids—hominines—hominids—hominoids—prosimians
    e. None of the above are correct.

11. The fossil nicknamed Lucy was a(n)
    a. early *Homo.*
    b. australopith.
    c. ardipithecine.
    d. modern human.

12. This genus is the direct ancestor of the genus *Homo.*
    a. *Ardipithecus*
    b. *Sahelanthropus*
    c. *Australopithecus*
    d. Denisovans
    e. None of these are correct.

32.5 Evolution of Modern Humans

13. Which species was probably the first of the genus *Homo* to migrate from Africa?
    a. *Homo erectus*
    b. *Homo habilis*
    c. Denisovans
    d. *Homo florensiensis*
    e. None of these are correct.

14. According to the fossil record, which of these species is the earliest of the *Homo* genus?
    a. *H. sapiens*
    b. *H. erectus*
    c. *H. habilis*
    d. *H. heidelbergensis*
    e. *H. ergaster*

15. According to the replacement model, *Homo sapiens*
    a. was the first to migrate from Africa.
    b. is the earliest species of the genus *Homo.*
    c. followed earlier *Homo* species out of Africa.
    d. originated in the New World.
    e. None of these are correct.

ENGAGE

Thinking Critically

1. Researchers have recently sequenced the genome of several vertebrate species, including the North American green anole and the African clawed frog. What information might these studies give us that could be applied to the development of new medical treatments for humans?

2. Bipedalism has many selective advantages, including the increased ability to spot predators and prey. However, bipedalism has one particular disadvantage—upright posture leads to a smaller pelvic opening, which makes giving birth to an offspring with a large head very difficult. This situation results in a higher percentage of deaths (of both mother and child) during birth in humans compared to other primates. How can you explain the selection for a trait, such as bipedalism, that has both positive and negative consequences for fitness?

3. In studying recent fossils of the genus *Homo,* such as Cro-Magnon, biologists have determined that modern humans have not undergone much biological evolution in the past 50,000 years. Rather, cultural anthropologists argue that cultural evolution has been far more important than biological evolution in the recent history of modern humans. What do they mean by this? Support your argument with some examples.

4. Some modern ethnic groups (white Europeans, Asians) have apparently inherited genes from Neandertals that may influence resistance to cold temperatures. What types of genes might these be?

©Frederic Coubet/Gallo Images/Getty Images

CHAPTER

33

Behavioral Ecology

The Benefits of Living in a Society

African lions often live in a group, or pride, containing between 2 and 18 closely related females. Living in a social unit increases the females' chances of survival. As a group, they can overcome large prey, which would not be possible as an individual. However, although the pride increases the success of hunting large animals, the food must be shared by the entire pride. The advantages and disadvantages of living in a society was a focus of the documentary *African Cats,* which helped to reveal the secrets of the lives of lions.

Within a pride, all of the females tend to be genetically related (mothers, daughters, sisters, aunts). They work together to protect their territory while helping raise each other's young. Female cubs will stay with the pride, while male cubs leave when they reach 2 to 4 years of age. These young males often form coalitions with related males. Once they reach maturity, the male coalition establishes a territory that over-laps with a pride.

Sometimes a pride contains young cubs that are not related to the local male coalition. In this case, the cubs are often killed. This triggers the females to enter estrus, allowing each of the new males to have mating opportunities with the females of the pride. Killing the original cubs and mating with the females increases the likelihood that the males will be able to produce cubs with which they will share their genes.

In this chapter, you will learn about animal behavior in general and gain a better understanding of the complexity of living in a social unit. We will explore whether behavior has a genetic or environmental basis. We will then examine how behavior and communication are necessary components of the struggle to survive.

As you read through this chapter, think about the following questions:

1. Which lion behaviors appear to have a genetic basis to them and which ones appear to have an environmental basis?

2. Which methods of communication are most likely employed by the lions to defend their territory?

CHAPTER OUTLINE

33.1 Nature Versus Nurture: Genetic Influences

33.2 Nature Versus Nurture: Environmental Influences

33.3 Animal Communication

33.4 Behaviors That Affect Fitness

BEFORE YOU BEGIN

Before beginning this chapter, take a few moments to review the following discussions:

Section 1.1 What is the relationship between a stimuli and behavior?

Section 23.3 How can the environment influence gene expression?

Section 27.3 What role does sexual selection and male competition play in the evolution of a species?

33.1 Nature Versus Nurture: Genetic Influences

Learning Outcomes

Upon completion of this section, you should be able to

1. Understand what is meant by the term *nature versus nurture*.
2. Describe an experiment that shows behavior can be genetically based.

Behavior represents the way an organism reacts to a stimulus or situation. This reaction may take many forms, including movement or changes in thought patterns. Recall from Section 1.1 that the ability to react to a stimulus is an important characteristic of life. Therefore, the nature of behavior plays an important role in our understanding of life.

The "nature versus nurture" question asks to what degree our genes (nature) and environmental influences (nurture) affect our behavior. Historically, scientists believed that behavior was either due to genetics or nature, but that pattern of thought has changed over the past few decades as new techniques and experiments have studied the interaction of genes and the environment. Behavior is now recognized as being multifactorial—meaning there are interactions between genes (or multiple genes) and environmental factors. In humans, the science of behavioral genetics often uses twin studies (see the Scientific Inquiry feature "Behavioral Genetics and Twin Studies") to explore these interactions.

While we usually discuss behavior in regard to animals because they respond more quickly to stimuli, behavior is now recognized to occur in any organism that responds to stimuli. In this section, we will examine a variety of experiments that have been conducted to help determine the degree to which genetics controls behavior.

Nurturing Behavior in Mice

In several scientific studies, investigators have found that maternal behavior in mice was dependent on the presence of a gene called *fosB*. Normally, when mothers first inspect their newborns, various sensory information from their eyes, ears, nose, and touch receptors travel to the hypothalamus. This incoming information causes *fosB* alleles to be activated, and a particular protein is produced. The protein begins a process during which cellular enzymes and other genes are activated. The end result is a change in the neural circuitry within the hypothalamus, which manifests itself in maternal nurturing behavior toward the young. Mice that do not engage in nurturing behavior were found to lack *fosB* alleles, and the hypothalamus failed to make any of the products or to activate any of the enzymes and other genes that lead to maternal nurturing behavior. Female mice with *fosB* alleles tended to retrieve their young and bring them back to them after they became separated (Fig. 33.1).

Food Choice in Garter Snakes

Several experiments have been conducted using the garter snake, *Thamnophis elegans,* to determine if food preference has a genetic basis. In California, inland garter snake populations are aquatic and commonly feed underwater on frogs and fish, whereas coastal populations are terrestrial and feed mainly on slugs. In the laboratory, inland adult snakes refused to eat slugs, while coastal snakes readily did so. Newborns resulting from matings between snakes

a.

b. *fosB* alleles present

c. *fosB* alleles not present

Figure 33.1 Maternal care in mice.   a. A mother mouse with *fosB* alleles spends time retrieving and crouching over its young, whereas mice that lack these alleles do not display these maternal behaviors. **b.** This typical mother mouse retrieves her young and crouches over them. **c.** This mouse, lacking *fosB*, does not retrieve her young and does not crouch over them. (b): ©Tom McHugh/Science Source; (c): ©Jean-Louis Klein & Marie-Luce Hubert/Science Source

from the two populations (inland and coastal) have an intermediate preference for slugs, suggesting a genetic basis for food choice.

Differences between slug acceptors and slug rejecters appear to be inherited, but what physiological difference is there between

Behavioral Genetics and Twin Studies

There are two types of twins—those that come from a single zygote (monozygotic, identical or paternal) and those that are the result of separate fertilization events (dizygotic or fraternal). Behavioral genetics often uses twin studies to examine the similarities and differences between monozygotic twins, dizygotic twins, and twins that have been raised apart, to attempt to address the nature or nurture debate. Many scientists attribute a person's outcome, and in our case, behavior, to two factors: nature and nurture. Nature (your genes) gives you traits for eye color, hair color, and blood type. Nurture is based on your lifestyle and environment, including diet, rearing, and education. The study of monozygotic twins (Fig. 33A) has demonstrated that even though the individuals share the same genetic information, they are not always completely identical with regard to their traits and behavior. These studies have led researchers to recognize that there is a bridge between nature and nurture in the form of epigenetics (Gk. *epi-*, "upon, over").

The specific chemical reactions, or epigenetic "tags," can come in different forms but are often associated with DNA methylation, in which a methyl group attaches to the cytosine base of DNA (Fig. 33B). With a methyl group attached, transcription cannot occur. The methyl group interferes with transcription factors and other proteins in the transcription machinery, thereby silencing or weakening a gene. Over time, the differences in these tags accumulate, making twins increasingly different from each other (Fig. 33C).

Epigenetics are heritable changes in gene expression without changing the DNA sequence. Chemical reactions due to environmental exposure influence how genes are regulated or turned off or on. This, in turn, may affect many aspects of an individual, including their behavior. Monozygotic twins present a unique opportunity to study epigenetics because they are clones resulting from a split in a single fertilized egg. Assuming a similar upbringing, their gradual differences over time can therefore be attributed to their disparate control of genes.

Epigenetics has important implications for behavior. The appearance of tags on genes helps scientists discover the cause of some conditions and illnesses that cannot be explained by DNA or genetic mutations alone. Identical twins discordant (different) for autism, psychiatric disorders, and dyslexia have been shown to have different DNA methylation on certain genes.

In addition, the epigenetic changes are reversible. A study using rats showed that rat pups that are licked and nurtured by their mothers become calm adults. Rat pups that are not nurtured are anxious. Injecting a calm rat with a drug that adds methyl groups creates an anxious rat. Conversely, injecting an anxious rat with a different drug that removes methyl groups creates a calm rat. In drug development, epigenetic medicines could be used to correct or reverse the particular effect of a tag.

Questions to Consider

1. What would a study of both monozygotic twins raised together and monozygotic twins raised apart tell you about environmental influences?
2. Epigenetics may be involved in the fine-tuning of gene expression. How might this affect the behavior of an individual?

DNA methylation is the addition of a methyl group (M) to the DNA base cytosine (C).

Figure 33B DNA methylation. DNA is methylated when a methyl group attaches to the cytosine nucleotide.

3-year-old twins

Yellow shows where the twins have epigenetic tags in the same place.

50-year-old twins

Red and green show where the twins have epigenetic tags in different places.

Figure 33C Epigenetic tags on chromosomes. One twin's epigenetic tags are dyed green, and the other twin's tags are dyed red. An overlap in green and red shows up as yellow. The 50-year-old twins have more epigenetic tags in different places than do the 3-year-old twins.

Figure 33A Identical twins. Identical twins come from a single fertilized egg that splits in two. Their genes are the same.

©Hero/Corbis/Glow Images

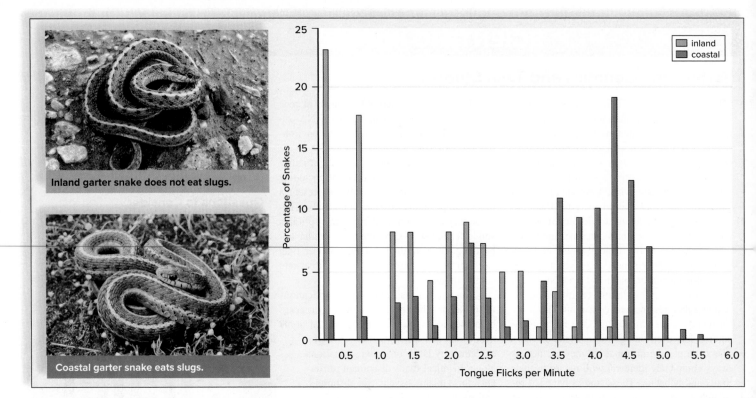

Figure 33.2 Feeding behavior in garter snakes.   The number of tongue flicks by inland and coastal garter snakes is measured in terms of their response to slug extract on cotton swabs. Coastal snakes tongue-flicked more than inland snakes, indicating that coastal snakes are more sensitive to the smell of slugs than inland snakes. (inland): ©John Serrao/Science Source; (coastal): ©Chris Mattison/Alamy

the two populations? A clever experiment answered this question. Snakes use tongue flicks to recognize their prey, and their tongues carry chemicals to an odor receptor in their mouth. Newborns will even flick their tongues at cotton swabs dipped in fluids of their prey. Swabs were dipped in slug extract, and the number of tongue flicks were counted for newborn inland and coastal snakes. Coastal snakes had a higher number of tongue flicks (indicating more interest in the smell) than inland snakes (Fig. 33.2). Thus, inland snakes apparently do not eat slugs because they are not sensitive to their smell. A genetic difference between the two populations of snakes has resulted in a physiological difference in their nervous systems. Although hybrids showed a great deal of variation in the number of tongue flicks, they were generally intermediate, as predicted by the genetic hypothesis.

Egg-Laying Behavior of Marine Snails

The nervous and endocrine systems are both responsible for the coordination of body systems. Is the endocrine system also involved in behavior? Experiments support endocrine system involvement. For example, the egg-laying behavior in the marine snail *Aplysia* involves a set sequence of movements. Following copulation, the animal extrudes long strings of more than a million eggs per egg case. It then takes the egg case string in its mouth, covers it with mucus, waves its head back and forth to wind the string into an irregular mass, and attaches the mass to a solid object, such as a rock. Scientists have isolated and analyzed an egg-laying hormone (ELH) that causes the snail to lay eggs even if it has not mated. ELH was found to be a small protein of 36 amino acids that diffuses into the

circulatory system and causes the smooth muscle cells of the reproductive duct to contract and expel the egg string. Using recombinant DNA techniques, the investigators isolated the entire *ELH* gene. The gene's product turned out to be a protein with 271 amino acids. The protein can be cleaved into as many as 11 possible products, and the ELH hormone is one of these. The hormone alone, or in conjunction with the gene's other products, is thought to control all the components of egg-laying behavior in *Aplysia*.

Check Your Progress 33.1

1. Identify how studies of animals suggest that behavior can be genetically based.
2. Explain how you would determine if a behavior was based primarily on genetics or on the environment.

33.2 Nature Versus Nurture: Environmental Influences

Learning Outcomes

Upon completion of this section, you should be able to

1. Identify an experiment that shows how behavior can be environmentally influenced.
2. Compare three different types of learned behavior.

Even though genetic inheritance serves as a basis for behavior, environmental influences (nurture) also affect behavior. For example, behaviorists originally believed that some behaviors were **fixed action patterns (FAPs)** elicited by a sign stimulus. But then they found that many behaviors improve with practice. In this context, **learning** is defined as a durable change in behavior brought about by experience.

Learning in Birds

Laughing gull (*Leucophaeus atricilla*) chicks' begging behavior appears to be a FAP, because it is always performed the same way in response to the parent's red bill (the sign stimulus). A chick directs a pecking motion toward the parent's bill, grasps it, and strokes it downward (Fig. 33.3*a*). Parents bring about the begging behavior by swinging their bill gently from side to side. After the chick responds, the parent regurgitates food onto the nest floor. If need be, the parent then encourages the chick to eat. This interaction between the chicks and their parents suggests that the begging behavior involves learning. To test this hypothesis, diagrammatic pictures of gull heads were painted on small cards. Then, eggs collected in the field were hatched in a dark incubator to eliminate visual recognition before the test. On the day of hatching, each chick was allowed to make about a dozen pecks at the model. The chicks were returned to the nest, and then each was retested. The tests showed that, on average, only one-third of the pecks by a newly hatched chick strike the model. But one day after hatching, more than half of the pecks are accurate, and two days after hatching, the accuracy reaches a level of more than 75% (Fig. 33.3*b, c*). Investigators concluded that improvement in motor skills, as well as visual experience, strongly affect the development of chick begging behavior—evidence that the behavior is, in part, learned.

Imprinting

Imprinting is considered a very simple form of learning, although it has a strong genetic component as well. Imprinting was first observed in birds when chicks, ducklings, and goslings followed the first moving object they saw after hatching. This object is ordinarily their mother, but investigators found that birds can imprint on any object, as long as it is the first moving object they see during a sensitive period of 2 to 3 days after hatching. The term *sensitive period* means that the behavior develops only during this time.

A chick imprinted on a red ball follows it around and chirps whenever the ball is moved out of sight. Social interactions between parent and offspring during the sensitive period seem key to normal imprinting. For example, female mallards cluck during the entire time imprinting is occurring, and it could be that vocalization before and after hatching is necessary for normal imprinting.

Social Interactions and Learning

White-crowned sparrows sing a species-specific song, but males of a particular region have their own dialect. Birds were caged to test the hypothesis that young white-crowned sparrows learn how to sing from older members of their species.

Three groups of birds were tested. Birds in the first group heard no songs at all. When grown, these birds sang a song, but it was not fully developed. Birds in the second group heard tapes of white-crowns singing. When grown, they sang in that dialect, as long as the tapes had been played during a sensitive period from about age 10 to 50 days. White-crowned sparrows' dialects (or other species' songs) played before or after this sensitive period had no effect on their song. Birds in a third group did not hear tapes and instead were given an adult tutor of a different species. These birds sang the song of the different species—no matter when the tutoring began—showing that social interactions apparently assist learning in birds.

a. Laughing gull adult and chick, *Leucophaeus atricilla*

b.

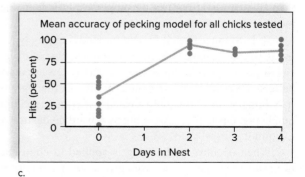

c.

Figure 33.3 Begging behavior in laughing gull chicks. a. At about 3 days of age, a laughing gull chick grasps the red bill of a parent, stroking it downward, and the parent then regurgitates food. **b.** The accuracy of a chick striking a test probe, painted red. **c.** Chick-pecking accuracy illustrated graphically. Note from these diagrams that a chick markedly improves its ability (within only 2 days) to peck a bill, a behavior that normally causes a parent to regurgitate food.

Associative Learning

A change in behavior that involves an association between two events is termed **associative learning.** For example, birds that get sick after eating a monarch butterfly no longer prey on monarch butterflies, even though they may be readily available. Or the smell of fresh-baked bread may entice you to eat even though you may have just eaten. Because you really enjoyed bread in the past, you associate the smell with those past experiences, which makes you hungry. Both classical conditioning and operant conditioning are examples of associative learning.

Classical Conditioning

In **classical conditioning,** the presentation of two different types of stimuli at the same time causes an animal to form an association between them. The best-known laboratory example of classical conditioning is an experiment done by the Russian psychologist Ivan Pavlov. First, Pavlov observed that dogs salivate when presented with food. He then rang a bell whenever the dogs were fed. Eventually, the dogs would salivate whenever the bell was rung, regardless of whether food was present (Fig. 33.4).

Classical conditioning suggests that an organism can be trained or conditioned to associate a specific response to a specific stimulus. Unconditioned responses are those that occur naturally, as when salivation follows the presentation of food. Conditioned responses are those that are learned, as when a dog learns to salivate when it hears a bell. Advertisements attempt to use classical conditioning. For example, commercials pair attractive people with a particular product in the hope that viewers will associate attractiveness with that product. This pleasant association may cause them to buy the product.

In a similar way, it's been suggested that holding children on your lap when reading to them facilitates an interest in reading. The belief is that they will associate a pleasant feeling with reading.

Operant Conditioning

During **operant conditioning,** a stimulus–response connection is strengthened. Most people know it is helpful to give an animal a reward, such as food or affection, when teaching them a trick. When we go to an animal show, it is quite obvious that trainers use operant conditioning. They present a stimulus, such as a hoop, and then give a reward (food) for the proper response (jumping through the hoop).

B. F. Skinner is well known for studying this type of learning in the laboratory. In Skinner's simplest type of experiment, a caged rat happens to press a lever and is rewarded with sugar pellets. Thereafter, the rat regularly presses the lever whenever it is hungry. In more sophisticated experiments, Skinner even taught pigeons to play ping-pong by rewarding desired responses to stimuli.

When operant conditioning is applied to child rearing, it's been suggested that parents who give a positive reinforcement for good behavior will be more successful than parents who punish behaviors they believe are undesirable.

saliva at sight of food (unconditioned response)

saliva at sound of bell only (conditioned response)

sound of bell (conditioned stimulus)

food (unconditioned stimulus)

apparatus to measure saliva

Figure 33.4 Classical conditioning.   Ivan Pavlov discovered classical conditioning by performing this experiment with dogs. A bell is rung when a dog is fed. Salivation is noted. Eventually, the dog salivates when the bell is rung even though no food is presented. Food is an unconditioned stimulus, and the sound of the bell is a conditioned stimulus that brings about the response—that is, salivation.

Check Your Progress 33.2

1. Identify examples of various types of learning.
2. Contrast the different types of associative learning.

33.3 Animal Communication

Learning Outcomes

Upon completion of this section, you should be able to

1. Summarize the various ways in which animals communicate.
2. Describe the advantages and disadvantages of chemical, auditory, visual, and tactile communication.

Animals exhibit a wide diversity of social behaviors. Some animal species are largely solitary and join with a member of the opposite sex only for the purpose of reproduction. Others pair, bond, and cooperate in raising offspring. Still others form a society in which members of the species are organized in a cooperative manner, extending beyond sexual and parental behavior. Social behavior among animals requires that they communicate with one another.

Communication is a signal by a sender that influences the behavior of a receiver. The communication can be purposeful, but does not have to be. Bats send out a series of sound pulses and listen for the corresponding echoes in order to find their way through dark caves and locate food at night. Some moths have an ability to hear these sound pulses, and they begin evasive tactics when they sense that a bat is near. The bats are not purposefully communicating with the moths. The bat sounds are simply a cue to the moths that danger is near.

The types of communication signals used by animals include chemical, auditory, visual, and tactile.

Chemical Communication

Chemical signals have the advantage of being effective both night and day. The term **pheromone** designates chemical signals in low concentration that are generally passed between members of the same species. For example, female moths secrete chemicals from special abdominal glands, which are detected by receptors on male antennae. The antennae are especially sensitive, and this ensures that only male moths of the same species will be able to detect them.

Cheetahs and other cats mark their territories by depositing urine, feces, and anal gland secretions at the boundaries (Fig. 33.5). Klipspringers (small antelope) use secretions from a gland below the eye to mark twigs and grasses of their territory.

Researchers are studying to what degree pheromones and hormones affect the behavior of mammals. Some researchers maintain that human behavior is influenced by undetectable pheromones wafting through the air. They have discovered that, like the mouse, humans have an organ in the nose, called the vomeronasal organ (VNO), which can detect not only odors, but also pheromones. The neurons from this organ lead to the hypothalamus, the part of the brain that controls the release of various hormones in the body. Perhaps this organ is even involved in how people choose their mates (see the Health feature "Mate Choice and Smelly T-Shirts" in Section 33.4).

Auditory Communication

Auditory (sound) communication has various advantages over other kinds of communication. It is faster than chemical communication, and is effective both night and day. Further, auditory communication can be modified not only by loudness but also by pattern, duration, and repetition. In an experiment with rats, a

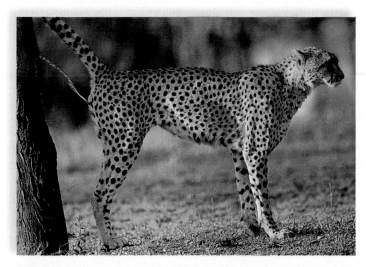

Figure 33.5 Use of a pheromone.   This male cheetah is spraying urine onto a tree to mark its territory. ©Gregory G. Dimijian/Science Source

researcher discovered that an intruder can avoid attack by increasing the frequency with which it makes an appeasement sound.

Male birds have songs for a number of different occasions, such as one song for distress, another for courting, and still another for marking territories. Sailors have long heard the songs of humpback whales transmitted through the hull of a ship. However, only recently has it been shown that the song has six basic themes, each with its own phrases, which can vary in length, be interspersed with sundry cries and chirps, and be heard for many miles. Interestingly, humpbacks' songs change from year to year, but all the humpbacks in an area learn and converge on singing the same song. The purpose of the song is probably sexual, serving to advertise the availability of the singer.

Language is the ultimate auditory communication. Only humans have the biological ability to produce a large number of different sounds and to put them together in many different ways. Nonhuman primates have different vocalizations, each having a definite meaning, such as when vervet monkeys give alarm calls (Fig. 33.6). Although chimpanzees can be taught to use an artificial

Figure 33.6 Auditory communication.   **a.** Vervet monkeys, *Cercopithecus aethiops,* are responding to an alarm call. Vervet monkeys can give different alarm calls according to whether a troop member sights an eagle or a leopard, for example. **b.** The frequency per second of the sound differs for each type of call. (running monkeys): ©Arco Images/GmbH/Alamy; (monkey): ©Flavio Vallenari/E+/Getty Images

language, they never progress beyond the capability level of a 2-year-old child. It also has been difficult to prove that chimps understand the concept of grammar or can use their language to reason. As such, most anthropologists agree that humans possess a communication ability unparalleled by most animals.

Visual Communication

Visual signals are most often used by species that are active during the day. Contests between males make use of threat postures to help prevent outright fighting, a behavior that might result in reduced fitness. A male baboon displaying full threat is an awesome sight that establishes his dominance and keeps peace within the baboon troop (Fig. 33.7).

Many animals use complex courtship behaviors and displays. The plumage of a male Raggiana Bird of Paradise allows him to put on a spectacular courtship dance to attract a female, giving her a basis on which to select a mate. Defense and courtship displays are exaggerated and always performed in the same way so that their meaning is clear. Male fireflies use a flash pattern to signal females of the same species (Fig. 33.8).

Visual communication allows animals to signal others of their intentions without the need to provide any auditory or chemical messages. The body language of students during a lecture provides an example. Some students lean forward in their seats and make eye contact with the instructor. They want the instructor to know they are interested and find the material of value. Others lean back in their chairs, look at the floor, or send text messages on their cell phones. These students indicate they are not interested in the material. Teachers can use students' body language to determine if they are effectively presenting the material, and make changes accordingly.

Other human behaviors also send visual clues. The hairstyle and dress of a person or the way he or she walks and talks are ways to send messages to others. Psychologists have long tried to understand

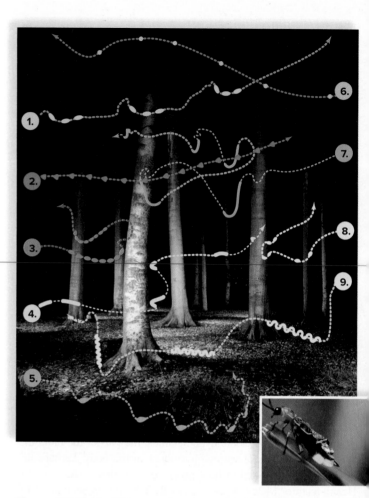

Figure 33.8 Fireflies use visual communication.   Each number represents the male flash pattern of a different species. The patterns are a behavioral reproductive isolation mechanism. (trees): ©Ghislain & Marie David de Lossy/Cultura/Getty Images; (firefly): ©James Jordan Photography/Getty Images

how visual clues can be used to better understand human emotions and behavior. Some studies have suggested that women are apt to dress in an appealing manner and be sexually inviting when they are ovulating. People who dress in black, move slowly, fail to make eye contact, and sit alone may be telling others that they are unhappy. Similarly, body language in animals is being used to suggest that they too have emotions, as discussed in the Bioethical feature "Do Animals Have Emotions?"

Tactile Communication

Tactile communication occurs when one animal touches another. For example, laughing gull chicks peck at the parent's beak to induce the parent to feed them (see Fig. 33.3). A male leopard nuzzles the female's neck to calm her and stimulate her willingness to mate. In primates, grooming—one animal cleaning the coat and skin of another—helps cement social bonds within a group.

Honeybees use a combination of communication methods, including tactile ones, to convey information about their environment. When a foraging bee returns to the hive, it performs a waggle dance that communicates the distance and direction to the food

Figure 33.7 A male olive baboon displaying full threat.
In olive baboons, males are larger than females and have enlarged canines. Competition between males establishes a dominance hierarchy for the distribution of resources. ©Raul Arboleda/AFP/Getty Images

SCIENCE IN YOUR LIFE ▶ BIOETHICAL

Do Animals Have Emotions?

In the light of recent YouTube video postings portraying the obvious expression of joy by dogs when their owners have returned from long absences, investigators have become more interested in determining to what extent animals have emotions. The body language of animals can be interpreted to suggest that they have feelings. When wolves reunite, they wag their tails to and fro, whine, and jump up and down. Many young animals play with one another or even by themselves, as when dogs chase their own tails. On the other end of the spectrum, upon the death of a friend or parent, chimps are apt to sulk, stop eating, and even die.

It seems reasonable to hypothesize that animals are "happy" when they reunite, "enjoy" themselves when they play, and are "depressed" over the loss of a close friend or relative. Most people would agree that an animal is feeling some type of emotion when it exhibits certain behaviors (Fig. 33D).

In the past, scientists found it expedient to collect data only about observable behavior and to ignore the possible mental state of an animal. B. F. Skinner, whose research method is described in this chapter, was able to condition animals to respond automatically to a particular stimulus. He and others never considered that animals might have feelings. But now, some scientists believe they have sufficient data to suggest that some animals express emotions such as fear, joy, embarrassment, jealousy, anger, love, sadness, and grief. One possible definition describes emotion as a psychological phenomenon that helps animals direct and manage their behavior.

Multiple instances of soldiers returning from combat have shown how dogs react to seeing their owners after they have been absent for extended periods of time. On many occasions, the dogs run in circles, jump up and down, and excitedly lick their owner's face. Cries and squeals of excitement can be heard coming from them continuously as they greet their owners.

The Gabriela Cowperthwaite documentary *Blackfish,* about captive killer whales at SeaWorld, shows a scene in which keepers remove a young killer whale from its mother in order to relocate the young whale to another facility. When the mother cannot locate her offspring within the enclosure, she begins to shake violently and scream a long-range-frequency call in an attempt to find her child.

Perhaps it would be reasonable to consider the suggestion of Charles Darwin, who said that animals are different in degree rather than in kind. This means that animals can feel love, but perhaps not to the degree humans can. When you think about it, it is unlikely that emotions first appeared in humans with no evolutionary homologies in animals.

Dopamine controls the reward and pleasure center in the brain, as well as regulating emotional responses. High levels of dopamine are found in the brain of rats when they play, with the levels increasing when the rats anticipate the opportunity to play. Field research is providing information that is helping researchers learn how animal emotions correlate with their behavior, just as emotions influence human behavior.

Questions to Consider

1. Do you believe animals have emotions? Why or why not? If not, what experiment(s) would convince you that animals do have emotions?
2. Pet psychology, an emerging field, is based on the premise that pets have feelings and emotions. Would you spend $50–100 per hour to take your pet to a psychologist?
3. If it can be shown that animals have emotions, should animals still be kept in captivity?

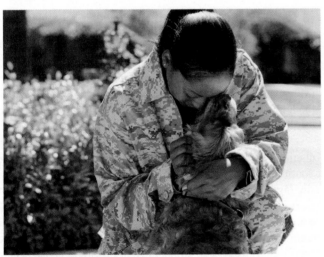

Is the dog happy to see his owner?

Does the mother feel love for her calf?

Figure 33D Emotions in animals.   (soldier): ©Image Source Trading, Ltd./Shutterstock; (whales): ©Bryce Flynn/Moment/Getty Images source (Fig. 33.9). Inside the hive, other foragers crowd around her and touch the forager with their antennae, and as the bee moves between the two loops of a figure 8, it buzzes noisily and shakes its entire body in so-called waggles. The angle of the straight run of the figure 8 to that of the direction of gravity is the same as that of the angle between the sun and the food source. In other words, a 40-degree angle to the left of vertical means that food is 40 degrees to the left of the sun. Bees can use the sun as a compass because they have a biological clock that allows them to compensate for the movement of the sun in the sky. (A biological clock is an internal means of telling time—for example, darkness outside stimulates many animals to sleep.) Outside the hive, the dance is done on a horizontal surface, and the straight-run part of the figure 8 indicates the direction of the food.

a.

b.

Figure 33.9 Communication among bees. **a.** Honeybees do a waggle dance to indicate the direction of food. **b.** If the dance is done outside the hive on a horizontal surface, the straight run of the dance will point to the food source. If the dance is done inside the hive on a vertical surface, the angle of the straightaway to that of the direction of gravity is the same as the angle of the food source to the sun. (a): ©Scott Camazine/Alamy

> **Check Your Progress 33.3**
>
> 1. Summarize the difference between tactile, visual, and auditory communication.
> 2. Identify ways in which communication is meant to affect the behavior of a receiver.
> 3. Describe a type of communication that would be the most effective in a dense forest.

33.4 Behaviors That Affect Fitness

> **Learning Outcomes**
>
> Upon completion of this section, you should be able to
> 1. Discuss why territoriality evolved in certain animal groups.
> 2. Explain why females should be choosier than males in selecting a mate.
> 3. Summarize the costs and benefits of living in a society.

Behavioral ecology assumes that most behavior is subject to natural selection. We have established that behaviors can have a genetic basis, and we would expect that certain behaviors more than others will lead to increased survival and number of offspring. Therefore, much of the behavior of organisms we observe today must have adaptive value.

Territoriality and Fitness

In order to gather food, animals often have a particular home range where they can be found during the course of the day. Animals may actively defend a portion of their home range for their exclusive use as a food source or as a mating area. This portion of the home range is called their territory and the behavior is called territoriality.

Territoriality is more likely to occur during times of reproduction. For example, gibbons live in the tropical rain forest of South and Southeast Asia, and territories are maintained by loud singing (Fig. 33.10). Males sing just before sunrise, and mated pairs sing duets during the morning. Males, but not females, show evidence of fighting to defend their territory in the form of broken teeth and scars. Obviously, defense of a territory has a certain cost; it takes energy to sing and fight off others. Also, you might get hurt. In order for territoriality to persist, it must have an adaptive value. Chief among the benefits of territoriality are to ensure a source of food, exclusive rights to one or more females, and to have a place to rear young and possibly to protect yourself from predators.

The territory has to be the right size for the animal. Too large a territory cannot be defended, and too small a territory may not

Figure 33.11 Foraging for food. When offered a choice of an equal number of each size of mussel, the shore crab, *Carcinus maenas,* prefers the intermediate size. This size provides the highest rate of net energy return. Net energy is determined by the energetic yield of the crab minus the energetic costs of breaking open the shell and digesting the mussel.

factors, such as escaping from predation. If an animal is killed and eaten, it has no chance at all of having offspring in the future. Thus, animals may have to avoid foraging when the threat of predation is high. Animals often face these types of trade-offs that cause them to modify their behavior. As another example, many animal species stop eating during peak reproductive periods because competition for mates is often fierce.

Reproductive Strategies and Fitness

Usually, primates are polygamous, and males monopolize multiple females. Because of gestation and lactation, females invest more in their offspring than do males and may not always be available for mating. Under these circumstances, it is adaptive for females to be concerned with a good food source. When food sources are clumped, females congregate in small groups. Because only a few females are expected to be receptive at a time, males will likely be able to defend these few from other males. Males are expected to compete with other males for the limited number of receptive females available (Fig. 33.12).

A limited number of primates are polyandrous. Tamarins are squirrel-sized New World monkeys that live in Central or South America. Tamarins live together in groups of one or more families in which one female mates with more than one male. The female normally gives birth to twins of such a large size that the father, and not the mother, carries them about. This may be the reason these animals are polyandrous. Polyandry also occurs when the environment does not have sufficient resources to support several young at a time.

Gibbons, as previously mentioned, are monogamous, which means they pair bond. Subsequently, both the male and female help with the rearing of the young. Males are active fathers, frequently grooming and handling infants. Monogamy is relatively rare in primates, which includes prosimians, monkeys, and apes (only about 18% are monogamous). In primates, monogamy occurs when males have limited mating opportunities, territoriality exists,

Figure 33.10 Male and female gibbons. Siamang gibbons, *Hylobates syndactylus,* are monogamous, and they both share the task of raising offspring. They also share the task of marking their territory by singing. As is often the case in monogamous relationships, the sexes are similar in appearance. Male is above and female is below. ©Nicole Duplaix/Getty Images

contain enough food. Cheetahs require a large territory to hunt for their prey and, therefore, they use urine to mark their territory (see Fig. 33.5). Hummingbirds are known to defend a very small territory because they depend on only a small patch of flowers as their food source.

Foraging for Food

Animals need to ingest food that will provide more energy than the effort expended acquiring the food. In one study, it was shown that shore crabs ate intermediate-sized mussels because the net energy gain was more than if they ate larger-sized mussels (Fig. 33.11). The large mussels take too much energy to open per the amount of energy they provide. The optimal foraging model states that it is adaptive for foraging behavior (i.e., searching for food) and food choice to be as energetically efficient as possible. Even though it can be shown that animals that take in more energy are more likely to have more offspring, animals often have to consider other

Figure 33.12 Hamadryas baboons.   Among Hamadryas baboons (*Papio hamadryas*), a male, which is silver-white and twice the size of a female, keeps and guards a harem of females with whom he mates exclusively. ©Thomas Dobner 2006/Alamy

Figure 33.13 Competition.   During the mating season, elk males of the species *Cervus elaphus* often find it necessary to engage in antler wrestling in order to have sole access to females in a territory. ©D. Robert & Lorri Franz/Corbis Documentary/Getty Images

and the male is fairly certain the offspring are his. In gibbons, females are evenly distributed in the environment, most likely because they are aggressive to one another. Studies have shown that females do attack a speaker when it plays female sounds in their territory.

Sexual Selection

Sexual selection is a form of natural selection that favors features that increase an animal's chances of mating. These features are adaptive in the sense that they lead to increased fitness. Sexual selection often results in female choice and male competition. Because females produce a limited number of eggs in their lifetime and generally provide the majority of the parental care, it is adaptive for them to be choosy about their mate. If they choose a mate that passes on features to a male offspring that will cause him to be chosen by females, their fitness has increased. Whether females choose features that are adaptive to the environment is in question. For example, peahens are likely to choose peacocks that have the most elaborate tails. Such a fancy tail could otherwise be detrimental to the male and make him more likely to be captured by a predator. In one study, an extra ornament was attached to a father zebra finch, and so the daughters of this bird underwent the process of imprinting. As a result, these females became more likely to choose a mate that also had the same artificial ornament.

While females can always be sure an offspring is theirs, males do not have this certainty. However, males produce a plentiful supply of sperm. The best strategy for males to increase their fitness, therefore, is to have as many offspring as possible. Competition may be required for them to gain access to females, and so ornaments, such as antlers, can enhance a male's ability to fight (Fig. 33.13). When bull elk compete, they issue a loud number of screams that give way to a series of grunts. If still necessary, the two bulls walk in parallel to show each other their physique. If this

doesn't convince one or the other to back off, the pair resorts to ramming each other with their antlers. Rarely is either bull actually hurt. Whereas a peacock cannot shed his tail, elk shed their antlers as soon as mating season is over.

Mating in Humans

A study of human mating behavior shows that the concepts of female choice and male competition apply to humans as well as to other animals. That is, mate choice behavior in men and women in most human cultures seems to be influenced by its fitness consequences. Of course, applying the principles of evolutionary biology to human behavior is not without controversy because we can't be reduced to being "preprogrammed" and still have the ability to make conscious choices. That said, understanding the evolutionary basis of animal behavior can provide some interesting insights into why people behave the way they do.

Human Males Compete

Consider that women, by nature, must invest more in having a child than men. After all, it takes almost 10 months to have a child, and pregnancy is followed by lactation when a woman may nurse her infant. Men, on the other hand, need only contribute sperm during a sex act that may require only a few minutes. The result is that men are generally more available for mating than are women. Because more men are available, they necessarily have to compete with others for the opportunity to mate.

Like many other animals, humans are dimorphic. Males tend to be larger and more aggressive than females, perhaps as a result of past sexual selection by females. As in other animals, males may pay a price due to high levels of testosterone, the energetic costs of male-male competition, and increased stress associated with finding mates. Human men tend to live, on average, 4 to 7 years less than females.

Females Choose

David Buss, an evolutionary psychologist at the University of Texas, conducted studies of female preference of a male mate across cultures in over 20 countries. His research, although somewhat controversial, suggests that the number-one trait females prefer in a male mate is his ability to obtain (financial) resources. Financial success means that men are more likely to provide females with the resources they need to raise their children, and thus increase the fitness (reproductive success) of the female.

Other recent studies have shown that symmetry in facial and body features is also important in female mate choice, possibly related to the "good genes" hypothesis. That is, females may choose males with "good genes" so that their offspring have good genes and a better chance at survival. As an example, symmetry in other animals is often a sign of good health and a strong immune system (e.g., animals with high parasite loads are often asymmetrical). Females may thus choose symmetrical males to pass these traits on to their offspring.

Men Also Have a Choice

Just as women choose men who can provide resources, men prefer youthfulness and attractiveness in females, signs that their partner can provide them with children. In one controversial study, men ages 8 to 80 across four continents preferred women with a waist-to-hip ratio (WHR) of 0.7, regardless of weight. Research showed that a WHR of 0.7 is optimal for conception; that is, for each increase of 10% in WHR, the odds of conception during each ovulation event decrease by 30%. Men responding to questionnaires prefer physical attributes in their female mates that biologists associate with a strong immune system and good health (e.g., symmetry), high estrogen levels (indicating fertility), and especially youthfulness. On average, men marry women 2.5 years younger than they are, but as men age, they tend to prefer women who are many years younger. Men are capable of reproduction for many more years than women. Therefore, by choosing younger women, older men can increase their fitness. Other factors are also involved in human mate choice, as explained in the Health feature "Mate Choice and Smelly T-Shirts."

Societies and Fitness

The principles of evolutionary biology can be applied to the study of social behavior in animals. Sociobiologists hypothesize that societies form when living in a society has a greater reproductive benefit than reproductive cost. An analysis of cost and benefit can help determine if this hypothesis is supported. Group living does have its benefits. It can help an animal avoid predators, rear

---

SCIENCE IN YOUR LIFE ▶ HEALTH

Mate Choice and Smelly T-Shirts

Mate choice has been studied extensively in humans, largely because of our curiosity about how evolutionary biology and behavior in other animals may give us insight into human behavior. And, of course, mate choice is one of the more fascinating human behaviors. In one unusual and recent study, scientists had several men wear T-shirts for a few nights while they slept, thereby infusing the T-shirts with their unique smell. The scientists then asked women to rate the attractiveness of the male without seeing the person—that is, based only on the smell of the T-shirt. It turns out that women tended to choose men whose major histocompatibility complex (MHC) alleles were different from their own. Recall from Section 13.3 that the MHC is associated with recognition of foreign antigens (e.g., viruses or bacteria) and enhances the immune system's ability to fight off infections. In choosing men with different (complementary) MHC alleles, females are maximizing the chance that their offspring have highly heterozygous MHC and potentially a more robust immune system. A more heterozygous MHC is associated with a better innate ability to recognize a more diverse group of antigens. In

fact, females in this study said unattractive shirt smells reminded them of their fathers, who, of course, have very similar MHC alleles. Where did scientists get the idea for such a strange experiment? It has already been well documented that mice preferentially choose mates with complementary MHC alleles based on smell.

Of course, while evolutionary and behavioral biology can provide insights into human mate choice, the process is clearly complex. Although males prefer young faithful females with physical attributes that indicate fertility, they also look for symmetry, intelligence, and a sense of humor. And while females prefer males with financial resources, they also list such qualities as dependability and emotional stability (indicators of good parenting), physical attractiveness (indicated by symmetry and testosterone markers, such as above-average strength and height, possibly to defend resources), sense of humor, and intelligence. The way a person smells also could play a factor, as indicated by the T-shirt study. Then there are unknowns—the context under which people meet, their professions, their likes/dislikes, etc.

Questions to Consider

1. Scientists have found a biological basis for infanticide. In lions, for example, males who kill the offspring of other males when they take over a pride then induce females to come into estrus, so they can mate with them. This increases male fitness. Should a biological basis for infanticide be used as a defense in human court cases where, say, a stepfather kills a stepchild?

2. As discussed in the paragraph just prior to this Health feature, some studies suggest that men prefer women with a waist-to-hip ratio of 0.7. Women of different weights can have this ratio because it is based on bone structure, rather than weight. Nonetheless, women may misinterpret this information, which can contribute to the epidemic of eating disorders, such as bulimia and anorexia. How can we better educate people about the scientific information on mate choice and the dangers of eating disorders?

3. Studies of human mate choice are generally complex, and often controversial. Should federal agencies fund these studies? Why or why not?

offspring, and find food. A group of impalas is more likely to hear an approaching predator than a solitary one. Many fish moving rapidly in different directions might distract a would-be predator. Weaver birds form giant colonies that help protect them from predators, but the birds may also share information about food sources. Primate members of the same baboon troop signal to one another when they have found an especially bountiful fruit tree. Having a larger number of individuals looking for food increases the chances of finding it.

Group living does have its disadvantages. When animals are crowded together into a small area, disputes can arise over access to the best feeding places and sleeping sites. Dominance hierarchies are one way to divide resources, but this puts subordinates at a disadvantage. Among red deer, sons increase the potential for the harem master to have a greater number of grandchildren. However, sons, being larger than daughters, need to be nursed more frequently and for a longer period of time. Subordinate females do not have access to enough food resources to adequately nurse sons and, therefore, they tend to rear daughters, not sons. Still, like the subordinate males in a baboon troop, subordinate females in a red deer harem may be better off in terms of fitness if they stay with a group instead of trying to survive on their own.

Living in close quarters exposes individuals to illness and parasites that can easily pass from one animal to another. Social behavior helps to offset some of the proximity disadvantages. For example, baboons and other types of social primates invest much time in grooming one another, and this most likely helps them remain healthy. Humans use extensive medical care to help offset the health problems that arise from living in densely populated cities around the world.

Sociobiology and Human Culture

Humans today rely on living in organized societies. Clearly, the benefits of social living must outweigh the costs because we have organized governments, with laws that tend to increase the potential for human survival. The **culture** of a human society involves a wide spectrum of customs, ranging from how to dress to forms of entertainment, marriage rituals, and types of food. Language and the use of tools are essential to human culture. Language is used to socialize children, teach them how to use tools, educate them, and train them in skills that will increase their chances of success. Technological skills, such as knowing how to use a computer or how to fix plumbing, are taught from an early age. Certainly, cultural evolution has surpassed biological evolution in the past few centuries. For example, medical advances, which are passed on through language, have increased the human life expectancy from 40 to 45 years a century ago to nearly 80 years today.

Why did human societies originate? Perhaps the earliest organized societies were composed of "hunter-gatherers." Which of our ancestors first became hunters is debated among paleontologists (biologists who study fossils) and sociologists. Nonetheless, scientists speculate that a predatory lifestyle may have encouraged the evolution of intelligence and the development of language. That is, cooperation, communication, and tools (such as weapons)

are necessary for hunting large animals. In hunter-gatherer societies, men tended to hunt, while women utilized wild and cultivated plants as a source of food. Sometimes, when animal food was scarce, hunter-gatherer societies relied on plant food cultivated by women. Cooperation increased their chances for survival.

Altruism Versus Self-Interest

Altruism can be generally described as a self-sacrificing behavior for the good of another member of the society. In an evolutionary sense, altruism may compromise the fitness of the altruist, while benefiting the fitness of the recipient. Are animals truly altruists, given that natural selection should eventually rid populations of altruists due to their decreased reproductive success? In humans, we can clearly think of examples—a volunteer firefighter dying while trying to save a stranger, a soldier losing his life for his country, a woman diving in front of a car to save an elderly person from being hit. But what about other animals?

In general, altruistic behavior in animals is explained by the concept of **kin selection.** Because close relatives share many of your genes, it may make sense to self-sacrifice to save them. For example, your siblings share 50% of your genes, so sacrificing your life for two brothers or sisters makes sense evolutionarily. **Inclusive fitness** refers to an individual's personal reproductive success, as well as that of his or her relatives, and thus to an individual's total genetic contribution to the next generation. The concepts of kin selection and inclusive fitness are well supported by research in complex animal societies. For example, in a colony of army ants (Fig. 33.14), the queen is inseminated only during her nuptial flight. Thereafter, she spends the rest of her life reproducing constantly, laying up to 30,000 eggs per day! The eggs hatch into three different sizes of sterile female workers whose jobs

Figure 33.14 Altruism and army ants.   A queen army ant has a large abdomen for egg production and is cared for by small ants, called nurses. The idea of inclusive fitness suggests that relatives, in addition to offspring, increase an individual's reproductive success. Therefore, sterile nurses are being altruistic when they help the queen produce offspring to whom they are closely related. ©Alexander Wild/www.alexanderwild.com

contribute to the benefit of the entire colony. The smallest workers (3 mm), called nurses, take care of the queen and the larvae by feeding and cleaning them. The intermediate-sized workers (3–12 mm), which constitute most of the population, go out on raids to collect food. The soldiers (14 mm), which have relatively huge heads and powerful jaws, surround raiding parties to protect them and the colony from attack by intruders. The worker ants engage in this apparently altruistic behavior because it has a reproductive advantage.

The queen ant is diploid (2n), but her mate is haploid. Thus, assuming the queen has had only one mate, her sister workers are more closely related to each other than normal siblings (75% versus 50%). Therefore, a worker can achieve higher inclusive fitness by helping her mother (the queen) produce additional sisters than by producing her own offspring, which would share only 50% of her genes. Thus, this behavior is not really altruistic. Rather, it is adaptive because it is likely due to the increase in inclusive fitness of the helpers relative to those that hypothetically might defect and breed on their own. Similar social patterns are observed in certain species of bees and wasps.

Reciprocal Altruism

In some bird species, offspring from a previous clutch of eggs may stay at the nest to help parents rear the next batch of offspring. In a study of Florida scrub jays, the number of fledglings produced by an adult pair doubled when they had helpers. Mammalian offspring are also observed to help their parents (Fig. 33.15). Among jackals in Africa, solitary pairs managed to rear an average of 1.4 pups, whereas pairs with helpers reared 3.6 pups. What are the benefits of staying behind to help? First, a helper is contributing to the survival of its own kin. Therefore, the helper actually gains a fitness benefit. Second, a helper is more likely than a nonhelper to inherit a parental territory—including other helpers. Helping, then, involves making a minimal, short-term reproductive sacrifice in order to maximize future reproductive potential. Therefore, helpers at the nest are also practicing a form of reciprocal altruism. Reciprocal altruism also occurs in animals that are not necessarily closely related. In this event, an animal helps or cooperates with another animal with no immediate benefit. However, the animal that was helped will repay the debt at some later time. Reciprocal altruism usually occurs in groups of animals that are mutually dependent. Cheaters in reciprocal altruism are recognized and not reciprocated in future events. Reciprocal altruism occurs in vampire bats that live in the tropics. Bats returning to the roost after a feeding activity share their blood meal with other bats in the roost. If a bat fails to share blood with one that had previously shared blood with it, the cheater bat will be excluded from future blood sharing.

Check Your Progress 33.4

1. Explain how territoriality is related to foraging for food.
2. Compare and contrast reproductive strategies and forms of sexual selection.
3. Describe the advantages and disadvantages of social living.

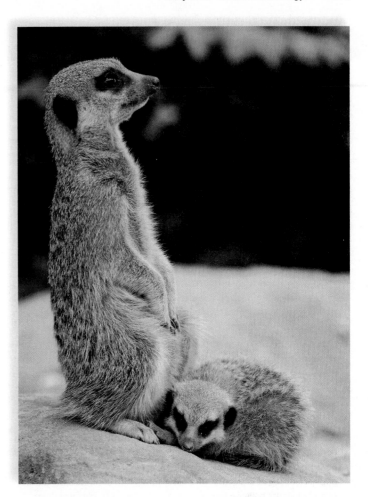

Figure 33.15 Inclusive fitness. A meerkat is acting as a babysitter for its younger sibling while their mother is away. Researchers point out that the helpful behavior of the older meerkat can lead to increased inclusive fitness. ©Biosphoto/Superstock

Conclusion

Female African lions form a society, called a pride, which provides benefits in foraging for food and protection of the young. The members of the pride are all closely related females (mothers, daughters, sisters, and aunts), which promotes altruism. However, this is not without a cost. While female cubs tend to stay in the pride their entire lives, the young males leave, and as they mature will form small groups of genetically related individuals. Upon reaching maturity, the males will establish a territory of their own that will coincide with that of a pride or prides of female lions. The males will often kill cubs that are not of their genetic line in order to get the females to enter estrus. They will then mate with the females in an attempt to pass on their genetics. While group life does have some disadvantages, the benefits must outweigh them in order for it to persist.

SUMMARIZE

33.1 Nature Versus Nurture: Genetic Influences

■ Investigators have long been interested in the degree to which nature (genetics) or nurture (environment) influences **behavior.**

■ Hybrid studies with lovebirds, as well as human twin studies, are consistent with the hypothesis that behavior has a genetic basis. Garter snake experiments indicate that the nervous system controls behavior. Egg-laying behavior in *Aplysia* has endocrine system involvement.

33.2 Nature Versus Nurture: Environmental Influences

■ Even some behaviors formerly thought to be **fixed action patterns (FAPs)** can sometimes be modified by **learning.** A red spot on the bill of laughing gulls initiates chick begging behavior. However, with experience, chick begging accuracy improves.

■ Other studies also support the involvement of learning in behaviors. **Imprinting** during a sensitive period causes birds to follow the first moving object they see.

■ Song learning in birds involves various elements—including learning during the sensitive period, as well as the effect of social interactions outside the period.

■ **Associative learning** includes **classical conditioning** and **operant conditioning.** In classical conditioning, the pairing of two different types of stimuli causes an animal to form an association between them (e.g., dogs salivating at the sound of a bell). In operant conditioning, animals learn behaviors because they are rewarded when they perform them.

33.3 Animal Communication

■ **Communication** is an action by a sender that affects the behavior of a receiver.

■ **Pheromones** are chemical signals passed between members of the same species.

■ Auditory communication includes language, which occurs in humans and other animals.

■ Visual communication allows signaling without auditory or chemical messages.

■ Tactile communication is especially associated with social behavior.

33.4 Behaviors That Affect Fitness

■ Traits that promote reproductive success are expected to have fitness advantages that outweigh their disadvantages.

■ Some animals are territorial and defend a territory that may have high-quality food and/or nesting sites.

■ **Sexual selection** is a form of natural selection that selects for traits that increase an animal's fitness.

■ Males, who produce many sperm, are expected to compete to inseminate as many females as possible.

■ Females, who produce few eggs, are expected to be selective about their mates. Females may choose mates based on increased survival of offspring or on traits that make their sons more attractive to other females.

■ **Culture** in human society has enabled us to dramatically increase the success of our species.

■ It is likely that mating behavior and mate choice in humans have also been shaped by evolutionary forces.

■ Living in a social group can have its advantages (e.g., avoiding predators, raising young, and finding food). It also has disadvantages (e.g., competition between members, spread of illness and parasites, and reduced reproductive potential).

■ When animals live in groups, the fitness (number of fertile offspring produced) benefits must outweigh the costs, or social behavior would not exist.

■ Sometimes animals exhibit **altruism** in order to increase their reproductive success. However, it is necessary to consider **inclusive fitness,** which includes personal reproductive success, as well as the reproductive success of relatives. **Kin selection** is important in social species because it increases the number of individuals who survive that share your genes. For example, social insects help their mother reproduce, but this behavior seems reasonable when we consider that siblings share 75% of their genes. Parental helpers in mammals and birds often inherit the parent's territory.

ASSESS

Testing Yourself

Choose the best answer for each question.

33. Nature Versus Nurture: Genetic Influences

1. An organism's response to a stimulus or situation is the definition of
   a. adaptation.
   b. learning.
   c. memory.
   d. behavior.
   e. epigenetics.

2. Which of the following is not an example of a genetically based behavior?
   a. Inland garter snakes do not eat slugs, whereas coastal populations do.
   b. Mice display a maternal caring instinct.
   c. Some bird species learn a begging behavior for food.
   d. Snails lay eggs in response to an egg-laying hormone.
   e. All of the above are genetic.

33.2 Nature Versus Nurture: Environmental Influences

3. How would the following graph differ if begging behavior in laughing gulls was a fixed action pattern?
   a. It would be a diagonal line with an upward incline.
   b. It would be a diagonal line with a downward incline.
   c. It would be a horizontal line.
   d. It would be a vertical line.
   e. None of these are correct.

4. Which of the following best describes classical conditioning?
   a. the gradual strengthening of stimulus–response connections that seemingly are unrelated
   b. a type of associative learning in which there is no contingency between response and reinforcer
   c. learning behavior in which an organism follows the first moving object it encounters
   d. learning behavior in which an organism exhibits a fixed action pattern from the time of birth

5. In white-crowned sparrows, social experience exhibits a very strong influence over the development of singing patterns. What observation led to this conclusion?
   a. Birds learned to sing only when they were trained by other birds.
   b. The window in which birds learn from other birds is wider than that in which birds learn from tape recordings.
   c. Birds could learn different dialects only from other birds.
   d. Birds that learned to sing from a tape recorder could change their song when they listened to another bird.

33.3 Animal Communication

For questions 6–9, match the type of communication in the key with its description. Answers may be used more than once or not at all.

Key:
   a. chemical communication
   b. auditory communication
   c. visual communication
   d. tactile communication

6. Aphids (insects) release an alarm pheromone when they sense they are in danger.
7. Male peacocks exhibit an elaborate display of feathers to attract females.
8. Ground squirrels give an alarm call to warn others of the approach of a predator.
9. Male silk moths are attracted to females by a sex attractant released by the female moth.
10. Sage grouses perform an elaborate courtship dance.

33.4 Behaviors That Affect Fitness

11. Which reproductive strategy is characterized by a female mating with multiple males?
    a. polygamy
    b. polyandry
    c. monogamy
    d. kin selection
    e. altruism

12. All of the following are benefits obtained through territoriality except
    a. access to mates.
    b. access to more food.
    c. access to more places to hide.
    d. access to more predators.
    e. increased space to raise more offspring.

ENGAGE

Thinking Critically

1. You are studying two populations of rats—one in New York and the other in Florida. Adult New York rats seem to prefer Swiss cheese, whereas Florida rats seem to prefer cheddar. Design an experiment to determine the extent to which this behavior may be genetically controlled versus environmentally influenced.

2. If altruistic behavior has usually evolved to improve the altruist's inclusive fitness by helping relatives, then why do you think humans perform altruistic acts for completely unrelated individuals (e.g., jumping into a pool to save a drowning person)?

CHAPTER

34

Population and Community Ecology

©Ethan Daniels/Shutterstock

CHAPTER OUTLINE

BEFORE YOU BEGIN

Before beginning this chapter, take a few moments to review the following discussions:

Section 6.1 What are the forms of energy?

Section 8.1 How does photosynthesis capture solar energy?

Section 33.4 How does a species' behavior determine the type of relationship it has with other species?

Exotic Species

Lionfish (*Pterois volitans*) are native to the Indo-Pacific region of the world. They were first discovered in the Atlantic as early as 1985 and by the early 2000s had established an Atlantic population that has since grown exponentially. With no natural predators in the Atlantic, they have been able to expand their range to encompass an area from New York to Bermuda, with populations spreading toward South America. With climate change increasing ocean temperatures, the lionfish range will likely extend farther north in the Atlantic.

Lionfish occupy the same ecological niche as economically important fish such as groupers and snappers. This leads to a significant amount of competition of resources between these species. Lionfish have also been shown to be able to reduce the population of prey fish along coral reefs by 90% in as little as 2 months.

The decrease in native prey species, along with the exponential growth of the lionfish population, is placing a significant amount of stress upon the coral reefs of the Atlantic and Caribbean ecosystems. Overconsumption of the prey species by lionfish leads to algae overgrowing coral, which eventually kills the coral. Our knowledge of population growth and regulation has led to strategies that are attempting to control the lionfish population to preserve the health and vitality of coral reef ecosystems in the Atlantic and Caribbean for generations to come.

As you read through this chapter, think about the following questions:

1. What type of factors will limit the growth of a population?
2. Do population growth models apply to humans the same way they apply to other species?
3. Do more-developed or less-developed countries have a greater impact upon the Earth's resources?

34.1 The Scope of Ecology

Learning Outcomes

Upon completion of this section, you should be able to
1. Identify which aspects of biology the study of ecology encompasses.
2. Recognize the different levels of ecological study.

Ecology is the study of the interactions of organisms with each other and with their physical environment. Ecology, like so many biological disciplines, is wide-ranging and involves several levels of study (Fig. 34.1).

At the most basic level, ecologists study how organisms are adapted to their environment. For example, they might study why a particular species of fish in a coral reef lives only within a narrow temperature range in warm tropical waters. Most organisms do not exist singly. Rather, they are part of a **population,** which is defined as all the organisms of the same **species** interacting with the environment in a particular area. At the population level of study, ecologists describe the size of populations over time. For example, an ecologist might study the relative sizes of parrotfish populations within a coral reef over time. A **community** consists of all the various populations at a particular locale. A coral reef community contains numerous populations of fish species, crustaceans, corals, and so forth. At the community level, ecologists study how various extrinsic factors (e.g., weather) and intrinsic factors (e.g., species' competition for resources) affect the size of these populations. An **ecosystem** encompasses a community of populations, as well as the nonliving environment. For example, energy flow and chemical cycling in a coral reef can affect the success of the organisms that inhabit it. Finally, the **biosphere** is that portion of the entire Earth's surface—air, water, land—where living organisms exist. Knowing the composition and diversity of an ecosystem, such as a coral reef, is important to the dynamics of the biosphere.

Modern ecology is not just descriptive; it primarily focuses on developing testable hypotheses. A central goal of modern ecology is to develop models that explain and predict the distribution and abundance of populations and species. Ultimately, ecology considers not just one particular area, but species' distributions throughout the biosphere. In this chapter, we particularly concentrate on the patterns of population growth, and how population growth is regulated. Then, at the community level, we examine the interactions between populations and how they change through time during the process of ecological succession.

Check Your Progress 34.1

1. Distinguish between a population and a community.
2. Describe the central goal of modern ecology.

Figure 34.1 Levels of ecological organization. The study of ecology encompasses levels of organization that proceed from the level of the individual organism to an entire ecosystem.

Organism ⟶ Species ⟶ Population ⟶ Community ⟶ Ecosystem

34.2 Patterns of Population Growth

Each population has a particular pattern of growth. The population size can change according to the **rate of natural increase (r)** or growth rate. Suppose, for example, a human population presently has a size of 1,000 individuals, the birthrate is 30 per year, and the death rate is 10 per year. The growth rate per year will be as follows:

$$30 - 10/1,000 = 0.02 = 2.0\% \text{ per year}$$

Note that this per capita rate of increase disregards both immigration and emigration, which for this example can be assumed to be equal and thus to cancel each other out. The highest possible per capita rate of increase for a population is called its **biotic potential** (Fig. 34.2). Whether the biotic potential is high or low depends on a variety of factors such as (1) reproductive potential of the current population, (2) availability of food, (3) presence or absence of disease, and (4) presence or absence of predators.

Suppose we are studying the growth of an insect population that is capable of infesting and taking over an area. Under these circumstances, **exponential growth** is expected. An exponential pattern of population growth results in a J-shaped curve (Fig. 34.3a). This pattern of population growth can be likened to compound interest at the bank: The more your money increases, the more interest you earn. If the insect population has 2,000 individuals and the per capita rate of increase is 20% per month, there will be 2,400 insects after one month, 2,880 after two months, 3,456 after three months, and so forth.

a.
b.

Figure 34.2 Biotic potential. A population's maximum growth rate under ideal conditions—that is, its biotic potential—is greatly influenced by the number of offspring produced in each reproductive event. **a.** Mice, which produce many offspring that quickly mature to produce more offspring, have a much higher biotic potential than (**b**) the rhinoceros, which produces only one or two offspring during infrequent reproductive events. (a): ©Daniel Heuclin/NHPA/Avalon; (b): ©Tracey Thompson/Corbis

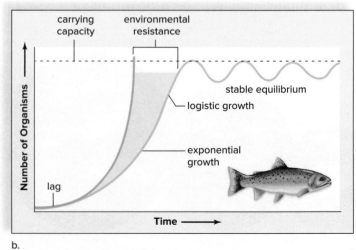

a.
b.

Figure 34.3 Patterns of population growth. **a.** Exponential growth results in a J-shaped growth curve because the growth rate is positive, and increasing, as in insects. **b.** Logistic growth results in an S-shaped growth curve because environmental resistance causes the population size to level off and be in a steady state at the carrying capacity of the environment, as in fish.

Notice that in exponential growth there are two distinct phases (Fig. 34.3*a*):

- *Lag phase.* Growth is slow because the population is small.
- *Exponential growth phase.* Growth is accelerating, and the population is exhibiting its biotic potential.

Usually, exponential growth cannot continue for long because of environmental resistance. **Environmental resistance** encompasses all those environmental conditions—such as limited food supply, accumulation of waste products, increased competition, or predation—that prevent populations from achieving their biotic potential. The amount of environmental resistance will increase as the population grows larger. This eventually causes the population growth to level off, resulting in an S-shaped pattern called **logistic growth** (Fig. 34.3*b*). The characteristics of logistic growth include:

- *Lag phase.* Growth is slow because the population is small.
- *Exponential growth phase.* Growth is accelerating, due to biotic potential.
- *Logistic growth phase.* The rate of population growth slows down.
- *Stable equilibrium phase.* Little if any growth takes place because births and deaths are about equal.

The stable equilibrium phase is said to occur at the **carrying capacity** of the environment. The carrying capacity is the number of individuals of a species that a particular environment can support.

Our knowledge of logistic growth has practical implications. The model predicts that exponential growth occurs only when population size is much lower than the carrying capacity. So, for example, if humans are using a fish population as a continuous food source, it would be best to maintain that population size in the exponential phase of growth when the birthrate is the highest. If we overharvest, the fish population will sink into the lag phase, and it will be years before exponential growth recurs. On the other hand, if we are trying to limit the growth of a pest, it is best to reduce the carrying capacity. Simply reducing the population size only encourages exponential growth to begin again. Farmers can reduce the carrying capacity for a pest by alternating rows of different crops within their fields instead of growing one type of crop throughout the entire field.

Survivorship

Population growth patterns assume that populations are made up of identically aged individuals. In real populations, however, individuals are in different stages of their life spans. Let us consider how many members of an original group of individuals born at the same time, called a **cohort,** are still alive after certain intervals of time. The result is a survivorship curve.

For the sake of discussion, three types of idealized survivorship curves are recognized (Fig. 34.4). The type I curve is characteristic of a population in which most individuals survive well past the midpoint, and death comes near the end of the maximum life span. On the other hand, the type III curve is typical for a population in which most individuals die very young. In the

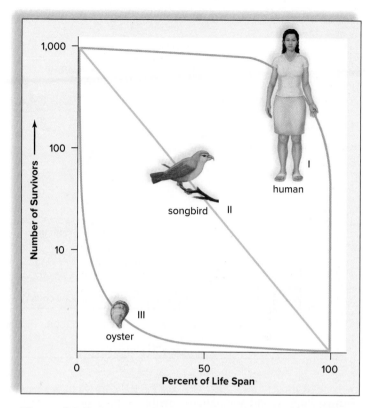

Figure 34.4 Survivorship curves.   Humans have a type I survivorship curve: The individual usually lives a normal life span, and then death is increasingly expected. Songbirds have a type II curve: The chances of surviving are the same for any particular age. Oysters have a type III curve: Most deaths occur during the free-swimming larva stage, but oysters that survive to adulthood usually live a normal life span.

type II curve, survivorship decreases at a constant rate throughout the life span. Some species, however, do not fit any of these curves exactly.

Much can be learned about the life history of a species by studying its survivorship curve. Only a few members of a population with a type III survivorship curve will actually contribute offspring to the next generation. Because death comes early for most members, only a few live long enough to reproduce.

Human Population Growth

Figure 34.5*a* illustrates human population growth. Growth in *less-developed countries* is still in the exponential phase and is not expected to level off until about 2040, while growth in *more-developed countries* has leveled off. The equivalent of a medium-sized city (roughly 220,000) is added to the world's population every day, and 80 million (equivalent to the combined populations of Argentina, Ecuador, and Peru) are added every year.

It has been increasingly difficult to predict patterns of future population growth, not only because the growth rate of the human population is decreasing, but also because economic factors and advances in medicine have an influence on the birthrate. Historically, growth rate was described by the **doubling time,** the length of time it takes for the population size to double. This was considered to be in the range of 35–60 years for humans, but most

b.

c.

Figure 34.5 World population growth. **a.** The graph shows world population size in the past, with estimates to 2150. People in (**b**) more-developed countries have a high standard of living and will contribute the least to world population growth, whereas people in (**c**) less-developed countries have a low standard of living and will contribute the most to world population growth. (b): ©Ariel Skelley/Blend Images LLC; (c): ©Henk Badenhorst/Getty Images

population experts recognize that it will probably take longer to reach 12–14 billion. Regardless, any increase in population size will put increased demands on our ability to produce and distribute resources. In the relatively near future, the world will need to increase the amount of food, jobs, water, energy, and so on just to maintain the present standard of living.

Many people are gravely concerned that the amount of time it takes to add an additional billion people to the world population has become shorter. The first billion was reached in 1800; the second billion in 1930; the third billion in 1960; and today there are over 7.4 billion. By contrast, **zero population growth,** in which the birthrate equals the death rate and population size remains steady, can be achieved only if the per capita rate of increase declines. Recent estimates by the United Nations suggest that the world's population may level off around 10 billion sometime around 2062.

More-Developed Versus Less-Developed Countries

The **more-developed countries (MDCs),** typified by countries in North America and Europe, are those in which population growth is low and people enjoy a good standard of living (Fig. 34.5b). The **less-developed countries (LDCs),** such as countries in Latin America, Africa, and Asia, are those in which population growth is expanding rapidly and the majority of people live in poverty (Fig. 34.5c).

The MDCs doubled their populations between 1850 and 1950 due to a decline in the death rate, the development of modern medicine, and improved socioeconomic conditions. However, the transition from an agricultural society to an urban society reduced the incentive for people to have large families (e.g., more people to help on the farm) because maintaining large families in urban areas is more costly than in rural areas, causing birthrates to decline. These factors, combined with the decreased death rate, caused populations in MDCs to experience only modest growth between 1950 and 1975. This sequence of events (i.e., decreased death rate followed by decreased birthrate) is termed a **demographic transition.**

Yearly growth of the MDCs as a whole has now stabilized at about 0.1%. The populations of several of the MDCs, including Germany, Greece, Italy, Hungary, and Japan, are not growing or are actually decreasing in size. In contrast, there is no leveling off in U.S. population growth due to the number of individuals that immigrate to the United States each year. In addition, an unusually large number of babies were born between 1947 and 1964 (called "baby boomers"). This means a large number of the children of this generation are still of reproductive age.

Although death rates began to decline steeply in the LDCs following World War II with the importation of modern medicine from the MDCs, birthrates remained high. The yearly growth of the LDCs peaked at 2.5% between 1960 and 1965. Since that time, a demographic transition has occurred: the decline in the death rates slowed, and birthrates fell. Yearly growth now averages 1.9%. Due to exponential growth, the population of LDCs may jump from around 5 billion to around 8 billion by 2050. Most of this growth will occur in Africa, Asia, and Latin America. Suggestions for greatly reducing the expected increase include the following:

- Establish and/or strengthen family planning programs. A decline in growth is seen in countries with good family planning programs supported by community leaders. Currently, 25% of women in sub-Saharan Africa say they would like to delay or stop childbearing, and yet they lack access to birth control. Similarly, 15% of women in Asia and Latin America have an unmet need for birth control.

- Use social progress to reduce the desire for large families. Many couples in LDCs presently desire as many as four to six children. But providing education, raising the status of women, reducing child mortality, and improving economic stability are desirable social improvements that could help decrease population growth.

- Delay the onset of childbearing. A delay in the onset of childbearing and wider spacing of births could cause a temporary decline in the birthrate and reduce the present reproductive rate.

Age Distributions

The **age-structure diagram** divides the population into three age groups: prereproductive, reproductive, and postreproductive (Fig. 34.6). The LDCs are experiencing a population momentum because they have more women entering the reproductive years than older women leaving them.

It is a misconception that if each couple has two children, referred to as **replacement reproduction,** zero population growth will occur. The population will continue to grow as long as there are more young women entering their reproductive years than there are older women leaving them, which is common among most of the LDCs. Living many years after reproduction also contributes to an increase in the population. This type of population growth results in *unstable age structure* (Fig. 34.6b). In LDCs, the more quickly replacement reproduction is achieved, the sooner zero population growth will occur. In MDCs, on the other hand, it is more typical that the number of people in the prereproductive class roughly equals the number in the reproductive age class. This causes a *stable age structure,* and population numbers will remain about the same for the foreseeable future (Fig. 34.6a).

a. More-developed countries (MDCs)

b. Less-developed countries (LDCs)

Figure 34.6 Age-structure diagrams.   The diagrams illustrate that (**a**) the MDCs are approaching stabilization, whereas (**b**) the LDCs will expand rapidly, resulting in unstable age structure.

34.3 Interactions Between Populations

Learning Outcomes

Upon completion of this section, you should be able to

1. Identify the difference between density dependence and density independence.
2. Describe how competition and predation can impact population growth.
3. List the three types of symbiosis, and discuss the characteristics of each.

Ecologists wish to determine the factors that regulate population growth. Life history patterns are characterized by how long it takes to reach reproductive maturity and the level of reproductive output

Check Your Progress 34.2

1. Explain why populations don't grow to their biotic potential.
2. Describe the three types of idealized population survivorship curves.
3. Recognize how differences in age structure explain differences in population growth rates between LDCs and MDCs.

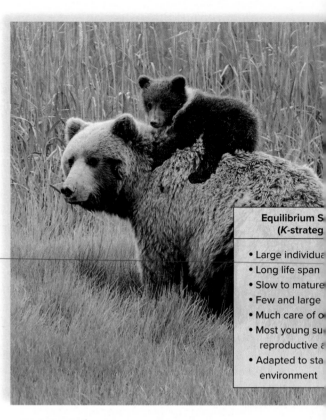

Opportunistic Species
(*r*-strategist)

- Small individuals
- Short life span
- Fast to mature
- Many offspring
- Little or no care of offspring
- Many offspring die before reproducing
- Early reproductive age

Equilibrium S
(*K*-strateg

- Large individua
- Long life span
- Slow to mature
- Few and large
- Much care of o
- Most young su reproductive a
- Adapted to sta environment

Figure 34.7 Life history patterns.   Dandelions are an opportunistic species (*r*-strategist) with the characteristics noted, and bears are an equilibrium species (*K*-strategist) with the characteristics noted. Often, the distinctions between these two possible life history patterns are not as clear-cut as they may seem. (photos) (dandelions): ©Elena Elisseeva/Alamy; (bears): ©Barrett Hedges/National Geographic/Getty Images

(Fig. 34.7). Populations that are small in size, mature early, and have a short life span exhibit an *opportunistic pattern*. They are also known as *r-strategist* species, which tend to produce many relatively small offspring and forego parental care in favor of a greater number of offspring. The more offspring, the more likely it is that some of them will survive to a reproductive age. Classic examples of opportunistic species are many insects and weeds.

In contrast, a population that remains pretty much at the carrying capacity and allocates energy to its growth and survival, as well as to the growth and survival of its offspring, follows an *equilibrium pattern*. These populations are known as *K-strategist* species, which tend to be fairly large, are slow to mature, and have a fairly long life span. They are specialists rather than generalists and tend to become extinct when their normal way of life is altered. The best examples of equilibrium species are found among birds and mammals. For example, the Florida panther, the largest mammal in the Florida Everglades, requires a very large range, and produces few offspring, which must be cared for. Currently, the Florida panther is unable to compensate for a reduction in its range due to human destruction of its habitat, and is therefore on the verge of extinction.

It is recognized that the environment contains both abiotic and biotic components. Abiotic factors, such as weather and natural disasters, are **density-independent factors,** meaning that the effects of the factor are the same for all sizes of populations. For example, fires don't necessarily kill a larger percentage of individuals as the population increases in size. On the other hand, biotic factors, such as competition, predation, and parasitism, are called **density-dependent**

factors. The effects of a density-dependent factor depend on the size of the population. The denser a population is, the faster a disease might spread, for example. Populations that have the opportunistic life history pattern tend to be regulated by density-independent factors, and those that have the equilibrium life history pattern tend to be regulated by density-dependent factors.

Competition

Competition, a density-dependent factor, occurs when members of different species try to utilize a resource (such as light, space, or nutrients) that is in limited supply. Understanding the effects of competition requires a grasp of the concept of ecological niche. The **ecological niche** is the role a species plays in the community, including the **habitat** it requires and its interactions with other organisms. The niche is essentially a species' total way of life, and thus includes the resources needed to meet its energy, nutrient, survival, and reproductive demands.

According to the **competitive exclusion principle,** no two species can occupy the same ecological niche at the same time if resources are limited. This principle is exemplified by an experiment involving two species of paramecia. When grown together in one test tube with limited resources, only one species will survive (Fig. 34.8). In another laboratory experiment, two species of paramecia can occupy the same tube if one species feeds on bacteria at the bottom of the tube and the other feeds on bacteria suspended in solution. The division of feeding niches, called **resource partitioning,** decreases competition between the two

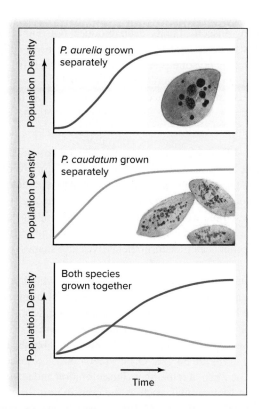

Figure 34.8 Competition between two laboratory populations of paramecia. When grown alone in pure culture (top two graphs), *Paramecium aurelia* and *Paramecium caudatum* exhibit logistic growth. When the two species are grown together in mixed culture (bottom graph), *P. aurelia* is the better competitor, and *P. caudatum* dies out. This experiment illustrates the competitive exclusion principle.

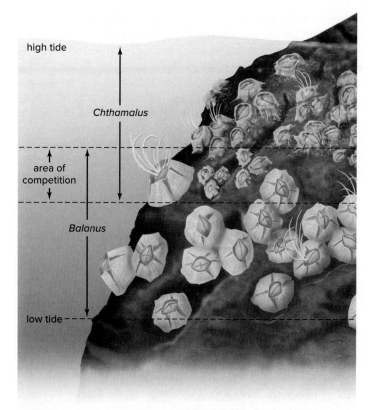

Figure 34.9 Competition between two species of barnacles. Competition prevents two species of barnacles from occupying as much of the intertidal zone as possible. Both exist in the area of competition between *Chthamalus* and *Balanus.* Above this area, only *Chthamalus* survives, and below it only *Balanus* survives.

species and allows occupancy of different niches and therefore survival.

On the Scottish coast, a small barnacle (*Chthamalus stellatus*) lives on the high part of the intertidal zone, and a large barnacle (*Balanus balanoides*) lives on the lower part (Fig. 34.9). Free-swimming larvae of both species attach themselves to rocks at any point in the intertidal zone, where they develop into the sessile adult forms. In the lower zone, the large *Balanus* barnacles seem to either force the smaller *Chthamalus* individuals off the rocks or grow over them. Competition is therefore restricting the range of *Chthamalus* on the rocks. *Chthamalus* is more resistant to drying out than is *Balanus.* Therefore, it has an advantage that permits it to grow in the upper intertidal zone.

It should be noted, however, that the greatest competition for resources will always come from within your own species, an occurrence known as *intraspecific competition.* Two individuals of the same species will have nearly identical requirements for survival, thus producing competition when resources are limited.

Predation

Predation occurs when one organism, called the **predator,** feeds on another, called the **prey.** In the broadest sense, predators include not only animals such as lions that kill zebras, but also filter-feeding blue whales that strain krill from ocean waters, and even white-tailed deer that feed upon a farmer's cornfield.

Predator–Prey Population Dynamics

Predators reduce the population density of prey, as shown by a laboratory study in which the protozoans *Paramecium caudatum* (prey) and *Didinium nasutum* (predator) were grown together in a culture medium. *Didinium* ate all the *Paramecium* and then died of starvation. In nature, we can find a similar example. When a gardener brought prickly-pear cactus to Australia from South America, the cactus spread out of control until millions of acres were covered with nothing but cacti. The cacti were brought under control when a moth from South America, whose caterpillar feeds only on the cactus, was introduced. Now, both cactus and moth are found at greatly reduced densities in Australia.

Mathematical formulas predict that predator and prey populations tend to cycle instead of maintaining a steady state. Cycling can occur when (1) the predator population overkills the prey, causing the predator population to decline in number, or (2) the prey population overshoots the carrying capacity and suffers a crash, causing the predator population to crash due to the lack of food. In either case, the result would be a series of peaks and valleys in population density of both species, with the predator population lagging slightly behind the prey.

A famous case of predator–prey cycles occurs between the snowshoe hare and the Canadian lynx, a type of small, predatory cat (Fig. 34.10). The snowshoe hare is a common herbivore in the coniferous forests of North America, where it feeds on terminal twigs of various shrubs and small trees. The Canadian lynx feeds on snowshoe hares, but also on ruffed grouse and spruce grouse, two types of birds. Investigators at first assumed that the lynx had brought about the decline of the hare population. But others noted that the decline in snowshoe hare abundance was accompanied by low growth and reproductive rates that could be signs of a food shortage. It appears that both explanations apply to the data. In other words, both a predator–hare cycle and a hare–food cycle have combined to produce an overall effect, which is observed in Figure 34.10. Although not shown on the graph, densities of the grouse populations also cycle, perhaps because the lynx switches to this food source when the hare population declines. Predators and prey do not normally exist as simple, two-species systems, and therefore, abundance patterns should be viewed with the complete community in mind.

Antipredator Defenses

While predators have evolved strategies to secure the maximum amount of food with minimal expenditure of energy, prey organisms have evolved strategies to escape predation. As is discussed in the Scientific Inquiry feature "Interactions and Invasive Species" the process of **coevolution** occurs when two species adapt in response to selective pressure imposed by the other. This phenomenon applies to predation, as well as to symbiotic interactions.

In plants, the sharp spines of the cactus, the pointed leaves of holly, and the tough, leathery leaves of the oak tree all discourage predation by insects. Plants even produce poisonous chemical compounds to deter predation. Animals have varied antipredator defenses such as poisonous secretions, concealment, fright, flocking together, and warning coloration (Fig. 34.11), as well as mimicry. A good example of concealment occurs with caddisfly larvae, which build protective cases out of sticks and sand, making themselves look like the stream bottom where they reside.

Mimicry   **Mimicry** occurs when one species resembles another species that has evolved to defend against predators or resembles an object in the environment. Mimicry can help a predator capture food or help a prey avoid capture. For example, angler fishes have lures that resemble worms for the purpose of bringing fish within reach. To avoid capture, some inchworms resemble twigs, and some caterpillars can transform themselves into shapes resembling snakes.

Batesian mimicry (named for Henry Bates, who discovered it) occurs when a prey species that is not harmful mimics another species that has a successful antipredator defense. Many examples of Batesian mimicry involve warning coloration. For example, stinging insects all tend to have black and yellow bands, like those of the wasp in Figure 34.12a. Once a predator has experienced the defense of the wasp, it remembers the coloration and tends to avoid animals that look similar, such as the nonstinging flower fly and longhorn beetle in Figure 34.12b, c, which can be considered Batesian mimics.

Figure 34.10 Predator–prey cycling of a lynx and a snowshoe hare.   Research indicates that the snowshoe hare population reaches a peak abundance before that of the lynx by a year or more. A study conducted from 1987 to 2009 shows a cycling of the lynx and snowshoe hare populations. (photo): ©Alan Carey/Science Source (graph): Source: Kluane Ecological Monitoring Project.

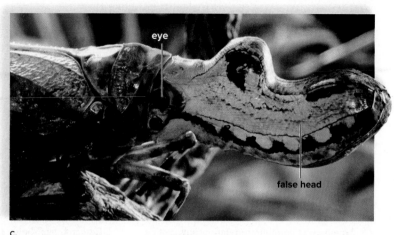

false eyespots

eye

false head

a.　　　　　　　　　b.　　　　　　　　　c.

Figure 34.11 Antipredator defenses. **a.** The skin secretions of poison-dart frogs are so poisonous that they were used by natives to make their arrows instant lethal weapons. The bright coloration of these frogs is called "warning coloration" because brightly colored species are generally toxic. **b.** The caterpillar of the eastern tiger swallowtail butterfly has false eyespots used to confuse a predator. **c.** The South American lantern fly has a large false head that resembles that of an alligator. This may frighten a predator into thinking it is facing a dangerous animal. (a): ©MedioImages/Superstock; (b): ©Scott Camazine/Science Source; (c): ©National Audubon Society/A. Cosmos Blank/Science Source

a.　　　　　　　　　b.　　　　　　　　　c.　　　　　　　　　d.

Figure 34.12 Mimicry among insects. **a.** A yellow jacket wasp uses stinging as its defense. **b.** A flower fly and (**c**) a longhorn beetle are Batesian mimics because they are incapable of stinging another animal, and yet they have the same appearance as the yellow jacket wasp. **d.** A bumblebee and the yellow jacket wasp are Müllerian mimics because they have a similar appearance, and both use stinging as a defense. (a): ©Rami Aapasuo/Alamy; (b): ©Pauline S. Mills/Getty Images; (c): ©blickwinkel/Alamy; (d): ©James H. Robinson/Science Source

Another type of mimicry occurs when species that resemble each other all have successful defenses. For example, many coral snake species have brilliant red, black, and yellow body rings and are venomous. Mimics that share the same protective defense are called Müllerian mimics, named after Fritz Müller, who discovered the phenomenon. The bumblebee in Figure 34.12*d* is a Müllerian mimic of the wasp in Figure 34.12*a*—that is, both insects have the same appearance and mode of defense.

Just as with other antipredator defenses, behavior plays a role in mimicry. Mimicry works better if the mimic acts like the model (the species it is mimicking) in addition to looking like it. For example, beetles that phenotypically resemble a wasp actively fly from place to place and spend most of their time in the same habitat as the wasp model.

Symbiosis

Symbiosis refers to close interactions between members of different species. Three types of symbiotic relationships have traditionally been defined—parasitism, commensalism, and mutualism.

Table 34.1 lists these three categories in terms of their benefits to one species or the other. Of the three types, only mutualism may increase the population size of both species. However, some ecologists now consider it difficult to try to classify symbiotic relationships into these categories, because the amount of harm or good the individuals of two species do to one another depends on what the investigator chooses to measure. Although the following discussion describes the traditional classification system, bear in mind that symbiotic relationships do not always fall neatly into these three categories.

TABLE 34.1 Symbiotic Relationships

	Species 1	**Species 2**
Parasitism*	Benefited	Harmed
Commensalism	Benefited	No effect
Mutualism	Benefited	Benefited

* Can be considered a type of predation.

Interactions and Invasive Species

A symbiotic relationship is one in which there is a close association between two species. Mutualism, commensalism, parasitism, and predation are examples of symbiotic relationships that occur within communities. As communities go through succession, populations evolve in response to each other. If they don't, one or both species may go extinct.

However, a significant disruption in the dynamics of community relationships can develop if an invasive species is introduced into a new community. Native predators and prey have not evolved in the presence of the invasive species. This results in the native predators being unable to control the population size of the invasive species. The prey of the invasive species will not have evolved any type of defensive mechanism against the invasive species, often resulting in the prey population being severely depleted.

The emerald ash borer (*Agrilus planipennis*) is an attractive beetle 1/2 inch long by 1/8 inch wide that has metallic green wing covers and a coppery red or purple abdomen. The beetles were first discovered in southeastern Michigan during the summer of 2002 and by August 3, 2017, had spread across the entire eastern, central, and southern United States (Fig. 34A).

It is believed that the beetles were brought, accidentally, to the United States in the wood of packing crates from Asia. Adult beetles cause little damage to the overall health of the ash trees, but the larval stage can cause enough damage to kill the trees.

Research has shown that the emerald ash borer can have a 1-, 2-, or 3-year life cycle. Adults usually emerge between May and late June, leaving a small D-shaped hole in the bark of the tree. They mate shortly after emergence and females deposit their eggs beneath the bark of nearby ash trees. The larvae will go through four different stages of development, overwinter in a small chamber beneath the bark, and then emerge in the spring. It is during the stages of development that the larvae burrow their way through the phloem layers of the tree, creating a serpentine pattern of tunnels that disrupt the tree's flow of water and nutrients. The upper leaves and branches of the tree die first, and usually within 1–3 years, most trees have died.

The emerald ash borer is sometimes confused with the native cottonwood borer or the white-spotted pine sawyer, but both native species have natural predators in the United States that help keep their population in check. The problem with the emerald ash borer is that there are no natural predators in the United States, so its population has been able to grow exponentially. It is this uncontrolled population growth and predation that has led to the loss of hundreds of millions of ash trees. As a result of the emerald ash borer being accidentally introduced into the United States, the symbiotic relationships that have evolved over the course of millions of years in our forests are in danger of being destroyed.

Questions to Consider

1. What happens to a species if it cannot coevolve along with the species it is interacting with?
2. What will happen to the native borer and sawyer species if the emerald ash borer population continues to grow exponentially?

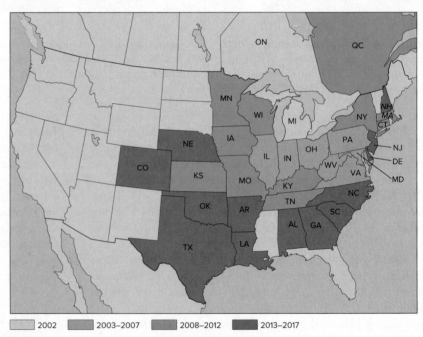

2002   2003–2007   2008–2012   2013–2017

Figure 34A Emerald ash borer. **a.** The emerald ash borer is an invasive species that has (**b**) spread to most of the states in the central and eastern United States. ©Monique van Someren/Moment/Getty Images

Parasitism

Parasitism is a symbiotic relationship in which the **parasite** derives nourishment from another organism, called the *host*. Therefore, the parasite benefits and the host is harmed. Parasites occur in all kingdoms of life. Bacteria (e.g., strep infection), protists (e.g., malaria), fungi (e.g., athlete's foot), plants (e.g., mistletoe), and animals (e.g., tapeworm) all include parasitic species. The effects of parasites on the health of the host can range from slightly weakening them to eventually killing them over time.

In addition to providing nourishment, some host organisms also provide their parasites with a place to live and reproduce, as well as a mechanism for dispersing offspring to new hosts. Many parasites have both a primary and secondary host. The secondary host may be a vector that transmits the parasite to the next primary host. As an example, consider the deer ticks *Ixodes dammini* and *I. ricinus* in the eastern and western United States, respectively. Deer ticks are arthropods that go through a number of stages (egg, larva, nymph, adult). They are so named because adults feed and mate on white-tailed deer in the fall. The female lays her eggs on the ground, and when the eggs hatch in the spring, they become larvae that feed primarily on white-footed mice. If a mouse is infected with the bacterium *Borrelia burgdorferi,* the larvae become infected also. The larvae overwinter and molt the next spring to become nymphs that can, by chance, take a blood meal from a human. At this time, the tick may pass the bacterium on to a human, who subsequently comes down with Lyme disease, characterized by arthritic-like symptoms and often a "bull's-eye" rash around the site of the tick bite. The nymphs develop into adults, and the cycle begins again.

Commensalism

Commensalism is a symbiotic relationship between two species in which one species is benefited and the other is neither benefited nor harmed. Often, one species provides a home and/or transportation for the other species, as when barnacles attach themselves to whales.

Clownfishes live within the waving mass of tentacles of sea anemones. Because most fishes avoid the venomous tentacles of the anemones, clownfishes are protected from predators. If clownfishes attract other fishes on which the anemone can feed, this relationship borders on mutualism. Other examples of commensalism may also be considered mutualistic. For example, birds called cattle egrets benefit from feeding near cattle because the cattle flush insects and other animals from the vegetation as they graze (Fig. 34.13).

Mutualism

Mutualism is a symbiotic relationship in which both members of the association benefit, though not necessarily equally. An example is the relationship between plants and their animal pollinators. When herbivores, such as insects, feed on pollen, they gain a meal, while the plants increase their reproductive chances. Ants form mutualistic relationships with both plants and insects. In tropical America, the bullhorn acacia tree is adapted to provide a home for ants of the species *Pseudomyrmex ferruginea.* Unlike other acacias, this species has swollen thorns with a hollow interior,

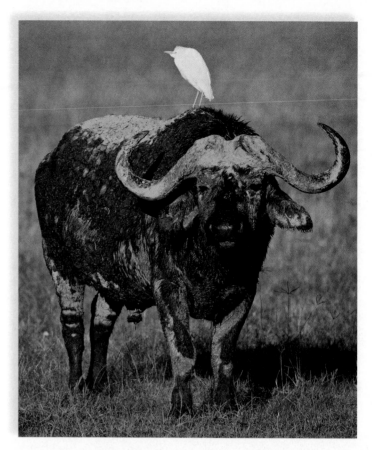

Figure 34.13 Egret symbiosis. Cattle egrets eat insects off and around various animals, such as this African cape buffalo. ©James Hager/Getty Images

where ant larvae can grow and develop. In addition to housing the ants, the acacias provide them with food. The ants constantly protect the plant from the caterpillars of moths and butterflies by swarming and stinging them, and from other plants that might shade the plant. When the ants on experimental trees were poisoned, the trees died.

Cleaning symbiosis is a relationship in which the individuals being cleaned are often vertebrates. Crustaceans, fish, and birds act as cleaners and are associated with a variety of vertebrate clients. Large fish in coral reefs line up at cleaning stations and wait their turn to be cleaned by small fish that often enter the mouths of the large fish (Fig. 34.14). It's been suggested that cleaners may be exploiting the relationship by feeding on host tissues, as well as on ectoparasites. On the other hand, cleaning could ultimately lead to net gains in client fitness by ridding them of parasites.

Check Your Progress 34.3

1. Describe an example of mutualism.
2. Identify examples of populations that would exhibit opportunistic growth.
3. Explain how resource partitioning can allow for the coexistence of species that have similar niches.
4. Recognize the various interactions that will exist between various species within a community.

Figure 34.14 Cleaning symbiosis. A cleaner wrasse, *Labroides dimidiatus*, in the mouth of a spotted sweetlip, *Plectorhincus chaetodonoides*, is feeding off parasites. Does this association improve the health of the sweetlip, or is the sweetlip being exploited? Investigation is underway. ©Bill Wood/Avalon

34.4 Ecological Succession

Learning Outcomes

Upon completion of this section, you should be able to
1. Compare the various types of succession.
2. Choose the correct sequence of events that occur during ecological succession.

Communities are composed of all the interacting populations in an area that are regulated by biotic interactions and ecological disturbances. **Ecological succession** is a change in a community's composition that is directional and follows a continuous pattern of colonization by new species. On land, *primary succession* is the establishment of a plant community in a newly formed area where there is no soil formation. Primary succession typically follows a major disturbance such as a volcanic eruption or a glacial retreat. *Secondary succession* is the return of a community to its natural vegetation following a disturbance, as when a cultivated cornfield returns to its natural state.

Pioneer species are the first species to begin the process of secondary succession of disturbed habitats. Succession progresses through a series of stages, as shown in Figure 34.15. Note that, in this case, the process began with grasses followed by shrubs, and then a mixture of shrubs and trees, until finally there were only trees.

Succession can also occur in aquatic communities, as when lakes and ponds sometimes undergo stages whereby they become filled in and eventually disappear.

Models of Succession

Ecologists have developed various hypotheses to explain succession and predict future events. The *climax-pattern model* of succession says that particular areas will always lead to the same type of community, called a **climax community.** This model is based on the observation that climate plays a significant role in determining whether a desert, a grassland, or a particular type of forest results. Therefore, coniferous forests are expected to occur in northern latitudes, deciduous forests in temperate zones, and tropical rain forests in the tropics. The climax-pattern model of succession now recognizes that the exact composition of a community need not always be the same. That is, while we might expect to see a coniferous forest as opposed to a tropical rain forest in northern latitudes, the exact mix of plants and animals after each stage of succession can vary.

Figure 34.15 Secondary succession in a forest. Secondary succession is a process that begins in areas that already have soil (not on bare rock as primary succession does). This example of secondary succession occurred in a large conifer plantation in central New York State. Notice that certain species are common to the particular stages. However, the process of regrowth shows approximately the same series of stages as secondary succession would in a former cornfield. (Arrows indicate the passage of time.)

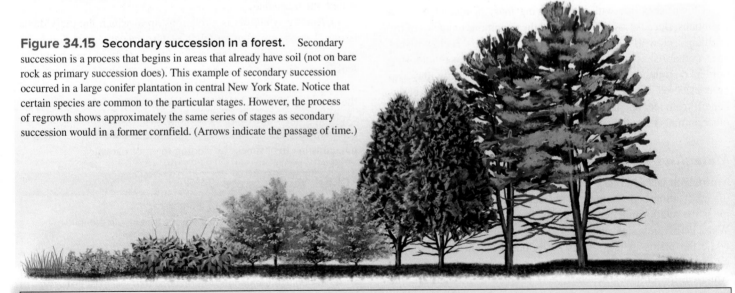

grasses → low shrubs → high shrubs → shrub/tree mix → low trees → high trees → climax community

The *facilitation model* of succession can be observed as shown in the example in Figure 34.15. It's possible that each successive community prepares the way for the next, and this is why grass–shrub–forest development occurs sequentially.

On the other hand, the *inhibition model* says that colonists hold on to their space and inhibit the growth of other plants until the colonists die or are damaged. Still another possible model, the *tolerance model,* predicts that different types of plants can colonize an area at the same time. Sheer chance determines which seeds arrive first, and successional stages may simply reflect the length of time it takes individuals to mature. This alone could account for the grass–shrub–forest development seen in Figure 34.15. In reality, the models we have mentioned are not mutually exclusive, and succession is recognized as a complex process.

Check Your Progress 34.4

1. Identify the events that would lead to succession.
2. Compare and contrast the various models used to explain succession.

Conclusion

Lionfish were accidentally introduced into the Atlantic in the mid 1980s. Since then, their population has exploded and now extends as far north as New York and are spreading as far south as South America. Lionfish can be found across a wide range of marine habitats, ranging from hard bottoms, mangroves, and sea grass beds to coral reefs and artificial habitats like shipwrecks in water as shallow as 1 foot or as deep as 1,000 feet.

Due to the ecological role of these fish, their biotic potential, and a lack of native predators, lionfish are on track to utterly disrupt coral reef ecosystems across the Atlantic and Caribbean.

Concerns include the loss of commercially important species like groupers and snappers, as well as the impact lionfish may have upon the ecotourism industry of the Atlantic and Caribbean. Understanding population ecology is critical to determining the impact this species will have upon the other species that live in the same community.

SUMMARIZE

34.1 The Scope of Ecology

- Ecologists study biotic and abiotic interactions of organisms at several levels: individual, **species, population, community, ecosystem,** and **biosphere. Ecology** is the study of the interactions between organisms and their physical environment.

34.2 Patterns of Population Growth

- The **rate of natural increase (r)** is calculated by subtracting the number of deaths from the number of births and dividing by the number of individuals in the population.

- Two patterns of population growth are possible. Exponential growth results in a J-shaped curve because, as the population increases in size, so does the number of new members.

- Under ideal circumstances, exponential growth occurs due to a population's **biotic potential.** However, exponential growth cannot continue indefinitely because **environmental resistance** opposes biotic potential, and logistic population growth occurs.

- Under **logistic growth,** an S-shaped growth curve results when the population stops growing near the **carrying capacity.**

- Populations tend to have one of three types of survivorship curves, depending on whether most individuals live out the normal life span (type I), die at a constant rate regardless of age (type II), or die early (type III). To best follow population growth, a **cohort,** or group of individuals born at the same time, is tracked throughout their life.

- Human population can be appreciated by looking at the **doubling time,** or the length of time it takes for the population size to double. **Zero population growth** can be achieved when the birthrate equals the death rate.

- **More-developed countries** (MDCs), such as the United States and most of western Europe, are those in which the population growth is low and people have a good standard of living. A **demographic transition** occurs when a decreased death rate is followed by a decreased birthrate, causing the population growth of countries to slow.

- The human population is expanding exponentially, mostly within **less-developed countries** (LDCs) in Africa, Asia, and Latin America. Support for family planning, human development, and delayed childbearing could help lessen the expected increase.

- **Age-structure diagrams** help divide the population into three age groups: prereproductive, reproductive, and postreproductive. **Replacement reproduction** occurs when each couple has two children, but it does not lead to zero population growth.

34.3 Interactions Between Populations

- Opportunistic species rapidly produce many young, generally lack parental care, and rely on rapid dispersal to new, unoccupied environments. Their population size is regulated by **density-independent factors.**

- Equilibrium species produce few young, which often require parental care, and their population size is typically regulated by **density-dependent factors,** such as **competition** and **predation.**

- The **competitive exclusion principle** states that no two species can occupy the same **ecological niche** at the same time when resources are limiting. With limited resources, either **resource partitioning** occurs or one species goes locally extinct. Each species will require a specific **habitat** in order to meet its basic needs.

- **Coevolution** occurs when two species adapt in response to selective pressure imposed by the other.

- Predator–prey interactions can cause **prey** populations to decline and remain at relatively low densities, cause decline of **predator** populations, or cycling of both predator and prey population densities.

- Prey defenses take many forms, such as concealment, use of fright, and warning coloration. Batesian **mimicry** occurs when one species has the warning coloration but lacks the defense of another species. Müllerian mimicry occurs when two species with the same warning coloration have a similar defense.
- In **parasitism,** the **parasite** benefits and the host is harmed. Parasites often utilize more than one host. In **commensalism,** neither party is harmed, and one species often provides a home and/or transportation for another species. In **mutualism,** both partners benefit. Examples of mutualistic relationships include flowers and their pollinators, ants and acacia trees, and cleaning **symbiosis.**

34.4 Ecological Succession

- A change in community composition over time is called **ecological succession.**
- The climax-pattern model of succession says that particular areas will always lead to the same type of community.
- The facilitation model implies that some community members prepare the landscape for new species during succession. The inhibition model suggests that some species prevent colonization, and the tolerance model suggests that species colonize simultaneously.
- A **climax community** is stable and associated with a particular geographic area.

ASSESS

Testing Yourself

Choose the best answer for each question.

34.1 The Scope of Ecology

1. Place the following levels of organization in order, from lowest to highest.
   a. population, organism, community, ecosystem
   b. community, ecosystem, population, organism
   c. organism, community, population, ecosystem
   d. population, ecosystem, organism, community
   e. organism, population, community, ecosystem

2. Which of these levels of ecological study involves an interaction between a community of organisms and their environment?
   a. organisms
   b. populations
   c. communities
   d. ecosystems
   e. All of these are correct.

34.2 Patterns of Population Growth

3. Assume that a deer population contains 500 animals. The birthrate is 105 animals per year, and the death rate is 100 animals per year. The rate of natural increase per year is
   a. 5%.
   b. 4%.
   c. 3%.
   d. 2%.
   e. 1%.

4. What type of survivorship curve would you expect for a plant species in which most seedlings die?
   a. type I
   b. type II
   c. type III

5. Label this logistic growth curve:

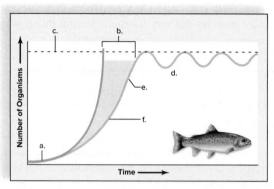

6. If an age-structure diagram indicates that there are more people in the prereproductive stage of life than in the reproductive stage, then replacement reproduction, over the long run, will result in
   a. zero population growth.
   b. an increase in population size.
   c. a decrease in population size.

34.3 Interactions Between Populations

7. Which of the following defines the role of a species in a community, including its habitat and interaction with other species?
   a. ecological niche
   b. competitive exclusion
   c. competition level
   d. mimicry
   e. None of these are correct.

8. A colorful, nonpoisonous frog that mimics a poison-dart frog is exhibiting
   a. Batesian mimicry.
   b. Müllerian mimicry.

For problems 9–11, indicate the type of symbiosis illustrated by each example. Some answers may be used more than once.

Key:
   a. parasitism
   b. commensalism
   c. mutualism

9. Single-celled algae live in the tissues of coral animals. The algae provide food for the coral, while the coral provides a stable home for the algae.

10. Flowering plants reward visiting insects with nectar. The insects, in turn, carry pollen to other flowers.

11. Small wasps lay eggs on other insects. The eggs hatch into larvae that feed on the insects and kill them.

34.4 Ecological Succession

12. Mosses growing on bare rock will eventually help to create soil. These mosses are involved in _____ succession.
   a. primary
   b. secondary
   c. tertiary

13. Assume that a farm field is allowed to return to its natural state. By chance, the field is first colonized by native grasses, which begin the succession process. This is an example of which model of succession?
   a. climax-pattern
   b. tolerance
   c. facilitation
   d. inhibition

ENGAGE

Thinking Critically

1. You are a farmer fighting off a crop insect pest. You know from evolutionary biology and natural selection that using the same pesticide causes resistance to evolve in the insect population. From what you have learned about population growth and regulation, what are some strategies you might use to control the insect population?

2. You find two species of insects, both using bright coloration and similar color patterns as an antipredator defense. Design an experiment to tell whether this is an example of Müllerian or Batesian mimicry.

3. Upland game hunters in Illinois have been noticing a decrease in the rabbit and pheasant populations over the past 5 years. They have lobbied the Department of Natural Resources and the state legislature to allow them to shoot red-tailed hawks, one of the rabbit's and pheasant's main predators. Explain why shooting hawks will not ultimately solve the problem of a decreased rabbit and pheasant population.

4. If predators tend to reduce the population densities of prey, what prevents predators from reducing prey populations to such low levels that they drive themselves extinct?

C H A P T E R

35

Nature of
Ecosystems

BEFORE YOU BEGIN

Before beginning this chapter, take a few moments to review the following discussions:

Section 2.1 Which elements are basic to all life?

Section 2.3 How do the properties of water influence the biogeochemical cycles?

Section 34.3 What role does the predator–prey cycle play in the energy flow of an ecosystem?

The Wolves of Yellowstone

By 1926, the last known wolf pack had been eliminated from Yellowstone National Park. The extinction of the wolves produced a multitude of ecological effects that would ripple through the entire park for the next 70 years. Fewer wolves meant less predation upon the elk and other large herbivores. An increase in herbivores meant more pressure upon their food sources. Cottonwoods, aspen, and streamside vegetation soon became depleted due to the overabundance of elk.

Overbrowsing of aspen trees by the elk and beavers produced a ripple effect throughout the stream ecosystem. This decreased the stability of the stream bank, which in turn caused increased erosion and sediment load in the streams. Fish, insects, birds, and all the wildlife associated with the streams were negatively impacted.

Since the reintroduction of wolves into Yellowstone in 1995, their numbers have grown to 108 wolves, distributed across 11 packs, in 2016. With elk making up the majority of the wolves' prey base, the elk herd has seen a decrease from 19,000 in 1993 to 5,000 in 2016. The decrease in the elk population has resulted in greater stream bank stabilization, along with widespread aspen stand recovery. Aspen forests support more biodiversity than other forest types found in Yellowstone.

In this chapter, you will learn about how energy flows through the members of biotic communities and how abiotic chemicals and nutrients cycle through ecosystems.

As you read through this chapter, think about the following questions:

1. How do trophic levels shape the nature of an ecosystem?

2. How does energy move from producers to consumers in an ecosystem?

35.1 The Biotic Components of Ecosystems

An **ecosystem** is the level of biological organization that includes both abiotic and biotic components (see Fig. 1.2). The **abiotic** components include the nonliving aspects, such as sunlight, inorganic nutrients, and water availability; and conditions, such as type of soil, average temperature, and wind speed. The **biotic** components of an ecosystem are the various populations of organisms that form the community.

Populations Within an Ecosystem

The populations within an ecosystem are categorized according to their food source. **Autotrophs** (i.e., "self-feeders") require only inorganic nutrients and an outside energy source to produce organic nutrients for their own use as food. Because they produce food for themselves and other members of the community, they are called **producers** (Fig. 35.1a). Photosynthetic organisms, also called photoautotrophs, such as algae and plants produce most of the organic nutrients for the biosphere.

Some bacteria are chemoautotrophs. They reduce carbon dioxide to an organic compound by using energetic electrons derived from inorganic compounds, such as hydrogen sulfides, hydrogen gas, or ammonium ions. Chemoautotrophs have been found to support communities at hydrothermal vents along deep-sea oceanic ridges.

Heterotrophs (i.e., "other feeders") need an outside source of organic nutrients. Because they consume food, they are called **consumers. Herbivores** are animals that feed directly on plants or algae (Fig. 35.1b). **Carnivores** feed on other animals. Snakes that feed on rabbits are carnivores, and so are the hawks that feed on these snakes (Fig. 35.1c). In this example, herbivorous rabbits are primary consumers, snakes are secondary consumers, and the hawks are tertiary consumers. Sometimes tertiary consumers are called top predators. **Omnivores** are animals (including humans) that feed on both plants and animals.

Decomposers are heterotrophic bacteria and some species of protists and fungi (such as molds and mushrooms) that break down nonliving organic matter (Fig. 35.1d). They perform a very valuable service because they release inorganic nutrients taken up by plants back into the environment. **Detritus** is partially decomposed matter in the water or soil. Detritivores are found in all ecosystems. Earthworms, termites, maggots, and some beetles are soil detritivores.

a. Producers

b. Herbivores

c. Carnivores

d. Decomposers

Figure 35.1 Biotic components. **a.** Diatoms, a type of algae, and green plants are producers. **b.** Caterpillars and rabbits are herbivores. **c.** Snakes and hawks are carnivores. **d.** Bacteria and mushrooms are decomposers. (a) (left): ©Ed Reschke; (right): ©Hermann Eisenbeiss/Science Source; (b) (left): ©McGraw-Hill Education/Ken Cavanagh, photographer; (right): ©Gerald C. Kelley/Science Source; (c) (left): ©Derrick Hamrick/ImageBroker/Getty Images; (right): ©Tze-hsin Woo/Getty Images; (d) (left): ©Image Source/Getty Images; (right): ©Denise McCullough

Energy Flow and Chemical Cycling

Every ecosystem is characterized by two fundamental phenomena: energy flow and chemical cycling. Energy flow begins when producers absorb solar energy, and chemical cycling begins when producers take in inorganic nutrients from the physical environment (Fig. 35.2). Thereafter, producers make organic nutrients (food) directly for themselves and indirectly for the other populations of the ecosystem. Most ecosystems cannot exist without a continual supply of solar energy. Chemicals cycle when inorganic nutrients are returned to the producers from the atmosphere or soil.

Only a portion of the organic nutrients made by autotrophs is passed on to heterotrophs because plants use some of the organic molecules to fuel their own cellular respiration. Similarly, only a small percentage of nutrients taken in by heterotrophs is available to higher-level consumers. Figure 35.3 shows why. A certain amount of the food eaten by a herbivore is eliminated as feces that are recycled by decomposers. Metabolic wastes are excreted as urine. Of the assimilated energy, a large portion is utilized during cellular respiration to provide energy to the organism and thereafter becomes heat. Only the remaining food, which is converted into increased body weight (or additional offspring), becomes available to carnivores.

In Section 6.1, you learned that energy flow is described by the laws of thermodynamics: (1) energy cannot be created or destroyed; and (2) in every energy transformation, some energy is lost as heat. These principles explain why ecosystems are dependent on a continual outside source of energy, such as sunlight, and why only a part of the original energy from producers is available to consumers.

Figure 35.3 Energy balances.   Only about 10% of the food energy taken in by a herbivore is passed on to carnivores. Much of the rest is lost as heat to the environment. A large portion goes to detritivores via defecation, excretion, and death, and another large portion is used for cellular respiration. (photo): ©George D. Lepp/Science Source

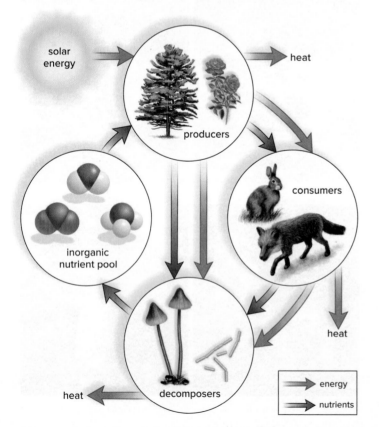

Figure 35.2 Energy flow and nutrient cycling.   Nutrients cycle, but energy flows through an ecosystem. As energy transformations repeatedly occur, all the energy derived from the sun eventually dissipates as heat.

Check Your Progress 35.1

1. Identify the nutritional differences between producers, consumers, and decomposers.
2. Explain why most energy fails to be converted to a usable form when one organism eats another.

35.2 Energy Flow

Learning Outcomes

Upon completion of this section, you should be able to
1. Recognize the difference between a food chain and a food web.
2. Explain the energy flow among populations through food webs and ecological pyramids.

The interconnecting paths of energy flow within ecosystems are represented by diagrams called **food webs.** Figure 35.4*a* is a **grazing food web.** Grazing food webs do not begin with grazers; instead they begin with producers like the trees in a forest. Caterpillars feed on leaves in the trees; and mice, rabbits, and deer feed on leaves at or near the ground. Birds, chipmunks, and mice eat fruits and nuts, but are omnivores because they also feed on caterpillars. These herbivores and omnivores all provide food for a number of different carnivores.

Figure 35.4 Grazing and detrital food webs. Food webs are descriptions of who eats whom. **a.** Tan arrows illustrate possible grazing food webs. For example, birds, which feed on nuts, may be eaten by a hawk. Autotrophs such as the tree are producers (first trophic level); the first series of animals are primary consumers (second trophic level); and the next group of animals are secondary consumers (third trophic level). **b.** Green arrows illustrate possible detrital food webs, which begin with detritus, the waste products and remains of dead organisms. A large portion of these remains are from the grazing food web illustrated in (**a**). The organisms in the detrital food web are sometimes consumed by animals in the grazing food web, as when robins feed on earthworms. Thus, the grazing food web and the detrital food web are connected.

Figure 35.4*b* is a **detrital food web,** which begins with detritus. Detritus is food for soil organisms such as earthworms and beetles. These animals are, in turn, fed on by salamanders and shrews. Because the members of the detrital food web may become food for aboveground carnivores, the detrital and grazing food webs are joined.

In this particular forest, the organic matter lying on the forest floor and mixed into the soil contains over twice as much energy as the leaves of living trees. Therefore, more energy in a forest may be funneling through the detrital food web than through the grazing food web.

Trophic Levels

You can see that Figure 35.4*a* would allow us to link organisms one to another in a straight line, according to who eats whom. Diagrams that show a single path of energy flow such as this are called **food chains.** For example, in the grazing food web, we could find this **grazing food chain:**

leaves → caterpillars → sparrows → hawks

And in the detrital food web (Fig. 35.4*b*), we could find this **detrital food chain:**

detritus → earthworms → shrews

A **trophic level** is composed of all the organisms that feed at a particular link in a food chain. In the grazing food web in Figure 35.4*a*, going from left to right, the trees are primary producers (first trophic level). The first series of animals, the herbivores, are primary consumers (second trophic level). The group of animals at the far right, the carnivores, are secondary consumers (third trophic level).

Ecological Pyramids

The shortness of food chains can be attributed to the loss of energy between trophic levels. In general, only about 10% of the energy of one trophic level is available to the next trophic level. Therefore, if a herbivore population consumes 1,000 kg of plant material, only about 100 kg is converted to herbivore tissue, 10 kg to first-level carnivores, and 1 kg to second-level carnivores. This 10% rule explains why few carnivores can be supported in a food web. The large energy losses that occur between successive trophic levels are sometimes depicted as an **ecological pyramid** (Fig. 35.5).

Biomass is the number of organisms multiplied by their weight. Thinking in terms of biomass rather than simply the number of organisms can be useful, because in terms of only numbers, one oak tree may have hundreds of caterpillars. Although it appears that there are more herbivores than autotrophs, the biomass of autotrophs is much greater. Similarly, the biomass of herbivores is greater than that of carnivores.

However, in aquatic ecosystems, such as lakes and open seas where algae are the main producers, the herbivores may have a greater biomass than the producers. The reason is that, over time, the algae reproduce rapidly, but they are also consumed at a high rate. Any pyramids like this one, which have more herbivores than producers, are called inverted pyramids:

Ecologists are hesitant to use pyramids to describe ecological relationships for all situations. Detrital food chains are rarely included in pyramids, but they may represent a large portion of energy in many ecosystems, making conclusions based on above-ground pyramids potentially inaccurate.

Check Your Progress 35.2

1. Explain why ecosystems generally support few carnivores.

2. Explain how biomass relates to the structure of ecological pyramids.

35.3 Global Biogeochemical Cycles

Learning Outcomes

Upon completion of this section, you should be able to

1. Explain what is meant by a biogeochemical cycle.

2. Identify the steps of the water cycle, the phosphorus cycle, the nitrogen cycle, and the carbon cycle.

3. Identify how human activities can alter each of the biogeochemical cycles.

Because the pathways by which chemicals circulate through ecosystems involve both biotic and geological components, they are known as **biogeochemical cycles.** For each element, chemical cycling may involve (1) a **reservoir**—a source normally unavailable to producers, such as fossilized remains, rocks, and deep-sea sediments; (2) an **exchange pool**—a source from which organisms generally take chemicals, such as the atmosphere or soil; and

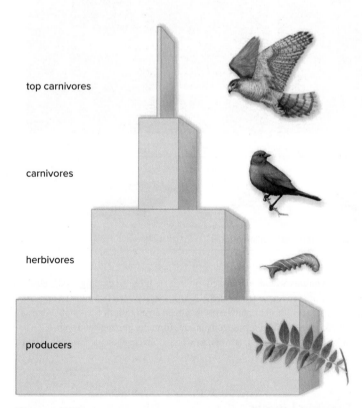

Figure 35.5 Ecological pyramid.   An ecological pyramid reflects the loss of energy from one trophic level to the next. Energy is lost as herbivores feed on producers, carnivores feed on herbivores, and top carnivores feed on carnivores. This loss of energy results in decreasing biomass and numbers of organisms at higher trophic levels relative to lower trophic levels (levels not to scale).

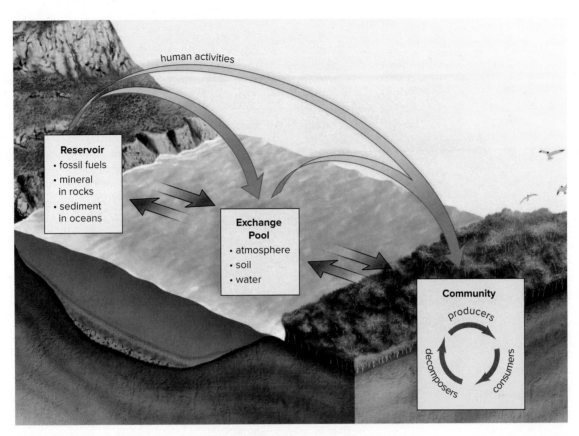

Figure 35.6 Model for chemical cycling.   Chemical nutrients cycle between these components of ecosystems. Reservoirs, which include fossil fuels, minerals in rocks, and sediments in oceans, are normally relatively unavailable sources. However, exchange pools, such as those in the atmosphere, soil, and water, are available sources of chemicals for the biotic community. When human activities (purple arrows) remove chemicals from reservoirs and pools and make them available to the biotic community, pollution can result.

(3) the biotic community—through which chemicals move along food chains, perhaps never entering an exchange pool (Fig. 35.6).

With the exception of water, which exists in gas, liquid, and solid forms (and sometimes as plasma), there are two types of bio-geochemical cycles. In a gaseous cycle, exemplified by the carbon and nitrogen cycles, the element is withdrawn from and returns to the atmosphere as a gas. In a sedimentary cycle, exemplified by the phosphorus cycle, the element is absorbed from soil by plant roots, passed to heterotrophs, and eventually returned to the soil by decomposers.

The diagrams in the next few sections show how nutrients flow between terrestrial and aquatic ecosystems. In the nitrogen and phosphorus cycles, these nutrients run off from a terrestrial to an aquatic ecosystem, and in this way enrich the aquatic ecosystem. However, too much of these nutrients can lead to excessive growth of algae. Decaying organic matter in aquatic ecosystems can be a source of nutrients for intertidal inhabitants such as fiddler crabs. Seabirds feed on fish but deposit guano (droppings) on land, and in that way phosphorus from the water is deposited on land. Anything put into the environment in one ecosystem can find its way to another ecosystem.

As evidence of this, scientists find the soot from urban areas and pesticides from agricultural fields in the snow and animals of the Arctic.

The Water Cycle

The **water (hydrologic) cycle** is described in Figure 35.7. During the water cycle, fresh water is distilled from salt water. First, evaporation occurs. During **evaporation,** a liquid (in this case, water) changes from a liquid state to a gaseous state. The sun's rays cause fresh water to evaporate from seawater, and the salts are left behind. Next, condensation occurs. During **condensation,** a gas is changed to a liquid. The amount of water evaporating from the oceans exceeds the amount of precipitation that falls back into the ocean. Often, this excess moves over and falls on land. Water also evaporates from land, from plants (transpiration), and from bodies of fresh water. Because land lies above sea level, gravity eventually returns all fresh water to the sea. In the meantime, water is contained within standing waters (lakes and ponds), flowing water (streams and rivers), and groundwater.

Figure 35.7 The water cycle.   Evaporation from the ocean exceeds precipitation, so there is a net movement of water vapor onto land, where precipitation results in the eventual flow of surface water and groundwater back to the sea. On land, transpiration by plants contributes to evaporation.

Some of the water from **precipitation** (e.g., rain, snow, sleet, hail, and fog) sinks, or percolates, into the ground and saturates the earth to a certain level. The top of the saturation zone is called the groundwater table, or simply, the water table. Because water infiltrates through the soil and rock layers, sometimes groundwater is also located in **aquifers,** rock layers that contain water and release it in appreciable quantities to wells or springs. Aquifers are recharged when rainfall and melted snow percolate into the soil.

Human Activities

In some parts of the United States, especially the arid West and southern Florida, withdrawals from aquifers exceed any possibility of recharge. This is called "groundwater mining." In these locations, the groundwater level is dropping, and residents may run out of groundwater, at least for irrigation purposes, within a few years.

Fresh water, which makes up only about 3% of the world's supply of water, is called a renewable resource because a new supply is always being produced as a result of the water cycle. But it is possible to run out of fresh water when the available supply is not adequate or has become polluted so it is not usable. As the Ecology feature "The California Drought" explains, the consequences of these droughts may extend well past the regional level.

The Phosphorus Cycle

In the phosphorus cycle, phosphorus moves from rocks on land to the oceans, where it gets trapped in sediments. Phosphorus then moves back onto land following a geological upheaval. You can verify this by following the appropriate arrows in Figure 35.8. However, on land, the very slow weathering of rocks makes phosphate ions (PO_4^{3-} and HPO_4^{2-}) available to plants, which take up phosphate from the soil.

Producers use phosphate to form a variety of molecules, including phospholipids, ATP, and the nucleotides that become a part of DNA and RNA. Animals incorporate some of the phosphate into teeth, bones, and shells that take many years to decompose. Eventually, however, phosphate ions become available to producers once again. Because the available amount of phosphate is already being utilized within food chains, phosphate is usually a limiting inorganic nutrient for plants—that is, their growth is limited by the amount of available phosphorus.

Some phosphate runs off into aquatic ecosystems, where algae acquire phosphate before the excess becomes trapped in sediments. Phosphate in marine sediments does not become available to producers on land again until a geological upheaval exposes sedimentary rocks to weathering once more. Phosphorus does not enter the atmosphere; therefore, the phosphorus cycle is called a sedimentary cycle.

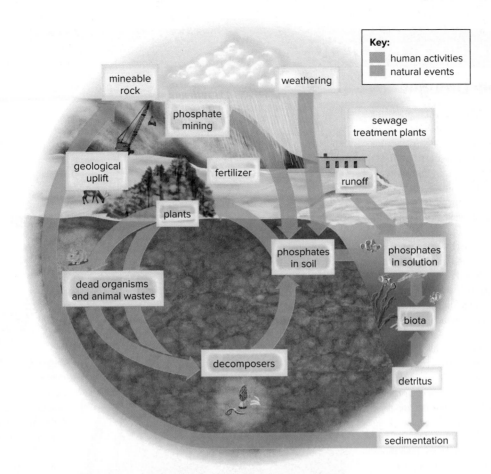

Figure 35.8 The phosphorus cycle. Purple arrows represent human activities; green arrows represent natural events.

Human Activities

Humans boost the amount of available phosphate by mining phosphate ores and using them to make fertilizers, animal feed supplements, and detergents. Fertilizers usually contain three basic ingredients: nitrogen, phosphorus, and potassium. Some laundry detergents still contain approximately 35–75% sodium triphosphate, but many companies are phasing out the phosphates altogether. The amount of phosphate in animal feed varies.

Animal wastes from livestock feedlots, fertilizers from lawns and cropland, and untreated and treated sewage discharged from cities all add excess phosphate to nearby waters. The end result is **cultural eutrophication** (over-enrichment), which can lead to an algal bloom, as indicated by green scum floating on the water or excessive mats of filamentous algae. When the algae die off, decomposers use up all available oxygen for cellular respiration. The result can be a massive fish kill.

The Nitrogen Cycle

Even though nitrogen gas (N_2) makes up about 78% of the atmosphere, it is unavailable for plants to use. Therefore, nitrogen is also a limiting inorganic nutrient for plants.

Nitrogen Fixation

In the nitrogen cycle, **nitrogen fixation** occurs when nitrogen gas (N_2) is converted to ammonium ions (NH_4^+), a form plants can use (Fig. 35.9). Some cyanobacteria in aquatic ecosystems and some free-living bacteria in soil are able to fix atmospheric nitrogen in this way. Other nitrogen-fixing bacteria live in nodules on the roots of legumes (see Chapter 9). They make organic compounds containing nitrogen available to the host plants so the plant can form proteins and nucleic acids.

Nitrification

Plants can also use nitrates as a source of nitrogen. The production of nitrates during the nitrogen cycle is called nitrification. Nitrification can occur in two ways: (1) Nitrogen gas is converted to nitrate (NO_3^-) in the atmosphere when cosmic radiation, meteor trails, and lightning provide the high energy needed for nitrogen to react with oxygen. (2) Ammonium ions in the soil from decomposition are converted to nitrate by soil bacteria in a two-step process. First, nitrite-producing bacteria convert ammonium to nitrite (NO_2^-), and then nitrate-producing bacteria convert nitrite to nitrate. These two groups of bacteria, called the nitrifying bacteria, are chemoautotrophs.

The California Drought

As of April 7, 2017, California governor Edmund G. Brown ended the drought state of emergency for most of the state. Governor Brown credits the water conservation efforts and an abundance of winter rain and snow as the reason for lifting the state of emergency. From 2011 to 2017, California experienced one of the worst multiple-year droughts in its history (Fig. 35A*a*).

There are two sources of water for California—surface water in rivers and streams and water pumped from underground aquifers. The majority of California's water (over 80%) comes from surface water sources, which originates as spring snow melts in the inland mountain ranges. Between 2011 and 2017, the snowpack in nearby mountains had been considerably lower than normal, thus reducing the amount of surface water available.

The aquifers under California offer only a temporary solution because they do not have large enough reserves to compensate for long periods of drought. In California, the aquifers are also deep underground and may take thousands, or even millions, of years to replenish.

Even though the dry conditions of the drought have been mostly eliminated, the damage it caused will remain for many years to come. Farm production in many regions of California has been reduced, over 100 million trees have been killed, wildlife has been harmed, and the drinking water supplies for many rural communities have been disrupted. These consequences will challenge California for decades.

The drought that spanned 2012 to 2016 contained the driest 4-year period in the state's history, along with 2015 recording the smallest Sierra-Cascades snowpack on record. 2014, 2015, and 2016 set records for the three warmest years in terms of statewide average temperatures. What worries scientists most are the computer models, which suggest that the worst is yet to come. When predictions of the current levels of climate change are factored into the models, the data suggest that there is an 80% chance that California and the remainder of the Southwest may experience a "megadrought" (an event lasting more than 35 years) before 2080.

Solutions for the Future

There are only two solutions—decrease water demand or increase the water supply. The population of California has increased by almost 20 million people in the past 40 years, and is

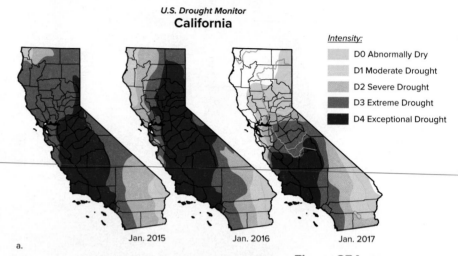

U.S. Drought Monitor
California

Intensity:
- D0 Abnormally Dry
- D1 Moderate Drought
- D2 Severe Drought
- D3 Extreme Drought
- D4 Exceptional Drought

Jan. 2015   Jan. 2016   Jan. 2017

a.

b.

Figure 35A The water crisis in California. **a.** The extent of the drought. **b.** A desalinization plant in southern California. (a): *Source:* The U.S. Drought Monitor is jointly produced by the National Drought Mitigation Center at the University of Nebraska- Lincoln, the United States Department of Agriculture, and the National Oceanic and Atmospheric Administration. Maps courtesy of NDMC-UNL.; (b): ©Gregory Bull/Associated Press

predicted to continue to grow in the future. Mandatory water restrictions frequently help, but these measures often specifically target residential use. As is the case in many parts of the country, most of the water is used by agriculture and industry. In California, the major user is agriculture. California grows the majority of the United States' fruits and vegetables, and thus any disruption has a ripple effect in the food supply.

Increasing the water supply is also difficult. Aquaducts, canals, and pipelines are expensive, and take considerable time to build. Desalinization plants (Fig. 35A*b*), which remove the salt from ocean water, require a significant amount of energy, and often supply water for only a small percentage of the population. For example, the new desalinization plant in San Diego, the largest in the world, will supply around only 10% of the local population with water.

Water usage is a complex problem that requires efforts across all aspects of our society to solve. One single solution will not adequately

address the entire problem. New technologies in the areas of water desalinization may help, as will more research in how reservoirs are recharged in the water cycle. Californians responded to the drought with a significant effort to reduce and conserve their water supplies. Urban water use, across the state, was reduced by nearly 25 percent. California also enacted the Sustainable Groundwater Management Act to improve the management of its water resources. While this drought has come to an end, most scientists agree that there will be more in the future. Water conservation must become a standard way of life for not only Californians, but all people.

Questions to Consider

1. Do you think that the government should be able to require citizens to enact water conservation measures? If so, what type of measures should be required?
2. Should federal tax dollars be allocated to California to help manage their water supply, or should the funds come from California?

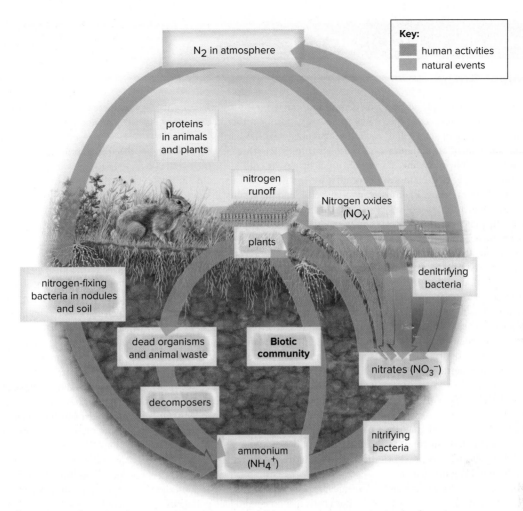

Figure 35.9 The nitrogen cycle.   Purple arrows represent human activities; green arrows represent natural events.

Denitrification

Denitrification is the conversion of nitrate back to nitrogen gas, which will enter the atmosphere. Denitrifying bacteria living in the anaerobic mud of lakes, bogs, and estuaries carry out this process as a part of their own metabolism. In the nitrogen cycle, denitrification would counterbalance nitrogen fixation except for human activities.

Human Activities

Human activities nearly double the nitrogen fixation rate when fertilizers are produced from N_2. Fertilizer, which also contains phosphate, runs off into lakes and rivers and results in algal overgrowth and fish kill.

Fertilizer use also results in the release of nitrous oxide (N_2O), a greenhouse gas, component of acid rain, and contributor to ozone shield depletion.

The Carbon Cycle

In the carbon cycle, plants in both terrestrial and aquatic ecosystems take up carbon dioxide (CO_2) from the air through photosynthesis. They incorporate carbon into food that is used by autotrophs and heterotrophs alike (Fig. 35.10). When all organisms, including plants, respire, a portion of this carbon is returned to the atmosphere as carbon dioxide.

In aquatic ecosystems, the exchange of carbon dioxide with the atmosphere is primarily indirect. However, there is a small quantity of free carbon dioxide in the water. Carbon dioxide from the air combines with water to produce bicarbonate ions (HCO_3^-), a source of carbon for algae that make up the base of most aquatic ecosystems. Similarly, when aquatic organisms respire, the carbon dioxide they give off becomes bicarbonate ion. The amount of bicarbonate in the water is in equilibrium with the amount of carbon dioxide in the air. Thus far, the ocean has acted as a buffer, absorbing much of the excess CO_2, but scientists are concerned about how much the ocean can actually hold.

Reservoirs Hold Carbon

Living and dead organisms contain organic carbon and serve as one of the reservoirs for the carbon cycle. The world's biotic components, particularly trees, contain 800 billion tons of organic carbon, and an additional 1,000–3,000 billion metric tons are estimated to be held in the remains of plants and animals in the soil. Before decomposition can occur, some of these remains are subjected to physical processes that transform them into coal, oil, and natural gas. We call these materials the **fossil fuels.** Most of

Photochemical Smog

Smog is well known to most people, especially those who live in large cities, as a type of air pollution that is a mixture of smoke and fog. While breathing any type of air pollution can be damaging to the respiratory system, a particular type of smog that may be even more harmful is called photochemical smog. Photochemical smog arises when primary pollutants react with one another under the influence of sunlight to form a more deadly combination of chemicals. For example, two primary pollutants, nitrogen oxides (NO_x) and volatile organic compounds (VOCs), including hydrocarbons from fossil fuel use, react with one another in the presence of sunlight to produce nitrogen dioxide (NO_2), ozone (O_3), and PAN (peroxyacetylnitrate). Exposure to these chemicals can result in respiratory distress, eye irritation, headache, and certain types of cancer.

Large, industrialized cities with warm, sunny climates—such as Los Angeles; Sydney, Australia; Mexico City; and Buenos Aires, Argentina—are particularly susceptible to photochemical smog. If the city is surrounded by hills, a thermal inversion may aggravate the situation. Normally, warm air near the ground rises, so that pollutants are dispersed and carried away by air currents. But sometimes, during a thermal inversion, smog gets trapped near the Earth by a blanket of warm air (Fig. 35B). This may occur when a cold front brings in cold air, which settles beneath a warm layer. The trapped pollutants cannot disperse, and the results are dangerous to a person's respiratory health. Even healthy adults experience a reduction in lung capacity when exposed to photochemical smog for long periods or during vigorous outdoor activities. Repeated exposures to high concentrations of ozone are associated with respiratory problems, such as an increased rate of lung infections and permanent lung damage. Children, elderly people, asthmatics, and individuals with emphysema or other similar disorders are particularly at risk.

Even though federal legislation is in place to bring air pollution under control, more than half the people in the United States live in cities polluted by too much smog. In the long run, pollution prevention is usually easier and cheaper than pollution cleanup. Some prevention suggestions are as follows:

- Encourage use of public transportation and burn fuels that do not produce pollutants.
- Increase recycling in order to reduce the amount of waste that is incinerated.
- Reduce energy use so that power plants need to provide less.
- Use renewable energy sources, such as solar, wind, or water power.
- Require industries to meet clean-air standards.

Questions to Consider

1. One of the most significant sources of the pollutants that contribute to photochemical smog is exhaust from automobiles. Hydrogen cell cars run on hydrogen and electricity and emit only water. Do you think that the government should provide strong incentives, such as tax breaks, for people who pay extra for these cars?
2. Why do you think exposure to ozone causes more serious respiratory problems in younger and older persons than in healthy adults?
3. How often do you use public transportation? Whether you live in a big city or not, would you be willing to use public transportation if it added, for example, an hour to the total time you spend each day traveling to school or work? Why or why not?

a. Ground-level ozone formation

b. Normal pattern

c. Thermal inversion

Figure 35B   Thermal inversion.   **a.** The chemical reactions that produce photochemical smog. **b.** Normally, pollutants escape into the atmosphere when warm air rises. **c.** During a thermal inversion, a layer of warm air (warm inversion layer) overlies and traps pollutants in cool air below. (photo) (a):

Figure 35.10 The carbon cycle.   Carbon cycles between atmospheric and geological reservoirs. Purple arrows represent human activities; green arrows represent natural events.

the fossil fuels were formed during the Carboniferous period, 286 to 360 MYA, when an exceptionally large amount of organic matter was buried before decomposing. Another reservoir is the inorganic carbonate that accumulates in limestone and in calcium carbonate shells of many marine organisms.

Human Activities

The transfer rates of carbon dioxide due to photosynthesis and cellular respiration are just about even. However, due to human activities, more carbon dioxide is being released into the atmosphere than is being removed. In 1850, atmospheric CO_2 was at about 280 parts per million (ppm); today, it is over 400 ppm. This increase is largely due to the burning of fossil fuels and the destruction of forests to make way for farmland and pasture. Today, the amount of carbon dioxide released into the atmosphere is about twice the amount that remains in the atmosphere. It's believed that most of this dissolves in the ocean.

The increased amount of carbon dioxide (and other gases) in the atmosphere is causing a rise in temperature called **global warming.** These gases allow the sun's rays to pass through, but they absorb and

radiate heat back to Earth, a phenomenon called the **greenhouse effect.** Global warming is contributing to **climate change,** which is causing significant changes to the Earth, as discussed in Section 36.1. The Ecology feature "Climate Change and Carbon Emissions" discusses worldwide efforts to cut back on greenhouse gases.

Carbon dioxide is not the only greenhouse gas. The following gases also play a role in the greenhouse effect:

- Methane (CH_4): A single molecule of methane has 21 times the warming potential of a molecule of carbon dioxide, making it a powerful greenhouse gas. Methane is a natural by-product of the decay of organic material, but is also released by landfills and the production of coal, natural gas, and oil.
- Nitrous oxide (N_2O): Nitrous oxide is released from the combustion of fossil fuels and as gaseous waste from many industrial activities.
- Hydrofluorocarbons: These were initially produced to reduce the levels of ozone-depleting compounds in the upper atmosphere. Unfortunately, although present in very small quantities, they are potent greenhouse gases.

SCIENCE IN YOUR LIFE ▶ ECOLOGY

Climate Change and Carbon Emissions

Over 97% of actively publishing climate scientists support the consensus that climate change is occurring. Scientists are currently collecting and interpreting environmental indicators that help us understand how and why the Earth's climate is changing. Changes in the average temperature of a region, precipitation patterns, sea levels, and greenhouse gas concentrations are all indicators that our climate is changing. These data support the scientific view that the climate-warming trends we have been experiencing over the past century are the result of human activities.

Since 1901, the average temperature across the United States has risen overall, with 2000–2009 being the warmest decade on record worldwide. 2016 was the warmest year on record, with 16 of the 17 warmest years on record having occurred since 2001.

Average precipitation rates have also increased by 6% over the past century, all while 30–60% of the United States experienced drought conditions. Since 1990, the United States has experienced eight of the top ten years of extreme precipitation events in its history. In line with this, increases in sea surface temperatures have produced more active hurricane seasons: Since the mid-1990s, the Atlantic Ocean, the Caribbean, and the Gulf of Mexico have seen six of its ten most active hurricane seasons.

Sea levels worldwide have risen an average of 1 inch per decade due to the overall increase in the surface temperature of the world's oceans. Climate models suggest we will see a rise in sea levels of 3 to 4 feet over the next century. This is concerning, given that over half the human population lives within 60 miles of the coast. New York City ranges from 5 feet to 16 feet above sea level, while the Florida Keys are an average of 3 to 4 feet above sea level. Even if the rising waters don't produce flooding, many coastal areas will be exposed to increasingly severe storms and storm surges that could lead to significant economic losses. The 2017 hurricane season, for instance, is ranked among the top ten all-time most active hurricane seasons, as well as the most expensive in history.

Atmospheric carbon dioxide levels have steadily increased (Fig. 35C), with the production of electricity being the largest contributor of greenhouse gas emissions in the United States, followed by transportation. As of October 2017, carbon dioxide concentrations in the atmosphere were measured at 407.06 parts per million (ppm).

The Kyoto Protocol was adopted on December 11, 1997, and initiated in force on February 16, 2005. The goal of the protocol was to achieve stabilization and reduction of the greenhouse gas concentrations in the atmosphere. As of November 2009, 187 countries had ratified the protocol, with the goal of reducing emissions by an average of 5.2% by the year 2012. Unfortunately, the United States never ratified the agreement, even though it is responsible for approximately 19% of the global CO_2 emissions from fossil-fuel combustion and is the largest per-capita emission producer in the world.

The Copenhagen conference in 2009 ended without any type of binding agreement for long-term action against climate change. It did produce a collective commitment by many developed nations to raise $30 billion to help poor nations cope with and combat the effects of climate change.

The 2010 Cancun summit on climate change helped solidify this agreement. Because deforestation produces about 15% of global carbon emissions, many developing countries can now receive incentives to prevent the destruction of their rain forests.

The 2013 UN Climate Change Conference in Warsaw was successful in establishing ways to help developing nations reduce greenhouse emissions (due to deforestation) and create financial commitments to assist these nations. The conference also helped keep governments on track to a universal climate agreement to reduce emissions, starting in 2015.

Interestingly, there is some evidence that change is occurring. In 2014, for the first time in 40 years, there was no increase in the growth of carbon dioxide emissions. However, no growth does not indicate a reduction in carbon dioxide entering the atmosphere, nor does it lessen the potential problems associated with global climate change. But it may indicate that these conferences and agreements are having an effect.

Questions to Consider

1. Should the United States and other developed countries pay developing nations to preserve these nations' forests?
2. Do individuals have a personal responsibility to help prevent climate change, or is it a governmental responsibility?

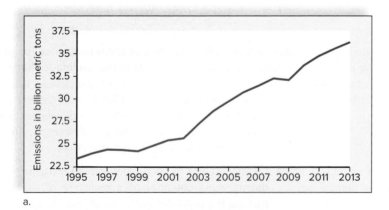

a.

Figure 35C **Global atmospheric carbon dioxide concentrations.** CO_2 emissions have increased significantly in the past hundred years, resulting in changes to the global climate patterns, including more severe droughts and erratic patterns of precipitation.

Figure 35.11 Climate change in the United States. The average temperature in the United States has steadily increased over the past two centuries, leading to more severe droughts and more erratic periods of precipitation that are altering the composition of many communities.

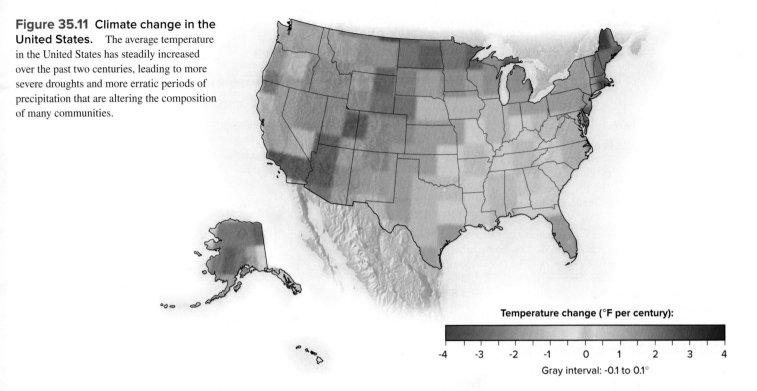

Temperature change (°F per century):

-4   -3   -2   -1   0   1   2   3   4

Gray interval: -0.1 to 0.1°

Figure 35.11 shows how the average temperature in the United States has steadily increased over the past two centuries. If the Earth's temperature continues to rise, more water will evaporate, forming more clouds. This sets up a positive feedback effect that could increase global warming still more. The global climate has already warmed about 0.6°C (1.1°F) since the Industrial Revolution.

Enhancements in computer science are allowing scientists to explore the majority of the variables that influence the global climate. Most climate scientists agree that the Earth's temperature may rise 1.5–4.5°C (2.0–8.1°F) by 2100 if greenhouse emissions continue at their current rates.

Check Your Progress 35.3

1. Identify the natural reservoirs for each of the biogeochemical cycles.
2. Provide examples of how human activities can disrupt the biogeochemical cycles.

Conclusion

In order to comprehend the role of wolves in ecosystems such as Yellowstone, scientists had to piece together information on the evolutionary history of the organisms in that community, with an understanding of how the ecosystem functions as a biological system. Coupled with this was some fascinating detective work, with experimental results suggesting that it was the demise of the wolves, not other ecological factors, that was responsible for the disruption of Yellowstone's ecosystem.

The story of the wolves is a positive one for science. By reintroducing wolves into Yellowstone in the late 1990s and closely monitoring the health of the wolf population, elk herds, and the recovery of the aspen forests, conservation ecologists have been able to document that the Yellowstone ecosystem is returning to its most "natural" state in over a hundred years.

SUMMARIZE

35.1 The Biotic Components of Ecosystems

- An ecosystem is composed of populations of organisms (**biotic** component) plus their physical environment (**abiotic** component).
- **Producers** are **autotrophs** that transform solar energy into food for themselves and all consumers. **Consumers** are **heterotrophs** that take in organic food. As **herbivores** feed on plants or algae, and **carnivores** feed on herbivores, energy is converted to heat. **Omnivores** are organisms that feed on both plants and animals. Feces, urine, and dead bodies become food for **decomposers.** Decomposers return some proportion of inorganic nutrients to autotrophs, and other portions are imported or exported among **ecosystems** in global cycles. **Detritus** is the partially decomposed matter in soil and water.

- Ecosystems are characterized by energy flow and chemical cycling. Energy is lost from the biosphere, but inorganic nutrients are not. They recycle within and among ecosystems. Eventually, all the solar energy that enters an ecosystem is converted to heat, and thus ecosystems require a continual supply of solar energy.

35.2 Energy Flow

- Ecosystems contain **food webs** in which the various organisms are connected by trophic relationships. In a **grazing food web, food chains** begin with a producer. In a **detrital food web,** food chains begin with detritus. The two food webs are joined when the same consumer links both a **grazing food chain** and a **detrital food chain.**

- A **trophic level** contains all the organisms that feed at a particular link in a food chain. **Ecological pyramids** show trophic levels stacked one on top of the other like building blocks. They are shaped like pyramids because most energy is lost from one trophic level to the next, and thus the number of species that can be sustained decreases. **Biomass** is the number of organisms multiplied by their weight.

35.3 Global Biogeochemical Cycles

- **Biogeochemical cycles** consist of reservoirs, exchange pools, and biotic communities. A **reservoir** pool contains elements available on a limited basis to living organisms, such as **fossil fuels,** sediments, and rocks. An **exchange pool,** such as the atmosphere, soil, and water, is a ready source of nutrients for living organisms.

- In the **water (hydrologic) cycle, evaporation** over the ocean is not compensated for by rainfall. Evaporation from terrestrial ecosystems includes transpiration from plants. **Condensation** occurs when a gas is changed into a liquid. **Precipitation** returns water to the land in the form of rain, snow, sleet, and fog. Rainfall over land results in bodies of fresh water plus groundwater, including **aquifers.** Eventually, all water returns to the oceans.

- In the phosphorus cycle, the biotic community recycles phosphorus back to the producers, and only limited quantities are made available by the weathering of rocks. Phosphates are mined for fertilizer production. When phosphates and nitrates enter lakes, ponds, and eventually the ocean, pollution and **cultural eutrophication,** or over-enrichment, occurs.

- In the nitrogen cycle, certain bacteria in water, soil, and within root nodules undergo **nitrogen fixation. Denitrification** is when various bacteria return nitrogen to the atmosphere. Human activities convert atmospheric nitrogen to fertilizer, which is broken down by soil bacteria. Humans also burn fossil fuels. In this way, nitrogen oxides can contribute to the formation of smog and acid rain, both of which are detrimental to animal and plant life.

- In the carbon cycle, organisms add as much carbon dioxide to the atmosphere as they remove. Shells in ocean sediments, organic compounds in living and dead organisms, and fossil fuels are reservoirs for carbon. Human activities, such as burning fossil fuels and trees, add carbon dioxide and other gases to the atmosphere. A buildup of these "greenhouse gases" allows the sun's rays to pass through, but they radiate the heat back to the Earth, creating a **greenhouse effect.** This is leading to **global warming** and causing **climate change.** A rise in sea level and a change in climate patterns is expected to follow.

ASSESS

Testing Yourself

Choose the best answer for each question.

35.1 The Biotic Components of Ecosystems

1. Label the populations in the ecosystem diagram below.

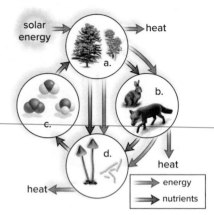

2. In an ecosystem, these organisms are responsible for converting solar energy to the stored energy found in organic compounds.
   a. herbivores
   b. decomposers
   c. producers
   d. carnivores
   e. None of these are correct.

3. Generally, energy _____ an ecosystem, and nutrients _____ an ecosystem.
   a. cycles within; cycle within
   b. cycles within; flow through
   c. flows through; cycle within
   d. flows through; flow through

35.2 Energy Flow

4. Of the total amount of energy that passes from one trophic level to another, about 10% is
   a. respired and becomes heat.
   b. passed out as feces or urine.
   c. stored as body tissue.
   d. recycled to autotrophs.
   e. All of these are correct.

5. This form of a food web begins with waste materials and the remains of dead organisms.
   a. aquatic
   b. detrital
   c. grazing
   d. atmospheric
   e. geological

6. The _____ of an ecosystem is equal to the weight of an organism times the number of this organism in the ecosystem.
   a. niche
   b. trophic level
   c. food web
   d. biomass
   e. biodiversity

35.3 Global Biogeochemical Cycles

For questions 7–12, match each characteristic to the cycles listed in the key. More than one answer can be used, and answers can be used more than once.

Key:

   a. water cycle
   b. carbon cycle
   c. nitrogen cycle
   d. phosphorus cycle
   e. none of the cycles
   f. all of the cycles

   7. Involves transpiration and precipitation
   8. Utilizes bacteria to make the compounds usable to plants
   9. Involves the participation of decomposers
   10. The atmosphere acts as a reservoir.
   11. Rocks are the reservoir in this cycle.
   12. Imbalances are contributing to global climate change.

ENGAGE

Thinking Critically

   1. A large forest has been removed by timber harvest (clear-cutting) and the land has not been replanted. After several years, humidity and rainfall in the area seem to have decreased. Use your knowledge of the water cycle to explain these observations.

   2. What impact might fungicides and pesticides have on detrital food webs? How do you think the nutrient cycles would be affected by the use of fungicides and pesticides?

   3. Why would an agricultural extension agent recommend planting a legume, such as clover, in a pasture?

   4. What types of things can you do at home to facilitate the cycling of nutrients such as carbon and nitrogen? What types of things can you do at home to conserve water or improve the quality of a nearby body of water?

36

Major Ecosystems of the Biosphere

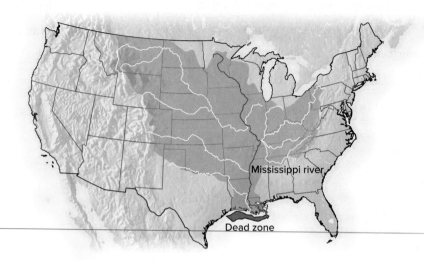

Mississippi river

Dead zone

BEFORE YOU BEGIN

Before beginning this chapter, take a few moments to review the following discussions:

Section 8.4 How do cellular processes like photosynthesis adapt to climate?

Section 34.3 What types of symbiotic relationships may allow species to survive in harsh environments?

Section 35.3 What is the difference between global warming and climate change?

Nitrates in the Water

Every year, large amounts of nitrogen- and phosphorus-based fertilizers are applied to farm fields across the Midwest to maximize crop yields. The problem is, when fertilizer is applied to a farm field, not all of the nutrients are retained within the field. Most fields have been modified to drain excess water after it rains. This can carry nitrogen and phosphorus into the nearby waterways, which are part of the Mississippi River Watershed. With climate change increasing the severity of precipitation events, we can expect to see more field runoff. The Mississippi River Watershed drains an area about 1.2 million square miles, including all or part of 31 different states. As the Mississippi empties into the Gulf of Mexico, the average flow rate is 600,000 cubic feet of water per second. Large amounts of nitrates are deposited into the Gulf every spring and summer.

The excess of nitrates entering the Gulf causes eutrophication. This occurs when there is an overstimulation of the growth of aquatic plants and algae due to an abundance of nutrients. The eventual decomposition of the aquatic plants and algae leads to hypoxia, a decrease in the dissolved oxygen levels within the water. Hypoxic water is what causes the fish and other aquatic organisms to leave the area or to die off, leading to a "dead" body of water.

As of July 2017, the Gulf of Mexico dead zone was the largest ever measured. It covered 8,776 square miles, or an area the size of New Jersey. Ocean currents eventually replace the hypoxic water with fresh water, but globally, dead zones are occurring with greater frequency and magnitude.

As you read through this chapter, think about the following questions:

1. Should nitrogen- and phosphorus-based fertilizers be banned from use in agriculture?

2. Which biomes are most closely associated with contributing to the dead zone in the Gulf of Mexico?

3. Is it possible to isolate the actions in one biome from the rest of the planet?

36.1 Climate and the Biosphere

> **Learning Outcomes**
>
> Upon completion of this section, you should be able to
> 1. Describe how solar radiation produces variations in the Earth's climate.
> 2. Explain how global air circulation patterns and physical geographic features are associated with the Earth's temperature and rainfall.

Climate refers to the prevailing weather conditions in a particular region. Climate is dictated by temperature and rainfall, which in turn are influenced by the following factors: (1) variations in the distribution of solar radiation due to the tilt of the Earth as it orbits about the sun; and (2) other effects, such as topography and proximity to water bodies.

Effect of Solar Radiation

Because the Earth is spherically shaped, the sun's rays are more direct at the equator and more spread out progressing toward the poles. Therefore, the tropical regions nearest the equator are warmer than the temperate regions farther away from the equator. In addition, Earth does not face the sun directly. Rather, it is on a slight tilt (about 23°) away from the sun. As the Earth orbits around the sun throughout the year, different parts of the planet are tilted toward or away from the sun, which determines the seasons. For example, during winter, the Northern Hemisphere is tilted away from the sun (Fig. 36.1). At the same time, the Southern Hemisphere is tilted toward the sun and it experiences summer.

Because the Earth completes one rotation on its axis per day and its surface consists of continents and oceans, the flow of warm and cold air forms three large circulation patterns in each hemisphere. The direction in which the air rises and cools determines the direction of the wind (Fig. 36.2). At the equator, the sun heats the air and evaporates water. The warm, moist air rises and loses most of its moisture as rain, resulting in the greatest amounts of

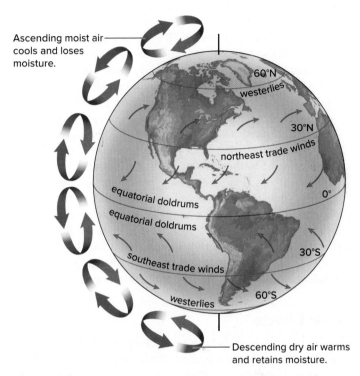

Ascending moist air cools and loses moisture.

Descending dry air warms and retains moisture.

Figure 36.2 Global wind circulation. Air ascends and descends as shown because Earth rotates on its axis. Also, the trade winds blow from the northeast to the west in the Northern Hemisphere, and blow from the southeast to the west in the Southern Hemisphere. The westerlies blow toward the east.

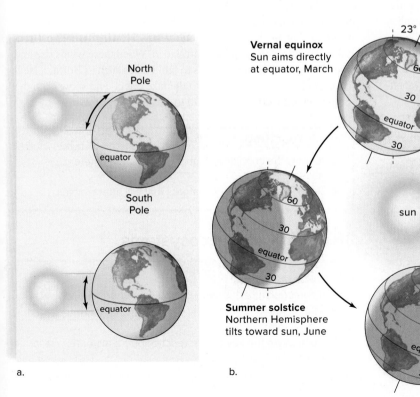

Vernal equinox Sun aims directly at equator, March

Winter solstice Northern Hemisphere tilts away from sun, December

Summer solstice Northern Hemisphere tilts toward sun, June

Autumnal equinox Sun aims directly at equator, September

Figure 36.1 Distribution of solar energy.
a. Because Earth is a sphere, beams of solar energy striking the planet near one of the poles are spread over a wider area than similar beams striking Earth at the equator.
b. The seasons of the Northern and Southern Hemispheres are due to the tilt of Earth on its axis as it rotates around the sun.

a.

b.

rainfall occurring nearest to the equator. The rising air flows toward the poles, but at about 30° north and south latitude, it sinks toward Earth's surface and reheats. As the dry air descends and warms once again, areas of high pressure are generated, which results in low rainfall. As a result, the great deserts of Africa, Australia, and the Americas occur at these latitudes. At about 60° north and south latitude, the warm air rises and cools as it does near the equator, producing a low-pressure area. Low pressure results in zones of high rainfall. Between 30° and 60° latitude, the strong wind patterns known as the westerlies occur in both the Northern and Southern Hemispheres. The westerlies move from west to east. The west coasts of the continents at these latitudes are wet, as in the U.S. Pacific Northwest, where a temperate (evergreen) rain forest is located. Weaker winds, called the polar easterlies, blow from east to west at latitudes higher than 60° in both hemispheres.

The direction of wind patterns, such as the easterlies and westerlies, is affected by the spinning of the Earth about its axis. That is, in the Northern Hemisphere, large-scale winds generally move clockwise, while in the Southern Hemisphere, they move counterclockwise. This explains, for example, why the northeast trade winds blow from the northeast toward the southwest and the southeast trade winds blow from the southeast toward the northwest (Fig. 36.2). Trade winds are so called because early sailors depended on them to power the movement of sailing ships.

Other Effects

The term *topography* refers to the physical features, or "the lay," of the land. One physical feature that affects climate is the presence of mountains. As air blows up and over a mountain range, it rises and cools, causing condensation to occur. One side of the mountain, called the windward side, receives more rainfall than the other side, called the leeward side. On the leeward side, the dry air descends, often producing a dry arid environment (Fig. 36.3). The difference between the windward side and the leeward side can be quite dramatic. In the Hawaiian Islands, for example, the windward side of the mountains typically receives more than 750 cm of rain a year while the leeward side, which is in a **rain shadow,** gets an average of only 50 cm of rain and is generally sunny. In the United States, the western side of the Sierra Nevada mountain range is lush, while the eastern side is a semidesert.

In contrast to landmasses, the oceans are slower to change temperature. This causes coasts to have a unique weather pattern that is not seen farther inland. During the day, the land warms more quickly than the ocean, and the air above the land rises. Then, a cool sea breeze blows in from the ocean. At night, the reverse happens and the breeze blows from the land to the sea.

In India and some other countries in southern Asia, the land heats more rapidly than the waters of the Indian Ocean during spring. The difference in temperature between the land and the ocean causes a gigantic circulation of air: Warm air rises over the land, and cooler air comes in off the ocean to replace it. As the warm air rises, it loses its moisture, creating a **monsoon** climate in which wet ocean winds blow onshore for almost half the year. During the monsoon season, rainfall is particularly heavy on the windward side of hills. Cherrapunji, in northern India, receives an annual average of 1,090 cm of rain a year because of its high altitude. The chief crop of India is rice, which starts to grow when

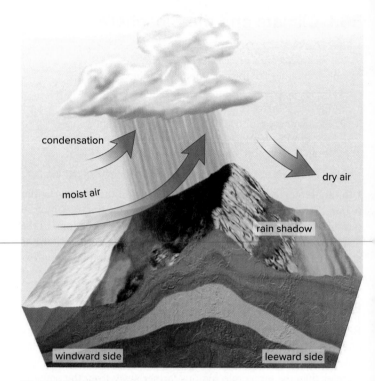

Figure 36.3 Formation of a rain shadow.   When winds from the sea cross a coastal mountain range, they rise and release their moisture as they cool the windward side of the mountain. The leeward side of a mountain is warmer and receives relatively little rain. Therefore, it is said to lie in a "rain shadow."

the monsoon rains begin. The weather pattern has reversed by November, when the land has become cooler than the ocean. Therefore, dry winds blow from the Asian continent across the Indian Ocean. In the winter, the air over the land is dry, the skies cloudless, and temperatures pleasant.

Other large bodies of water create major weather patterns. For example, in the United States, people often speak of the "lake effect," meaning that in the winter, arctic winds blowing over the Great Lakes become warm and moisture-laden. When these winds rise and lose their moisture, snow begins to fall. Places such as Buffalo, New York, get heavy snowfalls due to the lake effect, and snow is on the ground there for an average of 90 to 140 days every year.

Check Your Progress 36.1

1. Identify the conditions that account for the different seasons.
2. Recognize the various features that will affect rainfall.

36.2 Terrestrial Ecosystems

Learning Outcomes

Upon completion of this section, you should be able to

1. Identify the geographic distribution of the major terrestrial ecosystems.
2. Summarize the key characteristics of the major terrestrial biomes.

A major type of terrestrial ecosystem is called a **biome.** Biomes are characterized by particular climatic conditions, as well as the plants and animals living there. When terrestrial biomes are plotted according to their mean annual temperature and mean annual precipitation, a particular pattern results (Fig. 36.4*a*). The distribution of biomes is shown in Figure 36.4*b*. Even though this figure shows what appear to be clear demarcations, note that biomes gradually change from one type to another. To simplify our discussion in this chapter, we will group the biomes into general categories. Although we will be discussing each type of biome separately, keep in mind that all biomes are linked to form the biosphere, with each biome having connections to all the other terrestrial and aquatic ecosystems.

Figure 36.4 Pattern of biome distribution. **a.** Pattern of world biomes in relation to temperature and moisture. The dashed line encloses a wide range of environments in which either grasses or woody plants can dominate the area, depending on the soil type. **b.** The same type of biome can occur in different regions of the world, as shown on this global map.

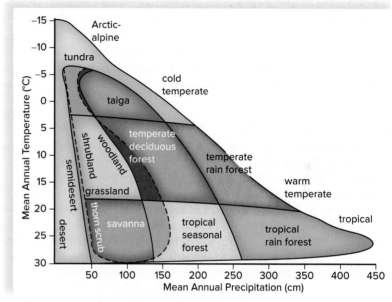

a. Biome pattern of temperature and precipitation

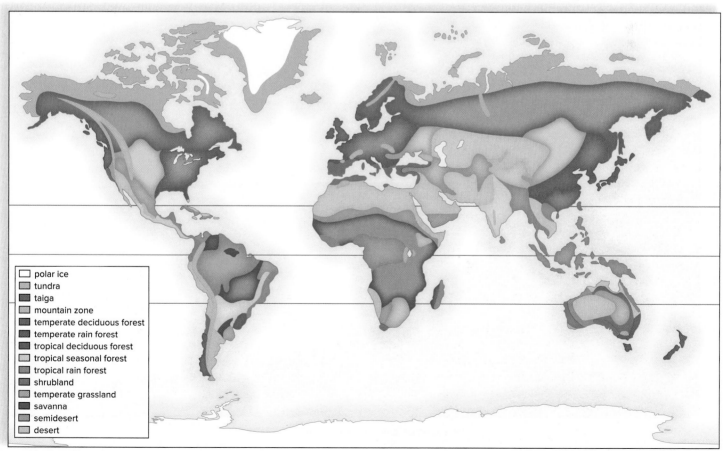

b. Distribution of biomes

The distribution of the biomes and their corresponding communities of organisms is determined principally by differences in climate due to solar radiation, water, and defining topographical features. Both latitude and altitude are responsible for temperature gradients. When traveling from the equator to the North Pole, it would be possible to observe first a tropical rain forest, followed by a temperate deciduous forest, a coniferous forest, and tundra. A similar type of sequence can also be seen when ascending a mountain (Fig. 36.5). The coniferous forest of a mountain is called a *montane coniferous forest,* and the tundra near the peak of a mountain is called an *alpine tundra.* However, when going from the equator to the South Pole, one would not reach a region corresponding to the coniferous forest and tundra of the Northern Hemisphere because of the absence of large landmasses in the Southern Hemisphere.

Tundra

The **tundra** biome, which encircles the arctic region just south of the ice-covered polar seas in the Northern Hemisphere, covers about 20% of Earth's land surface (Fig. 36.6a). As previously mentioned, a similar ecosystem, called the alpine tundra, occurs above the timberline on mountain ranges. The *arctic tundra* is cold and dark much of the year. Because rainfall amounts to only about 20 cm a year, the tundra could possibly be considered a desert. However, melting snow creates a landscape of pools and bogs in the summer. This is because only the topmost layer of earth thaws.

The **permafrost** beneath this layer is always frozen, and therefore, drainage is minimal.

Trees are not found in the tundra because the growing season is too short, their roots cannot penetrate the permafrost, and they cannot become anchored in the shallow, boggy soil during the brief summer. Instead, the ground is covered with short grasses and sedges, as well as numerous patches of lichens and mosses in summer (Fig. 36.6b). Dwarf woody shrubs, such as dwarf birch, flower and seed quickly during the short growing season.

Few animals live in the tundra year-round, but nearly all species have adaptations for living in extreme cold and short growing seasons. In winter, for example, the ptarmigan (a grouse) burrows in the snow during storms, and the musk ox conserves heat because of its thick coat and short, squat body. Other species, such as caribou (Fig. 36.6c) and reindeer, migrate to and from the tundra, as do the wolves that prey upon them. In the summer, the tundra is alive with numerous insects and birds, particularly shorebirds and waterfowl that migrate inland.

Coniferous Forests

Coniferous forests are found in three locations: in the **taiga,** which extends around the world in the northern part of North America and Eurasia; near mountaintops (where it is called a montane coniferous forest); and along the Pacific coast of North America, as far south as northern California (Fig. 36.7a).

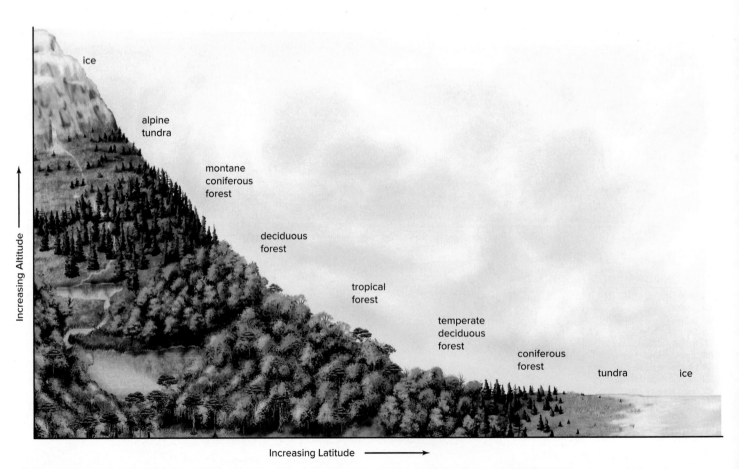

Figure 36.5 Climate and terrestrial biomes. Biomes change with altitude just as they do with latitude because vegetation is partly determined by temperature. Precipitation also plays a significant role, which is one reason grasslands, instead of tropical or deciduous forests, are sometimes found at the base of mountains.

a. Distribution of tundra

b. Tundra

c. Bull caribou

Figure 36.6 Tundra. **a.** In this biome, which is nearest the polar regions, the vegetation consists principally of lichens, mosses, grasses, and low-growing shrubs. **b.** The melting snow forms pools of water on the permanently frozen ground. **c.** Caribou will feed on lichens, grasses, and shrubs during the summer, then migrate south during the winter. (b): *Source: U.S. Fish & Wildlife Service*; (c): ©Roberta Olenick/All Canada Photos/Getty Images

a. Distribution of the taiga biomes

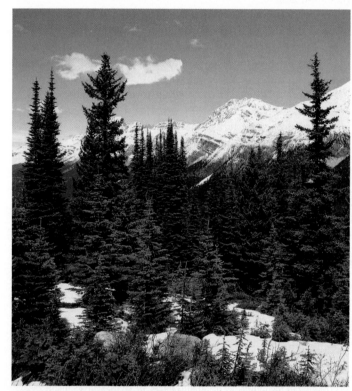

b. Spruce trees in the taiga biome

c. Bull moose

Figure 36.7 Coniferous forest. **a.** The taiga, which means "swampland," spans northern Europe, Asia, and North America. This biome has a dominant presence of (**b**) spruce trees and (**c**) moose, which frequent the ponds. (b): ©Creatas/Jupiterimages; (c): ©Jupiterimages/liquidlibrary/360/Getty Images

Taiga typifies the coniferous forest with its cone-bearing trees, such as spruce, fir, and pine (Fig. 36.7*b*). These trees are well adapted to the cold because both the leaves (reduced to needles) and bark have thick coverings. Also, the needlelike leaves can withstand the weight of heavy snow. There is a limited understory of plants, but the floor is covered by low-lying mosses and lichens. Birds harvest the seeds of the conifers, and bears, deer, moose, beaver, and muskrat live around the cool lakes and along the streams (Fig. 36.7*c*). Wolves prey on these larger mammals. A montane coniferous forest also harbors the wolverine and the mountain lion.

The coniferous forest that runs along the west coast of Canada and the United States is sometimes called a *temperate rain forest.* The plentiful rainfall and rich soil have produced some of the tallest conifer trees ever known, including the coastal redwoods. Small sections of this forest are considered old-growth forest because trees average over 150 years old, with some trees even as old as 800 years. It truly is an evergreen forest because mosses, ferns, and other plants often grow on tree trunks. Whether the limited portion of old-growth forest that remains should be protected from logging is an important and controversial conservation issue.

Temperate Deciduous Forests

Temperate deciduous forests are found south of the taiga in eastern North America, eastern Asia, and much of Europe (Fig. 36.8*a*). The climate in these areas is moderate, with relatively high rainfall (75–150 cm per year). The seasons are well defined, and the growing season ranges between 140 and 300 days. The trees, such as oak, beech, and maple, have broad leaves and are deciduous because they lose their leaves in fall and regrow them in spring.

The tallest trees form a canopy, an upper layer of leaves that are the first to receive sunlight and thereby create shade below (Fig. 36.8*b*). Even so, enough sunlight penetrates to provide energy for another layer of trees, called understory trees. Beneath these trees are shrubs and herbaceous plants that may flower in the spring before the trees have put forth their leaves. Mosses, lichens, and ferns can reside beneath the shrub layer. This stratification provides a variety of habitats for insects and birds. Ground life is also plentiful. Squirrels, cottontail rabbits, shrews, skunks, woodchucks, and chipmunks are small herbivores. These and ground birds such as turkeys, pheasants, and grouse are preyed on by various carnivores, such as the bobcat (Fig. 36.8*c*). White-tail deer and black bears have recently increased in number. In contrast to the taiga, a greater diversity of amphibians and reptiles occur in this biome because the winters are not as cold. Frogs and turtles generally prefer an aquatic existence, as do the beaver and muskrat.

Abundant fruits, nuts, and berries provide a supply of food for the winter. The leaves, after turning brilliant colors and falling to the ground, contribute to the rich layer of humus after decomposition. The minerals within the rich soil are washed far into the ground by spring rains, but the deep tree roots capture these nutrients and cycle them back through the forest system.

Tropical Forests

The most common type of tropical forest is the **tropical rain forest,** which is found in areas of South America, Africa, and the Indo-Malayan region near the equator (Fig. 36.9*a*). The temperature in a

Figure 36.8 Temperate deciduous forest.   **a.** Global distribution of the temperate deciduous forest. **b.** Trees, such as maples, beech, and oak, and mammals, such as this (**c**) bobcat, are found in this biome. (b): ©Bereczki Barna/Alamy; (c): ©Tom McHugh/Science Source

a. Distribution of temperate deciduous forest

b. Temperate deciduous forest

c. Bobcat

a. Distribution of tropical rain forests

b. An epiphyte growing on a broadleaf evergreen

c. Jaguar

**Figure 36.9
Tropical rain
forests.** **a.** The
distribution of tropical rain
forests. Broadleaf
evergreen trees, often with
(**b**) epiphyte attached, and
mammals such as this
(**c**) jaguar, are found in
tropical rain forests.
(b):©Ricochet Creative
Productions LLC; (c): ©Sandy
Windelspecht/Ricochet
Creative Productions LLC

tropical rain forest is always warm (between 20° and 25°C), and rainfall is plentiful (a minimum of 190 cm per year). As a result of these favorable climate conditions, this is the richest land biome in terms of species diversity. The diversity of species is enormous—a 10-km^2 area of tropical rain forest may contain 750 species of trees and 1,500 species of flowering plants.

A tropical rain forest has a complex structure, with many levels of life. Some broadleaf evergreen trees grow to heights of 15–50 m or more. These tall trees often have trunks buttressed at ground level to prevent them from toppling over. Lianas, or woody vines, often encircle rainforest trees as they grow. **Epiphytes** are plants that grow on other plants, but usually have roots of their own that absorb moisture and minerals leached from the canopy (Fig. 36.9*b*). Others catch rain and debris in hollows produced by overlapping leaf bases. The most common epiphytes are related to pineapples, orchids, and ferns.

Although some animals live on the ground (e.g., pacas, agoutis, peccaries, and armadillos), many also live in the trees (Fig. 36.9). Insect life is so abundant that the majority of species have not yet been identified. Termites play a vital role in the decomposition of woody plant material, and ants are found everywhere. The various birds, such as hummingbirds, parakeets, parrots, and toucans, are often beautifully colored, while amphibians and reptiles are well represented by many types of frogs, snakes, and lizards. Lemurs, sloths, and monkeys are well-known primates that feed on the fruits of the trees. The largest carnivores are the big cats—jaguars in South America and leopards in Africa and Asia (Fig. 36.9*c*).

Whereas the soil of a temperate deciduous forest biome is rich enough for agricultural purposes, the soil of a tropical rain forest biome is not. Nutrients are cycled directly from the litter to the plants again. Productivity is high because of warm and consistent temperatures, a year-long growing season, and the rapid recycling of nutrients from litter decomposition. To make up for the soil's low fertility, trees are sometimes cut and burned, so that the resulting ashes can provide enough nutrients for several harvests. This practice, called swidden agriculture, or slash-and-burn agriculture, can be successful if done on a small scale. Once the crops are harvested, the soil nutrients become depleted, consequently removing nutrients from the system. Erosion is high due to lack of tree roots and the heavy rainfall, causing additional nutrients to be washed away.

While we usually think of tropical forests as nonseasonal rain forests, *tropical deciduous forests* that have wet and dry seasons are found in India, Southeast Asia, West Africa, South and Central America, the West Indies, and northern Australia. They also have deciduous trees, and some of these forests also contain elephants, tigers, and hippopotamuses in addition to the animals mentioned previously.

Shrublands

Shrublands tend to occur along coasts that have dry summers and receive most of their rainfall during their winter (Fig. 36.10*a*). They are characterized by shrubs that have small but thick evergreen leaves, often coated with a waxy material that prevents loss of

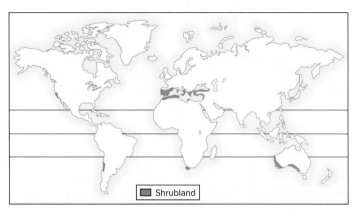

a. Distribution of shrubland

Figure 36.10 Shrublands.   **a.** Distribution of shrublands. **b.** Shrublands, such as this chaparral in California, are subject to raging fires, but the shrubs are adapted to quickly regrow. **c.** Greater roadrunners find a home in the chaparral. (b): ©Steven P. Lynch; (c): *Source:* U.S. Fish & Wildlife Service/ Gary Kramer, photographer

b. Shrubland

c. Roadrunner

moisture. These shrubs are adapted to withstand arid conditions and can also quickly regrow after a fire (Fig. 36.10*b*). In fact, the seeds of many species require the heat and scarring action of fire to induce germination. Other shrubs sprout from the roots after a fire. Typical shrubland species include coyotes, jackrabbits, gophers, and other rodents, as well as fire-adapted plant species such as chemise.

The dense shrubland that occurs in California is known as chaparral. This type of shrubland, called Mediterranean, lacks an understory and ground litter, and is highly flammable. Chaparrals are also found in South Africa and Australia. Typical animals of the chaparral include mule deer, rodents, lizards, and greater road-runners (Fig. 36.10*c*).

Grasslands

Grasslands occur where rainfall is greater than 25 cm but generally insufficient to support trees. For example, it is too dry for forests and too wet for deserts to form in temperate areas where rainfall is between 25 to 75 cm. Natural grasslands once covered more than 40% of Earth's land surface, but most areas that were once grasslands are now used to grow crops such as wheat, corn, and soybeans.

Grasses are well adapted to a changing environment and can tolerate some grazing, as well as flooding, drought, and sometimes fire. Where rainfall is high, large tall grasses that reach more than 2 m in height (e.g., pampas grass) can flourish. In drier areas, shorter grasses between 5 and 10 cm are dominant (e.g., grama grass). The growth of grasses is also seasonal. As a result, some

grassland animals such as bison migrate, whereas ground squirrels hibernate, when there is little grass for them to eat.

Temperate Grasslands

The *temperate grasslands* include the Russian *steppes,* the South American *pampas,* and the North American *prairies* (Fig. 36.11*a, b*). Large herds of bison—estimated at hundreds of thousands—once roamed the prairies, as did herds of pronghorn antelope (Fig. 36.11*c*). Now, small mammals, such as mice, prairie dogs, and rabbits, typically live below ground but usually feed above ground. Hawks, snakes, badgers, coyotes, and foxes feed on these mammals.

Savannas

Savannas, which are grasslands that contain some trees, occur in regions where a relatively cool, dry season is followed by a hot, rainy season (Fig. 36.12*a*). One tree that can survive the severe dry season is the flat-topped acacia, which sheds its leaves during a drought (Fig. 36.12*b*). The African savanna supports a tremendous variety and number of large herbivores (Fig. 36.12*c*). Elephants and giraffes are browsers that feed on tree vegetation. Antelopes, zebras, wildebeests, water buffalo, and rhinoceroses are grazers that feed on grasses. Any plant litter that is not consumed by grazers is attacked by a variety of smaller organisms, among them termites. Termites build towering mounds in which they tend fungal gardens that they use for food. The herbivores support a large population of carnivores. Lions and hyenas

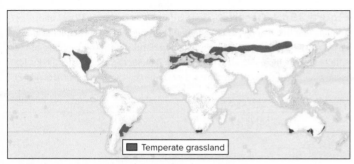

a. Distribution of temperate grasslands

b. Tall-grass prairie

c. American bison

Figure 36.11 Temperate grassland.   a. Distribution of temperate grasslands. **b.** Tall-grass prairies are seas of grasses dotted by pines and junipers. **c.** Bison, once abundant, are now being reintroduced into certain areas. (b): ©Jim Steinberg/Science Source; (c): ©Fuse/Getty Images

a. Distribution of savanna

b. Acacia trees of the savanna

c. Mammals of the savanna

Figure 36.12 The savanna.   a. The distribution of savanna. **b.** The African savanna varies from grassland to widely spaced shrubs and Acacia trees. **c.** This biome supports a large assemblage of animal species. (b): ©iStockphoto/Getty Images; (c): ©Danita Delimont/Alamy

sometimes hunt in packs, cheetahs hunt singly by day, and leopards hunt singly by night.

Deserts

As discussed in Section 36.1, **deserts** are usually found at latitudes of about 30°, in both the Northern and Southern Hemispheres (Fig. 36.13a). The winds that descend in these regions lack moisture, and annual rainfall is less than 25 cm. Days are hot because a lack of cloud cover allows the sun's rays to penetrate easily, but nights are cold because heat escapes easily into the atmosphere.

The Sahara—which stretches all the way from the Atlantic coast of Africa to the Arabian Peninsula—and a few other deserts have little or no vegetation. But most deserts have a variety of

plants (Fig. 36.13b). The best-known desert perennials in North America are the succulent, spiny-leafed cacti, which have stems that store water and carry on photosynthesis. Also common are nonsucculent shrubs, such as the many-branched sagebrush and the spiny-branched ocotillo that produces leaves during wet periods and sheds them during dry periods.

Some animals are adapted to the desert environment. For example, many desert animals are nocturnal. They are active at night when it is cooler. Reptiles and insects have waterproof outer

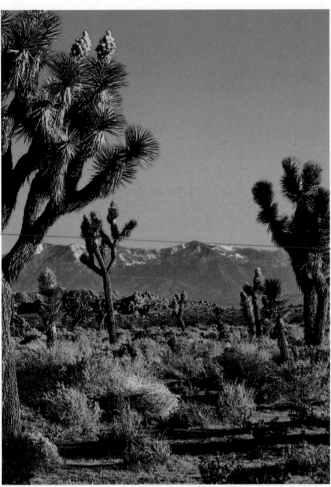

b. Plants of the desert

Figure 36.13 Deserts.   a. Distribution of desert. **b.** Plants and animals that live in a desert are adapted to arid conditions. The plants are either succulents, which retain moisture, or shrubs with woody stems and small leaves, which lose little moisture. **c.** Among the animal life, the kit fox is an example of a desert carnivore. (b): ©Corbis; (c): ©Kevin Schafer/Getty Images

a. Distribution of desert

c. Kit fox

coverings that conserve water. A desert has numerous insects, which pass through the stages of development from pupa to adult while there is rain. Reptiles, especially lizards and snakes, are a characteristic group of vertebrates found in deserts, but burrowing birds (e.g., burrowing owls) and rodents (e.g., the kangaroo rat) are also well known. Larger mammals, such as the kit fox (Fig. 36.13*c*), prey on the rodents, as do hawks.

Check Your Progress 36.2

1. Identify the features that would enable one biome to have a greater biodiversity than another.
2. Compare and contrast the biodiversity of the various land biomes.

36.3 Aquatic Ecosystems

Learning Outcomes

Upon completion of this section, you should be able to

1. Compare the characteristics of freshwater and saltwater ecosystems.
2. Recognize the differences between the pelagic and benthic divisions of the ocean.

Aquatic ecosystems are generally classified as freshwater (inland) or saltwater (usually marine). Brackish water, however, is a mixture of fresh and salt water. In this section, we describe lakes as examples of freshwater ecosystems and oceans as examples of saltwater ecosystems. Coastal ecosystems represent areas of brackish water.

Lakes

Lakes are bodies of fresh water often classified by their nutrient abundance. Oligotrophic (nutrient-poor) lakes are characterized by a small amount of organic matter and low productivity (Fig. 36.14*a*). Eutrophic (nutrient-rich) lakes typically have plentiful organic matter and high productivity (Fig. 36.14*b*). Such lakes are usually situated in naturally nutrient-rich regions or are enriched by agricultural or urban and suburban runoff. Oligotrophic lakes can become eutrophic through large inputs of nutrients, a process called **eutrophication.** Excess nutrient inputs, such as nitrogen fertilizer, can cause eutrophication, which can lead to fish kills.

In temperate environments, deep lakes are stratified during the summer and winter. In summer, lakes in the temperate zone have three layers of water that differ in temperature (Fig. 36.15). The surface layer, the epilimnion, is warm due to solar radiation; the thermocline is the middle layer, which decreases 1°C per meter of depth; and the lowest layer, the hypolimnion, is cold. These differences in temperature prevent mixing. The warmer,

a. Oligotrophic lake

b. Eutrophic lake

Figure 36.14 Types of lakes.   Lakes can be classified according to whether they are (**a**) oligotrophic (nutrient-poor) or (**b**) eutrophic (nutrient-rich). Eutrophic lakes tend to have large populations of algae and rooted plants, resulting in a large population of decomposers that use up much of the oxygen and leave little oxygen for fishes. (a): ©Roger Evans/Science Source; (b): ©McGraw-Hill Education/Pat Watson, photographer

Figure 36.15 Lake stratification in a temperate region.   Temperature profiles of a large oligotrophic lake in a temperate region vary with the season. During the spring and fall overturns, the deep waters receive oxygen from surface waters, and surface waters receive inorganic nutrients from deep waters.

less dense water of the epilimnion "floats" on top of the colder, more dense water of the hypolimnion.

As the season progresses, the epilimnion becomes nutrient-poor, while the hypolimnion begins to be depleted of oxygen. Phytoplankton (see Section 29.1), found in the sunlit epilimnion, use up nutrients during photosynthesis, and in turn release oxygen. Detritus naturally falls by gravity to the bottom of the lake where oxygen is used up as decomposition occurs. Decomposition in turn releases nutrients.

In the fall, as the epilimnion cools, and in the spring, as it warms, an overturn occurs. In the fall, the upper epilimnion waters become cooler than the hypolimnion waters. This causes the surface water to sink and the deep water to rise. The **fall overturn** continues until the temperature is uniform throughout the lake. At this point, wind helps circulate the water so that mixing occurs.

As winter approaches, the water cools further. Ice formation begins at the top, and the ice floats because it is less dense than cold water. Ice has an insulating effect, preventing further cooling of the water below. This permits aquatic organisms to live through the winter in the water beneath the surface of the ice.

In the spring, as the ice melts, the cooler water on top sinks below the warmer water below it. The **spring overturn** continues until the temperature is uniform through the lake. At this point, wind aids in the circulation of water as before. When the surface waters absorb solar radiation, thermal stratification occurs once more.

This vertical stratification and seasonal change of temperatures in a lake basin influence the seasonal distribution of fish and other aquatic life. For example, cold-water fish move to the deeper water in summer and inhabit the upper water in winter. In the fall and spring, just after mixing occurs, phytoplankton growth at the surface is most abundant.

Life Zones

In both fresh and salt water, microscopic **plankton,** which includes phytoplankton and also zooplankton (see Section 29.1), play important roles in aquatic ecosystems. Lakes and ponds can be divided into several life zones, as shown in Figure 36.16. Aquatic plants are rooted in the shallow *littoral zone* of a lake, and various microscopic organisms cling to these plants and to rocks. Some organisms, such as the water strider, live at the water-air interface and can literally walk on water. In the limnetic zone, small fishes,

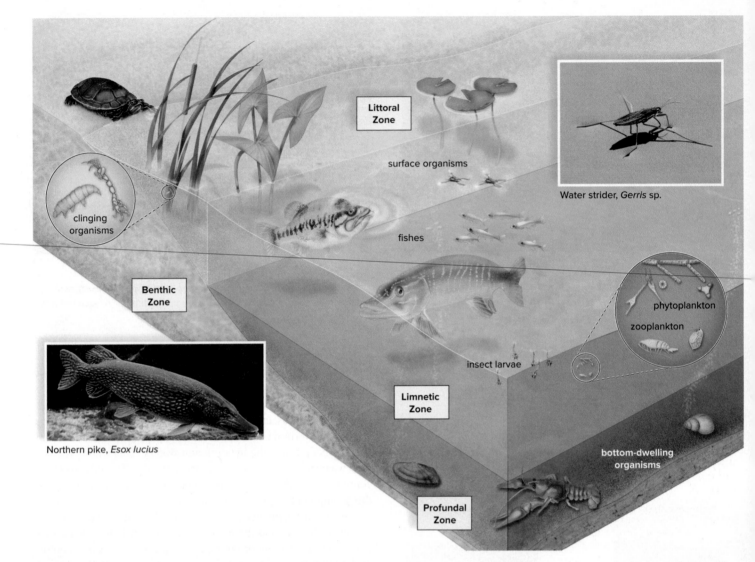

Figure 36.16 Zones of a lake. Rooted plants and clinging organisms live in the littoral zone. Phytoplankton, zooplankton, and fishes are in the sunlit limnetic zone. Water striders stand on the surface film of water with water-repellent feet. Crayfishes and molluscs are in the profundal zone, as well as the littoral zone. Pike are top carnivores prized by fishermen. (photos) (pike): ©abadonian/iStock/360/Getty Images; (water strider): ©Matti Suopajarvi/Getty Images

such as minnows and killifish, feed on plankton and also serve as food for large fishes. In the *profundal zone,* zooplankton and fishes such as whitefish feed on debris that falls from above. Pike species are an example of "lurking predators." They wait among vegetation around the margins of lakes and surge out to catch passing prey.

A few insect larvae are in the *limnetic zone,* but they are far more abundant in both the littoral and profundal zones. Midge larvae and ghost worms are common members of the benthos, animals that live on the bottom in the benthic zone. In a lake, benthos organisms include crayfishes, snails, clams, and various types of worms and insect larvae.

Coastal Ecosystems

Estuaries

An **estuary** is where fresh water and salt water meet and mix (Fig. 36.17). Coastal bays, tidal marshes, fjords (an inlet of

water between high cliffs), some deltas (a triangular-shaped area of land at the mouth of a river), and lagoons (a body of water separated from the sea by a narrow strip of land) are all examples of estuaries (Fig. 36.18).

Organisms living in an estuary must be able to withstand constant mixing of waters and rapid changes in salinity. Not many organisms are adapted to this environment, but those that can survive here find an abundance of nutrients. An estuary acts as a nutrient trap because the sea prevents the rapid escape of nutrients brought by a river. It has been estimated that well over half of all marine fishes develop in the protective environment of an estuary, which explains why estuaries are called the nurseries of the sea. Estuaries are also feeding grounds for many birds, fish, and shellfish because they offer a ready supply of food.

Salt marshes, dominated by salt marsh cordgrass, are often associated with estuaries. So are mudflats and mangrove swamps, where sediment and nutrients from the land collect.

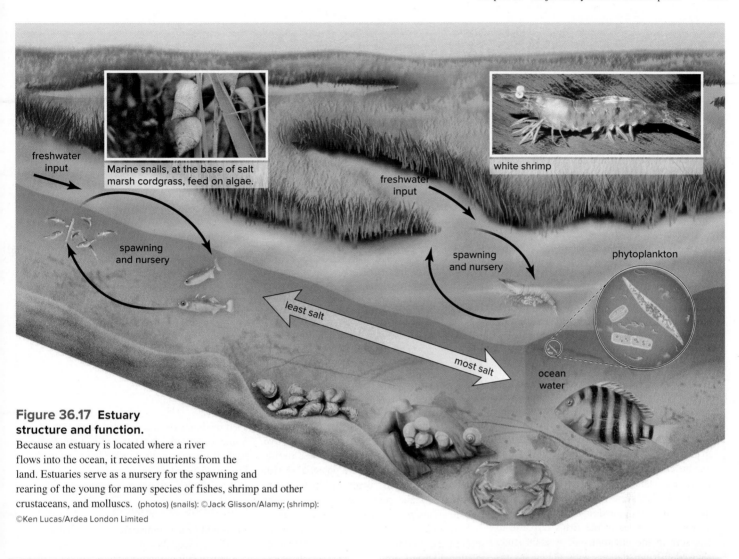

Figure 36.17 Estuary structure and function.
Because an estuary is located where a river flows into the ocean, it receives nutrients from the land. Estuaries serve as a nursery for the spawning and rearing of the young for many species of fishes, shrimp and other crustaceans, and molluscs. (photos) (snails): ©Jack Glisson/Alamy; (shrimp): ©Ken Lucas/Ardea London Limited

a. Mudflat

b. Mangrove swamp

Figure 36.18 Types of estuaries. Types of estuaries include (**a**) mudflats, which form at the mouth of rivers in temperate zones and are frequented by migrant birds, and (**b**) mangrove swamps, which skirt the coastlines of many tropical and subtropical lands. The tangled roots of mangrove trees trap sediments and nutrients that sustain many immature forms of sea life. (a): ©Marc Lester/Getty Images; (b): ©shakzu/iStock/360/Getty Images

Seashores

Seashores, which may be rocky or sandy, are constantly bombarded by the sea as the tides roll in and out (Fig. 36.19). The **littoral zone** lies between the high- and low-tide marks. The littoral zone of a rocky beach is divided into subzones. In the upper portion of the littoral zone, barnacles are attached so tightly to stone by their own secretions that their calcareous outer plates remain in place even after the enclosed shrimplike animal dies. In the midportion of the littoral zone, brown algae known as rockweed may overlie the barnacles. In the lower portions of the littoral zone, oysters and mussels attach themselves to the rocks by filaments called *byssal threads.* Also present are

SCIENCE IN YOUR LIFE ▶ ECOLOGY

The Fate of Prescription Medicine

What happens to prescriptions when they are flushed down the drain?

Multiple studies have found detectable concentrations of pharmaceuticals and organic wastewater chemicals in rivers and streams across the United States (Fig. 36A). U.S. Geological Survey scientists found 12 of 22 pharmaceuticals and 32 of the 47 organic wastewater chemicals looked for in the streams of the Boulder Creek Watershed, Colorado. Some of these chemicals included antibiotics used to treat bacterial infections, antimicrobial agents used in soaps, endocrine-active chemicals (such as those found in birth control pills), as well as caffeine. While few of the detectable chemicals exceeded water-quality standards, the scientists did note that no standards exist for a number of these compounds, raising concerns about what represents acceptable levels in the water supply.

While difficult to fully assess the complete ecological impact of these chemicals, a wide range of problems have been attributed to them. Native fish populations were found to exhibit endocrine disruption due to exposure to these contaminants. Endocrine disruptions can produce adverse effects on the development of the nervous and reproductive systems, as well as changes in the fish's response to stressors in the environment. A 2008 study showed fish populations with altered sex ratios, reduced gonad size, and disrupted ovarian and testicular development in streams containing a mixture of endocrine-active chemicals.

Figure 36A Effects of pharmaceuticals on fish. In some cases, male fish have developed ovaries as a result of exposure to endocrine-active chemicals. (pills): ©McGraw-Hill Education/Mark Dierker, photographer; (fish): ©Pat Clayton/Image Bank/Getty Images

Fish in contaminated streams were also found to contain both male and female reproductive organs (referred to as intersex). These results indicate that the reproductive potential of fish populations living in streams in which various pharmaceuticals and organic wastewater chemicals are present may be compromised.

In an attempt to decrease the contamination of the nation's waterways, the Office of National Drug Control Policy suggests working with local agencies such as pharmacies or hazardous waste collection sites that hold "take-back days" in which they collect un-

used medicines and dispose of them properly. Some states are also exploring a mail/ship back program to collect unused prescriptions. Proper disposal of prescriptions is the first step in keeping our nation's waterways safe.

Questions to Consider

1. Should each community be held liable for the water quality of the aquatic ecosystems to which it is connected?
2. What do you think should be the highest-priority chemicals that should be banned from entering aquatic ecosystems?

a. Sea stars at low tide

b. Shorebirds

Figure 36.19 Seashores. **a.** Some organisms of a rocky coast live in tidal pools. **b.** A sandy shore looks devoid of life except for the birds that feed there. (a): *Source:* NOAA Central Library; (b): ©Jeff Greenburg/Science Source snails called limpets and periwinkles. Periwinkles have a coiled shell and secure themselves by hiding in crevices or under sea-weeds, while limpets press their single, flattened cone tightly to a rock. Below the littoral zone, macroscopic seaweeds, which are the main photosynthesizers, anchor themselves to the rocks by holdfasts.

Organisms cannot attach themselves to shifting, unstable sands on a sandy beach. Therefore, nearly all the permanent beach residents dwell underground. They either burrow during the day and surface to feed at night, or they remain permanently within their burrows and tubes. Ghost crabs and sandhoppers (amphipods) burrow themselves above the high-tide mark and feed at night when the tide is out. Sandworms and sand (ghost) shrimp remain within their burrows in the littoral zone and feed on detritus whenever possible. Still lower in the sand, clams, cockles, and sand dollars can be found.

Oceans

Oceans cover approximately three-quarters of our planet. Climate is driven by the sun, but the oceans play a major role in redistributing heat in the biosphere. Water tends to be warm at the equator and much cooler at the poles because of the distribution of the sun's rays, as discussed earlier in this chapter (see Fig. 36.1*a*). Air takes on the temperature of the water below, and then warm air moves from the equator to the poles. In other words, the oceans have a major influence on wind patterns. (The landmasses also play a role, but water holds heat longer and remains cool longer than land during periods of changing temperature.)

When the wind blows strongly and steadily across a great expanse of ocean for a long time, friction from the moving air begins to drag the water along with it. Once the water has been set in motion, its momentum, aided by the wind, keeps it moving in a steady flow called a current. Because the ocean currents eventually strike land, they move in a circular path—clockwise in the Northern Hemisphere and counterclockwise in the Southern Hemisphere. As the currents flow, they move warm water from the equator to the poles. One such current, called the Gulf Stream, brings tropical Caribbean water to the east coast of North America and the higher latitudes of western Europe. Without the Gulf Stream, Great Britain, which has a relatively warm temperature, would be as cold as Greenland. In the Southern Hemisphere, another major ocean current warms the eastern coast of South America.

Also, in the Southern Hemisphere, the Humboldt Current flows toward the equator. The Humboldt Current carries phosphorus-rich cold water northward along the west coast of South America. During a process called **upwelling,** cold offshore winds cause cold nutrient-rich waters to rise and take the place of warm nutrient-poor waters. In South America, the enriched waters nourish an abundance of marine life that supports the fisheries of Peru and northern Chile. When the Humboldt Current is not as cool as usual, upwelling does not occur, stagnation results, the fisheries decline, and climate patterns change globally. This phenomenon is discussed in the Ecology feature "El Niño–Southern Oscillation."

Pelagic Division

It is customary to place the organisms of the oceans into either the pelagic division (open waters) or the benthic division (ocean floor). The geographic areas and zones of an ocean are shown in Figure 36.20, and the animals that inhabit the different zones are shown in Figure 36.21. The **pelagic division** includes the neritic province and the oceanic province.

Neritic Province The abundant sunlight and inorganic nutrients provide for a large concentration of organisms in the neritic province. Phytoplankton, consisting of suspended algae, is food not only for zooplankton but also for small fishes. These small fishes, in turn, are food for larger commercially valuable fishes.

Coral reefs are areas of high biological abundance and productivity found in shallow, nutrient-poor, tropical waters. Their

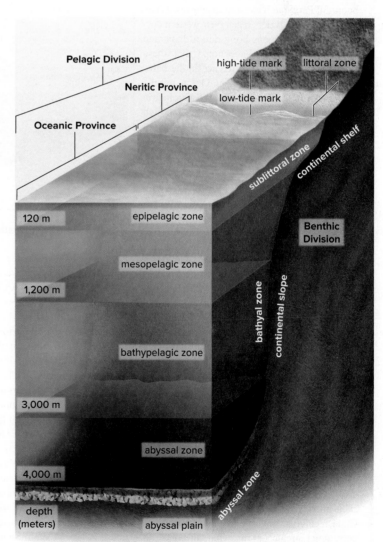

Figure 36.20 Marine environment. Organisms reside in the pelagic division (blue) where waters are divided as indicated. Organisms also reside in the benthic division (brown) in the surfaces and zones indicated.

El Niño–Southern Oscillation

Climate largely determines the distribution of life on Earth. Short-term variations in climate, which we call weather, also have a pronounced effect on living organisms. There is no better example than an El Niño. Originally, El Niño referred to a warming of the seas off the coast of Peru around Christmas—hence, the name El Niño, "the boy child," after Jesus.

Now scientists prefer the term *El Niño–Southern Oscillation (ENSO)* to denote a severe weather change brought on by an interaction between the atmosphere and ocean currents. Ordinarily, the southeast trade winds move along the coast of South America and turn west because of Earth's daily rotation on its axis. As the winds drag warm ocean waters from east to west, there is an upwelling of nutrient-rich cold water from the ocean's depths, resulting in a bountiful Peruvian harvest of anchovies. When the warm ocean waters reach their western destination, the monsoons bring rain to India and Indonesia. Scientists have noted that these events correlate with a difference in the barometric pressure over the Indian Ocean and the southeastern Pacific—that is, the barometric pressure is low over the Indian Ocean and high over the southeastern Pacific. But when a "southern oscillation" occurs and the barometric pressures switch, an El Niño begins.

During an El Niño, both the northeast and the southeast trade winds slacken. Upwelling no longer occurs, and the anchovy catch off the coast of Peru plummets. During a severe El Niño, waters from the east never reach the west, and the winds lose their moisture in the middle of the Pacific instead of over the Indian Ocean. The monsoons fail, and drought occurs in India, Indonesia, Africa, and Australia. Harvests decline, cattle must be slaughtered, and famine is likely in highly populated India and Africa, where funds to import replacement supplies of food are limited. In 2009, severe drought caused fires that killed hundreds of people and caused billions of dollars in damage in the state of Victoria, Australia.

A backward movement of winds and ocean currents may even occur so that the waters warm to more than 14°F above normal along the west coast of the Americas. This is a sign that a severe El Niño has occurred. The weather changes are dramatic in the Americas also. Southern California is hit by storms and even hurricanes, and the deserts of Peru and Chile receive so much rain that flooding occurs. A jet stream (strong wind currents) can carry moisture into Texas, Louisiana, and Florida, with flooding a near certainty. Or the winds can turn northward and deposit snow in the mountains along the west coast so that

La Niña
• Upwelling off the west coast of South America brings cold waters to the surface.
• Barometric pressure is high over the southeastern Pacific.
• Monsoons associated with the Indian Ocean occur.
• Hurricanes occur off the east coast of the United States.

El Niño
• Great ocean warming occurs off the west coast of the Americas.
• Barometric pressure is low over the southeastern Pacific.
• Monsoons associated with the Indian Ocean fail.
• Hurricanes occur off the west coast of the United States.

Figure 36B La Niña and El Niño.

flooding occurs in the spring. Some parts of the United States, however, benefit from an El Niño. The Northeast is warmer than usual, few if any hurricanes hit the east coast, and there is a lull in tornadoes throughout the Midwest. Altogether, a severe El Niño affects the weather over three-quarters of the globe.

Eventually, an El Niño dies out, and normal conditions return. The normal cold-water state off the coast of Peru is known as La Niña (the girl). Figure 36B contrasts the weather conditions of a La Niña with those of an El Niño. Since 1991, the sea surface has been almost continuously warm, and two record-breaking El Niños have occurred. What could be causing more of the El Niño state than the La Niña state? Some scientists argue that this environmental change is related to global warming, a rise in environmental temperature due to greenhouse gases in the atmosphere, as discussed in Chapter 35. Greenhouse gases, including carbon dioxide, are released by humans in mass quantities into the atmosphere by, for example, burning fossil fuels. Like the glass of a greenhouse,

these gases allow the sun's rays to pass through, but they trap the heat.

Questions to Consider

1. Recent research has suggested that global warming is contributing to the increased frequency and severity of El Niño events. This can cause severe drought in India and Africa, as well as greatly reduce fish catch in parts of the Southern Hemisphere. Should the responsibility for decreasing greenhouse gas emissions be greater for more-developed countries (MDCs) because they produce a greater amount of greenhouse gases per capita than less-developed countries (LDCs), even though El Niños have more severe consequences for LDCs?

2. Aside from the recent increases in gasoline prices, would you consider buying a more fuel-efficient car or spend more on plane flights to purchase "carbon credits" to do your part in decreasing the emission of greenhouse gases such as carbon dioxide? Why or why not?

chief constituents are stony corals, animals that have a calcium carbonate (limestone) exoskeleton, and calcareous red and green algae. Corals tend to form colonies derived from individual corals that have reproduced by means of budding. Corals provide a home for a microscopic alga called zooxanthella. The corals, which feed at night, and the algae, which photosynthesize during the day, are mutualistic and share materials and nutrients. The close relationship of corals and zooxanthellae is likely the reason coral reefs form only in shallow sunlit water—the algae need sunlight for photosynthesis.

A reef is densely populated with life. The large number of crevices and caves provide shelter for filter feeders (e.g., sponges, sea squirts, and fan worms) and for scavengers (e.g., crabs and sea urchins). There are many types of small, beautifully colored fishes. Parrotfishes feed directly on corals, and others feed on plankton or detritus. Small fishes become food for larger fishes, including snappers and other fish that are caught for human consumption. The barracuda, moray eel, and various shark species are top predators in coral reefs.

Oceanic Province The oceanic province lacks the inorganic nutrients of the neritic province and, therefore, does not have as high a concentration of phytoplankton, even though it is sunlit. Photosynthesizers are food for a large assembly of zooplankton, which then become food for herrings and bluefishes. These, in turn, are eaten by larger mackerels, tunas, and sharks. Flying fishes, which glide above the surface, are preyed upon by dolphins. Whales are other mammals found in the epipelagic zone (Fig. 36.21). Baleen whales strain krill (small crustaceans) from the water, and toothed sperm whales feed primarily on the common squid.

Animals in the mesopelagic zone are carnivores that are adapted to the absence of light, and tend to be translucent, red, or luminescent. There are luminescent shrimps, squids, and fishes, such as lantern and hatchet fishes.

The bathypelagic zone is in complete darkness except for an occasional flash of bioluminescent light. Carnivores and scavengers are found in this zone. Strange-looking fishes with distensible mouths and abdomens and small, tubular eyes feed on infrequent prey.

Figure 36.21 Ocean inhabitants.   Different organisms are characteristic of the epipelagic, mesopelagic, and bathypelagic zones of the pelagic division compared to the abyssal zone of the benthic division.

baleen whale

jellyfish

phytoplankton

zooplankton

squid

dolphin

shark

sea turtle

barracuda

ocean bonito

tuna

mackerel

prawn

lantern fish

giant squid

midshipman

sperm whale

viperfish

hagfish

anglerfish

deep-sea shrimp

tripod fish

gulper

glass sponges

brittle stars

sea cucumber

Epipelagic Zone (0–120 m)

Mesopelagic Zone (120–1,200 m)

Bathypelagic Zone (1,200–3,000 m)

Abyssal Zone (3,000 m–bottom)

Benthic Division

The **benthic division** includes organisms that live on or in the soil of the continental shelf (sublittoral zone), the continental slope (bathyal zone), and the abyssal plain (abyssal zone) (see Fig. 36.20).

Seaweed grows in the sublittoral zone, and can be found in batches on outcroppings as the water gets deeper. More diversity of life exists in the sublittoral and bathyal zones than in the abyssal zone. In the first two zones, clams, worms, and sea urchins are preyed upon by starfishes, lobsters, crabs, and brittle stars. Photosynthesizing algae occur in the sunlit sublittoral zone, but benthic organisms in the bathyal zone are dependent on the detritus that falls from the waters above.

The abyssal zone is inhabited by animals that live at the soil-water interface of the abyssal plain (Fig. 36.21). It once was thought that few animals exist in this zone because of the intense pressure and the extreme cold. Yet many invertebrates live there by feeding on debris floating down from the mesopelagic zone. Sea lilies rise above the seafloor; sea cucumbers and sea urchins crawl around on the sea bottom; and tube worms burrow in the mud.

The flat abyssal plain is interrupted by enormous underwater mountain chains called oceanic ridges. Along the axes of the ridges, crustal plates spread apart, and molten magma from Earth's core rises to fill the gap. At **hydrothermal vents,** seawater percolates through cracks and is heated to about 350°C, causing sulfate to react with water and form hydrogen sulfide (H_2S). Free-living or mutualistic chemoautotrophic bacteria use energetic electrons derived from hydrogen sulfide to reduce bicarbonate to organic compounds. These compounds support an ecosystem that includes huge tube worms and clams. It was a surprise to find such communities of organisms living so deep in the ocean, where light never penetrates. Life is possible so deep because, unlike photosynthesizers, chemoautotrophs are producers that do not require solar energy.

Check Your Progress 36.3

1. Compare and contrast the various freshwater ecosystems.
2. Explain the major characteristics of the two oceanic divisions.

Conclusion

Fertilizers have become a necessity in today's agricultural world. They help farmers increase crop yields that are becoming even more vital for feeding a growing population. As climate change causes more extreme weather events, we will see increased levels of nutrients running off from agricultural fields. This will continue to cause the eutrophication of aquatic ecosystems, larger dead zones, and increased environmental damage.

Every citizen in the United States plays a role in contributing to the Gulf of Mexico's dead zone, as well as solving the problem of eutrophication. It is estimated that the dead zone costs the U.S. seafood and tourism industry $82 million a year. With more than 50 cities and over 18 million people relying directly on the Mississippi for their water supply, it is vital that all Americans work together to maintain its health.

SUMMARIZE

36.1 Climate and the Biosphere

- **Climate** refers to the prevailing weather conditions in a particular region.
- The sun's rays are spread out over a larger area at the poles than the vertical rays at the equator because the Earth is spherical. Thus, surface temperatures decrease from the equator toward each pole.
- The planet is tilted on its axis, and the seasons change depending on which hemisphere is closest to the sun as Earth completes its annual rotation.
- Warm air rises near the equator, loses its moisture, and then descends at about 30° north and south latitude, and continuing toward the poles. When the air descends, it retains moisture, and therefore, the great deserts of the world are formed at 30° latitudes.
- Because Earth rotates on its axis daily, the winds blow in opposite directions above and below the equator.
- Air rising over coastal regions loses its moisture on the windward side as the elevation increases. This may cause **monsoons,** or extended periods of rain. Less rain falls on the leeward side, creating a **rain shadow.**

36.2 Terrestrial Ecosystems

- A **biome** is a major type of terrestrial ecosystem.
- Temperature and rainfall influence the distribution of biomes around the world.

- The **tundra** is the northernmost biome and consists largely of short grasses and sedges and dwarf woody plants. Most of the water in the soil is frozen year-round (i.e., **permafrost**) because of cold winters and short summers.
- The **taiga,** a **coniferous forest,** has less rainfall than other types of forests.
- The **temperate deciduous forest** has trees that gain and lose their leaves during spring and fall, respectively.
- **Tropical rain forests** are the most complex and productive of all biomes. Many of the plants in the tropics are **epiphytes,** meaning that they grow on other plants.
- **Shrublands** usually occur along coasts that have dry summers, and receive most of their rainfall in the winter.
- Among **grasslands,** the savanna, a tropical grassland, supports the greatest number of different types of large herbivores. Prairies, such as those in the mid United States, have a limited variety of vegetation and animal life.
- **Deserts** are characterized by rainfall of less than 25 cm a year. Plants, such as cacti, are succulents, and others are shrubs with thick leaves—both of which are adaptations to conserve water.

36.3 Aquatic Ecosystems

- In deep temperate zone lakes, water depth determines the temperature and the concentrations of nutrients and gases. A large input of nutrients into a body of water is called **eutrophication.** During **spring overturn** and **fall overturn,** nutrients from the bottom lake layers are redistributed while the entire body of water is cycled.

- **Lakes** and ponds have three life zones. Rooted plants and clinging organisms live in the littoral zone; **plankton** and fishes live in the sunlit limnetic zone; and bottom-dwelling organisms, such as crayfishes and molluscs, live in the profundal zone.

- Oceans play a major role in determining the climate of the land biomes. Ocean currents help move warm water from the equator toward the poles. During an **upwelling,** cold offshore winds cause cold nutrient-rich waters to rise and take the place of warm nutrient-poor waters.

- Marine ecosystems are divided into coastal ecosystems and the oceans.

- The coastal ecosystems, especially **estuaries,** are more productive than the oceans. Estuaries (e.g., salt marshes, mudflats, and mangrove forests) are near the mouth of a river and are called the nurseries of the sea. **Coral reefs** contain a tremendous diversity of species and play important roles in the health of the ocean.

- An ocean is divided into the **pelagic division** and the **benthic division.** The pelagic division (open waters) has three zones. The epipelagic zone receives adequate sunlight and supports the most life. The meso-pelagic and bathypelagic zones contain organisms adapted to minimum light and no light, respectively. The benthic division (ocean floor) includes organisms living on the continental shelf in the sublittoral zone, the continental slope in the bathyal zone, and the abyssal plain in the abyssal zone. Underwater **hydrothermal vents** are formed when seawater percolates through cracks and is heated by molten magma from the Earth's core. Life here is based upon the sulfate reacting with water to form hydrogen sulfide.

ASSESS

Testing Yourself

Choose the best answer for each question.

36.1 Climate and the Biosphere

1. The seasons are best explained by
    a. the distribution of temperature and rainfall in biomes.
    b. the tilt of the Earth as it orbits about the sun.
    c. the daily rotation of the Earth on its axis.
    d. the fact that the equator is warm and the poles are cold.

2. Which region on Earth receives the largest amount of direct solar radiation throughout the year?
    a. Northern Hemisphere above 60°
    b. Northern Hemisphere between 30° and 60°
    c. Southern Hemisphere below 60°
    d. Southern Hemisphere between 30° and 60°
    e. the equator, between 30° north and 30° south

3. Which biome is associated with the rain shadow of a mountain?
    a. desert   d. coniferous forest
    b. tropical rain forest   e. chaparral
    c. taiga

36.2 Terrestrial Ecosystems

4. Which of these influences the location of a particular biome?
    a. latitude   d. altitude
    b. average annual rainfall e. All of these are correct.
    c. average annual temperature

5. What is the geographic distribution of a temperate deciduous forest?
    a. the northern part of North America and Eurasia
    b. just south of the polar caps in the Northern Hemisphere
    c. eastern North America and eastern Asia
    d. near the equator in South America and Africa
    e. central United States and central Asia

6. Which of these pairs is mismatched?
    a. tundra—permafrost
    b. savanna—*Acacia* trees
    c. prairie—epiphytes
    d. coniferous forest—evergreen trees

7. The forest with a multilevel canopy is the
    a. tropical rain forest.
    b. coniferous forest.
    c. tundra.
    d. temperate deciduous forest.

36.3 Aquatic Ecosystems

8. An estuary acts as a nutrient trap because of the
    a. action of rivers and tides.
    b. depth at which photosynthesis can occur.
    c. amount of rainfall received.
    d. height of the water table.

9. Phytoplankton are more likely to be found in which life zone of a lake?
    a. limnetic zone
    b. profundal zone
    c. benthic zone
    d. All of these are correct.

10. El Niño
    a. prevents the ocean current from upwelling.
    b. causes stagnation of the nutrients.
    c. changes global climate patterns.
    d. decreases the local fisheries.
    e. All of these are correct.

ENGAGE

Thinking Critically

1. Climate change is predicted to have the least severe consequences in terms of temperature change nearest the equator and more severe consequences progressing into the more temperate zones. Why do you think this is?

2. Scientists have documented a phenomenon called "biodiversity gradients," which shows that the greatest amounts of species diversity exist nearest the equator and that diversity tends to diminish moving through increasing latitudes. Based on your knowledge of climate, topography, and general characteristics of terrestrial biomes, explain the phenomenon of biodiversity gradients.

3. Based on what you know about tropical forests versus temperate forests, why do you think temperate forest habitats are more suitable for long-term agriculture, despite the fact that they have a shorter growing season than tropical forests?

4. Pharmaceutical companies are "bioprospecting" the tropical rain forests. These companies are looking for naturally occurring compounds in plants or animals that can be used as drugs for a variety of diseases. The most promising compounds act as anti-bacterial or antifungal agents. Even discounting the fact that the higher density of species in tropical rain forests would produce a wider array of compounds than another biome, why would anti-bacterial and antifungal compounds be more likely to evolve in this biome?

C H A P T E R

37

Conservation Biology

CHINA

©Fuse/Getty Images

BEFORE YOU BEGIN

Before beginning this chapter, take a few moments to review the following discussions:

Table 27.1 What is the cause of the current extinction crisis?

Section 34.2 What factors are influencing human population growth?

Section 35.2 What role does biodiversity play in the energy flow of ecosystems?

Conservation Success Stories

In recent years, conservation efforts have been able to highlight a variety of success stories. European bison have returned to the wild in Russia since their extinction in the early twentieth century. Satellite technology is being used to track river dolphin populations in Brazil, and the tiger population in Bhutan Royal Manas National Park has doubled in the past 6 years.

The giant panda (*Ailuropoda melanoleuca*) population has risen nearly 17% to over 1,864 individuals in the past decade. This increase has moved pandas from endangered to vulnerable status according to the International Union Conservation of Nature (IUCN).

Giant pandas are native to the temperate broadleaf and mixed forests of southwestern China, where they are an important disperser of bamboo seeds. A major challenge is that the panda's habitat lies in the geographic and economic heart of China. Habitat loss due to economic development, along with poaching, were the main reasons that the giant panda was historically classified as an endangered species.

Being the iconic logo of the World Wildlife Fund, as well as a national treasure in China, has helped earn global support for panda conservation. The Chinese government has established over 50 protected panda reserves that cover 3.8 million acres of forest and are home to approximately 61% of the panda population. A decrease in poaching due to strict laws and greater public awareness has played a role in their conservation status. With continued international and national efforts, the giant panda can continue to play a significant role in the forests of southwestern China.

In this chapter, we will explore the science of conservation biology and investigate some of the practices by which scientists are attempting to conserve the biodiversity on the planet. We will also explore some of the threats to biodiversity, including those currently threatening the giant panda.

As you read through this chapter, think about the following questions:

1. What type of ecological consequences would occur if the giant panda went extinct?

2. Which is more important for the conservation of giant pandas, their direct or indirect value?

37.1 Conservation Biology and Biodiversity

Conservation biology is an interdisciplinary science with the explicit goal of protecting biodiversity and the Earth's natural resources. Conservation biology is considered *interdisciplinary* because it relies on many subdisciplines of biology for the development of basic scientific concepts to describe biodiversity. The application of these concepts, so that biodiversity can be sustainably managed and conserved for future generations, is critical to the future of our planet.

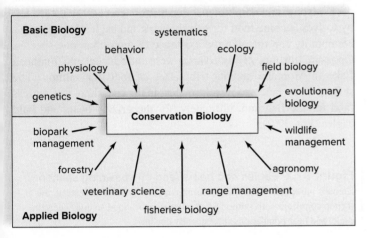

The application of basic scientific principles for conservation and management means that it is often necessary for conservation biologists to work with government officials at both the local and federal levels. Social scientists and economists are also involved because many conservation decisions have socioeconomic impacts in addition to biological ones. Public education is also critical for the success of conservation biology—educating people facilitates better-informed consumer decisions, such as product choices that minimize environmental impact.

Conservation biology is unique among the life sciences in that it embodies the following ethical principles: (1) biodiversity is desirable for the biosphere and therefore for humans; (2) human-induced extinctions are therefore undesirable; (3) complex interactions within ecosystems and communities support biodiversity and the maintenance of such interactions is therefore desirable; and (4) biodiversity generated by evolutionary change has intrinsic value, regardless of any practical benefit. Because species within communities often interact in subtle and complex ways, disruption to the ecosystems in which they live can have unpredictable and potentially dire consequences. For example, loss of a single, ecologically important species could result in large numbers of secondary extinctions. Thus, disruption of ecosystems or ecosystem functioning needs to be avoided.

Conservation biology has often been called a "crisis discipline" or a "discipline with a deadline" because we are facing a sixth mass extinction. Unlike the five previous mass extinctions that have occurred naturally throughout Earth's history, the current mass extinction is much more rapid. It is estimated that as many as 10–20% of all species may be extinct in the next 30 to 50 years, and some researchers suggest we may lose as many as 50% of all species by the year 2100. Thus, it is urgent we take steps to ensure that everyone understands the concept of biodiversity, the value of biodiversity, the causes of biodiversity loss, and the potential consequences of reduced biodiversity.

Biodiversity

At its simplest level, **biodiversity** is the variety of life on Earth. Biodiversity is commonly described in terms of the number of species found in a given area or ecosystem. To date, approximately 1.6–1.7 million species have been described and catalogued across the globe (Fig. 37.1). This may be only a small fraction of Earth's species, however. Recent scientific studies of biodiversity suggest there are probably around 8.7 million species (not counting bacteria or viruses) on the planet, but some think this number is much higher. Obviously, most species have yet to be discovered and described.

According to the U.S. Fish and Wildlife Service (FWS), there are over 500 animal species and 770 plant species in the United States that are in danger of extinction. Globally, over 30,000 identified species are threatened with extinction. An **endangered species** is one that faces immediate extinction throughout all or

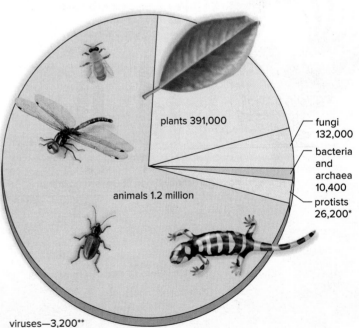

* Includes protozoans and chromists.
** Not considered by some to be living.

Figure 37.1 Number of catalogued species. Efforts are underway to catalogue all of the described species on the planet. Currently, between 1.6 to 1.7 million species have been catalogued. *Source:* www.catalogueoflife.org/.

most of its range. Examples of endangered species include the hawksbill sea turtle, California condor, and the snow leopard. Threatened species are likely to become endangered in the foreseeable future. Examples of **threatened species** include the Navajo sedge, northern spotted owl, and coho salmon.

To develop a meaningful understanding of life on Earth, we need to know more about species than just their total number. Although we most often think at the species level, ecologists and conservation biologists necessarily describe biodiversity at three other levels of biological organization: genetic diversity, ecosystem diversity, and landscape diversity.

Genetic diversity includes the number of different alleles, as well as the relative frequencies of those alleles in populations and species. It is genetic diversity that underlies the evolutionary potential of populations, or their capacity to adapt to future environmental change. Thus, populations with high genetic diversity are more likely to have some individuals that can survive changes in their community or ecosystem. Low genetic diversity made the 1846 potato blight in Ireland and the 1984 outbreak of citrus canker in Florida worse than they would have been if the populations were more genetically diverse. Higher genetic diversity would have likely meant that there would have been a higher proportion of individuals that, by chance, were resistant to the pathogens that decimated these crops.

Ecosystem diversity is dependent on the interactions between species and their abiotic environment in a particular area. An ecosystem may have one or several types of communities within it.

The composition of species within a community may differ dramatically among different community types. Consider the species you would find in a temperate rain forest versus a savannah. Variation in community composition increases the levels of biodiversity in the biosphere. Although past conservation efforts frequently focused on saving particular charismatic species, such as the California condor, the black-footed ferret, or the spotted owl, such species-specific methods may not be the most beneficial for all species in the ecosystem. Instead, a more effective approach is to conserve species that play critical roles in their ecosystem, such as those whose loss may result in many secondary extinctions.

The reintroduction of wolves in Yellowstone National Park has had a dramatic effect upon the health of the park. Without wolves, the elk population significantly increased, which then caused overgrazing of the aspen groves. Wolves have helped reduce the number of elk, thereby increasing the number of aspen and other trees found in floodplains that, in turn, increase the habitat for songbirds and beavers. Beavers have then built dams that have flooded areas, making them suitable for muskrats, ducks, otters, and amphibians. Further, elk carcasses that are left by wolves become food for grizzly bears and eagles. Disrupting a community can threaten the existence of more than one species. Opossum shrimp, *Mysis relicta,* were introduced into Flathead Lake in Montana and its tributaries as food for salmon. The shrimp ate so much zooplankton that, in the end, there was far less food for the fish and, ultimately, for the grizzly bears and bald eagles (Fig. 37.2).

Figure 37.2 Eagles and bears feed on spawning salmon. Humans introduced the opossum shrimp as prey for salmon. Instead, the shrimp competed with salmon for zooplankton as a food source. The salmon, eagle, and bear populations subsequently declined.

Landscape diversity involves a group of interacting ecosystems within a single landscape. For example, plains, mountains, and rivers may be fragmented to the point that they are connected only by patches (or corridors) that allow organisms to move from one ecosystem to another. Fragmentation of the landscape may reduce reproductive capacity by increasing predation risk during dispersal between suitable habitat patches or creating difficulty in finding mates.

Check Your Progress 37.1

1. Describe how conservation biology is supported by a variety of disciplines.

2. Explain why ecosystem-level conservation may be more important than species-level conservation.

37.2 Value of Biodiversity

Learning Outcomes

Upon completion of this section, you should be able to

1. Compare the direct and indirect values of biodiversity.

2. Describe the role that biodiversity plays in a natural ecosystem.

Conservation biology strives to reverse the trend toward the extinction of thousands of plants and animals. To achieve this goal, it is necessary to make all people aware that biodiversity is a resource of immense value.

Direct Value

Various individual species perform services for humans and contribute greatly to the value we should place on biodiversity. Figure 37.3 gives examples of the direct value of wildlife.

Medicinal Value

Most of the prescription drugs used in the United States, valued at over $200 billion, were originally derived from living organisms. The rosy periwinkle from Madagascar is an excellent example of a tropical plant that has provided us with useful medicines. Potent chemicals from this plant are now used to treat two forms of cancer: leukemia and Hodgkin disease. Because of these drugs, the survival rate for childhood leukemia has improved from 10% to 90%, and Hodgkin disease is usually curable. Although the value of saving a life cannot be calculated, it is still sometimes easier for us to appreciate the worth of a resource if it is explained in monetary terms. Thus, researchers estimate that, judging from the success rate in the past, hundreds of additional types of drugs are yet to be found in tropical rain forests, and the value of this resource to society is probably $147 billion.

You may already know that the antibiotic penicillin is derived from a fungus, and that certain species of bacteria produce the antibiotics tetracycline and streptomycin. These drugs have proven to be indispensable in the treatment of various sexually transmitted diseases (STDs). Leprosy is among those diseases for which, at

Wild species, like the rosy periwinkle, *Catharanthus roseus,* are sources of many medicines.

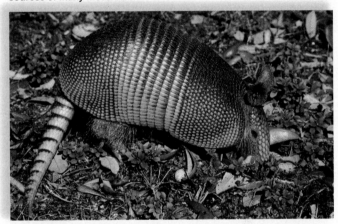

Wild species, like the nine-banded armadillo, *Dasypus novemcinctus,* play a role in medical research.

Wild species, like ladybugs, play a role in biological control of agricultural pests.

Figure 37.3 Direct value of wildlife. The direct services of wild species benefit humans immensely. It is sometimes possible to calculate the monetary value, which is always surprisingly large. (periwinkle): ©Steven P. Lynch; (armadillo): ©Steve Bower/Shutterstock; (ladybug): ©D. Hurst/Alamy

this time, there is no cure. The bacterium that causes leprosy will not grow in the laboratory, but scientists discovered that it grows naturally in the nine-banded armadillo. Having a source of the bacterium may make it possible to find a cure for leprosy. The blood of horseshoe crabs contains a substance called limulus amoebocyte lysate, which is used to ensure that medical devices, such as pacemakers, surgical implants, and prosthetic devices, are free of bacteria. Blood is taken from 250,000 crabs a year, which are then returned to the sea unharmed.

Agricultural Value

Crops such as wheat, corn, and rice are derived from wild plants that have been modified to be high producers. The same high-yield, genetically similar strains tend to be grown worldwide. When rice crops in Africa were being devastated by a virus, researchers grew wild rice plants from thousands of seed samples until they found one that contained a gene for resistance to the virus. These wild plants were then used in a breeding program to transfer the gene into high-yield rice plants. If this variety of wild rice had become extinct before it could be discovered, rice cultivation in Africa might have collapsed.

Biological pest controls—natural predators and parasites—are often preferable to using chemical pesticides. When a rice pest called the brown planthopper became resistant to pesticides, farmers began to use natural brown planthopper enemies instead. The economic savings were calculated at well over $1 billion. Similarly, cotton growers in Cañete Valley, Peru, found that pesticides were no longer working against the cotton aphid because they had evolved resistance. Research identified natural predators that are now being used to an ever-greater degree by cotton farmers. The savings have been enormous.

Most flowering plants are pollinated by animals such as bees, wasps, butterflies, beetles, birds, and bats. Honeybees play a major role in this process. As is explored in the Ecology feature "Colony Collapse Disorder," honeybee populations worldwide are decreasing, which presents a serious threat to modern agriculture.

SCIENCE IN YOUR LIFE ▶ ECOLOGY

Colony Collapse Disorder

Over $15 billion of United States agricultural production is directly associated with the presence of a viable honeybee population. While wild honeybees play a role in moving pollen from one plant to another, the domesticated honeybees are the ones that do most of the pollination. Populations of managed honeybees have come under serious threat over the past several decades due to a variety of pressures.

These pressures can be broken down into four main categories: pathogens, poor nutrition, pesticides, and sublethal pesticide effects. Any one of these, or more likely a combination of several, are responsible for the declining bee populations and occurrences of colony collapse disorder (CCD).

Since the 1980s, a variety of invasive and native pathogens have made their way into the bee populations. Viruses, bacteria, fungi, and mites play different roles in disrupting the bee populations. Varroa mites have been identified as one of the worst problems facing bee colonies today (Fig. 37A). They indirectly damage honeybees by sucking out their body fluids, allowing a wide range of other health issues to occur.

Honeybee nutrition is derived from the pollen and nectar gathered from a variety of different plants. When bees are used to pollinate monocultures, their diet lacks essential nutrients. This leads to weaker immune systems and less food availability to survive droughts and winter conditions.

The U.S. EPA has regulations to protect bees from pesticide misuse. Unfortunately, some individuals may not be aware of, understand, or be concerned about the regulations. Accidental overspraying, in which the wind carries a pesticide across the wrong

Figure 37A Our primary pollinator. Honeybee populations are decreasing as a result of colony collapse disorder (CCD). *Source:* USDA/Lila De Guzman, photographer

field, can occur, and the bees suffer the consequences.

Nicotine-based pesticides called neonicotinoids were developed in the mid-1990s due to their reduced toxicity to wildlife compared to other pesticides. However, the scientific data about the impact they cause are mixed. In some cases, they have had a sublethal effect when applied at or below approved doses, while in other cases they have not had any type of adverse impact.

Colony collapse disorder is defined as a colony in which there are no adult bees or dead bodies present in the hive, but there is a live queen, honey, and immature bees. Without the adult worker bees, the hive cannot sustain itself and it collapses. There has been no identification of a single factor that leads to CCD;

rather it is most likely a combination of factors that is leading to CCD and the decline in bee populations. Since 2010, the number of colonies lost due to CCD has been decreasing, but a high percentage of colonies are being lost due to other factors. It will take a combined effort from governmental, local, and private agencies to find ways to stabilize and, hopefully, increase honeybee populations in the United States.

Questions to Consider

1. What are some ways in which individuals can contribute to helping honeybee populations?
2. How could a decline in honeybee populations contribute to an ecological collapse within a community?

Consumptive Use Value

We have experienced much success in cultivating crops, keeping domesticated animals, growing trees on plantations, and so forth. But so far, aquaculture, the growing of fish and shellfish for human consumption, has contributed only minimally to human welfare. Instead, most freshwater and marine harvests depend on wild animals, such as fishes (e.g., trout, cod, and tuna) and crustaceans (e.g., lobsters, shrimps, and crabs). Obviously, these aquatic organisms are an invaluable biodiversity resource.

The environment provides all sorts of other products that are sold in the marketplace worldwide, including wild fruits and vegetables, skins, fibers, beeswax, and seaweed. Also, some people obtain their meat directly from the wild. In one study, researchers calculated that the economic value of wild pig in the diet of native hunters in Sarawak, East Malaysia, was approximately $40 million per year.

Similarly, many trees are still felled for their wood. Researchers have calculated that a species-rich forest in the Peruvian Amazon region is worth far more if the forest is used for fruit and rubber production than for timber production. Whereas fruit and the latex needed to produce rubber can be brought to market indefinitely, no more timber or other resulting products can be produced when trees are cut down.

Indirect Value

The wild species we have been discussing live in ecosystems. If we want to preserve them, many scientists argue it is more economical to save the ecosystems than the individual species. Ecosystems perform many services for modern humans. These services are said to be indirect because they are pervasive, and it is not easy to associate a direct dollar value to them. Human survival depends on the functions that natural ecosystems perform for us. Studies have suggested that the indirect value of ecosystem services surpasses that of the global gross national product—about $33 trillion per year!

Biogeochemical Cycles

You'll recall from Section 35.1 that ecosystems are characterized by energy flow and chemical cycling. The biodiversity within ecosystems contributes to the workings of the water, carbon, nitrogen, phosphorus, and other biogeochemical cycles. We are dependent on these cycles for fresh water, removal of carbon dioxide from the atmosphere, uptake of excess soil nitrogen, and provision of phosphate. When human activities upset the usual workings of biogeochemical cycles, the dire environmental consequences include the release of excess pollutants that are harmful to us. Unfortunately, technology is unable to artificially replicate any of the biogeochemical cycles.

Waste Disposal

Decomposers break down dead organic matter and other types of wastes into inorganic nutrients that are used by the producers within ecosystems. This function aids humans immensely because we dump millions of tons of waste material into natural ecosystems each year. If it were not for decomposition, waste would soon cover the entire surface of our planet. We can build sewage treatment plants, but they are expensive. In addition, few of them break down solid wastes completely to inorganic nutrients. It is less expensive and more efficient to water plants and trees with partially treated wastewater and let soil bacteria cleanse it completely.

Biological communities are also capable of breaking down and immobilizing pollutants like heavy metals and pesticides that humans release into the environment. A review of wetland functions in Canada assigned a value of $50,000 per hectare (2.5 acres or 10,000 m²) per year to the ability of natural areas to purify water and take up pollutants.

Provision of Fresh Water

Few terrestrial organisms are adapted to living in a salty environment—they need fresh water. The water cycle continually supplies fresh water to terrestrial ecosystems. Humans use fresh water in innumerable ways, including drinking and irrigation of their crops. Freshwater ecosystems, such as rivers and lakes, also provide us with fish and other types of edible organisms. Unlike other commodities, there is no substitute for fresh water. We can remove salt from seawater to obtain fresh water, but the cost of desalinization is about four to eight times the average cost of fresh water acquired via the water cycle.

Flood Prevention

Forests and other natural ecosystems exert a "sponge effect," thereby reducing flooding. They soak up water and then release it at a regular rate. When rain falls in a natural area, plant foliage and dead leaves lessen its impact, and the soil slowly absorbs it, especially if the soil has been aerated by organisms. The water-holding capacity of forests reduces the possibility of flooding. The value of a marshland outside Boston, Massachusetts, has been estimated at $72,000 per hectare per year based solely on its ability to reduce floods. Flooding in New Orleans by Hurricane Katrina in 2005 would likely have been much less severe if the natural wetlands and marshlands around the Gulf of Mexico had still been intact.

Prevention of Soil Erosion

Intact ecosystems naturally retain soil and prevent soil erosion. The importance of this ecosystem attribute is especially observed following deforestation. In Pakistan, the world's largest dam, the Tarbela Dam, is losing its storage capacity of 12 billion cubic meters (m³) many years sooner than expected because silt is building up behind the dam due to deforestation. At one time, the Philippines exported $100 million worth of oysters, mussels, and clams each year. Now, silt carried down rivers following deforestation is smothering the mangrove ecosystem that serves as a nursery for the sea. In general, most coastal ecosystems are not as bountiful as they once were because of deforestation, coastal development, and aquatic pollution.

Regulation of Climate

At the local level, trees provide shade, reduce the need for fans and air conditioners during the summer, and provide buffers from noise. In tropical rain forests, trees maintain regional rainfall, and without them the forests become arid.

Globally, forests ameliorate the climate because they take up carbon dioxide and release oxygen. The leaves of trees use carbon dioxide when they photosynthesize, and the bodies of the trees store carbon. When trees are cut and burned, carbon dioxide is released into the atmosphere. Carbon dioxide makes a significant contribution to global warming, which is expected to be stressful for many plants and animals.

Ecotourism

Many people prefer to vacation in natural areas. In the United States, nearly 100 million people spend nearly $5 billion each year on fees, travel, lodging, and food associated with ecotourism. Tourists are often interested in activities such as sport fishing, whale watching, boat riding, hiking, birdwatching, and the like. Some merely want to immerse themselves in the beauty and serenity of nature. Many underdeveloped countries in tropical regions, such as Belize and Costa Rica, are taking advantage of this by offering "ecotours" of the local biodiversity. Providing guided tours of natural ecosystems, such as tropical rain forests, is often more profitable than destroying them. Ecotourism also provides indirect value for wildlife. As an example, the tusks of an elephant are worth around $10,000 for the ivory, but the Kenyan government estimates that a single elephant is worth as much as $1.2 million in tourist dollars.

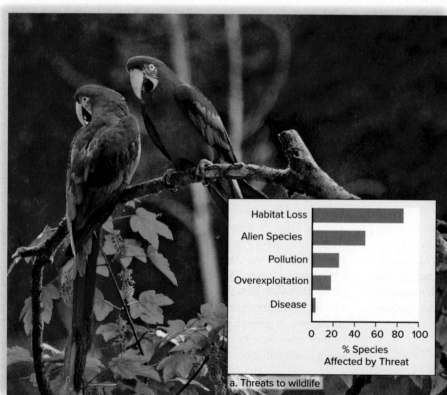

Figure 37.4 Habitat loss. **a.** In a study that examined records of imperiled U.S. plants and animals, habitat loss emerged as the greatest threat to wildlife. **b.** Macaws that reside in South American tropical rain forests are endangered for some of the reasons listed in the graph (**a**).
(photo) (b): ©Charan Reddy/EyeEm/Getty Images

b. Macaws

Check Your Progress 37.2

1. Identify the differences between consumptive use value and agricultural value of biodiversity.
2. Explain the difference between direct and indirect values of biodiversity.

37.3 Threats to Biodiversity

Learning Outcomes

Upon completion of this section, you should be able to

1. Classify the five causes of extinction.
2. Compare natural and human influenced causes of extinction.

We are presently in a biodiversity crisis—the number of extinctions (loss of species) expected to occur in the near future is unparalleled in Earth's history. To identify the role that humans are playing in modern extinctions, researchers examined the records of 1,880 threatened and endangered wild species in the United States. Habitat loss is the most significant factor that was involved in 85% of the cases (Fig. 37.4*a*). Exotic species had a hand in nearly 50%, pollution was a factor in 24%, overexploitation in 17%, and disease in 3%. Note that the percentages add up to more than 100% because most of these species are imperiled for more than one reason.

These five causes reflect the relative importance of the five leading causes for species' extinctions worldwide. Macaws illustrate that a combination of factors can lead to a species decline (Fig. 37.4*b*). Not only has their habitat been reduced by encroaching timber and mining companies, but macaws are also hunted for food and collected for the pet trade.

DNA technology may now help enforce laws to protect wildlife collected for the pet trade, as described in the Scientific Inquiry feature "Wildlife Conservation and DNA."

Habitat Loss

Habitat loss has occurred in all ecosystems, and human disruption of natural habitats is the most influential factor in biodiversity loss. Concern has now centered on tropical rain forests and coral reefs because they are particularly rich in species diversity. A sequence of events in Brazil offers a fairly typical example of the manner in which rain forest is converted to land uninhabitable for wildlife. The construction of a major highway into the forest first provided a way to reach the interior of the forest. Small towns and industries sprang up along the highway, and roads branching off the main highway gave rise to even more roads. The result was fragmentation of the once immense forest. The government offered subsidies to anyone willing to take up

SCIENCE IN YOUR LIFE ▶ SCIENTIFIC INQUIRY

Wildlife Conservation and DNA

After DNA analysis, scientists were amazed to find that some 60% of loggerhead turtles found drowned in Mediterranean fishing nets were from beaches in the southeastern United States. Because the unlucky creatures were a good representative sample of the turtles in the area, that meant more than half of the young turtles living in the Mediterranean Sea had hatched from nests on beaches in Florida, Georgia, and South Carolina (Fig. 37B*a*). Some 20,000 to 50,000 loggerheads die each year due to the Mediterranean fisheries, which may partly explain the decline in loggerheads nesting on southeastern U.S. beaches for the last 25 years.

Jaguars (*Panthera onca*) (Fig. 37B*b*) are the third-largest cats in the world behind lions and tigers. They are the largest cats in the Western Hemisphere. Their natural range extends from as far north as Mexico, to as far south as Argentina. Currently, they are listed as "Near Threatened" by the International Union for the Conservation of Nature (IUCN). Conservation of this top-level predator, which has an extensive range, requires support from all the countries that are home to jaguars.

Detailed genetic analysis of jaguar DNA has indicated that, whether they live in Mexico, Argentina, or anywhere in-between, they are all the same species. They are the only wide-ranging carnivore in the world that shows genetic continuity across their entire range. This genetic information led to the formation of the Jaguar Corridor Initiative (JCI), whose goal is to create a genetic corridor (a habitat corridor that allows gene flow) that links jaguar populations in all of the 18 countries in Latin America, from Mexico to Argentina, hopefully ensuring the survival of this species.

In what will become a classic example of how DNA analysis might be used to protect endangered species from future ruin, scientists from the United States and New Zealand carried out discreet experiments in a Japanese hotel room on whale sushi bought in local markets. Sushi, a staple of the Japanese diet, is rice and

Figure 37B DNA studies.   Conservation of the (**a**) loggerhead turtle and (**b**) jaguar is being aided by DNA analysis. (turtle): ©hotshotsworldwide/Getty Images; (jaguar): ©Photodisc/Getty Images

meat wrapped in seaweed. Armed with a miniature DNA sampling machine, the scientists found that, of the 16 pieces of whale sushi they examined, many were from whales that are endangered or protected under an international moratorium on whaling. "Their findings demonstrated the true power of DNA studies," says David Woodruff, a conservation biologist at the University of California, San Diego.

One sample was from an endangered humpback, four were from fin whales, one was from a northern minke, and another from a beaked whale. Stephen Palumbi, of Harvard University, says the technique could be used for monitoring and verifying catches. Until then, he says, "no species of whale can be considered safe."

Meanwhile, the U.S. Fish and Wildlife Service Forensics Laboratory in Ashland, Oregon, is already on the watch for wildlife crimes in the United States and 122 other countries that send samples to them for analysis. The lab has blood samples, for example, for all the wolves

being released into Yellowstone National Park. The lab has many cases currently pending in court that cannot be discussed. However, one story tells of the lab's first DNA-matching case. Shortly after the lab opened in 1989, California wildlife authorities contacted the director of the lab. They had seized the carcass of a trophy-sized deer from a hunter. They believed the deer had been shot illegally on a 3,000-acre preserve. The agents found a gut pile on the property but had no way to match it to the carcass. The hunter had two witnesses to deny the deer had been shot on the preserve. However, lab analysis made a perfect match between tissue from the gut pile and tissue from the carcass.

Questions to Consider

1. How might DNA analysis be used to prevent the poaching of endangered species?
2. What benefits would occur if there was an international database of DNA from endangered and threatened species?

residence in the forest, and the people who came cut and burned trees in patches to facilitate grazing. Tropical soils contain limited nutrients, and burning of trees releases additional nutrients that support cattle grazing for only about 3 years. Once the land was degraded, the farmers moved on to another portion of the forest to start over again.

Loss of habitat also affects freshwater and marine biodiversity. Coastal degradation is mainly due to the large concentration of people living on or near the coast. At least 40% of the world's

population lives within 100 km (60 mi) of a coastline, and this number is expected to increase. In the United States, over one-half of the population lives within 80 km (50 mi) of the coasts (including the Great Lakes). Coastal habitation leads to beach erosion and direct inputs of pollutants into freshwater or marine systems. Already, 60% of coral reefs have been destroyed or nearly so, with siltation resulting from beach erosion as a leading cause. It is possible that all coral reefs may disappear during the next 40 years unless our behaviors drastically change. Mangrove forest destruction is also a problem. Indonesia, with the most mangrove acreage, has lost 45% of its mangroves, and the percentage is even higher for other tropical countries. Wetland areas, estuaries, and seagrass beds are also being rapidly destroyed by the actions of humans.

Alien Species

Alien species, sometimes also called *exotic species,* are nonnative members of a community. Communities around the world are characterized by unique assemblages of species that have evolved together in an area. Introduction of exotic species can disrupt this balance by changing the interactions between species in a food web, as shown in Figure 37.2. In this example, opossum shrimp introduced in a lake in Montana added a trophic level that, in the end, meant less food for bald eagles and grizzly bears. Often, exotic species can directly compete with or prey upon native species, thereby reducing their abundance or causing localized extinction (Fig. 37.5). For these reasons, exotics are the second most important reason for biodiversity loss. Humans have introduced exotic species into new ecosystems by the following means:

■ *Colonization.* Europeans, in particular, brought various familiar species with them when they colonized new places. For example, the pilgrims brought the dandelion to the United States as a familiar salad green. The British brought foxes and stoats (a type of weasel) to New Zealand, and these predators have resulted in the loss of nearly 40% of all of New Zealand's bird species. These birds are easy prey because they evolved without mammalian predators, which were completely absent from New Zealand.

■ *Horticulture and agriculture.* Some exotics now taking over vast tracts of land have escaped from cultivated areas. Kudzu is a vine from Japan that the U.S. Department of Agriculture thought would help prevent soil erosion. The plant now covers much of the landscape in the South, including walnut, magnolia, and gum trees. Similarly, the water hyacinth was introduced to the United States from South America because

of its beautiful flowers. Today, it clogs up waterways and diminishes natural diversity.

■ *Accidental transport.* Global trade and travel accidentally bring many new species from one country to another. The zebra mussel from the Caspian Sea was accidentally introduced into the Great Lakes in 1988. It now forms dense beds that reduce biodiversity by outcompeting native mussels. Zebra mussels also decrease the amount of food available for higher levels in the food chain because they are such effective filter feeders. Zebra mussels cause millions in damage per year by clogging sewage and other pipes. Other exotics introduced into the United States include the Argentinian fire ant and the nutria, a large rodent found throughout the Southeast.

Exotics on Islands

Islands are particularly susceptible to environmental discord caused by the introduction of exotic species. Islands have unique assemblages of native species that are closely adapted to one another and cannot compete well against exotics. Myrtle trees, *Myrica faya,* introduced into the Hawaiian Islands from the Canary Islands, are symbiotic with a type of bacterium that is capable of nitrogen fixation. This feature allows the species to establish itself on nutrient-poor volcanic soil such as that found in Hawaii. Once established, myrtle trees halt normal ecological succession by outcompeting native plants on volcanic soil.

The brown tree snake has been introduced onto a number of islands in the Pacific Ocean. The snake eats eggs, nestlings, and adult birds. On Guam, it has reduced ten native bird species to the point of extinction. Mongooses introduced into the Hawaiian Islands to control rats have increased dramatically in abundance and also prey on native birds, causing some population declines (Fig. 37.5*b*).

Pollution

In the present context, **pollution** can be defined as any environmental change that adversely affects the lives and health of living

a. Kudzu

b. Mongoose

Figure 37.5 Exotic species.   a. Kudzu, a vine from Japan, was introduced in several southern states to control erosion. Today, kudzu has taken over and displaced many native plants. **b.** Mongooses were introduced into Hawaii to control rats, but they also prey on native birds. (a): ©Chuck Pratt/Avalon; (b): ©Chris Johns/National Geographic Image Collection organisms. Pollution has been identified as the third main cause of extinction. Pollution can also weaken organisms and make them more susceptible to disease. Biodiversity is particularly threatened by the following types of environmental pollution:

- *Acid deposition.* Sulfur dioxide from power plants and nitrogen oxides from auto exhaust are converted to acids when combined with atmospheric water vapor. These acids return to Earth during precipitation (rain or snow) or via dry deposition. This acid deposition decimates forests because it weakens trees and increases their susceptibility to disease and insects. In addition, because acid deposition can make the pH of water too low for organisms to survive, many lakes in northern states are now lifeless.

- *Eutrophication.* Lakes are also under stress due to overenrichment. When lakes receive excess nutrients due to runoff from agricultural fields and wastewater from sewage treatment, algae begin to grow in abundance. Death of algae leads to an abundance of decomposers that decreases the available oxygen and often leads to fish kills.

- *Ozone depletion.* The ozone shield protects Earth from harmful ultraviolet (UV) radiation. The release of chlorofluorocarbons (CFCs), in particular, into the atmosphere causes the shield to break down, leading to impaired growth of crops and trees and the death of plankton that sustain oceanic life. The immune systems and ability of all organisms to resist disease will most likely be weakened. Ozone depletion has also led to dramatic increases in skin cancer.

- *Organic chemicals.* Our modern society uses organic chemicals in all sorts of ways. Organic chemicals called nonylphenols are used in products ranging from pesticides to dishwashing detergents, cosmetics, plastics, and spermicides. Many of these chemicals can mimic the effects of hormones. Those that do are called endocrine-disruptors. For example, investigators exposed young salmon to nonylphenol. Although these fish are born in fresh water and mature in salt water, 20–30% of fish exposed to nonylphenol were unable to make the transition from fresh to salt water. Nonylphenols cause the pituitary to produce prolactin, a hormone that may prevent saltwater adaptation. Other endocrine-disrupting contaminants can possibly affect the endocrine system and thereby the reproductive potential of other animals, including humans.

- *Climate change.* The term **climate change** refers to recent changes in the Earth's climate. The major contributor to climate change is the phenomenon of global warming, which is an increase in Earth's temperature due to the increase of greenhouse gases, such as carbon dioxide and methane, in the atmosphere. Climate change is expected to have many detrimental effects, including the destruction of coastal wetlands due to a rise in sea levels (Fig. 37.6), a shift in suitable temperatures in locations where various species cannot live, and the death of coral reefs. An upward shift in temperatures could influence everything from growing seasons in plants to migratory patterns in animals. As temperatures rise, regions of suitable climate for various terrestrial species may shift toward the poles and higher elevations. Extinctions are expected because the present assemblages of species in eco-

a.

b.

Figure 37.6 Climate change and global warming. **a.** Mean global temperature is expected to rise due to the introduction of greenhouse gases into the atmosphere. **b.** An increase in the global temperature causes glaciers to melt, thus increasing the sea level and threatening coastal ecosystems. (b): ©Marco Simoni/Getty Images

systems will be disrupted as some species migrate northward (or southward in the Southern Hemisphere) following the environmental changes. Other species with limited mobility may not be able to migrate quickly enough. To remain in a favorable habitat, it's been calculated that the rate of beech tree migration would have to be 40 times faster than has ever been observed.

Overexploitation

Overexploitation occurs when the number of individuals taken from a wild population is so great that the population becomes severely reduced in numbers. A positive feedback cycle explains overexploitation: the smaller the population, the more commercially valuable its members, and the greater the incentive to capture the few remaining organisms. Poachers are very active in the collecting and sale of endangered and threatened species because it has become so lucrative. The overall international value of trading wildlife species is $20 billion, of which $8 billion is attributed to the illegal sale of rare species.

Markets for rare plants and exotic pets support both legal and illegal trade in wild species. Rustlers dig up rare cacti, such as the crested saguaros, and sell them to gardeners for as much as $15,000 each. Parrots are among birds taken from the wild for sale to pet owners. For every bird delivered alive, many more have died in the process. The same holds true for tropical fish, which often come from the coral reefs of Indonesia and the Philippines. Divers dynamite reefs or use plastic squeeze-bottles of cyanide to stun the fish within them. In the process, many fish and corals die.

The Convention on International Trade in Endangered Species (CITES) was an agreement established in 1973 to ensure that international trade of species does not threaten their survival. Today, over 35,600 species of plants and animals receive some level of protection from over 172 countries worldwide.

Poachers still hunt for hides, claws, tusks, horns, or bones of many endangered mammals. Because of its rarity, a single Siberian tiger is now worth more than $500,000—its bones are pulverized and used as a medicinal powder. The horns of rhinoceroses become ornate carved daggers, and their bones are also ground up to sell as a medicine. The ivory of an elephant's tusk is used to make art objects, jewelry, or piano keys. The fur of a Bengal tiger sells for as much as $100,000 in Tokyo.

The Food and Agricultural Organization of the United Nations tells us that we have now overexploited 11 of 15 major oceanic fishing areas. Fish are a renewable resource if harvesting does not exceed the ability of the population to reproduce. Our society uses larger and more efficient fishing fleets to decimate fishing stocks. Pelagic species such as tuna are captured by purse seine fishing, in which a very large net surrounds a school of fish, and then the net is closed in the same manner as a drawstring purse. Often, dolphins that swim above the schools of tuna are captured and killed in this type of net. Other fishing boats drag huge trawling nets, large enough to accommodate 12 jumbo jets, along the seafloor to capture bottom-dwelling fish (Fig. 37.7*a*). Only large fish are kept; undesirable small fish and sea turtles are discarded, dying, back into the ocean. Trawling has been called the marine equivalent of clear-cutting trees because after the net goes by, the sea bottom is devastated (Fig. 37.7*b*). Today's fishing practices don't allow fisheries to recover. Cod and haddock, once the most abundant bottom-dwelling fish along the northeast coast of the United States, are now often outnumbered by dogfish and skate.

Entire marine ecosystems can be disrupted by overfishing, as exemplified on the U.S. West Coast. When sea otters began to decline in numbers, investigators found they were being eaten by orcas (killer whales). Usually, orcas prefer seals and sea lions to sea otters, but they began eating sea otters when few seals and sea lions could be found. The decline in the seal and sea lion population was due to the decline in the perch and herring populations as a result of overfishing. Ordinarily, sea otters keep the population of sea urchins, which feed on kelp, under control. But with fewer sea otters around, the sea urchin population exploded and decimated the kelp beds. Thus, overfishing set in motion a chain of events that detrimentally altered the food web of an ecosystem.

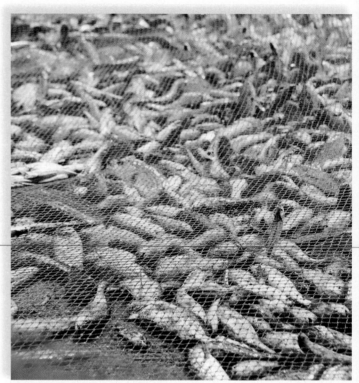

a. Fishing by use of a drag net

b. Result of drag net fishing

Figure 37.7 Trawling.   a. These fish were caught by dragging a net along the seafloor. **b.** Appearance of the seafloor after the net passed.
(a): ©StrahilDimitrov/iStock/360/Getty Images; (b): ©Peter Auster/University of Connecticut

Disease

Wildlife is subject to emerging diseases just as humans are. Due to the encroachment of humans on their habitat, wildlife have been exposed to new pathogens from domestic animals living nearby. For example, canine distemper was spread from domesticated dogs to lions in the African Serengeti, causing population declines. Avian influenza likely emerged from domesticated fowl (e.g., chicken) populations and has led to the deaths of millions of wild birds.

The significant effect of diseases on biodiversity is underscored by National Wildlife Health Center findings that almost half of sea otter deaths along the coast of California are due to infectious diseases. Scientists tell us the number of pathogens that cause disease is on the rise, and just as human health is threatened, so is that of wildlife. Extinctions due simply to disease may occur, and are currently implicated in the worldwide decline of amphibians (Fig. 37.8).

Figure 37.8 Amphibian at risk. The harlequin toad is nearly extinct due to infections by a fungal pathogen. Diseases are implicated in global population declines and extinctions of amphibians. ©Dr. Paul A. Zahl/ Science Source

---

Check Your Progress 37.3

1. Discuss two ways in which exotic species are introduced into new areas.
2. Identify the five main causes of extinction.
3. Recognize various reasons why pollution can impact biodiversity.

37.4 Habitat Conservation and Restoration

Learning Outcomes

Upon completion of this section, you should be able to

1. Describe the value of preserving biodiversity hotspots.
2. Distinguish between keystone species and flagship species.
3. Summarize the goals of habitat restoration.

Habitat Conservation

Because habitat loss is the leading cause of species' extinctions, conservation of habitat is of primary concern. One way to prioritize which habitats to conserve is to focus on those that contain the highest levels of biodiversity. Generally, biodiversity is highest at the tropics, and it declines toward each pole. Tropical rain forests and coral reefs are examples of ecosystems known for possessing high levels of biodiversity.

Some regions of the world are called **biodiversity hotspots** because they contain unusually large concentrations of species. Biodiversity in these hotspots accounts for about 44% of all known higher plant species and 35% of all terrestrial vertebrate species but covers only about 1.4% of Earth's land area. Hotspots may also contain large numbers of endemic species, or those not found anywhere else. An example of such a hotspot is the tropical forests of Madagascar, within which 93% of the primate species, 99% of the amphibian species, and over 80% of the plant species are endemic. The Cape region of South Africa, Indonesia, the coast of California, and the Great Barrier Reef of Australia are also considered biodiversity hotspots.

We can also focus our efforts on conserving habitat for **keystone species,** or those whose loss would result in a great number of secondary extinctions. Keystone species need not be the most abundant. Although they are relatively low in numbers, wolves can be considered a keystone species in Yellowstone National Park (see Chapter 35). Bats have also been designated keystone species in Old World tropical forests, because they are extremely important pollinators and dispersers of tree seeds. The loss of bats results in the failure of many tree species to reproduce and can lead to a loss of biodiversity. Other keystone species include grizzly bears (Fig. 37.9a), beavers in wetlands, and elephants in grasslands and forests.

Keystone species should not be confused with **flagship species,** or those that evoke an emotional response from humans. Flagship species are considered charismatic and are valued for their beauty, regal nature, or similarity to people's pets. Flagship species such as lions, tigers, dolphins, and the giant panda can motivate the public to conserve biodiversity.

Landscape Conservation and Reserve Design

Conservation often has to occur at the landscape level because sufficient habitat may not be available in a single place to sustain a viable population of a particular species. Grizzly bears, once numbering between 50,000 and 100,000 south of Canada, are now estimated at about 1,000 individuals in six small, subdivided populations. Grizzly bear conservation entails maintenance of a number of different types of ecosystems, including plains, mountains, and rivers. Other species also rely on multiple ecosystem types, such as amphibians that rely on wetlands and terrestrial habitat. Conserving a single one of these ecosystems would not be sufficient, nor would conservation of several ecosystems in isolation of one another. It is thus necessary to establish **conservation corridors** that allow animals to move safely between habitats that would otherwise be isolated. Corridors are often necessary because landscapes are subdivided due to urbanization, agriculture, and other aspects of human development. A corridor may be as small as an overpass for wildlife to cross a highway or a strip of forested habitat that is maintained along a stream after timber harvest. They can be as large as a patch of land several kilometers wide to allow for animals to follow migration routes.

Landscape conservation for a single species is often beneficial for other species that share the same habitats. For example, conservation of the northern spotted owl (Fig. 37.9b) helps protect many other species that inhabit temperate old growth forests. The last of the contiguous 48 states' harlequin ducks, westslope cutthroat trout, lynx, pine martins, wolverines, and great gray owls are found in areas occupied by grizzly bears. The geographic range of grizzly bears also overlaps with 40% of Montana's vascular plants of conservation concern.

a. Grizzly bear, *Ursus arctos horribilis*

b. Old-growth forest; northern spotted owl, *Strix occidentalis caurina* (inset)

Figure 37.9 Habitat preservation. When particular species are protected, other wildlife benefits. **a.** The Greater Yellowstone Ecosystem has been delineated in an effort to save grizzly bears, which need a very large habitat. **b.** Currently, the remaining portions of old-growth forests in the Pacific Northwest are not being logged in order to save the northern spotted owl (inset). (a): ©Deb Garside/Design Pics; (b) (forest): ©Timothy Epp/Shutterstock; (owl): ©Michael Sewell/Getty Images

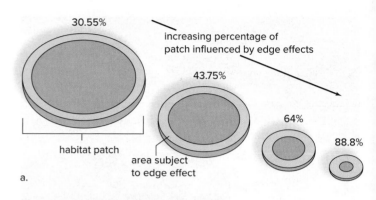

increasing percentage of patch influenced by edge effects

30.55%

43.75%

64%

88.8%

habitat patch

area subject to edge effect

a.

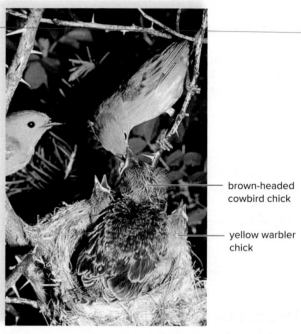

brown-headed cowbird chick

yellow warbler chick

b.

Figure 37.10 Edge effect. **a.** The smaller the patch, the greater the proportion that is subject to the edge effect. **b.** Cowbirds lay their eggs in the nests of songbirds (yellow warblers). A cowbird is bigger than a warbler nestling and will be able to acquire most of the food brought by the warbler parent. (b): ©Jeff Foott/Getty Images

Edge Effects

When conserving landscapes, it is necessary to consider **edge effects.** An edge reduces the amount of habitat typical of an ecosystem because the edges around a patch have a habitat slightly different from the interior of the patch. For example, forest edges are brighter, warmer, drier, and windier, with more vines, shrubs, and weeds than the forest interior. Also, Figure 37.10*a* shows that a small and a large patch of habitat have the same amount of edge. Therefore, the effective habitat shrinks as a patch gets smaller.

Many popular game animals, such as turkeys and white-tail deer, are more plentiful in the edge region of a particular area. However, today it is known that creating edges can be detrimental to wildlife because of habitat fragmentation.

Edge effects can also have a serious impact on population size. Songbird populations west of the Mississippi have been declining of late, and ornithologists have noticed that the nesting success of songbirds is quite low at the edge of a forest. The cause turns out to be the brown-headed cowbird, a social parasite of songbirds. Adult cowbirds prefer to feed in open agricultural areas, and they only briefly enter the forest when searching for a host nest in which to lay their eggs (Fig. 37.10*b*). Cowbirds are therefore benefited, while songbirds are disadvantaged, by the edge effect.

Reserve Design

Conservation reserves are those areas that are set aside with the primary goal of protecting biodiversity within them. As such, reserves should be largely protected from human activities, except perhaps ecotourism or limited scientific research. When considering the arguments made earlier in this section, reserves should contain sufficient amounts of habitat to sustain the biodiversity within them. This space should include multiple ecosystems that are connected by corridors to allow the movement of animals and the dispersal of plant species between habitat types. This space should also include enough area to account for edge effects, meaning that the absolute amount of land included in each reserve is likely higher than the usable habitat within it.

Globally, a network of more than 650 biosphere reserves has been designated by the United Nations. The maintenance of biosphere reserves is in the hands of the countries and territories in which they are located, and compliance is voluntary. Because the goals of biosphere reserves include preservation of local cultural values, as well as maintenance of biodiversity and the ability to foster sustainable human development, there has been more cooperation than was initially expected.

Each biosphere reserve is divided into three areas. First, there is the central core reserve, which allows only research, light tourism, and limited sustainable use for cultural purposes. Next, there is a buffer zone that surrounds the core where only low-impact human activities are allowed. A transition area then surrounds the buffer zone. This area is meant to support sustainable human development, tourism, and agriculture.

Although there has been widespread cooperation with the establishment of biosphere reserves, there are few that strictly follow the United Nations' model. People are often unwilling to modify their lifestyles to the extent necessary to achieve sustainability, and funding for biosphere maintenance is often limited. For example, funding levels may be insufficient for compensating landowners for adhering to sustainable development practices in buffer zones or transition areas.

Habitat Restoration

In cases where habitat has already been modified in an area to the extent that conservation and reserve formation may not be viable, or to reverse existing damage, habitat restoration is an alternative. **Restoration ecology** is a subdiscipline of conservation biology that seeks scientific ways to return ecosystems to their state prior to habitat degradation. Although habitat restoration is perceived as beneficial, there is some concern that the restored areas may not be functionally equivalent to the natural regions that were once there. Nonetheless, habitat restoration is increasing, and in the process, three principles have emerged thus far. First, it is best to begin as soon as possible before remaining fragments of the original habitat are lost. These fragments are sources of wildlife and seeds from which to restock the restored habitat. Second, once the natural histories of the species in the habitat are understood, it is best to use biological techniques that mimic natural processes to bring about restoration. This might take the form of using controlled burns to bring back grassland habitats, biological pest controls to rid the area of exotic species, or bioremediation techniques to clean up pollutants. Third, the goal is sustainable development, the ability of an ecosystem to maintain itself while providing services to humans. The Ecology feature "Emiquon Floodplain Restoration" presents an example of a successful conservation project in west-central Illinois.

> ### Check Your Progress 37.4
> 1. Identify ways in which landscape preservation is more valuable than ecosystem preservation.
> 2. Explain how edge effects influence nature reserve design.
> 3. List and explain the three principles of habitat restoration.

37.5 Working Toward a Sustainable Society

> ### Learning Outcomes
> Upon completion of this section, you should be able to
> 1. Compare renewable versus nonrenewable resources.
> 2. List two ways to transition to renewable energy resources.
> 3. Discuss how modern agriculture can be changed to minimize environmental impacts.

The majority of biodiversity loss is the result of human consumption of resources (Fig. 37.11). Some resources are nonrenewable, and some are renewable. **Nonrenewable resources** are limited in supply. For example, the amount of land, fossil fuels, and minerals is finite and can be exhausted. Even though better extraction methods can make more fossil fuels and minerals available, and efficient use and recycling can make the supply last longer, eventually these resources will run out. **Renewable resources** are not limited in supply. For example, certain forms of energy, such as solar and wind energy, can be replenished.

The following characteristics indicate that human society in its current form is not sustainable (Fig. 37.12*a*):

1. A considerable proportion of land, and therefore of natural ecosystems, is being altered for human purposes.
2. Our society primarily utilizes nonrenewable fossil fuel energy, which leads to global warming, acid precipitation, and smog.

Human population

Figure 37.11 Resources. Humans use land, water, food, energy, and minerals to meet their basic needs, such as a place to live, food to eat, and products that make their lives easier. (land): ©Budimir Jevtic/Shutterstock; (water): ©Evelyn Jo Johnson; (food): ©John Thoeming/McGraw-Hill Education; (energy): ©Geniuskys/Shutterstock; (minerals): ©TTstudio/Shutterstock

Emiquon Floodplain Restoration

Emiquon was once a vital component of the Illinois River system, which is a major tributary of the Mississippi River. It included a complex system of backwater wetlands and lakes that originally covered an area of approximately 14,000 acres (Fig. 37C). The vast diversity of native plants and animals found within Emiquon have supported human civilizations for over 10,000 years. Early cultures accessed the abundant inland fisheries and a variety of mussel species that could be found there. River otters, mink, beaver, and migrating birds, as well as a variety of prairie grasses on the floodplains, would have been used by native peoples (Fig. 37D).

In the early 1900s, the Illinois River was one of the most economically significant river ecosystems in the United States. It boasted the highest mussel abundance and productive commercial fisheries per mile of any stream in the United States. Each year during the seasonal flooding, the river would deposit nutrient-laden sediment onto the floodplain. This created a highly productive floodplain ecosystem. With the abundance of nutrients, a wide diversity of plant life would have been supported. This broad producer base would have been able to support a number of trophic levels, as well as a wide diversity of food webs.

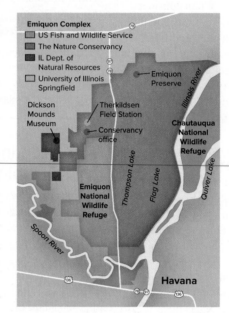

Figure 37C The Emiquon restoration project. Restoration plans call for allowing the river to return to some of its former range.

Throughout the twentieth century, modern cultures altered the natural ebb and flow of the Illinois River to better suit their needs.

Levees were constructed, resulting in the isolation of nearly half of the original floodplain from the river. Native habitats were converted to agricultural land. Corn and soybeans soon replaced the variety of prairie and wetland vegetation that were once a hallmark of the Emiquon floodplain. The river itself was also altered to better facilitate the export and import of a variety of commercial goods by boat. With the isolation of these areas from the seasonal flooding, a decrease in the amount of nutrients being deposited also occurred. This resulted in a continuous decline in the natural productivity of the floodplain.

Restoration Plan

The National Research Council has identified Emiquon as one of three large-floodplain river ecosystems in the United States that can be restored to some semblance of its original diversity and function. The floodplain restoration work at Emiquon is a key part of The Nature Conservancy's efforts to conserve the Illinois River.

The Emiquon Science Advisory Council is composed of over 40 scientists and managers who provide guidance for the restoration and managed connection between Emiquon's restored floodplain and the Illinois River. Once

a. Human society at present

b. Sustainable society

Figure 37.12 Current human society versus a sustainable society. **a.** At present, our "throwaway" society is characterized by high input of energy and raw materials, large output of waste materials and energy in the form of heat (red arrows), and minimal recycling (purple arrows). **b.** A sustainable society would be characterized by the use of only renewable energy sources, reuse of heat and waste materials, and maximal recycling of products (purple arrows). (photos) (a): ©lmantsu/Shutterstock; (b): ©Jeff Whyte/Shutterstock

Figure 37D Emiquon restoration.
Forty-nine black-crowned night-heron nests have been recorded at Emiquon since the restoration project began. Research at Emiquon is providing critical information about wetland restoration techniques. (heron): ©NHPA/Superstock; (researcher): ©Marilyn Kok, University of Illinois Springfield

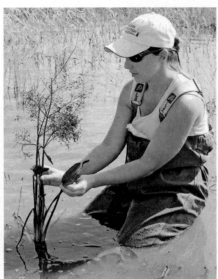

Springfield, and others to survey the community diversity. Since 2007, they have documented over 250 species of birds, including over 90% of Illinois endangered and threatened wetland bird species. In 2017, a 30-minute sample collected over 55,000 pounds of fish leaving Emiquon and heading out to the Illinois River.

In 2016, Emiquon completed a 6-million-dollar water control and research structure, along with seeing over 50,000 visitors through their visitor use areas. As usage continues to increase, so does the economic impact of Emiquon on local economies. New research into methods of controlling the spread of Asian carp are also being conducted. The knowledge gained from the research at Emiquon and The Nature Conservancy's work on the Illinois River and within the Upper Mississippi River system is helping influence floodplain restoration projects around the world.

Questions to Consider

1. In what ways might scientists assess whether the restoration plan was successful in its goal?
2. How might the Emiquon Floodplain Restoration plan be used as a model for other restoration projects?

established, this connection would allow for the seasonal flooding that it previously experienced. This seasonal flooding would help naturalize the flow of the river and restore the cyclical process of flooding and drying out. Flooding would provide more nutrients for wetland plants, as well as help to improve the water

quality. Access between the river and floodplain would once again be restored for various aquatic species, such as paddlefish and gar. Both species need a variety of habitats for reproduction and survival. The Conservancy is continuing to work with the Illinois Natural History Survey, the University of Illinois,

---

3. Even though fresh water is a renewable resource, we are using it faster than it can be replenished in aquifers and other sources.
4. Agriculture requires large inputs of nonrenewable fossil fuel energy, fertilizer, and pesticides, which create much pollution.
5. At least half of the agricultural yield in the United States goes toward feeding animals (it takes 10 lb of grain to produce 1 lb of meat).
6. Minerals are nonrenewable, and the mining, manufacture, and use of mineral products is responsible for a significant amount of environmental pollution.

To move toward a more sustainable society, we should move away from the use of nonrenewable resources to renewable ones. A **sustainable** society, like a sustainable ecosystem, should be able to provide the same goods and services for future generations that it provides for the current one. At the same time, biodiversity would be conserved.

A natural ecosystem can offer clues as to what a sustainable human society would be like. A natural ecosystem uses only renewable solar energy, and its materials cycle through herbivores, carnivores, and detritivores, and back to producers once again. It is

clear that if we want to develop a sustainable society, we too should use renewable energy sources and recycle materials (Fig. 37.12b). Section 37.4 discusses ways to conserve land and habitat for species. In addition to recycling and reuse of many of Earth's minerals (e.g., aluminum, copper, iron, and gold), which are nonrenewable, we should also consider alternative energy use, water conservation, and modifications to the way we conduct modern agriculture.

Energy

Presently, about 14% of the world's energy supply comes from nuclear power, and 61% comes from fossil fuels; both of these are finite, nonrenewable sources. Fossil fuels (oil, natural gas, and coal) are so named because they are derived from the compressed remains of plants and animals that died millions of years ago. Currently, shortages of fossil fuels such as oil have contributed to the dramatic increases in heating and gasoline costs. Comparatively speaking, each person in a more-developed country (MDC) uses approximately as much energy in 1 day as a person in a less-developed country (LDC) uses in 1 year. Increasing our reliance on renewable energy resources is a major step toward becoming a sustainable society.

Figure 37.13 Hydropower. Hydropower dams provide a clean form of energy but can be ecologically disastrous in other ways. ©Maxim Burkovskiy/Shutterstock

Renewable Energy Sources

Renewable types of energy include hydropower, geothermal energy, wind power, and solar energy. Hydroelectric plants convert the energy of falling water into electricity (Fig. 37.13). Worldwide, hydropower presently generates approximately 17% of all electricity utilized. In the United States, hydropower accounts for approximately 9% of total energy produced and 50% of the renewable energy used. Brazil, New Zealand, and Switzerland produce at least 75% of their electricity with hydropower, but Canada is the Western world's leading hydropower producer, accounting for 97% of its renewable energy generation. One way to reduce our reliance on nonrenewable resources such as fossil fuels is to increase our reliance on hydropower. However, much of the recent hydropower development has been due to construction of enormous dams that have detrimental environmental effects. Small-scale dams that generate less power per dam, but do not have the same environmental impact, are believed to be the more environmentally responsible choice.

Geothermal energy is produced because Earth has an internal source of heat. Elements such as uranium, thorium, radium, and plutonium undergo radioactive decay underground and then heat the surrounding rocks to hundreds of degrees Celsius. When the rocks are in contact with underground streams or lakes, huge amounts of steam and hot water are produced. This steam can be piped up to the surface to supply hot water for home heating or to run steam-driven turbogenerators. The California Geysers Project, for example, is one of the world's largest geothermal electricity generating complexes.

Wind power is expected to account for a significant percentage of our energy needs in the future (Fig. 37.14a). Despite the common belief that a huge amount of land is required for "wind farms" that produce commercial electricity, the actual amount of space for a wind farm compares favorably to the amount of land required by a coal-fired power plant or a solar thermal energy system. A community that generates its own electricity by using wind power can solve the problem of uneven energy production by selling electricity to a local public utility when an excess is available and buying electricity from the same facility when wind power is in short supply.

Solar energy is diffuse energy that must be (1) collected, (2) converted to another form, and (3) stored if it is to compete with other available forms of energy. Passive solar heating of a house is successful when the windows of the house face the sun, the building is well insulated, and heat can be stored in water tanks, rocks, bricks, or some other suitable material.

In a photovoltaic (solar) cell, a wafer of an electron-emitting metal is in contact with another metal that collects the electrons and passes them along into wires in a steady stream. Spurred by the oil shocks of the 1970s and in 2008, the U.S. government has been supporting the development of photovoltaic cells. As a result, the price of buying one has dropped from about $100 per watt (in the early 1970s) to around $0.74 (2013). The photovoltaic cells placed on roofs, for example, generate electricity that can be used inside a building and/or sold back to a power company (Fig. 37.14b).

Traditional cars have internal combustion engines that run on gasoline, but hybrid cars that are increasing in popularity due to rising fuel costs, run on both gasoline and electricity, increasing mileage per gallon. Now that we have better ways to capture solar energy, scientists are investigating the possibility of using solar energy to extract hydrogen from water via electrolysis. The hydrogen can then be used as a clean-burning fuel that produces water as a waste product.

a. Wind

b. Solar

c. Hydrogen

Figure 37.14 Other renewable energy sources. **a.** Wind power requires land on which to place enough windmills to generate energy. **b.** Photovoltaic cells on rooftops and **(c)** the use of hydrogen cars will reduce air pollution and dependence on fossil fuels. (a): ©Glen Allison/Getty Images; (b): ©Danita Delimont/Getty Images; (c): ©Associated Press

Increasingly, vehicles are powered by fuel cells that use hydrogen to produce electricity. Fuel cells are now powering buses in Vancouver and Chicago. Hydrogen cars are already in limited production (Fig. 37.14c). Hydrogen fuel can be produced locally or in central locations, using energy from photovoltaic cells. If in central locations, hydrogen can be piped to filling stations using the natural gas pipes already in place in the United States. The advantages of such a solar–hydrogen revolution would be at least twofold: (1) the world would no longer be dependent on its limited oil reserves, and (2) environmental problems, such as global warming, acid rain, and smog, would begin to lessen.

Water

In some areas of the world, people do not have ready access to clean and safe drinking water. It's considered a human right for people to have clean drinking water, but actually, most fresh water is utilized by industry and agriculture. Worldwide, 60% of all fresh water is used to irrigate crops! Domestically in MDCs, more water is usually used for bathing, flushing toilets, and watering lawns than for drinking and cooking.

Although the needs of the human population overall do not exceed the renewable supply of water, this is not the case in certain regions of the United States and the world. When needed, humans increase the supply of fresh water by damming rivers and withdrawing water from aquifers. Dams have drawbacks: (1) Reservoirs behind the dam lose water due to evaporation and seepage into underlying rock beds. The amount of water lost sometimes equals the amount dams make available! (2) The salt left behind by evaporation and agricultural runoff increases salinity and can make a river's water unusable farther downstream. (3) Over time, dams hold back less water because of sediment buildup. Sometimes a reservoir becomes so full of silt that it is no longer useful for storing water. (4) The alteration of a river or stream has a negative impact on the native wildlife.

To meet their freshwater needs, people are pumping vast amounts of water from underground reservoirs called **aquifers.** Aquifers hold about 1,000 times the amount of water that falls on land as precipitation each year. In the past 50 years, groundwater depletion has become a problem in many areas of the world. Removal of water is causing land **subsidence,** settling of the soil as it dries out. In California's San Joaquin valley, an area of more than 13,000 km² has subsided at least 30 cm due to groundwater depletion, and in the worst spot, the surface of the ground has dropped more than 9 m! Subsidence damages canals, buildings, and underground pipes. For more on the California aquifers, see the Ecology feature "The California Drought" in Section 35.3.

Conservation of Water

By 2025, two-thirds of the world's population may be facing serious water shortages. Some solutions for expanding water supplies have been suggested. Planting drought- and salt-tolerant crops would help a lot. Development of many such crops is already underway due to genetic engineering, as discussed in Section 26.2. Using drip irrigation delivers more water to crops and saves about 50% over traditional methods, while increasing crop yields. Although the first drip systems were developed in 1960, they are currently used on less than 1% of irrigated land. Most governments subsidize irrigation so heavily that farmers have little incentive to invest in drip systems or other water-saving methods.

Reusing water and adopting conservation measures could help the world's industries cut their water demands by more than half. For example, recycling washing machine water, shower water, or water used to wash cars through a filter before it is discarded as sewage could significantly reduce domestic water usage. Home yard irrigation should occur during dusk and dawn hours, as opposed to in the middle of the day when evaporation is at its highest. Purchasing and using dual-flush toilets can also save millions of gallons of water per year.

Agriculture

In 1950, the global human population numbered 2.5 billion, and at that time, there was only enough food to provide less than 2,000 calories per person per day. Today, current agricultural practices provide enough food to provide everyone on Earth a healthy diet consisting of 2,500 calories per day. However, one-sixth of the world's population (over 1 billion people) are currently considered malnourished due to lack of proper distribution and the redirection of grain to feed livestock. In addition, modern farming methods are environmentally destructive in the following ways. First, planting only a few genetic varieties, or monocultures (a genetically identical crop), means that a single destructive parasite or pathogen can cause huge crop losses. Second, modern farming relies on heavy use of fertilizers, pesticides, and herbicides. Pesticides reduce soil fertility because they kill off beneficial soil organisms, as well as pests. Pesticides also select for artificial resistance in insects, increasing eradication costs via increased pesticide use. In addition, fertilizers, pesticides, and herbicides all contribute to pollution. Third, modern agriculture uses significant amounts of fresh water through irrigation. Fourth, modern agriculture uses large amounts of fuels: Fertilizer production is energy intensive; irrigation pumps require energy to remove water from aquifers; large tractors and even airplanes are used to spread fertilizers, pesticides, and herbicides; and large machines are often used to harvest crops.

Several alternatives exist to employing modern agricultural methods. Polyculture, or planting of several varieties of crop in the field simultaneously can reduce the susceptibility of crops to pests or diseases (Fig. 37.15a). Polyculture also reduces the amount of herbicides necessary to kill competing weeds and can be used to replenish nutrients to topsoil. Crop rotation—where, for example, nitrogen-fixing crops, such as legumes, are alternated across harvest years with crops that utilize soil nitrogen, such as corn—can help reduce the use of nitrogen-containing fertilizers. Such multiuse farming techniques generally help increase the amount of organic matter and nutrients in the soil.

Organic farms are also increasing in number. Organic farms, as mandated by the U.S. Department of Agriculture, are those in which synthetic pesticides and herbicides are avoided or largely excluded from use. Organic farming has become increasingly profitable in recent years because people are more willing to purchase organic produce, despite the increased cost relative to nonorganic produce. Health concerns surrounding pesticide content in nonorganic produce, as well as better tasting food, have helped this trend.

One way that organic farmers have eliminated the need for pesticides is by using integrated pest management. This technique

a. Polyculture

b. Contour farming

c. Biological pest control

Figure 37.15 Methods that make farming more friendly to the environment. **a.** Polyculture reduces the ability of one parasite to wipe out an entire crop and reduces the need to use an herbicide to kill weeds. This farmer has planted alfalfa in between strips of corn, which also replenishes the nitrogen content of the soil (instead of adding fertilizers). Alfalfa, a legume, has root nodules that contain nitrogen-fixing bacteria. **b.** Contour farming with no-till conserves topsoil, because water has less tendency to run off. **c.** Instead of pesticides, it is possible to use natural predators. Here, ladybugs are feeding on aphids, an insect pest species. (a): ©David R. Frazier/Science Source; (b): ©Inga Spence/Alamy; (c): ©Perennou Nuridsany/Science Source

encourages the growth of competitive beneficial insects and uses biological pest control methods (also called "biocontrol"). Biocontrol has also helped reduce pesticide use in traditional farms as well. Such methods include using natural predators, such as spiders and ladybug beetles, to reduce the numbers of pests (Fig. 37.15c). Use of natural or engineered pathogens has also been suggested for biocontrol of pests, but there is concern that such diseases will escape crop areas and affect natural plants or wildlife.

Several types of farming now exist that reduce erosion and help to minimize topsoil loss, such as contour farming (Fig. 37.15b). Terrace farming involves converting steep slopes into steplike hills to minimize erosion, and some farmers plant "natural fences," such as rows of trees, around crops to prevent topsoil loss due to wind or other factors. These trees can also be used as other products; mature rubber trees provide us with rubber, and tagua nuts are an excellent substitute for ivory, for example. Cover crops, which are often a mixture of legumes and grasses, also help stabilize soil between rows of cash crops. Finally, avoiding farming on steep slopes also helps reduce erosion. Soil nutrients can be increased through composting, organic

farming techniques, or other self-renewable methods. In general, we should consider using precision farming (PF) techniques that rely on accumulated knowledge to reduce habitat destruction, while improving crop yields.

Urban Growth

More and more people are moving to cities. Growth of cities increases pollution via many sources, including automobile exhaust, runoff of pollutants on impervious surfaces (e.g., roads) and into waterways, noise pollution, and industrial and domestic wastes. Thus, the growth of cities involves careful planning to serve the needs of new arrivals, without overexpansion of the city.

Energy use in cities can be curtailed by providing a good public transportation system, preferably one that is energy efficient. Many U.S. cities are now encouraging carpooling by having HOV (high-occupancy vehicle) lanes on highways, as well as reduced toll costs for cars occupied by more than one person. Portland, Oregon, even has short-term electric car rental stations around the city, whereby people can rent cars for an evening and easily drop them off when they are finished. Maintaining a network of safe bicycle lanes also encourages people to ride their bikes to work.

A way to curtail urban sprawl, or extensive expansion of cities outward, is to build them upward. Several cities, such as Vancouver, Canada, are building many high-rise apartment buildings to accommodate more people in a smaller area. As new buildings are built, they should be as "green" as possible—by using solar or geothermal energy for heating, as well as being constructed out of sustainable or recycled materials. Space on top of buildings could be used to make "green roofs," whereby a garden of grasses, herbs, and vegetables is planted to assist temperature control, supply food, reduce rainwater runoff, and be visually appealing. As more people move into cities, city officials can focus on renovating older parts of the city, as opposed to spreading out further.

Modern cities can also have more planned green spaces, including plentiful walking and bicycle paths. In city parks, native species that attract bees and butterflies and require less water and fertilizers could be planted, as opposed to traditional grasses.

Sustainable cities can also improve storm-water management by using sediment traps for storm drains, artificial wetlands, and holding ponds. As new development occurs, cities can increase the use of porous surfaces for walking paths, parking lots, and roads. These surfaces reflect less heat, while soaking up rainwater runoff.

Check Your Progress 37.5

1. Identify which nonrenewable energy use is potentially environmentally harmful.
2. List ways in which we can conserve water to minimize negative environmental impacts.
3. Discuss two alternative agricultural practices that are environmentally friendly.

Conclusion

Despite pandas being a national treasure in China and their lack of natural predators, they are still under the threat of extinction. Conservation efforts have moved them from endangered to vulnerable status, but if continuous efforts are not made, they could easily slip back to the endangered status. Quite often, more value is placed on the economic importance of a species instead of their ecological importance. If it becomes economically more beneficial for China to develop panda habitat, it may cause many other species that live in the forests of southwestern China to also suffer.

SUMMARIZE

37.1 Conservation Biology and Biodiversity

- **Conservation biology** is an interdisciplinary science, with the goal of conserving Earth's biodiversity for the good of all species, that requires input from a variety of subdisciplines.

- **Biodiversity** is the variety of life that exists on Earth. Of the estimated 8 million species that live on Earth, nearly 75,000 of the described species are classified as **endangered species.** This means they face immediate extinction throughout all or most of their range. Thousands of species are listed as **threatened species,** meaning they may become endangered in the foreseeable future. Although biodiversity is often described as the total number of species, it is also considered at the genetic, ecosystem, and landscape levels.

- **Genetic diversity** helps ensure that populations can evolve in the face of future environmental change.

- **Ecosystem diversity** considers the functioning of communities within ecosystems and the interconnectedness of species within such communities.

- **Landscape diversity** includes all of the ecosystems that may interact in a particular area.

37.2 Value of Biodiversity

- A direct value of biodiversity includes the use of wild species for our medical needs, agricultural value, and consumptive purposes.

- The indirect value of biodiversity includes the workings of biogeochemical cycles, waste disposal, provision of fresh water, prevention of soil erosion, and regulation of climate.

- Ecotourism is an indirect value of biodiversity that is increasing as an important source of income for many less-developed countries (LDCs).

37.3 Threats to Biodiversity

- The five major causes of extinction, in decreasing order of importance, are habitat loss, introduction of **alien species, pollution, overexploitation,** and disease.

- Habitat loss has occurred in all parts of the biosphere, but concern has now centered on tropical rain forests and coral reefs, where biodiversity is especially high.

- Exotic species have been introduced into foreign ecosystems because of colonization, horticulture or agriculture, and accidental transport.

- Among the various causes of pollution (acid rain, eutrophication, ozone depletion, and organic chemicals), **climate change** and global warming are expected to cause the most instances of extinction.

- Overexploitation is exemplified by commercial fishing, which is so efficient that fisheries of the world are collapsing.

- Wildlife species are increasingly threatened by disease-causing pathogens.

37.4 Habitat Conservation and Restoration

- **Biodiversity hotspots,** such as tropical rain forests and coral reefs, are often high-priority conservation areas because they contain unusually large concentrations of species.

- **Keystone species** are often a target for conservation because their loss would result in many secondary extinctions. **Flagship species** are those that elicit an emotional response and often motivate the public to support conservation.

■ Habitat conservation should be considered at the landscape level, with the formation of **conservation corridors** that allow species to travel safely between ecosystem types.

■ **Edge effects** reduce the amount of suitable habitat and should be considered in reserve design.

■ **Conservation reserves** are areas set aside with the primary goal of protecting the biodiversity within them.

■ Biosphere reserves should consist of a core area with nearly no disturbance, a surrounding buffer area where only low-impact human activities are allowed, and an outer transition area meant to support sustainable human development, tourism, and agriculture.

■ **Restoration ecology** is the field of biology used to help return ecosystems to their state prior to habitat degradation.

■ Restoration ecology has three basic principles: (1) begin as soon as possible; (2) use biological techniques that mimic natural processes to bring about restoration; and (3) focus on sustainable development—the ability of an ecosystem to maintain itself, while providing services to humans.

37.5 Working Toward a Sustainable Society

■ The current biodiversity crisis is ultimately caused by human actions.

■ Today's society is considered unsustainable because we rely heavily on **nonrenewable resources** to maintain our existence.

■ A shift of focus toward **renewable resources** (wind, solar, geothermal, and hydropower) and **sustainable** use of water and agricultural practices will help reduce human impact.

■ Globally, many people do not have access to clean drinking water. In the southwestern United States, people are rapidly draining the **aquifers** to meet the growing demand for water. The removal of the water from the aquifers has led to land **subsidence,** settling of the soil as it dries out. In order to prevent problems, people need to become more aware of water conservation methods.

■ These methods include drip irrigation, planting drought- and salt-tolerant crops, and household water reuse and recycling methods.

■ Modern-day agriculture can employ new methods, such as biocontrol to reduce reliance on pesticides, and new forms of farming, such as contour farming, to reduce topsoil erosion.

ASSESS

Testing Yourself

Choose the best answer for each question.

37.1 Conservation Biology and Biodiversity

1. Eagles and bears feed on spawning salmon. If shrimp are introduced that compete with salmon for food,
   a. the salmon population will decline.
   b. the eagle and bear populations will decline.
   c. only the shrimp population will decline.
   d. all populations will increase in size.
   e. Both a and b are correct.

2. Which of the following is based on the interactions of species with their abiotic environment?
   a. genetic diversity
   b. landscape diversity
   c. ecosystem diversity
   d. biocultural diversity
   e. None of these are correct.

3. A species faced with an immediate extinction threat is said to be
   a. threatened.   c. alien.
   b. endangered.   d. extinct.

37.2 Value of Biodiversity

4. The most significant cause of the loss of biodiversity is
   a. habitat loss.   d. disease.
   b. pollution.   e. overexploitation.
   c. exotic species.

5. Which of these is not an indirect value of species?
   a. participates in biogeochemical cycles
   b. participates in waste disposal
   c. helps provide fresh water
   d. prevents soil erosion
   e. All of these are correct.

37.3 Threats to Biodiversity

6. Which of these is a true statement?
   a. Habitat loss is the most frequent cause of extinctions today.
   b. Exotic species are often introduced into ecosystems by accidental transport.
   c. Climate change may cause many extinctions but also expands the ranges of other species.
   d. Overexploitation of fisheries could lead to a complete collapse of the fishing industry.
   e. All of these statements are true.

7. Which of these is expected if the average global temperature increases?
   a. the inability of some species to migrate to cooler climates as environmental temperatures rise
   b. the possible destruction of coral reefs
   c. an increase in the number of parasites in the temperate zone
   d. a population decline for some species but an increase for others
   e. All of these are correct.

8. Complete the following graph by labeling each bar (a–e) with a cause of extinction, from the most influential to the least influential.

% Species
Affected by Threat

37.4 Habitat Conservation and Restoration

9. Edge effects
   a. result from species crowding at the edge of habitats.
   b. mean that no plants can grow well at the edge of a mountain.
   c. may increase wind and temperatures around the outside of habitats.
   d. increase competition between species at the edge of their territories.
   e. All of these are correct.

10. Biodiversity hotspots
    a. have few populations because the temperature is too hot.
    b. contain about 20% of Earth's species even though their area is small.
    c. are always found in tropical rain forests and coral reefs.
    d. are sources of species for the ecosystems of the world.
    e. All except a are correct.
11. Which of the following should not be considered when designing nature reserves?
    a. edge effects that may require an increase in size
    b. corridors to facilitate wildlife movement among habitat patches
    c. primarily the needs of a single, important species
    d. inclusion of multiple ecosystem types

37.5 Working Toward a Sustainable Socity

12. Which feature is part of a sustainable society?
    a. recycling and composting
    b. use of mass transit in urban environments
    c. integrated pest management to control crop damage
    d. preservation of wetlands
    e. All of these are correct.
13. Which feature is associated with the sustainability of a rural society?
    a. planting cover crops when farming
    b. using green roofs
    c. designing energy-efficient transportation systems
    d. making areas more pedestrian- and bike-friendly
    e. All of these are correct.

ENGAGE

BioNOW

Want to know how this science is relevant to your life? Check out the BioNOW video below.

■ Biodiversity

What was the effect of the exotic (alien) species on the biodiversity? Why do you think this was the case?

Thinking Critically

1. Visit www.myfootprint.org to determine the size of your ecological footprint and then discuss ways you can reduce that value.
2. What are some ways we can reduce our reliance on nonrenewable resources and move toward increased reliance on renewable resources?
3. Why should people be generally more concerned about pollutants in our meat than in our vegetables?
4. Discuss how removal of a keystone species can disrupt ecosystem functioning. Give an example.

Periodic Table of the Elements, and the Metric System

Periodic Table of the Elements

The Metric System

Unit and Abbreviation	Metric Equivalent	Approximate English-to-Metric Equivalents
Length		
nanometer (nm)	$= 10^{-9}$ m $(10^{-3}$ µm)	
micrometer (µm)	$= 10^{-6}$ m $(10^{-3}$ mm)	
millimeter (mm)	$= 0.001$ (10^{-3}) m	
centimeter (cm)	$= 0.01$ (10^{-2}) m	1 inch = 2.54 cm
		1 foot = 30.5 cm
meter (m)	$= 100$ (10^2) cm	1 foot = 0.30 m
	$= 1,000$ mm	1 yard = 0.91 m
kilometer (km)	$= 1,000$ (10^3) m	1 mi = 1.6 km
Weight (mass)		
nanogram (ng)	$= 10^{-9}$ g	
microgram (µg)	$= 10^{-6}$ g	
milligram (mg)	$= 10^{-3}$ g	
gram (g)	$= 1,000$ mg	1 ounce = 28.3 g
		1 pound = 454 g
kilogram (kg)	$= 1,000$ (10^3) g	= 0.45 kg
metric ton (t)	$= 1,000$ kg	1 ton = 0.91 t
Volume		
microliter (µl)	$= 10^{-6}$ l $(10^{-3}$ ml)	
milliliter (ml)	$= 10^{-3}$ l	1 tsp = 5 ml
	$= 1$ cm^3 (cc)	1 fl oz = 30 ml
	$= 1,000$ mm^3	
liter (l)	$= 1,000$ ml	1 pint = 0.47 l
		1 quart = 0.95 l
		1 gallon = 3.79 l
kiloliter (kl)	$= 1,000$ l	

Units of Temperature

°C	°F	
100	212	Water boils at standard temperature and pressure.
71	160	Flash pasteurization of milk
57	134	Highest recorded temperature in the United States, Death Valley, July 10, 1913
41	105.8	Average body temperature of a marathon runner in hot weather
37	98.6	Human body temperature
13.7	56.66	Human survival is still possible at this body temperature.
0	32.0	Water freezes at standard temperature and pressure.

To convert temperature scales:

$$°C = \frac{(°F - 32)}{1.8}$$

$$°F = 1.8(°C) + 32$$

Answer Key

CHAPTER 1
Check Your Progress

p. 6: 1. Characteristics are: (1): organization; (2) ways to acquire materials and energy; (3) reproduction and development; (4) responses to stimuli; (5) homeostasis; (6) capacity for adaptation to their environment. **2.** Levels of organization are cell, tissue, organ, organ system, organism, species, population, community, ecosystem, and biosphere. **3.** As an organism acquires an adaptation that makes it better able to live and reproduce, these traits are passed on to the next generation and the population evolves. **p. 8: 1.** Domains are the largest, most inclusive classification category. A supergroup is a classification system based on DNA similarities. Kingdoms are a historical classification system that is being replaced by the use of supergroups. **2.** Domain, Kingdom, Phylum, Class, Order, Family, Genus, Species. For eukaryotes, the supergroup level may be added between domain and kingdom. **3.** A hierarchical classification system allows increasing specificity about the organisms under study, which helps define their evolutionary relationships. **p. 12: 1.** An experimental variable is the one factor that is being tested in an experiment to determine what it contributes. **2.** The control group in an experiment is included to determine if the experiment is sensitive to the effect of the variable of interest. The test group experiences the variable and the control group does not. **3.** An observation is made and a hypothesis is constructed to possibly explain the observation. An experiment is designed to test the hypothesis, involving deductive reasoning and a prediction. The experiment is conducted, data collected and analyzed, and conclusions made about whether the hypothesis is supported by the data or not. **p. 14: 1.** New technology applies scientific knowledge to issues important to humans and can be used in investigations that lead to scientific discoveries. **2.** Biodiversity contributes to the health of ecosystems on which we depend for food, medicines, and raw materials. Ecosystems that are destroyed can no longer function well in water cycling, soil conservation, and chemical cycling. **3.** Emerging diseases appear in a population that has no immunity against them. Often, little is known about the causative agent and how to control it. Techniques for containing the emerging disease are often undeveloped. As Earth's climate warms, ecosystems are challenged by changes in water availability, temperature extremes, and altered habitats, all of which can reduce biodiversity.

Testing Yourself

1. c; **2.** b; **3.** b; **4.** c; **5.** c; **6.** a; **7.** b; **8.** d; **9.** c; **10.** d; **11.** c; **12.** d

Thinking Critically

1. Model organisms have been characterized and standardized. They are identical in their makeup and physiology. When analyzing the data from an experiment testing an experimental variable in a model organism, one can be more confident that the dependent variable is related to the experimental variable rather than some other factor. **2.** If life on Earth is unique in its characteristics, then the processes by which life emerged and evolved by natural selection are not universal. **3.** The experimental chemical should be studied in a model species. The experiment should have both a control group (that receives no drug, or the inert ingredients used in the drug's formulation) and a treatment group (that receives the drug). Both groups should experience identical conditions and have the same type of cancer. Data are collected on tumor development in the treatment group relative to the control group. If the treatment group shows greater tumor reduction than the control group, the drug may then be approved for trials on humans.

CHAPTER 2
Check Your Progress

p. 22: 1. Earth's crust has a different function and is constructed differently than living organisms. It is not a carbon-based life-form that exhibits the characteristics of life. **2.** See Figure 2.3. Ca has 20 protons and 20 neutrons in its nucleus, two electrons in the first orbital, eight electrons in the second orbital, eight electrons in the third orbital, and two electrons in the fourth orbital. **3.** High levels of radiation can cause burns, harm cells, and change DNA structure, with cancer being a possible result of exposure. However, radiation can be used beneficially to kill cancer cells. Also, radioisotopes are used extensively in medicine to image the body through X-rays, positron emission tomography, and computerized tomography. Bacteria and fungi present on medical equipment, in mail, and in food can be killed by exposure to radiation. **4.** Oxygen 16 has eight protons, eight electrons, and eight neutrons. Oxygen 18 differs by having ten neutrons (two more). **p. 25: 1.** Nitrogen gas (N_2) is a molecule because two atoms of nitrogen share a covalent bond. Carbon dioxide (CO_2) is a compound because two different elements are bonded together. However, CO_2 is also a molecule because if you broke apart the carbon and oxygen atoms, the properties of carbon dioxide would be lost. **2.** Carbon has six electrons, two in the first orbital, and four in the second orbital. Its incomplete outer orbital can accept or share four electrons, forming four covalent bonds. **3.** In nonpolar covalent bonds, there is equal sharing of electrons between the two atoms. In polar covalent bonds, the electrons are shared unequally, due in part to differences in the sizes of the two atoms involved. **p. 28: 1.** Heat capacity of water is the amount of energy needed to raise the temperature of 1 g of H_2O by 1°C, while heat of vaporization is the amount of energy needed to convert 1 g of the hottest water to a gas. **2.** Cohesion describes the ability of water molecules to cling to each other. Adhesion is a measure of water's ability to cling to other polar surfaces, such as the sides of a vessel. **3.** A solution at pH 3 contains 1×10^{-3} moles per liter of H^+ ions, while a solution at pH 6 contains 1×10^{-6} moles per liter of H^+ ions. 1×10^{-3} is a thousand times larger than 1×10^{-6}. **p. 29: 1.** Because they contain carbon and hydrogen and are present in living organisms. **2.** Hydrolysis reactions involve the addition of a molecule of water across a chemical bond, splitting the molecule in two. Dehydration reactions remove a molecule of water, joining two molecules. **p. 32: 1.** A ring structure of carbons with each bonded to an — H and an —OH group. **2.** Humans are able to break the bonds between the glucose molecules in starch, but not between the glucose molecules in cellulose. **p. 34: 1.** All lipids have hydrophobic characteristics and are insoluble in water. **2.** A double bond tends to make a fatty acid liquid because it introduces a "kink" in the chain that prevents the fatty acids from packing as tightly together, producing a solid. **3.** Triglycerides are used for storage of energy, insulation, and protection of organs. The presence of the steroid cholesterol in plasma membranes influences their fluidity. It is a precursor for the hormones testosterone and estrogen, as well as bile salts. **p. 37: 1.** Proteins have a variety of functions, including those of enzymes, structural molecules, hormones, transport molecules, and membrane channels. **2.** Each one has an amino group, an acidic group, and an R group bound to a central carbon atom. The R group varies in structure between one amino acid and another. **3.** Primary—linear sequence of amino acids; secondary—orientation of polypeptide chain, coiled or folded, determined by hydrogen bonding; tertiary—three-dimensional structure maintained by covalent, ionic, or hydrogen bonding; quaternary—structure of two or more polypeptides joined in a protein. **p. 39: 1.** DNA is a double-helix molecule that contains the nucleotides adenine, quanine, cytosine, and thymine, while RNA is a single-stranded molecule that contains the nucleotides adenine, guanine, cytosine, and uracil. **2.** It is stored in the sequence of the nucleotide bases: A, C, T, and G. **3.** It is stored in the chemical bonds of the last two phosphate groups.

Testing Yourself

1. b; **2.** c; **3.** d; **4.** c; **5.** a; **6.** b; **7.** d; **8.** e; **9.** a; **10.** c; **11.** a; **12.** c; **13.** c; **14.** b; **15.** b; **16.** d; **17.** d; **18.** c; **19.** d; **20.** a; **21.** b; **22.** c

Thinking Critically

1. The surface of the water is smooth because of the cohesion between water molecules. Water droplets cling to skin because water is cohesive. You are cooler because the water is evaporating and this removes a lot of heat from your skin because water has a high heat of evaporation. **2.** Protein function is dependent on the three-dimensional shape a protein molecule assumes. How a polypeptide chain folds into sheets or coils, how parts of the polypeptide chains interact with other parts, and how multiple chains of proteins interact among themselves all define the shape and ultimately the function of protein molecules. **3.** How reactive an element is depends on the number of electrons in its outer shell. Isotopes do not differ in their number of electrons.

CHAPTER 3
Check Your Progress

p. 46: 1. If humans were just one cell, the surface-area-to-volume ratio would be too small to allow for efficient movement of nutrients and oxygen into the cell and wastes, including carbon dioxide, out of the cell. **2.** The cell is the basic unit of life because all living organisms are made up of cells, and all cells come from preexisting cells. **3.** A large cell has a reduced surface-area-to-volume ratio, and the amount of diffusion of nutrients, wastes, and gases across the cell membrane is reduced. **p. 48: 1.** To form a barrier between the inside and outside of the cell and regulate what crosses that barrier. **2.** See Figure 3.3. **3.** See Table 3.1. **p. 56: 1.** Protecting the cell while remaining permeable. **2.** The endomembrane system transports molecules throughout the cell because the organelles that compose it are connected directly or by transport vesicles. **3.** Plant cells need to make their food. Chloroplasts allow them to capture solar energy to produce organic molecules. Plant cells need mitochondria to break down the organic molecules to produce ATP which is used for energy in their metabolism. **p. 58: 1.** All three are part of the cytoskeleton and are dynamic. Intermediate filaments play a structural role, while both actin filaments and microtubules have a structural and a movement role. Actin filaments are composed of actin monomers, microtubules are composed of tubulin monomers, and intermediate filaments are composed of various types of fibrous polypeptides. **2.** Centrioles function in the process of cell division, while both cilia and flagella are involved in locomotion. **3.** The dynein side arms on the microtubule doublets slide past each other using the energy of ATP. **p. 60: 1.** They are similar to bacteria in size and structure, are bound by a double membrane, contain a limited amount of genetic material, divide by splitting, have their own ribosomes (similar to prokaryotic ribosomes) and produce some proteins, and have ribosomal RNA sequences similar to prokaryotic rRNA sequences. **2.** Mitochondria were originally aerobic heterotrophic bacteria taken up by a eukaryotic cell and chloroplasts were cyanobacteria similarly taken up. Neither the bacteria nor cyanobacteria were destroyed, but instead successfully lived inside the host cell, contributing their metabolic functions to the whole.

Testing Yourself

1. c; 2. a; 3. d; 4. d; 5. c; 6. e; 7. d; 8. c; 9. b; 10. c; 11. a; 12. b; 13. d; 14. b

Thinking Critically

1. Locate a cell line that has the mechanism to secrete proteins and can be grown under laboratory conditions. Modify the genome of that cell line so it manufactures and secretes the drug protein so it can be harvested. 2. Cells, whether naturally occurring or synthesized, depend on diffusion and transport processes whose efficiency varies with cell size. 3. Knowing the basic structure and function of cells on Earth helps in recognizing similarities and differences in life-forms from other planets.

CHAPTER 4
Check Your Progress

p. 66: 1. Phospholipids compose a bilayer that separates the inside from the outside of the cell. Steroids in the bilayer regulate the fluidity of the membrane. Proteins embedded in the membrane contribute to its structure, the passage of molecules across the membrane, signaling pathways, cell recognition, and enzyme reactions. 2. See Figure 4.2. p. 74: 1. During diffusion, molecules move from an area of high concentration to an area of low concentration. Facilitated transport promotes the movement of these molecules down a concentration gradient across a plasma membrane. Carrier proteins reversibly bind to the molecule and speed up their passage. 2. A hypertonic environment has a lower concentration of water and a higher concentration of solutes than a hypotonic environment. Water will move from the hypotonic environment where it is present at a higher concentration, to the hypertonic environment across a semipermeable membrane by osmosis. 3. Both move molecules across the plasma membrane and require a carrier molecule. Facilitated transport does not use energy, while active transport does. Facilitated transport moves items down their concentration gradient, while active transport moves against the concentration gradient. 4. Allows the cell to target specific molecules to be brought in by endocytosis, thus increasing efficiency. p. 76: 1. Collagen and elastin fibers provide structure to the ECM. Fibronectin binds to integrin in the membrane and can signal the cell's cytoskeleton. Proteoglycans, shaped as bottle brushes, assist in cell signaling by regulating the passage of molecules through the ECM. 2. Adhesion junction—mechanically attaches adjacent cells; gap junction—allows cells to communicate through channels; tight junction—connects plasma membranes, creating a tight barrier. 3. The ECM of an animal cell is a meshwork of proteins and polysaccharides that can vary in amount and flexibility depending on the cell type. The plant cell wall, external to the plasma membrane, is composed of cellulose fibrils and pectin that allows flexibility during growth and eventually hardens into a rigid structure.

Testing Yourself

1. b; 2. e; 3. d; 4. c; 5. c; 6. b; 7. e; 8. d; 9. c; 10. c; 11. c

Thinking Critically

1. The receptor molecule changes its shape and that activates an internal signaling pathway, usually involving peripheral proteins and secondary messengers. 2. Lower concentrations of chloride outside of the cell results in a lower ionic concentration. Less water is attracted to the mucus outside the cells.

CHAPTER 5
Check Your Progress

p. 81: 1. See Figure 5.1. 2. It removes unwanted tissue and abnormal cells. p. 84: 1. Cells stop at the G_1 checkpoint if conditions are not favorable for cell division and/or there is DNA damage. Cells stop at the G_2 checkpoint if the DNA has not finished replicating and/or there is DNA damage. Cells stop at the M checkpoint if the chromosomes are not lined up correctly at the metaphase plate. 2. Growth factors are external signals that promote cell division of target cells. 3. Oncogenes encode proteins that continuously promote the cell cycle, leading to unregulated cell division. Tumor suppressor genes function to inhibit the cell cycle. If mutated, the proteins they express would not be active and the cell cycle would continue, possibly leading to cancer. p. 87: 1. One chromatid before replication; two chromatids after. 2. See Figure 5.7. 3. Both result in two daughter cells with identical genetic material. Animal cells form a cleavage furrow between the daughter nuclei,

which is constricted by the action of a band of actin filaments. Plant cells build a new cell wall between the daughter cells by fusing together vesicles produced by the Golgi apparatus. p. 91: 1. In order to divide the DNA content in half; because the DNA was already replicated before meiosis began. 2. See Figures 5.11 and 5.14. 3. During crossing-over, genetic material is exchanged between nonsister chromatids. This produces a different combination of genes. Independent assortment mixes the whole chromosomes donated by both parents. The end result of both processes is cells that do not have genetic material identical to either parent. p. 94: 1. See Table 5.1. 2. See Table 5.2. p. 96: 1. Sperm and egg cells are haploid. They function in sexual reproduction. 2. Four sperm; one egg.

Testing Yourself

1. a. G_0 stage—the cell is no longer active in the cell cycle. b. G_1 stage—the cell is growing by increasing the number of organelles. c. S stage—the DNA is being replicated. d. G_2 stage—the cell is preparing for cell division. e. Mitotic stage, or mitosis—the cell is undergoing nuclear division. f. Cytokinesis—the cell is undergoing cytoplasmic division. 2. d; 3. b; 4. c; 5. d; 6. e; 7. b; 8. c; 9. c; 10. c; 11. d; 12. b; 13. d; 14. d; 15. c

Thinking Critically

1. a. BPA could increase or decrease the production of cyclins, which would affect cell cycle progress through the checkpoints. b. Because BPA acts like a hormone, its effects are amplified within the cell. 2. Skin cells need to be able to divide to not only replace worn-out cells, but also to repair damage such as cuts and abrasions. 3. Sexual reproduction introduces genetic variation that may be an adaptive advantage to organisms responding to changing environments.

CHAPTER 6
Check Your Progress

p. 101: 1. Once energy is released as heat, it cannot be recaptured for recycling. 2. With digestion, you are converting chemical energy to kinetic energy with a loss of heat according to the 1st law, which says that energy cannot be created or destroyed but can be changed from one form to another; and according to the 2nd law, there will be a loss of usable energy every time energy is converted from one type to another. p. 104: 1. An anabolic reaction is more likely to be endergonic because it usually involves an investment of energy. 2. ATP, with its three phosphate groups, is analogous to a charged battery that can provide energy via a coupled reaction when one phosphate is cleaved off. ADP that is produced is like a discharged battery, which requires an input of energy to become ATP again. p. 109: 1. Enzymes are needed to reduce the activation energy of the reactions present in a biochemical reaction, thus allowing reactions to occur under conditions of the cell. A cell can convert an enzyme from an inactive form to an active one by the addition or removal of phosphate groups or by cleaving off parts of the protein. Cellular enzymes are subject to feedback inhibition. The presence of cofactors and coenzymes in the cell regulates enzyme activity. 2. Enzymes lower the activation energy of a reaction, allowing it to occur with lower inputs of energy. 3. How well the enzyme interacts with the reactants and the rate at which product is formed is determined by the three-dimensional shape of the protein molecule and its active site. p. 111: 1. They both involve the gain and loss of hydrogen atoms. In photosynthesis, water loses hydrogen atoms and carbon dioxide gains them. In cellular respiration, glucose loses hydrogen atoms and oxygen gains them. 2. See Figure 6.11.

Testing Yourself

1. c; 2. e; 3. b; 4. d; 5. b; 6. c; 7. d; 8. d; 9. b; 10. a; 11. d; 12. c

Thinking Critically

1. The energy of activation must still be overcome. Enzymes enable the reactions to occur at the conditions present within the body. 2. The main features that researchers would need to know about the vampire bat enzyme in order to synthesize it are the amino acids that make up the enzyme, the shape of the enzyme, and the optimal temperature and pH range that the enzyme functions best within. 3. The active site is the location on the enzyme in which the substrate fits in order to cause the reaction to occur. If a drug is produced so it binds to the active site of the enzyme, this will block the enzyme from interacting with the substrate, effectively shutting down the pathway.

CHAPTER 7
Check Your Progress

p. 117: 1. NAD^+ and FAD serve as coenzymes for the enzymes involved in cellular respiration. Both molecules receive electrons and become reduced (NADH and $FADH_2$) and transport the electrons to inside the mitochondria, where they function in the electron transport chain. 2. Glycolysis is an anaerobic process. The preparatory reaction, citric acid cycle, and electron transport chain are all aerobic and therefore dependent on the presence of oxygen. 3. See Figure 7.2. p. 117: 1. Glucose must be activated by the investment of 2 ATPs before the cascade of reactions oxidizing glyceraldehyde 3-phosphate, forming 4 ATPs, can happen. 2. For each molecule of glucose, the inputs are 2 ATPs and 2 NAD^+s. The outputs are 2 molecules of pyruvate, 2 NADHs, 2 ADPs, and 4 ATPs. There is a net gain of 2 ATPs. p. 121: 1. When muscles are working hard in a burst of activity, using up available oxygen. 2. During fermentation, pyruvate accepts electrons from NADH formed during glycolysis, and lactate is formed. The oxidized NAD^+ is available to pick up more hydrogen atoms from glycolysis. p. 126: 1. See Figure 7.2. 2. See Figure 7.9. 3. The NADH formed outside the mitochondria during glycolysis sometimes cannot cross into the mitochondria. Instead, NADH delivers its electrons to the electron transport chain using a process that costs one ATP/2 electrons. Because 2 NADH are formed during glycolysis, the total ATP count is reduced to 4 from 6.

Testing Yourself

1. b; 2. d; 3. a; 4. c; 5. e; 6. c; 7. d; 8. d; 9. d; 10. e; 11. a; 12. b; 13. c

Thinking Critically

1. The amount of sugar used in the initial reactions. 2. If the electron transport chain is the same in insects and humans, then Rotenone would inhibit the human chain also. In fact, rotenone is very toxic to humans. 3. The faster fatty acids are converted to acetyl CoA and enter the citric acid cycle, the greater the rate of respiration. These compounds may convert fatty acids to acetyl CoA more rapidly.

CHAPTER 8
Check Your Progress

p. 132: 1. The sun. 2. See Figure 8.2. 3. $6 CO_2 + 6 H_2O \rightarrow C_6H_{12}O_6 + 6 O_2$ with the input of solar energy and presence of pigments. 4. Light reactions require the input of light energy. They result in the generation of NADPH and ATP, which function in the Calvin cycle. The Calvin cycle reactions are independent of the input of light energy. They reduce CO_2 to produce carbohydrates. p. 135: 1. They reflect and do not absorb wavelengths of light that humans see as green. 2. In the noncyclic electron pathway, the excited electron from photosystem II passes down an electron transport chain to photosystem I, producing ATP. The excited electron from photosystem I is passed to $NADP^+$, making NADPH. Both ATP and NADPH are used in the Calvin cycle reactions. 3. See Figure 8.8. 4. H^+ ions present in the thylakoid space are at a higher concentration than in the stroma. The tendency is for the ions to move down this electrochemical gradient, releasing the stored energy to fuel ATP synthesis. p. 138: 1. Carbon dioxide is attached to RuBP by the enzyme RuBP carboxylase, and then converted to G3P in a two-step reaction involving ATP as the energy source and NADPH + H^+ as the source of the electrons for the reduction step. 2. It takes three turns to make one G3P, but for every three turns, five molecules of G3P are used to re-form three molecules of RuBP. 3. G3P can be converted into glucose, sucrose, starch, cellulose, fatty acids, or amino acids. p. 139: 1. Sugarcane, corn, succulent plants such as cacti 2. It allows the stomata to close during the day so there is less loss of water without affecting the plant's ability to fix carbon dioxide. p. 141: 1. See Figure 8.12. 2. Because the chemical pathways are not the reverse of each other. Different enzymes, cellular structures, and pigments are involved in the two processes.

Testing Yourself

1. a. thylakoid, b. oxygen, c. stroma, d. Calvin cycle, e. granum; 2. c; 3. e; 4. d; 5. a; 6. d; 7. e; 8. d; 9. c; 10. e; 11. c; 12. c

Thinking Critically

1. Ideally, biofuel production should be carbon neutral because the plant growth removes carbon dioxide to the

atmosphere, which is then released during combustion. However, production processes often result in a net addition of carbon dioxide to the atmosphere, but at a lower level than fossil fuels. **2.** Chlorophyll in the chloroplasts of the alga cannot absorb green wavelengths of light, so the middle area exposed to green light is not photosynthesizing and producing oxygen. The bacteria are heterotrophs and require oxygen to perform cellular respiration. Because each end of the alga is photosynthesizing under blue and red wavelengths, oxygen is being produced, which the bacteria are consuming. **3.** Carbon dioxide gas fixed from the atmosphere and water account for the increase in mass of the plant. During the Calvin cycle of photosynthesis, carbon is fixed to create molecules of G3P. These molecules are used to build plant structures like cell walls. Most of the mass of each cell in the plant is composed of water taken up by the plant roots.

CHAPTER 9

Check Your Progress

p. 148: 1. Meristematic tissue. **2.** Epidermal tissue: protection, conservation of water; Ground tissue: photosynthesis, carbohydrate storage; Vascular tissue: transport of water and nutrients. **3.** Ground tissue: parenchyma (progenitor cells), collenchyma (flexible support), sclerenchyma (support, some transport water); Vascular tissue: xylem (transports water and minerals from roots to leaves), phloem (transports sugar and other organic compounds throughout the plant).
p. 150: 1. The root system of a plant consists of the roots, while the shoot system consists of the leaves and stems of the plant. **2.** See Figure 9.1 for structure comparison. Roots: the root hairs, with their large surface area, allow for uptake of water and minerals. The branching of the roots allows them to stabilize the aboveground portion of the plant. Stems allow for continued growth and contain xylem and phloem for transportation of water and nutrients through the plant. Leaves are thin so that sunlight can reach the chloroplasts and gases can diffuse easily. The petiole allows for maximum exposure to the sun. **p. 150: 1.** See Figure 9.7. **2.** Monocots: corn, grass, palms; Eudicots: dandelions, oak trees, potatoes, kale. **p. 154: 1.** Via a Casparian strip, which requires water and minerals to pass through the cells to enter the xylem. **2.** The monocot root has a vascular cylinder surrounding a central pith, while the eudicot root has the phloem and xylem in the center of the root. **3.** Adventitious roots (stabilizing the shoot system), epiphytes (absorption of water), haustoria (absorption of water and nutrients), root nodules (obtaining nutrients), mycorrhizae (absorption of water and nutrients). **p. 159: 1.** See Figures 9.15 and 9.16. **2.** In the spring when water is abundant, secondary xylem contains wide vessel elements with thin walls. When water becomes restricted in the summer, the wood has fewer vessels and contains thick-walled vessels. **3.** Stolons in aboveground horizontal stems produce new plants when touching the soil. Cacti have aboveground vertical stems that store water. Many vines have stems with tendrils that twine around structures. Rhizomes are underground horizontal stems. Corms are bulbous underground stems that are dormant in winter. **p. 161: 1.** See Figure 9.21. **2.** Shade plants have broad, wide leaves; desert plants have reduced leaves with sunken stomata. Leaves tend to be thin and flat to allow maximum sunlight absorption. **p. 167: 1.** Cohesion—water molecules tend to cling together; adhesion—water molecules interact with the walls of the xylem vessels; both are required to keep the column of water together as it ascends the xylem channel. **2.** Evaporation of water from a leaf causes the whole channel of water to move upward, drawing in more water from the roots. **3.** After sugar is actively transported into the phloem sieve tubes, water follows by osmosis. This creates a pressure that causes the phloem contents to flow from the source to the sink.

Testing Yourself

1. b; **2.** d; **3.** a; **4.** c; **5.** c; **6.** c; **7.** c; **8.** c; **9.** c; **10.** b; **11.** b; **12.** b; **13.** a; **14.** c; **15.** d

Thinking Critically

1. Stems have nodes and internodes. The xylem and phloem in roots are arranged centrally, while xylem and phloem in stems are found in bundles, either dispersed throughout the stem or arranged in a ring toward the outside of the stem. **2.** Woody plants can increase in girth and grow taller to reach sunlight, but this costs energy. **3.** Both use active transport and osmosis to move substances into the vessels. Plants use passive means of transport, whereas humans use a pump. Plants separate water and minerals from organic

nutrients, while humans do not. Plants have two separate sets of vessels (xylem and phloem), with the xylem being arranged in a one-way pathway for transport, while humans have two connecting sets of vessels (arteries and veins) arranged in a circular pathway. Some of the vessels in plants are dead, while all of the vessels are alive in humans. **4.** At night, the large number of stomata open and are able to absorb moisture from coastal fog. During the day, the stomata close to prevent loss of moisture.

CHAPTER 10

Check Your Progress

p. 176: 1. The sporophyte generation produces microspores and megaspores. A microspore undergoes mitosis and becomes a pollen grain (male gametophyte), and the megaspore undergoes mitosis to become a microscopic embryo sac (female gametophyte). **2.** Anther: development of pollen grains. Filament: holds up anther. Stamen: male part of flower consisting of anther and filament. Pistil (or carpel): female part of flower, consisting of stigma, style, ovary, and ovule. Stigma: sticky knob for holding pollen. Style: holds up stigma. Ovary: enlarged base that holds one or more ovules. Ovule: houses megaspores and, when fertilized, is chamber where seed develops. **3.** Heterosporous means that a plant produces two types of spores: microspores which become a pollen grain (male gametophyte), and megaspores which become the embryo sac (female gametophyte). **4.** See Figure 10.5. **p. 179: 1.** In monocots, food molecules are absorbed from the endosperm. In eudicots, the nutrients are stored in the cotyledons and the endosperm disappears. **2.** Adaptions include: wooly hairs; plumes; wings, like on maple seeds; small size, as with orchids; parachutes of dandelions; and the bursting seed pods of touch-me-nots. **3.** Regulation of the timing of the germination process ensures that seeds are in the best environment for survival. **4.** Compare Figure 10.9 with Figure 10.10. **p. 181: 1.** Vegetative propagation and tissue culture. **2.** See Figure 10.11. **3.** The newly formed plant is often supported nutritionally by the parent plant until it is established **p. 186: 1.** Auxins promote shoot growth from the apical meristem or from the ancillary buds when the terminal bud is removed. Gibberellins cause elongation of cells. Cytokinins promote cell division. Abscisic acid initiates and maintains seed and bud dormancy, while controlling the opening of stomata under water stress. Ethylene functions in the process of abscission. Brassinosteroids are involved in plant growth and development. **2.** Plant growth tropisms include growth toward or away from a stimulus, such as light, gravity, and touch. **3.** Short-day plants flower during shorter daylight periods, while long-day plants flower during longer-light days. **4.** Because phytochrome conversion from inactive to active forms is the first step in a signaling pathway that results in flowering, regulating the conversion could lead to controlling plant flowering commercially.

Testing Yourself

1. c; **2. a.** diploid, **b.** anther, **c.** ovule, **d.** ovary, **e.** haploid, **f.** megaspore, **g.** pollen grain, **h.** embryo sac, **i.** sperm, **j.** seed, **3. a.** stigma, **b.** style, **c.** ovary, **d.** ovule, **e.** receptacle, **f.** peduncle, **g.** sepal, **h.** petal, **i.** filament, **j.** anther; **4.** b; **5.** c; **6.** e; **7.** a; **8.** c; **9.** d; **10.** b; **11.** e; **12.** a; **13.** c; **14.** e; **15.** f

Thinking Critically

1. Meiosis is used to produce gametes (sperm and eggs) in animals and to produce microspores and megaspores in flowering plants. Mitosis in animals leads to growth and development of the zygote. In plants, microspores and megaspores undergo mitosis to produce male and female gametophytes. **2.** A single, specialized pollinator species will ensure a greater probability of pollen transfer and fertilization to the correct species. **3.** You could measure the presence and distribution of abscisic acid in the stems of sunflowers as they bend throughout the day and look for correlations. **4.** Seeds are a product of sexual reproduction and contain a certain amount of genetic variability. While asexual propagation by tubers is cheaper, faster, and easier than by seeds, there is an inherent risk of not having genetic variability if the environment changes.

CHAPTER 11

Check Your Progress

p. 195: 1. See Figure 11.1. **2.** See Figure 11.2. **3.** See Figure 11.4. **4.** A neuron has a cell body, dendrites, and an axon. Neurons function in sensory input, integration of information,

and motor output. Neuroglia support and nourish neurons and participate in brain function. **p. 196: 1.** Thoracic, abdominal. **2.** Mucus protects the digestive, respiratory, urinary, and reproductive systems from bacteria and viruses. Serous fluid lubricates membranes that support internal organs and divide body cavities. Synovial fluid lubricates cartilage at the end of bones. **p. 198: 1.** Integumentary—skin; cardiovascular—heart; lymphatic and immune—lymph nodes; digestive—small intestine; respiratory—lungs; urinary—kidneys; nervous—spinal cord; musculoskeletal—skeleton; endocrine—pancreas; reproductive—ovaries. **2.** Integumentary, and lymphatic and immune. **p. 202: 1.** See Figure 11.8. **2.** They might not absorb enough UVB rays to produce adequate levels of vitamin D, which helps to keep bones strong. **3.** Nails—protect the distal parts of digits; hair follicles—produce hair; oil glands—secrete sebum. **p. 205: 1.** In negative feedback, a sensor detects a change in internal conditions, resulting in a response that brings the conditions back to normal. Positive feedback involves an ever-greater change in the same direction until the initial stimulus stops. **2.** The respiratory, digestive, and urinary systems interact to take in O_2 and nutrients needed by the body, and eliminate unused or waste products such as CO_2 and nitrates. They ensure that blood pH, blood volume, and glucose levels are maintained within strict ranges. **3.** Failure of the immune system results in infectious diseases such as shingles and candidiasis. Failure of the cardiovascular system results in heart failure or strokes. Failure of endocrine organs can result in diabetes or hypothyroidism.

Testing Yourself

1. c; **2.** a; **3.** e; **4. a.** ventral cavity, **b.** thoracic cavity, **c.** abdominal cavity, **d.** pelvic cavity, **e.** dorsal cavity, **f.** cranial cavity, **g.** vertebral canal, **h.** diaphragm; **5.** d; **6.** b; **7. a.** hair shaft, **b.** epidermis, **c.** dermis, **d.** subcutaneous layer, **e.** adipose tissue, **f.** sweat gland, **g.** hair follicle, **h.** arrector pili muscle, **i.** oil gland, **j.** sensory receptor, **k.** basal cells, **l.** sweat pore; **8.** c; **9.** b; **10.** b

Thinking Critically

1. Blood is like a tissue because it contains similarly specialized cells that perform a common body function. Blood cells are specialized to circulate throughout the body, bringing O_2 to the tissues (red blood cells), preventing the blood from being lost from the body (platelets), and protecting the body from infection (white blood cells). **2. a.** Negative feedback, because secretion of the pituitary gland hormone is inhibited by epinephrine. **b.** Positive feedback, because the stronger the signal becomes, the stronger the stimulus is to respond, until urination occurs. **c.** Negative feedback, because the various sensory receptors are detecting a change in a set point (blood volume), and the kidney is responding by increasing urine production. **3.** A loss of homeostasis may influence a number of parameters that keep the cells of the body alive (pH, temperature, nutrient concentration), thus leading to tissue and organ failure, and eventually death.

CHAPTER 12

Check Your Progress

p. 210: 1. Blood flow is controlled in arteries by contraction of smooth muscle in artery walls. Capillary blood flow is affected by the pressure of arterial supply plus the contraction of precapillary sphincters. Venous blood flow is affected by arterial and capillary blood flow, and valves that prevent blood from flowing backward. **2.** O_2, CO_2, glucose, amino acids. **p. 217: 1.** See Figure 12.3. **2.** See Figure 12.6. **3.** Diabetes; heart disease; liver disease; Alzheimer and Parkinson diseases. **4.** If the protein content of the blood was reduced, then the osmotic pressure of the blood would be lowered and less water would move from the interstitial fluid into the blood. **p. 221: 1.** Blood enters the right atrium, then the right ventricle, then travels through the pulmonary arteries to the lungs, and through the pulmonary veins back to the left atrium and then left ventricle. **2.** If the left ventricle was not contracting strongly, blood would back up into the lung, increasing blood pressure there, resulting in increased fluid leakage from blood vessels in the lung. **3.** The heartbeat "lub" sound is caused by the closure of the atrioventricular valves and the "dub" sound is caused by the semilunar valves closing. **4.** Intrinsic control of the heartbeat is needed to maintain regular coordinated contractions, while extrinsic control allows the heartbeat to change in response to different physiological demands. **p. 223: 1.** Pulmonary arteries. **2.** Blood flows to the lungs via the pulmonary arteries, and returns to the heart via the pulmonary veins. Blood flows to the body via the aorta, and returns via the superior and inferior venae cavae. **3.** Blood flows in one direction in arteries due to the pressure created from the pumping of the heart, and in veins mainly due to muscle contraction and the presence of valves. **p. 227: 1.** Atherosclerosis occurs when plaque builds up in the linings of arteries. The plaque can restrict blood flow through vessels, or break off and form an embolus. Hypertension can lead to stroke and heart attack. Heart valve disease results in the malfunction of the pumping of the heart. Strokes result when brain blood vessels burst or are blocked. Heart attacks occur when coronary arteries are blocked. Aneurysms happen when the artery wall bulges out and sometimes bursts. **2.** See Table 12.2. **3.** Failing hearts can be replaced with a transplant from a human donor, or with an artificial heart. A left ventricular assist device can help a failing heart function until it can be replaced. Potential problems include rejection of a transplanted heart by the recipient's immune system, and the mechanical failure of an artificial pump.

Testing Yourself

1. a; **2.** d; **3.** c; **4.** b; **5.** e; **6.** d; **7.** b; **8.** e; **9.** a; **10.** d; **11.** a; **12. a.** carotid artery, **b.** pulmonary vein, **c.** aorta, **d.** renal artery, **e.** pulmonary artery, **f.** superior vena cava, **g.** inferior vena cava, **h.** hepatic vein, **i.** hepatic portal vein, **j.** renal vein; **13.** s; **14.** c

Thinking Critically

1. Tissues that do not contain blood vessels include the corneas of the eyes, the outermost layer of the skin, and the cartilage that lines joints. These tissues presumably receive their required nutrients by diffusion from nearby vessels, and/or have lower metabolic requirements than more highly vascularized tissues. **2.** The answer will vary depending on age, but for someone with an average heart rate of 70 beats per minute, on their 20th birthday their heart will have beaten 70 × 60 minutes/hour × 24 hours/day × 365 days/year × 20 years, or 735,840,000 times. In terms of volume, 5.25 liters/minute × 60 minutes/hour × 24 hours/day × 365 days/year × 20 years = 55,188,000 liters (over 14 million gallons). **3.** Tracing "a" shows too many QRS complexes, indicating atrial fibrillation. The complexes in tracing "b" are missing the "T" waves, and the heartbeat is too slow, suggesting a problem with electrical transmission in the heart. A pacemaker is an electrical device that can be implanted in a patient's heart muscle and takes the place of a malfunctioning SA node to deliver a regular electrical impulse to the heart muscle. **4.** A total artificial heart (TAH) would have to be as incredibly durable as the heart is, to last even 20 years. The TAH also would need to be made of a material that would not stimulate the body to react against it in any way, including the immune and clotting systems.

CHAPTER 13

Check Your Progress

p. 233: 1. Excess interstitial fluid moves into the lymph vessels and is returned to the bloodstream. Fats from the small intestine move into the lymph vessels and are transported into the bloodstream. White blood cells in the lymph system function in protecting the body from disease. **2.** Lymph is the excess interstitial fluid that has been absorbed into lymphatic vessels. It is mostly water but also contains nutrients, electrolytes, oxygen, hormones, enzymes, and wastes. Blood contains erythrocytes, leukocytes, platelets, waste solutes, and products of cells. **3.** In the red bone marrow and the thymus, which are primary lymphoid organs, lymphocytes develop and mature. In the secondary lymph organs, such as the lymph nodes and the spleen, lymphocytes become activated. **p. 235: 1.** The skin and mucous membranes are physical barriers; a chemical barrier is the acid of the stomach. **2.** See Figure 13.3. **3.** Macrophages kill pathogens by engulfing them into a vesicle that has an acid pH, hydrolytic enzymes, and reactive oxygen compounds. NK cells induce cells that lack self-MHC-I molecules to undergo apoptosis (cell suicide). **4.** The complement system enhances phagocytosis of pathogens, activates inflammation, and kills pathogens by forming a membrane attack complex. **p. 239: 1.** Diversity in antigen receptors occurs because: 1. Genes for T- and B-cell receptors contain segments that code for many parts of the antigen receptor. 2. Enzymes in the T and B cells cut out these segments and combine them differently. 3. Mutations may be introduced as these segments are combined. 4. The two protein chains, containing the variation of steps 1–3, combine to produce a new receptor. **2.** The clonal selection theory states that B cells and T cells have cell surface receptors for only one specific antigen. When the cell contacts that

specific antigen, it is selected to undergo clonal expansion (divide) and differentiate into either memory cells or cells that actively fight infection. **3.** The five classes are IgG (activates complement, crosses the placenta), IgM (early response, activates complement), IgA (protects respiratory tract), IgD (activates B cells), and IgE (protects against parasites). **4.** Helper T cells recognize antigen fragments in combination with MCH molecules presented by antigen-presenting cells that have MHC class II proteins on their surface. Cytotoxic T cells recognize antigen-presenting cells with MHC class I proteins. Suppressor T cells inhibit these responses and act to control adaptive immunity. **p. 243: 1.** Three types include: attenuated (live but nonvirulent), genetically engineered (also called subunit vaccines), and DNA vaccines. **2.** The immune system of the newborn is immature, plus the newborn has not been exposed to any infectious agents, so it would be very susceptible to infectious diseases if it did not receive antibodies from its mother. **3.** The interaction of cytokines with their receptors could be blocked using antibodies that react with the cytokine, or possibly with the receptor. **4.** Monoclonal antibodies are extremely specific because they react with only one antigen. Because of their specificity and purity, reaction against the preparation is avoided. **p. 246: 1.** Immediate allergic responses are caused by IgE antibodies that have already been produced against the offending allergen; delayed-type responses take longer because they require sensitized T cells to secrete cytokines that cause inflammation. **2.** If a mother is Rh-positive, she will not produce anti-Rh antibodies that would react with the antigens of the fetus. **3.** Drugs that inhibit cytokine production may inhibit certain desirable immune responses that protect the individual. **p. 248: 1.** Muscular weakness occurs in patients with myasthenia gravis because antibodies attach to and disrupt neuromuscular junctions. In multiple sclerosis, T cells attack the myelin surrounding nerves and hamper nerve conduction. Antibody–antigen complexes form in people with systemic lupus erythematosus and these get deposited in organs, including the kidneys. Antibody–antigen complexes form in rheumatoid arthritis and these primarily affect the joints. **2.** A congenital immunodeficiency disease is severe combined immunodeficiency. AIDS is an acquired immunodeficiency disease.

Testing Yourself

1. c; **2.** b; **3.** a; **4.** c; **5.** c; **6.** c; **7.** a; **8.** a; **9.** e; **10. a.** antigen-binding site; **b.** light chain; **c.** heavy chain. V stands for variable regions and C stands for constant regions. **11.** a; **12.** d; **13.** d; **14.** d; **15.** c; **16.** b

Thinking Critically

1. Innate immune mechanisms are valuable mainly because they can react very quickly to a wide variety of pathogens. However, organisms lacking adaptive immunity are not able to mount specific antibody and T-cell responses targeted at one specific organism, nor are they able to develop memory cells that are able to respond faster to subsequent exposures to the same organism. **2.** The administration of antivenom antibodies is an example of passive immunity. The antibodies will go away over time. If a person is bitten again they need another injection of antivenom. **3.** ABO incompatibility between mother and fetus usually causes no problems because the antibodies the mother has against the nonself A or B antigen are normally of the IgM type, which is too large to cross the placenta. IgG antibodies can cross the placenta. Because around 85% of the population is Rh-positive, most females do not produce anti-Rh antibodies that could damage or kill the fetus; therefore, there is not a strong evolutionary selective pressure to eliminate the Rh molecule.

CHAPTER 14

Check Your Progress

p. 258: 1. Food passes through the mouth, pharynx, esophagus, stomach, small intestine, and large intestine. Mechanical digestion occurs in the mouth and stomach. Chemical digestion occurs in the stomach and small intestine. **2.** The small intestine processes food for digestion and absorbs nutrients. Its lower part, the ileum, is involved in immune responses. The large intestine absorbs water, salts, and some vitamins, and eliminates indigestible material as feces. **3.** Hormones that increase digestive activity include gastrin, secretin, and CCK. Inhibiting these might regulate weight but they can cause difficulties in digesting food, leading to excess gas or constipation. **p. 259: 1.** The pancreas secretes hormones (endocrine) such as insulin and glucagon into the blood, and also

secretes digestive enzymes (exocrine) into the small intestine by way of ducts. **2.** Salivary glands produce saliva, which moistens food and contains amylase, which begins the digestion of starch. The liver performs functions summarized in Table 14.2. The main function of the gallbladder is to store bile. The pancreas secretes bicarbonate, digestive enzymes, and the hormones insulin and glucagon. **3.** The gallbladder stores bile and people can live without this function. The liver has essential functions of detoxification, storage, synthesis, lipid regulation, and hemoglobin breakdown, all of which are necessary for survival. **p. 261: 1.** Carbohydrate digestion begins in the mouth, with salivary amylase, and continues in the small intestine, due to maltase and pancreatic amylase. Protein digestion begins in the stomach, where pepsin breaks proteins into peptides, and continues with trypsin and peptidases in the small intestine. Fats are mainly digested by pancreatic lipase in the small intestine. **2.** Carbohydrates are broken down to glucose; proteins to amino acids; fats to glycerol and fatty acids. **3.** The digestive system is divided into many compartments, each of which has a distinctive internal environment and pH. These differences contribute to regulation of digestive enzymes. **p. 269: 1.** The glycemic index (GI) of foods is an indicator of whether the blood glucose response to their ingestion is high or low. The advantage of consuming foods with a high GI is that they can provide a quick burst of energy, but that energy is usually short-lived, and over time, consuming too many high-GI foods can lead to type 2 diabetes and heart disease. **2.** A diet deficient in protein does not supply adequate amounts of essential amino acids used to synthesize necessary proteins. Too much protein must be deaminated in the liver, which requires abundant water for excretion and calcium loss in urine. **3.** Vitamins are organic molecules (other than carbohydrates, fats, and glucose) that act as coenzymes and cannot be synthesized by the body. Minerals are inorganic elements that serve a variety of functions. Major minerals such as calcium are in cells and body fluids and also can serve as structural components of tissues. Trace minerals are needed in smaller amounts and are parts of larger molecules. **4.** Dietary supplements can correct deficiencies in vitamin and mineral intake. However, in developed countries such deficiencies do not exist. In that case, the supplements are in excess of what is needed and can be present at toxic levels. The body eliminates water soluble nutrients in the urine, but lipid-soluble vitamins A and D are stored in body fat and can reach high concentrations. **p. 271: 1.** People with anorexia nervosa restrict their food intake, thinking of themselves as overweight no matter what they weigh. Those with bulimia nervosa eat more than they think is appropriate and rid themselves of the food by vomiting or using laxatives. Binge eating involves ingesting more food than needed, usually for emotional reasons. This most often leads to obesity. People with pica eat unusual substances for reasons not understood. **2.** Abnormally increased or decreased levels of hormones that influence appetite, such as gastrin, gastric inhibitory peptide (GIP), or leptin, can predispose a person to eating disorders. **p. 273: 1.** Most stomach ulcers are caused by the bacterium *Helicobacter pylori*. **2.** The pancreas is essential in regulating serum glucose levels by the secretion of insulin and glucagon. The pancreatic juice it secretes neutralizes stomach acid and provides digestive enzymes. The liver synthesizes bile and essential proteins, regulates glucose and cholesterol levels, metabolizes many waste products, and stores iron and many vitamins. Both organs are essential to life.

Testing Yourself

1. c; **2.** c; **3.** d; **4.** c; **5.** a; **6.** d; **7.** e; **8.** c; **9.** a; **10.** c; **11.** b; **12.** b; **13.** e; **14.** b; **15.** a; **16.** d; **17.** b

Thinking Critically

1. Digestion of complex carbohydrates begins in the mouth with salivary amylase, and continues in the small intestine with pancreatic amylase. Ultimately, monosaccharides such as glucose are absorbed from the small intestine into the hepatic portal vein, where some travel to the liver and some enter the general circulation via the hepatic veins and inferior vena cava. Cells transport glucose molecules across their plasma membranes and utilize them to produce ATP. **2.** Excess amino acids are transported to the liver where they are deaminated and urea is produced. This process uses water and can result in dehydration. Eating too much protein can also result in calcium loss in the urine. **3.** Only tube 4 containing pepsin, HCl, and water should result in the breakdown of the egg white. The other tubes lack a necessary component of this reaction mix.

CHAPTER 15

Check Your Progress

p. 280: 1. The upper respiratory tract consists of the nose and nasal cavities, pharynx, and larynx. The lower respiratory tract consists of the trachea, bronchial tree, and lungs. **2.** The trachea lies ventral to the esophagus. When food is swallowed, the epiglottis covers the tracheal opening (glottis) so the food slides over the epiglottis and into the esophagus. **3.** O_2 travels through the nasal cavity, pharynx, larynx, trachea, bronchial tree, and lungs, where it diffuses across the alveolar and capillary endothelial cells to the bloodstream. **4.** Pulmonary surfactants lower the surface tension of the coating of the alveoli, preventing their sides from collapsing upon themselves. **p. 283: 1.** Tidal volume is the amount of air that normally moves in and out of the lungs with each breath. Vital capacity is the maximum volume of air that can be moved in and out during a single breath. Expiratory reserve volume is the air that can be forcibly exhaled beyond the tidal volume. Residual volume is the air left in the lungs after a forced exhalation. **2.** Inspiration requires the contraction of the diaphragm and external intercostal muscles. Expiration is passive because it requires no muscle contractions, just the elastic recoil of the thoracic wall and lungs. **3.** The respiratory center in the brain automatically sends nerve impulses to the diaphragm and intercostal muscles. The vagus nerve carries inhibitory impulses from the lungs to the brain to stop the lungs from overstretching. The carotid bodies have chemoreceptors that monitor levels of O_2 in the blood. **p. 286: 1.** Hemoglobin's most essential function is to carry O_2 (as oxyhemoglobin) to the tissues, but it also carries a small amount of CO_2 (as carbaminohemoglobin) back to the lungs. Excess H^+ ions can be taken up by hemoglobin, which then becomes reduced hemoglobin. **2.** Arterial blood is brighter red than venous blood because the red blood cells in arterial blood contain oxyhemoglobin, which is bright red, compared to the red blood cells in venous blood that contain reduced hemoglobin, which is darker. Blood from a cut appears bright red because hemoglobin becomes oxyhemoglobin upon exposure to O_2 in the air. **3.** The exchange of gases in respiration is by diffusion. No energy is required, which indicates a passive process. **p. 291: 1.** Laryngitis is an inflammation of the larynx, while tonsillitis occurs when the tonsils become inflamed and enlarged. **2.** Smoking damages the tissues of the respiratory tract directly. Infections from bacteria, fungi, protozoans, and viruses become more likely if the lung tissue is damaged. **3.** Cystic fibrosis.

Testing Yourself

1. a. nasal cavity, **b.** nostril, **c.** pharynx, **d.** epiglottis, **e.** glottis, **f.** larynx, **g.** trachea, **h.** bronchus, **i.** bronchiole; **2. a.** sinus, **b.** hard palate, **c.** epiglottis, **d.** larynx, **e.** uvula, **f.** glottis; **3.** b; **4.** b; **5.** b; **6.** a; **7.** a; **8.** d; **9.** d; **10.** d; **11.** d

Thinking Critically

1. The respiratory system of birds is more efficient than that of mammals where inspired air passes into a "dead end" (the alveoli) and must be moved back out along the same path. Some amount of air always remains in the lungs. In the bird system, air passes in one direction through the air sacs. **2.** Enlarged tonsils and adenoids partially block the airway, restricting flow. **3.** Since CO binds to hemoglobin much more strongly than O_2, much less O_2 is delivered to the tissues and respiration is compromised. The specific cause of death would be asphyxiation. **4.** When a person breathes through a tracheostomy, the air does not pass through the upper respiratory tract, which normally warms and humidifies the air, as well as removing many impurities. Therefore, tracheostomy patients are prone to lower respiratory tract infections.

CHAPTER 16

Check Your Progress

p. 296: 1. Excretion of metabolic wastes, osmoregulation, maintaining acid-base balance, secretion of hormones. **2.** See Figure 16.1. **3.** With kidney disease, wastes will build up, salt and pH balance will be disrupted, and hormone secretion of the kidneys will be changed, all of which would interfere with homeostasis. **p. 300: 1.** The renal cortex contains the glomeruli, proximal convoluted tubules (PCTs), and distal convoluted tubules (DCTs). The renal medulla contains the loops of the nephron. **2.** The epithelium of the PCT has many microvilli, which increase the surface area for reabsorption. The DCT is composed of cuboidal epithelial cells that are not designed for reabsorption. **3.** All of these processes involve the movement of a substance across biological membranes. Diffusion and passive transport are dependent on concentration; active transport is not, but it requires energy. Active and passive transport usually require carrier molecules. **4.** Glucose is normally returned from the glomerular filtrate to the blood by reabsorption via carrier proteins at the proximal convoluted tubule. **p. 303: 1.** See Figure 16.7. Reabsorption of salt, establishment of a solute gradient, reabsorption of water. **2.** When blood volume and pressure is low, the juxtaglomerular apparatus secretes renin, which converts angiotensin I, which is then converted to angiotensin II that causes aldosterone secretion. **3.** Diuretics can inhibit the secretion of ADH, increase the glomerular filtration rate, decrease the tubular reabsorption of sodium, or inhibit active transport of sodium at the loop of the nephron or at the DCT. **4.** Kidneys regulate H^+, HCO_3^-, OH^-, and NH_4^+ to maintain blood pH 7.4. **p. 305: 1.** If kidneys cannot produce urine, then excretion of metabolic wastes is blocked, regulation of blood volume is compromised, maintenance of blood pH at 7.4 is impaired, and hormone secretion including renin, aldosterone, and erythropoietin is hindered. The lack of any of these functions is life threatening. **2.** Smoking increases the risk of bladder cancer fivefold due to the toxic by-products of cigarettes being secreted into the bladder. **3.** Females have a shorter, broader urethra that is closer to the anal opening and is more easily contaminated with bacteria. An enlarged prostate in males may cause difficult urination.

Testing Yourself

1. a. renal cortex, **b.** renal vein, **c.** ureter, **d.** renal pelvis; **2.** b; **3.** d; **4.** c; **5.** b; **6.** a; **7. a.** renal cortex, **b.** renal vein, **c.** ureter, **d.** renal pelvis; **8.** b; **9.** b; **10.** a; **11.** b; **12.** a

Thinking Critically

1. If one is not eating large amounts of salt, then the regulation of sodium levels in the blood and the symptom of high blood pressure would be dependent on kidney function rather than diet. If the release of aldosterone is being successfully regulated by the kidney's release of renin, then blood pressure should be under control. **2.** Angiotensin II is a powerful vasoconstrictor in addition to its stimulating the adrenal cortex to release aldosterone, which leads to the reabsorption of sodium ions. Blocking angiotensin II's production would lower blood pressure. **3.** Frequent urination with diabetes mellitus is due to the presence of excess glucose in the urine filtrate, which draws water into the urine by osmosis. With diabetes insipidus, the increased urine output is due to the absence of ADH.

CHAPTER 17

Check Your Progress

p. 310: 1. The three classes of neurons are sensory neurons, which take messages to the CNS; interneurons, which sum up messages from sensory neurons and other interneurons and communicate with motor neurons; and motor neurons, which take messages away from the CNS to effector organs, muscles, or glands. **2.** Most neurons contain dendrites, a cell body, and an axon. See Figure 17.2. **3.** Myelin is formed by Schwann cells in the PNS, and by oligodendrocytes in the CNS. **4.** Gray matter contains nonmyelinated nerve fibers. In white matter, the fibers are myelinated. The brain has gray matter on the surface and white matter in deeper tissue. That pattern is reversed in the spinal cord. **p. 314: 1.** The sodium–potassium pumps in neurons are always transporting Na^+ to the outside, and K^+ to the inside of the cell. **2.** During an action potential, the Na^+ gates open and Na^+ enters the cell, causing a depolarization. Then, the Na^+ gates close and the K^+ gates open, causing a repolarization (even a slight hyperpolarization). See Figure 17.4. **3.** See Figure 17.5. **4.** After the action potential passes one part of an axon, the sodium gates in that part are unable to open for a period of time, called the refractory period. This prevents action potentials from moving backward. A node of Ranvier is a gap between Schwann cells that make up the myelin sheath of an axon of a PNS neuron. Saltatory conduction is the "jumping" of action potentials as they spread from node to node. Synaptic integration is the summing up of all incoming excitatory and inhibitory messages by a neuron. **p. 319: 1.** The CNS is composed of the spinal cord and the brain. The PNS, which lies outside of the CNS, is composed of nerves and ganglia. Sensory fibers send information to the CNS, and motor fibers conduct information away from the CNS to tissues and organs. Ganglia are areas of nerves that are collections of cell bodies. **2.** The spinal cord is structured so that the brain can communicate with the peripheral nerves. This communication involves sensory and motor functions. **3.** Cerebrum: receives sensory information, integrates it, and commands voluntary motor responses. Diencephalon has two parts: hypothalamus serves as a link between the nervous and endocrine system, maintaining homeostasis; thalamus receives sensory inputs except for smell. Cerebellum: processes information about body position and maintains posture and balance. Brain stem: acts as a relay between the cerebrum and spinal cord or cerebellum; regulates breathing through the pons; regulates vital functions through the medulla oblongata; receives and sends signals between the higher brain centers and the spinal cord. **p. 322: 1.** The hippocampus within the limbic system is involved in learning and memory. It communicates with the prefrontal area of the brain in these functions. **2.** Short-term memory is useful in managing operations within the present by holding a thought in mind. Long-term memory involves semantic and episodic memory, as well as skill memory. It allows recall of the past and use of that information to shape future actions. **3.** Mice that lack a glutamate receptor in their hippocampus were unable to learn to run mazes. **4.** Wernicke's area is involved in the comprehension of speech. Broca's area is involved in the actual motor function of speech. **p. 326: 1.** See Figure 17.15. **2.** When you touch a hot stove, your hand withdraws before you feel the pain because the nerve pathway for a reflex arc travels directly through the spinal cord, without input from the brain. **3.** See Table 17.1. **4.** The neurological explanation for a stomachache after jogging might be that your brain directed more activity to the sympathetic system, and less to the parasympathetic. This could decrease peristalsis and direct blood away from the digestive system, leading to some discomfort. **p. 329: 1.** Continued exposure to a drug over time might lead to compensatory responses at synaptic clefts that result in a decreased sensitivity to the drug. For example, if the drug prevents release of the NT (see Fig. 17.18 part 2), the cell might synthesize more of the NT. **2.** Cocaine prevents synaptic uptake of dopamine. Heroin binds to endorphin receptors. Methamphetamine results in high amounts of dopamine released in the brain. Bath salts inhibit the reuptake of several neurotransmitters. **p. 332: 1.** Major symptoms of AD include loss of short-term memory, progressing to an inability to carry out basic functions. These symptoms can begin before age 50, but usually not before 65. **2.** With Parkinson disease, the dopamine-releasing neurons in the brain degenerate. With lower levels of dopamine, motor control decreases and involuntary tremors and muscular rigidity develop. **3.** Autoimmune diseases occur when the immune system attacks self-cells, tissues, or organs. In MS, the immune system attacks myelin, oligodendrocytes, and neurons. **4.** *H. influenzae* can cause disease if the immune status of the carrier is compromised by fatigue, poor health, viral infection, lack of exercise, or risky behaviors.

Testing Yourself

1. d; **2.** a; **3.** a; **4.** b; **5.** b; **6.** c; **7.** c; **8.** d; **9.** d; **10.** b; **11.** a; **12. a.** central canal, **b.** dorsal horn, **c.** white matter, **d.** cell body of interneuron, **e.** cell body of sensory neuron; **13.** e; **14.** d; **15.** d; **16.** d; **17.** d

Thinking Critically

1. A way scientists study the functions of specific brain areas is to note which brain functions are disrupted by brain tumors or other disorders like strokes that affect only a specific part of the brain. Other data have been obtained from brain surgeries in which the patient is kept awake, and different areas of the brain can be electrically stimulated to see which parts of the body (or higher functions) they control. **2.** The first reason that nicotine, cocaine, and methamphetamine would be poor choices as therapeutic drugs is the high incidence of addiction and other undesirable side effects. A second problem would be the adaptation of the body over time, so that less dopamine is produced in the brain, which would counteract any temporary beneficial effects for Parkinson disease. **3.** In order to develop rational treatments for any disease, it is very helpful, if not essential, to understand what causes the disease. In many brain diseases, these factors are not completely understood. In addition, the brain is less accessible to investigation than other organs, due to its protected location inside the skull, the blood-brain barrier, and its essential function for life. **4.** When acetylcholinesterase is inhibited, the effects of acetylcholine, which include muscle contraction, salivation, elevated heart rate, and increased blood pressure, are prolonged.

CHAPTER 18

Check Your Progress

p. 338: 1. Interoceptors are involved in regulating critical body functions like blood pressure, blood volume, and blood pH. **2.** Chemoreceptors—taste and smell; photoreceptors—vision; mechanoreceptors—hearing and balance; thermoreceptors—heat and cold. **3.** Without sensory adaptation, we would constantly be stimulated by all sorts of inputs, which could overwhelm the brain's ability to interpret. **p. 340: 1.** Somatic sensory receptors that are interoceptors include proprioceptors (such as muscle spindles and Golgi tendon organs), and pain receptors in the skin and internal organs. Somatic sensory receptors that are exteroceptors include cutaneous receptors like Meissner corpuscles, Krause end bulbs, Merkel disks, Pacinian corpuscles, Ruffini endings, and free nerve endings sensitive to temperature. **2.** See Figure 18.3. **3.** Someone who lacks muscle spindles would be susceptible to injuring their muscles and joints due to a lack of awareness of limb position and degree of muscle contraction. A person who lacked pain receptors would be without the benefit of the protective function of pain and would be very prone to injury. **p. 342: 1.** The tongue has taste cells and the nasal cavity has olfactory cells. Both are chemoreceptors to which molecules bind and nerve impulses are sent to the brain. The tongue has five types of taste cells distributed over the tongue, while there are about 1,000 different types of olfactory cells in the olfactory epithelium. **2.** Adaptations that allow some animals to have a more sensitive sense of smell would include having a greater number of olfactory receptors (and increased space in the nasal cavity for these receptors) and having more sensory nerve axons to receive signals from these receptors. **3.** The olfactory bulbs in the brain have direct connections to centers for memory in the limbic system. **p. 346: 1.** Conjunctiva, cornea, aqueous humor in the anterior compartment, through the iris, lens, vitreous humor in the posterior compartment, retina including the ganglion cell layer, bipolar cell layer, and finally the rod cell and cone cell layer where light is converted to nerve impulses. **2.** When the ciliary muscle is relaxed, the suspensory ligaments that attach to the lens are taut, which causes the lens to become flatter. When the ciliary muscle contracts, the suspensory ligaments are relaxed, allowing the lens to become more round. **3.** As many as 150 rod cells, but perhaps only one cone cell, synapse(s) on an individual ganglion cell, so that more information can be sent to the brain for an individual cone cell versus a rod cell. **4.** The optic chiasma is X-shaped, formed by the optic nerve fibers crossing over. The image sent to the brain is split because one side of the optic tract carries information about the opposite side of the visual field. The two sides of the visual field must communicate with each other in the brain for perception of the entire visual field. **p. 348: 1.** See Figure 18.11. Sound waves travel down the auditory canal to the tympanic membrane, and then to the bones of the middle ear, which are connected to the cochlea in the inner ear. It is in the cochlea that the sound waves are converted to neural impulses that are transmitted to the brain. **2.** The mechanoreceptors responsible for transducing sound waves into nerve impulses are located on hair cells on the basilar membrane of the spiral organ in the cochlea of the inner ear. **3.** Sound waves transmitted via the oval window cause the basilar membrane to vibrate. Near the tip of the spiral organ, the basilar membrane vibrates more in response to higher pitched sound, and at the base it responds to lower pitches. Different sensory nerves supply each part of the basilar membrane, and take information to different parts of the auditory cortex, where the pitch is interpreted. **4.** Loud sounds cause the basilar membrane to vibrate more rapidly, which is interpreted by the brain as volume. **p. 349: 1.** See Figure 18.13. **2.** The ampulla and cupula are associated with the semicircular canals; the otoliths, saccule, and utricle are associated with the vestibule. **3.** Having two systems for equilibrium allows the brain to receive information about rotational movement from the semicircular canals, and about the body's position at rest from the vestibule. Although it might be possible to distinguish between these with just one system, it would not be as efficient or sensitive. **p. 353: 1.** The sense of smell can protect us from dangerous foods, chemicals, and situations. Without it, one is more vulnerable to injury from such sources. **2.** Nearsightedness results when the eyeball is elongated; hence, the lens focuses in front of the retina. Farsightedness results when the eyeball is too short. Astigmatism is caused by a misshapen cornea or lens. **3.** In macular degeneration, the choroid vessels thicken and no longer function, which leads to the destruction of cone cells, resulting in blindness. The drainage system in the eyes

fails and fluid builds up in patients with glaucoma. The increased pressure in the eye destroys nerves first in the peripheral vision but can lead eventually to complete blindness. With cataracts, the lens of the eye becomes cloudy and eventually opaque. No light can get through and total lack of vision can result. **4.** The most common cause of hearing loss is exposure to loud noises (above 80 db). However, birth defects, infections, and head traumas are also causes of deafness.

Testing Yourself

1. a; 2. c; 3. d; 4. a; 5. b; 6. e; 7. b; 8. a. retina, **b.** choroid, **c.** sclera, **d.** optic nerve, **e.** fovea centralis, **f.** ciliary body, **g.** lens, **h.** iris, **i.** pupil, **j.** cornea; **9. d; 10. c; 11. a.** stapes, **b.** incus, **c.** malleus, **d.** tympanic membrane, **e.** semicircular canals, **f.** cochlea, **g.** auditory tube; **12. d; 13. c; 14. a; 15. a; 16. d; 17. a**

Thinking Critically

1. Without sensory adaptation to constant everyday stimuli, we would constantly be bombarded with and distracted by a huge amount of sensory data (about touch, temperature, pressure, etc.). In contrast, there are large risks for not responding to pain, as shown by the damage that occurs to people who lack normal pain perception. **2.** Once it reaches the retina, light must pass through layers of axons, ganglion cells, and bipolar cells before it reaches the rod or cone cells. From the right side of the left eyeball, nerve impulses travel via the optic nerve, through the optic chiasma, then through the right optic tract, until they synapse with neurons in the thalamus. Axons from thalamic neurons transmit this information to the visual area of the occipital cortex. **3.** The null hypothesis would be that there is no association between the density of taste buds on the tongue and obesity. The prediction based on this hypothesis is that obese people will have the same density of taste buds as normal weight people. **4.** smell, chemoreceptors in olfactory epithelium; sight, rods and cones in the retina; hearing, hair cells in spiral organ; gravity, hair cells in vestibule; motion, hair cells in semicircular canals.

CHAPTER 19

Check Your Progress

p. 359: 1. Compact bone is composed of tubular units and is highly organized, with central canals providing blood, nerves, and lymph. Spongy bone appears at the end of the long bones and is composed of thin plates arranged to form spaces. **2.** Hyaline cartilage is found at the ends of long bones (in joints), in the nose, ribs, larynx, and trachea. The discs between the vertebrae and in the knee are made of fibrocartilage, and elastic cartilage is found in the ear and epiglottis. **3.** Both ligaments and tendons are composed of dense fibrous connective tissue, but ligaments connect bone to bone, while tendons connect muscle to bone. **4.** During bone remodeling, osteoclasts break down bone, and osteoblasts rebuild it. **p. 366: 1.** See Figure 19.3. **2.** True ribs are attached to the sternum by costal cartilage. False ribs connect to the sternum via a common cartilage. Floating ribs do not attach to the sternum. **3.** The knee is a typical hinge joint, the elbow is a pivot joint, and the hip is a ball-and-socket joint. **p. 370: 1.** Smooth, cardiac, and skeletal. **2.** The origin of a muscle is its bone attachment that remains stationary, while the insertion is on the bone that moves. **3.** See Tables 19.1 and 19.2. **4.** Flexion involves movement that decreases the angle between bones; extension is the opposite of flexion; abduction moves a bone away from the body; adduction brings a limb toward the midline; rotation moves the part around an axis. **p. 376: 1.** Three unique components of muscle cells are the T tubules, the myofibrils, and myoglobin. **2.** During muscle contraction, myosin filaments use energy from ATP to pull actin filaments toward the center of the sarcomere. **3.** Neurons control muscle contraction by releasing neurotransmitters at the neuromuscular junction. These chemicals bind to receptors on the sarcolemma, stimulating impulses that spread via the T tubules. **4.** Creatine phosphate breakdown is the fastest way to provide ATP for muscle contraction, but it only supplies enough energy for a few seconds of intense activity. Cellular respiration provides most ATP used by a muscle cell by breaking down glycogen and fatty acids. Fermentation can supply ATP without requiring oxygen. **p. 378: 1.** Three stages of a muscle twitch are the latent, contraction, and relaxation periods. **2.** The index finger requires fine motor control and would have a higher ratio of motor nerve axons per muscle fibers than a muscle in your back. **3.** Domesticated birds no longer migrate by flying long

distances, so they have no need for the endurance associated with slow-twitch fibers. **p. 380: 1.** Untreated bone fractures can heal, especially if the individual is able to prevent the bone from moving, or if it occurs in a bone that does not move a lot (e.g., a rib). However, if two broken ends of a bone continue to move, an exaggerated bone callus usually forms, and the fracture may not heal well. **2.** Astronauts experience bone loss because the bones are not bearing weight. They would be remodeled over time by overactivity of the osteoclasts. **3.** Rheumatoid arthritis is an autoimmune disease in which the immune system attacks and damages the joints. Osteoarthritis is usually secondary to traumatic damage to a joint (either from an injury or over the lifetime of an older person).

Testing Yourself

1. b; 2. d; 3. b; 4. d; 5. a. coxal bones, **b.** patella, **c.** metatarsals, **d.** phalanges, **e.** humerus, **f.** ulna, **g.** radius, **h.** femur, **i.** fibula, **j.** tibia, **k.** tarsals; **6. a; 7. c; 8. e; 9. a.** cross-bridge, **b.** myosin, **c.** actin, **d.** Z line, **e.** H band, **f.** A band, **g.** I band; **10. d; 11. d; 12. a; 13. e; 14. b; 15. c**

Thinking Critically

1. a. With a loss of dystrophin, the force of muscle contraction cannot be conducted efficiently or in a coordinated manner to the connective tissue of the muscle. **b.** With a loss of dystrophin, the muscle is not exercised as much and atrophies over time. **2.** Aerobic respiration is preferable to fermentation because it generates much more ATP per molecule of glucose. Oxygen serves as a final electron acceptor in the electron transport chain present in mitochondria. **3.** Animal tissues contain high levels of amino acids, and the consumption of excess amino acids causes the urine to become more acidic, which may cause increased loss of calcium in the urine. The level of daily exercise may be another important factor, if Chinese people are more active than Americans.

CHAPTER 20

Check Your Progress

p. 388: 1. Hypothalamus, pituitary gland, thyroid gland, adrenal glands, thymus, pancreas. **2.** When an endocrine gland is controlled by negative feedback, it is sensitive to the condition it is regulating or the blood level of the hormone it is producing. **3.** Peptide hormones bind to receptors in the plasma membrane and do not enter the cell. A second messenger inside the cell is needed to initiate an enzyme cascade. **4.** It has been found that women prefer axillary odors from men with different MHC than themselves. **p. 391: 1.** Neurons in the hypothalamus produce hormones that pass through axons into the posterior pituitary where they are stored. The hypothalamus controls the anterior pituitary by producing hormones that travel to the anterior pituitary through a portal system. **2.** See Table 20.1. **3.** The release of oxytocin from the posterior pituitary is controlled by positive feedback in which the amount of oxytocin released increases as the stimulus continues. **4.** Diagnosis is difficult because of the many hormones controlled by the pituitary and the many organs and tissues affected. **p. 392: 1.** Levels of T_3 and T_4 are controlled by negative feedback. When levels of T_3 and T_4 rise, the anterior pituitary stops producing thyroid-stimulating hormone. **2.** Calcitonin decreases blood calcium by increasing calcium deposition in bones. A person with a calcitonin-producing thyroid tumor would be expected to have low blood calcium levels, leading to disrupted nerve conduction, muscle contraction, and blood clotting, as well as increased bone density. **3.** Parathyroid hormone increases the activity of osteoclasts, promotes reabsorption of calcium by the kidneys, causes blood phosphate to decrease, and activates vitamin D, so removal of the parathyroid glands would tend to cause decreased blood calcium and increased blood phosphate. **p. 394: 1.** Secretion of hormones by the adrenal medulla is under direct nervous control. Secretions of the adrenal cortex are controlled either by release of ACTH from the anterior pituitary (for glucocorticoids) or by the renin-angiotensin system (for mineralocorticoids). **2.** Epinephrine from the adrenal medulla helps the body deal with stress by activating the fight-or-flight response (increased heart rate, blood pressure, and energy level). Glucocorticoids from the adrenal cortex raise blood glucose levels and inhibit the body's inflammatory response. **3.** Aldosterone causes the kidneys to absorb sodium and excrete potassium, which generally raises blood pressure. **p. 395: 1.** The exocrine tissue of the pancreas produces and secretes juices that help in digestion in the small intestine. The endocrine cells of the pancreas produce the hormones glucagon, insulin, and somatostatin. **2.** See Figure 20.11. **3.** Somatostatin inhibits the release of growth hormone and suppresses glucagon and insulin secretion by negative feedback mechanism. **p. 398: 1.** Testosterone and estrogen stimulate the development of male and female sex characteristics and the production of gametes. **2.** Some of the worst side effects from taking anabolic steroids are liver dysfunction and cancer, kidney disease, heart damage, stunted growth, infertility, and impotency. **3.** Leptin signals satiety, a sense of not being hungry. **4.** Prostaglandins are not distributed in the blood, but instead act close to where they are produced within cells. **p. 401: 1.** ADH causes more water to be reabsorbed in the kidneys, decreasing urine volume. Inhibition of ADH secretion by alcohol results in increased amounts of urine and frequency of urination. **2.** In iodine deficiency, the thyroid gland cannot synthesize enough T_3 and T_4. The anterior pituitary responds by releasing large amounts of TSH, which stimulates the thyroid gland to enlarge. **3.** In Addison disease, antibodies play a role in destruction of the adrenal cortex by the immune system. In Graves disease, antibodies bind to the TSH receptor, stimulating production of T_3 and T_4 rather than destroying the gland. **4.** Type 1 diabetes is due to an insulin shortage and is treated by the administration of insulin. Type 2 results from an insensitivity to insulin, often specifically due to insufficient numbers of insulin receptors on cells. It is treated with medications, weight loss, and increased exercise.

Testing Yourself

1. a. pineal gland, **b.** hypothalamus, **c.** pituitary gland, **d.** thymus, **e.** pancreas, **f.** adrenal glands, **g.** thyroid gland; **2.** a; **3.** b; **4.** b; **5.** c; **6.** c; **7.** d; **8.** c; **9.** d; **10.** b; **11.** e; **12.** c; **13.** b; **14.** c; **15.** d

Thinking Critically

1. The nervous system allows for quick, almost immediate responses to stress or danger, and fine-tuning of all other bodily functions (plus all the other benefits of higher brain functions), while the endocrine system provides longer-lasting responses. **2.** If Cushing syndrome is caused by a problem (such as cancer) of the adrenal gland itself, ACTH levels would be very low, because of negative feedback action of cortisol on the hypothalamus/pituitary. **3.** Pheromones perceived as odors may affect sexual attraction, aggression, love, or almost any emotion, and we would not necessarily be aware of these effects.

CHAPTER 21
Check Your Progress

p. 409: 1. The seminiferous tubules in the testes produce sperm. The epididymis stores the sperm, and the vasa deferentia carry the sperm to the female via the urethra. The urethra also functions in urination. Interstitial cells, which lie between the seminiferous tubules, secrete androgens, most importantly testosterone. **2.** See Figure 21.3. **3.** GnRH stimulates the anterior pituitary to secrete the two gonadotropic hormones: FSH and LSH. FSH promotes the production of sperm. ICSH (LH hormone in males) controls the production of testosterone by interstitial cells. Testosterone controls the development and function of the testes and associated organs. It also brings about male secondary sex characteristics. **p. 411: 1.** The uterine tubes are severed and tied off. Movement of eggs down the uterine tubes is prevented. **2.** In males, the urethra and external genitalia (the penis) function in both reproduction and urination. In contrast, these two systems are separate in females. **3.** Contractions of the uterus may help to move sperm from the uterus to the uterine tubes. **p. 415: 1.** During the follicular phase, FSH from the anterior pituitary promotes the development of an ovarian follicle. Ovulation signals the end of the follicular phase. The major feature of the luteal phase is secretion of progesterone by the corpus luteum, which causes the uterine lining to thicken in order to be ready to receive a fertilized oocyte. **2.** Estrogen from an ovarian follicle stimulates the proliferative phase of the uterine cycle, during which the endometrium begins to thicken. Progesterone from the corpus luteum leads to the secretory phase of the uterine cycle, in which the endometrium thickens and uterine glands produce thick mucus. **3.** In pregnancy, the placenta produces human chorionic gonadotropin, which maintains progesterone production by the corpus luteum until the placenta can produce progesterone and estrogen, which inhibit the ovarian cycle and maintain the endometrium. See Figure 21.9 for an explanation of the cycle of a nonpregnant woman. During menopause, no eggs develop and the ovarian and uterine cycles cease. **p. 417: 1.** Methods that prevent

ovulation: oral contraception; contraceptive implants; contraceptive injections/patch; vaginal ring. Methods that prevent fertilization: diaphragm; cervical cap; male condom; female condom; coitus interruptus; spermicidal jellies, creams, or foams; natural family planning; vaginal film. **2.** The following methods physically block sperm from entering the uterus: condoms, diaphragm, and cervical cap. **3.** Birth control methods that prevent the contact of male ejaculate with female tissues reduce the transmission of STDs. **4.** Emergency contraception includes methods of birth control that prevent pregnancy after unprotected sex. **p. 425: 1.** STDs caused by bacteria: chlamydia, gonorrhea, syphilis; STDs caused by viruses: genital herpes, HPV, hepatitis. **2.** HIV can destroy CD4$^+$ T cells by directly infecting them, by inducing apoptosis in infected and uninfected cells, and by causing cytotoxic T cells to kill HIV-infected cells. **3.** Reverse transcriptase inhibitors block production of viral DNA from RNA; protease inhibitors prevent cleavage of long viral proteins; fusion inhibitors interfere with HIV entry into cells; integrase inhibitors prevent insertion of viral DNA into host DNA. **4.** Viruses are not affected by antibiotics. They can remain inside cells, protected from antibodies produced by the immune system. **p. 429: 1.** Viagra, Levitra, and Cialis increase the blood flow to the penis during sexual intercourse. **2.** Endometriosis is the presence of endometrial-like tissue outside of the uterine cavity. It can induce inflammation of the uterine lining, which is painful, and can result in infertility. **3.** Ovarian cancer is more frequently fatal than testicular cancer mainly because the ovaries are inside a woman's body, hidden from detection until the cancer has spread to other organs. **4.** Artificial insemination by donor requires the least medical intervention, because sperm is simply collected, concentrated, and placed into the vagina (or uterus) by a physician. In vitro fertilization mixes oocytes and sperm outside the body, and the embryos are transferred to the uterus of the female. Gamete intrafallopian transfer requires more intervention, because the oocytes and sperm are brought together in vitro, and then introduced into one of the woman's uterine tubes. Intracytoplasmic sperm injection requires the most intervention, because a single sperm must be injected into an oocyte.

Testing Yourself

1. a. seminal vesicle, **b.** ejaculatory duct, **c.** prostate gland, **d.** bulbourethral gland, **e.** anus, **f.** vas deferens, **g.** epididymis, **h.** testis, **i.** scrotum, **j.** foreskin, **k.** glans penis, **l.** penis, **m.** urethra, **n.** vas deferens, **o.** urinary bladder. **2.** c; **3.** d; **4. a.** uterine tube, **b.** ovary, **c.** uterus, **d.** urethra, **e.** clitoris, **f.** anus, **g.** vagina, **h.** cervix; **5.** b; **6.** b; **7.** c; **8.** b; **9.** c; **10.** d; **11.** e; **12.** c; **13.** c; **14.** b; **15.** a; **16.** d

Thinking Critically

1. Current exposure levels to chemicals that potentially could reduce sperm production in males could be measured and compared to previous levels. **2.** The main advantage of a monthly menstrual cycle is that the human female ovulates every month, which increases the number of opportunities for becoming pregnant. **3.** Although BPH may cause frequent or difficult urination, it usually does not block urination completely. Untreated prostate cancer, however, often spreads to other tissues, such as the bone marrow, where the cancer cells may interfere with normal functions. **4.** A potential mother who has low body fat is likely less well nourished and less able to nurse or otherwise provide for her offspring successfully.

CHAPTER 22
Check Your Progress

p. 437: 1. See Figure 22.1. **2.** The oocyte's membrane depolarizes and lifts off of the oocyte, preventing any further sperm from binding. Enzymes released by the cortical granules lead to the formation of the fertilization membrane, which results in the lifting of the zona pellucida from the surface of the oocyte. This prevents other sperm from binding to the zona pellucida. **3.** Cellular: All three lead to cleavage resulting in the formation of a multicellular embryo and the formation of a blastula. In the lancelet, the cleavage produces a ball of cells of equal size. In the frog, the cells are not of equal size, and in the chicken, cleavage is restricted to a layer of cells over the yolk. Tissue: All develop three germ layers: ectoderm, mesoderm, and endoderm. In the lancelet, mesoderm forms by outpocketing of the archenteron, while in the frog and chicken, mesoderm forms by migration of cells between the ectoderm and endoderm. **4.** See Table 22.1. **p. 441: 1.** Cellular differentiation results in cells becoming specialized in structure and function due to differential gene

expression. Morphogenesis produces the shape and form of the body first through the migration of cell types to produce the germ layers. **2.** A morphogen is the product of a morphogen gene and is present usually as a gradient within a developing organism. The gradients determine the shape of the organism. **3.** The homeobox sequence in the homeotic gene codes for a homeodomain composed of 60 amino acids. The homeodomain protein binds to DNA and controls gene expression. **p. 447: 1.** Umbilical blood vessels from the allantois; first blood cells from the yolk sac; fetal half of the placenta from the chorion. **2.** Months 1 and 2: week 1—fertilization, cleavage, blastocyst formation; week 2—implantation, yolk sac and amnion form, gastrulation occurs; week 3—the nervous system appears, development of the heart begins; weeks 4–5—development of the chorion and allantois, umbilical cord forms, limb buds appear, head enlarges, developing eyes, ears, and nose; weeks 6–8—head assumes normal position relative to body, all organ systems are developed. Months 3 and 4: cartilage replaces bone, head growth slows, eyelashes, eyebrows, head hair, fingernails, and nipples appear. Months 5 through 7: lanugo covered by vernix caseosa appears, eyelids are open. **3.** See Figure 22.15. The unique fetal structures are foramen ovale, ductus arteriosus, umbilical arteries, umbilical vein, and ductus venosus. **4.** The placenta is a structure of the developing embryo that is attached to the uterine wall. It permits the exchange of gases, nutrients, and wastes between the maternal and embryo circulatory systems. **p. 450: 1.** Progesterone causes smooth muscle to relax and arteries to expand, leading to low blood pressure. The renin-angiotensin-aldosterone mechanism leads to sodium and water retention. Blood volume, number of RBCs, and cardiac output increase. Blood flow to the kidneys, placenta, skin, and breasts rises. **2.** Stage 2. **3.** Colostrum is a thin, yellow, milky fluid, rich in proteins including antibodies, that is produced by breasts during the first two days after birth, before milk production begins. Benefits of breast-feeding include the transfer of antibodies from the mother to child, a lower rate of gastrointestinal diseases in the child, and post-pregnancy physiological benefits for the mother. **p. 454: 1.** Telomeres have a genetic influence on aging. The number of times a cell divides is controlled by the length of telomeres. The older the cell, the shorter the telomere. **2.** A high-calorie diet results in the generation of more free radicals, which can damage DNA and lead to cellular aging. Low-calorie diets have been associated with longer life spans. Also, the consumption of foods high in antioxidants counteracts the effects of free radicals and likely slows aging. **3.** atherosclerosis and elevated blood pressure, cardiovascular system; increased susceptibility to infections, immune system; pneumonia, respiratory system; enlarged prostate, excretory system; presbyopia and cataracts, sensory systems; osteoporosis, musculoskeletal system; insulin insensitivity, depression due to low testosterone, and increased heart disease in postmenopausal women, reproductive systems.

Testing Yourself

1. c; **2.** c; **3.** d; **4.** d; **5.** a; **6.** c; **7. a.** ovulation, **b.** fertilization, **c.** cleavage, **d.** morula formation, **e.** early blastocyst formation, **e.** implantation; **8.** e; **9.** e; **10.** e; **11.** d; **12.** c; **13.** e

Thinking Critically

1. The egg is what contributes the cytoplasm and the organelles to the developing embryo. Therefore, only the mother's mitochondria are present in the embryo. **2.** HCG is produced by the placenta and is excreted by the kidneys. **3.** The presence of FSH, which promotes the development of a follicle in the ovary, and a surge in LH, which causes ovulation, are measured in these tests. The tests are based on color-coded antibodies to the hormones.

CHAPTER 23
Check Your Progress

p. 465: 1. The phenotype of an individual is its appearance, which is determined by the genes it carries for traits carried in its genome. The genes present in the genome comprise its genotype. **2.** A testcross is conducted to determine if an individual that is expressing a dominant allele is heterozygous or homozygous for that gene. **3.** The law of segregation states that each individual has two factors for each trait and these factors separate during gamete formation such that each gamete receives only one factor. At fertilization, the zygote receives two factors for each trait. The law of independent assortment states that during gamete formation, each pair of factors separates independently of other factors and that all possible combinations of factors can occur. **4.** A phenotypic ratio of 3:1 results from a one-trait cross and a ratio of 9:3:3:1 results from a two-trait cross. **p. 468: 1.** See Figures 23.9 and 23.11. **2.** Autosomal recessive disorders result from the inheritance of two recessive alleles. The child is affected and is homozygous recessive, but neither parent is because they are heterozygous. In autosomal dominant disorders, both parents are affected but the child can be unaffected. **p. 471: 1.** With incomplete dominance, the heterozygote exhibits a phenotype that is intermediate between the homozygous recessive and the homozygous dominant. **2.** The mother has to be heterozygous for type A blood, and the second allele for type O blood would have come from the father. He could have been heterozygous for type A, type B, or type O blood. **p. 472: 1.** Environmental factors include nutritional status, diet, temperature, and exposure to sunlight. **2.** Identical twins have identical genomes and have developed from identical oocytes containing the same cytoplasmic composition. However, the expression of their genes is influenced by environmental inputs. Variation between identical twins is attributed to the inputs from the environment on their development.

Testing Yourself

1. d; **2.** e; **3.** b; **4.** a; **5.** a; **6.** b; **7.** b; **8.** b; **9.** d; **10.** d; **11.** d; **12.** d

Thinking Critically

1. Cross the fruit fly with the trait to a true breeding fruit fly without the trait and determine the phenotypic ratios. If the trait does not appear in the offspring or appears 25% of the time, then it is recessive. If the trait appears 75% of the time, then it is dominant. **2.** Track the appearance of that trait in identical twins raised in different environments. **3.** Even identical twins have differences in environmental influences. Beginning in the womb, each twin experiences their environment in different ways.

CHAPTER 24
Check Your Progress

p. 477: 1. Linked alleles are all on one chromosome and they tend to be inherited all together, thus they do not exhibit independent assortment. **2.** Two genes that are ten map units apart exhibit a 10% chance of crossing-over. **p. 480: 1.** Sex-linked alleles appear on the X chromosome and more rarely on the Y chromosome. Their inheritance is dependent on the distribution of the X and Y chromosomes in males and females. Recessive alleles on the X chromosome are expressed in males, while they are not in heterozygous females. **2.** 50% of female offspring and 50% of male offspring would express the recessive allele. **3.** Color blindness: A number of different genetic mechanisms can contribute to X-linked color blindness. Duchenne muscular dystrophy: caused by the absence of the protein dystrophin which leads to calcium entering the cell, which promotes an enzyme that dissolves muscle fibers. Fragile X syndrome: an abnormally high number of repeat sequences of CGG on the X chromosome. Hemophilia: either due to the absence or low level of clotting factor VIII or factor IX, resulting in a reduced ability of the blood to clot. **p. 484: 1.** In normal XX females, one of the X chromosomes becomes inactive and forms a Barr body, which indicates that only one functional X chromosome is needed in females, just like in males. Extra X chromosomes all become Barr bodies. In males, extra X chromosomes also become Barr bodies. Extra Y chromosomes do not seem to interfere with normal development, perhaps because there are so few genes on the Y chromosome. **2.** Turner syndrome and Kleinefelter syndrome can be caused by nondisjunction during oogenesis at meiosis I or at meiosis II. **p. 486: 1.** When chromosomes break and do not rejoin, the deletion from one chromosome can be added to the corresponding homologous chromosome, creating a duplication. **2.** If in the process of translocation the allele of interest remains whole, then it is possible there will be no loss of function. If, after an inversion, the alignment of the two homologous chromosomes can be achieved by forming a loop, it is possible the individual can be phenotypically normal.

Testing Yourself

1. c; **2.** c; **3.** b; **4.** a; **5.** b; **6.** d; **7.** c; **8.** a; **9.** e; **10.** d; **11.** c; **12.** a; **13.** b

Thinking Critically

1. Mosaicism in men with Klinefelter syndrome can happen during cell division after conception. A person with mosaicism might have tissues with two different phenotypes,

depending on whether the cell contains XY or XXY chromosomes, and which of the X chromosomes is inactivated in the XXY cells. Parents will need to prepare for the necessary added responsibilities of a child with a disability. They may need to arrange the finances for additional caregivers in the home and extensive medical care. **2.** Chromosome 1 is so large and carries so many genes that three copies are lethal. **3.** As long as the translocation is balanced, meaning that the entire amount of DNA is still present, and no alleles are disrupted, the person can express all of the genes normally and be phenotypically normal. However, when the translocated chromosomes go through meiosis and crossing-over, gametes without the entire amount of DNA present will be produced. This would result in infertility. **4.** The father must be color blind unless there is a genetic abnormality in the daughter (i.e., she has only one X chromosome). The husband is not the father.

CHAPTER 25
Check Your Progress

p. 492: 1. See Figure 25.1. Griffith concluded that some substance was passed from the dead S strain bacteria to the living R strain, and this substance had the capacity to transform the R strain. This was indicated to be genetic material. Later, Avery and others identified the genetic material as DNA. **2.** See Figure 25.2. **3.** See Figure 25.3. **p. 493: 1.** A new double strand of DNA is composed of one old strand and one newly synthesized strand. **2.** See Figures 25.4 and 25.5. **p. 499: 1.** mRNA carries the genetic information encoded in DNA from the nucleus to the ribosomes in the cytoplasm. At the ribosome composed of rRNA and proteins, mRNA assembles amino acids carried by tRNA into chains, which are eventually released to form functional proteins. **2.** The information stored in DNA is transcribed into mRNA which then, with the help of tRNA at the ribosomes made of rRNA, translates the information into protein structure. **3.** The genetic code contains more than one triplet codon for each amino acid. **p. 503: 1.** Different cells are subject to different controls of gene expression, such as X-inactivation and pretranscriptional, transcriptional, posttranscriptional, translational, and posttranslational control. **2.** In the absence of lactose, a repressor is bound to the promoter/operator complex and no lactase is produced. Lactose will bind to the repressor and prevent it from binding to the operator. The gene for lactase can then be expressed. **3.** See Figure 25.18. **p. 506: 1.** Gene mutations can happen when DNA replication produces errors, when a mutagen causes physical damage to DNA, and when transposons jump into a new location, disrupting the neighboring genes. **2.** A point mutation occurs when one DNA nucleotide is changed. A frameshift mutation occurs when a nucleotide is inserted or deleted. **3.** Cancer cells have high levels of mutagenesis and disregulated cell cycles. They avoid apoptosis and can metastasize.

Testing Yourself

1. a; **2.** b; **3.** c; **4.** c; **5.** a; **6.** c; **7.** b; **8.** c; **9.** a; **10.** c; **11.** c; **12.** a; **13.** e

Thinking Critically

1. Housekeeping genes are those that all cells require, such as those that encode for rRNA, genes encoding enzymes needed for energy metabolism, and genes for membrane transport proteins. **2.** Cancer cells express telomerase, allowing the cells that have damaged DNA to continue to divide. The cell loses the mechanism for cell death. **3.** In cases where a copy of a gene inherited from a parent is defective, the second gene inherited from the other parent must be altered before cancer develops. Multiple environmental exposures and habits can influence whether cancer develops or not.

CHAPTER 26
Check Your Progress

p. 514: 1. Using a restriction enzyme to cut open plasmid or other DNA, followed by use of DNA ligase to seal the foreign gene into the plasmid. **2.** See Figure 26.2. **3.** Each person has a unique number of short tandem repeats (STRs) at multiple loci. Separating these STRs by gel electrophoresis or by the technique of STR profiling allows the identification of each individual. **p. 518: 1.** Improved agriculture, production of pharmaceuticals, production of hormones, cleanup of environmental pollution. **2.** A cloned organism has the identical genetic makeup. Transgenic organisms carry a gene that has been inserted from another organism. **p. 520: 1.** In vivo and

ex vivo methods using viral vectors are used to infect cells and introduce corrected genes. They also infect people with viruses that help fight off cancer. **2.** RNA interference silences an existing gene to prevent gene expression; in vivo gene therapy provides an additional copy of a gene to enhance gene expression. **3.** An example of in vivo gene therapy is use of an adenovirus vector as an inhalant to treat cystic fibrosis patients. An example of ex vivo gene therapy is removal of dysfunctional genes from an SCID patient, using a virus to "inject" the correct gene into these cells, and re-inoculating the patient with the corrected cells. **p. 523: 1.** Genomics is the study of the complete genomic sequences of organisms. Proteomics studies the proteins produced from the genome within a cell. **2.** Comparative genomics can allow us to better understand gene function because unknown function in a focal organism that is close in genetic sequence to known genes in related organisms can often have a similar function. **3.** Bioinformatics, which involves computer technologies, special software, and statistics, is used to study the enormous amount of information obtained from genomics and proteomics.

Testing Yourself

1. e. **2.** c; **3.** c; **4.** d; **5.** d; **6.** c; **7.** b; **8.** c; **9.** c; **10.** e

Thinking Critically

1. Pros of viral gene therapy include treating or correcting potentially fatal or severely debilitating genetic disorders such as SCID. Gene therapy can also be used to treat difficult cancers. Cons include safety of using viruses and the fact that many patients have developed leukemia as a result of gene therapy. Many consequences of gene therapy are unknown. **2.** Genes for basic function, such as DNA replication, protein synthesis, and basic cellular processes such as respiration, should be similar.

CHAPTER 27
Check Your Progress

p. 529: 1. The four elements of Darwin's theory of natural selection are: (1) Individuals within a species exhibit genetically inheritable variations. (2) Organisms compete for limited resources. (3) Individual organisms have different reproductive success. (4) Organisms adapt to their environment over generations through natural selection. **2.** Changes to an organism's phenotype acquired during its lifetime are not heritable and therefore are not passed on to the next generation. **3.** Adaptations are evolved traits that result in an organism being better suited to its environment. **p. 536: 1.** Fossils found in the strata of sedimentary rocks of different geological eras, periods, and epochs show evolutionary links between modern organisms and ancient organisms. The range and distribution of plants and animals is studied in the field of biogeography and this information helps to understand how different communities begin whenever landmasses or water environments are separated. **2.** Breeders use artificial selection to increase the frequency of desired traits in a breed of dogs. The breed is evolving over time due to this. **3.** Chimpanzees, because they are more closely related evolutionarily. **p. 538: 1.** There must be no mutations, selection, genetic drift, or gene flow. The population must also have random mating. **2.** Allele frequencies will change if the Hardy-Weinberg conditions are not met. As an example, directional selection will decrease the frequency of the allele that is not favored. **p. 545: 1.** Mutations create genetic variation in a population. Genetic drift is the change in allelic frequency due to the random meeting of gametes during fertilization. Small populations are particularly susceptible to this process. Gene flow is the movement of alleles between populations and results in increased genetic diversity. Nonrandom mating, in which individuals select a mate based on desired traits, results in increased allelic frequencies of those traits. Natural selection results in individuals who are better adapted to the environment having more offspring than those less well-adapted. **2.** With natural selection, there must be variation in phenotypic traits. This variation must be heritable. There must be over-reproduction and a struggle for survival due to limited resources. There must be differential reproductive success, where some individuals produce more offspring than others. **3.** The parasite that causes malaria cannot live within the red blood cells of the heterozygote because, upon infection, the cell sickles and loses K+, resulting in the death of the parasite. Carrying the trait costs the individual. They are less fit, and in areas of the world without the malaria parasite, this would represent a disadvantage. **p. 547: 1.** See Figure 27.20. **2.** Punctuated equilibrium. **p. 550: 1.** Taxonomy involves identifying, naming, and classifying organisms into taxa. Phylogenetics studies the evolutionary relatedness of organisms and classifies organisms into clades that reflect their evolutionary descent. **2.** A phylogenetic tree is an estimation and visual representation of the evolutionary history of biodiversity. A cladogram is composed of branches called clades. Each clade contains the most recent ancestor and its descendants who share a derived trait. **3.** Domain Bacteria, domain Archaea, domain Eukarya.

Testing Yourself

1. b; **2.** b; **3.** b; **4.** c; **5.** e; **6.** e; **7.** d; **8.** c; **9.** f; **10.** b; **11.** e; **12.** b; **13.** b; **14.** b; **15.** b; **16.** b

Thinking Critically

1. To solve this, $q^2 = 0.0064$, therefore $q = 0.08$ and $p = 0.92$. $2pq = 2(0.08)(0.92) = 0.147 = 14.7\%$. **2.** HIV is hard to treat because it evolves resistance to drugs used to treat it very quickly. HIV's mutation rates are so high that millions of new strains evolve daily, some of which are resistant to drugs that an infected patient may be being treated with. Eventually, these strains will increase in frequency within a patient so much that a drug is no longer effective. The somewhat good news is that this process often takes years, and with proper and early treatment, HIV-infected patients can live an additional 15 to 20 years or more before the virus becomes fatal. **3.** The disease may be in higher allele frequency in South America because there is balancing selection. That is, heterozygote carriers of the disease may be resistant to some pathogen that is present in South America but not in North America. **4.** 16%. **5.** 10%.

CHAPTER 28
Check Your Progress

p. 555: 1. Pasteur demonstrated that sterile broth enclosed in a flask that was not open to the air did not spontaneously generate microbes. **2.** Microbes include bacteria, archaea, protists, fungi, and viruses. **3.** Your microbiota are all the microbes, the beneficial ones and the harmful ones, that live on or within you. **p. 559: 1.** The chemical evolution of biomolecules from inorganic compounds was the first step in biological evolution. **2.** Biomolecules, such as amino acids and nucleotides, had to be generated before polymers of these (DNA, RNA, proteins) could be constructed. Once polymers appeared, protobionts could evolve, leading eventually to the appearance of living cells. **3.** The "primordial soup" hypothesis explains the evolution of organic molecules from inorganic molecules by simulating the atmospheric environment containing CH_4, NH_3, H_2, and H_2O exposed to heat and electrical sparks. The "iron-sulfur world" hypothesis explains the appearance of organic molecules at thermal vents at the bottom of the oceans. Ammonia, CO, and hydrogen sulfide pass over iron and nickel sulfide minerals, which act as catalysts to form organic molecules. **4.** Which came first, the proteins or the genetic material? In the protein-first hypothesis, polypeptides formed and some of them had enzymatic activity, which eventually led to DNA formation. In the RNA-first hypothesis, RNA appeared first and could act as a substrate and an enzyme. DNA and proteins appeared later. Another hypothesis is that polypeptides and RNA evolved simultaneously, thus eliminating the paradox. **p. 561: 1.** Archaea and Bacteria lack a nucleus and membrane-bound organelles in the cytoplasm. Archaea shares features with Eukarya listed in Table 28.1. **2.** The plasma membranes of Archaea have a monolayer of lipids with branched side chains, compared to the lipid bilayer of bacterial and eukaryotic membranes. This contributes to the ability of many archaea to grow at very high temperatures. **3.** In addition to the membrane structural variations, archaeal enzymes must be able to function at high temperatures and/or salt concentrations that would inactivate the enzymes of most other organisms. **4.** Archaea living in the digestive tracts of cattle produce methane that contributes to global warming. Giving up beef in your diet would help to reduce this cause of global warming. **p. 566: 1.** The three basic bacterial shapes are bacillus (rod), coccus (spherical), and spirillum (spiral-shaped). **2.** In bacterial conjugation, actual physical contact (via a sex pilus) is required for two bacteria to share DNA. Transformation is the uptake of DNA from the environment by a bacterial cell, and in transduction DNA is transferred from one bacterium to another by viruses. **3.** Heterotrophic bacteria use organic compounds from their environment for energy. Chemoautotrophs reduce CO_2 to an organic compound using electrons derived from ammonia, hydrogen gas, and hydrogen sulfide, and sometimes minerals

such as iron. **4.** Antibiotics work by interfering with normal bacterial metabolism such as protein synthesis and cell wall synthesis. Bacteria can become resistant to antibiotics if a mutation occurs that alters the target of the drug (often an enzyme). Alternatively, some bacteria may acquire the ability to alter the antibiotic itself, so the drug is destroyed or inactivated. **p. 570: 1.** A typical enveloped virus has a protein capsid that contains its nucleic acid (DNA or RNA) and is surrounded by a lipid envelope in which protein spikes are embedded. **2.** Attachment: viral spikes bind to receptor molecules on host cell. Entry: viral envelope fuses with plasma membrane and capsid, and then viral genome enters cell. Replication: viral enzyme makes copies of its genome. Biosynthesis: more capsid and spike proteins are made on host ribosomes. Assembly: capsid forms around viral genome. Budding: new viruses containing some host plasma membrane and spikes are released from cell. **3.** Vaccines are available for measles, mumps, and rubella. Vaccines are not available for HIV, hepatitis C, and mononucleosis. **4.** Viruses, viroids, and prions are all acellular pathogens. Viroids' genomes are a single strand of circular RNA that is 1/10 the size of a viral genome. It codes for no proteins. Viroids cause disease in plants and not animals. Prions are proteinaceous molecules that are derived from normal proteins that have changed shape. They lead to neurological wasting diseases in animals, including humans.

Testing Yourself

1. a; **2.** d; **3.** a; **4.** d; **5.** d; **6.** d; **7.** b; **8.** d; **9.** b; **10.** b; **11.** a; **12.** b; **13.** b

Thinking Critically

1. Three reasons why bacteria are useful as a model genetic organism include: fast generation time, small size, and ease of generating genetic mutations. **2.** Viruses that use RNA as their genetic material would be more likely to contain their own enzyme than those that use DNA, because DNA viruses can use the enzymes that their host cells normally use to copy their cellular DNA. Because host cells do not have enzymes to make copies of RNA, RNA viruses must produce their own RNA-replicating enzymes. **3.** Antiviral drugs could interfere with any phase of viral replication including attachment, entry, replication, biosynthesis, assembly, and budding. For instance, the drug might mimic a molecule that normally binds to receptor proteins on the membrane and prevent the virus from binding. **4.** Cell wall synthesis; endospore production; mechanisms of protein synthesis unique to prokaryotes.

CHAPTER 29
Check Your Progress

p. 584: 1. The endosymbiotic theory proposes that mitochondria were the result of a nucleated cell engulfing an aerobic bacterial cell. Under the theory, chloroplasts originated when a nucleated cell with a mitochondrion engulfed a cyanobacteria. **2.** Protists carry out sexual reproduction when experiencing unfavorable conditions. This often results in cysts or spores which can survive in the environment until conditions improve, thus promoting dispersion to environments better suited for growth. **3.** Algae are typically photosynthetic, converting CO_2 into carbohydrates which are used as a food source. Protozoa are heterotrophs. Water molds and cellular slime molds are saprotrophs that feed on dead plant material. Cellular slime molds can also eat bacteria. **4.** Plasmodium causes malaria. *Trypanosoma* causes African sleeping sickness. Members of the genus *Entamoeba* cause amoebic dysentery. **p. 591: 1.** Fungi secrete enzymes and digest available food externally. The broken down food is then absorbed. Algae are primarily photosynthetic and require carbon dioxide, water, and a source of energy. Amoeboids surround algae, bacteria, or other protists, which they use as food, with their pseudopods, and digest the food in a food vacuole. **2.** Hyphae—individual filaments of a fungus. Mycelium—mass of hyphae composing the body of a fungus. Chitin—a component of the fungal cell wall. **3.** See Figure 29.20 for zygospore fungi reproduction; Figure 29.21 for sac fungi; Figure 29.22 for yeast; Figure 29.23 for club fungi. **4.** Superficial mycoses include athlete's foot and ringworm. Systemic mycoses include histoplasmosis and coccidioidomycosis.

Testing Yourself

1. a; **2.** b; **3.** c; **4.** b; **5.** a; **6.** d; **7.** b; **8.** d; **9.** b; **10.** c; **11.** c; **12.** b; **13.** e; **14.** a; **15.** d

Thinking Critically

1. The *Plasmodium* parasites complete part of their life cycle inside red blood cells. Two basic mechanisms have been proposed to explain malaria resistance in people who carry the sickle-cell gene: (1) the abnormal hemoglobin inhibits the parasite from completing some part of its life cycle, or (2) infected RBCs from sickle-cell carriers are more susceptible to being destroyed by the host immune system than infected RBCs from normal individuals. If an effective malaria vaccine becomes widely implemented, the natural resistance to malaria provided by the sickle-cell gene would become a less significant factor. Over time, the sickle-cell gene would become less prevalent in the population as the advantage of having only one copy of the gene diminished. **2.** A chemical to prevent termite infestations could be targeted at killing the bacteria like an antibiotic works, or it could prevent the bacteria from producing the enzyme, or it could interfere with the action of the enzyme. **3.** The outer layer of skin is composed of dead epithelial cells which are constantly being shed. Human skin is often exposed to soap and water. In order to attach and grow on the skin, the fungi need to be able to penetrate to the lower layers of the dermis and be able to withstand the usual methods of washing. The lungs are more hospitable to fungi, due to the lining being thin, moist, and protected.

CHAPTER 30
Check Your Progress

p. 598: 1. Protection of the embryo from desiccation developed in mosses. Vascular tissue appeared in the lycophytes. Ferns evolved megaphylls with their large surface to capture light energy. Seeds first appeared in the gymnosperms. Angiosperms have flowers and fruits. **2.** The sporophyte produces haploid spores by meiosis. Spores are reproductive cells that can produce gametophytes by mitosis. The gametophytes then produce gametes by mitosis that can fuse to form the diploid zygote, which develops into the sporophyte by mitosis. **p. 600: 1.** Nonvascular plants are limited by not having vascular tissue to transport water and nutrients. They are restricted to moist environments to grow because the gametophyte requires a film of water for the flagellated sperm to swim in to reach the egg. **2.** Nonvascular plants have developed rootlike, stemlike, and leaflike structures that help them to live on land. Because they are small and simple, they can inhabit microhabitats that other plants cannot. **3.** A liverwort has a flat-lobed thallus, whereas a moss has leafy green shoots. **p. 604: 1.** Lycophytes were the first plants to have vascular tissue and true roots, stems, and leaves. **2.** Microphylls are small leaves with a single vein containing xylem and phloem. Megaphylls are larger and their vascular tissue is branched. **3.** See Figure 30.10. **p. 610: 1.** Microspores develop in pollen sacs and become pollen grains (male gametophytes). Megaspores develop in the ovule and one out of four becomes an embryo sac (female gametophyte). **2.** Double fertilization in flowering plants results in the presence of endosperm, in addition to a zygote. Presence of an ovary leads to the production of seeds enclosed by a fruit. Animals are often used as pollinators. **3.** See Table 30.1. Monocots have one cotyledon (seed leaves that nourish the embryo) and eudicots have two.

Testing Yourself

1. a; **2.** c; **3. a.** sporophyte, **b.** meiosis, **c.** gametophyte, **d.** fertilization; **4.** c; **5.** a; **6.** e; **7.** e; **8.** b; **9.** b; **10.** b; **11.** c; **12. a.** stamen, **b.** carpel, **c.** receptacle, **d.** petals, **e.** sepals

Thinking Critically

1. Desiccation is overcome with the presence of complex tissues and organs, a waxy cuticle, and closing stomata at night or during water stress. Movement of fluids is accomplished with specialized vascular tissue. Genetic variability due to the alternation-of-generations life cycle can decrease deleterious effects of mutations caused by UV light. **2.** Angiosperms have protected seeds that can lay dormant before germination. They also have double fertilization, which provides endosperm as nourishment for the embryo. Angiosperms have flowers, which attract animal pollinators and allow for dispersal of pollen and fertilization of other plants. Finally, angiosperms have fruits, which are also attractive to animals and can allow long-distance dispersal of seeds to new habitats. **3.** Humans do not reach reproductive maturity until puberty. All cells, except eggs and sperm, are diploid. Haploid eggs or sperm are produced by meiosis in ovaries or testes. Fertilization occurs when a sperm cell meets an egg cell, forming a diploid zygote which grows into a fetus. In contrast, plants have an alternation between a diploid sporophyte generation and a haploid gametophyte generation. The sporophyte produces haploid reproductive spores by meiosis that can develop into a gametophyte by mitosis. The gametophyte produces gametes by mitosis. Sperm and egg fuse, forming a diploid zygote that undergoes mitosis and becomes the sporophyte.

CHAPTER 31
Check Your Progress

p. 617: 1. Unlike plants, animals are heterotrophs and must digest their food internally. Fungi secrete enzymes to break down food externally and then absorb the nutrients. **2.** Five characteristics animals have in common are: 1. movement or locomotion via muscle fibers; 2. multicellularity; 3. typically diploid adults; 4. usually undergo sexual reproduction and experience developmental stages; 5. usually heterotrophic. **3.** A mollusc is multicellular, has three tissue layers and a body cavity, undergoes protostome development, and has or had a trocophore larva. **p. 621: 1.** Sponges are multicellular, sessile filter feeders. They have an outer layer of flattened epidermal cells, a middle semifluid matrix with amoeboid cells, and an inner layer composed of collar cells. They are capable of reassembling from single cells into a complete organism. **2.** Cnidarians have true tissues and have an ectoderm and endoderm as embryonic germ layers. They exhibit radial symmetry as adults and demonstrate two body forms: a polyp and a medusa. See Figure 31.7. **p. 629: 1.** Flatworms, molluscs, and annelids all are protostomes that are bilaterally symmetrical, have three tissue layers, and are multicellular. **2.** Flatworms have no body cavity, an incomplete digestive tract, and no circulatory system. Molluscs have a true coelom, a complete digestive tract, and an open circulatory system. Annelids have a true coelom, a complete digestive tract, and a closed circulatory system. **3.** Molluscs have a visceral mass, a strong muscular foot, and a mantle. **p. 636: 1.** Both roundworms and arthropods periodically shed their outer covering, and they both have a complete digestive tract. **2.** Rigid, but jointed exoskeleton, segmentation, well-developed nervous system, wide variety of respiratory organs, and reduced competition through metamorphosis. **3.** Crustaceans have hard, calcified exoskeletons and typically have heads with compound eyes, and five pairs of appendages. **p. 638: 1.** Echinoderms and chordates are closely related because they are both deuterostomes. They share similar embryological development and the coelom forms from an outpocketing of the primitive gut. **2.** Echinoderms have endoskeletons composed of calcium, containing plates which have spines that protrude through their skin. They are most often radially symmetrical and have free-swimming larvae.

Testing Yourself

1. d; **2.** e; **3.** b; **4.** a; **5.** c; **6.** b; **7.** c; **8.** d; **9.** e; **10.** d; **11.** d; **12.** c; **13. a.** head, **b.** antenna, **c.** simple eye, **d.** compound eye, **e.** thorax, **f.** tympanum, **g.** abdomen, **h.** forewing, **i.** hindwing, **j.** ovipositor, **k.** spiracles; **14.** e; **15.** d; **16.** e

Thinking Critically

1. Cnidarians are aquatic and usually have a sessile polyp and motile medusa life stage. Cnidarians can reproduce both asexually and sexually. Bilaterally symmetrical species are typically motile and most often reproduce sexually. **2.** Multicellularity allows for specialization of cells, development of tissues, and the formation of organs. Separate functions are assumed by each, which increases specificity and efficiency of function. **3.** A parasite gains some benefit from the host such as nutrients, housing, or protection. In return, a parasite must compromise its structure, functions, or behaviors to comply with the host's body.

CHAPTER 32
Check Your Progress

p. 644: 1. Chordates have a notochord, nerve cord, pharyngeal pouches, and a postanal tail. **2.** Lancelets are shaped like a two-edged knife, while tunicates look like a sac with thick walls. **3.** Mammals have a bony skeleton, lungs, four limbs, an amniotic egg, and mammary glands. **p. 649: 1.** Fish are adapted to living in water, undergo external fertilization, and have a zygote that is a swimming larva. **2.** Amphibians have jointed appendages, four limbs, eyelids, ears, and a larynx.

3. The amphibian larval stage requires a water environment for survival. Metamorphosis of the larvae to the tetrapod adult that can breathe through lungs and its skin allows survival on land. **p. 653: 1.** Reptiles compared to mammals are covered by keratinized scales, do not have a diaphragm, and have an interventricular septum that varies as to its completeness. **2.** Placental mammals are unique because the extraembryonic membranes have been modified for internal development within the uterus. **3.** Primates are adapted to an arboreal life, have mobile limbs, and the hands and feet have five digits each. **p. 658: 1.** Bipedalism may have evolved because a dramatic change in climate caused forests to be replaced by grassland about 4 MYA. It is thought that hominids began to stand upright to keep the sun off their backs and to have better vision for detecting predators or prey in grasslands. However, recent fossil findings suggest bipedalism might have evolved as early as 7 MYA, while hominids still lived in forests. The first hominids may have walked upright on large branches as they collected fruit from overhead. The upright stance later made it easier for traveling on the ground and foraging among bushes. **2.** Bipedalism. **3.** Early species of *Homo* had a smaller brain size, had humanlike teeth, and could use tools. **p. 659: 1.** The replacement model, or out-of-Africa hypothesis, proposes that modern humans evolved from archaic humans only in Africa, and then modern humans migrated to Asia and Europe, where they replaced the archaic species about 100,000 years ago. In contrast, the multiregional continuity hypothesis proposes that modern humans arose from archaic humans in Africa, Asia, and Europe at roughly the same time. **2.** Neandertals lived between 200,000 and 28,000 years ago and were likely supplanted by modern humans (Cro-Magnons at the time). Neandertal brains were slightly larger (1,400 cc). They had massive brow ridges, a forward-sloping forehead, a receding lower jaw, and wide, flat noses. The bones of Neandertals were shorter and thicker than those of Cro-Magnons. Cro-Magnons had lighter bones, flat high foreheads, domed skulls housing brains of 1,350 cc, small teeth, and a distinct chin.

Testing Yourself

1. e; **2. a.** pharyngeal pouch, **b.** dorsal tubular nerve cord, **c.** notochord, **d.** postanal tail; **3.** c; **4.** c; **5.** c; **6.** d; **7.** d; **8.** a; **9.** e; **10.** a; **11.** b; **12.** c; **13.** a; **14.** c; **15.** d

Thinking Critically

1. Because we have a distant common ancestry with all vertebrates, we share many genes in common with them. Continued study of processes such as regeneration or immune response in amphibians could give us clues as to which human genes may be similar and potentially "turned on" or stimulated in such a way to benefit people by curing diseases or regenerating limbs. In general, genomic studies benefit humans because we are better off understanding genes associated with developmental processes and disease resistance. **2.** Bipedalism must have provided more of a fitness advantage than the cost of a smaller pelvis for birthing. Upright stance may have given hominids advantages when forests shifted to grasslands about 4 MYA. These advantages included keeping the sun off their backs, as well as better vision to detect predators and prey. Thus, being better able to avoid starvation and/or being eaten outweigh the costs of a more difficult birthing process. **3.** Cultural evolution encompasses human behavior and products, including technology. Language is also critical for cultural evolution because products and ideas are passed on from generation to generation, not by genetic inheritance but by verbal transmission. We learn about history from others, which plants are good for us and which ones are poisonous, and most recently, about medicines that can save our lives. In the past 100 years, the human life span has increased by more than 50% due to technology and culture. **4.** Genes that control insulating body fat, oxygen usage and metabolic rate, shorter stature, and a more boxlike body form.

CHAPTER 33
Check Your Progress

p. 666: 1. Identical twins that are separated at birth and exposed to different environments growing up will show similarities simply because they are genetically identical. Similar food preferences or other behaviors suggest there is a genetic basis, because their environments were mostly different. **2.** A study comparing the variability of specific behaviors of monozygotic twins raised in the same environment with those behaviors of monozygotic twins raised in different environments would help to determine if a behavior was based on

environmental factors. **p. 668: 1.** Examples of learning include laughing gull chicks learning begging behaviors, ducklings imprinting on their mother, and white-crowned sparrows learning a species-specific song. **2.** Classical conditioning presents a subject with two different stimuli and leads them to make an association between them (e.g., dogs and food and the sound of a bell). Operant conditioning involves strengthening a stimulus-response connection through such methods as rewards. **p. 672: 1.** Auditory communication is fast, and is not dependent on daylight. It can vary in loudness, pattern, duration, and repetition. Visual communication is most often used during the day though fireflies who emit their own light use this at night. It is a mechanism of expression of status, happiness, and intention without making physical contact. Tactile communication occurs when one organism touches another to provide information about the environment, express their emotional status or receptivity, or strengthen social bonds. **2.** Communication could warn a receiver, show receptivity to mating, mark a territory, tell a receiver where food is, and help in navigation. **3.** Auditory communication would be most effective in a forest. Visual would be least effective. **p. 677: 1.** Animals may defend a portion of their home range that has particularly good food resources. **2.** Females produce a limited number of eggs in their lifetime, whereas males produce essentially limitless amounts of sperm. Females are also usually the sex that provides more parental care, so they are choosy to ensure the highest-quality males in terms of fitness or parental care that the males may provide for the offspring. **3.** Advantages: avoiding predators, finding food, cooperative offspring rearing. Disadvantages: higher competition for mates and food; spread of diseases and parasites.

Testing Yourself

1. d; **2.** c; **3.** c; **4.** a; **5.** b; **6.** a; **7.** c; **8.** b; **9.** a; **10.** d; **11.** b; **12.** d

Thinking Critically

1. One experiment would be to catch rats from New York and Florida, breed them in the lab, and see if hand-raised newborns (without their parents) have the same preferences as the adults in natural populations. If offspring have no preference, an environmental or learned component would be suggested. This could further be tested by rearing one offspring group (composed of both New York and Florida rats) on one type of cheese and another offspring group on another type of cheese. A cheese preference test could be conducted later in life, and if rats that showed no preference at birth preferred the cheese they were raised on later in life, an environmental ("nurture") component is suggested. **2.** It is not exactly clear why altruistic behavior exists in humans. In the context of culture, someone might save someone else's life because it makes them feel good to perform such an act. Perhaps people do it so others would be indebted to them or for monetary gain. Perhaps some people perform altruistic behavior so that their societal status would increase (e.g., to be a "hero")—this may afford them more opportunities (e.g., jobs) that may allow them to improve their ability to obtain resources. Nonetheless, altruism in humans is a complicated issue, but because we are a social species, there may be some benefit in the context of living in a society.

CHAPTER 34
Check Your Progress

p. 681: 1. A population is all of the organisms of the same species within a specific area. A community is all of the populations present in one location. **2.** Ecology seeks to better understand how organisms interact with other organisms within their physical environment. **p. 685: 1.** To reach their biotic potential, a population would need to experience no environmental resistance. It would have no limitations on its reproductive potential, have unlimited food, be disease-free, and have no predators. This does not happen. **2.** Type I individuals have a normal life span and then die at an increasing rate after that. Type II individuals experience a constant rate of survival throughout their lives. Type III individuals have a decreased rate of survival early in their life, and the rate of survival increases during their life. **3.** LDCs tend to have a larger percentage of their population in the prereproductive group than MDCs. This produces an unstable age structure because that group will reach reproductive age and have babies, increasing the population. **p. 691: 1.** Mutualism is exemplified by insect pollinators that receive nectar from flowers and spread pollen to other flowers. **2.** Opportunistic growth patterns are exhibited by organisms such as insects and weeds that produce many small offspring with little parental input. **3.** By partitioning resources such as food into different niches, competition between species is reduced and survival is increased. **4.** Species within a community can experience many types of interactions, including predator–prey, parasitism, commensalism, and mutualism. **p. 693: 1.** The directional changes in a community's composition that is called succession occurs when the community is newly established and there is no soil, or when a major disturbance has disrupted the structure of the community. **2.** Climax-pattern model—areas will always lead to the climax community. Facilitation model—each successive community helps the development of the next community. Inhibition model—each species in a community resists being replaced by a successive species. Tolerance model—chance determines which species colonize an area. The order of succession may reflect the time it takes the species to mature.

Testing Yourself

1. e; **2.** d; **3.** e; **4.** c; **5. a.** lag phase, **b.** environmental resistance, **c.** carrying capacity, **d.** stable equilibrium, **e.** deceleration, **f.** exponential growth; **6.** b; **7.** a; **8.** a; **9.** c; **10.** c; **11.** a; **12.** a; **13.** b

Thinking Critically

1. Rotate use of different pesticides. Use biological controls. Grow different crops year to year. **2.** You would need to test if the two species have the same mode of defense. If so, it would be Müllerian mimicry. If only one species was harmful, then it would be Batesian mimicry **3.** Predator–prey systems do not exist as simple, two-species systems. The decline in pheasant and rabbit populations could in part be due to limited food supplies and resulting lower fertility rates. Eliminating the red-tailed hawk will not change food availability. As the supply of rabbits and pheasants becomes scarce, it is likely that the hawk would shift to another source of food. **4.** As a predator reduces the population of prey, it will become increasingly difficult to survive on the limited resource of that prey, and so the predator will seek out different sources of food to survive. This will happen before the prey becomes extinct.

CHAPTER 35

Check Your Progress

p. 698: 1. Producers are at the base of the food chain because they photosynthesize. They are autotrophs. Consumers need to eat other organisms for nutrients to survive and are heterotrophs. Decomposers consume dead organic matter and break it down. **2.** Most energy (90%) is lost as heat to the environment due to inefficiencies in consumption and digestion. **p. 700: 1.** Little biomass, in the form of carnivores, remains at the top trophic levels in an ecosystem because most energy is lost as heat when one organism eats another. **2.** Usually, the biomass of autotrophs is greater than that of herbivores, which is greater than that of carnivores. However, in aquatic ecosystems, herbivores can have a higher biomass than producers, which can be consumed at a high rate. **p. 709: 1.** See Figures 35.7, 35.8, 35.9, and 35.10. **2.** Humans burn fossil fuels, releasing carbon dioxide into the atmosphere, thus causing rises in global temperatures. Humans mine phosphorus and distribute it into the environment as fertilizers, animal feed supplements, and detergents, where it becomes available to the biotic community. Humans apply nitrogen as fertilizer, which runs off into lakes and rivers and is released as nitrous oxide into the atmosphere. Humans deplete sources of fresh water, creating scarcities.

Testing Yourself

1. a. producers, **b.** consumers, **c.** inorganic nutrient pool, **d.** decomposers; **2.** c; **3.** c; **4.** c; **5.** b; **6.** d; **7.** a; **8.** c; **9.** c, d; **10.** a, b, c; **11.** d; **12.** f

Thinking Critically

1. Removal of trees changes the balance of evaporation, condensation, and precipitation in an ecosystem. Removal of trees removes shade, causing increases in evaporation relative to condensation and precipitation. In addition, removal of trees causes erosion (due to reduction of tree roots that hold the soil together), which causes more water to run off into the ocean. This causes precipitation patterns to shift from overland, where the forest was, to over the ocean. **2.** Fungicides and pesticides would decrease the numbers of organisms involved in the detrital food web. When they are used, the amount of phosphate, ammonium, nitrite, and nitrate available in the soil and sediments would be reduced. **3.** Legumes contain nitrogen-fixing bacteria in the nodules on their roots. Growing them

increases the amount of nitrogen available to plants grown in that soil. **4.** Composting would recycle both carbon and nitrogen. Water can be conserved by shorter showers, less laundry washing, no lawn watering, and more efficient dishwashing. Decreasing the use of fertilizers or detergents would decrease the runoff into nearby bodies of water.

CHAPTER 36

Check Your Progress

p. 714: 1. The proximity of the portion of the Earth to the sun determines the different seasons. Because the Earth is tilted on its axis, during part of the year the southern hemisphere is farther from the sun than is the northern hemisphere, and vice versa. **2.** Solar energy heats up air and causes water to evaporate. As this air rises, it cools and the moisture condenses. It falls out as rain. Mountains affect rainfall. As air blows up and over a mountain range, it rises and cools and the moisture condenses as rain. **p. 722: 1.** Features in a biome that promote biodiversity include warm temperatures, plenty of rain, and a complex structure with many levels of living organisms. **2.** The tropical rain forest has the highest species diversity, while the tundra has the lowest. **p. 730: 1.** Lakes can range from nutrient-poor (oligotrophic) to nutrient-rich (eutrophic). Deep lakes are stratified during summer and winter into epilimnion, thermocline, and hypolimnion layers. Estuaries are where fresh water and saltwater meet and mix. **2.** The pelagic division, which is divided into the neritic and oceanic provinces, is characterized by organisms that live in the open waters under varied levels of sunlight and nutrient availability. The benthic division is characterized by organisms that live on or in the ocean floor.

Testing Yourself

1. b; **2.** e; **3.** a; **4.** e; **5.** c; **6.** c; **7.** d; **8.** a; **9.** a; **10.** e

Thinking Critically

1. Because the sun aims directly at the equator, it is already relatively warm and temperatures are relatively constant in tropical areas near the equator. As you move into the temperate zones (both north and south) away from the equator, the sun's rays get dispersed. Rising air flows toward the poles, sinks to the Earth's surface and reheats. With greenhouse gases trapping solar radiation, the warmer air will move toward the poles and away from the equator, making temperature increases in temperate areas larger than those in tropical areas nearest to the equator. **2.** Equatorial climates have the highest amount of rainfall, which supports high primary productivity. They also have the most stable climate in terms of temperature, so they can support many plant and animal species that have narrow temperature tolerances. As one moves more north or south from the equator, temperatures get hotter in the deserts, with large temperature fluctuations and little rainfall. This requires plant and animal species adapted for water conservation and those that can deal with large temperature fluctuations. Farther from the equator are taiga and tundra. Taiga has a very short growing season, because it is cold and the ground only thaws for a few months of the year. This makes primary productivity lower, thus fewer animals can be supported in the food web. Tundra is extremely cold and has permafrost, making ground cover ice and plant growth virtually impossible. Thus, there is very low diversity in areas of tundra—for example, marine life such as penguins that are aquatic for part of the year where they can feed. **3.** Although tropical forests have basically a year-round growing season, versus a growing season that ranges from 140 to 300 days in temperate forest habitats, temperate forests have more nutrient-rich soil. Decomposition rates are slower in temperate areas because of colder temperatures, thus nutrients are recycled more slowly. In tropical forests, nutrients are cycled rapidly from the soil directly back into the plants. When tropical forests are slashed and burned to initiate agriculture, nutrients are released into the soil, providing nutrients for several harvests. However, the soils become nutrient-poor rather quickly, requiring forest regrowth to occur before further harvesting can occur. **4.** The rich diversity and high density of species within a tropical rain forest increases the amount of competition between species. The evolution of antibacterial and antifungal compounds within a species might improve their success at survival and reproduction.

CHAPTER 37

Check Your Progress

p. 735: 1. Conservation biology relies on multiple disciplines in basic (e.g., physiology, genetics, ecology) and applied

(e.g., forestry, fisheries biology, agronomy) biology to achieve its goal of protecting biodiversity and Earth's natural resources. **2.** Ecosystem-level conservation will protect habitats for many communities and the species they contain, whereas species-level conservation may protect one species while not accounting for the needs of the whole community. **p. 738: 1.** Consumptive use value of biodiversity describes the direct value of a natural product, such as timber. Agricultural use value is the value of those products derived from wild plants and developed by humans for agriculture, such as crops. **2.** Direct value is the dollar value of product in the market. Indirect values, which are difficult to measure, are those benefits incurred as a result of the existence and functioning of ecosystems. **p. 743: 1.** Exotic species can be introduced by humans when they colonize a new area, by accidental transport, or by horticulture and agriculture. **2.** The five main causes of extinction are habitat loss, introduction of exotic species, pollution, overexploitation, and disease. **3.** Pollution can weaken immunity and lead to increased susceptibility to disease, as well as directly poison organisms. Pollution leads to global warming, which can have negative impacts on species' geographic ranges. **p. 745: 1.** Landscape preservation maintains multiple ecosystem types that are connected by corridors. This supports the populations of species such as grizzly bears, that require different types of ecosystems for survival. Maintaining just one ecosystem often cannot provide adequate habitat for species. **2.** Because edge effects essentially remove the amount of suitable habitat in an area, nature reserves should account for this and generally be designed to be larger. **3.** Three principles of habitat restoration include the following: (1) Begin as soon as possible to protect remaining habitat; (2) use biological techniques to mimic natural processes; and, (3) focus on sustainable development, allowing the ecosystem to maintain itself while providing services to humans. **p. 751: 1.** Most nonrenewable energy use is concentrated on fossil fuel burning, which releases greenhouse gases into the environment and contributes to global warming. **2.** We can plant salt- and drought-tolerant crops, use drip irrigation, reduce and recycle home use water, limit yard irrigation, and use water-efficient toilets. **3.** Crop rotation helps maintain nutrients in the soil (e.g., by alternating nitrogen-fixing crops, such as legumes). Both organic farming and biological pest control remove or reduce the use of pesticides and herbicides. Contour farming, terrace farming, planting cover crops, and natural fences all help reduce erosion.

Testing Yourself

1. e; **2.** c; **3.** b; **4.** a; **5.** e; **6.** e; **7.** e; **8. a.** habitat loss, **b.** exotic species, **c.** pollution, **d.** overexploitation, **e.** disease; **9.** c; **10.** d; **11.** c; **12.** e; **13.** a

Thinking Critically

1. Each person's answer will depend on the lifestyle they lead, but reducing use of fossil fuels, living in and maintaining a smaller home, reducing water consumption, and reducing meat consumption are ways to reduce ecological footprints. **2.** One simple way to reduce our reliance on nonrenewable resources is to reduce consumption by conserving electricity, water, and gas, for example. Another way is through recycling, which reuses nonrenewable resources. We can also switch from using nonrenewable energy resources, such as fossil fuels, to renewable energy sources, such as hydropower, geothermal energy, wind power, solar power, and hydrogen fuel. **3.** Biological magnification is the process by which pollutants increase in their concentration up the food chain. While a herbivore may eat a certain amount of biomass of plants, a carnivore eats several herbivores, thereby magnifying the concentration of pesticides in the plants. The farther an organism is up the food chain, the more the chemicals are magnified in the food. Thus, people who often are tertiary or even quaternary consumers should be more concerned about pesticide concentrations in their food than herbivores, because they eat plants directly. **4.** Extinction of a keystone species may cause secondary extinctions or change the food web dynamics of an ecosystem, as seen with wolves in Yellowstone National Park. Without wolves, elk become more numerous and alter aspen groves, their favorite food. Wolves have helped reduce the number of elk, and their carcasses become food for grizzly bears and eagles. Reduction of elk also results in increases in the number of aspen and other trees found in floodplains, which, in turn, increases the habitat for songbirds and beavers. Beavers' dams flood areas, thus making them suitable for muskrats, ducks, otters, and amphibians.

A

abiotic Nonliving; examples are water, gases, and minerals.

abscisic acid (ABA) Plant hormone that causes stomata to close and initiates and maintains dormancy.

abscission Dropping of leaves, fruits, or flowers from a plant.

acid Molecules tending to raise the hydrogen ion concentration in a solution and thus to lower its pH numerically.

acquired immunodeficiency syndrome (AIDS) Disease caused by the HIV virus that destroys helper T cells and macrophages of the immune system, thus preventing an immune response to pathogens; caused by sexual contact with an infected person, intravenous drug use, and transfusions of contaminated blood (rare).

acromegaly Condition caused by an overproduction of growth hormones as an adult.

actin One of two major proteins of muscle; makes up thin filaments in myofibrils of muscle fibers. *See also* myosin.

actin filament Component of the cytoskeleton; plays a role in the movement of the cell and its organelles; a protein filament in a sarcomere of a muscle, its movement shortens the sarcomere, yielding muscle contraction.

action potential Electrochemical changes that take place across the axon membrane; the nerve impulse.

active immunity Ability to produce antibodies due to the immune system's response to a microorganism or a vaccine.

active site Region of an enzyme where the substrate binds and where the chemical reaction occurs.

active transport Use of a plasma membrane carrier protein to move a molecule or ion from a region of lower concentration to one of higher concentration; it opposes equilibrium and requires energy.

acute bronchitis Inflammation of the bronchi in the lungs; often caused by a viral or bacterial infection.

adaptation Species' modification in structure, function, or behavior that makes a species more suitable to its environment.

adaptive immunity Type of immunity that is characterized by the response of lymphocytes to specific antigens.

adaptive radiation Rapid evolution of several species from a common ancestor into new ecological or geographic zones.

Addison disease Condition caused by an insufficient production of cortisol by the adrenal glands.

adenine (A) One of four nitrogen-containing bases in nucleotides composing the structure of DNA and RNA. Pairs with uracil (U) and thymine (T).

adenosine diphosphate (ADP) Nucleotide with two phosphate groups that can accept another phosphate group and become ATP.

adenosine triphosphate (ATP) Nucleotide with three phosphate groups. The breakdown of ATP into ADP + ℗ makes energy available for energy-requiring processes in cells.

adhesion The ability of water molecules to cling to, or be attracted to, a surface, such as a transport vessel in a plant or animal.

adhesion junction Junction between cells in which the adjacent plasma membranes do not touch but are held together by intercellular filaments attached to buttonlike thickenings.

adipose tissue Connective tissue in which fat is stored.

adrenal cortex Outer portion of the adrenal gland; secretes mineralocorticoids, such as aldosterone, and glucocorticoids, such as cortisol.

adrenal gland Gland that lies atop a kidney; the *adrenal medulla* produces the hormones epinephrine and norepinephrine, and the *adrenal cortex* produces the glucocorticoid and mineralocorticoid hormones.

adrenal medulla Inner portion of the adrenal gland; secretes the hormones epinephrine and norepinephrine.

adrenocorticotropic hormone (ACTH) Hormone secreted by the anterior lobe of the pituitary gland that stimulates activity in the adrenal cortex.

aerobic A chemical process that requires air (oxygen); phase of cellular respiration that requires oxygen.

African sleeping sickness Disease caused by the protist *Trypanosoma brucei*, which is transmitted by the tsetse fly.

afterbirth Another term for the placenta; the expulsion of the afterbirth represents the third stage of parturition.

age structure diagram In demographics, a display of the age groups of a population; a growing population has a pyramid-shaped diagram.

aging General term for the progressive physiological changes that occur to the human body over time.

aldosterone Hormone secreted by the adrenal cortex that regulates the sodium and potassium ion balance of the blood.

alien species Nonnative species that migrate or are introduced by humans into a new ecosystem; also called exotic species.

allantois Extraembryonic membrane that accumulates nitrogenous wastes in the eggs of reptiles, including birds; contributes to the formation of umbilical blood vessels in mammals.

allele Alternative form of a gene; alleles occur at the same locus on homologous chromosomes.

allergy Immune response to substances that usually are not recognized as foreign.

allopatric speciation Model that proposes that new species arise due to an interruption of gene flow between populations that are separated geographically.

alternation of generations Life cycle, typical of land plants, in which a diploid sporophyte alternates with a haploid gametophyte.

altruism Social interaction that has the potential to decrease the lifetime reproductive success of the member exhibiting the behavior.

alveoli (sing., **alveolus**) In humans, terminal, microscopic, grapelike air sac found in lungs.

Alzheimer disease Disease of the central nervous system (brain) that is characterized by an accumulation of beta amyloid protein and neurofibrillary tangles in the hippocampus and amygdala.

AM fungi Fungi with branching invaginations (arbuscular mycorrhizal or AM) used to invade plant roots.

amino acid Organic molecule composed of an amino group and an acid group; covalently bonds to produce peptide molecules.

ammonia Nitrogenous end product that takes a limited amount of energy to produce but requires much water to excrete because it is toxic.

amnion Extraembryonic membrane of birds, reptiles, and mammals that forms an enclosing, fluid-filled sac.

amniotic egg Egg that has an amnion, as seen during the development of reptiles (including birds) and mammals.

amoebic dysentery Condition caused by protistans belonging to the genus *Entamoeba;* characterized by severe diarrhea and fluid loss.

amoeboid Cell that moves and engulfs debris with pseudopods.

amphibian Member of vertebrate class Amphibia that includes frogs, toads, and salamanders; they are tied to a watery environment for reproduction.

amygdala Part of the limbic system of the brain; associated with emotional experiences.

amyotrophic lateral sclerosis (ALS) Neurodegenerative disease affecting the motor neurons in the brain and spinal cord, resulting in paralysis and death; also known as Lou Gehrig's disease.

anabolic steroid Synthetic steroid that mimics the effect of testosterone.

anabolism Chemical reaction in which smaller molecules (monomers) are combined to form larger molecules (polymers); anabolic metabolism.

anaerobic A chemical reaction that occurs in the absence of oxygen; an example is the fermentation reactions.

analogous structure Structure that has a similar function in separate lineages but differs in anatomy and ancestry.

anaphase Fourth phase of mitosis; chromosomes move toward the poles of the spindle.

anaphylactic shock Severe systemic form of anaphylaxis involving bronchiolar constriction, impaired breathing, vasodilation, and a rapid drop in blood pressure with a threat of circulatory failure.

androgen Male sex hormone (e.g., testosterone).

anemia Disease that is characterized by a decrease in the number of red blood cells (erythrocytes); may be genetic in origin or the result of another disease or condition in the body.

aneurysm Expansion in the walls of the blood vessel; often caused by atherosclerosis or hypertension.

angina pectoris Condition characterized by thoracic pain resulting from occluded coronary arteries; may precede a heart attack.

angiogenesis Formation of new blood vessels; rapid angiogenesis is a characteristic of cancer cells.

angiosperm Flowering land plant; the seeds are borne within a fruit.

animal Multicellular, heterotrophic eukaryote that undergoes development to achieve its final form. In general, animals are mobile organisms, characterized by the presence of muscular and nervous tissue.

annelid The segmented worms, such as the earthworm and the clam worm.

annual ring Layer of wood (secondary xylem) usually produced during one growing season.

anorexia nervosa Eating disorder characterized by a severe psychological fear of gaining weight.

anterior pituitary Portion of the pituitary gland that is controlled by the hypothalamus and produces six types of hormones, some of which control other endocrine glands.

anther Male structure of a flowering plant where pollen is produced; attached to the flower by the filament.

antheridium (pl., **antheridia**) Sperm-producing structures, as in the moss life cycle.

antibiotics Chemicals that interfere with the physiological activities, or structure, of bacteria.

antibody Protein produced in response to the presence of an antigen; each antibody combines with a specific antigen.

antibody-mediated immunity Specific mechanism of defense in which plasma cells derived from B cells produce antibodies that combine with antigens.

anticodon Three-base sequence in a transfer RNA molecule base that pairs with a complementary codon in mRNA.

antidiuretic hormone (ADH) Hormone, secreted by the posterior pituitary, that increases the permeability of the collecting ducts in a kidney.

antigen Foreign substance, usually a protein or a polysaccharide, that stimulates the immune system to react, such as to produce antibodies.

antigen-presenting cell (APC) Cell that displays an antigen to certain cells of the immune system so they can defend the body against that particular antigen.

anus Outlet of the digestive tube.

aorta In humans, the major systemic artery that takes blood from the heart to the tissues.

apical dominance Influence of a terminal bud in suppressing the growth of axillary buds.

apical meristem In vascular land plants, masses of cells in the root and shoot that reproduce and elongate as primary growth occurs.

apoptosis Programmed cell death; involves a cascade of specific cellular events leading to death and destruction of the cell.

appendicular skeleton Part of the vertebrate skeleton comprising the appendages, shoulder girdle, and hip girdle.

appendix (vermiform appendix) In humans, small, tubular appendage that extends outward from the cecum of the large intestine.

aquaporin Channel protein through which water can diffuse across a membrane.

aquifer Rock layers that contain water and release it in appreciable quantities to wells or springs.

arachnid Group of arthropods that includes spiders and scorpions.

archaea Prokaryotic organisms that are members of the domain Archaea.

archegonia (sing., **archegonium**) Egg-producing structures, as in the moss life cycle.

ardipithecines Common name for species of the genus *Ardipithecus,* an early, and now extinct, hominin.

arteriole Vessel that takes blood from an artery to capillaries.

artery Blood vessel that transports blood away from the heart.

arthritis Condition characterized by an inflammation of the joints; two common forms are osteoarthritis and rheumatoid arthritis

arthropod Invertebrates, with an exoskeleton and jointed appendages, such as crustaceans and insects.

artificial selection Intentional breeding of certain traits, or combinations of traits, over others to produce a desirable outcome.

ascocarp Reproductive structure of the sac fungi (phylum Ascomycota).

associative learning Acquired ability to associate two stimuli or between a stimulus and a response.

asthma Condition in which bronchioles constrict and cause difficulty in breathing.

astigmatism Uneven shape of the cornea or lens of the eye; causes a distortion in the light reaching the retina of the eye.

asymmetrical Lack of any symmetrical relationship in the morphology of an organism.

atherosclerosis Form of cardiovascular disease characterized by the accumulation of fatty materials (usually cholesterol) in the arteries.

atom Smallest particle of an element that displays the properties of the element.

atomic mass Average of atom mass units for all the isotopes of an atom.

atomic number Number of protons within the nucleus of an atom.

ATP synthase Complex of proteins in the cristae of mitochondria and thylakoid membrane of chloroplasts that produces ATP from the diffusion of hydrogen ions across a membrane.

atrial natriuretic hormone (ANH) Hormone secreted by the heart that increases sodium excretion by inhibiting the secretion of aldosterone.

atrioventricular valve Heart valve located between an atrium and a ventricle.

atrium Chamber; particularly an upper chamber of the heart, lying above a ventricle.

auditory canal Structure that funnels sounds from the outer ear to the middle ear.

auditory tube Also called the eustachian tube; connects the middle ear to the nasopharynx for the equalization of pressure.

australopithecine One of several species of *Australopithecus,* a genus that contains the first generally recognized humanlike hominins.

autoimmune disease Disease that results when the immune system mistakenly attacks the body's own tissues.

autonomic system Portion of the peripheral nervous system that regulates internal organs.

autosome Chromosome pairs that are the same between the sexes; in humans, all but the X and Y chromosomes.

autotroph Organism that can capture energy and synthesize organic molecules from inorganic nutrients.

auxin Plant hormone regulating growth, particularly cell elongation; also called indoleacetic acid (IAA).

axial skeleton Part of the vertebrate skeleton forming the vertical support or axis, including the skull, the rib cage, and the vertebral column.

axillary bud Bud located in the axil of a leaf.

axon Elongated portion of a neuron that conducts nerve impulses, typically from the cell body to the synapse.

B

B-cell receptor (BCR) Molecule on the surface of a B cell that binds to a specific antigen.

bacterium (pl., **bacteria**) Single-celled prokaryotic organisms that are members of the domain Bacteria. Lack a nucleus and membrane-bound organelles.

bark External part of a tree, containing cork, cork cambium, and phloem.

Barr body Dark-staining body in the cell nuclei of female mammals that contains a condensed, inactive X chromosome; named after its discoverer, Murray Barr.

basal nuclei Subcortical nuclei deep within the white matter that serve as relay stations for motor impulses and produce dopamine to help control skeletal muscle activities.

base Molecules tending to lower the hydrogen ion concentration in a solution and thus raise the pH numerically.

basidium Clublike structure in which nuclear fusion, meiosis, and basidiospore production occur during sexual reproduction of club fungi.

basophil White blood cell with a granular cytoplasm; able to be stained with a basic dye.

behavior Observable, coordinated responses to environmental stimuli.

benign Mass of cells derived from a single mutated cell that has repeatedly undergone cell division but has remained at the site of origin.

benign prostatic hyperplasia (BPH) Enlargement of the prostate gland that may cause irritation of the bladder and an urge to urinate more frequently.

benthic division Ocean or lake floor, extending from the high-tide mark to the deepest depths.

bicarbonate ion Ion that participates in buffering the blood, and the form in which carbon dioxide is transported in the bloodstream.

bilateral symmetry Body plan having two corresponding or complementary halves.

bile Secretion of the liver that is temporarily stored and concentrated in the gallbladder before being released into the small intestine, where it emulsifies fat.

binary fission Splitting of a parent cell into two daughter cells; serves as an asexual form of reproduction in bacteria.

biodiversity Total number of species, the variability of their genes, and the communities in which they live.

biodiversity hotspot Region of the world that contains unusually large concentrations of species.

biogeochemical cycle Circulating pathway of elements, such as carbon and nitrogen, involving exchange pools, storage areas, and biotic communities.

biogeography Study of the geographic distribution of organisms.

bioinformatics Area of scientific study that utilizes computer technologies to analyze large sets of data, typically in the study of genomics and proteomics.

biomass The number of organisms multiplied by their weight.

biome One of the biosphere's major communities, characterized in particular by certain climatic conditions and particular types of plants.

biomolecules Organic molecules such as proteins, nucleic acids, carbohydrates, and fats.

biosphere Zone of air, land, and water at the surface of the Earth, in which living organisms are found.

biotic Term that indicates life; a cell is an example of a biotic factor in the environment.

biotic potential Maximum population growth rate under ideal conditions.

bipedalism Walking erect on two feet; distinguishing characteristic of primates.

bird Endothermic reptile that has feathers and wings, is often adapted for flight, and lays hard-shelled eggs.

birth control pill Form of oral contraception that typically contains a combination of estrogen and progesterone.

bivalve Type of mollusc with a shell composed of two valves; includes clams, oysters, and scallops.

bladder stones Condition that causes inflammation of the bladder; may be caused by kidney stones, infections, or other conditions that restrict the flow of urine from the bladder.

blade Broad, expanded portion of a land plant leaf that may be single or compound leaflets.

blastocoel Fluid-filled cavity of a blastula.

blastocyst Early stage of human embryonic development that consists of a hollow, fluid-filled ball of cells.

blastopore Opening into the primitive gut formed at gastrulation.

blastula Hollow, fluid-filled ball of cells occurring during animal development prior to gastrula formation.

blind spot Region of the retina, lacking rods or cones, where the optic nerve leaves the eye.

blood Fluid circulated by the heart through a closed system of vessels; type of connective tissue.

blood pressure Force of blood pushing against the inside wall of blood vessels.

body cavity In vertebrates, defined regions of the body in which organs reside.

bone Connective tissue having protein fibers and a hard matrix of inorganic salts, notably calcium salts.

bony fishes Vertebrates belonging to the class of fish called Osteichthyes that have a bony, rather than cartilaginous, skeleton.

bottleneck effect Type of genetic drift; occurs when a majority of genotypes are prevented from participating in the production of the next generation as a result of a natural disaster or human interference.

brachiopod Invertebrate animal (lophotrochozoan) that possesses hard shells (or valves) on the upper and lower surfaces of the organism.

brain Ganglionic mass at the anterior end of the nerve cord; in vertebrates, the brain is located in the cranial cavity of the skull.

brain stem In mammals, portion of the brain consisting of the medulla oblongata, pons, and midbrain.

brassinosteroids Class of plant hormone that is involved in regulating the growth and development of the plant.

bronchi (sing., **bronchus**) In terrestrial vertebrates, branch of the trachea that leads to the lungs.

bronchiole In terrestrial vertebrates, small tube that conducts air from a bronchus to the alveoli.

brown algae Marine photosynthetic protists with a notable abundance of xanthophyll pigments; this group includes well-known seaweeds of northern rocky shores.

bryophytes Term used to identify plants that do not have vascular tissue. Includes the liverworts, hornworts, and mosses.

bryozoan Invertebrate animal (lophotrochozoan) that lives as a mass of interconnected individuals.

buffer Substance or group of substances that tend to resist pH changes of a solution, thus stabilizing its relative acidity and basicity.

bulbourethral glands Male sex glands that produce pre-ejaculate fluid that neutralizes acid in the urethra.

bulimia nervosa Eating disorder characterized by binge eating followed by purging of food (vomiting).

bundle sheath Layer of cells surrounding the vascular tissue in the leaf of a plant.

bursae (sing., **bursa**) Fluid-filled sacs that reduce friction between tendons and ligaments; often found near synovial joints such as the elbow and knee.

bursitis Inflammation of the bursae.

C

C$_4$ plant Plant that fixes carbon dioxide to produce a C$_4$ molecule that releases carbon dioxide to the Calvin cycle.

calcitonin Hormone secreted by the thyroid gland that increases the blood calcium level.

calorie Amount of heat energy required to raise the temperature of one gram of water 1°C.

Calvin cycle reaction Portion of photosynthesis that takes place in the stroma of chloroplasts and can occur in the dark. It uses the products of the light reactions to reduce CO$_2$ to a carbohydrate.

calyx The sepals collectively; the outermost flower whorl.

CAM Crassulacean-acid metabolism; a form of photosynthesis in succulent plants that separates the light-dependent and Calvin reactions by time.

capillary Microscopic blood vessel; gases and other substances are exchanged across the walls of a capillary between blood and tissue fluid.

capsid Protective protein containing the genetic material of a virus.

capsule A form of glycocalyx that consists of a gelatinous layer; found in blue-green algae and certain bacteria.

carbaminohemoglobin Hemoglobin carrying carbon dioxide.

carbohydrate Class of organic compounds that typically contain carbon, hydrogen, and oxygen in a 1:2:1 ratio; includes the monosaccharides, disaccharides, and polysaccharides.

carbon dioxide fixation Process by which carbon dioxide gas is attached to an organic compound; in photosynthesis, this occurs in the Calvin cycle reactions.

carbonic anhydrase Enzyme in red blood cells that speeds the formation of carbonic acid from water and carbon dioxide.

cardiac cycle One complete cycle of systole and diastole for all heart chambers.

cardiac muscle Striated, involuntary muscle tissue found only in the heart.

cardiac output Blood volume pumped by each ventricle per minute (not total output pumped by both ventricles).

cardiovascular system Organ system of humans that consists of the heart and blood vessels; transports blood, nutrients, gases, and wastes; assists in the defense against disease; helps control homeostasis.

carnivore Consumer in a food chain that eats other animals.

carotenoid An accessory photosynthetic pigment of plants and algae that are often yellow or orange in color; consists of two classes—the xanthophylls and the carotenes.

carpel Ovule-bearing unit that is a part of a pistil.

carrier protein Protein in the plasma membrane that combines with and transports a molecule or ion across the plasma membrane.

carrying capacity (**K**) Largest number of organisms of a particular species that can be maintained indefinitely by a given environment.

cartilage Connective tissue in which the cells lie within lacunae embedded in a flexible, proteinaceous matrix.

cartilaginous fishes Vertebrates that belong to the class of fish called the Chondrichthyes: possess a cartilaginous, rather than bony, skeleton; include sharks, rays, and skates.

cartilaginous joint Form of joint that is connected by hyaline cartilage.

Casparian strip Layer of impermeable lignin and suberin bordering four sides of root endodermal cells; prevents water and solute transport between adjacent cells.

catabolism Metabolic process that breaks down large molecules into smaller ones; catabolic metabolism.

cataracts Condition where the lens of the eye becomes opaque, preventing the transmission of light to the retina.

cecum Region of the large intestine, just below the small intestine, to which the appendix is attached.

cell The smallest unit of life that displays all the properties of life; composed of cytoplasm surrounded by a plasma membrane.

cell body Portion of a neuron that contains a nucleus and from which dendrites and an axon extend.

cell cycle An ordered sequence of events in eukaryotes that involves cell growth and nuclear division; consists of the stages G_1, S, G_2, and M.

cell plate Structure across a dividing plant cell that signals the location of new plasma membranes and cell walls.

cell recognition protein Glycoproteins in the plasma membrane that identify self and help the body defend itself against pathogens.

cell theory One of the major theories of biology, which states that all organisms are made up of cells; cells are capable of self-reproduction and come only from preexisting cells.

cell wall Cellular structure that surrounds a plant, protistan, fungal, or bacterial cell and maintains the cell's shape and rigidity; composed of polysaccharides.

cell-mediated immunity Specific mechanism of defense in which T cells destroy antigen-bearing cells.

cellular differentiation Process and developmental stages by which a cell becomes specialized for a particular function.

cellular respiration Metabolic reactions that use the energy from carbohydrate, fatty acid, or amino acid breakdown to produce ATP molecules.

cellular slime mold Free-living amoeboid cells that feed on bacteria and yeasts by phagocytosis and aggregate to form a plasmodium that produces spores.

cellulose Polysaccharide that is the major complex carbohydrate in plant cell walls.

central dogma Term used to describe the process by which information flows from DNA to RNA to protein in a cell.

central nervous system (CNS) Portion of the nervous system consisting of the brain and spinal cord.

centriole Cell structure, existing in pairs, that occurs in the centrosome and may help organize a mitotic spindle for chromosome movement during animal cell division.

centromere Constriction where sister chromatids of a chromosome are held together.

centrosome Central microtubule organizing center of cells. In animal cells, it contains two centrioles.

cephalization Having a well-recognized anterior head with a brain and sensory receptors.

cephalopod Type of mollusc in which the head is prominent and the foot is modified to form two arms and several tentacles; includes squids, cuttlefish, octopuses, and nautiluses.

cerebellum In terrestrial vertebrates, portion of the brain that coordinates skeletal muscles to produce smooth, graceful motions.

cerebral cortex Outer layer of cerebral hemispheres; receives sensory information and controls motor activities.

cerebral hemisphere Either of the two lobes of the cerebrum in vertebrates.

cerebrospinal fluid Fluid found in the ventricles of the brain, in the central canal of the spinal cord, and in association with the meninges.

cerebrum Largest part of the brain in mammals.

cervical cancer Form of cancer that is often caused by the human papillomavirus (HPV); vaccinations are available to reduce risk.

cervix Narrow end of the uterus, which leads into the vagina.

channel protein Protein that forms a channel to allow a particular molecule or ion to cross the plasma membrane.

chemical energy Energy associated with the interaction of atoms in a molecule.

chemiosmosis Process by which mitochondria and chloroplasts use the energy of an electron transport chain to create a hydrogen ion gradient that drives ATP formation.

chemoautotroph Organism able to synthesize organic molecules by using carbon dioxide as the carbon source and the oxidation of an inorganic substance (such as hydrogen sulfide) as the energy source.

chemoreceptor Sensory receptor that is sensitive to chemical stimulation—for example, receptors for taste and smell.

chitin Strong but flexible nitrogenous polysaccharide found in the exoskeleton of arthropods and in the cell walls of fungi.

chlamydia Sexually transmitted disease caused by *Chlamydia trachomatis;* one of the more common forms of STDs.

chlorophyll Green photosynthetic pigment of algae and plants that absorbs solar energy; occurs as chlorophyll *a* and chlorophyll *b*.

chloroplast Membrane-bound organelle in algae and plants with chlorophyll-containing membranous thylakoids; where photosynthesis takes place.

choanoflagellate Single-celled and colonial protists that are most closely related to the animals.

chordate Animals that have a dorsal tubular nerve cord, a notochord, pharyngeal gill pouches, and a postanal tail at some point in their life cycle; includes a few types of invertebrates (e.g., sea squirts and lancelets) and the vertebrates.

chorion Extraembryonic membrane functioning for respiratory exchange in birds and reptiles; contributes to placenta formation in mammals.

choroid Vascular, pigmented middle layer of the eyeball.

chromatid Following replication, a chromosome consists of a pair of sister chromatids, held together at the centromere; each chromatid is comprised of a single DNA helix.

chromatin Network of DNA strands and associated proteins observed within a nucleus of a cell.

chromosomal mutations Changes in the physical structure of a chromosome; includes deletions, duplications, inversions, and translocations.

chromosome The structure that transmits the genetic material from one generation to the next; composed of condensed chromatin; each species has a particular number of chromosomes that is passed on to the next generation.

chronic bronchitis Inflammation of the bronchi in the lungs that is usually caused by smoking or exposure to environmental contaminants.

chyme Thick, semiliquid food material that passes from the stomach to the small intestine.

chytrid Mostly aquatic fungi with flagellated spores that may represent the most ancestral fungal lineage.

cilia (sing., **cilium**) Short, hairlike projections from the plasma membrane, occurring usually in larger numbers.

ciliary muscle Within the ciliary body of the vertebrate eye, the ciliary muscle controls the shape of the lens.

ciliate Complex unicellular protist that moves by means of cilia and digests food in food vacuoles.

circadian rhythm Biological rhythm with a 24-hour cycle.

circumcision Surgical removal of the foreskin from the penis; female circumcision involves the partial or total removal of some part of the female genitalia.

cirrhosis Chronic, irreversible injury to liver tissue; commonly caused by frequent alcohol consumption.

citric acid cycle Cycle of reactions in mitochondria that begins with citric acid. This cycle breaks down an acetyl group and produces CO_2, ATP, NADH, and $FADH_2$; also called the Krebs cycle.

cladistics Method of systematics that uses derived characters to determine monophyletic groups and construct cladograms.

cladogram In cladistics, a branching diagram that shows the relationship among species in regard to their shared derived characters.

class One of the categories, or taxa, used to group species; the taxon above the order level.

classical conditioning Type of learning whereby an unconditioned stimulus that elicits a specific response is paired with a neutral stimulus so that the response becomes conditioned.

classification Process of naming organisms and assigning them to taxonomic groups (taxa).

cleavage Cell division without cytoplasmic addition or enlargement; occurs during the first stage of animal development.

cleavage furrow Indentation in the plasma membrane of animal cells during cell division; formation marks the start of cytokinesis.

climate Generalized weather patterns of an area, primarily determined by temperature and average rainfall.

climate change Recent changes in the Earth's climate; evidence suggests that this is primarily due to human influence, including the increased release of greenhouse gases.

climax community In ecology, community that results when succession has come to an end.

clitoris Organ of sexual arousal in a female.

clonal selection States that the antigen selects which lymphocyte will undergo clonal expansion and produce more lymphocytes bearing the same type of receptor.

cloning Production of identical copies. In organisms, the production of organisms with the same genes; in genetic engineering, the production of many identical copies of a gene.

closed circulatory system A type of circulatory system where blood is confined to vessels and is kept separate from the interstitial fluid.

clotting Also called coagulation, the response of the body to an injury in the vessels of the circulatory system; involves platelets and clotting proteins.

club fungi Fungi that produce spores in club-shaped basidia within a fruiting body; includes mushrooms, shelf fungi, and puffballs.

club moss Type of seedless vascular plant; also called ground pines.

cnidarian Invertebrates existing as either a polyp or medusa, with two tissue layers and radial symmetry.

cnidocytes Specialized stinging cells of a cnidarian; contain a toxin-filled capsule called a nematocyst.

cochlea Spiral-shaped structure of the vertebrate inner ear containing the sensory receptors for hearing.

codominance Inheritance pattern in which both alleles of a gene are equally expressed in a heterozygote.

codon Three-base sequence in messenger RNA that during translation directs the addition of a particular amino acid into a protein or directs termination of the process.

coelom Body cavity of an animal; the method by which the coelom is formed (or lack of formation) is an identifying characteristic in animal classification.

coenzyme Nonprotein organic molecule that aids the action of the enzyme to which it is loosely bound.

coevolution Mutual evolution in which each of two species exert selective pressures on the other species.

cofactor Nonprotein assistant required by an enzyme in order to function. Many cofactors are metal ions; others are coenzymes.

cohesion The ability of water molecules to cling to each other due to the process of hydrogen bonding.

cohesion-tension model Explanation for upward transport of water in xylem based upon transpiration-created tension and the cohesive properties of water molecules.

cohort Group of individuals having a statistical factor in common, such as year of birth, in a population study.

collagen fiber White fiber in the matrix of connective tissue giving flexibility and strength.

collar cells Choanocytes of poriferans (sponges); inner layer of flagellated cells.

collecting duct Duct within the kidney that receives fluid from several nephrons; the reabsorption of water occurs here.

collenchyma Plant tissue composed of cells with unevenly thickened walls; supports growth of stems and petioles.

color blindness Inability to detect specific wavelengths of light associated with color; red-green color blindness is the most common type.

color vision Ability of the eye to detect specific wavelengths of light; involves specialized cone cells that detect blue, green, and red wavelengths.

colostrum Substance produced during the early days of lactation; rich in proteins and antibodies.

columnar epithelium Type of epithelial tissue with cylindrical cells.

comb jelly Invertebrates that resemble jelly fishes and are the largest animals to be propelled by beating cilia.

commensalism Symbiotic relationship in which one species is benefited, and the other is neither harmed nor benefited.

communication Signal by a sender that influences the behavior of a receiver.

community Assemblage of species interacting with one another within the same environment.

compact bone Type of bone that contains osteons consisting of concentric layers of matrix and osteocytes in lacunae.

companion cell Plant cell in the vascular tissue of plants that metabolically supports the conducting cells of the phloem.

comparative genomics Study of genomes through the direct comparison of the genes and DNA sequences from multiple species.

competition Results when members of a species attempt to use a resource that is in limited supply.

competitive exclusion principle Theory that no two species can occupy the same niche in the same place and at the same time.

complement system Collective name for a series of proteins in the blood, some of which act as enzymes and activators, which bind to antibodies and may lead to rupture of a foreign cell.

complementary base pairing Hydrogen bonding between particular purines and pyrimidines; responsible for the structure of DNA, and some RNA, molecules.

compound Substance having two or more different elements in a fixed ratio.

condensation The process by which water vapor turns into liquid; at the cellular level, the removal of water between two molecules for the purpose of forming a chemical bond.

cone cell Photoreceptor in vertebrate eyes that responds to bright light and makes color vision possible.

conifer Member of a group of cone-bearing gymnosperm land plants that includes pine, cedar, and spruce trees.

coniferous forest Forested area generally found in areas with temperate climate and characterized by cone-bearing trees that are mostly evergreens with needle-shaped or scalelike leaves.

conjugation Transfer of genetic material from one cell to another.

conjunctiva Delicate membrane that lines the eyelid, protecting the sclera.

connective tissue Type of animal tissue that binds structures together, provides support and protection, fills spaces, stores fat, and forms blood cells; adipose tissue, cartilage, bone, and blood are types of connective tissue; living cells in a nonliving matrix.

conservation biology Discipline that seeks to understand the effects of human activities on species, communities, and ecosystems, and to develop practical approaches to preventing the extinction of species and the destruction of ecosystems.

conservation corridors Strips of land that allow animals to move safely between habitats that would otherwise be isolated.

conservation reserves Areas that are set aside with the primary goal of protecting biodiversity within them.

constipation Changes in the frequency of bowel movements that leads to the formation of hard, dry fecal material.

consumer Organism that feeds on another organism in a food chain generally; primary consumers eat plants, and secondary consumers eat animals.

continental drift The movement of the Earth's crust by plate tectonics, resulting in the movement of continents with respect to one another.

contraceptive Medications and devices that reduce the chance of pregnancy.

contraceptive implants Synthetic forms of progesterone that prevent ovulation by disrupting the ovarian cycle.

contraceptive injection This form of contraception contains synthetic forms of progesterone (or a combination of progesterone and estrogen) that prevent ovulation by disrupting the ovarian cycle.

contraceptive patch Applied to the skin, this form of contraception contains synthetic forms of progesterone that prevent ovulation by disrupting the ovarian cycle.

contractile vacuoles Structure found in some protistan cells, usually freshwater species, that are involved in osmoregulation.

control Sample that goes through all the steps of an experiment but does not contain the variable being tested; a standard against which the results of an experiment are checked.

convergent evolution Similarity in structure in distantly related groups, generally due to similar selective pressures in like environments.

coral reef Coral formations in shallow tropical waters that support an abundance of diversity.

cork Outer covering of the bark of trees; made of dead cells that may be sloughed off.

cork cambium Lateral meristem that produces cork.

cornea Transparent, anterior portion of the outer layer of the eyeball.

corolla The petals, collectively; usually the conspicuously colored flower whorl.

corpus luteum Follicle that has released an egg and increases its secretion of progesterone.

cortex In plants, ground tissue bounded by the epidermis and vascular tissue in stems and roots; in animals, outer layer of an organ, such as the cortex of the kidney or adrenal gland.

cortisol Glucocorticoid secreted by the adrenal cortex that responds to stress on a long-term basis; reduces inflammation and promotes protein and fat metabolism.

cotyledon Seed leaf for embryo of a flowering plant; provides nutrient molecules for the developing plant before photosynthesis begins.

coupled reactions Reactions that occur simultaneously; one is an exergonic reaction that releases energy, and the other is an endergonic reaction that requires an input of energy in order to occur.

covalent bond Chemical bond in which atoms share one pair of electrons.

cranial nerve Nerve that arises from the brain.

creatine phosphate High-energy compound found in muscular tissue; represents the fastest source of energy for muscle contraction.

cristae (sing., crista) Short, fingerlike projections formed by the folding of the inner membrane of mitochondria.

Cro-Magnon Common name for the first fossils to be designated *Homo sapiens*.

crossing-over Exchange of segments between nonsister chromatids of a bivalent during meiosis.

crustacean Member of a group of aquatic arthropods that contains, among others, shrimps, crabs, crayfish, and lobsters.

cuboidal epithelium Type of epithelial tissue with cube-shaped cells.

cultural eutrophication Over-enrichment caused by human activities, leading to excessive bacterial growth and oxygen depletion of a body of water.

culture Total pattern of human behavior; includes technology and the arts, and is dependent upon the capacity to speak and transmit knowledge.

Cushing syndrome Condition caused by the overproduction of ACTH, resulting in too much cortisol in the body.

cutaneous receptor Sensory receptor of the dermis that is activated by touch, pain, pressure, and temperature.

cuticle In plants, a waxy layer covering the epidermis of plants, that protects the plant against water loss and disease-causing organisms. In animals, an outer covering that protects and supports the organism.

cyanobacterium (pl., **cyanobacteria**) Photosynthetic bacterium that contains chlorophyll and releases oxygen; formerly called a blue-green alga.

cycad Type of gymnosperm with palmate leaves and massive cones; cycads are most often found in the tropics and subtropics.

cyclic electron pathway Alternate form of the light-dependent reactions in photosynthesis that does not produce NADPH.

cyst In protists and invertebrates, the resting structure that contains reproductive bodies or embryos.

cystic fibrosis (CF) Genetic disease caused by a defect in the *CFTR* gene, which is responsible for the formation of a transmembrane chloride ion transporter; causes the mucus of the body to be viscous.

cystitis Inflammation of the bladder.

cytokine Type of protein secreted by a T lymphocyte that attacks viruses, virally infected cells, and cancer cells.

cytokinesis Division of the cytoplasm following mitosis or meiosis.

cytokinin Plant hormone that promotes cell division; often works in combination with auxin during organ development in plant embryos.

cytoplasm Region of a cell between the nucleus, or the nucleoid region of a bacterium, and the plasma membrane; contains the organelles of the cell.

cytoplasmic segregation Process that parcels out the maternal determinants, which play a role in development, during mitosis.

cytosine (C) One of four nitrogen-containing bases in nucleotides composing the structure of DNA and RNA; pairs with guanine.

cytoskeleton Internal framework of the cell, consisting of microtubules, actin filaments, and intermediate filaments.

cytotoxic T cell T lymphocyte that attacks and kills antigen-bearing cells.

D

data Facts or information collected through observation and/or experimentation.

day-neutral plant Plant whose flowering is not dependent on day length—e.g., tomato and cucumber.

deciduous Land plant that sheds its leaves annually.

decomposer Organism, usually a bacterium or fungus, that breaks down organic matter into inorganic nutrients that can be recycled in the environment.

deductive reasoning The use of general principles to predict specific outcomes. Often uses "if… then" statements.

dehydration reaction Chemical reaction in which a water molecule is released during the formation of a covalent bond.

delayed allergic response Allergic response initiated at the site of the allergen by sensitized T cells, involving macrophages and regulated by cytokines.

deletion Change in chromosome structure in which the end of a chromosome breaks off or two simultaneous breaks lead to the loss of an internal segment; often causes abnormalities—e.g., cri du chat syndrome.

demographic transition Due to industrialization, a decline in the birthrate following a reduction in the death rate so that the population growth rate is lowered.

denatured Loss of a protein's or enzyme's normal shape so that it no longer functions; usually caused by a less than optimal pH and temperature.

dendrite Part of a neuron that sends signals toward the cell body.

dendritic cell Antigen-presenting cell of the epidermis and mucous membranes.

Denisovans A recently discovered member of the genus *Homo* that lived in Asia, and possibly Europe, around the same time as the Neandertals.

denitrification Conversion of nitrate to nitrogen gas by bacteria in soil.

dense fibrous connective tissue Type of connective tissue containing many collagen fibers packed together; found in tendons and ligaments, for example.

density-dependent factor Biotic factor, such as disease or competition, that affects population size in a direct relationship to the population's density.

density-independent factor Abiotic factor, such as fire or flood, that affects population size independent of the population's density.

dermis In mammals, thick layer of the skin underlying the epidermis.

desert Ecological biome characterized by a limited amount of rainfall; deserts have hot days and cool nights.

detrital food chain Straight-line linking of organisms according to who eats whom, beginning with detritus.

detrital food web Complex pattern of interlocking and crisscrossing food chains, beginning with detritus.

detritus Partially decomposed material; may be found in either soil or water.

deuterostome Group of coelomate animals in which the second embryonic opening is associated with the mouth; the first embryonic opening, the blastopore, is associated with the anus.

development Process of regulated growth and differentiation of cells and tissues.

diabetes insipidus Condition caused by a lack of antidiuretic hormone (ADH); characterized by excessive thirst and overproduction of urine.

diabetes mellitus Condition caused by an insulin imbalance in the body; type 1 diabetes results from not enough insulin being produced; type 2 diabetes is caused by the body (specifically adipose tissue) not responding to insulin in the blood.

diabetic retinopathy Complication of diabetes that causes the capillaries in the retina to become damaged, potentially causing blindness.

diaphragm In mammals, dome-shaped muscularized sheet separating the thoracic cavity from the abdominal cavity; contraceptive device that prevents sperm from reaching the egg.

diarrhea Excessively frequent and watery bowel movements.

diastole Relaxation period of a heart chamber during the cardiac cycle.

diatom Golden-brown alga with a cell wall in two parts, or valves; significant component of phytoplankton.

diffusion Movement of molecules or ions from a region of higher to lower concentration; it requires no energy and tends to lead to an equal distribution (equilibrium).

digestive system Organ system that consists of digestive organs (stomach, intestines, etc.) and accessory organs (liver, etc.); ingests and digests food; absorbs nutrients and eliminates wastes.

dihybrid cross Cross between parents that differ in two traits.

dinoflagellate Photosynthetic unicellular protist with two flagella—one whiplash and the other located within a groove between protective cellulose plates; a significant part of phytoplankton.

diploid (2n) Cell condition in which two of each type of chromosome are present.

directional selection Outcome of natural selection in which an extreme phenotype is favored, usually in a changing environment.

disaccharide Sugar that contains two monosaccharide units; e.g., maltose.

disease Abnormality in the body's physiology that typically upsets homeostasis.

disruptive selection Outcome of natural selection in which the two extreme phenotypes are favored over the average phenotype, leading to more than one distinct form.

distal convoluted tubule Final portion of a nephron that joins with a collecting duct; associated with tubular secretion.

diuretic Chemical that increases the flow of urine; an example is caffeine.

DNA (deoxyribonucleic acid) Nucleic acid polymer produced from covalent bonding of nucleotide monomers that contain the sugar deoxyribose; the genetic material of living organisms.

DNA fingerprint Process that examines differences in DNA patterns to identify the source of a DNA sample; also called DNA profiling

DNA ligase Enzyme that links DNA fragments; used during production of recombinant DNA to join foreign DNA to vector DNA.

DNA replication Synthesis of a new DNA double helix prior to mitosis and meiosis in eukaryotic cells and during prokaryotic fission in prokaryotic cells.

domain Largest of the categories, or taxa, used to group species; the three domains are Archaea, Bacteria, and Eukarya.

domain Archaea One of the three domains of life; contains prokaryotic cells that often live in extreme habitats and have unique genetic, biochemical, and physiological characteristics; its members are sometimes referred to as *archaea*.

domain Bacteria One of the three domains of life; contains prokaryotic cells that differ from archaea because they have their own unique genetic, biochemical, and physiological characteristics.

domain Eukarya One of the three domains of life, consisting of organisms with eukaryotic cells; includes protists, fungi, plants, and animals.

dominant allele Allele that exerts its phenotypic effect in the heterozygote; it masks the expression of the recessive allele.

dormancy In plants, a cessation of growth under conditions that seem appropriate for growth.

dorsal cavity One of two main body cavities in humans; contains the cranial cavity and vertebral canal.

dorsal root ganglion Mass of sensory neuron cell bodies located in the dorsal root of a spinal nerve.

double fertilization In flowering plants, one sperm nucleus unites with the egg nucleus, and a second sperm nucleus unites with the polar nuclei of an embryo sac.

double helix Double spiral; describes the three-dimensional shape of DNA.

doubling time Number of years it takes for a population to double in size.

Duchenne muscular dystrophy Genetic disorder that is characterized by a wasting of muscle tissue; displays an X-linked recessive pattern of inheritance.

ductus arteriosus Structure of the heart associated with fetal circulation; provides a pathway for returning any blood that enters the right ventricle back to the aorta.

duodenum First part of the small intestine, where chyme enters from the stomach.

duplication Change in chromosome structure in which a particular segment is present more than once in the same chromosome.

E

ecdysozoa Protostome characterized by periodic molting of the exoskeleton. Includes roundworms and arthropods.

echinoderm Invertebrates such as sea stars, sea urchins, and sand dollars; characterized by radial symmetry and a water vascular system.

ecological niche Role an organism plays in its community, including its habitat and its interactions with other organisms.

ecological pyramid Visual depiction of the biomass, number of organisms, or energy content of various trophic levels in a food web—from the producer to the final consumer populations.

ecological succession The gradual replacement of communities in an area following a disturbance (secondary succession) or the creation of new soil (primary succession).

ecology Study of the interactions of organisms with other organisms and with the physical and chemical environment.

ecosystem Biological community together with the associated abiotic environment; characterized by a flow of energy and a cycling of inorganic nutrients.

ecosystem diversity Variety of species in a particular locale, dependent on the species' interactions.

ectoderm Outermost primary tissue layer of an animal embryo; gives rise to the nervous system and the outer layer of the integument.

ectothermic Body temperature that varies according to the environmental temperature.

edge effect Phenomenon in which the edges around a landscape patch provide a slightly different habitat than the favorable habitat in the interior of the patch.

egg Also called an ovum; haploid cell that is usually fertilized by a sperm to form a diploid zygote.

elastic cartilage Type of cartilage composed of elastic fibers, allowing greater flexibility.

elastic fiber Yellow fiber in the matrix of connective tissue, providing flexibility.

electrocardiogram (ECG) Recording of the electrical activity associated with the heartbeat.

electron Negative subatomic particle, moving about in an energy level around the nucleus of an atom.

electron shell The average location, or energy level, of an electron in an atom. Often drawn as concentric circles around the nucleus.

electron transport chain (ETC) Process in a cell that involves the passage of electrons along a series of membrane-bound electron carrier molecules from a higher to lower energy level; the energy released is used for the synthesis of ATP.

electronegativity The ability of an atom to attract electrons toward itself in a chemical bond.

elements Substances that cannot be broken down into substances with different properties; composed of only one type of atom.

elephantiasis Disease caused by a roundworm called the filarial worm.

elongation Middle stage of translation in which additional amino acids specified by the mRNA are added to the growing polypeptide.

embryo Stage of a multicellular organism that develops from a zygote before it becomes free-living; in seed plants, the embryo is part of the seed.

embryo sac Female gametophyte (megagametophyte) of flowering plants.

embryonic development In humans, the first 2 months of development following fertilization, during which the major organs are formed.

embryonic disk During human development, flattened area during gastrulation from which the embryo arises.

emergency contraception Form of contraception that is taken following unprotected intercourse.

emerging disease Disease, such as SARS or MERS, that has not previously been detected in humans.

emphysema Disorder of the respiratory system, specifically the lungs, that is characterized by damage to the alveoli, thus reducing the ability to exchange gases with the external environment.

endangered species A species that is in peril of immediate extinction throughout all or most of its range (e.g., California condor, snow leopard).

endergonic reaction Chemical reaction that requires an input of energy; opposite of exergonic reaction.

endocrine gland Ductless organ that secretes hormone(s) into the bloodstream.

endocrine system Organ system involved in the coordination of body activities; uses hormones as chemical signals secreted into the bloodstream.

endocytosis Process by which substances are moved into the cell from the environment; includes phagocytosis, pinocytosis, and receptor-mediated endocytosis.

endoderm Innermost primary tissue layer of an animal embryo that gives rise to the linings of the digestive tract and associated structures.

endodermis Internal plant root tissue forming a boundary between the cortex and the vascular cylinder.

endometrioisis Condition in which endometrial tissue is located outside of the uterine cavity, causing pain and discomfort.

endometrium Mucous membrane lining the interior surface of the uterus.

endoplasmic reticulum (ER) System of membranous saccules and channels in the cytoplasm, often with attached ribosomes.

endosperm In flowering plants, nutritive storage tissue that is derived from the union of a sperm nucleus and polar nuclei in the embryo sac.

endospore Spore formed within a prokaryotic cell that protects the DNA within the cell from environmental stress.

endosymbiotic theory Explanation of the evolution of eukaryotic organelles by phagocytosis of prokaryotes.

energy Capacity to do work and bring about change; occurs in a variety of forms.

energy of activation Energy that must be added in order for molecules to react with one another.

entropy Measure of disorder or randomness in a system.

envelope Lipid covering of some viruses; located outside of the capsid.

environmental resistance Total of factors in the environment that limit the numerical increase of a population in a particular region.

enzymatic protein Protein that catalyzes a specific reaction; may be found in the plasma membrane or the cytoplasm of the cell.

enzyme Organic catalyst, usually a protein, that speeds a reaction in cells due to its particular shape.

enzyme inhibition Means by which cells regulate enzyme activity; may be competitive or noncompetitive inhibition.

eosinophil White blood cell containing cytoplasmic granules that stain with acidic dye.

epidermal tissue Exterior tissue, usually one cell thick, of leaves, young stems, roots, and other parts of plants.

epidermis In mammals, the outer, protective layer of the skin; in plants, tissue that covers roots, leaves, and stems of nonwoody organisms.

epididymis Location of sperm maturation in an adult human male; located in the testis.

epigenetic inheritance An inheritance pattern in which a nuclear gene has been modified but the changed expression of the gene is not permanent over many generations; the transmission of genetic information by means that are not based on the coding sequences of a gene.

epiglottis Structure that covers the glottis, the air-tract opening, during the process of swallowing.

epinephrine Hormone secreted by the adrenal medulla in times of stress; also referred to as adrenaline.

epiphyte Plant that takes its nourishment from the air because its placement in other plants gives it an aerial position.

episiotomy Surgical procedure during the process of birth (parturition) that involves a small incision to enlarge the vagina; allows for passage of the baby's head.

epithelial tissue Tissue that lines hollow organs and covers surfaces.

erectile dysfunction (ED) Also called impotence, this represents an inability to produce or maintain an erection.

erythrocyte Red blood cell; contains hemoglobin and carries oxygen from the lungs or gills to the tissues in vertebrates.

erythropoietin (EPO) Hormone produced by the kidneys that speeds red blood cell formation.

esophagus Muscular tube for moving swallowed food from the pharynx to the stomach.

essential amino acid One of eight amino acids (nine for children) that are unable to be produced by the human body and therefore must be obtained in the diet.

essential fatty acids Forms of fatty acids that are unable to be produced by the human body and therefore must be obtained in the diet.

estrogen Female sex hormone that helps maintain sexual organs and secondary sex characteristics.

estuary Portion of the ocean located where a river enters and fresh water mixes with salt water.

ethylene Plant hormone that causes ripening of fruit and is also involved in abscission.

eudicot Eudicotyledones; flowering plant group; members have two embryonic leaves (cotyledons), net-veined leaves, vascular bundles in a ring, flower parts in fours or fives and their multiples, and other characteristics.

euglenid Flagellated and flexible freshwater single-celled protist that usually contains chloroplasts and has a semirigid cell wall.

eukaryotic cell (eukaryote) Type of cell that has a membrane-bound nucleus and membranous organelles; found in organisms within the domain Eukarya.

eutrophication Enrichment of water by inorganic nutrients used by phytoplankton. Often, over-enrichment caused by human activities leads to excessive bacterial growth and oxygen depletion.

evaporation The process by which liquid water turns into gaseous form (water vapor).

evergreen Land plant that sheds leaves over a long period, so some leaves are always present.

evolution Genetic change in a species over time, resulting in the development of genetic and phenotypic differences that are the basis of natural selection; descent of organisms from a common ancestor.

exchange pool Location in a biogeochemical cycle from which organisms generally take a specific chemical (i.e., phosphorus).

excretion Elimination of metabolic wastes by an organism at exchange boundaries such as the plasma membrane of unicellular organisms and excretory tubules of multicellular animals.

exergonic reaction Chemical reaction that releases energy; opposite of endergonic reaction.

exocytosis Process in which an intracellular vesicle fuses with the plasma membrane so that the vesicle's contents are released outside the cell.

exon Segment of mRNA containing the protein-coding portion of a gene that remains within the mRNA after splicing has occurred.

experiment A test of a hypothesis that examines the influence of a single variable. Often involves both control and test groups.

experimental design Methodology by which an experiment will seek to support the hypothesis.

experimental variable Factor of the experiment being tested.

expiration Act of expelling air from the lungs; exhalation.

exponential growth Growth, particularly of a population, in which the increase occurs in the same manner as compound interest.

external respiration Exchange of oxygen and carbon dioxide between alveoli of the lungs and blood.

exteroceptor Sensory receptors of the peripheral nervous system that detect stimuli from outside the body.

extinct; extinction Total disappearance of a species or higher group.

extracellular matrix (ECM) Nonliving substance secreted by some animal cells; is composed of protein and polysaccharides.

extraembryonic membrane Membrane that is not a part of the embryo but is necessary to the continued existence and health of the embryo.

eyespot apparatus Structure found in some species of protists that is used to sense light and light intensity.

F

facilitated transport Passive transfer of a substance into or out of a cell along a concentration gradient by a process that requires a protein carrier.

FAD (flavin adenine dinucleotide) A coenzyme of oxidation-reduction that becomes $FADH_2$ as oxidation of substrates occurs in the mitochondria during cellular respiration, FAD then delivers electrons to the electron transport chain.

fall overturn Mixing process that occurs in fall in stratified lakes, whereby oxygen-rich top waters mix with nutrient-rich bottom waters.

familial hypercholesterolemia Genetic disorder that causes an accumulation of cholesterol in the blood due to defects in the LDL receptors on the cell surface.

family One of the categories, or taxa, used to group species; the taxon located above the genus level.

farsighted Condition in which an individual cannot focus on objects closely; caused by the focusing of the image behind the retina of the eye.

fat Organic molecule that contains glycerol and three fatty acids; energy storage molecule.

fate map Diagram used by developmental geneticists that tracks the differentiation of a cell in an embryo.

fatty acid Molecule that contains a hydrocarbon chain and ends with an acid group.

female condom Contraceptive device, usually made from polyurethane, that fits onto the female's cervix.

fermentation Anaerobic breakdown of glucose that results in a gain of two ATP and end products such as alcohol and lactate; occurs in the cytoplasm of cells.

fern Member of a group of land plants that have large fronds; in the sexual life cycle, the independent gametophyte produces flagellated sperm, and the vascular sporophyte produces windblown spores.

fertilization Fusion of sperm and egg nuclei, producing a zygote that develops into a new individual.

fetal development Period in human development from 3 to 9 months after conception that involves growth of the fetus and development of the internal organs.

fiber General term for indigestible plant material, typically complex carbohydrates such as cellulose.

fibrin Protein involved in blood clotting; acts to trap cells to seal wounds in the blood vessels.

fibroblast Cell found in loose connective tissue that synthesizes collagen and elastic fibers in the matrix.

fibrocartilage Cartilage with a matrix of strong collagenous fibers.

fibromyalgia Disorder associated with chronic, widespread pain; cause is not currently known.

fibrous joint Type of joint between two bones that is immovable; sutures are examples of fibrous joints.

fibrous root system In most monocots, a mass of similarly sized roots that cling to the soil.

filament End-to-end chains of cells that form as cell division occurs in only one plane; in plants, the elongated stalk of a stamen.

fimbria (pl., **fimbriae**) Small, bristlelike fiber on the surface of a bacterial cell that attaches bacteria to a surface; also fingerlike extension from the oviduct near the ovary.

first messenger Chemical signal, such as a peptide hormone, that binds to a plasma membrane receptor protein and alters the metabolism of a cell because a second messenger is activated.

fitness Ability of an organism to reproduce and pass its genes to the next fertile generation; measured against the ability of other organisms to reproduce in the same environment.

fixed action pattern (FAP) Innate behavior pattern that is stereotyped, spontaneous, independent of immediate control, genetically encoded, and independent of individual learning.

flagella (sing., **flagellum**) Long, slender extension used for locomotion by some bacteria, protozoans, and sperm.

flagellates Heterotrophic protozoans that propel themselves using one or more flagella.

flagship species Species that evoke a strong emotional response in humans; charismatic, cute, regal (e.g., lions, tigers, dolphins, pandas).

flatworms Invertebrates such as planarians and tapeworms with extremely thin bodies, a three-branched gastrovascular cavity, and a ladder type nervous system.

flower Reproductive organ of a flowering plant, consisting of several kinds of modified leaves arranged in concentric rings and attached to a modified stem called the receptacle.

fluid-mosaic model Model for the plasma membrane based on the changing location and pattern of protein molecules in a fluid phospholipid bilayer.

follicle Structure in the ovary of animals that contains an oocyte; site of oocyte production.

follicle-stimulating hormone (FSH) Hormone released by the anterior pituitary; in males, it promotes the production of sperm; in females, it promotes the development of the follicle in the ovary.

fontanels Membranous regions found in the skull of infants; permit the skull to move through the birth canal.

food chain The order in which one population feeds on another in an ecosystem, thereby showing the flow of energy from a detritivore (detrital food chain) or a producer (grazing food chain) to the final consumer.

food web In ecosystems, a complex pattern of interlocking and crisscrossing food chains.

foot Animal structure involved in locomotion.

foramen magnum Large opening at the base of the skull that allows for a connection between the spinal cord and brain stem.

foramen ovale Structural feature of the fetal heart that allows for blood to move directly between the right and left atrium.

foraminiferan A protist bearing a calcium carbonate test with many openings through which pseudopods extend.

formed elements Portion of the blood that consists of erythrocytes, leukocytes, and platelets (thrombocytes).

fossil Any past evidence of an organism that has been preserved in the Earth's crust.

fossil fuel Remains of once-living organisms that are burned to release energy, such as coal, oil, and natural gas.

founder effect Cause of genetic drift due to colonization by a limited number of individuals who, by chance, have different genotype and allele frequencies than the parent population.

fovea centralis Region of the retina consisting of densely packed cones; responsible for the greatest visual acuity.

fragile X syndrome Genetic condition caused by an abnormal number of nucleotide repeats; named after the appearance, and not physical characteristics, of the X chromosome.

free energy Energy in a system that is capable of performing work.

fruit Flowering plant structure consisting of one or more ripened ovaries that usually contain seeds.

functional genomics Study of gene function at the genome level. It involves the study of many genes simultaneously and the use of DNA microarrays.

functional group Specific cluster of atoms attached to the carbon skeleton of organic molecules that enters into reactions and behaves in a predictable way.

fungi (sing., **fungus**) Eukaryotic saprotrophic decomposer; the body is made up of filaments called hyphae that form a mass called a mycelium.

G

gallbladder Organ attached to the liver that serves to store and concentrate bile.

gallstones Hardened form of bile that may block the bile ducts and cause pancreatitis.

gamete Haploid sex cell; e.g., egg or sperm.

gametophyte Haploid generation of the alternation-of-generations life cycle of a plant; produces gametes that unite to form a diploid zygote.

ganglia (sing., **ganglion**) Collection or bundle of neuron cell bodies usually outside the central nervous system.

gap junction Junction between cells formed by the joining of two adjacent plasma membranes; it lends strength and allows ions, sugars, and small molecules to pass between cells.

gastropod Mollusc with a broad, flat foot for crawling (e.g., snails and slugs).

gastrovascular cavity Blind digestive cavity in animals that have a sac body plan.

gastrulation Formation of a gastrula from a blastula; characterized by an invagination to form cell layers of a caplike structure.

gene Unit of heredity existing as alleles on the chromosomes; in diploid organisms, typically two alleles are inherited—one from each parent.

gene cloning DNA cloning to produce many identical copies of the same gene.

gene flow Sharing of genes between two populations through interbreeding.

gene mutation Altered gene whose sequence of bases differs from the original sequence.

gene pool Total of the alleles of all the individuals in a population.

gene therapy Correction of a detrimental mutation by the insertion of DNA sequences into the genome of a cell.

genetic diversity Variety among members of a population.

genetic drift Mechanism of evolution due to random changes in the allelic frequencies of a population; more likely to occur in small populations or when only a few individuals of a large population reproduce.

genetic engineering Human-induced changes in the genome of an organism, often performed for the benefit of humans.

genetic equilibrium Description for a population in which the frequency of alleles for a given trait is not changing over time.

genetic profile An individual's genome, including any possible mutations.

genetically modified organisms (GMOs) Organisms whose genetic material has been altered or enhanced using DNA technology.

genital herpes Sexually transmitted disease caused by the herpes simplex virus (both type 1 and type 2).

genital warts Sexually transmitted disease caused by the human papillomavirus (HPV).

genome Sum of all the genetic information in a cell or organism.

genomics Area of study that examines the genome of a species or group of species.

genotype Genes of an organism for a particular trait or traits; often designated by letters—for example, *BB* or *Aa*.

genus One of the categories, or taxa, used to group species; contains those species that are most closely related through evolution.

germ layer Primary tissue layer of a vertebrate embryo—namely, ectoderm, mesoderm, or endoderm.

germinate Beginning of growth of a seed, spore, or zygote, especially after a period of dormancy.

gerontology Study of aging and the aging processes.

gibberellin Plant hormone promoting increased stem growth; also involved in flowering and seed germination.

gigantism Condition caused by the overproduction of growth hormone during childhood; individuals often also have diabetes mellitus and other health problems.

ginkgo Member of phylum Ginkgophyta; maidenhair tree.

gland Epithelial cell or group of epithelial cells that are specialized to secrete a substance.

glaucoma Condition in which the fluid in the eye (aqueous humor) accumulates, increasing pressure in the eye and damaging nerve fibers.

global warming Predicted increase in the Earth's temperature due to human activities that promote the greenhouse effect.

glomerular capsule Cuplike structure that is the initial portion of a nephron.

glomerular filtration Movement of small molecules from the glomerulus into the glomerular capsule due to the action of blood pressure.

glomerulus Capillary network within the glomerular capsule of a nephron.

glottis Opening for airflow in the larynx.

glucagon Hormone, produced by the pancreas, that stimulates the liver to break down glycogen, thus raising blood glucose levels.

glucocorticoid Type of hormone secreted by the adrenal cortex that influences carbohydrate, fat, and protein metabolism; *see also* cortisol.

glucose Six-carbon monosaccharide; used as an energy source during cellular respiration and as a monomer of the structural polysaccharides.

glycemic index Nutritional measurement of how the body responds to the glucose level of a food. Foods with a high glycemic index result in the release of high amounts of insulin.

glycogen Storage polysaccharide found in animals; composed of glucose molecules joined in a linear fashion but having numerous branches.

glycolipid Lipid in plasma membranes that contains an attached carbohydrate chain; assembled in the Golgi apparatus.

glycolysis Anaerobic breakdown of glucose that results in a gain of two ATP and the production of pyruvate; occurs in the cytoplasm of cells.

glycoprotein Protein in plasma membranes that has an attached carbohydrate chain; assembled in the Golgi apparatus.

gnetophyte Member of one of the four phyla of gymnosperms; Gnetophyta has only three living genera, which differ greatly from one another—e.g., *Welwitschia* and *Ephedra*.

Golgi apparatus Organelle consisting of sacs and vesicles that processes, packages, and distributes molecules about or from the cell.

gonad Organ that produces gametes; the ovary produces eggs, and the testis produces sperm.

gonadotropic hormone Substance, secreted by the anterior pituitary, that regulates the activity of the ovaries and testes; principally, follicle-stimulating hormone (FSH) and luteinizing hormone (LH).

gonadotropin-releasing hormone (GnRH) Hormone released by the hypothalamus that influences the secretion of hormones by the anterior pituitary, thus regulating the activity of the testes.

gonorrhea Sexually transmitted disease caused by the bacterium *Neisseria gonorrhoeae*.

gout Condition caused by the inability of the urinary system to remove uric acid from the body.

gradualistic model Model of genetic change in a species which suggests that change occurs at a slow, steady pace over time.

Gram stain Method of determining whether a culture of bacteria possess a layer of exposed peptidoglycan outside of their plasma membranes; historically used to classify bacteria as Gram-positive or Gram-negative, now used to determine the most effective antibiotic treatment.

grana (sing., **granum**) Stack of chlorophyll-containing thylakoids in a chloroplast.

grassland Biome characterized by rainfall greater than 25 cm/yr, grazing animals, and warm summers; includes the prairie of the U.S. Midwest and the African savanna.

Graves disease Disease associated with the overproduction of T_3 and T_4 by the thyroid; caused by antibodies that interact with TSH receptors in the thyroid.

gravitational equilibrium Maintenance of balance when the head and body are motionless.

gravitropism Growth response of roots and stems of plants to the Earth's gravity; roots demonstrate positive gravitropism, and stems demonstrate negative gravitropism.

gray crescent Region that forms during early cellular differentiation of the embryo; believed to contain unique chemical signals that are important during development.

gray matter Nonmyelinated axons and cell bodies in the central nervous system.

grazing food chain A flow of energy to a straight-line linking of organisms according to who eats whom.

grazing food web Complex pattern of interlocking and crisscrossing food chains that begins with populations of autotrophs serving as producers.

green algae Members of a diverse group of photosynthetic protists; contain chlorophylls *a* and *b* and have other biochemical characteristics like those of plants.

greenhouse effect Reradiation of solar heat toward the Earth, caused by an atmosphere that allows the sun's rays to pass through, but traps the heat in the same manner as the glass of a greenhouse.

ground tissue Tissue that constitutes most of the body of a plant; consists of parenchyma, collenchyma, and sclerenchyma cells that function in storage, basic metabolism, and support.

growth factor A hormone or chemical, secreted by one cell, that may stimulate or inhibit growth of another cell or cells.

growth hormone (GH) Substance secreted by the anterior pituitary; controls size of an individual by promoting cell division, protein synthesis, and bone growth.

guanine (G) One of four nitrogen-containing bases in nucleotides composing the structure of DNA and RNA; pairs with cytosine.

guard cell One of two cells that surround a leaf stoma; changes in the turgor pressure of these cells cause the stoma to open or close.

Guillain–Barré syndrome Disease caused by inflammation in the nervous system, resulting in demyelination of the nerve axons in the peripheral nervous system.

gymnosperm Type of woody seed plant in which the seeds are not enclosed by fruit and are usually borne in cones, such as those of the conifers.

H

habitat Place where an organism lives and is able to survive and reproduce.

hair cells Cells located in the inner ear that function as mechanoreceptors for hearing.

hair follicle Tubelike depression in the skin in which a hair develops.

halophile Type of archaea that lives in extremely salty habitats.

haploid (n) Cell condition in which only one of each type of chromosome is present.

Hardy-Weinberg equilibrium Mathematical law stating that the gene frequencies in a population remain stable if evolution does not occur due to nonrandom mating, selection, migration, and genetic drift.

heart Muscular organ whose contraction causes blood to circulate in the body of an animal.

heart attack Damage to the myocardium due to blocked circulation in the coronary arteries; myocardial infarction.

helper T cell Secretes lymphokines, which stimulate all kinds of immune cells.

hemocoel Residual coelom found in arthropods, which is filled with hemolymph.

hemodialysis Treatment for kidney failure in which the patient's blood is passed through an artificial kidney to remove waste material and toxins.

hemoglobin Iron-containing respiratory pigment occurring in vertebrate red blood cells and in the blood plasma of some invertebrates.

hemophilia Genetic disorder that is caused by a deficiency of a clotting factor in the blood.

hepatitis Inflammation of the liver. Viral hepatitis occurs in several forms.

herbivore Primary consumer in a grazing food chain; a plant eater.

hermaphroditic Type of animal that has both male and female sex organs.

heterotroph Organism that cannot synthesize needed organic compounds from inorganic substances and therefore must take in organic food.

heterozygous Possessing unlike alleles for a particular trait.

hexose Any monosaccharide that contains six carbons; examples are glucose and galactose.

hippocampus Region of the central nervous system associated with learning and memory; part of the limbic system.

histamine Substance, produced by basophils in blood and mast cells in connective tissue, that causes capillaries to dilate.

homeodomain Conserved DNA-binding region of transcription factors encoded by the homeobox of homeotic genes.

homeostasis Maintenance of normal internal conditions in a cell or an organism by means of self-regulating mechanisms.

homeotic genes Genes that control the overall body plan by controlling the fate of groups of cells during development.

homologous chromosome Member of a pair of chromosomes that are alike and come together in synapsis during prophase of the first meiotic division; a *homologue*.

homologous structure A structure that is similar in different types of organisms because these organisms descended from a common ancestor.

homozygous Possessing two identical alleles for a particular trait.

hormone Chemical messenger, produced in one part of the organism, that controls the activity of other tissues or organs in that organism.

horsetail A seedless vascular plant having only one genus (*Equisetum*) in existence today; characterized by rhizomes, scalelike leaves, strobili, and tough, rigid stems.

human chorionic gonadotropin (HCG) Gonadotropic hormone produced by the chorion that functions to maintain the uterine lining.

Human Genome Project (HGP) Initiative to determine the complete sequence of the human genome and to analyze this information.

human immunodeficiency virus (HIV) A retrovirus that is responsible for the disease AIDS.

Huntington disease Autosomal dominant genetic disorder that affects the nervous system; results in a progressive loss of neurons in the brain.

hyaline cartilage Cartilage whose cells lie in lacunae separated by a white translucent matrix containing very fine collagen fibers.

hybridization Interbreeding between two different species; typically prevented by prezygotic isolation mechanisms.

hydrogen bond Weak bond that arises between a slightly positive hydrogen atom of one molecule and a slightly negative atom of another molecule, or between parts of the same molecule.

hydrolysis reaction Splitting of a chemical bond by the addition of water, with the H^+ going to one molecule and the OH^- going to the other.

hydrophilic Type of molecule, often polar, that interacts with water by dissolving in water and/or by forming hydrogen bonds with water molecules.

hydrophobic Type of molecule that is typically nonpolar and therefore does not interact easily with water.

hydrostatic skeleton Fluid-filled body compartment that provides support for muscle contraction, resulting in movement; seen in cnidarians, flatworms, roundworms, and segmented worms.

hydrothermal vent Hot springs in the seafloor along ocean ridges, where heated seawater and sulfate react to produce hydrogen sulfide; here, chemosynthetic bacteria support a community of varied organisms.

hypertension Form of cardiovascular disease characterized by blood pressure over 140/95 (over 45 years of age) or 130/90 (under 45 years of age).

hyperthyroidism Caused by the oversecretion of hormones from the thyroid gland; symptoms include hyperactivity, nervousness, and insomnia.

hypertonic solution Higher solute concentration (less water) than the cytoplasm of a cell; causes cell to lose water by osmosis.

hypha (pl., **hyphae**) Filament of the vegetative body of a fungus.

hypothalamic-inhibiting hormone One of many hormones, produced by the hypothalamus, that inhibits the secretion of an anterior pituitary hormone.

hypothalamic-releasing hormone One of many hormones, produced by the hypothalamus, that stimulates the secretion of an anterior pituitary hormone.

hypothalamus In vertebrates, part of the brain that helps regulate the internal environment of the body—for example, heart rate, body temperature, and water balance.

hypothesis Supposition established by reasoning after consideration of available evidence; it can be tested by obtaining more data, often by experimentation.

hypothyroidism Caused by the under-secretion of hormones from the thyroid gland; symptoms include weight gain, lethargic behavior, and depression.

hypotonic solution Solution that contains a lower solute (more water) concentration than the cytoplasm of a cell; causes cell to gain water by osmosis.

I

ileum Region of the small intestine; connects the jejunum and large intestine.

immediate allergic response Allergic response that occurs within seconds of contact with an allergen; caused by the attachment of the allergen to IgE antibodies.

immune system System associated with protection against pathogens, toxins, and some cancerous cells. In humans, this is an organ system.

immunity Ability of the body to protect itself from foreign substances and cells, including disease-causing agents.

immunization Strategy for achieving immunity to the effects of specific disease-causing agents.

immunodeficiency disease Condition that is the result of the immune system's inability to protect the body.

immunoglobulin (Ig) Globular plasma protein that functions as an antibody.

imprinting Learning to make a particular response to only one type of animal or object.

inclusive fitness Fitness that results from personal reproduction and from helping nondescendant relatives reproduce.

incomplete dominance Inheritance pattern in which an offspring has an intermediate phenotype, as when a red-flowered plant and a white-flowered plant produce pink-flowered offspring.

incus Bone found in the middle ear that assists in the transmission of sound to the inner ear; also called the anvil.

independent assortment Alleles of unlinked genes segregate independently of each other during meiosis so that the gametes can contain all possible combinations of alleles.

induced fit model Change in the shape of an enzyme's active site that enhances the fit between the active site and its substrate(s).

induction Ability of a chemical or a tissue to influence the development of another tissue.

inductive reasoning Using specific observations and the process of logic and reasoning to arrive at general scientific principles.

infertility Failure of a couple to achieve pregnancy after one year of regular, unprotected intercourse.

inflammatory response Tissue response to injury that is characterized by redness, swelling, pain, and heat.

inheritance of acquired characteristics Lamarckian belief that characteristics acquired during the lifetime of an organism can be passed on to offspring.

initiation First stage of translation in which the translational machinery binds an mRNA and assembles.

innate immunity An immune response that does not require a previous exposure to the pathogen.

inner ear Portion of the ear consisting of a vestibule, semicircular canals, and the cochlea where equilibrium is maintained and sound is transmitted.

insect Type of arthropod. The head has antennae, compound eyes, and simple eyes; the thorax has three pairs of legs and often wings; and the abdomen has internal organs.

insertion When referring to a muscle, the connection of the muscle to the bone that moves.

inspiration Act of taking air into the lungs; inhalation.

insulin Hormone, released by the pancreas, that serves to lower blood glucose levels by stimulating the uptake of glucose by cells, especially muscle and liver cells.

integration Summing up of excitatory and inhibitory signals by a neuron or by some part of the brain.

integumentary system Organ system of humans that contains the skin, protects the body, synthesizes sensory input, assists in temperature regulation, and synthesizes vitamin D.

interferon Antiviral agent, produced by an infected cell, that blocks the infection of another cell.

interkinesis Period of time between meiosis I and meiosis II during which no DNA replication takes place.

intermediate filament Ropelike assemblies of fibrous polypeptides in the cytoskeleton that provide support and strength to cells; so called because they are intermediate in size between actin filaments and microtubules.

internal respiration Exchange of oxygen and carbon dioxide between blood and tissue fluid.

interneuron Neuron located within the central nervous system that conveys messages between parts of the central nervous system.

internode In vascular plants, the region of a stem between two successive nodes.

interoceptor Sensory receptors of the peripheral nervous system that detect stimuli from inside the body.

interphase Stages of the cell cycle (G_1, S, G_2) during which growth and DNA synthesis occur when the nucleus is not actively dividing.

interstitial cells Cells of the male reproductive system that secrete the androgens (such as testosterone).

interstitial fluid Fluid that surrounds the body's cells; consists of dissolved substances that leave the blood capillaries by filtration and diffusion.

intervertebral discs Fibrocartilage located between the vertebrae that act as shock absorbers.

intrauterine device (IUD) Contraceptive device that is inserted into the female's uterus; believed to work by altering the environment of the uterus.

intron Intervening sequence found between exons in mRNA; removed by RNA processing before translation.

inversion Change in chromosome structure in which a segment of a chromosome is turned around 180°; this reversed sequence of genes can lead to altered gene activity and abnormalities.

invertebrate Animal without a vertebral column or backbone.

ion Charged particle that carries a negative or positive charge.

ionic bond Chemical bond in which ions are attracted to one another by opposite charges.

iris Muscular ring that surrounds the pupil and regulates the passage of light through this opening.

isomer Molecules with the same molecular formula but a different structure, and therefore a different shape.

isotonic solution Solution that is equal in solute concentration to that of the cytoplasm of a cell; causes cell to neither lose nor gain water by osmosis.

isotope Atoms of the same element having the same atomic number but a different mass number due to a variation in the number of neutrons.

J

jawless fishes Type of fishes that lack jaws (agnathan); includes hagfishes and lampreys.

jejunum Region of the small intestine located between the duodenum and ileum.

joint Articulation between two bones of a skeleton.

K

karyotype Chromosomes arranged by pairs according to their size, shape, and general appearance in mitotic metaphase.

keystone species Species whose activities significantly affect community structure.

kidney stones Hardened granules that may form in the renal pelvis of the kidney; caused by factors such as diet, infections, or pH imbalance.

kidneys Paired organs of the vertebrate urinary system that regulate the chemical composition of the blood and produce a waste product called urine.

kin selection Indirect selection; adaptation to the environment due to the reproductive success of an individual's relatives.

kinetic energy Energy associated with motion.

kingdom One of the categories, or taxa, used to group species; the taxon above the phylum level.

L

lacteal Lymphatic vessel in an intestinal villus; aids in the absorption of fats.

lacuna Small pit or hollow cavity, as in bone or cartilage, where a cell or cells are located.

lake Body of fresh water, often classified by nutrient status, such as oligotrophic (nutrient-poor) or eutrophic (nutrient-rich).

lancelet Invertebrate chordate with a body that resembles a lancet and has the four chordate characteristics as an adult.

landscape diversity Variety of habitat elements within an ecosystem (e.g., plains, mountains, and rivers).

lanugo Fine, downlike covering of the fetus.

large intestine In vertebrates, portion of the digestive tract that follows the small intestine. In humans, consists of the cecum, colon, rectum, and anal canal.

laryngitis Inflammation of the larynx, usually resulting in an inability or difficulty in speaking.

larynx Cartilaginous organ located between the pharynx and the trachea; in humans, contains the vocal cords; sometimes called the voice box.

last universal common ancestor (LUCA) The first living organism on the planet, from which all life evolved.

law Universal principle that describes the basic functions of the natural world.

law of independent assortment Mendelian principle that explains how combinations of traits appear in gametes; *see also* independent assortment.

law of segregation Mendelian principle that explains how, in a diploid organism, alleles separate during the formation of the gametes.

laws of thermodynamics Two laws explaining energy and its relationships and exchanges. The first, also called the "law of conservation," says that energy cannot be created or destroyed but can only be changed from one form to another. The second says that energy cannot be changed from one form to another without a loss of usable energy.

leaf Lateral appendage of a stem, highly variable in structure, often containing cells that carry out photosynthesis.

leaf vein Vascular tissue within a leaf.

learning Relatively permanent change in an animal's behavior that results from practice and experience.

leech Blood-sucking annelid, usually found in fresh water, with a sucker at each end of a segmented body.

lens Transparent "disk" structure found in the vertebrate eye behind the iris; brings objects into focus on the retina.

leptin Hormone produced by adipose tissue that acts on the hypothalamus to signal satiety (fullness).

less-developed country (LDC) Country that is becoming industrialized; typically, population growth is expanding rapidly, and the majority of people live in poverty.

leukocyte White blood cell, of which there are several types, each having a specific function in pro-

tecting the body from invasion by foreign substances and organisms.

lichen Symbiotic relationship between certain fungi and either cyanobacteria or algae, in which the fungi possibly provide inorganic food or water and the algae or cyanobacteria provide organic food.

ligament Tough cord or band of dense fibrous tissue that binds bone to bone at a joint.

light reaction Portion of photosynthesis that captures solar energy and takes place in thylakoid membranes of chloroplasts; it produces ATP and NADPH.

lignin Chemical that hardens the cell walls of land plants.

limbic system In humans, functional association of various brain centers, including the amygdala and hippocampus; governs learning and memory and various emotions such as pleasure, fear, and happiness.

lineage Line of descent represented by a branch in a phylogenetic tree.

linkage group Term used to identify groups of alleles on a chromosome that tend to be inherited together.

lipase Fat-digesting enzyme secreted by the pancreas.

lipid Class of organic compounds that tends to be soluble in nonpolar solvents; includes fats and oils.

lipopolysaccharide Lipid-sugar molecules that are found on the exterior of some species of bacteria; presence indicates that the species is Gram-negative.

liposome Droplet of phospholipid molecules formed in a liquid environment.

littoral zone Shore zone between high-tide mark and low-tide mark; also, shallow water of a lake where light penetrates to the bottom.

liver Large, dark red internal organ that produces urea and bile, detoxifies the blood, stores glycogen, and produces the plasma proteins, among other functions.

liverwort Bryophyte with a lobed or leafy gametophyte and a sporophyte composed of a stalk and capsule.

lobe-finned fishes Type of fishes with limblike fins; also called the sarcopterygians.

locus Physical location of a trait (or gene) on a chromosome.

logistic growth Population increase that results in an S-shaped curve; growth is slow at first, steepens, and then levels off due to environmental resistance.

long-day plant Plant that flowers when day length is longer than a critical length; e.g., wheat, barley, clover, and spinach.

loop of the nephron Portion of a nephron between the proximal and distal convoluted tubules; functions in water reabsorption.

loose fibrous connective tissue Tissue composed mainly of fibroblasts widely separated by a matrix containing collagen and elastic fibers.

lophophoran A general term to describe several groups of lophotrochozoans that have a feeding structure called a lophophore.

lumen Cavity inside any tubular structure, such as the lumen of the digestive tract.

lung cancer Uncontrolled cell growth that affects any component of the respiratory system.

lung Internal respiratory organ containing moist surfaces for gas exchange.

luteinizing hormone (LH) Hormone released by the anterior pituitary; in males, it regulates the production of testosterone by the interstitial cells; in females, it promotes the development of the corpus luteum in the ovary; also called interstitial cell-stimulating hormone (ICSH) in males.

lymph Fluid, derived from interstitial fluid, that is carried in lymphatic vessels.

lymph node Mass of lymphatic tissue located along the course of a lymphatic vessel.

lymphatic capillary Smallest vessels of the lymphatic system; closed-ended; responsible for the uptake of fluids from the surrounding tissues.

lymphatic system Organ system consisting of lymphatic vessels and lymphatic organs; transports lymph and lipids, and aids the immune system.

lymphatic vessel Vessel of the lymphatic system that is responsible for transporting excess interstitial fluid, or lymph, from the tissues to the circulatory system.

lymphocyte Specialized white blood cell that functions in specific defense; occurs in two forms—T lymphocytes and B lymphocytes.

lymphoid organ Organ other than a lymphatic vessel that is part of the lymphatic system; the lymphatic organs are the lymph nodes, tonsils, spleen, thymus gland, and bone marrow.

lysosome Membrane-bound vesicle that contains hydrolytic enzymes for digesting macromolecules and bacteria; used to recycle worn-out cellular organelles.

M

macroevolution Large-scale evolutionary change, such as the formation of new species.

macrophage In vertebrates, large phagocytic cell derived from a monocyte, that ingests microbes and debris.

macular degeneration Condition in which the capillaries supplying the retina of the eye become damaged, resulting in reduced vision and blindness.

malaria Disease caused by the protozoan *Plasmodium vivax;* transmitted by the *Anopheles* mosquito.

male condom Contraceptive device, usually made of latex, that fits over the male penis.

malignant Term that is used to describe cancer cells meaning that it has the ability to move to other areas of the body and threaten life.

malleus Bone found in the middle ear that assists in the transmission of sound to the inner ear; also called the hammer.

Malpighian tubule Blind, threadlike excretory tubule attached to the gut of an insect.

maltase Enzyme produced in small intestine that breaks down maltose to two glucose molecules.

mammal Endothermic vertebrate characterized especially by the presence of hair and mammary glands.

mantle In molluscs, an extension of the body wall that covers the visceral mass and may secrete a shell.

Marfan syndrome Autosomal dominant genetic disorder of the connective tissue, specifically the fibrillin protein.

marsupial Member of a group of mammals bearing immature young nursed in a marsupium, or pouch—for example, kangaroo and opossum.

mass number Mass of an atom equal to the number of protons plus the number of neutrons within the nucleus.

mast cell Connective tissue cell that releases histamine in allergic reactions.

matrix Unstructured semifluid substance that fills the space between cells in connective tissues or inside organelles.

matter Anything that takes up space and has mass.

mechanical energy A type of kinetic energy associated with the position, or motion (such as walking or running), of an object.

mechanoreceptor Sensory receptor that responds to mechanical stimuli, such as pressure, sound waves, or gravity.

medulla oblongata In vertebrates, part of the brain stem that is continuous with the spinal cord; controls heartbeat, blood pressure, breathing, and other vital functions.

medusa Among cnidarians, bell-shaped body form that is directed downward and contains much mesoglea.

megaphyll Large leaf with several to many veins.

megaspore One of the two types of spores produced by seed plants; develops into a female gametophyte (embryo sac).

meiosis Type of nuclear division that reduces the chromosome number from 2n to n; daughter cells receive the haploid number of chromosomes in varied combinations; also called reduction division.

melanocyte Specialized cell in the epidermis that produces melanin, the pigment responsible for skin color.

melanocyte-stimulating hormone (MSH) Substance that causes melanocytes to secrete melanin in most vertebrates.

melatonin Hormone, secreted by the pineal gland, that is involved in biorhythms.

membrane-first hypothesis Proposes that the plasma membrane was the first component of the early cells to evolve.

memory Capacity of the brain to store and retrieve information about past sensations and perceptions; essential to learning.

memory B cell Forms during a primary immune response but enters a resting phase until a secondary immune response occurs.

memory T cell T cell that differentiates during an initial infection and responds rapidly during subsequent exposure to the same antigen.

meninges Protective membranous coverings around the central nervous system.

meningitis A condition that refers to inflammation of the brain or spinal cord meninges (membranes).

menopause Termination of the ovarian and uterine cycles in older women.

menstruation Periodic shedding of tissue and blood from the inner lining of the uterus in primates.

meristematic tissue Undifferentiated embryonic tissue in the active growth regions of plants.

mesoderm Middle primary tissue layer of an animal embryo that gives rise to muscle, several internal organs, and connective tissue layers.

mesoglea Transparent jellylike substance located between the endoderm and ectoderm of some sponges and cnidarians.

mesophyll Inner, thickest layer of a leaf, consisting of palisade and spongy mesophyll; the site of most of photosynthesis.

messenger RNA (mRNA) Type of RNA formed from a DNA template and bearing coded information for the amino acid sequence of a polypeptide.

metabolic pathway Series of linked reactions, beginning with a particular reactant and terminating with an end product.

metabolism The sum of the chemical reactions that occur in a cell.

metamorphosis Change in shape and form that some animals, such as insects, undergo during development.

metaphase Third phase of mitosis; chromosomes are aligned at the metaphase plate.

metastasis Spread of cancer from the place of origin throughout the body; caused by the ability of cancer cells to migrate and invade tissues.

methanogen Type of archaea that lives in oxygen-free habitats, such as swamps, and releases methane gas.

MHC (major histocompatibility complex) protein Protein marker that is a part of cell-surface markers anchored in the plasma membrane, which the immune system uses to identify "self."

micelle Single layer of fatty acids (or phospholipids) that orientate themselves in an aqueous environment.

microbiota Typical complement of bacterial species found on the human body; includes species found both internally and externally.

microevolution Change in gene frequencies between populations of a species over time.

microspore One of the two types of spores produced by seed plants; develops into a male gametophyte (pollen grain).

Microsporidia Phyla of fungi that are characterized as intracellular parasites of animals.

microtubule Small, cylindrical organelle composed of tubulin protein around an empty central core; present in the cytoplasm, centrioles, cilia, and flagella.

midbrain In mammals, the part of the brain located below the thalamus and above the pons.

middle ear Portion of the ear consisting of the tympanic membrane, the oval and round windows, and the ossicles, where sound is amplified.

mimicry Superficial resemblance of two or more species; a survival mechanism that avoids predation by appearing to be noxious.

mineral Naturally occurring inorganic substance containing two or more elements; certain minerals are needed in the diet.

mineralocorticoid Hormone secreted by the adrenal cortex that regulates salt and water balance, leading to increases in blood volume and blood pressure.

mitochondria (sing., **mitochondrion**) Membrane-bound organelle in which ATP molecules are produced during the process of cellular respiration.

mitosis The stage of cellular reproduction in which nuclear division occurs; process in which a parent nucleus produces two daughter nuclei, each having the same number and kinds of chromosomes as the parent nucleus.

model Simulation of a process that aids conceptual understanding until the process can be studied firsthand; a hypothesis that describes how a particular process could possibly be carried out.

molecular clock Idea that the rate at which mutational changes accumulate in certain genes is constant over time and is not involved in adaptation to the environment.

molecule Union of two or more atoms of the same element; also, the smallest part of a compound that retains the properties of the compound.

molluscs Invertebrates including squids, clams, snails, and chitons; characterized by a visceral mass, a mantle, and a foot.

monoclonal antibody One of many antibodies produced by a clone of hybridoma cells that all bind to the same antigen.

monocot Monocotyledone; flowering plant group; members have one embryonic leaf (cotyledon), parallel-veined leaves, scattered vascular bundles, flower parts in threes or multiples of three, and other characteristics.

monocyte Type of agranular leukocyte that functions as a phagocyte, particularly after it becomes a macrophage, which is also an antigen-presenting cell.

monohybrid cross Cross between parents that differ in only one trait.

monomer Small molecule that is a subunit of a polymer—e.g., glucose is a monomer of starch.

monosaccharide Simple sugar; a carbohydrate that cannot be broken down by hydrolysis—e.g., glucose; also, any monomer of the polysaccharides.

monosomy Chromosome condition in which a diploid cell has one less chromosome than normal; designated as 2n − 1.

monotreme Egg-laying mammal—e.g., duckbill platypus or spiny anteater.

monsoon Climate in India and southern Asia caused by wet ocean winds that blow onshore for almost half the year.

more-developed country (MDC) Country that is industrialized; typically, population growth is low, and the people enjoy a good standard of living overall.

morphogen genes Specific sequences of DNA that determine the relationship of structures during development; often expressed in a gradient across the embryo.

morphogenesis Emergence of shape in tissues, organs, or the entire embryo during development.

morula Spherical mass of cells resulting from cleavage during animal development prior to the blastula stage.

moss Bryophyte that is typically found in moist habitats.

motor (efferent) neuron Nerve cell that conducts nerve impulses away from the central nervous system and innervates effectors (muscle and glands).

motor molecule Protein that moves along either actin filaments or microtubules and translocates organelles.

motor unit Combination of nerve fibers and muscles fibers in a muscle.

mouth In humans, organ of the digestive tract where food is chewed and mixed with saliva.

mucous membrane Body membrane that lines the tubes of several organ systems in humans; secretes mucus that helps protect the body from infection.

multiple alleles Inheritance pattern in which there are more than two alleles for a particular trait; each individual has only two of all possible alleles.

multiple sclerosis (MS) Disease of the central nervous system characterized by the breakdown of myelin in the neurons; considered to be an autoimmune disease.

muscle tone Physiological condition by which, at any given time, some muscle fibers are always contracting in a muscle, even when the muscle is at rest.

muscle twitch Single contraction of a muscle fiber; typically lasts only a fraction of a second.

muscular dystrophy (MD) Disease that causes a weakening of the muscles; has a strong genetic component.

muscular tissue Type of animal tissue composed of fibers that shorten and lengthen to produce movements.

musculoskeletal system Name for the combined muscular and skeletal systems of humans; involved in movement and posture.

mutagen Chemical or physical agent that increases the chance of mutation.

mutation Change in the nucleotide structure of an organism's DNA.

mutualism Symbiotic relationship in which both species benefit in terms of growth and reproduction.

myasthenia gravis (MG) Autoimmune disease in which antibodies interfere with neuromuscular junctions, causing muscle weakness.

mycelium Tangled mass of hyphal filaments composing the vegetative body of a fungus.

mycorrhizae (sing., **mycorrhiza**) Mutualistic relationship between fungal hyphae and roots of vascular plants.

mycoses General name for fungal diseases of animals.

myelin sheath White, fatty material—derived from the membrane of neurolemmocytes—that forms a covering for nerve fibers.

myofibril Specific muscle cell organelle containing a linear arrangement of sarcomeres, which shorten to produce muscle contraction.

myosin Muscle protein making up the thick filaments in a sarcomere; it pulls actin to shorten the sarcomere, yielding muscle contraction. *See also* actin.

N

NAD$^+$ (nicotinamide adenine dinucleotide) Coenzyme in oxidation-reduction reactions that accepts electrons and hydrogen ions to become NADH + H$^+$ as oxidation of substrates occurs. During cellular respiration, NADH carries electrons to the electron transport chain in mitochondria.

NADP$^+$ (nicotinamide adenine dinucleotide phosphate) Coenzyme in oxidation-reduction reactions that accepts electrons and hydrogen ions to become NADPH + H$^+$. During photosynthesis, NADPH participates in the reduction of carbon dioxide to a carbohydrate.

nail Flattened epithelial tissue from the stratum lucidum of the skin; located on the tips of fingers and toes.

natural killer (NK) cell Lymphocyte that causes an infected or cancerous cell to burst.

natural selection Mechanism of evolutionary change caused by environmental selection of organisms most fit to reproduce; results in adaptation to the environment.

Neandertal Hominin with a sturdy build that lived during the last Ice Age in Europe and the Middle East; hunted large game and left evidence of being culturally advanced.

nearsighted (myopic) Condition in which an individual cannot focus on objects at a distance; caused by the focusing of the image in front of the retina of the eye.

negative feedback Mechanism of homeostatic response by which the output of a system suppresses or inhibits activity of the system.

nematocyst In cnidarians, a capsule that contains a threadlike fiber, the release of which aids in the capture of prey.

nephridium (pl., **nephridia**) Segmentally arranged, paired excretory tubule of many invertebrates, as in the earthworm.

nephron Microscopic kidney unit that regulates blood composition by glomerular filtration, tubular reabsorption, and tubular secretion.

nerve Bundle of long axons outside the central nervous system.

nerve cord Component of an invertebrate's nervous system; usually located ventrally in invertebrates; commonly called the spinal cord in vertebrates.

nerve fiber Axon; conducts nerve impulses away from the cell. Axons are classified as either myelinated or unmyelinated based on the presence or absence of a myelin sheath.

nerve impulse Electrical signal that conveys information along the length of a neuron.

nervous system Organ system of humans that includes the brain, spinal cord, sense organs (eyes, ears, etc.), and associated nerves. Receives, integrates, and stores sensory input; coordinates activity of other organ systems.

nervous tissue Tissue that contains nerve cells (neurons), which conduct impulses, and neuroglia, which support, protect, and provide nutrients to neurons.

neuroglia Nonconducting nerve cells that are intimately associated with neurons and that function in a supportive capacity.

neuromuscular junction Region where an axon bulb approaches a muscle fiber; contains a presynaptic membrane, a synaptic cleft, and a postsynaptic membrane.

neuron Nerve cell that characteristically has three parts: dendrites, cell body, and an axon.

neurotransmitter Chemical stored at the ends of axons that is responsible for transmission across a synapse.

neutron Neutral subatomic particle, located in the nucleus and assigned one atomic mass unit.

neutrophil Granular leukocyte that is the most abundant of the white blood cells; first to respond to infection.

nitrogen fixation Process whereby free atmospheric nitrogen is converted into compounds, such as ammonium and nitrates, usually by bacteria.

node In plants, the place where one or more leaves attach to a stem.

nodes of Ranvier Gaps in the myelin sheath around a nerve fiber.

noncyclic electron pathway Light-dependent photosynthetic pathway that is used to generate ATP and NADPH. Because the pathway is noncyclic, the electrons must be replaced by the splitting of water (photolysis).

nondisjunction Failure of the homologous chromosomes or sister chromatids to separate during either mitosis or meiosis; produces cells with abnormal chromosome numbers.

nonpolar covalent bond Bond in which the sharing of electrons between atoms is fairly equal.

nonrandom mating Mating among individuals on the basis of their phenotypic similarities or differences, rather than mating on a random basis.

nonrenewable resource Minerals, fossil fuels, and other materials present in essentially fixed amounts (within the human timescale) in our environment.

nonvascular plants Bryophytes, such as mosses and liverworts, that have no vascular tissue and either occur in moist locations or have special adaptations for living in dry locations.

norepinephrine Neurotransmitter of the postganglionic fibers in the sympathetic division of the autonomic system; also, a hormone produced by the adrenal medulla. Sometimes referred to as noradrenaline.

nose External structure of the respiratory system that is involved in the process of ventilation.

notochord Cartilage-like supportive dorsal rod in all chordates at some time in their life cycle; replaced by vertebrae in vertebrates.

nuclear envelope Double membrane that surrounds the nucleus in eukaryotic cells and is connected to the endoplasmic reticulum; has pores that allow substances to pass between the nucleus and the cytoplasm.

nuclear pore Opening in the nuclear envelope that permits the passage of proteins into the nucleus and ribosomal subunits out of the nucleus.

nucleoid Region of the cytoplasm in prokaryotic cells where DNA is located; it is not bound by a nuclear envelope.

nucleolus Dark-staining, spherical body in the nucleus that produces ribosomal subunits.

nucleoplasm Semifluid medium of the nucleus, containing chromatin.

nucleotide Monomer of DNA and RNA consisting of a 5-carbon sugar bonded to a nitrogenous base and a phosphate group.

nucleus Membrane-bound organelle within a eukaryotic cell that contains chromosomes and controls the structure and function of the cell.

O

obesity Overweight condition characterized as having a body mass index (BMI) of 30 or greater.

observation Initial step in the scientific method that often involves the recording of data from an experiment or natural event.

octet rule The observation that an atom is most stable when its outer shell is complete and contains eight electrons; an exception is hydrogen, which requires only two electrons in its outer shell to have a completed shell.

oil Triglyceride, usually of plant origin, that is composed of glycerol and three fatty acids and is liquid in consistency due to many unsaturated bonds in the hydrocarbon chains of the fatty acids.

oil gland Gland of the skin, associated with a hair follicle, that secretes sebum; sebaceous gland.

olfactory cell Modified neuron that is a sensory receptor for the sense of smell.

oligodendrocyte Type of glial cell that forms myelin sheaths around neurons in the CNS.

omnivore Organism in a food chain that feeds on both plants and animals.

oncogene Cancer-causing gene formed by a mutation in a proto-oncogene; codes for proteins that stimulate the cell cycle and inhibit apoptosis.

oocyte Immature egg that is undergoing meiosis. Upon completion of meiosis, the oocyte becomes an egg.

oogenesis Production of eggs in females by the process of meiosis and maturation.

open circulatory system Arrangement of internal transport in which blood bathes the organs directly, and there is no distinction between blood and interstitial fluid.

operant conditioning Learning that results from rewarding or reinforcing a particular behavior.

operon Group of structural and regulating genes that function as a single unit.

opportunistic infections Diseases caused by pathogens that are normally suppressed by the immune system but may become pathogenic if the immune system is compromised, as is the case with HIV/ AIDS.

order One of the categories, or taxa, used to group species; the taxon located above the family level.

organ Combination of two or more different tissues performing a common function.

organ of Corti Structure in the vertebrate inner ear that contains auditory receptors (also called spiral organ).

organ system Group of related organs working together; examples are the digestive and endocrine systems.

organelle Small, membranous structure in the cytoplasm having a specific structure and function.

organic molecule Molecule that always contains carbon and hydrogen, and often contains oxygen as well; organic molecules are associated with living organisms.

organism Level of biological organization that includes multiple organ systems; an individual of a species.

origin When referring to a muscle, the connection on the stationary bone.

osmoregulation Regulation of the water-salt balance to maintain a normal balance within internal fluids.

osmosis Diffusion of water through a selectively permeable membrane.

osmotic pressure Measure of the tendency of water to move across a selectively permeable membrane; visible as an increase in liquid on the side of the membrane with higher solute concentration.

ossicle The small bones of the vertebrate middle ear—malleus, incus, and stapes.

osteoarthritis (OA) Form of degenerative joint disease that is the result of the loss of cartilage in a synovial joint.

osteoblast Bone-forming cell.

osteoclast Cell that is responsible for bone resorption.

osteocyte Mature bone cell located within the lacunae of bone.

osteoporosis Condition characterized by a loss of bone density; associated with levels of sex hormones and diet.

otitis media Inflammation of the middle ear.

otolith Calcium carbonate granule associated with sensory receptors for detecting movement of the head; in vertebrates, located in the utricle and saccule.

outer ear Portion of the ear consisting of the pinna and the auditory canal.

oval window Structure of the middle ear that conducts sound from the middle ear to the inner ear.

ovarian cancer Form of cancer that affects the uterus; often difficult to diagnose due to a lack of symptoms.

ovarian cycle Monthly changes occurring in the ovary that determine the level of sex hormones in the blood.

ovary In flowering plants, the enlarged, ovule-bearing portion of the carpel that develops into a fruit; female gonad in animals that produces an egg and female sex hormones.

overexploitation When the number of individuals taken from a wild population is so great that the population becomes severely reduced in numbers.

ovulation Bursting of a follicle when a secondary oocyte is released from the ovary; if fertilization occurs, the secondary oocyte becomes an egg.

ovule In seed plants, a structure that contains the female gametophyte and has the potential to develop into a seed.

oxidation-reduction reaction A paired set of chemical reactions in which one molecule gives up electrons (oxidized) while another molecule accepts electrons (reduced); commonly called a redox reaction.

oxygen debt Amount of oxygen required to oxidize lactic acid produced anaerobically during strenuous muscle activity.

oxyhemoglobin Compound formed when oxygen combines with hemoglobin.

oxytocin Hormone, released by the posterior pituitary, that causes contraction of the uterus and milk letdown.

P

pacemaker Cells of the sinoatrial node of the heart; electrical device designed to mimic the normal electrical patterns of the heart.

pain receptors Sensory receptors that are sensitive to chemicals released by damaged cells; also called nociceptors.

pancreas Internal organ that produces digestive enzymes and the hormones insulin and glucagon.

pancreatic amylase Enzyme that digests starch to maltose.

pancreatic cancer Form of cancer that originates in the pancreas; one of the more fatal forms of cancer.

pancreatic islet Masses of cells that constitute the endocrine portion of the pancreas.

pancreatitis Inflammation of the pancreas; may be caused by infection, gallstones, or excessive use of alcohol.

Pap test Procedure that removes a few cells from the cervix to look for evidence of cervical cancer.

parasite Species that is dependent on a host species for survival, usually to the detriment of the host species.

parasitism Symbiotic relationship in which one species (the *parasite*) benefits in terms of growth and reproduction, to the detriment of the other species (the *host*).

parasympathetic division Division of the autonomic system that is active under normal conditions; uses acetylcholine as a neurotransmitter.

parathyroid gland Gland embedded in the posterior surface of the thyroid gland; it produces parathyroid hormone.

parathyroid hormone (PTH) Hormone secreted by the four parathyroid glands that increases the blood calcium level and decreases the phosphate level.

parenchyma Plant tissue composed of the least-specialized of all plant cells; found in all organs of a plant.

Parkinson disease (PD) Progressive deterioration of the central nervous system due to a deficiency in the neurotransmitter dopamine.

parsimony In systematics, the simplest solution in the analysis of evolutionary relationships.

parturition Process of giving birth; divided into three stages.

passive immunity Protection against infection, acquired by transfer of antibodies to a susceptible individual.

pathogen Disease-causing agent such as viruses, parasitic bacteria, fungi, and animals.

pattern formation Positioning of cells during development that determines the final shape of an organism.

pectoral girdle Portion of the vertebrate skeleton that provides support and attachment for the upper (fore) limbs; consists of the scapula and clavicle on each side of the body.

pedigree Chart of genetic relationship of family individuals across generations.

pelagic division Open portion of the ocean.

pellicle Structure that supports the cell membrane of some species of protists; used to maintain structure in a water environment.

pelvic girdle Portion of the vertebrate skeleton to which the lower (hind) limbs are attached; consists of the coxal bones.

pelvic inflammatory disease (PID) Condition in which the uterine tubes become inflamed; often caused by chlamydial infections.

penis Male copulatory organ; in humans, the male organ of sexual intercourse.

pentose Five-carbon monosaccharide. Examples are deoxyribose found in DNA and ribose found in RNA.

pepsin Enzyme secreted by gastric glands that digests proteins to peptides.

peptidase Intestinal enzyme that breaks down short chains of amino acids to individual amino acids that are absorbed across the intestinal wall.

peptide bond Type of covalent bond that joins two amino acids.

peptide hormone Type of hormone that is a protein, a peptide, a glycoprotein, or is derived from an amino acid.

perception Processing of sensory stimuli that occurs when the brain interprets information being received from the sensory receptors.

pericycle Layer of cells surrounding the vascular tissue of roots; produces branch roots.

periosteum Fibrous connective tissue that covers the surface of a long bone; contains blood vessels that service the cells within the bone.

peripheral nervous system (PNS) Nerves and ganglia that lie outside the central nervous system.

peristalsis Wavelike contractions that propel substances along a tubular structure such as the esophagus.

permafrost Permanently frozen ground, usually occurring in the tundra, a biome of Arctic regions.

peroxisome Enzyme-filled vesicle in which fatty acids and amino acids are metabolized to hydrogen peroxide that is broken down into harmless products.

petal A flower part that occurs just inside the sepals; often conspicuously colored to attract pollinators.

petiole The part of a plant leaf that connects the blade to the stem.

pH scale Measurement scale for hydrogen ion concentration. Based on the formula $-\log[H^+]$.

phagocytes Type of white blood cells that destroy pathogens using phagocytosis.

phagocytosis Process by which cells engulf large substances, forming an intracellular vacuole.

pharyngeal pouches Developmental characteristic of a chordate; these may develop into gills, or other structures such as auditory tubes and glands.

pharyngitis Inflammation of the pharynx; often caused by viruses or bacteria.

pharynx In vertebrates, common passageway for both food intake and air movement; located between the mouth and the esophagus.

phenotype Visible expression of a genotype—e.g., brown eyes or attached earlobes.

phenylketonuria (PKU) Autosomal recessive genetic disorder that causes a lack of the enzyme that metabolizes phenylalanine; the accumulation of phenylalanine causes problems with nervous system development and function.

pheromone Chemical messenger that works at a distance and alters the behavior of another member of the same species.

phloem Vascular tissue that conducts organic solutes in plants; contains sieve-tube members and companion cells.

phoronid Invertebrate animal (lophophoran) that possesses a protective tube made of chitin.

phospholipid Molecule that forms the bilayer of the cell's membranes; has a polar, hydrophilic head bonded to two nonpolar, hydrophobic tails.

photoperiod (photoperiodism) Relative lengths of daylight and darkness that affect the physiology and behavior of an organism.

photoreceptor Sensory receptor that responds to light stimuli.

photosynthesis Process, usually occurring within chloroplasts, that uses solar energy to reduce carbon dioxide to carbohydrate.

photosystem Photosynthetic unit where solar energy is absorbed and high-energy electrons are generated; contains a pigment complex and an electron acceptor; occurs as PS (photosystem) I and PS II.

phototropism Growth response of plant stems to light; stems demonstrate positive phototropism.

phylogenetics Area of systematic biology that uses evidence to construct phylogeny, or a hypothesis of the evolutionary relationships between species.

phylogeny Evolutionary history of a group of organisms.

phylum One of the categories, or taxa, used to group species; the taxon located above the class level.

phytochrome Photoreversible plant pigment that is involved in photoperiodism and other responses of plants, such as etiolation.

pineal gland Gland—either at the skin surface (fish, amphibians) or in the third ventricle of the brain (mammals)—that produces melatonin.

pinna External flap of the ear; serves to collect and funnel sound toward the middle ear.

pinocytosis Process by which vesicle formation brings macromolecules into the cell.

pith Parenchyma tissue in the center of some stems and roots.

pituitary dwarfism Condition caused when too little growth hormone is produced during childhood. Individuals with this condition have a small stature, but have body parts in normal proportions.

pituitary gland Small gland that lies just inferior to the hypothalamus; consists of the anterior and posterior pituitary, both of which produce hormones.

placenta Organ formed from the chorion and the uterine wall during the development of placental mammals; allows the embryo, and then the fetus, to acquire nutrients and rid itself of wastes; produces hormones that regulate pregnancy.

placental mammal Also called the eutherians; species that rely on internal development whereby the fetus exchanges nutrients and wastes with its mother via a placenta.

plankton Freshwater and marine organisms that are suspended on or near the surface of the water; includes phytoplankton and zooplankton.

plants Multicellular, photosynthetic eukaryotes that increasingly became adapted to live on land.

plasma In vertebrates, the liquid portion of blood; contains nutrients, wastes, salts, and proteins.

plasma cell Mature B cell that mass-produces antibodies.

plasma membrane Membrane surrounding the cytoplasm that consists of a phospholipid bilayer with embedded proteins; functions to regulate the entrance and exit of molecules to and from the cell.

plasmid Extrachromosomal ring of accessory DNA in the cytoplasm of prokaryotes.

plasmodesmata In plants, cytoplasmic connections in the cell wall that connect two adjacent cells.

plasmodial slime mold Free-living mass of cytoplasm that moves by pseudopods on a forest floor or in a field, feeding on decaying plant material by phagocytosis; reproduces by spore formation.

platelet Thrombocyte; component of blood that is necessary for blood clotting.

pneumonia Condition of the respiratory system characterized by the filling of the bronchi and alveoli with fluid; caused by a viral, fungal, or bacterial pathogen.

polar body Nonfunctional product of oogenesis produced by the unequal division of cytoplasm in females during meiosis. In humans, three of the four cells produced by meiosis are polar bodies.

polar covalent bond Bond in which the sharing of electrons between atoms is unequal.

pollen cone Reproductive structure of a gymnosperm that produces windblown pollen; microstrobili.

pollen grain In seed plants, structure that is derived from a microspore and develops into a male gametophyte.

pollen tube In seed plants, a tube that forms when a pollen grain lands on the stigma and germinates. The tube grows, passing between the cells of the stigma and the style to reach the egg inside an ovule, where fertilization occurs.

pollination In gymnosperms, the transfer of pollen from pollen cone to seed cone; in angiosperms, the transfer of pollen from anther to stigma.

pollinator Animal, typically an insect or bird, that transfers pollen between plants.

pollution Any environmental change that adversely affects the lives and health of living organisms.

polygenic inheritance Pattern of inheritance in which a trait is controlled by several allelic pairs.

polymer Macromolecule consisting of covalently bonded monomers; for example, a polypeptide is a polymer of monomers called amino acids.

polymerase chain reaction (PCR) Biotechnology technique that uses the enzyme DNA polymerase to produce billions of copies of a specific piece of DNA.

polyp Among cnidarians, body form that is directed upward and contains much mesoglea. In anatomy: small, abnormal growth that arises from the epithelial lining.

polypeptide Polymer of many amino acids linked by peptide bonds.

polyribosome String of ribosomes simultaneously translating regions of the same mRNA strand during protein synthesis.

polysaccharide Polymer made from carbohydrate monomers; the polysaccharides starch and glycogen are polymers of glucose monomers.

pons Portion of the brain stem above the medulla oblongata and below the midbrain; assists the medulla oblongata in regulating the breathing rate.

population Group of organisms of the same species occupying a certain area and sharing a common gene pool.

population genetics The study of gene frequencies and their changes within a population.

Porifera Invertebrates that are pore-bearing filter feeders whose inner body wall is lined by collar cells that resemble a unicellular choanoflagellate.

portal system Pathway of blood flow that begins and ends in capillaries, such as the portal system located between the small intestine and liver.

positive feedback Mechanism of homeostatic response in which the output of the system intensifies and increases the activity of the system.

postanal tail Characteristic of a chordate; tail extends past the anus of the digestive system.

posterior pituitary Portion of the pituitary gland that stores and secretes oxytocin and antidiuretic hormone produced by the hypothalamus.

postzygotic isolating mechanisms Processes that result in the production of sterile offspring between two different species.

potential energy Stored energy in a potentially usable form, as a result of location or spatial arrangement.

precipitation The process by which water leaves the atmosphere and falls to the ground as water, ice, or snow.

predation Interaction in which one organism (the predator) uses another (the prey) as a food source.

predator Organism that practices predation.

prediction Step of the scientific process that follows the formulation of a hypothesis and assists in creating the experimental design.

preparatory (prep) reaction Reaction that oxidizes pyruvate with the release of carbon dioxide; results in acetyl CoA and connects glycolysis to the citric acid cycle.

pressure-flow model Explanation for phloem transport; osmotic pressure following active transport of sugar into phloem produces a flow of sap from a source to a sink.

prey Organism that provides nourishment for a predator.

prezygotic isolating mechanism Anatomical, physiological, or behavioral difference between two species that prevents the possibility of mating.

primate Member of the order Primates; includes prosimians, monkeys, apes, and hominins, all of whom have adaptations for living in trees.

principle Theory that is generally accepted by an overwhelming number of scientists; also called a law.

prion Infectious particle consisting of protein only and no nucleic acid.

producer Photosynthetic organism at the start of a grazing food chain that makes its own food—e.g., green plants on land and algae in water.

product Substance that forms as a result of a reaction.

progesterone Female sex hormone that helps maintain sexual organs and secondary sex characteristics.

proglottid Segment of a tapeworm that contains both male and female sex organs and becomes a bag of eggs.

prokaryote Organism that lacks a nucleus and the membrane-bound organelles typically found in eukaryotes.

prokaryotic cell Cell that generally lacks a membrane-bound nucleus and organelles; the cell type within the domains Bacteria and Archaea.

prolactin Hormone secreted by the anterior pituitary that stimulates the production of milk from the mammary glands.

prometaphase Phase of cell division that occurs between prophase and metaphase and is characterized by attachment of the spindle fibers to the kinetochores of each sister chromatid.

promoter In an operon, a sequence of DNA where RNA polymerase binds prior to transcription.

prophase First phase of mitosis; characterized by the condensation of the chromatin; chromosomes are visible, but scattered in the nucleus.

proprioceptor Class of mechanoreceptor responsible for maintaining the body's equilibrium and posture; involved in reflex actions.

prostaglandin Hormone that has various and powerful local effects.

prostate cancer Cancer of the prostate gland; may be detected using a PSA test.

prostate gland Gland in male humans that secretes an alkaline, cloudy fluid that increases the motility of sperm.

protein Polymer of amino acids; often consisting of one or more polypeptides and having a complex three-dimensional shape.

protein-first hypothesis In chemical evolution, the proposal that protein originated before other macromolecules and made possible the formation of protocells.

proteomics Study of the complete collection of proteins that a cell or organism expresses.

protists The group of eukaryotic organisms that are not a plant, fungus, or animal. Protists are generally a microscopic, complex single cell; they evolved before other types of eukaryotes in the history of Earth.

proto-oncogene Gene that promotes the cell cycle and prevents apoptosis; may become an oncogene through mutation.

protobiont (protocell) In biological evolution, a possible cell forerunner that became a cell once it acquired genes.

proton Positive subatomic particle located in the nucleus and assigned one atomic mass unit.

protostome Group of coelomate animals in which the first embryonic opening (the blastopore) is associated with the mouth.

protozoan Heterotrophic, single-celled protist that moves by flagella, cilia, or pseudopodia.

proximal convoluted tubule Portion of a nephron following the glomerular capsule where tubular reabsorption of filtrate occurs.

pseudocoelom Body cavity lying between the digestive tract and body wall that is incompletely lined by mesoderm.

pteridophyte Ferns and their allies (horsetails and whisk ferns).

pulmonary artery Blood vessel that transports oxygen-poor blood from the heart to the lungs.

pulmonary circuit Circulatory pathway between the lungs and the heart.

pulmonary fibrosis Respiratory condition characterized by the buildup of connective tissue in the lungs; typically caused by inhalation of coal dust, silica, or asbestos.

pulmonary tuberculosis Respiratory infection caused by the bacterium *Mycobacterium tuberculosis*.

pulmonary vein Blood vessel that transports oxygen-rich blood from the lungs to the heart.

pulse Vibration felt in arterial walls due to expansion of the aorta following ventricular contraction.

punctuated equilibrium Model of evolutionary change in a species which suggests that there are long periods of little or no change, followed by brief periods of rapid speciation.

Punnett square Visual representation developed by Reginald Punnett that is used to calculate the expected results of simple genetic crosses.

pupil Opening in the center of the iris of the vertebrate eye.

purine Nucleotides with a double-ring structure; examples are adenine and guanine.

pyelonephritis Infection of the kidneys, often caused by an initial infection in one of the ureters.

pyrimidine Nucleotides with a single ring in their structure; examples are thymine, cytosine, and uracil.

R

radial symmetry Body plan in which similar parts are arranged around a central axis, like spokes of a wheel.

rain shadow Leeward side (side sheltered from the wind) of a mountainous barrier, which receives much less precipitation than the windward side.

rate of natural increase (r) Growth rate dependent on the number of individuals that are born each year and the number of individuals that die each year.

ray-finned bony fishes Group of bony fishes with fins supported by parallel bony rays connected by webs of thin tissue.

reactant Substance that participates in a reaction.

receptacle Area where a flower attaches to a floral stalk.

receptor protein Protein located in the plasma membrane or within the cell; binds to a substance that alters some metabolic aspect of the cell.

receptor-mediated endocytosis Selective uptake of molecules into a cell by vacuole formation after they bind to specific receptor proteins in the plasma membrane.

recessive allele Allele that exerts its phenotypic effect only in the homozygote; its expression is masked by a dominant allele.

recombinant DNA (rDNA) DNA that contains genes from more than one source.

red algae Marine photosynthetic protists with a notable abundance of phycobilin pigments; includes coralline algae of coral reefs.

red blood cell Erythrocyte; contains hemoglobin and carries oxygen from the lungs or gills to the tissues in vertebrates.

red bone marrow Vascularized, modified connective tissue that is sometimes found in the cavities of spongy bone; site of blood cell formation.

red tide A population bloom of dinoflagellates that causes coastal waters to turn red. Releases a toxin that can lead to paralytic shellfish poisoning.

redox reaction A paired set of chemical reactions in which one molecule gives up electrons (oxidized) while another molecule accepts electrons (reduced); also called an oxidation-reduction reaction.

reduced hemoglobin Globin chains within the hemoglobin molecule that have combined with hydrogen ions (H^+).

reflex action Automatic, involuntary response of an organism to a stimulus.

refractory period Time following an action potential when a neuron is unable to conduct another nerve impulse.

renewable resource Resource normally replaced or replenished by natural processes and not depleted by moderate use.

renin Enzyme released by the kidneys that leads to the secretion of aldosterone and a rise in blood pressure.

replacement model Hypothesis of the evolution of modern humans from archaic species; also called the out-of-Africa hypothesis.

replacement reproduction Population in which each person is replaced by only one child.

reproduction The process of producing a new individual of the same kind.

reproductive system Organ system in humans that includes the sex-specific organs (testes, ovaries, etc.); produces and transports gametes; in females, nurtures and gives birth to offspring.

reptile Terrestrial vertebrate with internal fertilization, scaly skin, and an egg with a leathery shell; includes snakes, lizards, turtles, crocodiles, and birds.

reservoir Location in a biogeochemical cycle where a chemical or resource is stored for long periods of time and is typically unavailable to living organisms.

resource partitioning Mechanism that increases the number of niches by apportioning the supply of a resource such as food or living space between species.

respiratory center Group of nerve cells in the medulla oblongata that sends out nerve impulses on a rhythmic basis, resulting in involuntary inspiration on an ongoing basis.

respiratory system Organ system of humans that includes the lungs and associated structures; involved in the exchange of gases; helps control pH.

responding variable Result or change that occurs when an experimental variable is utilized in an experiment.

resting potential Membrane potential of an inactive neuron.

restoration ecology Seeks scientific ways to return ecosystems to their state prior to habitat degradation.

restriction enzyme Bacterial enzyme that stops viral reproduction by cleaving viral DNA; used to cut DNA at specific points during production of recombinant DNA.

reticular activating system (RAS) Area of the brain that contains the reticular formation; acts as a relay for information to and from the peripheral nervous system and higher processing centers of the brain.

reticular connective tissue Form of connective tissue that supports the lymph nodes, spleen, thymus, and bone marrow.

reticular fiber Very thin collagen fiber in the matrix of connective tissue, highly branched and forming delicate supporting networks.

retina Innermost layer of the vertebrate eyeball, containing the photoreceptors—rod cells and cone cells.

retinal Light-absorbing molecule found in the photoreceptors of the eye; usually combined with opsin to form rhodopsin.

retinal detachment Condition characterized by the separation of the retina from the choroid layer of the eye.

retrovirus RNA virus, containing the enzyme reverse transcriptase, that carries out RNA/DNA transcription.

rheumatoid arthritis (RA) Autoimmune disease that causes inflammation of the joints.

rhizome Rootlike underground stem.

rhodopsin Light-absorbing molecule in rod cells and cone cells that contains a pigment and the protein opsin.

ribosomal RNA (rRNA) Structural form of RNA found in the ribosomes.

ribosome Site of protein synthesis in a cell; composed of proteins and ribosomal RNA (rRNA).

ribozyme RNA molecule that functions as an enzyme that can catalyze chemical reactions.

ringworm Condition of the skin caused by a fungal infection.

RNA (ribonucleic acid) Nucleic acid produced from covalent bonding of nucleotide monomers that contain the sugar ribose; occurs in many forms, including messenger RNA, ribosomal RNA, and transfer RNA.

RNA polymerase During transcription, an enzyme that creates an mRNA transcript by joining nucleotides complementary to a DNA template.

RNA-first hypothesis In chemical evolution, the proposal that RNA originated before other macromolecules and allowed the formation of the first cell(s).

rod cell Photoreceptor in vertebrate eyes that responds to dim light.

root Organ system of a plant that is responsible for anchoring the plant, absorbing water and minerals, and storing carbohydrates.

root cap Protective cover of the root tip, whose cells are constantly replaced as they are ground off when the root pushes through rough soil particles.

root hair Extension of a root epidermal cell that collectively increases the surface area for the absorption of water and minerals.

root nodule Structure on plant root that contains nitrogen-fixing bacteria.

root system Includes the main root and all of its lateral (side) branches.

rotational equilibrium Maintenance of balance when the head and body are suddenly moved or rotated.

rough ER (endoplasmic reticulum) Membranous system of tubules, vesicles, and sacs in cells; has attached ribosomes.

round window Structure of the middle ear that assists in the transmission of sound from the middle ear to the inner ear.

roundworm Invertebrates with nonsegmented cylindrical body covered by a cuticle that molts; some forms are free-living in water and soil, and many are parasitic.

RuBP carboxylase An enzyme that starts the Calvin cycle reactions by catalyzing attachment of the carbon atom from CO_2 to RuBP.

S

sac fungi Fungi that produce spores in fingerlike sacs called asci within a fruiting body; includes morels, truffles, yeasts, and molds.

saccule Saclike cavity in the vestibule of the vertebrate inner ear; contains sensory receptors for gravitational equilibrium.

salivary amylase In humans, enzyme in saliva that digests starch to maltose.

salivary gland In humans, gland associated with the mouth, that secretes saliva.

saltatory conduction Movement of nerve impulses from one node to another along a myelinated axon.

saprotroph Organism that secretes digestive enzymes and absorbs the resulting nutrients back across the plasma membrane.

sarcolemma Plasma membrane of a muscle fiber; also forms the tubules of the T system involved in muscular contraction.

sarcomere One of many units, arranged linearly within a myofibril, whose contraction produces muscle contraction.

sarcoplasmic reticulum Smooth endoplasmic reticulum of skeletal muscle cells; surrounds the myofibrils and stores calcium ions.

saturated fatty acid Fatty acid molecule that lacks double bonds between the carbons of its hydrocarbon chain. The chain bears the maximum number of hydrogens possible.

Schwann cell Cell that surrounds a fiber of a peripheral nerve and forms the myelin sheath.

scientific method Process by which scientists formulate a hypothesis, gather data by observation and experimentation, and come to a conclusion.

sclera White, fibrous, outer layer of the eyeball.

sclerenchyma Plant tissue composed of cells with heavily lignified cell walls; functions in support.

scolex Tapeworm head region; contains hooks and suckers for attachment to host.

scrotum Saclike structure of a male that houses the testes.

second messenger Chemical signal, such as cyclic AMP, that causes the cell to respond to the first messenger—a hormone bound to a plasma membrane receptor protein.

seed Mature ovule that contains an embryo, with stored food enclosed in a protective coat.

seed cone Reproductive structure of a gymnosperm that produces windblown seeds; megastrobili.

selectively permeable Property of the plasma membrane that allows some substances to pass, but prohibits the movement of others.

semen (seminal fluid) Thick, whitish fluid consisting of sperm and secretions from several glands of the male reproductive tract.

semicircular canal One of three half-circle-shaped canals of the vertebrate inner ear; contains sensory receptors for rotational equilibrium.

semilunar valve Valve resembling a half-moon, located between the ventricles and their attached vessels.

seminal vesicles Glands in male humans that secrete a viscous fluid that provides nutrition to the sperm cells.

seminiferous tubule Long, coiled structure contained within chambers of the testis, where sperm are produced.

senescence Sum of the processes involving aging, decline, and eventual death of a plant or plant part.

sensation Processing of sensory stimuli that involves detection of nerve impulses by the cerebral cortex of the brain.

sensory (afferent) neuron Nerve cell that transmits nerve impulses to the central nervous system after a sensory receptor has been stimulated.

sensory adaptation Change in the sensitivity of a receptor that usually makes the receptor less sensitive to its stimulus.

sensory receptor Structure that receives either external or internal environmental stimuli and is a part of a sensory neuron or transmits signals to a sensory neuron.

sensory transduction Process by which a sensory receptor converts an input to a nerve impulse.

sepal Outermost, leaflike covering of the flower; usually green in color.

serous membrane Body membrane in the thoracic and abdominal cavities that secretes a watery lubricant.

Sertoli cells Male reproductive cells that support, nourish, and regulate the sperm-producing cells.

sessile Organism that is permanently attached to a substrate, such as rock.

seta (pl., **setae**) A needlelike, chitinous bristle in annelids, arthropods, and others.

sex chromosomes Chromosomes that differ between the sexes; in humans, these represent the X and Y chromosomes.

sex-linked Trait controlled by a gene on a sex chromosome; often described as either X-linked or Y-linked.

sexual selection Changes in males and females, often due to male competition and female selectivity, leading to increased fitness.

shared derived traits Used in classification systems, such as cladistics, to determine the evolutionary relationship between two species.

shoot system Aboveground portion of a plant consisting of the stem, leaves, and flowers.

short tandem repeat (STR) profiling Procedure of analyzing DNA in which PCR and gel electrophoresis are used to create a banding pattern; these are usually unique for each individual; process used in DNA barcoding.

short-day plant Plant that flowers when day length is shorter than a critical length—e.g., cocklebur, poinsettia, and chrysanthemum.

shrubland Arid terrestrial biome characterized by shrubs and tending to occur along coasts that have dry summers and receive most of their rainfall in the winter.

sickle-cell disease Autosomal recessive genetic disorder that causes a malformation of hemoglobin molecules, causing red blood cells to form a sickle shape; also sometimes called sickle-cell anemia due to the symptoms of the disease.

sieve-tube member Member that joins with others in the phloem tissue of plants as a means of transport for nutrient sap.

sink In the pressure-flow model of phloem transport, the location (roots) from which sugar is constantly being removed. Sugar will flow to the roots from the source.

sinuses Spaces within the cranium that reduce the overall weight of the skull.

sinusitis Inflammation of the cranial sinuses in the head.

sister chromatid One of two genetically identical chromosomal units that are the result of DNA replication and are attached to each other at the centromere.

skeletal muscle Striated, voluntary muscle tissue that comprises the majority of the muscles in the human body; also called striated muscle.

skin Outer covering of the body; can be called the integumentary system because it contains organs such as sense organs.

skull Common name for facial bones and bones of the cranium.

sliding filament model An explanation for muscle contraction based on the movement of actin filaments in relation to myosin filaments.

small intestine In vertebrates, the portion of the digestive tract that precedes the large intestine. In humans, consists of the duodenum, jejunum, and ileum.

smooth ER (endoplasmic reticulum) Membranous system of tubules, vesicles, and sacs in eukaryotic cells; site of lipid synthesis; lacks attached ribosomes.

smooth muscle Nonstriated, involuntary muscle found in the walls of internal organs.

sodium–potassium pump Carrier protein in the plasma membrane that moves sodium ions out of, and potassium ions into, cells; important in the function of nerve and muscle cells in animals.

solute Substance that is dissolved in a solvent, forming a solution.

solution Fluid (the solvent) that contains a dissolved solid (the solute).

solvent Liquid portion of a solution that serves to dissolve a solute.

somatic cell Body cell; excludes cells that undergo meiosis and become sperm or eggs.

somatic system Portion of the peripheral nervous system containing motor neurons that control skeletal muscles.

somatostatin Hormone, produced by the pancreas, stomach, and small intestine, that inhibits the effects of growth hormones; also suppresses activity of hormones such as insulin and glucagon.

source In the pressure-flow model of phloem transport, the location (leaves) of sugar production. Sugar will flow from the leaves to the sink.

speciation Origin of new species due to the evolutionary process of descent with modification.

species Group of similarly constructed organisms capable of interbreeding and producing fertile offspring; organisms that share a common gene pool; the taxon at the lowest level of classification.

sperm Male gamete having a haploid number of chromosomes and the ability to fertilize an egg, the female gamete.

spermatogenesis Production of sperm in males by the process of meiosis and maturation.

spicule Skeletal structure of sponges composed of calcium carbonate or silicate.

spinal cord In vertebrates, the nerve cord that is continuous with the base of the brain and housed within the vertebral column.

spinal nerve Nerve that arises from the spinal cord.

spindle Collection of microtubules that assist in the orderly distribution of chromosomes during cell division.

Spiralia Classification of invertebrate animals that includes the lophotrochozoans and platyzoans.

spleen Large, glandular organ located in the upper left region of the abdomen; stores and filters blood.

spongy bone Type of bone that has an irregular, meshlike arrangement of thin plates of bone.

sporangium (pl., **sporangia**) Structure that produces spores.

spore Asexual reproductive or resting cell capable of developing into a new organism without fusion with another cell, in contrast to a gamete.

sporophyte Diploid generation of the alternation-of-generations life cycle of a plant; produces haploid spores that develop into the haploid generation.

sporozoan Spore-forming protist that has no means of locomotion and is typically a parasite with a complex life cycle; usually has both sexual and asexual phases.

spring overturn Mixing process that occurs in spring in stratified lakes, whereby oxygen-rich top waters mix with nutrient-rich bottom waters.

squamous epithelium Type of epithelial tissue that contains flat cells.

stabilizing selection Outcome of natural selection in which extreme phenotypes are eliminated and the average phenotype is conserved.

stamen In flowering plants, the portion of the flower that consists of a filament and an anther containing pollen sacs where pollen is produced.

stapes Bone found in the middle ear that assists in the transmission of sound to the inner ear; also called the stirrup.

starch Storage polysaccharide found in plants that is composed of glucose molecules joined in a linear fashion with few side chains.

stem Usually the upright, vertical portion of a plant that transports substances to and from the leaves.

stem cell Type of cell that acts as a source for other types of cells; capable of continuously dividing.

steroid Type of lipid molecule having a complex of four carbon rings—e.g., cholesterol, estrogen, progesterone, and testosterone.

steroid hormone Type of hormone that has a complex of 4-carbon rings, but each one has different side chains.

stigma In flowering plants, portion of the carpel where pollen grains adhere and germinate before fertilization can occur.

stolon Stem that grows horizontally along the ground and may give rise to new plants where it contacts the soil—e.g., the runners of a strawberry plant.

stomach In vertebrates, muscular sac that mixes food with gastric juices to form chyme, which enters the small intestine.

stomach ulcer Open sore in the lining of the stomach; frequently caused by the bacteria *Helicobacter pylori*.

stomata (sing., **stoma**) Small openings between two guard cells on the underside of leaf epidermis, through which gases pass.

stroke Condition resulting when an arteriole in the brain bursts or becomes blocked by an embolism; cerebrovascular accident.

stroma Region within a chloroplast that surrounds the grana; contains enzymes involved in the synthesis of carbohydrates during the Calvin cycle of photosynthesis.

style Elongated, central portion of the carpel between the ovary and stigma.

subsidence Occurs when a portion of Earth's surface gradually settles downward.

substrate-level ATP synthesis Process in which ATP is formed by transferring a phosphate from a metabolic substrate to ADP.

supergroup Systematic term used to classify the major groups of eukaryotes based on DNA similarities.

surface tension Force that holds moist membranes together due to the attraction of water molecules through hydrogen bonds.

surface-area-to-volume ratio Ratio of a cell's outside area to its internal volume; the relationship limits the maximum size of a cell.

sustainable Ability of a society or ecosystem to maintain itself while also providing services to humans.

suture Line of union between two nonarticulating bones, as in the skull.

sweat gland Skin gland that secretes a fluid substance for evaporative cooling; sudoriferous gland.

symbiosis Relationship that occurs when two different species live together in a unique way; it may be beneficial, neutral, or detrimental to one or both species.

sympathetic division Division of the autonomic system that is active when an organism is under stress; uses norepinephrine as a neurotransmitter.

sympatric speciation Origin of new species in populations that overlap geographically.

synapse Junction between neurons consisting of the presynaptic (axon) membrane, the synaptic cleft, and the postsynaptic (usually dendrite) membrane.

synapsis Pairing of homologous chromosomes during meiosis I.

synaptic cleft Small gap between presynaptic and postsynaptic cells of a synapse.

synovial joint Freely moving joint in which two bones are separated by a cavity.

synovial membrane Body membrane that lines synovial joints; secretes synovial fluid that lubricates the joint.

syphilis Sexually transmitted disease caused by the bacterium *Treponema pallidum.*

systematic biology Another name for the study of systematics, or the study of the evolutionary relationships between species.

systematics Study of the diversity of life for the purpose of understanding the evolutionary relationships between species.

systemic circuit Circulatory pathway of blood flow between the tissues and the heart.

systole Contraction period of the heart during the cardiac cycle.

T

T (traverse) tubules Portions of the sarcolemma (plasma membrane) of a muscle cell that interact with the sarcoplasmic reticulum of the cell.

T-cell receptor (TCR) Molecule on the surface of a T cell that can bind to a specific antigen fragment in combination with an MHC molecule.

taiga Terrestrial biome that is a coniferous forest extending in a broad belt across northern Eurasia and North America.

taproot Main axis of a root that penetrates deeply and is used by certain plants (such as carrots) for food storage.

taste bud Structure in the vertebrate mouth containing sensory receptors for taste; in humans, most taste buds are on the tongue.

taxon (pl., **taxa**) Group of organisms that fills a particular classification category.

taxonomist Scientist that investigates the identification and naming of new organisms.

taxonomy Branch of biology concerned with identifying, describing, and naming organisms.

Tay-Sachs disease Autosomal recessive genetic disorder that results in a deficiency in the enzyme hexosaminidase A; causes an accumulation of glycolipids in the lysosomes, resulting in a progressive loss of psychomotor functions.

technology Application of scientific information to solve or address a need of humans.

telomere Tip of the end of a chromosome that shortens with each cell division and may thereby regulate the number of times a cell can divide.

telophase Final phase of mitosis; daughter cells are located at each pole.

temperate deciduous forest Forest found south of the taiga; characterized by deciduous trees such as oak, beech, and maple, moderate climate, relatively high rainfall, stratified plant growth, and plentiful ground life.

tendon Strap of fibrous connective tissue that connects skeletal muscle to bone.

terminal bud Bud that develops at the apex of a shoot.

termination End of translation that occurs when a ribosome reaches a stop codon on the mRNA that it is translating, causing release of the completed protein.

testcross Cross, between an individual with a dominant phenotype and an individual with a recessive phenotype, to determine whether the dominant individual is homozygous or heterozygous.

testes Male gonads that produce sperm and the male sex hormones.

testicular cancer One of several forms of cancer that affect the testes of males; usually characterized by abnormal tenderness or lumps in one of the testicles.

testosterone Male sex hormone that helps maintain sexual organs and secondary sex characteristics.

tetanus Summation of muscle contractions to a level of maximum sustainability.

thalamus In vertebrates, the portion of the diencephalon that passes on selected sensory information to the cerebrum.

theory Concept, or a collection of concepts, widely supported by a broad range of observations, experiments, and data.

thermoacidophile Type of archaea that lives in hot, acidic, aquatic habitats, such as hot springs or near hydrothermal vents.

thermoreceptor Sensory receptor that detects heat.

threatened species Species that is likely to become an endangered species in the foreseeable future (e.g., bald eagle, gray wolf, Louisiana black bear).

three-domain system Classification system that places all organisms into one of three large domains—Bacteria, Archaea, and Eukarya.

thrombin Enzyme that is involved in blood clotting; acts on fibrinogen molecules to produce fibrin.

thrombocyte Platelet; component of blood that is necessary for blood clotting.

thrush Condition caused by an infection of the oral cavity with the fungi *Candida.*

thylakoid Flattened sac within a granum of a chloroplast; membrane contains chlorophyll; location where the light reactions of photosynthesis occur.

thymine (T) One of four nitrogen-containing bases in nucleotides composing the structure of DNA; pairs with adenine.

thymus Lymphoid organ involved in the development and functioning of the immune system; T lymphocytes mature in the thymus.

thyroid gland Large gland in the neck that produces several important hormones, including thyroxine, triiodothyronine, and calcitonin.

thyroid-stimulating hormone (TSH) Substance produced by the anterior pituitary that causes the thyroid to secrete thyroxine and triiodothyronine.

thyroxine (T_4) Hormone secreted from the thyroid gland that promotes growth and development; in general, it increases the metabolic rate in cells.

tight junction Junction between cells when adjacent plasma membrane proteins join to form an impermeable barrier.

tissue Group of similar cells combined to perform a common function.

tissue culture Process of growing tissue artificially, usually in a liquid medium in laboratory glassware.

tonsillitis Inflammation of the tonsils—lymphoid tissue located in the pharynx.

totipotent Cell that has the full genetic potential of the organism, including the potential to develop into a complete organism.

trachea (pl., **tracheae**) In insects, air tube located between the spiracles and the tracheoles. In tetrapod vertebrates, air tube (windpipe) that runs between the larynx and the bronchi.

tracheid In vascular plants, type of cell in xylem that has tapered ends and pits through which water and minerals flow.

tract Bundle of myelinated axons in the central nervous system.

trans fats Unsaturated fatty acid chains in which the configuration of the carbon–carbon double bonds is such that the hydrogen atoms are across from each other, as opposed to being on the same side (cis).

transcription First stage of gene expression; process whereby a DNA strand serves as a template for the formation of mRNA.

transcription factor In eukaryotes, protein required for the initiation of transcription by RNA polymerase.

transduction Exchange of DNA between bacteria by means of a bacteriophage.

transfer RNA (tRNA) Type of RNA that transfers a particular amino acid to a ribosome during protein synthesis. At one end, it binds to the amino acid, and at the other end, it has an anticodon that binds to an mRNA codon.

transformation Taking up of extraneous genetic material from the environment by bacteria.

transgenic organism An organism whose genome has been altered by the insertion of genes from another species.

transitional links Evidence of evolution, typically fossils, that bear a resemblance to two groups that in the present day are classified separately.

translation During gene expression, the process whereby ribosomes use the sequence of codons in mRNA to produce a polypeptide with a particular sequence of amino acids.

translocation Movement of a chromosomal segment from one chromosome to another nonhomologous chromosome, leading to abnormalities—e.g., Down syndrome.

transpiration Plant's loss of water to the atmosphere, mainly through evaporation at leaf stomata.

transposon DNA sequence capable of randomly moving from one site to another in the genome.

trichinosis Infection caused by the roundworm *Trichinella spiralis.*

triglyceride Neutral fat composed of glycerol and three fatty acids; typically involved in energy storage.

trisomy Chromosome condition in which a diploid cell has one more chromosome than normal; designated as $2n + 1$.

trochozoan Type of protostome that produces a trochophore larva; also has two bands of cilia around its middle.

trophic level Feeding level of one or more populations in a food web.

trophoblast Outer membrane surrounding the embryo in mammals; when thickened by a layer of mesoderm, it becomes the chorion, an extraembryonic membrane.

tropical rain forest Biome found near the equator, characterized by warm weather, plentiful rainfall, high biodiversity, and mainly tree-living animal life.

tropism In plants, a growth response toward or away from a directional stimulus.

trypsin Protein-digesting enzyme secreted by the pancreas.

tubal ligation Surgical form of reproductive sterilization that cuts and seals the uterine tubes in a female.

tubular reabsorption Movement of primarily nutrient molecules and water from the contents of the nephron into blood at the proximal convoluted tubule.

tubular secretion Movement of certain molecules from blood into the distal convoluted tubule of a nephron so that they are added to urine.

tumor suppressor gene Gene that codes for a protein that ordinarily suppresses the cell cycle; inactivity due to a mutation can lead to a tumor.

tundra Biome characterized by permanently frozen subsoil found between the ice cap and the tree line of regions of the Arctic, just south of the ice-covered polar seas in the Northern Hemisphere; also known as the arctic tundra. Alpine tundra, having similar characteristics, is found near the peak of a mountain.

tunicate Type of primitive invertebrate chordate.

tympanic membrane Membranous region that receives air vibrations in an auditory organ; in humans, the eardrum.

U

umbilical cord Cord connecting the fetus to the placenta, through which blood vessels pass.

unsaturated fatty acid Fatty acid molecule that contains double bonds between some carbons of its hydrocarbon chain; thus contains fewer hydrogens than a saturated hydrocarbon chain.

upwelling Upward movement of deep, nutrient-rich water along coasts; it replaces surface waters that move away from shore when the direction of the prevailing wind shifts.

uracil (U) Pyrimidine base that occurs in RNA, replacing thymine.

urea Main nitrogenous waste of terrestrial amphibians and most mammals.

ureter Tubular structure conducting urine from the kidney to the urinary bladder.

urethra Tubular structure that receives urine from the bladder and carries it to the outside of the body.

uric acid Main nitrogenous waste of insects, reptiles, and birds.

urinary bladder Organ where urine is stored.

urinary system Organ system of humans that includes the kidneys, urinary bladder, and associated structures; excretes metabolic wastes; maintains fluid balance; helps control pH.

urine Liquid waste product made by the nephrons of the vertebrate kidney through the processes of glomerular filtration, tubular reabsorption, and tubular secretion.

uterine cycle Cycle that runs concurrently with the ovarian cycle; it prepares the uterus to receive a developing zygote.

uterine tubes Transport the oocyte (egg) from the ovary to the uterus; location of fertilization; also called the oviducts or fallopian tubes.

uterus In mammals, expanded portion of the female reproductive tract through which eggs pass to the environment or in which an embryo develops and is nourished before birth.

utricle Cavity in the vestibule of the vertebrate inner ear; contains sensory receptors for gravitational equilibrium.

V

Vaccine Preparation that acts as a preventative measure against a disease by using antigens that promote immunity without causing the actual disease.

vacuole Membrane-bound sac, larger than a vesicle; usually functions in storage and can contain a variety of substances. In plants, the central vacuole fills much of the interior of the cell.

vagina Component of the female reproductive system that serves as the birth canal; receives the penis during copulation.

vaginal contraceptive ring Form of contraceptive containing estrogen and progesterone that is inserted into the vagina following menstruation.

valence shell The outer electron shell of an atom. Contains the valence electrons, which determine the chemical reactivity of the atom.

vas deferens Also called the ductus deferens; storage location for mature sperm before they pass into the ejaculatory duct.

vascular bundle In plants, primary phloem and primary xylem enclosed by a bundle sheath.

vascular cambium In plants, lateral meristem that produces secondary phloem and secondary xylem.

vascular cylinder In eudicots, the tissues in the middle of a root, consisting of the pericycle and vascular tissues.

vascular plant Plant that has xylem and phloem.

vascular tissue Transport tissue in plants, consisting of xylem and phloem.

vasectomy Surgical form of reproductive sterilization that cuts and seals the vasa deferentia in a male.

vector In genetic engineering, a means to transfer foreign genetic material into a cell—e.g., a plasmid.

vein Blood vessel that arises from venules and transports blood toward the heart.

vena cava Large systemic vein that returns blood to the right atrium of the heart in tetrapods; either the superior or inferior vena cava.

ventilation Process of moving air into and out of the lungs; breathing.

ventral cavity One of two main body cavities in humans; contains the thoracic, abdominal, and pelvic cavities.

ventricle Cavity in an organ, such as a lower chamber of the heart or the ventricles of the brain.

venule Vessel that takes blood from capillaries to a vein.

vernix caseosa White, greasy covering of the developing fetus; protects the fetus' skin from the amniotic fluid.

vertebral column Portion of the vertebrate endoskeleton that houses the spinal cord; consists of many vertebrae separated by intervertebral discs.

vertebrate Chordate in which the notochord is replaced by a vertebral column.

vertigo Equilibrium disorder that is often associated with problems in the receptors of the semicircular canals in the ear.

vesicle Small, membrane-bound sac that stores substances within a cell.

vessel element Cell that joins with others to form a major conducting tube found in xylem.

vestibule Space or cavity at the entrance to a canal, such as the cavity that lies between the semicircular canals and the cochlea.

vestigial structure Remnant of a structure that was functional in some ancestor but is no longer functional in the organism in question.

villus Small, fingerlike projection of the inner small intestinal wall.

viroid Infectious strand of RNA devoid of a capsid and much smaller than a virus.

virus Acellular parasitic agent consisting of an outer capsid of protein and an inner core of nucleic acid (DNA or RNA).

visceral mass Internal organs of an organism; typically includes components of the digestive, reproductive, and urinary systems (if present).

visual accommodation Ability of the eye to focus at different distances by changing the curvature of the lens.

vitamin Organic nutrient that is required in small amounts for metabolic functions. Vitamins are often part of coenzymes.

vocal cord In humans, fold of tissue within the larynx; creates vocal sounds when it vibrates.

vulva Common name for the external genital organs of a female.

W

water (hydrologic) cycle Interdependent and continuous circulation of water from the ocean, to the atmosphere, to the land, and back to the ocean.

water molds Filamentous organisms having cell walls made of cellulose; typically decomposers of dead freshwater organisms, but some are parasites of aquatic or terrestrial organisms.

white blood cell Leukocyte, of which there are several types, each having a specific function in protecting the body from invasion by foreign substances and organisms.

white matter Myelinated axons in the central nervous system.

wood Secondary xylem that builds up year after year in woody plants and becomes the annual rings.

X

X-linked Allele that is located on an X chromosome; not all X-linked genes code for sexual characteristics.

xylem Vascular tissue that transports water and mineral solutes upward through the plant body; it contains vessel elements and tracheids.

Y

yeast Unicellular fungus that has a single nucleus and reproduces asexually by budding or fission, or sexually through spore formation.

yolk sac One of the extraembryonic membranes that, in shelled vertebrates, contains yolk for the nourishment of the embryo, and in placental mammals is the first site for blood cell formation.

Z

zero population growth No growth in population size.

zygospore Thick-walled resting cell formed during sexual reproduction of zygospore fungi.

zygospore fungi Fungi, such as black bread mold, that reproduce by forming windblown spores in sporangia; sexual reproduction involves a thick-walled zygospore.

zygote Diploid cell formed by the union of two gametes; the product of fertilization.

Note: Page numbers in *italics* indicate material presented in figures and tables.

Malleus, 346, *346*
Malpighi, Marcello, 165
Malpighian tubules, 633, *635*
Maltase, 260
Malthus, Thomas, 529
Maltose, 30, *30*, 260
Mammals, *643*, 650–653, *651–653*
Mandible
 fish, *647*
 human, *360*, 361, *361*
Manganese, 268
Mantle, in molluscs, 622, 625
Manubrium, *363*, 364
Marfan syndrome, 468
Marijuana, 327–328
Marine diesel, 137
Marine snails, egg-laying behavior of, 666–667
Marsupials, 652, *652*
The Martian, 181
Mass, atomic, 19
Masseter, *369–370*
Mass extinctions, *532*, 533
Mass number, 18
Mast cells, 234
Mastoid sinuses, 361
Materials, acquisition by organisms, 4, *4*
Maternal determinants, 438
Mating, 540–541, 674–675
Matrix
 of connective tissue, 189
 of mitochondria, 56, *56*, 121
Matter, defined, 18
Maxillae, *360*, 361, *361*
McCarty, Maclyn, 489
McClintock, Barbara, 503, 521
MD (muscular dystrophy), 78, 380, *478*
MDCs (more-developed countries), *684*, 684–685, 747
MDMA, 329
Measles, 240, 569
Measles-mumps-rubella (MMR) vaccine, 240, 569
Mechanical digestion, 252
Mechanical energy, 100, *100*
Mechanoreceptors, 337, *337, 347, 348*
Medial, defined, 195
Medial condyle, *365*
Medial malleolus, *365*
Medicinal leeches, 626–627, *627*
Medicinal value of species, 735–736
Medicine. *See* Health and medicine
Medulla oblongata, 318
Medullary cavity, *358*
Medusa stage, 619
Megakaryocytes, 213
Megaphylls, 62, 597
Megaspores, 171, 172
Meiosis
 anaphase in, 90, *90, 92, 93–94*
 in animal cells, *90, 92*
 chromosomes and, 88–91, *89–92*
 comparison with mitosis, 92–94, *93–94*
 crossing-over and, 89, *90*, 477
 fertilization and, 89
 genetic variation and, 89
 importance of, 91
 independent assortment and, 89–90, *91*
 interkinesis in, *90*, 91
 meiosis I, 88–89, *89–90, 93–94*
 meiosis II, 89, *89*, 91, *92*
 metaphase in, 89–90, *90, 92, 93–94*
 occurrence of, 93–94
 overview, 88–89, *89*
 prophase in, 89, *90*, 91, *92, 93–94*
 telophase in, 90, *90, 92, 93–94*
Meissner corpuscles, 339, *340*
Melanin, 200
Melanocytes, 200
Melanocyte-stimulating hormone (MSH), *386*, 390
Melanoma, 65, 199, *199*
Melatonin, 318, *386*, 396, 397
Membrane-assisted transport, 72

Membrane attack complex, 235
Membrane-first hypothesis, 55
Membranes. *See also* Plasma membrane
 basement, 189
 basilar, 347
 evolution of, *55*, 557–559
 extraembryonic, 441, *441*
 fertilization, 44, *434*
 in human body, *195*, 195–196
 mucous, 196
 otolithic, 349
 serous, 196
 synovial, 196, 366
 tectorial, 347
 thylakoid, 134–135
 tympanic, 346, *346*
Memory, 319–321, *321*
Memory B cells, 236
Memory T cells, 239
Mendel, Gregor, *458*, 458–459
Mendel's laws, 458–465, *458–465*
Meniere's disease, 353
Meninges, 196, 314
Meningitis, 331
Menisci, 366
Menopause, 414, 454
Menstruation, 413, *413*
Menstruation disorders, 427
Meristematic tissue, 85, 145, *145*
Meristems, 154–155, *155*
Merkel disks, 339, *340*
MERS (Middle East respiratory syndrome), 14
Mesoderm, 435, *436*, 617
Mesoglea, 619
Mesopelagic zone, 729, *729*
Mesophyll, 160
Mesozoic era, *532*
Messenger RNA (mRNA), 39, 494–495, *494–495*
Metabolic pathways, 104–109
Metabolic wastes, excretion of, 295
Metabolism
 aerobic, 114, 115
 anaerobic, 114, 115
 athletes and, 114
 bacterial, *56*, 562–563
 defined, 4, 102, 555
 energy and, *103, 103–104*
 oxidation-reduction reactions and, 109–111, *110–111*
Metacarpal bones, *360, 364*, 365
Metamorphosis, 631, 634–635
Metaphase, 85, *87, 93–94*
Metaphase I, 89–90, *90, 93–94*
Metaphase II, 91, *92*
Metaphase plates, 85
Metaphysis, 357
Metastasis, 232, 506
Metatarsals, *360*, 365, *365*
Methamphetamine, 329
Methane, 555–557, 707
Methanogens, *56*, 561
Methicillin-resistant *Staphylococcus aureus* (MRSA), 526, 542, 564
MG (myasthenia gravis), 248, 332
MHC (major histocompatibility complex), 234, 237, 675
Mice, nurturing behavior in, 664, *664*
Micelles, 558
Microbes. *See also* Archaea; Bacteria; Eukarya
 decomposition and, 556
 defined, 55
 disease and, 554–555
 evolution of, 555–559, *557–559*
 number of, 55
 origin of, 555–559, *556–559*
 Pasteur's experiments with, 55, *55*
Microbiology, defined, 55
Microbiota, 262, 555
Microevolution, 536–538, *536–538*
Microfibrils, 76
Microglia, 195
Micronucleus, 578

Microphylls, 62
Microscopes, 44, 45, *45*
Microspores, 171, 172
Microsporidia, 584, 585
Microtubules, *49–50, 57*, 58, 84
Microvilli, 46, 189, 256
Micturition, 296
Midbrain, 318
Middle ear, *346*, 346–347
Middle East respiratory syndrome (MERS), 14
Middle lamella, 76
Miller, Stanely, 55
Miller–Urey experiment, 55, *55*
Millipede, *631*
Mimicry, 688–690, *690*
Mineralocorticoids, *386, 393*, 394
Minerals, 267–268, *267–268*
Miocene epoch, *532*
Mites, 635
Mitochondria
 aging and, 451
 in cell anatomy, *49*
 in cellular respiration, 122–123, *122–124*
 cellular respiration and, *110*, 110–111
 composition and function of, *48*
 energy and, 54, 55–56
 evolution of, 59–60
 in plant cells, *50*
 structure of, 55–56, *56*
Mitosis
 anaphase, 85, *87, 93–94*
 in animal cells, 84–85, *86–87*
 chromosomes and, 84–88, *85, 86–87*
 comparison with meiosis, 92–94, *93–94*
 defined, 81, 84
 metaphase, 85, *87, 93–94*
 occurrence of, 93–94
 overview, 84, *85*
 in plant cells, 85, *86–87*
 prometaphase, 85, *86*
 prophase, 84–85, *86, 93–94*
 telophase, 85, *87, 93–94*
Mitotic stage of cell cycle, 81
MMR (measles-mumps-rubella) vaccine, 240, 569
Mnemiopsis, *619*
Models, in science, 10
Molds
 slime, 580–581, *581*
 water, 576–577, *578*
Molecular clock, 548, 654
Molecules
 in compounds, 23–25, *23–25*
 defined, *3*, 22
 formation of, 2, 4
 hydrophilic, 26
 hydrophobic, 26
 motor, 57
 organic, 28–29, *29*
 passage into and out of cells, 67, *67*
 shape of, 24
 signaling, 65
Molluscs, *616, 622–627, 623*
Molting, 631
Molybdenum, 268
Mongooses, 740, *740*
Monkeys, *653*, 669, *669*
Monoclonal antibodies, *242*, 242–243
Monocots
 embryo development in, 176
 eudicots *vs.*, 607, *608*
 flowering of, 173, *173*
 seed germination in, 178–179, *179*
 structure of, 150, *150*, 152, *152*
Monocot stems, *156*
Monocotyledones, 607
Monocytes, 211, *212*, 234
Monoecious plants, 172
Monohybrid crosses, *460*, 461